Texts and Monographs in Physics

Springer
Berlin
Heidelberg
New York
Barcelona
Budapest
Hong Kong
London
Milan
Paris
Santa Clara
Singapore
Tokyo

Texts and Monographs in Physics

From Microphysics to Macrophysics I + II Methods and Applications of Statistical Physics By R. Balian

Variational Methods in Mathematical Physics A Unified Approach By P. Blanchard and E. Brüning

Quantum Mechanics: Foundations and Applications 3rd enlarged edition By A. Böhm

The Early Universe Facts and Fiction 3rd corrected and enlarged edition By G. Börner

Operator Algebras and Quantum Statistical Mechanics I + II 2nd edition By O. Bratteli and D. W. Robinson

Geometry of the Standard Model of Elementary Particles By A. Derdzinski

Random Walks, Critical Phenomena, and Triviality in Quantum Field Theory By R. Fernández, J. Fröhlich and A. D. Sokal

Quantum Relativity A Synthesis of the Ideas of Einstein and Heisenberg By D. R. Finkelstein

Quantum Mechanics I + II By A. Galindo and P. Pascual

The Elements of Mechanics By G. Gallavotti

Local Quantum Physics Fields, Particles, Algebras 2nd revised and enlarged edition By R. Haag

Elementary Particle Physics Concepts and Phenomena By O. Nachtmann

Inverse Schrödinger Scattering in Three Dimensions By R. G. Newton

Scattering Theory of Waves and Particles 2nd edition By R. G. Newton

Quantum Entropy and Its Use By M. Ohya and D. Petz

Generalized Coherent States and Their Applications By A. Perelomov

Essential Relativity Special, General, and Cosmological Revised 2nd edition By W. Rindler

Path Integral Approach to Quantum Physics An Introduction 2nd printing By G. Roepstorff

Advanced Quantum Theory and Its Applications Through Feynman Diagrams 2nd edition By M. D. Scadron

Finite Quantum Electrodynamics The Causal Approach 2nd edition By G. Scharf

From Electrostatics to Optics A Concise Electrodynamics Course By G. Scharf

Large Scale Dynamics of Interacting Particles By H. Spohn

General Relativity and Relativistic Astrophysics By N. Straumann

The Mechanics and Thermodynamics of Continuous Media By M. Šilhavý

The Dirac Equation By B. Thaller

The Theory of Quark and Gluon Interactions 2nd completely revised and enlarged edition By F. J. Ynduráin

Relativistic Quantum Mechanics and Introduction to Field Theory By F. J. Ynduráin

Supersymmetric Methods in Quantum and Statistical Physics By G. Junker

Miroslav Šilhavý

The Mechanics and Thermodynamics of Continuous Media

With 24 Figures

Springer

Miroslav Šilhavý

Academy of Sciences of the Czech Republic
Mathematical Institute
Žitná 25
115 67 Prague 1, Czech Republic

With 4 Tables

Library of Congress Cataloging-in-Publication Data

Šilhavý, M. (Miroslav), 1949– , The mechanics and thermodynamics of continuous media / M. Šilhavý. p. cm. – (Texts and monographs in physics, ISSN 0172-5998) Includes bibliographical references and index. ISBN 3-540-58378-5 (hardcover : alk. paper) 1. Mechanics, Analytic. 2. Mathematical physics. I. Title. II. Series. QA807.5.S55 1996 531–dc20 96-26659 CIP

ISSN 0172-5998
ISBN 3-540-58378-5 Springer-Verlag Berlin Heidelberg New York

Typesetting: Camera-ready copy from the author
Cover design: *design & production* GmbH, Heidelberg
SPIN: 10128606 55/3144-543210 - Printed on acid-free paper

Danušce a Mirovi

Preface

This book presents the nonlinear theories of continuum thermomechanics. Throughout I emphasize issues that are foundational in nature, and seek results common to materials of arbitrary symmetry. The central part of the book deals with thermoelastic bodies with heat conduction and viscosity, including the inviscid or ideal dissipationless bodies. A surprising variety of phenomena can be modeled within this framework. Moreover, the main ideas can be transferred into more complicated theories.

At present, the major challenge to the nonlinear thermoelasticity is posed by phase transformations with changes in symmetry. J. W. Gibbs' immensely influential treatise *On the equilibrium of heterogeneous substances* has provided a highly successful theory of phase transitions in fluids. Gibbs brought the view that the thermodynamics is not only the theory of heat, but also a theory of equilibrium, with the main tool the minimum principles. A large portion of the book is an extension of Gibbs' ideas to bodies of general symmetry by the methods of the calculus of variations. The interplay between the convexity properties of the stored energy functions, the resulting equations, and the physics of the phenomena is a leading theme.

I would like to thank many people for discussions held over the years of this long-term project, and reserve the chief place among them for Jan Kratochvíl and Ivan Samohýl. Earlier versions were written partly during my stays at the Institute for Mathematics and its Applications, University of Minnesota, Minneapolis, CNUCE, Istituto di C.N.R., Pisa, Helsinki Technological University, Espoo, and at Carnegie Mellon University, Pittsburgh, Pa. The hospitality of these institutions is gratefully acknowledged. Special thanks are due to I. Samohýl, W. Hrusa, and J. Vítek, who read earlier versions and suggested many improvements, and to Professor W. Beiglböck who contributed with valuable remarks. I also gratefully acknowledge the support of Grant No. 119110 of the Academy of Sciences of the Czech Republic during the final stage of the project.

Prague, September 1996 M. Šilhavý

Table of Contents

II Foundations

III Constitutive Theory

V Dynamics

Synopsis

To allow the existence of phase transformations, shock waves and other phenomena, the response of a body must be nonlinear. As this book is designed to treat bodies with a response of general symmetry, the deformation gradient must be used as the basic measure of deformation. The passage to specific symmetries – isotropic solids, fluids, or crystals – is made only to reach conclusions that do not hold generally, or to give the general assertions a more concrete form in the variables of the specific situation. The relationship between the response functions and the material behavior is established by isolating leading features of the response functions rather than by examining concrete models. Apart from the natural restrictions from the entropy inequality, frame indifference, and symmetry, the main unifying concepts are the convexity/nonconvexity properties of the thermodynamic potentials.

The understanding of these concepts is now much fuller than, say, at the time of TRUESDELL & NOLL [1965]. Since then, the closely related concepts of quasiconvexity and rank 1 convexity have proved to be basic for the mathematical theory of materials. They are related to the qualitative features of the equilibrium states, like their existence/nonexistence, stability/instability, uniqueness/nonuniqueness, the occurrence/nonoccurrence of phase boundaries, etc. Moreover, the violation of these and other 'mathematically desirable' features is now understood *not* as a mathematical pathology, but as a sign indicating (the possibility of) an interesting physical phenomenon, phase transition, observable large- or fine-scale instability or another 'catastrophic' feature. For fluids, the quasiconvexity and rank 1 convexity in the deformation gradient reduce to the convexity in the specific volume, and its relevance to the stability has been known since the times of Gibbs. Also the violation of the convexity of the energy, i.e., the occurrence of an interval where the pressure increases with the specific volume (the spinodal interval of the Van der Waals isotherm), is clearly necessary for a phase transition. The values of the specific volumes of the stably coexistent liquid and its vapor are determined from the equality of the chemical potentials of the phases.

The quasiconvexity and rank 1 convexity in the deformation gradient are the generalizations of the convexity in the specific volume for fluids, and the continuity of the normal component of the *Eshelby energy-momentum tensor* across the static phase interface is the generalization of the equality of the chemical potentials. The common origin of the quasiconvexity and of the continuity of the Eshelby tensor is in the minimum principles, e.g., the principle of minimum total stored energy. (The past attempts to generalize the convexity on an a priori basis, and the attempts to generalize the equality of chemical potentials by formal considerations, have proved unsuccessful.) At the regions where the rank 1 convexity prevails, the equations of mechanical equilibrium are elliptic. The occurrence of the spinodal region in fluids is translated into the violation of the rank 1 convexity on a subinterval of the line segment connecting the deformation gradients at the two sides of the interface ("generalized spinodal region"). These are unavoidable consequences of minima, valid for the response of a general symmetry, with numerous consequences. The theory of materials in equilibrium appears to be quite mature. Wide varieties of mathematical tools have been assembled, with major profit from and impact on the calculus of variations, partial differential equations and nonlinear functional analysis.

In dynamics, the thermal phenomena cannot be neglected. If the static theory admits states with several coexistent phases, the dynamic theory must describe the evolution of the phase interfaces, and in this respect it may be viewed as a broad generalization of the Stefan problem. Moreover, the energy function cannot be globally elliptic and the evolution equations change their type from hyperbolic to *elliptic* in the spinodal region. A violent nonuniqueness in the initial-value problem accompanies that, a nonuniqueness far more severe than the known nonuniqueness in the hyperbolic systems of nonlinear conservation laws. The results concerning this general approach are scarce. The present book describes a recent proposal for an additional kinetic equation for the speed of an evolving phase boundary. This is another interesting application of the Eshelby tensor.

The idealized dissipationless materials are described in a more detail. The ever-present discontinuities of solutions are shock waves, and the thermodynamics provides the entropy admissibility criterion, i.e., the increase of entropy across the shock. In fluids in the hyperbolic and genuinely nonlinear regimes, the shock waves and the entropy criterion are understood, at least on the constitutive level, since the late forties. The genuine nonlinearity means here that the pressure is a convex function of the specific volume at constant entropy, another remarkable constitutive restriction. The entropy criterion is known to secure the realistic behavior of shocks, most notably the uniqueness in the Riemann problem and Lax' inequalities. When the genuine nonlinearity fails but the hyperbolicity still holds, the entropy criterion is insufficient for that, and Liu's criterion must be employed, which may be interpreted as a strengthening of Lax' inequalities, or as a generalization of Oleinik's E-condition to systems of equations. Its relation to the entropy criterion is established, both for bodies of general material symmetry and for fluids. Naturally, the latter case leads to more perfect results.

Part I deals with the basic language of continuum mechanics. This includes tensor algebra and analysis, the geometry and kinematics of continuous bodies, and the balance equations. The constitutive equations characterizing particular materials are kept distinct form the balance equations, as is now common. The direct notation is used throughout. In the tensor analysis, the emphasis is on somewhat less standard questions like the differentiation of the eigenvalues of the stretch tensors (principal strains) with respect to the deformation gradient, differentiation of the square root of a tensor, etc. The kinematics deliberately avoids the classical analysis of the deformation in the manner of engineering elasticity. Not that it would be useless, but the just mentioned analysis of principal strains is more advantageous in the nonlinear range. The displacement vector and the infinitesimal strain tensor enter only at the stage of linearization. The shock waves and (coherent) phase fronts are sharp surfaces across which the deformation gradient and velocity have jump discontinuities (singular surfaces) but the actual position is continuous. The main consequence is the Hadamard lemma saying that the limiting values of the deformation gradient at the two sides of the interface are rank 1 connected. This opens the way to a number of essentially geometric topics, such as the mechanical theory of twinning, the austenite/martensite interface and others. The singular surfaces are assumed to bear no material structure: the surface tension, surface heating, and the surface concentration of mass are excluded. The transport theorems for processes with singular surfaces are proved. Then the balances of mass, momentum, energy, and entropy are introduced and their local forms for the bulk matter and for singular surfaces are derived.

Part II is somewhat independent of the rest. It deals with the foundations – the basic quantities of continuum thermodynamics, the total energy and entropy, are derived from a set of elementary axioms. For this, a state space formalism is introduced with a generality that covers also the memory phenomena and hysteresis. (This is also a basis for the general constitutive theory of materials.) There the reader will find statements of the first and the second laws of thermodynamics free from traditional ambiguities. The appealing programme of founding thermodynamics on the first and the second laws is achieved by simple but conceptually clear means. The theory resulting from Part II is the thermodynamics based on the Clausius–Duhem inequality, which is used throughout. There is no doubt about the appropriateness of this choice for the selected class of materials. Part II provides a strong support for this.

Part III describes the constitutive equations of viscous materials with heat conduction. The restrictions placed on the response functions by the Clausius–Duhem inequality, frame-indifference, and symmetry are derived. The only classes of symmetry to be dealt with explicitly are isotropic solids, fluids and crystals having the symmetry of the underlying Bravais lattice. Each of these classes has its own representation theorems. The approach to the representation theorems for isotropic solids is based here on the principal strains, principal stresses, and principal directions of strain rather than on the Rivlin–Ericksen representation theorem and the principal invariants. This sometimes provides a more direct way to fitting the experimental data on the empirical side and a somewhat better control over the convexity and ellipticity properties on the mathematical side. A discussion of the change-of-variables mechanism is given, emphasizing that each change of variables leads to a new quantity that can be useful for understanding certain aspects. Thus, for instance, Eshelby's energy momentum tensor is associated with the exchange of roles of the reference and actual configurations, besides being associated with the translational invariance in the reference configuration and phase transformations. In addition, a basis for the changes of the convexity properties under changes of variables is developed. Related to these are the thermodynamic coefficients – essentially the second derivatives of the thermodynamic potentials. These are the fourth-order tensors of elasticities (the elastic moduli), in isothermal and adiabatic versions, the coefficients of thermal expansion, the stress–temperature coefficients, the tensor of latent heats, and the scalar specific heats at constant deformation or stress. Some of these occur in the linearized equations in static and dynamic situations. The dynamic part of the response is treated similarly, which leads to the kinetic coefficients like the tensors of heat conductivity and viscosity etc. In addition, brief accounts are given of the classical linear irreversible thermodynamics and of its recent generalization–the extended irreversible thermodynamics.

Part IV is the theory of thermodynamic equilibrium. While Part II treats thermodynamics as the theory of heat and work, Part IV treats it as the theory of minima of integral functionals. The exposition starts with the thermodynamic background: in the conservative loading conditions, the canonical free energy decreases along processes. It follows that if an equilibrium state compatible with the external conditions is stable, the canonical free energy must take a minimum value among all states satisfying the kinematical constraints. The states that satisfy the extremum principles are examined assuming that they have the smoothness that allows a derivation of the Euler–Lagrange equations. This smoothness admits the singular surface; hence states of coexistent phases are included. The quasiconvexity along a minimizer is derived as a necessary condition for the minimum, as well as its consequence rank 1 convexity. Also the quasiconvexity at the boundary, equally important to the quasiconvexity itself, is shown to prevail on the free part of the boundary. While the quasiconvexity can be viewed as a condition for internal stability, the quasiconvexity at the boundary is a condition for the surface stability. It has been conjectured that its violation can cause surface wrinkling of certain metals.

The considerations are then applied to bodies of specified symmetry. For fluids this gives the classical thermostatics, treated in Part IV with the emphasis on the convexity/nonconvexity properties of the energy surface in the volume-entropy space. A rigorous proof of the Gibbs phase rule is given based on the Carathéodory theorem on the convex hull. The Gibbs function is shown to have singularities at the pressures and temperatures of phase transitions: Its supergradient must contain the specific volumes and entropies of all stable phases of the given pressure and temperature. Solids are treated first using the local approach, based on the linearization and implicit function theorem. This includes the linearized elasticity of stressed bodies and, as an extremely special but important case, the classical linear elasticity, i.e., the linearization about a stress-free state. For the general linearized elasticity, the positivity of the second variation is crucial for the existence and uniqueness of the solutions and for the absence/presence of bifurcations of the nonlinear equations. The general pointwise conditions on the tensor of elastic moduli related to the positivity of the second variation are discussed. These are the well-known strong ellipticity condition governing the behavior of the bulk matter, and the complementing and Agmon's conditions, controlling the surface phenomena. The latter appear to be less recognized explicitly, but the more enlightened portion of the elasticity literature treats the violation of the complementing condition under the name surface instabilities. The complementing and Agmon's conditions are discussed for isotropic states under tension or compression in some detail. A linear bifurcation analysis of a column under compression is presented as an example for the determination of parameters at which the uniform positivity of the second variation fails. Also an absence of instability under tension is explained for a model stored energy of the Blatz–Ko (special) type: at the tension, for which the principal forces are all nonnegative, the stored energy is convex. (In fact, one of this book's side goals is a rehabilitation of points of convexity, within limits. The folklore wrongly deems the convexity in the deformation gradient as 'contradicting the frame-indifference.')

The global theory of existence of solutions depends critically on the convexity properties of the stored energy and the related sequential lower semicontinuity of the total energy. The direct methods of the calculus of variations are the main working tools. For rubber-like materials the existence theory is based on a strengthened version of the quasiconvexity known as polyconvexity. The minimizers must be sought in the Sobolev spaces, and their quality depends dramatically on the exponent p in $W^{1,p}$. This is explained on the famous radial deformations with or without cavities. Also the devices to ensure reasonable versions of injectivity (invertibility) of deformations are reviewed in some detail. Unfortunately the book's size did not permit a systematic treatment, despite the author's opinion that a rather complete picture can be reconstructed from the existing literature. The polyconvexity, let alone rank 1 convexity, cannot be satisfied by crystalline bodies. In this case the total energy fails to be lower semicontinuous, and the properties of the minimizing sequences are not sufficiently fully reflected by the limiting macroscopic deformation. A finer device, Young's measure, must be used to encode the (spatial) oscillation phenomena occurring in the minimizing sequence. Physically, Young's measure describes the possible microstructures consistent with the given macroscopic deformation (in this context), and the set of all Young's measures may be viewed as an enlargement of the state space of the body.

Part V is about the dynamics. Besides the propagating phase boundaries mentioned above, this part deals with the dissipationless materials with shock waves, and with the resulting hyperbolic systems of equations. The current activities in the theory of hyperbolic systems involve one-dimensional models, susceptible to a fairly detailed analysis. Since the present book attempts to present general, three-dimensional bodies and the features common to responses of arbitrary symmetry, I found it necessary to present the entropy and Liu's criteria, Lax' inequalities, and the necessary background in this general setting. There appears to be no systematic treatment except for the isentropic case or fluids. The main problem is that the equations cannot be put in the form of a first-order system of conservation laws popular in fluid dynamics. Moreover, when the entropy is used as the independent thermal variable, which seems to be the only natural setting, the system does not have, even in the hyperbolic regime, a strictly (rank 1) convex mathematical entropy function: here the entropy is linear. The hyperbolicity, strict hyperbolicity, and genuine nonlinearity are formulated in this setting. The existence of the Hugoniot curve is proved locally, and the local equivalence of the entropy criterion with Lax' inequalities and Liu's criterion is established in the strictly hyperbolic, genuinely nonlinear regimes. The analogs of the centered waves of fluid dynamics are defined, and using this, the existence of the solution of the Riemann problem for nearly equilibrium initial data is proved through the implicit function theorem. Only then do I pass to fluids and show, using the classical analysis of Zemplén, Bethe, and Weyl, the global equivalence of the criteria for fluids satisfying the classical restrictions.

The presence of dissipation, e.g. the viscosity of the differential type, makes the problem of proving the existence of the time-evolution and examining its asymptotics easier. Despite that, the existing proofs deal only with one-dimensional models. Rather than these, this book includes only the proof of the existence of the time evolution of the model linearized at an equilibrium state of a solid. Even though the proof uses standard means, it is presented because in this case the ideal goal of the theory can be carried out to the end: the realistic conditions for the existence of evolution are given and the asymptotic properties, i.e., the trend to equilibrium, is proved when the second variation of the stored energy is positive definite. Moreover, efficient Liapunov functions can be calculated explicitly.

Part I

Balance Equations

1 Elements of Tensor Algebra and Analysis

The reader is assumed to have some familiarity with the elements of linear algebra. A convenient reference is HALMOS [1958]. A direct notation is used throughout, with the same conventions as in TRUESDELL & NOLL [1965] and GURTIN [1981]. The treatment below emphasizes the differentiation of functions of tensor arguments. Thus the functional calculus for symmetric tensors is introduced and a formula is given for its derivative; as a consequence of this, e.g., the formula for the derivative of the square root is derived. Also the derivatives of the eigenvalues, singular values and eigenvectors are calculated. Some of these results can be omitted on the first reading, as they will be used only sporadically until Chap. 8 on isotropic functions and Chap. 18 on the convexity properties thereof.

1.1 Vectors and Second-Order Tensors

Throughout, real linear spaces with inner products are used to describe the processes in continuous bodies. Only Cartesian tensors, of orders 0,1,2,3,4 and 6, will be employed. The tensors of order 0 and 1 are real numbers and vectors, respectively, and tensors of order 2 are linear transformations. Throughout the chapter, Vect is an n-dimensional real vector space with inner product (typically, $n = 3$) and the index integers $i, j, \ldots$ range between 1 and n unless stated otherwise.

1.1.1 The scalar (= inner) product of $\boldsymbol{u}, \boldsymbol{v} \in$ Vect is denoted $\boldsymbol{u} \cdot \boldsymbol{v}$; we write $|\boldsymbol{u}| := (\boldsymbol{u} \cdot \boldsymbol{u})^{1/2}$ for the length of $\boldsymbol{u}$ and denote $\boldsymbol{u}^2 = \boldsymbol{u} \cdot \boldsymbol{u} = |\boldsymbol{u}|^2$. In an orthonormal basis $\{\boldsymbol{e}_i\}$ we have $\boldsymbol{u} = \sum_{i=1}^{n} u_i \boldsymbol{e}_i$, where u_i are the components of $\boldsymbol{u}$. If $n = 3$, we assume that Vect is oriented, denote by $\boldsymbol{a} \times \boldsymbol{b}$ the cross product and write $[\boldsymbol{a}, \boldsymbol{b}, \boldsymbol{c}] := (\boldsymbol{a} \times \boldsymbol{b}) \cdot \boldsymbol{c}$ for every $\boldsymbol{a}, \boldsymbol{b}, \boldsymbol{c} \in$ Vect. A second-order tensor $\boldsymbol{A}$ is a linear transformation from Vect into Vect. The linear space of all second-order tensors is denoted by Lin. In an orthonormal basis $\{\boldsymbol{e}_i\}$ the tensor $\boldsymbol{A}$ is has the components $(\boldsymbol{A})_{ij} = A_{ij} := \boldsymbol{A}\boldsymbol{e}_j \cdot \boldsymbol{e}_i$ which can be arranged into an $n \times n$ matrix $A = [A_{ij}]$. The transpose $\boldsymbol{A}^{\mathrm{T}}$ is the unique tensor for which $\boldsymbol{A}^{\mathrm{T}}\boldsymbol{u} \cdot \boldsymbol{w} = \boldsymbol{u} \cdot \boldsymbol{A}\boldsymbol{w}$ for all $\boldsymbol{u}, \boldsymbol{w} \in$ Vect and one has $(\boldsymbol{A}^{\mathrm{T}})_{ij} = (\boldsymbol{A})_{ji}$. If $\boldsymbol{A}, \boldsymbol{B} \in$ Lin, then the product $\boldsymbol{A}\boldsymbol{B} \in$ Lin is defined as the composition i.e., $(\boldsymbol{A}\boldsymbol{B})\boldsymbol{u} = \boldsymbol{A}(\boldsymbol{B}\boldsymbol{u})$ for all $\boldsymbol{u} \in$ Vect; one has $(\boldsymbol{A}\boldsymbol{B})_{ik} = \sum_{j=1}^{n} A_{ij} B_{jk}$. If $\boldsymbol{A}$ is invertible, i.e., if $\boldsymbol{A}\boldsymbol{u} = \boldsymbol{0}$ only for $\boldsymbol{u} = \boldsymbol{0}$, then $\boldsymbol{A}^{-1}$ denotes the inverse of $\boldsymbol{A}$, i.e., the unique tensor for which $\boldsymbol{A}^{-1}\boldsymbol{A} = \boldsymbol{A}\boldsymbol{A}^{-1} = \boldsymbol{1}$, where $\boldsymbol{1} \in$ Lin is the identity tensor, given by $\boldsymbol{1}\boldsymbol{u} = \boldsymbol{u}$ for every $\boldsymbol{u} \in$ Vect. The tensor $\boldsymbol{1}$ has Kronecker's delta as its matrix in any basis.

The ***tensor product*** of two vectors $\boldsymbol{a}, \boldsymbol{b} \in$ Vect is a second-order tensor $\boldsymbol{a} \otimes \boldsymbol{b}$ such that $(\boldsymbol{a} \otimes \boldsymbol{b})\boldsymbol{u} = \boldsymbol{a}(\boldsymbol{b} \cdot \boldsymbol{u})$ for every $\boldsymbol{u} \in$ Vect. In components, $(\boldsymbol{a} \otimes \boldsymbol{b})_{ij} = a_i b_j$. Every tensor $\boldsymbol{A}$ can be expressed as

$$\boldsymbol{A} = \sum_{i,j=1}^{n} A_{ij}\boldsymbol{e}_i \otimes \boldsymbol{e}_j$$

in any orthonormal basis $\{\boldsymbol{e}_i\}$, where A_{ij} are its components. The ***rank*** $r(\boldsymbol{A})$ of a tensor $\boldsymbol{A}$ is the dimension of the range of the linear transformation $\boldsymbol{A}$. A tensor is of rank $r(\boldsymbol{A}) = n$ if and only if it is invertible; it is of rank ≤ 1 if and only if it is of the form $\boldsymbol{A} = \boldsymbol{a} \otimes \boldsymbol{b}$ for some $\boldsymbol{a}, \boldsymbol{b} \in \mathrm{Vect}$; if $\boldsymbol{a} \neq \boldsymbol{0}, \boldsymbol{b} \neq \boldsymbol{0}$, then $\boldsymbol{A} = \boldsymbol{a} \otimes \boldsymbol{b}$ is of rank 1. If $\boldsymbol{F}$ is an invertible tensor and $\boldsymbol{a}, \boldsymbol{b} \in \mathrm{Vect}$, then $\boldsymbol{F} + \boldsymbol{a} \otimes \boldsymbol{b}$ is invertible if and only if $1 + (\boldsymbol{F}^{-1}\boldsymbol{a} \cdot \boldsymbol{b}) \neq 0$ and if this is the case, then

$$(\boldsymbol{F} + \boldsymbol{a} \otimes \boldsymbol{b})^{-1} = \boldsymbol{F}^{-1} - (1 + \boldsymbol{F}^{-1}\boldsymbol{a} \cdot \boldsymbol{b})^{-1}(\boldsymbol{F}^{-1}\boldsymbol{a} \otimes \boldsymbol{F}^{-\mathsf{T}}\boldsymbol{b}). \tag{1.1.1}$$

1.1.2 Special Tensors Throughout the book, the following notation is used:

$\mathrm{Lin}^+ = \{\boldsymbol{L} \in \mathrm{Lin} : \det \boldsymbol{L} > 0\}$ = the set of all second-order tensors with positive determinant;

$\mathrm{Lin}^- = \{\boldsymbol{L} \in \mathrm{Lin} : \det \boldsymbol{L} < 0\}$ = the set of all second-order tensors with negative determinant;

$\mathrm{InvLin} = \{\boldsymbol{L} \in \mathrm{Lin} : \det \boldsymbol{L} \neq 0\}$ = the set of all invertible second-order tensors;

$\mathrm{Sym} = \{\boldsymbol{L} \in \mathrm{Lin} : \boldsymbol{L}^\mathsf{T} = \boldsymbol{L}\}$ = the set of all symmetric second-order tensors;

$\mathrm{Sym}^+ = \{\boldsymbol{L} \in \mathrm{Lin} : \boldsymbol{L}^\mathsf{T} = \boldsymbol{L}, \boldsymbol{L}\boldsymbol{v} \cdot \boldsymbol{v} > 0 \text{ for all } \boldsymbol{v} \in \mathrm{Vect}, \boldsymbol{v} \neq 0\}$ = the set of all positive definite symmetric second-order tensors,

$\mathrm{Skew} = \{\boldsymbol{W} \in \mathrm{Lin} : \boldsymbol{W}^\mathsf{T} = -\boldsymbol{W}\}$ = the set of all skew tensors;

$\mathrm{Orth} = \{\boldsymbol{Q} \in \mathrm{Lin} : \boldsymbol{Q}^\mathsf{T}\boldsymbol{Q} = \boldsymbol{Q}\boldsymbol{Q}^\mathsf{T} = \boldsymbol{1}\}$ = the set of all orthogonal tensors; Orth is a compact subgroup of InvLin; it is called the ***orthogonal group***;

$\mathrm{Orth}^+ = \{\boldsymbol{Q} \in \mathrm{Orth} : \det \boldsymbol{Q} = +1\}$ = the set of all proper orthogonal tensors, also called rotations; Orth^+ is called the ***proper orthogonal group***;

$\mathrm{Unim} = \{\boldsymbol{H} \in \mathrm{Lin} : |\det \boldsymbol{H}| = 1\}$ = the set of all unimodular tensors; Unim is called the ***unimodular group***;

$\mathrm{Sph} = \{\boldsymbol{n} \in \mathrm{Vect} : |\boldsymbol{n}| = 1\}$ = the unit sphere in Vect.

The symmetric and the skew parts of a tensor $\boldsymbol{H} \in \mathrm{Lin}$ are $\boldsymbol{E} \in \mathrm{Sym}$, $\boldsymbol{W} \in \mathrm{Skew}$, respectively, defined by

$$\boldsymbol{E} = \tfrac{1}{2}(\boldsymbol{H} + \boldsymbol{H}^\mathsf{T}), \quad \boldsymbol{W} = \tfrac{1}{2}(\boldsymbol{H} - \boldsymbol{H}^\mathsf{T}); \text{ hence } \boldsymbol{H} = \boldsymbol{E} + \boldsymbol{W}.$$

1.1.3 Derivatives Following e.g., DIEUDONNÉ [1960], TRUESDELL & NOLL [1965], and GURTIN [1981], we identify the derivatives with linear transformations: Let f : $\mathrm{D} \to \mathrm{Y}$ be a mapping between two normed vector spaces X, Y, with D an open subset of X. The derivative $Df(\boldsymbol{x}) = \partial_{\boldsymbol{x}} f(\boldsymbol{x})$ of f at $\boldsymbol{x} \in \mathrm{D}$ is a linear transformation from X into Y, its value on $\boldsymbol{u} \in \mathrm{X}$ is denoted $Df(\boldsymbol{x})[\boldsymbol{u}] = \partial_{\boldsymbol{x}} f(\boldsymbol{x})[\boldsymbol{u}]$. The phrase that a mapping f is differentiable at a point $\boldsymbol{x}$ will always mean that f has a Fréchet differential at that point. The second derivative (= the second (Fréchet) differential) of a mapping is then a bilinear form, etc. In the inner product spaces, the derivative of a real-valued function is identified with the corresponding vector via the representation theorem for linear forms and one writes $Df(\boldsymbol{x}) \cdot \boldsymbol{u} = \partial_{\boldsymbol{x}} f(\boldsymbol{x}) \cdot \boldsymbol{u}$ for $Df(\boldsymbol{x})[\boldsymbol{u}]$.

These conventions are illustrated on the following examples:

$$\tfrac{1}{2}\partial_{\boldsymbol{u}}\boldsymbol{u}^2[\boldsymbol{v}] = \boldsymbol{u} \cdot \boldsymbol{v}, \quad \boldsymbol{u}, \boldsymbol{v} \in \mathrm{Vect}; \text{ hence } \tfrac{1}{2}\partial_{\boldsymbol{u}}\boldsymbol{u}^2 = \boldsymbol{u},$$

by the representation theorem for linear forms on Vect. The derivative of a tensor-valued function of a tensor argument at a given point is a linear transformation from Lin to Lin. For instance, for every $\boldsymbol{A}, \boldsymbol{B} \in \mathrm{Lin}$,

$$\partial_{\boldsymbol{A}} \boldsymbol{A}[\boldsymbol{B}] = \boldsymbol{B}, \quad \partial_{\boldsymbol{A}} (\boldsymbol{A}^{\mathrm{T}}\boldsymbol{A})[\boldsymbol{B}] = \boldsymbol{B}^{\mathrm{T}}\boldsymbol{A} + \boldsymbol{A}^{\mathrm{T}}\boldsymbol{B}, \tag{1.1.2}$$

$$\partial_{\boldsymbol{A}} \boldsymbol{A}^{\mathrm{T}}[\boldsymbol{B}] = \boldsymbol{B}^{\mathrm{T}}, \quad \partial_{\boldsymbol{A}} \boldsymbol{A}^{m}[\boldsymbol{B}] = \sum_{k=0}^{m-1} \boldsymbol{A}^{k}\boldsymbol{B}\boldsymbol{A}^{m-k-1}, \quad (m \text{ a positive integer}), \tag{1.1.3}$$

$$\partial_{\boldsymbol{F}} \boldsymbol{F}^{-1}[\boldsymbol{B}] = -\boldsymbol{F}^{-1}\boldsymbol{B}\boldsymbol{F}^{-1}, \quad (\boldsymbol{F} \in \mathrm{InvLin}). \tag{1.1.4}$$

If G is an open subset of X and r a nonnegative integer or ∞, we denote by $C^r(G, \mathrm{Y})$ the set of all r times continuously differentiable mappings f from G to Y (infinitely differentiable if $r = \infty$). If the range is clear from the context, we abbreviate and write $C^r(G)$ for $C^r(G, \mathrm{Y})$; similar convention is used also for other spaces of mappings. We say that f is of class C^r if $f \in C^r(G)$. If F is *any* subset of X, we say that a mapping $f : F \to \mathrm{Y}$ is strongly r times continuously differentiable (and write $f \in C_s^r(F, \mathrm{Y})$) if f has an extension $\tilde{f}$ to an open set $\tilde{F}$ containing $\mathrm{cl}\, F$ such that $\tilde{f} \in C^r(\tilde{F}, \mathrm{Y})$. Clearly, if F is open, then $C_s^r(F) \subset C^r(F)$, but the inclusion may be strict: the mappings from $C_s^r(F)$ are locally bounded at each point of the boundary of F. If $\mathrm{cl}\, F = \mathrm{cl}\,\mathrm{int}\, F$ and $f \in C_s^r(F, \mathrm{Y})$, then the values of all partial derivatives of $\tilde{f}$ of orders $m \le r$ on $\mathrm{cl}\, F$ are uniquely determined by f, and we denote them unambiguously by $D^m f$. Indeed, if $f \in C_s^r(F, \mathrm{Y})$, and $\tilde{f}$ is a class C^r extension of f to an open set $\tilde{F}$ containing $\mathrm{cl}\, F$, then the derivatives of $\tilde{f}$ coincide with those of f on $\mathrm{int}\, F$, and as the derivatives of $\tilde{f}$ of orders $\le r$ are continuous, their values on $\mathrm{cl}\,\mathrm{int}\, F$ are the limits of the values of the derivatives of f on $\mathrm{int}\, F$. Note that the last is not true when the condition $\mathrm{cl}\, F = \mathrm{cl}\,\mathrm{int}\, F$ is violated: if, e.g., F consists of a finite number of isolated points in X, then any function $f : F \to \mathbb{R}$ is in $C_s^r(F)$, but the first derivatives of the extension $\tilde{f}$ at the points of F are arbitrary.

1.1.4 Determinant and Cofactor The reader is referred to HALMOS [1958, Sect. 53] for a coordinate-free definition of a determinant $\det \boldsymbol{A}$ of a second-order tensor $\boldsymbol{A} \in \mathrm{Lin}$. Let $\boldsymbol{A}$ be represented by the matrix $A = [A_{ij}]$ and let $\varepsilon_{i_1 \ldots i_n}$, $1 \le i_k \le n$, be the n-dimensional permutation symbol, i.e., $\varepsilon_{i_1 \ldots i_n} = 1(-1)$ if $(i_1, \ldots i_n)$ is an even (odd) permutation of $(1, \ldots, n)$ and $\varepsilon_{i_1 \ldots i_n} = 0$ otherwise. Then

$$\det \boldsymbol{A} = \frac{1}{n!} \sum \varepsilon_{i_1 \ldots i_n} \varepsilon_{j_1 \ldots j_n} A_{i_1 j_1} \cdots A_{i_n j_n} \tag{1.1.5}$$

where the sum is taken over all possible choices of $i_1, \ldots, i_n, j_1, \ldots, j_n = 1, \ldots, n$.

The characteristic polynomial of $\boldsymbol{A} \in \mathrm{Lin}$ is $p(\lambda) := \det(\lambda \mathbf{1} - \boldsymbol{A})$, $\lambda \in \mathbb{R}$; the principal invariants $I_k(\boldsymbol{A})$, $k = 0, 1, \ldots, n$, of $\boldsymbol{A}$ are defined as the coefficients of the characteristic polynomial:

$$\det(\lambda \mathbf{1} - \boldsymbol{A}) = \sum_{k=0}^{n} (-1)^k I_k(\boldsymbol{A}) \lambda^{n-k}, \quad \lambda \in \mathbb{R};$$

necessarily $I_0(\boldsymbol{A}) \equiv 1$, $I_n(\boldsymbol{A}) = \det \boldsymbol{A}$. If $\boldsymbol{F}$ is invertible and $\boldsymbol{a}, \boldsymbol{b} \in \mathrm{Vect}$, then

$$\det(\boldsymbol{F} + \boldsymbol{a} \otimes \boldsymbol{b}) = \det \boldsymbol{F} (1 + \boldsymbol{F}^{-1}\boldsymbol{a} \cdot \boldsymbol{b}). \tag{1.1.6}$$

The ***trace*** of $\boldsymbol{A} \in \mathrm{Lin}$ is $\mathrm{tr}\, \boldsymbol{A} := I_1(\boldsymbol{A})$; it is a linear form on Lin, the sum of the diagonal elements of $\boldsymbol{A}$ in any basis. The ***inner product on*** Lin is defined by $\boldsymbol{A} \cdot \boldsymbol{B} = \mathrm{tr}(\boldsymbol{A}\boldsymbol{B}^{\mathrm{T}})$ for any $\boldsymbol{A}, \boldsymbol{B} \in \mathrm{Lin}$; the length of $\boldsymbol{A}$ is $|\boldsymbol{A}| := (\boldsymbol{A} \cdot \boldsymbol{A})^{1/2}$. In any orthonormal basis,

$A \cdot B = \sum_{i,j=1}^{n} A_{ij} B_{ij}$. We shall use the following easily verifiable formulas without any reference:

$$A \cdot BC = AC^{\mathrm{T}} \cdot B = B^{\mathrm{T}} A \cdot C, \quad A, B, C \in \mathrm{Lin}.$$

Note also that $I_2(A) = \frac{1}{2}[(\mathrm{tr}\,A)^2 - \mathrm{tr}(A^2)]$.

The determinant and the principal invariants are infinitely differentiable scalar-valued functions of a tensor argument; their derivatives are given by

$$\partial_A I_k(A) = \sum_{j=0}^{k-1} (-A^{\mathrm{T}})^j I_{k-j-1}(A), \quad k = 1, \ldots, n, \quad A \in \mathrm{Lin}. \tag{1.1.7}$$

If A is invertible, then

$$\partial_A \det A = (\det A) A^{-\mathrm{T}}, \tag{1.1.8}$$

where $A^{-\mathrm{T}} = (A^{\mathrm{T}})^{-1} = (A^{-1})^{\mathrm{T}}$. (See TRUESDELL & NOLL [1965].) In particular,

$$\partial_A I_1(A) = \mathbf{1}, \quad \partial_A I_2(A) = \mathrm{tr}(A)\mathbf{1} - A^{\mathrm{T}}, \quad \partial_A I_n(A) = (\det A) A^{-\mathrm{T}} \tag{1.1.9}$$

for every $A \in \mathrm{Lin}$ in the first two formulas and for every $A \in \mathrm{InvLin}$ in the third formula. If $A, B, C \in \mathrm{Lin}$ then

$$\begin{aligned} \partial_A^2 I_1(A)[B, C] &= 0, \\ \partial_A^2 I_2(A)[B, C] &= \mathrm{tr}\,B\,\mathrm{tr}\,C - \mathrm{tr}(BC), \\ \partial_A^2 I_n(A)[B, C] &= (\det A)(\,\mathrm{tr}(A^{-1}B)\,\mathrm{tr}(A^{-1}C) - \mathrm{tr}(A^{-1}BA^{-1}C)) \end{aligned} \tag{1.1.10}$$

with $A \in \mathrm{InvLin}$ in the third formula.

The ***(tensor of) cofactor(s)*** of $A \in \mathrm{Lin}$ is defined by $\mathrm{cof}\,A := \partial_A \det A$. It is important in relation to changes of areas (see Subsect. 2.2.4), the definition of polyconvexity (see Sect. 17.5) and the concept of null Lagrangians and conservation laws (see Sects. 13.5, 13.6).

1.1.5 Proposition *Let $n \geq 2$. If $A \in \mathrm{Lin}$ then*

$$\mathrm{cof}\,A = \begin{cases} (\det A) A^{-\mathrm{T}} & \text{if } r(A) = n, \\ s_1 \cdots s_{n-1} Re \otimes e & \text{if } r(A) = n - 1, \\ \mathbf{0} & \text{if } r(A) \leq n - 2, \end{cases} \tag{1.1.11}$$

where $\{s_1, \ldots, s_{n-1}\}$ denotes the set of all positive eigenvalues of $(A^{\mathrm{T}}A)^{1/2}$ and $e \in$ Vect, $R \in$ Orth$^+$ are such that $Ae = \mathbf{0}$, $|e| = 1$, and $A = R(A^{\mathrm{T}}A)^{1/2}$. In particular, $\mathrm{cof}\,A = \mathbf{0} \Leftrightarrow r(A) \leq n - 2$.

See Sect. 1.2 for the definition of the square root of a positive semidefinite tensor. One of the eigenvalues of $(A^{\mathrm{T}}A)^{1/2}$ is 0 since $r(A) = n - 1$, and $\{s_1, \ldots, s_{n-1}\}$ denote the remaining eigenvalues. The existence of R, e with the above properties is part of the assertion, as well as the independence of $(1.1.11)_2$ of their choice.

Proof $(1.1.11)_1$ follows from (1.1.8). To prove $(1.1.11)_2$, let $r(A) = n - 1$ and e_i, $i = 1, \ldots, n$ be an orthonormal basis of eigenvectors of $(A^{\mathrm{T}}A)^{1/2}$, corresponding to s_i, $i = 1, \ldots, n - 1$ and $s_n = 0$, respectively. Set $e = e_n$. Let $f \in$ Vect be such that $|f| = 1$ and f is orthogonal to the range of A. Define, for each $\varepsilon > 0$, the tensor $A_\varepsilon = A + \varepsilon f \otimes e$. Then $r(A_\varepsilon) = n$,

$$(A_\varepsilon^{\mathrm{T}} A_\varepsilon)^{1/2} = (A^{\mathrm{T}}A)^{1/2} + \varepsilon e \otimes e, \quad (A_\varepsilon^{\mathrm{T}} A_\varepsilon)^{-1/2} = (A^{\mathrm{T}}A)^{-1/2} + \varepsilon^{-1} e \otimes e, \quad \det A_\varepsilon = s_1 \cdots s_{n-1}\varepsilon. \tag{1.1.12}$$

By the polar decomposition theorem 1.4.3, there exists an orthogonal tensor $\boldsymbol{R}$ such that $\boldsymbol{A}_\varepsilon = \boldsymbol{R}(\boldsymbol{A}_\varepsilon^{\mathsf{T}}\boldsymbol{A}_\varepsilon)^{1/2}$, the tensor $\boldsymbol{R}$ is easily seen to be independent of ε. Replacing $\boldsymbol{f}$ by $-\boldsymbol{f}$ if necessary, we may assume that $\boldsymbol{R}$ is a proper orthogonal tensor. Applying $(1.1.11)_1$ to $\boldsymbol{A}_\varepsilon = \boldsymbol{R}(\boldsymbol{A}_\varepsilon^{\mathsf{T}}\boldsymbol{A}_\varepsilon)^{1/2}$ and using $(1.1.12)_{2,3}$, we obtain

$$\operatorname{cof}\boldsymbol{A}_\varepsilon = s_1 \cdots s_{n-1}\varepsilon\boldsymbol{R}\big((\boldsymbol{A}^{\mathsf{T}}\boldsymbol{A})^{-1/2} + \varepsilon^{-1}\boldsymbol{e}\otimes\boldsymbol{e}\big) = s_1 \cdots s_{n-1}\varepsilon\boldsymbol{R}(\boldsymbol{A}^{\mathsf{T}}\boldsymbol{A})^{-1/2} + s_1 \cdots s_{n-1}\boldsymbol{R}\boldsymbol{e}\otimes\boldsymbol{e}.$$

Letting $\varepsilon \to 0$ and using the continuity of cof, we obtain $(1.1.11)_2$. Finally, $(1.1.11)_3$ is obtained from $(1.1.11)_2$ by letting one of the numbers s_i tend to 0. □

We have $\operatorname{cof}(\boldsymbol{AB}) = \operatorname{cof}\boldsymbol{A}\operatorname{cof}\boldsymbol{B}$, $\operatorname{cof}(\lambda\boldsymbol{A}) = \lambda^{n-1}\operatorname{cof}\boldsymbol{A}$. The first formula follows from $(1.1.11)_1$ for $\boldsymbol{A}, \boldsymbol{B}$ invertible, and it is extended by continuity to all tensors $\boldsymbol{A}, \boldsymbol{B}$. Also

$$\boldsymbol{A}^{\mathsf{T}}\operatorname{cof}\boldsymbol{A} = (\operatorname{cof}\boldsymbol{A})\boldsymbol{A}^{\mathsf{T}} = (\det\boldsymbol{A})\mathbf{1} \tag{1.1.13}$$

for any $\boldsymbol{A} \in \mathrm{Lin}$. Alternatively, one can define the cofactor as the unique continuous function from Lin to Lin which satisfies (1.1.13) for every $\boldsymbol{A} \in \mathrm{Lin}$.

If $\boldsymbol{A}$ is represented by $A = [A_{ij}]$ then $\operatorname{cof}\boldsymbol{A}$ is represented by $C = [C_{ij}]$ where $C_{ij} := (-1)^{i+j}\det M_{ij}$ and M_{ij} is the $(n-1)\times(n-1)$ matrix obtained from A by deleting the i-th row and j-th column. This follows from (1.1.5) by $(\operatorname{cof}\boldsymbol{A})_{ij} = \partial_{A_{ij}}\det A$. Equivalently,

$$(\operatorname{cof}\boldsymbol{A})_{i_1 j_1} = \frac{1}{(n-1)!}\sum \varepsilon_{i_1\ldots i_n}\varepsilon_{j_1\ldots j_n}A_{i_2 j_2}\cdots A_{i_n j_n} \tag{1.1.14}$$

where i_1, j_1 are fixed and the sum is taken over all choices of $i_2, \ldots, i_n, j_2, \ldots, j_n = 1, \ldots, n$. The formulas (1.1.5), (1.1.14) are but the first two members of a chain defining

$$M_{i_1\ldots i_p j_1\ldots j_p} = \frac{1}{(n-p)!}\sum \varepsilon_{i_1\ldots i_n}\varepsilon_{j_1\ldots j_n}A_{i_{p+1}j_{p+1}}\cdots A_{i_n j_n} \tag{1.1.15}$$

where the sum is taken over all choices of $i_{p+1}, \ldots, i_n, j_{p+1}, \ldots, j_n = 1, \ldots, n$. Each $M_{i_1\ldots i_p j_1\ldots j_p}$ is either 0 or a multiple by ± 1 of some minor of A of order $n-p$. A coordinate-free definition of the determinant and cofactor can be given using the exterior algebra: see, e.g., BISHOP & GOLDBERG [1968, Sect. 2.18, 2.19] (although no names are attached to the objects corresponding to the cofactors there). With the definition of the cofactor via the exterior algebra, the relation $\operatorname{cof}\boldsymbol{A} = \partial_{\boldsymbol{A}}\det\boldsymbol{A}$ would become a derived result. For our purposes, of more interest is the following characterization of the linear combinations of minors:

1.1.6 Proposition *Let k be a positive integer and f : Lin $\to \mathbb{R}$ a function of the form $f(\boldsymbol{A}) = \omega(\boldsymbol{A}, \ldots, \boldsymbol{A})$, where ω is a completely symmetric k-form on* Lin *such that the function*

$$(\boldsymbol{a}_1, \ldots, \boldsymbol{a}_k, \boldsymbol{b}_1, \ldots, \boldsymbol{b}_k) \mapsto \omega(\boldsymbol{a}_1\otimes\boldsymbol{b}_1, \ldots, \boldsymbol{a}_k\otimes\boldsymbol{b}_k), \qquad (\boldsymbol{a}_i, \boldsymbol{b}_j \in \mathrm{Vect})$$

is completely skew in $(\boldsymbol{a}_1, \ldots, \boldsymbol{a}_k)$ for every fixed $(\boldsymbol{b}_1, \ldots, \boldsymbol{b}_k)$ and completely skew in $(\boldsymbol{b}_1, \ldots, \boldsymbol{b}_k)$ for every fixed $(\boldsymbol{a}_1, \ldots, \boldsymbol{a}_k)$. The following assertions hold:
(1) *if $k > n$, then $f(\boldsymbol{A}) = 0$ identically;*
(2) *if $1 \le k \le n$, then in every basis, $f(\boldsymbol{A})$ is a linear combination of minors of $\boldsymbol{A}$ of order k.*
In particular, if $k = n$, then $f(\boldsymbol{A}) = c\det\boldsymbol{A}$ for some $c \in \mathbb{R}$ and all $\boldsymbol{A} \in$ Lin; *if $k = n-1$, then $f(\boldsymbol{A}) = \boldsymbol{B}\cdot\operatorname{cof}\boldsymbol{A}$ for some $\boldsymbol{B} \in$* Lin *and all $\boldsymbol{A} \in$* Lin; *if $k = 1$, then $f(\boldsymbol{A}) = \boldsymbol{B}\cdot\boldsymbol{A}$ for some $\boldsymbol{B} \in$* Lin *and all $\boldsymbol{A} \in$* Lin.

(A completely skew function is one which changes the sign if two arguments are interchanged.) This proposition, whose proof is omitted, will be used to derive the form of the null Lagrangians in Sect. 13.5. Also the following proposition is easy to prove:

1.1.7 Proposition *Let $r < n$ be a nonnegative integer. A tensor $\boldsymbol{A} \in$* Lin *is of rank $\le r$ if and only if all minors of $\boldsymbol{A}$ of order $r+1$ vanish. In particular, if $n = 3$ then $\boldsymbol{A}$ is of rank ≤ 1 if and only if* $\operatorname{cof}\boldsymbol{A} = \mathbf{0}$ *and if $n = 2$ then $\boldsymbol{A}$ is of rank ≤ 1 if and only if* $\det\boldsymbol{A} = 0$.

From $(1.1.10)_3$ one finds that for every $\boldsymbol{A} \in \mathrm{InvLin}$, $\boldsymbol{B} \in \mathrm{Lin}$,

$$\partial_{\boldsymbol{A}}\operatorname{cof}\boldsymbol{A}[\boldsymbol{B}] = \operatorname{cof}\boldsymbol{A}(\operatorname{tr}(\boldsymbol{B}\boldsymbol{A}^{-1})\mathbf{1} - \boldsymbol{B}^{\mathsf{T}}\boldsymbol{A}^{-\mathsf{T}}) = (\operatorname{tr}(\boldsymbol{B}\boldsymbol{A}^{-1})\mathbf{1} - \boldsymbol{A}^{-\mathsf{T}}\boldsymbol{B}^{\mathsf{T}})\operatorname{cof}\boldsymbol{A}. \tag{1.1.16}$$

If F is an invertible tensor and $a, b \in \mathrm{Vect}$, then

$$\operatorname{cof}(F + a \otimes b) = \operatorname{cof} F((1 + F^{-1}a \cdot b)\mathbf{1} - b \otimes F^{-1}a). \tag{1.1.17}$$

This follows from (1.1.1), (1.1.6) and $(1.1.11)_1$ if $1 + (F^{-1}a \cdot b) \neq 0$ and is extended by continuity to the general case. Consequently,

$$\operatorname{cof}(F + a \otimes b)b = (\operatorname{cof} F)b. \tag{1.1.18}$$

1.1.8 If $n = 3$, (1.1.5) gives

$$\det A = \sum_{i,j,k=1}^{n} \varepsilon_{ijk} A_{i1} A_{j2} A_{k3} = \tfrac{1}{6} \sum_{i,j,k,m,n,p=1}^{n} \varepsilon_{ijk}\varepsilon_{mnp} A_{im} A_{jn} A_{kp}. \tag{1.1.19}$$

For any $u, v, w \in \mathrm{Vect}$,

$$[Au, Av, Aw] = (\det A)[u, v, w], \tag{1.1.20}$$

with $[\cdot,\cdot,\cdot]$ introduced in Subsect. 1.1.1. Differentiation of (1.1.20) and rearrangements give

$$[Au, Av, w] = [u, v, (\operatorname{cof} A)^{\mathsf{T}} w], \qquad Au \times Av = (\operatorname{cof} A)(u \times v) \tag{1.1.21}$$

for every $u, v, w \in \mathrm{Vect}$. If A is invertible, a replacement $A \mapsto A^{-\mathsf{T}}$ leads to

$$A(u \times v) = (\det A) A^{-\mathsf{T}} u \times A^{-\mathsf{T}} v.$$

Furthermore,

$$I_2(A) = \operatorname{tr} \operatorname{cof} A, \qquad (\operatorname{cof} A)_{ij} = \tfrac{1}{2} \sum_{k,m,n,p=1}^{n} \varepsilon_{ikm}\varepsilon_{jnp} A_{kn} A_{mp}. \tag{1.1.22}$$

If $A = \operatorname{diag}(a_1, a_2, a_3)$ then $\operatorname{cof} A = \operatorname{diag}(a_2 a_3, a_3 a_1, a_1 a_2)$. This is $(1.1.11)_1$ when A is invertible; the general case is obtained by continuity.

1.2 Symmetric Tensors

For every unit vector $e \in \mathrm{Vect}$, the tensor $P := e \otimes e$ is a projection onto the line spanned by e. The spectral theorem for symmetric linear transformations on finite-dimensional spaces (Halmos [1958, Sect. 79]) therefore reads as follows.

1.2.1 Spectral Theorem for Symmetric Tensors *For every $A \in \mathrm{Sym}$ there exists an orthonormal basis $\{e_i\}$ and numbers $a_i \in \mathbb{R}$ such that*

$$A = \sum_{i=1}^{n} a_i e_i \otimes e_i. \tag{1.2.1}$$

The vectors e_i are the eigenvectors and a_i the corresponding eigenvalues of A. The set $\{a_i\}$ is the spectrum of A. We can assume that the numbers a_i as in (1.2.1) are ordered nonincreasingly, i.e., $a_1 \geq a_2 \geq \ldots \geq a_n$. If this is the case, then they are uniquely determined and we write $\hat{a}_i(A)$ for them; hence $\hat{a}_i : \mathrm{Sym} \to \mathbb{R}$. If $\hat{a}_i(A)$ are all distinct, then e_i are determined uniquely to within the change of signs, and the projections $P_i(A) := e_i \otimes e_i$ onto the eigenspaces of A are uniquely determined. Let us denote by $\mathrm{Sym}^\sharp$ the set of all $A \in \mathrm{Sym}$ for which $\hat{a}_i(A)$ are all distinct. This in an open dense subset of Sym (see below).

We shall now examine the differentiability of $\hat{a}_i$, $\boldsymbol{P}_i$ on $\mathrm{Sym}^\sharp$. This will be based on the proposition on the perturbation of a simple eigenvalue, below, to be used several times (e.g., to deduce the differentiability of the characteristic vectors and the differentiability of Hugoniot's curves). Let X, Y be finite-dimensional real vector spaces and $\boldsymbol{M}$ a function of class C^r, $r \geq 1$, on an open subset O of Y, with values in the set Lin(X) of all linear transformations from X to X. The proposition shows that locally any eigenvector of $\boldsymbol{M}(\boldsymbol{y})$ corresponding to a simple eigenvalue can be chosen to be an r-times continuously differentiable function of $\boldsymbol{y}$. Recall that an eigenvalue of a linear transformation $\boldsymbol{T}$ is said to be simple if it is a simple root (= root of multiplicity 1) of the characteristic polynomial of $\boldsymbol{T}$.

1.2.2 Proposition *Let X, Y, O, and $\boldsymbol{M}$ be as above. If $\boldsymbol{x} \in$ O and $\boldsymbol{e}$ is an eigenvector of $\boldsymbol{M}(\boldsymbol{x})$ with a simple eigenvalue $\lambda \in \mathbb{R}$, then the following assertions hold:*

(1) *there exist functions $\tilde{\lambda}$, $\tilde{\boldsymbol{e}}$, of class C^r, defined on a neighborhood O′ of $\boldsymbol{x}$, with values in $\mathbb{R}$ and X, respectively, such that $\tilde{\lambda}(\boldsymbol{x}) = \lambda$, $\tilde{\boldsymbol{e}}(\boldsymbol{x}) = \boldsymbol{e}$ and for every $\boldsymbol{y} \in$ O′, $\tilde{\boldsymbol{e}}(\boldsymbol{y})$ is an eigenvector of $\boldsymbol{M}(\boldsymbol{y})$ with the simple eigenvalue $\tilde{\lambda}(\boldsymbol{y})$; moreover, if O′ is chosen small enough, there exists an $\varepsilon > 0$ such that $\tilde{\lambda}(\boldsymbol{y})$ is the only eigenvalue of $\boldsymbol{M}(\boldsymbol{y})$ in $(\lambda - \varepsilon, \lambda + \varepsilon)$;*

(2) *if $\boldsymbol{M}(\boldsymbol{x}) = \boldsymbol{M}(\boldsymbol{x})^{\mathrm{T}}$ and $\tilde{\boldsymbol{e}}$ is normalized, i.e., $|\tilde{\boldsymbol{e}}(\boldsymbol{y})| = 1$, $\boldsymbol{y} \in$ O′, then for every $\boldsymbol{a} \in$ Y, $\partial_{\boldsymbol{x}}\tilde{\boldsymbol{e}}(\boldsymbol{x})[\boldsymbol{a}]$ is in the orthogonal complement $\boldsymbol{e}^\perp$ of $\boldsymbol{e}$, $\boldsymbol{M}(\boldsymbol{x}) - \lambda\boldsymbol{1}$ maps $\boldsymbol{e}^\perp$ onto $\boldsymbol{e}^\perp$ and*

$$\partial_{\boldsymbol{x}}\tilde{\lambda}(\boldsymbol{x})[\boldsymbol{a}] = \partial_{\boldsymbol{x}}\boldsymbol{M}(\boldsymbol{x})[\boldsymbol{a}]\boldsymbol{e} \cdot \boldsymbol{e}, \quad \partial_{\boldsymbol{x}}\tilde{\boldsymbol{e}}(\boldsymbol{x})[\boldsymbol{a}] = (\boldsymbol{M}(\boldsymbol{x}) - \lambda\boldsymbol{1})^{-1}(\boldsymbol{e} \otimes \boldsymbol{e} - \boldsymbol{1})\partial_{\boldsymbol{x}}\boldsymbol{M}(\boldsymbol{x})[\boldsymbol{a}]\boldsymbol{e}, \tag{1.2.2}$$

where $(\boldsymbol{M}(\boldsymbol{x}) - \lambda\boldsymbol{1})^{-1} : \boldsymbol{e}^\perp \to \boldsymbol{e}^\perp$ denotes the inverse of the restriction of $\boldsymbol{M}(\boldsymbol{x}) - \lambda\boldsymbol{1}$ to $\boldsymbol{e}^\perp$.

Proof (1) (cf. HANYGA [1985, p. 206]) Let $p(\boldsymbol{y}, t) = \det(\boldsymbol{M}(\boldsymbol{y}) - t\boldsymbol{1})$. Since λ is a simple root of $p(\boldsymbol{x}, \cdot)$, the derivative of the latter with respect to t satisfies $p'(\boldsymbol{x}, \lambda) \neq 0$. By the implicit function theorem (e.g., DIEUDONNÉ [1960, 10.2.1]) there exists a neighborhood O′ of $\boldsymbol{x}$ in O and a function $\tilde{\lambda}$: O′ $\to \mathbb{R}$ of class C^r such that $p(\boldsymbol{y}, \tilde{\lambda}(\boldsymbol{y})) = 0$ for every $\boldsymbol{y} \in$ O′. Moreover, there exists an $\varepsilon > 0$ such that $\tilde{\lambda}(\boldsymbol{y})$ is the only solution of $p(\boldsymbol{y}, t) = 0$ in the interval $(\lambda - \varepsilon, \lambda + \varepsilon)$. Using the definition of the cofactor and the chain rule, it is found that

$$p'(\boldsymbol{y}, t) = -\operatorname{cof}(\boldsymbol{M}(\boldsymbol{y}) - t\boldsymbol{1}) \cdot \boldsymbol{1} = -\operatorname{tr}(\operatorname{cof}(\boldsymbol{M}(\boldsymbol{y}) - t\boldsymbol{1})).$$

By $p'(\boldsymbol{x}, \lambda) \neq 0$ we have $\operatorname{tr}(\operatorname{cof}(\boldsymbol{M}(\boldsymbol{x}) - \lambda\boldsymbol{1})) = \operatorname{tr}(\operatorname{cof}(\boldsymbol{M}^{\mathrm{T}}(\boldsymbol{x}) - \lambda\boldsymbol{1})) \neq 0$ and hence $\operatorname{cof}(\boldsymbol{M}^{\mathrm{T}}(\boldsymbol{x}) - \lambda\boldsymbol{1})\boldsymbol{w} \neq \boldsymbol{0}$ for some $\boldsymbol{w} \in$ X. By continuity then $\tilde{\boldsymbol{e}}(\boldsymbol{y}) := \operatorname{cof}(\boldsymbol{M}^{\mathrm{T}}(\boldsymbol{y}) - \tilde{\lambda}(\boldsymbol{y})\boldsymbol{1})\boldsymbol{w} \neq \boldsymbol{0}$ for all $\boldsymbol{y}$ sufficiently close to $\boldsymbol{x}$. Applying (1.1.13) to $\boldsymbol{A} = \boldsymbol{M}^{\mathrm{T}}(\boldsymbol{y}) - \tilde{\lambda}(\boldsymbol{y})\boldsymbol{1}$, using $\det(\boldsymbol{M}^{\mathrm{T}}(\boldsymbol{y}) - \tilde{\lambda}(\boldsymbol{y})\boldsymbol{1}) = 0$, and multiplying by $\boldsymbol{w}$ we obtain $(\boldsymbol{M}(\boldsymbol{y}) - \tilde{\lambda}(\boldsymbol{y})\boldsymbol{1})\tilde{\boldsymbol{e}}(\boldsymbol{y}) = \boldsymbol{0}$ which shows that $\tilde{\boldsymbol{e}}(\boldsymbol{y})$ is an eigenvector of $\boldsymbol{M}(\boldsymbol{y})$ with the eigenvalue $\tilde{\lambda}(\boldsymbol{y})$. In particular, as $\boldsymbol{e}$ is a simple eigenvector of $\boldsymbol{M}(\boldsymbol{x})$ corresponding to the eigenvalue $\lambda = \tilde{\lambda}(\boldsymbol{x})$, we see that $\tilde{\boldsymbol{e}}(\boldsymbol{x})$ is a multiple of $\boldsymbol{e}$. We can choose a multiple of $\boldsymbol{w}$ such that $\boldsymbol{e} = \tilde{\boldsymbol{e}}(\boldsymbol{x})$. From the definition we see that $\tilde{\boldsymbol{e}}$ is of class C^r. (2): Let $\boldsymbol{a} \in$ Y and denote $\lambda' = \partial_{\boldsymbol{x}}\tilde{\lambda}(\boldsymbol{x})[\boldsymbol{a}]$, $\boldsymbol{M}' = \partial_{\boldsymbol{x}}\boldsymbol{M}(\boldsymbol{x})[\boldsymbol{a}]$, $\boldsymbol{e}' = \partial_{\boldsymbol{x}}\tilde{\boldsymbol{e}}(\boldsymbol{x})[\boldsymbol{a}]$. The normalization implies $\boldsymbol{e}' \cdot \boldsymbol{e} = 0$ i.e., $\boldsymbol{e}'$ is in $\boldsymbol{e}^\perp$. That $\boldsymbol{M}(\boldsymbol{x}) - \lambda\boldsymbol{1}$ maps bijectively $\boldsymbol{e}^\perp$ onto $\boldsymbol{e}^\perp$ follows from the symmetry of $\boldsymbol{M}(\boldsymbol{x})$ and the fact that $\boldsymbol{e}$ is a simple eigenvalue of $\boldsymbol{M}(\boldsymbol{x})$. Differentiating $(\boldsymbol{M}(\boldsymbol{y}) - \tilde{\lambda}(\boldsymbol{y}))\tilde{\boldsymbol{e}}(\boldsymbol{y}) = \boldsymbol{0}$ with respect to $\boldsymbol{y}$ at $\boldsymbol{y} = \boldsymbol{x}$ we obtain

$$(\boldsymbol{M}' - \lambda'\boldsymbol{1})\boldsymbol{e} + (\boldsymbol{M} - \lambda\boldsymbol{1})\boldsymbol{e}' = \boldsymbol{0}. \tag{1.2.3}$$

Forming the inner product with $\boldsymbol{e}$, using the normalization and $(\boldsymbol{M} - \lambda\boldsymbol{1})\boldsymbol{e}' \cdot \boldsymbol{e} = \boldsymbol{e}' \cdot (\boldsymbol{M} - \lambda\boldsymbol{1})\boldsymbol{e} = 0$ we obtain $(1.2.2)_1$. Moreover, (1.2.3) may be rewritten as $(\boldsymbol{1} - \boldsymbol{e} \otimes \boldsymbol{e})\boldsymbol{M}'\boldsymbol{e} + (\boldsymbol{M} - \lambda\boldsymbol{1})\boldsymbol{e}' = \boldsymbol{0}$, which implies $(1.2.2)_2$. □

1.2.3 Proposition *The functions $\hat{a}_i$ are Lipschitz continuous on* Sym. *The set* $\mathrm{Sym}^\sharp$ *is open and dense in* Sym *and $\hat{a}_i$, $\boldsymbol{P}_i$ are infinitely differentiable on it. Moreover, for every $\boldsymbol{A} \in \mathrm{Sym}^\sharp$ there is a neighborhood N on which the eigenvectors of any $\boldsymbol{B} \in N$ can be chosen to be infinitely differentiable functions of $\boldsymbol{B}$.*

Proof The Lipschitz continuity of $\hat{a}_i$ on Sym follows from the inequality (see HORN & JOHNSON [1986, Exercise in Sect. 6.3])

$$\sum_{i=1}^{n} (\hat{a}_i(\boldsymbol{A}) - \hat{a}_i(\boldsymbol{B}))^2 \leq |\boldsymbol{A} - \boldsymbol{B}|^2, \quad \boldsymbol{A}, \boldsymbol{B} \in \mathrm{Sym}.$$

To prove that $\mathrm{Sym}^\sharp$ is dense in Sym, we fix $\boldsymbol{A} \in$ Sym and use the spectral decomposition (1.2.1). In the basis $\{\boldsymbol{e}_i\}$, $\boldsymbol{A}$ is represented by a diagonal matrix with the elements a, and it suffices to note that this matrix is a limit of a sequence of diagonal matrices with mutually distinct elements. To prove that $\mathrm{Sym}^\sharp$ is open in Sym and that $\hat{a}_i, \boldsymbol{P}_i$ are infinitely differentiable on $\mathrm{Sym}^\sharp$, it suffices to apply Proposition 1.2.2 to the infinitely differentiable function $\boldsymbol{A} \mapsto \boldsymbol{A}$ from $\mathrm{Sym}^\sharp$ to itself. □

1.2.4 For every $\boldsymbol{A} \in \mathrm{Sym}^\sharp$, $\boldsymbol{B}, \boldsymbol{C} \in \mathrm{Sym}$,

$$\partial_{\boldsymbol{A}} \hat{a}_i = \boldsymbol{P}_i, \quad \partial^2_{\boldsymbol{A}} \hat{a}_i[\boldsymbol{B}, \boldsymbol{C}] = 2 \sum \frac{B_{ij} C_{ij}}{a_i - a_j}, \tag{1.2.4}$$

$$\partial_{\boldsymbol{A}} \boldsymbol{e}_i[\boldsymbol{B}] = \sum \frac{B_{ij}}{a_i - a_j} \boldsymbol{e}_j, \quad \partial_{\boldsymbol{A}} \boldsymbol{P}_i[\boldsymbol{B}] = \sum \frac{B_{ij}}{a_i - a_j} (\boldsymbol{e}_j \otimes \boldsymbol{e}_i + \boldsymbol{e}_i \otimes \boldsymbol{e}_j), \tag{1.2.5}$$

where the argument $\boldsymbol{A}$ of the derivatives has been omitted, $\{\boldsymbol{e}_i\}$ is any orthonormal basis of eigenvectors of $\boldsymbol{A}$ corresponding to $a_i := \hat{a}_i(\boldsymbol{A})$ and B_{ij}, C_{ij} are the components of $\boldsymbol{B}, \boldsymbol{C}$ in this basis; the sums are taken over all j, $j \neq i$. If for every $\boldsymbol{A} \in \mathrm{Sym}^\sharp$, $\boldsymbol{B} \in \mathrm{Sym}$ one defines the skew-symmetric ***twirl tensor***

$$\boldsymbol{\Omega}(\boldsymbol{A}, \boldsymbol{B}) := \sum \frac{B_{ij}}{a_j - a_i} \boldsymbol{e}_i \otimes \boldsymbol{e}_j,$$

where the sum is taken over all i, j, $i \neq j$, then (1.2.5) may be rewritten as

$$\partial_{\boldsymbol{A}} \boldsymbol{e}_i[\boldsymbol{B}] = \boldsymbol{\Omega}(\boldsymbol{A}, \boldsymbol{B}) \boldsymbol{e}_i, \quad \partial_{\boldsymbol{A}} \boldsymbol{P}_i[\boldsymbol{B}] = \boldsymbol{\Omega}(\boldsymbol{A}, \boldsymbol{B}) \boldsymbol{P}_i - \boldsymbol{P}_i \boldsymbol{\Omega}(\boldsymbol{A}, \boldsymbol{B}).$$

Equations (1.2.4) and (1.2.5) are direct consequences of the formulas (1.2.2) applied to the situation $\boldsymbol{x} \equiv \boldsymbol{A}$ and $\boldsymbol{M}(\boldsymbol{A}) = \boldsymbol{A}$. Alternatively, they are consequences of the independently proved results of Sect. 8.2 on the derivatives of general isotropic scalar- and tensor-valued functions, see Remark 8.2.4. The reader is referred to GUO & al. [1992] for a further discussion of the twirl tensors.

1.2.5 Functional Calculus (cf. HALMOS [1958, Sect. 82]) Let $f : D \to \mathbb{R}$ be a function on $D \subset \mathbb{R}$ and let $\boldsymbol{A} \in \mathrm{Sym}$ be a symmetric tensor with the spectral decomposition (1.2.1). If $\boldsymbol{A}$ has all its eigenvalues a_i in D, we define the value of f on $\boldsymbol{A}$ to be the symmetric tensor $f(\boldsymbol{A})$ given by

$$f(\boldsymbol{A}) = \sum_{i=1}^{n} f(a_i) \boldsymbol{e}_i \otimes \boldsymbol{e}_i.$$

This definition is independent of the possible arbitrariness of the spectral decomposition. Furthermore, denote by U_f the set of all $\boldsymbol{A} \in \mathrm{Sym}$ having all eigenvalues in D. The correspondence $(f, \boldsymbol{A}) \mapsto f(\boldsymbol{A})$ is called the ***functional calculus.*** It is easily verified that the set U_f and the function $\boldsymbol{A} \mapsto f(\boldsymbol{A})$ are isotropic (see Definition 8.1.1 below), i.e., $\boldsymbol{Q}\boldsymbol{A}\boldsymbol{Q}^{\mathrm{T}} \subset U_f$ and $f(\boldsymbol{Q}\boldsymbol{A}\boldsymbol{Q}^{\mathrm{T}}) = \boldsymbol{Q} f(\boldsymbol{A}) \boldsymbol{Q}^{\mathrm{T}}$ for any $\boldsymbol{A} \in U_f$, and $\boldsymbol{Q} \in \mathrm{Orth}$. This gives a powerful method for constructing isotropic functions. Some special cases are:

(1) For $f(x) = x^m$, $m = 0, 1, 2, \ldots$, $D = \mathbb{R}$, this definition leads to the usual definition of the powers of a tensor, with $U_f = \mathrm{Sym}$.

(2) For an arbitrary $\alpha \in \mathbb{R}$, let $f(x) = x^\alpha$ for $x \in D = \mathbb{R}^+ := [0, \infty)$. Then $U_f = \mathrm{Sym}^+$, and $f(\boldsymbol{A}) = \boldsymbol{A}^\alpha$ is a fractional power of a positive definite, symmetric tensor. One has

$$\boldsymbol{A}^\alpha \boldsymbol{A}^\beta = \boldsymbol{A}^{\alpha+\beta}, \quad (\boldsymbol{A}^\alpha)^\beta = \boldsymbol{A}^{\alpha\beta}. \tag{1.2.6}$$

(3) For $f(x) = e^x$, $x \in \mathbb{R}$, one obtains the exponential of a symmetric tensor, $D_f = \mathrm{Sym}$, and $f(\boldsymbol{A}) = e^{\boldsymbol{A}}$, with $e^{\boldsymbol{A}}$ given also by the sum of the corresponding power series. The latter definition is more general, though, as it also applies to every, not necessarily symmetric, tensor.

(4) For $f = \ln : \mathbb{R}^{++} := (0, \infty) \to \mathbb{R}$ one obtains the definition of a logarithm of a positive definite, symmetric tensor, $f(\boldsymbol{A}) = \ln \boldsymbol{A}$. One has

$$e^{\ln A} = A, \quad A \in \mathrm{Sym}^+, \qquad \ln e^B = B, \quad B \in \mathrm{Sym}.$$

The following three propositions deal with the the differentiation of the functional calculus. We denote by $\partial_A f(A)$ the derivative of the tensor-valued function $A \mapsto f(A)$ at $A \in U_f$ whenever it exists.

1.2.6 Proposition *Let $f : D \to \mathbb{R}$ be a function on $D \subset \mathbb{R}$. Then*

(1) *D is open if and only if U_f is open in* Sym;

(2) *the scalar-valued function f is of class C^r, $r = 0, \ldots, \infty$, on D if and only if the tensor-valued function $A \mapsto f(A)$ is of class C^r on U_f;*

(3) *if f is continuously differentiable then in the components in the basis of eigenvectors of A we have $\partial_A f(A)[B]_{ij} = H_{ij}(a)B_{ij}$ for every $B \in$* Sym, *where*

$$H_{ij}(a) = \begin{cases} \dfrac{f(a_i) - f(a_j)}{a_i - a_j} & \text{if } a_i \neq a_j, \\ f'(a_i) & \text{if } a_i = a_j \end{cases} \tag{1.2.7}$$

with a_i the eigenvalues of A. (f' is the derivative of f with respect to its scalar argument.)

The direct implication in Assertion (1) follows from the fact that the spectrum of a symmetric tensor A depends continuously on A, see Proposition 1.2.3, and the converse implication is immediate. The differentiability of the functional calculus and the formula for the derivative in Assertion (3) are known in matrix theory, see DONOGHUE [1974, Chap. VIII], and the elementwise multiplication of matrices $H_{ij}B_{ij}$ in the formula for the derivative is known as the Hadamard or Schur multiplication. Alternatively, see Proposition 8.1.9 and Remark 8.1.10(2) below for the independent proofs of (2) and (3).

Let us give an alternative view on the derivative of the functional calculus. For every $A \in$ Sym define the commutator of A by $\mathrm{C}(A) := \{B \in \mathrm{Sym} : AB = BA\}$; denote furthermore by $\mathrm{C}^{\perp}(A)$ its orthogonal complement in Sym and by $P(A)$ the orthogonal projection from Sym onto $\mathrm{C}(A)$. If A is given by (1.2.1) with $a_1 \geq a_2 \geq \ldots \geq a_n$, then denoting by $[B_{ij}]$ the matrix of a $B \in$ Sym in the basis of eigenvectors of A, we have

$$\mathrm{C}(A) = \{B \in \mathrm{Sym} : \text{ if } a_i \neq a_j \text{ for some } i,j \text{ then } B_{ij} = 0\},$$
$$\mathrm{C}^{\perp}(A) = \{B \in \mathrm{Sym} : \text{ if } a_i = a_j \text{ for some } i,j \text{ then } B_{ij} = 0\}.$$

Hence, if $B \in \mathrm{C}(A)$ then $[B_{ij}]$ is block-diagonal with each block extending over the maximal segment of i on which a_i is constant. Similarly, if $B \in \mathrm{C}^{\perp}(A)$ then $[B_{ij}]$ is off-block-diagonal, i.e., the elements of $[B_{ij}]$ on the main block-diagonal vanish. For a given $B \in$ Sym, the matrix of $P(A)B$ is obtained from that of B by replacing all off-block-diagonal elements by zeros; similarly the matrix of $(\mathbf{1} - P(A))B$ is obtained from that of B by replacing all block-diagonal elements by zeros. Here $\mathbf{1}$ is the identity linear transformation from Sym to Sym.

1.2.7 Proposition *If $f : D \to \mathbb{R}$ is continuously differentiable and $A \in D_f$, then*

(1) *$\partial_A f(A)[\cdot]$ is a symmetric linear transformation on* Sym *which leaves* $\mathrm{C}(A)$ *and* $\mathrm{C}^{\perp}(A)$ *invariant, i.e.,*

$$\partial_A f(A)[\mathrm{C}(A)] \subset \mathrm{C}(A), \quad \partial_A f(A)[\mathrm{C}^{\perp}(A)] \subset \mathrm{C}^{\perp}(A); \tag{1.2.8}$$

(2) *for every $C \in$* Sym *there exists a unique tensor $Q(f, A)[C] \in$* Sym *satisfying*

$$Q(f, A)[C] \in \mathrm{C}^{\perp}(A), \quad [A, Q(f, A)[C]] = [f(A), C]; \tag{1.2.9}$$

here $[D, E] := DE - ED$ for every $D, E \in$ Sym;

(3) *for every $B \in$* Sym,

$$\partial_A f(A)[B] = f'(A)P(A)B + Q(f, A)[(\mathbf{1} - P(A))B]$$

where $f'(A)$ is the value of the functional calculus for f' on A.

Hence $\partial_A f(A)[B]$ coincides with $f'(A)B$ on $C(A)$ and with $Q(f, A)[B]$ on $C^\perp(A)$. Alternatively, $\partial_A f(A)[\cdot]$ may be characterized as the unique linear transformation on Sym such that $\partial_A f(A)[B] = f'(A)B$ if $B \in C(A)$ and $\partial_A f(A)[B] \in C^\perp(A)$, $[A, \partial_A f(A)[B]] = [f(A), B]$ if $B \in C^\perp(A)$.

Proof We take the existence of $\partial_A f(A)$ for granted from Proposition 1.2.6. (1): Let $m : D_f \to \mathbb{R}$ be defined by $m(A) = \operatorname{tr} g(A)$, $A \in D_f$, where $g : D \to \mathbb{R}$ is any function satisfying $g' = f$. We have $m(A) = g(\hat{a}_1(A)) + \ldots + g(\hat{a}_n(A))$ and a differentiation using Subsect. 1.2.4 gives $\partial_A m(A) = f(A)$. Then $\partial_A f(A) = \partial_A^2 m(A)$ is symmetric by the interchangeability of the second partial derivatives. If $B \in C(A)$ then there exists a basis in which both A, B are represented by diagonal matrices diag a, diag b where a, b are n-tuples of eigenvalues. Then for every $\lambda \in \mathbb{R}$ sufficiently close to 0 $f(A+\lambda B)$ is represented by $\operatorname{diag}(f(a_1+\lambda b_1), \ldots, f(a_n+\lambda b_n))$ and a differentiation gives that $\partial_A f(A)[B]$ is represented by $\operatorname{diag}(f(a_1)b_1, \ldots, f(a_n)b_n)$, which is the matrix of $f'(A)B$. Clearly, $f'(A)B \in C(A)$, which proves $(1.2.8)_1$. Equation $(1.2.8)_2$ follows from $(1.2.8)_1$ and the fact that $\partial_A f(A)[\cdot]$ is a symmetric linear transformation on Sym. (2): Let A be given by (1.2.1) with $a_1 \ge a_2 \ge \ldots \ge a_n$, and define $Q(f, A)[C] \in$ Sym by

$$Q(f, A)[C]_{ij} = \begin{cases} \dfrac{f(a_i) - f(a_j)}{a_i - a_j} C_{ij} & \text{if } a_i \neq a_j, \\ 0 & \text{if } a_i = a_j. \end{cases}$$

Then $Q(f, A)[C]$ satisfies (1.2.9). To prove that $Q(f, A)[C]$ is unique, let $L, M \in$ Sym both satisfy (1.2.9) with $Q(f, A)[C]$ replaced by L, M respectively. From $(1.2.9)_2$ follows that $[A, L-M] = \mathbf{0}$, i.e., $L-M \in C(A)$. On the other hand, $L-M \in C^\perp(A)$ by $(1.2.9)_1$ and hence $L-M = \mathbf{0}$. (3): In view of the linearity of $\partial_A f(A)[\cdot]$, it suffices to prove the formula separately for $B \in C(A)$ and for $B \in C^\perp(A)$. The case $B \in C(A)$ has been proved in (1). By differentiating $Af(A) - f(A)A = \mathbf{0}$ in the direction $B \in$ Sym we obtain $[f(A), B] = [A, \partial_A f(A)[B]]$. If $B \in C^\perp(A)$ then $\partial_A f(A)[B] \in C^\perp(A)$ by $(1.2.8)_2$, and $[A, \partial_A f(A)[B]] = [f(A), B]$. A comparison with (2) gives $\partial_A f(A)[B] = Q(f, A)[B]$. □

Various expressions for the derivative of the functional calculus are given in CARLSON & HOGER [1986], SCHEIDLER [1991], WANG & DUAN [1991] and MAN & GUO [1993].

When f is real analytic, Cauchy's formula may be helpful in finding a component-independent formula for the derivative. In this case f has an analytic extension $\tilde{f}$ to an open subset $\tilde{D}$ of the complex plane $\mathbb{C}$ containing the eigenvalues of $A \in U_f$. Then, under the complexifications of the spaces, $f(A)$ is given by the resolvent formula

$$f(A) = \frac{1}{2\pi i} \int_C (z\mathbf{1} - A)^{-1} \tilde{f}(z)\, dz, \tag{1.2.10}$$

where C is any positively oriented simple curve in $\tilde{D}$ that encloses the eigenvalues of A. The differentiation under the integral sign in (1.2.10) is permissible and gives (see (1.1.4) or RUDIN [1973])

$$\partial_A f(A)[B] = \frac{1}{2\pi i} \int_C (z\mathbf{1} - A)^{-1} B (z\mathbf{1} - A)^{-1} \tilde{f}(z)\, dz, \quad B \in \text{Sym}.$$

In concrete cases the integration path can be chosen so as to simplify the expression and in particular as to eliminate the complex plane. This matter is not pursued further here.

1.2.8 Proposition *Let $\phi : D \to \mathbb{R}$ be a continuously differentiable monotonic function with $\phi' \neq 0$ everywhere on $D \subset \mathbb{R}$, let ψ be its inverse, and consider the functional calculi $\bar{Q} : U \mapsto \phi(U)$, $U \in D_\phi$, and $\bar{U} : Q \mapsto \psi(Q)$, $Q \in D_\psi$, on D_ϕ, $D_\psi \subset$ Sym specified in Subsect. 1.2.5. Then*

(1) *$\bar{Q}$ is a diffeomorphism from D_ϕ onto D_ψ and $\bar{U}$ is its inverse;*

(2) *if $U \in D_\phi$ and $Q = \phi(U)$, then in the basis of eigenvectors of U,*

$$\partial_U \bar{Q}(U)[B]_{ij} = H_{ij}(v) B_{ij}, \quad \partial_Q \bar{U}(Q)[B]_{ij} = K_{ij}(q) B_{ij}, \quad B \in \text{Sym},$$

where v_i, $q_i = \phi(v_i)$ are the eigenvalues of U, Q, respectively,

$$H_{ij}(v) \neq 0, \quad K_{ij}(q) \neq 0, \quad H_{ij} K_{ij}(q) = 1 \text{ for every } i, j$$

and

$$H_{ij}(v) = \begin{cases} \dfrac{\phi(v_i) - \phi(v_j)}{v_i - v_j} & \text{if } v_i \neq v_j, \\ \phi'(v_i) & \text{if } v_i = v_j, \end{cases} \qquad K_{ij}(q) = \begin{cases} \dfrac{\psi(q_i) - \psi(q_j)}{q_i - q_j} & \text{if } q_i \neq q_j, \\ \psi'(q_i) & \text{if } q_i = q_j. \end{cases}$$

Proof (1): We omit a simple proof that $\bar{Q}$ is one-to-one. That $\bar{U}, \bar{Q}$ are the inverses of one another follows directly from the definition of the functional calculus. Since under the above hypotheses ψ is continuously differentiable, Proposition 1.2.6 tells us that $\bar{Q}, \bar{U}$ are continuously differentiable. To check that the derivative of $\bar{Q}(U)$ is an invertible linear transformation, differentiate $\bar{Q}(\bar{U}(Q)) = Q$ to obtain $\partial_U \bar{Q}(U)[\partial_Q \bar{U}(Q)[B]] = B$ for every $B \in$ Sym. Hence the range of $\partial_U \bar{Q}(U)$ is Sym and $\partial_U \bar{Q}(U)$ is invertible. (2): The formulas for the derivatives follow from Proposition 1.2.6 and it is immediate to verify that H_{ij}, K_{ij} have all the claimed properties. □

As consequences of the above assertions, we have

1.2.9 Proposition *For every $\alpha \in \mathbb{R}$ the function $f : A \mapsto A^\alpha$ is infinitely differentiable on* Sym$^+$. *If $\alpha \neq 0$, then f maps* Sym$^+$ *bijectively onto* Sym$^+$ *and its inverse is $A \mapsto A^{1/\alpha}$. For $\alpha = \frac{1}{2}$ one writes $A^{\frac{1}{2}} = \sqrt{A}$ for the square root. One has*

$$\partial_A \sqrt{A}[B] = \int_0^\infty e^{-t\sqrt{A}} B e^{-t\sqrt{A}}\, dt, \tag{1.2.11}$$

for every $A \in$ Sym$^+$ and $B \in$ Sym. If A has the spectral decomposition (1.2.1) *with $a_i > 0$ then*

$$\partial_A \sqrt{A}[B]_{ij} = \frac{B_{ij}}{\sqrt{a_i} + \sqrt{a_j}}, \qquad 1 \le i, j \le n, \tag{1.2.12}$$

where B_{ij} are the components of B in the basis $\{e_i\}$.

Proof It suffices to prove only (1.2.11), (1.2.12) is a consequence. A differentiation of $\sqrt{A}\sqrt{A} = A$ in the direction $B \in$ Lin and the product rule give

$$\partial_A \sqrt{A}[B]\sqrt{A} + \sqrt{A}\partial_A \sqrt{A}[B] = B.$$

Denoting $X = \partial_A \sqrt{A}[B]$, $U = \sqrt{A}$, we have the Liapunov equation for X: $XU + UX = B$. For $U \in$ Sym$^+$ it is known to be uniquely solvable with the solution $X = \int_0^\infty e^{-tU} B e^{-tU}\, dt$, which gives (1.2.11). □

For dimensions $n = 2, 3$ there are explicit expressions for $\sqrt{A}$ in terms of powers of A of order $\le n - 1$ and principal invariants of A, see Marsden & Hughes [1983, Proposition 3.15 & Problem 3.3], Hoger & Carlson [1984], Ting [1985], and Sawyers [1986].

1.3 Skew and Orthogonal Tensors

We shall use the spectral theorem in the complex linear spaces to obtain a description of the structure of the skew and proper orthogonal tensors. Below, and at several other places throughout the book, we shall need to use the complex numbers in connection with vectors (we have already seen this in the preceding section when we used the Cauchy formula for tensors).

1.3.1 Complex Formalism (cf. Halmos [1958, Sect. 77]) We denote by Vect$^{\mathbb{C}}$ the complexification of the real vector space Vect. This is the set of all formal sums $a = b + ic$ where $b, c \in$ Vect and i is the imaginary unit. For the elements of Vect, the operations of addition and multiplications by complex numbers are defined in a natural way. The set Vect is naturally embedded into Vect$^{\mathbb{C}}$. For any $a = b + ic \in$ Vect$^{\mathbb{C}}$, ($b, c \in$ Vect) the symbol $a^* = b - ic$ denotes the complex conjugate vector. If $a = (a_1, \cdots, a_n)$ has complex components in a (real) orthonormal basis in Vect, then a^* has complex conjugate components: $a^* = (a_1^*, \cdots, a_n^*)$. The dot will denote the inner product of two complex vectors $a, b \in$ Vect$^{\mathbb{C}}$ defined to be *linear* in both factors, and not conjugate linear in the second. In components, $a \cdot b = a_1 b_1 + \cdots + a_n b_n$ with a_i, b_i the components of a, b. Hence the inner product defined in this way is no longer positive definite. However, the form $(\boldsymbol{a}, \boldsymbol{b}) \mapsto \boldsymbol{a} \cdot \boldsymbol{b}^*$ is a positive definite inner product in the sense of complex vectorspaces with inner product, i.e., $\boldsymbol{a} \cdot \boldsymbol{a}^* \geq 0$. We write $|\boldsymbol{a}| = \sqrt{\boldsymbol{a} \cdot \boldsymbol{a}^*}$ for the norm of $\boldsymbol{a} \in \mathrm{Vect}^{\mathbb{C}}$. We define the complex second-order tensors $\boldsymbol{H}$ to be linear mappings of the complex vector space $\mathrm{Vect}^{\mathbb{C}}$ into itself and denote by $\mathrm{Lin}^{\mathbb{C}}$ the complex linear space of all complex second-order tensors. Any real second order tensor $\boldsymbol{H} \in \mathrm{Lin}$ can be considered to be an element of $\mathrm{Lin}^{\mathbb{C}}$ by setting $\boldsymbol{H}(\boldsymbol{a}+\mathrm{i}\boldsymbol{b}) = \boldsymbol{H}\boldsymbol{a} + \mathrm{i}\boldsymbol{H}\boldsymbol{b}$, $\boldsymbol{a}, \boldsymbol{b} \in \mathrm{Vect}$. The symbols $\boldsymbol{H}^*$ and $\boldsymbol{H} \cdot \boldsymbol{L}$ for $\boldsymbol{H}, \boldsymbol{L} \in \mathrm{Lin}^{\mathbb{C}}$ are defined in a way analogous to the case of complex vectors. We write $|\boldsymbol{H}| = \sqrt{\boldsymbol{H} \cdot \boldsymbol{H}^*}$ for the norm of a complex tensor $\boldsymbol{H} \in \mathrm{Lin}^{\mathbb{C}}$, noting that $\boldsymbol{H} \cdot \boldsymbol{H}^* \geq 0$. Also the fourth-order complex tensors are defined analogously (see Sect. 1.6) and any real fourth-order tensor can be extended to a complex fourth-order tensor in a natural way.

1.3.2 Spectral Theorem for Skew Tensors *Let us set* $\boldsymbol{p} \wedge \boldsymbol{q} = \boldsymbol{p} \otimes \boldsymbol{q} - \boldsymbol{q} \otimes \boldsymbol{p}$, $\boldsymbol{p}, \boldsymbol{q} \in \mathrm{Vect}$. *Every* $\boldsymbol{W} \in \mathrm{Skew}$ *is of the form*

$$\boldsymbol{W} = \sum_{i=1}^{p} \omega_i \boldsymbol{p}_i \wedge \boldsymbol{q}_i \tag{1.3.1}$$

where $p \leq n/2$, $\omega_i \in \mathbb{R}$ *and* $\{\boldsymbol{p}_i, \boldsymbol{q}_i, i = 1, \ldots, p\}$ *is an orthonormal system in* Vect.

Proof Let us define a linear transformation $\boldsymbol{H}$ on $\mathrm{Vect}^{\mathbb{C}}$ by $\boldsymbol{H} = \mathrm{i}\boldsymbol{W}$. Then $\boldsymbol{H}$ is a hermitian linear transformation on $\mathrm{Vect}^{\mathbb{C}}$ and hence all its eigenvalues are real. One finds that if $\omega \in \mathbb{R}$ is an eigenvalue of $\boldsymbol{H}$ corresponding to the eigenvector $\boldsymbol{e} \in \mathrm{Vect}^{\mathbb{C}}$, then $-\omega$ is an eigenvalue corresponding to the eigenvector $\boldsymbol{e}^*$. Therefore, if $\omega \neq 0$ so that $\omega \neq -\omega$, the vectors $\boldsymbol{e}, \boldsymbol{e}^*$ must be orthogonal in the sense of the complex vector space $\mathrm{Vect}^{\mathbb{C}}$, i.e., $\boldsymbol{e} \cdot (\boldsymbol{e}^*)^* \equiv \boldsymbol{e} \cdot \boldsymbol{e} = 0$. Applying the spectral theorem (see HALMOS [1958, Sect. 79]) and the above observation to $\boldsymbol{H}$, one finds that the complete orthogonal system of eigenvectors for $\boldsymbol{H}$ can be chosen of the form $\{\boldsymbol{e}_i, \boldsymbol{e}_i^*, \boldsymbol{m}_j, 1 \leq i \leq p, 1 \leq j \leq n-2p\}$, where $\boldsymbol{e}_i$ are all eigenvectors of $\boldsymbol{H}$ corresponding to the positive eigenvalues ω_i, $\boldsymbol{e}_i^*$ are eigenvectors of $\boldsymbol{H}$ corresponding to $-\omega_i$, and $\boldsymbol{m}_j \in \mathrm{Vect}$ are real eigenvectors corresponding to the eigenvalue 0. Splitting $\boldsymbol{e}_i = \boldsymbol{p}_i + \mathrm{i}\boldsymbol{q}_i$, $\boldsymbol{p}_i, \boldsymbol{q}_i \in \mathrm{Vect}$, one finds that $\{\boldsymbol{p}_i, \boldsymbol{q}_i, \boldsymbol{m}_j\}$ is an orthogonal basis in Vect and $|\boldsymbol{p}_i| = |\boldsymbol{q}_i|$. One can therefore normalize to achieve $|\boldsymbol{p}_i| = |\boldsymbol{q}_i| = |\boldsymbol{m}_j| = 1$. The conditions $\boldsymbol{H}\boldsymbol{e}_i = \omega_i \boldsymbol{e}_i$, $\boldsymbol{H}\boldsymbol{m}_j = \boldsymbol{0}$ read

$$\boldsymbol{W}\boldsymbol{p}_i = \omega_i \boldsymbol{q}_i, \quad \boldsymbol{W}\boldsymbol{q}_i = -\omega_i \boldsymbol{p}_i, \quad \boldsymbol{W}\boldsymbol{m}_j = \boldsymbol{0}. \tag{1.3.2}$$

Since the linear transformation $\boldsymbol{W}'$ given by the right-hand side of (1.3.1) also satisfies (1.3.2), the proof is complete. □

1.3.3 Spectral Theorem for Proper Orthogonal Tensors *Let us set*

$$\boldsymbol{R}(\boldsymbol{p}, \boldsymbol{q}, \omega) = \boldsymbol{1} + (\cos\omega - 1)(\boldsymbol{p} \otimes \boldsymbol{p} + \boldsymbol{q} \otimes \boldsymbol{q}) + \sin\omega(\boldsymbol{p} \otimes \boldsymbol{q} - \boldsymbol{q} \otimes \boldsymbol{p}) \tag{1.3.3}$$

for any two perpendicular unit vectors $\boldsymbol{p}, \boldsymbol{q}$ *and* $\omega \in \mathbb{R}$. *Then* $\boldsymbol{R}(\boldsymbol{p}, \boldsymbol{q}, \omega) \in \mathrm{Orth}^+$. *Every* $\boldsymbol{R} \in \mathrm{Orth}^+$ *is of the form*

$$\boldsymbol{R} = \boldsymbol{R}_1 \cdots \boldsymbol{R}_p, \quad \boldsymbol{R}_i \equiv \boldsymbol{R}(\boldsymbol{p}_i, \boldsymbol{q}_i, \omega_i), \quad i = 1, \ldots, p, \tag{1.3.4}$$

where $p \leq n/2$ *and* $\{\boldsymbol{p}_i, \boldsymbol{q}_i, i = 1, \ldots, p\}$ *is an orthonormal system in* Vect.

A similar result holds for improper orthogonal tensors (see the proof below).

Proof (cf. HALMOS [1958, Sect. 81]) Let us consider $\boldsymbol{R}$ as a linear transformation on $\mathrm{Vect}^{\mathbb{C}}$. Then $\boldsymbol{R}$ is a unitary linear transformation on $\mathrm{Vect}^{\mathbb{C}}$ and hence all its eigenvalues $\lambda \in \mathbb{C}$ are of unit modulus, $|\lambda| = 1$. From the fact that $\boldsymbol{R}$ is a complexification of a real linear transformation it follows that if $\lambda \in \mathbb{C}$ is an eigenvalue corresponding to the eigenvector $\boldsymbol{e} \in \mathrm{Vect}^{\mathbb{C}}$, then λ^* is an eigenvalue corresponding to the eigenvector $\boldsymbol{e}^*$. Therefore, if $\lambda \neq \pm 1$ so that $\lambda \neq \lambda^*$, the vectors $\boldsymbol{e}, \boldsymbol{e}^*$ must be orthogonal in the sense of the complex vector space $\mathrm{Vect}^{\mathbb{C}}$. Applying the spectral theorem and the above observation to $\boldsymbol{R}$, one finds that the complete orthogonal system of eigenvectors for $\boldsymbol{H}$ can be chosen of the form $\{\boldsymbol{e}_i, \boldsymbol{e}_i^*, \boldsymbol{m}_j, \boldsymbol{n}_k, 1 \leq i \leq p, 1 \leq j \leq s, 1 \leq k \leq t\}$, where $\boldsymbol{e}_i$ are all eigenvectors of $\boldsymbol{R}$ corresponding to the eigenvalues λ_i with positive imaginary part, $\boldsymbol{e}_i^*$ are eigenvectors of $\boldsymbol{R}$ corresponding to λ_i^*, $\boldsymbol{m}_j \in \mathrm{Vect}$ are real eigenvectors corresponding to the eigenvalue 1, and $\boldsymbol{n}_k$ are real eigenvectors corresponding to the eigenvalue -1. As in the proof of Proposition 1.3.2 one finds that one can normalize to achieve $|\boldsymbol{p}_i| = |\boldsymbol{q}_i| = |\boldsymbol{m}_j| = |\boldsymbol{n}_k| = 1$. The conditions $\boldsymbol{R}\boldsymbol{e}_i = \lambda_i \boldsymbol{e}_i$ read

$$\boldsymbol{R}\boldsymbol{p}_i = \cos\omega_i \boldsymbol{p}_i - \sin\omega_i \boldsymbol{q}_i, \quad \boldsymbol{R}\boldsymbol{q}_i = \sin\omega_i \boldsymbol{p}_i + \cos\omega_i \boldsymbol{q}_i, \tag{1.3.5}$$

where $\omega_i \in \mathbb{R}$ is such that $\lambda_i = e^{i\omega_i}$. One has $1 = \det \boldsymbol{R} = (-1)^t$ from which t is even. One can therefore group the eigenvectors $\boldsymbol{n}_k$, $1 \le k \le t$ into pairs and denoting $\boldsymbol{p}_{p+1} = \boldsymbol{n}_1, \boldsymbol{q}_{p+1} = \boldsymbol{n}_2, \boldsymbol{p}_{p+2} = \boldsymbol{n}_3$ etc., one finds that for the so defined $\boldsymbol{p}_i, \boldsymbol{q}_i$, the equation (1.3.5) holds with $\omega_i = \pi$. Since the linear transformation $\boldsymbol{R}'$ given by the right-hand side of $(1.3.4)_1$ also satisfies (1.3.5), the proof is complete. □

1.3.4 Proposition *The tensor* $\boldsymbol{R}$ *is in* Orth$^+$ *if and only if it is of the form* $\boldsymbol{R} = e^{\boldsymbol{W}}$ *for some* $\boldsymbol{W} \in$ Skew. *(Here* $e^{\boldsymbol{W}}$ *is the exponential of a linear transformation* $\boldsymbol{W}$*, defined as the sum of the exponential power series, see, e.g.,* Rudin [1973].*) More precisely, if* $\boldsymbol{W}$ *is of the form* (1.3.1)*, then* $e^{\boldsymbol{W}}$ *is given by the right-hand side of* $(1.3.4)_1$ *and conversely.*

Proof Let $\boldsymbol{W}$ be as in (1.3.1) and set $\boldsymbol{e}_i = \boldsymbol{p}_i + i\boldsymbol{q}_i$, $1 \le i \le p$. Let $\boldsymbol{m}_j$, $1 \le j \le n - 2p$ be vectors such that $\{\boldsymbol{e}_i, \boldsymbol{e}_i^*, \boldsymbol{m}_j\}$ is an orthogonal basis in Vect$^{\mathbb{C}}$. It is found that both $e^{\boldsymbol{W}}$ and the $\boldsymbol{R}$ given by (1.3.4) act on the elements of $\{\boldsymbol{e}_i, \boldsymbol{e}_i^*, \boldsymbol{m}_j\}$ in the same way. □

1.3.5 Rotations in Three Dimensions In the three-dimensional space, the integer p from Propositions 1.3.2 and 1.3.3 is necessarily 0 or 1 and we conclude that every skew tensor in is of the form $\boldsymbol{W} = \omega \boldsymbol{p} \wedge \boldsymbol{q}$ where $\boldsymbol{p}, \boldsymbol{q}$ are perpendicular unit vectors and $\omega \in \mathbb{R}$. If $\boldsymbol{r} \in$ Vect is a unit vector such that $\{\boldsymbol{p}, \boldsymbol{q}, \boldsymbol{r}\}$ is a positively oriented orthonormal basis (the latter means $(\boldsymbol{p} \times \boldsymbol{q}) \cdot \boldsymbol{r} = 1$), then $\boldsymbol{W}\boldsymbol{a} = \omega \boldsymbol{a} \times \boldsymbol{r}$ for every $\boldsymbol{a} \in$ Vect, and to within the change of sign, $\boldsymbol{W}$ may be characterized as the $\boldsymbol{W} \in$ Skew for which $\boldsymbol{W}\boldsymbol{r} = \boldsymbol{0}$, and $|\boldsymbol{W}|^2 = 2\omega^2$. $\boldsymbol{W}$ is represented by

$$W = \begin{bmatrix} 0 & \omega & 0 \\ -\omega & 0 & 0 \\ 0 & 0 & 0 \end{bmatrix}, \quad W = \pm \begin{bmatrix} 0 & \omega_3 & -\omega_2 \\ -\omega_3 & 0 & \omega_1 \\ \omega_2 & -\omega_1 & 0 \end{bmatrix}$$

in the basis $\{\boldsymbol{p}, \boldsymbol{q}, \boldsymbol{r}\}$ and in a general orthonormal basis $\{\boldsymbol{e}_1, \boldsymbol{e}_2, \boldsymbol{e}_3\}$, respectively, where ω_i are the components of $\boldsymbol{\omega} := \omega \boldsymbol{r}$ and the + sign holds if $\{\boldsymbol{e}_1, \boldsymbol{e}_2, \boldsymbol{e}_3\}$ is positively oriented and − applies otherwise. Every $\boldsymbol{R} \in$ Orth$^+$ is of the form $\boldsymbol{R} = \boldsymbol{R}(\boldsymbol{p}, \boldsymbol{q}, \omega)$ for some perpendicular unit vectors $\boldsymbol{p}, \boldsymbol{q}$ and some $\omega \in \mathbb{R}$. If $\boldsymbol{r}$ is as above, then $\boldsymbol{r}$ and ω are the axis and the angle of rotation, respectively. Formula (1.3.3) reduces to

$$\boldsymbol{R}(\omega, \boldsymbol{r}) \equiv \boldsymbol{R}(\boldsymbol{p}, \boldsymbol{q}) = \boldsymbol{r} \otimes \boldsymbol{r} + (\boldsymbol{1} - \boldsymbol{r} \otimes \boldsymbol{r}) \cos\omega + \boldsymbol{W}_0 \sin\omega, \tag{1.3.6}$$

where $\boldsymbol{W}_0 \in$ Skew is given by $\boldsymbol{W}_0 \boldsymbol{a} = \boldsymbol{a} \times \boldsymbol{r}$, $\boldsymbol{a} \in$ Vect. Alternatively,

$$\boldsymbol{R}(\omega, \boldsymbol{r}) = \boldsymbol{r} \otimes \boldsymbol{r} + \boldsymbol{A}^+ \exp(i\omega) + \boldsymbol{A}^- \exp(-i\omega) \quad \text{where} \quad \boldsymbol{A}^\pm = \tfrac{1}{2}(\boldsymbol{1} - \boldsymbol{r} \otimes \boldsymbol{r} \mp i\boldsymbol{W}_0).$$

In components,

$$\boldsymbol{R}(\omega, \boldsymbol{r})_{ij} = r_i r_j + \cos\omega(\delta_{ij} - r_i r_j) \pm \sin\omega \sum_{k=1}^{3} \varepsilon_{ijk} r_k$$

with the + sign holding if $\{\boldsymbol{p}, \boldsymbol{q}, \boldsymbol{r}\}$ is a positive basis and − sign holding otherwise. In the basis $\boldsymbol{p}, \boldsymbol{q}, \boldsymbol{r}$ the tensor $\boldsymbol{R}$ is represented by

$$\boldsymbol{R} = \begin{bmatrix} \cos\omega & \pm\sin\omega & 0 \\ \mp\sin\omega & \cos\omega & 0 \\ 0 & 0 & 1 \end{bmatrix}. \tag{1.3.7}$$

If $\boldsymbol{m} \in$ Sph and $k = 1, 2, \ldots$, then a ***rotation of order*** k ***about*** $\boldsymbol{m}$ is a proper orthogonal tensor $\boldsymbol{R} = \boldsymbol{R}(\omega, \boldsymbol{m})$ with $\omega = 2\pi/k$. It satisfies $\boldsymbol{R}^k = \boldsymbol{1}$ and $\boldsymbol{R}^m \ne \boldsymbol{1}$ for $m = 1, \ldots, k-1$. Conversely, if the last sentence holds for some $\boldsymbol{R} \in$ Orth$^+$ and $k = 2, 3, \ldots$, then $\boldsymbol{R}$ is a rotation of order k about some axis $\boldsymbol{m}$, and to within the change of sign, $\boldsymbol{m}$ is determined as one of the two unit vectors $\boldsymbol{h}$ satisfying $\boldsymbol{R}\boldsymbol{h} = \boldsymbol{h}$. A ***reflection on the plane*** of normal $\boldsymbol{m} \in$ Sph is an improper orthogonal tensor $\boldsymbol{H} = \boldsymbol{1} - 2\boldsymbol{m} \otimes \boldsymbol{m}$.

1.3.6 *Remark* If dim Vect ≥ 3, then for every $\boldsymbol{R}$ of the form $\boldsymbol{R} = \boldsymbol{R}(\boldsymbol{p}, \boldsymbol{q}, \omega)$ as in (1.3.3) there exists a $\boldsymbol{Z} \in$ Orth$^+$ such that $\boldsymbol{R}^{-1} = \boldsymbol{Z}\boldsymbol{R}\boldsymbol{Z}^{-1}$. It suffices to take any unit vector $\boldsymbol{r} \in$ Vect orthogonal to $\boldsymbol{p}, \boldsymbol{q}$ and to set $\boldsymbol{Z} = \boldsymbol{1} - 2\boldsymbol{q} \otimes \boldsymbol{q} - 2\boldsymbol{r} \otimes \boldsymbol{r}$. In three dimensions, if $\boldsymbol{R}$ is of the form (1.3.7), then $\boldsymbol{Z}$ is represented by the diagonal matrix diag$(1, -1, -1)$.

The following proposition on the group homomorphisms from Orth to the additive group $\mathbb{R}$ will be used in Sect. 8.3 to prove a result on objective functions.

1.3.7 Proposition *Let* $c : \mathrm{Orth} \to \mathbb{R}$ *be a function satisfying*

$$c(\boldsymbol{RS}) = c(\boldsymbol{R}) + c(\boldsymbol{S}), \quad \boldsymbol{R}, \boldsymbol{S} \in \mathrm{Orth}. \tag{1.3.8}$$

Let either $\dim \mathrm{Vect} \geq 3$ *or* c *be bounded. Then* $c(\boldsymbol{R}) = 0$ *for every* $\boldsymbol{R} \in \mathrm{Orth}^+$ *and* $c(\boldsymbol{R}) = c_1$ *for some* $c_1 \in \mathbb{R}$ *and all* $\boldsymbol{R} \in \mathrm{Orth}$ *with* $\det \boldsymbol{R} = -1$.

If $\dim \mathrm{Vect} = 2$, then Orth^+ is parametrized by the angle of rotation in $\mathbb{R}^2$. Using Hamel's basis in $\mathbb{R}$, one can construct an additive 2π-periodic function from $\mathbb{R}$ to $\mathbb{R}$ which in turn leads to an unbounded solution c of (1.3.8).

Proof From (1.3.8) we obtain $c(\boldsymbol{1}) = 0$ and $c(\boldsymbol{R}^{-1}) = -c(\boldsymbol{R})$. Let us prove that c vanishes on Orth^+. Consider first the case $\dim \mathrm{Vect} \geq 3$. Then every $\boldsymbol{R} \in \mathrm{Orth}^+$ is a product of $\boldsymbol{R}_i$ as in Proposition 1.3.3 so that and it suffices to prove $c(\boldsymbol{R}) = 0$ for $\boldsymbol{R} := \boldsymbol{R}_i$. If $\boldsymbol{Z}$ is as in Remark 1.3.6, then $c(\boldsymbol{R}^{-1}) = c(\boldsymbol{ZRZ}^{-1}) = c(\boldsymbol{Z}) + c(\boldsymbol{R}) + c(\boldsymbol{Z}^{-1}) = c(\boldsymbol{R})$, where we have also taken into account that $c(\boldsymbol{1}) = 0$. A combination of $c(\boldsymbol{R}^{-1}) = c(\boldsymbol{R})$ and $c(\boldsymbol{R}^{-1}) = -c(\boldsymbol{R})$ gives $c(\boldsymbol{R}) = 0$. When c is bounded, one has $c(\boldsymbol{R}^n) = nc(\boldsymbol{R})$ for any positive integer n and any $\boldsymbol{R} \in \mathrm{Orth}^+$ so that if $c(\boldsymbol{R}) \neq 0$, the sequence $c(\boldsymbol{R}^n)$ is unbounded. This proves that c vanishes on Orth^+ in the second case. Let us now prove that c is constant on the set of all $\boldsymbol{R} \in \mathrm{Orth}$ with $\det \boldsymbol{R} = -1$. Let $\boldsymbol{J} \in \mathrm{Orth}$ with $\det \boldsymbol{J} = -1$ be arbitrary and set $c_1 := c(\boldsymbol{J})$. For every $\boldsymbol{R} \in \mathrm{Orth}$ with $\det \boldsymbol{R} = -1$ we have $\boldsymbol{RJ} \in \mathrm{Orth}^+$ and so $0 = c(\boldsymbol{RJ}) = c(\boldsymbol{R}) + c(\boldsymbol{J}) = c_1$. □

1.4 Invertible Tensors

For a general nonsymmetric tensor $\boldsymbol{F}$ no spectral decomposition is available, but it is possible to use the spectral decomposition of $\sqrt{\boldsymbol{FF}^{\mathrm{T}}}$ to infer indirectly the properties of $\boldsymbol{F}$. With the interpretation of $\boldsymbol{F} \in \mathrm{InvLin}$ as a deformation gradient (see Subsect. 2.2.3 below), this section actually starts the analysis of deformation, conventionally carried out for different measures of deformation, like the Cauchy-Green deformation tensors etc.

1.4.1 Singular Values Let $\boldsymbol{F} \in \mathrm{Lin}$. A number $\nu \geq 0$ and a vector $\boldsymbol{e}$ are said to be a ***singular value*** and a ***singular vector*** of $\boldsymbol{F}$ if they are an eigenvalue and an eigenvector of $\sqrt{\boldsymbol{FF}^{\mathrm{T}}}$. When ordered nonincreasingly, the singular values of $\boldsymbol{F}$ are uniquely determined and we denote them by $\hat{\nu}_i(\boldsymbol{F}) \equiv \hat{a}_i(\sqrt{\boldsymbol{FF}^{\mathrm{T}}}) = \hat{a}_i(\sqrt{\boldsymbol{F}^{\mathrm{T}}\boldsymbol{F}})$. When $\boldsymbol{F} \in \mathrm{InvLin}$ is interpreted as the deformation gradient then the singular values of $\boldsymbol{F}$ are called the principal stretches and the singular vectors the principal directions of deformation.

Denote by $\mathrm{InvLin}^\sharp$ the set of all $\boldsymbol{F} \in \mathrm{InvLin}$ which have all singular values different. If $\boldsymbol{F} \in \mathrm{InvLin}^\sharp$, then its singular vectors $\boldsymbol{e}_i$ corresponding to $\nu_i := \hat{\nu}_i(\boldsymbol{F})$ are determined to within a change of the signs, and the projections $\boldsymbol{P}_i = \boldsymbol{e}_i \otimes \boldsymbol{e}_i$ are unique. Using the definitions of $\hat{\nu}_i$, the formula $\boldsymbol{P}_i(\boldsymbol{F}) = \boldsymbol{P}_i(\sqrt{\boldsymbol{FF}^{\mathrm{T}}})$, Propositions 1.2.3, 1.2.9, equations $(1.1.2)_2$, (1.2.12), and Subsect. 1.2.4, we obtain the following result.

1.4.2 Proposition *In the above notation,*

(1) *the functions* $\hat{\nu}_i$ *are Lipschitz continuous on* Lin; *the functions* $\hat{\nu}_i$, $\boldsymbol{P}_i$ *are infinitely differentiable on* $\mathrm{InvLin}^\sharp$ *and satisfy* $\partial_{\boldsymbol{F}}\hat{\nu}_i = \boldsymbol{P}_i$;

(2) *if* $\boldsymbol{F} \in \mathrm{InvLin}^\sharp \cap \mathrm{Sym}^+$ *then for any* $\boldsymbol{B}, \boldsymbol{C} \in \mathrm{Lin}$,

$$\partial_{\boldsymbol{F}}\boldsymbol{e}_i[\boldsymbol{B}] = \sum \frac{\nu_j B_{ij} + \nu_i B_{ji}}{\nu_i^2 - \nu_j^2}\boldsymbol{e}_j = \sum \Big(\frac{E_{ij}}{\nu_i - \nu_j} - \frac{W_{ij}}{\nu_i + \nu_j}\Big)\boldsymbol{e}_j,$$

$$\partial_{\boldsymbol{F}}\boldsymbol{P}_i[\boldsymbol{B}] = \sum \frac{\nu_j B_{ij} + \nu_i B_{ji}}{\nu_i^2 - \nu_j^2}(\boldsymbol{e}_j \otimes \boldsymbol{e}_i + \boldsymbol{e}_i \otimes \boldsymbol{e}_j) = \sum \Big(\frac{E_{ij}}{\nu_i - \nu_j} + \frac{W_{ij}}{\nu_i + \nu_j}\Big)(\boldsymbol{e}_j \otimes \boldsymbol{e}_i + \boldsymbol{e}_i \otimes \boldsymbol{e}_j),$$

$$\partial_F^2 \hat{v}_i[B, C] = 2\sum \frac{E_{ij}D_{ij}}{v_i - v_j} + 2\sum \frac{W_{ij}X_{ij}}{v_i + v_j},$$

where the sums are taken over all $j, j \neq i$, E, W (D, X) are the symmetric and skew parts of B (C), respectively, and the subscripts denote the components in $\{e_i\}$.

Note. It suffices to calculate the derivatives at $F \in \mathrm{InvLin}^\sharp \cap \mathrm{Sym}^+$. The derivatives at a general $F \in \mathrm{InvLin}^\sharp$ are then obtained by using the transformation properties of the polar factors, below.

1.4.3 The Polar Decomposition Theorem (Autonne, 1902) *For every $F \in$ InvLin there exist unique tensors $U, V \in$ Sym$^+$ and a unique $R \in$ Orth such that*

$$F = RU = VR; \tag{1.4.1}$$

these tensors are given by

$$U = \sqrt{F^\mathsf{T} F}, \quad V = \sqrt{FF^\mathsf{T}}, \quad R = FU^{-1} = V^{-1}F. \tag{1.4.2}$$

Moreover, if $F \in$ Lin$^+$, then $R \in$ Orth$^+$.

See HALMOS [1958, Sect. 83] for the proof and HALMOS [1967, Problem 105] for the version when F is not invertible. The pairs R, U and V, R are called the ***right*** and ***left polar factors*** of F, respectively.

1.4.4 Proposition *The functions $F \mapsto U(F), V(F), R(F)$, defined on* InvLin *by the polar decomposition theorem, are infinitely differentiable and*

$$\partial_F U[A] = \int_0^\infty e^{-Ut}(A^\mathsf{T} F + F^\mathsf{T} A)e^{-Ut}\,dt, \tag{1.4.3}$$

$$\partial_F V[A] = \int_0^\infty e^{-Vt}(AF^\mathsf{T} + FA^\mathsf{T})e^{-Vt}\,dt, \tag{1.4.4}$$

$$\partial_F R[A] = R\int_0^\infty e^{-Ut}(R^\mathsf{T} A - A^\mathsf{T} R)e^{-Ut}\,dt = \int_0^\infty e^{-Vt}(AR^\mathsf{T} - RA^\mathsf{T})e^{-Vt}\,dt\,R \tag{1.4.5}$$

where $U = U(F)$, $V = V(F)$, $R = R(F)$. If $F \in$ Sym$^+$, then the component forms of these equations read

$$\partial_F U[A]_{ij} = \frac{v_i A_{ij} + v_j A_{ji}}{v_i + v_j}, \quad \partial_F V[A]_{ij} = \frac{v_j A_{ij} + v_i A_{ji}}{v_i + v_j}, \quad \partial_F R[A]_{ij} = \frac{A_{ij} - A_{ji}}{v_i + v_j}$$

where v_i are the eigenvalues of F and A_{ij} are the components of A in the basis of eigenvectors of F (the argument F has been omitted).

Proof The differentiability: The mapping $F \mapsto U(F)$ is a composition of $F \mapsto C(F) = F^\mathsf{T}F$ and $C \mapsto \sqrt{C}$, and both these are infinitely differentiable, cf. Proposition 1.2.9. The differentiability of V is proved similarly. By $(1.4.2)_3$ also R is infinitely differentiable, since the function $U \mapsto U^{-1}$ is infinitely differentiable on Sym$^+$. Let us now prove (1.4.3). By the chain rule, $\partial_F U(F)[A] = \partial_C \sqrt{C}[A^\mathsf{T}F + F^\mathsf{T}A]$, see $(1.1.2)_2$, and the derivative of the square root is in (1.2.11). This gives (1.4.3) and similarly (1.4.4). One can integrate the second term in (1.4.3) by parts:

$$\int_0^\infty e^{-Ut} F^\mathsf{T} A e^{-Ut}\,dt = -\int_0^\infty \frac{d}{dt}(e^{-Ut})R^\mathsf{T} A e^{-Ut}\,dt = R^\mathsf{T} A - \int_0^\infty e^{-Ut} R^\mathsf{T} A U e^{-Ut}\,dt.$$

From this,

$$\partial_F U(F)[A] = R^\mathsf{T} A + \int_0^\infty e^{-Ut}(A^\mathsf{T} R - R^\mathsf{T} A)Ue^{-Ut}\,dt. \tag{1.4.6}$$

Differentiating $F = RU$ with respect to F in the direction A and omitting the argument F we obtain $A = \partial_F R[A]U + R\partial_F U[A]$ from which $R^\mathsf{T}\partial_F R[A] = R^\mathsf{T} A U^{-1} - \partial_F U[A]U^{-1}$. A combination with (1.4.6) leads to $(1.4.5)_1$ and similarly to $(1.4.5)_2$. □

1.4.5 The Transformation of Polar Factors *For every* $F \in \mathrm{InvLin}$ *and* $Q \in \mathrm{Orth}$,

$$U(QF) = U(F), \quad V(QF) = QV(F)Q^{\mathrm{T}}, \quad R(QF) = QR(F),$$
$$U(FQ) = Q^{\mathrm{T}}U(F)Q, \quad V(FQ) = V(F), \quad R(FQ) = R(F)Q.$$

This follows directly from the definitions and the fact that $\sqrt{QAQ^{\mathrm{T}}} = Q\sqrt{A}Q^{\mathrm{T}}, Q \in \mathrm{Orth}$, $A \in \mathrm{Sym}^+$, see Proposition 1.2.6. Alternatively, one can check that the expressions on the right-hand sides of the above relations give the polar decompositions of the transformed tensors and invoke the uniqueness in the polar decomposition. □

1.5 Bravais Lattices

One of the most interesting recent extensions of the phenomenological theory of materials involve solid–solid phase transformations in nonlinear crystalline materials. By way of preparation, let us introduce some underlying concepts.

1.5.1 Bravais Lattices Let $\mathbb{Z}$ be the set of all integers. A ***Bravais lattice,*** or briefly a ***lattice,*** L, is a subset of Vect of the form

$$L = \{\sum_{i=1}^{n} m_i e_i : m_i \in \mathbb{Z}, 1 \le i \le n\} \tag{1.5.1}$$

where $\{e_i\}$ is a basis in Vect. Any collection $\{e_i\}$ of vectors such that (1.5.1) holds is called a basis of L and L is said to be generated by $\{e_i\}$. Denote by $\mathrm{GL}(\mathbb{Z}^n)$ the set of all $n \times n$ matrices $m = [m_{ij}]$ with integral elements ($m_{ij} \in \mathbb{Z}$) and $\det m = \pm 1$. This is a group: Clearly, if $m, p \in \mathrm{GL}(\mathbb{Z}^n)$, then $mp \in \mathrm{GL}(\mathbb{Z}^n)$. Furthermore, $m^{-1} = \pm \operatorname{cof} m^{\mathrm{T}}$ and (1.1.14), $(1.1.22)_2$ imply that $\operatorname{cof} m^{\mathrm{T}}$ has integral elements and hence $m^{-1} \in \mathrm{GL}(\mathbb{Z}^n)$. The group $\mathrm{GL}(\mathbb{Z}^n)$ is infinite: for example,

$$m_k := \begin{bmatrix} 1 & k & 0 \\ 0 & 1 & 0 \\ 0 & 0 & 1 \end{bmatrix}, \quad k \in \mathbb{Z},$$

is an infinite subgroup of $\mathrm{GL}(\mathbb{Z}^3)$. The basis of a lattice is nonunique.

1.5.2 Proposition *If L is a lattice with basis $\{e_i\}$, then an n-tuple of vectors $\{f_i\}$ is a basis for L if and only if*

$$f_i = \sum_{j=1}^{n} m_{ij} e_i \tag{1.5.2}$$

for some $m = [m_{ij}] \in \mathrm{GL}(\mathbb{Z}^n)$.

Proof Since $f_i \in L$, there must exist integers m_{ij} such that (1.5.2) holds. Since $\{f_i\}$ is a basis, there must be integers p_{ij} such that $e_i = \sum_{j=1}^{n} p_{ij} f_i$, and a combination with (1.5.2) gives that the matrices $m = [m_{ij}]$, $p = [p_{ij}]$ satisfy $mp = pm = I$ where I is the $n \times n$ identity matrix. Since m, p have integral elements, $\det m$, $\det p$ are integers and $\det m \det p = 1$. The last can hold only if $\det m = \pm 1$, $\det p = \pm 1$. The converse implication is proved similarly. □

If L is a lattice with basis $\{e_i\}$ then the ***dual lattice*** L^* is a lattice generated by the dual basis $\{e_i^*\}$, i.e., by the basis satisfying $e_i \cdot e_j^* = \delta_{ij}$. If L is a lattice and $F \in \mathrm{InvLin}$, then also $\bar{L} := FL = \{Fu : u \in L\}$ is a lattice. The ***symmetry group of a lattice*** is the set of all $H \in \mathrm{InvLin}$ that leave L unchanged: $G(L) := \{H \in \mathrm{InvLin} : HL = L\}$. From Proposition 1.5.2 one easily deduces that if $\{e_i\}$ is a basis for L, then $H \in$

$G(L)$ if and only if $\boldsymbol{He}_i = \sum_{j=1}^{n} m_{ij}\boldsymbol{e}_j$ for some $m = [m_{ij}] \in \mathrm{GL}(\mathbb{Z}^n)$. In the generally nonorthonormal basis $\{\boldsymbol{e}_i\}$ of Vect, the tensors $\boldsymbol{H} \in G(L)$ are represented by matrices $m = [m_{ij}] \in \mathrm{GL}(\mathbb{Z}^n)$; hence $G(L)$ has the same structure as $\mathrm{GL}(\mathbb{Z}^n)$. In particular, every $\boldsymbol{H} \in G(L)$ satisfies $\det \boldsymbol{H} = \pm 1$. Note that if $\boldsymbol{H} := \mathbf{1} + \boldsymbol{c} \otimes \boldsymbol{b}$ where $\boldsymbol{c} \in L$, $\boldsymbol{b} \in L^*$ and $\boldsymbol{c} \cdot \boldsymbol{b} = 0$, then $\boldsymbol{H} \in G(L)$ and

$$\boldsymbol{H}^k = \mathbf{1} + k\boldsymbol{c} \otimes \boldsymbol{b}, \quad k \in \mathbb{Z}. \tag{1.5.3}$$

It is easy to show that $G(L^*) = G(L)^{\mathsf{T}} = \{\boldsymbol{H}^{\mathsf{T}} : \boldsymbol{H} \in G(L)\}$. We have the following transformation rules under the change $L \to \boldsymbol{F}L$:

$$\boldsymbol{e}_i \to \boldsymbol{F}\boldsymbol{e}_i, \quad G(L) \to \boldsymbol{F}G(L)\boldsymbol{F}^{-1}, \tag{1.5.4}$$
$$L^* \to \boldsymbol{F}^{-\mathsf{T}}L^*, \quad \boldsymbol{e}_i^* \to \boldsymbol{F}^{-\mathsf{T}}\boldsymbol{e}_i^*, \quad G(L^*) \to \boldsymbol{F}^{-\mathsf{T}}G(L^*)\boldsymbol{F}^{\mathsf{T}}.$$

The following proposition is useful.

1.5.3 Proposition *Let $\boldsymbol{F}$ be an invertible tensor with the polar decomposition $\boldsymbol{F} = \boldsymbol{RU} = \boldsymbol{VR}$ and let $\boldsymbol{Q}$ be an orthogonal tensor such that $\bar{\boldsymbol{Q}} := \boldsymbol{FQF}^{-1}$ is an orthogonal tensor. Then*

$$\bar{\boldsymbol{Q}} = \boldsymbol{RQR}^{\mathsf{T}}, \quad \boldsymbol{UQ} = \boldsymbol{QU}, \quad \boldsymbol{V}\bar{\boldsymbol{Q}} = \bar{\boldsymbol{Q}}\boldsymbol{V}; \tag{1.5.5}$$

if, additionally, $\boldsymbol{Q} \neq \mathbf{1}$ is a rotation about axis $\boldsymbol{q} \in$ Sph through angle ω, then $\bar{\boldsymbol{Q}}$ is a rotation about axis $\bar{\boldsymbol{q}} := \boldsymbol{Rq}$ through angle ω; furthermore, $\boldsymbol{q}$ is an eigenvector of $\boldsymbol{U}$ and $\bar{\boldsymbol{q}}$ is an eigenvector of $\boldsymbol{V}$.

Proof The hypothesis implies that $\boldsymbol{Z} := \boldsymbol{R}^{\mathsf{T}}\bar{\boldsymbol{Q}}\boldsymbol{R} = \boldsymbol{UQU}^{-1}$ is an orthogonal tensor. The condition $\boldsymbol{Z}^{\mathsf{T}}\boldsymbol{Z} = \mathbf{1}$ and rearrangements give $\boldsymbol{U}^2 = \boldsymbol{Q}^{\mathsf{T}}\boldsymbol{U}^2\boldsymbol{Q}$; from the uniqueness of the square root then $\boldsymbol{U} = \boldsymbol{Q}^{\mathsf{T}}\boldsymbol{UQ}$ and hence $\boldsymbol{U}$ commutes with $\boldsymbol{Q}$; the rest of (1.5.5) are easy consequences of this. Let $\boldsymbol{Q} \neq \mathbf{1}$ be a rotation as in the statement of the proposition. Using $\bar{\boldsymbol{Q}} = \boldsymbol{RQR}^{\mathsf{T}}$ and $\boldsymbol{Qq} = \boldsymbol{q}$ we obtain $\bar{\boldsymbol{Q}} \neq \mathbf{1}$ and $\bar{\boldsymbol{Q}}\bar{\boldsymbol{q}} = \bar{\boldsymbol{q}}$; since the last equation characterizes the axis uniquely up to a change in sign, we see that $\bar{\boldsymbol{q}}$ is the axis of $\bar{\boldsymbol{Q}}$. Similarly, the commutation of $\boldsymbol{Q}, \boldsymbol{U}$ implies $\boldsymbol{QUq} = \boldsymbol{Uq}$ and thus $\boldsymbol{Uq}$ is a multiple of $\boldsymbol{q}$, i.e., $\boldsymbol{q}$ is an eigenvector of $\boldsymbol{U}$; that $\bar{\boldsymbol{q}}$ is an eigenvector of $\boldsymbol{V}$ is proved similarly. The assertion about the angle is obtained by combining (1.3.6) with $(1.5.5)_1$. □

Throughout the rest of this section, assume that $n \le 3$.

1.5.4 Point Groups and Crystallographic Systems The ***point group*** of a lattice L is

$$P(L) := G(L) \cap \mathrm{Orth}.$$

It is easy to prove that $P(L)$ is finite. According to the classical idea of symmetry of lattices, two lattices $L, \bar{L}$ are regarded as to have the same symmetry if their *points groups* are conjugate in InvLin, i.e., if there exists a $\boldsymbol{M} \in$ InvLin such that

$$P(\bar{L}) = \boldsymbol{M}P(L)\boldsymbol{M}^{-1}. \tag{1.5.6}$$

Proposition 1.5.3 implies that if (1.5.6) holds with some $\boldsymbol{M} \in$ InvLin, then also

$$P(\bar{L}) = \boldsymbol{R}P(L)\boldsymbol{R}^{\mathsf{T}}, \tag{1.5.7}$$

where $\boldsymbol{M} = \boldsymbol{RU}$ is the polar decomposition of $\boldsymbol{M}$; hence the point groups are also conjugate in Orth^+. The lattices $L, \bar{L}$ are said to belong to the same ***crystallographic system*** if (1.5.6) holds with some $\boldsymbol{M} \in$ InvLin or equivalently if (1.5.7) holds with some $\boldsymbol{R} \in$ Orth. The classical result says that in three dimensions, there are exactly 7 crystallographic systems, i.e., to within the conjugacy, only 7 groups can arise from the totality of all lattices as the point groups. In two dimensions, the number of crystallographic systems is 4. A finer classification is obtained if two lattices $L, \bar{L}$ are declared to be equivalent if there exists an $\boldsymbol{M} \in$ InvLin such that simultaneously $P(\bar{L}) = \boldsymbol{M}P(L)\boldsymbol{M}^{-1}$ and $\bar{L} = \boldsymbol{M}L$. Then some of the crystallographic systems split into subclasses called Bravais classes, often called Bravais lattices. For example, the cubic system splits into the simple, face centered, and body centered cubic lattices. In three dimensions, there are 14 Bravais classes and in two dimensions 5. The reader is referred to any treatise on crystallography, such as BRADLEY & CRACKNELL [1972], for the description of the crystallographic systems and the point groups. (See also SCHWARZENBERGER [1972].) We shall need here only the following special cases.

1.5.5 Example (Cubic and Tetragonal Point Groups) A subgroup G^c of Orth is said to be a ***cubic point group*** if there exists an orthogonal basis $\{\boldsymbol{e}_i\}$ with $|\boldsymbol{e}_1| = |\boldsymbol{e}_2| = |\boldsymbol{e}_3|$ such that $G^c = \{\boldsymbol{Q} \in \mathrm{Orth}^+ : \boldsymbol{Q}C = C\}$ where C is the cube $C = \{x_1\boldsymbol{e}_1 + x_2\boldsymbol{e}_2 + x_3\boldsymbol{e}_3 : |x_i| < 1, i = 1, 2, 3\}$. The basis $\{\boldsymbol{e}_i\}$ is determined by G^c uniquely to within changes of the signs of its elements, permutations, and the multiplication by a fixed positive number, and its elements are called the cubic axes of G^c. All cubic point groups are conjugate in Orth. Given a cubic point group G^c with cubic axes $(\boldsymbol{e}_1, \boldsymbol{e}_2, \boldsymbol{e}_3)$, a Bravais lattice L satisfies $P(L) = G^c$ if and only if its basis $\{\bar{\boldsymbol{e}}_i\}$ may be chosen such that one of the following three possibilities occurs:

(1) $\bar{\boldsymbol{e}}_i = \lambda\boldsymbol{e}_i$, $i = 1, 2, 3$, for some $\lambda > 0$; in this case $L \equiv L^{sc}$ is called the ***simple cubic lattice***;

(2) $\bar{\boldsymbol{e}}_1 = \lambda\boldsymbol{e}_1, \bar{\boldsymbol{e}}_2 = \lambda\boldsymbol{e}_2, \bar{\boldsymbol{e}}_3 = \lambda(\boldsymbol{e}_1 + \boldsymbol{e}_2 + \boldsymbol{e}_3)/2$ for some $\lambda > 0$; in this case $L \equiv L^{bcc}$ is called the ***body centered cubic lattice***;

(3) $\bar{\boldsymbol{e}}_i = \lambda(\boldsymbol{e}_j + \boldsymbol{e}_k)/2$, $i = 1, 2, 3$, for some $\lambda > 0$ where j, k are the complementing indices to i; in this case $L \equiv L^{fcc}$ is called the ***face centered cubic lattice.***

All simple, body centered, and face centered cubic lattices belong to the same crystallographic class; however, they are in distinct Bravais classes: for example, it is impossible to find an $\boldsymbol{M} \in$ InvLin such that $L^{sc} = \boldsymbol{M}L^{bcc}$ and simultaneously $\boldsymbol{M}G^c\boldsymbol{M}^{-1} = G^c$.

P^c has 48 elements; for each element $\boldsymbol{Q} \in P^c$ there exists a permutation $\pi = (\pi(1), \pi(2), \pi(3))$ of $\{1, 2, 3\}$ and a triple $\sigma = (\sigma_1, \sigma_2, \sigma_3)$, $\sigma_i = \pm 1$, such that

$$\boldsymbol{Q}\boldsymbol{e}_i = \sigma_i\boldsymbol{e}_{\pi(i)}, \quad i = 1, 2, 3,$$

where $(\boldsymbol{e}_1, \boldsymbol{e}_2, \boldsymbol{e}_3)$ are the cubic axes of G^c. Assume henceforth that $|\boldsymbol{e}_1| = |\boldsymbol{e}_2| = |\boldsymbol{e}_3| = 1$. We have $\det\boldsymbol{Q} = \operatorname{sgn}\pi \operatorname{sgn}\sigma$ where $\operatorname{sgn}\pi = \pm 1$ according to whether π is an even or odd permutation, respectively, and $\operatorname{sgn}\sigma = \sigma_1\sigma_2\sigma_3$. Of the 48 elements of P^c, 24 are proper orthogonal tensors, namely the identity and:

- rotations of order 4 about axes $\boldsymbol{e}_i$, $i = 1, 2, 3$, and their powers of order 2, 3;
- rotations of order 2 about the axes $\boldsymbol{n}_{ij} = (\boldsymbol{e}_i \pm \boldsymbol{e}_j)/\sqrt{2}$, $i < j$;
- rotations of order 3 about the axes $\boldsymbol{e}_\tau = (\boldsymbol{e}_1 + \tau_2\boldsymbol{e}_2 + \tau_3\boldsymbol{e}_3)/\sqrt{3}$, where $\tau = (\tau_2, \tau_3)$, $\tau_2 = \pm 1$, $\tau_3 = \pm 1$ and their squares.

A subgroup G^t of Orth is said to be a ***tetragonal point group*** if there exists an orthogonal basis $\{\boldsymbol{e}_i\}$ with $|\boldsymbol{e}_1| = |\boldsymbol{e}_2| \neq |\boldsymbol{e}_3|$ such that $G^c = \{\boldsymbol{Q} \in \mathrm{Orth}^+ : \boldsymbol{Q}P = P\}$ where P is the prism $P = \{x_1\boldsymbol{e}_1 + x_2\boldsymbol{e}_2 + x_3\boldsymbol{e}_3 : |x_i| < 1, i = 1, 2, 3\}$. The elements $(\boldsymbol{e}_1, \boldsymbol{e}_2, \boldsymbol{e}_3)$ are called the tetragonal axes of G^t. Given the tetragonal group G^c with axes $(\boldsymbol{e}_1, \boldsymbol{e}_2, \boldsymbol{e}_3)$, a lattice L satisfies $P(L) = G^c$ if and only if its basis $\{\bar{\boldsymbol{e}}_i\}$ may be chosen such that one of the following two possibilities occurs:

(1) $\bar{\boldsymbol{e}}_1 = \lambda\boldsymbol{e}_1, \bar{\boldsymbol{e}}_2 = \lambda\boldsymbol{e}_2, \bar{\boldsymbol{e}}_3 = \mu\boldsymbol{e}_3$ for some positive λ, μ such that $|\bar{\boldsymbol{e}}_1| = |\bar{\boldsymbol{e}}_2| \neq |\bar{\boldsymbol{e}}_3|$; in this case $L \equiv L^{st}$ is called the ***simple tetragonal lattice***;

(2) $\bar{\boldsymbol{e}}_i = (\tilde{\boldsymbol{e}}_j + \tilde{\boldsymbol{e}}_k)/2$, $i = 1, 2, 3$, where j, k are the complementing indices to i and $\tilde{\boldsymbol{e}}_1 = \lambda\boldsymbol{e}_1, \tilde{\boldsymbol{e}}_2 = \lambda\boldsymbol{e}_2, \tilde{\boldsymbol{e}}_3 = \mu\boldsymbol{e}_3$ for some positive λ, μ such that $|\tilde{\boldsymbol{e}}_1| = |\tilde{\boldsymbol{e}}_2| \neq |\tilde{\boldsymbol{e}}_3|$; in this case $L \equiv L^{fct}$ is called the ***face centered tetragonal lattice***.

P^t has 16 elements; for each element $\boldsymbol{Q} \in P^t$ there exists a unique permutation $\pi = (\pi(1), \pi(2))$ of $\{1, 2\}$ and a triple $\sigma = (\sigma_1, \sigma_2, \sigma_3)$, $\sigma_i = \pm 1$, such that

$$\boldsymbol{Q}\boldsymbol{e}_i = \sigma_i\boldsymbol{e}_{\pi(i)}, \quad i = 1, 2, \quad \boldsymbol{Q}\boldsymbol{e}_3 = \sigma_3\boldsymbol{e}_3,$$

where (e_1, e_2, e_3) are the tetragonal axes of G^t. Assume henceforth that $|e_1| = |e_2| = 1 \neq |e_3|$. Of the 16 elements of P^t, 8 are proper orthogonal tensors. These are the identity and:

- the rotation of order 4 about the axis $e_3/|e_3|$ and its powers of order 2, 3;
- two rotations of order 2 about the axes e_1, e_2;
- two rotations of order 2 about the axes $(e_1 \pm e_2)/\sqrt{2}$.

In phenomenological theories, the crystalline structure is reflected by the assignment of the symmetry group G of the response. The early elaborations of this idea involved linearized materials; in this setting the crystal structure implies specific forms of the tensor of elasticity. The modeling of phase transitions needs a fully nonlinear response, and the models under current study assume that the global symmetry group of the material is either the *whole* symmetry group $G(L)$ of some Bravais lattice or *only* the corresponding point group $P(L)$. The two approaches are related, as will be explained in Sect. 9.4. In both cases we are confronted with the change in symmetry due to deformation: the symmetry of the local response at the state of deformation gradient $\boldsymbol{F}$ and temperature θ is given by the ***local symmetry group*** $G_F := \boldsymbol{F}G\boldsymbol{F}^{-1} \cap \mathrm{Orth}$, where G is the global symmetry group. For example, the tensor of instantaneous elastic constants $\bar{\mathsf{B}}(\boldsymbol{F}, \theta)$ of a nonlinear material depends on $\boldsymbol{F}$ and θ and even if the global symmetry group is some specified group $G \subset \mathrm{InvLin}$ (such as the orthogonal group) the symmetry of $\bar{\mathsf{B}}(\boldsymbol{F}, \theta)$ will be G_F, which may be substantially smaller than G (see Sect. 10.5 for details). As an additional geometric motivation for G_F, notice that for $G = G(L)$ the construction of G_F produces the point group of the deformed lattice $\boldsymbol{F}L$ (see $(1.5.4)_2$), and for $G = \mathrm{Orth}$, G_F produces the set of all orthogonal tensors $\boldsymbol{Q}$ for which $\boldsymbol{QUQ}^{\mathsf{T}} = \boldsymbol{U}$; here $\boldsymbol{F} = \boldsymbol{RU}$ is the polar decomposition of $\boldsymbol{F}$ (see Proposition 1.5.3).

The following proposition describes a deformation-induced change from the cubic to tetragonal symmetries. We will return to it when examining the wells structure of the stored energy associated with cubic to tetragonal phase transitions. Similarly, Proposition 11.1.4 describes the deformation-induced anisotropy of isotropic materials.

1.5.6 Proposition *Let $\boldsymbol{F}$ be an invertible tensor with the polar decomposition $\boldsymbol{F} = \boldsymbol{RU}$, let P^c be a cubic point group with cubic axes (e_1, e_2, e_3). Then a necessary and sufficient condition that $\boldsymbol{F}P^c\boldsymbol{F}^{-1} \cap \mathrm{Orth}$ be a tetragonal point group is that $\boldsymbol{U}$ be of the form*

$$\boldsymbol{U} = \eta_1 \mathbf{1} + (\eta_2 - \eta_1) e_i \otimes e_i, \tag{1.5.8}$$

where i is some index in $\{1, 2, 3\}$ and $\eta_1 > 0$, $\eta_2 > 0$, $\eta_1 \neq \eta_2$. If this is the case, then $\boldsymbol{F}P^c\boldsymbol{F}^{-1} \cap \mathrm{Orth} \subset \boldsymbol{R}P^c\boldsymbol{R}^{\mathsf{T}}$.

Proof Suppose that $\boldsymbol{F}P^c\boldsymbol{F}^{-1} \cap \mathrm{Orth}$ is a tetragonal point group. Then also $\boldsymbol{U}P^c\boldsymbol{U}^{-1} \cap \mathrm{Orth}$ is a tetragonal point group P^t and we denote by $(\bar{e}_1, \bar{e}_2, \bar{e}_3)$ the tetragonal axes. If $\boldsymbol{Q} \in P^t$, then $\boldsymbol{U}^{-1}\boldsymbol{QU} \in P^c \subset \mathrm{Orth}$. By Proposition 1.5.3, the fact that $\boldsymbol{U}^{-1}\boldsymbol{QU}$ is an orthogonal tensor implies that $\boldsymbol{U}$ commutes with $\boldsymbol{Q}$ and hence $\boldsymbol{U}^{-1}\boldsymbol{QU} = \boldsymbol{Q}$; from this, $P^t \subset P^c$. Thus, in particular, the rotation of order 4 about $\bar{e}_3/|\bar{e}_3|$ is in P^c and hence $\bar{e}_3/|\bar{e}_3| = \pm e_i$ for some $i = 1, 2, 3$. Furthermore, by Proposition 1.5.3, each axis of rotation in P^t is an eigenvector of $\boldsymbol{U}$. The axes of rotations in P^t have been determined in Subsect. 1.5.5: they are $e_i = \pm\bar{e}_3/|\bar{e}_3|, \bar{e}_1, \bar{e}_2, (\bar{e}_1 \pm \bar{e}_2)/\sqrt{2}$ and the fact that the last three eigenvectors of $\boldsymbol{U}$ are not orthogonal implies that one eigenvalue of $\boldsymbol{U}$ has multiplicity 2. Hence $\boldsymbol{U}$ is of the asserted form, with $\eta_1 \neq \eta_2$ since otherwise we would have $\boldsymbol{U}P^c\boldsymbol{U}^{-1} \cap \mathrm{Orth} = P^c$. This completes the proof of the direct implication. The converse implication follows from the fact that by Proposition 1.5.3, $\boldsymbol{U}P^c\boldsymbol{U}^{-1} \cap \mathrm{Orth}$ is identical with those elements $\boldsymbol{Q}$ of P^c which commute with $\boldsymbol{U}$. □

1.6 Higher-Order Tensors

Higher-order tensors can always be interpreted in several ways, and we shall indicate here the interpretations to be used throughout the book.

1.6.1 A ***fourth-order tensor*** is a linear transformation A from Lin into itself. Its value on $\boldsymbol{H} \in \mathrm{Lin}$ is written $\mathsf{A}[\boldsymbol{H}] \in \mathrm{Lin}$. In indices, with the summation convention, if $\boldsymbol{K} = \mathsf{A}[\boldsymbol{H}]$, then $K_{ij} = A_{ijkl}H_{kl}$. This definition is natural when one deals with the derivatives of the second-order-tensor-valued functions with respect to a second-order tensor argument, for then the derivative is the linear part of the function, see $\partial_{\boldsymbol{F}}\boldsymbol{U}(\boldsymbol{F})$, $\partial_{\boldsymbol{F}}\boldsymbol{V}(\boldsymbol{F})$, $\partial_{\boldsymbol{F}}\boldsymbol{R}(\boldsymbol{F})$, $\partial_{\boldsymbol{F}}\boldsymbol{P}_i(\boldsymbol{F})$ in Propositions 1.4.4 and 1.4.2. Since Lin has an inner product, one can identify a fourth-order tensor A with a bilinear form B on Lin with values written as $\mathsf{B}[\boldsymbol{H}, \boldsymbol{K}]$, $\boldsymbol{H}, \boldsymbol{K} \in \mathrm{Lin}$, by the formula

$$\mathsf{B}[\boldsymbol{H}, \boldsymbol{K}] = \boldsymbol{H} \cdot \mathsf{A}[\boldsymbol{K}] = \mathrm{tr}(\mathsf{A}[\boldsymbol{K}]\boldsymbol{H}^{\mathrm{T}}), \qquad \boldsymbol{H}, \boldsymbol{K} \in \mathrm{Lin}.$$

This identification allows us to use the same letter for both the two objects, i.e., we write $\mathsf{A}[\boldsymbol{H}, \boldsymbol{K}] = \boldsymbol{H} \cdot \mathsf{A}[\boldsymbol{K}]$. The interpretation as a bilinear form on Lin is natural when one deals with the second derivatives of scalar-valued functions of a second-order tensor argument, for then the second derivative is the quadratic part of the Taylor expansion of the function (e.g., $\partial_{\boldsymbol{F}}^2 \hat{v}_i(\boldsymbol{F})$ in Proposition 1.4.2).

1.6.2 Operations with Fourth-Order Tensors We denote by $\mathbf{1}$ the unit fourth-order tensor defined by $\mathbf{1}\boldsymbol{H} = \boldsymbol{H}$ for every $\boldsymbol{H} \in \mathrm{Lin}$, with $\mathbf{1}_{ijkm} = \delta_{ik}\delta_{jm}$. The ***tensor product*** $\boldsymbol{M} \otimes \boldsymbol{N}$ of two second-order tensors is a fourth-order tensor defined by

$$(\boldsymbol{M} \otimes \boldsymbol{N})[\boldsymbol{H}] = \boldsymbol{M}(\boldsymbol{N} \cdot \boldsymbol{H}), \qquad (\boldsymbol{M} \otimes \boldsymbol{N})[\boldsymbol{H}, \boldsymbol{K}] = (\boldsymbol{M} \cdot \boldsymbol{K})(\boldsymbol{N} \cdot \boldsymbol{H}),$$

$\boldsymbol{H}, \boldsymbol{K} \in \mathrm{Lin}$. The ***major transposition*** of a fourth-order tensor A is the fourth-order tensor $\mathsf{A}^{\mathbb{T}}$ satisfying

$$\mathsf{A}^{\mathbb{T}}[\boldsymbol{H}, \boldsymbol{K}] = \mathsf{A}[\boldsymbol{K}, \boldsymbol{H}], \quad \mathsf{A}^{\mathbb{T}}[\boldsymbol{H}] \cdot \boldsymbol{K} = \boldsymbol{H} \cdot \mathsf{A}[\boldsymbol{K}], \qquad \boldsymbol{H}, \boldsymbol{K} \in \mathrm{Lin}.$$

A is said to have a ***major symmetry*** if $\mathsf{A}^{\mathbb{T}} = \mathsf{A}$. In indices this means that $A_{ijkl} = A_{klij}$. The symmetry in the first pair of indices means that $\mathsf{A}[\boldsymbol{H}]^{\mathrm{T}} = \mathsf{A}[\boldsymbol{H}]$, $\boldsymbol{H} \in \mathrm{Lin}$, while the symmetry in the second pair of indices means that $\mathsf{A}[\boldsymbol{H}^{\mathrm{T}}] = \mathsf{A}[\boldsymbol{H}]$, $\boldsymbol{H} \in \mathrm{Lin}$.

1.6.3 Third-Order Tensors A third order tensor D is a linear transformation from Vect into Lin, or a linear transformation E from Lin into Vect, or a trilinear form on Vect. A major transpose of a linear transformation D from Vect into Lin is a linear transformation $\mathsf{D}^{\mathbb{T}}$ form Lin into Vect such that

$$\mathsf{D}[\boldsymbol{v}] \cdot \boldsymbol{H} = \mathsf{D}^{\mathbb{T}}[\boldsymbol{H}] \cdot \boldsymbol{v}, \qquad \boldsymbol{v} \in \mathrm{Vect}, \quad \boldsymbol{H} \in \mathrm{Lin},$$

and a major transpose of a linear transformation E from Lin into Vect is a linear transformation $\mathsf{E}^{\mathbb{T}}$ form Vect into Lin such that

$$\mathsf{E}[\boldsymbol{H}] \cdot \boldsymbol{v} = \mathsf{E}^{\mathbb{T}}[\boldsymbol{v}] \cdot \boldsymbol{H}, \qquad \boldsymbol{v} \in \mathrm{Vect}, \quad \boldsymbol{H} \in \mathrm{Lin}.$$

Hence $(\mathsf{D}^{\mathrm{T}})^{\mathrm{T}} = \mathsf{D}$, $(\mathsf{E}^{\mathrm{T}})^{\mathrm{T}} = \mathsf{E}$. If F is a trilinear form on Lin, we denote by $\mathsf{F}[\boldsymbol{u}, \boldsymbol{v}, \boldsymbol{w}]$ its value on $\boldsymbol{u}, \boldsymbol{v}, \boldsymbol{w} \in \mathrm{Vect}$. The relationship to D, E as above is given by

$$\mathsf{F}[\boldsymbol{u}, \boldsymbol{v}, \boldsymbol{w}] = \mathsf{D}[\boldsymbol{u}] \cdot (\boldsymbol{v} \otimes \boldsymbol{w}) = \boldsymbol{u} \cdot \mathsf{E}[\boldsymbol{v} \otimes \boldsymbol{w}], \qquad \boldsymbol{u}, \boldsymbol{v}, \boldsymbol{w} \in \mathrm{Vect}.$$

(The formula of Proposition 1.4.2 for $\partial_{\boldsymbol{F}}\boldsymbol{e}_i(\boldsymbol{F})$ is an example of a third-order tensor identified with the linear transformation from Lin to Vect.)

2 Geometry and Kinematics of Continuous Bodies

Sects. 2.1 and 2.2 introduce inhomogeneous processes with moving singular surfaces across which various physical quantities may have jump discontinuities. The singular surfaces will be employed to model phase boundaries and shock waves. The Hadamard lemma says that the limiting values of the deformation gradient on the singular surface of a continuous (coherent) motion differ by a rank 1 tensor. Sects. 2.3–2.5 deal with consequences of this fact. Geometrical aspects include layered mixtures of twinned phases of solids, the construction of the austenite/martensite interface, and the structure of potential wells of the stored energy at the transition temperature. More algebraic results are the determination of the principal stretches of rank 1 perturbations of a given deformation gradient and the solution of the twinning equations. Many of these kinematical ideas underlie the energy stability considerations and constitutive inequalities to be treated in the subsequent parts of the book.

2.1 Processes with Singular Surfaces

2.1.1 Continuous Bodies A ***continuous body*** is a set B of points called material points. Let Eucl be an n-dimensional euclidean point space (= affine space) with the translation vector space Vect considered in the preceding chapter. A configuration of the body assigns to each material point its position in Eucl. It is convenient to choose one configuration as a reference configuration. If $\Omega \subset$ Eucl is the region occupied by B in this reference configuration, the material points of B can be labeled by the positions $\boldsymbol{p} \in \Omega$. It is assumed that the region Ω is an open (but not necessarily bounded or connected) subset of Eucl whose boundary $\partial\Omega$ is regular enough to ensure the validity of the divergence theorem. Appropriate assumptions about the regularity of $\partial\Omega$ will be made where they are needed; e.g., Ω can be a region with almost smooth boundary defined in Subsect. 2.6.2. We shall always assume that the exterior normal $\boldsymbol{n}$ to Ω is defined for almost every point $\boldsymbol{p}$ of $\partial\Omega$, where here 'almost every' means that that the set of point where $\boldsymbol{n}$ is not defined has the area measure 0. (Generally, given a measure M on a set X, we say that some property $P(\boldsymbol{x})$ holds for M-almost every $\boldsymbol{x} \in$ X, (abbreviated M-a.e.) if the measure M of the set of all $\boldsymbol{x} \in$ X where $P(\boldsymbol{x})$ is violated is 0.) In the subsequent treatment, the letter $\boldsymbol{n}$ is also used to denote the normal to the singular surface, and to avoid confusion, we shall sometimes write $\boldsymbol{n}^*$ or $\boldsymbol{n}^\Omega$ for the exterior normal to Ω. Each open region $\Omega \subset$ Eucl whose boundary has the chosen degree of regularity is called an ***admissible region***. Each bounded subset

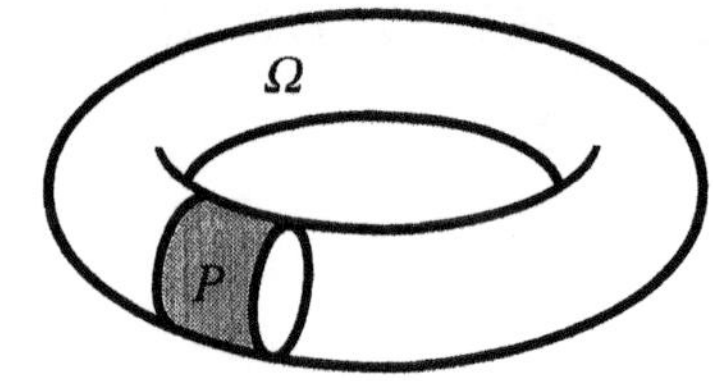

Fig. 2.1.1. A continuous body

$P \subset \Omega$ which is an admissible region is called a ***part*** of Ω. See Fig. 2.1.1, in which Ω has a boundary of class C^∞ and P has only a Lipschitz boundary.

2.1.2 Moving Surfaces If the body undergoes a phase transition or a shock wave, there is a time-dependent region of material points where physical quantities change extremely rapidly in space and time. For the description of such processes within phenomenological theories, fields with jump discontinuities across a moving two-dimensional surface (singular surface) can be used. We start with a description of a moving surface in the reference configuration.

Let $[t_1, t_2]$ be a time-interval, $t_1 < t_2$, Ω an admissible region and ϕ a continuously differentiable function defined on an open set containing the closure of $\Omega \times [t_1, t_2]$ such that $\nabla\phi \neq \mathbf{0}$ at every point where $\phi = 0$. Here $\nabla\phi$ and $\dot{\phi}$ are the referential gradient and the material time derivative of ϕ, i.e., the derivative of $\phi(\boldsymbol{p}, t)$ with respect to $\boldsymbol{p}$ and t, respectively. A ***moving surface*** in Ω on $[t_1, t_2]$ is a family $\{S_t, t \in [t_1, t_2]\}$ of closed oriented surfaces in Ω given by

$$S_t = \{\boldsymbol{p} \in \operatorname{cl}\Omega : \phi(\boldsymbol{p}, t) = 0\}, \quad t \in [t_1, t_2].$$

Here $\operatorname{cl}\Omega$ denotes the closure of Ω. The spacetime set U composed of the spacetime points on the singular surface, given by

$$U = \{(\boldsymbol{p}, t) \in \operatorname{cl}\Omega \times [t_1, t_2] : \phi(\boldsymbol{p}, t) = 0\}, \tag{2.1.1}$$

is a three-dimensional hypersurface; frequently it will be this set which will be referred to as the moving surface. Define the spatial regions Ω_t^+, Ω_t^-, $t \in [t_1, t_2]$, and the spacetime regions T^+, T^- by

$$\Omega_t^\pm = \{\boldsymbol{p} \in \Omega : \pm\phi(\boldsymbol{p}, t) > 0\}, \quad T^\pm = \{(\boldsymbol{p}, t) \in \Omega \times (t_1, t_2) : \pm\phi(\boldsymbol{p}, t) > 0\}. \tag{2.1.2}$$

The ***referential normal n*** to the surface and the ***referential speed*** U of the surface are defined by

$$\boldsymbol{n} = \nabla\phi / |\nabla\phi|, \quad U = -\dot{\phi} / |\nabla\phi|. \tag{2.1.3}$$

With this definition, the normal $\boldsymbol{n}$ points from Ω_t^- into Ω_t^+. If $\dot{\phi}(\boldsymbol{p}, t) \neq 0$, at some point of the singular surface, one can always choose ϕ in such a way that $\dot{\phi} < 0$ and hence $U > 0$ in some neighborhood of $(\boldsymbol{p}, t)$; in this case Ω_t^+ is the region ahead of the surface and Ω_t^- the region behind the surface. If $U(\boldsymbol{p}, t) = 0$ at some point and U changes its sign from negative to positive, then the surface stops and then starts to move in the opposite direction; in this case the above change is impossible. However, for considerations about a single point of the singular surface, one can always assume $U \geq 0$, and this will be mostly our convention. Also notice that to within the signs, the

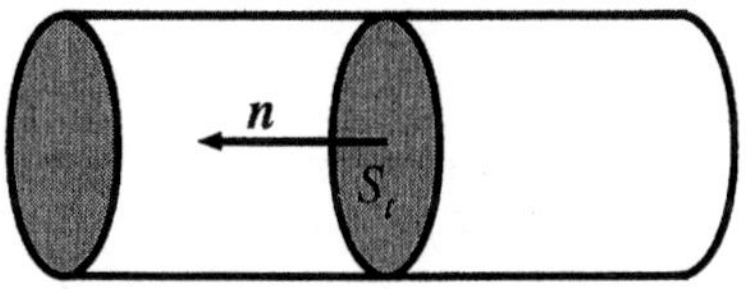

Fig. 2.1.2. A moving singular surface

definition of $\boldsymbol{n}$ and U are independent of the choice of the function ϕ used to describe the surface. See Fig. 2.1.2.

The definition does not imply that $T^{\pm}$, U, S_t or $\Omega_t^{\pm}$ are connected. If Ω is bounded, then U is a compact subset of a larger n-dimensional surface $\tilde{U}$ in Vect $\times$ $\mathbb{R}$, namely of the set of all $(\boldsymbol{p}, t)$ in the domain of ϕ satisfying $\phi(\boldsymbol{p}, t) = 0$; recall that it is assumed that ϕ is defined on some open set containing the closure of $\Omega \times (t_1, t_2)$. Hence the total area od S_t is finite for all times. The definition does not exclude that S_t be empty for some or all values of time $t \in [t_1, t_2]$. However, if $S_s \neq \emptyset$ for some $s \in (t_1, t_2)$, then $S_t \neq \emptyset$ for all t sufficiently close to s by the implicit function theorem. This prevents the singular surface from emerging or ceasing suddenly at some interior point of Ω : the concentric spheres of radii proportional to t starting to expand from some point of Ω at $t = 0$ are excluded. The only way for the surface to be empty for all $t \le s$ and nonempty for all $t > s$ is if it enters the body through the boundary. Also the singular surface cannot intersect itself. This rather restrictive notion of a singular surface can be generalized in several ways, e.g. by admitting that the description through a single ϕ holds only locally, and Sect. 2.6 introduces a broader notion of the singular surface. However, for the derivation of the divergence and transport theorems for fields with singular surface we shall use the definition as it is formulated above. This means that the regions ahead of and behind the singular surface are unambiguously defined on Ω globally.

2.1.3 Time-Dependent Fields Consider a long tube filled with a gas divided in the middle by a thin wall. The gas is in equilibrium in each of the two halves of the tube, but with different pressures. A sudden removal of the wall will start a process with one or more moving singular surfaces that separate regions where the process is smooth. The singular surfaces are known as shock waves and contact discontinuities, with jumps in the specific volume and other quantities, and the smooth portions of the process known as rarefaction waves. This is the Riemann problem in gas dynamics, see Sect. 23.5. An example involving a solid state is an impact problem in which a long steel bar is hit at one end by a hammer. This causes a jump in the value of the deformation gradient and the future evolution of the bar will again involve a shock wave. An example of a different kind is an equilibrium state of two coexistent phases separated by an immobile phase boundary, such as a solid–melt interface, and its dynamical counterpart, the process of melting or solidification. Of particular interest are states resulting from solid–solid martensitic transformations, in which the austenite phase coexists with a number of twinned variants of the martensite. The kinematics and the general balance equations for shock waves and (coherent) phase boundaries are identical. The distinction is constitutive–the response functions behave differently near phase transitions and differently in regions where the shock waves occur.

A time-dependent scalar-, vector- or tensor-valued field f with a singular surface on $\Omega \times (t_1, t_2)$ is any function defined on $\Omega \times (t_1, t_2) \setminus U$, where $U = \{S_t, t \in [t_1, t_2]\}$ is (a possibly empty) moving surface. The field f is said to be ***piecewise continuous*** (***piecewise continuously differentiable***) if the restriction of f to T^+ (see (2.1.2)) has a continuous (continuously differentiable, respectively) extension h^+ to a neighborhood

O^+ of the closure of T^+ and the restriction of f to T^- has a continuous (continuously differentiable) extension h^- to a neighborhood O^- of the closure of T^-.

2.1.4 The Values on the Singular Surface The extensions $h^\pm$ are not unique, but their values on U, being the limits of f from both sides of S_t, are uniquely determined by f. These values will be denoted by $f^\pm$; as a consequence,

$$f^\pm(\boldsymbol{p}, t) = \lim_{\delta \downarrow 0} f(\boldsymbol{p} \pm \delta \boldsymbol{n}, t), \quad f^\pm(\boldsymbol{p}, t) = \lim_{\delta \downarrow 0} f(\boldsymbol{p}, t \mp \delta)$$

for every $\boldsymbol{p} \in S_t \cap \Omega$. The ***jump*** $[f]$ and the ***mean value*** $\langle f \rangle$ of f on the singular surface are defined by

$$[f] = f^+ - f^-, \quad \langle f \rangle = \tfrac{1}{2}(f^+ + f^-)$$

so that $[f]$ and $\langle f \rangle$ are continuous functions defined on U. If f, g are two piecewise continuous fields with the same singular surface, then

$$[fg] = [f]\,\langle g \rangle + \langle f \rangle\,[g] \tag{2.1.4}$$

everywhere on the singular surface. (The similarity to the rule for the derivative of a product is not incidental: the jump $[f]$ determines the singular part of the distributional derivative of f, see Proposition 3.8.7.) The field f is said to be ***continuous and piecewise continuously differentiable*** if it is piecewise continuously differentiable and $[f] = 0$ identically on the singular surface.

Let the operator ∇ denote the ***referential gradient,*** i.e., a differentiation with respect to the reference position $\boldsymbol{p}$, and the superimposed dot the ***material time derivative,*** i.e., a derivative with respect to t holding $\boldsymbol{p}$ fixed:

$$\nabla f(\boldsymbol{p}, t) = \partial_{\boldsymbol{p}} f(\boldsymbol{p}, t), \quad \dot{f}(\boldsymbol{p}, t) = \partial_t f(\boldsymbol{p}, t),$$

whenever the derivatives exist. For a piecewise continuously differentiable field f one can define the values $\nabla f^\pm, \dot{f}^\pm$ on the singular surface as the values of ∇h^+, ∇h^- and $\partial_t h^+$, $\partial_t h^-$ at the given point, and consequently the jumps on the singular surface:

$$[\nabla f] = \nabla h^+ - \nabla h^-, \quad [\dot{f}] = \partial_t h^+ - \partial_t h^-.$$

2.1.5 The Boundary Values Recall that the exterior normal $\boldsymbol{n}^*$ to Ω is defined for A-a.e. point of the boundary, where A is the standard area measure on $\partial\Omega$. The boundary of the spacetime region $Q := \Omega \times (t_1, t_2)$ is $\partial Q = \partial\Omega \times [t_1, t_2] \cup \Omega \times \{t_1\} \cup \Omega \times \{t_2\}$ and clearly $\partial Q \subset \partial T^+ \cup \partial T^-$; hence at each point $(\boldsymbol{p}, t)$ of ∂Q at least one of the values $h^\pm(\boldsymbol{p}, t)$ is defined and completely determined by the behavior of f in $Q \setminus U$. On the sets $\Omega \times \{t_1\}$ and $\Omega \times \{t_2\}$ these values just extend f to the times $t = t_1, t_2$. On $\partial\Omega \times [t_1, t_2]$ the values $h^\pm(\boldsymbol{p}, t)$ enable one to define the ***boundary value*** f^b of f by

$$f^b(\boldsymbol{p}, t) = \begin{cases} h^+(\boldsymbol{p}, t) & \text{if } (\boldsymbol{p}, t) \in B^+ \cup D^+, \\ h^-(\boldsymbol{p}, t) & \text{if } (\boldsymbol{p}, t) \in B^- \cup D^-, \\ \text{undefined} & \text{if } (\boldsymbol{p}, t) \in D^0, \end{cases}$$

where

$$B^\pm = \{(\boldsymbol{p}, t) \in \partial T^\pm \setminus U : \boldsymbol{p} \in \partial\Omega\}, \quad D^\pm = \{(\boldsymbol{p}, t) \in U : \boldsymbol{p} \in \partial\Omega, \boldsymbol{n}^*(\boldsymbol{p}) = \mp \boldsymbol{n}(\boldsymbol{p}, t)\},$$

and D^0 is the set of all $(\boldsymbol{p}, t) \in U$ with $\boldsymbol{p} \in \partial\Omega$ such that $\boldsymbol{n}^*(\boldsymbol{p})$ is either undefined or $\boldsymbol{n}^*(\boldsymbol{p})$, $\boldsymbol{n}(\boldsymbol{p}, t)$ are linearly independent. Note that $\partial\Omega \times [t_1, t_2]$ can be written disjointly as

$$\partial\Omega \times [t_1, t_2] = B^+ \cup D^+ \cup B^- \cup D^- \cup D^0;$$

moreover, $H^n(D^0) = 0$ by Proposition 3.8.8(1), where H^n is the n-dimensional area measure (Hausdorff measure) in the $n+1$ dimensional space Vect $\times$ IR. Hence $f^b(\boldsymbol{p}, t)$ is defined for H^n-a.e. point $(\boldsymbol{p}, t) \in \partial\Omega \times [t_1, t_2]$, or, put otherwise, for every $t \in [t_1, t_2]$, $f^b(\boldsymbol{p}, t)$ is defined for A-a.e. point $\boldsymbol{p} \in \Omega$. Whenever there is no danger of confusion, we write f for f^b. The definition of the boundary values applies to $\nabla f, \dot{f}$ if f is piecewise continuously differentiable. See also Subsect. 3.1.1 and Fig. 3.1.1.

For each t, the set S_t is closed. However, *unless stated otherwise, throughout the rest of the book, by the singular surface we shall mean the part of S_t that is contained in the interior of Ω, and we shall use the same symbol S_t for it.*

The appropriate function space for processes with singular surfaces is the space BV of functions with bounded variation, see Sect. 3.8. In Part I, we shall mostly restrict ourselves to piecewise continuous and piecewise continuously differentiable processes as defined above. This is no serious loss of generality, for two reasons: (a) the structural result for functions of bounded variation says that for each such function there exists a countable family of surfaces across which the functions have jump discontinuities in the measure-theoretic sense; (b) the Riemann problem is in many cases solvable in the class of piecewise continuously differentiable functions.

2.1.6 Proposition (The Hadamard Lemma) *If f is a scalar-valued continuous and piecewise continuously differentiable field, then there exists a continuous scalar-valued function C on the singular surface U (see* (2.1.1)*) such that*

$$[\nabla f] = C\boldsymbol{n}, \quad [\dot{f}] = -UC \text{ on } U. \tag{2.1.5}$$

Conversely, if f is a piecewise continuously differentiable field, the singular surface U is connected, (2.1.5) *holds with some function C on U, and* $[f] = 0$ *at some point of U, then* $[f] = 0$ *everywhere on U.*

This is a kinematical compatibility condition; the quantity C is called the ***amplitude of the jump*** in derivatives of f.

Proof Let $h^\pm : O^\pm \to \mathbb{R}$ be the extensions of f from the definition of a piecewise continuously differentiable field. Then $U \subset O := O^+ \cap O^-$ and $h := h^+ - h^- : O \to \mathbb{R}$ is continuously differentiable. Since $[f] = 0$, h vanishes identically on $U \subset \text{Vect} \oplus \mathbb{R}$. At every point on U, the function h has therefore a relative minimum (and also maximum) with respect to the constraint $\phi(\boldsymbol{p}, t) = 0$. The theorem on Lagrange's multipliers applies and asserts the existence of a $\lambda = \lambda(\boldsymbol{p}, t) \in \mathbb{R}$ such that the four-dimensional gradient GRAD $= (\nabla, \dot{\ })$ of h satisfies GRAD $h = \lambda$ GRAD ϕ. Splitting into the spatial and time components and using $[\nabla f] = \nabla h$, $[\dot{f}] = \partial_t h$, we find $[\nabla f] = \lambda \nabla\phi$, $[\dot{f}] = \lambda\dot{\phi}$. From (2.1.3) it follows that (2.1.5) holds with $C = \lambda|\nabla\phi|$. To prove the converse part, let $(\boldsymbol{p}, t) \in U$ be the point where $[f] = 0$. For any $(\boldsymbol{q}, s) \in U$ there exists a continuously differentiable curve $\gamma : [0, 1] \to U$ connecting $(\boldsymbol{q}, s)$ with $(\boldsymbol{p}, t)$. Then $d/d\tau[f](\gamma(\tau)) = \text{GRAD}\, h(\gamma(\tau)) \cdot \dot{\gamma}(\tau) = C(\gamma(\tau))(\boldsymbol{n}, -U) \cdot \dot{\gamma}(\tau) = 0$. Integration gives $[f] = 0$ at $(\boldsymbol{q}, s)$. □

2.2 Motion and Deformation

In a chosen inertial reference frame, the ***classical spacetime*** can be identified with the cartesian product Eucl $\times$ IR of the euclidean space Eucl and the time axis IR. A typical element of Eucl $\times$ IR is $(\boldsymbol{x}, t)$ with $\boldsymbol{x}$ the position and t the time. If convenient, Eucl can be identified with $\mathbb{R}^n$. I refer to NOLL [1973], WANG & TRUESDELL [1973], and MARSDEN & HUGHES [1983] for a more detailed analysis of classical spacetimes. Changes of frame will be discussed in Chap. 6.

2.2.1 A ***motion*** χ of Ω on $[t_1, t_2]$ is a continuous and piecewise continuously differentiable field on $\Omega \times (t_1, t_2)$ with values in Eucl which satisfies Conditions (1)–(3) below. By continuity, $\chi(\boldsymbol{p}, t)$ is unequivocally defined even for $(\boldsymbol{p}, t)$ on the singular surface U; moreover, χ has a well-defined ***boundary value*** $\chi^b(\boldsymbol{p}, t) \equiv \chi(\boldsymbol{p}, t)$ for every $\boldsymbol{p} \in \partial\Omega$, $t \in [t_1, t_2]$. The value $\boldsymbol{x} = \chi(\boldsymbol{p}, t)$ is the position of the material point $\boldsymbol{p} \in \Omega$ at time $t \in [t_1, t_2]$. The partial mapping $\chi(\cdot, t)$ is the ***configuration*** at time t. The spatial set $r_t = \chi(\Omega, t) = \{\chi(\boldsymbol{p}, t) \in \text{Eucl} : \boldsymbol{p} \in \Omega\}$ is the current region occupied by the body at time t and the spacetime set

$$T = \{(\boldsymbol{x}, t) : t \in [t_1, t_2], \boldsymbol{x} = \chi(\boldsymbol{p}, t) \text{ for some } \boldsymbol{p} \in \Omega\} \tag{2.2.1}$$

is the spacetime tube of the body during the motion. If $S_t \subset \Omega$ is the singular surface at time t, then the set $s_t \equiv \chi(S_t, t) \equiv \{\chi(\boldsymbol{p}, t) \in \text{Eucl} : \boldsymbol{p} \in S_t\}$ is the ***spatial configuration of the singular surface*** at time t. By the definition of a piecewise continuously differentiable field in Subsect. 2.1.3, the restrictions of χ to $T^\pm$ have continuously differentiable extensions $\tilde{\chi}^\pm$ to open sets containing cl $T^\pm$. It is assumed that

(1) for every $t \in [t_1, t_2]$, $\chi(\cdot, t)$ maps Ω bijectively onto r_t; i.e., $\chi(\cdot, t)$ is a one-to-one mapping of Ω onto r_t;

(2) χ is twice continuously differentiable on $\Omega \times (t_1, t_2) \setminus U$;

(3) the extensions $\tilde{\chi}^\pm$ may be chosen such that either $\det \nabla\tilde{\chi}^\pm(\boldsymbol{p}, t) > 0$ for every $(\boldsymbol{p}, t) \in \text{cl } T^\pm$ or $\det \nabla\tilde{\chi}^\pm(\boldsymbol{p}, t) < 0$ for every $(\boldsymbol{p}, t) \in \text{cl } T^\pm$.

A ***deformation*** χ (equivalently, ***configuration***) of Ω is a motion of Ω that is independent of time; hence $\chi : \Omega \to \text{Eucl}$.

2.2.2 ***Remarks***

(1) Conditions (1)–(3) allow us to employ freely the standard change of variable formulas for volume and surface integrals and thus guarantee, e.g., the equivalence of the referential and spatial forms of balance equations. However, the existence theorems for nonlinear elasticity require spaces of deformations for which this equivalence may fail (see Sect. 21.4). These matters are currently under discussion and even though several proposals exist, an all-embracing theory seems to be outstanding. For this reason, and also to keep the presentation on an elementary level, we adopt (1)–(3) above as model assumptions.

(2) The class of continuous motions models adequately coherent shock waves and coherent phase boundaries. For the process of melting the continuity of motion is probably unnecessary; on the other hand, it may be satisfied because of the nature of the problem. Of the problems involving discontinuous motions we mention (a) fractures, (b) dislocations, (c) incoherent phase transitions, and (d) cavitation. I refer to Sect. 21.2 for a brief description of (d). In fracture mechanics, the discontinuity models a crack (see MAUGIN [1992, Chap. 7]). In the opening mode, the crack creates a hole, and there is no additional restriction on χ beyond the impenetrability of matter. In the shear plane and antiplane modes, no hole is created and the lips of the crack slip one relative another, which imposes a condition on the kind of discontinuity of χ. Volterra's or Somigliana's dislocations (see, e.g., ESHELBY [1956, Sect. 4], GURTIN [1972, Sect. 14], HANYGA [1985, Sect. 4.2], and HEAD & al. [1993]) of the continuum theory of dislocations are visualized by the cut-and-weld operations in the reference configuration. These lead to discontinuous χ which again do not open holes. Finally, also the theory of incoherent phase transformations (see GRINFELD [1991] and CERMELLI & GURTIN [1994] and the references therein) leads to motions in which the matter on one side of the singular surface slips relative to the other.

(3) The discontinuous motions which do not produce macroscopic holes may be described as follows. One can consider a piecewise continuously differentiable (but discontinuous) motion χ with a singular surface S_t having the property that for every $t \in [t_1, t_2]$ there exists a diffeomorphism $\boldsymbol{h}(\cdot, t)$ of S_t onto S_t such that

$$\chi^+(\boldsymbol{p}, t) = \chi^-(\boldsymbol{h}(\boldsymbol{p}, t), t) \quad \text{for every } \boldsymbol{p} \in S_t$$

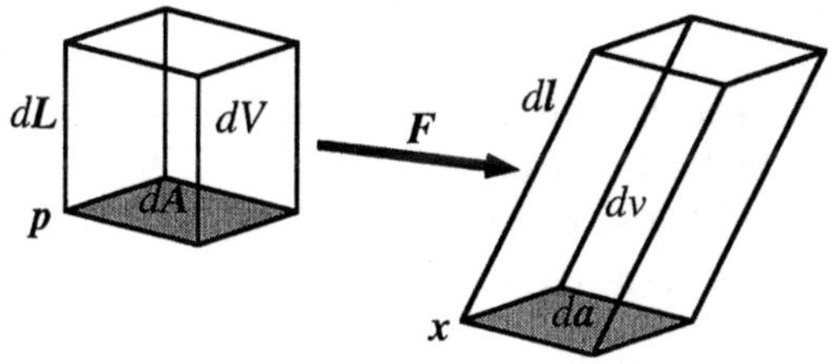

Fig. 2.2.1. Deformation of lines, areas, and volumes

where $\chi^{\pm}$ are the limits of χ on S_t. Kinematically, at every instant, the body is cut along S_t, then the points of the two sides of the cut are allowed to slip along S_t and finally they are glued together. I refer to CERMELLI & GURTIN [1994]; see also DEL PIERO & OWEN [1993, 1995] for a different approach.

(4) For a given Ω, the set of all continuous and continuously differentiable deformations can have a rather complicated topological structure. For example, there may exist two positively oriented deformations which cannot be connected by a continuous family of positively oriented deformations, etc. See HANYGA [1985, Subsects. 1.2.2–3] and ANTMAN [1995, Chap. XIII].

2.2.3 The First Spacetime Derivatives These are the ***deformation gradient*** $\boldsymbol{F}$ and the ***velocity*** $\boldsymbol{v}$, defined outside the singular surface by

$$\boldsymbol{F} = \nabla\chi, \quad \boldsymbol{v} = \dot{\chi}.$$

We denote by $\boldsymbol{F}^{\pm}$, $\boldsymbol{v}^{\pm}$ the values of these quantities on the singular surface and by $\boldsymbol{F}$, $\boldsymbol{v}$ the well-defined boundary values $\boldsymbol{F}^b$, $\boldsymbol{v}^b$ in the sense of Subsect. 2.1.5. The definition of motion implies that $\boldsymbol{F}, \boldsymbol{F}^{\pm}, \boldsymbol{F}^b$ are invertible tensors. One can apply the polar decomposition theorem 1.4.3 to $\boldsymbol{F} = \boldsymbol{F}(\boldsymbol{p}, t)$ to obtain $\boldsymbol{F} = \boldsymbol{RU} = \boldsymbol{VR}$, where $\boldsymbol{R}, \boldsymbol{U}, \boldsymbol{V}$ are the polar factors of $\boldsymbol{F}$ depending on $\boldsymbol{p}, t$. The tensor $\boldsymbol{R}$ is the ***rotation tensor*** and $\boldsymbol{U}, \boldsymbol{V}$ are the ***right*** and ***left stretch tensors***, respectively. The eigenvalues of $\boldsymbol{U}, \boldsymbol{V}$ coincide; they are called the ***principal stretches.*** The eigenvectors $\boldsymbol{f}_i$ of $\boldsymbol{U}$ are called the principal axes of strain in the undeformed state and the eigenvectors $\boldsymbol{e}_i = \boldsymbol{R}\boldsymbol{f}_i$ of $\boldsymbol{V}$ are called the principal axes of strain in the deformed state; alternatively, $\boldsymbol{f}_i$ and $\boldsymbol{e}_i$ are called the axes of the Lagrangian and Eulerian strain ellipsoid, respectively. We denote by $J, J^{\pm}$ the Jacobian determinants $J = \det \boldsymbol{F}$, $J^{\pm} = \det \boldsymbol{F}^{\pm}$. They are of the same sign and we denote

$$s = \operatorname{sgn} \det \boldsymbol{F} = 1 \text{ if } J > 0, \quad s = \operatorname{sgn} \det \boldsymbol{F} = -1 \text{ if } J < 0.$$

The deformation gradient $\boldsymbol{F}$ and the polar factors $\boldsymbol{U}, \boldsymbol{V}, \boldsymbol{R}$ are the primary measures of deformation. Many different measures of strain, such as the left and right Cauchy–Green deformation tensors, the Almansi tensor etc., are used. These are remnants of the linear theories and will not be used in the main exposition. (Occasionally the left and right Cauchy–Green deformation tensors will be used for comparison purposes.) A unified treatment of other deformation measures is briefly sketched in Sect. 10.4 together with the conjugate stress measures. Also the classical infinitesimal deformation tensor enters on a much later stage: in connection with linearization, cf. Sect. 10.5.

2.2.4 Deformation of Lines, Areas, and Volumes Let $(\boldsymbol{p}, t)$ be a point outside the singular surface of a motion χ and consider the oriented line, area and volume elements $d\boldsymbol{L}, d\boldsymbol{A}, dV$ at $(\boldsymbol{p}, t)$ (see Fig. 2.2.1). The deformation changes these infinitesimal objects into $d\boldsymbol{l}, d\boldsymbol{a}, dv$, respectively, situated at $\boldsymbol{x} = \chi(\boldsymbol{p}, t)$, where

$$d\boldsymbol{l} = \boldsymbol{F}\, d\boldsymbol{L}, \quad d\boldsymbol{a} = s \operatorname{cof} \boldsymbol{F}\, d\boldsymbol{A}, \quad dv = s \det \boldsymbol{F}\, dV. \tag{2.2.2}$$

The sign convention in $(2.2.2)_2$ with $s = \operatorname{sgn} \det \boldsymbol{F}$ is chosen so that the exterior normal to the boundary of the body maps into the exterior normal, and not the interior normal, even under transformations with negative determinant. Similarly, if $S_0 \subset \Omega$ is an oriented surface not intersecting the singular surface, and if $\boldsymbol{n}$ is its field of normals, then $\boldsymbol{\chi}(S_0, t)$ is an oriented surface with the normal

$$\boldsymbol{m} = |\boldsymbol{F}^{-\mathrm{T}}\boldsymbol{n}|^{-1}\boldsymbol{F}^{-\mathrm{T}}\boldsymbol{n} = |(\operatorname{cof}\boldsymbol{F})\boldsymbol{n}|^{-1}(\operatorname{cof}\boldsymbol{F})\boldsymbol{n}. \tag{2.2.3}$$

The formulas (2.2.2) are the symbolical expressions of the change of variables formulas for the line, surface, and volume integrals. Here and throughout, dV denotes the element of volume, i.e., the Lebesgue measure in Eucl and Vect; the reader is referred to RUDIN [1970] for the Lebesgue measure and for the change of variable formula for the Lebesgue measure under diffeomorphisms. For the general form, see Subsect. 21.2.3. The element of area is denoted by dA; the reader is referred to NEČAS [1967] for the area measure on Lipschitz surfaces (see also Proposition 3.8.10 and its proof). Alternatively, A may be interpreted as the $n-1$-dimensional Hausdorff measure. The proof of the change of variables formula for the surface integrals $(2.2.2)_2$ is harder to find in the mathematical literature; however, $(2.2.2)_2$ is an elementary consequence of the integration theory of exterior forms over differentiable manifolds, see e.g., BISHOP & GOLDBERG [1968]. The geometric measure theory provides a unified treatment of the change of variable formulas under minimal smoothness assumptions; I refer to FEDERER [1969], EVANS & GARIEPY [1992], and the forthcoming treatise GIAQUINTA, MODICA & SOUČEK [1996]. We shall return to these matters in Chap. 3.

2.2.5 The Spatial Description If f is a referential description of a time-dependent field with the singular surface S_t (see Sect. 2.1.3), then for a given motion one can express f as a function $\tilde{f}$ of the actual position $\boldsymbol{x}$ and time t where $f(\boldsymbol{p}, t) = \tilde{f}(\boldsymbol{x}, t)$ and $\boldsymbol{x} = \boldsymbol{\chi}(\boldsymbol{p}, t)$. The function $\tilde{f}$ is defined on the spacetime tube T (see (2.2.1)). The concepts of piecewise continuous, piecewise continuously differentiable and continuous and piecewise continuously differentiable fields in the spatial description are defined analogously to those for the referential description. The ***spatial gradient*** and the ***partial time derivative*** of f are denoted by $\operatorname{grad}\tilde{f}$ and $\partial\tilde{f}/\partial t$, respectively. Since for a given motion both f and $\tilde{f}$ express the same quantity, from now on we shall not distinguish f and $\tilde{f}$ graphically and denote consistently both of them by f. We write similarly $\operatorname{grad} f$ and $\partial f/\partial t$ for $\operatorname{grad}\tilde{f}$ and $\partial\tilde{f}/\partial t$ with the important convention that from now on the partial derivative symbol $\partial f/\partial t$ is reserved, unless otherwise stated, for the partial time derivative $\partial\tilde{f}/\partial t$ while $\dot{f}$ is used exclusively for the material time derivative of f. Both $\operatorname{grad} f$ and $\partial f/\partial t$ can be expressed as functions of $\boldsymbol{p}$ and t if convenient and conversely ∇f and $\dot{f}$ as functions of $(\boldsymbol{x}, t)$. The following well-known relations hold between the referential and spatial differential operations:

$$\nabla f = \boldsymbol{F}^{\mathrm{T}} \operatorname{grad} f, \quad \dot{f} = \frac{\partial f}{\partial t} + (\operatorname{grad} f) \cdot \boldsymbol{v}, \tag{2.2.4}$$

$$\operatorname{grad} f = \boldsymbol{F}^{-\mathrm{T}} \nabla f, \quad \frac{\partial f}{\partial t} = \dot{f} - \nabla f \cdot \boldsymbol{F}^{-1}\boldsymbol{v}. \tag{2.2.5}$$

The referential vector $-\boldsymbol{F}^{-1}\boldsymbol{v}$ can be called the referential velocity as it gives the rate of change of the referential positions of material points that occupy the same place during the motion.

The ***spatial gradient of velocity L***, ***stretching D***, and ***spin W*** are defined by

$$\boldsymbol{L} = \operatorname{grad} \boldsymbol{v}, \quad \boldsymbol{D} = \tfrac{1}{2}(\boldsymbol{L} + \boldsymbol{L}^{\mathrm{T}}), \quad \boldsymbol{W} = \tfrac{1}{2}(\boldsymbol{L} - \boldsymbol{L}^{\mathrm{T}}). \tag{2.2.6}$$

Roughly, $\boldsymbol{D}$ measures the rate of change of pure deformation, while $\boldsymbol{W}$ measures the rate of rotation. By $(2.2.5)_1$, $\boldsymbol{L} = \dot{\boldsymbol{F}}\boldsymbol{F}^{-1}$. If the velocity is known as a function of the actual position and time, $\boldsymbol{v} = \boldsymbol{v}(\boldsymbol{x}, t)$, then the deformation function $\boldsymbol{\chi} = \boldsymbol{\chi}(\boldsymbol{p}, t)$ can be reconstructed by solving the ordinary differential equation

$$\dot{\boldsymbol{\chi}}(\boldsymbol{p}, t) = \boldsymbol{v}(\boldsymbol{\chi}(\boldsymbol{p}, t), t), \quad \boldsymbol{p} \in \Omega, \quad t \in [t_1, t_2]$$

with the initial condition that, say, $\boldsymbol{\chi}(\boldsymbol{p}, t_1) = \boldsymbol{\chi}_0(\boldsymbol{p}), \boldsymbol{p} \in \Omega$, where $\boldsymbol{\chi}_0$ is a configuration of the body at time $t = t_1$, fixing the reference configuration. Every motion $\boldsymbol{\chi}$ satisfies Euler's identities

$$\dot{J} = J \operatorname{tr} \boldsymbol{L} = J \operatorname{tr} \boldsymbol{D} = J \operatorname{div} \boldsymbol{v}, \quad \frac{\partial J^{-1}}{\partial t} + \operatorname{div}(J^{-1}\boldsymbol{v}) = 0, \tag{2.2.7}$$

$$-u[J^{-1}] + [J^{-1}\boldsymbol{v}] \cdot \boldsymbol{m} = 0 \tag{2.2.8}$$

outside and on the singular surface, respectively. In (2.2.7), div denotes the ***spatial divergence,*** i.e., the divergence defined in the usual way through the differentiation with respect to the spatial position $\boldsymbol{x}$ and $u, \boldsymbol{m}$ are the spatial speed and the spatial normal to the singular surface, defined below.

2.2.6 The Singular Surface Applying the Hadamard Lemma 2.1.6 to each component of a motion $\boldsymbol{\chi}$, we obtain a continuous vector-valued function $\boldsymbol{a}$ on the singular surface such that

$$[\boldsymbol{F}] = \boldsymbol{a} \otimes \boldsymbol{n}, \quad [\boldsymbol{v}] = -U\boldsymbol{a}. \tag{2.2.9}$$

The quantity $\boldsymbol{a}$ is the ***amplitude of the jump of*** $\boldsymbol{\chi}$. From (2.2.9) and (1.1.18),

$$-U[\boldsymbol{F}] = [\boldsymbol{v}] \otimes \boldsymbol{n}, \quad [\operatorname{cof} \boldsymbol{F}]\boldsymbol{n} = \boldsymbol{0}. \tag{2.2.10}$$

These equations are further manifestations of the assumed coherency of motions. Since for an oriented area $\boldsymbol{b} \in$ Vect before the deformation, $\operatorname{cof} \boldsymbol{F}\boldsymbol{b}$ is the oriented area after the deformation, the equality $(2.2.10)_2$, i.e., $\operatorname{cof} \boldsymbol{F}^+\boldsymbol{n} = \operatorname{cof} \boldsymbol{F}^-\boldsymbol{n}$, says that the planar surface tangent to the singular surface is deformed in the same way on both sides of the singular surface (coherency). The image $s_t = \boldsymbol{\chi}(S_t, t)$ of a singular surface S_t under $\boldsymbol{\chi}$ is a moving singular surface in physical space. Its kinematics in the spatial description is completely analogous to that in the reference description. The singular surface s_t has a well-defined spatial normal $\boldsymbol{m}$ and spatial speed u. They are related to the referential quantities by

$$\boldsymbol{m} = (\boldsymbol{F}^{\pm})^{-\mathsf{T}}\boldsymbol{n}/|(\boldsymbol{F}^{\pm})^{-\mathsf{T}}\boldsymbol{n}|, \quad \boldsymbol{n} = (\boldsymbol{F}^{\pm})^{\mathsf{T}}\boldsymbol{m}/|(\boldsymbol{F}^{\pm})^{\mathsf{T}}\boldsymbol{m}|, \tag{2.2.11}$$

and

$$u = U|(\boldsymbol{F}^{\pm})^{\mathsf{T}}\boldsymbol{m}| + \boldsymbol{m} \cdot \boldsymbol{v}^{\pm}, \quad U = u^{\pm}|(\boldsymbol{F}^{\pm})^{-\mathsf{T}}\boldsymbol{n}|, \tag{2.2.12}$$

where

$$u^{\pm} = u - \boldsymbol{m} \cdot \boldsymbol{v}^{\pm} \tag{2.2.13}$$

are the relative velocities of the singular surface with respect to the material in front of the singular surface and behind the singular surface.

If f is a continuous and piecewise continuously differentiable scalar field, then the spatial version of the Hadamard lemma asserts the existence of a scalar field c on the singular surface such that

$$[\operatorname{grad} f] = c\boldsymbol{m}, \quad \left[\frac{\partial f}{\partial t}\right] = -uc$$

and the relation between this c and the C from Proposition 2.1.6 is

$$C = c\langle|\boldsymbol{F}^{\mathsf{T}}\boldsymbol{m}|\rangle + \langle\operatorname{grad} f\rangle \cdot [\boldsymbol{F}]\boldsymbol{n}, \quad c = C\langle|\boldsymbol{F}^{-\mathsf{T}}\boldsymbol{n}|\rangle + \langle\nabla f\rangle \cdot [\boldsymbol{F}^{-1}]\boldsymbol{m}.$$

These formulas are verified by computing C as $[\nabla f] \cdot \boldsymbol{n} = [\boldsymbol{F}^{\mathsf{T}} \operatorname{grad} f] \cdot \boldsymbol{n}$, using (2.1.4), (2.2.4) and (2.2.5).

2.2.7 Changes of the Stretch Tensors and of Axes of Deformation Consider a fixed material point p under the deformation and suppress the dependence of the quantities on p. The polar decomposition $F(t) = R(t)U(t) = V(t)R(t)$ determines uniquely U, V, R as functions of time. Differentiating $F = RU = VR$, taking the symmetric and skew parts, and using the fact that the tensor $Z := \dot{R}R^{\mathrm{T}}$ is skew, it is found that

$$\begin{aligned} D &= \tfrac{1}{2}R(\dot{U}U^{-1} + U^{-1}\dot{U})R^{\mathrm{T}} = \tfrac{1}{2}(\dot{V}V^{-1} + V^{-1}\dot{V}) + \tfrac{1}{2}(VZV^{-1} - V^{-1}ZV), \\ W &= Z + \tfrac{1}{2}R(\dot{U}U^{-1} - U^{-1}\dot{U})R^{\mathrm{T}} = \tfrac{1}{2}(\dot{V}V^{-1} - V^{-1}\dot{V}) + \tfrac{1}{2}(VZV^{-1} + V^{-1}ZV). \end{aligned} \tag{2.2.14}$$

For given U, V, R, D, W, these equations represent systems of equations for $\dot{U}$, $\dot{V}$ and Z. The plain disymmetry between the equations with $\dot{U}$ and $\dot{V}$ is worth noting. If $R(t) = \mathbf{1}$ at some particular t, then at that t, the components of $\dot{U}(t)$, $\dot{V}(t)$, $\dot{R}(t)$ in the basis of eigenvectors of $F(t) = U(t) = V(t)$ are given by

$$\dot{U}_{ij} = \frac{2v_iv_j}{v_i + v_j}D_{ij}, \quad \dot{V}_{ij} = \frac{v_i^2 + v_j^2}{v_i + v_j}D_{ij} + (v_j - v_i)W_{ij}, \quad \dot{R}_{ij} = \frac{v_j - v_i}{v_i + v_j}D_{ij} + W_{ij},$$

$1 \le i, j \le n$, where $v_i = v_i(t)$ are the ordered principal stretches. This follows either from the above equations or from the formulas for the derivatives of the polar factors with respect to F in Proposition 1.4.4 by using the chain rule. For instance, $\dot{U} = \partial_F U(F)[\dot{F}] = \partial_F U(F)[LF]$ and combining this with the expression for $\partial_F U(F)$ and $L = D + W$, one obtains the result. (In $n = 3$ dimensions, Guo [1984] gives explicit expressions for $\dot{U}$, $\dot{V}$, $\dot{R}$ in terms of V, R, L.) Also

$$\dot{v}_i = v_iD_{ii}, \quad \text{or equivalently} \quad \frac{d}{dt}\ln v_i = D_{ii}.$$

This follows from the expression for $\partial_F \hat{v}_i$ in Proposition 1.4.2. Suppose additionally that the eigenvalues of $F(t)$ are distinct. Then also the principal axes of strain, i.e., the normalized eigenvectors $e_i = e_i(t)$ of V and the normalized eigenvectors $f_i = f_i(t)$ of U, related by $Rf_i = e_i$, are uniquely determined to within the changes of signs. At the time t at which $R = \mathbf{1}$ one has

$$\dot{e}_i = (\Theta + W)e_i, \quad \dot{f}_i = \Omega e_i, \quad i = 1, \ldots, n,$$

where W is the spin tensor and Ω, Θ are time-dependent skew tensors given by

$$\Theta = \sum \frac{(v_i^2 + v_j^2)D_{ij}}{v_j^2 - v_i^2} e_i \otimes e_j, \quad \Omega = \sum \frac{2v_iv_jD_{ij}}{v_j^2 - v_i^2} e_i \otimes e_j,$$

where the sums are taken over all pairs i, j, $1 \le i, j \le n$, $i \neq j$. The formula for $\dot{e}_i$ follows from the formula for $\partial_F e_i$ in Proposition 1.4.2. The formula for $\dot{f}_i$ is obtained from $f_i = R^{\mathrm{T}}e_i$ and from the formulas for $\dot{e}_i$, $\dot{R}$ above. The formulas for this subsection form the basis of the so-called Hill's principal axis method. When combined with the results of Chap. 8, they can be used to evaluate the time-derivatives isotropic and objective isotropic scalar and tensor-valued functions in terms of D, W.

2.3 Compatibility of Deformations at the Interface

By the Hadamard lemma for deformations (2.2.9), the limiting values of deformation gradient on the singular surface differ by a tensor product $a \otimes n$. Two tensors $F, G \in$ Lin are said to be ***rank 1 connected*** (equivalently, F, G are said to be ***compatible,*** and G is said to be a ***rank 1 perturbation of*** F) if $G - F = a \otimes b$ for some $a, b \in$ Vect. The rest of this chapter is devoted to a closer examination of rank 1 connections. This section reviews piecewise homogeneous deformations in which the Hadamard lemma plays a significant role. Some of these deformations arise, physically, in ground states of crystalline bodies in martensitic transformations, and mathematically, in treating the constitutive inequalities (see Chap. 17). It will be also explained how the existence and form of the ground states depends on the structure of potential wells of the free energy function. The subsequent two sections deal with more algebraic aspects of rank 1 connections. We identify Eucl with Vect.

A deformation χ of Ω is said to be ***homogeneous*** if its singular surface is empty and $\nabla\chi \equiv \boldsymbol{F}$ is constant over Ω. When Ω is connected, then $\chi(\boldsymbol{p}) = \boldsymbol{F}\boldsymbol{p} + \boldsymbol{c}$ for some $\boldsymbol{c} \in$ Vect and all $\boldsymbol{p} \in \Omega$. A deformation χ of Ω with a singular surface $S \subset \Omega$ is said to be ***pairwise homogeneous*** if $\{\nabla\chi(\boldsymbol{p}) : \boldsymbol{p} \in \Omega \setminus S\}$ contains exactly two elements $\boldsymbol{F}^{\pm}$.

2.3.1 Proposition *If χ is a pairwise homogeneous deformation of a connected body Ω such that $[\boldsymbol{F}] \neq \boldsymbol{0}$ everywhere on the singular surface S, then*

(1) $\boldsymbol{F}^+ = \boldsymbol{F}^- + \boldsymbol{a} \otimes \boldsymbol{n}$ *for some* $\boldsymbol{a} \in$ Vect, $\boldsymbol{a} \neq \boldsymbol{0}$, *and* $\boldsymbol{n} \in$ Sph;

(2) *there exists a countable (finite if Ω is bounded) family of parallel hyperplanes H_i, $i = 1, \ldots,$ such that*

$$S \subset \Omega \cap \bigcup_{i=1}^{\infty} H_i; \qquad (2.3.1)$$

(3) *if Ω is convex, there exists a continuous and piecewise continuously differentiable function $\theta : \mathbb{R} \to \mathbb{R}$ with $\theta' \in \{0, 1\}$ and a $\boldsymbol{c} \in$ Vect such that*

$$\chi(\boldsymbol{p}) = \boldsymbol{F}^- \boldsymbol{p} + \boldsymbol{a}\theta(\boldsymbol{p} \cdot \boldsymbol{n}) + \boldsymbol{c}, \quad \boldsymbol{F}(\boldsymbol{p}) = \boldsymbol{F}^- + \boldsymbol{a} \otimes \boldsymbol{n}\theta'(\boldsymbol{p} \cdot \boldsymbol{n}), \quad \boldsymbol{p} \in \Omega; \qquad (2.3.2)$$

if H_i denote the discontinuity surfaces of $\boldsymbol{p} \mapsto \theta'(\boldsymbol{p} \cdot \boldsymbol{n})$ then the equality holds in (2.3.1).

Thus the deformation consists of parallel layers in Ω on each of which the deformation gradient is constant; these are separated by planar interfaces.

Proof (1): follows from the Hadamard lemma and the fact that Ω is connected. (2): Let S' is any connected component of S. Since the limiting values of $\nabla\chi$ are $\boldsymbol{F}^{\pm}$ at every point of the singular surface, necessarily the normal is equal either to $\boldsymbol{n}$ on S' or to $-\boldsymbol{n}$ on S'. An elementary argument then shows that S' is contained in some hyperplane of normal $\boldsymbol{n}$. By the general theorems, S can have at most a countable number of components, and hence (2.3.1) holds with some family H_i. If Ω is bounded, then the number of distinct hyperplanes H_i must be finite, for otherwise these planes would accumulate at some point of $\operatorname{cl}\Omega$, and this would contradict the fact that S is given globally as the solution of $\phi = 0$ with ϕ defined in a neighborhood of $\operatorname{cl}\Omega$ and $\nabla\phi \neq \boldsymbol{0}$ whenever $\phi = 0$. (3): Let S' be a connected component of S, H a hyperplane of normal $\boldsymbol{n}$ such that $S' \subset H$, $\boldsymbol{p} \in S'$ and $\boldsymbol{q} \in \Omega \cap H$. Our goal is to prove that $\boldsymbol{q} \in S'$. Since Ω is convex, the line segment with endpoints $\boldsymbol{p}, \boldsymbol{q}$ is contained in $\Omega \cap H$. If $\boldsymbol{q} \notin S'$, there must be a point $\boldsymbol{r}$ in the interior of the line segment which is on the relative boundary of S'. This point is then outside S' since S is the set of all solutions of $\phi = 0$. Hence $\boldsymbol{r}$ is a point of continuity of $\nabla\chi$. On the other hand, $\boldsymbol{r}$ is a limit of points of S' and hence a limit of points of discontinuity of $\nabla\chi$; since $[\nabla\chi]$ has a constant magnitude on S', this contradicts the continuity of $\nabla\chi$ at $\boldsymbol{r}$. This proves (2.3.1) with the equality sign. Let $\Omega^+ = \{\boldsymbol{p} \in \Omega \setminus S : \boldsymbol{F}(\boldsymbol{p}) = \boldsymbol{F}^+\}$. The above argument shows that each component of Ω^+ is of the form $\{\boldsymbol{p} \in \Omega : k < \boldsymbol{p} \cdot \boldsymbol{n} < l\}$ for some $k < l$. Consequently, the characteristic function χ of Ω^+ is expressible as $\chi(\boldsymbol{p}) = m(\boldsymbol{p} \cdot \boldsymbol{n})$, where $m : \mathbb{R} \to \{0, 1\}$. If θ is the primitive of m, i.e., $\theta' = m$, then clearly $(2.3.2)_2$ holds, and the integration gives $(2.3.2)_1$. □

The deformation to be now described provides a simple visualization of rank 1 convexity (see Definition 17.3.1). Let $\boldsymbol{n}$ be a unit vector and consider a unit cube Ω in Vect given by

$$\Omega = \{\boldsymbol{p} \in \text{Vect} : 0 < \boldsymbol{p} \cdot \boldsymbol{e}_i < 1, i = 1, \ldots, n\}$$

where $\{\boldsymbol{e}_i\}$ is an orthonormal basis in Vect such that $\boldsymbol{e}_1 = \boldsymbol{n}$. Let D_0 and D_1 denote the lower and the upper of the two faces of Ω perpendicular to $\boldsymbol{n}$, respectively, and set $D = D_0 \cup D_1$.

2.3.2 Proposition *Let $\bar{\boldsymbol{F}} \in \text{Lin}^+$ and let χ be a pairwise homogeneous deformation such that $\Omega \setminus S$ has exactly two components $\Omega^{\pm}$ and*

$$\chi(\boldsymbol{p}) = \bar{\boldsymbol{F}}\boldsymbol{p} \quad \text{for every } \boldsymbol{p} \in D. \qquad (2.3.3)$$

Then there exists an $\alpha \in (0, 1)$ and $\boldsymbol{a} \in$ Vect such that

$$\Omega^+ = \{p \in \Omega : 0 < p \cdot n < \alpha\}, \quad \Omega^- = \{p \in \Omega : \alpha < p \cdot n < 1\}, \tag{2.3.4}$$

and

$$F = \bar{F} + (1-\alpha)a \otimes n \text{ on } \Omega^+, \quad F = \bar{F} - \alpha a \otimes n \text{ on } \Omega^-. \tag{2.3.5}$$

Fig. 2.3.1 shows χ when $\bar{F} = 1$. The average deformation per unit cube is **1** and the points of the faces D_0, D_1 do not change their positions. The vector $\alpha(1-\alpha)a$ is a uniform displacement vector of the points in the plane $S = \{p \in \Omega : p \cdot n = \alpha\}$. The stored energy of the material is rank 1 convex at $\bar{F}$ if and only if the passage from the homogeneous state with deformation gradient $\bar{F}$ to the inhomogeneous state χ does not reduce the total stored energy of the cube.

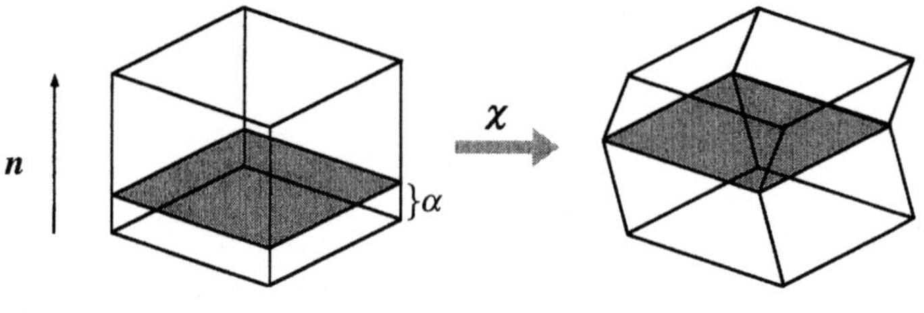

Fig. 2.3.1

Proof The integration of $\nabla\chi = F^{\pm}$ on $\Omega^{\pm}$ gives $\chi(p) = F^{\pm}p + c^{\pm}$ for every $p \in \Omega^{\pm}$, where $c^{\pm}$ are the integration constants. The boundary of at least one of $\Omega^{\pm}$ must intersect D_1 on a set of positive area and assume that this happens for Ω^+. The boundary condition (2.3.3) gives $F^+p + c^+ = \bar{F}p$ for every $p \in D_1 \cap \partial\Omega^+$. Writing this for $p + \gamma e$, where $e \cdot n = 0$ and $\gamma \in \mathbb{R}$ is sufficiently small, one deduces that $F^+e = \bar{F}e$ for every e with $e \cdot n = 0$. But this means that $F^+ = \bar{F} + a^+ \otimes n$ for some $a^+ \in$ Vect; similarly, $F^- = \bar{F} + a^- \otimes n$ for some $a^- \in$ Vect. Hence $F^+ = F^- + a \otimes n$, where $a = a^+ - a^-$. Combining this with the Hadamard lemma $F^+ = F^- + \bar{a} \otimes \bar{n}$, where $\bar{n}$ is the normal to S, we obtain $\bar{n} = n$, $\bar{a} = a$ (changing the orientation of S if necessary). This implies that S is contained in a plane of normal n, and hence $S = \{p \in \Omega : p \cdot n = \alpha\}$ for some α, $0 < \alpha < 1$; consequently (2.3.4) holds. Set $\bar{\chi}(p) = \bar{F}p$, $p \in \Omega$. Denoting by m the field of normal to $\partial\Omega$, we find by the divergence theorem and (2.3.3) that

$$\int_\Omega \nabla\chi(p)n \, dV = \int_{\partial\Omega} \chi(p)(m \cdot n) \, dA = \int_D \chi(p)(m \cdot n) \, dA = \int_D \bar{\chi}(p)(m \cdot n) \, dA = \int_\Omega \nabla\bar{\chi}(p)n \, dV = \bar{F}n. \tag{2.3.6}$$

Using $\nabla\chi = F^{\pm} = F + a^{\pm} \otimes n$ on $\Omega^{\pm}$ and (2.3.4), it is found that (2.3.6) leads to $\alpha a^+ + (1-\alpha)a^- = 0$. Combining with $a = a^+ - a^-$ we find that $a = a^-/\alpha$ and (2.3.5) follow. □

Rank 1 connections arise also in one dimensional deformations, used, e.g., to model the structure of the shock layers, see Sect. 23.8. A deformation χ of Vect is said to be ***one-dimensional*** if for every point $p \in$ Vect outside the singular surface,

$$F(p) \equiv \nabla\chi(p) = \phi(p \cdot n) \tag{2.3.7}$$

for some $n \in$ Sph and $\phi : \mathbb{R} \to$ Lin. Equation (2.3.2) shows that any pairwise homogeneous deformation of a convex region Ω is a restriction to Ω of some one-dimensional deformation. A one-dimensional deformation is said to be ***periodic*** if the function ϕ as in (2.3.7) is periodic. If p is the period, the value

$$\bar{F} := \frac{1}{p}\int_0^p \phi(s)\, ds$$

is called the ***mean deformation gradient.*** The following proposition shows that the deformation gradients realized in a one-dimensional deformation are mutually rank 1 connected.

2.3.3 Proposition

(1) *If* χ *is a one-dimensional deformation then there exists an* $\bar{F} \in \mathrm{Lin}$ *and a continuous and piecewise continuously differentiable function* $u : \mathbb{R} \to \mathrm{Vect}$ *such that*

$$\chi(p) = \bar{F}p + u(p \cdot n), \quad F(p) \equiv \nabla\chi(p) = \bar{F} + u'(p \cdot n) \otimes n \tag{2.3.8}$$

for every $p \in \mathrm{Vect} \setminus S$*; moreover, if* χ *is periodic, then* u *may be chosen periodic and* $\bar{F}$ *may be identified with the mean deformation gradient.*

(2) *Given* u *and* $\bar{F} \in \mathrm{Lin}^+$*, the mapping* χ *satisfies* $\det F(p) > 0$ *for every* $p \in \mathrm{Vect}$ *(including the values on the singular surface) if and only if* $1 + \bar{F}^{-1}u'(s) \cdot n > 0$ *for every* $s \in \mathbb{R}$*, including the limiting values of* $u'(s)$ *at points of discontinuity of* u'*. This also ensures that* χ *is globally invertible.*

Proof (1): Since the deformation gradient is continuously differentiable outside the singular surface as part of the definition of the deformation, the function ϕ as in (2.3.7) must be continuously differentiable in a neighborhood of each $s = p \cdot n$ for which p is not on a singular surface. Then $\nabla^2\chi(p) = \phi'(s) \otimes n$ and the interchangeability of the second partial derivatives, $\nabla^2\chi(p)[c, d] = \nabla^2\chi(p)[d, c]$, gives $\phi'(s)c(n \cdot d) = \phi'(s)d(n \cdot c)$. Setting $d = n$ we obtain $\phi'(s) = \phi'(s)n \otimes n$ and denoting the primitive of $\phi'(\cdot)n$ by ψ, then $(\phi - \psi \otimes n)' = 0$, i.e., $\phi = \psi \otimes n + \bar{F}$ for some integration constant $\bar{F} \in \mathrm{Lin}$. Denoting v the primitive of ψ, and defining $\bar{\chi}(p) = \bar{F}p + v(p \cdot n)$, $p \in \mathrm{Vect}$, it is easily found that $\nabla(\chi - \bar{\chi}) = 0$, i.e., $\chi(p) = \bar{F}p + v(p \cdot n) + c$ for some integration constant $c \in \mathrm{Vect}$, and defining finally $u = v + c$, one recovers the assertion. The proof of (2) is left to the reader. □

A ***periodic pairwise homogeneous deformation*** is a periodic one-dimensional deformation for which (2.3.8) holds with $u(s) = af(s)$ with a fixed $a \in \mathrm{Vect}$, $a \neq 0$ where $f : \mathbb{R} \to \mathbb{R}$ is a p-periodic continuous and piecewise continuously differentiable function such that

$$f'(s) = \begin{cases} 1 - \alpha & \text{if } 0 \le s < \alpha p, \\ -\alpha & \text{if } \alpha p \le s < p \end{cases} \tag{2.3.9}$$

for some $0 < \alpha < 1$. Then

$$F(p) = \begin{cases} \bar{F} + (1 - \alpha)a \otimes n & \text{if } 0 \le p \cdot n < \alpha p, \\ \bar{F} - \alpha a \otimes n & \text{if } \alpha p \le p \cdot n < p \end{cases}$$

and so on periodically. The deformation of $\Omega = \mathrm{Vect}$ consists of periodic layers separated by planar interfaces. The values $F^+ = \bar{F} + (1 - \alpha)a \otimes n$ and $F^- = \bar{F} - \alpha a \otimes n$ will be referred to as the phases; their concentrations are α and $1 - \alpha$, respectively and n is referred to as the normal of the deformation. The structure becomes finer and finer as $p \to 0$; for the weak limit of χ as $p \to 0$, see Example 21.1.2(2), and for the Young measure associated with χ, see Example 21.1.4(2).

Let us now consider deformations with intersecting singular surfaces. The deformation is a continuous and piecewise continuously differentiable field $\chi : \Omega \to \mathrm{Vect}$ in the sense of definition in Subsect. 2.6.6; moreover, we assume that the regions $Q^p \equiv \Omega^p$, $p = 1, \ldots, q$, as in Subsect. 2.6.6 may be chosen connected, that χ is twice continuously differentiable on $\Omega \setminus S$, and the inverse mapping exists. We denote by S the singular surface (see Subsect. 2.6.4). A deformation is said to be ***piecewise homogeneous*** if the restriction of χ to each of the regions Ω^p is of the form

$$\chi(p) = F_p p + c_p, \quad p \in \Omega^p, \quad p = 1, \ldots, q,$$

where $F_p \in \mathrm{Lin}^+$ and $c_p \in \mathrm{Vect}$ are constants. Clearly, the Hadamard lemma holds at the regular part S^r of the singular surface. In a way quite similar to that in the proof of

Proposition 2.3.1, it is possible to prove that if χ is piecewise homogeneous, then the singular surface is contained in a union of a finite number of (possibly intersecting) planes in Vect and hence each of the regions Ω^p is an intersection of Ω with a polyhedral region. At the points $\boldsymbol{p} \in S$ of intersection of several planes, there are several normals, limiting values of $\boldsymbol{F}$, and amplitudes. The Hadamard lemma applies to any two regions $\Omega^p, \Omega^s, p \neq s$, such that $\boldsymbol{p} \in \partial\Omega^p \cap \partial\Omega^s$ and $A(\partial\Omega^p \cap \partial\Omega^s) > 0$. JAMES [1986a,b] has given algorithms for treating the Hadamard lemma in these situations.

Piecewise homogeneous deformations are frequently observed in crystal structures but not in, say, isotropic rubber-like materials. This is related to the differences in the properties of the free energy functions of these two classes of materials. In particular, the occurrence/nonoccurrence of nontrivial piecewise homogeneous deformations with zero stress is related to the structure of the set of points of minimal stored energy.

2.3.4 Orbits and Wells Structure Let f be the stored energy of a material body at some temperature θ (see Subsect. 9.2.4) and denote by B the set of all $\boldsymbol{F} \in \mathrm{Lin}^+$ for which $f(\boldsymbol{F})$ takes the minimum value of f. (Assume that $B \neq \emptyset$.) By the frame-indifference (see Sect. 9.3), B is an objective set, i.e., if $\boldsymbol{F} \in B$ and $\boldsymbol{Q} \in \mathrm{Orth}^+$, then also $\boldsymbol{QF} \in B$, or equivalently, $O := \{\boldsymbol{QF} : \boldsymbol{Q} \in \mathrm{Orth}^+\} \subset B$. Each set O of this form, where $\boldsymbol{F} \in \mathrm{Lin}^+$ is fixed, is called an ***orbit.*** By the polar decomposition theorem, each orbit O intersects Sym^+ at exactly one point $\boldsymbol{U}$ which is called its ***center;*** then $O = \{\boldsymbol{QU} : \boldsymbol{Q} \in \mathrm{Orth}^+\}$. If O_1, O_2 are two orbits, then either $O_1 \cap O_2 = \emptyset$ or $O_1 = O_2$ according to whether the centers satisfy $\boldsymbol{U}_1 \neq \boldsymbol{U}_2$ or $\boldsymbol{U}_1 = \boldsymbol{U}_2$. From the objectivity, B is a union of distinct orbits; each orbit in B is called a ***potential well*** for the stored energy f. The number of potential wells is related to the material symmetry: if $\boldsymbol{H} \in G$ is a material symmetry (see Sect. 9.4 for the definition of the symmetry group G of a material) and O is a potential well, then also $O\boldsymbol{H} := \{\boldsymbol{FH} : \boldsymbol{F} \in O\}$ is a potential well. This well may, but need not, be distinct from O. It is easily verified that if in particular the symmetry $\boldsymbol{H} \equiv \boldsymbol{Q}$ is an orthogonal tensor, then the center of the orbit $O\boldsymbol{Q}$ is $\boldsymbol{Q}^{\mathrm{T}}\boldsymbol{UQ}$, hence $O\boldsymbol{Q}$ different from O if and only if $\boldsymbol{U} \neq \boldsymbol{Q}^{\mathrm{T}}\boldsymbol{UQ}$. In the context of phase transitions, the phase with the deformation $\boldsymbol{Q}^{\mathrm{T}}\boldsymbol{UQ}$ is said to be a ***variant*** of the phase with the deformation gradient $\boldsymbol{U}$.

The deformations (= states) χ which give the absolute minimum of the total stored energy are called the ***ground states*** for f. The problem of determining all ground states at the given temperature then amounts to determining all deformations χ of Ω such that

$$\nabla\chi(\boldsymbol{p}) \in B \tag{2.3.10}$$

for every $\boldsymbol{p} \in \Omega$, including the limiting values on the singular surface. Since the stress is a derivative of the stored energy with respect to $\boldsymbol{F}$, the ground states are stress-free and hence they trivially satisfy the equation of balance of forces and the Maxwell relation (see Sect. 17.4). The determination of the ground states is thus a geometric problem depending only on B, other details of f are irrelevant. Clearly, any homogeneous state of deformation gradient $\boldsymbol{F} \in B$ is a ground state. However, depending on the structure of B, there may be inhomogeneous ground states which may be preferable over the homogeneous ones; e.g., they may accommodate the boundary conditions. For the following discussion assume that B is an objective set and call each deformation satisfying (2.3.10) a ground state for B. Furthermore, assume that Ω is connected. Two potential wells (orbits) $O_1, O_2 \subset B$ are said to be ***rank 1 connected*** (equivalently, ***compatible***) if there is a pair $(\boldsymbol{F}_1, \boldsymbol{F}_2)$ of rank 1 connected tensors with $\boldsymbol{F}_1 \in O_1, \boldsymbol{F}_2 \in O_2$.

2.3.5 Martensitic Transformations; Wells Structure for Cubic/Tetragonal Change Martensitic transformations are temperature-induced, diffusionless solid-solid phase transformations involving a change in crystalline symmetry. When the binary alloy Indium–Thalium is cooled through a critical temperature it undergoes such a martensitic transformation from its cubic parent phase (austenite) to its tetragonal phase (martensite). Thus at the critical temperature θ, there is a homogeneous stress-free state corresponding to the austenite, and we can assume without any loss of generality that this state is described by the deformation gradient $\boldsymbol{F} = \mathbf{1}$; the states with $\boldsymbol{F} = \boldsymbol{Q} \in \mathrm{Orth}^+$ are obtained from $\boldsymbol{F} = \mathbf{1}$ by rotations, which leads to the orbit $A = \mathrm{Orth}^+$. There is also a homogeneous stress-free state corresponding to the martensite; this state together with the states obtainable from it by rotating the specimen, will form another orbit M, with center denoted by $\boldsymbol{U} \in \mathrm{Sym}^+$. Clearly $A \neq M$. We now make the assumption that the stored energy is minimal on the orbits A and M, i.e., $A, M \subset B$. Furthermore, we formalize the cubic symmetry of the austenite and the tetragonal symmetry of the martensite by the following assumptions: (1) the (global) symmetry group of the material is the cubic point group P^c of some lattice generated by an orthonormal basis $\{\boldsymbol{e}_1, \boldsymbol{e}_2, \boldsymbol{e}_3\}$ (see Subsect. 1.5.5), and (2) the local symmetry group at $\boldsymbol{U}$, defined by $G_U := \boldsymbol{U}G\boldsymbol{U}^{-1} \cap \mathrm{Orth}$, is a tetragonal point group, generated by some lattice (see Subsect. 1.5.5 for the tetragonal point group and the text following it for the local symmetry group). It turns out that these requirements imply that B contains two more distinct wells, and $\boldsymbol{U}$ is related to the cubic axes $\{\boldsymbol{e}_1, \boldsymbol{e}_2, \boldsymbol{e}_3\}$ by an explicit formula. Namely, by Proposition 1.5.6, $\boldsymbol{U}$ must be of the form (1.5.8) with some $i = 1, 2, 3$, and as explained above, for any $\boldsymbol{Q} \in P^c$, necessarily $M\boldsymbol{Q}$ is again an orbit in B, with center $\boldsymbol{Q}^{\mathrm{T}}\boldsymbol{U}\boldsymbol{Q}$. For most of $\boldsymbol{Q} \in P^c$, one has $\boldsymbol{Q}^{\mathrm{T}}\boldsymbol{U}\boldsymbol{Q} = \boldsymbol{U}$, in which case $M\boldsymbol{Q} = M$ is not a new orbit, but letting $\boldsymbol{Q}$ run over the whole of P^c, the set of all values of $\boldsymbol{Q}^{\mathrm{T}}\boldsymbol{U}\boldsymbol{Q}$ is found to consist of three tensors, namely (see (1.5.8)):

$$\boldsymbol{U}_i = \eta_1 \mathbf{1} + (\eta_2 - \eta_1)\boldsymbol{e}_i \otimes \boldsymbol{e}_i, \quad i = 1, 2, 3,$$

where $\eta_1 > 0, \eta_2 > 0, \eta_1 \neq \eta_2$. The local symmetry group $G_{U_i} := \mathrm{Orth} \cap \boldsymbol{U}_i G \boldsymbol{U}_i^{-1}$ is then the (tetragonal) point group of the lattice with basis $\{\bar{\boldsymbol{e}}_1, \bar{\boldsymbol{e}}_2, \bar{\boldsymbol{e}}_3\}$, where $\bar{\boldsymbol{e}}_k = \eta_1 \boldsymbol{e}_k$ if $k \neq i$ and $\bar{\boldsymbol{e}}_i = \eta_2 \boldsymbol{e}_i$. A minimal B with these properties thus consists of four potential wells: one well with center $\mathbf{1}$ corresponding to the parent phase (austenite) and three wells with centers $\boldsymbol{U}_1, \boldsymbol{U}_2, \boldsymbol{U}_3$, corresponding to the variants of the martensite:

$$B = A \cup M_1 \cup M_2 \cup M_3, \quad \text{where} \quad \begin{array}{lll} A = \mathrm{Orth}^+ & \ldots & \text{austenite well,} \\ M_i = \{\boldsymbol{Q}\boldsymbol{U}_i : \boldsymbol{Q} \in \mathrm{Orth}^+\} & \ldots & \text{martensite wells.} \end{array}$$

The tensors $\boldsymbol{U}_i$ are known as the transformation strains; the parameters η_1, η_2 can be determined experimentally. The tensors $\boldsymbol{U}_i$ are related by the symmetry transformations $\boldsymbol{R}_{ij}, \boldsymbol{H}_{ij} \in P^c$ as follows: for $i \neq j$,

$$\boldsymbol{U}_j = \boldsymbol{R}_{ij}^{\mathrm{T}} \boldsymbol{U}_i \boldsymbol{R}_{ij} = \boldsymbol{H}_{ij}^{\mathrm{T}} \boldsymbol{U}_i \boldsymbol{H}_{ij},$$

where $\boldsymbol{R}_{ij} = \boldsymbol{e}_i \otimes \boldsymbol{e}_j - \boldsymbol{e}_j \otimes \boldsymbol{e}_i + \boldsymbol{e}_k \otimes \boldsymbol{e}_k$, $i \neq k \neq j$, is the rotation of order 2 about $\boldsymbol{e}_k$, and $\boldsymbol{H}_{ij} = 2\boldsymbol{n}_{ij} \otimes \boldsymbol{n}_{ij} - \mathbf{1}$, $\boldsymbol{n}_{ij} = (\boldsymbol{e}_i \pm \boldsymbol{e}_j)/\sqrt{2}$ is the rotation of order 2 about $\boldsymbol{n}_{ij}$.

2.3.6 Proposition *Let χ be a ground state for an objective set B which is a union of a finite number of orbits. Then*

(1) *χ is piecewise homogeneous;*

(2) *if $\boldsymbol{p} \in S \cap \Omega$ and $\boldsymbol{F}^+(\boldsymbol{p}) \neq \boldsymbol{F}^-(\boldsymbol{p})$, then there are two distinct orbits O^+, O^- in B such that $\boldsymbol{F}^{\pm}(\boldsymbol{p}) \in O^{\pm}$;*

(3) *if B is just one orbit or if there are no rank 1 connections between different orbits of B, then χ is homogeneous.*

Hence the occurrence of an inhomogeneous ground state requires that B contains at least two compatible potential wells. For isotropic rubber-like materials, realistic stored energies lead to B of the form $B = \{\boldsymbol{R}\boldsymbol{w} : \boldsymbol{R} \in \mathrm{Orth}^+\}$, where $w > 0$ is some positive number (depending on θ). Thus, in this case, B is just one well and so each ground state is homogeneous. In contrast, we shall show in Example 2.5.5 that each two potential wells with the centers U_i as in Subsect. 2.3.5 are compatible, and thus there are inhomogeneous ground states.

Proof (1): Since B is a union of a finite number of orbits, and the orbits are connected, compact and disjoint, the distance of any two distinct orbits in B is positive. Let $\Omega^p, p = 1, \ldots, q$ be the partition of Ω associated with χ. Since Ω^p is connected and $\boldsymbol{F} \equiv \nabla\chi$ continuous on it, the set $\boldsymbol{F}(\Omega^p)$ must be contained in a single orbit O of B. If $\boldsymbol{U}$ is the center of the orbit O, then for every $\boldsymbol{p} \in \Omega^p$ there exists a unique $\boldsymbol{R}(\boldsymbol{p}) \in \mathrm{Orth}^+$ such that $\nabla\chi(\boldsymbol{p}) = \boldsymbol{R}(\boldsymbol{p})\boldsymbol{U}$. By the standard characterization of rigid deformations (see, e.g., GURTIN [1981, Theorem on p. 49], or CIARLET [1988, Thm. 1.8-1]), $\boldsymbol{R}$ is constant on Ω^p. Thus the deformation is homogeneous on Ω^p. (2): Suppose that $\boldsymbol{p} \in S \cap \Omega, \boldsymbol{F}^+(\boldsymbol{p}) \neq \boldsymbol{F}^-(\boldsymbol{p})$, and $\boldsymbol{F}^\pm \equiv \boldsymbol{F}^\pm(\boldsymbol{p}) \in O \subset B$, where O is an orbit. We have that $\boldsymbol{F}^+ - \boldsymbol{F}^- = \boldsymbol{a} \otimes \boldsymbol{n}$ and as again $\boldsymbol{F}^\pm = \boldsymbol{R}^\pm \boldsymbol{U}$ for some $\boldsymbol{R}^\pm \in \mathrm{Orth}^+$, we arrive at

$$\boldsymbol{R} = \mathbf{1} + \boldsymbol{b} \otimes \boldsymbol{m} \tag{2.3.11}$$

where $\boldsymbol{R} = (\boldsymbol{R}^-)^\mathsf{T}(\boldsymbol{R}^+)^\mathsf{T}, \boldsymbol{b} = (\boldsymbol{R}^-)^\mathsf{T}\boldsymbol{a}, \boldsymbol{m} = \boldsymbol{U}^{-1}\boldsymbol{n}$. Taking the determinant, and writing the condition $\boldsymbol{R}\boldsymbol{R}^\mathsf{T} = \mathbf{1}$ we obtain

$$\boldsymbol{b} \cdot \boldsymbol{m} = 0, \quad \boldsymbol{b} \otimes \boldsymbol{m} + \boldsymbol{m} \otimes \boldsymbol{b} + \boldsymbol{m}^2 \boldsymbol{b} \otimes \boldsymbol{b} = \mathbf{0}.$$

Multiplying the second of these equations by $\boldsymbol{m}$ and using the first equation, we obtain $\boldsymbol{m}^2\boldsymbol{b} = \mathbf{0}$ from which $\boldsymbol{b} = \mathbf{0}$ and (2.3.11) gives $\boldsymbol{R} = \mathbf{1}$, i.e., $\boldsymbol{R}^+ = \boldsymbol{R}^-$. Then $\boldsymbol{F}^+ = \boldsymbol{F}^-$, which is a contradiction. (3): A consequence of (1), (2). □

The rank 1 connections obtained so far were consequences of either the Hadamard lemma on the singular surface or of the interchangeability of the second derivatives of the deformation outside the singular surface. The following proposition addresses a question whether a periodic one-dimensional deformation can be tailored coherently to a homogeneous deformation along a planar interface. The answer involves a rank 1 connection between the mean deformation gradient of the periodic deformation and the deformation gradient of the homogeneous deformation. A scaled deformation of the type constructed in the proof below will be employed in Subsect. 2.3.8.

2.3.7 Proposition *Let* χ_0 : Vect $\to$ Vect *be a periodic deformation with the mean deformation gradient* $\bar{\boldsymbol{F}}$, *let* $\boldsymbol{G} \in \mathrm{Lin}^+$, $\boldsymbol{m} \in \mathrm{Sph}$ *and* $\beta > 0$. *Then the following conditions are equivalent:*

(1) *there exists a continuous and piecewise continuously differentiable function* χ : Vect $\to$ Vect *such that*

$$\chi(\boldsymbol{p}) = \begin{cases} \boldsymbol{G}\boldsymbol{p} & \text{if } \boldsymbol{p} \cdot \boldsymbol{m} < 0, \\ \chi_0(\boldsymbol{p}) & \text{if } \boldsymbol{p} \cdot \boldsymbol{m} > \beta; \end{cases} \tag{2.3.12}$$

and $|\nabla \boldsymbol{F}(\boldsymbol{p})| < c$ *for some* $c > 0$ *and all* $\boldsymbol{p} \in$ Vect;

(2) *there exists a* $\boldsymbol{b} \in$ Vect *such that*

$$\boldsymbol{G} = \bar{\boldsymbol{F}} + \boldsymbol{b} \otimes \boldsymbol{m}. \tag{2.3.13}$$

In particular, if χ_0 *is a periodic pairwise homogeneous deformation with the phases* $\boldsymbol{F}^\pm$, *of concentrations* α *and* $1 - \alpha$, *of amplitude* $\boldsymbol{a}$, *and of normal* $\boldsymbol{n}$, *then* (2.3.13) *reads*

$$\boldsymbol{G} = \boldsymbol{F}^- + \alpha \boldsymbol{a} \otimes \boldsymbol{n} + \boldsymbol{b} \otimes \boldsymbol{m}. \tag{2.3.14}$$

Here the plane H of normal $\boldsymbol{m}$ through the origin is the interface between the homogeneous deformation $\boldsymbol{G}$ and the periodic deformation χ_0. The actual process of tailoring these two deformation is realized in the strip of width β along H. A similar construction is also employed to prove that the quasiconvexity implies rank 1 convexity (see

Proof of Lemma 17.3.4). Notice that (2.3.13) involves a rank 1 connection only between $\boldsymbol{G}$ and the mean deformation gradient $\bar{\boldsymbol{F}}$; the individual deformation gradients realized in χ_0 need not be rank 1 connected to $\boldsymbol{G}$. In particular, when χ_0 is a periodic pairwise homogeneous deformation, then $\boldsymbol{F}^+, \boldsymbol{F}^-$ must be rank 1 connected, but there need not be rank 1 connections between $\boldsymbol{G}, \boldsymbol{F}^+$ or $\boldsymbol{G}, \boldsymbol{F}^-$. The proposition does not assert that the mapping χ is a deformation; it need not be invertible and it may even happen that $\det \nabla\chi < 0$ somewhere in the strip $0 < \boldsymbol{p} \cdot \boldsymbol{m} < \beta$. However, it will be clear from the proof that if β is large enough, then χ may be chosen both invertible and satisfying $\det \nabla\chi > 0$. The assumption that $\boldsymbol{F} = \nabla\chi$ be globally bounded is essential for the equivalence of (1) and (2); easy constructions show that without this assumption *any* two homogeneous deformations may be connected by some bridging function χ with unbounded $\nabla\chi$, no matter whether or not they are rank 1 connected.

Proof To simplify notation, assume that $\beta = 1$. The general case is reduced to this by an appropriate scaling. (1) $\Rightarrow$ (2) : Let $\boldsymbol{e} \in$ Vect be any vector such that $\boldsymbol{e} \cdot \boldsymbol{m} = 0$, let $r > 0$ and consider the parallelogram P (in the reference configuration Ω = Vect) with vertices $-\boldsymbol{m}, -\boldsymbol{m} + r\boldsymbol{e}, r\boldsymbol{e} + \boldsymbol{m}, \boldsymbol{m}$. Denote by ∂P its boundary with the orientation determined by the order in the list of vertices; then the line integral of $\nabla\chi$ along ∂P, being the integral of the gradient of the potential χ along a closed curve, must vanish, i.e.,

$$\int_{\partial P} \nabla\chi(\boldsymbol{p})\, d\boldsymbol{p} = \boldsymbol{0}. \tag{2.3.15}$$

Let us evaluate/estimate the integrals along the individual sides of P. We have

$$\int_{-\boldsymbol{m}}^{r\boldsymbol{e}-\boldsymbol{m}} \nabla\chi(\boldsymbol{p})\, d\boldsymbol{p} = r\boldsymbol{G}\boldsymbol{e}, \qquad \int_{r\boldsymbol{e}+\boldsymbol{m}}^{\boldsymbol{m}} \nabla\chi(\boldsymbol{p})\, d\boldsymbol{p} = -r\bar{\boldsymbol{F}}\boldsymbol{e} + \boldsymbol{u}(\boldsymbol{m} \cdot \boldsymbol{n}) - \boldsymbol{u}(\boldsymbol{m} \cdot \boldsymbol{n} + r\boldsymbol{n} \cdot \boldsymbol{e}) \tag{2.3.16}$$

where we have used (2.3.8) with $\boldsymbol{F}$ replaced by $\bar{\boldsymbol{F}}$; on the other hand,

$$\Big| \int_{r\boldsymbol{e}-\boldsymbol{m}}^{r\boldsymbol{e}+\boldsymbol{m}} \nabla\chi(\boldsymbol{p})\, d\boldsymbol{p} \Big| < 2c, \qquad \Big| \int_{\boldsymbol{m}}^{-\boldsymbol{m}} \nabla\chi(\boldsymbol{p})\, d\boldsymbol{p} \Big| < 2c \tag{2.3.17}$$

since the deformation gradient is bounded by c. If $\boldsymbol{n} \cdot \boldsymbol{e} = 0$ then the difference of $\boldsymbol{u}$ on the right-hand side of $(2.3.16)_2$ vanishes; if $\boldsymbol{n} \cdot \boldsymbol{e} \neq 0$, one can choose $r = kp/(\boldsymbol{n} \cdot \boldsymbol{e})$ where k is an integer and p is the period of $\boldsymbol{u}$; the difference of $\boldsymbol{u}$ vanishes again. In each case, (2.3.15) gives

$$r\boldsymbol{G}\boldsymbol{e} - r\bar{\boldsymbol{F}}\boldsymbol{e} + \int_{r\boldsymbol{e}-\boldsymbol{m}}^{r\boldsymbol{e}+\boldsymbol{m}} \nabla\chi(\boldsymbol{p})\, d\boldsymbol{p} + \int_{\boldsymbol{m}}^{-\boldsymbol{m}} \nabla\chi(\boldsymbol{p})\, d\boldsymbol{p} = \boldsymbol{0}.$$

By (2.3.17) then $|r\boldsymbol{G}\boldsymbol{e} - r\bar{\boldsymbol{F}}\boldsymbol{e}| \le 4c$ and dividing by $r > 0$ and letting $r \to \infty$ (in such a way that $r = kp/(\boldsymbol{n} \cdot \boldsymbol{e})$ if $\boldsymbol{n} \cdot \boldsymbol{e} \neq 0$) we obtain $|\boldsymbol{G}\boldsymbol{e} - \bar{\boldsymbol{F}}\boldsymbol{e}| = 0$ i.e., $\boldsymbol{G}\boldsymbol{e} = \bar{\boldsymbol{F}}\boldsymbol{e}$. But the validity of the last equation for each $\boldsymbol{e}$ such that $\boldsymbol{e} \cdot \boldsymbol{m} = 0$ is equivalent to the existence of a $\boldsymbol{b} \in$ Vect such that (2.3.13) holds.

(2) $\Rightarrow$ (1) : Let $\varphi : \mathbb{R} \to [0, 1]$ be an infinitely differentiable function which vanishes on $(-\infty, 0]$ and which is equal to 1 on $[1, \infty)$ and define χ by

$$\chi(\boldsymbol{p}) = \begin{cases} \boldsymbol{G}\boldsymbol{p} & \text{if } \boldsymbol{p} \cdot \boldsymbol{m} < 0, \\ \bar{\boldsymbol{F}}\boldsymbol{p} + \varphi(\boldsymbol{p} \cdot \boldsymbol{m})\boldsymbol{u}(\boldsymbol{p} \cdot \boldsymbol{n}) & \text{if } \boldsymbol{p} \cdot \boldsymbol{m} \ge 0. \end{cases}$$

Then clearly χ is continuous and piecewise continuously differentiable on each of the half-spaces $H^{\pm} = \{\pm\boldsymbol{p} \cdot \boldsymbol{m} > 0\}$. Let us check that χ is continuous across the plane $\boldsymbol{p} \cdot \boldsymbol{m} = 0$. The limit of χ from H^- is $\boldsymbol{G}\boldsymbol{p}$, while the limit of χ from H^+ is $\bar{\boldsymbol{F}}\boldsymbol{p}$, were we have used that $\varphi(\boldsymbol{p} \cdot \boldsymbol{m}) = 0$ if $\boldsymbol{p} \cdot \boldsymbol{m} = 0$. But the application of (2.3.13) to the $\boldsymbol{p}$ gives $\boldsymbol{G}\boldsymbol{p} = \bar{\boldsymbol{F}}\boldsymbol{p} + \boldsymbol{b}(\boldsymbol{m} \cdot \boldsymbol{p})$ which for $\boldsymbol{p} \cdot \boldsymbol{m} = 0$ gives the equality of the limits from $H^{\pm}$. Hence χ is continuous. That (2.3.12) holds is immediate. □

2.3.8 Austenite/Martensite Interface The deformations constructed in Proposition 2.3.7 lie in the foundations of the crystallographic theory of martensite due to WECHSLER, LIEBERMAN & READ [1953] and BOWLES & MACKENZIE [1954]. At the transition temperature θ, the set B of all minimum points of f contains a well A of the austenite and several wells M_i of the variants of martensite, see Subsect. 2.3.5 for the case of the cubic to tetragonal transformation. It often happens that that the wells M_i are mutually compatible, but incompatible with the well A of the austenite. (Such is the case of Subsect. 2.3.5, as shown in Example 2.5.5 below.) Let us therefore assume that B contains three different tensors $\boldsymbol{G}, \boldsymbol{F}^+, \boldsymbol{F}^-$ such that $\boldsymbol{F}^+, \boldsymbol{F}^-$ are rank 1 connected, namely $\boldsymbol{F}^+ = \boldsymbol{F}^- + \boldsymbol{a} \otimes \boldsymbol{n}$ and neither $\boldsymbol{G}, \boldsymbol{F}^+$ nor $\boldsymbol{G}, \boldsymbol{F}^-$ are rank 1 connected. Let Ω be a connected reference region of the body. Then the only ground states involving the wells of $\boldsymbol{G}, \boldsymbol{F}^\pm$ are the homogeneous states of deformation gradients $\boldsymbol{G}$ and $\boldsymbol{F}^\pm$, and the pairwise homogeneous states with deformation gradients $\boldsymbol{F}^\pm$ (and the states obtained form these by rotations). In particular, there is no ground state containing simultaneously the austenite $\boldsymbol{G}$ and the martensites $\boldsymbol{F}^\pm$. However, observations show that there are states involving both the austenite and the martensite, separated by some kind of interface. Namely, if (2.3.14) holds with some $\alpha, \boldsymbol{b}, \boldsymbol{m}$ then one can construct a sequence of states which involves a mixture of homogeneous deformations with all three deformation gradients $\boldsymbol{G}, \boldsymbol{F}^\pm$ and whose total energy is arbitrarily close to the minimum total energy (approximate ground states). Let $\boldsymbol{\chi}$ be a deformation as in Proposition 2.3.7(1), assume that $\boldsymbol{0} \in \Omega$, and consider a family of deformations $\boldsymbol{\chi}_\varepsilon$ given by

$$\boldsymbol{\chi}_\varepsilon(\boldsymbol{p}) = \varepsilon\boldsymbol{\chi}(\boldsymbol{p}/\varepsilon), \quad \boldsymbol{p} \in \Omega; \text{ hence } \boldsymbol{F}_\varepsilon(\boldsymbol{p}) := \nabla\boldsymbol{\chi}_\varepsilon(\boldsymbol{p}) = \nabla\boldsymbol{\chi}(\boldsymbol{p}/\varepsilon).$$

Set

$$\Omega_0 = \{\boldsymbol{p} \in \Omega : \boldsymbol{p}\cdot\boldsymbol{m} < 0\}, \ \Omega_{0,\varepsilon} = \{\boldsymbol{p} \in \Omega : 0 \le \boldsymbol{p}\cdot\boldsymbol{m} \le \beta\varepsilon\},$$
$$\Omega_\varepsilon = \{\boldsymbol{p} \in \Omega : \boldsymbol{p}\cdot\boldsymbol{m} > \beta\varepsilon\}.$$

On Ω_0, $\boldsymbol{\chi}_\varepsilon$ is homogeneous with the deformation gradient $\boldsymbol{G}$, on Ω_ε, $\boldsymbol{\chi}_\varepsilon$ is a periodic pairwise homogeneous deformation with the deformation gradients $\boldsymbol{F}^\pm$, of concentrations α, $1-\alpha$ and on the strip $\Omega_{0,\varepsilon}$ the deformation takes arbitrary values, but the deformation gradients are bounded by a constant independent of ε. As $\varepsilon \to 0$, the periodic layers $\boldsymbol{F}^+/\boldsymbol{F}^-/\boldsymbol{F}^+/\boldsymbol{F}^-, \ldots$ in the region Ω_ε become finer and finer and $V(\Omega_{0,\varepsilon}) \to 0$. Let $B_0 = \{\boldsymbol{G}, \boldsymbol{F}^+, \boldsymbol{F}^-\}$. Since $\{\boldsymbol{p} \in \Omega : \boldsymbol{F}_\varepsilon(\boldsymbol{p}) \notin B_0\} \subset \Omega_{0,\varepsilon}$, we have

$$V\{\boldsymbol{p} \in \Omega : \boldsymbol{F}_\varepsilon(\boldsymbol{p}) \notin B_0\} \to 0, \quad V\{\boldsymbol{p} \in \Omega : \boldsymbol{F}_\varepsilon(\boldsymbol{p}) \in B_0\} \to V(\Omega).$$

Hence, as $\varepsilon \to 0$, most of the points $\boldsymbol{p} \in \Omega$ have the deformation gradient in B_0, where the stored energy takes the minimum value; although the state $\boldsymbol{\chi}_\varepsilon$ is not a ground state, its total stored energy converges to the minimum value. Thus the material can reduce its total stored energy by exhibiting the fine-structure state $\boldsymbol{\chi}_\varepsilon$. Moreover, the region $\Omega_{0,\varepsilon}$ plays a role of a planar interface separating the homogeneous deformation $\boldsymbol{G}$ from the fine mixture of the deformations $\boldsymbol{F}^\pm$.

The amounts of the three phases $\boldsymbol{G}, \boldsymbol{F}^\pm$ satisfy

$$V\{\boldsymbol{p} \in \Omega : \boldsymbol{F}_\varepsilon(\boldsymbol{p}) = \boldsymbol{G}\} \to V(\Omega_0), \quad V\{\boldsymbol{p} \in \Omega : \boldsymbol{F}_\varepsilon(\boldsymbol{p}) = \boldsymbol{F}^+\} \to \alpha V(\Omega \setminus \Omega_0),$$
$$V\{\boldsymbol{p} \in \Omega : \boldsymbol{F}_\varepsilon(\boldsymbol{p}) = \boldsymbol{F}^-\} \to (1-\alpha)V(\Omega \setminus \Omega_0).$$

For the weak limit of $\boldsymbol{\chi}_\varepsilon$, see Example 21.1.2(4), and for the Young measure associated with $\boldsymbol{\chi}_\varepsilon$, see Example 21.1.4(4).

For the wells structure from Subsect. 2.3.5, (2.3.14) is explicitly solvable; in addition to the literature at the beginning of this subsection, the reader is referred to BALL & JAMES [1987], where also general conditions for the solvability of (2.3.14) are derived. The deformation from Proposition 2.3.7 is also the basic ingredient of the wedge-like microstructure in martensite. For this construction, and for a number of related results, including the shape-memory effect, see BHATTACHARYA

[1991, 1992, 1993], BALL & JAMES [1992], RUDDOCK [1994], BHATTACHARYA & al. [1994], and the references therein. A widely used related approach to the morphology of martensitic transformations was developed by KHACHATURYAN [1967, 1983], ROITBURD [1978], and KHACHATURYAN & SHATALOV [1969]. Also in that theory, the interface orientations are calculated by seeking rank 1 connections between energy wells, and layering is identified as a mechanism for energy reduction. The only substantial difference from the above approach is that the theory of Khachaturyan, Roitburd, & Shatalov is 'geometrically linear,' i.e., the material satisfies the principle of material frame indifference only approximately inasmuch as the stored energy depends on the deformation gradient only through the tensor of infinitesimal deformation $\boldsymbol{E}$. The rotations, represented above by the elements of Orth^+, are represented by skew tensors. This, and the use of $\boldsymbol{E}$, makes some algebraic problems easier. However, when, loosely speaking, large rotations are involved in the problem, the two theories can lead to significantly different predictions. The reader is referred to BALL & JAMES [1992] and BHATTACHARYA [1993] for detailed comparisons of the two theories, and for further references.

2.4 Rank 1 Connections

This and the following sections deal with the algebraic aspects of rank 1 connections. If $\boldsymbol{F}, \boldsymbol{G}$ are rank 1 connected, then also $\boldsymbol{F}^{\mathrm{T}}, \boldsymbol{G}^{\mathrm{T}}$ are rank 1 connected; by (1.1.1), $\boldsymbol{F}^{-1}, \boldsymbol{G}^{-1}$ are rank 1 connected provided $\boldsymbol{F}, \boldsymbol{G}$ are invertible. The following proposition presents some additional properties of rank 1 connected tensors; all of these properties are immediate and intuitively clear in view of the geometric compatibility of the underlying deformation. The proof is therefore left to the reader.

2.4.1 Proposition *Let $\boldsymbol{F}, \boldsymbol{G} \in \mathrm{Lin}^+$ be rank 1 connected and denote by H the orthogonal complement $\boldsymbol{n}^{\perp}$ of $\boldsymbol{n}$. Then*

(1) *$\boldsymbol{F}\boldsymbol{c} = \boldsymbol{G}\boldsymbol{c}$ for every $\boldsymbol{c} \in H$; equivalently, if $\boldsymbol{P} = \boldsymbol{1} - \boldsymbol{n} \otimes \boldsymbol{n}$ is the projection onto H, then $\boldsymbol{G}\boldsymbol{P} = \boldsymbol{F}\boldsymbol{P}$;*

(2) *$\boldsymbol{F}^{-\mathrm{T}}\boldsymbol{n}/|\boldsymbol{F}^{-\mathrm{T}}\boldsymbol{n}| = \boldsymbol{G}^{-\mathrm{T}}\boldsymbol{n}/|\boldsymbol{G}^{-\mathrm{T}}\boldsymbol{n}|$; if this common value is denoted by $\boldsymbol{m}$, then the hyperplane H' of normal $\boldsymbol{m}$ is the image of H under $\boldsymbol{F}$ and also under $\boldsymbol{G}$;*

(3) $\operatorname{cof} \boldsymbol{F}\boldsymbol{n} = \operatorname{cof} \boldsymbol{G}\boldsymbol{n}$.

Conversely, if $\boldsymbol{F}, \boldsymbol{G} \in \mathrm{Lin}^+$ and if there exists a hyperplane $H \subset \mathrm{Vect}$ such that $\boldsymbol{F}\boldsymbol{c} = \boldsymbol{G}\boldsymbol{c}$ for every $\boldsymbol{c} \in H$ then $\boldsymbol{F}, \boldsymbol{G}$ are rank 1 connected; if $\boldsymbol{n}$ is any of the two unit normals to H, then $\boldsymbol{G} = \boldsymbol{F} + \boldsymbol{a} \otimes \boldsymbol{n}$ for some $\boldsymbol{a} \in \mathrm{Vect}$.

A somewhat less straightforward is the description of the set of possible principal stretches of a rank 1 perturbation of a given tensor. We shall give two results concerning this.

2.4.2 Proposition *Let $\boldsymbol{F} \subset \mathrm{Lin}^+$ have the principal stretches $v_1 \geq v_2 \geq \ldots v_n$. Then $w_1 \geq w_2 \geq \ldots w_n$ are principal stretches of some rank 1 perturbation $\boldsymbol{G} \in \mathrm{Lin}^+$ of $\boldsymbol{F}$ if and only if*

$$w_1 \geq v_2, \quad v_1 \geq w_2 \geq v_3, \quad \ldots, \quad v_{n-1} \geq w_n. \tag{2.4.1}$$

For $n = 3$ these conditions read

$$w_1 \geq v_2, \quad v_1 \geq w_2 \geq v_3, \quad v_2 \geq w_3.$$

I refer to ROSAKIS [1990] for proof. Note that the above inequalities are symmetric under the exchange $w_i \leftrightarrow v_i$. A rank 1 perturbation can increase the largest principal stretch unlimitedly, but cannot reduce it below the second principal stretch of the original tensor; generally, the rank 1 perturbation can shift the i-th principal stretch in the interval between the preceding and the following principal stretches of the original. Related to this result is the classical result in the matrix theory that if $\boldsymbol{A}, \boldsymbol{B} \in$ Sym and $\boldsymbol{B} = \boldsymbol{A}+\boldsymbol{c}\otimes\boldsymbol{c}$ for some $\boldsymbol{c} \in$ Vect, then the nonincreasingly ordered eigenvalues a_i, b_i of $\boldsymbol{A}, \boldsymbol{B}$ satisfy the interlacing inequalities

$$b_1 \geq a_1 \geq b_2 \geq a_2 \geq \ldots \geq b_n \geq a_n. \tag{2.4.2}$$

(See, DONOGHUE [1974, Chap. 7] or HORN & JOHNSON [1986, Chap. 4].) In fact, the direct part of the proof of the above proposition may be derived from this, since if $\boldsymbol{G} = \boldsymbol{F} + \boldsymbol{a} \otimes \boldsymbol{n}$, then the square of the left stretch tensor of $\boldsymbol{G}$ is

$$(\boldsymbol{F} + \boldsymbol{a} \otimes \boldsymbol{n})(\boldsymbol{F} + \boldsymbol{a} \otimes \boldsymbol{n})^{\mathsf{T}} = \boldsymbol{F}\boldsymbol{F}^{\mathsf{T}} + (\boldsymbol{a} + \boldsymbol{F}\boldsymbol{n}) \otimes (\boldsymbol{a} + \boldsymbol{F}\boldsymbol{n}) - \boldsymbol{F}\boldsymbol{n} \otimes \boldsymbol{F}\boldsymbol{n}.$$

Thus the square of the left stretch tensor of $\boldsymbol{G}$ is obtained as a superposition of a change of the type $\boldsymbol{A} \mapsto \boldsymbol{B} = \boldsymbol{A} + \boldsymbol{c} \otimes \boldsymbol{c}$, which changes the eigenvalues according to (2.4.2), and a change of the type $\boldsymbol{B} \mapsto \boldsymbol{C} = \boldsymbol{B} - \boldsymbol{d} \otimes \boldsymbol{d}$, which changes the eigenvalues from b_i to c_i, where

$$b_1 \geq c_1 \geq b_2 \geq c_2 \geq \ldots \geq b_n \geq c_n.$$

The resulting change leads to (2.4.1).

The following special case is of particular interest.

2.4.3 Proposition *Let $n = 3$ and $v_1 \geq v_2 \geq v_3 > 0$.*

(1) *A necessary and sufficient condition for (v_1, v_2, v_3) to be an ordered set of principal stretches of some rank 1 perturbation $\boldsymbol{F} \in \mathrm{Lin}^+$ of the unit tensor $\mathbf{1}$ is that $v_2 = 1$; if this condition holds then a tensor $\boldsymbol{F} = \mathbf{1} + \boldsymbol{a} \otimes \boldsymbol{n}$ with $|\boldsymbol{n}| = 1$ has the required principal stretches if an only if*

$$|\boldsymbol{a}| = v_1 - v_3, \quad \boldsymbol{a} \cdot \boldsymbol{n} = v_1 v_3 - 1; \tag{2.4.3}$$

moreover, if $\boldsymbol{F} \neq \mathbf{1}$, then in a properly oriented orthonormal basis $\boldsymbol{e}_i$ of eigenvectors of $\boldsymbol{V} := \sqrt{\boldsymbol{F}\boldsymbol{F}^{\mathsf{T}}}$ corresponding to v_i,

$$\boldsymbol{a} = \tau(v_1 - v_3)(\sqrt{v_1^2 - 1}\, \boldsymbol{e}_1 - \sqrt{1 - v_3^2}\, \boldsymbol{e}_3),$$

$$\boldsymbol{n} = \tau(v_3 \sqrt{v_1^2 - 1}\, \boldsymbol{e}_1 + v_1 \sqrt{1 - v_3^2}\, \boldsymbol{e}_3),$$

where $\tau = 1/\sqrt{v_1^2 - v_3^2}$.

(2) *Any $\boldsymbol{U} \in \mathrm{Sym}^+$ whose ordered set of eigenvalues (v_1, v_2, v_3) satisfies $v_2 = 1$ is the right stretch tensor of some rank 1 perturbation $\boldsymbol{F} \in \mathrm{Lin}^+$ of $\mathbf{1}$.*

Proof (1): Assume that v_1, v_2, v_3 are the principal stretches of $\boldsymbol{F} = \mathbf{1} + \boldsymbol{a} \otimes \boldsymbol{n}$. Since

$$\boldsymbol{V}^2 = \boldsymbol{F}\boldsymbol{F}^{\mathsf{T}} = \mathbf{1} + \boldsymbol{a} \otimes \boldsymbol{n} + \boldsymbol{n} \otimes \boldsymbol{a} + \boldsymbol{a} \otimes \boldsymbol{a}, \tag{2.4.4}$$

any vector $\boldsymbol{d}$ perpendicular to $\boldsymbol{a}, \boldsymbol{n}$ satisfies $\boldsymbol{V}^2\boldsymbol{d} = \boldsymbol{d}$ which shows that one of the three eigenvalues of $\boldsymbol{V}$ is 1. Denoting the remaining two by v_1, v_3 and assuming that $v_1 \geq v_3$, we have to prove that $v_1 \geq 1 \geq v_3$. By (2.4.4),

$$\operatorname{tr} \boldsymbol{V}^2 = 3 + 2\boldsymbol{a} \cdot \boldsymbol{n} + \boldsymbol{a}^2 = 1 + v_1^2 + v_3^2, \quad \det \boldsymbol{V} = 1 + \boldsymbol{a} \cdot \boldsymbol{n} = v_1 v_3 \tag{2.4.5}$$

and a rearrangement gives (2.4.3). By Schwarz's inequality, $-|\boldsymbol{a}| \leq \boldsymbol{a} \cdot \boldsymbol{n} \leq |\boldsymbol{a}|$; inserting the values of $|\boldsymbol{a}|, \boldsymbol{a} \cdot \boldsymbol{n}$ calculated in (2.4.3) one obtains

$$v_3 - v_1 \leq v_1 v_3 - 1 \leq v_1 - v_3; \tag{2.4.6}$$

these inequalities give $v_1 \geq 1 \geq v_3$, which completes the proof of the necessity part. The sufficiency will be proved by exhibiting $\boldsymbol{F}$. Let us prove that there exists a pair $\boldsymbol{a} \in$ Vect, $\boldsymbol{n} \in$ Sph satisfying (2.4.3). By Schwarz's inequality, this amounts to checking (2.4.6), and this in turn is obtained from

$\nu_1 \geq 1 \geq \nu_3$ by reversing the steps in the proof above. The formulas (2.4.5) then show that $\boldsymbol{F} := \mathbf{1}+\boldsymbol{a}\otimes\boldsymbol{n}$ has the principal stretches (ν_1, ν_2, ν_3), which completes the proof of the sufficiency part. Finally, assuming $\boldsymbol{F} \neq \mathbf{1}$, we proceed to determining the form of $\boldsymbol{a}, \boldsymbol{n}$ in the basis of eigenvectors of $\boldsymbol{V}^2$. Hence, let $\nu_1 \geq 1 \geq \nu_3, \nu_1 - \nu_3 > 0$, let $\boldsymbol{F} = \mathbf{1} + \boldsymbol{a} \otimes \boldsymbol{n}$ with $|\boldsymbol{n}| = 1$ have the principal stretches $(\nu_1, 1, \nu_3)$ and write $\lambda_1 = \nu_1^2, \lambda_3 = \nu_3^2$. Recall that it has been demonstrated that the eigenvector $\boldsymbol{e}_2$ corresponding to the eigenvalue 1 may be chosen orthogonal to $\boldsymbol{a}, \boldsymbol{n}$; assuming that, the components of $\boldsymbol{a}, \boldsymbol{n}$ corresponding to $\boldsymbol{e}_2$ vanish. We have $\boldsymbol{V}^2 = \boldsymbol{F}\boldsymbol{F}^{\mathrm{T}} = \mathbf{1} + \boldsymbol{c} \otimes \boldsymbol{c} - \boldsymbol{n} \otimes \boldsymbol{n}$ where $\boldsymbol{c} = \boldsymbol{a} + \boldsymbol{n}$. In the basis of eigenvectors of $\boldsymbol{V}$ the equation $\boldsymbol{V}^2 - \boldsymbol{c} \otimes \boldsymbol{c} = \mathbf{1} - \boldsymbol{n} \otimes \boldsymbol{n}$ reads

$$\begin{bmatrix} \lambda_1 - c_1^2 & c_1 c_3 \\ c_1 c_3 & \lambda_3 - c_3^2 \end{bmatrix} = \begin{bmatrix} 1 - n_1^2 & n_1 n_3 \\ n_1 n_3 & 1 - n_3^2 \end{bmatrix}.$$

Taking the trace and determinant gives, using $n_1^2 + n_3^2 = 1$,

$$c_1^2 + c_3^2 = \lambda_1 + \lambda_3 - 1, \quad \lambda_3 c_1^2 + \lambda_1 c_3^2 = \lambda_1 \lambda_3. \tag{2.4.7}$$

Similarly, writing the normalization condition for $\boldsymbol{n}$, taking the determinant of $\boldsymbol{V}^2 + \boldsymbol{n}\otimes\boldsymbol{n} = \mathbf{1}+\boldsymbol{c}\otimes\boldsymbol{c}$, and using $(2.4.7)_1$ we obtain

$$n_1^2 + n_3^2 = 1, \quad \lambda_3 n_1^2 + \lambda_1 n_3^2 = \lambda_1 + \lambda_3 - \lambda_1 \lambda_3. \tag{2.4.8}$$

Solving the system (2.4.7), (2.4.8) and changing the orientation of the axes $\boldsymbol{e}_1, \boldsymbol{e}_3$ (if necessary) to reach $n_1 \geq 0, n_3 \geq 0$ we obtain

$$c_1 = \sigma_1 \sqrt{\frac{\lambda_1(\lambda_1 - 1)}{\lambda_1 - \lambda_3}}, \quad c_3 = \sigma_3 \sqrt{\frac{\lambda_3(1 - \lambda_3)}{\lambda_1 - \lambda_3}}, \quad n_1 = \sqrt{\frac{\lambda_3(\lambda_1 - 1)}{\lambda_1 - \lambda_3}}, \quad n_3 = \sqrt{\frac{\lambda_1(1 - \lambda_3)}{\lambda_1 - \lambda_3}},$$

where $\sigma_1, \sigma_3 = \pm 1$. To determine σ_1, σ_3, notice that $\det \boldsymbol{F} = 1 + \boldsymbol{a}\cdot\boldsymbol{n} = \boldsymbol{c}\cdot\boldsymbol{n} = \nu_1\nu_3$. This condition may be restated as $(\sigma_1 - 1)(\lambda_1 - 1) + (\sigma_3 - 1)(1 - \lambda_3) = 0$. Since the two terms in the last sum are nonpositive, they must vanish separately: $(\sigma_1 - 1)(\lambda_1 - 1) = (\sigma_3 - 1)(1 - \lambda_3) = 0$, which for $\lambda_1 - 1 > 0$, $1 - \lambda_3 > 0$ gives $\sigma_1 = \sigma_2 = 1$ whereas when any of $\lambda_1 - 1$, $1 - \lambda_3$ vanishes (which exhausts the remaining possibilities), then the value of σ_i does not influence the value of c_i. With this choice of signs, the definition $\boldsymbol{a} = \boldsymbol{c} - \boldsymbol{n}$ leads to the asserted formulas. (2): If $\nu_2 = 1$ then there exists a $\bar{\boldsymbol{F}}$ of the form $\bar{\boldsymbol{F}} = \mathbf{1} + \bar{\boldsymbol{a}} \otimes \bar{\boldsymbol{n}}$ whose principal stretches are (ν_1, ν_2, ν_3). This means that if $\bar{\boldsymbol{F}} = \bar{\boldsymbol{R}}\bar{\boldsymbol{U}}$ is the polar decomposition of $\bar{\boldsymbol{F}}$, then $\bar{\boldsymbol{U}}$ has the eigenvalues (ν_1, ν_2, ν_3) and since also $\boldsymbol{U}$ has these eigenvalues, there exists a $\boldsymbol{Q} \in \mathrm{Orth}^+$ such that $\boldsymbol{U} = \boldsymbol{Q}\bar{\boldsymbol{U}}\boldsymbol{Q}^{\mathrm{T}}$. Defining $\boldsymbol{F} := \boldsymbol{Q}\bar{\boldsymbol{F}}\boldsymbol{Q}^{\mathrm{T}}$, it is found first that $\boldsymbol{F} = \mathbf{1} + \boldsymbol{a} \otimes \boldsymbol{n}$ where $\boldsymbol{a} = \boldsymbol{Q}\bar{\boldsymbol{a}}$, $\boldsymbol{n} = \boldsymbol{Q}\bar{\boldsymbol{n}}$ and second that $\boldsymbol{F}^{\mathrm{T}}\boldsymbol{F} = \boldsymbol{Q}\bar{\boldsymbol{F}}^{\mathrm{T}}\bar{\boldsymbol{F}}\boldsymbol{Q}^{\mathrm{T}} = \boldsymbol{Q}\bar{\boldsymbol{U}}^2\boldsymbol{Q}^{\mathrm{T}} = \boldsymbol{U}^2$. Hence the right stretch tensor of $\boldsymbol{F}$ is $\boldsymbol{U}$. □

The preceding proposition will be now used to derive a simple test for the compatibility of wells.

2.4.4 Proposition *Let $n = 3$ and let O_1, O_2 be two distinct orbits with centers $\boldsymbol{U}_1, \boldsymbol{U}_2 \in \mathrm{Sym}^+$. Then*

(1) *for a given $\boldsymbol{F} \in O_1$ there are precisely zero, one or two rank 1 connections between $\boldsymbol{F}$ and O_2;*

(2) *there is at least one rank 1 connection (i.e., O_1, O_2 are compatible) if and only if the triple $\lambda_1 \geq \lambda_2 \geq \lambda_3 > 0$ of ordered eigenvalues of*

$$\boldsymbol{C} := \boldsymbol{U}_1^{-1}\boldsymbol{U}_2^2\boldsymbol{U}_1^{-1} \tag{2.4.9}$$

satisfies $\lambda_2 = 1$;

(3) *there exists exactly one rank 1 connection if and only if 1 is a double eigenvalue of $\boldsymbol{C}$, i.e., if and only if*

$$\boldsymbol{U}_2^2 = \boldsymbol{U}_1^2 + \delta \boldsymbol{U}_1\boldsymbol{e} \otimes \boldsymbol{U}_1\boldsymbol{e} \tag{2.4.10}$$

where $|\boldsymbol{e}| = 1$ and $\delta > -1, \delta \neq 0$;

(4) *if* $\boldsymbol{U}_2 = \boldsymbol{R}\boldsymbol{U}_1\boldsymbol{R}^{\mathrm{T}}$ *for some* $\boldsymbol{R} \in \mathrm{Orth}^+$ *(in particular, if* $\mathcal{O}_1$, $\mathcal{O}_2$ *are wells corresponding to two variants) then the number of rank 1 connections can be only zero or two.*

Of particular importance are the rank 1 connections arising in the context of (4): for a given $\boldsymbol{F} \in \mathcal{O}_1$, any rank 1 connected tensor $\boldsymbol{G} \in \mathcal{O}_2$ is called a twin of $\boldsymbol{F}$ on $\mathcal{O}_2$. Hence, any $\boldsymbol{F} \in \mathcal{O}_1$ has either exactly zero or exactly two twins on $\mathcal{O}_2$; in the latter case the two different twins $\boldsymbol{G}, \bar{\boldsymbol{G}} \in \mathcal{O}_2$ of $\boldsymbol{F}$ are called the reciprocal twins; Proposition 2.4.5 determines the relations between their normals and amplitudes. (Twins will be treated systematically in Sect. 2.5.)

Proof (2): If $\mathcal{O}_1$, $\mathcal{O}_2$ have a rank 1 connection, then $\boldsymbol{R}_2\boldsymbol{U}_2 - \boldsymbol{R}_1\boldsymbol{U}_1 = \boldsymbol{a} \otimes \boldsymbol{n}$ for some $\boldsymbol{R}_1, \boldsymbol{R}_2 \in \mathrm{Orth}^+$ and $\boldsymbol{a}, \boldsymbol{n} \in \mathrm{Vect}$. After premultiplication by $\boldsymbol{R}_1^{\mathrm{T}}$ and postmultiplication by $\boldsymbol{U}_1^{-1}$ we get the equivalent form

$$\boldsymbol{R}\boldsymbol{U}_2\boldsymbol{U}_1^{-1} = \mathbf{1} + \boldsymbol{b} \otimes \boldsymbol{m}, \tag{2.4.11}$$

where $\boldsymbol{R} = \boldsymbol{R}_1^{\mathrm{T}}\boldsymbol{R}_2$, $\boldsymbol{b} = \boldsymbol{R}_1^{\mathrm{T}}\boldsymbol{a}$, $\boldsymbol{m} = \boldsymbol{U}_1^{-1}\boldsymbol{n}$. Eliminating $\boldsymbol{R}$, we obtain equivalently

$$\boldsymbol{C} = (\mathbf{1} + \boldsymbol{m} \otimes \boldsymbol{b})(\mathbf{1} + \boldsymbol{b} \otimes \boldsymbol{m}) \tag{2.4.12}$$

where $1 + \boldsymbol{b} \cdot \boldsymbol{m} > 0$. By Proposition 2.4.3, the eigenvalues of $\boldsymbol{C}$ must satisfy $\lambda_2 = 1$. Conversely, if this condition is satisfied, then $\boldsymbol{C}$ may be written in the form (2.4.12) with some $\boldsymbol{b}, \boldsymbol{m}$; this implies

$$\boldsymbol{U}_2^2 = (\boldsymbol{U}_1 + \boldsymbol{n} \otimes \boldsymbol{b})(\boldsymbol{U}_1 + \boldsymbol{b} \otimes \boldsymbol{n}) \quad \text{where } \boldsymbol{n} = \boldsymbol{U}_1\boldsymbol{m}.$$

The last equation shows that the tensor $\bar{\boldsymbol{F}} := \boldsymbol{U}_1 + \boldsymbol{b} \otimes \boldsymbol{n}$ has the right stretch tensor $\boldsymbol{U}_2$ and hence $\boldsymbol{U}_1 + \boldsymbol{b} \otimes \boldsymbol{n} = \bar{\boldsymbol{R}}\boldsymbol{U}_2$ for some $\bar{\boldsymbol{R}} \in \mathrm{Orth}^+$, which gives the desired rank 1 connection of $\mathcal{O}_1$, $\mathcal{O}_2$. (1) & (3): Let us determine the number of rank 1 connected tensors of $\boldsymbol{U}_1$ on $\mathcal{O}_2$. If there are two different rank 1 connected tensors of $\boldsymbol{U}_1$ on $\mathcal{O}_2$, then (2.4.11) has two solutions $\boldsymbol{R}, \boldsymbol{b}, \boldsymbol{m}$ and $\bar{\boldsymbol{R}}, \bar{\boldsymbol{b}}, \bar{\boldsymbol{m}}$ with $\boldsymbol{R} \neq \bar{\boldsymbol{R}}$. Eliminating $\boldsymbol{U}_2\boldsymbol{U}_1^{-1}$, we see that this is equivalent to the existence of a solution of

$$\boldsymbol{Z}(\mathbf{1} + \bar{\boldsymbol{b}} \otimes \bar{\boldsymbol{m}}) = \mathbf{1} + \boldsymbol{b} \otimes \boldsymbol{m} \tag{2.4.13}$$

with $\boldsymbol{Z} = \boldsymbol{R}\bar{\boldsymbol{R}}^{\mathrm{T}}$ and $\boldsymbol{b} \otimes \boldsymbol{m} \neq \bar{\boldsymbol{b}} \otimes \bar{\boldsymbol{m}}$. Proposition 2.4.5 (below) shows that there can be at most one nontrivial solution of (2.4.13), which completes the proof of (1). By Proposition 2.4.5, there is no nontrivial solution of (2.4.13) if and only if $\boldsymbol{b}, \boldsymbol{m}$ are linearly dependent, and it is easy to see that this happens if and only if (2.4.10) holds. (4): Taking the trace of $\boldsymbol{U}_2 = \boldsymbol{R}\boldsymbol{U}_1\boldsymbol{R}^{\mathrm{T}}$ one finds that (2.4.10) cannot hold. □

2.4.5 Proposition *Let* $n = 3$ *and let* $\boldsymbol{m}, \bar{\boldsymbol{m}} \in \mathrm{Sph}$, $\boldsymbol{b}, \bar{\boldsymbol{b}} \in \mathrm{Vect}$, $\boldsymbol{Z} \in \mathrm{Orth}^+$ *satisfy* $1 + \boldsymbol{b} \cdot \boldsymbol{m} > 0$. *Then*

(1) *if* $\boldsymbol{b}, \boldsymbol{m}$ *are linearly dependent and* (2.4.13) *holds then* $\bar{\boldsymbol{m}} = \pm\boldsymbol{m}$, $\bar{\boldsymbol{b}} = \pm\boldsymbol{b}$;

(2) *if* $\boldsymbol{b}, \boldsymbol{m}$ *are linearly independent and* $\boldsymbol{b} \otimes \boldsymbol{m} \neq \bar{\boldsymbol{b}} \otimes \bar{\boldsymbol{m}}$ *then* (2.4.13) *holds if and only if*

$$\bar{\boldsymbol{b}} = \tau b^2(2(\boldsymbol{b} \cdot \boldsymbol{m} + 1)\boldsymbol{m} - \boldsymbol{b}), \quad \bar{\boldsymbol{m}} = \tau(b^2\boldsymbol{m} + 2\boldsymbol{b}), \tag{2.4.14}$$

where $\tau = \pm 1/|b^2\boldsymbol{m} + 2\boldsymbol{b}|$ *and* $|b^2\boldsymbol{m} + 2\boldsymbol{b}| \neq 0$.

Proof Suppose that (2.4.13) holds, and prove that

$$\boldsymbol{b} \cdot \boldsymbol{m} = \bar{\boldsymbol{b}} \cdot \bar{\boldsymbol{m}}, \quad b^2 = \bar{b}^2. \tag{2.4.15}$$

Taking the determinant, we obtain $(2.4.15)_1$. Set $\gamma = \boldsymbol{b} \cdot \boldsymbol{m} = \bar{\boldsymbol{b}} \cdot \bar{\boldsymbol{m}}$. The transpose of (2.4.13) reads

$$(\mathbf{1} + \bar{\boldsymbol{m}} \otimes \bar{\boldsymbol{b}})\boldsymbol{Z}^{\mathrm{T}} = \mathbf{1} + \boldsymbol{m} \otimes \boldsymbol{b}; \tag{2.4.16}$$

multiplying (2.4.13) from the left by (2.4.16), using $\boldsymbol{Z}^{\mathrm{T}}\boldsymbol{Z} = \mathbf{1}$, and rearranging, we obtain

$$\boldsymbol{m} \otimes \boldsymbol{b} + \boldsymbol{b} \otimes \boldsymbol{m} + b^2\boldsymbol{m} \otimes \boldsymbol{m} = \bar{\boldsymbol{m}} \otimes \bar{\boldsymbol{b}} + \bar{\boldsymbol{b}} \otimes \bar{\boldsymbol{m}} + \bar{b}^2\bar{\boldsymbol{m}} \otimes \bar{\boldsymbol{m}}. \tag{2.4.17}$$

Taking the trace and using $(2.4.15)_1$ we obtain $(2.4.15)_2$. If $\boldsymbol{b} = \delta\boldsymbol{m}$, then a simple analysis that is left to the reader leads to (1). (2): Set $\alpha = b^2 + 2\gamma$, $\beta = \sqrt{b^2 - \gamma^2}$. Since $\boldsymbol{b}$ is not a multiple of $\boldsymbol{m}$, Schwarz's inequality implies that $b^2 > \gamma^2$ and hence $\beta > 0$. This allows us to set

$$\boldsymbol{p} := (\boldsymbol{b} - \gamma\boldsymbol{m})/\beta, \quad \bar{\boldsymbol{p}} := (\bar{\boldsymbol{b}} - \gamma\bar{\boldsymbol{m}})/\beta$$

and clearly $\{\boldsymbol{m}, \boldsymbol{p}\}$ and and $\{\bar{\boldsymbol{m}}, \bar{\boldsymbol{p}}\}$ are pairs of orthogonal unit vectors. Equation (2.4.17) reads

$$\alpha m \otimes m + \beta(m \otimes p + p \otimes m) = \alpha \bar{m} \otimes \bar{m} + \beta(\bar{m} \otimes \bar{p} + \bar{p} \otimes \bar{m}). \tag{2.4.18}$$

It follows that the span of $\{m, p\}$ coincides with the span of $\{\bar{m}, \bar{p}\}$. We denote it by T and define an orthogonal transformation $Q : T \to T$ by $Qm = \bar{m}$, $Qp = \bar{p}$. Then (2.4.18) implies

$$Q(\alpha m \otimes m + \beta(m \otimes p + p \otimes m))Q^{\mathrm{T}} = \alpha m \otimes m + \beta(m \otimes p + p \otimes m) \quad \text{on } T. \tag{2.4.19}$$

If Q is the matrix of Q in the basis $\{m, p\}$, then (2.4.19) reads

$$QMQ^{\mathrm{T}} = M, \quad \text{where } M = \begin{bmatrix} \alpha & \beta \\ \beta & 0 \end{bmatrix} \tag{2.4.20}$$

is the matrix of $\alpha m \otimes m + \beta(m \otimes p + p \otimes m)$. Let us first assume that $\det Q = 1$ so that

$$Q = \begin{bmatrix} c & -s \\ s & c \end{bmatrix} \quad \text{for some } c, s \in \mathbb{R} \text{ with } c^2 + s^2 = 1.$$

Writing $(2.4.20)_1$ in the form $QM = MQ$, calculating the products and comparing the elements of the so obtained matrices, it is found that $s\alpha = s\beta = 0$ and hence $s = 0$ by $\beta > 0$. This implies that $Q = \pm I$, where I is the 2×2 identity matrix. Consequently $\bar{m} = \pm m$, $\bar{b} = \pm b$ which is excluded by the hypothesis $b \otimes m \neq \bar{b} \otimes \bar{m}$. If $\det Q = -1$, then

$$Q = \begin{bmatrix} c & s \\ s & -c \end{bmatrix} \quad \text{for some } c, s \in \mathbb{R} \text{ with } c^2 + s^2 = 1.$$

Equation $(2.4.20)_1$ this time leads to $c = \pm\alpha/\theta$ $s = \pm 2\beta/\theta$, where $\theta = \sqrt{4\beta^2 + \alpha^2}$. But then $\bar{m} = cm + sp$, $\bar{p} = sm - cp$, and using the definitions of $p, \bar{p}$, the formulas for c, s, and noting that $\theta = |b^2 m + 2b|$, we obtain (2.4.14). □

2.5 Twins

A pair of tensors $F, G \in \mathrm{Lin}^+$ is said be a ***twin*** (and ***G is said to be a twin of F***) if F, G are rank 1 connected and the principal stretches of F, G coincide. We write

$$G = F(1 + c \otimes n) \tag{2.5.1}$$

where $c \in \mathrm{Vect}$, $n \in \mathrm{Sph}$, and call c the amplitude of the twin. The condition that F, G have the same principal stretches is equivalent to $G = OFH$ for some $O, H \in \mathrm{Orth}^+$ and hence

$$OFH = F(1 + c \otimes n). \tag{2.5.2}$$

We shall see that any twin is either of type I or type II, where, by definition, a ***type I twin*** satisfies

$$c = 2(F^{-1}F^{-\mathrm{T}}n/|F^{-\mathrm{T}}n|^2 - n) \tag{2.5.3}$$

and a ***type II twin*** satisfies

$$n = 2(c/|c|^2 - F^{\mathrm{T}}Fc/|Fc|^2). \tag{2.5.4}$$

A ***compound*** twin is a twin that is simultaneously of types I and II. A ***pure type I twin*** is a type I twin that is not a type II twin. A ***pure type II twin*** is defined analogously. Writing $F = RU$ for the polar decomposition, we find that (2.5.2) reads

$$QUH = U(1 + c \otimes n) \tag{2.5.5}$$

where $Q = R^{\mathrm{T}}OR \in \mathrm{Orth}^+$. Throughout the section, assume that $U \in \mathrm{Sym}^+$, $n \in \mathrm{Sph}$ are given and the goal is to determine all c such that $U(1 + c \otimes n)$ and U are twins. Let $n^{\perp}$ denote the orthogonal complement of n and $P = 1 - n \otimes n$ the orthogonal projection onto $n^{\perp}$. The following decomposition of U^2 into the parts orthogonal and parallel to n will be useful.

2.5.1 Proposition *If* $U \in \mathrm{Sym}^+$ *then*

(1) *there exists a unique triple* A, x, β *where* $A \in \mathrm{Sym}$, $x \in \mathrm{Vect}$, $\beta > 0$, *such that*

$$U^2 = A + n \otimes x + x \otimes n + \beta n \otimes n \tag{2.5.6}$$

and $An = 0$, $An^\perp = n^\perp$, $x \in n^\perp$; *namely,*

$$A = PU^2P, \quad x = PU^2n = U^2n - |Un|^2 n, \quad \beta = |Un|^2; \tag{2.5.7}$$

(2) U^{-2} *may be written similarly as*

$$U^{-2} = B + n \otimes y + y \otimes n + \gamma n \otimes n \tag{2.5.8}$$

where $B = PU^{-2}P$, $y = PU^{-2}n = U^{-2}n - |U^{-1}n|^2 n$, *and* $\gamma = |U^{-1}n|^2$;

(3) *the restriction* T *of* A *to* $n^\perp$ *is an invertible linear transformation from* $n^\perp$ *onto* $n^\perp$ *and* $\beta - A^{-1}x \cdot x > 0$; *if* A^{-1} *denotes the linear transformation from* Vect *to* Vect *that is equal to the inverse of* T *on* $n^\perp$ *and to* 0 *on the span of* n, *then*

$$\gamma = (\beta - A^{-1}x \cdot x)^{-1}, \quad B = A^{-1} + \gamma A^{-1}x \otimes A^{-1}x, \quad y = -\gamma A^{-1}x. \tag{2.5.9}$$

(4) n *is not an eigenvector of* U *if and only if* $x \neq 0$.

Throughout the rest of the section, we use the notation of this proposition. In the matrix form,

$$U^2 = \begin{bmatrix} A & x \\ x^{\mathrm{T}} & \beta \end{bmatrix}, \quad U^{-2} = \begin{bmatrix} A^{-1} + \gamma A^{-1}x \otimes A^{-1}x & -\gamma A^{-1}x \\ -\gamma x^{\mathrm{T}} A^{-1} & (\beta - A^{-1}x \cdot x)^{-1} \end{bmatrix}.$$

Note that T is positive definite and if its eigenvalues are denoted $t_1^2 \geq t_2^2 \geq \ldots \geq t_{n-1}^2$, $t_i > 0$, then they satisfy the interlacing inequalities

$$v_1 \geq t_1 \geq v_2 \geq \ldots \geq v_{n-1} \geq t_{n-1} \geq v_n,$$

where v_i are the principal stretches. This follows from HORN & JOHNSON [1986, Thms. 4.3.8 & 4.3.10]; however, these inequalities will not be employed in the following treatment.

Proof (1): We have $1 = P + n \otimes n$ and hence

$$U^2 = (P + n \otimes n)U^2(P + n \otimes n) = PU^2P + n \otimes PU^2n + PU^2n \otimes n + n \otimes nU^2n \otimes n.$$

With the identification (2.5.7) this formula reduces to (2.5.6). This completes the proof of the existence. The proof of the uniqueness is left to the reader. (2): Same as (1). (3): A straightforward calculation shows that if U^{-2} is given by (2.5.8) with B, y, γ as in (2.5.9), then $U^2U^{-2} = U^{-2}U = 1$. (4): Immediate. □

Throughout the rest of the section, assume that $n = 3$.

2.5.2 Proposition U *and* $U(1 + c \otimes n)$ *are twins if and only if the following relations hold:*

$$c \cdot n = 0, \quad 2Uc \cdot Un + |Uc|^2 = 0, \quad 2U^{-1}c \cdot U^{-1}n - c^2|U^{-1}n|^2 = 0. \tag{2.5.10}$$

Equations (2.5.10) *may be equivalently rewritten as*

$$c \cdot n = 0, \quad |\sqrt{A}(c - w)|^2 = |\sqrt{A}w|^2, \quad (c - w)^2 = w^2, \tag{2.5.11}$$

where $w = -A^{-1}x$.

Proof Suppose that (2.5.5) holds. Taking the determinant of (2.5.5), we obtain $(2.5.10)_1$. Let $\bar{U} \in \mathrm{Sym}^+$ be the unique tensor such that

$$\bar{U}^2 = U(\mathbf{1} + c \otimes n)(\mathbf{1} + n \otimes c)U. \tag{2.5.12}$$

By (2.5.5), $\bar{U}^2 = QU^2Q^{\mathrm{T}}$, $\bar{U}^{-2} = QU^{-2}Q^{\mathrm{T}}$, and consequently

$$\operatorname{tr} \bar{U}^2 = \operatorname{tr} U^2, \quad \operatorname{tr} \bar{U}^{-2} = \operatorname{tr} U^{-2}. \tag{2.5.13}$$

On the other hand, by (2.5.12) and (1.1.6)

$$\bar{U}^2 = U^2 + Uc \otimes Un + Un \otimes Uc + Uc \otimes Uc,$$
$$\bar{U}^{-2} = U^{-2} - U^{-1}n \otimes U^{-1}c - U^{-1}c \otimes U^{-1}n + c^2 U^{-1}n \otimes U^{-1}n.$$

Conditions (2.5.13) then imply $(2.5.10)_{2,3}$. Conversely, suppose that (2.5.10) hold, and define $\bar{U} \in \mathrm{Sym}^+$ by (2.5.12). Equation $(2.5.10)_1$ implies that

$$\det \bar{U} = \det U, \quad \det \bar{U}^2 = \det U^2 \tag{2.5.14}$$

and equations $(2.5.10)_{2,3}$ imply (2.5.13). From the general formula $I_2(A) = \det A \operatorname{tr} A^{-1}$ and from (2.5.13), (2.5.14) we infer that the principal invariants of $\bar{U}^2$, U^2 are the same and hence $\bar{U} = QUQ^{\mathrm{T}}$ for some $Q \in \mathrm{Orth}^+$. But by the polar decomposition theorem, $U(\mathbf{1}+c\otimes n) = \bar{U}R$ for some $R \in \mathrm{Orth}$. Consequently, $U(\mathbf{1}+c\otimes n) = QUQ^{\mathrm{T}}R$. From the condition $(2.5.10)_1$, $\det R = 1$. Consequently, (2.5.5) holds with $H = Q^{\mathrm{T}}R$. To prove (2.5.11), note that for $c \in n^{\perp}$, (2.5.6) gives

$$U^2c = Ac + (x \cdot c)n, \quad |Uc|^2 = Ac \cdot c, \quad Uc \cdot Un = x \cdot c$$

so that $(2.5.10)_2$ reduces to $2x \cdot c + Ac \cdot c = 0$, which is equivalent to $(2.5.11)_2$. Similarly, (2.5.8) combined with Proposition 2.5.1(3) gives

$$U^{-2}c = Bc + (y \cdot c)n, \quad U^{-1}c \cdot U^{-1}n = y \cdot c = -|U^{-1}n|^2 A^{-1}x \cdot c,$$

and hence $(2.5.10)_3$ reduces to $2A^{-1}x \cdot c + c^2 = 0$ which is equivalent to $(2.5.11)_3$. □

In the following proposition the term *solution* means any *nonzero* solution c of (2.5.10) (for given U and n). We denote by q the number of solutions.

2.5.3 Proposition *q can take only the values 0, 1, 3 and ∞. We have $q \geq 1$ if and only if n is not an eigenvector of U (equivalently, $x \neq 0$). Assuming this,*

(1) *$q = 1$ if and only if T has two distinct eigenvalues and x is its eigenvector; then*

$$c = 2(U^{-2}n/|U^{-1}n|^2 - n) = -2|x|^2 x/|Ux|^2; \tag{2.5.15}$$

alternatively, c is the unique solution $c \in n^{\perp}, c \neq 0$ of

$$n = 2(c/c^2 - U^2c/|Uc|^2); \tag{2.5.16}$$

and hence c leads to a compound twin;

(3) *$q = 3$ if and only if T has two distinct eigenvalues and x is not its eigenvector; the solutions are*

$$c_0 = 2(U^{-2}n/|U^{-1}n|^2 - n) \quad \text{and} \quad c_i = -2(x \cdot e_i)e_i/|Ue_i|^2, \tag{2.5.17}$$

$i = 1, 2$, where e_i are the normalized perpendicular eigenvectors of T; alternatively, c_i may be characterized as the only two solutions $c \in n^{\perp}, c \neq 0$ of (2.5.16); moreover c_0 leads to a pure type I twin and c_i lead to pure type II twins;

(3) *$q = \infty$ if and only if T has two identical eigenvalues, i.e., if $PU^2P = \lambda P$ for some $\lambda > 0$; the solutions are identical with the nonzero solutions c of*

$$c \cdot x + \tfrac{1}{2}\lambda c^2 = 0, \quad c \cdot n = 0; \tag{2.5.18}$$

and also with the nonzero solutions $c \in n^{\perp}$ of (2.5.16); hence the arising twins are type II twins; of them exactly one is compound, namely that with c given by (2.5.15).

This gives a complete solution of the twinning equations (2.5.5). Notice that the identity tensor $\boldsymbol{U} = \mathbf{1}$ has no nontrivial twin. The proposition also implies that a twin with the amplitude $\boldsymbol{c}$ is compound if and only if $\boldsymbol{x}$ is an eigenvector of $\boldsymbol{A}$.

Proof If $\boldsymbol{n}$ is an eigenvector of $\boldsymbol{U}$, then by Proposition 2.5.1(4) $\boldsymbol{x} = \mathbf{0}$; then also $\boldsymbol{w} := -\boldsymbol{A}^{-1}\boldsymbol{x} = \mathbf{0}$ and hence (2.5.11) degenerate to $\boldsymbol{c} = \mathbf{0}$; consequently, there is no nonzero solution of (2.5.10). Conversely, clearly $\boldsymbol{c} = 2\boldsymbol{w}$ is always a solution of (2.5.10); hence the absence of a nonzero solution of (2.5.10) means that $\boldsymbol{w} = \mathbf{0}$, $\boldsymbol{x} = \mathbf{0}$ and consequently that $\boldsymbol{n}$ is an eigenvector of $\boldsymbol{U}$.

For the rest of the proof assume that $\boldsymbol{n}$ is not an eigenvector of $\boldsymbol{U}$ and hence $\boldsymbol{x} \neq \mathbf{0}$. Let $\boldsymbol{e}_i \in \boldsymbol{n}^{\perp}$, $\lambda_i > 0$, $i = 1, 2$, be the normalized perpendicular eigenvectors and eigenvalues of $\boldsymbol{T}$, respectively. If c_i, w_i are the components of $\boldsymbol{c}$, $\boldsymbol{w}$ in the basis $\{\boldsymbol{e}_1, \boldsymbol{e}_2\}$, then $(2.5.11)_{2,3}$ read

$$\left.\begin{aligned} \lambda_1(c_1 - w_1)^2 + \lambda_2(c_2 - w_2)^2 &= \lambda_1 w_1^2 + \lambda_2 w_2^2, \\ (c_1 - w_1)^2 + (c_2 - w_2)^2 &= w_1^2 + w_2^2. \end{aligned}\right\} \tag{2.5.19}$$

In the plane $\mathbb{R}^2$ with coordinates (c_1, c_2), the first of these represents an ellipsis with the center $\boldsymbol{w}$ and the axes of direction of the coordinate axes and of magnitudes $a_1^2 = w_1^2 + w_2^2\lambda_2/\lambda_1$, $a_2^2 = w_1^2\lambda_1/\lambda_2 + w_2^2$. The second equation represents a concentric circle of radius $|\boldsymbol{w}|$ and center $\boldsymbol{w} \in \boldsymbol{n}^{\perp}$. Clearly, the points

$$(0, 0), \quad (0, 2w_2), \quad (2w_1, 2w_2), \quad (2w_1, 0) \tag{2.5.20}$$

are solutions of (2.5.19), and surely the points $(0, 0)$ and $(2w_1, 2w_2)$ are distinct. The other two may degenerate to $(0, 0)$ or to $(2w_1, 2w_2)$. Given the constraints that the circle and the ellipsis are concentric and that there are two distinct points common to both of them, it is clear that the circle and the ellipsis have either exactly two, or exactly four, or infinitely many common points. These possibilities correspond to the following three exhaustive conditions on λ_i, w_i : (1): $\lambda_1 \neq \lambda_2$ and one of the numbers w_i, $i = 1, 2$, vanishes; (2): $\lambda_1 \neq \lambda_2$ and none of the numbers w_i, $i = 1, 2$, vanishes; (3): $\lambda_1 = \lambda_2$. We take these possibilities in turn.

(1): $\lambda_1 \neq \lambda_2$ and one of the numbers w_i, $i = 1, 2$, vanishes. Let us first prove that this is equivalent to asserting that (2.5.19) have exactly two distinct solutions. If the conditions on λ_i, w_i hold, we may assume without any loss of generality that $w_2 = 0$. Then (2.5.19) read

$$(c_1 - w_1)^2 + \lambda_2 c_2^2/\lambda_1 = w_1^2, \quad (c_1 - w_1)^2 + c_2^2 = w_1^2,$$

and subtracting the second equation from the first we obtain $(\lambda_2/\lambda_1 - 1)c_2^2 = 0$. As $\lambda_2/\lambda_1 - 1 \neq 0$, necessarily $c_2 = 0$ and consequently $(0, 0)$, $(2w_1, 0)$ are the only two solutions of (2.5.19). Conversely, if the system (2.5.19) has exactly two solutions, then two pairs of the points in (2.5.20) must coincide, which can happen only if one of the numbers w_i, $i = 1, 2$, vanishes. Also necessarily $\lambda_1 \neq \lambda_2$ for otherwise the circle and the ellipsis would coincide and there would be infinitely many solutions of (2.5.19). Moreover, one of the numbers w_i, $i = 1, 2$, vanishes if and only if $\boldsymbol{w} = -\boldsymbol{A}^{-1}\boldsymbol{x} = -\boldsymbol{T}^{-1}\boldsymbol{x}$ is an eigenvector of $\boldsymbol{T}$ and this is equivalent to $\boldsymbol{x}$ being an eigenvector of $\boldsymbol{T}$. Finally we prove that the solution $\boldsymbol{c}$ may be characterized as in Item (1) of the proposition. By (2.5.8), $\boldsymbol{U}^{-2}\boldsymbol{n} = \boldsymbol{y} + \gamma\boldsymbol{n}$, $|\boldsymbol{U}^{-1}\boldsymbol{n}|^2 = \gamma$, and consequently

$$2(\boldsymbol{U}^{-2}\boldsymbol{n}/|\boldsymbol{U}^{-1}\boldsymbol{n}|^2 - \boldsymbol{n}) = 2((\boldsymbol{y} + \gamma\boldsymbol{n})/\gamma - \boldsymbol{n}) = 2\boldsymbol{y}/\gamma = -2\boldsymbol{A}^{-1}\boldsymbol{x} = 2\boldsymbol{w}$$

which proves $(2.5.15)_1$. If the eigenvalue of $\boldsymbol{T}$ corresponding to the eigenvector $\boldsymbol{x}$ is $\lambda > 0$, then from (2.5.6), $\boldsymbol{U}^2\boldsymbol{x} = \lambda\boldsymbol{x} + x^2\boldsymbol{n}$ and hence $|\boldsymbol{U}\boldsymbol{x}|^2 = \lambda x^2$. Consequently, $\boldsymbol{c} = -2\boldsymbol{A}^{-1}\boldsymbol{x} = -2\lambda^{-1}\boldsymbol{x} = -2x^2\boldsymbol{x}/|\boldsymbol{U}\boldsymbol{x}|^2$ which is $(2.5.15)_2$. By (2.5.6),

$$\boldsymbol{U}^2\boldsymbol{c} = \boldsymbol{A}\boldsymbol{c} + (\boldsymbol{x} \cdot \boldsymbol{c})\boldsymbol{n} = \lambda\boldsymbol{c} + (\boldsymbol{x} \cdot \boldsymbol{c})\boldsymbol{n}, \quad |\boldsymbol{U}\boldsymbol{c}|^2 = \lambda c^2, \quad \boldsymbol{U}\boldsymbol{c} \cdot \boldsymbol{U}\boldsymbol{n} = \boldsymbol{x} \cdot \boldsymbol{c}.$$

By $(2.5.10)_2$ then $\boldsymbol{x} \cdot \boldsymbol{c} = -\frac{1}{2}|\boldsymbol{U}\boldsymbol{c}|^2 = -\frac{1}{2}\lambda c^2$. Using these relations, one finds that

$$2(\boldsymbol{c}/c^2 - \boldsymbol{U}^2\boldsymbol{c}/|\boldsymbol{U}\boldsymbol{c}|^2) = 2(\boldsymbol{c}/c^2 - (\lambda\boldsymbol{c} + (\boldsymbol{x} \cdot \boldsymbol{c})\boldsymbol{n})/(\lambda c^2)) = -2\boldsymbol{n}(\boldsymbol{x} \cdot \boldsymbol{c})/(\lambda c^2) = \boldsymbol{n}$$

and hence $\boldsymbol{c}$ satisfies (2.5.16). To see that there is exactly one solution of (2.5.16), notice that any solution of (2.5.16) satisfies (2.5.10). Indeed, a multiplication of (2.5.16) by $\boldsymbol{n}$ gives $(2.5.10)_2$ and a multiplication of (2.5.16) by $\boldsymbol{U}^{-2}\boldsymbol{n}$ gives $(2.5.10)_3$. But in the present case, there is only one nonzero solution of (2.5.10), namely the $\boldsymbol{c}$ given by (2.5.15).

(2): $\lambda_1 \neq \lambda_2$ and none of the numbers w_i, $i = 1, 2$, vanishes. This condition is equivalent to the existence of four distinct solution of (2.5.19), namely those listed in (2.5.20). This is proved by an analysis that is slightly more complicated than the corresponding portion of the proof in (1), and the details are omitted. A calculation identical with that in the proof of $(2.5.15)_1$ above shows that $c_0 = 2w$ is given by $(2.5.17)_1$. The points $(2w_1, 0)$ and $(0, 2w_2)$ may be written invariantly as $c_i = 2(w \cdot e_i)e_i$, $i = 1, 2$ which give $(2.5.17)_2$. A calculation identical with that performed in proving (2.5.16) shows that c_i, $i = 1, 2$, satisfy (2.5.16). As in the proof of (1), every solution of (2.5.16) satisfies (2.5.10) and an appeal to the fact that the above c_0, c_1, c_2 are the only nonzero solutions of (2.5.10) and the easily verifiable fact that x does not satisfy (2.5.16) leads to the assertion that c_1, c_2 are the only solutions of (2.5.16).

(3): Finally, there are infinitely many common points if and only if the circle and the ellipsis coincide. This happens if and only if A is on $n^\perp$ a multiple of P by a scalar λ (say). Then (2.5.10) reduces to (2.5.18). The easy details are left to the reader. □

2.5.4 Proposition *If F, G is a twin with $F \neq G$ and if U is the right stretch tensor of F, then*

(1) *the twin F, G is either of type I or of type II and its amplitude is as described in Proposition 2.5.3(1)–(3);*

(2) *the tensors $O, H \in \mathrm{Orth}^+$ in (2.5.2) may be chosen as 180° rotations*

$$O = 2o \otimes o - 1, \quad H = 2h \otimes h - 1, \quad O^2 = H^2 = 1, \tag{2.5.21}$$

where the axes of rotation are determined as follows:

$$\text{for a type I twin:} \quad o = \pm F^{-T}n/|F^{-T}n|, \quad h = \pm n; \tag{2.5.22}$$

$$\text{for a type II twin:} \quad o = \pm Fc/|Fc|, \quad h = \pm c/|c|; \tag{2.5.23}$$

(3) *any other pair $\bar{O}, \bar{H}$ of tensors satisfying (2.5.2) is given by*

$$\bar{O} = ORS^T R^T, \quad \bar{H} = SH,$$

where O, H are as in (2), S is any proper orthogonal tensor such that $SU = US$, and $F = RU$ is the polar decomposition of F.

Proof (1) follows from Proposition 2.5.3 and the polar decomposition theorem. (2): This is just an inspection using (2.5.3), (2.5.4) and (2.5.21). The details are left to the reader. (3): If $\bar{O}, \bar{H}$ satisfy (2.5.2) and if $S = \bar{H}H^T$, then (2.5.2) written for $\bar{O}, \bar{H}$ and O, H imply $R^T\bar{O}^T OR = USU^{-1}$. Eliminating the tensor $R^T\bar{O}^T OR$ we obtain, after some rearrangements, $SU^2 = U^2 S$. This implies $SU = US$ and the rest is a calculation that is left to the reader. □

2.5.5 Example (Twins in the Tetragonal Martensite) Let us show that the wells with centers U_1, U_2, U_3 at the variants of the tetragonal phases of the martensite as in Subsect. 2.3.5 are compatible. It is enough to determine the rank 1 connections between the wells U_1, U_2, the connections between other pairs of wells follow by symmetry. Thus we are looking for $R \in \mathrm{Orth}^+$, $c \in \mathrm{Vect}$, $n \in \mathrm{Sph}$ such that

$$RU_2 = U_1(1 + c \otimes n). \tag{2.5.24}$$

Eliminating R, we obtain

$$U_2^2 = (1 + n \otimes c)U_1^2(1 + c \otimes n) \tag{2.5.25}$$

where $U_i^2 = \eta_1^2 1 + (\eta_2^2 - \eta_1^2)e_i \otimes e_i$, $i = 1, 2, 3$. Since $(1 + n \otimes c)^{-1} = 1 - n \otimes c$, (2.5.25) implies

$$(1 - n \otimes c)U_2^2 = U_1^2(1 + c \otimes n), \quad U_2^2 - U_1^2 = U_1^2 c \otimes n + n \otimes U_2^2 c. \tag{2.5.26}$$

Multiplying $(2.5.26)_2$ by c and using $c \cdot n = 0$ we obtain

$$(U_2^2 - U_1^2)c = |U_2 c|^2 n. \tag{2.5.27}$$

In the basis $\{e_1, e_2, e_3\}$, $U_2^2 - U_1^2 = \mathrm{diag}(\eta_1^2 - \eta_2^2, \eta_2^2 - \eta_1^2, 0)$ from which the range of $U_2^2 - U_1^2$ is equal to the span of $\{e_1, e_2\}$. Since $|U_2 c| \neq 0$, it follows that $n \in \mathrm{span}\{e_1, e_2\}$; multiplying $(2.5.26)_2$ by n, it is found that also $c \in \mathrm{span}\{e_1, e_2\}$. The component form of (2.5.27) therefore reads

$$(\eta_1^2 - \eta_2^2)c_1 = |U_2 c|^2 n_1, \quad (\eta_1^2 - \eta_2^2)c_2 = -|U_2 c|^2 n_2. \tag{2.5.28}$$

Multiplying $(2.5.28)_1$ by n_1, $(2.5.28)_2$ by n_2, summing, and using $\boldsymbol{c} \cdot \boldsymbol{n} = 0$, $\boldsymbol{n}^2 = 1$, one finds that

$$n_1^2 = n_2^2 = \tfrac{1}{2}. \tag{2.5.29}$$

Furthermore, (2.5.28) implies $\boldsymbol{c} = k(n_1, -n_2, 0)$ for some $k \neq 0$. Inserting this form of $\boldsymbol{c}$ and using (2.5.29) it is found that $k = 2(\eta_1^2 - \eta_2^2)/(\eta_1^2 + \eta_2^2)$. Assuming without any loss of generality that $n_1 = +1/\sqrt{2}$, one concludes finally that

$$\begin{gathered} \boldsymbol{n} = \frac{1}{\sqrt{2}}(\boldsymbol{e}_1 \pm \boldsymbol{e}_2), \quad \boldsymbol{c} = \frac{\sqrt{2}(\eta_1^2 - \eta_2^2)}{\eta_1^2 + \eta_2^2}(\boldsymbol{e}_1 \mp \boldsymbol{e}_2), \\ \boldsymbol{a} = \frac{\sqrt{2}(\eta_1^2 - \eta_2^2)}{\eta_1^2 + \eta_2^2}(\eta_2 \boldsymbol{e}_1 \mp \eta_1 \boldsymbol{e}_2), \end{gathered} \tag{2.5.30}$$

where $\boldsymbol{n}, \boldsymbol{c}$ are as in (2.5.24) and $\boldsymbol{a} = \boldsymbol{U}_1 \boldsymbol{c}$ so that $\boldsymbol{R}\boldsymbol{U}_2 - \boldsymbol{U}_1 = \boldsymbol{a} \otimes \boldsymbol{n}$. To determine $\boldsymbol{R}$, we note that by Subsect. 2.3.5, we have

$$\boldsymbol{U}_2 = \boldsymbol{H}\boldsymbol{U}_1\boldsymbol{H}, \quad \boldsymbol{H} = 2\boldsymbol{n} \otimes \boldsymbol{n} - \boldsymbol{1}, \quad \boldsymbol{H} = \boldsymbol{H}^{\mathrm{T}},$$

where $\boldsymbol{n}$ is as in $(2.5.30)_1$. Further, using the above calculated $\boldsymbol{c}, \boldsymbol{n}$, it is easily found that $\boldsymbol{U}_1, \boldsymbol{U}_2$ is a compound twin, and in particular a type I twin. One may thus consistently choose $\boldsymbol{O}$ with $\boldsymbol{o} = \boldsymbol{U}_1^{-1}\boldsymbol{n}/|\boldsymbol{U}_1^{-1}\boldsymbol{n}|$ (see (2.5.22) and (2.5.21)) to find

$$\boldsymbol{O}\boldsymbol{U}_1\boldsymbol{H} = \boldsymbol{U}_1(\boldsymbol{1} + \boldsymbol{c} \otimes \boldsymbol{n}), \quad \text{where } \boldsymbol{O} = 2\boldsymbol{o} \otimes \boldsymbol{o} - \boldsymbol{1}, \quad \boldsymbol{o} = \frac{\eta_1 \boldsymbol{e}_1 \pm \eta_2 \boldsymbol{e}_2}{\sqrt{\eta_1^2 + \eta_2^2}}.$$

The rotation $\boldsymbol{R}$ from (2.5.24) is $\boldsymbol{R} = \boldsymbol{O}\boldsymbol{H}$, and a calculation gives

$$\boldsymbol{R} = \boldsymbol{e}_3 \otimes \boldsymbol{e}_3 + (\boldsymbol{1} - \boldsymbol{e}_3 \otimes \boldsymbol{e}_3)c + \boldsymbol{W}_0 s,$$

where

$$\boldsymbol{W}_0 = \boldsymbol{e}_1 \otimes \boldsymbol{e}_2 - \boldsymbol{e}_2 \otimes \boldsymbol{e}_1, \quad c = \pm\frac{2\eta_1\eta_2}{\eta_1^2 + \eta_2^2}, \quad s = \frac{\eta_2^2 - \eta_1^2}{\eta_1^2 + \eta_2^2}.$$

A comparison with (1.3.6) shows that $\boldsymbol{R}$ is a rotation about axis $\boldsymbol{e}_3$ through angle ω whose sine and cosine are s, c, respectively. The choices of $\pm$ in the above formulas thus give two different (reciprocal) twins of $\boldsymbol{U}_1$ on the orbit of $\boldsymbol{U}_2$. (Alternatively, one may determine the eigenvalues (ν_1, ν_2, ν_3) of $\boldsymbol{U}_2\boldsymbol{U}_1^{-1}$ and apply Proposition 2.4.3 to arrive at the above results.)

2.6 Appendix: Piecewise Smooth Objects

In this section we introduce a class of piecewise continuously differentiable fields f with intersecting singular surfaces. Briefly, for each piecewise continuously differentiable field f, the spacetime region $Q = \Omega \times (t_1, t_2)$ will be divided into a finite union of regions Q^p whose closures cover cl Q, and the restriction of f to each Q^p will be required to have a continuously differentiable extension to an open set containing cl Q^p. Less clear is the question of the appropriate restrictions on the sets Q^p. There appears to be no formal treatment of the piecewise continuously differentiable fields in the literature except in the one-dimensional case; the reason may become clear after the perusal of this paragraph. Namely, if one requires that the resulting class be a linear space, then the class of regions Q^p entering the definition must be closed under intersection. The reason is that the singular surface of the sum $f + g$ is generally the union of the singular surfaces of f, g, and the partition for $f + g$ is obtained from the partitions Q^p, R^q for f, g as the collection of the intersections $P^{p,q} = Q^p \cap R^q$. There appears to be no naturally defined class of regions with this property, except for the class of polygonal regions, which is too narrow, and the class of sets of finite perimeter (or classes derived from it). Even when the boundaries of Q^p, R^q are of class C^∞, the boundary of $Q^p \cap R^q$ can be rather strongly irregular; irregular to the extent that the Gauss–Green theorem does not hold with the usual topological boundary and the geometric exterior normal (see Example 3.8.9). One has to pass to the measure-theoretic counterparts of these, and to the notions of the geometric measure theory. One is thus quickly led to the functions of bounded variation, which is indeed the only a natural setting for these matters (see Sect. 3.8). Nevertheless, it appears worthwhile to present an elementary exposition of piecewise continuously differentiable fields with intersecting singular surfaces, bearing in mind that the resulting class is not a linear space.

Throughout the section, let W, X be finite-dimensional real inner product spaces and let n be the dimension of W. Later, we shall identify W with Vect or with Vect $\times$ $\mathbb{R}$. If $\boldsymbol{x} \in$ W and $r > 0$, we denote by $B(\boldsymbol{x}, r) = \{\boldsymbol{y} \in \mathrm{W} : |\boldsymbol{y} - \boldsymbol{x}| < r\}$ the open ball of radius r and center $\boldsymbol{x}$ in W, where $|\cdot|$ is the Euclidean norm on W. An affine orthogonal transformation $\boldsymbol{T} : \mathbb{R}^n \to \mathrm{W}$ is a mapping of the form $\boldsymbol{T}x = \boldsymbol{O}x + \boldsymbol{b}$ where $\boldsymbol{O}$ is a linear isometry from $\mathbb{R}^n$ to W and $\boldsymbol{b} \in$ W a constant.

2.6.1 Proposition *Let M be a subset of* W *and k an integer such that $0 < k < n$. The following conditions are equivalent:*

(1) *for every $\boldsymbol{x} \in M$ there exists a neighborhood O of $\boldsymbol{x}$ and a class C^1 mapping $\phi : O \to \mathbb{R}^{n-k}$ with $D\phi$ of rank $n-k$ on O such that*

$$M \cap O = \phi^{-1}(0);$$

(2) *for every $\boldsymbol{x} \in M$ there exists a neighborhood O of $\boldsymbol{x}$ and a class C^1 mapping $\psi : P \to$ W on a neighborhood P of 0 in $\mathbb{R}^k$ with $D\psi$ of rank k on P such that*

$$M \cap O = \psi(P) \text{ and } \boldsymbol{x} = \psi(0);$$

(3) *for every $\boldsymbol{x} \in M$ there exists a neighborhood O of $\boldsymbol{x}$, a class C^1 mapping $\omega : P \to \mathbb{R}^{n-k}$ on a neighborhood P of 0 in $\mathbb{R}^k$ and an affine orthogonal transformation $\boldsymbol{T}$ from $\mathbb{R}^n$ to* W *which maps the origin in $\mathbb{R}^n$ to $\boldsymbol{x}$ such that*

$$M \cap O = \boldsymbol{T}(\{(z, \omega(z)) : z \in P\}) \text{ and } \omega(0) = 0, \quad D\omega(0) = 0.$$

Any subset M of W satisfying the above three equivalent conditions is called a ***smooth k-dimensional surface*** in W; the descriptions in (1), (2), (3) above are the ***implicit, parametric,*** and ***graph*** descriptions of M near $\boldsymbol{x}$. Smooth surfaces need not be connected. Let $\boldsymbol{x} \in M$ be a point of a smooth k-dimensional surface and let ϕ, ψ and ω, $\boldsymbol{T}$ be an implicit, parametric and graph descriptions of M near $\boldsymbol{x}$. The ***tangent space*** to M at $\boldsymbol{x} \in M$ is the linear subspace $T_{\boldsymbol{x}}(M)$ of W of all $\boldsymbol{t} \in$ W satisfying the following three equivalent conditions: (a) $D\phi(\boldsymbol{x})[\boldsymbol{t}] = \boldsymbol{0}$ (i.e., $T_{\boldsymbol{x}}(M)$ is the kernel of $D\phi(\boldsymbol{x})$); (b) $\boldsymbol{t} = D\psi(0)[t]$ for some $t \in \mathbb{R}^k$ (i.e., $T_{\boldsymbol{x}}(M)$ is the range of $D\psi(0)$); (c) $\boldsymbol{t} = \boldsymbol{O}(t, 0)$ for some $t \in \mathbb{R}^k$, where $\boldsymbol{O}$ is the linear isometry occurring in the definition of an affine orthogonal transformation above. The ***normal space*** to M at $\boldsymbol{x} \in M$ is the linear subspace $N_{\boldsymbol{x}}(M)$ of W given by the following three equivalent conditions: (a) $N_{\boldsymbol{x}}(M)$ is the range of $D\phi(\boldsymbol{x})^{\mathrm{T}}$ (which is a linear transformation from $\mathbb{R}^{n-k}$ to W), (b) $N_{\boldsymbol{x}}(M)$ is the kernel of $D\psi(0)^{\mathrm{T}}$ (which is a linear transformation from W to $\mathbb{R}^k$); (c) $N_{\boldsymbol{x}}(M)$ is the set of all $\boldsymbol{n} \in$ W such that $\boldsymbol{n} = \boldsymbol{O}(0, n)$ for some $n \in \mathbb{R}^{n-k}$. Clearly, $T_{\boldsymbol{x}}(M)$, $N_{\boldsymbol{x}}(M)$ have dimensions k and $n-k$, respectively, and $T_{\boldsymbol{x}}(M) \oplus N_{\boldsymbol{x}}(M) = \mathrm{W}$.

A ***hypersurface*** in W is any $n-1$ dimensional surface. A mapping $\boldsymbol{n} : M \to$ W is said to be an ***orientation*** of M if $\boldsymbol{n}$ is continuous and $\boldsymbol{n}(\boldsymbol{x}) \in N_{\boldsymbol{x}}(M)$, $|\boldsymbol{n}(\boldsymbol{x})| = 1$ for every $\boldsymbol{x} \in M$. An ***oriented smooth hypersurface*** in W is a pair $(M, \boldsymbol{n})$ where M is a smooth hypersurface and $\boldsymbol{n}$ is an orientation of M.

2.6.2 Regions with Almost Smooth Boundary Let $Q \subset$ W be an open set. A point $\boldsymbol{x} \in \partial Q$ is said to be a ***point of smoothness*** of ∂Q if there exists an $r > 0$ such that $\partial Q \cap B(\boldsymbol{x}, r)$ is a hypersurface. The set Q is said to be ***locally on one side of*** ∂Q at $\boldsymbol{x} \in \partial Q$ if for every $r > 0$ there exists a neighborhood $D \subset B(\boldsymbol{x}, r)$ of $\boldsymbol{x}$ such that $D \setminus \partial Q$ has exactly two (nonempty) connected components $D \cap Q$ and $D \cap \operatorname{int}(\mathrm{W} \setminus Q)$. A point $\boldsymbol{x} \in \partial Q$ is said to be a ***regular point*** of ∂Q if it is a point of smoothness of Q and Q is locally on one side of ∂Q at $\boldsymbol{x}$. We denote by $\partial^r Q$ the set of all regular points of ∂Q. It can be shown that if $\boldsymbol{x} \in \partial^r Q$, then for every $r > 0$ there exists a local implicit description $\phi : O \to \mathbb{R}$ such that $O \subset B(\boldsymbol{x}, r)$ and $Q \cap O = \{\boldsymbol{x} \in O : \phi(\boldsymbol{x}) < 0\}$. We then define $\boldsymbol{n}(\boldsymbol{x}) = D\phi(\boldsymbol{x})/|D\phi(\boldsymbol{x})|$ and call $\boldsymbol{n}$ the (geometric) ***exterior normal*** to Q (or to ∂Q) at $\boldsymbol{x}$. It is easy to see that $\boldsymbol{n}$ is the same for all local implicit descriptions of the type considered above. The mapping $\boldsymbol{x} \mapsto \boldsymbol{n}(\boldsymbol{x})$, $\boldsymbol{x} \in \partial^r Q$ is continuous and gives $\partial^r Q$ a natural orientation. An open subset Q of W is said to have an ***almost smooth boundary*** if

(1) for any $\boldsymbol{x} \in \partial Q$, Q is locally on one side of ∂Q at $\boldsymbol{x}$;

(2) $H^{n-1}(\partial Q \setminus \partial^r Q) = 0$, $H^{n-1}(\partial Q) < \infty$;

(3) $Q = \operatorname{int} \operatorname{cl} Q$.

Here H^{n-1} is the $n-1$-dimensional Hausdorff measure in W (the 'surface' measure; see Sect. 3.8). Hence the exterior normal is defined for H^{n-1}-a.e. point $x \in \partial Q$. An open subset Q of W is said to have a ***smooth boundary*** if it is a set with almost smooth boundary and $\partial Q = \partial^r Q$. We shall often use the term region with almost smooth boundary for open sets with almost smooth boundary. Clearly, open balls and solid spherical shells are regions with smooth boundary, while open cubes are regions with almost smooth boundary: the regular part of the boundary consists of the faces, while the edges and corners are not the regular points of the boundary. Their $n-1$-dimensional Hausdorff measure as they are covered by a finite number of smooth curves, for example. The regions with almost smooth boundary may have cusps as well.

2.6.3 Example *Let $O \subset$ W be an open set, let $\phi : O \to \mathbb{R}$ be of class C^1, and suppose that there exists a closed set $F \subset O$ such that $\phi > 0$ on $O \setminus F$. Suppose furthermore that $D\phi(x) \neq \mathbf{0}$ whenever $\phi(x) = 0$. Then the set $Q = \{x \in O : \phi(x) < 0\}$ is an open subset of* W *with smooth boundary and $\partial Q = \{x \in O : \phi(x) = 0\}$.*

The proof is left to the reader. The condition $D\phi \neq \mathbf{0}$ whenever $\phi = 0$ excludes isolated points x where $\phi(x) = 0$; the set F excludes "the possibility of the boundary ∂Q leaving the domain of definition of ϕ through the boundary of O."

2.6.4 Singular Surfaces Let Q be a region with a almost smooth boundary. A closed subset U of cl Q is called a ***singular surface*** in Q if there exists a finite number of pairwise disjoint regions with almost smooth boundary $Q^p \subset Q$, $p = 1, \ldots, q$, such that

$$Q \setminus U = \bigcup_{p=1}^{q} Q^p, \quad U \cap Q = Q \cap \Big(\bigcup_{p=1}^{q} \partial Q^p\Big), \quad U = \mathrm{cl}(U \cap Q). \tag{2.6.1}$$

One can show that cl $Q = \cup_{p=1}^{q}$ cl Q^p; hence for every $x \in$ cl Q there exists at least one region Q^p such that $x \in$ cl Q^p; each such a region is called a ***region adjacent*** to x. The number of regions adjacent to $x \in$ cl Q is denoted by $n(x)$ and is called the number of phases at x. A point $x \in U$ is called a ***regular point of the singular surface*** if $x \in Q$ and if there exists an $r > 0$ such that $B(x, r) \cap U$ is a smooth hypersurface. We denote by U^r the set of all regular points of U and note that U^r is a smooth hypersurface.

An ***orientation of a singular surface*** U is any orientation n of U^r. An ***oriented singular surface*** is a pair (U, n) where U is a singular surface and n is an orientation of U. Each singular surface has at least one orientation n. To see it, it suffices to realize that $U^r \subset \cup_{p=1}^{q} \partial^r Q^p$ where Q^p are the regions from the definition of the singular surface. Hence, one defines n as the exterior normal to Q^1 on $U \cap \partial^r Q^1$; then one defines n as the exterior normal to ∂Q^2 on $U \cap \partial^r Q^2 \setminus U \cap \partial^r Q^1$ etc., which eventually defines n on the whole of U^r. Clearly, if n is an orientation, then $-n$ is also an orientation, but generally, since U^r need not be connected, there are orientations different from the above two. We shall always assume that U has an assigned orientation n. The assignment of n influences the following important convention: If x is a regular point of U, then there exists a local implicit description $\phi : O \to \mathbb{R}$ of U^r near x such that $n(x) = D\phi(x)$; we then define, for every $r > 0$ sufficiently small, the sets $Q^+ = Q^+(x, r)$, $Q^- = Q^-(x, r)$ by $Q^\pm = \{y \in B(x, r) : \pm\phi(y) > 0\}$. If $r > 0$ is sufficiently small, then $Q^\pm$ belong to two distinct adjacent regions Q^p, Q^s to x and we shall later define the limits $f^\pm(x)$ of a piecewise continuous function at x as the limits of f at x relative to $Q^\pm$. Note that the normal n then points from Q^- to Q^+.

It can be proved that the singular surface has the following properties.

2.6.5 Proposition *Let U be a singular surface in Q and $x \in$ cl Q. Then*

(1) *$n(x) = 1$ if and only if $x \in$ cl $Q \setminus U$ and $n(x) \geq 2$ if and only if $x \in U$;*

(2) *if $x \in U^r$, then $n(x) = 2$;*

(3) *$n(x) = 2$ for H^{n-1}-a.e. point $x \in U$; more precisely, $H^{n-1}(U \setminus U^r) = 0$.*

2.6.6 Piecewise Smooth Mappings Let Q be a region with a almost smooth boundary and let r be a nonnegative integer or ∞. An X-valued field f on Q with a singular surface is any mapping $f : Q \setminus U \to$ X where U is some singular surface in Q. A field f with a singular surface is said to be r-times piecewise continuously differentiable on Q if the singular surface U can be chosen in such a way that the restriction of f to Q^p is in $C^r_s(Q^p)$ for every $p = 1, \ldots, q$, where Q^p are the regions from the definition of a singular surface. (See Subsect. 1.1.3 for the definition of C^r_s.) If $r = 0$, then f is called a piecewise continuous field; if $r = 1$ then f is called piecewise continuously differentiable.

Thus, for every fixed component Q^p, the restriction of f to the open set Q^p has an extension $\tilde{f} = \tilde{f}_p$ to an open set $\tilde{Q}^p$ containing the closure of Q^p which is r times continuously differentiable in the classical sense. The derivatives of $\tilde{f}_p$ of orders $\le r$ coincide with those of f at every point $\boldsymbol{x} \in Q^p$. For any $\boldsymbol{x} \in \partial Q^p$ we denote unambiguously

$$D^m f_p(\boldsymbol{x}) := D^m \tilde{f}_p(\boldsymbol{x}) = \lim_{\substack{z \to x \\ z \in Q^p}} D^m f(z), \quad 0 \le m \le r;$$

for $m = 0$ we write $f_p(\boldsymbol{x})$. For every $\boldsymbol{x} \in \partial Q \cup U$ this assigns the values of f and of its derivatives up to order r also to points $\boldsymbol{x} \in \partial Q \cup U$, where originally f was not defined. If $\boldsymbol{x} \in \partial Q \setminus U$, then there exists exactly one Q^p adjacent to $\boldsymbol{x}$ (see Proposition 2.6.5) and we write unambiguously $D^m f(\boldsymbol{x})$ in this case. If $\boldsymbol{x}$ is a regular point of U, and $Q^\pm = Q^\pm(\boldsymbol{x}, r)$ are the regions introduced above, we write $D^m f^\pm(\boldsymbol{x})$ for the two values $D^m f_p(\boldsymbol{x})$, $D^m f_s(\boldsymbol{x})$ where Q^p, Q^s are the two regions adjacent to $\boldsymbol{x}$ such that $Q^+ \subset Q^p$, $Q^- \subset Q^s$; for $m = 0$ we write $f^\pm(\boldsymbol{x})$. Notice that if f is piecewise continuous on Q, then it is bounded on $Q \setminus U$, if f is piecewise continuously differentiable, then both f and Df are bounded on $Q \setminus U$; if f is continuous and piecewise continuously differentiable, then it is locally Lipschitz continuous.

2.6.7 Evolving Surfaces Let dim Vect $= n$. For any subset M of Vect $\times$ IR and any time $t \in$ IR we denote by M_t the section of M by the instant Vect $\times \{t\}$:

$$M_t := \{\boldsymbol{p} \in \text{Vect} : (\boldsymbol{p}, t) \in M\};$$

M_t is referred to as an ***instantaneous configuration*** of M at time t. An (oriented) surface is any (oriented) smooth $n-1$-dimensional surface in Vect (= hypersurface in Vect). If $n = 3$, then the surfaces are the surfaces in the usual sense. A subset U of Vect $\times$ IR is called an ***evolving spatial surface*** if it is a smooth hypersurface in Vect $\times$ IR and for every $\boldsymbol{x} \equiv (\boldsymbol{p}, t) \in M$, any of the two unit normals $N = (N_s, N_t) \in$ Vect $\oplus$ IR to U at $\boldsymbol{x} = (\boldsymbol{p}, t)$ satisfies $N_s \neq \boldsymbol{0}$. This means that that the tangent space has never a purely space-like character. Thus the hypersurface Vect $\times \{0\}$ in Vect $\times$ IR is not an evolving spatial surface. It is easy to prove that if U is an evolving spatial surface in Vect then for every $t \in$ IR, the instantaneous configuration $S_t \equiv U_t$ is a (possibly empty) surface in Vect. If U has an orientation $N = (N_s, N_t)$, $|N| = 1$, then one can define a field $\boldsymbol{n}$ by $\boldsymbol{n} := N_s / |N_s|$. For every $t \in$ IR, the restriction of $\boldsymbol{n}$ to the instantaneous configuration $S_t \equiv U_t$ is an orientation of S_t. Moreover, the scalar field $U := -N_t / |N_s|$ is the speed of S_t. Given an orientation and a point $(\boldsymbol{p}, t)$ of an evolving surface U one can always find a local implicit description $\phi(\boldsymbol{q}, s) = 0$ such that $\boldsymbol{n} = \nabla\phi / |\nabla\phi|$, $U = -\dot{\phi} / |\nabla\phi|$ on $U \cap O$, and this establishes the contact with the singular surface introduced in Sect. 2.1, where the description by ϕ was assumed global.

Let $\Omega \subset$ Vect be a region with a almost smooth boundary and $t_1 < t_2$. An ***evolving singular surface*** on $\Omega \times (t_1, t_2)$ is a closed subset U of cl $\Omega \times [t_1, t_2]$ which is a singular surface in $\Omega \times (t_1, t_2)$ in the sense of the definition in Subsect. 2.6.4 and the set U^r of regular points of U is an evolving spatial surface.

Bibliographical Note

The formula for $\dot{\boldsymbol{e}}_i$ in Subsect. 2.2.7 is stated equivalently using a different language in BIOT [1965] and in HILL [1970a] (who calls it 'the rotation of the Eulerian strain ellipsoid;') the formula for $\dot{\boldsymbol{f}}_i$ is equivalent to the 'rotation of the Lagrangian strain ellipsoid' due to HILL [1970a]. The formulas for $\dot{\boldsymbol{e}}_i, \dot{\boldsymbol{f}}_i$ are not employed below: they are present only implicitly in a related formula for the derivatives of isotropic functions to be given in Sects. 8.4 and 8.5. See also OGDEN [1972b (Appendix), 1984].)

Proposition 2.4.3 is essentially due to KHACHATURYAN [1983]; see also WECHSLER, LIEBERMAN & READ [1953] and BALL & JAMES [1987]. Proposition 2.5.3 appears to be new; the treatments of twins in ERICKSEN [1981] and GURTIN [1983] are made under the additional assumption that the tensor $\boldsymbol{O}$ as in (2.5.2) is a 180° rotation about some axis, which is here a consequence of Proposition 2.5.3. That $\boldsymbol{O}$ may be chosen a 180° rotation is also mentioned without proof in ERICKSEN [1985], who also considers more general twins of the form (2.5.2) with $\boldsymbol{H} \in$ Unim such that $\boldsymbol{H}^N = \boldsymbol{1}$ for some integer $N > 1$; his result says that then an appropriate change of the reference configuration reduces this more general case to that in the main text above. PITTERI [1985b] analyses the case $\boldsymbol{H} \in$ Unim, $\boldsymbol{H}^2 = 1$. As yet different view on twinning is expressed in ZANZOTTO [1992]. Proposition 2.4.5 is due to JAMES [1986b]. Most of the results of Sect. 2.3 can be given weak formulations for deformations from Sobolev spaces; see BALL & JAMES [1987, 1992], and FONSECA [1989b]. The characterization of rigid deformations employed in the proof of Proposition 2.3.6 is of particular interest. That a twice continuously differentiable deformation χ of a connected region Ω with $\nabla\chi(\boldsymbol{p}) \in \mathrm{Orth}^+$ for every $\boldsymbol{p} \in \Omega$ is a rigid deformation can be proved by a standard computational proof, suggested in CIARLET [1988, Exercise 1.14]. The conclusion holds if χ is merely continuously differentiable as shown in CIARLET [1988, Thm. 1.8-1]. A still more general version is due to RESHETNYAK [1967], who shows that the conclusion remains valid even when χ is in $W^{1,\infty}(\Omega)$. Finally, KINDERLEHRER [1988] showed that if ν_p is the Young measure associated with a sequence of deformation gradients and ν_p is supported in Orth^+ for almost every $\boldsymbol{p} \in \Omega$, then ν_p is a constant Dirac measure concentrated at some $\boldsymbol{R} \in \mathrm{Orth}^+$ and the resulting weak limit is a rigid deformation. (See also JAMES & KINDERLEHRER [1989].) The construction of a periodic pairwise homogeneous deformation is often called a simple layering construction; its refinements, like the one involving layers in layers, are used in the relaxation and homogenization theories (in addition to the context of Sect. 2.3). These methods are also employed in the study of the problem of the characterization of the set of gradient Young measures. The reader is referred to the literature cited in Chap. 21.

3 Balance Equations

This chapter presents the basic formal aspects of balance equations of continuum thermodynamics for fields with singular surfaces. Sections 3.1–3.2 deal with unspecified quantities, and treat in detail such features as the divergence theorem for fluxes with jump discontinuities, the passage from the reference to the spatial descriptions, and transport theorem for fields with singular surface. Several forms of a general balance equation are given and their relationships are explained in Sect. 3.3. After these preparations, the balances of concrete thermodynamic quantities are presented in Sections 3.4–3.7. The appealing elegance makes these balance equations acceptable without much motivation, and this is indeed the approach taken in this chapter. Only in the four chapters which follow will these concepts along with their balance equations be derived from simpler statements.

3.1 Extensive Quantities: Fluxes

The fluxes of scalar quantities through oriented surfaces in continuous bodies are represented by the fields of flux vectors. It is possible to prove the existence of the flux vector from a set of elementary axioms on the global fluxes through the surfaces. The classical Cauchy's stress theorem based on the tetrahedron argument (see, e.g., GURTIN [1981]) is an elementary example of such a derivation. I refer to NOLL [1959, 1973], GURTIN & MARTINS [1976], ANTMAN & OSBORN [1979], ZIEMER [1983], and ŠILHAVÝ [1985b, 1991] for more advanced developments. Here we take the existence of the flux vector for granted.

Throughout this chapter, the ***body*** Ω is assumed to be a region with almost smooth boundary (see Subsect. 2.6.2); a ***part*** of the body is any bounded region with almost smooth boundary such that $\operatorname{cl} P \subset \Omega$. The singular surface is assumed to be a (possibly empty) evolving surface defined in Subsect. 2.1.2. We shall consider piecewise continuously differentiable fields in the referential or spatial descriptions, defined in Sect. 2.1 and Subsect. 2.2.5, respectively. In the case of the spatial description, such fields are defined on a ***spacetime tube*** of some motion (see (2.2.1)) of Ω on $[t_1, t_2]$. If $t \in (t_1, t_2)$, then a bounded set $p \subset \mathrm{Eucl}$ with almost smooth boundary is said to be a ***spatial control volume*** at time t if

$$\operatorname{cl} p \subset T_t := \{\boldsymbol{x} \in \mathrm{Eucl} : (\boldsymbol{x}, t) \in T\}. \tag{3.1.1}$$

It follows that for all s sufficiently close to t, $\operatorname{cl} p \subset T_s$.

3.1.1 The Values of a Piecewise Continuous Field on a Surface Let f be a piecewise continuous time-dependent field on Ω. Along the singular surface S_t, f has two different values $f^{\pm}$. Since S_t has null Lebesgue measure, for the volume integration this ambiguity causes no problems. For the surface integration, one has to define the value $f^S(\boldsymbol{p}, t)$ taken by f at a point $\boldsymbol{p}$ of an oriented surface $S = (S, \boldsymbol{n})$. To this end, we fix time t and introduce subsets of S as follows:

$$S^{\pm} = \{\boldsymbol{p} \in S \cap S_t : \boldsymbol{n}(\boldsymbol{p}) = \mp \boldsymbol{n}_t(\boldsymbol{p})\},$$
$$S^0 = S \cap S_t \setminus (S^+ \cup S^-) = \{\boldsymbol{p} \in S \cap S_t : \boldsymbol{n}(\boldsymbol{p}), \boldsymbol{n}_t(\boldsymbol{p}) \text{ linearly independent}\},$$

where $\boldsymbol{n}$ is the normal to S and $\boldsymbol{n}_t$ is the normal to S_t at time t; see Fig. 3.1.1, in which $S = \partial P$ and S^0 is empty. Then

$$S = (S \setminus S_t) \cup S^+ \cup S^- \cup S^0 \text{ and } A(S^0) = 0, \tag{3.1.2}$$

where the union in $(3.1.2)_1$ is disjoint; $(3.1.2)_2$ is proved in Proposition 3.8.8(1) below. We set

$$f^S(\boldsymbol{p}, t) := \begin{cases} f(\boldsymbol{p}, t) & \text{if } \boldsymbol{p} \in S \setminus S_t \\ f^{\pm}(\boldsymbol{p}, t) & \text{if } \boldsymbol{p} \in S^{\pm}, \text{ respectively,} \end{cases}$$

and f^S is undefined on S^0. This convention is chosen so as to make, for every part $P \subset \Omega$, the values $f^{\partial P}$ equal (essentially) to the limit of f from the interior of P. (If P is a part of Ω and $\bar{f}$ the restriction of f to P, then $\bar{f}^{\partial P}$ coincides with the boundary values of $\bar{f}$ on P defined in Subsect. 2.1.3.) With the above definition, a change of orientation from $S = (S, \boldsymbol{n})$ to $-S := (S, -\boldsymbol{n})$ changes the values of f^S, namely,

$$f^S - f^{-S} = \begin{cases} 0 & \text{on } S \setminus S_t \\ \pm [f] & \text{on } S^{\pm}, \text{ respectively.} \end{cases}$$

Consequently, if $\boldsymbol{Q}$ is a piecewise continuous vector-valued field, then

$$\int_S \boldsymbol{Q}^S \cdot \boldsymbol{n}\, dA - \int_S \boldsymbol{Q}^{-S} \cdot \boldsymbol{n}\, dA = -\int_{S \cap S_t} [\boldsymbol{Q}] \cdot \boldsymbol{n}_t\, dA$$

where the integral on the right is unequivocally defined as an integral of the function $[\boldsymbol{Q}] : S_t \to \text{Vect}$. The same convention applies to the spatial configuration of the singular surfaces and to second order tensor-valued functions. When integrating a piecewise continuous field f over an oriented surface S, we often simplify the notation and write $\int_S f\, dA$ for $\int_S f^S\, dA$.

3.1.2 Piola's Transformation Let $\boldsymbol{Q}$ be a piecewise continuously differentiable vector field in the reference description with the singular surface S_t. The interpretation of $\boldsymbol{Q}$ as of a flux vector is that if S is any oriented surface in the reference configuration with the normal $\boldsymbol{n}$, then the integral

$$\mathsf{F}(S, t) = \int_S \boldsymbol{Q} \cdot \boldsymbol{n}\, dA$$

is the rate at which the quantity under consideration flows through S. For the given quantity F, the vector field $\boldsymbol{Q}$ is called the ***referential flux vector field of*** F. For a motion $\boldsymbol{\chi}$ with the same singular surface S_t the ***spatial flux vector field*** $\boldsymbol{q}$ is defined by the basic requirement that

$$\mathsf{F}(S, t) = \int_S \boldsymbol{Q} \cdot \boldsymbol{n}\, dA = \int_{\boldsymbol{\chi}(S,t)} \boldsymbol{q} \cdot \boldsymbol{m}\, da$$

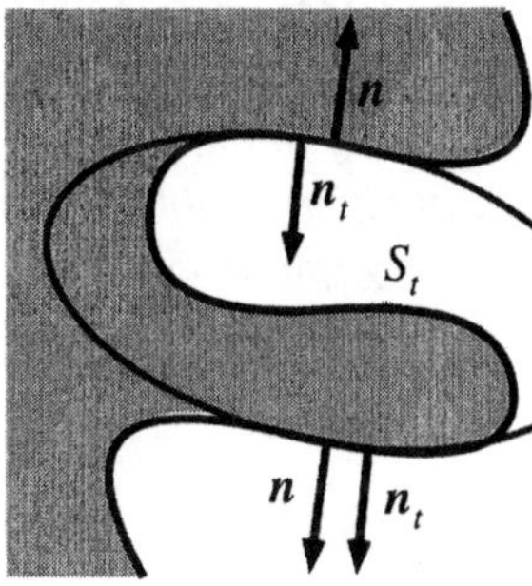

Fig. 3.1.1

holds for every S, where $\boldsymbol{m}$ is the normal to the image $\chi(S, t)$ of S under the motion. The change of variables formula for surface integrals (see Subsect. 2.2.4) leads to

$$\boldsymbol{Q} = s(\operatorname{cof} \boldsymbol{F})^{\mathsf{T}} \boldsymbol{q}, \tag{3.1.3}$$

where $s := \operatorname{sgn} \det \boldsymbol{F} = \pm 1$. It is observed that $\boldsymbol{Q}$ is piecewise continuously differentiable if and only if $\boldsymbol{q}$ is. The passage from $\boldsymbol{Q}$ to $\boldsymbol{q}$ and back via (3.1.3) is the ***Piola transformation.***

Throughout the chapter, it is understood without explicit mention that the singular surface of the motion χ coincides with that of the quantity under consideration; if more than one quantity is present, then all of them have the same singular surface, and if both the referential and spatial descriptions are used simultaneously, then the spatial and referential configurations of the singular surface are related by $s_t = \chi(S_t, t)$.

3.1.3 Proposition *If χ is a motion and if $\boldsymbol{Q}, \boldsymbol{q}$ are piecewise continuously differentiable vector fields related by* (3.1.3), *then*

$$\operatorname{Div} \boldsymbol{Q} = s \det \boldsymbol{F} \operatorname{div} \boldsymbol{q} \tag{3.1.4}$$

outside the singular surface and

$$[\boldsymbol{Q}] \cdot \boldsymbol{n} = s|(\operatorname{cof} \boldsymbol{F})\boldsymbol{n}|[\boldsymbol{q}] \cdot \boldsymbol{m} \tag{3.1.5}$$

on the singular surface, where $(\operatorname{cof} \boldsymbol{F})\boldsymbol{n}$ *denotes the common value* $(\operatorname{cof} \boldsymbol{F})^{+}\boldsymbol{n}$ − $(\operatorname{cof} \boldsymbol{F})^{-}\boldsymbol{n}$, *cf.* $(2.2.10)_2$.

In (3.1.4) Div denotes the referential divergence, i.e., the divergence defined through the differentiation with respect to the referential position $\boldsymbol{p}$. Equation (3.1.4) is the ***Piola identity.*** It can be verified by a direct computation using coordinates, the important intermediate step being the identity

$$\operatorname{Div} \operatorname{cof} \boldsymbol{F} = \boldsymbol{0}$$

outside the singular surface. Equation (3.1.5) follows immediately.

3.1.4 Proposition (The divergence theorem with a singular surface)

(1) *If $\boldsymbol{Q}$ is a piecewise continuously differentiable vector field on $\Omega \times (t_1, t_2)$, then for every part $P \subset \Omega$ and every $t \in [t_1, t_2]$,*

$$\int_{\partial P} \boldsymbol{Q} \cdot \boldsymbol{n}\, dA = \int_{P} \operatorname{Div} \boldsymbol{Q}\, dV + \int_{P \cap S_t} [\boldsymbol{Q}] \cdot \boldsymbol{n}\, dA, \tag{3.1.6}$$

where S_t is the singular surface of $\boldsymbol{Q}$ at time t.

(2) *If $\boldsymbol{q}$ be a piecewise continuously differentiable vector field on a spacetime tube T then for every $t \in (t_1, t_2)$ and every spatial control volume $p \subset$ Eucl at time t, we have*

$$\int_{\partial p} \boldsymbol{q} \cdot \boldsymbol{m}\, da = \int_{p} \operatorname{div} \boldsymbol{q}\, dv + \int_{p \cap s_t} [\boldsymbol{q}] \cdot \boldsymbol{m}\, da$$

where s_t is the spatial configuration of the singular surface.

In the integral on the left in (3.1.6), $\boldsymbol{Q}$ stands for $\boldsymbol{Q}^{\partial P}$. Equation (3.1.6) is obtained formally by applying the divergence theorem for continuously differentiable fields to the regions $P \cap \Omega^+$ and $P \cap \Omega^-$ and adding the results. Here $\Omega^\pm = \Omega_t^\pm$ are the regions in front of and behind the singular surface introduced in Subsect. 2.1.2. However, $P \cap \Omega^\pm$ need not be sets with almost smooth boundary; one has to use the measure-theoretic boundary and the measure-theoretic exterior normal rather than the topological boundary and geometric normal in the formulation of the Gauss–Green theorem for $P \cap \Omega^\pm$. These concepts are introduced in Sect. 3.8 in the context of sets of finite perimeter, of which $P \cap \Omega^\pm$ are special cases. The crucial equalities (3.1.7), (3.1.8) for the measure-theoretic boundaries of $P \cap \Omega^\pm$ will be derived from the results established there. Let us emphasize, however, that the above final result (1) uses the geometric normal and the topological boundary; only the derivation needs the measure-theoretic notions.

Proof Only (1) will be proved; the proof of (2) is similar. Fix time t. Write $\Omega^\pm$ for $\Omega_t^\pm$, $\boldsymbol{n}_t$ for the normal to S_t at time t and $\boldsymbol{n}$ for the exterior normal to ∂P. Since the function ϕ occurring in the definition of the singular surface has a continuously differentiable extension to $\operatorname{cl} \Omega$, it follows that the sets $\Omega^\pm$ can be written as an intersection of two open sets with almost smooth boundary. Hence, by Propositions 3.8.6 and 3.8.5, $\Omega^\pm$ are sets of finite perimeter. Moreover, if O is any open subset of Ω such that $\operatorname{cl} O \subset \Omega$, then $O \cap \partial \Omega^\pm = O \cap S_t$. Let now P be a part of Ω and set $P^+ := P \cap \Omega^+$, $P^- := P \cap \Omega^-$; again $P^\pm$ are sets of finite perimeter by Proposition 3.8.5 and the first task is to determine the measure-theoretic boundaries of $P^\pm$. Define $S^\pm := \{\boldsymbol{p} \in S_t \cap \partial P : \boldsymbol{n}(\boldsymbol{p}) = \mp \boldsymbol{n}_t(\boldsymbol{p})\}$, where ∂P is the topological and simultaneously a measure-theoretic boundary of P (see Subsect. 3.8.1 below). Then to within sets of null areas,

$$\partial^* P^\pm = (\partial P \cap \Omega^\pm) \cup (S_t \cap P) \cup S^\pm, \quad \partial P = (\partial P \cap \Omega^+) \cup (\partial P \cap \Omega^-) \cup S^+ \cup S^-. \tag{3.1.7}$$

The formula for the measure-theoretic boundaries $\partial^* P^\pm$ of $P^\pm$ is an application of $(3.8.7)_1$ from Proposition 3.8.5; the formula for ∂P follows from the set-theoretic identity

$$\partial P = (\partial P \cap \Omega^+) \cup (\partial P \cap \Omega^-) \cup S^+ \cup S^- \cup S^0,$$

where S^0 is the set of those $\boldsymbol{p} \in S_t \cap \partial P$ for which either $\boldsymbol{n}(\boldsymbol{p})$ does not exist or $\boldsymbol{n}(\boldsymbol{p})$ and $\boldsymbol{n}_t(\boldsymbol{p})$ are linearly independent. Combining the assumption that $\boldsymbol{n}$ exists for A-a.e. point of ∂P with Proposition 3.8.8(1) below, we obtain that $A(S^0) = 0$ and $(3.1.7)_2$ follows. Let us denote by $\boldsymbol{n}^\pm$ the measure-theoretic exterior normal to $P^\pm$. Then

$$\boldsymbol{n}^\pm = \boldsymbol{n} \text{ on } \partial P \cap \Omega^\pm, \quad \boldsymbol{n}^\pm = \mp \boldsymbol{n}_t \text{ on } P \cap S_t, \quad \boldsymbol{n}^\pm = \boldsymbol{n} \text{ on } S^\pm \tag{3.1.8}$$

as a consequence of $(3.8.8)_1$. Let $\boldsymbol{Q}^\pm$ be the smooth extensions of $\boldsymbol{Q}$ to a neighborhood of the closure of $\Omega^\pm$ from the definition of piecewise continuous and piecewise smooth fields. By the Gauss–Green theorem for continuously differentiable fields (3.8.2), and by $(3.1.7)_1$, (3.1.8),

$$\int_{P^\pm} \operatorname{Div} \boldsymbol{Q}\, dV = \int_{\partial P^\pm} \boldsymbol{Q}^\pm \cdot \boldsymbol{n}^\pm\, dA = \int_{\partial P \cap \Omega^\pm} \boldsymbol{Q} \cdot \boldsymbol{n}\, dA \mp \int_{S_t \cap P} \boldsymbol{Q}^\pm \cdot \boldsymbol{n}_t\, dA + \int_{S^\pm} \boldsymbol{Q}^\pm \cdot \boldsymbol{n}\, dA$$

Adding these equations, we obtain

$$\int_P \operatorname{Div} Q\,dV = \int_{\partial P\cap\Omega^+} Q\cdot n\,dA + \int_{\partial P\cap\Omega^-} Q\cdot n\,dA + \int_{S^+} Q^+\cdot n\,dA + \int_{S^-} Q^-\cdot n\,dA - \int_{S_t\cap P} [Q]\cdot n_t\,dA.$$

On the other hand, by $(3.1.7)_2$ and the definition in Subsect. 3.1.1 of the boundary values of Q on ∂P, the sum of the first four terms on the right-hand side gives $\int_{\partial P} Q\cdot n\,dA$ and (3.1.6) follows. □

3.2 Extensive Quantities: Densities and Transport Theorems

An extensive quantity B associates with every part P the content of B in P. Only the extensive quantities will be considered that have a piecewise continuous density with respect to the referential or (which is equivalent) the spatial volume; concentrations on surfaces, lines or at points are excluded.

3.2.1 Volume Densities Let χ be a motion with a singular surface S_t. Consider a scalar quantity with the ***referential density*** B which is piecewise continuously differentiable with the same singular surface. For every part P and time t the integral

$$\mathsf{B}(P,t) = \int_P B\,dV \tag{3.2.1}$$

is the ***content*** of the quantity in part P at t. Here dV is the volume element in the reference configuration. In terms of the ***spatial density*** b,

$$\mathsf{B}(P,t) = \int_{\chi(P,t)} b\,dv$$

for every part P, where dv is the element of volume in the actual configuration and $B = |J|b$. It is easily seen that B is piecewise continuously differentiable if and only if b is.

3.2.2 Proposition *If B is a piecewise continuously differentiable scalar field on $\Omega\times(t_1, t_2)$, then for every part $P\subset\Omega$ the function $t\mapsto\int_P B\,dV$ is absolutely continuous on $[t_1, t_2]$ and*

$$\frac{d}{dt}\int_P B\,dV = \int_P \dot{B}\,dV - \int_{P\cap S_t} U[B]\,dA \tag{3.2.2}$$

for almost every $t\in[t_1, t_2]$.

Using Lebesgue's theorem, it is not hard to see that the volume integral on the right-hand side of (3.2.2) is a continuous function of time, but the surface integral can have jumps in t as the integration range $P\cap S_t$ can jump suddenly, for instance from the empty set to a set of positive area. Hence the derivative on the left-hand side of (3.2.2) is not continuous. The absolute continuity asserted by the proposition can probably be made stricter, because the derivative stems from integrals of piecewise continuous fields, but the absolute continuity completely suffices. There appears to be no immediately available known theorem of analysis dealing with this situation except in the one-dimensional case. The proof below applies the Gauss–Green theorem to the spacetime region delimited by the singular surface, and the integration with respect to time seems to be the only way to obtain absolute continuity and hence the existence of the derivative.

Proof Let $R := P\times(u, v)\subset \mathrm{Eucl}\times\mathbb{R}$ where u, v are times such that $t_1\le u\le v\le t_2$; then

$$\partial R = R_u \cup R_v \cup (\partial P \times [u, v]) \tag{3.2.3}$$

where $R_u = P \times \{u\}$, $R_v = P \times \{v\}$, and the union on the right-hand side of (3.2.3) is disjoint. Let L denote the $n+1$-dimensional Lebesgue measure in Eucl $\times \mathbb{R}$ and H^n the n-dimensional Hausdorff measure in Eucl $\times \mathbb{R}$, let $M = (M_s, M_t) \in$ Vect $\times \mathbb{R}$ be the exterior normal to ∂R and $N = (N_s, N_t) \in$ Vect $\times \mathbb{R}$ the unit normal to U. Set $R^\pm = R \cap T^\pm$ where $T^\pm$ are defined by $(2.1.2)_2$. By Proposition 3.8.5, to within a set of null H^n-measure, the measure-theoretic boundary of $R^\pm$ is $\partial^* R^\pm = (\partial R \cap T^\pm) \cup (U \cap R) \cup S^\pm$ where $S^\pm = \{(p, t) \in \partial R \cap U : M(p, t) = \mp N(p, t)\}$. The Gauss–Green theorem (3.8.2) gives

$$\int_{R^\pm} \dot{B}^\pm \, dL = \int_{\partial R \cap T^\pm} B^\pm M_t \, dH^n \pm \int_{R \cap U} B^\pm N_t \, dH^n + \int_{S^\pm} B^\pm M_t \, dH^n, \tag{3.2.4}$$

where $B^\pm$ are the extensions of the restrictions of B from the definition of a piecewise continuously differentiable field. We have $M_s \neq \mathbf{0}$ only for points from $\partial P \times (u, v)$ while $N_s \neq \mathbf{0}$ everywhere on U, from which necessarily $S^\pm \subset \partial P \times (u, v)$. As $M_t = 0$ on $\partial P \times [u, v]$, it follows that the third integral on the right-hand side of (3.2.4) vanishes, and the integration range of the first reduces to $(R_u \cup R_v) \cap T^\pm$. Adding the so simplified equations (3.2.4), we obtain

$$\int_R \dot{B} \, dL = \int_{R_u \cup R_v} B M_t \, dH^n - \int_{U \cap R} [B] N_t \, dH^n. \tag{3.2.5}$$

Combining with

$$\int_{R_u \cup R_v} B M_t \, dL = \int_{R_v} B \, dV - \int_{R_u} B \, dV, \qquad \int_{U \cap R} [B] M_t \, dH^n = -\int_u^v \int_{P \cap S_t} U[B] \, dA \, dt,$$

(see Proposition 3.8.10), we are led to

$$\int_{R_u} B \, dV - \int_{R_v} B \, dV = \int_u^v \Big(\int_P \dot{B} \, dV + \int_{P \cap S_t} U[B] \, dA \Big) \, dt.$$

As this must hold for every u, v, we see that $t \mapsto \int_P B \, dV$ is expressible as an integral of an integrable function of t and hence it is an absolutely continuous function and its derivative equals the integrand for a.e. t. □

3.2.3 Proposition *Let b a piecewise continuously differentiable scalar field on a spacetime tube T and $p \subset$ Eucl a spatial control volume at time $t_0 \in (t_1, t_2)$. Then for all t sufficiently close to t_0, $\mathrm{cl}\, p \subset T_t$, the function $t \mapsto \int_p b \, dv$ is absolutely continuous and*

$$\frac{d}{dt} \int_p b \, dv = \int_p \frac{\partial b}{\partial t} \, dv - \int_{p \cap s_t} u[b] \, da \tag{3.2.6}$$

for almost every t.

Apart from obvious modifications due to the use of the spatial description, this proposition is mathematically just a restatement of Proposition 3.2.2 in a different notation. Physically, however, this corresponds to a different situation. Namely, (3.2.6) gives the time-rate of the contents of the quantity b in a fixed control volume in physical space, whereas (3.2.2) gives the change of the content of the quantity in a fixed part of the body, i.e., in a fixed *material* region.

3.2.4 Proposition (The transport theorem with singular surface) *Let χ be a motion and b a piecewise continuously differentiable scalar field on the spacetime tube T of χ (see (2.2.1)). Then for every part $P \subset \Omega$ the function $t \mapsto \int_{\chi(P,t)} b \, dv$ is absolutely continuous and*

$$
\begin{aligned}
\frac{d}{dt}\int_{\chi(P,t)} b\,dv &= \int_{\chi(P,t)} \left(\frac{\partial b}{\partial t} + \operatorname{div}(bv)\right) dv - \int_{\chi(P,t)\cap s_t} (u[b] - [bv]\cdot m)\,da \\
&= \int_{\chi(P,t)} \frac{\partial b}{\partial t}\,dv + \int_{\partial\chi(P,t)} bv\cdot m\,da - \int_{\chi(P,t)\cap s_t} u[b]\,da
\end{aligned}
\tag{3.2.7}
$$

for almost every t.

Proof The integral on the left-hand side of (3.2.7) can be expressed as $s\int_P Jb\,dV$, where s is the sign of J. Assume for simplicity that $s = 1$. Proposition 3.2.2 tells us that

$$
\frac{d}{dt}\int_{\chi(P,t)} b\,dv = \int_P (Jb)^{\cdot}\,dV - \int_{P\cap S_t} U[Jb]\,dA. \tag{3.2.8}
$$

By Euler's identity $(2.2.7)_1$ and the formula $(2.2.4)_2$, the volume integral on the right-hand side of (3.2.8) can be rewritten as

$$
\int_P (\dot{J}b + J\dot{b})\,dV = \int_P J\left(b\operatorname{div} v + \frac{\partial b}{\partial t} + (\operatorname{grad} b)\cdot v\right) dV = \int_{\chi(P,t)} \left(\frac{\partial b}{\partial t} + \operatorname{div}(bv)\right) dv. \tag{3.2.9}
$$

With the help of (2.2.12), the integrand of the surface integral in (3.2.8) can be gradually transformed in the following way

$$
\begin{aligned}
U[Jb] &= UJ^+b^+ - UJ^-b^- = b^+u^+J^+|(F^+)^{-\mathrm{T}}n| - b^-u^-J^-|(F^-)^{-\mathrm{T}}n| \\
&= b^+u^+|(\operatorname{cof} F^+)n| - b^-u^-|(\operatorname{cof} F^-)n| = (u[b] - [bv]\cdot m)|(\operatorname{cof} F^-)n|,
\end{aligned}
$$

where we have used $|(\operatorname{cof} F^+)n| = |(\operatorname{cof} F^-)n|$ (see $(2.2.10)_2$) and (2.2.13). Hence, using $(2.2.2)_2$,

$$
\int_{P\cap S_t} U[Jb]\,dA = \int_{\chi(P,t)\cap s_t} (u[b] - [bv]\cdot m)\,da. \tag{3.2.10}
$$

Equations (3.2.9) and (3.2.10) give $(3.2.7)_1$ and an appeal to the divergence theorem 3.1.3 allows us to transform the two terms with bv into the surface integral occurring in $(3.2.7)_2$. □

3.3 Extensive Quantities: Balance Equations

A balance equation expresses the time derivative of an extensive quantity contained in a volume in terms of its flux through the boundary and the external source of the quantity. Several equivalent forms exist, differing according to whether the balancing is made from the standpoint of a fixed material volume through the boundary of which there is no flux of matter, or with respect to fixed volume in the physical space, through the boundary of which the material points can move (material versus control volume forms). Another possibility is to use the densities with respect to the referential volume or the densities with respect to the actual volume. We shall analyze only scalar balance equations. The generalization to the vector or tensor balance equation is made component-wise. The hypotheses on the densities and the flux vectors will be exactly those of the preceding two sections.

If T is an open subset of Eucl × $\mathbb{R}$, denote by $C_0^\infty(T)$ the set of all infinitely differentiable scalar-valued functions w on Eucl × $\mathbb{R}$ with compact support supp w such that supp $w \subset T$, i.e., infinitely differentiable functions which vanish near the boundary of T.

3.3.1 Proposition (The material form) *Let B, Q, Z be time-dependent fields with the singular surface S_t, B scalar-valued, Q vector-valued, both piecewise continuously differentiable, and Z scalar-valued, piecewise continuous. Then the following conditions are equivalent:*

(1) *for every part $P \subset \Omega$ and almost every t,*

$$\frac{d}{dt}\int_P B\,dV + \int_{\partial P} \boldsymbol{Q}\cdot\boldsymbol{n}\,dA = \int_P Z\,dV; \tag{3.3.1}$$

(2) *we have*

$$\dot{B} + \operatorname{Div}\boldsymbol{Q} = Z, \quad -U[B] + [\boldsymbol{Q}]\cdot\boldsymbol{n} = 0 \tag{3.3.2}$$

everywhere outside and on the singular surface, respectively;

(3) *for every $w \in C_0^\infty(\Omega \times (t_1, t_2))$,*

$$\int_{t_1}^{t_2}\Big(\int_\Omega (B\dot{w} + \boldsymbol{Q}\cdot\nabla w + wZ)\,dV\Big)dt = 0. \tag{3.3.3}$$

The field B is the reference volume density of the balanced quantity, $\boldsymbol{Q}$ its referential flux vector and Z the reference volume density of the rate of supply of the balanced quantity from the exterior. Conditions (1), (2), and (3) are the integral, the local, and the weak forms of the balance equation, respectively. The weak form can also be formulated using the test functions w which do not vanish at the boundary of Ω and at times t_1 and t_2. Then the surface integrals must be added to (3.3.3).

Proof (1) $\Rightarrow$ (2) : We use Proposition 3.2.1 to evaluate the time-derivative and the divergence theorem 3.1.4 to transform the surface integral as suggested by the formulas in these propositions. The result is, after simple rearrangements,

$$\int_P (\dot{B} + \operatorname{Div}\boldsymbol{Q} - Z)\,dV + \int_{P\cap S_t} (-U[B] + [\boldsymbol{Q}]\cdot\boldsymbol{n})\,dA = 0. \tag{3.3.4}$$

For a given part P, this must be satisfied for almost every t. If P is chosen so as not to intersect the singular surface for a while, then the surface integral in (3.3.4) vanishes and the volume integral is a continuous function of time and hence (3.3.4) holds for all such times in this case, not just for a.e. t. The freedom in the choice of such parts and the continuity of the integrand of the volume integral then implies $(3.3.2)_1$. Equation (3.3.4) then reduces to

$$\int_{P\cap S_t} (-U[B] + [\boldsymbol{Q}]\cdot\boldsymbol{n})\,dA = 0 \tag{3.3.5}$$

and again as P is arbitrary and (3.3.5) holds for a.e. t, a similar reasoning leads to $(3.3.2)_2$. (2) $\Rightarrow$ (1) : This proof is obtained by reversing the steps in the preceding part of the proof. (2) $\Leftrightarrow$ (3) is obtained by multiplying $(3.3.2)_1$, $(3.3.2)_2$ by w, adding the results, integrating over $\Omega \times [t_1, t_2]$ and using the spacetime version of the divergence theorem for fields with jump discontinuities. The details of this part of the proof are omitted. □

3.3.2 Proposition (The control-volume form) *Let b, $\boldsymbol{q}'$, z fields with singular surface s_t defined on a spacetime tube T, b scalar-valued, $\boldsymbol{q}'$ vector-valued, both piecewise continuously differentiable, and z scalar-valued, piecewise continuous. Then the following conditions are equivalent:*

(1) *for every spatial control volume p at time t_0,*

$$\frac{d}{dt}\int_p b\,dv + \int_{\partial p} \boldsymbol{q}'\cdot\boldsymbol{m}\,da = \int_p z\,dv;$$

for almost every t for which $p \subset T_t$ (see (3.1.1));

(2) *we have*

$$\frac{\partial b}{\partial t} + \operatorname{div}\boldsymbol{q}' = z, \quad -u[b] + [\boldsymbol{q}']\cdot\boldsymbol{m} = 0 \tag{3.3.6}$$

everywhere on and outside the singular surface, respectively;

(3) *for every* $w \in C_0^\infty(T)$,

$$\int_T (b\frac{\partial w}{\partial t} + \boldsymbol{q}' \cdot \operatorname{grad} w + wz) dv\, dt = 0. \tag{3.3.7}$$

The field b is the spatial volume density of the balanced quantity, $\boldsymbol{q}'$ its total flux vector and z is the spatial volume density of the rate of supply of the balanced quantity. The balancing is made from the point of view of a fixed control volume in the physical space. As we shall see in Proposition 3.3.4, besides the true flux, $\boldsymbol{q}'$ includes also the transport of the balanced quantity through the boundary of the control volume. Otherwise the above proposition is mathematically just a restatement of Proposition 3.3.1 using different letters, and the proof, based on Propositions 3.1.4(2) and 3.2.3, is omitted.

3.3.3 Proposition (The mixed form) *Let* χ *be a motion with the spacetime tube* T. *Let* b, $\boldsymbol{q}$, z *be time-dependent fields with singular surface,* b *scalar-valued,* $\boldsymbol{q}$ *vector-valued, both piecewise continuously differentiable, and* z *scalar-valued, piecewise continuous. Then the following conditions are equivalent:*

(1) *for every part* $P \subset \Omega$ *and almost every* t,

$$\frac{d}{dt}\int_{\chi(P,t)} b\, dv + \int_{\partial\chi(P,t)} \boldsymbol{q} \cdot \boldsymbol{m}\, da = \int_{\chi(P,t)} z\, dv; \tag{3.3.8}$$

(2) *we have*

$$\frac{\partial b}{\partial t} + \operatorname{div}(b\boldsymbol{v} + \boldsymbol{q}) = z, \qquad -u[b] + [b\boldsymbol{v} + \boldsymbol{q}] \cdot \boldsymbol{m} = 0 \tag{3.3.9}$$

everywhere outside and on the singular surface, respectively;

(3) *for every* $w \in C_0^\infty(T)$,

$$\int_T (b\frac{\partial w}{\partial t} + (b\boldsymbol{v} + \boldsymbol{q}) \cdot \operatorname{grad} w + wz)\, dv\, dt = 0. \tag{3.3.10}$$

Proof The steps are similar to those in the proof of Proposition 3.3.1. One has only to use the transport theorem 3.2.4 to compute the time-derivative in (3.3.8). □

3.3.4 Proposition (Relations among the balance equations) *Consider the objects* χ, S_t, s_t, B, b, $\boldsymbol{Q}$, $\boldsymbol{q}$, $\boldsymbol{q}'$, Z, z *as in Propositions* 3.3.1–3.3.3, *and suppose that they are related by*

$$B = sJb, \quad \boldsymbol{Q} = s(\operatorname{cof} \boldsymbol{F})^{\mathsf{T}}\boldsymbol{q}, \quad \boldsymbol{q}' = \boldsymbol{q} + b\boldsymbol{v}$$

outside the singular surface ($s = \operatorname{sgn}\det \boldsymbol{F}$, $J = \det \boldsymbol{F}$). *Then all of the conditions in Propositions* 3.3.1–3.3.3 *are equivalent and the following relations hold:*

$$\dot{B} = sJ(\dot{B} + \operatorname{div} \boldsymbol{v}) = sJ(\frac{\partial b}{\partial t} + \operatorname{div}(b\boldsymbol{v})), \quad \operatorname{Div} \boldsymbol{Q} = sJ \operatorname{div} \boldsymbol{q} \tag{3.3.11}$$

outside the singular surface and

$$-U[B] = s|(\operatorname{cof} \boldsymbol{F})\boldsymbol{n}|(-u[b] + [b\boldsymbol{v}] \cdot \boldsymbol{m}), \quad [\boldsymbol{Q}] \cdot \boldsymbol{n} = s|(\operatorname{cof} \boldsymbol{F})\boldsymbol{n}|[\boldsymbol{q}] \cdot \boldsymbol{m} \tag{3.3.12}$$

on the singular surface. Moreover, also the following differential form of the balance equation holds:

$$J^{-1}\dot{B} + \operatorname{div} \boldsymbol{q} = z. \tag{3.3.13}$$

We note that the total flux $\boldsymbol{q}'$ occurring in the control-volume form is the sum of the transport (or convection) $b\boldsymbol{v}$ and of the true flux $\boldsymbol{q}$ from the mixed form 3.3.3. The differential form (3.3.13) falls somewhat outside the systemization given above. Note that of all of the differential forms given in this section (3.3.13) is the only one whose left-hand side is not in the form of the full four-dimensional spacetime divergence. Related to this is the fact that there is no natural form of the jump condition corresponding to (3.3.13); the differential form must be combined with the jump conditions in 3.3.1(2), 3.3.2(2) or 3.3.3(2). It should also be noted that the reference volume density B is somewhat preferable to the spatial density, b, since, as we shall see in Sect. 3.4, B is proportional to the mass-density and thus B is more invariant than b.

Proof Relation $(3.3.11)_1$ is obtained by differentiating the product $B = sJb$ and using the Euler identity $(2.2.7)_1$ in combination with $(2.2.4)_2$. $(3.3.11)_2$ and $(3.3.12)_2$ are the identities from Proposition 3.1.3 concerning the Piola transformation. $(3.3.12)_1$ is proved by computing the jump of $B = sJb$ with the help of the identity for the jump of products (2.1.4) and the transformation formulas (2.2.11) and (2.2.12) for the kinematical variables for the singular surface in the reference and the current configurations. The differential form of the balance equation (3.3.13) follows from $(3.3.2)_1$ with the help of $(3.3.11)_2$. To prove the equivalence of the conditions given in Propositions 3.3.1–3.3.3, it is enough to establish the equivalence of the local forms 3.3.1(2), 3.3.2(2) or 3.3.3(2). But this is immediate in view of the formulas just proved. □

3.3.5 Example (Balance of the referential volume) Consider the balance of the quantity $\mathsf{V}(P) = \int_P dV$. Its referential density B is 1 and as $d\mathsf{V}(P)/dt = 0$, a comparison with (3.3.1) shows that this quantity satisfies the integral form (3.3.1) with $\boldsymbol{Q} \equiv 0$, $z \equiv 0$. Using the formulas in Proposition 3.3.4, one finds that the quantities $b, \boldsymbol{q}, \boldsymbol{q}', z$ are $b = sJ^{-1}$, $\boldsymbol{q} = 0$, $\boldsymbol{q}' = sJ^{-1}\boldsymbol{v}$, $z = 0$ and the equivalent spatial local forms 3.3.2(2) and 3.3.3(2) read

$$\frac{\partial sJ^{-1}}{\partial t} + \operatorname{div}(sJ^{-1}\boldsymbol{v}) = 0, \quad -u[sJ^{-1}] + [sJ^{-1}\boldsymbol{v}] \cdot \boldsymbol{m} = 0.$$

The constant factor $s = \pm 1$ can be canceled and we arrive at $(2.2.7)_2$ and (2.2.8), i.e., at one of the forms of Euler's identity. This example is closely connected with the conservation of mass, cf. Subsect. 3.4.3.

3.4 Mass

Throughout the rest of the chapter we assume that the motions have the smoothness specified in Subsect. 2.2.1, that the fields of internal energy, entropy and temperature, to be introduced below, are piecewise continuously differentiable and the body force and the rate of supply of heat are piecewise continuous.

It is postulated that the mass of parts of the body can be expressed through the ***referential density of mass*** $\rho_\circ > 0$, $\rho_\circ : \Omega \to \mathbb{R}^{++}$ a continuous function. The mass $\mathsf{M}(P)$ of the part $P \subset \Omega$ is

$$\mathsf{M}(P) = \int_P \rho_\circ \, dV.$$

3.4.1 The Spatial Density of Mass The ***spatial density of mass*** ρ during a motion χ is defined by the requirement that

$$\mathsf{M}(P) = \int_{\chi(P,t)} \rho \, dv$$

for every part $P \subset \Omega$. The continuity of $\rho_\circ$ does not preclude the spatial density from having jump discontinuities. The change of variables formula gives

$$\rho = \rho_\circ / |\det \boldsymbol{F}| = \rho_\circ / |J|$$

outside the singular surface; on the singular surface this formula assigns two values $\rho^\pm = \rho_\circ / |\det \boldsymbol{F}^\pm|$. The referential density $\rho_\circ$ is independent of time, which expresses the conservation of mass. Hence $B \equiv \rho_\circ$ satisfies the balance equation of Proposition 3.3.1 with the flux $\boldsymbol{Q} \equiv \boldsymbol{0}$ and the source $Z \equiv 0$. The spatial density ρ is then the spatial density b of B in the sense of Sect. 3.3. By Proposition 3.3.4 the local control-volume form of the equation of balance of mass is

$$\frac{\partial \rho}{\partial t} + \operatorname{div}(\rho \boldsymbol{v}) = 0, \quad -u[\rho] + [\rho \boldsymbol{v}] \cdot \boldsymbol{m} = 0 \tag{3.4.1}$$

outside and on the singular surface, respectively. We recognize $\rho \boldsymbol{v}$ to be the ***transport of mass;*** there is no true flux of mass. The ***specific volume*** is defined by $v := \rho^{-1} = |\det \boldsymbol{F}|/\rho_\circ$ outside the singular surface and $v^\pm := (\rho^\pm)^{-1} = |\det \boldsymbol{F}^\pm|/\rho_\circ$ on the singular surface. Using $(2.2.4)_2$, equation $(3.4.1)_1$ can be given the form $\dot{\rho} + \rho \operatorname{div} \boldsymbol{v} = 0$ outside the singular surface.

For a singular surface moving with referential speed U the quantity $J := \rho_\circ U$, of physical dimension mass per length^2 × time, is the amount of mass which passes through unit referential area of the singular surface per unit time. The ***flux of mass*** through the singular surface is

$$j := \rho_\circ U / |\operatorname{cof} \boldsymbol{F}^+ \boldsymbol{n}| = \rho_\circ U / |\operatorname{cof} \boldsymbol{F}^- \boldsymbol{n}|. \tag{3.4.2}$$

Since $|\operatorname{cof} \boldsymbol{F}^+ \boldsymbol{n}| = |\operatorname{cof} \boldsymbol{F}^- \boldsymbol{n}|$ is the actual area of the element of singular surface whose referential area is 1 (see Subsect. 2.2.4 and equation $(2.2.10)_2$), j is the amount of mass passing through the unit actual area of the singular surface per unit time.

3.4.2 Proposition *On the singular surface of a motion* $\boldsymbol{\chi}$,

$$j = \rho^+ u^+ = \rho^- u^-, \quad j = u^+/v^+ = u^-/v^-, \tag{3.4.3}$$

where $u^\pm$ *are the relative speeds of the singular surface with respect to the motion ahead of and behind the singular surface defined by* (2.2.13) *and* $v^\pm$ *are the two specific volumes on the singular surface. Also*

$$j[v] + [\boldsymbol{v}] \cdot \boldsymbol{m} = 0, \quad j = \langle \rho \rangle \, u - \langle \rho \boldsymbol{v} \rangle \cdot \boldsymbol{m}. \tag{3.4.4}$$

The convention $U \geq 0$ *is equivalent to* $j \geq 0$.

$j > 0$ means that the amount of mass ahead of the singular surface diminishes.

Proof To prove (3.4.3), it suffices to eliminate U between the definition (3.4.2) and $(2.2.12)_2$. The first formula of (3.4.4) follows from $jv^\pm = u^\pm = u - \boldsymbol{v}^\pm \cdot \boldsymbol{m}$, by subtracting the formula with $-$ from the formula with $+$. To obtain the second formula, note that we have $\langle \rho \rangle \, u - \langle \rho \boldsymbol{v} \rangle \cdot \boldsymbol{m} = \frac{1}{2}(\rho^+ u + \rho^- u - \rho^+ \boldsymbol{v}^+ \cdot \boldsymbol{m} - \rho^- \boldsymbol{v}^- \cdot \boldsymbol{m}) = \frac{1}{2}(\rho^+ u^+ + \rho^- u^-) = j.$ □

3.4.3 The Specific Quantities In Sect. 3.2 the referential and spatial densities B, b of an extensive quantity were introduced. As the concept of mass is now available, we also introduce the ***specific value*** $\beta := B/\rho_\circ = b/\rho$; i.e., the amount of the quantity per unit mass. The content of B in part P is

$$\int_{\chi(P,t)} \rho \beta \, dv = \int_P \rho_\circ \beta \, dV.$$

Since $\rho_\circ$ is independent of time, one has $\rho_\circ\dot{\beta} = \dot{B}$ and the referential forms of the balance equations can be trivially modified to replace B by β. For instance, the equations of Condition (2) of Proposition 3.3.1 can be written as

$$\rho_\circ\dot{\beta} + \operatorname{Div}\boldsymbol{Q} = \rho_\circ\zeta, \quad -U\rho_\circ[\beta] + [\boldsymbol{Q}]\cdot\boldsymbol{n} = 0$$

outside and on the singular surface, respectively, with $\zeta = z/\rho_\circ$ the specific rate of supply of the balanced quantity. It is assumed that $\rho_\circ$ is a continuous function of $\boldsymbol{p}$; one therefore has $[\rho_\circ\beta] = \rho_\circ[\beta]$.

3.4.4 Proposition *Let χ be a motion with a singular surface and β a continuous and piecewise continuously differentiable function defined on the spacetime tube* (2.2.1) *with the same singular surface. Then*

(1) *we have*

$$\rho\dot{\beta} = \frac{\partial\rho\beta}{\partial t} + \operatorname{div}(\rho\beta\boldsymbol{v}), \quad j[\beta] = u[\rho\beta] - [\rho\beta\boldsymbol{v}]\cdot\boldsymbol{m} \tag{3.4.5}$$

outside and on the singular surface, respectively;

(2) *for every part $P \subset \Omega$ and almost every t,*

$$\frac{d}{dt}\int_{\chi(P,t)}\rho\beta\,dv = \int_{\chi(P,t)}\rho\dot{\beta}\,dv - \int_{\chi(P,t)\cap s_t} j[\beta]\,da. \tag{3.4.6}$$

Proof (1): Equation $(3.4.5)_1$ is a calculation using $(3.4.1)_1$. To prove $(3.4.5)_2$, we apply the product rule (2.1.4) to calculate the jumps $[\rho\beta]$, $[\rho\beta\boldsymbol{v}]$, with $\rho\beta\boldsymbol{v}$ viewed as a product of $\rho\boldsymbol{v}$ and β. Rearranging the terms, we obtain

$$u[\rho\beta] - [\rho\beta\boldsymbol{v}]\cdot\boldsymbol{m} = (u[\rho] - [\rho\boldsymbol{v}]\cdot\boldsymbol{m})\langle\beta\rangle + (u\langle\rho\rangle - \langle\rho\boldsymbol{v}\rangle\cdot\boldsymbol{m})[\beta] = j[\beta],$$

since the first round bracket vanishes by $(3.4.1)_2$ and the second round bracket is j by $(3.4.4)_2$. (2): Apply $(3.2.7)_1$ to $b := \rho\beta$ and use (1). □

3.5 Linear and Angular Momenta

As a preliminary to the discussion of the equations of balance of linear and angular momenta, the structure of forces acting in a continuous body is briefly reviewed.

3.5.1 The Contact and Body Forces It is postulated that the contact forces acting in the body Ω are described by a stress tensor and that the external volume forces acting upon the body have a volume density. Specifically, associated with a process is

(1) a time-dependent field $\boldsymbol{S}$ on $\Omega \times [t_1, t_2]$ with values in Lin such that for any two parts P_1, P_2 which have no points in common, the force which the part P_2 exerts on P_1 is given by

$$\boldsymbol{f}(P_1, P_2, t) = \int_S \boldsymbol{Sn}\,dA, \tag{3.5.1}$$

where $S = \partial P_1 \cap \partial P_2$ is a surface of contact and $\boldsymbol{n}$ is the unit normal to S pointing from P_1 into P_2;

(2) a time-dependent field $\boldsymbol{b}$ on $\Omega \times [t_1, t_2]$ with values in Vect such that for any part P the force acting on P from the exterior of the body is

$$\int_P \rho_\circ\boldsymbol{b}\,dV.$$

The tensor S is called the ***referential*** or the ***Piola–Kirchhoff stress tensor*** and b is the ***specific body force.*** Using the Piola transformation described in Sect. 3.1, it is possible to write (3.5.1) as

$$f(P_1, P_2, t) = \int_{s_t} \boldsymbol{Tm}\, da,$$

where T is the ***spatial*** or the ***Cauchy stress tensor,*** related to S by

$$\boldsymbol{T} = s\boldsymbol{S}(\operatorname{cof}\boldsymbol{F})^{-1}, \quad \boldsymbol{S} = s\boldsymbol{T}\operatorname{cof}\boldsymbol{F}, \quad (s = \operatorname{sgn}\det\boldsymbol{F})$$

and $s_t = \partial\chi(P_1, t) \cap \partial\chi(P_2, t)$ is the spatial configuration of the surface of contact. These formulas are the application of the scalar formula (3.1.3) to "each of the components" of the force, or, alternatively, one can start from the requirement that for every elementary surface $\boldsymbol{Sn}\, dA = \boldsymbol{Tm}\, da$ where the quantities $\boldsymbol{n}$, dA, $\boldsymbol{m}$, da are related by $\boldsymbol{m}\, da = s \operatorname{cof} \boldsymbol{Fn}\, dA$, cf. Subsect. 2.2.4. In the subsequent discussion of the balance equations self-explanatory interpretations are omitted.

3.5.2 Balance of Linear Momentum The material form of balance of linear momentum reads

$$\frac{d}{dt}\int_P \rho_0 \boldsymbol{v}\, dV = \int_{\partial P} \boldsymbol{Sn}\, dA + \int_P \rho_0 \boldsymbol{b}\, dV \tag{3.5.2}$$

for every part and almost every time. Equivalently,

$$\frac{d}{dt}\int_{\chi(P,t)} \rho \boldsymbol{v}\, dv = \int_{\partial\chi(P,t)} \boldsymbol{Tm}\, da + \int_{\chi(P,t)} \rho \boldsymbol{b}\, dv. \tag{3.5.3}$$

The local form is

$$\rho_0 \dot{\boldsymbol{v}} = \operatorname{Div}\boldsymbol{S} + \rho_0 \boldsymbol{b}, \quad -U\rho_0[\boldsymbol{v}] = [\boldsymbol{S}]\boldsymbol{n} \tag{3.5.4}$$

outside and on the singular surface, respectively. By $(2.2.10)_1$,

$$U^2 \rho_0 [\boldsymbol{F}]\boldsymbol{n} = [\boldsymbol{S}]\boldsymbol{n}.$$

Using the spatial density, (3.5.4) can be rewritten as (see (3.3.11), (3.3.12) and (3.4.2))

$$\rho \dot{\boldsymbol{v}} = \operatorname{div}\boldsymbol{T} + \rho \boldsymbol{b}, \quad -j[\boldsymbol{v}] = [\boldsymbol{T}]\boldsymbol{m}. \tag{3.5.5}$$

Applying the prescription of Sect. 3.3 to each of the three components of $\rho\boldsymbol{v}$ one finds that the ***transport of linear momentum*** is $\rho\boldsymbol{v} \otimes \boldsymbol{v}$. The control-volume form of balance of linear momentum reads

$$\frac{d}{dt}\int_p \rho \boldsymbol{v}\, dV = \int_p (\boldsymbol{T} - \rho\boldsymbol{v} \otimes \boldsymbol{v})\boldsymbol{m}\, da + \int_p \rho \boldsymbol{b}\, dv$$

for every spatial control volume p and almost every time for which $p \subset T_t$. The local forms are

$$\frac{\partial \rho \boldsymbol{v}}{\partial t} + \operatorname{div}(\rho\boldsymbol{v} \otimes \boldsymbol{v} - \boldsymbol{T}) = \rho \boldsymbol{b}, \quad -u[\rho\boldsymbol{v}] + [\rho\boldsymbol{v} \otimes \boldsymbol{v} - \boldsymbol{T}]\boldsymbol{m} = \boldsymbol{0} \tag{3.5.6}$$

outside and on the singular surface, respectively. Forming a scalar product of $(3.5.5)_2$ with $\boldsymbol{m}$ and using $(3.4.4)_1$ we obtain

$$j^2[v] = [\boldsymbol{T}]\boldsymbol{m} \cdot \boldsymbol{m}. \tag{3.5.7}$$

3.5.3 Balance of Angular Momentum The material form reads

$$\frac{d}{dt}\int_P \rho_\circ(\boldsymbol{x}-\boldsymbol{x}_0)\times \boldsymbol{v}\,dV = \int_{\partial P}(\boldsymbol{x}-\boldsymbol{x}_0)\times \boldsymbol{Sn}\,dA + \int_P \rho_\circ(\boldsymbol{x}-\boldsymbol{x}_0)\times \boldsymbol{b}\,dV$$

for every part, almost every time and every point $\boldsymbol{x}_0 \in$ Eucl; the quantity $\boldsymbol{x}=\boldsymbol{\chi}$ is the actual position of the material point. The local form is

$$\rho_\circ((\boldsymbol{x}-\boldsymbol{x}_0)\times \boldsymbol{v})^{\cdot} = \mathrm{Div}\,((\boldsymbol{x}-\boldsymbol{x}_0)\times \boldsymbol{S}) + \rho_\circ(\boldsymbol{x}-\boldsymbol{x}_0)\times \boldsymbol{b}$$

outside the singular surface and

$$-U\rho_\circ[(\boldsymbol{x}-\boldsymbol{x}_0)\times \boldsymbol{v}] = [(\boldsymbol{x}-\boldsymbol{x}_0)\times \boldsymbol{Sn}]$$

on the singular surface. Since the motion $\boldsymbol{x}=\boldsymbol{\chi}$ is continuous, $\boldsymbol{x}-\boldsymbol{x}_0$ can be taken out of the square brackets and one finds that this equation follows from the jump condition for the balance of linear momentum.

3.5.4 Proposition

(1) *The balance of angular momentum implies the balance of linear momentum.*

(2) *The equation of balance of angular momentum is satisfied if and only if the equation of balance of linear momentum holds and*

$$\boldsymbol{SF}^{\mathsf T} = \boldsymbol{FS}^{\mathsf T};\ \text{equivalently}\ \boldsymbol{T}=\boldsymbol{T}^{\mathsf T} \tag{3.5.8}$$

everywhere during the process.

Proof (1) Writing the equation of balance of angular momentum with the point $\boldsymbol{x}_0$ as above and with the point $\boldsymbol{x}_0$ replaced by $\boldsymbol{x}_0+\boldsymbol{w}$, where $\boldsymbol{w}$ is arbitrary vector, and subtracting the corresponding equations, one obtains

$$\frac{d}{dt}\int_P \rho_\circ \boldsymbol{w}\times \boldsymbol{v}\,dV = \int_{\partial P} \boldsymbol{w}\times \boldsymbol{Sn}\,dA + \int_P \rho_\circ \boldsymbol{w}\times \boldsymbol{b}\,dV.$$

Taking $\boldsymbol{w}$ out of the integrals and using the arbitrariness of $\boldsymbol{w}$, one obtains (3.5.2).

(2) See, e.g., GURTIN [1981, Sect. 14]. □

3.6 Energy

In complete analogy to the assumptions about the structure of forces, the following subsection briefly summarizes the assumptions about the fluxes of heat in a continuous body.

3.6.1 Contact Heating and External Supply of Heat It is postulated that the contact fluxes of heat in the body Ω are described by a heat flux vector and that the external supply of heat to the body has a volume density. Accordingly, associated with a process is

(1) a time-dependent field $\boldsymbol{q}$ on $\Omega\times[t_1,t_2]$ with values in Vect such that for any two parts P_1, P_2 which have no points in common, the rate at which P_2 supplies heat to P_1 is

$$h(P_1,P_2,t) = \int_S \boldsymbol{q}\cdot\boldsymbol{n}\,dA, \tag{3.6.1}$$

where $S=\partial P_1\cap\partial P_2$ is a surface of contact of P_1 and P_2 and $\boldsymbol{n}$ is the unit normal to S pointing from P_1 into P_2;

(2) a time-dependent scalar field r on $\Omega \times [t_1, t_2]$ such that for any part P the rate of the heat supply to P from the exterior of the body is

$$\int_P \rho_\circ r \, dV.$$

The vector $\boldsymbol{q}$ is called the ***referential*** or the ***Piola–Kirchhoff heat flux vector*** and r is the ***specific rate of supply of heat.*** Using the Piola transformation (Sect. 3.1), it is possible to write (3.6.1) as

$$h(P_1, P_2, t) = \int_{s_t} \boldsymbol{h} \cdot \boldsymbol{m} \, da,$$

where $\boldsymbol{h}$ is the ***spatial*** or the ***Cauchy heat flux vector,*** related to $\boldsymbol{q}$ by

$$\boldsymbol{h} = s(\operatorname{cof} \boldsymbol{F})^{-\mathsf{T}} \boldsymbol{q}, \quad \boldsymbol{q} = s(\operatorname{cof} \boldsymbol{F})^{\mathsf{T}} \boldsymbol{h}$$

and $s_t = \partial \chi(P_1, t) \cap \partial \chi(P_2, t)$ is the spatial configuration of the surface of contact. The formulas for $\boldsymbol{h}$, $\boldsymbol{q}$ follow by an application of (3.1.3).

3.6.2 The Material Form of Balance of Energy Introducing the specific internal energy e, we state the material form of balance of energy as

$$\frac{d}{dt} \int_P \rho_\circ (e + \tfrac{1}{2} v^2) \, dV = \int_{\partial P} (\boldsymbol{v} \cdot \boldsymbol{S} \boldsymbol{n} - \boldsymbol{q} \cdot \boldsymbol{n}) \, dA + \int_P \rho_\circ (\boldsymbol{v} \cdot \boldsymbol{b} + r) \, dV \tag{3.6.2}$$

for every part and almost every time. This can be restated in the spatial form analogous to (3.5.3), which is omitted. The ***total energy***

$$\int_P \rho_\circ (e + \tfrac{1}{2} v^2) \, dV$$

is written as the sum of the internal and kinetic energies. In Chap. 5 the total energy will be derived from a more basic statement of the first law. In Chap. 6 the splitting of the total energy into the internal and kinetic parts will be seen to reflect the different transformation laws for the external and internal forces under changes of frame. The local material form of balance of energy is

$$\rho_\circ (e + \tfrac{1}{2} v^2)^{\boldsymbol{\cdot}} = \operatorname{Div}(\boldsymbol{S}^{\mathsf{T}} \boldsymbol{v} - \boldsymbol{q}) + \rho_\circ (\boldsymbol{v} \cdot \boldsymbol{b} + r) \tag{3.6.3}$$

outside the singular surface and

$$-U \rho_\circ [e + \tfrac{1}{2} v^2] = [\boldsymbol{S}^{\mathsf{T}} \boldsymbol{v} - \boldsymbol{q}] \boldsymbol{n} \tag{3.6.4}$$

on the singular surface. Using the spatial densities, (3.6.3) can be rewritten as

$$\rho (e + \tfrac{1}{2} v^2)^{\boldsymbol{\cdot}} = \operatorname{div}(\boldsymbol{T} \boldsymbol{v} - \boldsymbol{h}) + \rho (\boldsymbol{b} \cdot \boldsymbol{v} + r).$$

Using the balance of linear and angular momenta, (3.6.3) and (3.6.4) can be given the reduced forms:

$$\rho_\circ \dot{e} = \boldsymbol{S} \cdot \dot{\boldsymbol{F}} - \operatorname{Div} \boldsymbol{q} + \rho_\circ r, \quad U \rho_\circ [e] = U \langle \boldsymbol{S} \rangle \cdot [\boldsymbol{F}] + [\boldsymbol{q}] \cdot \boldsymbol{n} \tag{3.6.5}$$

outside and on the singular surface, respectively. Equation $(3.6.5)_1$ is obtained by computing the time derivative of $v^2/2$ and combining it with (3.5.4). Equation $(3.6.5)_2$ is obtained from (3.6.4) by calculating the products in square brackets by (2.1.4), subtracting $\langle \boldsymbol{v} \rangle$ times $(3.5.4)_2$, and using $(2.2.10)_1$. Equivalently, (3.6.5) can be written as

$$\rho \dot{e} = \boldsymbol{T} \cdot \boldsymbol{L} - \operatorname{div} \boldsymbol{h} + \rho r, \quad -j[e] = \langle \boldsymbol{T} \rangle [\boldsymbol{v}] \cdot \boldsymbol{m} - [\boldsymbol{h}] \cdot \boldsymbol{m}, \tag{3.6.6}$$

where $\boldsymbol{L} = \dot{\boldsymbol{F}} \boldsymbol{F}^{-1}$ is the spatial gradient of velocity; $(3.6.6)_2$ is obtained from $(3.6.5)_2$ by (3.3.11), (3.3.12) and (3.4.2).

3.6.3 The Control-Volume Form of Balance of Energy According to the general scheme of the passage from the material to the control volume forms of the balance equations explained in Sect. 3.3, the transport of total energy is $\rho(e+\frac{1}{2}v^2)v$. Therefore the control-volume form reads

$$\frac{d}{dt}\int_p \rho(e+\tfrac{1}{2}v^2)\,dv = \int_{\partial p} (Tv - h - \rho(e+\tfrac{1}{2}v^2)v)\cdot m\,da + \int_p \rho(v\cdot b + r)\,dv \tag{3.6.7}$$

for every spatial control volume p and almost every time for which $p \subset T_t$. The corresponding local forms are

$$\frac{\partial \rho(e+\frac{1}{2}v^2)}{\partial t} + \operatorname{div}(\rho(e+\tfrac{1}{2}v^2)v - Tv + h) = \rho r$$

outside the singular surface and

$$-u[\rho(e+\tfrac{1}{2}v^2)] + [\rho(e+\tfrac{1}{2}v^2)v - Tv + h]\cdot m = 0 \tag{3.6.8}$$

on the singular surface. By (3.4.5) we can rewrite (3.6.8) as

$$-j[e+\tfrac{1}{2}v^2] - [Tv]\cdot m + [h]\cdot m = 0.$$

3.7 Entropy

The existence of the absolute temperature and entropy is postulated for general "nonequilibrium" processes. It is assumed that these quantities satisfy the Clausius–Duhem inequality. In Chap. 7 the existence of the entropy and absolute temperature scale will be derived from the statements of the second law.

3.7.1 The Material Form of the Clausius-Duhem Inequality We introduce the ***specific entropy*** η and the positive ***absolute temperature*** $\theta > 0$ such that for every process the material form of the Clausius–Duhem inequality holds:

$$\frac{d}{dt}\int_P \rho_\circ \eta\,dV \geq -\int_{\partial P} \frac{q\cdot n}{\theta}\,dA + \int_P \rho_\circ \frac{r}{\theta}\,dV \tag{3.7.1}$$

for every part and almost every time. The integral on the left-hand side of (3.7.1) is the ***total entropy*** of part P. The local material form is

$$\rho_\circ \dot\eta \geq -\operatorname{Div}\frac{q}{\theta} + \rho_\circ \frac{r}{\theta}, \quad -U\rho_\circ[\eta] \geq -\left[\frac{q}{\theta}\right]\cdot n \tag{3.7.2}$$

outside and on the singular surface, respectively. Using the spatial densities, (3.7.2) can be rewritten as

$$\rho\dot\eta \geq -\operatorname{div}\frac{h}{\theta} + \rho\frac{r}{\theta}, \quad -j[\eta] \geq -\left[\frac{h}{\theta}\right]\cdot m. \tag{3.7.3}$$

Using the balance of energy, one can give the Clausius–Duhem inequality and the jump condition (3.7.2) the reduced forms

$$\rho_\circ \theta\dot\eta \geq \rho_\circ \dot e - S\cdot\dot F + q\cdot G/\theta, \tag{3.7.4}$$

outside the singular surface, where $G = \nabla\theta$ is the referential gradient of temperature, and

$$-U\rho_\circ \langle\theta\rangle [\eta] \geq -U(\rho_\circ[e] - \langle \boldsymbol{S}\rangle \cdot [\boldsymbol{F}]) + [\theta] \langle \boldsymbol{q}/\theta\rangle \cdot \boldsymbol{n} \tag{3.7.5}$$

on the singular surface. To prove (3.7.4), one inserts

$$\mathrm{Div}(\boldsymbol{q}/\theta) = (\mathrm{Div}\,\boldsymbol{q})/\theta - \boldsymbol{q} \cdot \boldsymbol{G}/\theta^2,$$

into (3.7.2) and eliminates $\mathrm{Div}\,\boldsymbol{q}$ from the resulting equation by (3.6.5). To prove (3.7.5), one inserts

$$[\boldsymbol{q}] = [\theta(\boldsymbol{q}/\theta)] = \langle\theta\rangle [\boldsymbol{q}/\theta] + [\theta] \langle \boldsymbol{q}/\theta\rangle$$

into (3.6.5) and eliminates $[\boldsymbol{q}/\theta]$ from (3.7.2). Equivalently, (3.7.4), (3.7.5) can be written in the spatial forms:

$$\rho\theta\dot{\eta} \geq \rho\dot{e} - \boldsymbol{T} \cdot \boldsymbol{L} + \boldsymbol{h} \cdot \boldsymbol{g}/\theta,$$

$$-j\langle\theta\rangle [\eta] \geq -j[e] - \langle \boldsymbol{T}\rangle \boldsymbol{m} \cdot [\boldsymbol{v}] + [\theta] \langle \boldsymbol{h}/\theta\rangle \cdot \boldsymbol{m}$$

where $\boldsymbol{g} = \mathrm{grad}\,\theta$ is the spatial gradient of temperature.

According to the general scheme, the control volume form of the Clausius–Duhem inequalityreads

$$\frac{d}{dt}\int_p \rho\eta\, dv \geq -\int_{\partial p} \left(\frac{\boldsymbol{h}}{\theta} + \rho\eta\boldsymbol{v}\right) \cdot \boldsymbol{m}\, da + \int_p \rho\frac{r}{\theta}\, dv$$

for every spatial control volume p and almost every time for which $p \subset T_t$. The corresponding local forms are

$$\begin{aligned} &\frac{\partial\rho\eta}{\partial t} + \mathrm{div}\left(\rho\eta\boldsymbol{v} + \frac{\boldsymbol{h}}{\theta}\right) \geq \rho\frac{r}{\theta}, \\ &-u[\rho\eta] + \left[\rho\eta\boldsymbol{v} + \frac{\boldsymbol{h}}{\theta}\right] \cdot \boldsymbol{m} \geq 0 \end{aligned} \tag{3.7.6}$$

outside and on the singular surface.

3.7.2 The Free Energy In terms of the free energy $f := e - \theta\eta$, the Clausius–Duhem inequality can be restated as the ***dissipation inequality***

$$\rho_\circ\dot{f} - \boldsymbol{S} \cdot \dot{\boldsymbol{F}} + \rho_\circ\eta\dot{\theta} \leq -\boldsymbol{q} \cdot \boldsymbol{G}/\theta, \tag{3.7.7}$$

$$-U(\rho_\circ[f] - \langle \boldsymbol{S}\rangle \cdot [\boldsymbol{F}] + \rho_\circ \langle\eta\rangle [\theta]) \leq -\langle \boldsymbol{q}/\theta\rangle \cdot \boldsymbol{n}[\theta] \tag{3.7.8}$$

outside and on the singular surface, respectively. We see that the energy was eliminated from the Clausius–Duhem inequality and the time-derivative of the entropy was replaced by the time-derivative of the absolute temperature. For this reason the form (3.7.7) of the entropy inequality is particularly useful when the temperature is used as an independent thermal variable while the original form (3.7.4) is useful when either the entropy or the internal energy is used as the independent thermal variable. When accompanied by the corresponding change of variables, the passage $e \mapsto f$ is an instance of the Legendre transformation; see Sect. 10.1.

3.7.3 The Eshelby Tensor Throughout the book, the ***referential Eshelby tensor***

$$\boldsymbol{P} := f\mathbf{1} - \rho_\circ^{-1}\boldsymbol{F}^{\mathrm{T}}\boldsymbol{S} = f\mathbf{1} - \rho^{-1}\boldsymbol{F}^{\mathrm{T}}\boldsymbol{T}\boldsymbol{F}^{-\mathrm{T}}, \tag{3.7.9}$$

will play an important role. Its physical dimension is that of the specific energy. The Eshelby tensor is generally nonsymmetric; in fact, using (3.5.8), it is found that $\boldsymbol{P}\boldsymbol{C} = \boldsymbol{C}\boldsymbol{P}^{\mathrm{T}}$ where $\boldsymbol{C} = \boldsymbol{F}^{\mathrm{T}}\boldsymbol{F}$ is the right Cauchy–Green deformation tensor. It will be seen later that $\boldsymbol{P}$ is symmetric in equilibrium states of isotropic solids and fluids, and that in the latter case, $\boldsymbol{P} = g\mathbf{1}$, where $g = f + pv$ is the Gibbs function of the fluid. The Clausius–Duhem inequality on the singular surface can be restated in terms of $\boldsymbol{P}$ as

$$\rho_\circ U\left(\boldsymbol{n}\cdot[\boldsymbol{P}]\boldsymbol{n} + \tfrac{1}{2}U^2\left[|\boldsymbol{F}\boldsymbol{n}|^2\right]\right) - (\langle \boldsymbol{q}/\theta\rangle \cdot \boldsymbol{n} - U\rho_\circ\langle\eta\rangle)[\theta] \geq 0. \tag{3.7.10}$$

Indeed, using (2.1.4), $(2.2.10)_1$ and (3.5.4), we find

$$\boldsymbol{n}\cdot[\boldsymbol{F}^\mathsf{T}\boldsymbol{S}]\boldsymbol{n} = \langle\boldsymbol{S}\rangle\cdot[\boldsymbol{F}] + \rho_\circ U^2\langle\boldsymbol{F}\rangle\,\boldsymbol{n}\cdot[\boldsymbol{F}]\boldsymbol{n}$$

and

$$\langle\boldsymbol{F}\rangle\,\boldsymbol{n}\cdot[\boldsymbol{F}]\boldsymbol{n} = \tfrac{1}{2}\left[|\boldsymbol{F}\boldsymbol{n}|^2\right].$$

Hence

$$\boldsymbol{n}\cdot[\boldsymbol{P}]\boldsymbol{n} = [f] - \rho_\circ^{-1}\langle\boldsymbol{S}\rangle\cdot[\boldsymbol{F}] - \tfrac{1}{2}U^2\left[|\boldsymbol{F}\boldsymbol{n}|^2\right].$$

A combination with (3.7.8) gives (3.7.10). For further properties of $\boldsymbol{P}$, see Sect. 10.3. Related to $\boldsymbol{P}$, but *different*, is the *symmetric **tensor of chemical potential***

$$\boldsymbol{K} := f\mathbf{1} - \rho_\circ^{-1}\boldsymbol{F}\boldsymbol{S}^\mathsf{T} = f\mathbf{1} - \rho^{-1}\boldsymbol{T} = \boldsymbol{F}^{-\mathsf{T}}\boldsymbol{P}\boldsymbol{F}^\mathsf{T}$$

by BOWEN [1967, 1976]. $\boldsymbol{K}$ is important in the theory of diffusing mixtures; see SAMOHÝL [1982, 1996b] and TRUESDELL [1984] for more details.

3.7.4 Interpretations of the Clausius–Duhem Inequality The local form (3.7.2) of the Clausius–Duhem inequality has three different statuses:

(1) We shall see that $(3.7.2)_1$ (or equivalently (3.7.7)) acts as a restriction on the response functions: these must be chosen as to satisfy $(3.7.2)_1$ "identically" in every smooth process compatible with the balance and constitutive equations; see Sect. 9.2.

The interpretation of $(3.7.2)_2$ (or equivalently (3.7.8) or (3.7.10)) depends whether the singular surface describes a shock wave (roughly, a discontinuity propagating at a speed comparable with the speed of sound) or a moving phase boundary (a discontinuity that propagates with a speed much smaller than the speed of sound).

(2) For a shock wave, $(3.7.2)_2$ acts as an admissibility criterion: only those solutions with singular surfaces are admitted (declared as physical) which in addition to the balance and constitutive equations, also satisfy $(3.7.2)_2$. See Sect. 17.4 and Chaps. 23 and 24.

(3) For a phase boundary, an *additional constitutive equation* must be prescribed to govern the speed of the singular surface for the problem to become well-posed. This constitutive equation must be such that $(3.7.2)_2$ holds identically. See Subsect. 22.2.1.

The different character of $(3.7.2)_2$ in (2), (3) is explained by the different character of the evolution equations in the region of state parameters where shock waves occur (where in the absence of dissipation the evolution equations are hyperbolic) and in the region of phase transitions (which are accompanied by subregions where the evolution equations are elliptic).

Another application of the Clausius–Duhem inequality will be given in Chaps. 13, 15, where it will be shown to provide natural Liapunov functions for bodies in equilibrium environments. When combined with the ideas of stability, it leads eventually to the extremum (variational) principles of thermodynamics.

3.8 Appendix: The Gauss–Green Theorem

The purpose of this section is to give a brief review of some results about sets of finite perimeter and functions of bounded variation. There are several detailed book expositions of this material: FEDERER [1969], GIUSTI [1984], VOL'PERT & HUDJAEV [1985], EVANS & GARIEPY [1992] and GIAQUINTA, MODICA & SOUČEK [1996]. In a standard way, it will be furthermore shown that the piecewise smooth objects introduced in Sect. 2.6 are special cases of the concepts discussed here. Explicit formulas for the measure-theoretic and topological boundaries of the intersection of two sets will be derived, and an example at the end of this section will show that the topological boundary of the intersection of two regions with C^∞ boundary can be substantially larger than its measure-theoretic boundary.

Let W be a finite-dimensional real inner product space of dimension n as in Sect. 2.6. We denote by $L^n = L$ the Lebesgue measure in W and by H^{n-1} the Hausdorff $n-1$-dimensional measure. Let $\omega_n > 0$ be the volume of the unit ball in W; then $L(B(x, r)) = \omega_n r^n$. If Q is a subset of W, we denote by $Q^c := \mathrm{W} \setminus Q$ its complement.

3.8.1 Sets of Finite Perimeter If $Q \subset \mathrm{W}$ is a Lebesgue measurable set and $\boldsymbol{x} \in \mathrm{W}$, we define $\theta(Q, \boldsymbol{x})$ by

$$\theta(Q, \boldsymbol{x}) := \lim_{r \to 0} L(Q \cap B(\boldsymbol{x}, r)) / \omega_n r^n \tag{3.8.1}$$

whenever the limit exists, and call $\theta(Q, \boldsymbol{x})$ the ***density of Q at $\boldsymbol{x}$***. The phrase $\theta(Q, \boldsymbol{x})$ exists means that the limit in (3.8.1) exists, and if $\alpha \in \mathbb{R}$, the statement $\theta(Q, \boldsymbol{x}) = \alpha$ means that $\theta(Q, \boldsymbol{x})$ exists and is equal to α. We say that $\boldsymbol{x} \in \mathrm{W}$ is a ***point of density*** of Q if $\theta(Q, \boldsymbol{x}) = 1$; we say that $\boldsymbol{x} \in \mathrm{W}$ is a ***point of rarefaction*** of Q if $\theta(Q, \boldsymbol{x}) = 0$. We denote by Q_* the set of all points of density of Q. A well-known result (Lebesgue's differentiation theorem) implies that Q_* differs from Q at most by a set of null Lebesgue measure, i.e., $L(Q_* \setminus Q) = L(Q \setminus Q_*) = 0$. The ***measure-theoretic boundary*** $\partial^* Q$ of Q is the set of all points $\boldsymbol{x} \in \mathrm{W}$ which are neither the points of density of Q nor the points of rarefaction of Q. Thus $\partial^* Q$ consists of all points $\boldsymbol{x} \in \mathrm{W}$ for which either $\theta(Q, \boldsymbol{x})$ exists and is different from 0, 1 or $\theta(Q, \boldsymbol{x})$ does not exist. Moreover, it follows that $L(\partial^* Q) = 0$ and that $\partial^* Q \subset \partial Q$. However, the difference $\partial Q \setminus \partial Q_*$ can be very large, as Example 3.8.9 shows. A Lebesgue measurable set Q is said to be a ***set of finite perimeter*** if $H^{n-1}(\partial^* Q) < \infty$. Clearly, if Q is a set of finite perimeter, then its complement Q^c is a set of finite perimeter and $\partial^* Q^c = \partial^* Q$. If $\boldsymbol{n} \in \mathrm{Sph}$ and $\boldsymbol{x} \in \mathrm{W}$, let us denote by $H(\boldsymbol{x}, \boldsymbol{n}) = \{\boldsymbol{y} \in \mathrm{W} : (\boldsymbol{y} - \boldsymbol{x}) \cdot \boldsymbol{n} < 0\}$ the open half-space with the boundary of normal $\boldsymbol{n}$ through $\boldsymbol{x}$. The unit vector $\boldsymbol{n}$ is said to be the ***measure-theoretic exterior normal*** to Q at $\boldsymbol{x} \in \mathrm{W}$ if

$$\theta(Q \cap H(\boldsymbol{x}, -\boldsymbol{n})) = 0 \quad \text{and} \quad \theta(Q^c \cap H(\boldsymbol{x}, \boldsymbol{n})) = 0.$$

If the measure-theoretic exterior normal to Q at $\boldsymbol{x}$ exists, it is uniquely determined. Moreover, since $(Q \cap H(\boldsymbol{x}, \boldsymbol{n})) \cup (Q^c \cap H(\boldsymbol{x}, \boldsymbol{n})) = H(\boldsymbol{x}, \boldsymbol{n})$, it follows from the easily verifiable fact that $\theta(H(\boldsymbol{x}, \boldsymbol{n}), \boldsymbol{x}) = \frac{1}{2}$ that if $\boldsymbol{x}$ has a normal at $\boldsymbol{x}$, then $\theta(Q, \boldsymbol{x}) = \frac{1}{2}$ and similarly that $\theta(Q^c, \boldsymbol{x}) = \frac{1}{2}$. Hence if the measure-theoretic exterior normal to Q at $\boldsymbol{x}$ exists, then $\boldsymbol{x} \in \partial^* Q$. Conversely, it can be proved that if Q has finite perimeter, then the measure-theoretic exterior normal to Q exists for H^{n-1}-a.e. point $\boldsymbol{x} \in \partial^* Q$, and the set of all $\boldsymbol{x}$ at which the normal exists is denoted by $\partial^{**} Q$, and called the ***reduced boundary*** of Q. Thus $\partial^{**} Q \subset \partial^* Q$ and $H^{n-1}(\partial^* Q \setminus \partial^{**} Q) = 0$. Thus for a set of finite perimeter Q, the measure-theoretic exterior normal is defined on $\partial^{**} Q$; it is convenient to extend the function $\boldsymbol{n}$ by $\mathbf{0}$ outside $\partial^{**} Q$. The resulting function is thus defined everywhere on W. Clearly, if $\boldsymbol{n}$ is a measure-theoretic exterior normal to Q, then $-\boldsymbol{n}$ is a measure-theoretic exterior normal to Q^c.

The ***Gauss–Green theorem for smooth functions*** says that if Q is a set of finite perimeter, then the measure-theoretic exterior normal $\boldsymbol{n}$ to Q is H^{n-1}-integrable and

$$\int_Q \nabla f \, dL = \int_{\partial^{**} Q} f \boldsymbol{n} \, dH^{n-1} \tag{3.8.2}$$

for every $f \in C_0^1(\mathrm{W})$. The class of all sets of finite perimeter is the largest class of sets for which the Gauss–Green theorem holds. More precisely, if Q is a Lebesgue measurable set in W such that there exists a vector-valued measure Borel measure $\boldsymbol{s}$ on W (of finite total variation) such that

$$\int_Q \nabla f \, dL = \int_{\mathrm{W}} f \, d\boldsymbol{s} \tag{3.8.3}$$

for every $f \in C_0^1(\mathrm{W})$, then Q is a set of finite perimeter and $\boldsymbol{s} = \boldsymbol{n} H^{n-1}$, where $\boldsymbol{n}$ is the measure-theoretic exterior normal to Q (with $\boldsymbol{n} = \boldsymbol{0}$ outside $\partial^{**} Q$). For the definition of a signed (i.e., real-valued) Borel measure on $\mathbb{R}$, see Sect. 7.1; in the present context one has just to replace the Borel subsets of $\mathbb{R}$ by the Borel subsets of W; and a vector-valued measure is just an n-tuple of real-valued measures. In the formula $\boldsymbol{s} = \boldsymbol{n} H^{n-1}$, the product of a Borel measurable and μ-integrable function h with a Borel measure μ is a Borel measure $\nu = h\mu$ given by

$$\nu(A) = (h\mu)(A) = \int_A h(\boldsymbol{x})\, d\mu(\boldsymbol{x})$$

for every Borel set $A \subset \mathrm{W}$.

3.8.2 Definitions Let Q be an open subset of W. A function $f \in L^1(Q)$ is said to be a function of ***bounded variation*** if there exists a vector-valued Borel measure $\boldsymbol{M}$ on W (of finite total variation) such that

$$\int_Q f\, D\phi\, dL = -\int_Q \phi\, d\boldsymbol{M} \tag{3.8.4}$$

for every $\phi \in C_0^\infty(Q)$. Here $L^1(Q)$ is the space of all L-summable functions on Q, see Sect. 20.2. Two functions $f, g \in L^1(Q)$ are regarded as identical if $f(\boldsymbol{x}) = g(\boldsymbol{x})$ for L-a.e. point $\boldsymbol{x} \in Q$. The measure $\boldsymbol{M}$ in (3.8.4) is uniquely determined and called the ***distributional derivative*** of f, and we shall sometimes write Df for $\boldsymbol{M}$. The set of all functions of bounded variation in Q is denoted by $\mathrm{BV}(Q)$. If $f \in \mathrm{BV}(Q)$, then for H^{n-1}-a.e. point $\boldsymbol{x} \in Q$ there exists the limit

$$\bar{f}(\boldsymbol{x}) = \lim_{r \to 0} \frac{1}{\omega_n r^n} \int_{B(\boldsymbol{x},r)} f\, dL,$$

and $\bar{f}(\boldsymbol{x})$ is called the mean value of f a $\boldsymbol{x}$ whenever the limit exists. The Lebesgue differentiation theorem implies that $\bar{f}(\boldsymbol{x}) = f(\boldsymbol{x})$ for L-a.e. point $\boldsymbol{x} \in \mathrm{W}$ whenever $f \in L^1(Q)$. That the mean value $\bar{f}(\boldsymbol{x})$ is defined for H^{n-1}-a.e. point $\boldsymbol{x}$ and not just for L-a.e. point is a consequence of $f \in \mathrm{BV}(Q)$. The function $\bar{f}$ is often called the ***precise representative*** of f; notice that two functions differing on a set of null L measure have the same precise representative.

The characteristic function χ_Q of a set $Q \subset \mathrm{W}$ is defined by $\chi_Q(\boldsymbol{x}) = 1$ if $\boldsymbol{x} \in Q$ and $\chi_Q(\boldsymbol{x}) = 0$ if $\boldsymbol{x} \notin Q$. Considerations about (3.8.3) say that Q is a set of finite perimeter with $L(Q) < \infty$ if and only if $\chi_Q \in \mathrm{BV}(\mathrm{W})$; moreover

$$D\chi_Q = -\boldsymbol{n} H^{n-1}, \quad \bar{\chi}_Q = \theta(Q, \cdot), \tag{3.8.5}$$

where again $\boldsymbol{n} H^{n-1}$ denotes the product of the Borel function $\boldsymbol{n}$ with the Borel measure H^{n-1}, and recall that $\boldsymbol{n} = \boldsymbol{0}$ outside $\partial^{**} Q$.

3.8.3 Proposition *Let Q be a bounded set of finite perimeter and $f \in \mathrm{BV}(\mathrm{W})$. Then*

(1) *for H^{n-1}-a.e. point $\boldsymbol{x} \in \partial^* Q$ there exist the inner trace f^b and the outer trace f^e of f on $\partial^* Q$, given by*

$$f^b(\boldsymbol{x}) = \lim_{r \to 0} \frac{2}{\omega_n r^n} \int_{Q \cap B(\boldsymbol{x},r)} f\, dL, \quad f^e(\boldsymbol{x}) = \lim_{r \to 0} \frac{2}{\omega_n r^n} \int_{Q^c \cap B(\boldsymbol{x},r)} f\, dL;$$

they satisfy

$$\bar{f}(\boldsymbol{x}) = \tfrac{1}{2}(f^b(\boldsymbol{x}) + f^e(\boldsymbol{x}))$$

for H^{n-1}-a.e. point $\boldsymbol{x} \in \partial^* Q$;

(2) ***(the Gauss–Green theorem for BV functions)*** *if $\boldsymbol{M}$ is the distributional derivative of f on W, and if f^b is H^{n-1}-integrable on $\partial^* Q$, then*

$$\boldsymbol{M}(Q_*) = \int_{\partial^* Q} f^b \boldsymbol{n}\, dH^{n-1},$$

where $\boldsymbol{n}$ is the measure-theoretic exterior normal to Q;

(3) *if f^b, f^e are H^{n-1}-integrable and if D is any Borel subset of $\partial^* Q$, then*

$$\boldsymbol{M}(D) = \int_D (f^e - f^b)\boldsymbol{n}\, dH^{n-1}.$$

The following result is a rule for the differentiation of a product of two functions of bounded variation.

3.8.4 Proposition *If $f, g \in \mathrm{BV}(W) \cap L^\infty(W)$, then $fg \in \mathrm{BV}(W) \cap L^\infty(W)$, and the distributional derivatives satisfy*

$$D(fg) = \bar{f}\, Dg + \bar{g}\, Df, \tag{3.8.6}$$

where $\bar{f}, \bar{g}$ are the mean values of f, g. The products in (3.8.6) are the products of a Borel measurable function with a Borel measure defined above.

Here $L^\infty(W)$ is the set of all L-measurable and essentially bounded functions. One of the advantages of the measure-theoretic boundary and the measure-theoretic exterior normal is that they behave well under the intersection and union of the sets of finite perimeter; the topological boundary behaves less regularly, see Proposition 3.8.8.

3.8.5 Proposition *Let P, Q be sets of finite perimeter with the measure-theoretic exterior normals $\boldsymbol{m}, \boldsymbol{n}$, respectively. Then $P \cap Q$ and $P \cup Q$ are sets of finite perimeter and moreover:*
(1) *to within H^{n-1}-null sets,*

$$\begin{aligned} \partial^*(P \cap Q) &= (\partial^* P \cap Q_*) \cup (P_* \cap \partial^* Q) \cup S, \\ \partial^*(P \cup Q) &= (\partial^* P \cap (Q^c)_*) \cup ((P^c)_* \cap \partial^* Q) \cup S, \end{aligned} \tag{3.8.7}$$

where $S = \{x \in \partial^{**} P \cap \partial^{**} Q : \boldsymbol{m}(x) = \boldsymbol{n}(x)\}$;
(2) *the measure-theoretic exterior normals $\boldsymbol{o}, \boldsymbol{p}$ to $P \cap Q$ and $P \cup Q$ are given by*

$$\boldsymbol{o} = \chi_{P_* \cap \partial^* Q}\, \boldsymbol{n} + \chi_{Q_* \cap \partial^* P}\, \boldsymbol{m} + \chi_S\, \boldsymbol{q}, \quad \boldsymbol{p} = -\chi_{(P^c)_* \cap \partial^* Q}\, \boldsymbol{n} - \chi_{(Q^c)_* \cap \partial^* P}\, \boldsymbol{m} - \chi_S\, \boldsymbol{q} \tag{3.8.8}$$

for H^{n-1}-a.e. point of W, where $\boldsymbol{q} := \boldsymbol{m} = \boldsymbol{n}$ on S.

Two H^{n-1}-measurable sets $A, B \subset W$ are said to be equal to within H^{n-1}-null sets if $H^{n-1}(B \setminus A) = H^{n-1}(A \setminus B) = 0$. Notice that the sets on the right-hand sides of (3.8.7) are disjoint. This proposition will be proved by applying the above product rule for BV functions. Another possibility is to employ directly the definitions, but this is lengthier and less illuminative.

Proof Let us restrict ourselves to sets P, Q of finite Lebesgue measure, i.e., $L(P) < \infty$, $L(Q) < \infty$; the generalization to sets not satisfying these conditions is easy. Since $\chi_R = \chi_P \chi_Q$ and $\chi_P, \chi_Q \in \mathrm{BV}(W)$, the product rule Proposition 3.8.4 implies that $\chi_R \in \mathrm{BV}(W)$, i.e., that the intersection of P, Q is a set of finite perimeter (and of finite Lebesgue measure). Moreover, if $\boldsymbol{o}$ is the measure-theoretic exterior normal to R (equal to $\boldsymbol{0}$ outside $\partial^{**} R$), then by $(3.8.5)_1$, $D\chi_R = -\boldsymbol{o}\, H^{n-1}$ and the same applies to the characteristic functions of P, Q, whose precise representatives are given by $(3.8.5)_2$. By (3.8.6),

$$\boldsymbol{o} H^{n-1} = \theta(P, \cdot)\, \boldsymbol{n}\, H^{n-1} + \theta(Q, \cdot)\, \boldsymbol{m}\, H^{n-1};$$

consequently,

$$\boldsymbol{o} = \theta(P, \cdot)\, \boldsymbol{n} + \theta(Q, \cdot)\, \boldsymbol{m} \quad \text{for } H^{n-1} \text{ a.e. point of } W. \tag{3.8.9}$$

Now $\theta(P, \cdot) = 1$ on P_*, $\theta(P, \cdot) = 0$ on $(P^c)_*$, and $\theta(P, \cdot) = \frac{1}{2}$ for H^{n-1}-a.e. point of $\partial^* P$. Consequently

$$\theta(P, \cdot) = \chi_{P_*} + \tfrac{1}{2} \chi_{\partial^* P} \quad \text{for } H^{n-1}\text{-a.e. point of } W$$

and similarly for $\theta(P, \cdot)$. As also clearly $\boldsymbol{m} = \chi_{\partial^* P}\, \boldsymbol{m}$ for H^{n-1}-a.e. point of W, equation (3.8.9) may be rewritten as

$$\boldsymbol{o} = \chi_{P_*} \cdot \chi_{\partial^* Q}\, \boldsymbol{n} + \chi_{Q_*} \cdot \chi_{\partial^* P}\, \boldsymbol{m} + \tfrac{1}{2} \chi_{\partial^* P} \cdot \chi_{\partial^* Q} (\boldsymbol{m} + \boldsymbol{n}) \quad \text{for } H^{n-1}\text{-a.e. point of } W. \tag{3.8.10}$$

Observe that for H^{n-1}-a.e. point $\boldsymbol{x} \in W$, at most one of the three terms in the right-hand side of (3.8.10) is different from $\boldsymbol{0}$. Since the range of $\boldsymbol{o}$ is $\{0\} \cup \mathrm{Sph}$, it follows that the values of each of the three terms on the right-hand side of (3.8.10) are in $\{0\} \cup \mathrm{Sph}$ for H^{n-1}-a.e. point in W. This is clearly true for the first two terms. For the third term this implies that for H^{n-1}-a.e. point $\boldsymbol{x}$ of $\partial^* P \cap \partial^* Q$ we have $\frac{1}{2}(\boldsymbol{m}(\boldsymbol{x}) + \boldsymbol{n}(\boldsymbol{x})) \in \{0\} \cup \mathrm{Sph}$. Since for any two unit vectors $\boldsymbol{m}, \boldsymbol{n}$ we have $\frac{1}{2}(\boldsymbol{m} + \boldsymbol{n}) \in \{0\} \cup \mathrm{Sph}$ if and only if either $\boldsymbol{m} = \boldsymbol{n}$ or $\boldsymbol{m} = -\boldsymbol{n}$, it follows that for H^{n-1}-a.e. point $\boldsymbol{x}$ of $\partial^* P \cap \partial^* Q$ either $\boldsymbol{m}(\boldsymbol{x}) = \boldsymbol{n}(\boldsymbol{x})$ or $\boldsymbol{m}(\boldsymbol{x}) = -\boldsymbol{n}(\boldsymbol{x})$. In the first case the third term in the right-hand side of (3.8.10) contributes with $\boldsymbol{m} = \boldsymbol{n}$ and in the second case with $\boldsymbol{0}$. Hence, (3.8.10) may be rewritten as

$$\boldsymbol{o} = \chi_{P_*} \cdot \chi_{\partial^* Q}\, \boldsymbol{n} + \chi_{Q_*} \cdot \chi_{\partial^* P}\, \boldsymbol{m} + \chi_S\, \boldsymbol{q} \quad \text{for } H^{n-1}\text{-a.e. point of } W \tag{3.8.11}$$

which implies $(3.8.8)_1$. Since to within a H^{n-1}-null set, the measure-theoretic boundary of R is the set of all x for which $o(x) \neq \mathbf{0}$, (3.8.11) also implies $(3.8.7)_1$. Equations $(3.8.7)_2$ and $(3.8.8)_2$ then follow from the just proved equations by applying de Morgan's formulas. □

Let us now relate the measure-theoretic boundary and normal to the topological boundary and the geometric normal.

3.8.6 Proposition

(1) *If Q is an open set in W and $x \in \partial^r Q$ then $x \in \partial^{**} Q$ and the measure-theoretic exterior normal to Q at x coincides with the geometric exterior normal to Q at x;*

(2) *if Q is a region with almost smooth boundary, then Q is a set of finite perimeter.*

Proof It suffices to prove only (1). We can assume that $x = \mathbf{0}$ and let m be the geometric exterior normal to Q at $\mathbf{0}$. Let $\{e_i, i = 1, \dots, n\}$ be an orthonormal basis such that $e_n = m$. If the components of an $y \in$ W relative to this basis are denoted by $y = (z, y) \in \mathbb{R}^n$, $z \in \mathbb{R}^{n-1}$, $y \in \mathbb{R}$, we can identify the objects in W with the corresponding objects in the coordinate space $\mathbb{R}^n$. Then the set $K_r :=$ $Q \cap H(\mathbf{0}, -m) \cap B(\mathbf{0}, r)$ is given by $K_r = \{(z, y) \in \mathbb{R}^n : |(z, y)| < r,\ 0 < y < \omega(z)\}$ where ω is a function defined on a neighborhood of 0 in $\mathbb{R}^{n-1}$ which gives a local parametric description of ∂Q, with $\nabla\omega(0) = 0$. This is derived using the fact that Q is locally on one side of its boundary. Since $\nabla\omega(0) = 0$, for every $\varepsilon > 0$ there exists a $\delta > 0$ such that $|\omega(z)| < \varepsilon|z|$ whenever $|z| < \delta$. Let $r < \delta$. Then

$$K_r \subset \{(z, y) \in \mathbb{R}^n : |z| < r,\ 0 < y < \varepsilon|z|\} \equiv L_r,$$

and the Lebesgue measure of the last set is easy to evaluate using the polar coordinates in $\mathbb{R}^{n-1}$ with the central distance of $z \in \mathbb{R}^{n-1}$ denoted by ρ. Namely,

$$L^n(L_r) = (n-1)\omega_{n-1} \int_0^r \varepsilon\rho \cdot \rho^{n-2}\, d\rho = (n-1)\omega_{n-1}\varepsilon r^n/n,$$

where $(n-1)\omega_{n-1}$ is the area of the $n-2$-dimensional unit sphere in $\mathbb{R}^{n-1}$. But then

$$L^n(K_r)/\omega_n r^n \le (n-1)\omega_{n-1}\varepsilon/(n\omega_n),$$

which proves that $\theta(Q \cap H(x, -m)) = 0$. The equation $\theta(Q^c \cap H(x, n)) = 0$ is proved similarly. □

The following proposition shows that the piecewise continuously differentiable fields in the sense of Definition 2.6.6 are functions of bounded variation, evaluates their distributional derivatives, and states the Gauss-Green theorem for them.

3.8.7 Proposition *Let Q be an open subset of W with almost smooth boundary, and let f be a real-valued piecewise continuously differentiable function with a singular surface U. Then*

(1) *for every subset P of Q with almost smooth boundary,*

$$\int_{\partial P} f^{\partial P} n^*\, dH^{n-1} = \int_P Df\, dL + \int_{P\cap U} [f] n\, dH^{n-1};$$

here Df is the classical derivative of f defined outside U, ∂P is the topological boundary of P, n is an orientation of U, $[f]$ is the jump of f (defined for H^{n-1}-a.e. point of U), and $f^{\partial P}$ is the boundary value of f on ∂P defined in Subsect. 2.1.5 and 3.1.1;

(2) *$f \in \mathrm{BV}(Q)$; the distributional derivative M of f is given by*

$$M(D) = \int_D Df\, dL + \int_{D\cap U} [f] n\, dH^{n-1} \tag{3.8.12}$$

for every Borel subset $D \subset Q$;

(3) *if $\tilde f$ is the extension of f from Q to the whole of W by 0 outside Q, then $\tilde f \in \mathrm{BV}(\mathrm{W})$ and the distributional derivative $\tilde M$ of $\tilde f$ is given by*

$$\tilde M(D) = \int_{D\cap Q} Df\, dL + \int_{D\cap U} [f] n\, dH^{n-1} - \int_{\partial Q\cap D} f n^*\, dH^{n-1}$$

for every Borel subset D of Q; in the last integral, ∂Q is the topological boundary of Q, n^* is the (geometric) exterior normal to Q, and f is the boundary value of f on ∂Q;

(4) *if $P \subset Q$ is a region with almost smooth boundary, then $f^b = f^{\partial P}$ for H^{n-1}-a.e. point of ∂P; here f^b is the inner trace of f on ∂P defined in Proposition 3.8.3(1) and $f^{\partial P}$ the boundary value of f on ∂P defined in Subsects. 2.1.3 and 3.1.1.*

From (2) we see that apart from the absolutely continuous part of the derivative, expressed by the volume integral in (3.8.12), there is also the singular part, associated with the jump of f on the singular surface. This part may be interpreted as the multiple of the Dirac measure concentrated on U. The formal proof of this proposition is omitted. It is only noted that the assertion that $f \in \mathrm{BV}(Q)$ is obtained applying directly the definition of BV. Namely, one divides Q into the regions Q^p from the definition of the singular surface and applies the classical Gauss-Green theorem to Q^p to find that

$$\int_{Q^p} f\, D\phi\, dL = -\int_{Q^p} Df\, \phi\, dL + \int_{\partial Q^p} f\, \phi \boldsymbol{n}^p\, dH^{n-1}$$

where $\boldsymbol{n}^p$ is the exterior normal to Q^p. Summing over all Q^p and using the Gauss-Green theorem to Q in the same way, combining with the assumption that $\phi = 0$ on ∂Q, one obtains that

$$\int_Q f\, D\phi\, dL = -\int_Q Df\, \phi\, dL - \int_U f\, \phi \boldsymbol{n}\, dH^{n-1},$$

and the right-hand side of this equation is just $\int_Q \phi\, d\boldsymbol{M}$, where $\boldsymbol{M}$ is as in Item (2). Item (3) is proved similarly. Item (4) may be verified using directly the definition and Proposition 3.8.6. Knowing that $f \in \mathrm{BV}(Q)$, we can now appeal to the Gauss-Green theorem for BV-functions to establish Item (1).

The following proposition shows that the topological boundary of the intersection of two sets with almost smooth boundary may be generally larger than the measure-theoretical boundary, and identifies precisely the difference between the two. The results (2)–(4) are stated for illustrative purposes and will be used only in the subsequent example.

3.8.8 Proposition *Let P, Q be open sets with almost smooth boundary with the exterior normals $\boldsymbol{m}$, $\boldsymbol{n}$, respectively. Then the following assertions hold:*

(1) *the set* $S^0 := \{\boldsymbol{x} \in \partial^r P \cap \partial^r Q : \boldsymbol{m}(\boldsymbol{x}), \boldsymbol{n}(\boldsymbol{x}) \text{ linearly independent}\}$ *is a smooth $n-2$-dimensional surface in* W *and hence* $H^{n-1}(S^0) = 0$;

(2) *if* $E := \{\boldsymbol{x} \in \partial^r P \cap \partial^r Q : \boldsymbol{m}(\boldsymbol{x}) = -\boldsymbol{n}(\boldsymbol{x})\}$, *then*

$$E \cap \mathrm{cl}(\partial P \cap Q) = E \cap \mathrm{cl}(P \cap \partial Q); \tag{3.8.13}$$

(3) *if T denotes the set on the two sides of* (3.8.13) *and* $S := \{\boldsymbol{x} \in \partial^r P \cap \partial^r Q : \boldsymbol{m}(\boldsymbol{x}) = \boldsymbol{n}(\boldsymbol{x})\}$, *then to within sets of H^{n-1}-null measure,*

$$\partial(P \cap Q) = (\partial P \cap Q) \cup (P \cap \partial Q) \cup S \cup T; \tag{3.8.14}$$

(4) *to within sets of H^{n-1}-null measure,* $\partial(P \cap Q) \setminus \partial^*(P \cap Q) = T$.

Notice that the sets on the right-hand side of (3.8.14) are disjoint. We shall see that it may happen that $H^{n-1}(T) > 0$ even when P, Q have boundaries of class C^∞.

Proof (1): Let $\boldsymbol{x} \in \partial P \cap \partial Q$ and let ϕ_1, ϕ_2 be the local implicit descriptions of $\partial P, \partial Q$, near $\boldsymbol{x}$, respectively, defined on O_1, O_2. Let $\phi := (\phi_1, \phi_2)$ on $O = O_1 \cap O_2$. Then the rank of $\nabla\phi(\boldsymbol{x})$ is 2 by the definition of S^0 and hence S^0 is an $n-2$-dimensional surface by Proposition 2.6.1. That its H^{n-1}-measure is 0 can be seen in a number of ways. For instance, ∂P can be covered by a countable number of parametric descriptions $\psi : N \subset \mathbb{R}^{n-1} \to$ W. Then $\psi^{-1}(S^0)$ is a hypersurface in $\mathbb{R}^{n-1}$ and its $n-1$-dimensional Lebesgue measure is 0 by Fubini's theorem. (2): If $\boldsymbol{x} \in E$, then since the normals to $\partial^r P, \partial^r Q$ are opposite and P, Q are locally on one side of their boundaries, one can find an affine isometry $\boldsymbol{T}$ and a neighborhood O of $\boldsymbol{x}$ such that

$$\boldsymbol{T}^{-1}(O \cap P) = \{(z, y) \in P \times (-\varepsilon, \varepsilon) : y > \psi(z)\},\ \boldsymbol{T}^{-1}(O \cap Q) = \{(z, y) \in P \times (-\varepsilon, \varepsilon) : y < \phi(z)\},$$

where P is some neighborhood of 0 in $\mathbb{R}^{n-1}$, $\varepsilon > 0$, ψ, ϕ are continuously differentiable functions on P such that $\psi(0) = \phi(0) = 0$, $D\psi(0) = D\phi(0) = 0$, and $\boldsymbol{T}^{-1}(\boldsymbol{x}) = (0, 0)$. If $\boldsymbol{x} \in E \cap \mathrm{cl}(\partial P \cap Q)$, there exists a sequence (z^k, y^k) such that $(z^k, y^k) \to (0, 0)$ as $k \to \infty$ and $y^k = \psi(z^k)$, $y^k < \phi(z^k)$. But then the sequence $(z^k, \phi(z^k))$ is in $P \cap \partial Q$ and $(z^k, \phi(z^k)) \to (0, 0)$ as $k \to \infty$, which shows that $\boldsymbol{x} \in E \cap \mathrm{cl}(P \cap \partial Q)$. (3): For any two open sets (see DIEUDONNÉ [1960, Problem 3c in Sect. III.8]),

$$(\partial P \cap Q) \cup (P \cap \partial Q) \subset \partial(P \cap Q) \subset (\partial P \cap Q) \cup (P \cap \partial Q) \cup (\partial P \cap \partial Q). \tag{3.8.15}$$

Since $\partial(P \cap Q)$ is closed, it follows from the first inclusion and the definition of T that

$$T \subset \mathrm{cl}(\partial P \cap Q) \cup \mathrm{cl}(P \cap \partial Q) \subset \partial(P \cap Q). \tag{3.8.16}$$

Let us now prove that

$$(\partial P \cap Q) \cup (P \cap \partial Q) \cup S \cup T \subset \partial(P \cap Q). \tag{3.8.17}$$

In view of $(3.8.15)_1$ and (3.8.16), it suffices to prove that $S \subset \partial(P \cap Q)$. Since the normals at any $\boldsymbol{x} \in S$ are the same, there exist local graph descriptions of P, Q such that

$$\boldsymbol{T}^{-1}(O \cap P) = \{(z, y) \in P \times (-\varepsilon, \varepsilon) : y > \psi(z)\}, \ \boldsymbol{T}^{-1}(O \cap Q) = \{(z, y) \in P \times (-\varepsilon, \varepsilon) : y > \phi(z)\},$$

where the objects O, P, ψ, ϕ, ε are of the same nature as in the proof of (2), which makes the inclusion $\boldsymbol{x} \in \partial(P \cap Q)$ obvious, completing the proof of (3.8.17). Let us now prove conversely that

$$\partial(P \cap Q) \subset (\partial P \cap Q) \cup (P \cap \partial Q) \cup S \cup T \cup N \tag{3.8.18}$$

where

$$N = S^0 \cup (\partial P \setminus \partial^r P) \cup (\partial Q \setminus \partial^r Q) \ \text{and} \ H^{n-1}(N) = 0.$$

The last equality follows from Item (1) and the facts that P, Q are sets with almost smooth boundary. We shall prove (3.8.18) by showing that if $\boldsymbol{x} \in \partial(P \cap Q)$ and $\boldsymbol{x} \notin \partial P \cap Q$, $\boldsymbol{x} \notin P \cap \partial Q$, $\boldsymbol{x} \notin S$, $\boldsymbol{x} \notin N$, then $\boldsymbol{x} \in T$. By $\boldsymbol{x} \notin \partial P \cap Q$, $\boldsymbol{x} \notin P \cap \partial Q$ and the second inclusion in (3.8.15) we have $\boldsymbol{x} \in \partial P \cap \partial Q$, and as $\boldsymbol{x} \notin N$, $\boldsymbol{x} \notin S$ necessarily $\boldsymbol{x} \in E$. Hence assume that $\boldsymbol{x} \in \partial(P \cap Q)$, $\boldsymbol{x} \in E$ and prove that $\boldsymbol{x} \in T$. Using the local graph descriptions as in the proof of (2), the condition $\boldsymbol{x} \in \partial(P \cap Q)$ implies that there exists a sequence of points in $P \cap Q$ converging to $\boldsymbol{x}$. In the graph descriptions, this means that there exists a sequence (z^k, y^k) such that $\psi(z^k) < y^k < \phi(z^k)$ and $(z^k, y^k) \to (0, 0)$ as $k \to \infty$. But then the sequence $(z^k, \psi(z^k))$ is in $\partial P \cap Q$ and $(z^k, \psi(z^k)) \to (0, 0)$ as $k \to \infty$. This means that $\boldsymbol{x} \in T$. (4): A consequence of (3.8.14) and $(3.8.7)_1$. □

3.8.9 Example *There exists a* $\phi \in C_0^\infty(\mathbb{R})$ *such that if*

$$P := \{(x, y) \in \mathbb{R}^2 : y < \phi(x)\}, \quad Q := \{(x, y) \in \mathbb{R}^2 : y > 0\}, \ \text{and} \ R := P \cap Q,$$

then $H^1(\partial R \setminus \partial^* R) > 0$.

Thus, P, Q are sets with smooth boundary, but R is not: the topological boundary of R is substantially larger than its measure-theoretic boundary, which cannot happen for sets with almost smooth boundary (see Proposition 3.8.6). It will be seen from the proof that the reason of this failure is not so much a loss of smoothness, since ∂R is contained in a union of two smooth curves ∂P, ∂Q, but rather the fact that R fails to be locally on one side of ∂R at the points of $\partial R \setminus \partial^* R$ in the sense of Subsect. 2.6.2. *Remark.* The set G to be constructed in the proof below gives an example of an open set whose boundary has positive Lebesgue measure, and the function ϕ gives an example of an infinitely differentiable function whose support has the boundary of positive Lebesgue measure.

Proof Let ε be such that $0 < \varepsilon < 1$ and let L denote the Lebesgue measure in $\mathbb{R}$. Let x_i be a sequence in (0, 1) which is dense in [0, 1] and let $\varepsilon_i > 0$ be a sequence such that $2\sum_{i=1}^\infty \varepsilon_i < \varepsilon$ and $I_i := (x_i - \varepsilon_i, x_i + \varepsilon_i) \subset (0, 1)$. Set $G = \bigcup_{i=1}^\infty I_i$, $F := [0, 1] \setminus G$. Then G is open, dense in [0, 1], $L(G) \le \sum_{i=1}^\infty L(I_i) = 2\sum_{i=1}^\infty \varepsilon_i < \varepsilon$, F is a closed set, $L(F) \ge 1 - \varepsilon > 0$ and as $x_i \notin F$, the interior of F is empty. The density of G implies $\operatorname{cl} G = [0, 1]$ and hence $\partial G = \operatorname{cl} G \setminus G = [0, 1] \setminus G = F$. Let $\psi : \mathbb{R} \to [0, 1]$ be a C^∞ function such that $\psi(x) = 0$ if $|x| \ge 1$ and $\psi(x) > 0$ if $|x| < 1$. Let c_i be a sequence of positive numbers such that

$$C_k := \sum_{i=1}^\infty c_i \varepsilon_i^{-k} < \infty \ \text{for every} \ k = 0, 1, \ldots \tag{3.8.19}$$

and set

$$\phi(x) = \sum_{i=1}^\infty c_i \psi((x - x_i)/\varepsilon_i), \quad x \in \mathbb{R}. \tag{3.8.20}$$

The elements of this series are majorized by those of the series in (3.8.19) with $k = 0$; likewise the elements of the series obtained from (3.8.20) by formal differentiation of order k are majorized by those of the series in (3.8.19); hence all the series are uniformly convergent and ϕ is a C^∞ function. Moreover, $\phi(x) > 0$ if and only if $x \in G$. Since G is dense in [0, 1], it follows that $\operatorname{supp}\phi = [0, 1]$. Clearly, P, Q are open sets with smooth boundary in the sense of the definition given in Subsect. 2.6.2 and

$$\partial P = \partial^{**} P = \{(x, y) \in \mathbb{R}^2 : y = \phi(x)\}, \quad \partial Q = \partial^{**} Q = \{(x, y) \in \mathbb{R}^2 : y = 0\}.$$

In the notation of Propositions 3.8.5 and 3.8.8, $S = \emptyset$, $E = F\times\{0\}$, $T = F\times\{0\}$. Thus, by Proposition 3.8.8(4), to within H^1-null sets, $\partial R \setminus \partial^* R = T = F\times\{0\}$ and hence $H^1(\partial R\setminus\partial^* R) = L(F) > 0$. □

Recall from Subsect. 2.6.7 the notion of an evolving smooth surface U which is a special case of an n-dimensional surface in the n+1-dimensional space Vect×$\mathbb{R}$. We denote by H^n the n-dimensional area measure on U, i.e., the restriction of the n-dimensional Hausdorff measure in Vect × $\mathbb{R}$ to U.

3.8.10 Proposition *Let U be an evolving smooth surface with unit normal $N = (N_s, N_t) \in$ Vect × $\mathbb{R}$. Let $\varphi : U \to \mathbb{R}$ be and H^n-summable function. Then for L^1-a.e. $t \in \mathbb{R}$, the function $\varphi(\cdot, t) : S_t \to \mathbb{R}$ is A-summable and*

$$\int_U \varphi\, dH^n = \int_{-\infty}^{\infty}\int_{S_t} \frac{\varphi}{|N_s|}\, dA\, dt = \int_{-\infty}^{\infty}\int_{S_t} \varphi\sqrt{1+U^2}\, dA\, dt; \tag{3.8.21}$$

consequently,

$$\int_U \varphi N\, dH^n = \int_{-\infty}^{\infty}\int_{S_t} \frac{\varphi N}{|N_s|}\, dA\, dt, \tag{3.8.22}$$

and splitting this into the spatial and temporal components,

$$\int_U \varphi N_s\, dH^n = \int_{-\infty}^{\infty}\int_{S_t} \varphi \boldsymbol{n}\, dA\, dt, \qquad \int_U \varphi N_t\, dH^n = -\int_{-\infty}^{\infty}\int_{S_t} \varphi U\, dA\, dt. \tag{3.8.23}$$

Proof It is enough to verify the formulas locally, i.e., only for functions φ which are nonzero in an arbitrarily small neighborhood of a general point $(\boldsymbol{q}, s) \in U$. The definition of an evolving surface implies that the surface has a local parametrization of the following form: There exists an $\varepsilon > 0$, and a mapping $\psi : P\times(s-\varepsilon, s+\varepsilon) \to$ Vect, where P is an open subset of $\mathbb{R}^{n-1}$, such that $U\cap O = \{(\boldsymbol{p}, t) \in \text{Vect}\times\mathbb{R} : \boldsymbol{p} = \psi(y, t)$ for some y such that $(y, t) \in P\times(s-\varepsilon, s+\varepsilon)\}$. We write $\psi = (\psi_1, \ldots, \psi_n)$ and define, for convenience, $\psi_{n+1}(y, t) = t$, $y_n = t$. Then

$$\int \varphi N_i\, dH^n = \int \varphi \varepsilon_{ii_1\cdots i_n}\psi_{i_1,1}\cdots\psi_{i_n,n}\, dy_1\cdots dy, \tag{3.8.24}$$

where $\varepsilon_{ii_1\cdots i_n}$ is $n+1$-dimensional permutation symbol, the summation convention is used, and a comma followed by a subscript k denotes the differentiation with respect to y_k. Let $i < n+1$ in (3.8.24). Then $\varepsilon_{ii_1\cdots i_n}$ is different from 0 only when exactly one of the indices $i_1, \ldots, i_n$ is $n+1$; moreover, if this happens for some index i_k with $k \neq n$, then the contribution is still 0, because then $\psi_{i_k,k}$ is 0. Hence the only nonzero contributions to the sum in (3.8.24) are those with $i_n = n+1$, and then $\psi_{i_n,n} = 1$. Hence if $i < n+1$,

$$\int \varphi N_i\, dH^n = \int \varphi\varepsilon_{ii_1\cdots i_{n-1}}\psi_{i_1,1}\cdots\psi_{i_{n-1},n-1}\, dy_1\cdots dy_{n-1}\, dt,$$

and by Fubini's theorem, this is exactly the right-hand side of $(3.8.23)_1$. Replacing φ in $(3.8.23)_1$ by $\varphi N_s/|N_s|^2$, and recalling that $N_s \neq \boldsymbol{0}$, $\boldsymbol{n} = N_s/|N_s|$, we obtain the first equality in (3.8.21). The second is a consequence of $U = -N_t/|N_s|$ and $|N_s|^2 + |N_t|^2 = 1$. Finally, (3.8.22) is a consequence of $(3.8.23)_1$ with φ replaced by φN and $(3.8.23)_2$ its temporal part. □

Bibliographical Note

The Clausius–Duhem inequality first appears in TRUESDELL & TOUPIN [1960], although particular cases were given earlier by Clausius and Duhem. For some time after 1960 there was no agreement about its meaning. Formally the equations of linear momentum and energy and the appropriate constitutive equations form a system of four scalar equations for four unknown functions – the three components of position and one temperature – and they should suffice to determine the processes in given external conditions. The situation was also complicated by the fact that the thermodynamic inequalities for processes like the Clausius–Duhem inequality were not clearly separated from the variational inequalities expressing the stability (the minimization of energy at fixed entropy etc.) The interpretation (1) above was first clearly pronounced by COLEMAN & NOLL [1963] in a fundamental paper. This paper showed the consistency of (1) and proved that this interpretation places concrete restrictions on the constitutive equations of bodies. Subsequent research has revealed that (1) is extremely powerful in applications to diverse classes of bodies.

In the approach adopted, the jumps bear no structure, in particular, surface tension effects are excluded. For a theory of singular surfaces which can exchange mass and other quantities with the bulk regions, including the surface tension effects, see GURTIN [1988], ANGENENT & GURTIN [1989], GURTIN & STRUTHERS [1990], GURTIN [1993a,b, 1994, 1995], CERMELLI & GURTIN [1994] and the references therein. Yet another approach to singular surfaces is to view them as regions of finite width with steep changes of the associated quantities, see Subsect. 23.8.6 and Sect. 24.5. For the literature relevant to Interpretation (3) above, see Sect. 22.2. For Subsect. 3.7.3, cf. KNOWLES [1979] and, in the isothermal case, TRUSKINOVSKY [1987]. A tensor closely related to $\boldsymbol{P}$, namely $\boldsymbol{P}' = \boldsymbol{P} - \rho_0^{-1}\boldsymbol{S} = f - \rho_0^{-1}(\boldsymbol{F} - \mathbf{1})\boldsymbol{S}$, was introduced by Eshelby (see ESHELBY [1951, 1956, 1975]) in the continuum theory of dislocations under the name the energy–momentum tensor. See Subsect. 14.2.5 for more information about $\boldsymbol{P}$, for remarks on the continuum theory of dislocations and further references. The relevance of $\boldsymbol{P}$ to the phase equilibria and the literature are to be found in Sect. 17.4.

Part II

Foundations

4 Material Bodies

An appealing feature of thermodynamics is that its structure can be derived from conceptually simple "empirical" statements summarizing and generalizing the experience with the behavior of real bodies. From the first and second laws we shall establish the existence and meaning of concepts and assertions like the interconvertibility of work and heat, the energy, the absolute temperature, and the entropy. The proofs elucidate similarities between the two laws. Most of Part II does not need the specific structure of continuous bodies, although for simplicity the derivation is carried only for them. Only one idea to be presented in Part II depends on the structure of continuous bodies – the derivation of Cauchy's equations of motion.

The first chapter of this part gives the definition of a material body for the purpose of the foundations. The concepts of mass, absolute temperature, energy, and entropy do not occur on the list of primitive concepts, and consequently the balance equations cannot, and need not, be formulated. The idea of state function is central to the definition of a material body, and this idea underlies also all of the general constitutive theory.

4.1 State Space

Classical thermodynamics uses the concept of state in an informal way, which creates a good deal of confusion it its foundations. The situation is better in applications since the idea of state is usually clear from the context. In modern treatments of continuum mechanics and thermodynamics the concepts of state and state functions are replaced by constitutive functionals expressing the dependent quantities as functionals of histories of independent variables, see NOLL [1958]. While the history formalism is useful for many applications, it is not general enough for the foundations, where one still wishes to speak about states, state functions and cyclic processes. Versions of the state-space formalism have been developed by NOLL [1972], PERZYNA & KOSIŃSKI [1973], COLEMAN & OWEN [1974, 1977], GURTIN [1975], ŠILHAVÝ [1980, 1987], KRATOCHVÍL & ŠILHAVÝ [1982], SERRIN [1986], RICOU [1986, 1988] and others. The version used in this book is not identical with any of them, but the essential contents are the same. Also note that these formalism fall within the scheme of the abstract dynamical systems theory. See, e.g., KALMAN, FALB & ARBIB [1969], ZADEH & DESOER [1969] and WILLEMS [1972].

We shall associate with each body a set Σ, the ***state space,*** and another set, Π, the ***class of processes.*** The elements σ of Σ are the ***states*** of the body and the elements

π of Π are the ***processes.*** Each process is a function $\pi : [0, d_\pi] \to \Sigma$, defined on the closed interval $[0, d_\pi]$ with values in Σ. The number $d_\pi > 0$ depends on the process; it is the duration of the process. The value $\pi(t)$ is the state of the body at time $t \in [0, d_\pi]$. The value $\pi^i = \pi(0)$ is the initial state of the process and the value $\pi^f = \pi(d_\pi)$ is the final state of the process. The specification of Σ and Π contains the basic information about the nature of the body. The states will be interpreted globally: they describe the instantaneous situation over the whole body. Concrete examples of pairs (Σ, Π) will be given at the end of Sect. 4.2. The interpretation leads to two natural requirements on Σ and Π :

4.1.1 Definition Let (Σ, Π) be a pair consisting of a set Σ and a set Π of functions $\pi : [0, d_\pi] \to \Sigma$. The elements π of Π are called processes. The pair (Σ, Π) is called a ***system*** if the following two conditions are satisfied:
(1) if $\pi_1, \pi_2 \in \Pi$ are two processes and $\pi_1^f = \pi_2^i$, then Π contains also a process $\pi_1 * \pi_2$, of duration $d_{\pi_1} + d_{\pi_2}$, given by

$$(\pi_1 * \pi_2)(t) = \begin{cases} \pi_1(t) & \text{if } t \in [0, d_{\pi_1}], \\ \pi_2(t - d_{\pi_1}) & \text{if } t \in [d_{\pi_1}, d_{\pi_1} + d_{\pi_2}]; \end{cases} \tag{4.1.1}$$

(2) if $\pi \in \Pi$ and $0 \le t_1 < t_2 \le d_\pi$, then Π contains also a process $\pi_{[t_1, t_2]}$, of duration $t_2 - t_1$, given by $\pi_{[t_1,t_2]}(t) = \pi(t + t_1)$, $t \in [0, t_2 - t_1]$.

The process $\pi_1 * \pi_2$ is the ***continuation of*** π_1 ***with*** π_2 and $\pi_{[t_1,t_2]}$ is the $[t_1, t_2]$-***segment*** of π, see Fig. 4.1.1.

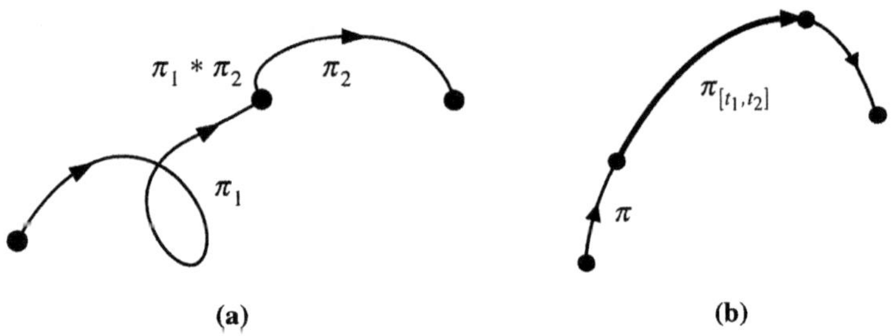

Fig. 4.1.1. Continuation and segment

4.1.2 Definitions Let (Σ, Π) be a the system.
(1) A process $\pi \in \Pi$ is said to be ***cyclic*** or to be a ***cycle*** if $\pi^i = \pi^f$.
(2) A process is said to ***start from (terminate in)*** $\sigma \in \Sigma$ if $\pi^i = \sigma$ ($\pi^f = \sigma$).
(3) A state $\sigma \in \Sigma$ is said to be ***stationary*** if for each $d > 0$ there exists a process $\pi \in \Pi$ of duration d such that $\pi(t) = \sigma$ for every $t \in [0, d_\pi]$.
(4) A system (Σ, Π) is said to satisfy the ***accessibility axiom*** if for every two states $\sigma_1, \sigma_2 \in \Sigma$ there exists a process $\pi \in \Pi$ such that $\sigma_1 = \pi^i$, $\sigma_2 = \pi^f$.

We shall also use the terms 'static state' and 'rest state' for the stationary states. The term 'equilibrium state' is reserved for the stationary states with a uniform temperature over the body, cf. Sect. 13.2 and 14.1. The state is stationary if the system can stay in it for arbitrarily long time, provided the external conditions are properly adjusted. The accessibility is important for the construction of the thermodynamic potentials like the energy or entropy. It is a sort of connectedness of the state space,

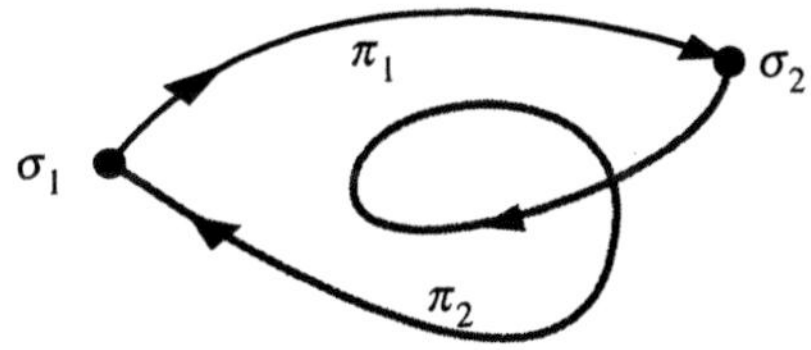

Fig. 4.1.2. Accessibility

the 'connecting' meaning 'connecting by means of real processes,' see Fig. 4.1.2. Practically, this means that whatever process the body undergoes, it can always return back to the initial state. This has nothing to do with reversibility, since the way back can be completely different from the way in the original direction. Of course, appropriate forces and sources of heat must be applied to realize the process. In principle, real materials can always be brought back to their virgin state by means of melting the body, etc. For this reason, I do not consider the accessibility axiom to be any serious restriction from the physical point of view. Another question is whether the mathematical models designed to describe only certain aspects of the behavior of the body satisfy the accessibility axiom. It turns out that elastic, viscous, and ideally plastic materials possess accessibility, at least locally (i.e., at a given material point). Topological complications arise with the global version formulated above. For bodies of certain shape there can be two configurations which cannot be connected by a continuous family of positively oriented globally invertible configurations. Speaking again about the local accessibility, materials with fading memory have only approximate accessibility and materials with internal state variables and plastic materials with hardening do not have even approximate accessibility. The significance of the accessibility for the constitutive theory was first fully recognized by NOLL [1972] and for the foundations of thermodynamics by COLEMAN & OWEN [1974]. The latter reference also examines the approximate accessibility for various types of models of material behavior. Note also that the accessibility condition can be used to generate a rich supply of cyclic processes as Fig. 4.1.2 shows.

4.2 Local State Functions; Material Bodies

4.2.1 Definition Let (Σ, Π) be a system, $\Omega \subset$ Eucl an admissible region and M a set. Any function $f : \Sigma \times \Omega \to M$, associating with any pair $(\sigma, \boldsymbol{p}) \in \Sigma \times \Omega$ an element $f(\sigma, \boldsymbol{p})$ of M, is called a ***local state function*** on (Σ, Π, Ω) with values in M. If M is a finite dimensional vector space, a state function f is said to be of class C^r, $(r = 0, 1, \dots, \infty)$ if for any $\pi \in \Pi$ the function f_π given by (4.2.1) below, has an extension of class C^r to an open set containing the closure of $\Omega \times [0, d_\pi]$.

In accord with the global interpretation of states, each state $\sigma \in \Sigma$ determines a field $f(\sigma, \cdot)$ of the quantity f over the body; $f(\sigma, \boldsymbol{p})$ is its value at $\boldsymbol{p}$ when the body is in the state σ. Each process $\pi \in \Pi$ determines the ***temporal evolution*** f_π of f, defined on $\Omega \times [0, d_\pi]$ by

$$f_\pi(\boldsymbol{p}, t) = f(\pi(t), \boldsymbol{p}), \quad (\boldsymbol{p}, t) \in \Omega \times [0, d_\pi], \tag{4.2.1}$$

see Fig. 4.2.1. When there is no danger of confusion, we write f for f_π. Equation (4.2.1) is a general constitutive equation. We shall use the terms "local state function," "state function," "constitutive function," "constitutive functional" for f interchangeably.

4.2.2 Proposition *Let f be a local state function of class C^r, $(r = 0, 1, \dots, \infty)$ on (Σ, Π, Ω) and let D be any (possibly mixed) spacetime derivative with respect to $\boldsymbol{p}$ and t of order at most r (e.g., if $r = 2$, D can be $\partial_p, \partial_t, \partial_p\partial_t$, etc.). Then there exists*

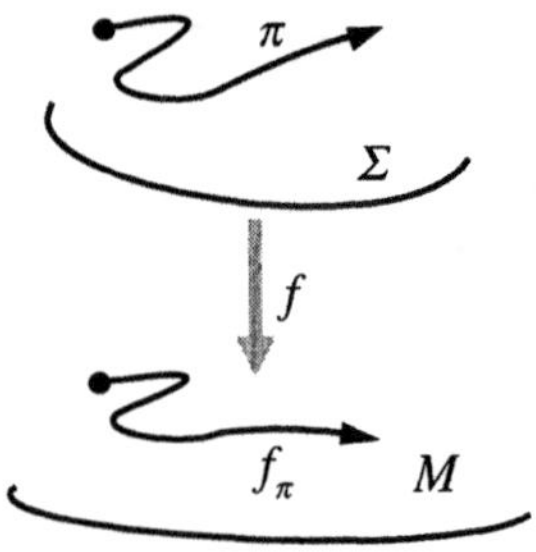

Fig. 4.2.1. A local state function

a local state function f_D on (Σ, Π, Ω) such that for every process $\pi \in \Pi$ and every $(\boldsymbol{p}, t) \in \Omega \times [0, d_\pi]$ one has

$$Df_\pi(\boldsymbol{p}, t) = f_D(\pi(t), \boldsymbol{p}).$$

In particular, if $r \geq 1$, then there exist local state functions f_G and f_T such that

$$\nabla f_\pi(\boldsymbol{p}, t) = f_G(\pi(t), \boldsymbol{p}), \quad \dot{f}_\pi(\boldsymbol{p}, t) = f_T(\pi(t), \boldsymbol{p}).$$

I refer to Šilhavý [1987, Proposition 8.1] for the proof. The proposition asserts that not only the values of the state function in question, but also its spacetime derivatives up to order r are completely determined by the state. This might seem somewhat paradoxical in the case of time derivatives, since their determination seemingly requires a knowledge of the evolution of the state in a small interval of time near t. The resolution is that the validity of the proposition depends on the assumption that the derivatives of f_π are continuous. When the derivatives are piecewise continuous only, the conclusion of the proposition is false. We shall see soon in examples that in order to satisfy the requirement that the derivatives be continuous at the point where one process is followed by another, it is necessary to include the time derivatives of the state parameters in the definition of state. See 4.2.5 below.

The concepts of systems and state functions are now illustrated on examples.

4.2.3 Classical Systems I If the system has n degrees of freedom and $\boldsymbol{x} = (x_1, \ldots, x_n)$ denote the external parameters from an open subset O of $\mathbb{R}^n$, then one may set

$$\Sigma = O, \quad \Pi = C^0(O), \tag{4.2.2}$$

where, in this section only, $C^0(O)$ denotes the set of all continuous functions $\pi : [0, d_\pi] \to O$, and it is understood that $d_\pi > 0$ may be different for different elements of $C^0(O)$. Hence, states σ are identified with the elements $\boldsymbol{x}$ of O. (For homogeneous elastic fluids $n = 2$ and states can be identified with the pairs $\sigma = (v, \theta)$ where $v > 0$ is the specific volume and $\theta > 0$ the absolute temperature; $O = (0, \infty) \times (0, \infty)$.) The pair (Σ, Π) defined by (4.2.2) is a system, as is easily verified. If Ω is a reference configuration, then the most general state function f on (Σ, Π, Ω) is any function of the two variables $(\sigma, \boldsymbol{p}), \boldsymbol{p} \in \Omega$. Reasonable state functions shall associate homogeneous fields with homogeneous states; such will be the case if f is independent of $\boldsymbol{p}$ and we shall consider only such state functions, i.e., $f : \Sigma \to \mathbb{R}$. If f is continuous, then f, viewed as a local state function, is of class C^0 in the sense of Definition 4.2.1. If f is nonconstant, then the state function f can *never* be of class C^r with $r \geq 1$ no matter how smooth f, because the processes π need not be smooth. At first glance a solution to this problem might seem to be in restricting the class of processes, e.g., one could require that the processes be k times continuously differentiable, i.e., set

$$\Sigma = O, \quad \Pi = C^k(O), \quad k \geq 1, \tag{4.2.3}$$

where for any $k = 1, \ldots, \infty$, $C^k(O)$ denotes the set of all functions $\boldsymbol{x} : [0, d_{\boldsymbol{x}}] \to O$, such that $\boldsymbol{x}$ is k times continuously differentiable on $(0, d_{\boldsymbol{x}})$ and the derivatives of order $\leq k$ have limits at the endpoints of $[0, d_{\boldsymbol{x}}]$. However, (Σ, Π) is not a system, because it violates the continuation axiom 4.1.1(1): if $\pi_1, \pi_2 \in \Pi$ are two continuously differentiable processes such that $\pi_1^f = \pi_2^i$, then $\pi_1 * \pi_2$ need not be continuously differentiable.

4.2.4 Classical Systems II Another possibility is to set

$$\Sigma = O, \quad \Pi = PC^1(O), \tag{4.2.4}$$

where $PC^1(O)$ denotes the set of all continuous and piecewise continuously differentiable functions $\pi : [0, d_\pi] \to O$. This gives a system (Σ, Π). Any continuous function $f : O \to \mathbb{R}$ gives rise to a class C^0 local state function. If f is continuously differentiable, then generally the corresponding local state function is still not of class C^1, since the jumps in the derivatives of the processes π induce jumps in the derivatives of f_π. Nevertheless, the choices (4.2.3) seem to be the easiest possibility to ensure that (Σ, Π) be a system and the classical derivatives of the processes be defined for almost every time. Later we shall employ this system to illustrate the construction of the state functions in thermodynamics. The way to arrange things so as to have class C^1 state functions is described in the following example.

4.2.5 Classical Systems III The idea is to include the time derivatives of the external parameters $\boldsymbol{x} = (x_1, \ldots, x_n)$ in the definition of state. If $\boldsymbol{x} \in C^1(O)$ and if these time derivatives are denoted by $\boldsymbol{a} = (a_1, \ldots, a_n)$, then they must satisfy the compatibility equations

$$\boldsymbol{a}(t) = \dot{\boldsymbol{x}}(t), \quad t \in [0, d_\pi] \tag{4.2.5}$$

with the superimposed dot denoting the ordinary derivative if t is an interior point of $[0, d_\pi]$ and the limits of the derivatives if t is any of the endpoints of $[0, d_\pi]$. If

$$\Sigma = O \times \mathbb{R}^n, \quad \Pi = \{\pi = (\boldsymbol{x}, \boldsymbol{a}) : \boldsymbol{x} \in C^1(O), \boldsymbol{a} \in C^0(\mathbb{R}^n) \text{ and (4.2.5) holds}\},$$

then (Σ, Π) is a system: the inclusion of the time derivatives in the state variables ensures that if $\pi_1 = (\boldsymbol{x}_1, \boldsymbol{a}_1)$, $\pi_2 = (\boldsymbol{x}_2, \boldsymbol{a}_2) \in \Pi$ are two processes such that $\pi_1^f = \pi_2^i$, then the path $\boldsymbol{x}_1 * \boldsymbol{x}_1$ is continuously differentiable even at the point where the processes are concatenated. Here we see the first example from the many where the class Π is far from the set of all (smooth) curves in the state space. Any homogeneous local state function is a function on Σ. If f is continuous on Σ, then f, considered as a local state function, is of class C^0 in the sense of Definition 4.2.1. If f depends only on $\boldsymbol{x}$ and f is continuously differentiable, then f, viewed as a local state function, is of class C^1. For a process $\pi = (\boldsymbol{x}, \boldsymbol{a}) \in \Pi$ the derivative $\dot{f}_\pi(t)$ is given by

$$\dot{f}_\pi(t) = \partial_{\boldsymbol{x}} f(\boldsymbol{x}(t)) \cdot \dot{\boldsymbol{x}}(t) = \partial_{\boldsymbol{x}} f(\boldsymbol{x}(t)) \cdot \boldsymbol{a}(t);$$

we see that the derivative $\dot{f}_\pi(t)$ is completely determined by the instantaneous state. This illustrates Proposition 4.2.2. In fact this proposition says that in order to have nontrivial state functions of class C^1, the derivatives of the external parameters *must* be included among the state parameters either directly or by way of some equivalent information.

Since the general state functions $f(\sigma) = f(\boldsymbol{x}, \boldsymbol{a})$ depend on the time derivatives $\dot{\boldsymbol{x}}$, such systems can be also used to model the memory of the differential type to describe the phenomena of viscosity and friction.

4.2.6 Systems with Internal Variables The memory is represented by a set $\boldsymbol{\alpha} = (\alpha_1, \ldots, \alpha_m)$ of scalar internal state variables. The state space Σ is

$$\Sigma = O \times \mathbb{R}^m,$$

where O is an open subset of $\mathbb{R}^n$, the range of external parameters. The evolution law for the internal parameters is the ordinary differential equation of the form

$$\dot{\boldsymbol{\alpha}}(t) = \boldsymbol{f}(\boldsymbol{x}, \boldsymbol{\alpha}), \tag{4.2.6}$$

where $\boldsymbol{f} : \Sigma \to \mathbb{R}^m$ is a given continuous function. The class of processes is taken to be

$$\Pi = \{\pi = (\boldsymbol{x}, \boldsymbol{\alpha}) : \boldsymbol{x} \in PC^1(O), \boldsymbol{\alpha} \in C^1(\mathbb{R}^m) \text{ and (4.2.6) holds}\}.$$

Again (Σ, Π) is a system. Π is not the set of all possible curves in the state space as we have the evolution law (4.2.6). Homogeneous state functions are functions defined on Σ. A state $\sigma \in \Sigma$ is stationary if and only if $\boldsymbol{f}(\sigma) = \boldsymbol{0}$. Standard theorems on ordinary differential equations (see, e.g., Coddington & Levinson [1955]) may be invoked to ensure that (4.2.6) is locally solvable and hence that Π is nonempty. Additional hypotheses can be imposed on $\boldsymbol{f}$ to ensure the global solvability. Systems of this class are frequently used to model rate-dependent hysteresis phenomena; e.g., rate-dependent plasticity.

For the purpose of Part II we lay down the following definition.

4.2.7 Definition A ***material body*** is a collection of 9 objects

$$(\Omega, \Sigma, \Pi, \hat{\chi}, \hat{\theta}, \hat{S}, \hat{b}, \hat{q}, \hat{r}) \tag{4.2.7}$$

of the following nature:

(1) Ω is an admissible region;

(2) (Σ, Π) is a system;

(3) $\hat{\chi}, \hat{\theta}, \hat{S}, \hat{b}, \hat{q}, \hat{r}$ are local state functions of class C^∞ on (Σ, Π, Ω), with values in the sets to be determined by the interpretation given below.

Here Ω is the reference configuration of the body, Σ its state space and Π its class of processes. The local state functions deliver the following quantities:

$\hat{\chi}$ configuration;
$\hat{\theta}$ temperature;
$\hat{S}$ referential stress;
$\hat{b}$ referential volume density of the body force;
$\hat{q}$ referential heat flux vector;
$\hat{r}$ referential density of the rate of heat supply.

The assumption that the local state functions are of class C^∞ is made for simplicity and for Part II only. Recall that the processes π correspond to appropriately chosen external forces and external sources of heat; these are assumed to be at the experimentalist's disposal. This assumption is in no contradiction with the necessity of considering weak solutions when dealing with the processes corresponding to *prescribed* external forces and sources of heat. For example, if the model of viscous bodies with heat conduction of Part III has the *response* functions of class C^∞, then the class of C^∞ processes π is extremely large: corresponding to arbitrarily chosen motion and temperature field (χ, θ) of class C^∞ there exists a π such that $\hat{\chi}_\pi = \chi, \hat{\theta}_\pi = \theta$. The sources $\hat{b}_\pi, \hat{r}_\pi$ are calculated from the equations of balance of momentum and energy; they are of class C^∞ as a consequence of the smoothness of the response functions.

The six quantities in (3) are the primitive for Part II; actually just their combinations *work*, *heat*, and the *heating measure*, to be defined below, would suffice. A slight deviation from the notation of Part I is that here θ denotes the *empirical* temperature (to be explained in Sect. 7.1) and $\hat{b}, \hat{r}$ denote the *reference volume densities* of the body force and the rate of heat supply, while in the rest of the book they denote the specific values of these quantities, differing just by a constant factor – the reference density of mass. We will abbreviate and use the term 'body' for 'material body' and the letter Ω for the collection (4.2.7). The time evolutions $\hat{\chi}_\pi, \hat{\theta}_\pi, \hat{S}_\pi, \hat{b}_\pi, \hat{q}_\pi, \hat{r}_\pi$, determined for a given process π by the state functions in (3), will be denoted χ, θ, S, b, q, r.

5 The First Law of Thermodynamics

Section 5.1 gives two statements of the first law. On our way from them to the energy and its balance equation we shall encounter Joule's relation of proportionality of work and heat in cyclic processes. Only the total energy occurs in this chapter; its splitting into the invariant internal energy and the kinetic energy is based on the invariance of the theory. This will be discussed in Chap. 6.

5.1 Work and Heat

For a process $\pi \in \Pi$ and a part $P \subset \Omega$ the ***work*** $w(\pi, P)$ ***done by*** P ***in*** π is

$$w(\pi, P) = -\int_0^d \Big(\int_{\partial P} \boldsymbol{v} \cdot \boldsymbol{Sn}\, dA + \int_P \boldsymbol{v} \cdot \boldsymbol{b}\, dV\Big)\, dt,$$

and the ***net heat*** $q(\pi, P)$ ***gained by*** P ***in*** π is

$$q(\pi, P) = \int_0^d \Big(-\int_{\partial P} \boldsymbol{q} \cdot \boldsymbol{n}\, dA + \int_P r\, dV\Big)\, dt.$$

These quantities behave additively in two ways: **(1)** for two processes that take place simultaneously in two different bodies (or parts of the same body) and **(2)** for one process followed by another in the same body. These properties will be used extensively in the proof of Joule's relation and in the construction of the energy function. We now give two not entirely equivalent versions of the first law of thermodynamics. The following two sections establish their relationship and consequences. Assume that a body Ω is given.

5.1.1 The First Law, Version I *For every cyclic process* π *and every part* P,

$$q(\pi, P) = 0 \Rightarrow w(\pi, P) = 0.$$

That is, whenever the net heat gained by the part P in a cycle vanishes, so also does the work done in that cycle.

5.1.2 The First Law, Version II *For every cyclic process* π *and every part* P,

$$w(\pi, P) > 0 \Leftrightarrow q(\pi, P) > 0.$$

A positive amount of work can be obtained from a cyclically operating part P if and only if a positive net heat is delivered to that part.

5.2 Joule's Relation

The first step in establishing the consequences of the first law (in either of the versions given in the preceding section) is the universal proportionality of work and heat in cyclic processes.

5.2.1 Definition A body Ω is said to satisfy ***Joule's relation*** with the coefficient of proportionality $J \in \mathbb{R}$ if for every cyclic process π and every part P,

$$w(\pi, P) = Jq(\pi, P). \tag{5.2.1}$$

The constant J is Joule's constant or the mechanical equivalent of a unit of heat and (5.2.1) is the interconvertibility of work and heat.

It is convenient to represent a process π by the work–heat pair $(w(\pi, P), q(\pi, P))$. If Ω is a body, $\mathbb{P}(\Omega)$ denotes the set of all its work–heat pairs of cyclic processes; the pair $(w, q) \in \mathbb{R}^2$ belongs to $\mathbb{P}(\Omega)$ if and only if there exists a cyclic process π of Ω and a part P of Ω such that $(w, q) = (w(\pi, P), q(\pi, P))$. If C is a collection (i.e., a set) of bodies, then $\mathbb{P}(C)$ denotes the union of all $\mathbb{P}(\Omega)$, $\Omega \in C$. The elements $(w, q) \in \mathbb{P}(C)$ are called the ***cyclic work–heat pairs*** of C.

5.2.2 Assumptions Let C be a collection of bodies.

(1) $\mathbb{P}(C)$ is additive, i.e., $(w_1, q_1), (w_2, q_2) \in \mathbb{P}(C) \Rightarrow (w_1, q_1) + (w_2, q_2) \in \mathbb{P}(C)$.
(2) $\mathbb{P}(C)$ contains elements (w^+, q^+) and (w^-, q^-) such that $q^+ > 0$ and $q^- < 0$.
(3) $\mathbb{P}(C)$ contains elements (w^+, q^+) and (w^-, q^-) such that $q^+ > 0, q^- < 0$ and $\alpha(w^+, q^+), \alpha(w^-, q^-) \in \mathbb{P}(C)$ for every $\alpha \in (0, 1]$.

If two cyclic processes, represented by (w_1, q_1) and (w_2, q_2), take place simultaneously on two different bodies Ω_1, Ω_2, then they constitute a cyclic process of the union $\Omega_1 \oplus \Omega_2$ of these two bodies and this process is represented by $(w_1, q_1)+(w_2, q_2)$. Without giving a formal definition of $\Omega_1 \oplus \Omega_2$, we may state Condition (1) as

$$\Omega_1, \Omega_2 \in C \Rightarrow \Omega_1 \oplus \Omega_2 \in C,$$

and this in turn means that $\Omega_1 \oplus \Omega_2$ obeys the same general laws as Ω_1 and Ω_2 do. The consistency of this interpretation requires that the two bodies do not have common material points and that the two processes have the same duration. There are many ways to satisfy these requirements. For instance, if each material body in C has a copy made of distinct material points but otherwise of identical properties, then the first requirement may be satisfied by passing to a suitable copy of one of the two bodies. Similarly, if one of the processes is rate independent in the sense that it can be executed with higher or lower speed without changing the work–heat pair then the second requirement may be satisfied by passing to a faster/slower process, etc. Assumption (2) expresses the existence of cyclic processes with positive and negative net gains of heat; Assumption (3) says, roughly, that these processes may be chosen to be homogeneous: Indeed, if (w^+, q^+) corresponds to a homogeneous cyclic process of some body Ω from C, then the pair $\alpha(w^+, q^+)$, where $0 < \alpha \le 1$, corresponds to the same process on the part $P \subset \Omega$ of referential volume $\mathsf{V}(P) = \alpha \mathsf{V}(\Omega)$.

5.2.3 Proposition *Let C satisfy Assumptions 5.2.2(1),(3). Then every body from C satisfies Version I of the first law if and only if every body from C satisfies Joule's relation with a universal coefficient of proportionality $J \in \mathbb{R}$.*

The universality means that the coefficient J is the same for all bodies from C. It is not asserted that J is positive.

Proof Only the implication that Version I of the first law implies the existence of J needs a proof. Version I says that the set $\mathbb{P}(C)$ does not intersect the "forbidden lines" F^+ and F^- given by $F^{\pm} = \{(w, q) \in \mathbb{R}^2 : q = 0, \pm w > 0\}$, see Fig. 5.2.1(a), while the Joule relation with a universal constant $J \in \mathbb{R}$ says that $\mathbb{P}(C)$ is contained in some nonhorizontal straight line through the origin in $\mathbb{R}$. Let us give a geometrical proof. Let (w^+, q^+) and (w^-, q^-) be the elements of $\mathbb{P}(C)$ from Assumption 5.2.2(3). Denote by $H^\pm$ the open half-lines starting at (0, 0) and going through $(w^\pm, q^\pm)$ respectively. It is asserted that both H^+ and H^- belong to $\mathbb{P}(C)$. Indeed the open segment between (0, 0) and (w^+, q^+) belongs to $\mathbb{P}(C)$ by Assumption 5.2.2(3). Since (w^+, q^+) belongs to $\mathbb{P}(C)$ too, we deduce from the additivity of $\mathbb{P}(C)$ (Assumption 5.2.2(1)), that also the segment between (w^+, q^+) and $2(w^+, q^+)$ belongs to $\mathbb{P}(C)$, and so on. Next it is claimed that the half-lines $H^\pm$ have the same slope. Indeed, if this were not to be true, $H^\pm$ would make a nonstraight angle, and as with every two points in $H^\pm \subset \mathbb{P}(C)$ their sum also belongs to $\mathbb{P}(C)$, it would follow that all points inside the angle belong to $\mathbb{P}(C)$, and this angle would necessarily intersect one of the two forbidden lines (see Fig. 5.2.1(b)), thus violating Version I of the first law. Hence the set $H^+ \cup H^- \cup \{(0,0)\}$ is a straight line $L := \{(w, q) : w = Jq\}$, for some $J \in \mathbb{R}$. Finally it is claimed that $\mathbb{P}(C) \subset L$. Suppose in contradiction that $(w, q) \notin L$ for some point of $\mathbb{P}(C)$. The additivity of $\mathbb{P}(C)$ implies that the shifted line $(w, q)+L$ belongs to $\mathbb{P}(C)$, but if (w, q) is not in L, then the shifted line necessarily intersects one of the forbidden lines F^+, F^-, see Fig. 5.2.2(a). □

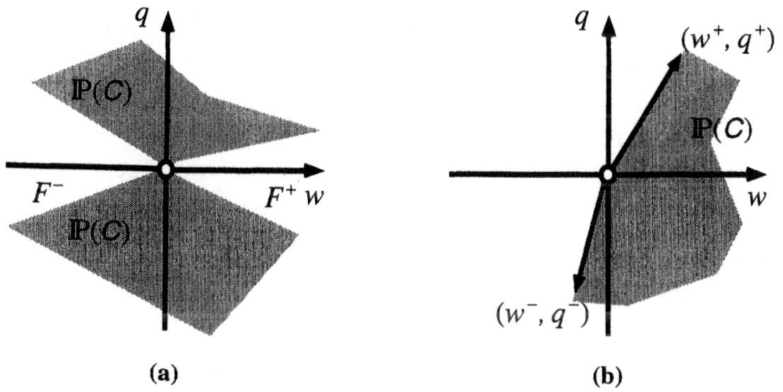

Fig. 5.2.1

5.2.4 Proposition *Let C satisfy Assumptions 5.2.2(1),(2). Then every body from C satisfies Version II of the first law if and only if every body from C satisfies Joule's relation with a positive universal coefficient of proportionality $J > 0$.*

Proof Version II of the first law says that the set $\mathbb{P}(C)$ does not intersect the "forbidden quadrants" $Q_2 := \{(w, q) \in \mathbb{R}^2 : w < 0, q > 0\}$, $Q_4 := \{(w, q) \in \mathbb{R}^2 : w > 0, q < 0\}$, while the existence of $J > 0$ says that the set $\mathbb{P}(C)$ is contained in some straight line through the origin and lying in the first and the third open quadrants. The geometrical idea behind the proof is similar to that in the proof of the preceding proposition, see Fig. 5.2.2(b). Let (w^+, q^+) and (w^-, q^-) be the elements of $\mathbb{P}(C)$ from Assumption 5.2.2(2). It is asserted that the points (w^+, q^+) and (w^-, q^-) belong to a straight line through the origin. Indeed if this were to be false, the points (w^+, q^+) and (w^-, q^-) would make a nonstraight angle, and as with every two points in $\mathbb{P}(C)$ their sum also belongs to $\mathbb{P}(C)$, it would follow by induction that all lattice points inside the angle, i.e., the set $M = \{n_1(w^+, q^+) + n_2(w^-, q^-) : n_1 \ge 0, n_2 \ge 0, (n_1, n_2) \ne 0, n_1, n_2 \text{ integers}\}$ would belong to $\mathbb{P}(C)$. But the set M would necessarily intersect one of the two forbidden quadrants, thus violating Version II of the first law. Hence the points (w^+, q^+) and (w^-, q^-) belong to some straight line $L := \{(w, q) : w = Jq\}$, which gives us the value of the constant J. Finally, the proof is completed by asserting that for every $(w, q) \in \mathbb{P}(C)$ we have $(w, q) \in L$. But this is proved by the argument with the lattice points as above, now applied to (w, q) and (w^+, q^+) if $q < 0$ and to (w, q) and (w^-, q^-) if $q \ge 0$. □

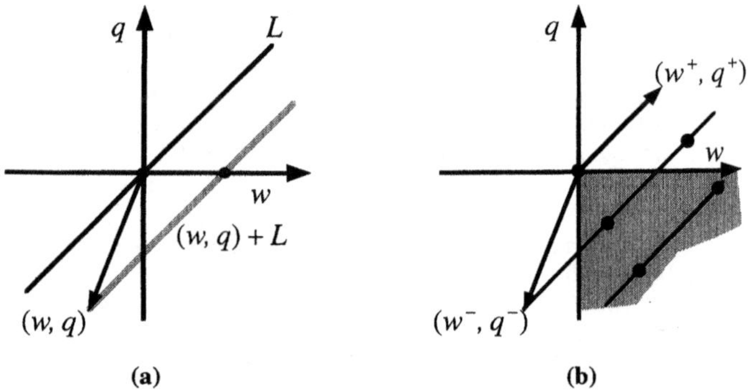

Fig. 5.2.2

5.3 Energy. The Equation of Balance of Energy

For cyclic processes the work done and the heat gained are equal provided suitable units are used. For noncyclic processes the latter two quantities may differ but it is possible to introduce a state function called the energy such that its changes balance the differences between the heat gained and work done.

5.3.1 Proposition *Let Ω be a body satisfying the accessibility axiom. Then Ω satisfies Joule's relation with $J = 1$ if and only if there exists a smooth scalar state function u on (Σ, Π, Ω) such that*

$$\dot{u} = \mathrm{Div}(\boldsymbol{S}^{\mathsf{T}}\boldsymbol{v} - \boldsymbol{q}) + \boldsymbol{b} \cdot \boldsymbol{v} + r \tag{5.3.1}$$

for every process π. The function u, if it exists, is unique to within a constant depending on $\boldsymbol{p}$, i.e., if u_1, u_2 are two functions such that each of them satisfies (5.3.1), *then there exists a smooth function $c : \Omega \to \mathbb{R}$ such that*

$$u_2(\sigma, \boldsymbol{p}) = u_1(\sigma, \boldsymbol{p}) + c(\boldsymbol{p}) \tag{5.3.2}$$

for every $(\sigma, \boldsymbol{p}) \in \Sigma \times \Omega$.

In accord with our convention, in (5.3.1) the letter u denotes the evolution u_π of u in the process π. Any scalar state function u satisfying (5.3.1) is called the ***total energy*** for the body Ω, and when the body is in the state σ, $u(\sigma, \boldsymbol{p})$ is the reference volume density of the total energy at $\boldsymbol{p} \in \Omega$. In the next chapter the total energy will be shown to be the sum of the internal and kinetic energies with different transformation laws. For any part P,

$$\mathsf{U}(\sigma, P) := \int_P u(\sigma, \cdot)\, dV$$

is the total energy of P in the state σ. The global version of (5.3.1) is

$$\mathsf{U}(\pi^f, P) - \mathsf{U}(\pi^i, P) = q(\pi, P) - w(\pi, P). \tag{5.3.3}$$

The energy will be constructed by a generalization of the familiar path-independence argument from the theory of potential vector fields (to which it reduces in the case of reversible systems, cf. Example 5.3.2 below). The general proof has nothing to do

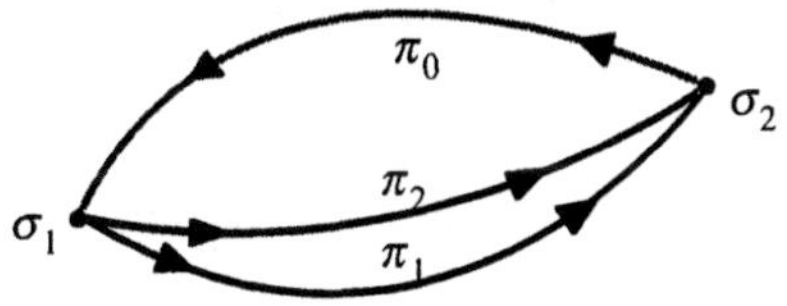

Fig. 5.3.1. Path independence

with the reversibility. Arguments like the one given below are more or less explicitly described in the treatments of classical thermodynamics, see, e.g. FERMI [1937].

Proof We set $g(\pi, \boldsymbol{p}) = \int_0^d (\operatorname{Div}(\boldsymbol{S}^{\mathsf{T}}\boldsymbol{v} - \boldsymbol{q}) + \boldsymbol{b} \cdot \boldsymbol{v} + r)\, dt$ for every process π and every $\boldsymbol{p} \in \Omega$, so that $\int_P g(\pi, \cdot)\, dV = q(\pi, P) - w(\pi, P)$ for every part P. Joule's relation with $J = 1$ is $\int_P g(\pi, \cdot)\, dV = 0$ for every cyclic process and every part P; the arbitrariness of P and the smoothness assumptions give that

$$g(\pi, \boldsymbol{p}) = 0 \tag{5.3.4}$$

for every cyclic process and every $\boldsymbol{p} \in \Omega$. We shall now prove the following auxiliary assertion: *if π_1 and π_2 are two processes such that $\pi_1^i = \pi_2^i$ and $\pi_1^f = \pi_2^f$, then*

$$g(\pi_1, \boldsymbol{p}) = g(\pi_2, \boldsymbol{p}) \tag{5.3.5}$$

for each $\boldsymbol{p} \in \Omega$. Indeed, by the accessibility axiom there exists a process π_0 starting at $\pi_1^f = \pi_2^f$ and terminating at $\pi_1^i = \pi_2^i$; we then construct the combined processes $\pi_1 * \pi_0$, $\pi_2 * \pi_0$. As they are cyclic, (5.3.4) gives $g(\pi_1 * \pi_0, \boldsymbol{p}) = g(\pi_2 * \pi_0, \boldsymbol{p}) = 0$. The additivity of g on combined processes yields $g(\pi_1, \boldsymbol{p}) + g(\pi_0, \boldsymbol{p}) = g(\pi_2, \boldsymbol{p}) + g(\pi_0, \boldsymbol{p}) = 0$ and (5.3.5) follows. We now choose an arbitrary reference state σ_0 and define u by

$$u(\sigma, \boldsymbol{p}) = g(\pi, \boldsymbol{p}), \tag{5.3.6}$$

for every $(\sigma, \boldsymbol{p}) \in \Sigma \times \Omega$, where π is any process starting at σ_0 and terminating at σ. The auxiliary assertion tells us that the right-hand side of (5.3.6) is independent of the choice of π and hence the definition is unambiguous. It will be now proved that u satisfies

$$u(\pi^f, \boldsymbol{p}) - u(\pi^i, \boldsymbol{p}) = g(\pi, \boldsymbol{p}) \tag{5.3.7}$$

for every process π and every $\boldsymbol{p} \in \Omega$. The accessibility provides a process π_0 starting at σ_0 and terminating at π^i. The process $\pi_0 * \pi$ then starts at σ_0 and terminates at π^f. The definition of u gives $u(\pi^i, \boldsymbol{p}) = g(\pi_0, \boldsymbol{p})$, $u(\pi^f, \boldsymbol{p}) = g(\pi_0 * \pi, \boldsymbol{p}) = g(\pi_0, \boldsymbol{p}) + g(\pi, \boldsymbol{p})$, and (5.3.7) follows. The application (5.3.7) to the segment $\pi_{[0,t]}$ of the process π (Property (2) in Definition 4.1.1) yields

$$u(\boldsymbol{p}, t) \equiv u(\pi(t), \boldsymbol{p}) = \int_0^t (\operatorname{Div}(\boldsymbol{S}^{\mathsf{T}}\boldsymbol{v} - \boldsymbol{q}) + \boldsymbol{b} \cdot \boldsymbol{v} + r)\, ds + u(\pi^i, \boldsymbol{p})$$

where the integrand is considered as a function of the time-like variable s. From this relation one learns that u is smooth and the differentiation with respect to t gives (5.3.1). This proves that Joule's relation implies the equation of balance of energy. If conversely the equation of balance of energy holds, then for cyclic process the left-hand side of (5.3.3) vanishes and Joule's relation follows. Finally, to prove the uniqueness assertion, let u_z, $z = 1, 2$, be two energy functions for Ω and set $\mathsf{U}_z(\sigma, P) = \int_P u_z(\sigma, \cdot)\, dV$. The global equations of balance of energy for the two energy functions are

$$\mathsf{U}_z(\pi^f, P) - \mathsf{U}_z(\pi^i, P) = q(\pi, P) - w(\pi, P), \quad z = 1, 2. \tag{5.3.8}$$

If σ_0, σ are any two states, the accessibility axiom provides a process π starting at σ_0 and terminating at σ. The equations of balance of energy (5.3.8) then lead to the equality $\mathsf{U}_2(\sigma, P) - \mathsf{U}_2(\sigma_0, P) = \mathsf{U}_1(\sigma, P) - \mathsf{U}_1(\sigma_0, P)$ a local version of which reads $u_2(\sigma, \boldsymbol{p}) - u_2(\sigma_0, \boldsymbol{p}) = u_1(\sigma, \boldsymbol{p}) - u_1(\sigma_0, \boldsymbol{p})$ and so (5.3.2) holds with $c(\boldsymbol{p}) = u_2(\sigma_0, \boldsymbol{p}) - u_1(\sigma_0, \boldsymbol{p})$. □

5.3.2 Example Consider the system (Σ, Π) of classical thermodynamics with n independent external parameters defined in Subsect. 4.2.4. Let for every process $\pi \in \Pi$ the work done and the heat gained be defined by path integrals

$$w(\pi) = \int_\pi w, \quad q(\pi) = \int_\pi q, \quad \text{where } w = \sum_{i=1}^n w_i dx_i, \quad q = \sum_{i=1}^n q_i dx_i$$

and the differential forms w, q are continuous on Σ. The system (Σ, Π) satisfies the accessibility axiom if and only if Σ is connected. Joule's relation asserts that $\int_\pi (q-w) = 0$ for every closed path π, i.e., for every cyclic process. This means that $\int_\pi (q-w)$ is path-independent. Proposition 5.3.1 or the classical result in the theory of path-integrals provide a globally defined, continuously differentiable function U on Σ such that

$$\mathsf{U}(\pi^f) - \mathsf{U}(\pi^i) = \int_\pi (q - w) \tag{5.3.9}$$

for every process; this is equivalent to

$$\frac{\partial \mathsf{U}}{\partial x_i} = q_i - w_i, \quad i = 1, \ldots, n. \tag{5.3.10}$$

The function U is unique to within an additive constant. If Σ has several connected components, then in the present special context Joule's relation still gives the existence of a globally defined function U satisfying both (5.3.9) and (5.3.10). The uniqueness assertion in this case reads: Any two functions satisfying (5.3.9) (or (5.3.10)) differ by a function which is constant on every connected component of Σ. If the coefficients q, w are continuously differentiable, then the interchangeability of the second partial derivatives of U give the cross relations

$$\frac{\partial(q_j - w_j)}{\partial x_i} = \frac{\partial(q_i - w_i)}{\partial x_j}, \quad i, j = 1, \ldots, n. \tag{5.3.11}$$

If Σ is simply connected, then conversely (5.3.11) leads to the existence of the potential U and hence to Joule's relation.

Bibliographical Note

The idea that the axioms of thermodynamics need not directly postulate the proportionality of work and heat in cyclic processes appears in Truesdell & Bharatha [1977] who derived Joule's relation for two-variable *reversible* systems from a set of axioms close in spirit to Carnot's ideas. Their axioms do not involve the first laws used here. The above treatment follows Šilhavý [1980, 1982]. The geometrical proofs close to those given above are presented in Šilhavý [1982]. Pitteri [1982] generalizes Truesdell & Bharatha [1977] to many-variable reversible systems.

A related development is due to Ricou [1986, 1988]. His formulation applies to noncyclic processes and its distinctive feature is that it gives an energy function that is bounded from below, a property that is satisfied in all sound physical theories. An essential point in Ricou's papers is that his systems need not satisfy the accessibility requirement. His formulation gives only the energy *inequality*, which first appears, without motivation, in Fosdick & Serrin [1975], who pointed out that the essential features of the structure of continuum thermodynamics are preserved when the equation of balance of energy is replaced by the energy inequality. Šilhavý [1980] showed that the energy inequality follows from the implication of the type of the first law used above. This point of view was extended by Serrin [1986] and by the quoted works of Ricou. Ricou's motivation in favor of energy inequality is based on peculiar symmetries between the collection of processes starting form σ and terminating at σ.

As has already been pointed out, the construction of energy in Sect. 5.3 is standard. Using the formalized language of systems and processes similar to, but more general than the present one, a modern version of the construction of the energy was given by Coleman & Owen [1974] under the approximate accessibility assumption; an improved version of the result is in Šilhavý [1985a].

6 The Principle of Material Frame Indifference

The principles of invariance and symmetry have played an important role ever since the early days of science. This chapter presents what many believe is an appropriate spacetime invariance principle for the continuum thermodynamics: the principle of material frame indifference, also called the principle of objectivity. This principle establishes a link between the general thermodynamic assertions such as the equation of balance of energy and the boundedness of energy from below on the one hand and the specific *continuum* thermodynamic concepts on the other. Among the latter concepts will be the mass, its positiveness, the equations of balance of linear and angular momenta, and the concepts of internal and kinetic energies. Apart from this theoretical role, the principle of material frame indifference is also important in the constitutive theory where it restricts the form of the constitutive functions. This feature of the principle of material frame indifference will be explained in Part III; here we shall concentrate on the fundamental aspects.

6.1 Formulation

6.1.1 The Changes of Frame A mapping i : Eucl$\to$Eucl is said to be an ***isometry*** if $|i(\boldsymbol{x})-i(\boldsymbol{y})| = |\boldsymbol{x}-\boldsymbol{y}|$ for every $\boldsymbol{x}, \boldsymbol{y} \in$ Eucl. A ***change of frame*** is a family $i = \{i_t, t \in \mathbb{R}\}$ of isometries. It is of the form

$$i_t(\boldsymbol{x}) = \boldsymbol{R}(t)(\boldsymbol{x}-\boldsymbol{x}_0)+\boldsymbol{x}_0'(t), \quad \boldsymbol{x} \in \text{Eucl}, \ t \in \mathbb{R}, \tag{6.1.1}$$

where $\boldsymbol{R}(\cdot) : \mathbb{R} \to$ Orth, and $\boldsymbol{x}_0'(\cdot) : \mathbb{R} \to$ Eucl are functions and $\boldsymbol{x}_0 \in$ Eucl a point. We shall assume that $\boldsymbol{R}, \boldsymbol{x}_0'$ are infinitely differentiable. A composition $i \circ j$ of two changes of frame $i = \{i_t, t \in \mathbb{R}\}$, $j = \{j_t, t \in \mathbb{R}\}$ is a change of frame $k = \{k_t, t \in \mathbb{R}\}$ given by

$$k_t(\boldsymbol{x}) = (i \circ j)_t(\boldsymbol{x}) = i_t(j_t(\boldsymbol{x})), \quad \boldsymbol{x} \in \text{Eucl}, \ t \in \mathbb{R}.$$

The set of all infinitely differentiable changes of frame is a group under the composition. Two changes of frame $i = \{i_t, t \in \mathbb{R}\}$, $j = \{j_t, t \in \mathbb{R}\}$ are said to be equal near time $d \in \mathbb{R}$ if there exists a $\delta > 0$ such that $i_t(\boldsymbol{x}) = j_t(\boldsymbol{x})$, $\boldsymbol{x} \in$ Eucl for every $t \in \mathbb{R}$ with $|t-d| < \delta$. A change of frame $i = \{i_t, t \in \mathbb{R}\}$ is said to be equal to the identity near time d if i and the identity change of frame are equal near d, i.e., if there exists a $\delta > 0$ such that $i_t(\boldsymbol{x}) = \boldsymbol{x}$, $\boldsymbol{x} \in$ Eucl, for all t such that $|s-t| < \delta$.

Let m be an integer, $m \ge 1$. We denote by Vect^m the cartesian product Vect$\times \cdots \times$ Vect with m copies of Vect. If π is any process of the body and $(\boldsymbol{p}, t) \in \Omega \times [0, d_\pi]$, then the vector of ***generalized velocities*** $\boldsymbol{w} = \boldsymbol{w}(\boldsymbol{p}, t)$ is the collection of the material time derivatives of the motion up to order m, i.e.,

$$\boldsymbol{w} = \left(d\chi/dt, d^2\chi/dt^2, \dots, d^m\chi/dt^m\right),$$

where here d^p/dt^p denotes the material time derivative of order p at $(\boldsymbol{p}, t)$. If $\boldsymbol{w} = (\boldsymbol{w}_0, \dots, \boldsymbol{w}_{m-1}) \in \text{Vect}^m$ and $\boldsymbol{R} \in \text{Orth}$, we define $\boldsymbol{R}\boldsymbol{w} := (\boldsymbol{R}\boldsymbol{w}_0, \dots, \boldsymbol{R}\boldsymbol{w}_{m-1})$.

6.1.2 Assumptions

(1) (***The principle of material frame indifference***) There exists an integer $m \geq 1$ and a smooth function $\boldsymbol{g} : \Omega \times \text{Vect}^m \to \text{Vect}$ such that for every $\pi \in \Pi$ and every change of frame i of the form (6.1.1) there exists a unique $\pi' \in \Pi$ of duration d_π, denoted $\pi' \equiv i \circ \pi$, satisfying

$$\chi' = \boldsymbol{R}(\chi - \boldsymbol{x}_0) + \boldsymbol{x}_0', \quad \theta' = \theta, \tag{6.1.2}$$

$$\boldsymbol{S}' = \boldsymbol{R}\boldsymbol{S}, \quad q' = q, \quad r' = r, \tag{6.1.3}$$

$$\boldsymbol{b}' = \boldsymbol{R}\boldsymbol{b} + \boldsymbol{g}(\boldsymbol{p}, \boldsymbol{w}' - \boldsymbol{R}\boldsymbol{w}) \tag{6.1.4}$$

for every $(\boldsymbol{p}, t) \in \Omega \times [0, d_\pi]$, where the quantities on both sides of the equations are evaluated at $(\boldsymbol{p}, t)$, and those associated with π' are distinguished by primes.

(2) (***Group property***) If π is a process and i, j two changes of frame, then $(i \circ j) \circ \pi = i \circ (j \circ \pi)$.

The unique process $i \circ \pi$ is called the ***transform*** of π under i. Subsection 6.1.2(1) is the active interpretation of the principle of material frame indifference: the processes π' and π are two different processes observed from the same frame. The passive interpretation would be to view the process π' to be 'the same' process observed from a frame moving relative to the inertial frame according to (6.1.1). The transformation laws are imposed only on the quantities postulated in Definition 4.2.7 of a material body. The transformation law for energy will be derived later in this chapter. In Chap. 7 we shall also establish the objectivity of entropy. Equations $(6.1.3)_1$ asserts that the contact forces behave 'like a vector.' That is, if P_1, P_2 are two disjoint parts which touch, then the contact force (see Subsect. 3.5.1) $\boldsymbol{f}' = \boldsymbol{f}'(P_1, P_2, t)$ between P_1, P_2 associated with π' is related to the contact force $\boldsymbol{f} = \boldsymbol{f}(P_1, P_2, t)$ associated with π by $\boldsymbol{f}' = \boldsymbol{R}\boldsymbol{f}$. By $(6.1.3)_{2,3}$, the rate of contact communication of heat and the rate of the external heat supply are invariant. Equation (6.1.4) says that the difference of the new body force form the original body force transformed into the 'new reference frame' depends, at the given material point, on the difference of the new velocity and the original velocity transformed into the new reference system and on similarly constructed differences of the time derivatives of the velocity up to some fixed order. The quantity $\boldsymbol{g}$ on the right-hand side of (6.1.4) is the apparent force, and later it will be shown that $\boldsymbol{g}$ necessarily reduces to the classical expression for the apparent force in terms of the centrifugal force, Coriolis force, etc. The difference in the transformation laws for the contact and body forces is due to the difference in their origins: the contact forces act between parts of the body, whose relative motions are unaffected by the change of frame, while the body forces are caused by the exterior of the body, and the relative motion of the body with respect to its exterior is different in the new frame. The group character (2) of the transformation rule for the processes forces the function $\boldsymbol{g}$ to be linear and isotropic in its vector arguments:

6.1.3 Proposition *There exist smooth scalar fields* $\lambda_0, \dots, \lambda_{m-1}$ *on* Ω *such that*

$$\boldsymbol{g}(\boldsymbol{p}, \boldsymbol{w}) = \lambda_0(\boldsymbol{p})\boldsymbol{w}_0 + \dots + \lambda_{m-1}(\boldsymbol{p})\boldsymbol{w}_{m-1} \tag{6.1.5}$$

for every $\boldsymbol{w} = (\boldsymbol{w}_0, \dots, \boldsymbol{w}_{m-1}) \in \mathrm{Vect}^m$ *and* $\boldsymbol{p} \in \Omega$.

With this knowledge, the transformation law (6.1.4) can be rewritten as

$$\boldsymbol{b}' - \boldsymbol{g}(\boldsymbol{p}, \boldsymbol{w}') = \boldsymbol{R}(\boldsymbol{b} - \boldsymbol{g}(\boldsymbol{p}, \boldsymbol{w}))$$

which suggests the interpretation that $\boldsymbol{g}(\boldsymbol{p}, \boldsymbol{w})$ is the quantity which must be subtracted from $\boldsymbol{b}$ to obtain a frame-indifferent quantity. $\boldsymbol{g}(\boldsymbol{p}, \boldsymbol{w})$ is then the acceleration force.

Proof In the proof, the material point $\boldsymbol{p}$ will play a role of a fixed parameter and we therefore suppress it and write $\boldsymbol{g}(\boldsymbol{w})$ for $\boldsymbol{g}(\boldsymbol{p}, \boldsymbol{w})$. The form (6.1.5) will be shown to be a consequence of the following auxiliary assertion: *for every* $\boldsymbol{u}, \boldsymbol{z} \in \mathrm{Vect}^m$ *and every* $\boldsymbol{R} \in \mathrm{Orth}$

$$\boldsymbol{g}(\boldsymbol{R}\boldsymbol{u} + \boldsymbol{z}) = \boldsymbol{R}\boldsymbol{g}(\boldsymbol{u}) + \boldsymbol{g}(\boldsymbol{z}). \tag{6.1.6}$$

To prove (6.1.6), let π be any process and i, j changes of frame of the forms

$$i_t(\boldsymbol{x}) = \boldsymbol{R}(\boldsymbol{x} - \boldsymbol{x}_0) + \boldsymbol{e}(t), \quad j_t(\boldsymbol{x}) = \boldsymbol{x} + \boldsymbol{c}(t), \quad \boldsymbol{x} \in \mathrm{Vect}, \ t \in \mathbb{R}, \tag{6.1.7}$$

where $\boldsymbol{e} : \mathbb{R} \to \mathrm{Eucl}$ and $\boldsymbol{c} : \mathbb{R} \to \mathrm{Vect}$ are infinitely differentiable functions such that

$$\frac{d\boldsymbol{e}(0)}{dt} = \boldsymbol{z}_0, \ \dots, \ \frac{d^m\boldsymbol{e}(0)}{dt^m} = \boldsymbol{z}_{m-1}, \quad \frac{d\boldsymbol{c}(0)}{dt} = \boldsymbol{u}_0, \ \dots, \ \frac{d^m\boldsymbol{c}(0)}{dt^m} = \boldsymbol{u}_{m-1}, \tag{6.1.8}$$

where $\boldsymbol{u} = (\boldsymbol{u}_0, \dots, \boldsymbol{u}_{m-1})$, $\boldsymbol{z} = (\boldsymbol{z}_0, \dots, \boldsymbol{z}_{m-1})$. A composition $k := i \circ j$ is

$$k_t(\boldsymbol{x}) = \boldsymbol{R}(\boldsymbol{x} - \boldsymbol{x}_0) + \boldsymbol{R}\boldsymbol{c}(t) + \boldsymbol{e}(t), \quad \boldsymbol{x} \in \mathrm{Vect}, \ t \in \mathbb{R}.$$

We consider the transforms $\pi' = j \circ \pi$, $\pi'' = i \circ \pi'$. We denote by $\boldsymbol{b}, \boldsymbol{b}', \boldsymbol{b}''$ the body forces and by $\boldsymbol{w}, \boldsymbol{w}', \boldsymbol{w}''$ the generalized velocities at $\boldsymbol{p} \in \Omega$, $t = 0$ corresponding to the processes π, π', π''. The application of the transformation rule (6.1.4) to the changes $\pi \mapsto \pi'$, $\pi' \mapsto \pi''$ gives

$$\boldsymbol{b}' - \boldsymbol{b} = \boldsymbol{g}(\boldsymbol{w}' - \boldsymbol{w}), \quad \boldsymbol{b}'' - \boldsymbol{R}\boldsymbol{b}' = \boldsymbol{g}(\boldsymbol{w}'' - \boldsymbol{R}\boldsymbol{w}'). \tag{6.1.9}$$

By Assumption 6.1.2(2) $\pi'' = i \circ (j \circ \pi) = (i \circ j) \circ \pi = k \circ \pi$ and the application of the transformation rule (6.1.4) to the change $\pi \mapsto k \circ \pi$ gives

$$\boldsymbol{b}'' - \boldsymbol{R}\boldsymbol{b} = \boldsymbol{g}(\boldsymbol{w}'' - \boldsymbol{R}\boldsymbol{w}). \tag{6.1.10}$$

From (6.1.9) and (6.1.10)

$$\boldsymbol{g}(\boldsymbol{w}'' - \boldsymbol{R}\boldsymbol{w}') - \boldsymbol{R}\boldsymbol{g}(\boldsymbol{w}' - \boldsymbol{w}) = \boldsymbol{g}(\boldsymbol{w}'' - \boldsymbol{R}\boldsymbol{w}), \tag{6.1.11}$$

and, from the definitions of the generalized velocities and (6.1.7) and (6.1.8) $\boldsymbol{w}'' - \boldsymbol{R}\boldsymbol{w} = \boldsymbol{R}\boldsymbol{u} + \boldsymbol{z}$, $\boldsymbol{w}' - \boldsymbol{w} = \boldsymbol{u}$, $\boldsymbol{w}'' - \boldsymbol{R}\boldsymbol{w}' = \boldsymbol{z}$. Insertion of these values of the differences of the generalized velocities into (6.1.11) gives (6.1.6). From (6.1.6) we see that the function $\boldsymbol{g}$ is additive (take $\boldsymbol{R} = \boldsymbol{1}$) and isotropic (take $\boldsymbol{z} = \boldsymbol{0}$). The additivity of $\boldsymbol{g}$ and the assumed continuity imply that $\boldsymbol{g}$ is linear in $\boldsymbol{w}$ (actually this can be deduced form (6.1.6) even without the assumption of continuity). The familiar representation theorem for linear isotropic functions of vector arguments (see TRUESDELL & NOLL [1965]) then gives the asserted form of $\boldsymbol{g}$. Invoking the assumed smoothness, we see that the fields λ must be smooth over Ω. □

6.1.4 Proposition *If π is a process, $t \in [0, d_\pi]$ a time, and i, j changes of frame that are equal near t, then*

$$(i \circ \pi)(t) = (j \circ \pi)(t). \tag{6.1.12}$$

The transformations of processes depend locally (in time) on the changes of frame in the sense that the state of the transformed process at time t depends only on the values of the change of frame near t.

Proof Assume for simplicity that t is an interior point of $[0, d_\pi]$. The boundary points are treated similarly. Since i, j are equal near t, there exists a $\delta > 0$ such that $i_s(\boldsymbol{x}) = j_s(\boldsymbol{x})$, $\boldsymbol{x} \in \mathrm{Eucl}$, for every s with $|s - t| < \delta$. Assume that δ is small enough that $t - \delta, t + \delta \in [0, d_\pi]$. The segments $(i \circ \pi)_{[t-\delta, t+\delta]}$ and $(j \circ \pi)_{[t-\delta, t+\delta]}$ are again processes, cf. 4.1.1(2). Since i, j are equal on the interval $[t - \delta, t + \delta]$, the transformation laws (6.1.2)–(6.1.4) for the transforms of the process $\pi_{[t-\delta, t+\delta]}$ under i and j are identical and as the transforms are completely determined by the transformation laws (6.1.2)–(6.1.4) (cf. the uniqueness assertion in 6.1.2(1)), the segments $(i \circ \pi)_{[t-\delta,t+\delta]}$ and $(j \circ \pi)_{[t-\delta,t+\delta]}$ are equal each to other. Hence also the transforms $(i \circ \pi)$, $(j \circ \pi)$ are the same on the interval $[t-\delta, t+\delta]$, which implies (6.1.12). □

6.2 The Transformation Law for Work; Mass

In order to arrive at the concept of mass and the equations of balance of linear and angular momenta, it will be assumed throughout the rest of the chapter that the body satisfies the equation of balance of energy (5.3.1) with a reference density of the total energy u bounded below:

6.2.1 Assumptions

(1) For every process, every part, and every time,

$$\frac{d}{dt}\int_P u\,dV = \int_{\partial P} (\boldsymbol{S}^{\mathsf{T}}\boldsymbol{v} - \boldsymbol{q}) \cdot \boldsymbol{n}\,dA + \int_P (\boldsymbol{b} \cdot \boldsymbol{v} + r)\,dV; \tag{6.2.1}$$

(2) there exists a continuous integrable function u_0 such that

$$\int_P \hat{u}(\sigma, \cdot)\,dV \ge \int_P u_0\,dV \tag{6.2.2}$$

for every state σ and every part P;

(3) the set Π contains processes of arbitrary duration, i.e., for every $d > 0$ there exists a process $\pi \in \Pi$ such that $d_\pi = d$;

(4) for every state σ there exists a process π with $\pi^f = \sigma$ and a process τ with $\tau^i = \sigma$.

6.2.2 Proposition *For any process π, any part P, and any change of frame i with the representation* (6.1.1) *we have*

$$q(i \circ \pi, P) = q(\pi, P), \tag{6.2.3}$$

$$w(i \circ \pi, P) = w(\pi, P) - \int_0^{d_\pi} (\boldsymbol{V} \cdot \boldsymbol{f} + \boldsymbol{\Omega} \cdot \boldsymbol{m} + G)\,dt \tag{6.2.4}$$

where

$$\boldsymbol{f} = \int_{\partial P} \boldsymbol{S}\boldsymbol{n}\,dA + \int_P \boldsymbol{b}\,dV, \quad \boldsymbol{m} = \int_{\partial P} (\boldsymbol{x} - \boldsymbol{x}_0) \times \boldsymbol{S}\boldsymbol{n}\,dA + \int_P (\boldsymbol{x} - \boldsymbol{x}_0) \times \boldsymbol{b}\,dV,$$

$$G = \int_P \boldsymbol{g}(\boldsymbol{p}, \boldsymbol{w}' - \boldsymbol{R}\boldsymbol{w}) \cdot \boldsymbol{v}'\,dV,$$

$\boldsymbol{v}'$ and $\boldsymbol{w}'$ are the velocity and the generalized velocity of the transformed process, $\boldsymbol{V} = \boldsymbol{R}^{\mathsf{T}}\dot{\boldsymbol{x}}_0'$, and $\boldsymbol{\Omega}$ is the axial vector of the skew tensor $\mathbf{W} = \dot{\boldsymbol{R}}\boldsymbol{R}^{-1}$, $\Omega_i = \frac{1}{2}\varepsilon_{ijk}W_{jk}$.

Formulas (6.2.3) and (6.2.4) are direct consequences of the postulated transformation laws and their easy verification is omitted. Heat is objective (invariant) but the work is not; consequently, the total energy, balancing the difference of work and heat, is not objective. However, work must be objective in special circumstances. Since the cyclic processes are transformed into cyclic processes (at least for special changes of frame, see below), the work, being equal to the objective heat by Joule's relation, must be objective there, too. We shall see presently that this requirement and another one derived from the boundedness from below of total energy has a number of consequences on the form (6.1.5) of the acceleration force $\boldsymbol{g}$.

6.2.3 Proposition

(1) *For every process π and every part P there exists a constant $C(\pi, P) < \infty$ such that*

$$w(i \circ \pi, P) \leq C(\pi, P) \tag{6.2.5}$$

for every change of frame i which is equal to the identity near $t = 0$.

(2) *For every process π, every part P, and every change of frame i which is equal to the identity near $t = 0$ and $t = d_\pi$,*

$$w(i \circ \pi, P) = w(\pi, P). \tag{6.2.6}$$

Proof (1): Define $C(\pi, P) := q(\pi, P) + \mathsf{U}(\pi^i, P) - C(P)$, where

$$\mathsf{U}(\sigma, P) = \int_P \hat{u}(\sigma, \cdot)\, dV, \quad C(P) = \int_P u_0\, dV,$$

with u_0 the lower bound occurring in (6.2.2). We shall now prove that with this $C(\pi, P)$ the inequality (6.2.5) holds. Denote $\bar{\pi} = i \circ \pi$; the integrated form of the equation of balance of energy for $\bar{\pi}$ reads

$$\mathsf{U}(\bar{\pi}^f, P) - \mathsf{U}(\bar{\pi}^i, P) = q(\bar{\pi}, P) - w(\bar{\pi}, P). \tag{6.2.7}$$

Since i is equal to the identity near the origin $t = 0$, Proposition 6.1.4 gives $\bar{\pi}^i = \pi^i$ and so $\mathsf{U}(\bar{\pi}^i, P) = \mathsf{U}(\pi^i, P)$; the boundedness from below of energy (6.2.2) provides $\mathsf{U}(\bar{\pi}^f, P) \geq C(P)$. The combination of the last two relations with (6.2.7) and (6.2.3) yields (6.2.5). (2) Proposition 6.1.4 implies that $\bar{\pi}^i = \pi^i, \bar{\pi}^f = \pi^f$ and so the changes of the total energy in the processes $\bar{\pi}$ and π are the same. By the equation of balance of energy then also the difference of the work and heat in the two processes must be the same, and since heat is objective (cf. (6.2.3)), this implies that the amounts of work in the two processes must be the same. □

A function $\boldsymbol{c} : \mathbb{R} \to \mathrm{Vect}$ is said to vanish near $s \in \mathbb{R}$ if there exists a $\delta > 0$ such that $\boldsymbol{c}(t) = \boldsymbol{0}$ for every $t \in (s - \delta, s + \delta)$.

6.2.4 Lemma *Let $\boldsymbol{c} : \mathbb{R} \to \mathrm{Vect}$ be infinitely differentiable and let $\boldsymbol{p} \in \Omega$. Then for every $d > 0$ we have:*

(1) *if $\boldsymbol{c}$ vanishes near 0, then*

$$\int_0^d \boldsymbol{g}(\boldsymbol{p}, d\boldsymbol{c}/dt, \ldots, d^m\boldsymbol{c}/dt^m) \cdot (d\boldsymbol{c}/dt)\, dt \geq 0; \tag{6.2.8}$$

(2) *if $\boldsymbol{c}$ vanishes near 0, and near d, then*

$$\int_0^d \boldsymbol{g}(\boldsymbol{p}, d\boldsymbol{c}/dt, \ldots, d^m\boldsymbol{c}/dt^m) \cdot (d\boldsymbol{c}/dt)\, dt = 0. \tag{6.2.9}$$

Proof Relations (6.2.8) and (6.2.9) will be derived form (6.2.5) and (6.2.6), respectively. Let π be an arbitrary process of duration d (see Assumption 6.2.1(3)). Let i be a change of frame given by $i_t(\boldsymbol{x}) = \boldsymbol{x} + \beta\boldsymbol{c}(t)$, $\boldsymbol{x} \in \mathrm{Vect}$, $t \in \mathbb{R}$, where $\boldsymbol{c}$ is a vector-valued function on $\mathbb{R}$ and $\beta > 0$ a number. If $\boldsymbol{c}$ vanishes near 0, then i is equal to the identity near 0 and a combination of the transformation law (6.2.4) with inequality (6.2.5) gives

$$\begin{aligned} C(\pi, P) - w(\pi, P) \geq &-\beta \int_0^d \int_P \boldsymbol{g}(\boldsymbol{p}, d\boldsymbol{c}/dt, \ldots, d^m\boldsymbol{c}/dt^m) \cdot \boldsymbol{v}\, dV\, dt \\ &-\beta^2 \int_0^d \int_P \boldsymbol{g}(\boldsymbol{p}, d\boldsymbol{c}/dt, \ldots, d^m\boldsymbol{c}/dt^m) \cdot (d\boldsymbol{c}/dt)\, dV\, dt \\ &-\beta \int_0^d \Big(\int_{\partial P} \boldsymbol{S}\boldsymbol{n}\, dA + \int_P \boldsymbol{b}\, dV \Big) \cdot (d\boldsymbol{c}/dt)\, dt, \end{aligned} \tag{6.2.10}$$

where we have used the linearity of $\boldsymbol{g}(\boldsymbol{p}, \cdot)$. Dividing by β^2 and letting $\beta \to \infty$ then gives

$$-\int_0^d \int_P g(p, dc/dt, \ldots, d^m c/dt^m) \cdot (dc/dt)\, dV\, dt \le 0;$$

and as P is arbitrary, we have (6.2.8). Equality (6.2.9) is obtained similarly: combining (6.2.4) with (6.2.6) one obtains (6.2.10) with the inequality sign replaced by the equality and the left-hand side replaced by 0. □

We now use (6.2.8) and (6.2.9) to obtain the final form of the apparent force.

6.2.5 Proposition *There exists a smooth, nonnegative field ρ_o on Ω such that*

$$g(p, w) = \rho_o(p) w_1 \tag{6.2.11}$$

for every $p \in \Omega$ and every $w = (w_0, \ldots, w_{m-1}) \in \mathrm{Vect}^m$. In other words, the transformation law for the body force (6.1.4) *has the form*

$$b' = Rb + \rho_o(a' - Ra)$$

where a and a' are the accelerations of π and $\pi' = i \circ \pi$ respectively.

The scalar field ρ_o is the density of mass of the body in the reference configuration. ρ_o is time- and process-independent quantity; mass is conserved. The independence of the transformation law (6.1.4) of the reference configuration implies the correct transformation law for ρ_o under the change of the reference configuration.

Proof of Proposition 6.2.5 Let $\beta > 0$ and replace the function c and the time d in (6.2.8) by the function $\bar{c}$ and time $\bar{d}$ given by $\bar{c}(t) = c(\beta t)$, $t \in \mathbb{R}$, $\bar{d} = d/\beta$. Inequality (6.2.8) provides

$$\int_0^{\bar{d}} g(p, \beta dc/dt, \ldots, \beta^m d^m c/dt^m) \cdot (\beta dc/dt)\, dt \ge 0,$$

where the argument of the time derivatives of c is βt. After a substitution $t \mapsto \beta t$,

$$\int_0^d g(p, \beta dc/dt, \ldots, \beta^m d^m c/dt^m) \cdot (dc/dt)\, dt \ge 0,$$

where now the argument of the time derivatives of c is t. Again, dividing by β^m and letting $\beta \to \infty$, we obtain, using the linearity of $g(p, \cdot)$,

$$\int_0^d g(p, 0, \ldots, 0, d^m c/dt^m) \cdot (dc/dt)\, dt \ge 0,$$

i.e., (cf. (6.1.5)),

$$\lambda_{m-1}(p) \int_0^d d^m c/dt^m \cdot (dc/dt)\, dt \ge 0. \tag{6.2.12}$$

This must be satisfied by every function c vanishing near 0. It is now claimed without proof (see ŠILHAVÝ [1989, p. 14]) that for $m > 2$, the integral on the right-hand side of (6.2.12) can take arbitrary values on the set of all functions vanishing near 0. Consequently, if $m > 2$, necessarily $\lambda_{m-1}(p) = 0$. Proceeding inductively, we conclude by this argument that $\lambda_k(p) = 0$ for all k such that $2 \le k \le m-1$. With this knowledge, the integral in (6.2.8) reduces to

$$\int_0^d g(p, dc/dt, \ldots, d^m c/dt^m) \cdot (dc/dt)\, dt = \lambda_0(p) \int_0^d \dot{c}(t)^2\, dt + \tfrac{1}{2}\lambda_1(p)\dot{c}(d)^2 \tag{6.2.13}$$

for every function vanishing near 0. Inequality (6.2.8) then implies $\lambda_0(p) \ge 0$, $\lambda_1(p) \ge 0$. Finally, the equality (6.2.9) for every c vanishing near 0 and d and (6.2.13) give $\lambda_0(p) = 0$ and the proof is complete if one identifies ρ_o with λ_1. □

6.3 Cauchy's Equations of Motion; Internal Energy

Within the present scheme the Cauchy equations of motion and the splitting of the total energy into the invariant internal energy and the kinetic energy of the usual form have status of derived assertions. The boundedness from below of the total energy is important for the derivation.

6.3.1 Proposition *For every process π, every part P, every point $x_0 \in$ Eucl and every $t \in [0, d_\pi]$ the Cauchy equations of motion hold:*

$$\frac{d}{dt}\int_P \rho_\circ \boldsymbol{v}\, dV = \int_{\partial P} \boldsymbol{Sn}\, dA + \int_P \boldsymbol{b}\, dV, \tag{6.3.1}$$

and

$$\frac{d}{dt}\int_P \rho_\circ (\boldsymbol{x}-\boldsymbol{x}_0)\times \boldsymbol{v}\, dV = \int_{\partial P} (\boldsymbol{x}-\boldsymbol{x}_0)\times \boldsymbol{Sn}\, dA + \int_P (\boldsymbol{x}-\boldsymbol{x}_0)\times \boldsymbol{b}\, dV. \tag{6.3.2}$$

Proof Let $\boldsymbol{c} : \mathbb{R} \to$ Vect be a smooth function vanishing near 0 and satisfying $\dot{\boldsymbol{c}}(d) = 0$, where $d = d_\pi$. The change of frame i defined by $i_t(\boldsymbol{x}) = \boldsymbol{x} + \boldsymbol{c}(t)$, $\boldsymbol{x} \in$ Vect, is then equal to the identity near 0; since i is equal to the identity near 0, Proposition 6.2.3(1) applies, which in combination with the transformation rule (6.2.4) and the form of $\boldsymbol{g}$ given by Proposition 6.2.5 implies

$$\int_0^d \dot{\boldsymbol{c}}(t)\cdot\Big(\int_{\partial P} \boldsymbol{Sn}\, dA + \int_P (\boldsymbol{b}-\rho_\circ \boldsymbol{a})\, dV\Big)\, dt \le C(\pi, P) - w(\pi, P).$$

As the left-hand side is linear in $\boldsymbol{c}$, one must have actually the equation

$$\int_0^d \dot{\boldsymbol{c}}(t)\cdot\Big(\int_{\partial P} \boldsymbol{Sn}\, dA + \int_P (\boldsymbol{b}-\rho_\circ \boldsymbol{a})\, dV\Big)\, dt = 0 \tag{6.3.3}$$

and this must be satisfied by every smooth function $\boldsymbol{c}$ on $\mathbb{R}$ vanishing near 0 and satisfying $\dot{\boldsymbol{c}}(d) = 0$. An elementary limiting procedure shows that (6.3.3) must be satisfied by every continuously differentiable function $\boldsymbol{c}$ satisfying $\boldsymbol{c}(0) = \boldsymbol{0}$, and this in turn implies that the expression in the square brackets in (6.3.3) vanishes. This completes the proof of (6.3.1). The equation of balance of angular momentum (6.3.2) is proved similarly by using time-dependent rotations about the point $\boldsymbol{x}_0$. □

The ***internal energy*** $\hat{e}$ is a local state function on (Σ, Π, Ω) defined by

$$\hat{e}(\sigma, \boldsymbol{p}) = \hat{u}(\sigma, \boldsymbol{p}) - \tfrac{1}{2}\rho_\circ(\boldsymbol{p})\boldsymbol{v}(\sigma, \boldsymbol{p})^2$$

for every state σ and every point $\boldsymbol{p}$. The internal and kinetic energies of the part P in the state σ are

$$\mathsf{E}(\sigma, P) = \int_P \hat{e}(\sigma, \cdot)\, dV, \quad \mathsf{K}(\sigma, P) = \tfrac{1}{2}\int_P \rho_\circ \boldsymbol{v}(\sigma, \cdot)^2\, dV; \tag{6.3.4}$$

hence $\mathsf{U}(\sigma, P) = \mathsf{K}(\sigma, P) + \mathsf{E}(\sigma, P)$. The local forms of (6.3.1), (6.3.2), and (6.2.1) read

$$\rho_\circ \dot{\boldsymbol{v}} = \operatorname{Div} \boldsymbol{S} + \boldsymbol{b}, \quad \boldsymbol{SF}^{\mathsf{T}} = \boldsymbol{FS}^{\mathsf{T}}, \quad \dot{e} = \boldsymbol{S}\cdot\dot{\boldsymbol{F}} - \operatorname{Div} \boldsymbol{q} + r. \tag{6.3.5}$$

(See Part I.)

6.3.2 Proposition

(1) *The internal energy is objective, i.e., for every process π and every change of frame i we have*

$$\hat{e}((i \circ \pi)(t), \cdot) = \hat{e}(\pi(t), \cdot), \tag{6.3.6}$$

for every part P and every $t \in [0, d_\pi]$.

(2) *The internal energy is bounded from below by* u_0.

Since the transformation law for the velocity in the expression for the kinetic energy is known, we have also obtained (implicitly) the transformation law for the total energy.

Proof (1): Using $(6.3.5)_3$ and the transformation laws postulated in the principle of material frame indifference, one learns that the right-hand side of $(6.3.5)_3$ is objective and hence also the time derivative of e is objective, i.e.,

$$\dot{e}' = \dot{e} \tag{6.3.7}$$

for every process π and every change of frame i, where e' and e are the evolutions of the internal energy corresponding to the processes $i \circ \pi$ and π, respectively. Equation (6.3.7) implies $e'(\boldsymbol{p}, t) = e(\boldsymbol{p}, t) + c(\boldsymbol{p})$. Now if i additionally is equal to the identity near 0, then the initial states of the original and the transformed processes coincide (Proposition 6.1.4) and so $e'(\boldsymbol{p}, 0) = e(\boldsymbol{p}, 0)$; the integration constant $c(\boldsymbol{p})$ then necessarily vanishes and (6.3.7) reduces to (6.3.6). To prove (6.3.6) for the general changes of frame, it suffices to reduce the general situation to the one when the partial result just obtained can be applied. This can be done as follows. The partial result is applied to the process $\pi_0 * \pi$, where π_0 is a process with $\pi_0^f = \pi^i$, whose existence is guaranteed by Assumption 6.2.1(4). If j is any change of frame which is equal to the identity near 0, and which coincides with i on the time interval of the process π in the combined process $\pi_0 * \pi$, then the evolutions of the internal energy during the processes $j \circ (\pi_0 * \pi)$ and $\pi_0 * \pi$ coincide with the evolutions of the internal energy during $i \circ \pi$ and π. The result then holds generally. (2): The continuity and the arbitrariness of P implies the local form of (6.2.2):

$$\hat{u}(\sigma, \boldsymbol{p}) \geq u_0(\boldsymbol{p}). \tag{6.3.8}$$

for every state σ and every $\boldsymbol{p}$. Let σ and $\boldsymbol{p}$ be given, and let π be the process with $\pi^i = \sigma$ whose existence is guaranteed by Assumption 6.2.1(4). For the given $\boldsymbol{p}$ there exists a change of frame i such that the initial velocity at $\boldsymbol{p}$ of the transformed process $i \circ \pi$ vanishes, $\boldsymbol{v}((i \circ \pi)^i, \boldsymbol{p}) = \boldsymbol{0}$ and hence,

$$\hat{e}((i \circ \pi)^i, \boldsymbol{p}) = \hat{u}((i \circ \pi)^i, \boldsymbol{p}). \tag{6.3.9}$$

On the other hand, by the objectivity of e,

$$\hat{e}((i \circ \pi)^i, \boldsymbol{p}) = \hat{e}(\pi^i, \boldsymbol{p}) = e(\sigma, \boldsymbol{p}), \tag{6.3.10}$$

and a combination of (6.3.9) with (6.3.10) yields $\hat{e}(\sigma, \boldsymbol{p}) = \hat{u}((i \circ \pi)^i, \boldsymbol{p})$ and so the inequality (6.3.8) applied to $(i \circ \pi)^i$ yields $\hat{e}(\sigma, \boldsymbol{p}) = \hat{u}((i \circ \pi)^i, \boldsymbol{p}) \geq u_0(\boldsymbol{p})$. □

If the boundedness from below of the total energy is omitted from the list of postulates, more general equations of balance of linear and angular momenta follow (see ŠILHAVÝ [1989]). In view of the balance of linear momentum, the transformation law for the body force can now be omitted: it follows from the former. In view of the reduced equation of balance of energy, the equation postulating the objectivity of the external heat supply can be replaced by the postulation of the objectivity of the internal energy. Later, when the concept of entropy is derived, we shall be able to prove the objectivity of entropy.

Bibliographical Note

The recognition of the role of the principle of material frame indifference in continuum mechanics is due to Noll. The material of the last three sections is taken from ŠILHAVÝ [1989]. (See also ŠILHAVÝ [1987, 1992].) The earlier approaches to the derivation of the balance of linear and angular momenta are due to NOLL [1963, 1973], GREEN & RIVLIN [1964a,b] and SERRIN [1972]. They differ from the approach above in that they presuppose the concept of mass and either the classical expression for the acceleration force or the splitting of the total energy into the invariant internal energy and the kinetic energy of the classical form. In SERRIN [1990] another approach is described which does not presuppose the concept of mass.

7 The Second Law of Thermodynamics

The second law specifies conditions under which a cyclically operating body can do positive net work. The conditions are stated in terms of the exchange of heat of the body with its environment. In this respect the second law is similar to the first. The difference is that while for the formulation of the first law the single net gain of heat of the body suffices, the second law requires a much finer concept which distinguishes amounts of heat absorbed or emitted at different temperatures. The concepts of heating measure and accumulation function, to be introduced in Sect. 7.1, serve this purpose. It will be seen that the framework of heating measures enables one to remove the traditional ambiguities in the statements of the second law.

The program of deriving the consequences of the second law culminates in Sect. 7.6 with the introduction of the entropy satisfying the Clausius–Duhem inequality. The important intermediate step is the Clausius inequality for cyclic processes, which also includes the introduction of the absolute temperature scale.

7.1 Empirical Temperature. The Heating Measure

According to our assumption in Definition 4.2.7, each body has a well-defined empirical temperature θ given by a local state function. Hence the temperature field is associated with any state and any process. The empirical temperature is specified by the laws it satisfies. This is the second law of thermodynamics and its consequences. Sometimes, the "zeroth law" is used to introduce the empirical temperature. There is no need for such an axiom. It is also often argued that the concept of temperature is meaningless outside thermodynamic equilibrium. Every concept has a limited range of applicability, including the temperature. But there are many nonequilibrium situations where the temperature still has a clear meaning, such as the dissipation of heat in flows of viscous fluids, or the temperature profiles in rigid heat conductors, and many others. These situations are not covered by the classical thermodynamics of homogeneous reversible systems, but they are covered by the present approach. The entropy function resulting from the versions of the second law to be given below may in general depend on the nonequilibrium variables. For example, in the case of the Maxwell–Cattaneo heat conduction, the entropy depends on the heat flux vector (see Proposition 7.6.5 below).

To approach the heating measure, we first briefly sketch some notions of measure theory. Further information is in RUDIN [1970, Chaps. 1, 2, 6 & 7].

7.1.1 Measures The concept of measure is designated to describe the distributions (in the physical space or elsewhere) of additive quantities such as volume, mass and electrical charge. A measure associates with each set of points a number –the amount of the quantity in that set. We restrict our attention to measures defined on sets of real numbers (to be identified with the empirical temperatures). For technical reasons, the measures to be considered will be defined only on Borel sets of real numbers. Without giving a formal definition we note that the class of all Borel subsets of $\mathbb{R}$ is extremely large: it contains all open and closed sets and all countable unions and intersections of them; in particular, all open, closed, and semi-open intervals. The reader not familiar with the Borel sets can think of a Borel set as of any 'reasonable' set of real numbers.

If $I \subset \mathbb{R}$ is an open interval, then $\mathbb{B}(I)$ denotes the collection of all Borel subsets of I. A ***signed Borel measure,*** or briefly a ***measure,*** on I, is a function $Q : \mathbb{B}(I) \to \mathbb{R}$ which associates with each Borel set $N \subset I$ a number $Q(N)$ in such a way that if $N_1, N_2, \ldots \subset I$ is a sequence of pairwise disjoint Borel sets (i.e., $N_i \cap N_j = \emptyset$ for $i \neq j$), then $Q(\bigcup_{i=1}^{\infty} N_i) = \sum_{i=1}^{\infty} Q(N_i)$, with the series on the right-hand side absolutely convergent as part of the definition. In particular, if M, N are two disjoint Borel sets, then we have the additivity $Q(M \cup N) = Q(M) + Q(N)$. The value of Q on a one-point set $\{\theta\}, \theta \in I$, is denoted here by $Q[\theta] := Q(\{\theta\})$. The set of all measures on I is a linear space under obviously defined operations of a sum $Q_1 + Q_2$ of two measures and a product λQ of a measure with a real number.

The ***distribution function*** $\hat{Q} : I \to \mathbb{R}$ of a measure Q on $I = (a, b) \subset \mathbb{R}$ is defined by

$$\hat{Q}(\theta) = Q((a, \theta]), \quad \theta \in I. \tag{7.1.1}$$

$\hat{Q}$ is a function of bounded variation satisfying simple additional restrictions not to be formulated here, and each function of bounded variation satisfying these restrictions determines a unique Borel measure on I which satisfies (7.1.1), i.e., Borel measures are in one-to-one correspondence with the distribution functions.

7.1.2 Examples

(1) ***Absolutely continuous measures*** Let $f : I \to \mathbb{R}$ be a Lebesgue integrable function (i.e., $\int_I |f| \, d\theta < \infty$) and define a measure Q on I by

$$Q(N) = \int_N f \, d\theta \quad \text{for each } N \in \mathbb{B}(I). \tag{7.1.2}$$

Each measure of this form is termed absolutely continuous with respect to the Lebesgue measure, and f is its density. The distribution function is

$$\hat{Q}(\theta) = \int_a^\theta f(\theta') \, d\theta', \quad \theta \in I.$$

(2) ***Dirac's measure at*** $\omega \in I$ This is a measure δ_ω on I defined by

$$\delta_\omega(N) = 0 \text{ if } \omega \notin N \text{ and } \delta_\omega(N) = 1 \text{ if } \omega \in N.$$

The distribution function is $\hat{\delta}_\omega(\theta) = H(\theta - \omega)$ where H is the Heaviside step function,

$$H(\theta) = 0 \text{ if } \theta < 0 \text{ and } H(\theta) = 1 \text{ if } \theta \geq 0.$$

Dirac's measure is an example of a singular measure. Any linear combination of the measures considered in (1), (2) is again a measure. By the Lebesgue decomposition theorem (see RUDIN [1970, Sect. 6.9]), each measure on I is a sum of an absolutely continuous measure as in (1) and a singular measure. The latter is defined by the requirement that it vanishes outside a Borel set of the Lebesgue measure 0, and a prototype of such a measure is a finite or "countable" linear combination of Dirac's measures concentrated at various points. Hence by linearly combining the measures given in Examples (1) and (2) one obtains essentially *any* Borel measure. The singular measures cannot be avoided, for we shall see that the heating measures of isothermal processes are multiples of Dirac's measure.

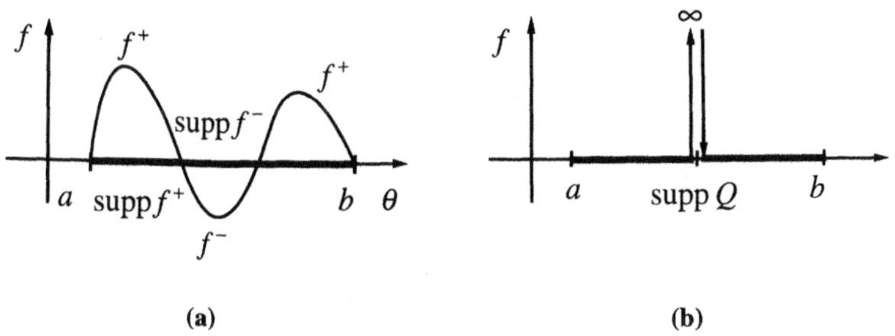

Fig. 7.1.1

Fig. 7.1.1(a) depicts the density of an absolutely continuous measure; Dirac's measure has no density in the traditional sense, yet this density is depicted in Fig. 7.1.1(b).

A measure Q in I is said to be ***nonnegative*** if $Q(N) \geq 0$ for every $N \in \mathbb{B}(I)$; this happens if and only if its distribution function is monotonically nondecreasing. Every measure can be written as the difference of two nonnegative measures.

7.1.3 Jordan's Decomposition Theorem *Let Q be a measure on I and define functions Q^+, Q^- and $|Q|$ on $\mathbb{B}(I)$ by*

$$Q^{\pm}(N) = \sup\{\pm Q(E) : E \in \mathbb{B}(I), E \subset N\},$$
$$|Q|(N) = \sup\left\{\sum |Q(E_i)| : E_i \in \mathbb{B}(I), E_i \text{ pairwise disjoint}, \bigcup E_i = N\right\}$$

for every $N \in \mathbb{B}(I)$ where the sum and the union are taken over all $i = 1, \ldots$ Then Q^+, Q^- and $|Q|$ are nonnegative Borel measures on I and

$$Q = Q^+ - Q^-, \quad |Q| = Q^+ + Q^-. \tag{7.1.3}$$

If $\bar{Q}^+$, $\bar{Q}^-$ is any pair of nonnegative measures on I satisfying

$$Q = \bar{Q}^+ - \bar{Q}^-, \tag{7.1.4}$$

then

$$\bar{Q}^+ \geq Q^+ \quad \bar{Q}^- \geq Q^-. \tag{7.1.5}$$

Equality $(7.1.3)_1$ is the ***Jordan decomposition*** of Q; the measures Q^+, Q^- and $|Q|$ are called the ***positive, negative,*** and ***total variations*** of Q, respectively. (The value $|Q|(N)$ should not be confused with $|Q(N)|$.) Inequalities (7.1.5) express the minimality of the Jordan decomposition: note that every measure can be written as the difference of infinitely many pairs of nonnegative measures. Indeed, once the Jordan decomposition is known, one can create further decompositions as in (7.1.4) by adding the same arbitrary nonnegative measure Q' to Q^+, Q^-: $\bar{Q}^+ = Q^+ + Q'$, $\bar{Q}^- = Q^- + Q'$. The Jordan decomposition corresponds to writing the distribution function as the difference of two monotonically nondecreasing functions –the positive and the negative variations of $\hat{Q}$: $\hat{Q} = \hat{Q}^+ - \hat{Q}^-$. For the absolutely continuous measures of Example 7.1.2(1) the Jordan decomposition corresponds to decomposing the density f into the difference of the positive and the negative parts of f, see Fig. 7.1.1(a).

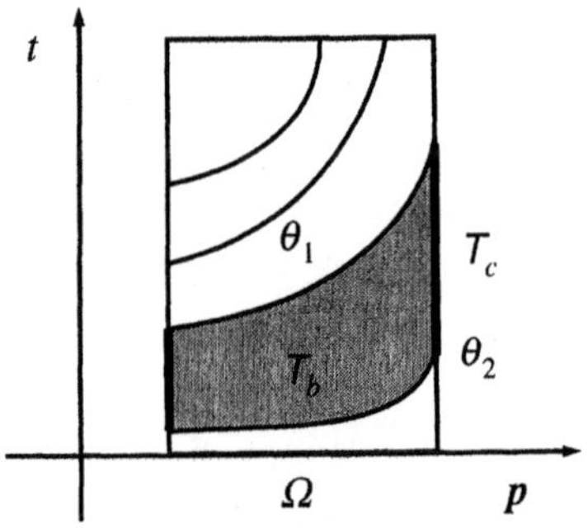

Fig. 7.1.2

7.1.4 Proposition *For any measure Q on I there exists the smallest relatively closed subset* $\operatorname{supp} Q$ *of I such that $Q(N) = 0$ for every $N \in \mathbb{B}(I)$ satisfying $N \cap \operatorname{supp} Q = \emptyset$.*

The set $\operatorname{supp} Q$ is called the ***support*** of Q and the measure vanishes outside it. The support of the Dirac measure concentrated at ω is the one-point set $\{\omega\}$; the support of the absolutely continuous measure (7.1.2) is, roughly speaking, the closure of the set of all points θ where $f(\theta) \neq 0$, see Figures 7.1.1(a),(b).

7.1.5 The Heating Measure of a Process We now identify the open interval $I \subset \mathbb{R}$ with the set of all possible empirical temperatures. Consider a process π of a body Ω with the evolutions of the heat flux vector, the rate of the external heat supply and the empirical temperature $\boldsymbol{q}$, r, and θ. It is assumed that the range of θ is contained in I. For every Borel set N of empirical temperatures it is possible to determine the amount of heat the body gained at temperatures from the set N during the process. One has to determine the points $(\boldsymbol{p}, t)$ where and when the body had the temperature in N, and then to sum the elementary gains of heat at all these points. This can be done also for any part $P \subset \Omega$. Denote

$$T_c := \{(\boldsymbol{p}, t) \in \partial P \times [0, d] : \theta(\boldsymbol{p}, t) \in N\}, \quad T_b := \{(\boldsymbol{p}, t) \in P \times [0, d] : \theta(\boldsymbol{p}, t) \in N\}.$$

Thus T_c is the set of all spacetime points "on the boundary" of part P where the temperature is in N and T_b is the set of all spacetime points "in the interior" of part P where the temperature is in N. Fig. 7.1.2 describes T_c, T_b for a one-dimensional body and for the set N of the form $N = [\theta_0, \theta_1]$. The lines in the rectangle $P \times [0, d]$ are the level sets of temperature in their temporal evolution. The set T_c consists of the bold lines on the boundary of the rectangle $P \times [0, d]$ while T_b is the shaded area in $P \times [0, d]$. The heat gained by P in π at temperatures from $N \subset I$ is

$$Q(\pi, P, N) = -\int_{T_c} \boldsymbol{q} \cdot \boldsymbol{n} \, dA \, dt + \int_{T_b} r \, dV \, dt. \tag{7.1.6}$$

For every fixed π, P, the set function $N \mapsto Q(\pi, P, N)$ is a measure on I. Indeed the additivity with respect to N is to be expected on physical grounds from the additivity of heat; the σ-additivity is easy to check formally by using the definition of $Q(\pi, P, N)$. The measure $Q(\pi, P, \cdot)$ is called the ***heating measure*** corresponding to π, P and is denoted by $Q \equiv Q(\pi, P)$. The corresponding distribution function $\hat{Q}(\pi, P, \theta)$ is called the

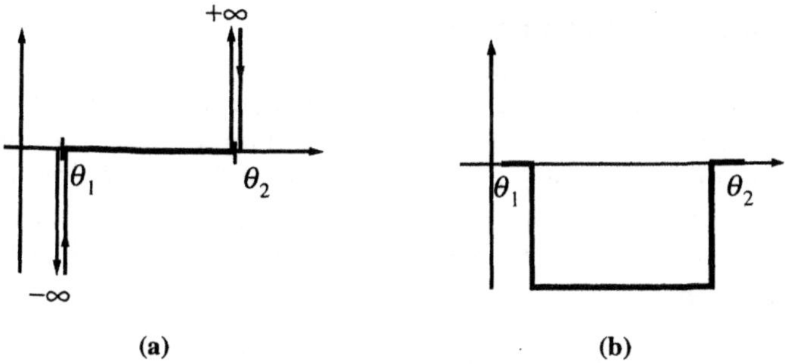

Fig. 7.1.3

accumulation function corresponding to π, P. For every $\theta \in I$, the number $\hat{Q}(\pi, P, \theta)$ is the heat gained by P at or below the temperature θ during π. One has

$$\hat{Q}(\pi, P, \theta) = \int_0^d \Big(- \int_{\partial P} H(\theta - \theta(\boldsymbol{p}, t)) \boldsymbol{q} \cdot \boldsymbol{n} \, dA + \int_P H(\theta - \theta(\boldsymbol{p}, t)) \cdot r \, dV \Big) \, dt,$$

where H is the Heaviside function.

7.1.6 Examples

(1) ***Adiabatic process*** A process π is said to be adiabatic on P if $r = 0$ on $P \times [0, d]$ and $\boldsymbol{q} \cdot \boldsymbol{n} = 0$ on $\partial P \times [0, d]$. Then $Q(\pi, P) = 0$.

(2) ***Isothermal process*** A process π is said to be isothermal of temperature $\omega \in I$ on P if $\theta = \omega$ on $P \times [0, d]$. Then $Q(\pi, P) = q(\pi, P)\delta_\omega$ and $\hat{Q}(\pi, P, \theta) = q(\pi, P) H(\theta - \omega)$.

(3) ***The body in contact with two bodies of different temperatures*** Consider a heat conducting block bounded by two parallel faces which are in contact with two bodies of different temperatures $\theta_1, \theta_2, \theta_1 < \theta_2$. The remaining faces are thermally insulated and $r = 0$. If the block is in a rest state, then the rate of heat gained on the face θ_2 equals the rate of heat emitted by the face θ_1. If $c > 0$ is this common value, then the heating measure corresponding to the rest process of duration d at this rest state is $Q(\pi, P) = cd(\delta_{\theta_2} - \delta_{\theta_1})$. Fig. 7.1.3(a) depicts $Q(\pi, P)$ and Fig. 7.1.3(b) $\hat{Q}(\pi, P, \theta)$.

7.1.7 Proposition

(1) *If $\pi_1, \pi_2 \in \Pi$ are processes with $\pi_1^f = \pi_2^i$ and $P \subset \Omega$, then*

$$Q(\pi_1 * \pi_2, P) = Q(\pi_1, P) + Q(\pi_2, P).$$

(2) *If $\pi \in \Pi$ is a process and P_1, P_2 two parts whose interiors have no points in common and if $P_1 \cup P_2$ is a part, then*

$$Q(\pi, P_1 \cup P_2) = Q(\pi, P_1) + Q(\pi, P_2).$$

The additive character (1), (2) of the heating measure under continuation of processes and joining parts is completely analogous to the properties of $w(\pi, P)$ and $q(\pi, P)$. The derivation of the consequences of the second law depends vitally on (1), (2). Proof of Proposition 7.1.7 is just a verification and is omitted.

The heating measure also has the following kind of cancellation property. Suppose a certain amount of heat is added to a body at the temperature θ_0 during some time interval and later the same amount of heat is emitted at the same temperature. As far as the heating measure of the process is concerned, the two amounts of heat cancel and do not contribute to the heating measure. This follows either directly from the definition of the heating measure or from Assertion (1) of Proposition 7.1.7. In particular, if no other exchange of heat occurs during the process, then its heating measure will vanish despite the fact that the body *exchanges* heat with its environment during the process. Hence, the vanishing of the heating measure does not imply that the process is adiabatic.

7.1.8 Thermodynamic Characteristics One of the key features of the second law is the lack of symmetry between the absorption and the emission of heat. We therefore wish to clearly separate these two effects and to associate appropriate objects with them. We shall use the Jordan decomposition in Proposition 7.1.3 and denote by $Q^\pm \equiv Q^\pm(\pi, P)$ the positive and negative variations of $Q \equiv Q(\pi, P)$, respectively. For every Borel set $N \subset I$, the number $Q^+(N) \geq 0$ is the heat absorbed by P in π at temperatures from N and $Q^-(N) \geq 0$ is the heat emitted by P in π at temperatures from N. The numbers

$$q^+ \equiv q^+(\pi, P) = Q^+(I), \quad q^- \equiv q^-(\pi, P) = Q^-(I)$$

are the ***heat absorbed*** and the ***heat emitted*** by P in π, respectively. Clearly,

$$q(\pi, P) = q^+(\pi, P) - q^-(\pi, P), \quad q^+(\pi, P) \geq 0, \quad q^-(\pi, P) \geq 0.$$

The support $\operatorname{supp} Q$ (see Proposition 7.1.4) is the set of empirical temperatures on which P exchanges heat during π. Since we assume that the temperature field is smooth and its range is in I, $\operatorname{supp} Q$ and $\operatorname{supp} Q^\pm$ are compact sets contained in I. $\operatorname{supp} Q^+$ is the set of all empirical temperatures on which heat is absorbed by P in π and $\operatorname{supp} Q^-$ is the set of all empirical temperatures on which heat is emitted by P in π. If $\operatorname{supp} Q^+ \neq \emptyset$, i.e., if $q^+ > 0$, then the number

$$\theta^+ \equiv \theta^+(\pi, P) = \max \operatorname{supp} Q^+$$

is referred to as ***the maximum empirical temperature at which heat is absorbed*** by P in π; if $\operatorname{supp} Q^- \neq \emptyset$, i.e., if $q^- > 0$, then the number

$$\theta^- \equiv \theta^-(\pi, P) = \min \operatorname{supp} Q^-$$

is referred to as ***the minimum empirical temperature at which heat is emitted*** by P in π.

The introduction of the quantities q^+, q^-, θ^+ and θ^- is the main point of this section. We shall use them to give the classical statements of the second law unambiguous meaning. They also appear in the efficiency estimates. The thermodynamics literature contains also several other nonequivalent definitions of the heat absorbed, emitted, etc., see DAY [1972b], DAY & ŠILHAVÝ [1977], FOSDICK & SERRIN [1975], ŠILHAVÝ [1977], TRUESDELL & BHARATHA [1977], GREEN & NAGHDI [1978]. The relationships between the definitions are discussed by ŠILHAVÝ [1980, 1982, 1983] and especially MAN [1989]. Replacing the definitions adopted here by others would strongly change the meaning of the statements of the second law to be given in the following section and consequently the validity of the results presented below. The possibility of giving several inequivalent mathematical meanings to each verbal statement of the second law most sharply demonstrates that the verbal statements themselves are by no means "operationally clear" as is sometimes claimed by those who are unwilling to accept any endeavor to improve the foundations of thermodynamics.

We conclude this section with a lemma on the integration of Borel functions with respect to the heating measure. The definition of Borel functions will not be given, but it is noted that the class of Borel functions is quite large and contains "any reasonable function". Since the heating measure is defined in terms of $\boldsymbol{q}$, r, θ, the integral $\int_I \bar{g}\, dQ(\pi, P)$ can be expressed in terms of $\boldsymbol{q}$, r, θ, too. If $\bar{g}$ is a bounded Borel function on I, one defines a local state function $\hat{g}$ by

$$\hat{g}(\sigma, \boldsymbol{p}) = \bar{g}(\theta(\sigma, \boldsymbol{p})), \quad \sigma \in \Sigma, \ \boldsymbol{p} \in \Omega. \tag{7.1.7}$$

7.1.9 Lemma *In the above situation,*

$$\int \bar{g}\, dQ(\pi, P) = \int_0^d \Big(- \int_{\partial P} g\boldsymbol{q} \cdot \boldsymbol{n}\, dA + \int_P gr\, dV \Big)\, dt \tag{7.1.8}$$

for any process π *and any part* P.

In accordance with the notation adopted, the letter g on the right-hand side of (7.1.8) denotes the evolution of the local state function $\hat{g}$ (7.1.7) in π. The integration range of the integral on the left-hand side of (7.1.8) is I. Here and in the sequel, any integral with unspecified range is an integral over I.

Proof *Step 1:* If $\bar{g} = \chi_N$ is the characteristic function of a Borel set $N \subset I$, i.e., if $\chi_N(\theta) = 0$ for $\theta \notin N$ and $\chi_N(\theta) = 1$ for $\theta \in N$, then the validity of (7.1.8) follows trivially from the definition of the heating measure. *Step 2:* If $\bar{g}$ is a simple function, i.e., a linear combination of the characteristic functions of Borel sets, then (7.1.8) is obtained by linearly combining equations (7.1.8) for the characteristic functions. *Step 3:* If $\bar{g}$ is any bounded Borel function, then there exists a sequence $\bar{g}_n$, $n = 1, \ldots,$ of simple functions such that $\bar{g}_n$ converges uniformly to $\bar{g}$ on I as $n \to \infty$. The temporal evolutions $g_n(\boldsymbol{p}, t) = \bar{g}_n(\theta(\boldsymbol{p}, t))$ also converge to the temporal evolution $g(\boldsymbol{p}, t) = \bar{g}(\theta(\boldsymbol{p}, t))$, uniformly on $\Omega \times [0, d]$. A limit with the equality (7.1.8) for the functions g_n, $\bar{g}_n$ then establishes (7.1.8) generally. □

7.2 Statements of the Second Law

While the first law says that for cyclic processes work and heat are universally proportional and hence in this sense equivalent, the second law expresses the more subtle fact that these two quantities are of unequal status in nature: it is easier to convert work into heat than to perform the opposite process. Treatises on classical thermodynamics contain many verbal statements indicating this fact in more or less suggestive ways, and associate them with the names Carnot, Clausius, Kelvin, and Planck. Below we shall formalize some of them. The unequal status of work and heat is optically manifested by the fact that it suffices to characterize the work by a single number while heat needs the heating measure or the quantities derived therefrom. Let Ω be a body. Given a process $\pi \in \Pi$ and a part $P \subset \Omega$, let us write $Q, Q^{\pm}, q^{\pm}, \theta^{\pm}$ for the heating measure and the other characteristics associated with (π, P) in the preceding section; write also w, q for $w(\pi, P)$, $q(\pi, P)$.

7.2.1 Carnot's Statement *For every cyclic process* π *and every part* P,

$$w > 0 \Rightarrow q^- > 0.$$

That is, to produce a positive amount of work in a cyclic process, the body must not only absorb heat (this follows from the first law); it must also emit some (perhaps very small) amount of it. This is the main content of the second law.

7.2.2 Kelvin's Statement *For every cyclic process* π *and every part* P,

$$w > 0 \Rightarrow \operatorname{supp} Q \ \text{ contains at least two temperatures.}$$

Stated negatively, this says that if the body exchanges heat with its environment in a cyclic process at just one temperature or if the process is adiabatic, then the body cannot produce positive net work.

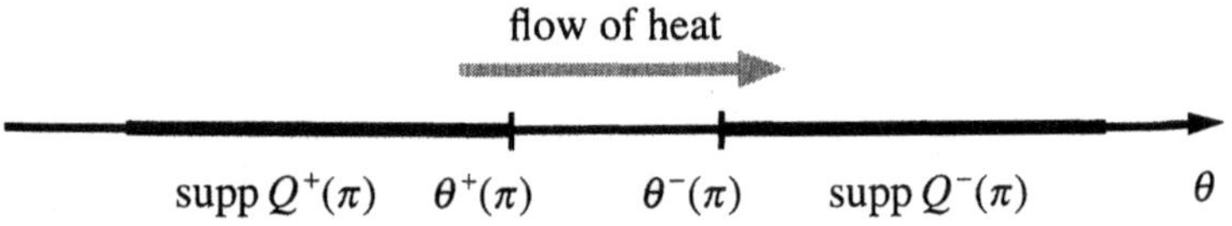

Fig. 7.2.1

7.2.3 Clausius' Statement, Version I *For every cyclic process π and every part P,*

$$w > 0 \Rightarrow q^+ > 0,\ q^- > 0 \text{ and } \theta^- < \theta^+.$$

Since $w > 0$, one has necessarily $q^+ > 0$ by the first law. Hence a process violating 7.2.3 would have either $q^- = 0$ or, if this is not the case, $\theta^- \geq \theta^+$. The impossibility of having $q^- = 0$ has already been discussed. The other possibility of violating 7.2.3, $\theta^- \geq \theta^+$, would mean that all temperatures at which heat is absorbed are below every temperature at which heat is emitted. In a process of this kind the heat absorbed at lower temperatures would be partly transferred to higher temperatures (in amount q^-) and partly transformed into positive work: see Fig. 7.2.1. This process is impossible, too.

7.2.4 Clausius' Statement, Version II *For every cyclic process π and every part P,*

$$w > 0,\ q^+ > 0 \text{ and } q^- > 0 \Rightarrow \theta^- < \theta^+.$$

A violation of 7.2.4 would mean that $w > 0$, $q^+ > 0$, $q^- > 0$ and $\theta^- \geq \theta^+$. As in the explanation of 7.2.3, such a process is impossible.

There are some immediate logical relations among the above statements of the second law. Namely, it is easily checked that under the validity of the first law, Carnot's version implies Kelvin's version and that generally Clausius' Version I implies Clausius' Version II and Carnot's version. The subsequent sections contain the analysis of the second laws under additional structural axioms, whereupon the logical relations simplify greatly.

7.3 Ideal Systems

As in the case of the first law, to derive the consequences of the second law, we shall consider a collection C of bodies with certain structural properties. The derivation relies vitally on the sufficiently rich supply of idealized reversible processes in C. Two ways of incorporating the reversibility will be considered: (a) the collection C will be assumed to contain a perfectly reversible body such as the reversible fluid to be defined below; or (b) the bodies in C are irreversible, but the heating measures of reversible processes will be obtained as limiting elements of the heating measures of processes of bodies from C. In this section the possibility (a) will be explored. The results obtained thereby will serve as a motivation for the hypotheses to be made in (b) in Sect. 7.4.

7.3.1 Notation Let $I \subset \mathbb{R}$ be an open interval of empirical temperatures. Denote by $\mathbb{M}(I)$ the linear space of all Borel measures with compact support contained in I. Every heating measure is in $\mathbb{M}(I)$ but $\mathbb{M}(I)$ contains measures which can never be obtained as a heating measure of a process. Namely the representation (7.1.6) imposes certain (very mild) restrictions on $Q(\pi, P)$. (The intrinsic description of the measures

$Q \in \mathbb{M}(I)$ which have the representations (7.1.6) appears to be a complicated problem.) We denote therefore by $\mathbb{M}_h(I)$ the set of those $Q \in \mathbb{M}(I)$ for which there exists a regular region $P \subset \Omega$ and three smooth functions $\boldsymbol{q}$, r, θ on $\Omega \times [0, d]$ such hat Q has the representation (7.1.6). The objects need not be associated with any body from the collection C to be dealt with.

If Ω is a body, then $\mathbb{D}(\Omega)$ denotes the set of all measures $Q \in \mathbb{M}_h(I)$ of the form $Q = Q(\pi, P)$ where π is a cyclic process and P a part of Ω. The elements of $\mathbb{D}(\Omega)$ are called ***cyclic heating measures***of Ω. If C is a collection (= a set) of bodies, then $\mathbb{D}(C)$ denotes the union of all $\mathbb{D}(\Omega)$, $\Omega \in C$. The elements of $\mathbb{D}(C)$ are called ***cyclic heating measures*** of C.

7.3.2 Proposition *The set* $\mathbb{M}_h(I)$ *is a linear space.*

It will be also seen that $\mathbb{M}_h(I)$ is dense in $\mathbb{M}(I)$.

Proof If $Q_1, Q_2 \in \mathbb{M}_h(I)$, then they can be represented by functions $(\boldsymbol{q}_\alpha, r_\alpha, \theta_\alpha)$, on $\Omega_\alpha \times [0, d_\alpha]$, $\alpha = 1, 2$. In view of the obvious invariance of the representations with respect to translations in space and with respect to changes of scales of time, we can assume that $\Omega_1 \cap \Omega_2 = \emptyset$, and $d_1 = d_2$. The measure $Q_1 + Q_2$ is then represented by functions $(\boldsymbol{q}, r, \theta)$ on $(\Omega_1 \cup \Omega_2) \times [0, d_2]$ given by

$$m = m_1 \text{ on } \Omega_1 \times [0, d_2] \text{ and } m = m_2 \text{ on } \Omega_2 \times [0, d_2]$$

where m stands for $\boldsymbol{q}$, r, θ. Hence $Q_1 + Q_2 \in \mathbb{M}_h(I)$. As far as the homogeneity is concerned, is suffices to note that if Q is represented by the triple $(\boldsymbol{q}, r, \theta)$ on $\Omega \times [0, d]$, then the measure λQ is represented by $(\lambda \boldsymbol{q}, \lambda r, \theta)$ on $\Omega \times [0, d]$. □

We now proceed to a detailed analysis of the set of cyclic heating measures of a reversible fluid Δ, our prototype of a perfectly reversible body. The main results are Propositions 7.3.7 and 7.3.8 which show that under certain mild qualitative assumptions, $\mathbb{D}(\Delta)$ is surprisingly large. Let $\Delta \subset$ Eucl be a region with sufficiently smooth boundary. A deformation-temperature path of duration $d > 0$ on Δ is a pair (χ, θ) of functions on $\Delta \times [0, d]$ such that χ is a motion of class C^∞ and θ a scalar field of class C^∞ with values in I. The path (χ, θ) is said to be ***cyclic*** if it can be extended to a pair of functions of class C^∞ on $\Omega \times \mathbb{R}$, periodic with respect to time, of period d. In the following definition, we assume that the reference region Δ is such that for every positive smooth function v on $\Delta \times [0, d]$, $(d > 0)$ there exists a smooth deformation χ on $\Delta \times [0, d]$ such that $v = \det \nabla \chi$ identically on $\Delta \times [0, d]$. This is satisfied, e.g., if Δ is a cube.

7.3.3 Definition A body $(\Delta, \Sigma, \Pi, \hat{\theta}, \hat{\chi}, \hat{S}, \hat{\boldsymbol{b}}, \hat{\boldsymbol{q}}, \hat{r})$ is said to be a ***reversible fluid*** if it satisfies the following conditions:

(1) for every deformation-temperature path (χ, θ) on Δ of duration $d > 0$ there exists a unique process $\pi \in \Pi$ such that (χ, θ) are the motion and the temperature field corresponding to π; if (χ, θ) is cyclic, then π is cyclic. Henceforth we identify processes with deformation-temperature paths;

(2) $\hat{\boldsymbol{q}} \equiv \boldsymbol{0}$ for every state;

(3) there exist smooth functions l_v, c_v on $\mathbb{R}^{++} \times I$ such that the rate of external supply of heat $r \equiv \hat{r}$ during any process π is given by

$$r = l_v(v, \theta)\dot{v} + c_v(v, \theta)\dot{\theta}, \tag{7.3.1}$$

where $v = v(\boldsymbol{p}, t)$ is the specific volume; the functions l_v, c_v satisfy

$$l_v \neq 0, \quad (\partial_\theta l_v - \partial_v c_v)/l_v > 0, \quad c_v/l_v > 0 \tag{7.3.2}$$

on $\mathbb{R}^{++} \times I$;

(4) if r and θ are smooth functions on $\Omega \times [0, d]$ with values in $\mathbb{R}$ and I, respectively, and if v_0 is a smooth positive field on Δ, then there exists a unique positive function v on $\Omega \times [0, d]$ such that (7.3.1) holds identically and $v(\boldsymbol{p}, 0) = v_0, \boldsymbol{p} \in \Omega$.

A construction of the state space and the class of processes satisfying the above assumptions is omitted, but the systems satisfying all of the above requirements really exist. A systematic analysis of the constitutive equations of thermoelastic fluids possibly with heat conduction and viscosity which include the reversible fluids will be presented in Part III. The function c_v in (3) is the ***specific heat at constant volume*** and l_v is the ***latent heat with respect to volume.*** Inequality $(7.3.2)_3$ will be shown to imply that on every adiabat the volume is a decreasing function of the empirical temperature. Inequality $(7.3.2)_2$ will be shown to imply that the work done in a homogeneous cyclic Carnot process does not vanish (cf. Proof of Proposition 7.4.5 below) and that the absolute temperature is a monotone function of the empirical temperature (cf. Proposition 7.5.3). Condition (4) is the assumption of the unique solvability of the ordinary differential equation (7.3.1) with respect to v.

7.3.4 Proposition *$\mathbb{D}(\Delta)$ is a linear subspace of $\mathbb{M}_h(I)$.*

Proof We shall make use of the following three special properties of Δ which are easy to verify: **(1)** *reversibility:* if π is a process then the process $\bar{\pi}$ corresponding to traversing the same deformation-temperature path in the opposite direction will give the minus the original heating measure; **(2)** *rate-independence:* if π is a process, then the process π' in which the same deformation-temperature path is traversed at different rates gives the same heating measure as the original process; **(3)** *scaling property:* a mapping ϕ : Eucl $\to$ Eucl is called a λ-rescaling if it is of the form $\phi(\boldsymbol{p}) = \lambda(\boldsymbol{p} - \boldsymbol{q}) + \boldsymbol{r}, \boldsymbol{p} \in$ Eucl, where $0 < \lambda \le 1$ is a constant and $\boldsymbol{q}, \boldsymbol{r}$ are constant points of Eucl. Let $P, R \subset \Delta$ be two parts and $\pi = (\chi, \theta)$, $\tau = (\psi, \zeta)$ two processes of duration d. The process τ is called a λ-rescaling of π from P to R if there exists a λ-rescaling ϕ such that $R = \phi(P)$ and $\psi(\phi(\boldsymbol{p}), t) = \phi(\chi(\boldsymbol{p}), t), \zeta(\phi(\boldsymbol{p}), t) = \theta(\boldsymbol{p}, t)$ for every $\boldsymbol{p} \in P$ and every $t \in [0, d]$. If this is the case, then $Q(\tau, R) = \lambda^3 Q(\pi, P)$. Using these properties, two preliminary results will be proved.

(a) if $Q \in \mathbb{D}(\Delta)$ and n is an integer, then $nQ \in \mathbb{D}(\Delta)$. Indeed, let Q be the heating measure of a cyclic process π on part P. If $n > 0$, the nQ is the heating measure of the n-times repeated process π. If $n = 0$, then $nQ = 0$ is the heating measure of any rest process. If $n < 0$, then nQ is the heating measure of the $-n$-times repeated time reversal of π.

(b) If $Q_1, Q_2 \in \mathbb{D}(\Delta)$ and $\beta > 0$ sufficiently small, then $\beta(Q_1 + Q_2) \in \mathbb{D}(\Delta)$. By definition, the measures Q_α, $(\alpha = 1, 2)$ are realized as the heating measures of cyclic processes π_α on parts P_α. By the rate-independence (2) we can assume that the durations of these two processes are the same. The regions P_1, P_2 may overlap in general. This difficulty is overcome by the scaling property (3). If $\lambda > 0$ is small enough, then the parts P_1, P_2 can be mapped onto smaller parts $R_1, R_2 \subset \Delta$ by two possibly different λ-rescalings (with the same λ) in such a way that $\operatorname{cl} R_1 \cap \operatorname{cl} R_2 = \emptyset$. Then we map the deformation-temperature path (χ_1, θ_1) corresponding to π_1 from P_1 onto R_1 and the deformation-temperature path (χ_2, θ_2) corresponding to π_2 from P_2 onto R_2 by the same rescalings. The two paths on R_1, R_2 obtained in this way can be extended into a cyclic path on Δ. The cyclic process τ corresponding to this path is then the λ-rescaling of π_1 from P_1 to R_1 and at the same time the λ-rescaling of π_2 from P_2 to R_2. In view of Proposition 7.1.7(2) and of the scaling property (3) then $Q(\tau, R_1 \cup R_2) = Q(\tau, R_1) + Q(\tau, R_2) = \lambda^3 Q(\pi, P_1) + \lambda^3 Q(\pi, P_2)$ and setting $\lambda = \sqrt[3]{\beta}$ we see that (b) holds.

Proof of the additivity of $\mathbb{D}(\Delta)$: Let $Q_1, Q_2 \in \mathbb{D}(\Delta)$. Choose $n > 0$ large enough to satisfy (b) with $\beta = n^{-1}$. Then $n^{-1}(Q_1 + Q_2) \in \mathbb{D}(\Delta)$ and hence $Q_1 + Q_2 = n \cdot n^{-1}(Q_1 + Q_2) \in \mathbb{D}(\Delta)$ in view of (a). Proof of the homogeneity of $\mathbb{D}(\Delta)$: Let $Q \in \mathbb{D}(\Delta)$ and $\lambda \subset \mathbb{R}$. If $\lambda > 0$, then $\lambda n^{-1} Q \in \mathbb{D}(\Delta)$ for a sufficiently large positive integer n in view of (b). By (a) then $\lambda Q = n \cdot \lambda n^{-1} Q \in \mathbb{D}(\Delta)$. If $\lambda < 0$ then $-\lambda n^{-1} Q \in \mathbb{D}(\Delta)$ for a sufficiently large positive integer n by (b). By (a) then $\lambda Q = (-n)(-\lambda n^{-1} Q) \in \mathbb{D}(\Delta)$. □

7.3.5 Lemma *For every $Q \in \mathbb{M}_h(I)$ there exists a process π of Δ and a part $P \subset \Delta$ such that $Q = Q(\pi, P)$.*

It is not asserted that π is cyclic.

Proof By definition, every $Q \in \mathbb{M}_h(I)$ has a representation by a triple of functions $(\boldsymbol{q}, r, \theta)$ on $\Omega\times[0, d]$ as in (7.1.6). Assume first that $\boldsymbol{q} \equiv \boldsymbol{0}$. By using the obvious changes of scales, we can assume that $\Omega \subset \Delta$. We now extend the functions r and θ on $\Omega \times [0, d]$ to any functions $\bar{r}, \bar{\theta}$ on $\Delta \times [0, d]$ with the range of $\bar{\theta}$ in I. By the solvability condition (4) in the definition of the reversible fluid, there exists a process π of Δ such that $\bar{\theta}$ is the temperature field during π and $\bar{r}$ is the external supply of heat during π. Then obviously $Q = Q(\pi, \Omega)$. If $Q \in \mathbb{M}_h(I)$ is represented by a general triple of functions $(\boldsymbol{q}, r, \theta)$ on Ω, it can be shown that Q can also be represented by $(\boldsymbol{q}', r', \theta')$ on some Ω' with $\boldsymbol{q}' \equiv \boldsymbol{0}$, i.e., the general case can be reduced to the above particular case. The reduction is based on the idea of replacing the boundary $\partial\Omega$ by a thin three-dimensional shell along the boundary and by replacing the surface influx $-\boldsymbol{q}\cdot\boldsymbol{n}$ across $\partial\Omega$ by appropriate rate r of external body supply defined on the shell. The details are omitted. □

To facilitate the proof of Proposition 7.3.7 we introduce the important concepts of adiabat and of Carnot process.

7.3.6 Definitions

(1) An ***adiabat*** of Δ is any maximal smooth parametrized curve A in $\mathbb{R}^{++}\times I$ such that $l_v dv + c_v d\theta = 0$ along A. An ***isotherm*** of Δ is any line l in $\mathbb{R}^{++}\times I$ of the form $l = \{(v, \theta_0) : v > 0\}$ for some fixed θ_0.

(2) A ***homogeneous cyclic Carnot process*** π of Δ at the operating temperatures $\theta, \omega \in I, \theta < \omega$, is a cyclic process π such that (a) the specific volume field $v(\boldsymbol{p}, t)$ and the temperature filed $\theta(\boldsymbol{p}, t)$ during π are uniform over Δ, i.e., $v(\boldsymbol{p}, t) = v(t), \theta(\boldsymbol{p}, t) = \theta(t), (\boldsymbol{p}, t) \in \Delta \times [0, d_\pi]$; (b) the path

$$P = \{(v(t), \theta(t)) \in \mathbb{R}^{++}\times I, t \in [0, d_\pi]\} \tag{7.3.3}$$

is simple and consists of two connected portions of different isotherms at temperatures θ and ω, respectively, and of two connected portions of different adiabats connecting the two isotherms; (c) the fluid Δ expands at the temperature ω, i.e., $v(t)$ is a monotonically nondecreasing function of t during the isothermal portion of the process at temperature ω.

If the parametrizations of the two adiabats are $v_1(\theta), v_2(\theta), v_1(\theta) < v_2(\theta)$, then for any pait P,

$$Q(\pi, P) = V(P)(c(\omega)\delta_\omega - c(\theta)\delta_\theta), \tag{7.3.4}$$

where $V(P)$ is the referential volume of $P \subset \Delta$ and

$$c(\theta) = \int_{v_1(\theta)}^{v_2(\theta)} l_v(v, \theta)\, dv, \qquad \theta \in I. \tag{7.3.5}$$

7.3.7 Proposition

(1) *For every $\theta \in I$ and $Q \in \mathbb{M}_h(I)$ there exists a $c \in \mathbb{R}$ such that*

$$Q + c\delta_\theta \in \mathbb{D}(\Delta), \tag{7.3.6}$$

i.e.,

$$\mathbb{D}(\Delta) + \mathbb{R}\cdot\delta_\theta = \mathbb{M}_h(I),$$

where $\mathbb{R}\cdot\delta_\theta := \{c\delta_\theta : c \in \mathbb{R}\}$ is the straight line through 0 and δ_θ;

(2) *For any $\theta, \omega \in I, \theta \neq \omega$, there is a $Q \in \mathbb{D}(\Delta)$ with $\operatorname{supp} Q = \{\theta, \omega\}$ and $Q(I) \neq 0$. (Here $\{\theta, \omega\}$ is the set consisting of the two points θ, ω.)*

Hence $\mathbb{D}(\Delta)$ is extremely large, namely, it is a subspace of $\mathbb{M}_h(I)$ of codimension of at most 1.

Proof (1): Assertion (1) says that corresponding to every $Q \in \mathbb{M}_h(I), \theta_0 \in I$ there exists a constant $c \in \mathbb{R}$, a cyclic process π of Δ and a part P of Δ such that $Q = Q(\pi, P) + c\delta_{\theta_0}$. The cyclic process π will be constructed in the form $\pi = \pi_1 * \pi_2 * \pi_3 * \pi_4$, where the processes π_α, $\alpha = 1, \ldots, 4$, are determined as follows. The process π_1 is the process satisfying $Q = Q(\pi_1, P)$ for some $P \subset \Delta$. The existence is guaranteed by Lemma 7.3.5. The process π_1 need not be cyclic. We have, at every point $\boldsymbol{p} \in \Delta$, the initial and final values of specific volume and temperature: $(v^i(\boldsymbol{p}), \theta^i(\boldsymbol{p}))$, $(v^f(\boldsymbol{p}), \theta^f(\boldsymbol{p}))$. Our task is to complete π_1 into a cycle. The process π_2 is defined to be an adiabatic process ($r = 0$, see Example 7.1.5(1) and recall that $\boldsymbol{q} \equiv \boldsymbol{0}$) which proceeds, at every $\boldsymbol{p} \in \Delta$, along the adiabat through $(v^f(\boldsymbol{p}), \theta^f(\boldsymbol{p}))$ until it reaches the temperature θ_0. The existence of this process follows from Definition 7.3.3(4): We take any smooth function θ_1 on $\Delta \times [0, d]$, $(d > 0)$ such that $\theta_1(\boldsymbol{p}, 0) = \theta^f(\boldsymbol{p}), \theta_1(\boldsymbol{p}, d) = \theta_0, \boldsymbol{p} \in \Delta$. Corresponding to this function, there exists a function v_1 on $\Delta \times [0, d]$ such that $v_1(\boldsymbol{p}, 0) = v^f(\boldsymbol{p}), \boldsymbol{p} \in \Delta$, and $l_v(v_1, \theta_1)\dot{v}_1 + c_v(v_1, \theta_1)\dot{\theta}_1 = 0$ identically on $\Delta \times [0, d]$. (One applies (4) with $r \equiv 0$.) The process π_4 is the time reversal of a process π_5 which is constructed as an adiabatic process which starts at $(v^i(\boldsymbol{p}), \theta^i(\boldsymbol{p}))$ and proceeds, at every $\boldsymbol{p} \in \Delta$, along the adiabat through $(v^i(\boldsymbol{p}), \theta^i(\boldsymbol{p}))$ until it reaches the temperature θ_0. The existence of π_5 again follows from Definition 7.3.3(4). Finally π_3 is an isothermal process of temperature θ_0 which connects, at every $\boldsymbol{p} \in \Delta$, the adiabat through $(v^f(\boldsymbol{p}), \theta^f(\boldsymbol{p}))$ with the adiabat through $(v^i(\boldsymbol{p}), \theta^i(\boldsymbol{p}))$. To construct π_3, consider a deformation temperature path (χ_3, θ_0) on $\Delta \times [0, d]$ with θ_0 a spatial-temporal constant (the given θ_0) and with χ chosen such that the corresponding specific volume v_3 satisfies $v_3(\boldsymbol{p}, 0) = v_2^f(\boldsymbol{p}), v_3(\boldsymbol{p}, d) = v_4^i(\boldsymbol{p}), \boldsymbol{p} \in \Delta$.

The rate-independence of Δ allows one to choose the parametrization of π_α, $\alpha = 1, \ldots, 4$, in such a way that the resulting continuation will give a smooth cyclic path. The 'corners' at the points of continuation do not preclude the existence of such a smooth parametrization. The process π is cyclic by Definition 7.3.3(1). Since π_2 and π_4 are adiabatic, we have $Q(\pi_2, P) = Q(\pi_4, P) = 0$ and since π_3 is isothermal at temperature θ_0, we have $Q(\pi_3, P) = c\delta_{\theta_0}$ where $c = q(\pi_3, P)$. Combining this with $Q = Q(\pi_1, P)$ we obtain

$$Q(\pi, P) = Q(\pi_1, P) + Q(\pi_2, P) + Q(\pi_3, P) + Q(\pi_4, P) = Q + c\delta_{\theta_0}.$$

(2): The measure Q with the prescribed support is a heating measure of any cyclic homogeneous Carnot process of Δ with the operating temperatures θ, ω. Let us now show that $(7.3.2)_2$ implies that the net gain of heat $Q(I)$ in a Carnot process satisfies $Q(I) \neq 0$. Denoting by P the path of the process in the v, θ-space, we have $Q(I) = \int_P (l_v dv + c_v d\theta)$ and using Green's theorem this can be written as $Q(I) = \int_O (\partial_\theta l_v - \partial_v c_v)\, dv d\theta$, where O is the region enclosed by P. But as O is connected and by $(7.3.2)_1$ the integrand different from 0, it follows that the integrand is either positive everywhere on O or negative everywhere on O. In both cases the integral is different from 0. □

7.3.8 Proposition (An Alternative) *Either*

(1) $\mathbb{D}(\Delta) = \mathbb{M}_h(I)$ *or*

(2) *there exists a smooth function* $T : I \to \mathbb{R}^{++}$ *such that*

$$\mathbb{D}(\Delta) = \{Q \in \mathbb{M}_h(I) : \int dQ/T = 0\}.$$

Assertion (2) is the Clausius equality for cyclic processes of Δ. We shall see in Sect. 7.5 that for reversible bodies it is equivalent to the second law of thermodynamics. The proposition gives a remarkable alternative for Δ. Either the set of all cyclic heating measures contains all of $\mathbb{M}_h(I)$ or, if it is not the case, then the Clausius equality, and hence the second law, holds for Δ. It must be emphasized that the alternative is completely independent of any postulation of the second law.

Proof Hence, assume that

$$\mathbb{D}(\Delta) \neq \mathbb{M}_h(I) \tag{7.3.7}$$

and prove (2). Let us first show that for every $Q \in \mathbb{M}_h(I)$ and $\theta \in I$ the number c as in (7.3.6) is unique. Suppose on the contrary that for some $R \in \mathbb{M}_h(I)$ and $\theta \in I$ there are two different numbers c, d such that

$$R + c\delta_\theta \in \mathbb{D}(\Delta), \quad R + d\delta_\theta \in \mathbb{D}(\Delta). \tag{7.3.8}$$

Since $\mathbb{D}(\Delta)$ is a linear subspace (Proposition 7.3.4), it follows that $(c-d)\delta_\theta \in \mathbb{D}(\Delta)$ and as $c-d \neq 0$, it follows that $\delta_\theta \in \mathbb{D}(\Delta)$ and hence also

$$-a\delta_\theta \in \mathbb{D}(\Delta) \tag{7.3.9}$$

for every $a \in \mathbb{R}$. By Proposition 7.3.7 for every $Q \in \mathbb{M}_h(I)$ there exists an $a \in \mathbb{R}$ such that $Q + a\delta_\theta \in \mathbb{D}(\Delta)$. As $\mathbb{D}(\Delta)$ is a linear space, we have $Q = (Q + a\delta_\theta) - a\delta_\theta \in \mathbb{D}(\Delta)$. That is, $Q \in \mathbb{D}(\Delta)$ for every $Q \in \mathbb{M}_h(I)$, in contradiction with (7.3.7). Hence really the number c as in (7.3.6) is uniquely determined by Q and θ_0. Let us now fix $\theta_0 \in I$ and define a functional F on $\mathbb{M}_h(I)$ by the requirement that

$$Q + F(Q)\delta_{\theta_0} \in \mathbb{D}(\Delta) \tag{7.3.10}$$

for every $Q \in \mathbb{M}_h(I)$. The above uniqueness assertion shows that F is well-defined. The linearity of $\mathbb{D}(\Delta)$ implies that F is a linear functional and for every $Q \in \mathbb{M}_h(I)$,

$$F(Q) = 0 \Leftrightarrow Q \in \mathbb{D}(\Delta) \tag{7.3.11}$$

i.e., that the kernel of F is exactly $\mathbb{D}(\Delta)$. Set $f(\theta) = F(\delta_\theta)$, $\theta \in I$, and prove that

$$f(\theta) \neq 0 \quad \text{for every} \quad \theta \in I. \tag{7.3.12}$$

Indeed, if $f(\theta) = 0$ for some $\theta \in I$, then $\delta_\theta = \delta_\theta + f(\theta)\delta_\theta = \delta_\theta + F(\delta_\theta)\delta_\theta \in \mathbb{D}(\Delta)$ in view of (7.3.10), and the inclusion $\delta_\theta \in \mathbb{D}(\Delta)$ was shown to lead to a contradiction with (7.3.7). Let π be a cyclic homogeneous Carnot process with operating temperatures $\theta, \omega \in I$, $\theta < \omega$. In view of (7.3.4) and of the linearity of F,

$$F(Q(\pi, P)) = V(P)c(\omega)F(\delta_\omega) - V(P)c(\theta)F(\delta_\theta) = V(P)c(\omega)f(\omega) - V(P)c(\theta)f(\theta),$$

where $c(\theta)$ is given by (7.3.5). Since π is cyclic, we have $F(Q(\pi, P)) = 0$ by (7.3.11) and the above computation gives

$$c(\theta)f(\theta) = c(\omega)f(\omega). \tag{7.3.13}$$

Let us now fix the temperature ω and the two adiabats of the Carnot process. Varying the temperature θ we obtain a family of Carnot processes. By 7.3.3(3) either $l_v > 0$ on $\mathbb{R}^{++} \times I$ or $l_v < 0$ on $\mathbb{R}^{++} \times I$ and hence either $c(\theta) > 0$ on I or $c(\theta) < 0$ on I. (7.3.13) can now be rewritten as $f(\theta) = c(\omega)f(\omega)/c(\theta)$. It follows that, firstly, f is of the same sign on I and, secondly, that f is smooth on I (invoke (7.3.5) and use the fact that $v_1(\theta)$, $v_2(\theta)$ and l_v are smooth). Taking into account also that (7.3.12) holds, we now define $T : I \to \mathbb{R}^{++}$ by $T = 1/|f|$. T is smooth. Equality (7.3.13) reads $c(\theta)/T(\theta) + c(\omega)/T(\omega) = 0$ which can be rewritten as $\int_P ((l_v/T)\,dv + (c_v/T)\,d\theta) = 0$ where P is the path (7.3.3). By Green's theorem, $\int_O (\partial_\theta(l_v/T) - \partial_v(c_v/T))\,dv\,d\theta = 0$ where O is the region bounded by P. The arbitrariness of the Carnot path P gives that the integrand vanishes, i.e.,

$$\partial_\theta(l_v/T) = \partial_v(c_v/T). \tag{7.3.14}$$

The differential form $\alpha := (l_v dv + c_v d\theta)/T$ is therefore closed. As $\mathbb{R}^{++} \times I$ is simply connected, α has a potential η, i.e., $d\eta = \alpha$. By (7.3.1) then $r(p, t)/T(\theta(p, t)) = d\eta(\theta(p, t), v(p, t))/dt$ for every process π. Hence if π is cyclic,

$$\int_0^d r(p, t)/T(\theta(p, t))\,dV\,dt = 0$$

for every $p \in \Delta$. By Lemma 7.1.9 this can be rewritten as $\int dQ(\pi, P)/T = 0$ which proves that $\mathbb{D}(\Delta) \subset \{Q \in \mathbb{M}_h(I) : \int dQ/T = 0\}$. Conversely, suppose that $\int dQ/T = 0$ and prove that $Q \in \mathbb{D}(\Delta)$. Let $\theta \in I$. By Proposition 7.3.7 there exists a c such that $R := Q + c\delta_\theta \in \mathbb{D}(\Delta)$. Then by the preceding part of the proof $\int dR/T = 0$ which means $\int dQ/T + c/T(\theta) = 0$. But $\int dQ/T = 0$ and hence $c = 0$. Hence $R = Q$ is in $\mathbb{D}(\Delta)$. □

7.4 The Collection of Bodies

This section is devoted to the possibility (b) mentioned at the beginning of Sect. 7.3. No kind of strictly reversible system is postulated. It is not expected that the set $\mathbb{D}(C)$ will contain a large collection of reversible measures, i.e., measures $Q \in \mathbb{D}(C)$ for which $-Q \in \mathbb{D}(C)$. Rather, the set $\mathbb{D}(C)$ is first augmented with limiting elements which are interpreted as "ideal" cyclic processes. The union of $\mathbb{D}(C)$ with the set of limits of families of measures from $\mathbb{D}(C)$ is identified with the closure of $\mathbb{D}(C)$ in the topology to be defined below. It is realistic to assume that even if $\mathbb{D}(C)$ contains no reversible heating measures, the closure of will contain many.

We denote by $C^\infty(I)$ the linear space of all infinitely differentiable real-valued functions defined on I. The pair $\mathbb{M}_h(I)$, $C^\infty(I)$ is a pair of linear spaces in duality (see BOURBAKI [1966]) with the pairing $\langle \cdot, \cdot \rangle : \mathbb{M}_h(I) \times C^\infty(I) \to \mathbb{R}$ given by

$$\langle Q, f \rangle = \int f \, dQ.$$

The space $\mathbb{M}_h(I)$ is endowed with the coarsest locally convex topology on $\mathbb{M}_h(I)$ that renders the functionals $Q \mapsto \langle Q, f \rangle$, $Q \in \mathbb{M}_h(I)$ continuous for every $f \in C^\infty(I)$. The choice of the topology can be justified on phenomenological grounds, see ŠILHAVÝ [1980,1982].

7.4.1 Proposition

(1) *Every continuous linear functional F on $\mathbb{M}_h(I)$ is of the form*

$$F(Q) = \int f \, dQ, \quad Q \in \mathbb{M}_h(I) \tag{7.4.1}$$

where $f \in C^\infty(I)$.

(2) *If $\mathbb{F}$ is a closed linear subspace of $\mathbb{M}_h(I)$ and $Q_0 \notin \mathbb{F}$, then there exists a continuous linear functional F on $\mathbb{M}_h(I)$ such that*

$$F(Q_0) \neq 0 \text{ and } \mathbb{F} \subset \{Q \in \mathbb{M}_h(I) : F(Q) = 0\}.$$

See BOURBAKI [1966] for a duality theory from which the proof follows. Assertion (2) is a version of the Hahn-Banach theorem. The structural properties imposed on the collection C are analogous to those for the first law.

7.4.2 Assumptions

(1) $\mathbb{D}(C)$ is additive.

(2) $\mathbb{D}(C)$ contains a closed linear subspace $\mathbb{E}$ such that for every $\theta, \omega \in I, \theta \neq \omega$, there exists a $Q \in \mathbb{E}$ with $\operatorname{supp} Q = \{\theta, \omega\}$ and $Q(I) \neq 0$.

(3) C contains a reversible fluid.

Assumption (1) reflects the additivity of heating measures (cf. Proposition 7.1.7) and the structural property

$$\Omega_1, \Omega_2 \in C \Rightarrow \Omega_1 \oplus \Omega_2 \in C.$$

The motivation is completely analogous to the additivity of the set of work–heat pairs $\mathbb{P}(C)$ in Sect. 5.2. The ideal reversible processes enter through (2). The set $\mathbb{E}$ is identified with the set of ideal elements of $\mathbb{D}(C)$ which correspond to homogeneous, reversible, cyclic, processes. The property $Q \subset \mathbb{E} \Rightarrow Q \subset \mathbb{E}$ expresses the reversibility; the property that $\mathbb{E}$ is closed means that the limits of the elements of $\mathbb{E}$ are already in $\mathbb{E}$; the inclusion $\mathbb{E} \subset \mathbb{D}(C)$ says that the ideal processes satisfy the same laws as the real ones. Finally, the additivity of $\mathbb{E}$ is a general requirement reflecting the additivity of the heating measures and the structure of the collection C; the motivation is the same as in Sect. 5.2 and the homogeneity of $\mathbb{E}$ can be derived from the fact that the processes considered are homogeneous, again in analogy with the discussion in Sect. 5.2. Finally, the existence of the process with $\operatorname{supp} Q = \{\theta, \omega\}$ is the requirement that there are sufficiently many reversible homogeneous cyclic Carnot processes. Proposition 7.4.5, below, says that (3) $\Rightarrow$ (2). The advantage of (3) is that it is very concrete and free of the use of the topology on $\mathbb{M}_h(I)$.

7.4.3 Proposition (An Alternative) *Let $\mathbb{E}$ be the closed linear subspace with the properties described in Assumption* 7.4.2(2). *Then either*
(1) $\mathbb{E} = \mathbb{M}_h(I)$ *or*
(2) *there exists a smooth function $T : I \to \mathbb{R}^{++}$ such that*

$$\mathbb{E} = \{Q \in \mathbb{M}_h(I) : \int dQ/T = 0\}. \tag{7.4.2}$$

If $\mathbb{E}$ does not contain all of $\mathbb{M}_h(I)$, then every element from $\mathbb{E}$ satisfies the Clausius equality for reversible cyclic processes and hence also the second law.

Proof Assume that

$$\mathbb{E} \neq \mathbb{M}_h(I) \tag{7.4.3}$$

and prove (2). We start with proving that under (7.4.3) there exists a nonzero linear continuous functional F on $\mathbb{M}_h(I)$ such that

$$\mathbb{E} = \{Q \in \mathbb{M}_h(I) : F(Q) = 0\}. \tag{7.4.4}$$

Indeed, since $\mathbb{E}$ is a closed linear subspace and (7.4.3) holds, by Proposition 7.4.1(2) there exists a nonzero continuous linear functional F such that

$$\mathbb{E} \subset \{Q \in \mathbb{M}_h(I) : F(Q) = 0\}. \tag{7.4.5}$$

If the inclusion is strict, i.e., if (7.4.4) does not hold, then there exists a R such that $F(R) = 0$ and $R \notin \mathbb{E}$. By Proposition 7.4.1(2) there exists another continuous linear functional G on $\mathbb{M}_h(I)$ such that $G(R) \neq 0$ and $\mathbb{E} \subset \{Q \in \mathbb{M}_h(I) : G(Q) = 0\}$. By Proposition 7.4.1(1) the functionals F, G have representations $F(Q) = \int f\, dQ$, $G(Q) = \int g\, dQ$, $Q \in \mathbb{M}_h(I)$, where $f, g \in C^\infty(I)$. Let $\tau, \omega \in I$, $\tau \neq \omega$. By Assumption 7.4.2(2) there is a measure Q in $\mathbb{E}$ such that $\operatorname{supp} Q = \{\tau, \omega\}$. Then

$$0 = F(Q) = f(\tau)Q[\tau] + f(\omega)Q[\omega], \quad 0 = G(Q) = g(\tau)Q[\tau] + g(\omega)Q[\omega], \tag{7.4.6}$$

Since $\operatorname{supp} Q = \{\tau, \omega\}$, necessarily $Q[\tau] \neq 0$, $Q[\omega] \neq 0$. We shall show that $f(\theta) \neq 0$, $g(\theta) \neq 0$ for every $\theta \in I$. Indeed, if $f(\tau) = 0$ were to hold for some $\tau \in I$, then $(7.4.6)_1$ would imply $f(\omega)Q[\omega] = 0$ and so $f(\omega) = 0$. Since Ω is arbitrary, this would give $f \equiv 0$ identically on I, in contradiction with $F \neq 0$. The same applies to g. Equations (7.4.6) imply that $g(\omega) = g(\tau)f(\omega)/f(\tau)$ and as τ, Ω are arbitrary, g is a multiple of f, i.e., $g(\theta) = kf(\theta)$ for some $k \in \mathbb{R}$ and every $\theta \in I$. Hence $G(Q) = kF(Q)$, $Q \in \mathbb{M}_h(I)$. But $F(R) = 0$ and $G(R) \neq 0$; this contradiction shows that (7.4.4) holds. Since f is smooth and different from 0 everywhere, either $f > 0$ everywhere on I or $f < 0$ everywhere on I. Defining $T : I \to \mathbb{R}^{++}$ by $T = 1/|f|$, we see that the assertion of the proposition holds. □

7.4.4 Proposition *Let $\mathbb{E}$ be the closed linear subspace with the properties described in Assumption* 7.4.2(2). *If the alternative* (2) *in Proposition* 7.4.3 *holds and $\theta \in I$ is given, then every $Q \in \mathbb{D}(C)$ can be written uniquely in the form $Q = c\delta_\theta - R$ where $R \in \mathbb{E}$ and $c \in \mathbb{R}$; moreover $c\delta_\theta \in \mathbb{D}(C)$ and c is of the same sign as $\int dQ/T$. In particular, $\mathbb{E} + \mathbb{R} \cdot \delta_\theta = \mathbb{M}_h(I)$.*

This has the following meaning. The measures Q, R are heating measures of some cyclic processes π, τ on some parts P, M of certain bodies $\Omega, \Lambda \in \mathbb{D}(C)$. Moreover the process τ is reversible since $R \in \mathbb{E}$. The proposition says that for the given π, P, Ω one always finds a reversible process τ and part M such that process consisting in running simultaneously the processes π, τ on the union of P, M will exchange heat with the environment only at θ.

Proof We define the number c and the measure R by $c = T(\theta) \int dQ/T$, $R = Q - c\delta_\theta$. Then $\int dR/T = 0$ and so $R \in \mathbb{E}$ by Proposition 7.4.3(2). Since $\mathbb{E}$ is a linear subspace, we have $-R \in \mathbb{E}$ and as $\mathbb{D}(C)$ is additive (Assumption 7.4.2(1)) and $Q \in \mathbb{D}(C)$, we have $c\delta_\theta = Q - R \in \mathbb{D}(C)$. □

We conclude this section by observing that the set $\mathbb{D}(\Delta)$ of cyclic heating measures of a reversible fluid has the properties imposed above as postulates on $\mathbb{E}$.

7.4.5 Proposition *$\mathbb{D}(\Delta)$ is a closed linear subspace of $\mathbb{M}_h(I)$; if $\Delta \in$ C, then Assumption* 7.4.2(2) *holds with* $\mathbb{E} = \mathbb{D}(\Delta)$.

Proof In view of the alternative 7.3.8 we easily learn that in both cases $\mathbb{D}(\Delta)$ is a closed linear subspace; by Proposition 7.3.7(2) it has all properties required in Assumption 7.4.2(2). □

7.5 The Absolute Temperature Scale. The Clausius Inequality

The second law of thermodynamics in any of its versions given in Sect. 7.2 leads to the existence of the absolute temperature scale satisfying the Clausius inequality.

7.5.1 Definition Let Ω be a body, $I \subset \mathbb{R}$ the range of empirical temperatures and $T : I \to \mathbb{R}^{++}$ a smooth function. The body Ω is said to satisfy the ***Clausius inequality with the absolute temperature scale*** T if

$$\int \frac{dQ(\pi, P)}{T} \le 0. \tag{7.5.1}$$

for every cyclic process π and every part $P \subset \Omega$. Given the absolute temperature scale T, the ***absolute temperature*** is the local state function $\hat{T} = T \circ \hat{\theta}$. By Lemma 7.1.9, (7.5.1) can be rewritten as

$$\int_0^{d_\pi} \Big(-\int_{\partial P} \frac{\boldsymbol{q} \cdot \boldsymbol{n}}{T}\, dA + \int_P \frac{r}{T}\, dV\Big)\, dt \le 0. \tag{7.5.2}$$

7.5.2 Proposition *Let C satisfy either Assumptions* 7.4.2(1),(2) *or* 7.4.2(1),(3). *Then*

(1) *Carnot's and Kelvin's statements of the second law are equivalent and hold if and only if each body from C satisfies the Clausius inequality with some universal absolute temperature scale T;*

(2) *Versions I and II of Clausius' statement of the second law are equivalent and hold if and only if each body from C satisfies the Clausius inequality some universal absolute temperature scale T that is an increasing function of the empirical temperature;*

(3) *the equivalent Versions I and II of Clausius' statement of the second law imply the equivalent Carnot and Kelvin statements of the second law;*

(4) *the absolute temperature scale, if it exists, is unique to within a constant positive multiple.*

The universality means that T is the same for all bodies from C. Neither the Carnot nor the Kelvin statement of the second law generally guarantee that the absolute temperature scale is an increasing function of the empirical temperature since they do not involve inequalities between the empirical temperatures, which would fix the order given by "hotter than". The Clausius versions are free of this ambiguity.

Proof By Proposition 7.4.5, Assumption 7.4.2(3) is stronger than Assumption 7.4.2(2) and so it suffices to give a proof under the weaker Assumption 7.4.2(1),(2). To prove Assertion (1), denote by (C), (K) the Carnot and Kelvin statements, respectively, and by (T) the existence of the absolute temperature scale T as in (1). It suffices to prove that (C) $\Rightarrow$ (K) $\Rightarrow$ (T) $\Rightarrow$ (C). (T) $\Rightarrow$ (C) : If π is a cyclic process and P a part of some body from C such that $w(\pi, P) > 0$, then necessarily $q^+(\pi, P) > 0$ by the first law and since $\int dQ(\pi, P)/T \le 0$ and T is positive, the measure Q cannot be nonnegative. That is, $q^-(\pi, P) > 0$. (C) $\Rightarrow$ (K) : The violation of (K) means the existence of a cyclic process π

and part P of some body from C such that $w(\pi, P) > 0$ and $Q(\pi, P) = c\delta_\theta$ for some $c \in \mathbb{R}$ and $\theta \in I$. By the first law then $c > 0$ and so $Q(\pi, P)$ is a nonnegative measure which implies $q^+(\pi, P) > 0$ and $q^-(\pi, P) = 0$; this is a violation of (C). (K) $\Rightarrow$ (T) : Assume that (K) holds. In view of Assumption 7.4.2(2) we have the alternative described in Proposition 7.4.3. By (K) $\mathbb{E} \nsupseteq \mathbb{M}_h(I)$. The alternative says that there exists a smooth function $T : I \to \mathbb{R}^{++}$ such that (7.4.2) holds. Let $Q \in \mathbb{D}(C)$ and prove that

$$\int \frac{dQ}{T} \le 0. \tag{7.5.3}$$

Take arbitrary $\theta \in I$. By Proposition 7.4.4 then there exists a $c \in \mathbb{R}$ of the same sign as $\int dQ/T$ such that $c\delta_\theta \in \mathbb{D}(C)$. By (K) necessarily $c \le 0$ for otherwise the process π and part P of some body from C for which $Q(\pi, P) = c\delta_\theta$ would have $w(\pi, P) > 0$ and $\operatorname{supp} Q(\pi, P)$ a one-point set. This completes the proof of (1).

To prove (2), denote by (C-I),(C-II) Versions I, II of the Clausius statement, respectively, and by (T^+) the existence of the increasing absolute temperature scale such that the Clausius inequality holds for every body from C. We shall prove (C-I) $\Rightarrow$ (C-II) $\Rightarrow$ (T^+) $\Rightarrow$ (C-I). The implication (C-I) $\Rightarrow$ (C-II) is immediate.

(C-II) $\Rightarrow$ (T^+) : By (C-II), $\mathbb{E} \nsupseteq \mathbb{M}_h(I)$. In view of Assumption 7.4.2(2) and the alternative Proposition 7.4.3 there exists a smooth function $T : I \to \mathbb{R}^{++}$ such that (7.4.2) holds. Let us prove that (7.5.3) holds for every $Q \in \mathbb{D}(C)$. Suppose, in contradiction, that $\int dQ/T > 0$ for some $Q \in \mathbb{D}(C)$. Let $\theta \in I$ be arbitrary. By Proposition 7.4.4 there exists a $c > 0$ such that $c\delta_\theta \in \mathbb{D}(C)$. Let $\omega \in I$ be any temperature such that $\omega > \theta$. By Assumption 7.4.2(2) there exists a $S \in \mathbb{E}$ such that $\operatorname{supp} S = \{\theta, \omega\}$. As $\int dS/T = 0$, necessarily $S[\theta]$ and $S[\omega]$ are of different signs. In view of the reversibility we can assume that $S[\omega] < 0$. Since $c\delta_\theta \in \mathbb{D}(C)$ and $S \in \mathbb{E}$, we have $W := c\delta_\theta + \lambda S \in \mathbb{D}(C)$ for every $\lambda \in \mathbb{R}$ and if $\lambda > 0$ is sufficiently small, then also

$$\operatorname{supp} W = \{\theta, \omega\}, \quad W[\theta] > 0, \quad W[\omega] < 0, \quad W(I) > 0. \tag{7.5.4}$$

By the definition of $\mathbb{D}(C)$ there exists a cyclic process π and a part P of some body from C such that $Q(\pi, P) = W$. Relations (7.5.4) imply $w(\pi, P) = W(I) > 0$, $q^+(\pi, P) = W[\theta] > 0$, $q^-(\pi, P) = -W[\omega] > 0$, $\theta^+(\pi, P) = \theta$, $\theta^-(\pi, P) = \omega$, and as $\omega > \theta$, the process π contradicts (C-II). Hence really (7.5.3) holds for every $Q \in \mathbb{D}(C)$. The only thing that now remains to be proved is that T is an increasing function. Let $\theta, \omega \in I, \theta < \omega$. By assumption 7.4.2(2) there is a $Q \in \mathbb{E} \subset \mathbb{D}(C)$ such that $\operatorname{supp} Q = \{\theta, \omega\}$. Without any loss of generality we can assume that $Q(I) > 0$. The equality $\int dQ/T = 0$ implies that $Q[\theta]$ and $Q[\omega]$ are of different signs and applying (C-II) to the cyclic process π represented by the measure Q tells us that $\theta^+(\pi, P) > \theta^-(\pi, P)$. This gives $Q[\omega] > 0$, and $Q[\theta] < 0$. But as $Q(I) > 0$, we also know that $Q[\omega] > -Q[\theta] > 0$. Combining this with $\int dQ/T = 0$ gives $T(\theta) < T(\omega)$.

(T^+) $\Rightarrow$ (C-I) : Let (T^+) hold and let π be a cyclic process and P any part. Then, setting $T^\pm = T(\theta^\pm(\pi, P))$ and omitting the argument (π, P) in the heating measure and in other characteristics of (π, P), we have

$$0 \ge \int \frac{dQ}{T} = \int \frac{dQ^+}{T} - \int \frac{dQ^-}{T} \ge \int \frac{dQ^+}{T^+} - \int \frac{dQ^-}{T^-} = \frac{q^+}{T^+} - \frac{q^-}{T^-}, \tag{7.5.5}$$

since T is increasing and $\theta \le \theta^+$ on $\operatorname{supp} Q^+$ and $\theta \le \theta$ on $\operatorname{supp} Q^-$. The above inequality provides the classical efficiency estimate

$$w = q^+ - q^- \le (1 - T^-/T^+)q^+.$$

If $w > 0$, the bracket in front of q^+ must be positive and taking into account that T is an increasing function, it follows that $\theta^+ > \theta^-$ and hence (C-I) holds. This completes the proof (2).

Noting that (3) is a consequence of (1) and (2), we proceed to the proof of the uniqueness of the absolute temperature scale as asserted in (4). Let T_1, T_2 be two absolute temperature scales. Invoking the process Q from Assumption 7.4.2(2) and writing down the Clausius inequalities for Q and the functions T_1, T_2 one arrives at $T_1(\theta) = T_1(\omega)T_2(\theta)/T_2(\omega)$ and so $T_1(\theta) = kT_2(\theta)$, $\theta \in I$ where $k = T_1(\omega)/T_2(\omega)$ and the proof is complete. □

As it has been pointed out above, the only difference between the Carnot and Kelvin versions of the second law, on the one hand, and the two versions of the Clausius statement of the second law, on the other, is that the latter two always guarantee that the absolute temperature scale is increasing while for the Carnot and Kelvin versions of the second law this is not generally true. However, the difference is purely formal, since the existence of one model system with "nice" properties will make the four empirical statements equivalent:

7.5.3 Proposition *Let C satisfy Assumption* 7.4.2(1) *and contain a reversible fluid. Then the Carnot, the Kelvin and the two Clausius statements of the second law are equivalent and the absolute temperature is an increasing function of the empirical temperature.*

Proof From the analysis of the reversible fluid in the proof of Proposition 7.3.8 we know that if the absolute temperature exists, then it satisfies (7.3.14) which gives $\partial_\theta T = T(\partial_\theta l_v - \partial_v c_v)/l_v$, and so by $(7.3.2)_2$ T is increasing. That is, the only possible absolute temperature scale is increasing. Hence, by Assertion (1) of Proposition 7.5.2 the Carnot or the Kelvin statement of the second law necessarily leads to the existence of an increasing absolute temperature scale and hence, by Assertion (2) of Proposition 7.5.2 also the two Clausius versions of the second law hold. □

We are now going to show that if the collection C contains an ideal gas, then the absolute temperature T coincides with the ideal gas temperature. Recall that we assume, through the whole chapter, that the first law holds. Hence the total energy is available. If also Cauchy's equations hold, which we now assume, then also the internal energy e is available, satisfying the reduced equation of balance of energy

$$\dot{e} = -p\dot{v} + r \tag{7.5.6}$$

in every process. A reversible fluid is said to be an ***ideal gas*** if its range of empirical temperatures is $I = \mathbb{R}^{++}$ and the following conditions hold: **(1)** the Cauchy stress $\boldsymbol{T}$ is given by $\boldsymbol{T}(\sigma, \boldsymbol{p}) = -p(v, \theta)\mathbf{1}$, where

$$p(v, \theta) = R'\theta/v \tag{7.5.7}$$

with $R' > 0$ a constant; **(2)** the body satisfies (7.5.6) with the internal energy function $\hat{e}$ of the form $\hat{e} = \varphi(\hat{\theta}(\sigma, \boldsymbol{p}))$, where φ is a smooth function.

7.5.4 Proposition *Let C satisfy Assumption* 7.4.2(1) *and contain an ideal gas* Δ. *If T is an absolute temperature scale for C, then $T(\theta) = k\theta$ for every $\theta \in \mathbb{R}^{++}$ and some positive constant k.*

Proof Inserting the postulated form of $\hat{e}$ and (7.5.7), (7.3.1) into (7.5.6), we obtain $c_v(v, \theta) = \partial_\theta\varphi(\theta)$, $l_v(v, \theta) = R'\theta/v$. The general equation (7.3.14) then leads to $\partial_\theta(\theta/T(\theta)) = 0$. □

Once we have derived the existence of an increasing absolute temperature scale, we can use it also as the empirical temperature, i.e., we can pass form the original empirical temperature θ to the new empirical temperature $\bar{\theta} = T(\theta)$. Such a change does not affect the validity of the empirical statements of the second law. This is seen from the rules of change of the thermodynamic characteristics of processes which read

$$\bar{q}^{\pm}(\pi, P) = q^{\pm}(\pi, P), \quad \bar{\theta}^{\pm}(\pi, P) = T(\theta^{\pm}(\pi, P)),$$
$$Q(\pi, P) \mapsto \bar{Q}(\pi, P) \text{ where } \bar{Q}(\pi, P)(\bar{N}) = Q(\pi, P)(T^{-1}(\bar{N}))$$

for every Borel set $N \subset T(I)$. In what follows we understand that the empirical temperature has been chosen in such a way as to coincide with the absolute temperature and use the letter θ for the absolute temperature. Hence the Clausius inequality reads

$$\int \frac{dQ(\pi, P)}{\theta} \leq 0 \tag{7.5.8}$$

for every cyclic process π and every part P. By Lemma 7.1.9, (7.5.8) can be rewritten as

$$\int_0^{d_\pi} \Big(-\int_{\partial P} \frac{\boldsymbol{q} \cdot \boldsymbol{n}}{\theta}\, dA + \int_P \frac{r}{\theta}\, dV \Big)\, dt \leq 0. \tag{7.5.9}$$

7.5.5 Proposition *Let Ω be a body satisfying the Clausius inequality, π its cyclic process and $P \subset \Omega$ a part. If $q^+ > 0$, then also $q^- > 0$, and*

$$w/q^+ \leq 1 - \theta^-/\theta^+.$$

This is the classical efficiency estimate. Its proof has been given in the proof of Proposition 7.5.2.

7.6 The Entropy. The Clausius–Duhem Inequality

The goal of this chapter is achieved in this section by the construction of entropy. We shall assume throughout that the range of the temperature θ is $\mathbb{R}^{++}$.

7.6.1 Definition A body Ω is said to satisfy the Clausius–Duhem inequality if there exists a local state function $\hat{\eta}$ on Ω (not necessarily smooth) such that **(1)** for every process $\pi \in \Pi$ and every $\boldsymbol{p} \in \Omega$,

$$\hat{\eta}(\pi^f) - \hat{\eta}(\pi^i) \geq \int_0^{d_\pi} \Big(-\operatorname{Div} \frac{\boldsymbol{q}}{\theta} + \frac{r}{\theta} \Big)\, dt \tag{7.6.1}$$

and **(2)** $\hat{\eta}(\sigma, \cdot)$ is integrable over Ω for every $\sigma \in \Sigma$. Every state function $\hat{\eta}$ satisfying **(1)**, **(2)** is called the ***entropy*** function for the body.

Clearly, if $\hat{\eta}$ is an entropy function and $c : \Omega \to \mathbb{R}$ an integrable field over Ω, then $\bar{\eta}$, given by

$$\bar{\eta}(\sigma, \boldsymbol{p}) = \hat{\eta}(\sigma, \boldsymbol{p}) + c(\boldsymbol{p}), \quad \sigma \in \Sigma,\ \boldsymbol{p} \in \Omega, \tag{7.6.2}$$

is again an entropy function. In this respect the entropy is similar to the energy. There are, however, two important differences. First, there are bodies which have two entropies $\hat{\eta}$, $\bar{\eta}$ whose difference is nonconstant, i.e., which are not related as in (7.6.2). Second, it can happen that the body does not have a smooth entropy. These two peculiarities are related to the fact that for the entropy we have only an inequality. Below we shall give two different constructions of the entropy function which are not generally equivalent and the degree of regularity guaranteed for them will be much lower than the smoothness. For this reason we have not incorporated the smoothness assumption into the above definition of entropy, and related to this is the fact that for the entropy we have postulated the time-integrated form of the Clausius–Duhem inequality. On the other hand, for many classes of bodies it is possible to prove that the entropy is unique to within a constant and that it is smooth. The quantity

$$\mathsf{H}(\sigma, P) = \int_P \hat{\eta}(\sigma, \cdot)\, dV$$

is the entropy of part P in σ. Property (2) in Definition 7.6.1 guarantees that $H(\sigma, P)$ is well-defined. The integral version of the Clausius–Duhem inequality is

$$\mathsf{H}(\pi^f, P) - \mathsf{H}(\pi^i, P) \geq \int_0^{d_\pi} \Big(-\int_{\partial P} \frac{\boldsymbol{q} \cdot \boldsymbol{N}}{\theta} \, dA + \int_P \frac{r}{\theta} \, dV \Big) \, dt. \tag{7.6.3}$$

For smooth entropy functions, (7.6.1) can be given a local form

$$\dot{\eta} \geq -\operatorname{Div} \frac{\boldsymbol{q}}{\theta} + \frac{r}{\theta}$$

for every process π and every $(\boldsymbol{p}, t) \in \Omega \times [0, d_\pi]$. This is obtained by the application of (7.6.1) to the segment of the process π between $t_1 < t_2$, dividing both sides of the resulting inequality by $t_2 - t_1 > 0$ and letting $t_2 \to t_1$.

7.6.2 Proposition *Let Ω be a body satisfying the accessibility axiom* 4.2.7(4). *Then Ω satisfies the Clausius inequality if and only if it satisfies the Clausius–Duhem inequality.*

Proof We set

$$s(\pi, \boldsymbol{p}) = \int_0^{d_\pi} \Big(-\operatorname{Div} \frac{\boldsymbol{q}}{\theta} + \frac{r}{\theta} \Big) \, dt$$

for every process π and every $\boldsymbol{p} \in \Omega$; the Clausius–Duhem inequality then reads

$$\hat{\eta}(\pi^f, \boldsymbol{p}) - \hat{\eta}(\pi^i, \boldsymbol{p}) \geq s(\pi, \boldsymbol{p}). \tag{7.6.4}$$

On the other hand, using the divergence theorem and interchanging the orders of the integrations in the Clausius inequality (7.5.2), one obtains $\int_P s(\pi, \boldsymbol{p}) \, dV \leq 0$ for every cyclic process π and every part P; the arbitrariness of P implies that the body satisfies the Clausius inequality if and only if

$$s(\pi, \boldsymbol{p}) \leq 0 \tag{7.6.5}$$

for every cyclic process π and every $\boldsymbol{p} \in \Omega$. To show that the Clausius–Duhem inequality implies the Clausius inequality, it suffices to note that for cyclic processes the difference on the left-hand side of (7.6.4) vanishes.

To prove the converse implication, let us choose an arbitrary reference state σ_0 and define, for any $\sigma \in \Sigma$ and $\boldsymbol{p} \in \Omega$,

$$\underline{\eta}(\sigma, \boldsymbol{p}) = \sup \{ s(\pi, \boldsymbol{p}) : \pi \in \Pi : \pi^i = \sigma_0, \pi^f = \sigma \}. \tag{7.6.6}$$

The idea of (7.6.6) is that all processes are considered connecting the reference state σ_0 with the state σ whose entropy is to be computed and the supremum is taken of the set of all Clausius integrals $s(\pi, \boldsymbol{p})$ corresponding to the totality of all such processes. The accessibility axiom 4.2.7(4) guarantees that there is at least one process connecting σ_0 with σ, so that the supremum (7.6.6) is taken over a nonempty set of numbers

$$\{ s(\pi, \boldsymbol{p}) : \pi \in \Pi : \pi^i = \sigma_0, \pi^f = \sigma \}. \tag{7.6.7}$$

Since the numbers $s(\pi, P)$ as in (7.6.7) are finite we have $\underline{\eta}(\sigma, \boldsymbol{p}) > -\infty$. However, it is not a priori clear that $\underline{\eta}(\sigma, \boldsymbol{p}) < \infty$, for the set (7.6.7) of finite numbers can be unbounded above. That this is not the case follows from the Clausius inequality (7.6.5), as we shall see presently. Indeed, invoking the accessibility axiom once again, we find a process π_0 with $\pi_0^i = \sigma$, $\pi_0^f = \sigma_0$. Then if π is as in (7.6.7), the process $\pi * \pi_0$ is cyclic and so $s(\pi, \boldsymbol{p}) + s(\pi_0, \boldsymbol{p}) = s(\pi * \pi_0, \boldsymbol{p}) \leq 0$, i.e., $s(\pi, \boldsymbol{p}) \leq -s(\pi_0, \boldsymbol{p})$. Hence the set (7.6.7) is bounded above by the definite bound $-s(\pi_0, \boldsymbol{p})$ which is independent of the processes π in (7.6.7). Hence $\underline{\eta}$ is a finite-valued state function on Ω. Furthermore, the definition (7.6.6) shows that $\underline{\eta}(\sigma, \cdot)$, being the supremum of smooth functions $s(\pi, \cdot)$, is lower semicontinuous and hence measurable. To prove that $\underline{\eta}(\sigma, \cdot)$ is integrable over Ω, it suffices to notice that $s(\pi_1, \boldsymbol{p}) \leq \underline{\eta}(\sigma, \boldsymbol{p}) \leq -s(\pi_0, \boldsymbol{p})$ where π_0 is as above and π_1 is any process with $\pi_1^i = \sigma_0$, $\pi_1^f = \sigma$. Since $s(\pi_1, \boldsymbol{p})$, $-s(\pi_0, \boldsymbol{p})$ have smooth extensions onto a set containing the closure of Ω, their integrals over Ω are finite and hence also the integral of $\underline{\eta}(\sigma, \cdot)$ over Ω is finite. This proves that $\underline{\eta}$ satisfies Condition (2) in Definition 7.6.1.

Let π be any process and $\boldsymbol{p} \in \Omega$ and prove that $\underline{\eta}$ satisfies the Clausius–Duhem inequality

$$\underline{\eta}(\pi^f, \boldsymbol{p}) - \underline{\eta}(\pi^i, \boldsymbol{p}) \geq s(\pi, \boldsymbol{p}) \tag{7.6.8}$$

for every $p \in \Omega$. By the accessibility axiom there exists a process π_2 such that $\pi_2^i = \sigma_0$, $\pi_2^f = \pi^i$. Then $\pi_2 * \pi$ is again a process and the definition (7.6.6) applied to the state $\sigma = \pi^f$ gives $s(\pi_2 * \pi, p) \le \underline{\eta}(\pi^f, p)$ i.e.,

$$s(\pi_2, p) + s(\pi, p) \le \underline{\eta}(\pi^f, p) \tag{7.6.9}$$

since $\pi_2 * \pi$ is a process connecting σ_0 with π^f. Varying now π_2 over the set of all processes connecting σ_0 with π^i, using the definition of $\underline{\eta}(\pi^i, p)$ and (7.6.9), we obtain $\underline{\eta}(\pi^i, p) \le \underline{\eta}(\pi^f, p) - s(\pi, p)$ and (7.6.8) follows. □

The construction of the entropy is a generalization of the generalized path-independence argument used in the proof of Proposition 5.3.1 to construct the total energy: The difference $g(\pi, p)$ of the work and heat in the case of the first law is replaced by the Clausius integral $s(\pi, P)$, and instead of the path-independence property (5.3.5) of g which gives the unique value $u(\sigma, p) = g(\pi, p)$ over the set of all processes connecting σ_0 with σ, in the definition of $\underline{\eta}(\sigma, p)$ we must consider the supremum of $s(\pi, p)$ over the same set of processes. It is also noted without proof that the definition

$$\overline{\eta}(\sigma, p) = \inf\{-s(\pi, p) : \pi \in \Pi, \pi^i = \sigma, \pi^f = \sigma_0\},$$

$(\sigma, p) \in \Sigma \times \Omega$ gives another possible entropy function for the body. In general it can happen that $\overline{\eta}$ is different from $\underline{\eta}$, i.e., their difference can be nonconstant. The above definitions lead to $\underline{\eta}(\sigma_0, p) \le 0$, $\overline{\eta}(\sigma_0, p) \ge 0$ in the reference state σ_0, and although it is not a formal consequence of the definitions, in all practical cases one has in fact $\underline{\eta}(\sigma_0, p) = \overline{\eta}(\sigma_0, p) = 0$ for all $p \in \Omega$. It is then easy to show that every entropy function η' for the body which vanishes at σ_0, is between the above two entropies, $\underline{\eta}(\sigma, p) \le \eta'(\sigma, p) \le \overline{\eta}(\sigma, p)$ for every $\sigma \in \Sigma$, and every $p \in \Omega$. That is, $\underline{\eta}, \overline{\eta}$ are the least and the greatest elements of the set of all entropy functions that vanish at σ_0.

The idea that the Clausius inequality leads to the existence of the entropy function, defined in and away from equilibrium, is due to DAY [1969, 1972a]. His original construction is given within the formalism of histories and response functionals defined on histories. COLEMAN & OWEN [1974] generalized Day's construction to systems with the state space satisfying the approximate accessibility axiom, described the greatest and the least entropies of the system, and analyzed concrete classes of material from this point of view. The reader is referred to COLEMAN & OWEN [1974] for the application of the theory to materials with genuine memory, such as materials with fading memory or viscoelastic materials. COLEMAN & OWEN [1975] gave an explicit example of a system having many different entropies (see also LUCCHESI [1993] and LUCCHESI & ŠILHAVÝ [1993]). Later COLEMAN & OWEN [1977] generalized the construction of the entropy to systems satisfying even weaker accessibility conditions. OWEN [1982] gave a construction of the local entropy functions for global bodies under the assumptions weaker than the assumptions considered here. WILLEMS [1972] and RICOU [1986, 1988] gave a construction of the entropy using a stronger version of the assumption on the Clausius integral. Their construction will be described in more detail in Sect. 7.7.

7.6.3 Example (The reversible system of classical thermodynamics) Let (Σ, Π) be the system as in Subsect. 4.2.4 and 5.3.2. Assume that $\Sigma \subset \mathbb{R}^n$ is connected so that (Σ, Π) satisfies the accessibility axiom. Given the differential form $q = \sum_{i=1}^n q_i dx_i$ for the net gain of heat and the state function $\theta : \Sigma \to \mathbb{R}^{++}$ for the absolute temperature, the Clausius integral $s(\pi)$ reduces to

$$s(\pi) = \int_\pi \frac{q}{\theta} \tag{7.6.10}$$

along the path π in Σ. The Clausius inequality reads $s(\pi) \le 0$. The application to the time reversal $\bar{\pi}$ of π, given by $\bar{\pi}(t - d_\pi) = \pi(d_\pi - t)$, $t \in [0, d_\pi]$, provides $s(\bar{\pi}) \le 0$. As $s(\bar{\pi}) = -s(\pi)$ by the reversibility, one finds that actually $s(\pi) = 0$. That is, the Clausius inequality reduces to the equality for cyclic processes. The line integral (7.6.10) is path independent and the classical result asserts the existence of a smooth function $\eta : \Sigma \to \mathbb{R}$ such that $d\eta = q/\theta$ with η the entropy for (Σ, Π). Also the Clausius–Duhem inequality reduces to the equality. Moreover, it can be shown that every entropy function is smooth and the difference of two entropy functions is constant on Σ.

The following example and the subsequent proposition show that the entropy constructed in Proposition 7.6.2 generally depends "on the nonequilibrium variables," cf. (7.6.15) below.

7.6.4 Example (Maxwell–Cattaneo heat conduction) A rigid heat conductor is a body Ω for which the motions are constrained by the condition $\chi(\boldsymbol{p}, t) = \boldsymbol{p}$ for all $\boldsymbol{p} \in \Omega$ and all times during the process. The equation of balance of linear momentum (3.5.4) is understood to determine the body force $\boldsymbol{b}$ necessary to render the body immobile. Assuming $\rho_o = 1$, the reduced equation of balance of energy reads as

$$\dot{e} = -\operatorname{Div} \boldsymbol{q} + r \tag{7.6.11}$$

where e is the specific internal energy. The process is completely determined by the evolutions of temperature $\theta = \theta(\boldsymbol{p}, t)$, the referential heat flux $\boldsymbol{q} = \boldsymbol{q}(\boldsymbol{p}, t)$ and the body supply of heat $r = r(\boldsymbol{p}, t)$. The ***Maxwell–Cattaneo rigid heat conductor*** is a rigid heat conductor whose processes $(\theta, \boldsymbol{q}, r)$ obey the equation of balance of energy (7.6.11) and the constitutive equations

$$\boldsymbol{T}(\theta)\dot{\boldsymbol{q}} + \boldsymbol{q} = -\boldsymbol{K}(\theta)\boldsymbol{G}, \qquad e = \bar{e}(\theta, \boldsymbol{q}) \tag{7.6.12}$$

at every point $(\boldsymbol{p}, t)$ during the process, where $\boldsymbol{G} = \nabla\theta$, $\boldsymbol{K}, \boldsymbol{T}$ are tensor-valued functions on $(0, \infty)$ with $\boldsymbol{K}(\theta)$ an invertible tensor for each $\theta > 0$ and $\bar{e}$ is a scalar-valued function on $(0, \infty) \times \text{Vect}$. $\boldsymbol{T}$ is called a tensor of relaxation times. Assume that $\boldsymbol{K}, \boldsymbol{T}, \bar{e}$ are twice continuously differentiable over their domains.

The state space Σ is the set of all pairs $\sigma = (\theta, \boldsymbol{q})$ of continuously differentiable fields on Ω. The class Π of all processes consists of all pairs $\pi = (\theta, \boldsymbol{q})$ where $\theta, \boldsymbol{q}$ are time-dependent fields defined on $\Omega \times [0, d_\pi]$ such that there exist $t_0 = 0 < t_1 < \ldots < t_n = d_\pi$ with the property that the restrictions of $\theta, \boldsymbol{q}$ to $\Omega \times [t_i, t_{i+1}]$ are continuously differentiable for each $i = 1, \ldots, n-1$ and (7.6.12) hold whenever the derivative occurring in them exist. It is not hard to see that (Σ, Π) is a system that satisfies the accessibility axiom.

Converting the surface integral in (7.5.9) to the volume integral, using $\operatorname{Div}(\boldsymbol{q}/\theta) = (\operatorname{Div} \boldsymbol{q})/\theta - \boldsymbol{q} \cdot \boldsymbol{G}/\theta^2$ and eliminating r, $\boldsymbol{G}$ via (7.6.11) and (7.6.12), respectively, we obtain the alternative form of the Clausius inequality

$$\int_0^{d_\pi} \int_P (\dot{e}/\theta - (\mathbf{Z}\dot{\boldsymbol{q}} \cdot \boldsymbol{q} + \boldsymbol{K}^{-1}\boldsymbol{q} \cdot \boldsymbol{q})/\theta^2)\, dV\, dt \le 0 \tag{7.6.13}$$

for every part P and every cyclic process π, where $\boldsymbol{Z} := \boldsymbol{K}^{-1}\boldsymbol{T}$.

7.6.5 Proposition *The Maxwell–Cattaneo rigid heat conductor satisfies the Clausius inequality* (7.6.13) *for cyclic processes if and only if the following two assertions hold simultaneously:*
(1) *for every $\theta > 0$, the tensor $\mathbf{K}(\theta)$ is positive definite and the tensor $\mathbf{Z}(\theta)$ is symmetric;*
(2) *$\bar{e}$ is of the form*

$$\bar{e}(\theta, \boldsymbol{q}) = \bar{e}_0(\theta) - \tfrac{1}{2}\theta^2 \frac{d}{d\theta}(\mathbf{Z}(\theta)/\theta^2)\boldsymbol{q} \cdot \boldsymbol{q}, \qquad \theta > 0, \qquad \boldsymbol{q} \in \text{Vect}, \tag{7.6.14}$$

where $\bar{e}_0 : (0, \infty) \to \mathbb{R}$ is a twice continuously differentiable function.
If these two conditions hold, then the system has an entropy function $\hat{\eta}$ given by

$$\hat{\eta}(\sigma, \boldsymbol{p}) = \bar{\eta}(\theta(\boldsymbol{p}), \boldsymbol{q}(\boldsymbol{p})), \qquad \sigma = (\theta, \boldsymbol{q}) \in \Sigma, \qquad \boldsymbol{p} \in \Omega,$$

where $\bar{\eta}$ is of the form

$$\bar{\eta}(\theta, \boldsymbol{q}) = \bar{\eta}_0(\theta) - \tfrac{1}{2}\frac{d}{d\theta}(\mathbf{Z}(\theta)/\theta)\boldsymbol{q} \cdot \boldsymbol{q}, \qquad \theta > 0, \qquad \boldsymbol{q} \in \text{Vect}, \tag{7.6.15}$$

and $\bar{\eta}_0$ is a continuously differentiable function on $(0, \infty)$ such that $\theta\partial_\theta\bar{\eta}_0 = \bar{e}_0$ on $(0, \infty)$. The entropy is determined uniquely in the sense that if $\hat{\eta}'$ is another entropy function, then there exists a constant field $c : \Omega \to \mathbb{R}$ such that $\hat{\eta}'(\sigma, \boldsymbol{p}) = \hat{\eta}(\sigma, \boldsymbol{p}) + c(\boldsymbol{p})$ for every $\sigma \in \Sigma$ and every $\boldsymbol{p} \in \Omega$. The free energy $\bar{f} := \bar{e} - \theta\bar{\eta}$ is given by

$$\bar{f}(\theta, \boldsymbol{q}) = \bar{f}_0(\theta) + \tfrac{1}{2}\mathbf{Z}(\theta)\boldsymbol{q} \cdot \boldsymbol{q}/\theta, \qquad \theta > 0, \qquad \boldsymbol{q} \in \text{Vect},$$

where $\bar{f}_0 = \bar{e}_0 - \theta\bar{\eta}_0$.

Proof (cf. Coleman, Fabrizio & Owen [1986]) Let Ω satisfy the Clausius inequality for cyclic processes. Since (Σ, Π) satisfies the accessibility axiom, by Proposition 7.6.2 Ω has an entropy function $\hat{\eta}$. Choose $\boldsymbol{p} \in \Omega$ arbitrarily and define a function $\bar{\eta} : (0, \infty) \times \text{Vect} \to \mathbb{R}$ by

$$\bar{\eta}(\theta, \boldsymbol{q}) = \hat{\eta}(\sigma, \boldsymbol{p}) \qquad (\theta, \boldsymbol{q}) \in (0, \infty) \times \text{Vect},$$

where $\sigma \in \Sigma$ is a homogeneous state with the uniform fields $\bar\theta, \bar{\boldsymbol q}$ of values $\theta, \boldsymbol q$ respectively. Let us consider a homogeneous process $\pi = (\theta, \boldsymbol q)$ in which $\theta, \boldsymbol q$ are functions of t alone. Then the Clausius–Duhem inequality (7.6.1) at $\boldsymbol p$ reads

$$\bar\eta(\theta^f, \boldsymbol q^f) - \bar\eta(\theta^i, \boldsymbol q^i) \ge \int_0^{d_\pi} (\dot e/\theta - (\boldsymbol Z\dot{\boldsymbol q}\cdot\boldsymbol q + \boldsymbol K^{-1}\boldsymbol q\cdot\boldsymbol q)/\theta^2)\,dt. \tag{7.6.16}$$

Let $\alpha > 0$ and replace the process π by a process $\bar\pi = (\bar\theta, \bar{\boldsymbol q})$ of duration d_π/α, given by $\bar\theta(t) = \theta(\alpha t)$, $\bar{\boldsymbol q}(t) = \boldsymbol q(\alpha t)$, $t \in [0, d_\pi/\alpha]$. Writing (7.6.16) for $\bar\pi$ and making the change of variables $s = \alpha t$, we obtain

$$\bar\eta(\theta^f, \boldsymbol q^f) - \bar\eta(\theta^i, \boldsymbol q^i) \ge \int_0^{d_\pi} (\dot e/\theta - (\boldsymbol Z\dot{\boldsymbol q}\cdot\boldsymbol q + \alpha\boldsymbol K^{-1}\boldsymbol q\cdot\boldsymbol q)/\theta^2)\,ds.$$

As $\alpha > 0$ is arbitrary, we let $\alpha \to 0$ to obtain

$$\bar\eta(\theta^f, \boldsymbol q^f) - \bar\eta(\theta^i, \boldsymbol q^i) \ge \int_0^{d_\pi} (\dot e/\theta - \boldsymbol Z\dot{\boldsymbol q}\cdot\boldsymbol q/\theta^2)\,dt, \tag{7.6.17}$$

where we have changed the notation of time from s back to t. This inequality must hold for every homogeneous process π. Replacing π by $\pi' = (\theta', \boldsymbol q')$ defined by $\theta'(t) = \theta(d_\pi - t)$, $\boldsymbol q'(t) = \boldsymbol q(d_\pi - t)$, $t \in [0, d_\pi]$ we obtain the same inequality except that $\ge$ is replaced by $\le$ and hence (7.6.17) holds with the equality sign. Evaluating the $\dot e$, we obtain

$$\bar\eta(\theta^f, \boldsymbol q^f) - \bar\eta(\theta^i, \boldsymbol q^i) = \int_0^{d_\pi} \left(\partial_\theta\bar e\cdot\dot\theta/\theta + (\partial_{\boldsymbol q}\bar e/\theta - \boldsymbol Z^{\mathrm T}\boldsymbol q/\theta^2)\cdot\dot{\boldsymbol q}\right)dt. \tag{7.6.18}$$

This shows that the vector field $(X, Y) := \left(\partial_\theta e/\theta, \partial_{\boldsymbol q} e/\theta - \boldsymbol Z^{\mathrm T}\boldsymbol q/\theta^2\right)$ has a potential $\bar\eta$, which, by a standard argument, is continuously differentiable and satisfies

$$\partial_\theta\bar\eta = \partial_\theta\bar e/\theta, \quad \partial_{\boldsymbol q}\bar\eta = \partial_{\boldsymbol q}\bar e/\theta - \boldsymbol Z^{\mathrm T}\boldsymbol q/\theta^2.$$

Since $\bar e, \boldsymbol Z$ are twice continuously differentiable, this in turn shows that $\bar\eta$ is twice continuously differentiable and the symmetry of $\partial^2_{\boldsymbol q}\bar\eta$, $\partial^2_{\boldsymbol q}\bar e$ gives that $\boldsymbol Z^{\mathrm T}$ is symmetric and hence also $\boldsymbol Z$ is symmetric. The condition $\partial_\theta Y = \partial_{\boldsymbol q} X$ gives

$$\partial_{\boldsymbol q}\bar e/\theta = -\theta^2\frac{d}{d\theta}(\boldsymbol Z/\theta^2)\boldsymbol q.$$

An integration leads to (7.6.14). Finally, the application of (7.6.16) to a homogeneous process π in which $\theta, \boldsymbol q$ are arbitrary constants independent of time reduces to $\boldsymbol K^{-1}\boldsymbol q\cdot\boldsymbol q \ge 0$ and hence $\boldsymbol K^{-1}$ is positive semidefinite which in combination with the assumed invertibility of $\boldsymbol K$ gives that $\boldsymbol K$ is positive definite. This proves that if the system satisfies the Clausius–Duhem inequality, then (1), (2) hold. Conversely, given that (1), (2) hold, one defines $\bar\eta$ by (7.6.15) where $\bar\eta_0$ is such that $\theta\partial_\theta\bar\eta_0 = \partial_\theta\bar e_0$. It is then a matter of calculation to verify that $\bar\eta$ is a potential for the vector field (X, Y) and hence (7.6.18) holds in every homogeneous process. Combining this with the positive definiteness of $\boldsymbol K^{-1}$ asserted in (1), one sees that (7.6.16) holds in every homogeneous process. This means that (7.6.16) holds for every $\boldsymbol p \in \Omega$ in any (not necessarily homogeneous) process. This in turn yields that the system has an entropy, which is equivalent to asserting that the system satisfies the Clausius inequality for cyclic processes. The proof that (1) and (2) imply the Clausius inequality is complete. The above argument also shows that if $\hat\eta'$ is any entropy function and $\boldsymbol p$ any point in Ω then

$$\hat\eta'(\pi^f, \boldsymbol p) - \hat\eta'(\pi^i, \boldsymbol p) = \int_0^{d_\pi} \left(\partial_\theta\bar e\cdot\dot\theta/\theta + (\partial_{\boldsymbol q}\bar e/\theta - \boldsymbol Z\boldsymbol q/\theta^2)\cdot\dot{\boldsymbol q}\right)dt.$$

for any process $\pi \in \Pi$. This is easily seen to imply the uniqueness assertion about $\hat\eta'$. Finally, the form of f is immediate. □

For a generalization of the Maxwell–Cattaneo model, see Sect. 12.4. The section is concluded with the following two general assertions.

7.6.6 Proposition *If Ω satisfies the accessibility axiom* 4.2.7(4), *the principle of material frame indifference in the form of Assumptions* 6.1.2, *and the Clausius–Duhem inequality, then it has at least one objective entropy.*

The proof is omitted.

7.6.7 Proposition *Let Ω be a body satisfying the equation of balance of energy* (5.3.3) *and the Clausius–Duhem inequality* (7.6.3), *π a process and P a part. If $q^+ > 0$, $q^- > 0$, then*

$$w \le \theta^- \Delta \mathsf{H} - \Delta \mathsf{U} + (1 - \theta^-/\theta^+) q^+, \tag{7.6.19}$$

where Δ denotes the change of the state functions in the process π.

For cyclic processes (7.6.19) reduces to the classical efficiency estimate. Inequality (7.6.19) is stated in TRUESDELL [1973] using different definitions of $q^\pm$, $\theta^\pm$. With the above definitions, it is due to MAN [1989] who used it to obviate the Clausius–Duhem inequality in determining the restrictions on constitutive relations and discussed its relationship to the classical statements of the second law. See also DAY [1972b], ŠILHAVÝ [1977], DAY & ŠILHAVÝ [1977].

Proof We have $\Delta \mathsf{U} = -w + q^+ - q^-$, and (see (7.5.5)) $\Delta \mathsf{H} \ge q^+/\theta^+ - q^-/\theta^-$. The elimination of q^- between these two relations leads to (7.6.19). □

7.7 Notes and Complements

Table 7.7.1 summarizes the consequences of the two laws and reflects the steps in their derivation. (A comment is due here perhaps on a common anthropocentric feature of the basic statements of the two laws, namely the concern in specifying conditions for a body to do a positive work. Only a man wants the things to do work for him.)

As the table shows, the derivation of the consequences of the two laws is analogous. Recall that in Sects. 5.2 and 7.3 we associated with each process a geometrical representative – the work–heat

Table 7.7.1. The two laws in juxtaposition

cyclic processes		
	first law	second law
statement	$w > 0 \Leftrightarrow q > 0$	$w > 0 \Rightarrow q^- > 0$
derived object	J	$T(\theta)$
normalizations	$J = 1$	$T(\theta) = \theta, \theta > 0$
relation	$w = q$	$\int \frac{dQ}{\theta} \le 0$

general processes		
	first law	second law
derived object	U	H
relation	$\Delta \mathsf{U} = q - w$	$\Delta \mathsf{H} \ge \int \frac{dQ}{\theta}$

pair in the case of the first law and the heating measure in the case of the second law. Denote by W the space of the representatives. Both the two laws can be given the form

$$C \cap F = \emptyset, \tag{7.7.1}$$

where $C \subset W$ is the set of representatives of all cyclic processes from the collection C and F is a "forbidden set" F depending on the law in question. This was shown to lead to Joule's constant J and the absolute temperature scale $T(\cdot)$. If $H := \{(w, q) : w = Jq\}$ in the case of the first law and $H := \{Q : \int dQ/\theta \le 0\}$ in the case of the second law, then $H \cap F = \emptyset$, $C \subset H$, which is the best way to satisfy (7.7.1). The objects $(1, -J)$ and $1/T(\cdot)$ then represent the normal to ∂H, which is their common geometrical meaning. It is instructive to simplify the universe by assuming that the empirical temperature can take only two values θ_1 and θ_2, $I = \{\theta_1, \theta_2\}$. The heating measure is then determined by the gains of heat Q_1, Q_2 on the temperatures θ_1, θ_2, respectively. The space $\mathbb{M}_h(I)$ of measures in $I = \{\theta_1, \theta_2\}$ is the plane $\mathbb{R}^2$ of pairs (Q_1, Q_2). The forbidden set for the second law is

$$F = \{(Q_1, Q_2) \in \mathbb{R}^2 : (Q_1, Q_2) \neq 0, Q_1 \ge 0, Q_2 \ge 0\}$$

which is the closed first quadrant from which the origin is removed. The derivation of the existence of the absolute temperature scale $T(\theta)$, $\theta = \theta_1$ or $\theta = \theta_2$, can then be based on the use of the lattice points in a complete analogy with the derivation of J in the proof of Proposition 5.2.4. I refer to KRATOCHVÍL & ŠILHAVÝ [1982] for further details.

SERRIN [1978, 1979, 1986] has developed a similar approach to the one presented in Sects. 7.1–7.6. The formal difference is that Serrin uses the equivalent language of accumulation functions (see Subsect. 7.1.6) rather than the heating measures. His statement of the second law is:

The condition $\hat{Q}(\pi, P, \cdot) \ge 0$ can occur for a cyclic process π and part P only in the exceptional case $\hat{Q}(\pi, P, \cdot) = 0$.

Here $\hat{Q}$ is the accumulation function. Under the first law this is equivalent to:

If π is a cyclic process and P any part then

$$w(\pi, P) > 0 \Rightarrow \text{ there exists a } \theta_0 \in I \text{ such that } \hat{Q}(\pi, P, \theta_0) < 0. \tag{7.7.2}$$

Let us compare (7.7.2) with the Carnot statement of the second law from Sect. 7.2. Let $I = (a, b)$, $-\infty \le a < b \le \infty$, be the interval of empirical temperatures. The Carnot statement of the second law says that for any cyclic process π and any part P with $w(\pi, P) > 0$ there exists a Borel set $N \subset I$ such that

$$Q(\pi, P, N) < 0. \tag{7.7.3}$$

On the other hand, (7.7.2) says that not only such a set N exists, but it also specifies its form $N = (a, \theta_0]$ for some $\theta_0 \in I$. Because of this additional information, (7.7.2) is stronger than the Carnot statement of the second law. Statement (7.7.2) is actually stronger than any of the versions of the second law in Sect. 7.2. It will suffice to prove that (7.7.2) implies Version I of the Clausius statement of the second law, because this is the strongest of all statements in Sect. 7.2. Let π be a cyclic process and P a part such that $w(\pi, P) > 0$. Statement (7.7.2) gives a θ_0 such that $Q(\pi, P, (a, \theta_0]) < 0$. It follows that

$$q^-(\pi, P) > 0 \text{ and } \operatorname{supp} Q^-(\pi, P) \cap (a, \theta_0] \neq \emptyset; \tag{7.7.4}$$

consequently,

$$\theta^-(\pi, P) \le \theta_0. \tag{7.7.5}$$

On the other hand, as $w(\pi, P) = Q(\pi, P, I) > 0$ and $Q(\pi, P, (a, \theta_0]) < 0$, necessarily $Q(\pi, P, (\theta_0, b)) > 0$ and it follows that

$$q^+(\pi, P) > 0 \text{ and } \operatorname{supp} Q^+(\pi, P) \cap (\theta_0, b) \neq \emptyset; \tag{7.7.6}$$

hence

$$\theta_0 < \theta^+(\pi, P). \tag{7.7.7}$$

Relations (7.7.5) and (7.7.7) give

$$\theta^-(\pi, P) < \theta^+(\pi, P). \tag{7.7.8}$$

But Version I of the Clausius statement of the second law asserts exactly $(7.7.4)_1$, $(7.7.6)_1$, and (7.7.8). Being stronger, Serrin's version of the second law yields an easier derivation of the consequences. Granted appropriate structural axioms on C, it results in establishing the existence of the absolute temperature scale such that the Clausius inequality holds for every body from C. Hence, under the structural assumptions, (7.7.2) is actually equivalent to the versions given in Sect. 7.2. MAN [1980] has shown how to modify Serrin's approach in order that a strengthened Kelvin's law imply the Clausius inequality. COLEMAN, OWEN & SERRIN [1981] generalized the approach to systems with approximate cycles. For further developments and remarks, see MONLEÓN-PRADAS [1994], MAN [1995] and SERRIN [1995].

JOU & CASAS-VÁZQUEZ [1992] propose an experiment to check the reality of the nonequilibrium temperature.

RICOU [1986, 1988] gave a statement of the second law for noncyclic processes which implies that for every $\sigma \in \Sigma, \boldsymbol{p} \in \Omega$, the set

$$\{s(\pi, \boldsymbol{p}) : \pi \in \Pi, \pi^f = \sigma\}$$

is bounded from above. Using this, he exhibited an entropy function η that is nonnegative. His definition is

$$\eta(\sigma, \boldsymbol{p}) = \sup\{s(\pi, \boldsymbol{p}) : \pi \in \Pi, \pi^f = \sigma\},$$

for every σ and $\boldsymbol{p}$. Observe that this is independent of any accessibility assumption on the state space. Ricou's construction is essentially identical with the one given by WILLEMS [1972] for general dynamical systems. In connection with the positivity of the entropy it should be mentioned that the third law of thermodynamics is usually taken to assert that, at least for special systems, there exists an entropy function such that for every $\boldsymbol{p} \in \Omega$ and $\boldsymbol{F} \in \mathrm{Lin}^+$ the infimum

$$\inf\{\eta(\sigma, \boldsymbol{p}) : \sigma \in \Sigma, \nabla\chi(\sigma, \boldsymbol{p}) = \boldsymbol{F}\}$$

is finite and independent of $\boldsymbol{F}$ and $\boldsymbol{p}$. The choice of the additive constant in the entropy then leads to an entropy function for which the infimum is equal to 0. Comparing this with the positivity of the entropy as a result of Ricou's work, we see that it already foreshadows the third law.

FEINBERG & LAVINE [1983], working independently, also realized that the geometric treatment of the second law similar to the one presented in Sections 7.3 and 7.5 might provide a basis for the construction of the absolute temperature. The important conceptual difference of their theory is that they consider the heating measures with respect to states of the system rather than with respect to the empirical temperature. This allows them to remove that empirical temperature from the list of primitive notions. The construction of the entropy proceeds along lines entirely different from the one given in Sect. 7.6. In fact their proof of the existence of entropy can be given simultaneously with the proof of the absolute temperature.

Another line of the research in the foundations of thermodynamics returns to Carnot's original ideas. For homogeneous elastic fluids, TRUESDELL & BHARATHA [1977] and TRUESDELL [1979] analyzed Carnot's general axiom and explored its consequences.

Finally it is necessary to comment on the Carathéodory line to the foundations of thermodynamics, CARATHÉODORY [1909], see also ARENS [1963], COOPER [1967], WALTER [1978] and the references therein. The main limitation of the Carathéodory approach is that it provides only the absolute temperature in equilibrium, i.e., it works only for the reversible systems of classical thermodynamics. For this reason, the approach definitely cannot provide a basis for the nonequilibrium temperature and entropy. Apart from this, even when restricted to homogeneous reversible systems, the Carathéodory approach contains a number of conceptual and technical difficulties which will not be analyzed here. I refer to TRUESDELL [1986] and WALTER [1978] for the criticism.

For the history of thermodynamics, see TRUESDELL [1980] and MENDOZA [1961]. The reader can consult OWEN [1984] for an elementary exposition of thermodynamics based on statements of the first and the second laws close to those given in this book.

Part III

Constitutive Theory

8 Isotropic Functions

The constitutive equations (constitutive relations) differentiate between particular materials and complement the system of balance equations to determine the behavior of a body. On a general level the constitutive equations have been introduced in Sect. 4.1; this part is devoted to the constitutive equations of *elastic materials with heat conduction and viscosity*. Special cases are the Navier–Stokes–Fourier fluids, Kelvin–Voigt solids, thermoelastic materials, and ideal dissipationless "adiabatic materials" (in which the viscosity and heat conduction are absent). In the presence of heat conduction and viscosity, Carathéodory's approach to thermodynamics is not applicable as the mechanical power and the rate of heating are not expressible as differential forms in the "external parameters". Historically, it was this class of materials to which the "nonequilibrium" thermodynamics was first applied, and Sect. 12.2 is devoted to a brief discussion of this approach.

The exposition starts in this chapter with the representation theorems for scalar- and tensor-valued functions of a tensor argument, with the emphasis on the derivatives thereof. These are necessary for calculating, e.g., the elastic moduli at a general, nonisotropic state of an isotropic solid. Some readers may prefer to pass directly to Chap. 9, where the proper subject of the constitutive theory starts, and to use the present chapter for reference as needed.

8.1 Isotropic Tensor-Valued Functions

Throughout the chapter the dimension dim Vect $= n$ is arbitrary unless stated otherwise, and the indices $i, j, k, \ldots$ range from 1 to n.

8.1.1 Definitions A subset U of Sym is said to be ***isotropic*** if for every $\boldsymbol{A} \in U$ and every $\boldsymbol{Q} \in$ Orth also $\boldsymbol{Q}\boldsymbol{A}\boldsymbol{Q}^{\mathsf{T}} \in U$. A scalar-valued function $h : U \to \mathbb{R}$ is said to be ***isotropic*** if its domain is isotropic and $h(\boldsymbol{Q}\boldsymbol{A}\boldsymbol{Q}^{\mathsf{T}}) = h(\boldsymbol{A})$ for every $\boldsymbol{A} \in U$ and every $\boldsymbol{Q} \in$ Orth. A tensor-valued function $\boldsymbol{G} : U \to$ Lin is said to be ***isotropic*** if its domain is isotropic and

$$\boldsymbol{G}(\boldsymbol{Q}\boldsymbol{A}\boldsymbol{Q}^{\mathsf{T}}) = \boldsymbol{Q}\boldsymbol{G}(\boldsymbol{A})\boldsymbol{Q}^{\mathsf{T}} \tag{8.1.1}$$

for every $\boldsymbol{A} \in U$ and every $\boldsymbol{Q} \in$ Orth. For every orthonormal basis $\{\boldsymbol{e}_i\}$, denote by $\Delta(\boldsymbol{e}_1, \ldots, \boldsymbol{e}_n)$ the set of all symmetric tensors that are represented by diagonal matrices in the basis $\{\boldsymbol{e}_i\}$.

The isotropic functions can be represented either by the principal invariants and powers of the arguments, as in the Rivlin-Ericksen representation theorem 11.1.2, or by the eigenvectors and symmetric functions of the eigenvalues, as in Proposition 8.1.5 (the principal axis method). These two types of representations are completely equivalent. However, the principal axis method gives a better control over the differentiability, invertibility and convexity properties. For instance, an isotropic scalar-valued function h of $\boldsymbol{A} \in \mathrm{Sym}$ can be expressed as a function $\bar{f} = \bar{f}(I_i, \ldots, I_n)$ of the principal invariants I_i of $\boldsymbol{A}$ and also as a symmetric function $f = f(a_1, \ldots, a_n)$ of the eigenvalues of $\boldsymbol{A}$. The expression for $\partial_{\boldsymbol{A}}^2 h$ is given by an insight-providing formula (8.2.3) in terms of $f(a_1, \ldots, a_n)$, but by a complicated formula in terms of $\bar{f}(I_i, \ldots, I_n)$, cf. (11.1.5). Moreover, the passage from h to $\bar{f}$ is accompanied by a possible loss of differentiability (see SERRIN [1959], TRUESDELL & NOLL [1965, p. 28]; cf. also MAN [1994]). There is also a practical problem: for the Ogden stored energy functions (see Sect. 11.3), which use the (possibly noninteger) powers of the principal stretches, there are no closed-form formulas in terms of the principal invariants. In this book, where the emphasis is on the theoretical aspects, the representations of in eigenvalues and eigenvectors are used more frequently. The reader is referred to any standard treatment of continuum mechanics, such as TRUESDELL & NOLL [1965, Chap. B], for the representation theorems in powers and principal invariants, to SPENCER [1971] for the polynomial case, and to RYCHLEWSKI & ZHANG [1991] for a review of more recent developments.

In this section we discuss the representation theorems for isotropic tensor-valued functions; the scalar-valued functions are treated in the Sect. 8.2. The key observation is the following.

8.1.2 Proposition *Let* $\boldsymbol{G} : U \to \mathrm{Lin}$ *be a tensor-valued isotropic function and* $\boldsymbol{A} \in U$. *Then* $\boldsymbol{G}(\boldsymbol{A})$ *is a symmetric tensor and every eigenvector of* $\boldsymbol{A}$ *is an eigenvector of* $\boldsymbol{G}(\boldsymbol{A})$. *If* $\boldsymbol{u}, \boldsymbol{v}$ *are two eigenvectors of* $\boldsymbol{A}$ *corresponding to the same eigenvalue of* $\boldsymbol{A}$, *then they correspond to the same eigenvalue of* $\boldsymbol{G}(\boldsymbol{A})$. *Moreover,* $\boldsymbol{A}\boldsymbol{G}(\boldsymbol{A}) = \boldsymbol{G}(\boldsymbol{A})\boldsymbol{A}$.

Hence the requirement of isotropy forces the function to have its values in the set of symmetric tensors.

Proof (cf. TRUESDELL & NOLL [1965, p. 32]) Let $\boldsymbol{e}$ be an eigenvector of $\boldsymbol{A} \in U$, set $\boldsymbol{D} = \boldsymbol{G}(\boldsymbol{A})$ and define an orthogonal tensor $\boldsymbol{Q}$ by $\boldsymbol{Q}\boldsymbol{e} = -\boldsymbol{e}$, $\boldsymbol{Q}\boldsymbol{f} = \boldsymbol{f}$ if $\boldsymbol{f} \cdot \boldsymbol{e} = 0$. Then $\boldsymbol{A} = \boldsymbol{Q}\boldsymbol{A}\boldsymbol{Q}^{\mathrm{T}}$ and hence $\boldsymbol{G}(\boldsymbol{A}) = \boldsymbol{Q}\boldsymbol{G}(\boldsymbol{A})\boldsymbol{Q}^{\mathrm{T}}$, i.e., $\boldsymbol{Q}\boldsymbol{D} = \boldsymbol{D}\boldsymbol{Q}$. Therefore $\boldsymbol{Q}(\boldsymbol{D}\boldsymbol{e}) = \boldsymbol{D}\boldsymbol{Q}\boldsymbol{e} = -\boldsymbol{D}\boldsymbol{e}$. Since the only vectors transformed by $\boldsymbol{Q}$ into the opposite are multiples of $\boldsymbol{e}$, it follows that $\boldsymbol{D}\boldsymbol{e} = \mu\boldsymbol{e}$. If $\boldsymbol{e}, \boldsymbol{f}$ are two eigenvectors of $\boldsymbol{A}$ corresponding to the same eigenvalue, then also their arbitrary linear combination $\boldsymbol{g}$ is an eigenvector of $\boldsymbol{A}$ with the same eigenvalue. Hence by the previous part of the proof $\boldsymbol{e}, \boldsymbol{f}, \boldsymbol{g}$ are eigenvectors of $\boldsymbol{G}(\boldsymbol{A})$ and using that the linear combination $\boldsymbol{g}$ can be chosen arbitrary, it is easily found that the eigenvalues of $\boldsymbol{G}(\boldsymbol{A})$ corresponding to $\boldsymbol{e}, \boldsymbol{f}$ must be the same. Finally $\boldsymbol{A}\boldsymbol{G}(\boldsymbol{A}) = \boldsymbol{G}(\boldsymbol{A})\boldsymbol{A}$ follows from the fact that $\boldsymbol{A}, \boldsymbol{G}(\boldsymbol{A})$ share the same basis of eigenvectors. □

8.1.3 Lemma

(1) *If* $U \subset \mathrm{Sym}$ *is an open set, then the set*

$$U^{\natural} = \{\boldsymbol{A} \in U : \text{ the eigenvalues of } \boldsymbol{A} \text{ are mutually distinct}\}$$

is an open, dense subset of U. *If* U *is isotropic, then also* $U^{\natural}$ *is isotropic.*

(2) *If* $\boldsymbol{A} \in \mathrm{Sym}$ *is a symmetric tensor with mutually distinct eigenvalues and* $\boldsymbol{B} \in \mathrm{Lin}$ *is such that* $\boldsymbol{B}\boldsymbol{A} = \boldsymbol{A}\boldsymbol{B}$ *then* $\boldsymbol{B}$ *is symmetric.*

Proof (1): That $U^{\natural}$ is open and dense follows from Proposition 1.2.3. The isotropy of $U^{\natural}$ is clear. (2): Let $\{\boldsymbol{e}_i\}$ be an orthonormal basis of eigenvectors of $\boldsymbol{A}$ with a_i the corresponding eigenvalues. From $\boldsymbol{B}\boldsymbol{A} = \boldsymbol{A}\boldsymbol{B}$ and the fact that the eigenvalues of $\boldsymbol{A}$ are mutually distinct one deduces, in the standard way, that for any $i = 1, \ldots, n$, $\boldsymbol{B}\boldsymbol{e}_i = b_i\boldsymbol{e}_i$ with $b_i \in \mathbb{R}$. But then $\boldsymbol{B}$ is symmetric. Alternatively, one can write $\boldsymbol{A}\boldsymbol{B} = \boldsymbol{B}\boldsymbol{A}$ in the matrix form in the basis of eigenvectors of $\boldsymbol{A}$ to learn that $\boldsymbol{B}$ is symmetric. □

8.1.4 Symmetric Functions and Sets Denote by $\mathbb{M}^{n \times n}$ the set of all $n \times n$ matrices with real elements. A $P \in \mathbb{M}^{n \times n}$ is a ***permutation matrix*** if $P_{ij} = \delta_{i\pi(j)}$, where π is any permutation of $\{1, \ldots, n\}$. If $x = (x_1, \ldots, x_n) \in \mathbb{R}^n$, we define $Px \in \mathbb{R}^n$ by $(Px)_i = \sum_{j=1}^n P_{ij}x_j$. Any permutation matrix is orthogonal, i.e., $P^{\mathrm{T}} = P^{-1}$ and $\det P = 1$

if the permutation π is even and $\det P = -1$ if the permutation π is odd. A subset D of $\mathbb{R}^n$ is said to be ***symmetric*** if for every $a = (a_1, \dots, a_n) \in D$ and every permutation matrix P also $Pa \in D$. Let f be a function defined on a subset D of $\mathbb{R}^n$ with values in $\mathbb{R}$ ($\mathbb{R}^n$, $\mathbb{M}^{n \times n}$). The function f is said to be ***symmetric*** if its domain is symmetric and $f(Pa) = f(a)$ ($f(Pa) = Pf(a), f(Pa) = Pf(a)P^{\mathrm{T}}$) for every $a \in D$ and every permutation matrix P. Clearly, if $f : D \to \mathbb{R}$ is twice continuously differentiable and symmetric, then ∇f and $\nabla^2 f$ are symmetric. For a given $U \subset$ Sym denote by D the set of all $a = (a_1, \dots, a_n) \in \mathbb{R}^n$ for which there exists an $\boldsymbol{A} \in U$ with the spectrum equal to $\{a_1, \dots, a_n\}$. This is automatically a symmetric set. If U is isotropic, this can be stated equivalently as follows. D is the set of all n-tuples $(a_1, \dots, a_n) \in \mathbb{R}^n$ such that for some (and hence every) orthonormal basis $\{\boldsymbol{e}_i\}$ of Vect, the tensor

$$\boldsymbol{A} = \sum_{i=1}^{n} a_i \boldsymbol{e}_i \otimes \boldsymbol{e}_i \tag{8.1.2}$$

is in U. Conversely, for a given subset D of $\mathbb{R}^n$, denote by $U \subset$ Sym the set of all $\boldsymbol{A} \in$ Sym with the spectrum of the form $\{a_1, \dots, a_n\}$ for some $a = (a_1, \dots, a_n) \in D$. The set U is automatically isotropic. Hence the above relations establish a one-to-one relation between isotropic subsets U of Sym and symmetric subsets of D of $\mathbb{R}^n$. If U is open in Sym, then D is open in $\mathbb{R}^n$.

8.1.5 Representation Theorem I *Let* $\boldsymbol{G} : U \to$ Sym *be a function on an isotropic subset of* Sym *and let D be related to U as in Subsect.* 8.1.4. *Then* $\boldsymbol{G}$ *is isotropic if and only if there exists a symmetric function* $g = (g_1, \dots, g_n) : D \to \mathbb{R}^n$ *such that for any* $\boldsymbol{A} \in U$ *with the spectral decomposition* (8.1.2) *we have*

$$\boldsymbol{G}(\boldsymbol{A}) = \sum_{i=0}^{n} g_i(a) \boldsymbol{e}_i \otimes \boldsymbol{e}_i \tag{8.1.3}$$

where $a = (a_1, \dots, a_n)$ *is any (not necessarily ordered) n-tuple of eigenvalues.*

Proof Assume that $\boldsymbol{G}$ is isotropic. Let $a \in D$, let $\{\boldsymbol{e}_i\}$ be any orthonormal basis and let $\boldsymbol{A}$ be given by (8.1.2). By Proposition 8.1.2, $\boldsymbol{e}_i$ are the eigenvectors of $\boldsymbol{G}(\boldsymbol{A})$ and define $g_i(a)$ as the corresponding eigenvalues. This definition is independent of the choice of $\{\boldsymbol{e}_i\}$ and (8.1.3) holds. Let P be any permutation matrix with elements $P_{ij} = \delta_{i\pi(j)}$, where π is any permutation of $\{1, \dots, n\}$. To prove $g(Pa) = Pg(a)$, set $b = (b_1, \dots, b_n) := Pa$ so that $b_{\pi(i)} = a_i$ and let $\{\boldsymbol{f}_i\}$ be an orthonormal basis given by $\boldsymbol{f}_{\pi(i)} = \boldsymbol{e}_i$. Then $\boldsymbol{A}$, originally given by (8.1.2), is also given by $\boldsymbol{A} = \sum_{i=1}^{n} b_i \boldsymbol{f}_i \otimes \boldsymbol{f}_i$ and $g_i(b)$ is the eigenvalue of $\boldsymbol{G}(\boldsymbol{A})$ corresponding to $\boldsymbol{f}_i$. A comparison gives $g_{\pi(i)}(b) = g_i(a)$. This completes the proof of the direct implication. The converse is immediate. □

8.1.6 Corollary *Let* $\boldsymbol{G}$: Sym $\to$ Sym *be linear and isotropic. Then there exist constants* $c_0, c_1 \in \mathbb{R}$ *such that*

$$\boldsymbol{G}(\boldsymbol{A}) = c_0(\operatorname{tr}\boldsymbol{A})\mathbf{1} + c_1\boldsymbol{A}, \quad \boldsymbol{A} \in \text{Sym}.$$

Proof Working in a fixed orthonormal basis, it is found that the function $g : \mathbb{R}^n \to \mathbb{R}^n$ from Proposition 8.1.5 is linear and hence $g(a) = Ma$ for all a and some $M \in \mathbb{M}^{n \times n}$. The requirement that $g(Pa) = Pg(a)$ leads to

$$M = PMP^{\mathrm{T}} \tag{8.1.4}$$

for every permutation matrix P. The general form of M satisfying (8.1.4) is $M = c_0 J + c_1 I$ where $c_0 = M_{12}, c_1 = M_{11} - M_{12}$, I is the $n \times n$ identity matrix and

$$J = \begin{bmatrix} 1 & 1 & \dots & 1 \\ \vdots & & & \\ 1 & 1 & \dots & 1 \end{bmatrix}.$$

This is obtained by making appropriate choices of P in (8.1.4) and the details are omitted. Hence $g(a) = c_0(\Sigma_{i=1}^n a_i)(1, 1, \ldots, 1) + c_1 a$ and (8.1.3) gives the result. □

Let us proceed to derivatives of isotropic tensor-valued functions.

8.1.7 Differential Consequences of Isotropy *Let $\mathbf{G} : U \to \mathrm{Sym}$ be a continuously differentiable isotropic function and $\mathbf{A} \in U$. Then*

(1) *for every $\mathbf{Q} \in \mathrm{Orth}$, $\mathbf{B} \in \mathrm{Sym}$,*

$$\partial_{\mathbf{A}} \mathbf{G}(\mathbf{Q}\mathbf{A}\mathbf{Q}^{\mathsf{T}})[\mathbf{Q}\mathbf{B}\mathbf{Q}^{\mathsf{T}}] = \mathbf{Q}\partial_{\mathbf{A}} \mathbf{G}(\mathbf{A})[\mathbf{B}]\mathbf{Q}^{\mathsf{T}};$$

(2) *for every $\mathbf{W} \in \mathrm{Skew}$,*

$$\partial_{\mathbf{A}} \mathbf{G}(\mathbf{A})[\mathbf{A}\mathbf{W} - \mathbf{W}\mathbf{A}] = \mathbf{G}(\mathbf{A})\mathbf{W} - \mathbf{W}\mathbf{G}(\mathbf{A}); \tag{8.1.5}$$

(3) *for every $\mathbf{B} \in \mathrm{Sym}$ and every $\mathbf{W} \in \mathrm{Skew}$ such that $\mathbf{A}\mathbf{W} = \mathbf{W}\mathbf{A}$,*

$$\partial_{\mathbf{A}} \mathbf{G}(\mathbf{A})[\mathbf{B}\mathbf{W} - \mathbf{W}\mathbf{B}] = \partial_{\mathbf{A}} \mathbf{G}(\mathbf{A})[\mathbf{B}]\mathbf{W} - \mathbf{W}\partial_{\mathbf{A}} \mathbf{G}(\mathbf{A})[\mathbf{B}].$$

Proof (1): This is just the differentiation of (8.1.1) with respect to $\mathbf{A}$. (2): Write (8.1.1) with $\mathbf{Q} = \mathbf{Q}(t) = e^{\mathbf{W}t}$, $t \in \mathbb{R}$ (see 1.3.4) and a differentiate at $t = 0$. (3): If $\mathbf{A}\mathbf{W} = \mathbf{W}\mathbf{A}$, then $e^{\mathbf{W}t}\mathbf{A}e^{-\mathbf{W}t} = \mathbf{A}$ and (1) gives $\partial_{\mathbf{A}} \mathbf{G}(\mathbf{A})[e^{\mathbf{W}t}\mathbf{B}e^{-\mathbf{W}t}] = e^{\mathbf{W}t}\partial_{\mathbf{A}} \mathbf{G}(\mathbf{A})[\mathbf{B}]e^{\mathbf{W}t}$; a differentiation at $t = 0$ gives the result. □

8.1.8 Lemma *Let A be an $n \times n$ diagonal matrix with diagonal elements a_i and let B be a symmetric matrix with vanishing diagonal elements: $B_{ii} = 0$, $i = 1, \ldots, n$. Set $J = \mathrm{diag}(1, 2, \ldots, n)$. Then there exist skew matrices W^+, W^- such that*

$$B = AW^+ - W^+A + JW^- - W^-J, \quad W^-A - AW^- = 0;$$

the off-diagonal elements of $W^{\pm}$ are

$$W_{ij}^+ = \begin{cases} B_{ij}/(a_i - a_j) & \text{if } a_i \neq a_j, \\ 0 & \text{if } a_i = a_j, \end{cases} \qquad W_{ij}^- = \begin{cases} 0 & \text{if } a_i \neq a_j, \\ B_{ij}/(i-j) & \text{if } a_i = a_j. \end{cases}$$

This is just a calculation.

To evaluate the derivatives of the isotropic tensor functions in the representation (8.1.3), let $g = (g_1, \ldots, g_n) : D \to \mathbb{R}^n$ be a symmetric continuously differentiable function on an open subset of $\mathbb{R}^n$, and write $g_{i,j}(a)$ for $\partial_{a_j} g_i(a)$. Let i, j be indices such that $i \neq j$, and define a function $H_{ij} : D \to \mathbb{R}$ by

$$H_{ij}(a) = \begin{cases} \dfrac{g_i(a) - g_j(a)}{a_i - a_j} & \text{if } a_i \neq a_j, \\ g_{i,i}(a) - g_{i,j}(a) & \text{if } a_i = a_j, \end{cases} \tag{8.1.6}$$

for every $a \in D$; denote by $H(a)$ the $n \times n$ matrix with the diagonal elements 0 and the off-diagonal elements $H_{ij}(a)$. It is easily seen that H is a continuous, symmetric-matrix-valued function H on D. Moreover, $H(Pa) = PH(a)P^{\mathsf{T}}$ for every permutation matrix P and every $a \in D$.

8.1.9 Proposition *Let $\mathbf{G} : U \to \mathrm{Sym}$ be an isotropic function and let D and g be as in Proposition 8.1.5. Then $\mathbf{G}$ is of class C^r, $r = 0, \ldots, \infty$, if and only if g is of class C^r. If $\mathbf{G}$ is of class C^1 and $\mathbf{A} \in U$ has the spectral decomposition (8.1.2), then*

$$\partial_{\mathbf{A}} \mathbf{G}(\mathbf{A})[\mathbf{B}]_{ij} = \begin{cases} \sum_{k=1}^{n} g_{i,k}(a)B_{kk} & \text{if } i = j, \\ H_{ij}(a)B_{ij} & \text{if } i \neq j, \end{cases} \tag{8.1.7}$$

for every i, j and every $\mathbf{B} \in \mathrm{Sym}$ with components B_{ij} in the basis $\{\mathbf{e}_i\}$ of eigenvectors of $\mathbf{A}$. In other words,

$$\partial_A \boldsymbol{G}(\boldsymbol{A})[\boldsymbol{B}] = \sum_{i,j=1}^{n} g_{i,j}(a)B_{jj}\boldsymbol{e}_i \otimes \boldsymbol{e}_i + \sum H_{ij}(a)B_{ij}\boldsymbol{e}_i \otimes \boldsymbol{e}_j. \tag{8.1.8}$$

Here and throughout the chapter, the symbol $\sum$ with unspecified limits denotes a sum over all $i, j, i \neq j$.

Proof Let $\{\boldsymbol{e}_i\}$ be any orthonormal basis. If $\boldsymbol{A} \in \Delta(\boldsymbol{e}_1, \ldots, \boldsymbol{e}_n) \cap U$ has a spectral decomposition (8.1.2), then $\boldsymbol{G}(\boldsymbol{A}) = \sum_{i=1}^n g_i(a)\boldsymbol{e}_i \otimes \boldsymbol{e}_i$. If $\boldsymbol{G}$ is of class C^r on U, then its restriction to $\Delta(\boldsymbol{e}_1, \ldots, \boldsymbol{e}_n) \cap U$ is of class C^r and as there is one-to-one correspondence between the elements $\boldsymbol{A}$ of $\Delta(\boldsymbol{e}_1, \ldots, \boldsymbol{e}_n) \cap U$ and elements $a \in D$, we see that g is of class C^r on D. Hence the differentiability of $\boldsymbol{G}$ implies the same degree differentiability of g. Conversely, if g is of class C^r, then $\boldsymbol{G}$ is of class C^r on $U^\sharp$, since on $\mathrm{Sym}^\sharp$ both the eigenvalues and eigenvectors are infinitely differentiable by Proposition 1.2.3. When some of the eigenvalues of $\boldsymbol{A}$ coincide, this argument does not work and the reader is referred to SYLVESTER [1985] (see also BALL [1984]) for the proof in this case.

Let us now prove (8.1.7). We can assume that $\boldsymbol{A}$ is represented by a diagonal matrix and $\boldsymbol{B}$ by a symmetric matrix B. We can split B into the sum of two matrices C, D where C is diagonal with diagonal elements of B and D is the off-diagonal part of B. By the linearity of the derivative, it suffices to consider separately the cases when B is diagonal and when the diagonal elements of B vanish. If B is diagonal, we have $\boldsymbol{G}(\boldsymbol{A}+\alpha\boldsymbol{B}) = \sum_{i=1}^n g_i(a_i+\alpha b_i)\boldsymbol{e}_i \otimes \boldsymbol{e}_i$ for all $\alpha \in \mathbb{R}$ sufficiently close to 0, where b_i are the diagonal elements of $\boldsymbol{B}$. A differentiation gives $\partial_A \boldsymbol{G}(\boldsymbol{A})[\boldsymbol{B}] = \sum_{i,j=1}^n g_{i,j}(a)B_{jj}\boldsymbol{e}_i \otimes \boldsymbol{e}_i$, i.e.,

$$\partial_A \boldsymbol{G}(\boldsymbol{A})[\boldsymbol{B}]_{ij} = \sum_{k=1}^{n} g_{i,k}(a)B_{kk} \ \text{ if } i=j \text{ and } \partial_A \boldsymbol{G}(\boldsymbol{A})[\boldsymbol{B}]_{ij} = 0 \text{ if } i \neq j.$$

If the diagonal elements of B vanish, then B can be written in the form in Lemma 8.1.8. Let $\boldsymbol{W}^\pm, \boldsymbol{J}$ be the tensors represented by the matrices $W^\pm, J$. Then $\boldsymbol{W}^-\boldsymbol{A} - \boldsymbol{A}\boldsymbol{W}^- = \boldsymbol{0}$ and

$$\partial_A \boldsymbol{G}(\boldsymbol{A})[\boldsymbol{B}] = \boldsymbol{G}(\boldsymbol{A})\boldsymbol{W}^+ - \boldsymbol{W}^+\boldsymbol{G}(\boldsymbol{A}) + \partial_A \boldsymbol{G}(\boldsymbol{A})[\boldsymbol{J}]\boldsymbol{W}^- - \boldsymbol{W}^-\partial_A \boldsymbol{G}(\boldsymbol{A})[\boldsymbol{J}] \tag{8.1.9}$$

by 8.1.7(2),(3). Since $\boldsymbol{J}$ is diagonal, the above particular case can be used to calculate $\partial_A \boldsymbol{G}(\boldsymbol{A})[\boldsymbol{J}]$, and the result is again a diagonal matrix $D = \mathrm{diag}(p_1, \ldots, p_n)$ with $p_i = \sum_{k=1}^n g_{i,k} \cdot k$, where here and in the sequel the argument a of g_i and its derivatives has been omitted. The matrix form of (8.1.9) is

$$\partial_A \boldsymbol{G}(\boldsymbol{A})[\boldsymbol{B}]_{ij} = (g_i - g_j)W_{ij}^+ + (p_i - p_j)W_{ij}^-.$$

This is different from 0 only if $i \neq j$. If $i \neq j$ is such that $a_i \neq a_j$, then $W_{ij}^- = 0$ and using the expression for W_{ij}^+ from the lemma, we are directly led to $H_{ij}(a)B_{ij}$ with $H_{ij}(a)$ given by (8.1.6) for the case $a_i \neq a_j$. If $i \neq j$ is such that $a_i = a_j$, then $W_{ij}^+ = 0$ and it is found that

$$p_i - p_j = \sum_{k=1}^{n} g_{i,k} \cdot k - \sum_{k=1}^{n} g_{j,k} \cdot k = (g_{i,i} - g_{i,j})(i-j). \tag{8.1.10}$$

To prove $(8.1.10)_2$, we have to take into account that g is symmetric and hence also ∇g is symmetric. If P is a permutation matrix exchanging the indices i, j, then $Pa = a$ and consequently $\nabla g(a) = P\nabla g(a)P^{\mathrm{T}}$. In components this means $g_{i,i}(a) = g_{j,j}(a)$, $g_{i,j}(a) = g_{j,i}(a)$ and $g_{i,k}(a) = g_{j,k}(a)$ for every k such that $k \neq i, k \neq j$. From the last we see that all the terms in the two sums in (8.1.10) are the same except those where $k = i$ or $k = j$. Hence $p_i - p_j = i(g_{i,i} - g_{j,i}) + j(g_{i,j} - g_{j,j})$ and we obtain (8.1.10). Combining this result with the expression for W_{ij}^- in Lemma 8.1.8, we are led to $H_{ij}(a)B_{ij}$ with $H_{ij}(a)$ given by (8.1.6) for the case $a_i = a_j$. □

8.1.10 *Remarks*

(1) For $\boldsymbol{A}$ of the form $\boldsymbol{A} = \alpha\boldsymbol{1}$, $\alpha \in \mathbb{R}$, (8.1.8) takes the form

$$\partial_A \boldsymbol{G}(\boldsymbol{A})[\boldsymbol{B}] = c_0(\mathrm{tr}\,\boldsymbol{B})\boldsymbol{1} + c_1\boldsymbol{B}, \quad \boldsymbol{B} \in \mathrm{Sym},$$

where $c_0 = g_{12}(a)$, $c_1 = g_{11}(a) - g_{12}(a)$, $a = (\alpha, \ldots, \alpha)$. For the proof it suffices to note that at this a we have $g_{ii}(a) = g_{11}$ for all i and $g_{ij}(a) = g_{12}(a)$ for all $i, j, i \neq j$.

(2) The isotropic functions of the functional calculus Subsect. 1.2.5 fall within the scope of Proposition 8.1.5: For $\boldsymbol{A} \mapsto \boldsymbol{G}(\boldsymbol{A}) := f(\boldsymbol{A})$ the mapping g in Proposition 8.1.5 is $g(a_1, \ldots, a_n) = (f(a_1), f(a_2), \ldots, f(a_n))$. The differentiability assertion and the expression for the derivative in Proposition 8.1.5 imply those in Proposition 1.2.6.

8.2 Isotropic Scalar-Valued Functions

8.2.1 Representation Theorem II *Let $h : U \to \mathbb{R}$ be a function on an isotropic subset of* Sym *and let D be the set defined in Subsect.* 8.1.4. *Then h is isotropic if and only if there exists a symmetric function f on D such that $h(\boldsymbol{A}) = f(a_1, \ldots, a_n)$ for every $\boldsymbol{A}$ in U, where a_i are the eigenvalues of $\boldsymbol{A}$.*

Proof Assume that h is isotropic. Define f by $f(a) = h(\boldsymbol{A})$ for any $a \in D$, where $\boldsymbol{A}$ is given by (8.1.2) in a given orthonormal basis $\{\boldsymbol{e}_i\}$. Since h is isotropic, the definition is independent of the choice of the basis and so f is well-defined. The symmetry of f follows in the same way as in the corresponding part of the proof of 8.1.5. The converse implication is immediate. □

8.2.2 Proposition *Let $f : U \to \mathbb{R}$ be a continuously differentiable function on an isotropic subset $U \subset$ Sym. Then the following conditions are equivalent:*

(1) *f is an isotropic scalar-valued function;*

(2) *$\partial_{\boldsymbol{A}} f$ is an isotropic tensor-valued function;*

(3) *for every $\boldsymbol{A} \in U$,*

$$\partial_{\boldsymbol{A}} f(\boldsymbol{A})\boldsymbol{A} = \boldsymbol{A}\partial_{\boldsymbol{A}} f(\boldsymbol{A}). \tag{8.2.1}$$

Here the derivative $\partial_{\boldsymbol{A}} f(\boldsymbol{A})$ is a symmetric second order tensor. Implication (3) $\Rightarrow$ (2) shows that if a tensor-valued function $\boldsymbol{G}$ is a derivative of a scalar-valued (not necessarily isotropic) function and if $\boldsymbol{G}(\boldsymbol{A})$ commutes with $\boldsymbol{A}$ for every $\boldsymbol{A}$, then $\boldsymbol{G}$ is an isotropic-tensor-valued function. This is false without the assumption that $\boldsymbol{G}$ be the derivative of something.

Proof (1) $\Rightarrow$ (2) is immediate. (2) $\Rightarrow$ (3) follows from Proposition 8.1.2. (3) $\Rightarrow$ (1) : Let $\boldsymbol{A} \in U$, $\boldsymbol{Q} \in$ Orth, and let us calculate the difference $f(\boldsymbol{A}) - f(\boldsymbol{Q}\boldsymbol{A}\boldsymbol{Q}^{\mathsf{T}})$. Without any loss of generality we can assume that $\det \boldsymbol{Q} = 1$, since otherwise we pass to $\boldsymbol{Q}\boldsymbol{J}$ where $\boldsymbol{J} \in$ Orth with $\det \boldsymbol{J} = -1$ is such that $\boldsymbol{J}\boldsymbol{A}\boldsymbol{J}^{\mathsf{T}} = \boldsymbol{A}$. If $\det \boldsymbol{Q} = 1$, then there exists a $\boldsymbol{W} \in$ Skew such that $\boldsymbol{Q} = e^{\boldsymbol{W}}$ (see Subsect. 1.3.4). Set $\phi(t) = f(\boldsymbol{A}(t))$, $t \in [0, 1]$ where $\boldsymbol{A}(t) = e^{t\boldsymbol{W}}\boldsymbol{A}e^{-t\boldsymbol{W}}$. The function ϕ is differentiable and

$$\frac{d\phi(t)}{dt} = \partial_{\boldsymbol{A}} f(\boldsymbol{A}(t)) \cdot (\boldsymbol{W}\boldsymbol{A}(t) - \boldsymbol{A}(t)\boldsymbol{W}) = (\partial_{\boldsymbol{A}} f(\boldsymbol{A}(t))\boldsymbol{A}(t) - \boldsymbol{A}(t)\partial_{\boldsymbol{A}} f(\boldsymbol{A}(t))) \cdot \boldsymbol{W} = 0$$

by (8.2.1). Hence $\phi(0) = \phi(1)$. □

8.2.3 Proposition *Let $h : U \to \mathbb{R}$ be an isotropic function on an open subset of* Sym. *Let D be the set defined in Subsect.* 8.1.4 *and f the function occurring in* 8.2.1. *Then h is r-times continuously differentiable if and only if f is r-times continuously differentiable. If $r \ge 1$, then*

$$\partial_{\boldsymbol{A}} h(\boldsymbol{A}) = \sum_{i=1}^{n} f_{,i}(a)\boldsymbol{e}_i \otimes \boldsymbol{e}_i, \quad \boldsymbol{A} \in U, \tag{8.2.2}$$

where $\boldsymbol{A}$ has the spectral decomposition (8.1.2). *If $r \ge 2$, then*

$$\partial_{\boldsymbol{A}}^2 h(\boldsymbol{A})[\boldsymbol{B}, \boldsymbol{C}] = \sum_{i,j=1}^{n} f_{,ij}(a)B_{ii}C_{jj} + \sum H_{ij}(a)B_{ij}C_{ij}, \tag{8.2.3}$$

for every $\boldsymbol{B}, \boldsymbol{C} \in$ Sym, *where B_{ij}, C_{ij} are the components of $\boldsymbol{B}, \boldsymbol{C}$ in the basis of eigenvectors of $\boldsymbol{A}$ and $H_{ij}(a)$ is given by* (8.1.6) *with $g_i = f_{,i}$. Recall the convention about the sums in Proposition* 8.1.9.

Proof This is just a combination of Proposition 8.2.2 with Proposition 8.1.9. □

8.2.4 ***Remark*** Perhaps the most important isotropic functions are the ordered eigenvalues $\hat{a}_i(\boldsymbol{A})$ of $\boldsymbol{A} \in \mathrm{Sym}$, see Sect. 1.2. Each $\hat{a}_i$ is expressible as a symmetric function m_i of an unordered n-tuple of eigenvectors of $\boldsymbol{A}$ as asserted by Proposition 8.2.1. The function m_i is described in Example 18.1.6(1), below. At this moment, it suffices to know that $m_i(a) = a_i$ whenever $a \in \mathbb{R}^n$ and $a_1 > a_2 > \ldots > a_n$. Hence $m_{i,j}(a) = \delta_{ij}$, $m_{i,jk} = 0$ for all such a. If we apply Proposition 8.2.3 to $\hat{a}_i$ restricted to $U := \mathrm{Sym}^\sharp$ with $f = m_i$, we obtain the formulas in Subsect. 1.2.4.

8.3 Objective Functions

8.3.1 Definitions A subset U of Lin is said to be ***objective*** if for every $\boldsymbol{A} \in U$ and every $\boldsymbol{Q} \in \mathrm{Orth}$ also $\boldsymbol{QA} \in U$. A scalar-valued function $h : U \to \mathbb{R}$ is said to be ***objective*** if its domain is objective and $h(\boldsymbol{QA}) = h(\boldsymbol{A})$ for every $\boldsymbol{A} \in U$ and every $\boldsymbol{Q} \in \mathrm{Orth}$. A tensor-valued function $\boldsymbol{G} : U \to \mathrm{Lin}$ is said to be ***objective*** if its domain is objective and $\boldsymbol{G}(\boldsymbol{QA}) = \boldsymbol{QG}(\boldsymbol{A})$ for every $\boldsymbol{A} \in U$ and every $\boldsymbol{Q} \in \mathrm{Orth}$.

The following proposition relates the objectivity of h to the objectivity properties of $\partial_{\boldsymbol{F}} h$. In mechanical terms, Condition (2) will be seen to be equivalent to the transformation law of the referential stress under changes of frame and Condition (3) will be seen to be equivalent to the symmetry of the Cauchy stress.

8.3.2 Proposition *Let $h : U \to \mathbb{R}$ be a continuously differentiable function on an objective subset U if* InvLin. *Consider the following conditions:*

(1) *h is objective;*

(2) *$\partial_{\boldsymbol{F}} h$ is objective;*

(3) $\partial_{\boldsymbol{F}} h(\boldsymbol{F})\boldsymbol{F}^{\mathrm{T}} = \boldsymbol{F}(\partial_{\boldsymbol{F}} h(\boldsymbol{F}))^{\mathrm{T}}$ *for every $\boldsymbol{F} \in U$.*

Then (1) $\Rightarrow$ (2), (1) $\Rightarrow$ (3). *If $U \cap \mathrm{Sym}^+$ is connected and $h(\boldsymbol{JF}) = h(\boldsymbol{F})$ for some $\boldsymbol{J} \in \mathrm{Orth}$ with* $\det \boldsymbol{J} = -1$ *and all $\boldsymbol{F} \in U$ then the above three assertions are equivalent.*

Proof (1) $\Rightarrow$ (2) is immediate. (1) $\Rightarrow$ (3) : We apply $h(\boldsymbol{RF}) = h(\boldsymbol{F})$ to $\boldsymbol{R} = e^{t\boldsymbol{W}}$, $t \in \mathbb{R}$, $\boldsymbol{W} \in \mathrm{Skew}$, and differentiate at $t = 0$ to obtain

$$\partial_{\boldsymbol{F}} h(\boldsymbol{F}) \cdot \boldsymbol{WF} = 0, \tag{8.3.1}$$

which is rewritten as $\boldsymbol{F}^{\mathrm{T}} \partial_{\boldsymbol{F}} h(\boldsymbol{F}) \cdot \boldsymbol{W} = 0$. Since $\boldsymbol{W}$ is an arbitrary skew tensor, Assertion (3) follows. Let us now assume that $U \cap \mathrm{Sym}^+$ is connected and $h(\boldsymbol{JF}) = h(\boldsymbol{F})$ for some $\boldsymbol{J} \in \mathrm{Orth}$ with $\det \boldsymbol{J} = -1$ and all $\boldsymbol{F} \in U$. (2) $\Rightarrow$ (1) : From the assumption that $U \cap \mathrm{Sym}^+$ is connected, from the polar decomposition theorem and the fact that Orth has exactly two connected components we infer that U has exactly two components, one contained in Lin^+ and the other in Lin^-. Let U^+ be the component in Lin^+ and let $\boldsymbol{R} \in \mathrm{Orth}$. (2) implies $\partial_{\boldsymbol{F}} h(\boldsymbol{RF}) - \boldsymbol{R}\partial_{\boldsymbol{F}} h(\boldsymbol{F}) = 0$, which says that the derivative of the function $\boldsymbol{F} \mapsto h(\boldsymbol{RF}) - h(\boldsymbol{F})$ vanishes on U^+, and as U^+ is connected, it follows that this function is constant of value $c = c(\boldsymbol{R})$, i.e.,

$$h(\boldsymbol{RF}) = h(\boldsymbol{F}) + c(\boldsymbol{R}) \tag{8.3.2}$$

for every $\boldsymbol{F} \in U^+$, $\boldsymbol{R} \in \mathrm{Orth}$. Replacing $\boldsymbol{R}$ by $\boldsymbol{RS}$ with $\boldsymbol{R}, \boldsymbol{S} \in \mathrm{Orth}$, we obtain $h(\boldsymbol{RSF}) = h(\boldsymbol{F}) + c(\boldsymbol{RS})$. On the other hand, using (8.3.2) twice, we find $h(\boldsymbol{RSF}) = h(\boldsymbol{F}) + c(\boldsymbol{R}) + c(\boldsymbol{S})$, i.e., $c(\boldsymbol{RS}) = c(\boldsymbol{R}) + c(\boldsymbol{S})$. Moreover, the function c is continuous and hence bounded since Orth is compact. By Lemma 1.3.7, c has the form asserted there. Writing down (8.3.2) for $\boldsymbol{R} = \boldsymbol{J}$ we obtain $h(\boldsymbol{JF}) = h(\boldsymbol{F}) + c_1$ and the additional assumption in the proposition says that $c_1 = 0$; hence $c \equiv 0$ on Orth. Equation (8.3.2) reduces to

$$h(\boldsymbol{RF}) = h(\boldsymbol{F}) \tag{8.3.3}$$

for every $\boldsymbol{F} \in U^+$ and every $\boldsymbol{R} \in \mathrm{Orth}$. Clearly (8.3.3) then extends to every $\boldsymbol{F} \in U$. (3) $\Rightarrow$ (1) : Any $\boldsymbol{R} \in \mathrm{Orth}^+$ can be written as $\boldsymbol{R} = e^{\boldsymbol{W}}$ for some $\boldsymbol{W} \in \mathrm{Skew}$. If we replace $\boldsymbol{F}$ by $e^{t\boldsymbol{W}}\boldsymbol{F}$, $t \in \mathbb{R}$, in (8.3.1), we obtain $dh(e^{t\boldsymbol{W}}\boldsymbol{F})/dt = 0$ and the integration gives $h(\boldsymbol{RF}) = h(\boldsymbol{F})$ for all $\boldsymbol{R} \in \mathrm{Orth}^+$ and $\boldsymbol{F} \in U$. The proof is completed by showing that under the additional hypotheses, $h(\boldsymbol{RF}) = h(\boldsymbol{F})$ holds for every $\boldsymbol{R} \in \mathrm{Orth}$. □

Next we give a representation theorem for objective scalar-valued functions. Its main part is the assertion about the derivative.

8.3.3 Representation Theorem III *Let $h : U \to \mathbb{R}$ be a function on an objective subset of* InvLin.

(1) *Then h is objective if and only if it has a representation $h(\boldsymbol{F}) = m(\boldsymbol{U})$ where $\boldsymbol{U} = \sqrt{\boldsymbol{F}^{\mathsf{T}}\boldsymbol{F}}$ and $m : V \to \mathbb{R}$ is a function on $V = U \cap \mathrm{Sym}^+$.*
(2) *If h has the representation as in* (1), *then m is a restriction of h to $U \cap \mathrm{Sym}^+$. The set U is open in* Lin *if and only if V is open in* Sym. *The function h is of class C^r, $1 = 0, \ldots, \infty$, if and only if m is of class C^r.*
(3) *If h is continuously differentiable, then*

$$\partial_{\boldsymbol{F}} h(\boldsymbol{F}) = 2 \int_0^\infty \boldsymbol{F} e^{-\boldsymbol{U}t} \partial_{\boldsymbol{U}} m(\boldsymbol{U}) e^{-\boldsymbol{U}t}\, dt \tag{8.3.4}$$

for every $\boldsymbol{F} \in U$. If $\boldsymbol{F} \in V$ is symmetric, then the components of $\partial_{\boldsymbol{F}} h(\boldsymbol{F})$ in the basis of eigenvectors of $\boldsymbol{F} = \boldsymbol{U}$ are

$$\partial_{\boldsymbol{F}} h(\boldsymbol{F})_{ij} = \frac{2 v_i \partial_{\boldsymbol{U}} m(\boldsymbol{U})_{ij}}{v_i + v_j}, \tag{8.3.5}$$

where v_i are the eigenvalues of $\boldsymbol{F}$; hence the symmetric part of $\partial_{\boldsymbol{F}} h(\boldsymbol{U})$ coincides with $\partial_{\boldsymbol{U}} m(\boldsymbol{U})$.

Even when $\boldsymbol{F} \in \mathrm{Sym}^+$, the derivative of h does not coincide with that of m. The equation $\partial_{\boldsymbol{F}} h(\boldsymbol{U}) = \partial_{\boldsymbol{U}} m(\boldsymbol{U})$ holds only if $\partial_{\boldsymbol{U}} m(\boldsymbol{U})$ commutes with $\boldsymbol{U}$. The latter is true if m is isotropic, but not generally.

Proof (1) is a direct application of the polar decomposition theorem. (2) & (3): Equation (8.3.4) is obtained by a calculation using (1.4.3), and (8.3.5) is its consequence. □

8.4 Objective-Isotropic Tensor-Valued Functions

The main result of the section is the formula for the derivative of an objective-isotropic tensor-valued function in Proposition 8.4.5. This gives also the form of the second differential of an objective-isotropic scalar-valued function, which is helpful in examining the convexity properties of isotropic thermodynamic potentials.

8.4.1 Definitions A subset U of Lin is said to be ***objective-isotropic*** if for every $\boldsymbol{F} \in U$ and every $\boldsymbol{Q}, \boldsymbol{R} \in \mathrm{Orth}$ also $\boldsymbol{QFR} \in U$. A scalar-valued function $h : U \to \mathbb{R}$ is said to be ***objective-isotropic*** if its domain is objective-isotropic and $h(\boldsymbol{QFR}) = h(\boldsymbol{F})$ for every $\boldsymbol{F} \in U$ and every $\boldsymbol{Q}, \boldsymbol{R} \in \mathrm{Orth}$. A tensor-valued function $\boldsymbol{G} : U \to \mathrm{Lin}$ is said to be ***referentially objective-isotropic*** if its domain is objective-isotropic and

$$\boldsymbol{G}(\boldsymbol{QFR}) = \boldsymbol{Q}\boldsymbol{G}(\boldsymbol{F})\boldsymbol{R} \tag{8.4.1}$$

for every $\boldsymbol{F} \in U$ and every $\boldsymbol{Q}, \boldsymbol{R} \in \mathrm{Orth}$. Clearly, if $\boldsymbol{G} : U \to \mathrm{Lin}$ is an objective-isotropic tensor-valued function then the restriction of $\boldsymbol{G}$ to $U \cap \mathrm{Sym}$ is isotropic. Hence the results on isotropic functions can be used to derive the properties of objective-isotropic functions.

8.4.2 We are now going to discuss the representation theorems for objective-isotropic functions defined on subsets of InvLin. Let

$$\mathbb{R}^n_{++} = \{x = (x_1, \ldots, x_n) \in \mathbb{R}^n : x_i > 0, i = 1, \ldots, n\}.$$

For a given set $U \subset$ InvLin denote by D the set of all $v = (v_1, \ldots, v_n) \in \mathbb{R}^n_{++}$ for which there exists an $\boldsymbol{F} \in U$ such that the set $\{v_1, \ldots, v_n\}$ is a collection of singular values of $\boldsymbol{F}$ (taken in arbitrary order, see Sect. 1.4). This is automatically a symmetric set. If U is objective-isotropic, this definition can be stated equivalently as follows. D is the set of all n-tuples $(v_1, \ldots, v_n) \in \mathbb{R}^n_{++}$ such that for some (and hence every) orthonormal basis $\{\boldsymbol{e}_i\}$ of Vect, the tensor $\boldsymbol{V} := \sum_{i=1}^n v_i \boldsymbol{e}_i \otimes \boldsymbol{e}_i$ is in U or equivalently U is the set of all eigenvalues of the elements of $U \cap \mathrm{Sym}^+$. Conversely, for a given subset D of $\mathbb{R}^n_{++}$ denote by U the subset of InvLin consisting of all $\boldsymbol{F}$ with $(\hat{v}_1(\boldsymbol{F}), \ldots, \hat{v}_n(\boldsymbol{F})) \in D$. The set U is automatically isotropic. Hence the above relations establish a one-to-one relation between objective-isotropic subsets U of InvLin and symmetric subsets of D of $\mathbb{R}^n_{++}$. If U is open in InvLin, then D is an open subset of $\mathbb{R}^n$ and conversely.

8.4.3 Representation Theorem IV *Let $\boldsymbol{G} : U \to$ Lin be a tensor-valued function on an objective-isotropic subset of* InvLin *and let D be related to U as in Subsect.* 8.4.2. *Then $\boldsymbol{G}$ is referentially objective-isotropic if and only if there exists a symmetric function $g = (g_1, \ldots, g_n) : D \to \mathbb{R}^n$ such that for any $\boldsymbol{F} \in U$ with the singular vectors $\boldsymbol{e}_i$ (see Subsect.* 1.4.1) *we have*

$$\boldsymbol{G}(\boldsymbol{F}) = \sum_{i=0}^{n} g_i(v)\boldsymbol{e}_i \otimes \boldsymbol{R}^{\mathrm{T}}\boldsymbol{e}_i = \Big(\sum_{i=0}^{n} g_i(v)\boldsymbol{e}_i \otimes \boldsymbol{e}_i\Big)\boldsymbol{R} \tag{8.4.2}$$

where $v = (v_1, \ldots, v_n)$ and $\boldsymbol{R}$ are the singular values and the rotation tensor of $\boldsymbol{F}$.

This is completely analogous to 8.1.5 and the proof is omitted.

8.4.4 Proposition *Let $\boldsymbol{G} : U \to$ Sym be a continuously differentiable objective-isotropic function. Then the following assertions hold:*

(1) ***differential consequences of the objectivity-isotropy:***

$$\partial_{\boldsymbol{F}}\boldsymbol{G}(\boldsymbol{F})[\boldsymbol{W}\boldsymbol{F}] = \boldsymbol{W}\boldsymbol{G}(\boldsymbol{F}), \quad \partial_{\boldsymbol{F}}\boldsymbol{G}(\boldsymbol{F})[\boldsymbol{F}\boldsymbol{W}] = \boldsymbol{G}(\boldsymbol{F})\boldsymbol{W}$$

for every $\boldsymbol{F} \in U$ and $\boldsymbol{W} \in$ Skew.

(2) *If $\boldsymbol{F} \in \mathrm{Sym}^+$ and $\boldsymbol{W} \in$ Skew then there exists a unique $\boldsymbol{Z} \in$ Lin such that $\boldsymbol{Z}\boldsymbol{F} + \boldsymbol{F}\boldsymbol{Z} = \boldsymbol{W}$; this $\boldsymbol{Z}$ satisfies $\boldsymbol{Z} \in$ Skew and in the basis of eigenvectors of $\boldsymbol{F}$,*

$$Z_{ij} = W_{ij}/(v_i + v_j), \quad 1 \le i, j \le n.$$

(3) *Let $\boldsymbol{H}$ be the restriction of $\boldsymbol{G}$ to $U \cap \mathrm{Sym}^+$. Then $\boldsymbol{G}$ is continuously differentiable if and only if $\boldsymbol{H}$ is continuously differentiable; if this is the case, then for every $\boldsymbol{F} \in U \cap \mathrm{Sym}^+$, $\boldsymbol{B} \in$ Lin,*

$$\partial_{\boldsymbol{F}}\boldsymbol{G}(\boldsymbol{F})[\boldsymbol{B}] = \partial_{\boldsymbol{F}}\boldsymbol{H}(\boldsymbol{F})[\boldsymbol{E}] + \boldsymbol{Z}\boldsymbol{G}(\boldsymbol{F}) + \boldsymbol{G}(\boldsymbol{F})\boldsymbol{Z}, \tag{8.4.3}$$

where $\boldsymbol{E}$, $\boldsymbol{W}$ are the symmetric and skew parts of $\boldsymbol{B}$ and $\boldsymbol{Z}$ is the solution of $\boldsymbol{Z}\boldsymbol{F} + \boldsymbol{F}\boldsymbol{Z} = \boldsymbol{W}$.

Assertion (3) reduces the calculation of the derivative of $\boldsymbol{G}$ to the derivative of the isotropic function $\boldsymbol{H}$. Moreover, (8.4.3) gives a decomposition of $\partial_{\boldsymbol{F}}\boldsymbol{G}(\boldsymbol{F})[\boldsymbol{B}]$ into the symmetric and skew parts: $\partial_{\boldsymbol{F}}\boldsymbol{H}(\boldsymbol{F})[\boldsymbol{E}]$ is symmetric and $\boldsymbol{Z}\boldsymbol{G}(\boldsymbol{F}) + \boldsymbol{G}(\boldsymbol{F})\boldsymbol{Z}$ skew.

Proof (1): Differentiate $\boldsymbol{G}(e^{\boldsymbol{W}t}\boldsymbol{F}) = e^{\boldsymbol{W}t}\boldsymbol{G}(\boldsymbol{F})$ and $\boldsymbol{G}(\boldsymbol{F}e^{\boldsymbol{W}t}) = \boldsymbol{G}(\boldsymbol{F})e^{\boldsymbol{W}t}$ at $t = 0$. (2) is immediate. (3): We have, for any $\boldsymbol{F} \in U, \boldsymbol{G}(\boldsymbol{F}) = \boldsymbol{H}(\boldsymbol{V})\boldsymbol{R}$, where $\boldsymbol{F} = \boldsymbol{V}\boldsymbol{R}$ is the polar decomposition; using the fact that both $\boldsymbol{V}, \boldsymbol{R}$ are infinitely differentiable functions of $\boldsymbol{F}$ (Proposition 1.4.4), we obtain the assertion about the differentiability. To prove the formula, let us first consider the case when $\boldsymbol{B} = \boldsymbol{E} \in$ Sym. Then $\boldsymbol{F}+\lambda\boldsymbol{B} \in \mathrm{Sym}^+$ for all λ sufficiently close to 0 and hence $\boldsymbol{G}(\boldsymbol{F}+\lambda\boldsymbol{B}) = \boldsymbol{H}(\boldsymbol{F}+\lambda\boldsymbol{B})$. A differentiation gives (8.4.3) in this special case. On the other hand, if $\boldsymbol{B} = \boldsymbol{W} \in$ Skew, then $\boldsymbol{W}$ can be written uniquely as $ZF + FZ$ by (2). 8.4.4 gives $\partial_F G(F)[W] = ZG(F) + G(F)Z$. Combining these special cases, we obtain the general result. □

To evaluate the derivatives of the objective-isotropic tensor functions in the representation (8.4.2), assume that $g = (g_1, \ldots, g_n) : D \to \mathbb{R}^n$, $D \subset \mathbb{R}^n_{++}$, is a continuously differentiable symmetric function on an open subset of $\mathbb{R}^n$. Let i, j be indices, $i \neq j$. In addition to the function H_{ij}, defined by (8.1.6), define a function $L_{ij} : D \to \mathbb{R}$ by

$$L_{ij}(v) = \frac{g_i(v) + g_j(v)}{v_i + v_j}. \tag{8.4.4}$$

8.4.5 Proposition *Let $G : U \to$ Lin be an objective-isotropic function on an open subset of* InvLin *and let D and g be as in Proposition* 8.4.3. *Then G is of class C^r, $r = 0, \ldots, \infty$, if and only if g is of class C^r. If G is of class C^1, then*

$$\partial_F G(F)[B]_{ij} = \begin{cases} \sum_{k=1}^{n} g_{i,k}(v) B_{kk} & \text{if } i = j, \\ H_{ij}(v) E_{ij} + L_{ij}(v) W_{ji} & \text{if } i \neq j, \end{cases} \tag{8.4.5}$$

for every $F \in U \cap \mathrm{Sym}^+$, every i, j and every $B \in$ Lin. Here E, W are the symmetric and the skew parts of B and the subscripts denote the components in an arbitrary orthonormal basis of eigenvectors of F. In other words,

$$\partial_F G(F)[B] = \sum_{i,j=1}^{n} g_{i,j}(v) B_{jj} e_i \otimes e_i + \sum (H_{ij}(v) E_{ij} + L_{ij}(v) W_{ji}) e_i \otimes e_j. \tag{8.4.6}$$

where the second sum is taken over all i, j, $i \neq j$.

This is just a combination of 8.4.4 with 8.1.9. It is enough to evaluate the derivative at a positive-definite symmetric argument F, since the formula $\partial_F G(FQ)[BQ] = \partial_F G(F)[B]Q$ relates the derivative at a general F to that when F is positive definite symmetric. An alternative form of (8.4.5) is

$$\partial_F G(F)[B]_{ij} = \begin{cases} \sum_{k=1}^{n} g_{i,k}(v) B_{kk} & \text{if } i = j, \\ K_{ij}(v) B_{ij} + M_{ij}(v) B_{ji} & \text{if } i \neq j, \end{cases} \tag{8.4.7}$$

where

$$K_{ij}(v) = \frac{v_i g_i(v) - v_j g_j(v)}{v_i^2 - v_j^2}, \quad M_{ij}(v) = \frac{v_j g_i(v) - v_i g_j(v)}{v_i^2 - v_j^2} \quad \text{if } v_i \neq v_j \tag{8.4.8}$$

and

$$\left.\begin{aligned} K_{ij}(v) &= \tfrac{1}{2}(g_{i,i}(v) - g_{i,j}(v) + g_i(v)/v_i), \\ M_{ij}(v) &= \tfrac{1}{2}(g_{i,i}(v) - g_{i,j}(v) - g_i(v)/v_i), \end{aligned}\right\} \text{if } v_i = v_j. \tag{8.4.9}$$

8.4.6 ***Remarks***

(1) For F of the form $F = \alpha \mathbf{1}$, $\alpha > 0$, (8.4.6) takes the form

$$G(F)[B] = c_0 (\operatorname{tr} E) \mathbf{1} + c_1 E + c_2 W, \quad B \in \text{Lin},$$

where $c_0 = g_{12}(a)$, $c_1 = g_{11}(a) - g_{12}(a)$, $c_2 = g_1(a)/\alpha$, $a = (\alpha, \ldots, \alpha)$ and E, W are the symmetric and the skew parts of B, respectively.

(2) *Let $G : U \to$ Lin be an objective-isotropic function on an open subset of* InvLin *that is linear in the sense that it is a restriction of a linear function on* Lin. *Then there exists an $\alpha \in \mathbb{R}$ such that*

$$G(F) = \alpha F, \quad F \in U.$$

Proof Let us denote the linear extension of G to Lin again by G. Clearly G is of class C^∞ on U and

$$\partial_F G(F)[B] = G(B) \tag{8.4.10}$$

for every $F \in U$ and $B \in \text{Lin}$. On the other hand, differentiating (8.4.1) and using (8.4.10) we obtain

$$M(QBR) = QM(B)R \tag{8.4.11}$$

and so in particular the restriction of M to Sym is an isotropic function, and by Subsect. 8.1.6 it has the representation

$$M(B) = c_0(\text{tr}\, B)\mathbf{1} + c_1 B, \quad B \in \text{Sym}. \tag{8.4.12}$$

Writing $F = VR$ and combining (8.4.11), (8.4.12), we obtain

$$M(F) = c_0(\text{tr}\, V)R + c_1 F, \quad F \in \text{Lin}, \quad G(F) = M(F) = c_0(\text{tr}\, V)R + c_1 F, \quad F \in U.$$

Both the terms in the second expression for G are objective-isotropic; the second term is also linear, so that the first term must also be linear. We shall now show that the latter is true only if $c_0 = 0$. In other words, the function G_0 defined as $F \mapsto (\text{tr}\, V)R$, $F \in U$ (V, R the polar decomposition of F) is not linear. Applying the representation theorem 8.4.3, it is found that $g_i(v) = m := \sum_{m=1}^n v_m$ for every i, $1 \le i \le n$ and so $g_i(v) = g_j(v)$ for every i, j. By (8.4.5) then the off-diagonal element of $\partial_F G_0(F)[B]$ is

$$\partial_F G_0(F)[B]_{ij} = m(v)\frac{2W_{ij}}{v_i + v_j}, \quad i \neq j, \quad W_{ij} \equiv \tfrac{1}{2}(B_{ij} - B_{ji}).$$

This term explicitly depends on the singular values of F and hence is nonconstant and so G_0 is nonlinear. □

8.5 Objective-Isotropic Scalar-Valued Functions

The formula for the second derivative of an objective-isotropic scalar-valued function to be given below will be used several times in dealing with the convexity properties of the stored energy functions of isotropic solids. For example, it will provide a simple derivation of the Baker–Ericksen inequalities.

8.5.1 Representation Theorem V *Let $h : U \to \mathbb{R}$ be a function on an objective-isotropic subset U of* InvLin. *Let D be the set defined in Subsect.* 8.4.2. *Then h is objective-isotropic if and only if there exists a symmetric function f on D such that $h(F) = f(v_1, \dots, v_n)$ for every F in U, where v_i are the singular values of F.*

This is completely analogous to Proposition 8.2.1 and the proof is therefore omitted. The following proposition shows, among other things, that the derivative of an objective-isotropic scalar-valued function is an objective-isotropic tensor-valued function; hence the results of the preceding section can be applied.

8.5.2 Proposition *Let $h : U \to \mathbb{R}$ be a continuously differentiable function on an objective-isotropic set $U \subset$* InvLin. *Consider the following four conditions:*

(1) *h is objective-isotropic;*

(2) *$\partial_A h$ is referentially objective-isotropic;*

(3) *for every $A \in U$,*

$$\partial_A h(A)A^{\mathsf T} = A\partial_A h(A)^{\mathsf T}, \quad \partial_A h(A)^{\mathsf T} A = A^{\mathsf T}\partial_A h(A); \tag{8.5.1}$$

(4) *for every $A \in U$,*

$$\partial_A h(A)A^{\mathsf T}A = AA^{\mathsf T}\partial_A h(A). \tag{8.5.2}$$

Then (1) ⇒ (2), (1) ⇒ (3), (3) ⇔ (4). *If* $U \cap \mathrm{Sym}^+$ *is connected and* $h(\boldsymbol{A}) = h(\boldsymbol{JA})$ *for some* $\boldsymbol{J} \in \mathrm{Orth}$ *with* $\det \boldsymbol{J} = -1$ *and every* $\boldsymbol{A} \in U$ *then the above four conditions are equivalent.*

Condition (3) or equivalently (4) are the differential forms of the condition for the function to be objective-isotropic.

Proof Other parts of the proof being similar to those of Proposition 8.3.2 and noting that (3) ⇒ (4) is a direct calculation, we only prove (4) ⇒ (3). A multiplication of (8.5.2) by $\boldsymbol{A}^{\mathsf{T}}$ from the left and a multiplication of (8.5.2) by $\boldsymbol{A}^{\mathsf{T}}$ from the right gives

$$(\boldsymbol{A}^{\mathsf{T}}\partial_{\boldsymbol{A}} h(\boldsymbol{A}))(\boldsymbol{A}^{\mathsf{T}}\boldsymbol{A}) = (\boldsymbol{A}^{\mathsf{T}}\boldsymbol{A})(\boldsymbol{A}^{\mathsf{T}}\partial_{\boldsymbol{A}} h(\boldsymbol{A})), \quad (\partial_{\boldsymbol{A}} h(\boldsymbol{A})\boldsymbol{A}^{\mathsf{T}})(\boldsymbol{A}\boldsymbol{A}^{\mathsf{T}}) = (\boldsymbol{A}\boldsymbol{A}^{\mathsf{T}})(\partial_{\boldsymbol{A}} h(\boldsymbol{A})\boldsymbol{A}^{\mathsf{T}}),$$

which says that $\boldsymbol{A}^{\mathsf{T}}\partial_{\boldsymbol{A}} h(\boldsymbol{A})$ commutes with $\boldsymbol{A}^{\mathsf{T}}\boldsymbol{A}$ and $\partial_{\boldsymbol{A}} h(\boldsymbol{A})\boldsymbol{A}^{\mathsf{T}}$ with $\boldsymbol{A}\boldsymbol{A}^{\mathsf{T}}$. If the eigenvalues of $\boldsymbol{A}^{\mathsf{T}}\boldsymbol{A}$ and $\boldsymbol{A}\boldsymbol{A}^{\mathsf{T}}$ are mutually distinct, then Lemma 8.1.3(2) implies that $\boldsymbol{A}^{\mathsf{T}}\partial_{\boldsymbol{A}} h(\boldsymbol{A})$ and $\partial_{\boldsymbol{A}} h(\boldsymbol{A})\boldsymbol{A}^{\mathsf{T}}$ are symmetric, and this gives (8.5.1). Finally the continuity of the derivatives and the density asserted in Lemma 8.1.3(1) then extends the validity of (8.5.1) for all $\boldsymbol{A} \in U$. □

8.5.3 Proposition *Let* $m : U_0 \subset \mathrm{Sym}^+ \to \mathbb{R}$ *be an isotropic continuously differentiable function and let* $h : U \to \mathbb{R}$ *be defined on the domain*

$$U = \{\boldsymbol{F} \in \mathrm{InvLin} : \boldsymbol{U} \in U_0\} = \{\boldsymbol{F} \in \mathrm{InvLin} : \boldsymbol{V} \in U_0\}$$

by the formula $h(\boldsymbol{F}) = m(\boldsymbol{U}) = m(\boldsymbol{V})$. *(Here* $\boldsymbol{U}, \boldsymbol{V}, \boldsymbol{R}$ *are the polar factors of* $\boldsymbol{F}$.*) Then* h *is a continuously differentiable objective-isotropic function and*

$$\partial_{\boldsymbol{F}} h(\boldsymbol{F}) = \boldsymbol{R}\partial_{\boldsymbol{A}} m(\boldsymbol{U}) = \partial_{\boldsymbol{A}} m(\boldsymbol{V})\boldsymbol{R} \tag{8.5.3}$$

for every $\boldsymbol{F} \in U$, *where* $\partial_{\boldsymbol{A}} m$ *is a symmetric tensor-valued function delivering the derivative of* m *with respect to its symmetric-tensor argument.*

This reduces the calculation of the derivative of h with respect to the argument $\boldsymbol{F}$ varying over a subset of general, nonsymmetric tensors, to calculating the derivative of the restriction of h to symmetric tensors with respect to its symmetric-tensor-valued argument.

Proof By Proposition 8.5.2 the derivative $\partial_{\boldsymbol{F}} h$ is referentially objective-isotropic and hence its restriction to $U \cap \mathrm{Sym}$ is isotropic. Then Proposition 8.1.2 tells us that $\partial_{\boldsymbol{F}} h(\boldsymbol{A})$ is symmetric whenever $\boldsymbol{A} \in U \cap \mathrm{Sym}$. But $\partial_{\boldsymbol{A}} m$ is isotropic as well and as $h(\boldsymbol{A}+\lambda\boldsymbol{B}) = m(\boldsymbol{A}+\lambda\boldsymbol{B})$ for every $\lambda \in \mathbb{R}$ and $\boldsymbol{B} \in \mathrm{Sym}$, we have $\partial_{\boldsymbol{F}} h(\boldsymbol{A}) \cdot \boldsymbol{B} = \partial_{\boldsymbol{A}} m(\boldsymbol{A}) \cdot \boldsymbol{B}$. As this must be true for every $\boldsymbol{B} \in \mathrm{Sym}$, the symmetry of $\partial_{\boldsymbol{F}} h(\boldsymbol{A})$ and $\partial_{\boldsymbol{A}} m(\boldsymbol{A})$ allows us to conclude that $\partial_{\boldsymbol{F}} h(\boldsymbol{A}) = \partial_{\boldsymbol{A}} m(\boldsymbol{A})$, for every $\boldsymbol{A} \in U \cap \mathrm{Sym}$. But then for a general $\boldsymbol{F} \in U$, we have $\boldsymbol{F} = \boldsymbol{RU} = \boldsymbol{VR}$ and the referential objectivity and isotropy of $\partial_{\boldsymbol{F}} h$ gives $\partial_{\boldsymbol{F}} h(\boldsymbol{F}) = \partial_{\boldsymbol{F}} h(\boldsymbol{RU}) = \boldsymbol{R}\partial_{\boldsymbol{F}} h(\boldsymbol{U}) = \boldsymbol{R}\partial_{\boldsymbol{A}} m(\boldsymbol{U})$, which proves $(8.5.3)_1$. Equation $(8.5.3)_2$ is proved similarly. □

8.5.4 Proposition *Let* $h : U \to \mathbb{R}$ *be an objective-isotropic function on an open subset* U *of* InvLin. *Let* f, D *be as in Proposition* 8.5.1. *Then* h *is of class* C^r, $r = 0, \ldots, \infty$, *if and only if* f *is of class* C^r. *If* $r \ge 1$ *and* $\boldsymbol{F} \in U$, *then*

$$\partial_{\boldsymbol{F}} h(\boldsymbol{F}) = \sum_{i=1}^{n} f_{,i}(\boldsymbol{v})\boldsymbol{e}_i \otimes \boldsymbol{R}^{\mathsf{T}}\boldsymbol{e}_i, \quad \boldsymbol{F} \in U, \tag{8.5.4}$$

where $\boldsymbol{v}$ *and* $\boldsymbol{R}$ *are an* m*-tuple of singular values and the rotation tensor of* $\boldsymbol{F}$ *and* $\{\boldsymbol{e}_i\}$ *is any orthonormal basis of singular vectors of* $\boldsymbol{F}$ *(see Subsect.* 1.4.1*). If* $r \ge 2$ *and* $\boldsymbol{F} \in U \cap \mathrm{Sym}^+$, *then*

$$\partial_{\boldsymbol{F}}^2 h(\boldsymbol{F})[\boldsymbol{B}, \boldsymbol{C}] = \sum_{i,j=1}^{n} f_{,ij}(\boldsymbol{v})B_{ii}C_{jj} + \sum (H_{ij}(\boldsymbol{v})E_{ij}D_{ij} + L_{ij}(\boldsymbol{v})W_{ij}X_{ij}), \tag{8.5.5}$$

for every $\boldsymbol{B}, \boldsymbol{C} \in \mathrm{Lin}$, *where* E_{ij}, D_{ij} (W_{ij}, X_{ij}) *are the components of the symmetric (skew) parts of* $\boldsymbol{B}, \boldsymbol{C}$, *in the basis of eigenvectors of* $\boldsymbol{F}$ *and* $H_{ij}(v)$, $L_{ij}(v)$ *are given by* (8.1.6), (8.4.4) *with* $g_i = f_{,i}$. *Recall the convention about the sums in Proposition* 8.1.9. *Alternatively,*

$$\partial_{\boldsymbol{F}}^2 h(\boldsymbol{F})[\boldsymbol{B}, \boldsymbol{C}] = \sum_{i,j=1}^{n} f_{,ij}(v) B_{ii} C_{jj} + \sum (K_{ij}(v) B_{ij} C_{ij} + M_{ij}(v) B_{ij} C_{ji},) \tag{8.5.6}$$

where K_{ij}, M_{ij} *are given by* (8.4.8) *and* (8.4.9).

This is a direct consequence of Proposition 8.4.5.

Let h and $\boldsymbol{F}$ be as in Proposition 8.5.4. If $\partial_{\boldsymbol{F}}^2 h(\boldsymbol{F})$ is interpreted as a linear transformation on Lin (see Subsect. 1.6.1), then the components of $\partial_{\boldsymbol{F}}^2 h(\boldsymbol{F})[\boldsymbol{B}]$ are given by the right-hand side of (8.4.5). Since $\partial_{\boldsymbol{F}}^2 h(\boldsymbol{F})$ has the major symmetry as a consequence of the interchangeability of the second partial derivatives, $\partial_{\boldsymbol{F}}^2 h(\boldsymbol{F})$ is a symmetric linear transformation on Lin. By the spectral theorem, it has a complete orthonormal set of eigenvectors. Here eigenvectors are second-order tensors, i.e., a nonzero $\boldsymbol{H} \in \mathrm{Lin}$ is said to be an eigenvector of $\mathsf{B} := \partial_{\boldsymbol{F}}^2 h(\boldsymbol{F})$ and $\lambda \in \mathbb{R}$ its eigenvalue if $\mathsf{B}[\boldsymbol{H}] = \lambda \boldsymbol{H}$. The following proposition describes the set of all eigenvectors of $\partial_{\boldsymbol{F}}^2 h(\boldsymbol{F})$. Recall that a subspace S of a vector space X is said to be invariant under a linear transformation $M : \mathrm{X} \to \mathrm{X}$ if $MS \subset S$.

8.5.5 Proposition *Let* h *and* $\boldsymbol{F} \in U \cap \mathrm{Sym}$ *be as in Proposition* 8.5.4, *with* h *twice continuously differentiable and let* $\{\boldsymbol{e}_i\}$ *be any orthonormal basis of eigenvectors of* $\boldsymbol{F}$. *Then*

(1) *the following subspaces of* Lin *are invariant under* $\mathsf{B} := \partial_{\boldsymbol{F}}^2 h(\boldsymbol{F})$: Sym; Skew; *the set of all* $\boldsymbol{S} \in \mathrm{Sym}$ *represented in* $\{\boldsymbol{e}_i\}$ *by diagonal matrices; the set of all* $\boldsymbol{M} \in \mathrm{Sym}$ *represented in* $\{\boldsymbol{e}_i\}$ *by off-diagonal matrices;*

(2) (a) *if* $w = (w_1, \ldots, w_n) \in \mathbb{R}^n$ *is any eigenvector of the Hessian* $n \times n$ *matrix* $[f_{,ij}]$, *and* $\lambda \in \mathbb{R}$ *is the corresponding eigenvalue, then*

$$\boldsymbol{H} := \sum_{i=1}^{n} w_i \boldsymbol{e}_i \otimes \boldsymbol{e}_i$$

is an eigenvector of B *and* λ *is the corresponding eigenvalue;* (b) *if* $i \neq j$ *then* $\boldsymbol{e}_i \otimes \boldsymbol{e}_j + \boldsymbol{e}_j \otimes \boldsymbol{e}_i$ *is an eigenvector with the eigenvalue* $H_{ij}(v)$; (c) *if* $i \neq j$ *then* $\boldsymbol{e}_i \otimes \boldsymbol{e}_j - \boldsymbol{e}_j \otimes \boldsymbol{e}_i$ *is an eigenvector with the eigenvalue* $L_{ij}(v)$. *Altogether, this gives a basis consisting of* n^2 *orthogonal eigenvectors of* B.

This is just an inspection of the right-hand side of (8.4.5).

8.5.6 Example Let dim Vect = 3 and consider the functions J_i, $i = 1, 2, 3$, defined on InvLin by $J_i(\boldsymbol{F}) = I_i(\boldsymbol{V}) = I_i(\boldsymbol{U})$, $i = 1, 2, 3$, where $\boldsymbol{U}, \boldsymbol{V}$ are the polar factors of $\boldsymbol{F}$ and I_i the principal invariants. These functions are infinitely differentiable,

$$\partial_{\boldsymbol{F}} J_1(\boldsymbol{F}) = \boldsymbol{R}, \quad \partial_{\boldsymbol{F}} J_2(\boldsymbol{F}) = \mathrm{tr}(\boldsymbol{U})\boldsymbol{R} - \boldsymbol{R}\boldsymbol{U} = \mathrm{tr}(\boldsymbol{V})\boldsymbol{R} - \boldsymbol{V}\boldsymbol{R}, \quad \partial_{\boldsymbol{F}} J_3(\boldsymbol{F}) = (\det \boldsymbol{F})\boldsymbol{F}^{-\mathrm{T}} \tag{8.5.7}$$

for every $\boldsymbol{F} \in \mathrm{InvLin}$ and

$$\begin{aligned} \partial_{\boldsymbol{F}}^2 J_1(\boldsymbol{F})[\boldsymbol{B}, \boldsymbol{C}] &= 2 \sum \frac{1}{v_i + v_j} W_{ij} X_{ij}, \\ \partial_{\boldsymbol{F}}^2 J_2(\boldsymbol{F})[\boldsymbol{B}, \boldsymbol{C}] &= \mathrm{tr}\, \boldsymbol{B}\, \mathrm{tr}\, \boldsymbol{C} - \mathrm{tr}(\boldsymbol{B}\boldsymbol{C}^{\mathrm{T}}) + 2\, \mathrm{tr}\, \boldsymbol{V} \sum \frac{1}{v_i + v_j} W_{ij} X_{ij}, \\ \partial_{\boldsymbol{F}}^2 J_3(\boldsymbol{F})[\boldsymbol{B}, \boldsymbol{C}] &= (\det \boldsymbol{V})(\mathrm{tr}(\boldsymbol{V}^{-1}\boldsymbol{B})\, \mathrm{tr}(\boldsymbol{V}^{-1}\boldsymbol{C}) - \mathrm{tr}(\boldsymbol{V}^{-1}\boldsymbol{B}\boldsymbol{V}^{-1}\boldsymbol{C})) \end{aligned} \tag{8.5.8}$$

for every $\boldsymbol{F} \in \mathrm{Sym}^+$, where W_{ij}, X_{ij} are the components of the skew parts of $\boldsymbol{B}, \boldsymbol{C}$ in the basis of eigenvectors of $\boldsymbol{F}$. The first derivatives follow from Proposition 8.5.3. Equations (8.5.8) are obtained from the component expressions in Proposition 1.4.4.

Bibliographical Note

The formula (8.1.7) occurs independently in BOWEN & WANG [1970, 1971] and CHADWICK & OGDEN [1971b]. The latter reference establishes it for a somewhat more general case of tensor functions for which the principal directions of $\boldsymbol{F}$ and $\boldsymbol{G}(\boldsymbol{F})$ coincide. A formula of this type is implicit in HILL

[1970a]. Yet another derivation is in BALL [1984], however, without the limiting case when some of the eigenvalues coincide. The references BOWEN & WANG [1970, 1971] and CHADWICK & OGDEN [1971b] also give formulas for $\partial_F^2 G$. The formula (8.5.5) is due to BALL [1984] for the case of distinct singular values. The differentiability assertion in Proposition 8.1.9 for $r = 0, 1, \infty$ is also due to BALL [1984]; the general r is due to SYLVESTER [1985].

9 Constitutive Equations

The main principles used in determining the form of the constitutive equations are the Clausius–Duhem inequality, the principle of material frame indifference, the idea of symmetry, and linearization. The main tools are the Coleman–Noll procedure for finding the restrictions placed by the Clausius–Duhem inequality, the polar decomposition theorem, the reduced constitutive equations, and representation theorems for isotropic tensor functions. The methods are systematic, simple, and standard.

One of the complications in thermodynamics is the several possible choices of independent variables for use in the constitutive equations. The present chapter uses the temperature and its gradient as the independent thermal variables. The changes of variables mechanism for thermodynamics will be treated in Chap. 10. Throughout this part we shall deal with smooth processes and exclude jumps.

9.1 Response Functions

9.1.1 Viscous Bodies with Heat Conduction A body with the reference configuration Ω is said to be a ***viscous body with heat conduction*** if for every process the quantities $e, \eta, \boldsymbol{S}$ and $\boldsymbol{q}$ are related to the motion and the temperature fields by the constitutive equations of the form:

$$e = \bar{e}(\boldsymbol{F}, \theta, \boldsymbol{G}, \dot{\boldsymbol{F}}), \quad \eta = \bar{\eta}(\boldsymbol{F}, \theta, \boldsymbol{G}, \dot{\boldsymbol{F}}), \quad \boldsymbol{S} = \bar{\boldsymbol{S}}(\boldsymbol{F}, \theta, \boldsymbol{G}, \dot{\boldsymbol{F}}), \quad \boldsymbol{q} = \bar{\boldsymbol{q}}(\boldsymbol{F}, \theta, \boldsymbol{G}, \dot{\boldsymbol{F}}), \quad (9.1.1)$$

where $\bar{e}, \bar{\eta}, \bar{\boldsymbol{S}}, \bar{\boldsymbol{q}}$ are given functions–the ***response functions*** also called ***constitutive functions***. Here $\boldsymbol{G} = \nabla\theta$ and $e, \eta, \boldsymbol{S}, \boldsymbol{q}, \boldsymbol{F}, \theta, \boldsymbol{G}, \dot{\boldsymbol{F}}$ are evaluated at $(\boldsymbol{p}, t)$. The response functions $\bar{e}, \bar{\eta}, \bar{\boldsymbol{S}}, \bar{\boldsymbol{q}}$ are defined on a common domain of the form $\bar{U}\times\text{Vect}\times\text{Lin}$, where $\bar{U}$ is an open subset of $\text{InvLin} \times \mathbb{R}^{++}$; unless otherwise stated, the response functions are assumed to be infinitely differentiable. It is understood that all deformation-temperature paths (χ, θ) for which $(\boldsymbol{F}, \theta)$ is in $\bar{U}$ (at every material point and throughout the process) give rise to a possible process of the body in which $e, \eta, \boldsymbol{S}, \boldsymbol{q}$ are determined through (9.1.1). The body force and the external supply of heat are adjusted so as to satisfy the equations of balance of linear momentum and energy. Every such process is an ***admissible process;*** it is completely determined by the evolutions of deformation and of temperature (χ, θ).

Using $f = e - \theta\eta$, $\boldsymbol{T} = s\boldsymbol{S}(\text{cof}\,\boldsymbol{F})^{-1}$, $\boldsymbol{h} = s(\text{cof}\,\boldsymbol{F})^{-\mathrm{T}}\boldsymbol{q}$, we see that $f, \boldsymbol{T}, \boldsymbol{h}$ are also given by constitutive equations analogous to (9.1.1),

$$f = \bar{f}(\boldsymbol{F}, \theta, \boldsymbol{G}, \dot{\boldsymbol{F}}), \quad \boldsymbol{T} = \bar{\boldsymbol{T}}(\boldsymbol{F}, \theta, \boldsymbol{G}, \dot{\boldsymbol{F}}), \quad \boldsymbol{h} = \bar{\boldsymbol{h}}(\boldsymbol{F}, \theta, \boldsymbol{G}, \dot{\boldsymbol{F}}), \quad (9.1.2)$$

where the constitutive functions $\bar{f}, \bar{T}, \bar{h}$ are determined by those in (9.1.1) in the obvious way. (Recall that $s = \operatorname{sgn} \det \boldsymbol{F}$.) Using the relations $\boldsymbol{L} = \dot{\boldsymbol{F}}\boldsymbol{F}^{-1}, \boldsymbol{g} = \boldsymbol{F}^{-\mathrm{T}}\boldsymbol{G}$ for $\boldsymbol{L} = \operatorname{grad} \boldsymbol{v}, \boldsymbol{g} = \operatorname{grad} \theta$, we see that (9.1.1) and (9.1.2) can also be replaced by $e = \bar{e}(\boldsymbol{F}, \theta, \boldsymbol{g}, \boldsymbol{L})$ etc., where the constitutive functions in terms of $\boldsymbol{F}, \theta, \boldsymbol{g}, \boldsymbol{L}$ are denoted by the same symbols as those in terms of $\boldsymbol{F}, \theta, \boldsymbol{G}, \dot{\boldsymbol{F}}$.

9.1.2 The Equilibrium and the Dynamic Responses A process in which θ and $\boldsymbol{F}$ are constants in space and time is called an equilibrium process. The equilibrium stress response functions $\bar{\boldsymbol{S}}_e, \bar{\boldsymbol{T}}_e$ are defined on $\bar{U}$ by

$$\bar{\boldsymbol{S}}_e(\boldsymbol{F}, \theta) = \bar{\boldsymbol{S}}(\boldsymbol{F}, \theta, \boldsymbol{0}, \boldsymbol{0}), \quad \bar{\boldsymbol{T}}_e(\boldsymbol{F}, \theta) = \bar{\boldsymbol{T}}(\boldsymbol{F}, \theta, \boldsymbol{0}, \boldsymbol{0})$$

and the dynamic stress response functions $\bar{\boldsymbol{S}}_d, \bar{\boldsymbol{T}}_d$ are defined on $\bar{U} \times \mathrm{Vect} \times \mathrm{Lin}$ by

$$\bar{\boldsymbol{S}}_d(\boldsymbol{F}, \theta, \boldsymbol{G}, \dot{\boldsymbol{F}}) = \bar{\boldsymbol{S}}(\boldsymbol{F}, \theta, \boldsymbol{G}, \dot{\boldsymbol{F}}) - \bar{\boldsymbol{S}}_e(\boldsymbol{F}, \theta),$$
$$\bar{\boldsymbol{T}}_d(\boldsymbol{F}, \theta, \boldsymbol{G}, \dot{\boldsymbol{F}}) = \bar{\boldsymbol{T}}(\boldsymbol{F}, \theta, \boldsymbol{G}, \dot{\boldsymbol{F}}) - \bar{\boldsymbol{T}}_e(\boldsymbol{F}, \theta)$$

so that $\boldsymbol{S} = \boldsymbol{S}_d + \boldsymbol{S}_e, \boldsymbol{T} = \boldsymbol{T}_d + \boldsymbol{T}_e$. The equation of balance of angular momentum (3.5.8) splits into

$$\boldsymbol{S}_e\boldsymbol{F}^{\mathrm{T}} = \boldsymbol{F}\boldsymbol{S}_e^{\mathrm{T}}, \quad \boldsymbol{S}_d\boldsymbol{F}^{\mathrm{T}} = \boldsymbol{F}\boldsymbol{S}_d^{\mathrm{T}}, \text{ equivalently } \boldsymbol{T}_e = \boldsymbol{T}_e^{\mathrm{T}}, \quad \boldsymbol{T}_d = \boldsymbol{T}_d^{\mathrm{T}}. \tag{9.1.3}$$

9.1.3 Special Cases

(1) *Thermoelastic materials* They are characterized by $e = \bar{e}(\boldsymbol{F}, \theta, \boldsymbol{G}), \eta = \bar{\eta}(\boldsymbol{F}, \theta, \boldsymbol{G}), \boldsymbol{S} = \bar{\boldsymbol{S}}(\boldsymbol{F}, \theta, \boldsymbol{G}), \boldsymbol{q} = \bar{\boldsymbol{q}}(\boldsymbol{F}, \theta, \boldsymbol{G})$. There are no "memory effects," and the dissipation is weak: it is represented only by the conduction of heat and (possibly) shock waves or moving phase boundaries. The theory of thermal stresses and Fourier's theory of conduction of heat fall within this framework.

(2) *'Adiabatic materials'* These are characterized by $e = \bar{e}(\boldsymbol{F}, \theta), \eta = \bar{\eta}(\boldsymbol{F}, \theta), \boldsymbol{S} = \bar{\boldsymbol{S}}(\boldsymbol{F}, \theta), \boldsymbol{q} \equiv \boldsymbol{0}$. The dissipation is present only through the shock waves (see Chaps. 22–24). The dynamic (adiabatic) elasticity and the classical fluid dynamics without conduction of heat fall within this class. Except at shocks, every material point undergoes an adiabatic process in which the specific entropy is conserved. Classical thermodynamics applies to this case (and only to this case).

9.1.4 Material Inhomogeneities Instead of the constitutive equations (9.1.1) one can consider constitutive equations with response functions $\bar{e}$, etc., depending on the material point $\boldsymbol{p} \in \Omega$ in a parametric way, e.g.,

$$e = \bar{e}(\boldsymbol{F}, \theta, \boldsymbol{G}, \dot{\boldsymbol{F}}; \boldsymbol{p}).$$

Such bodies are called materially inhomogeneous. The dependence of the response on $\boldsymbol{p}$ can arise if a materially homogeneous body is viewed from an inhomogeneous reference configuration. For example consider a rubber torus which in its toroidal configuration Ω is materially homogeneous. If a deformed configuration is used as the reference configuration instead, then the response functions will depend on $\boldsymbol{p}$. There are bodies which do not have any materially homogeneous configuration. This can arise if different parts of the body consist of different materials, like an isotropic body with an nonisotropic inclusion. However, there are also bodies whose material points are all made of the same material, but which still do not have a global materially homogeneous configuration. Such bodies are called materially uniform inhomogeneous bodies. Consider again the rubber torus in an originally materially homogeneous configuration. Suppose that the torus is cut along the plane perpendicular to its circumferential central line, one side of the cut twisted by a nonzero angle about the central line, and then the two sides glued together. The resulting body will be materially uniform but inhomogeneous. Materially inhomogeneous bodies are used in the continuum theory of dislocations (see Remark 14.2.5(5)). All the considerations of Part III extend with obvious modifications to materially inhomogeneous bodies.

9.2 Consequences of the Clausius–Duhem Inequality

9.2.1 The Dissipation Principle The constitutive functions must be such that the Clausius–Duhem inequality holds in every smooth admissible process.

The reader is referred to Subsect. 3.7.4 for the discussion of this dissipation principle. Here we shall take the technical side of it and derive its consequences. Since we use the temperature as the independent variable, the dissipation inequality

$$\rho_o \dot{f} \le \boldsymbol{S} \cdot \dot{\boldsymbol{F}} - \rho_o \eta \dot{\theta} - \boldsymbol{q} \cdot \boldsymbol{G}/\theta \tag{9.2.1}$$

is the appropriate form of the Clausius–Duhem inequality as it contains the time-derivative of θ rather than that of e.

9.2.2 Proposition *A viscous body with heat conduction satisfies the dissipation principle if and only if the following three conditions hold:*

(1) *the constitutive functions $\bar{f}$, $\bar{\eta}$, $\bar{e}$ are independent of the "nonequilibrium" variables $\boldsymbol{G}$ and $\dot{\boldsymbol{F}}$ (or equivalently of $\boldsymbol{g}$ and $\boldsymbol{L}$): $\bar{f}(\boldsymbol{F}, \theta, \boldsymbol{G}, \dot{\boldsymbol{F}}) = \bar{f}(\boldsymbol{F}, \theta)$, etc., for every $(\boldsymbol{F}, \theta, \boldsymbol{G}, \dot{\boldsymbol{F}}) \in \bar{U} \times \mathrm{Vect} \times \mathrm{Lin}$;*

(2) *the equilibrium thermostatic relations*

$$\bar{\boldsymbol{S}}_e(\boldsymbol{F}, \theta) = \rho_o \partial_{\boldsymbol{F}} \bar{f}(\boldsymbol{F}, \theta), \quad \bar{\boldsymbol{T}}_e(\boldsymbol{F}, \theta) = \rho \partial_{\boldsymbol{F}} \bar{f}(\boldsymbol{F}, \theta) \boldsymbol{F}^{\mathrm{T}}, \tag{9.2.2}$$

$$\bar{\eta}(\boldsymbol{F}, \theta) = -\partial_\theta \bar{f}(\boldsymbol{F}, \theta), \quad \bar{e}(\boldsymbol{F}, \theta) = -\theta^2 \partial_\theta (\bar{f}(\boldsymbol{F}, \theta)/\theta), \tag{9.2.3}$$

hold for each $(\boldsymbol{F}, \theta) \in \bar{U}$; in differentials, $\rho_o d\bar{f} = \bar{\boldsymbol{S}}_e \cdot d\boldsymbol{F} - \rho_o \bar{\eta} d\theta$;

(3) *the internal dissipation inequality*

$$\bar{\boldsymbol{S}}_d(\boldsymbol{F}, \theta, \boldsymbol{G}, \dot{\boldsymbol{F}}) \cdot \dot{\boldsymbol{F}} - \bar{\boldsymbol{q}}(\boldsymbol{F}, \theta, \boldsymbol{G}, \dot{\boldsymbol{F}}) \cdot \boldsymbol{G}/\theta \ge 0, \tag{9.2.4}$$
$$\bar{\boldsymbol{T}}_d(\boldsymbol{F}, \theta, \boldsymbol{g}, \boldsymbol{L}) \cdot \boldsymbol{L} - \bar{\boldsymbol{h}}(\boldsymbol{F}, \theta, \boldsymbol{g}, \boldsymbol{L}) \cdot \boldsymbol{g}/\theta \ge 0$$

holds for each $(\boldsymbol{F}, \theta, \boldsymbol{G}, \dot{\boldsymbol{F}}) \in \bar{U} \times \mathrm{Vect} \times \mathrm{Lin}$.

This is the theorem of COLEMAN & NOLL [1963]. By (2), the entropy and the equilibrium stress are completely determined by the free energy, which is then a "potential" for the equilibrium stress and entropy. The form of the thermostatic relations in different variables is discussed in Chap. 10 and the forms for fluids and isotropic solids in Chap. 11. Inequality (9.2.4) generally *does not* split into the nonnegativity of the power of viscous stresses $\bar{\boldsymbol{S}}_d(\boldsymbol{F}, \theta, \boldsymbol{G}, \dot{\boldsymbol{F}}) \cdot \dot{\boldsymbol{F}} \ge 0$ and Fourier's inequality $\bar{\boldsymbol{q}}(\boldsymbol{F}, \theta, \boldsymbol{G}, \dot{\boldsymbol{F}}) \cdot \boldsymbol{G} \le 0$. The dynamic response functions are treated in Chap. 12.

Proof (Coleman and Noll's procedure). Assume that the body satisfies the Clausius–Duhem inequality in every smooth process. Inserting the constitutive equations into the dissipation inequality (9.2.1), one obtains

$$(\rho_o \partial_{\boldsymbol{F}} \bar{f} - \bar{\boldsymbol{S}}) \cdot \dot{\boldsymbol{F}} + \rho_o (\partial_\theta \bar{f} + \bar{\eta}) \cdot \dot{\theta} + \rho_o \partial_{\boldsymbol{G}} \bar{f} \cdot \dot{\boldsymbol{G}} + \rho_o \partial_{\dot{\boldsymbol{F}}} \bar{f} \cdot \ddot{\boldsymbol{F}} + \boldsymbol{q} \cdot \boldsymbol{G}/\theta \le 0 \tag{9.2.5}$$

at every point of the body and for every time during the process. The constitutive functions and their derivatives are evaluated at $(\boldsymbol{F}, \theta, \boldsymbol{G}, \dot{\boldsymbol{F}})$. Using the fact that every deformation-temperature path can be realized in a process, the values $(\boldsymbol{F}, \theta, \boldsymbol{G}, \dot{\boldsymbol{F}}, \dot{\theta}, \dot{\boldsymbol{G}}, \ddot{\boldsymbol{F}})$ can be chosen arbitrarily (subject to the restriction that $(\boldsymbol{F}, \theta)$ be in $\bar{U}$). Since the left-hand side of (9.2.5) depends on $\dot{\theta}, \dot{\boldsymbol{G}}, \ddot{\boldsymbol{F}}$ linearly, the coefficients in front of these variables must vanish: $\partial_\theta \bar{f} + \bar{\eta} = 0$, $\partial_{\boldsymbol{G}} \bar{f} = 0$, $\partial_{\dot{\boldsymbol{F}}} \bar{f} = 0$. These equations imply that $\bar{f}$ is independent of $\boldsymbol{G}$ and $\dot{\boldsymbol{F}}$ and that $(9.2.3)_1$ holds. Hence also $\bar{\eta}$ is independent of $\boldsymbol{G}$ and $\dot{\boldsymbol{F}}$. From $\bar{f} = \bar{e} - \theta \bar{\eta}$ also $\bar{e}$ is independent of $\boldsymbol{G}$ and $\dot{\boldsymbol{F}}$ and $(9.2.3)_2$ holds. Inequality (9.2.5) reduces to

$$(\rho_o \partial_{\boldsymbol{F}} \bar{f} - \bar{\boldsymbol{S}}) \cdot \dot{\boldsymbol{F}} + \boldsymbol{q} \cdot \boldsymbol{G}/\theta \le 0, \tag{9.2.6}$$

where now $\bar{S}$, q depend on $(F, \theta, G, \dot{F})$ and $\partial_F \bar{f}$ on (F, θ). Referring to the arbitrariness of these values, the same inequality must hold with $G, \dot{F}$ replaced by $\alpha G, \alpha\dot{F}$, where α is a positive number. The resulting inequality is

$$\alpha(\rho_o \partial_F \bar{f} - \bar{S}) \cdot \dot{F} + \alpha q \cdot G/\theta \le 0, \tag{9.2.7}$$

where $\bar{S}$, $\bar{q}$ are now evaluated at $(F, \theta, \alpha G, \alpha\dot{F})$ and $\partial_F \bar{f}$ at (F, θ). Dividing (9.2.7) by α, letting α tend to 0, and using the continuity of $\bar{S}$, $\bar{q}$, we obtain

$$(\rho_o \partial_F \bar{f}(F, \theta) - \bar{S}(F, \theta, 0, 0)) \cdot \dot{F} + \bar{q}(F, \theta, 0, 0) \cdot G/\theta \le 0.$$

The linearity of the left-hand side in $\dot{F}$, G then yields

$$\rho_o \partial_F \bar{f}(F, \theta) - \bar{S}(F, \theta, 0, 0) = 0, \quad \bar{q}(F, \theta, 0, 0) = 0. \tag{9.2.8}$$

By the definition of $\bar{S}_e$, $(9.2.8)_1$ can be rewritten as $(9.2.2)_1$ and referring to the definition of $\bar{S}_d$, we see that (9.2.7) can be rewritten as (9.2.4). The rest of the relations in the proposition are consequences of the partial results obtained so far; this proves that the dissipation principle implies (1)–(3). Conversely, reversing the order of the arguments shows that (1)–(3) imply the dissipation postulate. □

Throughout the rest of the book it is assumed that the body satisfies the dissipation postulate.

Let us define the specific production of entropy γ in a smooth process by

$$\rho_o \gamma = \rho_o \dot{\eta} + \operatorname{Div} \frac{q}{\theta} - \rho_o \frac{r}{\theta}. \tag{9.2.9}$$

By the Clausius–Duhem inequality, $\gamma \ge 0$ for every process of the body.

9.2.3 Proposition (1) *For every process, the Gibbs equations*

$$\rho_o \theta \dot{\eta} = \rho_o \dot{e} - S_e \cdot \dot{F}, \quad \rho \theta \dot{\eta} = \rho \dot{e} - T_e \cdot L, \tag{9.2.10}$$

$$\rho_o \dot{f} = S_e \cdot \dot{F} - \rho_o \eta \dot{\theta}, \quad \rho \dot{f} = T_e \cdot L - \rho \eta \dot{\theta}, \tag{9.2.11}$$

hold; (2)

$$\gamma = S_d \cdot \dot{F}/\rho_o \theta - q \cdot G/\rho_o \theta^2 = T_d \cdot L/\rho\theta - h \cdot g/\rho\theta^2. \tag{9.2.12}$$

The production of entropy equals, to within a factor $1/\rho_o \theta$, the expression occurring in the internal dissipation inequality. We shall have to say more about the production of entropy in Chap. 12.

Proof Equations (9.2.10) and (9.2.11) are immediate consequences of the thermostatic relations (Item (2) of Proposition 9.2.2). To prove the expressions for the production of entropy, one first eliminates the entropy from (9.2.9) via $(9.2.10)_1$ and then the reduced equation of balance of energy (3.6.5) is used to arrive at (2). □

In the proof of Proposition 9.2.2 we have also proved that

$$\bar{q}(F, \theta, 0, 0) = 0, \quad \bar{h}(F, \theta, 0, 0) = 0. \tag{9.2.13}$$

The process is said to be locally equilibrium at a given spacetime point if $\dot{F} = 0$, $G = 0$. Hence for locally equilibrium processes the heat flux vanishes. By definition, also

$$\bar{S}_d(F, \theta, 0, 0) = 0, \quad \bar{T}_d(F, \theta, 0, 0) = 0.$$

9.2.4 Isothermal Processes For every $\theta > 0$, the function $f \equiv W := \bar{f}(\cdot, \theta) : \bar{U}_\theta \to \mathbb{R}$ defined on $\bar{U}_\theta = \{F \in \text{InvLin} : (F, \theta) \in \bar{U}\}$ by $f(F) = \bar{f}(F, \theta)$ is called the ***stored energy*** at θ. The isothermal theory results when the processes are constrained to have a fixed temperature θ. Then by (9.2.13) $q = 0$ and $S = S(F, \dot{F})$, where $S(F, \dot{F}) = \bar{S}(F, \theta, 0, \dot{F})$ and the equilibrium stress $S(F, 0)$ is given by $S(F, 0) = \rho_o \partial_F f(F)$. In particular, the stored energy determines completely the response of the body in equilibrium situations.

9.3 Frame Indifference

We shall employ here the "practical" face of the principle of material frame indifference, the fundamental aspects were treated in Chap. 6.

9.3.1 Changes of Kinematical Variables Recall the form of the change of frame (6.1.1). Let us consider a process determined by the motion χ and the temperature field θ and another process of the same body, which in the same reference frame is given by χ' and θ', where

$$\chi'(p,t) = R(t)(\chi(p,t) - x_0) + x_0'(t), \quad \theta'(p,t) = \theta(p,t). \tag{9.3.1}$$

It is useful to list the changes of the main kinematical variables under changes of frame:

$$\begin{gathered} F \to F' = RF, \quad U \to U' = U, \quad G \to G' = G, \quad g \to g' = Rg, \\ \dot{F} \to \dot{F}' = R\dot{F} + ZRF, \quad L \to R' = RLR^{\mathsf{T}} + Z, \\ D \to D' = RDR^{\mathsf{T}}, \quad W \to W' = RWR^{\mathsf{T}} + Z, \end{gathered}$$

where

$$Z = \dot{R}R^{-1}. \tag{9.3.2}$$

9.3.2 The Principle of Material Frame Indifference Let (6.1.1) be a change of frame with smooth functions $R(\cdot)$, $x_0'(\cdot)$. For a given admissible process (cf. Subsect. 9.1.1) determined by χ and θ the admissible process determined by χ' and θ', given by (9.3.1), is also possible; the evolutions of energy, entropy, stress, and heat flux, given in both processes by the same constitutive functions, are related as follows:

$$e' = e, \quad \eta' = \eta, \quad S' = RS, \quad q' = q. \tag{9.3.3}$$

The last two equations can be restated as

$$T' - RTR^{\mathsf{T}}, \quad h' = Rh. \tag{9.3.4}$$

There was some discussion in the literature as to whether the *proper* orthogonal tensors ($\det R = +1$) or the general orthogonal tensors R should occur in the principle of material frame indifference. This has some consequences for the form of the constitutive equations. In this text it is assumed that the statement with the general orthogonal tensors holds. In specific situations, e.g., when the domain of the constitutive functions contains the deformation gradients with positive determinant only, one must employ the form of the principle of material frame indifference with the R restricted to proper orthogonal tensors only.

Since the quantities associated with the primed and unprimed processes are given by the same constitutive functions, the transformation laws impose restrictions on the form of the constitutive functions. In the derivations below we use the constitutive functions already restricted by the dissipation principle, e.g., that $\bar{e}$, $\bar{\eta}$, and $\bar{f}$ are independent of $\dot{F}$, G. For convenience we shall use the variables L and g in the constitutive functions rather than $\dot{F}$, G.

9.3.3 The Restrictions on the Response Functions Applying the constitutive equations to both the unprimed and primed processes and combining the transformation laws derived in Subsect. 9.3.1 with those postulated in Subsect. 9.3.2, we find that the body satisfies the principle of material frame indifference if and only if for every $(F, \theta) \in \bar{U}$ and for every $R \in \mathrm{Orth}$ also $(RF, \theta) \in \bar{U}$ and the following relations hold:

$$\bar{m}(\boldsymbol{RF}, \theta) = \bar{m}(\boldsymbol{F}, \theta), \tag{9.3.5}$$

where $\bar{m}$ stands for $\bar{e}, \bar{f}, \bar{\eta}$,

$$\bar{\boldsymbol{S}}_e(\boldsymbol{RF}, \theta) = \boldsymbol{R}\bar{\boldsymbol{S}}_e(\boldsymbol{F}, \theta), \quad \bar{\boldsymbol{T}}_e(\boldsymbol{RF}, \theta) = \boldsymbol{R}\bar{\boldsymbol{T}}_e(\boldsymbol{F}, \theta)\boldsymbol{R}^{\mathrm{T}} \tag{9.3.6}$$

and

$$\begin{aligned} \bar{\boldsymbol{S}}_d(\boldsymbol{RF}, \theta, \boldsymbol{Rg}, \boldsymbol{RLR}^{\mathrm{T}} + \boldsymbol{Z}) &= \boldsymbol{R}\bar{\boldsymbol{S}}_d(\boldsymbol{F}, \theta, \boldsymbol{g}, \boldsymbol{L}), \\ \bar{\boldsymbol{q}}(\boldsymbol{RF}, \theta, \boldsymbol{Rg}, \boldsymbol{RLR}^{\mathrm{T}} + \boldsymbol{Z}) &= \bar{\boldsymbol{q}}(\boldsymbol{F}, \theta, \boldsymbol{g}, \boldsymbol{L}) \end{aligned} \tag{9.3.7}$$

where $\boldsymbol{g} \in \mathrm{Vect}$, $\boldsymbol{L} \in \mathrm{Lin}$, and $\boldsymbol{Z} \in \mathrm{Skew}$ are arbitrary. It is necessary to take into account that $\boldsymbol{Z}$ as in (9.3.2) is skew and conversely that every skew tensor $\boldsymbol{Z}$ can be realized in some change of frame. This, in turn, is related to the fact that time-dependent orthogonal tensors $\boldsymbol{R}(\cdot)$ are admitted in changes of frame.

Representation theorems of the type Proposition 8.3.3 may be invoked to find that the essential part of the dependence of the equilibrium quantities $\bar{e}, \bar{f}, \bar{\eta}, \bar{\boldsymbol{S}}_e, \bar{\boldsymbol{T}}_e$ on $\boldsymbol{F}$ is realized through the right stretch tensor $\boldsymbol{U}$ while the dependence on the rotation tensor $\boldsymbol{R}$ is absent in scalar quantities or is trivial for the vector and tensor quantities. When $\boldsymbol{U}$ is used as the independent variable of the free energy, then the stress relation takes the form implied by (8.3.4) or (8.3.5). One can also use the Cauchy–Green tensor $\boldsymbol{C} = \boldsymbol{F}^{\mathrm{T}}\boldsymbol{F}$ in place of $\boldsymbol{U}$.

Equation (9.3.7) implies that the nonequilibrium response depends on the spatial gradient of velocity $\boldsymbol{L}$ only through its symmetric part $\boldsymbol{D}$:

9.3.4 Proposition *If the body satisfies the principle of material frame indifference, then the following two conditions hold:*

(1) *for every* $(\boldsymbol{F}, \theta, \boldsymbol{g}, \boldsymbol{L}) \in \bar{U} \times \mathrm{Vect} \times \mathrm{Lin}$ *the response depends on* $\boldsymbol{L}$ *only through its symmetric part* $\boldsymbol{D}$

$$\bar{z}(\boldsymbol{F}, \theta, \boldsymbol{g}, \boldsymbol{L}) = \bar{z}(\boldsymbol{F}, \theta, \boldsymbol{g}, \boldsymbol{D}),$$

where $\bar{z}$ *stands for* $\bar{\boldsymbol{S}}_d, \bar{\boldsymbol{T}}_d, \bar{\boldsymbol{q}}, \bar{\boldsymbol{h}}$, *and the functions on the right-hand side of these equations are the restrictions of the functions occurring on the left-hand side to the domain* $\bar{U} \times \mathrm{Vect} \times \mathrm{Sym}$,

(2) *these restrictions satisfy additionally*

$$\begin{aligned} \bar{\boldsymbol{S}}_d(\boldsymbol{RF}, \theta, \boldsymbol{Rg}, \boldsymbol{RDR}^{\mathrm{T}}) &= \boldsymbol{R}\bar{\boldsymbol{S}}_d(\boldsymbol{F}, \theta, \boldsymbol{g}, \boldsymbol{D}), \\ \bar{\boldsymbol{q}}(\boldsymbol{RF}, \theta, \boldsymbol{Rg}, \boldsymbol{RDR}^{\mathrm{T}}) &= \bar{\boldsymbol{q}}(\boldsymbol{F}, \theta, \boldsymbol{g}, \boldsymbol{D}), \end{aligned}$$

for every $(\boldsymbol{F}, \theta, \boldsymbol{g}, \boldsymbol{D}) \in \bar{U} \times \mathrm{Vect} \times \mathrm{Sym}$ *and* $\boldsymbol{R} \in \mathrm{Orth}$.

Let us also mention the relationship between the frame indifference of the free energy and the symmetry and frame indifference of the equilibrium stress. The equilibrium stress and entropy are derivatives of the free energy and hence the frame indifference of the latter implies that of the former two. The converse implication requires integration, and an orthogonal tensor-dependent additive constant will emerge initially. However, as an application of Proposition 8.3.2 to $m \equiv f(\cdot, \theta)$ shows, this constant actually vanishes thanks to the structure of the proper orthogonal group (see Remark 1.3.6 and Proposition 1.3.7). Further, Proposition 8.3.2 also shows that the objectivity of the free energy is equivalent to the symmetry of the equilibrium Cauchy stress $(9.1.3)_1$ and $(9.1.3)_3$. In Chap. 6 the symmetry of the total stress was shown to be a consequence of the frame indifference. The result just stated is another realization of this relationship; however, it applies only to the equilibrium stress.

9.4 The Symmetry Group

That the symmetry of the material provides further restrictions on the constitutive equations has been known since Cauchy's derivation of the tensor of elastic constants of an isotropic solid. In order to deal with more complicated types of symmetries, the symmetry group of the material is introduced.

9.4.1 Changes of the Reference Configuration Consider a viscous body with the reference configuration Ω and with the response functions $\bar{e}$, $\bar{\eta}$, $\bar{\boldsymbol{S}}$, $\bar{\boldsymbol{q}}$. The response functions depend on the reference configuration. To determine how they transform, consider a change of the reference configuration, i.e., let us relabel the reference positions $\boldsymbol{p}$ via a smooth invertible mapping $z : \Omega \to$ Eucl to obtain a new reference configuration $\Omega' = z(\Omega)$. In the reference configurations Ω, Ω', a given material point has positions $\boldsymbol{p}, \boldsymbol{p}'$ where $\boldsymbol{p}' = z(\boldsymbol{p})$. Consequently, the same process is described by different motions $\boldsymbol{\chi}, \boldsymbol{\chi}'$ and temperature fields θ, θ', respectively, related by

$$\boldsymbol{\chi}'(\boldsymbol{p}', t) = \boldsymbol{\chi}(\boldsymbol{p}, t), \quad \theta'(\boldsymbol{p}', t) = \theta(\boldsymbol{p}, t), \quad \text{with } \boldsymbol{p}' = z(\boldsymbol{p}). \tag{9.4.1}$$

A calculation gives that under the transformation $\boldsymbol{\chi} \to \boldsymbol{\chi}', \theta \to \theta'$, the kinematical quantities change as

$$\boldsymbol{F} \to \boldsymbol{F}' = \boldsymbol{F}\boldsymbol{H}^{-1}, \quad \boldsymbol{G} \to \boldsymbol{G}' = \boldsymbol{H}^{-\mathrm{T}}\boldsymbol{G} \quad \boldsymbol{g} \to \boldsymbol{g}' = \boldsymbol{g},$$
$$\dot{\boldsymbol{F}} \to \dot{\boldsymbol{F}}' = \dot{\boldsymbol{F}}\boldsymbol{H}^{-1}, \quad \boldsymbol{L} \to \boldsymbol{L}' = \boldsymbol{L}, \quad \boldsymbol{D} \to \boldsymbol{D}' = \boldsymbol{D}, \quad \boldsymbol{W} \to \boldsymbol{W}' = \boldsymbol{W},$$
$$\operatorname{cof} \boldsymbol{F}' = \operatorname{cof} \boldsymbol{F} \operatorname{cof} \boldsymbol{H}^{-1}, \quad \det \boldsymbol{F}' = \det \boldsymbol{F} \det \boldsymbol{H}^{-1},$$
$$dV \to dV' = s \det \boldsymbol{H}\, dV, \quad \boldsymbol{n}dA \to \boldsymbol{n}'\, dA' = s \operatorname{cof} \boldsymbol{H}\boldsymbol{n}\, dA,$$

where $\boldsymbol{H} = \nabla z$ and $s = \operatorname{sgn} \det \boldsymbol{H}$. Furthermore, the spatial volume and area are independent of the reference configuration, $dv \to dv' = dv$, $\boldsymbol{m}\, da \to \boldsymbol{m}'\, da' = \boldsymbol{m}\, da$.

We now proceed to derive the transformation rules for densities and fluxes of physical quantities. These are associated with parts of the body and material surfaces; in the reference configurations Ω, Ω', a given part is described by subsets $P \subset \Omega$, $P' \subset \Omega'$ related by $P' = z(P)$ and a given material surface by S, S' where $S' = z(S)$. The total quantities corresponding to P, P' and the total fluxes corresponding to S, S' must take the same values. Thus the referential densities of mass ρ_o, ρ_o' must satisfy $\rho_o\, dV = \rho_o'\, dV'$, where dV and dV' are related as above, from which

$$\rho_o \to \rho_o' = s\rho_o \det \boldsymbol{H}^{-1}.$$

Similarly, the referential densities of internal energy, entropy, and free energy $\rho_o e, \rho_o' e'$, etc., must satisfy $\rho_o' e'\, dV' = \rho_o e\, dV$, etc., from which

$$e \to e' = e, \quad \eta \to \eta' = \eta, \quad f \to f' = f.$$

The changes of the referential stress and the referential heat flux are determined from $\boldsymbol{S}'\boldsymbol{n}'\, dA' = \boldsymbol{S}\boldsymbol{n}\, dA$, $\boldsymbol{q}' \cdot \boldsymbol{n}'\, dA' = \boldsymbol{q} \cdot \boldsymbol{n}\, dA$; this leads to

$$\boldsymbol{S} \to \boldsymbol{S}' = s\boldsymbol{S} \operatorname{cof} \boldsymbol{H}^{-1}, \quad \boldsymbol{q} \to \boldsymbol{q}' = s \operatorname{cof} \boldsymbol{H}^{-\mathrm{T}}\boldsymbol{q}, \quad (s = \operatorname{sgn} \det \boldsymbol{H}), \tag{9.4.2}$$

while the invariance of the spatial description of surfaces leads to $\boldsymbol{T} \to \boldsymbol{T}' = \boldsymbol{T}$, $\boldsymbol{h} \to \boldsymbol{h}' = \boldsymbol{h}$, which can be also derived from (9.4.2) using the relations between $\boldsymbol{T}, \boldsymbol{S}$ and $\boldsymbol{h}, \boldsymbol{q}$.

In the new reference configuration Ω' the processes will be described by the constitutive equations of the general form (9.1.1), but with different response functions, which will be distinguished by a prime. They are defined so as to deliver the values of the primed quantities related to the unprimed quantities by the transformation laws above, where the arguments of the constitutive functions are changed as well:

$$\bar{m}'(\boldsymbol{F}',\theta) = \bar{m}(\boldsymbol{F}'\boldsymbol{H},\theta), \tag{9.4.3}$$

where $\bar{m}$ stands for $\bar{e}, \bar{f}, \bar{\eta}$, and

$$\bar{\boldsymbol{S}}_e'(\boldsymbol{F}',\theta) = s\bar{\boldsymbol{S}}_e(\boldsymbol{F}'\boldsymbol{H},\theta)\operatorname{cof}\boldsymbol{H}^{-1}, \quad (s = \operatorname{sgn}\det\boldsymbol{H}), \tag{9.4.4}$$

for the equilibrium response functions; they are defined on the domain

$$\bar{U}' = \{(\boldsymbol{F}',\theta) = (\boldsymbol{F}\boldsymbol{H}^{-1},\theta) : (\boldsymbol{F},\theta) \in \bar{U}\}.$$

The nonequilibrium response functions are given by

$$\bar{\boldsymbol{S}}_d'(\boldsymbol{F}',\theta,\boldsymbol{g},\boldsymbol{L}) = s\,\bar{\boldsymbol{S}}_d(\boldsymbol{F}'\boldsymbol{H},\theta,\boldsymbol{g},\boldsymbol{L})\operatorname{cof}\boldsymbol{H}^{-1},$$
$$\bar{\boldsymbol{q}}'(\boldsymbol{F}',\theta,\boldsymbol{g},\boldsymbol{L}) = s\operatorname{cof}\boldsymbol{H}^{-\mathrm{T}}\bar{\boldsymbol{q}}(\boldsymbol{F}'\boldsymbol{H},\theta,\boldsymbol{g},\boldsymbol{L})$$

on $\bar{U}' \times \text{Vect} \times \text{Lin}$.

9.4.2 On the Continuity of States in Different Phases Throughout the book, the point of view is taken that, ideally, a *single set* of constitutive equations describes the substance at all values of the state parameters $\boldsymbol{F}, \theta, \dot{\boldsymbol{F}}, \boldsymbol{G}$, including the values corresponding to different phases of the material. The van der Waals equation of state (9.4.10) provides an example of a response function which describes the liquid and gas phases of a fluid in a single analytical expression. In more realistic cases, the constitutive equations are tailored from pieces appropriate to different regimes of the body, with substantially different properties, and in particular with different symmetry of the response. An example is a solid in equilibrium with its melt, see Subsect. 9.4.6. The definition of the symmetry group to be given below will allow for such changes of symmetry when passing from one region of the state parameters to another.

9.4.3 Definition Consider a body with the response functions $\bar{f}, \bar{\boldsymbol{S}}, \bar{\boldsymbol{q}}$ and let $\bar{U}_0 \subset \bar{U}$. The ***symmetry group of the material*** on U_0 is the set $\mathsf{G}(\bar{U}_0)$ of all $\boldsymbol{H} \in \text{InvLin}$ such that for every $(\boldsymbol{F},\theta) \in \bar{U}_0$ also $(\boldsymbol{F}\boldsymbol{H},\theta) \in \bar{U}_0$ and

$$\bar{f}(\boldsymbol{F}\boldsymbol{H},\theta) = \bar{f}(\boldsymbol{F},\theta), \tag{9.4.5}$$
$$\bar{\boldsymbol{S}}(\boldsymbol{F}\boldsymbol{H},\theta,\boldsymbol{g},\boldsymbol{L}) = s\bar{\boldsymbol{S}}(\boldsymbol{F},\theta,\boldsymbol{g},\boldsymbol{L})\operatorname{cof}\boldsymbol{H}, \tag{9.4.6}$$
$$\bar{\boldsymbol{q}}(\boldsymbol{F}\boldsymbol{H},\theta,\boldsymbol{g},\boldsymbol{L}) = s\operatorname{cof}\boldsymbol{H}^{\mathrm{T}}\bar{\boldsymbol{q}}(\boldsymbol{F},\theta,\boldsymbol{g},\boldsymbol{L}), \tag{9.4.7}$$

for every $(\boldsymbol{F},\theta) \in \bar{U}_0, \boldsymbol{g} \in \text{Vect}, \boldsymbol{L} \in \text{Lin}$. The ***(global) symmetry group*** G of the material is its symmetry group on $\bar{U}$, i.e., $\mathsf{G} \equiv \mathsf{G}(\bar{U})$. The material is said to be

(1) a ***solid*** on $\bar{U}_0$ if in some reference configuration $\mathsf{G}(\bar{U}_0) \subset \text{Orth}$;

(2) ***isotropic*** on $\bar{U}_0$ if in some reference configuration $\mathsf{G}(\bar{U}_0) \supset \text{Orth}$; equivalently, this can be stated as that the material has at least one isotropic state;

(3) ***isotropic solid*** on $\bar{U}_0$ if in some reference configuration $\mathsf{G}(\bar{U}_0) = \text{Orth}$;

(4) a ***fluid*** on $\bar{U}_0$ if in some reference configuration $\mathsf{G}(\bar{U}_0) \supset \text{Unim}$; it follows from (9.4.8) that $\mathsf{G}(\bar{U}_0) \supset \text{Unim}$ in every reference configuration.

Thus the symmetry group of the material is the set of all changes of the reference configuration for which the response functions remain unchanged. Note that (9.4.5) implies similar equations for $\bar{\eta}$, $\bar{e}$ via the entropy relation; equation (9.4.6) splits into analogous equations for the equilibrium and dynamic parts of the stress and moreover (9.4.6) and (9.4.7) can be stated equivalently as

$$\bar{T}(FH, \theta, g, L) = \bar{T}(F, \theta, g, L), \quad \bar{h}(FH, \theta, g, L) = \bar{h}(F, \theta, g, L).$$

The sets G, $G(\bar{U}_0)$ are subgroups of InvLin. The symmetry group itself depends on the reference configuration. Under the change of the reference configuration $p \mapsto p' = z(p)$ the symmetry group changes according to the rule

$$G \to G' = HGH^{-1} \tag{9.4.8}$$

where G, G' correspond to subsets $\bar{U}_0$, $\bar{U}_0'$ of the state parameters related by the change H.

For realistic materials one can expect that the global symmetry group satisfies $G \subset$ Unim. Indeed, if G contains an element H with $\det H \neq 1$, then either H^2 or H^{-2} has the determinant positive and less than 1. Denoting such an element by K and using that G is a group, one finds that $K^n \in G$ for every positive integer n and $\det K^n = (\det K)^n \to 0$ as $n \to \infty$. Hence the free and internal energies remain constant under changes of deformation in which the body is compressed to arbitrarily small volume, which is unrealistic.

We shall see in Sect. 10.5 that the symmetry of the material constants (such as the tensor of elasticities, the thermal expansion tensor, etc.) at a point (F, θ, L, g) is determined by the local symmetry group:

9.4.4 Definitions Let $\bar{U}_0 \subset \bar{U}$. The ***local symmetry group*** G_F of the material at the state (F, θ, L, g) relative to $\bar{U}_0$ is the set $G_F \equiv G_F(\bar{U}_0) := FG(\bar{U}_0)F^{-1} \cap \text{Orth}$. The state (F, θ, L, g) is said to be ***isotropic*** if $G_F = \text{Orth}$.

The local symmetry group changes from state to state. By (9.4.8), FGF^{-1} is the ***instantaneous global symmetry group,*** i.e., the symmetry group in the reference configuration given by the current deformation F. The local symmetry group G_F is the part of FGF^{-1} contained in Orth. Alternatively,

$$G_F = \{Q \in \text{Orth} : QF = FH \ \text{ for some } \ H \in G\}.$$

Indeed, if Q is in the set on the right-hand side, then $Q = FHF^{-1} \in FGF^{-1} \cap \text{Orth}$ and conversely. Thus G_F is the set of all orthogonal transformations for which the 'rotated' state QF is also a symmetry-related state FH. By the principle of material frame indifference, the state QF has the same properties as F. In other words, the local symmetry group is the set of elements of the instantaneous global symmetry group FGF^{-1} which leave the state F unchanged (up to a rotation). Yet another local concept of symmetry is the symmetry group of the elastic moduli, see Sect. 10.5.

9.4.5 Crystalline Materials As mentioned in Sect. 1.5, there are two possibilities to reflect the crystalline structure of the material: either to assume that the global symmetry group is $G = G(L)$, i.e., the nonorthogonal symmetry group of some Bravais lattice L, which may be identified with the atomic lattice of the material in its reference configuration, or to assume that $G = P(L) \subset G(L)$ is the orthogonal point group of L. The first assumption is much more restrictive than the second, and the resulting theory has several singular features to be now mentioned. The free energy satisfies

$$\bar{f}(FH, \theta) = \bar{f}(F, \theta), \quad (F, \theta) \in \bar{U}, \quad H \in G(L). \tag{9.4.9}$$

(PARRY [1976] has given a representation theorem for the stored energy functions satisfying (9.4.9).) If the material is in the state with the deformation gradient $\boldsymbol{F} \in \text{InvLin}$, then $\boldsymbol{F}L$ can be identified with the atomic lattice in the deformed state (the Cauchy–Born rule). The local symmetry group is $G_{\boldsymbol{F}} = \boldsymbol{F}G(L)\boldsymbol{F}^{-1} \cap \text{Orth} = G(\boldsymbol{F}L) \cap \text{Orth}$, i.e., the *point group* of the deformed lattice $\boldsymbol{F}L$, see (1.5.4). For most of the deformation gradients, $G_{\boldsymbol{F}}$ will be the point group of the triclinic system, i.e., $G_{\boldsymbol{F}} = \{\mathbf{1}, -\mathbf{1}\}$: it can be proved that the set of all $\boldsymbol{F} \in \text{InvLin}$ for which $G_{\boldsymbol{F}} = \{\mathbf{1}, -\mathbf{1}\}$ is open and dense in InvLin. The higher (ideal) crystallographic symmetry, $G_{\boldsymbol{F}} \neq \{\mathbf{1}, -\mathbf{1}\}$, occurs only at exceptional values of $\boldsymbol{F}$. Note that a crystal is not a solid according to Definition 9.4.3(1). Rather, the occurrence of a group $G(L) \subset \text{Unim}$, whose major part lies outside Orth, indicates certain flow features similar to those of fluids and plastic materials. Indeed, the special deformations $\boldsymbol{F} = \boldsymbol{H} \in G(L)$ have the effect of slipping the crystal planes without changing L and without changing the response. (Further fluid-like features will be presented in Sects. 17.8 and 21.6.) However, since $G(L)$ is discrete, the deformations $\boldsymbol{F} = \boldsymbol{H}$ can never be small. Hence locally, in the neighborhood of a given deformation, the crystal behaves like a solid. PITTERI [1984] (see also PARRY [1976] and BALL & JAMES [1992]) showed that every state $(\boldsymbol{F}, \theta)$ has a neighborhood $\bar{U}_0$ on which the symmetry group $G(\bar{U}_0)$ equals the point group of the deformed lattice $\boldsymbol{F}L$. This establishes the relationship between the theories with $G = G(L)$ and $G = P(L)$.

There are two main mathematical differences between the free energy functions of, say, isotropic rubber-like materials, and those of crystals with $G = G(L)$. The first is that the free energies of crystals cannot be coercive, i.e., from (9.4.9) follows that there are sequences $\boldsymbol{F}_n$ such that $|\boldsymbol{F}_n| \to \infty$ and $\bar{f}(\boldsymbol{F}_n, \theta)$ is bounded. The second is that, as we shall see in Sect. 17.8, $\bar{f}(\cdot, \theta)$ cannot be quasiconvex. Both these have the dramatic consequences: Generally, there are no configurations χ minimizing the total canonical free energy strictly, states (χ, θ) in the traditional sense can only approach the minimum free energy to a certain degree. Typically, such states will be fine mixtures of homogeneous configurations and the better the degree of approximation of the minimum, the finer the structure needed. The notion of Young's measure is an appropriate tool for dealing with such problems. (See Sect. 21.6.)

The theories of crystalline bodies based on the assumption $G = P(L)$ were mentioned in Sect. 1.5 (in particular Proposition 1.5.6 and the text preceding it), Examples 2.3.5 and 2.5.5, and Subsect. 2.3.8.

9.4.6 Changes in Phase and Symmetry Anticipating, let us give an example of a stored energy function which exhibits a phase transformation from a solid to a fluid phase. Let U_Δ be an open subset of $(0, \infty) \times (0, \infty)$, let $f_0, \bar{\mu} : U_\Delta \to \mathbb{R}$ be smooth functions and consider a material with the free energy $\bar{f} : \bar{U} \to \mathbb{R}$, where

$$\bar{U} = \{(\boldsymbol{F}, \theta) \in \text{InvLin} \times (0, \infty) : (|\det \boldsymbol{F}|, \theta) \in U_\Delta\},$$
$$\bar{f}(\boldsymbol{F}, \theta) = f_0(v, \theta) + \bar{\mu}(v, \theta)v^{\frac{1}{3}}(\operatorname{tr} \boldsymbol{V}^2 - \tfrac{1}{3}(\operatorname{tr} \boldsymbol{V})^2), \quad (\boldsymbol{F}, \theta) \in \bar{U},$$

where $v = |\det \boldsymbol{F}|$ and $\boldsymbol{V}$ is the left stretch tensor of $\boldsymbol{F}$. The entropy $\bar{\eta}$ and equilibrium stress $\bar{\boldsymbol{S}}_e$ are given by the thermostatic relations (9.2.2) and (9.2.3); set $\rho_\circ = 1$ and $\bar{\boldsymbol{q}} = \boldsymbol{0}$, $\bar{\boldsymbol{S}}_d = \boldsymbol{0}$ identically. Then $\bar{\boldsymbol{S}} = \bar{\boldsymbol{S}}_e$ and

$$\bar{\boldsymbol{S}}(\boldsymbol{F}, \theta) = -q(\boldsymbol{F}, \theta)v\boldsymbol{F}^{-\mathsf{T}} + 2\bar{\mu}(v, \theta)v^{\frac{1}{3}}(\boldsymbol{F} - \tfrac{1}{3}(\operatorname{tr} \boldsymbol{V})\boldsymbol{R}),$$
$$\bar{\boldsymbol{T}}(\boldsymbol{F}, \theta) = -q(\boldsymbol{F}, \theta)\mathbf{1} + 2\bar{\mu}(v, \theta)v^{-\frac{2}{3}}(\boldsymbol{V}^2 - \tfrac{1}{3}(\operatorname{tr} \boldsymbol{V})\boldsymbol{V}),$$

where

$$q(\boldsymbol{F}, \theta) = p(v, \theta) - \partial_v(\bar{\mu}(v, \theta)v^{\frac{1}{3}})(\operatorname{tr} \boldsymbol{V}^2 - \tfrac{1}{3}(\operatorname{tr} \boldsymbol{V})^2), \quad p(v, \theta) = -\partial_v f_0(v, \theta),$$

and $\boldsymbol{R}$ is the rotation tensor of $\boldsymbol{F}$. In particular, if $\boldsymbol{F} = w\mathbf{1}$, $w > 0$, then

$$\bar{\boldsymbol{T}}(\boldsymbol{F}, \theta) = -p(v, \theta)\mathbf{1}, \quad v = w^3.$$

This material is isotropic, i.e., its global symmetry group satisfies $G \supset \text{Orth}$. In Sect. 11.2, we shall see that its isothermal Lamé moduli in the isotropic state $(w\mathbf{1}, \theta)$ are

$\bar{\lambda}(w^3, \theta)$, $\bar{\mu}(w^3, \theta)$ where $\bar{\mu}$ is the function given above and $\bar{\lambda}$ is given by $(11.2.9)_1$. If $\bar{\mu} = 0$ identically, then the material is a fluid, and if $\bar{\mu}$ is everywhere different form 0, then the material is an isotropic solid.

However, it is possible to take $\bar{\mu}$ such that it vanishes identically on some region of v, θ and is strictly positive on its complement (and is as smooth as one wishes). More specifically, let there exist a function $\phi : (0, \infty) \to (0, \infty)$ such that

$$\bar{\mu}(v, \theta) > 0 \text{ if } (v, \theta) \in U_\Delta \text{ and } 0 < v < \phi(\theta) \text{ and } \bar{\mu}(v, \theta) = 0 \text{ otherwise.}$$

Then on

$$\bar{U}_s := \{(\boldsymbol{F}, \theta) \in \bar{U} : |\det \boldsymbol{F}| < \phi(\theta)\}$$

the material behaves like an isotropic solid, i.e., $G(\bar{U}_s) = \text{Orth}$, and on

$$\bar{U}_f := \{(\boldsymbol{F}, \theta) \in \bar{U} : |\det \boldsymbol{F}| \geq \phi(\theta)\}$$

the material behaves like a fluid, i.e., $G(\bar{U}_f) = \text{Unim}$, while the global symmetry group is Orth. We see that the abrupt change in symmetry at the points $(\boldsymbol{F}, \theta)$ satisfying $|\det \boldsymbol{F}| = \phi(\theta)$ is in no contradiction with the smoothness of the response functions. In particular, the choice

$$f_0(v, \theta) = -R\theta \ln(v - b) + a/v + m(\theta), \qquad v > b,$$

where a, b, R are positive constants and m arbitrary function, gives the van der Waals pressure

$$p(v, \theta) = R\theta/(v - b) - a/v^2, \tag{9.4.10}$$

see Fig. 19.4.1. Let θ be sufficiently low so that $p(\cdot, \theta)$ be nonmonotone. Let us consider the specific volumes $v^+ = v^+(\theta)$, $v^- = v^-(\theta)$ determined by the equality of pressures $p(v^+, \theta) = p(v^-, \theta)$ and by Maxwell's rule $g(v^+, \theta) = g(v^-, \theta)$, $v^+ < v^-$, where $g(v, \theta) = f_0(v, \theta) + p(v, \theta) \cdot v$. (In the notation of Fig. 19.4.1, $v^+ = v_1$, $v^- = v_2$.) For each sufficiently low θ, there is a pair $v^+(\theta)$, $v^-(\theta)$ with the above properties. Let further ϕ be chosen to satisfy $v^+(\theta) < \phi(\theta) < v^-(\theta)$. Let $\boldsymbol{F}^+(\theta) = (v^+)^{1/3}\mathbf{1}$ and let $\boldsymbol{F}^-(\theta)$ be a tensor differing from $\boldsymbol{F}^+(\theta)$ by a rank 1 tensor and satisfying $|\det \boldsymbol{F}^-(\theta)| = v^-$. Then the "local states" $(\boldsymbol{F}^+(\theta), \theta)$ and $(\boldsymbol{F}^-(\theta), \theta)$ describe solid and fluid phases of the body, respectively. As the results of Subsect. 17.4 will justify, these phases can coexist in a stable global equilibrium state which describes a solid in equilibrium with its melt.

In addition to the restrictions coming from the Clausius–Duhem inequality, frame indifference, and (changes of) material symmetry, the main further clues for constructing realistic free energy functions are

(1) the convexity properties, to be studied extensively in Chaps. 17 and 18;
(2) the wells structure of $\bar{f}$, coming from the existence of the preferred states of isothermal minima of $\bar{f}$ at which $\boldsymbol{S} = \mathbf{0}$, see Sect. 2.3;
(3) the natural conditions that $\bar{f}(\boldsymbol{F}, \theta) \to \infty$ as $\det \boldsymbol{F} \to \mathbf{0}$ and if possible, $\bar{f}(\boldsymbol{F}, \theta) \to \infty$ as $|\boldsymbol{F}| \to \infty$.

9.5 Supply-Free Processes

The Coleman–Noll procedure from Sect. 9.2 relies on the assumption that the motion and the temperature field of an admissible process can be chosen arbitrarily. The external force and the external supply of heat needed to execute the process are

$$\boldsymbol{b} = \ddot{\boldsymbol{\chi}} - \operatorname{Div} \boldsymbol{S}, \qquad r = \dot{e} - \operatorname{Div} \boldsymbol{q} + \boldsymbol{S} \cdot \dot{\boldsymbol{F}},$$

(setting $\rho_o = 1$ for simplicity), and hence $\boldsymbol{b}$, r in turn must be at the experimentalist's disposal. In practical situations, however, the force $\boldsymbol{b}$ reduces to the gravity and hence $\boldsymbol{b}$ is completely determined (and in many cases negligible) and the external supply of heat is absent, $r = 0$. I shall pause in this section to show that under mild additional assumptions the thermodynamic consequences from Sect. 9.2 can be derived using only supply-free processes, i.e., those with $\boldsymbol{b} \equiv \boldsymbol{0}$, $r \equiv 0$. For them the equations of balance of linear momentum and energy and the Clausius–Duhem inequality read

$$\ddot{\boldsymbol{\chi}} = \operatorname{Div} \boldsymbol{S}, \quad \dot{e} = \boldsymbol{S} \cdot \dot{\boldsymbol{F}} - \operatorname{Div} \boldsymbol{q}, \quad \dot{\eta} \geq -\operatorname{Div} \frac{\boldsymbol{q}}{\theta}. \tag{9.5.1}$$

A local supply-free process is an infinitely differentiable process on an open subset of $\Omega \times \mathbb{R}$ that satisfies the constitutive equations and the balance equations (9.5.17) on its domain of definition.

9.5.1 The Dissipation Principle for Local Supply-Free Processes The Clausius–Duhem inequality $(9.5.1)_3$ holds for every local supply-free process.

In combination with the constitutive equations (9.1.1) the balance equations and the Clausius–Duhem inequality take the forms

$$\ddot{\boldsymbol{\chi}} = \partial_{\boldsymbol{F}} \bar{\boldsymbol{S}} \cdot \nabla \boldsymbol{F} + \partial_\theta \bar{\boldsymbol{S}} \cdot \boldsymbol{G} + \partial_{\boldsymbol{G}} \bar{\boldsymbol{S}} \cdot \nabla \boldsymbol{G} + \partial_{\dot{\boldsymbol{F}}} \bar{\boldsymbol{S}} \cdot \nabla \dot{\boldsymbol{F}}, \tag{9.5.2}$$

$$\begin{aligned} &\partial_{\boldsymbol{F}} \bar{e} \cdot \dot{\boldsymbol{F}} + \partial_\theta \bar{e} \cdot \dot{\theta} + \partial_{\boldsymbol{G}} \bar{e} \cdot \dot{\boldsymbol{G}} + \partial_{\dot{\boldsymbol{F}}} \bar{e} \cdot \ddot{\boldsymbol{F}} \\ &\qquad = \boldsymbol{S} \cdot \dot{\boldsymbol{F}} - (\partial_{\boldsymbol{F}} \bar{\boldsymbol{q}} \cdot \nabla \boldsymbol{F} + \partial_\theta \bar{\boldsymbol{q}} \cdot \boldsymbol{G} + \partial_{\boldsymbol{G}} \bar{\boldsymbol{q}} \cdot \nabla \boldsymbol{G} + \partial_{\dot{\boldsymbol{F}}} \bar{\boldsymbol{q}} \cdot \nabla \dot{\boldsymbol{F}}), \end{aligned} \tag{9.5.3}$$

$$\begin{aligned} &\partial_{\boldsymbol{F}} \bar{\eta} \cdot \dot{\boldsymbol{F}} + \partial_\theta \bar{\eta} \cdot \dot{\theta} + \partial_{\boldsymbol{G}} \bar{\eta} \cdot \dot{\boldsymbol{G}} + \partial_{\dot{\boldsymbol{F}}} \bar{\eta} \cdot \ddot{\boldsymbol{F}} \\ &\qquad \geq -(\partial_{\boldsymbol{F}} \boldsymbol{\phi} \cdot \nabla \boldsymbol{F} + \partial_\theta \boldsymbol{\phi} \cdot \boldsymbol{G} + \partial_{\boldsymbol{G}} \boldsymbol{\phi} \cdot \nabla \boldsymbol{G} + \partial_{\dot{\boldsymbol{F}}} \boldsymbol{\phi} \cdot \nabla \dot{\boldsymbol{F}}). \end{aligned} \tag{9.5.4}$$

Here for brevity the 'dot' notation has been employed to denote the contractions of products of tensors; thus, e.g., $\partial_{\boldsymbol{F}} \bar{\boldsymbol{S}} \cdot \nabla \boldsymbol{F}$ denotes a vector with components $(\partial_{\boldsymbol{F}} \bar{\boldsymbol{S}} \cdot \nabla \boldsymbol{F})_i = \partial_{F_{jB}} S_{iA} F_{jAB}$, (the explicit form of the other terms being similar and determined by the chain rule employed to evaluate the divergences) and the abbreviation

$$\boldsymbol{\phi} = \bar{\boldsymbol{q}} / \theta \tag{9.5.5}$$

has been used. The constitutive functions and their derivatives are evaluated at $(\boldsymbol{F}, \theta, \boldsymbol{G}, \dot{\boldsymbol{F}})$. We observe that (9.5.2)–(9.5.4) contain the quantities

$$\boldsymbol{F}, \theta, \boldsymbol{G}, \ddot{\boldsymbol{\chi}}, \nabla \boldsymbol{F} = \nabla^2 \boldsymbol{\chi}, \nabla \boldsymbol{G} = \nabla^2 \theta, \dot{\boldsymbol{F}}, \dot{\theta}, \dot{\boldsymbol{G}}, \ddot{\boldsymbol{F}}, \nabla \dot{\boldsymbol{F}}; \tag{9.5.6}$$

of these, the following are contained linearly:

$$\ddot{\boldsymbol{\chi}}, \nabla \boldsymbol{F}, \nabla \boldsymbol{G}, \dot{\theta}, \dot{\boldsymbol{G}}, \ddot{\boldsymbol{F}}, \nabla \dot{\boldsymbol{F}}. \tag{9.5.7}$$

We have to guarantee that the balance equations $(9.5.1)_{1,2}$ have enough local solutions; otherwise the Clausius–Duhem inequality $(9.5.1)_3$ could be satisfied vacuously.

9.5.2 Assumptions

(1) ***Local solvability assumption*** Let $\boldsymbol{p} \in \Omega$ and $t \in \mathbb{R}$ be arbitrary and let

$$\boldsymbol{F}_0, \theta_0, \boldsymbol{G}_0, \ddot{\boldsymbol{\chi}}_0, \nabla \boldsymbol{F}_0, \nabla \boldsymbol{G}_0, \dot{\boldsymbol{F}}_0, \dot{\theta}_0, \dot{\boldsymbol{G}}_0, \ddot{\boldsymbol{F}}_0, \nabla \dot{\boldsymbol{F}}_0 \tag{9.5.8}$$

be any set of quantities (not fields) of the same tensorial nature as the respective quantities in (9.5.6) with $(\boldsymbol{F}_0, \theta_0) \in \bar{U}$. Suppose that they satisfy equations (9.5.2), (9.5.3). Then there exists a local supply-free process $(\boldsymbol{\chi}, \theta)$ on a neighborhood of $(\boldsymbol{p}, t)$, such that the quantities in (9.5.6) evaluated for this process at $(\boldsymbol{p}, t)$ are equal to (9.5.8), i.e.,

$$\begin{aligned} &\nabla \boldsymbol{\chi} = \boldsymbol{F}_0, \theta = \theta_0, \nabla \theta = \boldsymbol{G}_0, \ddot{\boldsymbol{\chi}} = \ddot{\boldsymbol{\chi}}_0, \nabla^2 \ddot{\boldsymbol{\chi}} = \nabla \boldsymbol{F}_0, \nabla^2 \theta = \nabla \boldsymbol{G}_0, \\ &\nabla \dot{\boldsymbol{\chi}} = \dot{\boldsymbol{F}}_0, \dot{\theta} = \dot{\theta}_0, \nabla \dot{\theta} = \dot{\boldsymbol{G}}_0, \nabla \ddot{\boldsymbol{\chi}} = \ddot{\boldsymbol{F}}_0, \nabla \dot{\boldsymbol{\chi}} = \nabla \dot{\boldsymbol{F}}_0. \end{aligned}$$

(2) For every $(\boldsymbol{F}_0, \theta_0, \boldsymbol{G}_0, \dot{\boldsymbol{F}}_0) \in \operatorname{InvLin} \times \mathbb{R}^{++} \times \operatorname{Vect} \times \operatorname{Lin}$ with $(\boldsymbol{F}_0, \theta_0) \in \bar{U}$, every $(\boldsymbol{p}, t)$ and every $s \in \mathbb{R}$ there exists a local supply-free process $(\boldsymbol{\chi}, \theta)$ such that at $(\boldsymbol{p}, t)$

$$\nabla \boldsymbol{\chi} = \boldsymbol{F}_0, \theta = \theta_0, \nabla \theta = \boldsymbol{G}_0, \nabla \dot{\boldsymbol{\chi}} = \dot{\boldsymbol{F}}_0 \text{ and } s = \operatorname{Div} \boldsymbol{q}.$$

The "initial conditions" at $(\boldsymbol{p}, t)$ in (1) must be contrasted with the usual Cauchy problem, where the initial data are specified at time t over the whole body. The Cauchy–Kowalewskyaja theorem can be employed to prove the local solvability when the system is noncharacteristic and the response functions are analytic. Assumptions of the type (1) have been formulated by I.-S. LIU [1972] for thermoelastic materials with viscosity, by COLEMAN & DILL [1971] for immobile electromagnetic materials with memory, and by OLVER [1983, 1985; 1986, Sect. 2.6] within the context of general differential equations. Since the external supply of heat r is absent, the divergence of the heat flux vector is the only way heat can be communicated to the material elements of the body. Without assuming that it does not vanish identically one can hardly expect to obtain any sound thermodynamic consequences. Assumption (2) is equivalent to requiring that at least one of the symmetrized derivatives

$$\tfrac{1}{2}(\partial_{G_A} q_B + \partial_{G_B} q_A), \quad \tfrac{1}{2}(\partial_{F_{iA}} q_B + \partial_{F_{iB}} q_A), \quad \tfrac{1}{2}(\partial_{\dot F_{iA}} q_B + \partial_{\dot F_{iB}} q_A),$$

be nonzero, for these symmetrized derivatives occur in the expression for the divergence of $\boldsymbol{q}$ in front of the gradients $\nabla \boldsymbol{G}$, $\nabla \boldsymbol{F}$, $\nabla \dot{\boldsymbol{F}}$ obtained by applying the chain rule. Recall that these gradients have the symmetries following from the interchangeability of the partial derivatives.

9.5.3 Lemma *If Assumptions* 9.5.2 *hold, then for every*

$$\boldsymbol{F}_0, \theta_0, \boldsymbol{G}_0, \dot{\boldsymbol{F}}_0, \dot\theta_0, \dot{\boldsymbol{G}}_0, \ddot{\boldsymbol{F}}_0$$

with $(\boldsymbol{F}_0, \theta_0) \in \bar U$ *and every given* $(\boldsymbol{p}, t)$ *there exists a local supply-free process* (χ, θ) *such that at* $(\boldsymbol{p}, t)$,

$$\nabla \chi = \boldsymbol{F}_0, \theta = \theta_0, \nabla\theta = \boldsymbol{G}_0, \nabla\dot\chi = \dot{\boldsymbol{F}}_0, \dot\theta = \dot\theta_0, \nabla\dot\theta = \dot{\boldsymbol{G}}_0, \nabla\ddot\chi = \ddot{\boldsymbol{F}}_0.$$

That is, the values of the derivatives $\dot\theta_0, \dot{\boldsymbol{G}}_0, \ddot{\boldsymbol{F}}_0$ may be chosen independently of $\boldsymbol{F}_0, \theta_0, \boldsymbol{G}_0, \dot{\boldsymbol{F}}_0$.

Proof Define the number s by

$$s = \bar{\boldsymbol{S}} \cdot \dot{\boldsymbol{F}}_0 - \partial_{\boldsymbol{F}} \bar e \cdot \dot{\boldsymbol{F}}_0 - \partial_\theta \bar e \cdot \dot\theta_0 - \partial_{\boldsymbol{G}} \bar e \cdot \dot{\boldsymbol{G}}_0 - \partial_{\dot{\boldsymbol{F}}} \bar e \cdot \ddot{\boldsymbol{F}}_0$$

where the partial derivatives are evaluated at $(\boldsymbol{F}_0, \theta_0, \boldsymbol{G}_0, \dot{\boldsymbol{F}}_0)$. By the assumption of the nontriviality of the divergence of the heat flux there are quantities $\nabla\boldsymbol{G}_0, \nabla\boldsymbol{F}_0, \nabla\dot{\boldsymbol{F}}_0$ such that

$$s = \partial_{\boldsymbol{F}} \bar{\boldsymbol{q}} \cdot \nabla\boldsymbol{F}_0 + \partial_\theta \bar{\boldsymbol{q}} \cdot \boldsymbol{G}_0 + \partial_{\boldsymbol{G}} \bar{\boldsymbol{q}} \cdot \nabla\boldsymbol{G}_0 + \partial_{\dot{\boldsymbol{F}}} \bar{\boldsymbol{q}} \cdot \nabla\dot{\boldsymbol{F}}_0.$$

Define $\ddot\chi_0$ by (9.5.2) with the quantities $\boldsymbol{F}, \theta, \ldots$ replaced by $\boldsymbol{F}_0, \theta_0, \ldots$ respectively. At this stage we have a collection of quantities (9.5.8) and they satisfy (9.5.2) and (9.5.3) by construction. The local solvability assumption then gives the desired process. □

9.5.4 Reduction to a Problem in Linear Algebra Let us denote by z the variables (9.5.7) which are contained linearly in (9.5.2)–(9.5.4) and by $\boldsymbol{w}$ the rest of the variables in (9.5.6), i.e.,

$$z = (\ddot\chi, \nabla\boldsymbol{F}, \nabla\boldsymbol{G}, \dot\theta, \dot{\boldsymbol{G}}, \ddot{\boldsymbol{F}}, \nabla\dot{\boldsymbol{F}}), \quad \boldsymbol{w} = (\boldsymbol{F}, \theta, \boldsymbol{G}, \dot{\boldsymbol{F}}). \tag{9.5.9}$$

Let Z be the linear space of all z as in $(9.5.9)_1$, Y = Vect ⊕ ℝ, and O the set of all quadruples $\boldsymbol{w}$ as in $(9.5.9)_2$. Equations (9.5.2) and (9.5.3) can be written as a single equation

$$\boldsymbol{A}(\boldsymbol{w})z + \boldsymbol{B}(\boldsymbol{w}) = \boldsymbol{0}, \tag{9.5.10}$$

where $\boldsymbol{A}(\boldsymbol{w})$ is a linear mapping from the space Z into the space Y given by the coefficients in front of the variables z in (9.5.2), (9.5.3) and $\boldsymbol{B}(\boldsymbol{w})$ an element of Y denoting the absolute terms in (9.5.2), (9.5.3). The Clausius–Duhem inequality can be written as

$$\alpha(\boldsymbol{w}) \cdot z + \beta(\boldsymbol{w}) \geq 0, \tag{9.5.11}$$

where $\alpha(\boldsymbol{w})$ is an element of Z and $\beta(\boldsymbol{w})$ a number, again identified with the linear and the absolute terms in (9.5.4), respectively. The local solvability assumption 9.5.2(1) says that every pair $(\boldsymbol{w}, z) \in$ O × Z can be realized at some point $(\boldsymbol{p}, t)$ in a local supply-free process. The supply-free form of the dissipation principle then demands that $(9.5.1)_3$ hold for every $(\boldsymbol{w}, z)$ satisfying (9.5.10). Fixing $\boldsymbol{w} \in$ O, and denoting $\boldsymbol{A} \equiv \boldsymbol{A}(\boldsymbol{w})$, $\boldsymbol{B} \equiv \boldsymbol{B}(\boldsymbol{w})$, $\alpha \equiv \alpha(\boldsymbol{w})$, $\beta \equiv \beta(\boldsymbol{w})$, the exploitation of the consequences of the supply-free version of the dissipation principle reduces to a problem of finding conditions under which

$$\alpha \cdot z + \beta \geq 0 \tag{9.5.12}$$

holds for every $z \in$ Z satisfying

$$\boldsymbol{A}z + \boldsymbol{B} = \boldsymbol{0}. \tag{9.5.13}$$

The following proposition gives a complete answer.

9.5.5 Proposition *Let Z, Y be finite-dimensional real linear spaces with scalar product, $\boldsymbol{A}$ a linear mapping from Z into Y, and $\alpha \in$ Z, $\boldsymbol{B} \in$ Y, $\beta \in \mathbb{R}$. Suppose that there exists at least one $z \in$ Z such that* (9.5.13) *holds. Then the following three conditions are equivalent:*
(1) *for every $z \in$ Z satisfying* (9.5.13) *the inequality* (9.5.12) *holds;*
(2) *there exists an element $\boldsymbol{\Lambda} \in$ Y such that*

$$\alpha \cdot z + \beta - \boldsymbol{\Lambda} \cdot (\boldsymbol{A}z + \boldsymbol{B}) \geq 0 \tag{9.5.14}$$

for every $z \in$ Z;
(3) *there exists an element $\boldsymbol{\Lambda}$ in Y such that*

$$\alpha = \boldsymbol{A}^{\mathsf{T}}\boldsymbol{\Lambda}, \quad \beta - \boldsymbol{\Lambda} \cdot \boldsymbol{B} \geq 0. \tag{9.5.15}$$

The components of $\boldsymbol{\Lambda}$ will play the role of Lagrange's multipliers.

Proof For the original proof, see I.-S. LIU [1972]. The proposition is also a consequence of the known theorems on linear inequalities: It suffices to identify the spaces Z, Y with $\mathbb{R}^m$, $\mathbb{R}^n$ and to apply ROCKAFELLAR [1970, Thm. 22.3] to the system of inequalities $\boldsymbol{A}z+\boldsymbol{B} \geq \boldsymbol{0}$, $\boldsymbol{A}z+\boldsymbol{B} \leq \boldsymbol{0}$, where the inequalities are interpreted as inequalities between the components of vectors in $\mathbb{R}^n$. In the present context, $\boldsymbol{\Lambda}$ is also closely related to the notion of a characteristic of a conservation law for a general system of differential equations, see OLVER [1983, 1985; 1986, Sect. 4.3]. □

We can now employ Proposition 9.5.5 in our problem. First, it is necessary to verify that for every $\boldsymbol{w}$ as in $(9.5.9)_2$ there exists a z_0 as in $(9.5.9)_1$ satisfying (9.5.10). For this it suffices to apply Lemma 9.5.3 to $(\boldsymbol{F}_0, \theta_0, \boldsymbol{G}_0, \dot{\boldsymbol{F}}_0)$ identified with the components of $\boldsymbol{w}$ (see $(9.5.9)_2$) and $s = 0$. The values of z (see $(9.5.9)_1$) then satisfy (9.5.10). Employing the implication (1) ⇒ (2) for every $\boldsymbol{w} = (\boldsymbol{F}, \theta, \boldsymbol{G}, \dot{\boldsymbol{F}})$ gives the existence of a $\boldsymbol{\Lambda} = \boldsymbol{\Lambda}(\boldsymbol{F}, \theta, \boldsymbol{G}, \dot{\boldsymbol{F}}) \in \mathrm{Vect} \oplus \mathbb{R}$ such that (9.5.14) holds. We have $\boldsymbol{\Lambda} = (\boldsymbol{\Lambda}^v, \Lambda^e)$ where $\boldsymbol{\Lambda}^v$ is a vector and Λ^e a number, both depending on $\boldsymbol{w}$. Inequality (9.5.14) reads

$$\dot{\eta} + \mathrm{Div}\,\frac{\boldsymbol{q}}{\theta} - \Lambda^e(\dot{e} - \boldsymbol{S} \cdot \dot{\boldsymbol{F}} + \mathrm{Div}\,\boldsymbol{q}) - \boldsymbol{\Lambda}^v \cdot (\ddot{\chi} - \mathrm{Div}\,\boldsymbol{S}) \geq 0.$$

This can be written explicitly (see (9.5.2)–(9.5.5))

$$\begin{aligned}
&-\boldsymbol{\Lambda}^v \cdot \ddot{\chi} + (\partial_{\boldsymbol{F}}\bar{\eta} - \Lambda^e\partial_{\boldsymbol{F}}\bar{e} - \Lambda^e\bar{\boldsymbol{S}}) \cdot \dot{\boldsymbol{F}} \\
&\quad + (\partial_\theta\bar{\eta} - \Lambda^e\partial_\theta\bar{e}) \cdot \dot{\theta} + (\partial_{\boldsymbol{G}}\bar{\eta} - \Lambda^e\partial_{\boldsymbol{G}}\bar{e}) \cdot \dot{\boldsymbol{G}} + (\partial_{\dot{\boldsymbol{F}}}\bar{\eta} - \Lambda^e\partial_{\dot{\boldsymbol{F}}}\bar{e}) \cdot \ddot{\boldsymbol{F}} \\
&\quad + (\partial_{\boldsymbol{F}}\phi - \Lambda^e\partial_{\boldsymbol{F}}\bar{\boldsymbol{q}} + \boldsymbol{\Lambda}^v\partial_{\boldsymbol{F}}\bar{\boldsymbol{S}}) \cdot \nabla\boldsymbol{F} + (\partial_\theta\phi - \Lambda^e\partial_\theta\bar{\boldsymbol{q}} + \boldsymbol{\Lambda}^v\partial_\theta\bar{\boldsymbol{S}}) \cdot \boldsymbol{G} \\
&\quad + (\partial_{\boldsymbol{G}}\phi - \Lambda^e\partial_{\boldsymbol{G}}\bar{\boldsymbol{q}} + \boldsymbol{\Lambda}^v\partial_{\boldsymbol{G}}\bar{\boldsymbol{S}}) \cdot \nabla\boldsymbol{G} + (\partial_{\dot{\boldsymbol{F}}}\phi - \Lambda^e\partial_{\dot{\boldsymbol{F}}}\bar{\boldsymbol{q}} + \boldsymbol{\Lambda}^v\partial_{\dot{\boldsymbol{F}}}\bar{\boldsymbol{S}}) \cdot \nabla\dot{\boldsymbol{F}} \geq 0.
\end{aligned} \tag{9.5.16}$$

The point of Proposition 9.5.5 is that the variables in this inequality can be considered as independent, i.e., it must hold for *every* set of variables involved in it. The variables $\ddot{\chi}$, $\nabla\boldsymbol{F}$, $\nabla\boldsymbol{G}$, $\dot{\theta}$, $\dot{\boldsymbol{G}}$, $\ddot{\boldsymbol{F}}$, $\nabla\dot{\boldsymbol{F}}$ occur linearly, and therefore the coefficients in front of them must vanish. Hence $\boldsymbol{\Lambda}^v = \boldsymbol{0}$ and the nullity of the coefficients in front of $\dot{\theta}$, $\dot{\boldsymbol{G}}$, and $\ddot{\boldsymbol{F}}$ gives

$$\partial_\theta\bar{\eta} = \Lambda^e\partial_\theta\bar{e}, \quad \partial_{\dot{\boldsymbol{G}}}\bar{\eta} = \Lambda^e\partial_{\dot{\boldsymbol{G}}}\bar{e}, \quad \partial_{\dot{\boldsymbol{F}}}\bar{\eta} = \Lambda^e\partial_{\dot{\boldsymbol{F}}}\bar{e}. \tag{9.5.17}$$

In dealing with the coefficients for the variables $\nabla\boldsymbol{G}$, $\nabla\boldsymbol{F}$, $\nabla\dot{\boldsymbol{F}}$, it is necessary to take into account the symmetry of the second partial derivatives; by (9.5.5) here the nullity gives only

$$(\theta^{-1} - \Lambda^e)\partial_{\boldsymbol{G}}\bar{\boldsymbol{q}} \cdot \nabla\boldsymbol{G} = 0, \; (\theta^{-1} - \Lambda^e)\partial_{\boldsymbol{F}}\bar{\boldsymbol{q}} \cdot \nabla\boldsymbol{F} = 0, \; (\theta^{-1} - \Lambda^e)\partial_{\dot{\boldsymbol{F}}}\bar{\boldsymbol{q}} \cdot \nabla\dot{\boldsymbol{F}} = 0 \tag{9.5.18}$$

for every $\nabla\boldsymbol{G}$, $\nabla\boldsymbol{F}$, $\nabla\dot{\boldsymbol{F}}$ having the required symmetries. By Assumption 9.5.2(2) , at least one of the numbers behind the round brackets in (9.5.18) is nonzero for some choice of $\nabla\boldsymbol{G}$, $\nabla\boldsymbol{F}$, $\nabla\dot{\boldsymbol{F}}$ and hence $\Lambda^e = \theta^{-1}$. With(9.5.17) and (9.5.18), inequality (9.5.16) simplifies to

$$\partial_{\boldsymbol{F}}\bar{f} \cdot \dot{\boldsymbol{F}} + \boldsymbol{q} \cdot \boldsymbol{G}/\theta - \boldsymbol{S} \cdot \dot{\boldsymbol{F}} \geq 0 \tag{9.5.19}$$

where $\bar{f} = \bar{e} - \theta\bar{\eta}$ is the free energy. Furthermore, with $\Lambda = \theta^{-1}$ relations (9.5.17) reduce to $\partial_\theta\bar{f} = -\bar{\eta}$, $\partial_{\boldsymbol{G}}\bar{f} = \boldsymbol{0}$, $\partial_{\dot{\boldsymbol{F}}}\bar{f} = \boldsymbol{0}$. To summarize, we have proved that the free energy is independent of $\boldsymbol{G}$, $\dot{\boldsymbol{F}}$; that the entropy relation holds; and that (9.5.19) holds. Inequality (9.5.19) is identical in form with (9.2.6) obtained from the Clausius–Duhem inequality for general processes. Hence also its consequences are the same. This implies that the equilibrium stress relation (9.2.2) and the residual dissipation inequality (9.2.4) hold. Hence we have proved the following result:

9.5.6 Proposition *If Assumptions* 9.5.2 *hold, then the supply-free version of the dissipation principle is equivalent to the Assertions* (1)–(3) *of Proposition* 9.2.2.

It is possible to give a proof of the theorem without reference to Proposition 9.5.5. Namely, one can eliminate the derivative of the temperature from (9.5.4) by using (9.5.3) and then refer to Assumptions 9.5.2 to find that all of the variables in the so-obtained inequality are arbitrary, and comparing the coefficients to obtain the conclusion.

Closely related to the exploitation of the consequences of the Clausius–Duhem inequality for supply-free processes is the exploitation of the entropy inequality due to MÜLLER [1985]. Motivated by the kinetic theory of gases, Müller replaces the Clausius–Duhem inequality by an entropy inequality which for supply-free processes takes the form

$$\dot{\eta} + \operatorname{Div} \boldsymbol{\phi} \geq 0, \tag{9.5.20}$$

in which $\boldsymbol{\phi}$ is the entropy flux given by an independent constitutive equation, which for viscous materials with heat conduction reads

$$\boldsymbol{\phi} = \hat{\boldsymbol{\phi}}(\boldsymbol{F}, \theta, \boldsymbol{G}, \dot{\boldsymbol{F}}). \tag{9.5.21}$$

It is assumed that (9.5.20) holds for every local supply-free process and the goal is the same as in the case of the Clausius–Duhem inequality, namely, to find the consequences of (9.5.20) on the constitutive functions. Of course, some condition must be added to guarantee that the functions η, $\boldsymbol{\phi}$ be nontrivial, because identical zeros satisfy (9.5.20). The derivation of the consequences of (9.5.20) is identical with the one given above up to the point where the identification (9.5.5) has been used. In particular, one has (9.5.17). However, the rest of the analysis is much harder than in the case of the Clausius–Duhem inequality, and it leads to definite results only for isotropic materials or fluids. In this case it is possible to prove that Λ^e is a function of θ only and so the entropy flux is given by

$$\boldsymbol{\phi} = \Lambda^e \boldsymbol{q}. \tag{9.5.22}$$

The quantity θ is in this case interpreted as the empirical temperature and Λ^e as the reciprocal of the absolute temperature, called the coldness by Müller. With this identification the consequences of the entropy inequality (9.5.20) are identical with those obtained on the basis of the Clausius–Duhem inequality.

MÜLLER [1985] (see also MÜLLER & RUGGERI [1993] and Bibliographical Note to Chap. 12) gives a detailed overview of the application of the entropy inequality (9.5.20) to numerous particular cases. Here I restrict myself only to brief remarks.

(1) If the constitutive equations (9.1.1), (9.5.21) are replaced by the more general ones in which the constitutive quantities are allowed to depend on $(\boldsymbol{F}, \theta, \boldsymbol{G}, \dot{\boldsymbol{F}}, \dot{\theta})$, it is possible to prove that for isotropic materials or fluids

$$\Lambda^e = \Lambda^e(\theta, \dot{\theta}).$$

Using additional assumptions, it is asserted that the function Λ^e is universal and (9.5.22) holds again. As Müller notes, the experimental determination of the universal function $\Lambda^e(\theta, \dot{\theta})$ remains an open problem (MÜLLER [1985]).

(2) For elastic mixtures the theory gives the proper form of the entropy flux which in this case is not equal to $\boldsymbol{q}/\theta$.

(3) When applied to fluids in electromagnetic fields, and using the Ampère–Lorentz model of electromagnetism of deformable media, the magnetization vector is shown to lead to a correction term to the entropy flux, so that here again the entropy flux is not $\boldsymbol{q}/\theta$.

(4) Generally, the entropy inequality (9.5.20) gives weaker restrictions than the Clausius–Duhem inequality; the entropy flux "yields" the structure of the rest of the balance equations of the theory.

Bibliographical Note

Since the publication of COLEMAN & NOLL [1963], the Coleman–Noll procedure has been applied to many different classes of materials. Of basic importance are the classic papers COLEMAN [1964a,b] on the thermodynamics of materials with fading memory. The reader is referred to TRUESDELL & NOLL [1965, Chapters C.III, C.IV, & C.V] for a classification of materials and for treatments of special materials. Further developments from various standpoints and bibliographies are to be found in DAY [1972a], OWEN [1984], TRUESDELL [1984], KRAWIETZ [1986], SERRIN [1986], MAUGIN [1992],

BESSELING & VAN DER GIESSEN [1993]. No attempt is made here to make the bibliography complete. For more about the materials with $G = G(L)$, see, ERICKSEN [1970, 1979, 1980, 1989, 1992], PARRY [1976], PITTERI [1984, 1985a,b], FONSECA [1987,1989b], KINDERLEHRER [1988], CHIPOT & KINDERLEHRER [1988], BALL & JAMES [1987, 1992], FOSDICK & HERTOG [1990], and ZANZOTTO [1988, 1992]. For a related theory of defective crystals, see DAVINI [1986], DAVINI & PARRY [1989, 1991, 1992], FONSECA & PARRY [1992a,b] and DACOROGNA & FONSECA [1992]. The requirement that $f(\boldsymbol{F}) \to \infty$ as $\det \boldsymbol{F} \to 0$ is due to BALL [1977]; its analog in the theory of rods and shells is due to ANTMAN [1970–1972, 1995]. The properties of the response at extreme deformations are examined in PODIO-GUIDUGLI & VERGARA CAFFARELLI [1991].

10 The Equilibrium Response

The equilibrium response is a traditional domain of classical thermodynamics. Although we shall speak about the equilibrium response of viscous bodies with heat conduction, the results apply to a much wider class of bodies in equilibrium, such as viscoelastic bodies with memory. Two main topics to be discussed are the changes of variables and the second derivatives of the thermodynamic potentials, presented here for bodies of arbitrary symmetry. The specialization to isotropic bodies is carried out in the next chapter.

We shall consider the changes of thermal variables, the passage from the referential to the spatial descriptions, and the passage to generalized measures of strain. Via the equilibrium thermodynamic relations (9.2.2) and (9.2.3), each of these leads to a new quantity. Thus, by changes of thermal variables, the energy and entropy will gain a status of potentials for the stress and the conjugate thermal variable. The passage from the referential to the spatial descriptions is accompanied by the Eshelby tensor (also called the energy momentum tensor) and the passage to the generalized measures of strain leads to the conjugate stresses. The second derivatives of the thermodynamic potentials are related to measurable coefficients such as the specific and latent heats, elastic constants, and the thermal expansion tensor. A rather complete discussion of their properties is given. As we shall speak exclusively about the equilibrium stresses $\boldsymbol{S}_e$, $\boldsymbol{T}_e$ in this chapter, we write $\boldsymbol{S}$, $\boldsymbol{T}$ for $\boldsymbol{S}_e$, $\boldsymbol{T}_e$. Relations $(9.2.2)_1$, $(9.2.3)_1$ then take the form

$$\bar{\boldsymbol{S}} = \rho_0 \partial_{\boldsymbol{F}} \bar{f}, \quad \bar{\eta} = -\partial_\theta \bar{f},$$

which we assume to hold throughout the chapter together with the frame indifference. Only an occasional use of the balance equations will be made.

10.1 The Legendre Transformation

A substantial part of changes of variables in thermodynamics is a special case of the Legendre transformation, a procedure for passing from one or several variables to conjugate variables. In thermodynamics partial Legendre transformations are needed, in which some variables are replaced with the conjugate variables while others are unchanged. It is well-known that the full Legendre transformation preserves convexity properties and in this section the counterpart of this for partial Legendre transformation will be proved (cf. Proposition 10.1.4). The full Legendre transformation coincides with the Fenchel transformation (see Subsect. 16.4) when the function is very strictly convex and continuously differentiable.

10.1.1 Partial Changes of Variables Let U, R be two finite-dimensional real inner product spaces, and consider the space $X = U \times R$. Its elements will be written in the form $\boldsymbol{x} = (\boldsymbol{p}, \boldsymbol{r})$ or $\boldsymbol{y} = (\boldsymbol{q}, \boldsymbol{r})$ with $\boldsymbol{p}, \boldsymbol{q} \in U$, $\boldsymbol{r} \in R$. A ***partial change of variables*** is a mapping of the form

$$(\boldsymbol{p}, \boldsymbol{r}) \mapsto (\boldsymbol{q}, \boldsymbol{r}), \quad \boldsymbol{q} = \bar{\boldsymbol{q}}(\boldsymbol{p}, \boldsymbol{r}), \quad (\boldsymbol{p}, \boldsymbol{r}) \in \bar{O} \subset X, \tag{10.1.1}$$

where $\bar{\boldsymbol{q}} : \bar{O} \to U$. If the mapping has an inverse, then the inverse is of the form

$$(\boldsymbol{q}, \boldsymbol{r}) \mapsto (\boldsymbol{p}, \boldsymbol{r}), \quad \boldsymbol{p} = \tilde{\boldsymbol{p}}(\boldsymbol{q}, \boldsymbol{r}), \quad (\boldsymbol{q}, \boldsymbol{r}) \in \tilde{O} \subset X, \tag{10.1.2}$$

where $\bar{\boldsymbol{q}}, \tilde{\boldsymbol{p}}$ satisfy

$$\boldsymbol{p} = \tilde{\boldsymbol{p}}(\bar{\boldsymbol{q}}(\boldsymbol{p}, \boldsymbol{r}), \boldsymbol{r}), \quad (\boldsymbol{p}, \boldsymbol{r}) \in \bar{O}, \quad \boldsymbol{q} = \bar{\boldsymbol{q}}(\tilde{\boldsymbol{p}}(\boldsymbol{q}, \boldsymbol{r}), \boldsymbol{r}), \quad (\boldsymbol{q}, \boldsymbol{r}) \in \tilde{O}. \tag{10.1.3}$$

In this situation, the function $\bar{\boldsymbol{q}}$ is said to be ***invertible in p*** and $\tilde{\boldsymbol{p}}$ is called the ***partial inverse*** of $\bar{\boldsymbol{q}}$. The ***full change of variables p*** $\mapsto \bar{\boldsymbol{q}}(\boldsymbol{p})$ is the special case when $R = \{\boldsymbol{0}\}$. If both $\tilde{\boldsymbol{p}}, \bar{\boldsymbol{q}}$ are continuously differentiable, we obtain from (10.1.3) that

$$\partial_q \tilde{\boldsymbol{p}}(\boldsymbol{q}, \boldsymbol{r}) \partial_p \bar{\boldsymbol{q}}(\boldsymbol{p}, \boldsymbol{r}) = \mathbf{1}, \quad \partial_p \bar{\boldsymbol{q}}(\boldsymbol{p}, \boldsymbol{r}) \partial_q \tilde{\boldsymbol{p}}(\boldsymbol{q}, \boldsymbol{r}) = \mathbf{1}, \tag{10.1.4}$$

where $\boldsymbol{q} = \bar{\boldsymbol{q}}(\boldsymbol{p}, \boldsymbol{r}), \boldsymbol{p} = \tilde{\boldsymbol{p}}(\boldsymbol{q}, \boldsymbol{r})$; the partial derivatives are linear transformations from U to U and $\mathbf{1}$ is the identity transformation from U to U. In this case the change (10.1.1) is a diffeomorphism. As a consequence of (10.1.4),

$$\partial_p \bar{\boldsymbol{q}}(\boldsymbol{p}, \boldsymbol{r}) \text{ is nonsingular (= invertible)} \tag{10.1.5}$$

at every $(\boldsymbol{p}, \boldsymbol{r}) \in \bar{O}$. Conversely, the implicit function theorem 20.9.1 implies that if (10.1.5) holds at some point, then $\bar{\boldsymbol{q}}$ has a partial inverse in a neighborhood of that point and the change (10.1.1) is a diffeomorphism on that neighborhood.

If $\bar{\boldsymbol{q}}$ is invertible in $\boldsymbol{p}$ and if $\bar{h} : \bar{O} \to W$ is a smooth mapping to a linear space W (e.g., $W = \mathbb{R}$), one can define another smooth mapping $\tilde{h} : \tilde{O} \to W$ by $\tilde{h}(\boldsymbol{q}, \boldsymbol{r}) = \bar{h}(\tilde{\boldsymbol{p}}(\boldsymbol{q}, \boldsymbol{r}), \boldsymbol{r})$, $(\boldsymbol{q}, \boldsymbol{r}) \in \tilde{O}$. In standard presentations of thermodynamics, it is customary to make no graphic distinction between $\bar{h}$ and $\tilde{h}$; to avoid the ambiguity concerning the partial derivatives of these functions, the "subscript" notation is introduced:

$$\left(\frac{\partial h}{\partial p}\right)_r = \frac{\partial \bar{h}}{\partial p}, \quad \left(\frac{\partial h}{\partial r}\right)_p = \frac{\partial \bar{h}}{\partial r}, \quad \left(\frac{\partial h}{\partial q}\right)_r = \frac{\partial \tilde{h}}{\partial q}, \quad \left(\frac{\partial h}{\partial r}\right)_q = \frac{\partial \tilde{h}}{\partial r}.$$

Sometimes this notation is useful, especially when combined with the technique of Jacobians (see, e.g., McClellan [1980, Chap. 11]), but it will not be used in this book.

10.1.2 Partial Legendre Transformation Let $f : \bar{O} \to \mathbb{R}$ be a continuously differentiable function on an open subset of X and define the conjugate variable $\bar{\boldsymbol{q}}$ by $\bar{\boldsymbol{q}} = \partial_p f$. We say that f has a partial Legendre transformation $g : \tilde{O} \to \mathbb{R}$ if the partial change of variables (10.1.1) has a continuous inverse (10.1.2) on an open set $\tilde{O} \subset X$; the Legendre transformation (or the conjugate function) g of f is then defined by

$$g(\boldsymbol{q}, \boldsymbol{r}) = \boldsymbol{q} \cdot \tilde{\boldsymbol{p}}(\boldsymbol{q}, \boldsymbol{r}) - f(\tilde{\boldsymbol{p}}(\boldsymbol{q}, \boldsymbol{r}), \boldsymbol{r}) \tag{10.1.6}$$

for every $(\boldsymbol{q}, \boldsymbol{r}) \in \tilde{O}$. If f, g are sufficiently differentiable (see below for precise conditions), then a differentiation shows that the inverse change of variables (10.1.2) is given by $\tilde{\boldsymbol{p}} = \partial_q g$ and $\partial_r g(\boldsymbol{q}, \boldsymbol{r}) = -\partial_r f(\tilde{\boldsymbol{p}}(\boldsymbol{q}, \boldsymbol{r}), \boldsymbol{r})$. The relations then gain a symmetric form

$$f(\boldsymbol{p},\boldsymbol{r}) + g(\boldsymbol{q},\boldsymbol{r}) = \boldsymbol{p}\cdot\boldsymbol{q}, \tag{10.1.7}$$

$$\bar{\boldsymbol{q}}(\boldsymbol{p},\boldsymbol{r}) = \partial_p f(\boldsymbol{p},\boldsymbol{r}), \quad \bar{\boldsymbol{p}}(\boldsymbol{q},\boldsymbol{r}) = \partial_q g(\boldsymbol{q},\boldsymbol{r}), \tag{10.1.8}$$

$$\partial_r f(\boldsymbol{p},\boldsymbol{r}) + \partial_r g(\boldsymbol{q},\boldsymbol{r}) = \boldsymbol{0}, \tag{10.1.9}$$

$$\boldsymbol{q} = \bar{\boldsymbol{q}}(\boldsymbol{p},\boldsymbol{r}), \quad \boldsymbol{p} = \bar{\boldsymbol{p}}(\boldsymbol{q},\boldsymbol{r}). \tag{10.1.10}$$

The partial Legendre transformation of g is f. No information is lost in the passage from f to g, the function f can be reconstructed from g even when the relation between the original and the new variables is forgotten; it suffices to perform one more Legendre transformation. We shall frequently omit the term "partial" when speaking about Legendre transformation. The full Legendre transformation is a special case when $\mathrm{R} = \{\boldsymbol{0}\}$.

10.1.3 Proposition *If $f : \bar{\mathrm{O}} \to \mathbb{R}$ is twice continuously differentiable on $\bar{\mathrm{O}} \subset \mathrm{X}$ and the Hessian $\partial_p^2 f(\boldsymbol{p},\boldsymbol{r})$ is invertible at some $(\boldsymbol{p},\boldsymbol{r}) \in \bar{\mathrm{O}}$ then there exists a neighborhood of $(\boldsymbol{p},\boldsymbol{r})$ such that the restriction of f to this neighborhood has a twice continuously differentiable partial Legendre transformation g and* (10.1.8), (10.1.9) *hold on this neighborhood.*

Proof We have $\partial_p^2 f(\boldsymbol{p},\boldsymbol{r}) = \partial_p \bar{\boldsymbol{q}}(\boldsymbol{p},\boldsymbol{r})$ and hence $\bar{\boldsymbol{q}}$ is continuously differentiable by the hypothesis on f. By the local inverse theorem, the change of variables (10.1.1) has an inverse (10.1.2) in some neighborhood of $(\boldsymbol{p},\boldsymbol{r})$ and $\bar{\boldsymbol{p}}$ is continuously differentiable, with $\partial_q \bar{\boldsymbol{p}}$ invertible throughout its domain. From the definition, g is continuously differentiable and a differentiation gives $(10.1.8)_2$, (10.1.9). From these relations, $\partial_q g$, $\partial_r g$ are continuously differentiable. □

The above proposition does not cover all possible situations when the Legendre transformation is defined. A different type of sufficient conditions for the global existence of the Legendre transformation comes from the convexity conditions (see Propositions 16.4.6 and 16.4.7). Under these conditions, a continuous differentiability of f suffices and the nonsingularity of the Hessian $\partial_p^2 f(\boldsymbol{p},\boldsymbol{r})$ is not needed; one obtains only a continuous differentiability of g, with the changes of variables (10.1.1), (10.1.2) only continuous. It can happen that one of the two Hessians $\partial_p^2 f(\boldsymbol{p},\boldsymbol{r})$, $\partial_q^2 g(\boldsymbol{q},\boldsymbol{r})$ is undefined and the other is singular.

10.1.4 Proposition *In the situation of Proposition* 10.1.3,

$$d^2 f[\boldsymbol{w}_1;\boldsymbol{w}_2] = \partial_q^2 g[d\bar{\boldsymbol{q}}[\boldsymbol{w}_1]; d\bar{\boldsymbol{q}}[\boldsymbol{w}_2]] - \partial_r^2 g[\boldsymbol{v}_1;\boldsymbol{v}_2], \tag{10.1.11}$$

and

$$d^2 g[\boldsymbol{w}_1,\boldsymbol{w}_2] = \partial_p^2 f[d\bar{\boldsymbol{p}}[\boldsymbol{w}_1]; d\bar{\boldsymbol{p}}[\boldsymbol{w}_2]] - \partial_r^2 f[\boldsymbol{v}_1;\boldsymbol{v}_2], \tag{10.1.12}$$

for every $\boldsymbol{w}_1 = (\boldsymbol{u}_1,\boldsymbol{v}_1)$, $\boldsymbol{w}_2 = (\boldsymbol{u}_2,\boldsymbol{v}_2) \in \mathrm{X}$, where the differentials are evaluated at the points related by the Legendre transformation.

The total or partial second differentials are bilinear forms. Their definite signs reflect convexity properties. We shall use this proposition when transforming the convexity properties from one thermodynamic potential to another. A remarkable feature of (10.1.11), (10.1.12), and the one particularly important for an understanding of the constitutive inequalities, is the diagonal form of the right-hand sides: the mixed derivatives $\partial_p\partial_r f$ and $\partial_q\partial_r g$ do not occur. For a full Legendre transformation $\boldsymbol{p} \mapsto \boldsymbol{q} := \partial_p f(\boldsymbol{p})$, the formulas reduce to

$$d^2 f[\boldsymbol{u}_1;\boldsymbol{u}_2] = \partial_q^2 g[d\bar{\boldsymbol{q}}[\boldsymbol{u}_1]; d\bar{\boldsymbol{q}}[\boldsymbol{u}_2]], \quad d^2 g[\boldsymbol{u}_1;\boldsymbol{u}_2] = \partial_p^2 f[d\bar{\boldsymbol{p}}[\boldsymbol{u}_1]; d\bar{\boldsymbol{p}}[\boldsymbol{u}_2]].$$

Proof It is enough to prove, e.g., (10.1.12), the other is symmetric. A differentiation of the definition relation (10.1.6) with respect to $(\boldsymbol{p}, \boldsymbol{r})$ along the increment $(\boldsymbol{u}_1, \boldsymbol{v}_1)$ gives

$$dg[\boldsymbol{u}_1, \boldsymbol{v}_1] = \boldsymbol{u}_1 \cdot \tilde{\boldsymbol{p}}(\boldsymbol{q}, \boldsymbol{r}) - \partial_{\boldsymbol{r}} f(\tilde{\boldsymbol{p}}(\boldsymbol{q}, \boldsymbol{r}), \boldsymbol{r})[\boldsymbol{v}_1]. \tag{10.1.13}$$

The second differentiation of (10.1.13) with respect to $(\boldsymbol{p}, \boldsymbol{r})$ along the increment $(\boldsymbol{u}_1, \boldsymbol{v}_2)$ gives, after simple rearrangements,

$$d^2 g[\boldsymbol{u}_1, \boldsymbol{v}_1; \boldsymbol{u}_2, \boldsymbol{v}_2] = (\boldsymbol{u}_1 - \partial_{\boldsymbol{r}} \bar{\boldsymbol{q}}[\boldsymbol{v}_1]) \cdot d\tilde{\boldsymbol{p}}[\boldsymbol{u}_2, \boldsymbol{v}_2] - \partial_{\boldsymbol{r}}^2 f[\boldsymbol{v}_1, \boldsymbol{v}_2]. \tag{10.1.14}$$

Now the definition of $\bar{\boldsymbol{q}}, \tilde{\boldsymbol{p}}$ says that $\boldsymbol{q} = \bar{\boldsymbol{q}}(\tilde{\boldsymbol{p}}(\boldsymbol{q}, \boldsymbol{r}), \boldsymbol{r})$ and a differentiation along $(\boldsymbol{u}_1, \boldsymbol{v}_1)$ provides, after a rearrangement, $\boldsymbol{u}_1 - \partial_{\boldsymbol{r}} \bar{\boldsymbol{q}}[\boldsymbol{v}_1] = \partial_{\boldsymbol{r}}^2 f[d\tilde{\boldsymbol{p}}[\boldsymbol{u}_1, \boldsymbol{v}_1]]$ which in combination with (10.1.14) leads to (10.1.12). □

10.1.5 Corollary *If in the situation of the preceding proposition* $\bar{\mathrm{O}}, \tilde{\mathrm{O}}$ *are convex, then f is convex if and only if g is separately convex in* $\boldsymbol{q}$ *and concave in* $\boldsymbol{r}$*. In particular, in the case of the full Legendre transformation, f is convex if and only if g is convex.*

This is interesting, because generally the separate convexity in two variables does not imply the joint convexity.

10.2 Changes of Thermal Variables

In Chap. 9 the equilibrium response was expressed in terms of $(\boldsymbol{F}, \theta)$ as independent variables in the constitutive equations. For general bodies, there are two other natural choices of independent variables: $(\boldsymbol{F}, \eta)$ and $(\boldsymbol{F}, e)$, and the passage form θ to η will be seen to be a partial Legendre transformation. For fluids additionally the passage from the specific volume to the pressure is often possible; see Sect. 11.4.

The system of notation: The constitutive function for a given quantity f will be denoted by $\bar{f}, \tilde{f}$, or $\hat{f}$ according to whether it is expressed as function of $(\boldsymbol{F}, \theta)$, $(\boldsymbol{F}, \eta)$, or $(\boldsymbol{F}, e)$ respectively, i.e., $f = \bar{f}(\boldsymbol{F}, \theta) = \tilde{f}(\boldsymbol{F}, \eta) = \hat{f}(\boldsymbol{F}, e)$. All the processes and the constitutive functions considered in this chapter will be assumed to be infinitely differentiable.

10.2.1 The Entropy as Independent Variable Consider a body with the response functions for the free energy, equilibrium stress, and entropy $\bar{f}, \bar{\boldsymbol{S}}, \bar{\eta}$ on $\bar{U}$. We say that ***the body admits the entropy as independent variable*** if the function $(\boldsymbol{F}, \theta) \mapsto \bar{\eta}(\boldsymbol{F}, \theta)$ has a continuously differentiable partial inverse with respect to θ in the sense of Subsect. 10.1.1. We denote the partial inverse by $\tilde{\theta} : \tilde{U} \to \mathrm{IR}^{++}$ with the elements of $\tilde{U} \subset \mathrm{InvLin} \times \mathrm{IR}$ denoted $(\boldsymbol{F}, \eta)$. The local invertibility condition (10.1.5) is

$$\partial_\theta \bar{\eta} \neq 0 \text{ which implies } \partial_\eta \tilde{\theta} \neq 0. \tag{10.2.1}$$

Being continuous, the derivatives must be of the same sign on every connected component of $\bar{U}, \tilde{U}$. By the frame indifference and the structure of Orth, the set $\bar{U}$ has an even number of components; if it has exactly two components, then by $\bar{\eta}(-\boldsymbol{F}, \theta) = \bar{\eta}(\boldsymbol{F}, \theta)$, equation $(10.2.1)_1$ implies either

$$\partial_\theta \bar{\eta} > 0 \text{ everywhere on } \bar{U} \text{ or } \partial_\theta \bar{\eta} < 0 \text{ everywhere on } \bar{U}. \tag{10.2.2}$$

We shall see in Sect. 10.8 (namely in $(10.8.2)_1$) that $\partial_\theta \bar{\eta}$ is of the same sign as the specific heat at constant deformation; experience shows that there is no substance for which $(10.2.2)_1$ would be violated. Hence the local invertibility condition is practically always satisfied in the form $(10.2.2)_1$. However, $(10.2.2)_1$ is not laid down as a permanent assumption, for it is the aim of the subsequent discussion to show the relationship of $(10.2.2)_1$ to the stability.

The partial Legendre transformation of $\bar{f}$ with respect to θ is $-e$; the conjugate of θ is $-\eta$. We denote by $\tilde{e}, \tilde{\boldsymbol{S}}, \tilde{\boldsymbol{T}}$ the energy and stresses expressed as functions of $(\boldsymbol{F}, \eta)$, according to the general scheme of notation. Proposition 10.1.3, or a direct computation using the differentials, gives the thermostatic relations in entropy: If the body admits the entropy as independent variable, then

$$\tilde{\boldsymbol{S}} = \rho_o \partial_{\boldsymbol{F}} \tilde{e}, \quad \tilde{\theta} = \partial_\eta \tilde{e}, \quad \tilde{\boldsymbol{T}} = \rho \partial_{\boldsymbol{F}} \tilde{e} \boldsymbol{F}^{\mathrm{T}}, \tag{10.2.3}$$

on $\tilde{U}$. In differentials,

$$\rho_o d\tilde{e} = \tilde{\boldsymbol{S}} \cdot d\boldsymbol{F} + \rho_o \tilde{\theta} d\eta, \quad \rho d\tilde{e} = \tilde{\boldsymbol{T}} \boldsymbol{F}^{-\mathrm{T}} \cdot d\boldsymbol{F} + \rho \tilde{\theta} d\eta. \tag{10.2.4}$$

10.2.2 The Energy as Independent Variable The use of $(\boldsymbol{F}, e)$ as independent variables opens a new branch of thermodynamic potentials, with the physical dimension of energy divided by temperature, the "Massieu–Planck branch." This branch is obtained by writing the Clausius–Duhem inequality in the form

$$\rho_o \dot{\eta} \geq \rho_o \dot{e}/\theta - \boldsymbol{S} \cdot \dot{\boldsymbol{F}}/\theta + \boldsymbol{q} \cdot \boldsymbol{G}/\theta^2$$

and viewing the specific energy as independent thermal variable. A body described by the response functions $\bar{f}, \bar{\boldsymbol{S}}, \bar{e}, \bar{\eta}$ on $\bar{U}$ is said to ***admit the energy as independent variable*** if the function $\bar{e}(\boldsymbol{F}, \theta)$ has a continuously differentiable partial inverse with respect to θ in the sense of Subsect. 10.1.1. We denote by $\hat{\theta} : \hat{U} \to \mathbb{R}^{++}$ the partial inverse. Remarks quite analogous to those for $\partial_\theta \bar{\eta}$ apply here to the derivative $\partial_\theta \bar{e}$. If the body admits the energy as independent variable and if we consider $\hat{\boldsymbol{S}}, \hat{\boldsymbol{T}}, \hat{\eta}$ expressed as functions of $(\boldsymbol{F}, e)$, then

$$\hat{\boldsymbol{S}}/\hat{\theta} = -\rho_o \partial_{\boldsymbol{F}} \hat{\eta}, \quad 1/\theta = \partial_e \hat{\eta}, \quad \hat{\boldsymbol{T}}/\hat{\theta} = -\rho \partial_{\boldsymbol{F}} \hat{\eta} \boldsymbol{F}^{\mathrm{T}} \tag{10.2.5}$$

on $\hat{U}$. In differentials,

$$\rho_o d\hat{\eta} = -\hat{\boldsymbol{S}} \cdot d\boldsymbol{F}/\hat{\theta} + \rho_o de/\hat{\theta}, \quad \rho d\hat{\eta} = -\hat{\boldsymbol{T}} \boldsymbol{F}^{-\mathrm{T}} \cdot d\boldsymbol{F}/\hat{\theta} + \rho de/\hat{\theta}. \tag{10.2.6}$$

The entropy taken as function of $(\boldsymbol{F}, e)$ is a potential for $-\hat{\boldsymbol{S}}/\hat{\theta}, 1/\hat{\theta}$. The passage from $(\boldsymbol{F}, \theta)$ to $(\boldsymbol{F}, e)$ is an exchange of roles of the potential and the variable. The local changes *between the energy and entropy* are always possible, because the local invertibility conditions $\partial_\eta \tilde{e} \neq 0, \partial_e \hat{\eta} \neq 0$ are always satisfied by $(10.2.3)_2$ and $(10.2.5)_2$ and the positivity of the absolute temperature.

A Legendre transformation of $\hat{\eta}$ with respect to e gives the conjugate variable $1/\theta$ and the conjugate function $-f/\theta$. The quantity $y := f/\theta$ is the Massieu function. It is well-known that the dissipation inequality for y does not contain the entropy:

$$\rho_o \dot{y} \leq \boldsymbol{S} \cdot \dot{\boldsymbol{F}}/\theta - \rho_o e\dot{\theta}/\theta^2 - \boldsymbol{q} \cdot \boldsymbol{g}/\theta^2.$$

10.2.3 *Remark* The Legendre transformation of $\bar{f}$ with respect to $\boldsymbol{F}$, i.e., the function

$$m(\boldsymbol{S}, \theta) := \bar{f}(\boldsymbol{F}, \theta) - \rho_o^{-1} \bar{\boldsymbol{S}}(\boldsymbol{F}, \theta) \cdot \boldsymbol{F}$$

cannot be generally defined as a function of $\boldsymbol{S}, \theta$, since the stress relation $\boldsymbol{S} = \bar{\boldsymbol{S}}(\boldsymbol{F}, \theta)$ need not be invertible in $\boldsymbol{F}$. (The reader is referred to Sect. 18.4 for the discussion of the invertibility of the stress relation in isotropic elasticity.) The attempts to introduce m usually arise from the desire to generalize the classical Gibbs function of fluids (see Sect. 19.4). However, for fluids the above definition gives $m = f + 3pv$ while the Gibbs function is $g = f + pv$. The main motivation for the Gibbs function in the context of fluids is that its continuity across the phase boundary expresses the conditions of thermodynamic equilibrium. The theory of equilibrium in solids based on the continuity of m is wrong except in special circumstances, and the appropriate generalization of the equality of the Gibbs function is the continuity of the normal component of the Eshelby tensor across the phase boundary.

10.3 The Eshelby Tensor. The Spatial Description

The Eshelby tensor, also called the energy momentum tensor, is important for phase transitions, continuum theory of dislocations and fracture mechanics. Its also related to the referential translational symmetry by Noether's theorem, see Subsect. 14.5.6 and Proposition 14.5.7.

10.3.1 The Isothermal Eshelby Tensor For a body with the response functions $\bar{f}$, $\bar{S}$ on $\bar{U}$ we define the ***(isothermal) referential Eshelby tensor*** $\bar{P}$ by (cf. Subsect. 3.7.3)

$$\bar{P} = \bar{f}\mathbf{1} - \rho_o^{-1}F^{\mathsf{T}}\bar{S} = \bar{f}\mathbf{1} - F^{\mathsf{T}}\partial_F\bar{f} = -(\det F)F^{\mathsf{T}}\partial_F(f/\det F);$$

the ***spatial Eshelby tensor*** $\bar{Q}$ is the inverse Piola transformation of $\bar{P}$, i.e.,

$$\bar{Q} = |J|^{-1}\bar{P}F^{\mathsf{T}} = |J|^{-1}(\bar{f}F^{\mathsf{T}} - \rho_o^{-1}F^{\mathsf{T}}\bar{S}F^{\mathsf{T}}).$$

The physical dimension of the Eshelby tensor is that of specific energy. In Sect. 11.4 we shall see that for fluids the Eshelby tensor reduces to the Gibbs chemical potential.

The underlying reason for the usefulness of the Eshelby tensor in fracture mechanics is the balance of energy in a *control volume*. This, in turn, is related to the interchange of the roles of the reference and actual configurations. In the rest of this section we examine this operation.

10.3.2 The Inverse Motion The ***inverse motion*** $\mu := \chi^{-1}$ (see Subsect. 2.2.1(1)) gives the reference position $p = \mu(x, t)$ in terms of actual position x at time t. The inverse velocity V and the inverse deformation gradient Z are $V = \partial_t\mu = -F^{-1}v$, $Z = \partial_x\mu = F^{-1}$. We have already alluded to the fact that $-F^{-1}v$ is the velocity in the reference configuration in Subsect. 2.2.5.

Consider the spatial volume densities $\bar{f}^{\sharp} := \rho\bar{f}$, $\bar{\eta}^{\sharp} := \rho\bar{\eta}$, the spatial Eshelby tensor $\bar{Q}^{\sharp}$ and the Cauchy stress tensor $\bar{T}^{\sharp}$ expressed as functions of Z and θ,

$$\bar{f}^{\sharp}(Z, \theta) = |\det Z|\rho_o\bar{f}(Z^{-1}, \theta), \quad \bar{\eta}^{\sharp}(Z, \theta) = |\det Z|\rho_o\bar{\eta}(Z^{-1}, \theta),$$
$$\bar{Q}^{\sharp}(Z, \theta) = \bar{Q}(Z^{-1}, \theta), \quad \bar{T}^{\sharp}(Z, \theta) = \bar{T}(Z^{-1}, \theta) = |\det Z|\bar{S}(Z^{-1}, \theta)Z^{-\mathsf{T}}$$

defined on $\bar{U}^{\sharp} = \{(F^{-1}, \theta) : (F, \theta) \in \bar{U}\}$. In terms of Z, the thermostatic relations read

$$\bar{Q}^{\sharp} = \rho_o^{-1}\partial_Z\bar{f}^{\sharp}, \quad \bar{\eta}^{\sharp} = \partial_\theta\bar{f}^{\sharp}, \quad \bar{T}^{\sharp} = \bar{f}^{\sharp}\mathbf{1} - Z^{\mathsf{T}}\partial_Z\bar{f}^{\sharp} \tag{10.3.1}$$

for every $(Z, \theta) \in \bar{U}^{\sharp}$. This is obtained by using (1.1.4) and (1.1.8). *Under the exchange of the reference and actual configurations the roles of the stress and the Eshelby tensors are exchanged:* the rule leading to the referential stress in the referential description leads to the spatial Eshelby tensor in the spatial description and the rule leading to the referential Eshelby tensor in the referential description leads to the spatial stress tensor in the spatial description. An additional remarkable fact is that the quasiconvexity, rank 1 convexity, Legendre–Hadamard condition, and polyconvexity are preserved under the operation $f \mapsto f^{\sharp}$; see Sect. 17.6.

10.3.3 Balancing Energy in a Control Volume Using Referential Description Let $p \subset$ Eucl be a fixed, time-independent spatial region and denote by P_t its pre-image in the reference configuration,

$$P_t = \mu(p, t) = \{\mu(x, t) : x \in p\} = \{p \in \Omega : \chi(p, t) \in p\}.$$

The change of variables mechanism for volume and surface integrals explained in Sections 3.2, 3.1 allows one to rewrite the control volume balance of energy (3.6.7) in the form

$$\frac{d}{dt}\int_{P_t}\rho_o(e + \tfrac{1}{2}(FV)^2)\,dV = \int_{\partial P_t}(\rho_o V \cdot \tilde{P}n - q \cdot n)\,dA + \int_{P_t}(V \cdot B + r)\rho_o\,dV. \tag{10.3.2}$$

where $\tilde{P}$ is a tensor given by $\tilde{P} = \frac{1}{2}v^2\mathbf{1} + e\mathbf{1} - \rho_o^{-1}F^{\mathsf{T}}S$ and $B = -F^{\mathsf{T}}b$. The mechanical power term $V \cdot \tilde{P}n$ in the surface integral is analogous to the mechanical power term in the material form of the balance equation of energy, with the stress tensor replaced, however, by the tensor closely related to the Eshelby tensor. In fracture mechanics (10.3.2) is (implicitly) used to compute the dissipation associated with a crack propagating with constant velocity in an immobile body. I refer to Subsect. 14.2.5(5) for references.

The discussion of $\boldsymbol{P}$ continues in Subsect. 14.5.1, where it is shown that the equations of mechanical equilibrium and the Eshelby conservation law (including the jump condition, see Subsect. 17.4) interchange their roles when the deformation is replaced by the inverse deformation in the variational conditions of thermodynamic equilibrium. This is interesting because the jump condition for the Eshelby conservation law does not follow from the conditions of thermodynamic equilibrium.

10.4 The Generalized Stress and Strain Measures

In addition to $\boldsymbol{U}, \boldsymbol{V}, \boldsymbol{R}$, many different measures of strain, such as the left and right Cauchy–Green deformation tensors, the Almansi tensor, etc., are popular. All these derived measures are of the form $\boldsymbol{Q} = \phi(\boldsymbol{U})$ or $\boldsymbol{M} = \phi(\boldsymbol{V})$ where $\boldsymbol{U}, \boldsymbol{V}$ are the polar factors of $\boldsymbol{F}$, ϕ is a suitable scalar-valued function, and $\phi(\boldsymbol{U}), \phi(\boldsymbol{V})$ are the values of the functional calculus of Subsect. 1.2.5 on $\boldsymbol{U}, \boldsymbol{V}$. Let us briefly review the form of the thermostatic relations in the generalized measures of strain. Examples of various choices of ϕ will be given below.

The rate of work of internal forces per unit referential volume during a process is $W = \boldsymbol{S} \cdot \dot{\boldsymbol{F}} = |\det \boldsymbol{F}|\boldsymbol{T} \cdot \boldsymbol{D}$, where $\boldsymbol{S}, \boldsymbol{T}$ are the Piola–Kirchhoff and Cauchy stresses, $\boldsymbol{F}$ the deformation gradient and $\boldsymbol{D}$ the stretching. If we define the Kirchhoff stress $\boldsymbol{A} = |\det \boldsymbol{F}|\boldsymbol{T}$ then $W = \boldsymbol{A} \cdot \boldsymbol{D}$. It turns out that for every generalized measure of strain $\bar{\boldsymbol{Q}}$, the rate of internal work is expressible in the form $W = \boldsymbol{P} \cdot \dot{\boldsymbol{Q}}$, where $\boldsymbol{P}$ is a suitably defined stress tensor, the stress conjugate to $\bar{\boldsymbol{Q}}$:

10.4.1 Proposition *Let $\boldsymbol{Q} = \phi(\boldsymbol{U})$ be a measure of strain with ϕ subject to the restrictions stated in Proposition* 1.2.8. *For every local process with the time evolutions of the Piola–Kirchhoff stress and deformation gradient $\boldsymbol{S} = \boldsymbol{S}(t)$, $\boldsymbol{F} = \boldsymbol{F}(t)$, $\boldsymbol{F} = \boldsymbol{R}\boldsymbol{U}$, we have $W = \boldsymbol{P} \cdot \dot{\boldsymbol{Q}}$, where $\boldsymbol{P} = \boldsymbol{P}(t)$ is a symmetric tensor given by*

$$\boldsymbol{P} = \tfrac{1}{2}\partial_{\boldsymbol{Q}}\psi(\boldsymbol{Q})[\boldsymbol{R}^{\mathsf{T}}\boldsymbol{S} + \boldsymbol{S}^{\mathsf{T}}\boldsymbol{R}] = \tfrac{1}{2}(\partial_{\boldsymbol{U}}\phi(\boldsymbol{U}))^{-1}[\boldsymbol{R}^{\mathsf{T}}\boldsymbol{S} + \boldsymbol{S}^{\mathsf{T}}\boldsymbol{R}].$$

*Let us call $\boldsymbol{P}$ the **stress conjugate to** $\boldsymbol{Q}$.*

Proof Define $\boldsymbol{X} := \tfrac{1}{2}(\boldsymbol{R}^{\mathsf{T}}\boldsymbol{S} + \boldsymbol{S}^{\mathsf{T}}\boldsymbol{R})$. Using $\dot{\boldsymbol{F}} = \dot{\boldsymbol{R}}\boldsymbol{U} + \boldsymbol{R}\dot{\boldsymbol{U}}$ it is found that $W = \boldsymbol{X} \cdot \dot{\boldsymbol{U}}$ and combining with $\dot{\boldsymbol{U}} = \partial_{\boldsymbol{Q}}\phi(\boldsymbol{Q})[\dot{\boldsymbol{Q}}]$ gives the result. □

Let us consider a material with the free energy $\bar{f} = \bar{f}(\boldsymbol{F}, \theta) = \bar{f}(\boldsymbol{U}, \theta)$. If $\bar{\boldsymbol{Q}} = \phi(\boldsymbol{U})$ is a generalized measure of strain, then by Proposition 1.2.8, it is possible to express the free energy, entropy and the conjugate equilibrium stress $\boldsymbol{P}$ as functions of $\boldsymbol{Q}$ and θ; i.e., $\bar{g}(\boldsymbol{Q}, \theta) = \bar{f}(\boldsymbol{U}, \theta)$, $\bar{\bar{\eta}}(\boldsymbol{Q}, \theta) = \bar{\eta}(\boldsymbol{U}, \theta)$, $\boldsymbol{P} = \bar{\boldsymbol{P}}(\boldsymbol{Q}, \theta)$ for $\boldsymbol{Q}, \boldsymbol{U}$ related by $\boldsymbol{Q} = \bar{\boldsymbol{Q}}(\boldsymbol{U})$. The thermostatic relations read

$$\bar{\boldsymbol{P}} = \rho_{\mathrm{o}}\partial_{\boldsymbol{Q}}\bar{g}, \quad \bar{\bar{\eta}} = -\partial_{\theta}\bar{g}.$$

This is immediate from the Gibbs equations written in the forms $\rho_{\mathrm{o}}\dot{f} = \boldsymbol{P} \cdot \dot{\boldsymbol{Q}} - \rho_{\mathrm{o}}\dot{\eta}$. Completely analogous relations hold if the entropy is used as independent thermal variable.

10.4.2 Examples

(1) If $\phi(v) = v$, then $\boldsymbol{Q} = \boldsymbol{U}$ and the conjugate stress is $\boldsymbol{X} = \tfrac{1}{2}(\boldsymbol{R}^{\mathsf{T}}\boldsymbol{S} + \boldsymbol{S}^{\mathsf{T}}\boldsymbol{R})$; called Jaumann's or Biot's stress.

(2) If $\phi(v) = v^2$, then $\boldsymbol{Q} = \boldsymbol{U}^2 = \boldsymbol{F}^{\mathsf{T}}\boldsymbol{F}$ is the right Cauchy–Green deformation tensor. The conjugate stress is $\boldsymbol{P} = \tfrac{1}{2}|\det \boldsymbol{F}|\boldsymbol{F}^{-1}\boldsymbol{T}\boldsymbol{F}^{-\mathsf{T}}$.

(3) Let $\phi = \phi_m$, $m \in \mathbb{R}$, be defined by

$$\phi_m(v) = \begin{cases} (v^{2m} - 1)/2m & \text{if } m \neq 0, \\ \ln v & \text{if } m = 0, \end{cases} \qquad v > 0. \tag{10.4.1}$$

The resulting $\boldsymbol{Q} = \phi_m(\boldsymbol{U})$ are the Seth–Hill strain measures. Included among them is Green's deformation tensor ($m = 2$), Almansi's tensor ($m = -2$), nominal strain ($m = 1$), and the logarithmic strain ($m = 0$). The conjugate stresses $\boldsymbol{P}$ are given by the formula in Proposition 10.4.1 with the above $\phi = \phi_m$. The general expressions for $\boldsymbol{P}$ simplify only for isotropic materials, see Subsect. 11.1.3. The conjugate stresses are of limited applicability, because they do not enter directly the balance equations.

The reader is referred also to GUO & MAN [1992], MAN & GUO [1993], HOGER [1987], and LEHMAN & GUO [1991].

10.5 Isothermal Elastic Constants

The remaining part of the chapter is devoted to the second derivatives of thermodynamic potentials, which are directly measurable quantities and which also occur in the explicit forms of the balance equations (see Parts IV and V). This section and Sect. 10.7 are devoted to the the second derivatives of potentials with respect to $\boldsymbol{F}$, i.e., to the tensor of $\bar{\boldsymbol{A}}$ of "elastic constants" (which are not constants at all). $\bar{\boldsymbol{A}}$ is often also called the tensor of elastic moduli, the tensor of stiffness coefficients or the tensor of incremental elasticities, and $\bar{\boldsymbol{B}}$ (see below) the tensor of instantaneous moduli. The mixed second derivative with respect to $\boldsymbol{F}$ and the thermal variable are related to the latent heats and the thermal expansion tensor; they are treated in Sects. 10.6 and 10.8. The second derivatives with respect to the thermal variables are related to the specific heats as explained in Sect. 10.8. Let us call the elements $(\boldsymbol{F}, \theta)$ of $\bar{U}$ the states. For the interpretation of the second derivatives in dynamical situations we shall assume that the dynamical part of the response, $\bar{\boldsymbol{S}}_d, \bar{\boldsymbol{q}}$ vanishes. Practically, the contributions of the dynamical parts of the quantities will be negligible if the processes are slow and have small gradients of temperature ("quasi-static" processes of classical thermodynamics).

10.5.1 Definitions For a body with response functions $\bar{f}, \bar{\boldsymbol{S}}$ on $\bar{U}$ and for every $(\boldsymbol{F}, \theta) \in \bar{U}$ the fourth-order tensor

$$\bar{\boldsymbol{A}} = \partial_{\boldsymbol{F}}\bar{\boldsymbol{S}} = \rho_{\circ}\partial^2_{\boldsymbol{F}}\bar{f}$$

evaluated at $(\boldsymbol{F}, \theta)$ is called the ***referential tensor of isothermal elasticities*** at $(\boldsymbol{F}, \theta)$. The ***spatial tensor of isothermal elasticities*** $\bar{\boldsymbol{B}}$ at $(\boldsymbol{F}, \theta)$ is defined by

$$\bar{\boldsymbol{B}}[\boldsymbol{H}, \boldsymbol{K}] = |J|^{-1}\bar{\boldsymbol{A}}[\boldsymbol{HF}, \boldsymbol{KF}], \qquad \boldsymbol{H}, \boldsymbol{K} \in \mathrm{Lin}.$$

If, for a given $(\boldsymbol{F}, \theta)$, the reference configuration is identified with the actual configuration, i.e., if one uses the change of the reference configuration described in Sect. 9.4 with $\boldsymbol{H} = \boldsymbol{F}$, then the spatial tensor of isothermal elasticities coincides with the referential tensor of isothermal elasticities evaluated in the new reference configuration. The ***Cauchy tensor of isothermal elasticities*** $\bar{\boldsymbol{C}}$ at $(\boldsymbol{F}, \theta)$ is defined by

$$\bar{\boldsymbol{C}}[\boldsymbol{H}] = \partial_{\boldsymbol{F}}\bar{\boldsymbol{T}}[\boldsymbol{HF}], \qquad \boldsymbol{H} \in \mathrm{Lin},$$

where $\bar{\boldsymbol{T}}$ is Cauchy stress in terms of $\boldsymbol{F}, \theta$. The ***reduced spatial tensor of isothermal elasticities*** $\bar{\boldsymbol{B}}_{\circ}$ at $(\boldsymbol{F}, \theta)$ is defined by

$$\bar{\boldsymbol{B}}_{\circ}[\boldsymbol{H}] = \bar{\boldsymbol{B}}[\boldsymbol{H}] - \boldsymbol{H}\bar{\boldsymbol{T}}, \qquad \boldsymbol{H} \in \mathrm{Lin}.$$

The tensor $\bar{\boldsymbol{A}}$ is invertible as a linear transformation from Lin into Lin if and only if $\bar{\boldsymbol{B}}$ is invertible as a linear transformation from Lin into Lin. If this is the case, we introduce the referential and the spatial tensors of ***compliances*** $\bar{\boldsymbol{Z}}$ and $\bar{\boldsymbol{S}}$ by $\bar{\boldsymbol{Z}} = \bar{\boldsymbol{A}}^{-1}, \bar{\boldsymbol{S}} = \bar{\boldsymbol{B}}^{-1}$. We shall see in Subsect. 10.5.3(4) that the referential and spatial tensors of compliances do not exist in stress-free states. In states of nonvanishing stress the compliances can exist, as seen from Proposition 18.4.4, dealing with isotropic materials. Since the Cauchy stress tensor is symmetric, the Cauchy tensor of elasticities has its range in Sym. If $\bar{\boldsymbol{C}}$ maps Sym onto Sym we say that the ***tensor of Cauchy compliances exists*** and define it as a linear transformation $\bar{\mathsf{G}}$ from Sym into Sym given by $\bar{\mathsf{G}} = \bar{\mathsf{C}}^{-1}$, where the inverse on the right is understood to be the inverse of $\bar{\mathsf{C}}$ restricted to Sym.

In Sect. 10.7 we shall introduce the tensor of adiabatic elasticities defined by the same partial derivatives as the isothermal elasticities, but keeping the entropy, rather than the temperature, fixed. The basic distinction between the isothermal and adiabatic elasticities is that the former are relevant to static situations and the latter to dynamic situations.

To state the consequences of the material symmetry in a differential form, let us introduce the following notion.

10.5.2 The Lie Algebra of the Symmetry Group Let G be a subgroup of InvLin. The ***Lie algebra*** of G is the set Lia(G) of all $\boldsymbol{A} \in$ Lin for which there exists a continuously differentiable function $\gamma : (-\varepsilon, \varepsilon) \to G$ such that $\gamma(0) = \mathbf{1}$, $\dot{\gamma}(0) = \boldsymbol{A}$. The following properties follow from the definition:

(1) Lia(G) *is a linear subspace of* Lin;
(2) *if* $\boldsymbol{A} \in$ Lia(G) *and* $\boldsymbol{H} \in G$, *then* $\boldsymbol{HAH}^{-1} \in$ Lia(G);
(3) *if* $\boldsymbol{A}, \boldsymbol{B} \in$ Lia(G), *then* $[\boldsymbol{A}, \boldsymbol{B}] := \boldsymbol{AB} - \boldsymbol{BA} \in$ Lia(G).

For every $\boldsymbol{A} \in$ Lia(G), $\boldsymbol{A}$ small, the element $\mathbf{1}+\boldsymbol{A}$ is approximately in G. The summation in Lia(G) corresponds to the multiplication in G: $(\mathbf{1}+\boldsymbol{A})(\mathbf{1}+\boldsymbol{B}) \cong \mathbf{1}+\boldsymbol{A}+\boldsymbol{B}$. For every $\boldsymbol{F} \in$ InvLin we have Lia($\boldsymbol{F}G\boldsymbol{F}^{-1}$) $= \boldsymbol{F}$ Lia(G)$\boldsymbol{F}^{-1}$. Some examples are:

$$\mathrm{Lia(InvLin)} = \mathrm{Lin}, \quad \mathrm{Lia(Orth)} = \mathrm{Lia(Orth^+)} = \mathrm{Skew},$$
$$\mathrm{Lia(Unim)} = \{\boldsymbol{A} \in \mathrm{Lin} : \mathrm{tr}\,\boldsymbol{A} = 0\}$$

and Lia(G) $= \{\mathbf{0}\}$ if G is a discrete group (such as the point groups of crystal classes).

10.5.3 Proposition *Consider a material with a symmetry group G. Let* $(\boldsymbol{F}, \theta) \in \bar{U}$, *and consider the elasticities* $\bar{\mathsf{A}}, \bar{\mathsf{B}}, \bar{\mathsf{C}}$ *the Piola–Kirchhoff stress* $\bar{\boldsymbol{S}}$ *and the Cauchy stress* $\bar{\boldsymbol{T}}$ *at* $(\boldsymbol{F}, \theta)$. *The following assertions hold (for every* $\boldsymbol{H} \in$ Lin, $\boldsymbol{W} \in$ Skew):

(1) (relations among $\bar{\mathsf{A}}$, $\bar{\mathsf{B}}$ and $\bar{\mathsf{C}}$)

$$\bar{\mathsf{B}}[\boldsymbol{H}] = |J|^{-1}\bar{\mathsf{A}}[\boldsymbol{HF}]\boldsymbol{F}^{\mathrm{T}}, \quad \bar{\mathsf{C}}[\boldsymbol{H}] = \bar{\mathsf{B}}[\boldsymbol{H}] - \bar{\boldsymbol{T}}\,\mathrm{tr}\,\boldsymbol{H} + \bar{\boldsymbol{T}}\boldsymbol{H}^{\mathrm{T}};$$

(2) (consequences of the symmetry of stress)

$$\bar{\mathsf{A}}[\boldsymbol{H}]\boldsymbol{F}^{\mathrm{T}} = \boldsymbol{F}\bar{\mathsf{A}}[\boldsymbol{H}]^{\mathrm{T}} + \boldsymbol{H}\bar{\boldsymbol{S}}^{\mathrm{T}} - \bar{\boldsymbol{S}}\boldsymbol{H}^{\mathrm{T}},$$
$$\bar{\mathsf{B}}[\boldsymbol{H}] - \bar{\mathsf{B}}[\boldsymbol{H}]^{\mathrm{T}} + \boldsymbol{H}\bar{\boldsymbol{T}} - \bar{\boldsymbol{T}}\boldsymbol{H}^{\mathrm{T}}, \quad \bar{\mathsf{C}}[\boldsymbol{H}] = \bar{\mathsf{C}}[\boldsymbol{H}]^{\mathrm{T}};$$

(3) (consequences of the second law)

$$\bar{\mathsf{A}}^{\mathbb{T}} = \bar{\mathsf{A}}, \quad \bar{\mathsf{B}}^{\mathbb{T}} = \bar{\mathsf{B}},$$
$$\bar{\mathsf{C}}^{\mathbb{T}}[\boldsymbol{H}] = \bar{\mathsf{C}}[\boldsymbol{H}] + \bar{\boldsymbol{T}}\,\mathrm{tr}\,\boldsymbol{H} - (\bar{\boldsymbol{T}} \cdot \boldsymbol{H})\mathbf{1} + \boldsymbol{H}^{\mathrm{T}}\bar{\boldsymbol{T}} - \bar{\boldsymbol{T}}\boldsymbol{H}^{\mathrm{T}};$$

(4) (consequences of frame-indifference)

$$\bar{\mathsf{A}}[\boldsymbol{WF}] = \boldsymbol{W}\bar{\boldsymbol{S}}, \quad \bar{\mathsf{B}}[\boldsymbol{W}] = \boldsymbol{W}\bar{\boldsymbol{T}}, \quad \bar{\mathsf{C}}[\boldsymbol{W}] = \boldsymbol{W}\bar{\boldsymbol{T}} - \bar{\boldsymbol{T}}\boldsymbol{W};$$

(5) (combined consequences of frame-indifference and symmetry) *for every* $\boldsymbol{Q} \in G_{\boldsymbol{F}} \equiv \boldsymbol{F}G\boldsymbol{F}^{-1} \cap$ Orth,

$$\boldsymbol{Q}\bar{\boldsymbol{T}}\boldsymbol{Q}^{\mathrm{T}} = \bar{\boldsymbol{T}}, \quad \bar{\mathsf{B}}[\boldsymbol{QHQ}^{\mathrm{T}}] = \boldsymbol{Q}\bar{\mathsf{B}}[\boldsymbol{H}]\boldsymbol{Q}^{\mathrm{T}}, \quad \bar{\mathsf{C}}[\boldsymbol{QHQ}^{\mathrm{T}}] = \boldsymbol{Q}\bar{\mathsf{C}}[\boldsymbol{H}]\boldsymbol{Q}^{\mathrm{T}};$$

(6) (differential consequences of material symmetry) *for every* $A \in \mathrm{Lia}(FGF^{-1})$,

$$\bar{A}[AF] = \bar{S}\,\mathrm{tr}A - \bar{S}F^{\mathsf{T}}A^{\mathsf{T}}F^{-\mathsf{T}}, \quad \bar{B}[A] = \bar{T}\,\mathrm{tr}A - \bar{T}A^{\mathsf{T}}, \quad \bar{C}[A] = 0,$$
$$\bar{T}\cdot A = 0, \quad \bar{S}F^{\mathsf{T}}\cdot A = 0,$$

and for every $\Omega \in \mathrm{Skew} \cap \mathrm{Lia}(FGF^{-1})$, $\bar{T}\Omega = \Omega\bar{T}$.

Proof (1): The first relation is just the definition of $\bar{B}$ when it is interpreted as linear transformations on Lin. To prove the second relation (of (1)), one writes down $\bar{T}(F,\theta) = |J|^{-1}\bar{S}(F,\theta)F^{\mathsf{T}}$ and differentiates with respect to F using $(1.1.3)_1$ and (1.1.8). Replacing H by HF one obtains the relation. (2): the first relation (of (2)) is proved by differentiating $\bar{S}(F,\theta)F^{\mathsf{T}} = F\bar{S}(F,\theta)^{\mathsf{T}}$ with respect to F. The second relation is obtained form the first by combining with (1). The last relation is obtained directly by differentiating $\bar{T}(F,\theta) = \bar{T}(F,\theta)^{\mathsf{T}}$. (3): the first relation is the interchangeability of the second partial derivatives of $\bar{f}$ and the second is its direct consequence. To prove the third, take arbitrary elements $H, K \in \mathrm{Lin}$ and use (1) and $\bar{B} = \bar{B}^{\mathbb{T}}$ in the following computation:

$$\bar{C}[H]\cdot K = (\bar{B}[H] - \bar{T}(\mathrm{tr}\,H) + \bar{T}H^{\mathsf{T}})\cdot K = \bar{B}[H,K] - (\bar{T}\cdot K)\,\mathrm{tr}\,H + \bar{T}H^{\mathsf{T}}\cdot K$$
$$= \bar{B}[K,H] - (1\cdot H)(\bar{T}\cdot K) + K^{\mathsf{T}}\bar{T}\cdot H = (\bar{B}[K] - (\bar{T}\cdot K)1 + K^{\mathsf{T}}\bar{T})\cdot H.$$

Using $\bar{C}[H]\cdot K = \bar{C}^{\mathbb{T}}[K]\cdot H$, we deduce that $\bar{C}^{\mathbb{T}}[K] = \bar{B}[K] - (\bar{T}\cdot K)1 + K^{\mathsf{T}}\bar{T}$. Replacing K by H and subtracting the resulting equation from the second relation of (1) we obtain the result. (4): Differentiate $\bar{S}(e^{tW}F,\theta) = e^{tW}\bar{S}(F,\theta)$, $\bar{T}(e^{tW}F,\theta) = e^{tW}\bar{T}(F,\theta)e^{-tW}$. (5): Let $Q \in FGF^{-1} \cap \mathrm{Orth}$; then $Q^{\mathsf{T}} = Q^{-1} \in FGF^{-1} \cap \mathrm{Orth}$ and therefore $F^{-1}Q^{\mathsf{T}}F \in G$. By the definition of the symmetry group,

$$\bar{S}(GF^{-1}Q^{\mathsf{T}}F,\theta) = \bar{S}(G,\theta)F^{\mathsf{T}}Q^{\mathsf{T}}F^{-\mathsf{T}}, \quad \bar{T}(GF^{-1}Q^{\mathsf{T}}F,\theta) = \bar{T}(G,\theta) \tag{10.5.1}$$

for every $(G,\theta) \in \bar{U}$. By the principle of material frame indifference in combination with (10.5.1) then

$$\bar{S}(QGF^{-1}Q^{\mathsf{T}}F,\theta) = Q\bar{S}(G,\theta)F^{\mathsf{T}}Q^{\mathsf{T}}F^{-\mathsf{T}}, \quad \bar{T}(QGF^{-1}Q^{\mathsf{T}}F,\theta) = Q\bar{T}(G,\theta)Q^{\mathsf{T}}. \tag{10.5.2}$$

Setting $G = F$ in the last formula we obtain the first relation of (5). Differentiating (10.5.2) with respect to G in the direction $K \in \mathrm{Lin}$ we arrive at

$$\bar{A}[QKF^{-1}Q^{\mathsf{T}}F] = Q\bar{A}[K]F^{\mathsf{T}}Q^{\mathsf{T}}F^{-\mathsf{T}}, \quad \partial_F\bar{T}[QKF^{-1}Q^{\mathsf{T}}F] = Q\partial_F\bar{T}[K]Q^{\mathsf{T}},$$

which by the definitions of $\bar{B}$ and $\bar{C}$ can be rewritten as $\bar{B}[QKF^{-1}Q^{\mathsf{T}}] = Q\bar{B}[KF^{-1}]Q^{\mathsf{T}}$, $\bar{C}[QKF^{-1}Q^{\mathsf{T}}] = Q\bar{C}[KF^{-1}]Q^{\mathsf{T}}$. Writing K in the form $K = HF$ we obtain the remaining two relations. (6): Let $A \in \mathrm{Lia}(FGF^{-1})$ and let $\gamma : (-\varepsilon,\varepsilon) \to FGF^{-1}$ be a continuously differentiable function such that $\gamma(0) = 1$, $\dot{\gamma}(0) = A$. Then $\bar{\gamma} := F^{-1}\gamma F$ satisfies $\bar{\gamma}(t) \in G$ for $t \in (-\varepsilon,\varepsilon)$ and

$$\bar{\gamma}(0) = 1, \quad \dot{\bar{\gamma}}(0) = F^{-1}AF. \tag{10.5.3}$$

Then $\bar{T}(F\bar{\gamma}(t),\theta) = \bar{T}(F,\theta)$ and a differentiation at $t = 0$, relation (10.5.3) and the definition of G give the third relation in (6). The first two relations then follow by using (1). To prove the fourth and fifth relations, it suffices to differentiate $\bar{f}(F\bar{\gamma}(t),\theta) = \bar{f}(F,\theta)$ at $t = 0$. To prove $\bar{T}\Omega = \Omega\bar{T}$, we note that by (4) and the just proved part of (6), $\bar{C}[\Omega] = \Omega\bar{T}$ and $\bar{C}[\Omega] = \bar{T}\Omega$. □

Recall from Definition 9.4.4 that G_F is the local symmetry group in the state (F,θ). By (5) the form of the elastic constants is determined by G_F, and for a given F the local symmetry group can be very small, no matter how large G. Proposition 11.1.4 describes the dependence of G_F on F in isotropic materials. The decrease of the local symmetry group due to deformation is known as the deformation-induced anisotropy.

Alternatively, one can define the ***symmetry group*** $G(\bar{B})$ ***of elastic constants*** at (F,θ) to be the set of all $Q \in \mathrm{Orth}$ such that the relations of Proposition 10.5.3(5) hold. This gives a local symmetry group entirely independent of the global symmetry group G; Proposition 10.5.3(5) then says that $G(\bar{B}) \supset FGF^{-1} \cap \mathrm{Orth}$.

10.5.4 Proposition *The tensor* $\bar{\mathsf{B}}_{\mathrm{o}}$ *has the following properties as a consequence of Assertions* (1)–(6) *of Proposition* 10.5.3*:*

(1) *it has a major and minor symmetries:*

$$\bar{\mathsf{B}}_{\mathrm{o}}^{\mathbb{T}} = \bar{\mathsf{B}}_{\mathrm{o}}, \quad \bar{\mathsf{B}}_{\mathrm{o}}[H] = \bar{\mathsf{B}}_{\mathrm{o}}[H]^{\mathrm{T}}, \quad \bar{\mathsf{B}}_{\mathrm{o}}[H] = \bar{\mathsf{B}}_{\mathrm{o}}[H^{\mathrm{T}}];$$

(2) *for every* $Q \in G_F \equiv FGF^{-1} \cap \mathrm{Orth}$,

$$\bar{\mathsf{B}}_{\mathrm{o}}[QHQ^{\mathrm{T}}] = Q\bar{\mathsf{B}}_{\mathrm{o}}[H]Q^{\mathrm{T}};$$

(3) *for every* $A \in \mathrm{Lia}(FGF^{-1})$

$$\bar{\mathsf{B}}_{\mathrm{o}}[A] = \bar{T}\,\mathrm{tr}\,A - A\bar{T} - \bar{T}A^{\mathrm{T}}.$$

Conversely, given a tensor $\bar{\mathsf{B}}_{\mathrm{o}}$ *satisfying* (1)–(3) *above and defining* $\bar{\mathsf{A}}, \bar{\mathsf{B}}, \bar{\mathsf{C}}$ *by*

$$\bar{\mathsf{B}}[H] = \bar{\mathsf{B}}_{\mathrm{o}}[H] + H\bar{T}, \quad \bar{\mathsf{C}}[H] = \bar{\mathsf{B}}_{\mathrm{o}}[H] + H\bar{T} + \bar{T}H^{\mathrm{T}} - \bar{T}\,\mathrm{tr}\,H,$$
$$\bar{\mathsf{A}}[H] = |J|\bar{\mathsf{B}}[HF^{-1}]F^{-\mathrm{T}} + HF^{-1}\bar{S}, \quad H \in \mathrm{Lin},$$

then $\bar{\mathsf{A}}, \bar{\mathsf{B}}, \bar{\mathsf{C}}$ *will satisfy Assertions* (1)–(6) *of Proposition* 10.5.3.

Let us turn to an important special case.

10.5.5 The Classical Linear Elasticity (e.g., GURTIN [1972]) An element $(F, \theta) \in \bar{U}$ is said to be a ***natural*** (= ***stress-free***) ***state*** if $\bar{T}(F, \theta) = \mathbf{0}$. The classical linear elasticity (of aeolotropic bodies, i.e., of bodies of general symmetry) deals with the linearized response in the natural state. Then

(1) $\bar{\mathsf{B}} = \bar{\mathsf{B}}_{\mathrm{o}} = \bar{\mathsf{C}}$ and hence there is just one tensor of elasticities and the linearized Cauchy stress coincides with the linearized Piola–Kirchhoff stress.

(2) By $\bar{\mathsf{B}} = \bar{\mathsf{B}}_{\mathrm{o}}$, *Assertions* (1)–(3) *of Proposition* 10.5.4 *hold with* $\bar{\mathsf{B}}_{\mathrm{o}}$ *replaced by* $\bar{\mathsf{B}}$ *or* $\bar{\mathsf{C}}$. If $n = 3$, the dimension of the space of all fourth-order tensors that have major and minor symmetries is 21, whereas the dimension of the space of all fourth-order tensors is 81.

(3) The material symmetry places further restrictions. The treatises on classical elasticity describe the forms of the tensors of elasticities in the natural state when G_F is equal to various symmetry groups; $\bar{\mathsf{B}}$ then contains a limited number of scalar parameters. See, e.g., GURTIN [1972, Sect. 26] for the forms of $\bar{\mathsf{B}}$ for the crystal classes and GREEN & ZERNA [1954, Sect. 5.4] for orthotropic, transversely isotropic, and isotropic cases (the latter three symmetries will be also treated below for general stressed states). However, not every crystallographic point group can arise as a symmetry group of the tensor of elasticities $G(\bar{\mathsf{B}})$: e.g., the point group G_h of the hexagonal system is excluded because when $\bar{\mathsf{A}}$ is invariant with respect to the elements of G_h then it is automatically invariant with respect to the transversely isotropic group. This is a special case of ***Hermann's theorem*** which says that if a tensor of order m has an axis of symmetry (see Subsect. 1.3.5) of order $k > m$, then it is invariant with respect to all rotations about this axis. Hence, for the fourth-order tensor of elasticities, all axes of symmetry of order higher than 4 imply transverse isotropy. To within the conjugacy in the orthogonal group, there are exactly twelve subgroups of Orth that can arise as the symmetry groups of elastic constants, see ZHONG & DEL PIERO [1987].

(4) In the natural states the equation $T(F, \theta) = \mathbf{0}$ can be consistently interpreted as a consequence of the minimum of stored energy at F; the latter then implies that $\bar{\mathsf{B}}$ is positive-semidefinite. Because $\bar{\mathsf{B}}$ vanishes on the skew tensors, $\bar{\mathsf{B}}$ cannot be (strictly) positive definite on Lin, but its restriction to Sym can be positive definite, which is a standard assumption of the classical elasticity. For the stressed states, the positive-definiteness properties of $\bar{\mathsf{B}}$ are more complicated, see Chap. 18 for the general states of isotropic bodies.

(5) Since $\bar{\mathsf{B}}$ has the major symmetry, it is a symmetric linear transformations from Lin into itself (and its restriction to Sym a symmetric linear transformations from Sym into itself). Hence it possesses the spectral decomposition. In particular, a nonzero tensor $H \in \mathrm{Lin}$ (or $\in \mathrm{Sym}$) is an eigenstate of $\bar{\mathsf{B}}$ and $\lambda \in \mathbb{R}$ its eigenvalue if $\bar{\mathsf{B}}[H] = \lambda H$; $\bar{\mathsf{B}}$ then has an orthonormal basis of eigenstates in Lin (or in Sym), which for $n = 3$ has 9 (or 6) elements. The idea goes back to Kelvin and has been recently applied by RYCHLEWSKI [1984, 1995], where also examples for specific symmetries and further references are found.

10.5.6 The Tensors of Elasticity in Stressed States

(1) By Proposition 10.5.4, we have

$$\bar{\mathsf{B}}[\boldsymbol{H}] = \bar{\mathsf{B}}_{\circ}[\boldsymbol{E}] + \boldsymbol{H}\bar{\boldsymbol{T}},$$
$$\bar{\mathsf{C}}[\boldsymbol{H}] = \bar{\mathsf{B}}_{\circ}[\boldsymbol{E}] + \boldsymbol{E}\bar{\boldsymbol{T}} + \bar{\boldsymbol{T}}\boldsymbol{E} - \bar{\boldsymbol{T}}\,\mathrm{tr}\,\boldsymbol{E} + \boldsymbol{W}\bar{\boldsymbol{T}} - \bar{\boldsymbol{T}}\boldsymbol{W},$$

where $\boldsymbol{E}$, $\boldsymbol{W}$ are the symmetric and skew parts of $\boldsymbol{H} \in \mathrm{Lin}$, respectively. Hence unless the state is stress-free, the spatial tensor of elasticities depends on the skew part of $\boldsymbol{H}$. As will be shown in Chap. 20, it is the tensor $\bar{\mathsf{A}}$ or $\bar{\mathsf{B}}$ which enters the linearized equations of elasticity. Hence in general these linearized equations will contain $\boldsymbol{W}$ but by the above this dependence is of a perfectly controlable form. The only nontrivial information about the state is contained in $\bar{\mathsf{B}}_{\circ}$. (Also the Cauchy tensor of elasticities depends on $\boldsymbol{W}$ unless $\bar{\boldsymbol{T}} = -p\mathbf{1}$, $p \in \mathbb{R}$.)

(2) By Proposition 10.5.4, the tensor of elasticities $\bar{\mathsf{B}}$ splits into the sum of $\bar{\mathsf{B}}_{\circ}$ plus $\boldsymbol{H} \mapsto \boldsymbol{H}\boldsymbol{T}$. For a given symmetry $G_{\boldsymbol{F}}$, $\bar{\mathsf{B}}_{\circ}$ will have the form of $\bar{\mathsf{B}}$ in the natural states known from the classical elasticity (of course with different values of the parameters), and to this one adds the term $\boldsymbol{H}\boldsymbol{T}$ to obtain the $\bar{\mathsf{B}}$ in the general state of specified symmetry.

(3) In Sect. 11.1, the tensors of elasticities of isotropic solids will be shown to have either the isotropic, or the transversely isotropic, or the orthotropic symmetry, and their form will be described there. The elasticities of fluids are always isotropic; for their forms, see Sect. 11.4.

(4) For planar anisotropic elasticity ($n = 2$) in a natural state, OLVER [1988a] deals with the effect on $\bar{\mathsf{B}}$ of the changes of variables $\boldsymbol{p} \mapsto \boldsymbol{A}\boldsymbol{p}, \boldsymbol{x} \mapsto \boldsymbol{A}^{-1}\boldsymbol{x}$, where $\boldsymbol{A} \in \mathrm{InvLin}$. He shows that an appropriate change of variables and an addition of a null Lagrangian (which does not alter the equilibrium equations, see Sects. 13.5 and 13.6) reduces a general strongly elliptic (see Sect. 17.3) elasticity tensor $\bar{\mathsf{B}}$ to an orthotropic canonical elasticity tensor, which contains just two dimensionless scalar parameters. See OLVER [1988b] for an application of the result in analysing conservation laws and OLVER [1988c] for a theory applicable to stressed states.

10.6 The Thermal Coefficient of Stress

In this section the coefficients related to the mixed second derivatives of the free energy are examined.

10.6.1 Definitions The *spatial* and the ***referential stress-temperature tensors*** (the spatial and referential thermal coefficients of stress) are the tensors $\bar{\mathsf{N}}$, $\bar{\mathsf{M}}$ defined by

$$\bar{\mathsf{N}} = \partial_\theta \bar{\boldsymbol{T}}, \quad \bar{\mathsf{M}} = \partial_\theta \bar{\boldsymbol{S}} = \rho_\circ \partial_\theta \partial_{\boldsymbol{F}} \bar{f} = |J| \bar{\mathsf{N}} \boldsymbol{F}^{-\mathrm{T}}. \tag{10.6.1}$$

By considerations similar to those in the preceding section one finds that for every $(\boldsymbol{F}, 0) \in \bar{U}$, every $\boldsymbol{Q} \in \boldsymbol{F}G\boldsymbol{F}^{-1} \cap \mathrm{Orth}$ and every $\boldsymbol{A} \in \mathrm{Lin}(\boldsymbol{F}G\boldsymbol{F}^{-\mathrm{T}})$, we have

$$\boldsymbol{Q}\bar{\mathsf{N}}\boldsymbol{Q}^{\mathrm{T}} = \bar{\mathsf{N}}, \quad \bar{\mathsf{N}} \cdot \boldsymbol{A} = 0.$$

Both $\bar{\mathsf{M}}$ and $\bar{\mathsf{N}}$ measure the change of the stresses obtained by raising the temperature by 1 keeping the deformation fixed. An object dual to this is the thermal expansion tensor.

The thermal expansion tensor, to be defined below, is designed to measure the relative change of deformation in a process in which the temperature is raised by 1 and the *Cauchy* stress is kept fixed. Let $(\boldsymbol{F}(t), \theta(t))$ be the deformation-temperature path during such a process. Differentiating $\boldsymbol{T}(t) = \bar{\boldsymbol{T}}(\boldsymbol{F}(t), \theta(t)) = \mathrm{const}$ with respect to t and assuming that $\dot{\theta} = 1$ we obtain

$$\bar{\mathsf{C}}[\boldsymbol{L}] + \bar{\mathsf{N}} = \mathbf{0}, \tag{10.6.2}$$

where $\boldsymbol{L} = \dot{\boldsymbol{F}}\boldsymbol{F}^{-1}$ is the relative change of deformation in this process. Although physically the existence of such a process is natural, the existence of a solution $\boldsymbol{L}$ of (10.6.2) is not guaranteed. On the other hand, we shall show that if (10.6.2) has a solution $\boldsymbol{L}$, then it has a solution which satisfies the additional requirements (10.6.4).

10.6.2 Proposition *Let* $(\boldsymbol{F}, \theta) \in \bar{U}$ *be a point and suppose that there exists a* $\boldsymbol{L}$ *satisfying* (10.6.2). *Then there exists a second order tensor* $\bar{\boldsymbol{P}}$ *satisfying*

$$\bar{\boldsymbol{C}}[\bar{\boldsymbol{P}}] + \bar{\boldsymbol{N}} = \boldsymbol{0} \tag{10.6.3}$$

and

$$\boldsymbol{Q}\bar{\boldsymbol{P}}\boldsymbol{Q}^{\mathrm{T}} = \bar{\boldsymbol{P}}, \quad \bar{\boldsymbol{P}} \cdot \boldsymbol{A} = 0 \tag{10.6.4}$$

for every $\boldsymbol{Q} \in G_F \equiv \boldsymbol{F}G\boldsymbol{F}^{-1} \cap \mathrm{Orth}$, *and every* $\boldsymbol{A} \in \mathrm{Lia}(\boldsymbol{F}G\boldsymbol{F}^{-1})$.

The tensor $\bar{\boldsymbol{P}}$ satisfying (10.6.3) and (10.6.4), if it exists, is the ***thermal expansion tensor***. If the tensor of Cauchy compliances $\bar{\boldsymbol{G}} = \bar{\boldsymbol{C}}^{-1}$ exists in the given state, then $\bar{\boldsymbol{P}} = -\bar{\boldsymbol{G}}[\bar{\boldsymbol{N}}]$.

Proof We start form the solution $\boldsymbol{L}$ of (10.6.2). Using the symmetries of $\bar{\boldsymbol{C}}, \bar{\boldsymbol{N}}$ one sees that for every $\boldsymbol{Q} \in \boldsymbol{F}G\boldsymbol{F}^{-1} \cap \mathrm{Orth}$ also $\boldsymbol{Q}\boldsymbol{L}\boldsymbol{Q}^{\mathrm{T}}$ satisfies this equation. One can find a solution $\bar{\boldsymbol{P}}$ of (10.6.3) satisfying $(10.6.4)_1$ as the average of all $\boldsymbol{Q}\boldsymbol{L}\boldsymbol{Q}^{\mathrm{T}}$ over the compact group $\boldsymbol{F}G\boldsymbol{F}^{-1} \cap \mathrm{Orth}$, i.e., as the integral of $\boldsymbol{Q}\boldsymbol{L}\boldsymbol{Q}^{\mathrm{T}}$ with respect to the normalized Haar measure (see RUDIN [1973, Chap. 5]) on $\boldsymbol{F}G\boldsymbol{F}^{-1} \cap \mathrm{Orth}$. The details are easy to work out. Having such a $\bar{\boldsymbol{P}}$ one can improve it to satisfy also $(10.6.4)_2$ as follows. Since $\boldsymbol{Q}\,\mathrm{Lia}(\boldsymbol{F}G\boldsymbol{F}^{-1})\boldsymbol{Q}^{\mathrm{T}} \subset \mathrm{Lia}(\boldsymbol{F}G\boldsymbol{F}^{-1})$ for every $\boldsymbol{Q} \in \boldsymbol{F}G\boldsymbol{F}^{-1} \cap \mathrm{Orth}$, one finds that the orthogonal complement $\mathrm{Lia}(\boldsymbol{F}G\boldsymbol{F}^{-1})^{\perp}$ of $\mathrm{Lia}(\boldsymbol{F}G\boldsymbol{F}^{-1})$ in Lin has the same property: $\boldsymbol{Q}\,\mathrm{Lia}(\boldsymbol{F}G\boldsymbol{F}^{-1})^{\perp}\boldsymbol{Q}^{\mathrm{T}} \subset \mathrm{Lia}(\boldsymbol{F}G\boldsymbol{F}^{-1})^{\perp}$ for every $\boldsymbol{Q} \in \boldsymbol{F}G\boldsymbol{F}^{-1} \cap \mathrm{Orth}$. Then the projection of $\bar{\boldsymbol{P}}$ onto $\mathrm{Lia}(\boldsymbol{F}G\boldsymbol{F}^{-1})^{\perp}$ will satisfy both the two relations in (10.6.4). Moreover, since this projection differs from the original by an element of $\mathrm{Lia}(\boldsymbol{F}G\boldsymbol{F}^{-1})$, it follows from Proposition 10.5.3(6) that it will also satisfy (10.6.3). □

10.7 Adiabatic Elastic Constants

The adiabatic elasticities are constructed from the internal energy in the same way as the isothermal elasticities are constructed from the free energy. They are important in dynamical situations.

10.7.1 Definition Suppose that the body admits the entropy as independent variable and consider the response functions $\tilde{e}, \tilde{\boldsymbol{S}}$. For every $(\boldsymbol{F}, \eta) \in \tilde{U}$ the fourth-order tensor

$$\tilde{\boldsymbol{A}} = \partial_{\boldsymbol{F}}\tilde{\boldsymbol{S}} = \rho_{\circ}\partial_{\boldsymbol{F}}^2\tilde{e}$$

is called the ***referential tensor of adiabatic elasticities*** at $(\boldsymbol{F}, \eta)$. The ***spatial tensor of adiabatic elasticities*** $\tilde{\boldsymbol{B}}$ at $(\boldsymbol{F}, \eta)$ is defined by

$$\tilde{\boldsymbol{B}}[\boldsymbol{H}, \boldsymbol{K}] = |J|^{-1}\tilde{\boldsymbol{A}}[\boldsymbol{H}\boldsymbol{F}, \boldsymbol{K}\boldsymbol{F}]$$

for every $\boldsymbol{H}, \boldsymbol{K} \in \mathrm{Lin}$. The ***Cauchy tensor of adiabatic elasticities*** is defined by

$$\tilde{\boldsymbol{C}}[\boldsymbol{H}] = \partial_{\boldsymbol{F}}\tilde{\boldsymbol{T}}[\boldsymbol{H}\boldsymbol{F}].$$

for every $\boldsymbol{H} \in \mathrm{Lin}$. The ***referential*** and ***the spatial stress-entropy tensors*** $\tilde{\boldsymbol{M}}, \tilde{\boldsymbol{N}}$ at $(\boldsymbol{F}, \eta)$ are defined by

$$\tilde{\boldsymbol{M}} = \partial_\eta\tilde{\boldsymbol{S}} = \partial_\eta\partial_{\boldsymbol{F}}\tilde{e}, \quad \tilde{\boldsymbol{N}} = \partial_\eta\tilde{\boldsymbol{T}}.$$

The adiabatic elasticities have the properties completely analogous to those stated in Propositions 10.5.3 and 10.5.4. The following proposition establishes the relations between the adiabatic and isothermal elasticities.

10.7.2 Proposition *If the body admits the entropy as independent variable, then at the corresponding points of* $\bar{U}, \tilde{U}$,

$$\bar{A} = \tilde{A} - \frac{\theta}{\rho_o \bar{c}_F} \bar{M} \otimes \bar{M}, \quad \bar{B} = \tilde{B} - \frac{\theta}{\rho \bar{c}_F} \bar{N} \otimes \bar{N}, \quad \bar{C} = \tilde{C} - \frac{\theta}{\rho \bar{c}_F} \bar{N} \otimes \bar{N}, \tag{10.7.1}$$

where $\bar{c}_F \equiv -\theta \partial_\theta^2 \bar{f}$ *is the specific heat at constant deformation (see Sect.* 10.8), *and* $\bar{c}_F \neq 0$ *by* $(10.2.1)_1$.

Proof An application of Proposition 10.1.4 to the passage from the temperature to the entropy gives that for every $(G_1, v_1), (G_2, v_2) \in \mathrm{Lin} \times \mathbb{R}$,

$$d^2\bar{f}[G_1, v_1; G_2, v_2] = \partial_F^2 \tilde{e}[G_1, G_2] - \partial_\eta^2 \tilde{e} \cdot d\bar{\eta}[G_1, v_1] \cdot d\bar{\eta}[G_2, v_2], \tag{10.7.2}$$

$$d^2\tilde{e}[G_1, v_1; G_2, v_2] = \partial_F^2 \bar{f}[G_1, G_2] - \partial_\theta^2 \bar{f} \cdot d\tilde{\theta}[G_1, v_1] \cdot d\tilde{\theta}[G_2, v_2]. \tag{10.7.3}$$

The second differentials and the second partial derivatives are interpreted as bilinear forms as usual. (The above formulas will be used several times in the following.) To prove $(10.7.1)_1$ it suffices to set $v_1 = v_2 = 0$ in (10.7.2), to write down explicitly the differentials of $\bar{\eta}$ and to compare the coefficients on both sides of the equation. The definitions of $\bar{A}, \tilde{A}, \bar{M}$ then lead immediately to $(10.7.1)_1$. The rest is a consequence of $(10.7.1)_1$ and Proposition 10.5.3(1). □

10.7.3 *Remark* Real materials have $\bar{c}_F > 0$ (this will be discussed in Sect. 17.3) and from (10.7.1) we see that then $\tilde{A} \geq \bar{A}, \tilde{B} \geq \bar{B}$ in the sense that

$$\tilde{A}[H, H] \geq \bar{A}[H, H], \quad \tilde{B}[H, H] \geq \bar{B}[H, H]$$

for every $H \in \mathrm{Lin}$. This fact implies, among other things, that the correct calculation of the speed of sound waves using the adiabatic elasticities gives larger values than the (incorrect) calculation using isothermal elasticities. (See Sect. 23.3.)

10.8 Specific and Latent Heats; Calorimetry

Recall our convention for the present chapter that for the purpose of interpretation we assume that the dynamical part of the response functions vanish: $\bar{S}_d = \mathbf{0}, \bar{q} = \mathbf{0}$. In this situation the only way to communicate the heat to the body is the rate of the external supply of heat r as in (3.6.3).

10.8.1 Calorimetry; the Deformation as the Mechanical Variable *The rate of the external heat supply is given by*

$$r = \bar{c}_F \cdot \dot{\theta} + \bar{L}_F \cdot D, \tag{10.8.1}$$

where

$$\bar{c}_F = \theta \partial_\theta \bar{\eta} = -\theta \partial_\theta^2 \bar{f} = \partial_\theta \bar{e} \tag{10.8.2}$$

*is the **specific heat at constant deformation** and*

$$\bar{L}_F = -\theta \partial_F \partial_\theta \bar{f} F^{\mathsf{T}} = -\theta \bar{N}/\rho \tag{10.8.3}$$

*is the **latent heat with respect to deformation**. The tensor* $\bar{L}_F$ *is symmetric. If the body admits the entropy as independent thermal variable, then*

$$\bar{c}_F = \tilde{\theta}/\partial_\eta \tilde{\theta} = \tilde{\theta}/\partial_\eta^2 \tilde{e}, \quad \bar{L}_F = -\tilde{\theta} \partial_\eta \partial_F \tilde{e} F^{\mathsf{T}}/\partial_\eta^2 \tilde{e}. \tag{10.8.4}$$

The constant $\bar{c}_F$ is the heat that must be supplied to unit mass of the body in order to raise its temperature by 1 keeping the deformation fixed. $\bar{L}_F$ is the measure of the heat that must be supplied to the body to achieve the unit relative deformation maintaining the temperature fixed. As it was mentioned in the preceding text, the Clausius–Duhem inequality says nothing about the sign of $\bar{c}_F$; later the positiveness of $\bar{c}_F$ will be related to the stability of the body, cf. Sect. 17.3.

Proof From $\bar{S}_d = \mathbf{0}, \bar{q} = \mathbf{0}$, (9.2.10) and (3.6.5) we learn that $r = \theta\dot{\eta}$. Combining this with the constitutive equation $\eta = \bar{\eta}(F, \theta)$ we obtain

$$r = \theta(\partial_F \bar{\eta} \cdot \dot{F} + \partial_\theta \bar{\eta} \cdot \dot{\theta}) = -\theta \partial_F \partial_\theta \bar{f} F^{\mathrm{T}} \cdot L - \theta \partial_\theta^2 \bar{f} \cdot \dot{\theta} = \bar{c}_F \cdot \dot{\theta} + \bar{L}_F \cdot D$$

with $\bar{c}_F$ given by $(10.8.2)_{1,2}$ and $\bar{L}_F$ given by $(10.8.3)_1$. Recalling the symmetry of the Cauchy stress we see from this formula that $\bar{L}_F$ is symmetric and therefore L can be replaced by D. Equation $(10.8.2)_3$ is obtained by differentiating the identity $\bar{e} = \bar{f} + \theta\bar{\eta}$. Equation $(10.8.3)_2$ follows immediately from the definition (10.6.1) of $\bar{N}$. To prove $(10.8.4)_1$ it suffices to note that $\bar{c}_F = \theta\partial_\theta\bar{\eta} = \tilde{\theta}/\partial_\eta\tilde{\theta}$ from the relation $\bar{\eta}(F, \tilde{\theta}(F, \eta)) = \eta$. Similarly,

$$\bar{L}_F = -\theta\partial_\theta\partial_F \bar{f} F^{\mathrm{T}} = -\theta\partial_\theta \bar{S} F^{\mathrm{T}}/\rho_\circ = -\theta\partial_\eta \tilde{S} F^{\mathrm{T}} \partial_\theta \bar{\eta}/\rho_\circ = -\theta\partial_\eta\partial_F \tilde{e} F^{\mathrm{T}}/\partial_\eta\tilde{\theta}. \quad \square$$

10.8.2 Calorimetry; the Stress as the Mechanical Variable *Suppose that for every point* $(F, \theta) \in \bar{U}$ *there exists a second-order tensor* $\bar{L}_T$ *satisfying*

$$\bar{C}^{\mathrm{T}}[\bar{L}_T] = \bar{L}_F \tag{10.8.5}$$

and set

$$\bar{c}_T = \bar{c}_F + \frac{\rho}{\theta}\bar{C}[\bar{L}_T, \bar{L}_T]. \tag{10.8.6}$$

Then

(1) *for every process,*

$$r = \bar{c}_T \cdot \dot{\theta} + \bar{L}_T \cdot \dot{T}. \tag{10.8.7}$$

(2) *If* $\bar{L}_T$ *as in* (10.8.5) *exists, then it can be chosen symmetric and to satisfy*

$$Q\bar{L}_T Q^{\mathrm{T}} = \bar{L}_T \tag{10.8.8}$$

for every $Q \in G_F \equiv FGF^{-1} \cap$ Orth. *If the Cauchy tensors of compliances exists at the given point, then* $\bar{L}_T = \bar{G}^{\mathrm{T}}[\bar{L}_F]$.

(3) *If* X $\subset$ Sym *is any subspace that contains* $\bar{L}_F$ *and is invariant under* $\bar{C}$ *(i.e.,* $\bar{C}[\mathrm{X}] \subset \mathrm{X}$*) then it is also invariant under* $\tilde{C}$ *and the determinants of the restrictions of* $\bar{C}, \tilde{C}$ *to* X *(still denoted* $\det\bar{C}$ *and* $\det\tilde{C}$*, respectively) satisfy*

$$\bar{c}_T/\bar{c}_F = \det\tilde{C}/\det\bar{C} \tag{10.8.9}$$

provided the denominators do not vanish.

The quantity c_T is the ***specific heat at constant stress*** and $\bar{L}_T$ is the ***latent heat with respect to stress***. $\bar{c}_T$ is the heat that must be supplied to unit mass of the body in order to execute a process in which the temperature is raised by 1 and the Cauchy stress remains constant. $\bar{L}_T$ is the measure of the heat that must be supplied to the body in a process in which the temperature is constant and the Cauchy stress is changed by a unit. The solvability condition (10.8.5) is similar to that discussed in connection with the thermal expansion tensor. Since the ranges of $\bar{C}$ and $\tilde{C}$ are in Sym, the subspace X := Sym contains $\bar{L}_F$ and is invariant under $\bar{C}$. However, it may happen that $\bar{C}$ is singular on Sym, leaving (10.8.9) meaningless with this choice of X. Nevertheless, there may exist an X as is (3) such that $\bar{C}$ is nonsingular. For example, for fluids, the range of $\bar{C}$ is contained in $\mathbb{R}\mathbf{1}$, the space of all tensors of the form $c\mathbf{1}$ where $c \in \mathbb{R}$. Hence X = $\mathbb{R}\mathbf{1}$ is the only choice of an invariant subspace on which $\bar{C}$ can be nonsingular, and one has to apply (3) with this choice to obtain a meaningful result. For isotropic states of isotropic solids, both X = Sym and X = $\mathbb{R}\mathbf{1}$ are invariant, and $\bar{\boldsymbol C}$ can be nonsingular on both of these. Naturally, (10.8.9) gives the same value of $\bar c_T/\bar c_F$ with these two choices, which is explained in the remark following the proof. Sects. 11.1 and 11.4 give the concrete forms of (10.8.9) for isotropic solids and fluids.

Proof (1): From (10.8.5), (10.8.3) we have $\bar{\boldsymbol C}[\bar{\boldsymbol L}_T, \bar{\boldsymbol L}_T] = \bar{\boldsymbol C}^{\mathrm{T}}[\bar{\boldsymbol L}_T] \cdot \bar{\boldsymbol L}_T = -\theta \bar{\boldsymbol N} \cdot \bar{\boldsymbol L}_T/\rho$ and hence, by (10.8.6), $\bar c_T = \bar c_F - \bar{\boldsymbol N} \cdot \bar{\boldsymbol L}_T$. Then

$$\bar c_T \cdot \dot\theta + \bar{\boldsymbol L}_T \cdot \dot{\boldsymbol T} = \bar c_F \cdot \dot\theta - \bar{\boldsymbol L}_T \cdot \bar{\boldsymbol N} \cdot \dot\theta + \bar{\boldsymbol L}_T \cdot (\bar{\boldsymbol C}[\boldsymbol L] + \bar{\boldsymbol N}\dot\theta) = \bar c_F \cdot \dot\theta + \bar{\boldsymbol L}_F \cdot \boldsymbol D$$

and a reference to (10.8.1) completes the proof. (2): The possibility of choosing $\bar{\boldsymbol L}_T$ to satisfy additionally the condition (10.8.8) is similar to the same additional condition on the thermal expansion tensor and the proof is omitted. (3): Using the expression $(10.8.4)_1$ for $\bar c_F$, (10.8.3) for $\bar{\boldsymbol L}_F$ and (10.8.5) for $\bar{\boldsymbol L}_T$ we find that $(10.7.1)_3$ can be rewritten as

$$\tilde{\boldsymbol C} = \bar{\boldsymbol C} + \rho \bar{\boldsymbol L}_F \otimes \bar{\boldsymbol L}_F/\theta \bar c_F = (\mathbf 1 + \rho \bar{\boldsymbol L}_F \otimes \bar{\boldsymbol L}_T/\theta \bar c_F)\bar{\boldsymbol C},$$

where $\mathbf 1$ is the unit tensor on Lin. From the hypotheses of (3) then follows that X is invariant under $\tilde{\boldsymbol C}$. Restricting both sides of the last equation to X, taking the determinant and using

$$\det(\mathbf 1 + \rho \bar{\boldsymbol L}_F \otimes \bar{\boldsymbol L}_T/\theta \bar c_F) = 1 + \rho \bar{\boldsymbol L}_F \cdot \bar{\boldsymbol L}_T/\theta \bar c_F = 1 + \rho \bar{\boldsymbol C}[\bar{\boldsymbol L}_T, \bar{\boldsymbol L}_T]/\theta \bar c_F = \bar c_T/\bar c_F$$

(see (1.1.6) and (10.8.6)) leads to (10.8.9). □

10.8.3 ***Remark*** If $\bar{\boldsymbol G}$ exists, then $(10.7.1)_3$ can be written as

$$\bar{\boldsymbol G} \circ \tilde{\boldsymbol C} = \mathbf 1 + \theta \bar{\boldsymbol G}[\bar{\boldsymbol N}] \otimes \bar{\boldsymbol N}/\rho \bar c_F, \tag{10.8.10}$$

where the tensors $\bar{\boldsymbol G}$, $\tilde{\boldsymbol C}$ are interpreted as linear transformations on Sym and $\bar{\boldsymbol G} \circ \tilde{\boldsymbol C}$ is their composition. Formula (10.8.10) implies that $\bar{\boldsymbol G}[\bar{\boldsymbol N}]$ is an eigenvector of $\bar{\boldsymbol G} \circ \tilde{\boldsymbol C}$ with the eigenvalue $1 + \theta(\bar{\boldsymbol G}[\bar{\boldsymbol N}] \cdot \bar{\boldsymbol N})/\rho \bar c_F = \bar c_T/\bar c_F$, of multiplicity 1, and every $\boldsymbol H \in$ Sym with $\bar{\boldsymbol N} \cdot \boldsymbol H = 0$, $\boldsymbol H \neq \mathbf 0$ is an eigenvector of $\bar{\boldsymbol G} \circ \tilde{\boldsymbol C}$ with the eigenvalue 1, of multiplicity 5. Since the determinant of a linear transformation is the product of its eigenvalues, it follows that $\det(\bar{\boldsymbol G} \circ \tilde{\boldsymbol C}) = \bar c_T/\bar c_F$. Using $\det \bar{\boldsymbol G} = 1/\det \bar{\boldsymbol C}$, we obtain (10.8.9), with now det denoting the determinant of the restrictions of the linear transformations onto Sym. The occurrence of the multiple eigenvalue 1 explains the independence of (10.8.9) of the choice of X.

Table 10.8.1 summarizes the main thermodynamic coefficients.

Table 10.8.1. Thermodynamic coefficients

coefficient	proportional to	'reciprocal object'
spatial elasticities $\bar{\boldsymbol D}$	$\partial_F^2 \bar f$	spatial compliances $\bar{\boldsymbol S} = \bar{\boldsymbol D}^{-1}$
Cauchy elasticities $\bar{\boldsymbol C}$	$\partial_F \bar{\boldsymbol T}$	Cauchy compliances $\bar{\boldsymbol G} = \bar{\boldsymbol C}^{-1}$
spatial thermal coefficient of stress $\bar{\boldsymbol N}$, latent heat with respect to deformation $\bar{\boldsymbol L}_F$	$\partial_\theta \partial_F \bar f$	thermal expansion tensor $\bar{\boldsymbol P} = -\bar{\boldsymbol G}[\bar{\boldsymbol N}]$, latent heat with respect to stress $\bar{\boldsymbol L}_T = \bar{\boldsymbol G}^{\mathrm{T}}[\bar{\boldsymbol L}_F]$
specific heat at constant deformation $\bar c_F$	$\partial_\theta^2 \bar f$	specific heat at constant stress $\bar c_T$

10.9 Approximate Equilibrium Response

Throughout the section we consider a fixed point $(\boldsymbol F, \theta) \in \bar U$ and make the expansions of the response functions about this point. The symbols

$$\bar{\boldsymbol A},\ \bar{\boldsymbol B},\ \bar{\boldsymbol C},\ \bar{\boldsymbol M},\ \bar{\boldsymbol N},\ \bar c_F \tag{10.9.1}$$

will denote the values of the indicated objects at (F, θ); the values of $\bar{f}, \bar{S}, \bar{T}, \bar{e}$, at (F, θ) will be denoted by $\bar{f}_0, \bar{S}_0, \bar{T}_0, \bar{e}_0$. We further denote, for a given $(G, \omega) \in \bar{U}$, by H the second-order tensor defined by $G = (1+H)F$ and by E and W its symmetric and the skew parts, respectively, $H = E + W$. The tensor $1 + H = GF^{-1}$ is the deformation gradient relative to the reference configuration given by F : the tensor E is then the classical ***infinitesimal deformation tensor*** of linear elasticity and W the classical ***infinitesimal rotation tensor.*** Usually, when the linearization is made, the reference configuration is chosen as to coincide with the configuration about which the linearization is made. Then $F = 1$. Denote also $\vartheta = \omega - \theta$, $J = \det F$ and assume that $J > 0$.

10.9.1 Linear Expressions To within errors of order $o(|H| + |\vartheta|)$ we have

$$\bar{S}(G, \omega) = \bar{S}_0 + \bar{A}[G - F] + \bar{M}\vartheta,$$
$$\rho_\circ \bar{\eta}(G, \omega) = \rho_\circ \bar{c}_F \vartheta/\theta - \bar{M} \cdot (G - F) = \rho_\circ \bar{c}_F \vartheta/\theta - J\bar{N} \cdot E,$$
$$\rho_\circ \bar{e}(G, \omega) = (\bar{S}_0 - \theta \bar{M}) \cdot (G - F) + \rho_\circ \bar{c}_F \vartheta,$$
$$\bar{T}(G, \omega) = \bar{T}_0 + \bar{C}[H] + \bar{N}\vartheta = \bar{T}_0 + \bar{C}[E] + W\bar{T}_0 - \bar{T}_0 W + \bar{N}\vartheta,$$

where in the expressions for $\bar{\eta}$ and $\bar{e}$ the absolute terms have been omitted. Note that the knowledge of the coefficients (10.9.1) permits one to establish for energy only an expression linear in H, ϑ. The quadratic expression would also require knowledge of some "non-canonical" second derivatives, i.e., the derivatives that do not occur in the linearized equations of balance of momentum and energy (cf. Sect. 25.1). If $F = 1$, the expression for $\bar{S}$ can also be written in the form

$$\bar{S}(G, \omega) = \bar{T}_0 + \bar{B}[H] + \bar{N}\vartheta = \bar{T}_0 + \bar{B}[E] + W\bar{T}_0 + \bar{N}\vartheta,$$

and the difference between the linearized expressions for $\bar{S}$ and $\bar{T}$ is

$$\bar{S}(G, \omega) - \bar{T}(G, \omega) = \bar{T}_0 \operatorname{tr} E - \bar{T}_0 E + \bar{T}_0 W.$$

If, additionally, $\bar{T}_0 = 0$, then the difference vanishes and the referential stress is therefore symmetric to within the above error. We see from the above expressions that the response generally is not determined by the infinitesimal deformation tensor E; the infinitesimal rotation tensor is also needed. Only if the reference state is natural is the response completely determined by E.

10.9.2 Quadratic Expressions To within the errors of order $o(|H|^2 + |\vartheta|^2)$ we have the expression

$$\rho_\circ \bar{f} = \bar{S}_0 \cdot (G - F) - \rho_\circ \bar{\eta}_0 \vartheta + \tfrac{1}{2}\bar{A}[G - F, G - F] + \bar{M} \cdot (G - F) \cdot \vartheta - \tfrac{1}{2}\rho_\circ \bar{c}_F \vartheta^2/\theta$$
$$= J\bar{T}_0 \cdot E - \rho_\circ \bar{\eta}_0 \vartheta + \tfrac{1}{2}J\bar{B}[H, H] + J\bar{N} \cdot E \cdot \vartheta - \tfrac{1}{2}\rho_\circ \bar{c}_F \vartheta^2/\theta,$$

where $\bar{f} \equiv \bar{f}(G, \omega)$ and the absolute term is omitted. If q denotes the quantity defined by $\rho_\circ q(G, \omega) = \rho_\circ \bar{e}(G, \omega) - \rho_\circ \theta \bar{\eta}(G, \omega) = \rho_\circ \bar{f}(G, \omega) + \rho_\circ \vartheta \bar{\eta}(G, \omega)$, i.e., the density of the canonical free energy, then to within the errors of order $o(|H|^2 + |\vartheta|^2)$ we have

$$\rho_\circ q(G, \omega) = \bar{S}_0 \cdot (G - F) + \tfrac{1}{2}\bar{A}[G - F, G - F] + \tfrac{1}{2}\rho_\circ \bar{c}_F \vartheta^2/\theta. \qquad (10.9.2)$$

Note that although the knowledge of $\bar{A}$, $\bar{M}$, $\bar{c}_F$ permits one to establish only the linear expressions for $\bar{e}$ and $\bar{\eta}$, for the combination $\bar{e} - \theta\bar{\eta}$ we have a quadratic expression. Moreover, (10.9.2) is block-diagonal in the sense that the mixed term $(G - F)\vartheta$ does not appear, in contrast to the expression for $\bar{f}$.

In closing this section, it should be noted that the linearized theory for which the above approximate expressions would be exact is inconsistent – it violates the Clausius–Duhem inequality. Also the insertion of the above approximate expressions into the balance equations does not result in a linear system of equations. The equation of balance of energy will be nonlinear. The correct linearization procedure is described in Sect. 25.1 and Sect. 25.6.

Bibliographical Note

The Legendre transformation is traditionally used in analytical mechanics, see, e.g., ABRAHAM & MARSDEN [1978, Sect. 3.6], who consider in detail the case when the partial change of variables (10.1.1) is a C^∞ diffeomorphism. This implies that the Hessian is nondegenerate, and no convexity conditions have to be imposed on f. With the conjugate variable only continuous, matters become more complicated; the main problem is the differentiability of the conjugate function. Notice that a mechanical application of the chain rule seemingly requires that the conjugate variable be continuously differentiable, and not only continuous, in order to obtain the differentiability of the conjugate function, i.e., *only the existence* of the conjugate variable for the conjugate function. This is actually not the case, and one can profit from the theorems on the differentiability of convex functions. There is, therefore, a group of results which guarantee that the conjugate function is continuously differentiable when the original function is (only) continuously differentiable and has some convexity properties. I refer to Propositions 16.4.6 and 16.4.7 for results of this type for the full Legendre transformation, and also to the result of ROCKAFELLAR [1970, Sect. 26] guaranteeing that the Legendre and Fenchel transformations coincide globally. A more general result on a partial Legendre transformation is in PUCCI & SERRIN [1994].

The exchange of the roles of the stress tensor and the Eshelby tensor discussed in Sect. 10.3 is due to CHADWICK [1975] and ESHELBY [1975].

The generalized strain measures were introduced by SETH [1964] and HILL [1970a]. The use of the functional calculus as in Propositions 1.2.8 and 10.4.1 is new.

Earlier systematic treatments of the elastic constants and other thermodynamic coefficients include TRUESDELL & TOUPIN [1960, Sects. 246–249], TRUESDELL & NOLL [1965, Sects. 45 & 82], CHADWICK & OGDEN [1971a,b], MCLELLAN [1980, Chaps. 13–18], HILL [1981] and OGDEN [1984, Chap. 6]. The importance of $\bar{\boldsymbol{B}}_0$ has been emphasized by MAN & CARLSON [1994]. The equations involving the tensors of elasticities are often written in the rate form, e.g., the argument $\boldsymbol{H}$ in $\bar{\boldsymbol{B}}[\boldsymbol{H}]$ is identified with the spatial gradient of velocity $\boldsymbol{H} = \dot{\boldsymbol{F}}\boldsymbol{F}^{-1}$. The transformation properties 10.5.3(2), (4) are often written using various objective strain rates, such as the Jaumann time derivative, etc. BIOT [1965] was the first to systematically use the equations of linearized elasticity in stressed states with the appropriate tensors of elasticities. However, his derivation of the equations is entirely different.

The definition in Subsect. 10.5.2 identifies the Lie algebra with the tangent space (see ALEXEEV & al. [1979, Sect. 2.3]) of the group G at $\mathbf{1}$. When G happens to be a Lie group, this definition coincides with the usual one (see BISHOP & GOLDBERG [1968]). Proposition 10.5.3(5) on the local symmetry group appears to be new.

11 The Equilibrium Response of Isotropic Bodies

There are two classes of isotropic bodies: isotropic solids and fluids. In the thermodynamics of isotropic solids, there is a new issue in comparison with the isothermal case: the natural (unstressed) state depends on temperature. This calls for a treatment in which the stressed isotropic states are given the same status as the unstressed ones, and it is given below. For instance, Lamé's moduli are defined for general (possibly stressed) isotropic states. A compatibility relation for them is derived, and a free energy function is constructed fitting given ('experimentally determined') dependencies of the Lamé moduli on temperature and specific volume.

11.1 Response Functions for Isotropic Solids

Recall that a body is an isotropic solid if in a properly chosen reference configuration its symmetry group is Orth. Let us use this reference configuration and choose the temperature as the independent thermal variable. If m stands for the internal energy, free energy, or entropy, the constitutive equations read

$$m = m(\boldsymbol{F}, \theta), \quad \boldsymbol{S} = \bar{\boldsymbol{S}}(\boldsymbol{F}, \theta), \quad \boldsymbol{T} = \bar{\boldsymbol{T}}(\boldsymbol{F}, \theta). \tag{11.1.1}$$

The response functions must be such that for every $(\boldsymbol{F}, \theta) \in \bar{U}$ and every $\boldsymbol{Q}, \boldsymbol{R} \in \mathrm{Orth}$ also $(\boldsymbol{QFR}, \theta) \in \bar{U}$ and

$$m(\boldsymbol{QFR}, \theta) = m(\boldsymbol{F}, \theta), \quad \bar{\boldsymbol{S}}(\boldsymbol{QFR}, \theta) = \boldsymbol{Q}\bar{\boldsymbol{S}}(\boldsymbol{F}, \theta)\boldsymbol{R}, \quad \bar{\boldsymbol{T}}(\boldsymbol{QFR}, \theta) = \boldsymbol{Q}\bar{\boldsymbol{T}}(\boldsymbol{F}, \theta)\boldsymbol{Q}^{\mathrm{T}}. \tag{11.1.2}$$

Hence, for every fixed θ, the functions $m(\cdot, \theta)$, $\bar{\boldsymbol{S}}(\cdot, \theta)$, $\bar{\boldsymbol{T}}(\cdot, \theta)$ are objective isotropic functions as defined in Sect. 8.4.

11.1.1 Representation in Principal Stretches Let $(\boldsymbol{F}, \theta) \in \bar{U}$. Write $\boldsymbol{F} = \boldsymbol{RU} = \boldsymbol{VR}$ for the polar decomposition of $\boldsymbol{F}$ and denote by v_1, v_2, v_3 the principal stretches of $\boldsymbol{F}$ arranged in a nonincreasing order. By (11.1.2) and the representation theorem 8.5.1, $\bar{f}(\boldsymbol{F}, \theta) = f(v_1, v_2, v_3, \theta)$, $\bar{\eta}(\boldsymbol{F}, \theta) = \eta(v_1, v_2, v_3, \theta)$, with $f(\cdot, \theta)$, $\eta(\cdot, \theta)$ symmetric functions on the domain described in Subsect. 8.4.2. By (11.1.2), $\bar{\boldsymbol{S}}(\boldsymbol{F}, \theta) = \bar{\boldsymbol{S}}(\boldsymbol{V}, \theta)\boldsymbol{R}$, $\bar{\boldsymbol{T}}(\boldsymbol{F}, \theta) = \bar{\boldsymbol{T}}(\boldsymbol{V}, \theta)$ and the restrictions of $\bar{\boldsymbol{S}}$, $\bar{\boldsymbol{T}}$ to $\bar{U} \cap (\mathrm{Sym}^+ \times \mathbb{R})$ are isotropic tensor-valued functions. By Lemma 8.1.1, $\bar{\boldsymbol{S}}(\boldsymbol{V}, \theta)$ is symmetric and commutes with $\boldsymbol{V}$ (this is not true for materials satisfying only the principle of material frame indifference). Also $\bar{\boldsymbol{T}}(\boldsymbol{V}, \theta)$ is symmetric and commutes with $\boldsymbol{V}$. Then $\boldsymbol{V}$, $\bar{\boldsymbol{S}}(\boldsymbol{V}, \theta)$, $\bar{\boldsymbol{T}}(\boldsymbol{V}, \theta)$ admit simultaneous spectral decompositions,

$$\boldsymbol{V} = \sum_{i=1}^{3} v_i \boldsymbol{e}_i \otimes \boldsymbol{e}_i, \quad \bar{\boldsymbol{S}}(\boldsymbol{V}, \theta) = \sum_{i=1}^{3} s_i \boldsymbol{e}_i \otimes \boldsymbol{e}_i, \quad \bar{\boldsymbol{T}}(\boldsymbol{V}, \theta) = \sum_{i=1}^{3} t_i \boldsymbol{e}_i \otimes \boldsymbol{e}_i,$$

where $s_i = s_i(v_1, v_2, v_3, \theta)$, $t_i = t_i(v_1, v_2, v_3, \theta)$ are called the ***principal forces*** and ***principal stresses*** respectively. The relation $\boldsymbol{S} = (\det \boldsymbol{V})\boldsymbol{T}\boldsymbol{V}^{-1}$ gives $s_i = v_1 v_2 v_3 t_i v_i^{-1}$ (no summation). The thermostatic relations read

$$s_i = \rho_{\text{o}} \partial_{v_i} f, \quad i = 1, 2, 3, \qquad \eta = -\partial_\theta f. \tag{11.1.3}$$

11.1.2 Representation in B Since $\bar{\boldsymbol{T}}(\boldsymbol{F}, \theta)$ is determined by $\boldsymbol{V}$ and θ and since there is a one-to-one correspondence between $\boldsymbol{V}$ and the left Cauchy–Green tensor $\boldsymbol{B} = \boldsymbol{V}^2 = \boldsymbol{F}\boldsymbol{F}^{\mathrm{T}}$, we have a representation $\bar{\boldsymbol{T}}(\boldsymbol{F}, \theta) = \bar{\boldsymbol{T}}_1(\boldsymbol{B}, \theta)$ with $\bar{\boldsymbol{T}}_1$ an isotropic function on $U_2 \subset \mathrm{Sym} \times \mathbb{R}$. The Rivlin–Ericksen representation theorem (see TRUESDELL & NOLL [1965, Sect. 12]) says that

$$\bar{\boldsymbol{T}}_1(\boldsymbol{B}, \theta) = \bar{\varphi}_0 \mathbf{1} + \bar{\varphi}_1 \boldsymbol{B} + \bar{\varphi}_2 \boldsymbol{B}^2,$$

where $\bar{\varphi}_k$, $k = 0, 1, 2$, are isotropic functions of $\boldsymbol{B}, \theta$. The functions $\bar{\varphi}_k$ can be expressed as functions of the principal invariants I_1, I_2, I_3 of $\boldsymbol{B}$ (and of the temperature). The relation between $\bar{\boldsymbol{S}}$ and $\bar{\boldsymbol{T}}$ gives

$$\boldsymbol{S}(\boldsymbol{F}, \theta) = \varphi_0 \boldsymbol{F}^{-\mathrm{T}} + \varphi_1 \boldsymbol{F} + \varphi_2 \boldsymbol{F}\boldsymbol{F}^{\mathrm{T}}\boldsymbol{F}, \tag{11.1.4}$$

where φ_k, $k = 0, 1, 2$, are functions of I_1, I_2, I_3, θ. Also $\bar{f}, \bar{\eta}$ can be expressed as functions of I_1, I_2, I_3, θ. The thermostatic relations take the forms

$$\begin{aligned} \bar{\boldsymbol{S}}(\boldsymbol{F}, \theta) &= 2\rho_{\text{o}}(I_3 \partial_{I_3} \bar{f} \boldsymbol{F}^{-\mathrm{T}} + (\partial_{I_1} \bar{f} + I_1 \partial_{I_2} \bar{f})\boldsymbol{F} - \partial_{I_2} \bar{f} \boldsymbol{F}\boldsymbol{F}^{\mathrm{T}}\boldsymbol{F}), \\ \bar{\boldsymbol{T}}(\boldsymbol{F}, \theta) &= 2\rho(I_3 \partial_{I_3} \bar{f} \mathbf{1} + (\partial_{I_1} \bar{f} + I_1 \partial_{I_2} \bar{f})\boldsymbol{B} - \partial_{I_2} \bar{f} \boldsymbol{B}^2), \\ \bar{\eta}(\boldsymbol{F}, \theta) &= -\partial_\theta \bar{f}(I_1, I_2, I_3, \theta). \end{aligned} \tag{11.1.5}$$

11.1.3 The Generalized Stress and Strain Measures For isotropic materials the results of Sect. 10.4 specialize as follows. The functions $\bar{g}, \bar{\boldsymbol{P}}, \bar{\bar{\eta}}$ are isotropic functions of $\boldsymbol{Q}$ and hence

$$\bar{g}(\boldsymbol{Q}, \theta) = g(q_1, \ldots, q_n, \theta), \qquad \boldsymbol{P}(\boldsymbol{Q}, \theta) = \sum_{i=1}^{n} p_i(q_1, \ldots, q_n, \theta) \boldsymbol{f}_i \otimes \boldsymbol{f}_i,$$

where $\boldsymbol{f}_i$ are the eigenvectors of $\boldsymbol{U}$ and q_i are the eigenvalues of $\boldsymbol{Q}$. By Proposition 10.4.1, $p_i(q, \theta) = s_i(v, \theta)/\phi'(v_i)$ where s_i are the principal forces. The stress relation reads $p_i = g_{,i}$. The stress conjugate to $\boldsymbol{Q} = \boldsymbol{U}$ (Example 10.4.2(1)) is $\bar{\boldsymbol{X}}(\boldsymbol{F}, \theta) = \boldsymbol{S}(\boldsymbol{U}, \theta) = \boldsymbol{R}^{\mathrm{T}}\boldsymbol{S}(\boldsymbol{F}, \theta)$. With the Seth–Hill measures of strain (Example 10.4.2(3))

$$p_i(q, \theta) = v_i^{1-2m} s_i(v, \theta), \qquad \boldsymbol{P}(\boldsymbol{Q}, \theta) = \boldsymbol{U}^{1-2m}\boldsymbol{S}(\boldsymbol{U}, \theta), \qquad (\boldsymbol{Q} = \phi(\boldsymbol{U}))$$

and in particular, for the logarithmic strain $\boldsymbol{Q} = \ln \boldsymbol{U}$, $\boldsymbol{P}(\boldsymbol{Q}, \theta) = |\det \boldsymbol{F}| \bar{\boldsymbol{T}}(\boldsymbol{U}, \theta) = |\det \boldsymbol{F}| \boldsymbol{R}^{\mathrm{T}} \bar{\boldsymbol{T}}(\boldsymbol{F}, \theta)\boldsymbol{R}$, the Kirchhoff stress evaluated at $\boldsymbol{U}$.

Representations completely analogous to those of Subsects. 11.1.1–11.1.3 apply when the entropy is used as independent variable.

We now discuss the forms of the local symmetry groups $G_{\boldsymbol{F}} = \boldsymbol{F}\,\mathrm{Orth}\,\boldsymbol{F}^{-1} \cap$ Orth of states of isotropic bodies. Recall (see 10.5.3(5)) that it is $G_{\boldsymbol{F}}$, and not G, which governs the symmetry of the tensor of elasticities and of other thermodynamic coefficients at the given state.

11.1.4 Proposition *Let* $\boldsymbol{F} = \boldsymbol{R}\boldsymbol{U} = \boldsymbol{V}\boldsymbol{R} \in$ InvLin. *Then*

$$G_{\boldsymbol{F}} = \{\boldsymbol{Q} \in \mathrm{Orth} : \boldsymbol{Q}\boldsymbol{V} = \boldsymbol{V}\boldsymbol{Q}\} \tag{11.1.6}$$

and exactly one of the following three possibilities holds:

(1) ***(isotropy)*** $G_{\boldsymbol{F}} =$ Orth; *this happens if and only if* $\boldsymbol{V} = w\mathbf{1}$ *for some* $w \in \mathbb{R}$, $w > 0$, *i.e., if the principal stretches of* $\boldsymbol{F}$ *coincide;*

(2) ***(transverse isotropy)*** $G_{\boldsymbol{F}} = \{\boldsymbol{Q} \in \mathrm{Orth} : \boldsymbol{Q}\boldsymbol{e} = \boldsymbol{e} \text{ or } \boldsymbol{Q}\boldsymbol{e} = -\boldsymbol{e}\}$ *for some unit vector* $\boldsymbol{e} \in$ Vect; *this happens if and only if* $\boldsymbol{V} = \alpha\boldsymbol{e} \otimes \boldsymbol{e} + \beta(\mathbf{1} - \boldsymbol{e} \otimes \boldsymbol{e})$ *for some* $\alpha > 0, \beta > 0, \alpha \neq \beta$, *i.e., if exactly two of the principal stretches coincide;*

(3) ***(orthotropy)*** *there exists an orthonormal basis* $\{\boldsymbol{e}_i, i = 1, 2, 3\}$ *such that*

$$G_F = \{\boldsymbol{Q} \in \mathrm{Orth} : \boldsymbol{Q}\boldsymbol{e}_i = \sigma_i \boldsymbol{e}_i \ \text{ for } i = 1, 2, 3 \text{ and some } \sigma_i = \pm 1\};$$

this happens if and only if the principal stretches of $\boldsymbol{F}$ *are mutually distinct and* $\boldsymbol{e}_i$, $i = 1, 2, 3$, *are eigenvectors of* $\boldsymbol{V}$.

Proof The formula (11.1.6) follows from Proposition 1.5.3. To prove that exactly one of (1)–(3) holds, notice that the forms of $\boldsymbol{V}$ stated in (1)–(3) are exhaustive and disjoint. Hence one has to calculate G_F through (11.1.6). Case (1) being clear, we note that in Case (2) $\boldsymbol{e}$ is a simple eigenvector of $\boldsymbol{V}$ with the eigenvalue α. Referring to (11.1.6), we note that $\boldsymbol{VQ} = \boldsymbol{QV}$ implies $\boldsymbol{VQe} = \alpha\boldsymbol{Qe}$, i.e., $\boldsymbol{Qe}$ is an eigenvector of $\boldsymbol{V}$ with the eigenvalue α and hence $\boldsymbol{Qe} = \gamma\boldsymbol{e}$ for some $\gamma \in \mathbb{R}$, $\gamma \neq 0$. By the orthogonality of $\boldsymbol{Q}$, necessarily $\gamma = \pm 1$. Also the converse is true. In Case (3), every tensor $\boldsymbol{Q}$ commuting with $\boldsymbol{V}$ is symmetric by Lemma 8.1.3(2). The orthogonality of $\boldsymbol{Q}$ then gives $\boldsymbol{Q}^2 = \mathbf{1}$ and the claimed form of G_F follows. □

To each of the Cases (1)–(3) there corresponds a specific form of the tensor of elasticities and other coefficients (of increasing complexity when going from (1) to (3)). The general form of the referential elasticities $\bar{\mathsf{A}}$ is obtained by combining Definition 10.5.1 with Proposition 8.4.5; there is no simplification of the formula of that proposition in the present context. By Proposition 10.5.4, the essential information about the spatial tensor of elasticities $\bar{\mathsf{B}}$ is contained in the reduced spatial tensor of elasticities $\bar{\mathsf{B}}_\circ$ which has the major and the two minor symmetries; its form is as follows.

11.1.5 Proposition *Let* $(\boldsymbol{F}, \theta) \in \bar{U} \cap (\mathrm{Sym}^+ \times \mathbb{R})$, *let* $\boldsymbol{F}$ *have the principal stretches* v_i *and the orthonormal basis of eigenvectors* $\{\boldsymbol{e}_i, i = 1, 2, 3\}$. *If* $\bar{\boldsymbol{T}}$ *is the Cauchy stress at* $(\boldsymbol{F}, \theta)$, *then*

(1) *the spatial tensor of elasticities* $\bar{\mathsf{B}}$ *at* $(\boldsymbol{F}, \theta)$ *is given by*

$$\bar{\mathsf{B}}[\boldsymbol{H}] = \bar{\mathsf{B}}_\circ[\boldsymbol{E}] + \boldsymbol{H}\bar{\boldsymbol{T}} \ \text{ where } \boldsymbol{E} = \tfrac{1}{2}(\boldsymbol{H} + \boldsymbol{H}^{\mathrm{T}}), \quad \boldsymbol{H} \in \mathrm{Lin},$$

and

$$\bar{\mathsf{B}}_\circ[\boldsymbol{E}] = \sum_{i,j=1}^{3} a_{ij} E_{ii} \boldsymbol{e}_j \otimes \boldsymbol{e}_j + \sum b_{ij} E_{ij} \boldsymbol{e}_i \otimes \boldsymbol{e}_j, \quad \boldsymbol{E} \in \mathrm{Sym}. \tag{11.1.7}$$

Here E_{ij} *are the components of* $\boldsymbol{E}$ *in the basis* $\{\boldsymbol{e}_i, i = 1, 2, 3\}$,

$$a_{ij} = \rho(v_i v_j f_{,ij} - v_i f_{,i} \delta_{ij}), \quad i, j = 1, 2, 3, \quad b_{ij} = 2\rho v_i v_j M_{ij}, \quad i \neq j,$$

the partial derivatives of f *are evaluated at* v, $\rho = \rho_\circ/(v_1 v_2 v_3)$ *is the actual density,* $M_{ij} = M_{ij}(v)$ *is given by* (8.4.8), (8.4.9) *with* $g_i = f_{,i}$ *and the sum with unspecified limits denotes the sum over all* $i, j = 1, 2, 3$, $i \neq j$, *as always;*

(2) *the spatial stress-temperature tensor* $\bar{\mathsf{N}}$ *and the specific heat at constant deformation* $\bar{c}_F$ *at* $(\boldsymbol{F}, \theta)$ *are given by*

$$\bar{\mathsf{N}} = \rho \sum_{i=1}^{n} v_i f_{,i\theta} \boldsymbol{e}_i \otimes \boldsymbol{e}_i, \quad \bar{c}_F = -\theta f_{,\theta\theta}.$$

These expressions simplify when three or two principal stretches coincide (isotropy or transverse isotropy). The isotropic case will be dealt with in detail in the following section. In the transversely isotropic case, i.e., when $v_1 = v_2 \neq v_3$, we have

$$a_{11} = a_{22}, \quad a_{23} = a_{13}, \quad b_{12} = b_{13}$$

as a consequence of $f_{,1} = f_{,2}, f_{,11} = f_{,22}, f_{,13} = f_{,23}$ and $\bar{\mathsf{B}}_\circ$ simplifies accordingly. The expression for $\bar{\mathsf{B}}_\circ$ can be given a coordinate-free form (SYNGE [1956b]; see also

FEDOROV [1968, Chap. 6], WALPOLE [1984], PODIO-GUIDUGLI & VIRGA [1987] and CHADWICK [1989]). However, the 5 coefficients in this expression are explicitly determined by the first and second derivatives of f at v. The details are left to the reader as an exercise.

With unspecified coefficients $a_{ij} = a_{ji} \in \mathbb{R}$, $i, j = 1, 2, 3$, $b_{ij} = b_{ji} \in \mathbb{R}$, $i \neq j$, equation (11.1.7) provides the most general form of an elasticity tensor of an orthotropic body in a natural state, i.e., of a body whose symmetry group is $G := G_F$ as in Proposition 11.1.5(3). In other words, if a fourth-order tensor $\bar{\mathsf{B}}_o$ has the major and the two minor symmetries and satisfies $\bar{\mathsf{B}}_o[\boldsymbol{Q}\boldsymbol{H}\boldsymbol{Q}^{\mathrm{T}}] = \boldsymbol{Q}\bar{\mathsf{B}}_o[\boldsymbol{H}]\boldsymbol{Q}^{\mathrm{T}}$ for every $\boldsymbol{Q} \in G_F$, $\boldsymbol{H} \in \mathrm{Lin}$, then $\bar{\mathsf{B}}_o[\boldsymbol{H}] = \bar{\mathsf{B}}_o[\boldsymbol{E}]$ and the latter is given by (11.1.7). The details of the proof are left to the reader as an exercise (see GREEN & ZERNA [1954, Sect. 5.4(b)]).

11.2 Isotropic States

When specialized to isotropic states (see Definition 9.4.4), the restrictions on the thermodynamic coefficients derived in Sects. 10.5–10.8 give a number of scalar coefficients which will now be reviewed. These coefficients are eventually determined by the free energy function, but such expressions will be given only later.

11.2.1 The Isothermal Elasticities Let us start with the Cauchy elasticity tensor. Its properties may be summarized as follows:

$$\bar{\mathsf{C}}[\boldsymbol{Q}\boldsymbol{H}\boldsymbol{Q}^{\mathrm{T}}] = \boldsymbol{Q}\bar{\mathsf{C}}[\boldsymbol{H}]\boldsymbol{Q}^{\mathrm{T}}, \quad \bar{\mathsf{C}}[\boldsymbol{H}] = \bar{\mathsf{C}}[\boldsymbol{H}]^{\mathrm{T}}, \quad \bar{\mathsf{C}}[\boldsymbol{W}] = \boldsymbol{0}, \quad \bar{\mathsf{C}}^{\mathbb{T}} = \bar{\mathsf{C}},$$

for every $\boldsymbol{H} \in \mathrm{Lin}$, $\boldsymbol{Q} \in \mathrm{Orth}$, $\boldsymbol{W} \in \mathrm{Skew}$. By Corollary 8.1.6 then

$$\bar{\mathsf{C}}[\boldsymbol{H}] = \bar{\lambda}(\mathrm{tr}\,\boldsymbol{E})\mathbf{1} + 2\bar{\mu}\boldsymbol{E} \tag{11.2.1}$$

for every $\boldsymbol{H} \in \mathrm{Lin}$, where $\boldsymbol{E}$ is the symmetric part of $\boldsymbol{H}$ and where $\bar{\lambda}$ and $\bar{\mu}$ are constants, the ***isothermal Lamé's coefficients.*** This definition applies to general isotropic states, not necessarily with vanishing stress. The spatial elasticities are given by (see Proposition 10.5.3(1))

$$\bar{\mathsf{B}}[\boldsymbol{H}] = (\bar{\lambda} - \bar{p})(\mathrm{tr}\,\boldsymbol{E})\mathbf{1} + (2\bar{\mu} + \bar{p})\boldsymbol{E} - \bar{p}\boldsymbol{W} \tag{11.2.2}$$

where $\bar{p}$ is the pressure in the isotropic state occurring in $\boldsymbol{T} = -\bar{p}\mathbf{1}$, $\boldsymbol{S} = -w^2\bar{p}\mathbf{1}$. For isotropic states the relation between the referential and the spatial elasticities takes the form $\bar{\mathsf{A}} = w\bar{\mathsf{B}}$ so that

$$\bar{\mathsf{A}}[\boldsymbol{H}] = w(\bar{\lambda} - \bar{p})(\mathrm{tr}\,\boldsymbol{E})\mathbf{1} + w(2\bar{\mu} + \bar{p})\boldsymbol{E} - \bar{p}w\boldsymbol{W}. \tag{11.2.3}$$

The ***isothermal compressibility*** is $\bar{K} := 3/(3\bar{\lambda} + 2\bar{\mu})$.

The Lamé moduli $\bar{\lambda}, \bar{\mu}$ are usually defined only for natural states ($p = 0$) and their notation is firmly established in the literature. For stressed (pressured) isotropic states, there are several inequivalent possibilities to define $\bar{\lambda}, \bar{\mu}$ in such a way that they reduce to the standard ones when $\bar{p} = 0$. Here the definition was chosen which leads to the form of the Cauchy tensor of elasticities in the stressed isotropic states formally identical to that for a natural state (but of course with different numerical values of $\bar{\lambda}, \bar{\mu}$). This convention is identical with that in MURDOCH [1978, Appendix]. Another possibility is that one insists that the reduced spatial tensor of elasticities $\bar{\mathsf{B}}_o$ (see Definition 10.5.1) in the stressed state be formally the same as that at $\bar{p} = 0$. That is, one can define, for a state $(\boldsymbol{F}, \theta) = (w\mathbf{1}, \theta)$, the quantities $\bar{\lambda}_0$, $\bar{\mu}_0$ as coefficients in

$$\bar{\mathsf{B}}_o[\boldsymbol{H}] = \bar{\lambda}_0(\mathrm{tr}\,\boldsymbol{E})\mathbf{1} + 2\bar{\mu}_0\boldsymbol{E}, \quad \boldsymbol{H} \in \mathrm{Lin}.$$

Then

$$\bar{\boldsymbol{B}}[\boldsymbol{H}] = \bar{\lambda}_0(\operatorname{tr}\boldsymbol{E})\mathbf{1} + (2\bar{\mu}_0 - \bar{p})\boldsymbol{E} - \bar{p}\boldsymbol{W}, \quad \bar{\boldsymbol{C}}[\boldsymbol{H}] = (\bar{\lambda}_0 + \bar{p})(\operatorname{tr}\boldsymbol{E})\mathbf{1} + 2(\bar{\mu}_0 - \bar{p})\boldsymbol{E},$$

i.e., $\bar{\lambda}_0 = \bar{\lambda} - \bar{p}$, $\bar{\mu}_0 = \bar{\mu} + \bar{p}$. This convention is identical with that of TRUESDELL & NOLL [1965, eq. (68.24)], due to Poincaré (1892).

Instead of the Lamé coefficients $\bar{\lambda}, \bar{\mu}$ (at $\bar{p} = 0$), the ***Young modulus*** $E \equiv \beta$ and ***Possion's ratio*** ν are often employed, especially in engineering. Moreover, $\bar{\lambda}, \bar{\mu}, E, \nu, \bar{K}$ have simple physical meanings for elementary deformations. I refer to GURTIN [1972, Sect. 22] for the definitions of E, ν, for the conversion table, the tensors of compliances and the descriptions of deformations elucidating the meaning of these coefficients when $\bar{p} = 0$.

11.2.2 Second-Order-Tensor Coefficients The spatial stress–temperature tensor $\bar{\boldsymbol{N}}$ is written in the form

$$\bar{\boldsymbol{N}} = -m\mathbf{1}, \tag{11.2.4}$$

where the scalar m is the ***thermal coefficient of stress.*** By (10.8.3) the latent heat at constant deformation $\bar{\boldsymbol{L}}_F$ is given by $\bar{\boldsymbol{L}}_F = -(\theta m/\rho)\mathbf{1}$. The thermal expansion tensor $\bar{\boldsymbol{P}}$, if it exists, is a multiple of the identity tensor satisfying $\bar{\boldsymbol{C}}[\bar{\boldsymbol{P}}] + \bar{\boldsymbol{N}} = \mathbf{0}$. The equation is uniquely solvable if and only if $3\bar{\lambda} + 2\bar{\mu} \neq 0$ and the solution is

$$\bar{\boldsymbol{P}} = \tfrac{1}{3}\alpha\mathbf{1} \quad \text{with} \quad \alpha = -3m/(3\bar{\lambda} + 2\bar{\mu});$$

the scalar α is called the ***coefficient of volumetric thermal expansion.*** The latent heat with respect to stress is a spherical tensor $\bar{\boldsymbol{L}}_T$ solving the equation $\bar{\boldsymbol{C}}^{\mathrm{T}}[\bar{\boldsymbol{L}}_T] = \bar{\boldsymbol{L}}_F$. This equation is uniquely solvable in the class of spherical tensors if and only if the equation for $\bar{\boldsymbol{P}}$ is uniquely solvable and the solution is

$$\bar{\boldsymbol{L}}_T = -(\theta m/((3\bar{\lambda} + 2\bar{\mu})\rho))\mathbf{1}.$$

By (10.8.6)

$$\bar{c}_T - \bar{c}_F = 3\theta m^2/(3\bar{\lambda} + 2\bar{\mu}) = 3(3\bar{\lambda} + 2\bar{\mu})\alpha^2\theta/\rho.$$

11.2.3 The Adiabatic Elasticities The adiabatic elasticities naturally have the same form as the isothermal ones, but with different coefficients, namely,

$$\tilde{\boldsymbol{C}}[\boldsymbol{H}] = \tilde{\lambda}(\operatorname{tr}\boldsymbol{E})\mathbf{1} + 2\tilde{\mu}\boldsymbol{E}, \quad \tilde{\boldsymbol{B}}[\boldsymbol{H}] = (\tilde{\lambda} - p)(\operatorname{tr}\boldsymbol{E})\mathbf{1} + (2\tilde{\mu} + p)\boldsymbol{E} - p\boldsymbol{W}, \tag{11.2.5}$$

and

$$\tilde{\boldsymbol{A}}[\boldsymbol{H}] = w(\tilde{\lambda} - p)(\operatorname{tr}\boldsymbol{E})\mathbf{1} + w(2\tilde{\mu} + p)\boldsymbol{E} - pw\boldsymbol{W}$$

for every $\boldsymbol{H} \in \mathrm{Lin}$, where $\tilde{\lambda}$ and $\tilde{\mu}$ are the ***adiabatic Lamé coefficients.*** Using the definition of $\bar{c}_F$, $(10.7.1)_3$ reduces to $\bar{\boldsymbol{C}} = \tilde{\boldsymbol{C}} - \theta\bar{\boldsymbol{N}} \otimes \bar{\boldsymbol{N}}/\rho\bar{c}_F$, and from $(11.2.4)_2$,(11.2.1), and (11.2.5) we infer that $\bar{\lambda} = \tilde{\lambda} - \theta m^2/(\rho\bar{c}_F)$, $\tilde{\mu} = \bar{\mu}$. The ***adiabatic compressibility*** is $\tilde{K} := 3/(3\tilde{\lambda} + 2\tilde{\mu})$. Real materials have $\bar{c}_F > 0$ and we see that $\tilde{\lambda} \geq \bar{\lambda}$. From (10.8.9) we obtain $\bar{c}_T/\bar{c}_F = \bar{K}/\tilde{K}$.

Consider the free energy $\bar{f}(\boldsymbol{F}, \theta) = f(v_1, v_2, v_3, \theta)$ as in 11.1.1. The isotropic states are represented by the arguments of f of the form (w, w, w, θ); such arguments will be called the diagonal arguments. The derivatives evaluated at the diagonal arguments have the following symmetries:

$$f_{,1} = f_{,2} = f_{,3}, \quad f_{,11} = f_{,22} = f_{,33}, \quad f_{,12} = f_{,13} = f_{,23} \tag{11.2.6}$$

and the derivatives $f_{,1}, f_{,11}$ and $f_{,12}$ will be used to denote the common values of the derivatives in (11.2.6), which seemingly displays a dissymmetry in the formulas below. The coefficients $\bar{p}, m, \bar{c}_F, \bar{\lambda}, \bar{\mu}$ of Sect. 11.2 are defined for every isotropic state

$(\boldsymbol{F},\theta) = (w\mathbf{1},\theta) \in U$; so they are functions of w, θ. It is more convenient to consider them as functions of the specific volume $v = \rho_\circ^{-1} w^3$ and θ. Let us denote by $U_\Delta \subset \mathbb{R}^{++} \times \mathbb{R}^{++}$ the set of all pairs (v,θ) such that (w,w,w,θ) is in the domain of f where $w = (\rho_\circ v)^{\frac{1}{3}}$. We write $\bar{c}_v$ for $\bar{c}_{\boldsymbol{F}}$ when the latter is associated with an isotropic state.

11.2.4 Proposition *For every $(v,\theta) \in U_\Delta$,*

$$\bar{p} = -\rho_\circ w^{-2} f_{,1}, \quad m = -\rho_\circ w^{-2} f_{,1\theta}, \quad \bar{c}_v = -\theta f_{,\theta\theta}, \tag{11.2.7}$$

$$w(\bar{\lambda} - \bar{p}) = \rho_\circ f_{,12}, \quad w(2\bar{\mu} + \bar{p}) = \rho_\circ (f_{,11} - f_{,12}), \tag{11.2.8}$$

where the derivatives on the right are evaluated at $(w,w,w,\theta) \in U_1$ with $w = (\rho_\circ v)^{\frac{1}{3}}$. Everywhere on U_Δ,

$$3\bar{\lambda} + 2\bar{\mu} = -3v\partial_v \bar{p}, \quad \partial_v \bar{c}_v = \theta \partial_\theta^2 \bar{p}. \tag{11.2.9}$$

For a given isotropic state, there are four independent scalar thermodynamic coefficients related to second derivatives of the free energy; these can be chosen to be $m, \bar{c}_v, \bar{\lambda}, \bar{\mu}$. All other coefficients are expressed through these coefficients.

Proof Assume $\rho_\circ = 1$. Equation $(11.2.7)_1$ is a consequence of the definition of the pressure for isotropic states and of the thermostatic relations $(11.1.3)_1$. Equation $(11.2.7)_2$ is a consequence of the former. Equations (11.2.8) follow from Remark 8.4.6(1). To prove $(11.2.9)_1$, write the pressure relation $(11.2.7)_1$ with the explicit arguments as $-w^2 \bar{p}(w^3,\theta) = f_{,1}(w,w,w,\theta)$. A differentiation using the symmetries of the partial derivatives at the diagonal arguments gives (now again omitting the arguments) $-2w\bar{p} - 3w^4 \partial_v \bar{p} = f_{,11} + 2f_{,12}$ which in combination with $(11.2.8)_1$ and $(11.2.8)_2$ gives $(11.2.9)_1$. To prove $(11.2.9)_2$, it suffices to differentiate $(11.2.7)_1$ and $(11.2.7)_3$ in the obvious way. □

A natural state of a solid is an isotropic state $(\boldsymbol{F},\theta)$ for which $\boldsymbol{S} = \boldsymbol{T} = \mathbf{0}$. Assume that, for a given temperature, it is unique to within a rotation and denote by $\bar{v}(\theta)$ its unique specific volume. In terms of the pressure function $\bar{p}$, the value $\bar{v}$ is a solution of the equation $\bar{p}(\bar{v}(\theta),\theta) = 0$ and a differentiation with respect to θ and the use of (11.2.7) and $(11.2.9)_1$ give the equation $\bar{v}'(\theta) = 3m\bar{v}/(3\bar{\lambda} + 2\bar{\mu})$.

The quantities $\bar{p}, \bar{c}_v, \bar{\lambda}, \bar{\mu}$ can be determined experimentally. It will be shown that there always exists an isotropic material that fits the given $\bar{p}, \bar{c}_v, \bar{\lambda}, \bar{\mu}$ provided only the compatibility relation (11.2.9) holds. (In fact it will be clear from the proof that there are infinitely many such materials. Obviously, $\bar{p}, \bar{c}_v, \bar{\lambda}, \bar{\mu}$ are insufficient to determine the response in the whole range of deformation gradients.) Let φ be a function defined on $(0,\infty)^3$ by

$$\begin{aligned}\varphi(v_1,v_2,v_3) &= \tfrac{1}{3}(v_1v_2v_3)^{\frac{1}{3}}((v_1-v_2)^2 + (v_2-v_3)^2 + (v_3-v_1)^2)\\ &= (v_1v_2v_3)^{\frac{1}{3}}(\operatorname{tr}\boldsymbol{V}^2 - \tfrac{1}{3}(\operatorname{tr}\boldsymbol{V})^2).\end{aligned}$$

On the diagonal arguments,

$$\varphi = 0, \quad \varphi_{,i} = 0, \quad i = 1,2,3, \quad \varphi_{,11} = \tfrac{4}{3}w, \quad \varphi_{,12} = -\tfrac{2}{3}w. \tag{11.2.10}$$

We shall employ this function in the proof of the following

11.2.5 Proposition *Let $U_\Delta \subset \mathbb{R}^{++} \times \mathbb{R}^{++}$ be open and simply connected and let $\bar{p}, \bar{c}_v, \bar{\lambda}, \bar{\mu}$ be functions on U_Δ satisfying* (11.2.9). *Then there exists a function $f = f(v_1,v_2,v_3,\theta)$ such that* (11.2.7) *and* (11.2.8) *hold. More specifically, there exists a function $f_0 : U_\Delta \to \mathbb{R}$ such that*

$$\bar{p} = -\partial_v f_0, \quad \bar{c}_v = -\theta \partial_\theta^2 f_0$$

and then

$$\bar{f}(\boldsymbol{F}, \theta) = f(v_1, v_2, v_3, \theta) = f_0(v, \theta) + \rho_o^{-1} \bar{\mu}(v, \theta)(\rho_o v)^{\frac{1}{3}} (\operatorname{tr} \boldsymbol{V}^2 - \tfrac{1}{3}(\operatorname{tr} \boldsymbol{V})^2)$$

where $\boldsymbol{V}$ *and* v_1, v_2, v_3 *are the left stretch tensor and the principal stretches of* $\boldsymbol{F}$ *and* $v = \rho_o^{-1} v_1 v_2 v_3 = \rho_o^{-1} \det \boldsymbol{F}$.

This means that there are no other relations among $\bar{p}$, $\bar{c}_v$, $\bar{\lambda}$, $\bar{\mu}$ beyond the compatibility relations (11.2.9).

Proof Assume $\rho_o = 1$. We shall first determine the free energy function f_0 on the isotropic states, and then add to it a free energy function which vanishes on the isotropic states, but which will lead to the prescribed Lamé moduli. Writing $(11.2.9)_2$ in the form $\partial_v(\bar{c}_v/\theta) = \partial_\theta(\partial_\theta \bar{p})$, we see that there exists a continuously differentiable function η_0 on U_Δ such that $\partial_\theta \eta_0 = \bar{c}_v/\theta, \partial_v \eta_0 = \partial_\theta \bar{p}$. Substituting the value $\bar{c}_v = \theta \partial_\theta \eta_0$ into $(11.2.9)_2$, we obtain $\partial_\theta(\partial_v \eta_0 - \partial_\theta \bar{p}) = 0$ and it follows that there exists a function $\xi : \mathbb{R}^{++} \to \mathbb{R}$ such that $\partial_v \eta_0 - \partial_\theta \bar{p} = \xi(v)$. Denote by ω the primitive function to ξ, and define $\bar{\eta} : U_\Delta \to \mathbb{R}$ by $\bar{\eta}(v, \theta) = \eta_0(v, \theta) - \omega(v)$. Then $\partial_v \bar{\eta} = \partial_\theta \bar{p}$ and hence there exists a function $f_0 : U_\Delta \to \mathbb{R}$ such that

$$\bar{p} = -\partial_v f_0, \quad \bar{\eta} = -\partial_\theta f_0. \tag{11.2.11}$$

This function is seen to satisfy

$$c_v = -\theta \partial_\theta^2 f_0. \tag{11.2.12}$$

Now let φ be the function defined above. It is claimed that the free energy $f = f(v_1, v_2, v_3, \theta)$, defined by

$$f(v_1, v_2, v_3, \theta) = f_0(v_1 v_2 v_3, \theta) + \bar{\mu}(v_1 v_2 v_3, \theta)\varphi(v_1, v_2, v_3),$$

satisfies $(11.2.7)_{1,3}$ and (11.2.8) with the prescribed values of the pressure, specific heat, and Lamé constants. By (11.2.10), we have the following relations on the diagonal arguments:

$$f_{,1} = w^2 \partial_v f_0, \quad f_{,\theta} = \partial_\theta f_0, \tag{11.2.13}$$

$$f_{,11} = v w \partial_v^2 f_0 + \tfrac{4}{3} w \bar{\mu}, \quad f_{,12} = w \partial_v f_0 + v w \partial_v^2 f_0 - \tfrac{2}{3} w \bar{\mu}. \tag{11.2.14}$$

A combination of (11.2.13) with (11.2.11) and (11.2.12) show that the right-hand sides of $(11.2.7)_1$ and $(11.2.7)_3$ give the prescribed pressure function $\bar{p}$ and c_v. From (11.2.14) it follows that

$$f_{,11} - f_{,12} = -w \partial_v f_0 + 2\bar{\mu} = w\bar{p} + 2w\bar{\mu}, \tag{11.2.15}$$

and hence the expression on the right-hand side of $(11.2.8)_2$ equals the required value $w(2\bar{\mu} + \bar{p})$. $(11.2.14)_2$ can be rewritten as

$$f_{,12} = -w\bar{p} - v w \partial_v \bar{p} - \tfrac{2}{3} w \bar{\mu}. \tag{11.2.16}$$

The derivative $\partial_v \bar{p}$ is eliminated by using the compatibility relation $(11.2.9)_1$ for the prescribed functions $\bar{p}$, $\bar{\lambda}$, $\bar{\mu}$ with the result

$$f_{,11} = -w\bar{p}_0 + \tfrac{1}{3} w \cdot (3\bar{\lambda} + 2\bar{\mu}) - \tfrac{2}{3} w \bar{\mu} = w(\bar{\lambda} - \bar{p});$$

hence in this case too the right-hand side of $(11.2.8)_1$ gives the desired values of the prescribed functions. □

11.2.6 The Anisotropy Degree of Elastic Materials The practical aspects of the linear elasticity, such as the concrete engineering problems, universal & singular solutions, wave propagation, etc., are most accessible to an explicit analysis in the case of the linear elasticity in the natural isotropic state. For anisotropic materials, or isotropic materials in anisotropic states, the problems usually do not admit explicit solutions. However, if the anisotropy is not too large, one may expect that the solutions will be close to those for an isotropic material. This leads to the idea to associate with a general anisotropic material $\bar{\boldsymbol{B}}$ an isotropic material that is closest to it. In a crude form, this can be done as follows. Since the space of all isotropic elasticity tensors in (11.2.2) (with $\bar{\lambda}$, $\bar{\mu}$, $\bar{p}$ arbitrary) is a linear subspace E of the space X of all fourth-order tensors, one can project orthogonally a general $\bar{\boldsymbol{B}}$ onto E. Here the notion of orthogonality is understood in the sense of the natural scalar product on X, defined by $\bar{\boldsymbol{B}} \cdot \bar{\boldsymbol{C}} = B_{ijkm} C_{ijkm}$ with the summation convention over repeated indices. For a given

$\bar{\boldsymbol{B}}$, the projection $\bar{\boldsymbol{B}}_E$ is given by an explicit formula, i.e., by explicit formulas for $\bar{\lambda}, \bar{\mu}, \bar{p}$ in terms of the components of $\bar{\boldsymbol{B}}$. Moreover, the distance $d(\bar{\boldsymbol{B}})$ of $\bar{\boldsymbol{B}}$ from $\bar{\boldsymbol{B}}_E$, which is also the distance of $\bar{\boldsymbol{B}}$ from E, is a measure of the degree of anisotropy of $\bar{\boldsymbol{B}}$. Thus ideally, rather than solving the given hard problem for $\bar{\boldsymbol{B}}$, one solves it for $\bar{\boldsymbol{B}}_E$ and hopes that if $d(\bar{\boldsymbol{B}})$ is small, then the so obtained solution will be close to the unknown solution for $\bar{\boldsymbol{B}}$. The idea goes back to Voigt, and for its refined form and an application to wave propagation, see FEDOROV [1968, Chap. 5]. A different measure of anisotropy has been proposed in RYCHLEWSKI & ZHANG [1989], where also further references are found.

11.3 Free Energies of Isotropic Solids

In this section we shall briefly review some free and stored energy functions frequently employed in the literature. Assume that $\rho_o = 1$.

11.3.1 Ogden Materials (OGDEN [1972a,b]) Let $\bar{f}$ be a free energy defined on InvLin $\times \mathbb{R}^{++}$ by

$$\bar{f}(\boldsymbol{F}, \theta) = \sum a_i \psi(\alpha_i) + \sum b_j \chi(\beta_j) + \Gamma(v_1 v_2 v_3, \theta)$$

where a_i, b_j are functions of θ, α_i, β_j are constants, the summation is taken over a finite number of indices i, j, Γ is a given function, and

$$\psi(\alpha) = v_1^\alpha + v_2^\alpha + v_3^\alpha = \operatorname{tr} \boldsymbol{V}^\alpha, \quad \chi(\alpha) = (v_1 v_2)^\alpha + (v_2 v_3)^\alpha + (v_3 v_1)^\alpha = \det \boldsymbol{V}^\alpha \operatorname{tr} \boldsymbol{V}^{-\alpha}$$

for every $\alpha \in \mathbb{R}$. We shall mostly restrict ourselves to the case when $\Gamma(v, \theta) = \sum c_k v^{-\gamma_k}$ where c_k are functions of θ, γ_k are constants and the sum is taken over a finite set of indices. By (8.2.2), the stresses and the entropy are

$$\bar{\boldsymbol{S}}(\boldsymbol{F}, \theta) = \sum \alpha_i a_i \boldsymbol{V}^{\alpha_i - 1} \boldsymbol{R} + \sum \beta_j b_j ((\operatorname{tr} \boldsymbol{V}^{\beta_j}) \mathbf{1} - \boldsymbol{V}^{\beta_j}) \boldsymbol{V}^{\beta_j - 1} \boldsymbol{R} - \sum \gamma_k c_k (\det \boldsymbol{F})^{-\gamma_k} \boldsymbol{F}^{-\mathrm{T}},$$

$$\bar{\boldsymbol{T}}(\boldsymbol{F}, \theta) = (\det \boldsymbol{V})^{-1} \big(\sum \alpha_i a_i \boldsymbol{V}^{\alpha_i} + \sum \beta_j b_j ((\operatorname{tr} \boldsymbol{V}^{\beta_j}) \mathbf{1} - \boldsymbol{V}^{\beta_j}) \boldsymbol{V}^{\beta_j} - \sum \gamma_k c_k (\det \boldsymbol{V})^{-\gamma_k} \mathbf{1} \big),$$

$$\bar{\eta}(\boldsymbol{F}, \theta) = -\sum \partial_\theta a_i \psi(\alpha_i) - \sum \partial_\theta b_j \chi(\beta_j) - \sum \partial_\theta c_k (v_1 v_2 v_3)^{-\gamma_k}$$

where $\boldsymbol{F} = \boldsymbol{V}\boldsymbol{R}$ is the polar decomposition of $\boldsymbol{F}$.

11.3.2 Natural Requirements There are several reasonable requirements on $\bar{f}$ that restrict the ranges of the constants and functions occurring in the expressions for $\bar{f}$ above.

(1) $\bar{f}(\boldsymbol{F}, \theta) \geq 0$ for every $(\boldsymbol{F}, \theta) \in$ InvLin $\times \mathbb{R}^{++}$. A sufficient condition is that $a_i \geq 0, b_j \geq 0, c_k \geq 0$ for all i, j, k and all $\theta > 0$.

(2) $\bar{f}(\boldsymbol{F}, \theta) \to \infty$ as $|\det \boldsymbol{F}| \to 0$. A sufficient condition is that $\gamma_k \geq 0$ for every $k, \theta > 0$ and at least one k satisfies $c_k > 0, \gamma_k > 0$ for all $\theta > 0$.

(3) $\bar{f}(\boldsymbol{F}, \theta) \to \infty$ as $|\boldsymbol{F}| \to \infty$. A sufficient condition is that $a_i \geq 0, b_j \geq 0, c_k \geq 0$ for all i, j, k and all $\theta > 0$ and at least one i satisfies $\alpha_i > 0, a_i > 0$ for all $\theta > 0$.

(4) For every $\theta > 0$, the function $\bar{f}(\cdot, \theta)$ is polyconvex (see Subsect. 17.5 for a definition). This requirement guarantees that a good existence theory is available for the material (see Chap. 21). A sufficient condition is that $a_i \geq 0, b_j \geq 0, c_k \geq 0$ for all i, j, k and all $\theta > 0$, that $\alpha_i \geq 1, \beta_j \geq 1$, for all i, j, and for every k either $\gamma_k \geq 0$ or $\gamma_k \leq -1$.

(5) $\eta(\boldsymbol{F}, \theta) \geq 0$ for every $(\boldsymbol{F}, \theta) \in$ InvLin $\times \mathbb{R}^{++}$. A sufficient condition is $\partial_\theta a_i \leq 0, \partial_\theta b_j \leq 0, \partial_\theta c_k \leq 0$ for all i, j, k and all $\theta > 0$.

We now give some special cases. In these, we restrict ourselves to the isothermal situation at temperature θ, write $\bar{f} = \bar{f}(\cdot, \theta)$, $\bar{\boldsymbol{S}} = \bar{\boldsymbol{S}}(\cdot, \theta)$ with $\bar{f}, \bar{\boldsymbol{S}}$ defined on InvLin. Clearly, if $\bar{f}$ is a given stored energy, then for the model purposes the stored energy $a\bar{f}$, where $a > 0$ is a constant, leads to the same properties. Hence, wherever possible, we simplify the notation by choosing the value $a > 0$ suitably. Let $\Delta := |\det \boldsymbol{F}| = v_1 v_2 v_3$, $\boldsymbol{F} \in$ InvLin.

11.3.3 Special Cases

(1) ***Hadamard–Green material***

$$\bar{f}(\boldsymbol{F}) = \tfrac{1}{2}(v_1^2 + v_2^2 + v_3^2) + \tfrac{1}{2} b((v_1 v_2)^2 + (v_2 v_3)^2 + (v_3 v_1)^2) + h(\Delta),$$

$$\bar{f}(\boldsymbol{F}) = \tfrac{1}{2}|\boldsymbol{F}|^2 + \tfrac{1}{2} b |\operatorname{cof} \boldsymbol{F}|^2 + h(\Delta),$$

$$\bar{\boldsymbol{S}}(\boldsymbol{F}) = \boldsymbol{F} + b(|\operatorname{cof} \boldsymbol{F}|^2 \boldsymbol{F}^{-\mathrm{T}} - \operatorname{cof} \boldsymbol{F}\, \boldsymbol{F}^{-1} \operatorname{cof} \boldsymbol{F}) + \Delta h'(\Delta) \boldsymbol{F}^{-\mathrm{T}},$$

where b is a constant and h a twice continuously differentiable scalar-valued function on $(0, \infty)$. The "entropy" version of this material, i.e., the material in which the *internal* energy at fixed entropy is given by the expression above, leads to particularly simple properties of waves, see Sect. 23.4.

(2) *Hadamard material*

$$\bar{f}(\boldsymbol{F}) = \tfrac{1}{2}\boldsymbol{F}\cdot\boldsymbol{F} + h(\Delta), \quad \bar{\boldsymbol{S}}(\boldsymbol{F}) = \boldsymbol{F} + \Delta h'(\Delta)\boldsymbol{F}^{-\mathrm{T}},$$

where h is a twice continuously differentiable scalar-valued function on $(0, \infty)$. If $h'(1) = -1$, then $\bar{\boldsymbol{S}}(\boldsymbol{1}) = \boldsymbol{0}$. The referential tensor of elastic moduli at $\boldsymbol{F}$ is calculated using (1.1.8) with the result

$$\bar{\boldsymbol{A}}[\boldsymbol{H}] = \boldsymbol{H} + (\Delta h'(\Delta))'\Delta\,\mathrm{tr}(\boldsymbol{F}^{-1}\boldsymbol{H})\boldsymbol{F}^{-\mathrm{T}} - \Delta h'(\Delta)\boldsymbol{F}^{-\mathrm{T}}\boldsymbol{H}^{\mathrm{T}}\boldsymbol{F}^{-\mathrm{T}}, \quad \boldsymbol{H} \in \mathrm{Lin};$$

equivalently

$$\bar{\boldsymbol{A}}[\boldsymbol{H},\boldsymbol{H}] = |\boldsymbol{H}|^2 + (\Delta h'(\Delta))'\Delta\left[\mathrm{tr}(\boldsymbol{F}^{-1}\boldsymbol{H})\right]^2 - \Delta h'(\Delta)\,\mathrm{tr}\left[(\boldsymbol{F}^{-1}\boldsymbol{H})^2\right], \quad \boldsymbol{H} \in \mathrm{Lin}.$$

(3) *Blatz–Ko materials* (BLATZ & KO [1962])

$$\bar{f}(\boldsymbol{F}) = \delta(\tfrac{1}{2}\,\mathrm{tr}\,\boldsymbol{V}^2 + \Delta^{-\gamma}/\gamma) + (1-\delta)(\tfrac{1}{2}\,\mathrm{tr}\,\boldsymbol{V}^{-2} + \Delta^{\gamma}/\gamma)$$

where $\boldsymbol{V}$ is the left stretch tensor of $\boldsymbol{F}$ and $0 \le \delta \le 1$, $\gamma > 0$ are constants. Then

$$\bar{\boldsymbol{S}}(\boldsymbol{F}) = \delta(\boldsymbol{F} - \Delta^{-\gamma}\boldsymbol{F}^{-\mathrm{T}}) + (\delta - 1)(\boldsymbol{F}^{-\mathrm{T}}\boldsymbol{F}^{-1}\boldsymbol{F}^{-\mathrm{T}} - \Delta^{\gamma}\boldsymbol{F}^{-\mathrm{T}}),$$

$$\bar{\boldsymbol{T}}(\boldsymbol{F}) = \delta(\boldsymbol{B} - \Delta^{-\gamma}\boldsymbol{1})/\Delta + (\delta - 1)(\boldsymbol{B}^{-1} - \Delta^{\gamma}\boldsymbol{1})/\Delta,$$

see Subsect. 11.3.1. Hence $\bar{\boldsymbol{S}}(\boldsymbol{1}) = \boldsymbol{0}$. The special cases with $\delta = 1$ and $\delta = 0$, which will be referred to as the ***Blatz–Ko special I*** and ***Blatz–Ko special II,*** respectively, were shown by BLATZ & KO [1962] to fit their experimental data on solid polyurethane rubbers and foamed polyurethane elastomers, respectively. Later we shall see that Blatz–Ko special I is a polyconvex free energy (see Sect. 17.5) whereas Blatz–Ko special II was shown by KNOWLES & STERNBERG [1975] to violate the strong ellipticity condition (see Sect. 17.3) at every $\boldsymbol{F}$ for which the ratios of the principal stretches are either too small or too large.

(4) We shall consider a slight generalization of the Blatz–Ko special I as follows:

$$\bar{f}(\boldsymbol{F}) = \mathrm{tr}\,\boldsymbol{V}^{\alpha}/\alpha + \Delta^{-\gamma}/\gamma, \quad \bar{\boldsymbol{S}}(\boldsymbol{F}) = \boldsymbol{V}^{\alpha-1}\boldsymbol{R} - \Delta^{-\gamma}\boldsymbol{F}^{-\mathrm{T}} \tag{11.3.1}$$

where $\alpha \ge 1$, $\gamma > 0$ are constants. Let us also denote by $f : (\mathbb{R}^{++})^3 \to \mathbb{R}$ the function

$$f(v_1, v_2, v_3) = (v_1^{\alpha} + v_2^{\alpha} + v_3^{\alpha})/\alpha + (v_1 v_2 v_3)^{-\gamma}/\gamma \tag{11.3.2}$$

so that

$$\bar{f}(\boldsymbol{F}) = f(v_1, v_2, v_3). \tag{11.3.3}$$

11.4 Response Functions of Fluids

For fluids the symmetry group is G = Unim in any reference configuration and we have

$$m(\boldsymbol{QFH}, \theta) = m(\boldsymbol{F}, \theta), \quad \bar{\boldsymbol{S}}(\boldsymbol{QFH}, \theta) = \boldsymbol{Q}\bar{\boldsymbol{S}}(\boldsymbol{F}, \theta)\boldsymbol{H}^{-\mathrm{T}}, \quad \bar{\boldsymbol{T}}(\boldsymbol{QFH}, \theta) = \boldsymbol{Q}\bar{\boldsymbol{T}}(\boldsymbol{F}, \theta)\boldsymbol{Q}^{\mathrm{T}}$$

for every $(\boldsymbol{F}, \theta) \in \bar{U}$, every $\boldsymbol{Q} \in$ Orth, and $\boldsymbol{H} \in$ Unim, where m stands for $\bar{e}, \bar{f}$ and $\bar{\eta}$. These equations lead to the following straightforward result.

11.4.1 The Representation Theorem for Fluids *The functions $\bar{f}, \bar{\boldsymbol{S}}, \bar{\boldsymbol{T}}$ are expressible as follows:*

$$\bar{f}(\boldsymbol{F}, \theta) = \bar{f}(v, \theta), \quad \boldsymbol{S}(\boldsymbol{F}, \theta) = -\bar{p}(v, \theta)|J|\boldsymbol{F}^{-\mathrm{T}}, \quad \boldsymbol{T}(\boldsymbol{F}, \theta) = -\bar{p}(v, \theta)\boldsymbol{1}, \tag{11.4.1}$$

where $v = |J|/\rho_{\circ} = 1/\rho$ is the specific volume and $\bar{f}, \bar{p}$ on the right-hand sides of (11.4.1) are scalar-valued functions of (v, θ) defined on $\bar{V} \subset \mathbb{R}^2$ specified below. The thermodynamic relations take the forms

$$\bar{p} = -\partial_v \bar{f}, \quad \bar{\eta} = -\partial_{\theta}\bar{f}.$$

Alternatively,

$$\tilde{e}(\boldsymbol{F}, \eta) = \tilde{e}(v, \eta), \quad \tilde{\theta}(\boldsymbol{F}, \eta) = \tilde{\theta}(v, \eta), \quad \tilde{\boldsymbol{T}}(\boldsymbol{F}, \eta) = -\tilde{p}(v, \eta)\mathbf{1}, \tag{11.4.2}$$

where the functions on the right-hand side of these equations are defined on $\tilde{V} \subset \mathbb{R}^2$ *specified below.*

The form of the thermostatic relations in entropy is given in the following subsection. The relation of the domains $\bar{U}$, $\tilde{U}$ of the functions on the left-hand sides of the above equations to the domains $\bar{V}$, $\tilde{V}$ of the functions on the right-hand sides is as follows:

$$\tilde{V} = \{(v, \eta) \in \mathbb{R}^{++} \times \mathbb{R} : v = |\det \boldsymbol{F}|/\rho_\circ \ \text{ for some } \ (\boldsymbol{F}, \eta) \in \tilde{U}\}, \tag{11.4.3}$$

and conversely

$$\tilde{U} = \{(\boldsymbol{F}, \eta) \in \text{Lin} \times \mathbb{R} : (|\det \boldsymbol{F}|/\rho_\circ, \eta) \in \tilde{V}\}; \tag{11.4.4}$$

similar relations hold for $\bar{U}$ and $\bar{V}$. For the proof of the thermostatic relations it is enough to invoke (1.1.8).

11.4.2 Thermodynamic Potentials In many situations, although not always, the pressure function $\bar{p}(v, \theta)$ or the pressure function $\tilde{p}(v, \eta)$ is invertible in v. Then additional thermodynamic potentials can be introduced by performing partial Legendre transformations, called the Gibbs free energy and the enthalpy. We here briefly summarize the thermostatic relations in terms of the available potentials in a symmetrical way. For the purpose of the discussion we assume the universal invertibility allowing the passage from any pair of variables v, θ, η, p to any other, at least locally. We also use the same letter for a given quantity expressed as a function of any pair of variables, and indicate the variables by $e = e(v, \eta)$ etc.:
the ***energy***

$$e = e(v, \eta), \quad p = -\partial_v e, \ \theta = \partial_\eta e;$$

the ***free energy*** or the ***Helmholtz function***

$$f = f(v, \theta) = e - \theta\eta, \quad p = -\partial_v f, \ \eta = -\partial_\theta f;$$

the ***Gibbs free energy*** or the ***chemical potential*** or the ***Gibbs function***

$$g = g(p, \theta) = e - \theta\eta + pv, \quad v = \partial_p g, \ \eta = -\partial_\theta g;$$

the ***enthalpy***

$$h = h(p, \eta) = e + pv, \quad v = \partial_p h, \ \theta = \partial_\eta h.$$

Each of these four potentials is related to any other by a partial or full Legendre transformation (to within the signs). The natural variables and the relations can be read off the following:

$$de = -pdv + \theta d\eta, \ \ df = -pdv - \eta d\theta, \ \ dg = vdp - \eta d\theta, \ \ dh = vdp + \theta d\eta.$$

Each of the potentials is useful in different situations. Thus for instance the equality of the Gibbs potentials is related to the phase equilibrium, the enthalpy is useful in considerations in fluid dynamics etc. This will be clarified in the subsequent treatment.

For fluids, every state is isotropic and hence all of the tensorial thermodynamic coefficients are necessarily isotropic tensors. Using considerations similar to those for the isotropic solids we arrive at the following relations.

11.4.3 Elasticities The referential, spatial and the Cauchy tensors of isothermal elasticities are given by

$$\bar{\mathsf{A}}[\boldsymbol{H}] = -\rho_{\circ} v \partial_v(\bar{p}v)\, \mathrm{tr}(\boldsymbol{H}\boldsymbol{F}^{-1})\boldsymbol{F}^{-\mathrm{T}} + \rho_{\circ} v \bar{p} \boldsymbol{F}^{-\mathrm{T}}\boldsymbol{H}^{\mathrm{T}}\boldsymbol{F}^{-\mathrm{T}},$$

and

$$\bar{\mathsf{B}}[\boldsymbol{H}] = -\partial_v(\bar{p}v)(\mathrm{tr}\,\boldsymbol{H})\mathbf{1} + \bar{p}\boldsymbol{H}^{\mathrm{T}}, \quad \bar{\mathsf{C}}[\boldsymbol{H}] = -v\partial_v\bar{p}(\mathrm{tr}\,\boldsymbol{H})\mathbf{1}$$

for every $\boldsymbol{H} \in \mathrm{Lin}$. This is obtained from the definition 10.5.1 by a calculation using $(1.1.10)_3$. Observe that the above elasticities are special cases of those in Subsect. 11.2.1 with the shear modulus $\mu = 0$. The adiabatic elasticities $\tilde{\mathsf{A}}, \tilde{\mathsf{B}}, \tilde{\mathsf{C}}$ are given by the expressions above simply with $\bar{p}$ replaced by $\tilde{p}$.

11.4.4 Second-Order-Tensor Coefficients All of these are spherical tensors, i.e., the multiples, by scalar coefficients, of the identity tensor. The scalar coefficients are given names that will now be introduced. The thermal coefficient of stress is

$$\bar{\mathsf{N}} = -\partial_\theta \bar{p}\mathbf{1},$$

and the scalar $\beta = \partial_\theta \bar{p}$ is called the ***thermal coefficient of pressure.*** The thermal expansion tensor is defined if and only if $\partial_v \bar{p} \neq 0$ and is given by

$$\bar{\mathsf{P}} = -\tfrac{1}{3}\alpha\mathbf{1} \ \text{ where } \ \alpha = \partial_\theta\bar{p}/(v\partial_v\bar{p});$$

α is the coefficient of ***volumetric thermal expansion.*** The specific heat $\bar{c}_{\boldsymbol{F}}$ at constant deformation is called the ***specific heat at constant volume*** and is denoted by c_v. The latent heat $\bar{\mathsf{L}}_{\boldsymbol{F}}$ with respect to deformation is a tensor given by

$$\bar{\mathsf{L}}_{\boldsymbol{F}} = v\theta\partial_\theta\bar{p}\mathbf{1};$$

the scalar coefficient $l_v = \theta\partial_\theta\bar{p}$ is the ***latent heat with respect to volume.*** Equation (10.8.1) reads $r = c_v\dot{\theta} + l_v\dot{v}$. The latent heat with respect to stress is defined if and only if $\partial_v\bar{p} \neq 0$ and is given by

$$\bar{\mathsf{L}}_{\boldsymbol{T}} = -\tfrac{1}{3}l_p\mathbf{1} \ \text{ where } \ l_p = \theta\partial_\theta\bar{p}/\partial_v\bar{p};$$

The scalar l_p is the ***latent heat with respect to pressure.*** If it is defined, then so is the specific heat $\bar{c}_{\boldsymbol{T}}$ at constant stress; in the context of fluids it is called the ***specific heat at constant pressure*** and is denoted by c_p. Equation (10.8.7) reads $r = c_p\dot{\theta} + l_p\dot{p}$. Relation (10.8.6) gives

$$c_p - c_v = -\theta(\partial_\theta\bar{p})^2/\partial_v\bar{p}.$$

The coefficients $\bar{K} = -1/\partial_v\bar{p}$ and $\tilde{K} = -1/v\partial_v\tilde{p}$ are the isothermal compressibility and adiabatic compressibilities, respectively. Relation (10.8.9) gives

$$c_p/c_v = \bar{K}/\tilde{K} = \partial_v\tilde{p}/\partial_v\bar{p},$$

and relations (10.7.1) reduce to

$$\partial_v\tilde{p} = \partial_v\bar{p} - \theta(\partial_\theta\bar{p})^2/c_v. \qquad (11.4.5)$$

At the given value (v, θ) there are exactly three independent coefficients; the others can be expressed through them.

The experimental data usually involve information about the *state equation* $p = \bar{p}(v, \theta)$ and the specific heat $c_v = c_v(v, \theta)$. The following proposition, a direct specialization of Proposition 11.2.5, pertains to this.

11.4.5 Proposition *Let $\bar{p}$, c_v be functions on $\bar{V}$ satisfying*

$$\partial_v \bar{c}_v = \theta \partial_\theta^2 \bar{p}$$

and assume that $\bar{V}$ is simply connected. Then there exists a function $f = f(v, \theta)$ such that

$$\bar{p} = -\partial_v f, \quad c_v = -\theta \partial_\theta^2 f.$$

Bibliographical Note

The stored energy functions Subsect. 11.3.1 were further examined by BALL [1977] from the standpoint of the existence theory. For additional references and the discussion of the correlation with the experimental data, see TRELOAR [1975], OGDEN [1979, 1984 1986, 1987, 1992b], and BEATTY [1987, Sects. 8–11]. For further stored energy functions, see also BURGESS & LEVINSON [1972] and CIARLET [1988], with additional references. An interesting characterization of the Blatz–Ko materials was given by BEATTY & STALNAKER [1986].

VUKALOVICH & NOVIKOV [1948] (in Russian) list 150 analytical state equations proposed in the period 1664–1944 (with references). For a review of both analytical and empirical state equations, see KOLASIŃSKA [1986]. Historical remarks are found in TRUESDELL & BHARATHA [1977] and PARTINGTON [1949, Sect. 45]. RUMER & RYVKIN [1977, Sect. 11] define thermodynamic coefficients for fluids by $(\partial\lambda/\partial\mu)_\nu$ where λ, μ, ν stand for any of the quantities p, v, θ, η. With this definition, there are 12 thermodynamic coefficients, of which only 3 are independent, while others are expressible in terms of them by explicit rules.

12 The Dynamic Response

For the nonequilibrium response the second law takes the form of the internal dissipation inequality (9.2.4) and the combination of the frame indifference with the symmetry leads to explicit representations of the nonequilibrium response functions. As these depend on more vectorial and tensorial variables (namely on $\boldsymbol{F}, \theta, \boldsymbol{D}, \boldsymbol{g}$ or on $\boldsymbol{F}, \theta, \dot{\boldsymbol{F}}, \boldsymbol{G}$), the representation theorems are necessarily more complicated. Since there are essentially no new ideas in them, and since some parts of the development are closely parallel to those concerning the equilibrium response, the material will be presented only briefly. The rest of the chapter deals with the nonequilibrium response from a broader point of view. First, the classical linear irreversible thermodynamics, based on Onsager's relations, will be briefly described in view of its obvious historical importance. Then a possible nonlinear generalization of the Onsager relations, namely, the assumption of the existence of the convex dissipation potential, will be presented. Finally, the relaxation models including those studied by the extended linear irreversible thermodynamics will be examined from the point of view of the thermodynamics based on the Clausius–Duhem inequality.

Throughout the chapter it is assumed that the residual dissipation inequality and the principle of material frame indifference hold and that the Cauchy stress is symmetric.

12.1 Linearization, Kinetic Coefficients

We shall start form the response functions

$$\boldsymbol{T}_d = \bar{\boldsymbol{T}}_d(\boldsymbol{F}, \theta, \boldsymbol{L}, \boldsymbol{g}) = \bar{\boldsymbol{T}}_d(\boldsymbol{F}, \theta, \boldsymbol{D}, \boldsymbol{g}), \quad \boldsymbol{h} = \bar{\boldsymbol{h}}(\boldsymbol{F}, \theta, \boldsymbol{L}, \boldsymbol{g}) = \bar{\boldsymbol{h}}(\boldsymbol{F}, \theta, \boldsymbol{D}, \boldsymbol{g}) \tag{12.1.1}$$

and recall, as was shown in Sect. 9.3, that the nonequilibrium response depends on the spatial gradient of velocity $\boldsymbol{L}$ only through its symmetric part $\boldsymbol{D}$. The response functions can therefore be equally considered as functions on $\bar{U} \times \mathrm{Lin} \times \mathrm{Vect}$ or as functions on $\bar{U} \times \mathrm{Sym} \times \mathrm{Vect}$, where $\bar{U}$ is the domain of all possible values of $(\boldsymbol{F}, \theta)$.

12.1.1 Kinetic Coefficients *To within errors of order $o(|\boldsymbol{D}|+|\boldsymbol{g}|)$ the nonequilibrium response functions at $(\boldsymbol{F}, \theta)$ have the forms*

$$\bar{\boldsymbol{T}}_d(\boldsymbol{F}, \theta, \boldsymbol{D}, \boldsymbol{g}) = \mathsf{V}(\boldsymbol{F}, \theta)[\boldsymbol{D}] + \mathsf{D}(\boldsymbol{F}, \theta)[\boldsymbol{g}], \tag{12.1.2}$$

$$\bar{\boldsymbol{h}}(\boldsymbol{F}, \theta, \boldsymbol{D}, \boldsymbol{g}) = -\boldsymbol{K}(\boldsymbol{F}, \theta)[\boldsymbol{g}] - \mathsf{E}(\boldsymbol{F}, \theta)[\boldsymbol{D}], \tag{12.1.3}$$

where the values of V, D, E, *and* $\boldsymbol{K}$ *at* $(\boldsymbol{F}, \theta)$ *are*

$$\mathsf{V} = \partial_D \bar{\boldsymbol{T}}_d, \quad \mathsf{D} = \partial_g \bar{\boldsymbol{T}}_d, \quad \mathsf{E} = -\partial_D \bar{\boldsymbol{h}}, \quad \boldsymbol{K} = -\partial_g \bar{\boldsymbol{h}},$$

and the derivatives are evaluated at $(\boldsymbol{F}, \theta, \boldsymbol{0}, \boldsymbol{0})$.

This is the Taylor expansion using $\bar{\boldsymbol{T}}_d(\boldsymbol{F}, \theta, \boldsymbol{0}, \boldsymbol{0}) = \boldsymbol{0}$ and $\bar{\boldsymbol{h}}(\boldsymbol{F}, \theta, \boldsymbol{0}, \boldsymbol{0}) = \boldsymbol{0}$. The fourth-order tensor V is the ***tensor of spatial viscosities.*** The coefficient D describes the cross effect – the influence of the gradient of temperature on the value of the nonequilibrium stress. The second-order tensor K is the ***spatial thermal conductivity.*** The third-order tensor E again describes the cross effect – the influence of the gradient of velocity on the heat flux. Collectively, $\mathsf{V}, \mathsf{D}, \mathsf{K}, \mathsf{E}$ are the ***spatial kinetic coefficients.*** It is convenient to arrange them in the matrix

$$\mathsf{Q} = \begin{bmatrix} \mathsf{V} & \mathsf{D} \\ \frac{1}{\theta}\mathsf{E} & \frac{1}{\theta}\mathsf{K} \end{bmatrix}, \tag{12.1.4}$$

representing a linear transformation on V := Sym ⊕ Vect.

12.1.2 Properties of Kinetic Coefficients *The kinetic coefficients at a given point* $(\boldsymbol{F}, \theta)$ *satisfy*

(1) *the symmetry relations*

$$\mathsf{V}[\boldsymbol{L}] = \mathsf{V}[\boldsymbol{L}^{\mathsf{T}}] = \mathsf{V}[\boldsymbol{L}]^{\mathsf{T}}, \quad \mathsf{D}[\boldsymbol{g}] = \mathsf{D}[\boldsymbol{g}]^{\mathsf{T}}, \quad \mathsf{E}[\boldsymbol{L}] = \mathsf{E}[\boldsymbol{L}^{\mathsf{T}}]$$

for every $\boldsymbol{L} \in \mathrm{Lin}$, $\boldsymbol{g} \in \mathrm{Vect}$;

(2) *the linearized entropy production inequality*

$$\mathsf{V}[\boldsymbol{D}] \cdot \boldsymbol{D} + \mathsf{D}[\boldsymbol{g}] \cdot \boldsymbol{D} + \frac{1}{\theta}\mathsf{K}[\boldsymbol{g}] \cdot \boldsymbol{g} + \frac{1}{\theta}\mathsf{E}[\boldsymbol{D}] \cdot \boldsymbol{g} \geq 0 \tag{12.1.5}$$

for every $\boldsymbol{D} \in \mathrm{Sym}$, $\boldsymbol{g} \in \mathrm{Vect}$ (*i.e., the matrix* Q *is positive semidefinite*);

(3) *the material symmetry relations*

$$\mathsf{V}[\boldsymbol{Q}\boldsymbol{D}\boldsymbol{Q}^{\mathsf{T}}] = \boldsymbol{Q}\mathsf{V}[\boldsymbol{D}]\boldsymbol{Q}^{\mathsf{T}}, \quad \mathsf{D}[\boldsymbol{Q}\boldsymbol{g}] = \boldsymbol{Q}\mathsf{D}\boldsymbol{Q}^{\mathsf{T}},$$
$$\mathsf{E}[\boldsymbol{Q}\boldsymbol{D}\boldsymbol{Q}^{\mathsf{T}}] = \boldsymbol{Q}\mathsf{E}[\boldsymbol{D}], \quad \mathsf{K}[\boldsymbol{Q}\boldsymbol{g}] = \boldsymbol{Q}\mathsf{K}[\boldsymbol{g}]$$

for every $\boldsymbol{D} \in \mathrm{Sym}$, $\boldsymbol{g} \in \mathrm{Vect}$ *and every* $\boldsymbol{Q} \in \mathsf{G}_{\boldsymbol{F}} \equiv \boldsymbol{F}\mathsf{G}\boldsymbol{F}^{-1} \cap \mathrm{Orth}$ *where* G *is the symmetry group of the material.*

Relations (1), (3) lead to explicit forms of the kinetic coefficients for states of specified symmetry $\mathsf{G}_{\boldsymbol{F}}$, very similar to those of the tensor of elasticities. In particular, for isotropic states, i.e., when $\mathsf{G}_{\boldsymbol{F}}$ = Orth, the cross coefficients D, E vanish and V, K have the form given by the right-hand sides of (12.1.6) below. Likewise, one can derive the forms of the kinetic coefficients for the crystal classes, etc.

Proof (1) follows from the fact that $\mathsf{V}[\boldsymbol{L}]$ and $\mathsf{E}[\boldsymbol{L}]$ depend on $\boldsymbol{L}$ only through its symmetric part $\boldsymbol{D}$ and from the symmetry of the Cauchy stress. (2) is obtained by inserting the approximate expressions (12.1.2), (12.1.3) with the remainders into the dissipation inequality. (3) is completely analogous to Subsect. 10.5.3(5).

The properties 12.1.2 deal with the kinetic coefficients at a single point $(\boldsymbol{F}, \theta)$. To obtain the global forms of the dynamic response functions for a material of a specified symmetry, one has to invoke the representation theorems appropriate for the given symmetry. But since the dynamic response depends on more tensor variables, the representation theorems are complicated, even in the isotropic case. Some simplification is reached by passing to the linear materials.

12.1.3 Linear Materials A viscous material with heat conduction is said to be linear if T_d and h depend linearly on g and D, i.e., if (12.1.2) and (12.1.3) are the exact. (Equivalently, S_d, q depend linearly on G, $\dot{F}$.) Note that unlike the case of the equilibrium response, where the linear dependence of the stress on the deformation gradient is inconsistent with the frame indifference, the linear dynamic response (in D, g) is in a perfect agreement with the frame indifference. The linear fluids with heat conduction and viscosity are often called the Navier–Stokes–Fourier fluids. For general viscous fluids with heat conduction the material symmetry relations read

$$\bar{T}_d(FH, \theta, g, D) = \bar{T}_d(F, \theta, g, D), \quad \bar{h}(FH, \theta, g, D) = \bar{h}(F, \theta, g, D)$$

for every $(F, \theta) \in \bar{U}$, $D \in$ Sym, $g \in$ Vect, and $H \in$ Unim. Combining these with the frame indifference (Proposition 9.3.4(2)), we obtain the familiar form of the response of Navier–Stokes–Fourier fluids:

$$\bar{T}_d(F, \theta, g, D) = \lambda(\operatorname{tr} D)\mathbf{1} + 2\mu D, \quad \bar{h}(F, \theta, g, D) = -\kappa g \tag{12.1.6}$$

where λ, μ, κ are scalar functions of v and θ satisfying

$$3\lambda + 2\mu \geq 0, \quad \mu \geq 0, \quad \kappa \geq 0. \tag{12.1.7}$$

The functions λ, μ are the coefficients of viscosity and κ is the coefficient of thermal conductivity. Inequalities $(12.1.7)_{1,2}$ are equivalent to the assertion that the power of viscous forces is nonnegative and $(12.1.7)_3$ is the Fourier inequality.

12.2 Linear Irreversible Thermodynamics. Onsager's Relations

Onsager's relations form an independent postulate which does not follow from the second law. They were originally motivated by Onsager on microscopic grounds, and they are largely confirmed experimentally.

12.2.1 Formulation For viscous bodies with heat conduction Onsager's relations postulate that for every $(F, \theta) \in \bar{U}$ the matrix Q (see (12.1.4)) is symmetric. That is,

$$V^{\mathbb{T}} = V, \quad \frac{1}{\theta} E^{\mathbb{T}} = D, \quad K^{\mathrm{T}} = K,$$

where the superscript $\mathbb{T}$ denotes the major transpose, see Subsect. 1.6.1. In indices,

$$V_{ijkl} = V_{klij}, \quad \frac{1}{\theta} E_{ijk} = D_{jki}, \quad K_{ij} = K_{ji}.$$

For general states of fluids and for isotropic states of solids the Onsager relations are automatically satisfied as a consequence of symmetry alone. WANG [1984, Appendix 7A] discusses the symmetry of the thermal conductivity tensor for various classes of symmetry.

The linear irreversible thermodynamics, which we now briefly describe, uses Onsager's relations as the main ingredient. Apart from these, the linear irreversible thermodynamicsis based on the so-called postulate of local thermodynamic equilibrium and on the concepts of thermodynamic forces and fluxes.

12.2.2 Local Thermodynamic Equilibrium Linear irreversible thermodynamics takes for granted the concepts of entropy and absolute temperature from classical thermodynamics. The postulate of local thermodynamic equilibrium is the assumption that the internal energy, the entropy, and the free energy are functions of (F, θ) only and that they satisfy the Gibbs equation

$$\rho\theta\dot{\eta} = \rho\dot{e} - T_e \cdot D \tag{12.2.1}$$

where $T_e = T_e(F, \theta)$ is the equilibrium stress as defined in Sect. 9.1. Recall that in the thermodynamics based on the Clausius–Duhem inequality the postulate of the local thermodynamic equilibrium is a consequence of the constitutive assumptions, see Proposition 9.2.2. A combination of (12.2.1) with the reduced equation of balance of energy $(3.6.5)_1$ gives

$$\rho\dot{\eta} = -\operatorname{div}\frac{h}{\theta} + \rho\frac{r}{\theta} + \rho\gamma,$$

where γ, the production of entropy, is given by (9.2.12), with T_d defined by $T_d = T - T_e$ just as in Sect. 9.1. Next it is argued, on the grounds of the second law, that γ must be nonnegative.

There are many useful classes of materials which violate the postulate of local thermodynamic equilibrium. The most notable among them are materials with fading memory (see the bibliographical note to this chapter for references), where the entropy and other thermodynamic functionals depend on the whole history of deformation and temperature, and materials with relaxation to be described in Sect. 12.4. Here we mention another simple example.

12.2.3 Example (Fluids of Grade 2) A compressible ***fluid of grade 2*** is defined by the constitutive equations

$$\bar{f} = \bar{f}(F, \theta, \dot{F}, \ddot{F}, G), \quad \bar{\eta} = \bar{\eta}(F, \theta, \dot{F}, \ddot{F}, G),$$

$$\bar{T} = \bar{T}(F, \theta, \dot{F}, \ddot{F}, G) = -p_0\mathbf{1} + \lambda(\operatorname{tr} D)\mathbf{1} + 2\mu D + \alpha_1 A_2 + \alpha_2 A_1^2,$$

$$h = \bar{h}(F, \theta, \dot{F}, \ddot{F}, G),$$

where A_1, A_2 are the first two Rivlin–Ericksen tensors, $A_1 = 2D$, $A_2 = \dot{A}_1 + A_1 L + L^{\mathrm{T}} A_1$, and the functions $p_0, \lambda, \mu, \alpha_1, \alpha_2$ depend on the specific volume and temperature. The coefficients λ, μ are the classical viscosities, and the coefficients α_1, α_2, will be called the normal stress coefficients here. Many features of real fluids cannot be modeled within the theory of classical materials with heat conduction and viscosity, and the second grade fluids provide a better model in certain situations. There is a great amount of experimental data on the normal stress coefficients. Let us restrict ourselves, for simplicity, to the case when the heat flux h vanishes identically. It turns out that the material is thermodynamically compatible if and only if

(1) λ, μ satisfy $(12.1.7)_{1,2}$;

(2) $\alpha_1 + \alpha_2 = 0$;

(3) $\alpha_1(v, \theta) = \hat{\alpha}_1(\theta)/v$, where $\hat{\alpha}_1$ is a function of θ alone.

The free energy and entropy are then given by

$$\bar{f} = \bar{f}_0(v, \theta) + \hat{\alpha}_1(\theta)|D|^2, \quad \bar{\eta} = \bar{\eta}_0(v, \theta) - \partial_\theta\hat{\alpha}_1(\theta)|D|^2,$$

where f_0 is such that $p_0 = -\partial_v f_0$ and $\bar{\eta}_0 := -\partial_\theta \bar{f}_0$. These restrictions are derived by applying the Coleman-Noll's procedure (see Proof of Proposition 9.2.2).

12.2.4 Forces and Fluxes For a viscous material with heat conduction, the thermodynamic force X and the thermodynamic flux J are defined by

$$X = (D/\theta\rho, -g/\theta^2\rho), \quad J = (T_d, h); \text{ hence } X, J \in \mathrm{V}. \tag{12.2.2}$$

Then $\gamma = J \cdot X$ and the constitutive assumptions (12.1.1) are restated as $J = \bar{J}(F, \theta, X)$. The entropy production inequality $\gamma \geq 0$ requires that

$$J(F, \theta, X) \cdot X \geq 0 \tag{12.2.3}$$

for every $(F, \theta) \in \bar{U}$ and $X \in \mathrm{V}$. We have $J(F, \theta, \mathbf{0}) = \mathbf{0}$. The absolute term being $\mathbf{0}$, the linearization of $J(F, \theta, X)$ with respect to X is

$$J(F, \theta, X) = L(F, \theta)[X] \tag{12.2.4}$$

where $L(F, \theta)$ is a linear transformation on V, the ***matrix of kinetic coefficients***. A comparison gives $L = \rho\theta Q$ and hence

$$L = \rho\theta \begin{bmatrix} V & D \\ \frac{1}{\theta}E & \frac{1}{\theta}K \end{bmatrix}. \tag{12.2.5}$$

Onsager's relations assert that L is symmetric.

Within the more general contexts (such as the chemically reacting mixtures of thermoelectric phenomena) the concepts of forces and fluxes are introduced to identify the matrix of kinetic coefficients. The general Onsager relations are the postulate that the matrix of kinetic coefficients is symmetric. However, there is no satisfactory definition of forces and fluxes, and the matrix of kinetic coefficients can be symmetric with one choice and nonsymmetric with another (TRUESDELL [1984, Lecture 7]). Even within the restricted context of thermomechanics different authors define the forces and the fluxes differently, but these choices lead to equivalent sets of reciprocal relations. This is partially explained by the result (MEIXNER [1943]), which says that the passage from $\boldsymbol{X}$, $\boldsymbol{J}$ to $\boldsymbol{X}'$, $\boldsymbol{J}'$ given by

$$\boldsymbol{X}' = \boldsymbol{M}(\boldsymbol{F}, \theta)^{-\mathrm{T}}\boldsymbol{X}, \quad \boldsymbol{J}' = \boldsymbol{M}(\boldsymbol{F}, \theta)\boldsymbol{J} \tag{12.2.6}$$

where $\boldsymbol{M}(\boldsymbol{F}, \theta)$ is a nonsingular linear transformation on V does not affect the symmetry of the matrix of kinetic coefficients. Indeed, in terms of $\boldsymbol{X}'$, $\boldsymbol{J}'$ the production of entropy is $\gamma = \boldsymbol{J}' \cdot \boldsymbol{X}'$ and the linear law (12.2.4) implies $\boldsymbol{J}'(\boldsymbol{F}, \theta, \boldsymbol{X}') = \boldsymbol{L}'(\boldsymbol{F}, \theta)[\boldsymbol{X}']$ where $\boldsymbol{L}'(\boldsymbol{F}, \theta) = \boldsymbol{M}^{\mathrm{T}}(\boldsymbol{F}, \theta)\boldsymbol{L}(\boldsymbol{F}, \theta)\boldsymbol{M}(\boldsymbol{F}, \theta)$. Hence $\boldsymbol{L}'$ is symmetric if and only if $\boldsymbol{L}$ is. However, there are simple changes of forces and fluxes not covered by (12.2.6), see TRUESDELL [1984].

12.3 Dissipation Potential

A possible generalization of Onsager's relations to the nonlinear dynamical response (12.1.1) is as follows.

12.3.1 The Dissipation Potential The dynamic response of the body is said to have a ***dissipation potential*** if there exists a function $\psi = \psi(\boldsymbol{F}, \theta, \boldsymbol{L}, \boldsymbol{g})$, defined on the common domain of the dynamic response functions, such that

$$\bar{\boldsymbol{T}}_d = \rho\theta\partial_{\boldsymbol{L}}\psi, \quad \bar{\boldsymbol{h}} = -\rho\theta^2\partial_{\boldsymbol{g}}\psi \tag{12.3.1}$$

throughout the domain. Equivalently, there exists a function $\phi = \phi(\boldsymbol{F}, \theta, \dot{\boldsymbol{F}}, \boldsymbol{G})$ such that

$$\bar{\boldsymbol{S}}_d = \rho_{\mathrm{o}}\theta\partial_{\dot{\boldsymbol{F}}}\phi, \quad \bar{\boldsymbol{q}} = -\rho_{\mathrm{o}}\theta^2\partial_{\boldsymbol{G}}\phi$$

everywhere on $\bar{U} \times \mathrm{Lin} \times \mathrm{Vect}$. The functions ψ and ϕ, if they exist, are related by the substitution rule $\phi(\boldsymbol{F}, \theta, \dot{\boldsymbol{F}}, \boldsymbol{G}) = \psi(\boldsymbol{F}, \theta, \dot{\boldsymbol{F}}\boldsymbol{F}^{-1}, \boldsymbol{F}^{-\mathrm{T}}\boldsymbol{G})$; they are called the ***dissipation potentials.*** The production of entropy is then given by

$$\gamma = \partial_{\boldsymbol{L}}\psi(\boldsymbol{F}, \theta, \boldsymbol{L}, \boldsymbol{g}) \cdot \boldsymbol{L} + \partial_{\boldsymbol{g}}\psi(\boldsymbol{F}, \theta, \boldsymbol{L}, \boldsymbol{g}) \cdot \boldsymbol{g}, \tag{12.3.2}$$

which motivates the occurrence of the coefficients in front of the derivatives in (12.3.1). Since $\boldsymbol{T}_d$ is symmetric, ψ necessarily depends on $\boldsymbol{L}$ only through its symmetric part $\boldsymbol{D}$, i.e., $\psi(\boldsymbol{F}, \theta, \boldsymbol{L}, \boldsymbol{g}) = \psi(\boldsymbol{F}, \theta, \boldsymbol{L}^{\mathrm{T}}, \boldsymbol{g})$, or equivalently, $\psi(\boldsymbol{F}, \theta, \dot{\boldsymbol{F}}, \boldsymbol{G}) = \psi(\boldsymbol{F}, \theta, \boldsymbol{F}^{-\mathrm{T}}\dot{\boldsymbol{F}}^{\mathrm{T}}\boldsymbol{F}, \boldsymbol{G})$ throughout the domains. The relationship to the Onsager relations is that if the dissipation potential exists, then

$$\begin{bmatrix} \partial_{\boldsymbol{L}}\bar{\boldsymbol{T}} & \partial_{\boldsymbol{g}}\bar{\boldsymbol{T}} \\ -\frac{1}{\theta}\partial_{\boldsymbol{L}}\bar{\boldsymbol{h}} & -\frac{1}{\theta}\partial_{\boldsymbol{g}}\bar{\boldsymbol{h}} \end{bmatrix} = \rho\theta \begin{bmatrix} \partial^2_{\boldsymbol{L}}\psi & \partial_{\boldsymbol{g}}\partial_{\boldsymbol{L}}\psi \\ \partial_{\boldsymbol{L}}\partial_{\boldsymbol{g}}\psi & \partial^2_{\boldsymbol{g}}\psi \end{bmatrix},$$

and the interchangeability of the second partial derivatives implies that the above matrices are symmetric. In particular the matrix $\boldsymbol{Q}$ (see (12.1.4)) is symmetric. Conversely, if the matrix $\boldsymbol{Q}$ is symmetric and the response (12.1.1) is linear in $\boldsymbol{L}, \boldsymbol{g}$, then the dissipation potential ψ as in (12.3.1) exists; it suffices to set

$$\psi = \frac{1}{2\theta}\rho\Big(\mathsf{V}[\boldsymbol{D}] \cdot \boldsymbol{D} + \mathsf{D}[\boldsymbol{g}] \cdot \boldsymbol{D} + \frac{1}{\theta}\mathsf{K}[\boldsymbol{g}] \cdot \boldsymbol{g} + \frac{1}{\theta}\mathsf{E}[\boldsymbol{D}] \cdot \boldsymbol{g}\Big).$$

12.3.2 Proposition *Let the dynamic response have a dissipation potential* ψ.

(1) *If the production of entropy is nonnegative in every process, then* $\psi(\boldsymbol{F}, \theta, \cdot, \cdot)$ *has a minimum at* $(\boldsymbol{0}, \boldsymbol{0})$ *for every* $(\boldsymbol{F}, \theta) \in \bar{U}$; *hence, if* ψ *is normalized so as to satisfy* $\psi(\boldsymbol{F}, \theta, \boldsymbol{0}, \boldsymbol{0}) = 0$ *for every* $(\boldsymbol{F}, \theta) \in \bar{U}$, *then necessarily* $\psi \ge 0$;

(2) *if* ψ *is positive, convex in* $\boldsymbol{L}, \boldsymbol{g}$ *and normalized, then the production of entropy is nonnegative in every process; in fact,*

$$\gamma \ge \psi \ge 0. \tag{12.3.3}$$

Proof (1): If $\gamma \geq 0$ in every process, then (12.3.2) provides

$$\partial_L \psi(\boldsymbol{F}, \theta, \boldsymbol{L}, \boldsymbol{g}) \cdot \boldsymbol{L} + \partial_{\boldsymbol{g}} \psi(\boldsymbol{F}, \theta, \boldsymbol{L}, \boldsymbol{g}) \cdot \boldsymbol{g} \geq 0$$

and it suffices to fix $(\boldsymbol{F}, \theta)$ and integrate this inequality along any ray starting at $(\boldsymbol{L}, \boldsymbol{g}) = (\boldsymbol{0}, \boldsymbol{0})$ and ending at a general $(\boldsymbol{L}, \boldsymbol{g})$. (2): The differential form of the convexity condition (see Sect. 16.2) implies that

$$0 = \psi(\boldsymbol{F}, \theta, \boldsymbol{0}, \boldsymbol{0}) \geq \psi(\boldsymbol{F}, \theta, \boldsymbol{L}, \boldsymbol{g}) - \partial_L \psi(\boldsymbol{F}, \theta, \boldsymbol{L}, \boldsymbol{g}) \cdot \boldsymbol{L} - \partial_{\boldsymbol{g}} \psi(\boldsymbol{F}, \theta, \boldsymbol{L}, \boldsymbol{g}) \cdot \boldsymbol{g}$$

and hence (12.3.3). □

With the identification of Sect. 12.2.4, the existence of the dissipation potential requires that there exist a function ψ depending on $\boldsymbol{F}, \theta$ and $\boldsymbol{X}$, denoted $\psi = \psi(\boldsymbol{X})$ such that

$$\boldsymbol{J}(\boldsymbol{X}) = \partial_{\boldsymbol{X}} \psi(\boldsymbol{X})$$

for every $\boldsymbol{X}$ and $\boldsymbol{F}, \theta$. Note that the existence of the dissipation potential and the additional requirements on ψ in Proposition 12.3.2(2) are clearly invariant under the passage $\boldsymbol{X}, \boldsymbol{J} \to \boldsymbol{X}', \boldsymbol{J}'$ of the format (12.2.6). The following result, due to EDELEN [1973], bears a relation to the matters just discussed.

12.3.3 Proposition *Let $\boldsymbol{J} = \boldsymbol{J}(\boldsymbol{X})$ be a continuously differentiable vector field on a vector space* X *with scalar product. Then $\boldsymbol{J}$ has a unique decomposition $\boldsymbol{J}(\boldsymbol{X}) = \boldsymbol{J}_0(\boldsymbol{X}) + \boldsymbol{J}_1(\boldsymbol{X})$, $\boldsymbol{X} \in$ X, where the vector field $\boldsymbol{J}_1$ satisfies $\boldsymbol{J}_1(\boldsymbol{X}) \cdot \boldsymbol{X} = 0$, $\boldsymbol{X} \in$ X, and the vector field $\boldsymbol{J}_0$ has a potential, i.e., $\boldsymbol{J}_0(\boldsymbol{X}) = \partial_{\boldsymbol{X}} \psi_0(\boldsymbol{X})$ for some function ψ_0 on* X.

In other words, it is possible to subtract a vector field $\boldsymbol{J}_1$ which does not contribute to the scalar product $\boldsymbol{J} \cdot \boldsymbol{X}$ in such a way that the resulting vector field $\boldsymbol{J}_0$ has a potential. This implies that one can decompose the dynamic response (12.1.1) into

$$\bar{\boldsymbol{T}}_d(\boldsymbol{F}, \theta, \boldsymbol{L}, \boldsymbol{g}) = \bar{\boldsymbol{T}}_{d,0}(\boldsymbol{F}, \theta, \boldsymbol{D}, \boldsymbol{g}) + \bar{\boldsymbol{T}}_{d,1}(\boldsymbol{F}, \theta, \boldsymbol{D}, \boldsymbol{g}),$$
$$\bar{\boldsymbol{h}}(\boldsymbol{F}, \theta, \boldsymbol{L}, \boldsymbol{g}) = \bar{\boldsymbol{h}}_0(\boldsymbol{F}, \theta, \boldsymbol{D}, \boldsymbol{g}) + \bar{\boldsymbol{h}}_1(\boldsymbol{F}, \theta, \boldsymbol{D}, \boldsymbol{g})$$

in such a way that $\bar{\boldsymbol{T}}_{d,0}, \bar{\boldsymbol{h}}_0$ have the dissipation potential and $\bar{\boldsymbol{T}}_{d,1}, \bar{\boldsymbol{h}}_1$ do not contribute to the production of entropy and dissipation.

12.4 Relaxation Models. The Extended Linear Irreversible Thermodynamics

The relaxation is the dependence of $\boldsymbol{T}, \boldsymbol{h}$ on time on the interval $[0, \infty)$ even when $\boldsymbol{F}, \theta$, and $\boldsymbol{g}$ are held constant on $[0, \infty)$. This can happen only when at time $t = 0$ the material element is in a non-equilibrium state, and it is natural to assume that as $t \to \infty$, the values $\boldsymbol{T}(t), \boldsymbol{h}(t)$ approach the relaxed values $\boldsymbol{T}_e, \boldsymbol{h}_e$ of these quantities corresponding to the situation when the material is held on the given $\boldsymbol{F}, \theta, \boldsymbol{g}$ for all times. Clearly, the relaxation cannot be described by the materials with heat conduction and viscosity of Sect. 9.1. Two related ways to describe the relaxation are the models with the constitutive equations that involve functionals of the history of $\boldsymbol{F}, \theta, \boldsymbol{g}$ up to the present time (materials with fading memory; see the bibliographical note to this chapter for references) and the models in which the excess $\boldsymbol{T}_d = \boldsymbol{T} - \boldsymbol{T}_e$ of $\boldsymbol{T}$ over the relaxed value and $\boldsymbol{h}$ are governed by evolution equations involving the time derivatives of $\boldsymbol{T}_d, \boldsymbol{h}$ (relaxation models). In this section we shall give a brief description of the relaxation models. These have been recently studied extensively by the extended thermodynamics (see the bibliographical note to this chapter). One of the purposes of this section is to show that, contrary to the occasional claims, these models are *completely compatible with the Clausius–Duhem inequality.* Another objective is to show that they may be considered as an extension of the classical linear irreversible thermodynamics, with, however, *the postulate of local thermodynamic equilibrium abandoned.*

12.4.1 Constitutive Equations of the Relaxation Models A process of a body Ω made of relaxation material is a collection of C^∞ fields $\chi, \theta, e, \eta, \boldsymbol{T}, \boldsymbol{h}, \boldsymbol{b}, r$ defined on $\Omega \times [t_1, t_2]$ satisfying the equations of balance of linear momentum and energy and the following constitutive equations (at every $(\boldsymbol{p}, t) \in \Omega \times [t_1, t_2]$):

$$e = \bar{e}(\boldsymbol{F}, \theta, \boldsymbol{J}), \quad \eta = \bar{\eta}(\boldsymbol{F}, \theta, \boldsymbol{J}), \quad \boldsymbol{T} = \boldsymbol{T}_e + \boldsymbol{T}_d, \quad \boldsymbol{T}_e = \bar{\boldsymbol{T}}_e(\boldsymbol{F}, \theta, \boldsymbol{J}), \tag{12.4.1}$$

$$\mathsf{T}(\boldsymbol{F}, \theta)[\mathring{\boldsymbol{J}}] + \boldsymbol{J} = \mathsf{L}(\boldsymbol{F}, \theta)[\boldsymbol{X}]. \tag{12.4.2}$$

Here $\boldsymbol{X}$, $\boldsymbol{J}$ are the thermodynamic force and the thermodynamic flux defined in (12.2.2) and

$$\mathring{\boldsymbol{J}} := (\dot{\boldsymbol{T}}_d - \boldsymbol{W}\boldsymbol{T}_d + \boldsymbol{T}_d\boldsymbol{W}, \dot{\boldsymbol{h}} - \boldsymbol{W}\boldsymbol{h}) \text{ for } \boldsymbol{J} = (\boldsymbol{T}_d, \boldsymbol{h})$$

(with $\boldsymbol{W}$ the spin tensor) is the Jaumann derivative of $\boldsymbol{J}$, $\bar{f}, \bar{\eta}, \bar{\boldsymbol{T}}_e$ are response functions defined on $\bar{U} \times \mathrm{V}$ with values in $\mathbb{R}$, $\mathbb{R}$, Sym, respectively, and T, L are functions on $\bar{U}$ with values in the set of all linear transformations on V, called the (matrix of) relaxation times and the (matrix of) kinetic coefficients, respectively. Assume that all the response functions are infinitely differentiable. The Jaumann derivative, rather than the ordinary time derivative, occurs in (12.4.2) for the theory to meet the principle of material frame indifference, which also needs appropriate invariance requirements on T, L not to be discussed here. To simplify the treatment, we shall replace (12.4.2) by the simpler equation

$$\mathsf{T}(\boldsymbol{F}, \theta)[\dot{\boldsymbol{J}}] + \boldsymbol{J} = \mathsf{L}(\boldsymbol{F}, \theta)[\boldsymbol{X}], \tag{12.4.3}$$

with noting that (12.4.2) is treated similarly. Defining the free energy $f := e - \theta\eta$, we have the constitutive equation $f = \bar{f}(\boldsymbol{F}, \theta, \boldsymbol{J})$, where $\bar{f}$ is the response function for the free energy, defined on $\bar{U} \times \mathrm{V}$.

12.4.2 Special Cases

(1) When $\mathsf{T} \equiv \mathbf{0}$, (12.4.3) reduces to $\boldsymbol{J} = \mathsf{L}[\boldsymbol{X}]$, i.e. to (12.2.4), and inserting this value into (12.4.1), one obtains the constitutive equations of the form (9.1.1), i.e., of the material with heat conduction and viscosity. However, when T is different from $\mathbf{0}$, then the evolution equations resulting from combining the constitutive equations (12.4.1), (12.4.3) with the balance equations are generally hyperbolic, whereas the evolution equations of viscous materials with heat conduction are generally parabolic.

(2) When the material is immobile, i.e., when $\boldsymbol{F} = \mathbf{1}$ identically, then (12.4.3) reduces to the Maxwell–Cattaneo model discussed in Subsect. 7.6.4 and Proposition 7.6.5. One has $\boldsymbol{q} \equiv \boldsymbol{h}$ and assuming $\rho_0 = 1$, one sees that a consistent identification gives $\boldsymbol{J} \equiv \boldsymbol{h}$ and $\boldsymbol{Z} \equiv \theta^2\boldsymbol{H}$. The hyperbolic character of the heat equation leads to a finite speed of propagation of the heat signals, in contrast to the parabolic Fourier law gives infinite speeds of propagation. The finite speed of propagation of heat pulses is generally called *second sound*. It was first observed in liquid Helium-2 (see LANDAU & LIFSCHITZ [1978]) and later in dielectric crystals, such as Helium-3, Helium-4, NaF, and Bi, in appropriate ranges of temperature.

(3) In the extremely special case when there is a single relaxation time, i.e., when $\mathsf{T}[\dot{\boldsymbol{J}}] = \tau\dot{\boldsymbol{J}}$, for every $\dot{\boldsymbol{J}} \in \mathrm{V}$ and some constant $\tau \in \mathbb{R}$, (12.4.3) reduces to

$$\tau\dot{\boldsymbol{J}} + \boldsymbol{J} = \mathsf{L}[\boldsymbol{X}].$$

(4) When $\mathsf{T}[\dot{\boldsymbol{J}}] = (\tau_1\dot{\boldsymbol{T}}_d, \tau_2\dot{\boldsymbol{h}})$, for every $\dot{\boldsymbol{J}} = (\dot{\boldsymbol{T}}_d, \dot{\boldsymbol{h}}) \in \mathrm{V}$, where τ_1, τ_2 are constants, then (12.4.3) gives

$$\tau_1\dot{\boldsymbol{T}}_d + \boldsymbol{T}_d = \mathsf{V}[\boldsymbol{D}] + \mathsf{D}[\boldsymbol{g}], \quad \tau_2\dot{\boldsymbol{h}} + \boldsymbol{h} = -\mathsf{K}[\boldsymbol{g}] - \mathsf{E}[\boldsymbol{D}], \tag{12.4.4}$$

where the identification (12.2.5) has been used. Equation $(12.4.4)_1$ is a special case of the popular rheological models (see JOU, CASAS-VÁZQUEZ & LEBON [1993, Chap. 7] or MAUGIN [1992, Chap. 2]).

The subsequent discussion deals with the following special case of the relaxation materials. A ***quadratic material*** is a relaxation material satisfying the additional requirements that **(1)**

$$\bar{f}(\boldsymbol{F}, \theta, -\boldsymbol{J}) = \bar{f}(\boldsymbol{F}, \theta, \boldsymbol{J}), \quad \bar{\eta}(\boldsymbol{F}, \theta, -\boldsymbol{J}) = \bar{\eta}(\boldsymbol{F}, \theta, \boldsymbol{J}), \quad \bar{\boldsymbol{T}}_e(\boldsymbol{F}, \theta, -\boldsymbol{J}) = \bar{\boldsymbol{T}}_e(\boldsymbol{F}, \theta, \boldsymbol{J})$$

for every $(\boldsymbol{F}, \theta, \boldsymbol{J}) \in V$, and **(2)** $\mathsf{L}(\boldsymbol{F}, \theta)$, $\mathsf{T}(\boldsymbol{F}, \theta)$ are invertible linear transformations from V onto itself for every $(\boldsymbol{F}, \theta) \in \mathrm{InvLin} \times (0, \infty)$. Define a linear transformation $\mathsf{H}(\boldsymbol{F}, \theta)$ on V by

$$\mathsf{H}(\boldsymbol{F}, \theta)[\boldsymbol{J}] = \mathsf{L}(\boldsymbol{F}, \theta)^{-1}\left[\mathsf{T}(\boldsymbol{F}, \theta)[\boldsymbol{J}]\right], \quad \boldsymbol{J} \in \mathrm{V},$$

and the major transposition $\boldsymbol{H}^{\mathbb{T}}$ of $\boldsymbol{H}$ by $\boldsymbol{H}^{\mathbb{T}}[\boldsymbol{I}] \cdot \boldsymbol{J} = \boldsymbol{I} \cdot \boldsymbol{H}^{\mathbb{T}}[\boldsymbol{J}]$ for every $\boldsymbol{I}, \boldsymbol{J} \in \mathrm{V}$. Let us agree to say that $\boldsymbol{H}$ has a major symmetry if $\boldsymbol{H} = \boldsymbol{H}^{\mathbb{T}}$ and that $\boldsymbol{L}$ is positive definite if $\boldsymbol{L}[\boldsymbol{X}] \cdot \boldsymbol{X} > 0$ for every nonzero $\boldsymbol{X} \in \mathrm{V}$.

12.4.3 Proposition *The quadratic material satisfies the dissipation principle* 9.2.1 *if and only if the following assertions hold:*

(1) *for every* $(\boldsymbol{F}, \theta) \in \bar{U}$*, the tensor* $\boldsymbol{L}(\boldsymbol{F}, \theta)$ *is positive definite and the tensor* $\boldsymbol{H}(\boldsymbol{F}, \theta)$ *has the major symmetry;*

(2) $\bar{f}$ *is of the form*

$$\bar{f}(\boldsymbol{F}, \theta, \boldsymbol{J}) = \bar{f}_0(\boldsymbol{F}, \theta) + \tfrac{1}{2}\theta \boldsymbol{H}(\boldsymbol{F}, \theta)[\boldsymbol{J}] \cdot \boldsymbol{J}, \quad (\boldsymbol{F}, \theta, \boldsymbol{J}) \in \bar{U} \times \mathrm{V}, \tag{12.4.5}$$

where $\bar{f}_0$ *is a scalar-valued function on* $\bar{U}$;

(3) $\bar{\boldsymbol{T}}_e = \rho \partial_{\boldsymbol{F}} \bar{f} \boldsymbol{F}^{\mathsf{T}}$, $\bar{\eta} = -\partial_\theta \bar{f}$, *i.e.,*

$$\bar{\boldsymbol{T}}_e = \rho \partial_{\boldsymbol{F}} \bar{f}_0(\boldsymbol{F}, \theta) \boldsymbol{F}^{\mathsf{T}} + \tfrac{1}{2}\rho\theta\partial_{\boldsymbol{F}}(\boldsymbol{H}(\boldsymbol{F}, \theta)[\boldsymbol{J}] \cdot \boldsymbol{J})\boldsymbol{F}^{\mathsf{T}},$$

$$\bar{\eta} = -\partial_\theta \bar{f}_0(\boldsymbol{F}, \theta) - \tfrac{1}{2}\partial_\theta(\theta \boldsymbol{H}(\boldsymbol{F}, \theta)[\boldsymbol{J}] \cdot \boldsymbol{J})$$

for every $(\boldsymbol{F}, \theta, \boldsymbol{J}) \in \bar{U} \times \mathrm{V}$.

When these three conditions are satisfied, then $\bar{e}$ *is of the form*

$$\bar{e} = \bar{e}_0(\boldsymbol{F}, \theta) - \tfrac{1}{2}\theta^2 \partial_\theta \boldsymbol{H}[\boldsymbol{J}] \cdot \boldsymbol{J}, \quad (\boldsymbol{F}, \theta, \boldsymbol{J}) \in \bar{U} \times \mathrm{V},$$

where $\bar{e}_0 = \bar{f}_0 - \theta\partial_\theta \bar{f}_0$.

This generalizes Proposition 7.6.5: In the identifications 12.4.2(2) the expressions for the thermodynamic potentials coincide.

Proof Assume that the material satisfies the dissipation principle. Consider an arbitrary process. Using the chain rule and defining $\boldsymbol{J}_e(\boldsymbol{F}, \theta, \boldsymbol{J}) := (\bar{\boldsymbol{T}}_e(\boldsymbol{F}, \theta, \boldsymbol{J}) - \rho\partial_{\boldsymbol{F}} \bar{f}(\boldsymbol{F}, \theta, \boldsymbol{J})\boldsymbol{F}^{\mathsf{T}}, \boldsymbol{0})$, one can rewrite the dissipation inequality $\rho\dot{f} \le \boldsymbol{T} \cdot \boldsymbol{D} - \rho\eta\dot{\theta} - \boldsymbol{h} \cdot \boldsymbol{g}/\theta$ as

$$0 \le -\rho(\partial_\theta \bar{f} + \bar{\eta}) \cdot \dot{\theta} - \rho\partial_{\boldsymbol{J}} \bar{f} \cdot \dot{\boldsymbol{J}} + \rho\theta \boldsymbol{J}_e \cdot \boldsymbol{X} + \rho\theta \boldsymbol{J} \cdot \boldsymbol{X}, \tag{12.4.6}$$

where $\partial_\theta \bar{f}$, $\bar{\eta}$, $\boldsymbol{J}_e$, $\partial_{\boldsymbol{J}} \bar{f}$ are evaluated at $(\boldsymbol{F}, \theta, \boldsymbol{J})$. Writing (12.4.3) in the form $\boldsymbol{X} = \boldsymbol{H}[\dot{\boldsymbol{J}}] + \boldsymbol{L}^{-1}[\boldsymbol{J}]$ and inserting the so calculated $\boldsymbol{X}$ into (12.4.6), one obtains

$$0 \le -\rho(\partial_\theta \bar{f} + \bar{\eta}) \cdot \dot{\theta} - \rho\partial_{\boldsymbol{J}} \bar{f} \cdot \dot{\boldsymbol{J}} + \rho\theta \boldsymbol{J}_e \cdot (\boldsymbol{H}[\dot{\boldsymbol{J}}] + \boldsymbol{L}^{-1}[\boldsymbol{J}]) + \rho\theta \boldsymbol{J} \cdot (\boldsymbol{H}[\dot{\boldsymbol{J}}] + \boldsymbol{L}^{-1}[\boldsymbol{J}]). \tag{12.4.7}$$

This inequality involves only the variables

$$\boldsymbol{F}, \theta, \boldsymbol{J}, \dot{\theta}, \dot{\boldsymbol{J}}, \tag{12.4.8}$$

and it is a simple matter to realize that for a given point $\boldsymbol{p} \in \Omega$ and a given time t there is a process in which the variables (12.4.8) at $(\boldsymbol{p}, t)$ take arbitrarily prescribed values, subject only to the restrictions that they belong to the domains of the response functions. The linearity in $\dot{\boldsymbol{J}}, \dot{\theta}$ then implies that (12.4.7) decouples into

$$0 = \partial_\theta \bar{f} + \bar{\eta}, \quad 0 = -\partial_{\boldsymbol{J}} \bar{f} \cdot \dot{\boldsymbol{J}} + \theta \boldsymbol{J}_e \cdot \boldsymbol{H}[\dot{\boldsymbol{J}}] + \theta \boldsymbol{J} \cdot \boldsymbol{H}[\dot{\boldsymbol{J}}] \tag{12.4.9}$$

and

$$0 \le \boldsymbol{J}_e \cdot \boldsymbol{L}^{-1}[\boldsymbol{J}] + \boldsymbol{J} \cdot \boldsymbol{L}^{-1}[\boldsymbol{J}]. \tag{12.4.10}$$

Replacing $\boldsymbol{J}$ by $-\boldsymbol{J}$ in $(12.4.9)_2$ and using $\boldsymbol{J}_e(\boldsymbol{F}, \theta, -\boldsymbol{J}) = \boldsymbol{J}_e(\boldsymbol{F}, \theta, \boldsymbol{J})$, $\partial_{\boldsymbol{J}} \bar{f}(\boldsymbol{F}, \theta, -\boldsymbol{J}) = -\partial_{\boldsymbol{J}} \bar{f}(\boldsymbol{F}, \theta, \boldsymbol{J})$ (the material is quadratic), we obtain

$$0 = \partial_{\boldsymbol{J}} \bar{f} \cdot \dot{\boldsymbol{J}} + \theta \boldsymbol{J}_e \cdot \boldsymbol{H}[\dot{\boldsymbol{J}}] - \theta \boldsymbol{J} \cdot \boldsymbol{H}[\dot{\boldsymbol{J}}],$$

where $\boldsymbol{J}_e$, $\partial_{\boldsymbol{J}} \bar{f}$ stand for $\boldsymbol{J}_e(\boldsymbol{F}, \theta, \boldsymbol{J})$ and $\partial_{\boldsymbol{J}} \bar{f}(\boldsymbol{F}, \theta, \boldsymbol{J})$ respectively. A combination with $(12.4.9)_2$ gives $0 = -\partial_{\boldsymbol{J}} \bar{f} \cdot \dot{\boldsymbol{J}} + \theta \boldsymbol{J} \cdot \boldsymbol{H}[\dot{\boldsymbol{J}}]$, $0 = \theta \boldsymbol{J}_e \cdot \boldsymbol{H}[\dot{\boldsymbol{J}}]$, i.e., by the arbitrariness of $\dot{\boldsymbol{J}}$,

$$\partial_{\boldsymbol{J}} \bar{f} = \theta \boldsymbol{H}^{\mathbb{T}}[\boldsymbol{J}], \quad \boldsymbol{H}^{\mathbb{T}}[\boldsymbol{J}_e] = \boldsymbol{0}. \tag{12.4.11}$$

By $(12.4.11)_1$ and the interchangeability of the partial derivatives, $\boldsymbol{H}^{\mathbb{T}}$ is symmetric and hence also $\boldsymbol{H}$ is symmetric. The integration of $(12.4.11)_1$ leads to (2). Since $\boldsymbol{H} \equiv \boldsymbol{H}^{\mathbb{T}}$ is nonsingular, $(12.4.11)_2$ implies $\boldsymbol{J}_e \equiv \boldsymbol{0}$, which, together with $(12.4.9)_1$ gives (3). Finally, by $\boldsymbol{J}_e \equiv \boldsymbol{0}$ (12.4.10) reduces to $0 \le \boldsymbol{J} \cdot \boldsymbol{L}^{-1}[\boldsymbol{J}]$ and hence $\boldsymbol{L}$ is positive definite. With the already proved symmetry of $\boldsymbol{H}$ this gives

(1). This completes the proof of the direct implication. Reversing the steps in the above proof leads to the converse implication. □

A simple calculation that is left to the reader gives the following.

12.4.4 Proposition *For a quadratic material satisfying the dissipation principle the following equations hold in every process:*

(1) ***the generalized Gibbs equation***

$$\rho\theta\dot{\eta} = \rho\dot{e} + \boldsymbol{T}_e \cdot \boldsymbol{D}; \tag{12.4.12}$$

(2) *the specific production of entropy, defined by* (9.2.9), *is given by*

$$\gamma = \boldsymbol{J} \cdot \boldsymbol{L}^{-1}[\boldsymbol{J}] = \boldsymbol{J} \cdot (\boldsymbol{X} - \boldsymbol{H}[\dot{\boldsymbol{J}}]).$$

Equation (12.4.12) is formally identical with the Gibbs equation (12.2.1), with, however, η, $\boldsymbol{T}_e$ given by the response functions that depend not only on the equilibrium variables $\boldsymbol{F}, \theta$, but also on the nonequilibrium variable $\boldsymbol{J}$; hence the relaxation models give the Gibbs equation *without* the local equilibrium postulate.

We now sketch the construction of the evolution equation (12.4.3) within the scheme of the extended irreversible thermodynamics, cf. Jou, Casas-Vázquez & Lebon [1988,1993]. This approach may be viewed as an extension of the classical irreversible thermodynamics. Following the logic of the linear irreversible thermodynamics explained in Sect. 12.2, one may take the generalized Gibbs equation (12.4.12) as a postulate (i.e., a definition of temperature θ and equilibrium stress $\boldsymbol{T}_e$). Defining $\boldsymbol{J}$ as above and assuming that $\bar{f}$ is of the quadratic form (12.4.5) with $\boldsymbol{H}$ having the major symmetry, one finds that the production of entropy defined by (9.2.9) is given by

$$\gamma = (\boldsymbol{J} \cdot \boldsymbol{X} - \boldsymbol{H}[\boldsymbol{J}] \cdot \dot{\boldsymbol{J}}) = (\boldsymbol{X} - \boldsymbol{H}[\dot{\boldsymbol{J}}]) \cdot \boldsymbol{J} \geq 0$$

with the last inequality being the consequence of the dissipation inequality. The production of entropy is again a product of "forces times fluxes"

$$\gamma = \boldsymbol{X}' \cdot \boldsymbol{J},$$

with the force $\boldsymbol{X}'$ now defined by $\boldsymbol{X}' = \boldsymbol{X} - \boldsymbol{H}[\dot{\boldsymbol{J}}]$. The linear philosophy then suggests that $\boldsymbol{J}$ is given by the constitutive equation

$$\boldsymbol{J} = \boldsymbol{L}[\boldsymbol{X}'] = \boldsymbol{L}[\boldsymbol{X}] - \boldsymbol{L}\big[\boldsymbol{H}[\dot{\boldsymbol{J}}]\big],$$

where $\boldsymbol{L}$ is a matrix of kinetic coefficients. Finally, defining $\boldsymbol{T}$ by $\boldsymbol{T}[\dot{\boldsymbol{J}}] = \boldsymbol{L}\big[\boldsymbol{H}[\dot{\boldsymbol{J}}]\big]$, one recovers (12.4.3).

Bibliographical Note

Historically, the first materials studied within the framework of the linear thermodynamics of irreversible processes were viscous materials with heat conduction (Eckart [1940, 1948] and Meixner [1943]; see De Groot & Mazur [1962] for additional references). The generalization of Onsager's relation in Sect. 12.3 is due to Ziegler [1968], and the idea of the dissipation potential goes back to Rayleigh. For an approach based on a systematic use of the dissipation potential and the associated convexity methods, see Maugin [1992] and the references given there.

The incompressible fluids of grade 2 are studied by Dunn & Fosdick [1974], and also in this case the free energy depends on $\boldsymbol{D}$. Another example violating the principle of local equilibrium are materials with fading memory; see Coleman & Noll [1960,1961], Coleman [1964a,b], Truesdell & Noll [1965, Sects. 38–41], Day [1972a], Renardy, Hrusa, & Nohel [1987], and Gurtin & Hrusa [1991].

For extended thermodynamics, see Carrassi & Morro [1972], Lebon [1978, 1987] and Jou, Casas-Vázquez & Lebon [1988, 1993], where the emphasis is on the relations to the linear irreversible thermodynamics, rheological models, and fluctuations. Another group of references is Müller [1967; 1985, Chap. 13], Liu & Müller [1983] and Müller & Ruggeri [1993] where the theory is developed within the framework of Müller's entropy inequality with Lagrange multipliers (see Sect. 9.5), with however, an extended set of balance equations for 13 (or more) independent fields, with fluxes and sources depending on them. This approach, as well as the Maxwell–Cattaneo model, is motivated by the kinetic theory of gases, which leads to a similar structure of equations via approximations to the Boltzmann equation. – Note here that under various circumstances (12.4.2) can be given a form of a conservation law $\dot{\boldsymbol{J}}$+Div $\boldsymbol{M} = \boldsymbol{Z}$ where $\boldsymbol{J}$,i.e., $\boldsymbol{T}_d$ and $\boldsymbol{h}$ play roles of balanced quantities, the flux $\boldsymbol{M}$ is a pair consisting of the third- and second-order tensors and $\boldsymbol{Z}$ is a source term.

Other papers related to the Maxwell–Cattaneo heat equation and finite propagation of heat pulses are GURTIN & PIPKIN [1968], COLEMAN, FABRIZIO & OWEN [1982, 1986], COLEMAN, HRUSA & OWEN [1986], MORRO [1987], BRANDON & HRUSA [1988], ÖNCÜ & MOODIE [1992], and COLEMAN & LAI [1994].

Part IV

Thermodynamic Equilibrium

13 The Environment

Equilibrium states are states of zero velocity and uniform temperature, and, as a consequence, of zero production of entropy. This distinguishes them from the more general steady states, in which the production of entropy is nonzero due to gradients of temperature and velocity. This fundamental difference leads to many implied differences between equilibrium states and steady states. Most importantly, there are universal variational principles for determining equilibrium states – the extremum principles of thermodynamics – which do not have any counterpart for the steady states (despite many claims in the literature). The extremum principles are related to the dynamical stability of equilibrium states. Closely connected to the stability are the static constitutive inequalities – additional properties of the equilibrium response functions which do *not* follow (at least not directly) from the second law of thermodynamics. Basic assertions about the equilibrium states of coexistent phases and the proofs of the existence of equilibrium states are also based on extremum principles.

An ideal goal of continuum thermodynamics is to determine the processes of the body in a given environment, or to determine steady or equilibrium states in that environment. The environment is specified by prescribing the boundary conditions and the external body forces (assuming that the external volumetric supplies of heat are absent). One can then examine the existence and the qualitative properties of the solutions of the corresponding equations. In this part of the book only the equilibrium environment will be considered, i.e., an environment which eventually stops the dissipation and brings the body into an equilibrium state.

13.1 States and Processes

Throughout this part of the book we consider a viscous body with heat conduction Ω, specified by the response functions $\bar{f}$, $\bar{\eta}$, $\bar{S}$, $\bar{q}$, as in Sect. 9.1, that satisfy the Clausius–Duhem inequality and the principle of material frame indifference(see Sects. 9.2 and 9.3). Unless otherwise stated, it is assumed that the body also admits the entropy as independent thermal variable (see Sect. 10.2). Let $\tilde{U} \subset \mathrm{InvLin} \times \mathbb{R}$ denote the set of all possible pairs $(\boldsymbol{F}, \eta)$, i.e., the domain of $\tilde{e}$.

13.1.1 States A ***state*** σ of Ω is any triple $\sigma = (\boldsymbol{\chi}, \eta, \boldsymbol{v})$ of (time-independent) fields over Ω (where $\boldsymbol{\chi}$ is interpreted as a deformation and η, $\boldsymbol{v}$ as the fields of entropy and velocity over Ω) such that

(1) $\boldsymbol{\chi}$ is a deformation of Ω (see Subsect. 2.2.1), η is piecewise continuous over Ω and continuously differentiable outside the singular surface, the values of $(\nabla\boldsymbol{\chi}, \eta)$

outside the singular surface as well as the limiting values on the singular surface and on $\partial\Omega$ are in $\tilde{U} \cap (\mathrm{Lin}^+ \times \mathbb{R})$;

(2) $\boldsymbol{v}$ is continuously differentiable outside the singular surface, the fields $\boldsymbol{S}$, $\boldsymbol{q}$, given outside the singular surface by

$$\boldsymbol{S} = \bar{\boldsymbol{S}}(\boldsymbol{F}, \theta, \nabla\boldsymbol{v}, \nabla\theta), \quad \boldsymbol{q} = \bar{\boldsymbol{q}}(\boldsymbol{F}, \theta, \nabla\boldsymbol{v}, \nabla\theta), \quad \theta = \tilde{\theta}(\boldsymbol{F}, \eta), \quad \boldsymbol{F} = \nabla\boldsymbol{\chi}, \qquad (13.1.1)$$

are piecewise continuously differentiable.

The set of all states is denoted by Σ. The state $\sigma \in \Sigma$ is said to be a ***rest state*** if $\boldsymbol{v} = \boldsymbol{0}$ identically. Any rest state $\sigma = (\boldsymbol{\chi}, \eta, \boldsymbol{0}) \in \Sigma$ can be identified with $\sigma \simeq (\boldsymbol{\chi}, \eta)$. We denote by $\Sigma^{\#}$ the set of all pairs $\sigma = (\boldsymbol{\chi}, \eta)$ which satisfy Condition (1) above and identify the elements of $\Sigma^{\#}$ with the rest states. An ***equilibrium state*** state σ is any rest state $\sigma = (\boldsymbol{\chi}, \eta) \in \Sigma^{\#}$ such that the temperature θ, calculated by $(13.1.1)_{3,4}$, is uniform.

13.1.2 Processes A ***process*** is a triple $\pi = (\boldsymbol{\chi}, \eta, \boldsymbol{v})$ of (possibly time-dependent) fields defined either on $\Omega \times [t_1, t_2]$ with t_1, t_2 finite or on $\Omega \times [t_1, \infty)$ such that

(1) $\boldsymbol{\chi}$ is a motion of Ω, η is piecewise continuous and continuously differentiable outside the singular surface; the values of $(\nabla\boldsymbol{\chi}, \eta)$ outside the singular surface as well as the limiting values on the singular surface and on $\partial\Omega$ are in $\tilde{U} \cap (\mathrm{Lin}^+ \times \mathbb{R})$;

(2) outside the singular surface, the consistency condition $\boldsymbol{v} = \dot{\boldsymbol{\chi}}$ holds;

(3) the fields $\boldsymbol{S}$, $\boldsymbol{q}$, given outside the singular surface by (13.1.1), are piecewise continuously differentiable;

(4) on the singular surface, the jump condition for the entropy $(3.7.2)_2$ holds.

The value $\pi(t) = (\boldsymbol{\chi}(\cdot, t), \eta(\cdot, t), \boldsymbol{v}(\cdot, t)) \in \Sigma$ is the state of the body at time t. We denote by Π the set of all processes, defined either on $\Omega \times [t_1, t_2]$ or on $\Omega \times [t_1, \infty)$. For adiabatic bodies 9.4.1(2), the requirement (3) is a consequence of (1), (2), while for genuinely viscous bodies, it is necessary to impose additional smoothness on $\boldsymbol{\chi}$, η to satisfy (3).

13.2 Heating Environments

A wide range of external conditions can accompany the equations of thermomechanics, depending on whether the body is solid or fluid, whether it has a viscosity, etc. We shall restrict ourselves to the most classical conditions. The reader will have no difficulty in modifying the considerations below to more general environments. Many of the scalar equations presented below can be replaced by inequalities of appropriate sign, thus also adding the dissipation to the external conditions. We shall divide the external conditions into the heating and loading parts, and treat each of them separately.

13.2.1 The Standard Heating Let T, Q be two disjoint complementary subsets of $\partial\Omega$ (i.e., $T \cap Q = \emptyset$, $T \cup Q = \partial\Omega$) and $\theta_0 > 0$ a constant. The ***standard heating environment*** consists of prescribing

$$r(\boldsymbol{p}, t) = 0, \quad \boldsymbol{p} \in \Omega, \qquad \theta(\boldsymbol{p}, t) = \theta_0, \quad \boldsymbol{p} \in T, \qquad \boldsymbol{q}(\boldsymbol{p}, t) \cdot \boldsymbol{n}(\boldsymbol{p}) = 0, \quad \boldsymbol{p} \in Q,$$

for all t during the process. If the area of T is positive, θ_0 is called the ambient temperature. This kind of heating environment is suitable primarily for solid bodies in permanent contact with two different substances, one capable of easily communicating heat to the body and the other a thermal insulator.

Special cases are:

(1) ***Adiabatic isolation***

$$r(p, t) = 0, \quad p \in \Omega, \quad q(p, t) \cdot n(p) = 0, \quad p \in \partial\Omega$$

for all t. For adiabatic isolation the ambient temperature is not defined.

(2) ***Prescribed boundary temperature***

$$r(p, t) = 0, \quad p \in \Omega, \quad \theta(p, t) = \theta_0, \quad p \in \partial\Omega.$$

More general conditions than 13.2.1 would be to prescribe a spatially varying temperature on T and a possibly nonzero heat flux on Q. There is no equilibrium state compatible with such conditions: they cause gradients of temperature inside the body. Such conditions are outside the scope of the equilibrium theory.

13.2.2 The Confinement Condition We consider a motion of the body constrained to a fixed, a priori known spatial region $R \subset$ Eucl with smooth boundary, interpreted as a rigid container or canister. The set R can be unbounded and the body need not fill R completely. The choice $R =$ Eucl is not excluded. If $r_t \equiv \chi(\Omega, t)$ is the actual region occupied by the body, we require that

$$\chi(\Omega, t) \subset R. \tag{13.2.1}$$

Let us divide the actual boundary $\partial\chi(\Omega, t)$ into two *a priori unknown* parts, the surface of contact $C_t := \{p \in \partial\Omega : \chi(p, t) \in \partial R\}$ and the free part $F_t := \partial\Omega \setminus C_t$. As a consequence of (13.2.1) we have the slip condition

$$v(p, t) \cdot m = 0 \text{ if } p \in C_t \tag{13.2.2}$$

for smooth motions, where v is the velocity and m the normal to $\partial\chi(\Omega, t)$ at $\chi(p, t)$. (This normal is just the opposite of the normal to ∂R at $\chi(p, t)$ whenever $p \in C_t$.)

13.2.3 The Contact Heating In the situation 13.2.2, there are circumstances in which there is a difference in the communication of heat through C_t, and F_t. For instance, if the container is made of a good conductor of heat and if it is kept at the temperature θ_0 while the free boundary is in touch with an insulating gas, then one can lay down the following conditions of ***contact heating:***

$$r(p, t) = 0, \quad p \in \Omega, \quad \theta(p, t) = \theta_0, \quad p \in C_t, \quad q(p, t) \cdot n(p) = 0, \quad p \in F_t,$$

for all t during the process.

Consider the following state functions (for every state $\sigma = (\chi, \eta, v) \in \Sigma$): the entropy and the internal and kinetic energies

$$\mathsf{H}(\sigma) = \int_\Omega \eta\rho_\circ \, dV, \quad \mathsf{E}(\sigma) = \int_\Omega \tilde{e}(F, \eta)\rho_\circ \, dV, \quad \mathsf{K}(\sigma) = \int_\Omega \tfrac{1}{2}\rho_\circ v^2 \, dV,$$

respectively, and the ***canonical free energy*** at θ_0 (for equilibrium heating environment of ambient temperature θ_0),

$$\mathsf{F}(\sigma) = \mathsf{K}(\sigma) + \mathsf{E}(\sigma) - \theta_0 \mathsf{H}(\sigma).$$

Recall from Sect. 5.1 the work $w(\pi) \equiv w(\pi, \Omega)$ done by the body in the process π and the heat $q(\pi) \equiv q(\pi, \Omega)$ absorbed by it. The equilibrium character of the heating environment leads to the following result.

13.2.4 The Maximum Work *For every process* $\pi = (\chi, \eta, v)$ *compatible with the heating environment* 13.2.1 *or* 13.2.3, *the work* $w(\pi)$ *done by the body satisfies*

$$w(\pi) \le -[\mathsf{F}(\pi^f) - \mathsf{F}(\pi^i)]. \tag{13.2.3}$$

The right-hand side of (13.2.3) is independent of the path in the state space and provides thus an upper bound for the work. This result is independent of any assumption about the character of the loading environment. If the process is cyclic, then (13.2.3) reduces to the assertion that the work must be nonpositive, thus giving an alternative meaning to Kelvin's statement of the second law.

Proof We shall give the proof only for the environment 13.2.1; the case 13.2.3 is similar. The Clausius–Duhem inequality reads

$$\mathsf{H}(\pi^f) - \mathsf{H}(\pi^i) \ge \int_{t_1}^{t_2} \int_{\partial\Omega} \frac{q \cdot n}{\theta}\, dA\, dt = \frac{1}{\theta_0} \int_{t_1}^{t_2} \int_{\partial\Omega} q \cdot n\, dA\, dt = \frac{1}{\theta_0} q(\pi),$$

using 13.2.1, and the equation of balance of energy is $\mathsf{E}(\pi^f) + \mathsf{K}(\pi^f) - \mathsf{E}(\pi^i) - \mathsf{K}(\pi^i) = q(\pi) - w(\pi)$. A combination leads to (13.2.3). □

Proposition 13.2.4 can be slightly refined, namely, it is possible to calculate the difference between the left-and right-hand sides of (13.2.3) in terms of the dissipation in the process.

13.2.5 Consider a process π not necessarily compatible with any of the environments defined above. The ***total production of entropy*** in the body at time t is defined to be

$$\Gamma(t) = \frac{d}{dt} \int_{\Omega} \rho_\circ \eta\, dV + \int_{\partial\Omega} \frac{q \cdot n}{\theta}\, dA - \int_{\Omega} \rho_\circ \frac{r}{\theta}\, dV \tag{13.2.4}$$

for every time for which the derivative of the entropy exists. Using the results of Sects. 3.2 and 3.1, one obtains

$$\Gamma(t) = \int_{\Omega} \rho_\circ \gamma\, dV + \int_{S_t} \left(-U\rho_\circ [\eta] + \left[\frac{q}{\theta}\right] \cdot n \right) dA \tag{13.2.5}$$

where γ is the specific production of entropy for which Proposition 9.2.3 gives the expression $\gamma = T_d \cdot D/\rho\theta - h \cdot g/\rho\theta^2 = S_d \cdot \dot{F}/\rho_\circ\theta - q \cdot G/\rho_\circ\theta^2$. The Clausius–Duhem inequality then takes the form

$$\dot{\mathsf{H}}(t) = \int_{\partial\Omega} \frac{q \cdot n}{\theta}\, dA - \int_{\Omega} \rho_\circ \frac{r}{\theta}\, dV + \Gamma(t) \tag{13.2.6}$$

for a.e. t during the process, where $\Gamma(t) \ge 0$ is given by (13.2.5).

13.2.6 Proposition *For every process* $\pi = (\chi(\cdot), \eta(\cdot), v(\cdot))$ *compatible with the heating environment* 13.2.1 *or* 13.2.3, *the canonical free energy is an absolutely continuous function of time and for a.e. time t during the process,*

$$\dot{\mathsf{F}}(t) = \mathsf{A}(t) - \theta_0 \Gamma(t), \tag{13.2.7}$$

where

$$\mathsf{A}(t) = \int_{\partial\Omega} v \cdot Sn\, dA + \int_{\Omega} v \cdot b\rho_\circ\, dV$$

is the power of external forces exerted on the body and $\Gamma(t)$ *is given by* (13.2.5). *Hence,*

$$w(\pi) = -[\mathsf{F}(\pi^f) - \mathsf{F}(\pi^i)] - \theta_0 \int_{t_1}^{t_2} \Gamma\, dt. \tag{13.2.8}$$

This is proved in the same way as Proposition 13.2.4 using (13.2.6).

13.3 Loading Environments

The mechanical interaction of the body with its environment to be described in this section will be more general than the equilibrium heating conditions defined above. The format will be general enough to cover the most common boundary conditions for both solids and fluids. Namely, we shall consider general, configuration- and deformation-gradient-dependent surface tractions and body forces. To simplify the notation and to avoid repeated hypotheses, we let

$$\boldsymbol{q}(\boldsymbol{p}, t) := (\boldsymbol{F}(\boldsymbol{p}, t), \boldsymbol{\chi}(\boldsymbol{p}, t), \boldsymbol{p})$$

for a given deformation χ, every $\boldsymbol{p} \in \operatorname{cl}\Omega$ and every time t a time during the process. Also, define

$$A := \mathrm{Lin}^+ \times R,$$

where $R \subset \mathrm{Eucl}$ will be specified below for each environment separately, and let $\boldsymbol{s}, \boldsymbol{b}_0$ be given continuous vector-valued functions on $A \times S, A \times \Omega$, respectively. The functions $\boldsymbol{s}, \boldsymbol{b}_0$ and the set S are parts of the specifications of the environments.

13.3.1 The Displacement-Traction Environment Let D, S be two disjoint complementary subsets of $\partial\Omega$ and set $R := \mathrm{Eucl}$. The ***displacement–traction environment*** consists in imposing the following conditions on the processes of the body:

$$\left.\begin{aligned} \boldsymbol{\chi}(\boldsymbol{p}, t) &= \boldsymbol{d}(\boldsymbol{p}) \text{ if } \boldsymbol{p} \in D, \\ \boldsymbol{S}(\boldsymbol{p}, t)\boldsymbol{n}(\boldsymbol{p}) &= \boldsymbol{s}(\boldsymbol{q}(\boldsymbol{p}, t)) \text{ if } \boldsymbol{p} \in S, \\ \boldsymbol{b}(\boldsymbol{p}, t) &= \boldsymbol{b}_0(\boldsymbol{q}(\boldsymbol{p}, t)) \text{ if } \boldsymbol{p} \in \Omega, \end{aligned}\right\} \quad (13.3.1)$$

for all times during the process. Here $\boldsymbol{d}$ is a given continuous vector-valued function on D.

Recall that throughout this part of the book it is assumed that $\det \boldsymbol{F} > 0$ in each process. We impose the following *regularity assumption:* S is relatively open in $\partial\Omega$ and for every $(\boldsymbol{F}, \boldsymbol{x}, \boldsymbol{p}) \in A \times (\Omega \cup S)$ there exists a twice continuously differentiable mapping χ : Eucl $\to$ Eucl such that $\chi = \boldsymbol{d}$ on D, $\boldsymbol{F} = \nabla\chi(\boldsymbol{p})$, $\boldsymbol{x} = \chi(\boldsymbol{p})\, \nabla\chi(\bar{\boldsymbol{p}}) \in \mathrm{Lin}^+$ for every $\bar{\boldsymbol{p}} \in \Omega \cup S$, and χ is one-to-one on $\operatorname{cl}\Omega$. This assumption ensures that there is a large supply of motions of Ω compatible with the kinematical constraint $(13.3.1)_1$; this will be employed in Proposition 13.3.7.

The ***pure displacement environment*** is the special case of the above when $S = \emptyset$; $(13.3.1)_1$ is then the Dirichlet boundary condition. The ***pure traction environment*** is the special case $D = \emptyset$. If $S \neq \emptyset \neq D$, the environment is called a ***mixed displacement–traction environment.*** The pure displacement environment is appropriate not only for solids, but also for fluids when the fluid adheres to the boundary of a canister $\bar{R}$ and cannot slip along $\partial\bar{R}$. (In this case $\boldsymbol{d} : \partial\Omega \to \partial\bar{R}$.)

13.3.2 The Bilateral Contact Environment Let D_i, $i = 1, \ldots, s$, S be pairwise disjoint subsets of $\partial\Omega$ whose union is $\partial\Omega$, and let P_i, $i = 1, \ldots, s$, be pairwise disjoint smooth oriented surfaces in Eucl. Set $R = \mathrm{Eucl}$. The ***bilateral contact environment*** consists in imposing the following conditions:

$$\left.\begin{aligned} \boldsymbol{\chi}(\boldsymbol{p}, t) \in P_i \text{ and } (\mathbf{1} - \boldsymbol{m} \otimes \boldsymbol{m})\boldsymbol{S}(\boldsymbol{p}, t)\boldsymbol{n}(\boldsymbol{p}) &= \mathbf{0} \text{ if } \boldsymbol{p} \in D_i \text{ for some } i, \\ \boldsymbol{S}(\boldsymbol{p}, t)\boldsymbol{n}(\boldsymbol{p}) &= \boldsymbol{s}(\boldsymbol{q}(\boldsymbol{p}, t)) \text{ if } \boldsymbol{p} \in S, \\ \boldsymbol{b}(\boldsymbol{p}, t) &= \boldsymbol{b}_0(\boldsymbol{q}(\boldsymbol{p}, t)) \text{ if } \boldsymbol{p} \in \Omega \end{aligned}\right\} \quad (13.3.2)$$

for all times during the process. The vector $\boldsymbol{m} \equiv \boldsymbol{m}(\boldsymbol{p}, t)$ is the normal to the actual boundary $\partial\chi(\Omega, t)$ considered as function of $\boldsymbol{p}$, t, and $\mathbf{1} - \boldsymbol{m} \otimes \boldsymbol{m}$ is the projection onto the tangent space to $\partial\chi(\Omega, t)$. Note that the kinematical constraint $(13.3.2)_{1,1}$ implies the slip condition

$$v(p, t) \cdot m = 0 \text{ if } p \in D_i \text{ for some } i = 1, \ldots, s. \tag{13.3.3}$$

The portions D_i of $\partial\Omega$ and the smooth surfaces P_i are known a priori. The motions are constrained by the conditions that the material points from D_i never leave the surface P_i, but the body can slip freely along these surfaces. This is a constraint weaker than the displacement condition $(13.3.1)_1$. On the other hand the tangential component of the surface traction on D_i must be $\mathbf{0}$. This implies that the slip is frictionless. The surface traction Sn is therefore of the form $Sn = \alpha m$. If $\alpha < 0$, the surface P_i exerts a pressure and a lubricated smooth surface of a rigid body can be imagined to realize such an interaction. If $\alpha > 0$ the interaction is harder to realize, but ANTMAN [1995, Chap. XIII.11] proposes to use lubricated magnets. Note that the case $S = \emptyset$ is not excluded. This arises, for example, when one considers motions of fluids in a canister $\bar{R}$ such that the fluid adheres to $\partial\bar{R}$ and can slip along it. This corresponds to $s = 1$, $D_1 = \partial\Omega$, $P_1 = \partial\bar{R}$.

We impose the following *regularity assumptions:* (a) S is relatively open in $\partial\Omega$. (b) For every $(F, x, p) \in A \times \mathrm{cl}\,\Omega$ such that $x \in P_j$ if $p \in D_j$ for some $j = 1, \ldots, s$, there exists a twice continuously differentiable mapping χ : Eucl $\to$ Eucl such that $\chi \in P_i$ on D_i, $i = 1, \ldots, s$, $F = \nabla\chi(p)$, $x = \chi(p)$ and $\nabla\chi(\bar{p}) \in \mathrm{Lin}^+$ for every $\bar{p} \in \Omega \cup S$. In addition χ is one-to-one on $\mathrm{cl}\,\Omega$. (c) Each D_i is a relative closure (in $\partial\Omega$) of its relative interior.

13.3.3 The Unilateral Contact Environment Consider a motion in a region $R \subset$ Eucl as in Subsect. 13.2.2, with (13.2.1) holding at every t. Set $S = \partial\Omega$. The ***unilateral contact environment*** consists in imposing the following conditions:

$$\left.\begin{aligned} (\mathbf{1} - m \otimes m)S(p, t)n(p) &= \mathbf{0} \text{ if } p \in C_t, \\ S(p, t)n(p) &= s(q(p, t)) \text{ if } p \in F_t, \\ b(p, t) &= b_0(q(p, t)) \text{ if } p \in \Omega \end{aligned}\right\} \tag{13.3.4}$$

for all times during the process of the body. Here m the normal to $\partial\chi(\Omega, t)$ at $\chi(p, t)$.

The body is allowed to slip along the surface of the container, but the surface traction has zero tangential component. The latter condition is automatically satisfied in inviscid fluids where $T = -p\mathbf{1}$, and the slip condition (13.2.2) is common in these contexts. We impose the regularity assumption that $\partial\Omega$ is class C^1. Recall that (13.2.1) implies the slip condition (13.2.2).

Further environments can be produced by combining the three environments above; however, in this book, by a ***loading environment*** we shall mean only the environments of the three types 13.3.1, 13.3.2 or 13.3.3 above. For the contact environments 13.3.2 or 13.3.3, the power of surface tractions on the surface of contact vanishes at every point. (This follows from the slip conditions (13.3.3), (13.2.2) and the assumed normality of the surface tractions.) More general definitions would be to constraint the surface tractions as to make the power nonnegative.

We shall limit ourselves to the conservative loading environments. Let us first discuss the form of the potential energy and its time-derivative; only then we proceed to the definition of a conservative loading environment.

13.3.4 Potential Energy It is assumed that the potential energy W is of the format

$$\mathsf{W}(\chi) = \int_\Omega \psi(F, \chi, p)\rho_o \, dV + \int_S \phi(\chi, p) \, dA, \tag{13.3.5}$$

where $F = \nabla\chi$ and both F and χ are evaluated at p, t. Here ψ, ϕ are scalar-valued functions on $\mathrm{Lin}^+ \times \mathrm{cl}\, R \times \Omega$, $\mathrm{cl}\, R \times S$, with S, R as in the definition of the three types of loading environments above. It is assumed that ψ is twice continuously differentiable in the sense that it has a twice continuously differentiable extension to an open set containing the closure of its domain, and that ϕ is continuously differentiable in the same sense. The potential energy is defined for all deformations (configurations) χ such that

(1) χ is a deformation of Ω with $\det \nabla\chi > 0$ outside the singular surface;
(2) χ is of class C^2 outside the singular surface,
(3) $\chi(\Omega) \subset R$.

In the subsequent treatment we consider motions that satisfy Conditions (1)–(3) of Subsect. 2.2.1. A ***kinematically admissible motion*** for a loading environment is a motion χ that satisfies the kinematical constraints $(13.3.1)_1$, $(13.3.2)_{1,1}$ or (13.2.1), in the case of a displacement–traction, bilateral, or unilateral loading environment, respectively.

13.3.5 Proposition *Let ψ, ϕ be as above. If χ is a motion on $[t_1, t_2]$ satisfying* (13.2.1) *then the function $t \mapsto \mathsf{W}(\chi(\cdot, t))$ is absolutely continuous on $[t_1, t_2]$ and for almost every $t \in [t_1, t_2]$,*

$$\begin{aligned}\frac{d}{dt}\mathsf{W}(\chi(\cdot, t)) = &\int_\Omega (\rho_\circ\partial_x\psi - \operatorname{Div}(\rho_\circ\partial_F\psi)) \cdot \dot{\chi}\, dV \\ &- \int_{S_t} (U\rho_\circ[\psi] + \rho_\circ[\partial_F\psi^{\mathsf T}\dot{\chi}] \cdot \boldsymbol{n})\, dA \\ &+ \int_{\partial\Omega} \rho_\circ\partial_F\psi\boldsymbol{n} \cdot \dot{\chi}\, dA + \int_S \partial_x\phi \cdot \dot{\chi}\, dA.\end{aligned} \tag{13.3.6}$$

Proof Assume $\rho_\circ = 1$. The function $\tilde{\psi}(\boldsymbol{p}, t) := \psi(\boldsymbol{F}(\boldsymbol{p}, t), \chi(\boldsymbol{p}, t), \boldsymbol{p})$ is piecewise continuously differentiable. Hence Proposition 3.2.2 can be invoked to assert that the volume integral in (13.3.5) is an absolutely continuous function of time with

$$\frac{d}{dt}\int_\Omega \tilde{\psi}\, dV = \int_\Omega (\partial_F\psi \cdot \dot{\boldsymbol{F}} + \partial_x\psi \cdot \dot{\chi})\, dV - \int_{S_t} U[\psi]\, dA.$$

Then the identity $\partial_F\psi \cdot \dot{\boldsymbol{F}} = \operatorname{Div}(\partial_F\psi^{\mathsf T}\dot{\chi}) - \operatorname{Div}(\partial_F\psi) \cdot \dot{\chi}$ and the divergence theorem 3.1.4 are employed to convert the volume integral of $\operatorname{Div}(\partial_F\psi^{\mathsf T}\dot{\chi})$ into surface integrals. This leads to

$$\frac{d}{dt}\int_\Omega \tilde{\psi}\, dV = \int_\Omega (\partial_x\psi \cdot \dot{\chi} - \operatorname{Div}(\partial_F\psi) \cdot \dot{\chi})\, dV - \int_{S_t} (U[\psi] + [\partial_F\psi^{\mathsf T}\dot{\chi}] \cdot \boldsymbol{n})\, dA + \int_{\partial\Omega} \partial_F\psi\boldsymbol{n} \cdot \dot{\chi}\, dA.$$

Finally, the integrand in the surface integral in (13.3.5) is an absolutely continuous function of time, with the derivative given by the chain rule. One can use these facts to show that the surface integral is an absolutely continuous function of time, and its time derivative is obtained by differentiating under the integral sign. □

13.3.6 Conservative Environments A loading environment is said to be a ***conservative*** if there exists a potential energy (13.3.5) such that for every kinematically admissible motion χ,

$$\frac{d}{dt}\mathsf{W}(\chi(\cdot, t)) = -\int_\Omega \boldsymbol{b}_0(\boldsymbol{q}) \cdot \boldsymbol{v}\rho_\circ\, dV - \int_S \boldsymbol{s}(\boldsymbol{q}) \cdot \boldsymbol{v}\, dA \tag{13.3.7}$$

for almost every t during the motion, with S in (13.3.7) replaced by F_t in the case of the unilateral environment 13.3.3. The functions ψ, ϕ are called the volume and surface densities of potential energy, respectively. (Notice that (13.3.7) is postulated to hold for any *kinematically admissible motion* and not just for the *processes* of the body satisfying the constitutive equations and the balance equations with the prescribed external forces. Indeed even to prove that the latter class is nonempty would require to prove an existence theorem for the corresponding initial-value problem.) A

conservative environment is a pair consisting of a heating and conservative loading environment. If both the environments are unilateral it is assumed that the set R is the same for both. A process compatible with the environment is a process that satisfies all constraints imposed by the environments in question.

For every process compatible with a conservative loading environment we have

$$w(\pi) = \mathsf{W}(\chi^f) - \mathsf{W}(\chi^i) \tag{13.3.8}$$

where χ^i and χ^f are the initial and the final configurations in this process. For this it suffices to realize that the right-hand side of (13.3.7) differs from the power of external forces by $\int_D \boldsymbol{Sn} \cdot \boldsymbol{v}\, dA$, $\int_{\partial\Omega\setminus S} \boldsymbol{Sn} \cdot \boldsymbol{v}\, dA$ in the case of displacement–traction and bilateral contact environments, respectively, and by $\int_{C_t} \boldsymbol{Sn} \cdot \boldsymbol{v}\, dA$ in the case of the unilateral contact environment. All these integrals vanish as a consequence of the conditions imposed on the processes by the environments.

13.3.7 Proposition *A functional* W *from* (13.3.5) *is a potential energy for a loading environment if and only if the following three conditions hold simultaneously:*

(1) *for every* $(\boldsymbol{F}, \boldsymbol{x}, \boldsymbol{p}) \in A \times \Omega$, *the divergence of* $\rho_o \partial_{\boldsymbol{F}} \psi(\boldsymbol{F}, \boldsymbol{x}, \boldsymbol{p})$, *calculated formally by the chain rule, is independent of* $\nabla \boldsymbol{F}$ *and*

$$\rho_o \boldsymbol{b}_0 = \mathrm{Div}(\rho_o \partial_{\boldsymbol{F}} \psi) - \rho_o \partial_{\boldsymbol{x}} \psi; \tag{13.3.9}$$

(2) *for every* $\boldsymbol{F}^+, \boldsymbol{F}^-, \boldsymbol{x}, \boldsymbol{p}$ *such that* $(\boldsymbol{F}^\pm, \boldsymbol{x}, \boldsymbol{p}) \in A \times \Omega$ *and* $[\boldsymbol{F}]$ *is of rank 1,*

$$[\partial_{\boldsymbol{F}} \psi][\boldsymbol{F}]^{\mathsf{T}} = \boldsymbol{0}, \quad [\psi] - \langle \partial_{\boldsymbol{F}} \psi \rangle \cdot [\boldsymbol{F}] = 0; \tag{13.3.10}$$

(3) *for every* $(\boldsymbol{F}, \boldsymbol{x}, \boldsymbol{p}) \in A \times S$,

$$\boldsymbol{s} = -\rho_o \partial_{\boldsymbol{F}} \psi\, \boldsymbol{n} - \partial_{\boldsymbol{x}} \phi; \tag{13.3.11}$$

and moreover, (a) *for a bilateral contact environment,*

$$(\boldsymbol{1} - \boldsymbol{m} \otimes \boldsymbol{m}) \partial_{\boldsymbol{F}} \psi\, \boldsymbol{n} = \boldsymbol{0} \tag{13.3.12}$$

at every $(\boldsymbol{F}, \boldsymbol{x}, \boldsymbol{p}) \in A \times \partial\Omega$ *such that* $\boldsymbol{p} \in D_i$ *for some* $i = 1, \dots, s$; (b) *for a unilateral contact environment*

$$(\boldsymbol{1} - \boldsymbol{m} \otimes \boldsymbol{m})(\rho_o \partial_{\boldsymbol{F}} \psi\, \boldsymbol{n} + \partial_{\boldsymbol{x}} \phi) = \boldsymbol{0} \tag{13.3.13}$$

at every $(\boldsymbol{F}, \boldsymbol{x}, \boldsymbol{p}) \in A \times S$ *such that* $\boldsymbol{x} \in \partial R$.

Proof Assume that $\rho_o = 1$. Let the environment be conservative. Consider first the displacement–traction environment. A combination of (13.3.6) and (13.3.7) gives that the following equation must be satisfied for every motion compatible with the constraint $(13.3.1)_1$ and a.e. time during the motion:

$$\int_\Omega (\partial_{\boldsymbol{x}} \psi + \boldsymbol{b}_0 - \mathrm{Div}(\partial_{\boldsymbol{F}} \psi)) \cdot \dot{\chi}\, dV + \int_S (\partial_{\boldsymbol{F}} \psi\, \boldsymbol{n} + \partial_{\boldsymbol{x}} \phi + \boldsymbol{s}) \cdot \dot{\chi}\, dA - \int_{S_t} (U[\psi] + [\partial_{\boldsymbol{F}} \psi^{\mathsf{T}} \dot{\chi}] \cdot \boldsymbol{n})\, dA = 0 \tag{13.3.14}$$

Let us now fix a $t_0 \in [t_1, t_2]$ at which (13.3.14) holds and let $\boldsymbol{u}$ be any smooth time-independent velocity field which vanishes in a neighborhood of D. Define a new motion $\bar{\chi}$ by $\bar{\chi} = \chi + (t - t_0)\boldsymbol{u}$, for $t \in [t_0 - \varepsilon, t_0 + \varepsilon]$, $\boldsymbol{p} \in \Omega$. If $\varepsilon > 0$ is sufficiently small, $\bar{\chi}$ is invertible in $\boldsymbol{p}$ and $\det \nabla \bar{\chi} > 0$ during $[t_0 - \varepsilon, t_0 + \varepsilon]$. Since $\boldsymbol{u}$ vanishes near D, $\bar{\chi}$ satisfies the boundary condition $(13.3.1)_1$. Since $\boldsymbol{u}$ is smooth, the singular surface of $\bar{\chi}$ coincides with that of χ. At $t = t_0$ we have $\bar{\chi} = \chi$, $\dot{\bar{\chi}} = \dot{\chi} + \boldsymbol{u}$. We now write down (13.3.14) for the motion $\bar{\chi}$ at time $t = t_0$, and subtract from it equation (13.3.14) for the original motion. The result is

$$\int_{\Omega} (\partial_{\boldsymbol{x}}\psi + \boldsymbol{b}_0 - \operatorname{Div}(\partial_{\boldsymbol{F}}\psi)) \cdot \boldsymbol{u}\, dV + \int_{S} (\partial_{\boldsymbol{F}}\psi\, \boldsymbol{n} + \partial_{\boldsymbol{x}}\phi + \boldsymbol{s}) \cdot \boldsymbol{u}\, dA - \int_{S_t} [\partial_{\boldsymbol{F}}\psi]\boldsymbol{n} \cdot \boldsymbol{u}\, dA = 0,$$

where the expressions in front of $\boldsymbol{u}$ are calculated on the original motion χ. (Since t_0 was chosen such that (13.3.14) holds at t_0, equation (13.3.14) holds at t_0 also for $\bar{\chi}$ as well.) Since $\boldsymbol{u}$ is arbitrary and independent of χ, the last equation implies that at t_0,

$$-\operatorname{Div}(\partial_{\boldsymbol{F}}\psi) + \partial_{\boldsymbol{x}}\psi + \boldsymbol{b}_0 = \boldsymbol{0}, \quad \partial_{\boldsymbol{F}}\psi\, \boldsymbol{n} + \partial_{\boldsymbol{x}}\phi + \boldsymbol{s} = \boldsymbol{0}, \quad [\partial_{\boldsymbol{F}}\psi]\boldsymbol{n} = \boldsymbol{0}, \tag{13.3.15}$$

outside the singular surface, on S and on the singular surface, respectively. Hence, by the continuity of the expressions in (13.3.15), these equations hold everywhere outside the singular surface, on S and on the singular surface, respectively, for all times in $[t_1, t_2]$. By $(13.3.15)_{1,2}$ we see that the integral over the singular surface S_t in (13.3.14) must vanish. With $(13.3.15)_3$, (2.1.4), (2.2.9) this may be written as

$$\int_{S_t} U([\psi] - \langle \partial_{\boldsymbol{F}}\psi \rangle \cdot [\boldsymbol{F}])\, dA = 0 \tag{13.3.16}$$

Let now t_0 be such that (13.3.19) holds, let V be a smooth scalar-valued function on Ω which vanishes near $\partial\Omega$, and define a new motion $\bar{\chi}$ by $\bar{\chi}(\boldsymbol{p}, t) = \chi(\boldsymbol{p}, V(\boldsymbol{p})(t-t_0))$ for $t \in [t_0-\varepsilon, t_0+\varepsilon]$, $\boldsymbol{p} \in \Omega$. This is a motion and it again satisfies the kinematical constrains $(13.3.1)_1$. The singular surface $\bar{S}_t$ with normal $\bar{N}$ and speed $\bar{U}$ of $\bar{\chi}$ will be generally different from that for χ. However, at $t = t_0$, the instantaneous configurations of S_t, $\bar{S}_t$ will be identical and moreover $\bar{\boldsymbol{N}} = \boldsymbol{n}$, $\bar{U} = VU$, $[\bar{F}] = [F]$. Equation (13.3.16) for $t = t_0$ and the motion $\bar{\chi}$ reads

$$\int_{S_t} V \cdot U([\psi] - \langle \partial_{\boldsymbol{F}}\psi \rangle \cdot [\boldsymbol{F}])\, dA = 0.$$

Consequently, the arbitrariness of V and $(13.3.15)_3$ give

$$[\partial_{\boldsymbol{F}}\psi][\boldsymbol{F}]^{\mathrm{T}} = \boldsymbol{0}, \quad [\psi] - \langle \partial_{\boldsymbol{F}}\psi \rangle \cdot [\boldsymbol{F}] = 0 \tag{13.3.17}$$

with the second equation at every point where $U \neq 0$. Equations $(13.3.15)_{1,2}$, (13.3.17) are exactly the assertions of (1)–(3), except that it must the ensured that the arguments in these equations can be chosen arbitrarily. For this we shall employ the regularity assumption in 13.3.1 to make various choices of the motion χ. First, let $(\boldsymbol{F}, \boldsymbol{x}, \boldsymbol{p}) \in A \times (\Omega \cup S)$ and take the motion χ independent of time with χ as in the regularity assumption. By noting that $\boldsymbol{G} := \nabla \boldsymbol{F}(\boldsymbol{p}) = \nabla^2\chi(\boldsymbol{p})$ can be chosen to have an arbitrary value (symmetric in the last two indices) without changing the values $\nabla\chi(\boldsymbol{p})$ and $\chi(\boldsymbol{p})$, we see that $(13.3.15)_{1,2}$ imply the statements (1), (3). In the same way, the regularity assumption may be used to assert that given $\boldsymbol{F}^{\pm}, \boldsymbol{x}, \boldsymbol{p}$ as in (2), a motion with a singular surface obeying $(13.3.1)_1$ can be found such that $\boldsymbol{F}^{\pm}$ are realized at some point of the singular surface where $U \neq 0$. By (13.3.17), this proves (2). Conversely, to prove that (1), (2), and (3) imply that W is a potential energy, one uses the relations in these conditions to show that (13.3.14) holds for every motion, which in turn implies that W is a potential energy.

Next consider a bilateral contact environment and suppose that it is conservative. Conditions (1), (2) and (13.3.11) of (3) are proved in the same way as above. Let us now prove (a). With (1), (2) and (13.3.11), the equation (13.3.7) reduces to

$$\int_{D} \rho_0 \partial_{\boldsymbol{F}}\psi\, \boldsymbol{n} \cdot \boldsymbol{v}\, dA = 0 \tag{13.3.18}$$

where $D = \cup_{i=1}^{s} D_i$. The regularity assumptions (b), (c) can be used to formally show that the velocity $\boldsymbol{v}$ on D can be chosen arbitrarily subject to the condition that $\boldsymbol{v} \cdot \boldsymbol{m} = 0$ on D. Using this arbitrariness, one sees that (13.3.18) implies that $\partial_{\boldsymbol{F}}\psi\, \boldsymbol{n}$ must be parallel with $\boldsymbol{m}$, which is (13.3.12). That conversely (1), (2) and (13.3.11), (13.3.12) imply that the environment is conservative is proved by reversing the arguments above.

Finally, consider the unilateral contact environment. If the environment is conservative, then (1), (2) and (13.3.11) of (3) are obtained as above by considering the processes in which no contact with ∂R takes place. With this knowledge, (13.3.7) reduces to

$$\int_{C_t} (\rho_0 \partial_{\boldsymbol{F}}\psi\, \boldsymbol{n} + \partial_{\boldsymbol{x}}\phi) \cdot \boldsymbol{v}\, dA = 0. \tag{13.3.19}$$

The smoothness of $\partial\Omega$, ∂R can be invoked to ensure that one has enough freedom to choose $\boldsymbol{v}$ arbitrarily subject to $\boldsymbol{v} \cdot \boldsymbol{m} = 0$. One then infers from (13.3.19) that $\rho_0 \partial_{\boldsymbol{F}}\psi\, \boldsymbol{n} + \partial_{\boldsymbol{x}}\phi$ must be perpendicular to the tangent space to $\chi(\Omega, t)$ at a given point of the surface of contact, which is (13.3.13). Conversely, if (1), (2), (3) hold, then the reversal of the steps yields that the environment is conservative, with the essential use of (13.3.13). □

13.3.8 ***Remarks***

(1) The expression

$$\delta_x(\rho_\circ\psi) := \rho_\circ\partial_x\psi - \operatorname{Div}(\rho_\circ\partial_F\psi) \tag{13.3.20}$$

is the ***variational derivative*** of $\rho_\circ\psi$.

(2) Equations (13.3.10) imply that the potential ψ does not contribute to the balance of energy on the singular surface.

(3) Proposition 13.3.7 is only a first step towards determining the form of the potential energy. In Sect. 13.5 and 13.7 it will be shown that $(13.3.10)_1$, $(13.3.10)_2$ and the requirement that the divergence of $\rho_\circ\partial_F\psi(\boldsymbol{F}, \boldsymbol{\chi}, \boldsymbol{p})$ be independent of $\nabla\boldsymbol{F}$ are mutually equivalent and hold if and only if $\rho_\circ\psi$ has the form (13.7.1) below.

(4) As a consequence of (3)(b), if a unilateral loading environment is conservative, then

$$(\mathbf{1} - \boldsymbol{m} \otimes \boldsymbol{m})\boldsymbol{s}(\boldsymbol{F}, \boldsymbol{x}, \boldsymbol{p}) = \mathbf{0}$$

if $(\boldsymbol{F}, \boldsymbol{x}, \boldsymbol{p}) \in \mathsf{A} \times \mathsf{S}$ and $\boldsymbol{x} \in \partial\mathsf{R}$.

(5) The potentials ψ, ϕ enjoy a large degree of nonuniqueness. See Sect. 13.7.

13.3.9 Example (Dead loadings) Any of the loading environments of the types 13.3.1–13.3.3 is said to be of a ***dead loading type*** if the functions $\boldsymbol{s}$, $\boldsymbol{b}_0$ are independent of $\boldsymbol{F}, \boldsymbol{x}$, i.e.,

$$\boldsymbol{s}(\boldsymbol{F}, \boldsymbol{x}, \boldsymbol{p}) = \boldsymbol{s}(\boldsymbol{p}), \quad \boldsymbol{b}_0(\boldsymbol{F}, \boldsymbol{x}, \boldsymbol{p}) = \boldsymbol{b}_0(\boldsymbol{p})$$

for every $(\boldsymbol{F}, \boldsymbol{x}, \boldsymbol{p})$ in $\mathsf{A} \times \mathsf{S}$ and $\mathsf{A} \times \Omega$, respectively. A ***dead loading environment in the restricted sense*** is a displacement–traction environment of a dead loading type. A displacement–traction environment or the bilateral contact environment of the dead loading type is conservative with the potential energy

$$\mathsf{W}(\boldsymbol{\chi}) = -\int_\Omega \boldsymbol{b}_0 \cdot \boldsymbol{\chi}\rho_\circ\, dV - \int_\mathsf{S} \boldsymbol{s}(\boldsymbol{p}) \cdot \boldsymbol{\chi}\, dA.$$

The unilateral contact environment of a dead loading type is conservative only if $\boldsymbol{s} \equiv \mathbf{0}$ identically. Indeed the consequence of conservativity stated Remark 13.3.8(4) reads

$$|\boldsymbol{F}^{-\mathsf{T}}\boldsymbol{n}(\boldsymbol{p})|^2\boldsymbol{s}(\boldsymbol{p}) = \boldsymbol{F}^{-\mathsf{T}}\boldsymbol{n}(\boldsymbol{p}) \otimes \boldsymbol{F}^{-\mathsf{T}}\boldsymbol{n}(\boldsymbol{p})\boldsymbol{s}(\boldsymbol{p}).$$

for every $(\boldsymbol{F}, \boldsymbol{x}, \boldsymbol{p}) \in \mathsf{A} \times \mathsf{S}$ with $\boldsymbol{x} \in \partial\mathsf{R}$. Since $\boldsymbol{x}$ does not occur in the last equation, one can use the arbitrariness of $\boldsymbol{F}$ to infer that $\boldsymbol{s} \equiv \mathbf{0}$.

13.3.10 Example (A spatially-dependent pressure loading) Any of the loading environments 13.3.1–13.3.3 is said to be a ***spatially dependent pressure loading*** if $\boldsymbol{b}_0$ is independent of $\boldsymbol{F}, \boldsymbol{x}$ and $\boldsymbol{s}$ has the form

$$\boldsymbol{s}(\boldsymbol{F}, \boldsymbol{x}, \boldsymbol{p}) = -\pi(\boldsymbol{x}) \operatorname{cof} \boldsymbol{F}\boldsymbol{n}(\boldsymbol{p}), \quad (\boldsymbol{F}, \boldsymbol{x}, \boldsymbol{p}) \in \mathsf{A} \times \mathsf{S},$$

where π is a given function on R. In terms of the Cauchy stress, the traction condition on S reads

$$\boldsymbol{T}(\boldsymbol{x}, t)\boldsymbol{m}(\boldsymbol{x}, t) = -\pi(\boldsymbol{x})\boldsymbol{m}(\boldsymbol{x}, t) \ \text{ where } \ \boldsymbol{x} = \boldsymbol{\chi}(\boldsymbol{p}, t).$$

This is an example of a *live loading*. The pressure environment arises when the body is immersed in a fluid of a given, possibly spatially dependent pressure, like a bubble of a gas in a liquid, or a drop of a liquid in a gas. If $\boldsymbol{b}_0$ is conservative, i.e., if

$$\boldsymbol{b}_0(\boldsymbol{x}) = -\operatorname{grad} \psi_b(\boldsymbol{x}), \quad \boldsymbol{x} \in R \tag{13.3.21}$$

for some continuously differentiable function ψ_b on R, the spatially-dependent pressure loading is conservative and ψ, ϕ may be chosen to be

$$\psi(\boldsymbol{F}, \boldsymbol{x}, \boldsymbol{p}) = \psi_b(\boldsymbol{x}) + \rho_\circ^{-1}\pi(\boldsymbol{x}) \det \boldsymbol{F}, \quad \phi \equiv 0 \tag{13.3.22}$$

for every $\boldsymbol{F} \in \mathrm{Lin}^+$, $\boldsymbol{x} \in R$ and $\boldsymbol{p} \in \Omega$. This is proved by verifying Conditions of Proposition 13.3.7. Indeed that $\delta_{\boldsymbol{x}}(\rho_\circ \psi)$ is independent of $\nabla \boldsymbol{F}$ follows from $\operatorname{Div} \operatorname{cof} \boldsymbol{F} = \boldsymbol{0}$; this identity is also the main step in the proof of (13.3.9). The jump conditions of Proposition 13.3.7(2) follow from the jump conditions for determinant and cofactor. Finally, Proposition 13.3.7(3) follows from $\partial_{\boldsymbol{F}} \det \boldsymbol{F} = \operatorname{cof} \boldsymbol{F}$. (Alternatively, one may invoke the general results about the potentials to be given in Sect. 13.7 to verify the above statements.) The contribution to the potential energy from π is

$$\mathsf{W}_\pi(\chi) = \int_\Omega \pi \det \boldsymbol{F}\, dV = \int_{\chi(\Omega)} \pi(\boldsymbol{x})\, dv.$$

If π is constant, then $\mathsf{W}_\pi(\chi) = \pi \mathsf{V}(\chi(\Omega))$, where $\mathsf{V}(\chi(\Omega))$ is the actual volume of the body.

13.3.11 Example (A referentially-dependent pressure loading) Any of the loading environments 13.3.1–13.3.3 is said to be a ***referentially dependent pressure loading*** if $\boldsymbol{b}_0$ is independent of $\boldsymbol{F}, \boldsymbol{x}$ and s has the form

$$s(\boldsymbol{F}, \boldsymbol{x}, \boldsymbol{p}) = -\pi(\boldsymbol{p}) \operatorname{cof} \boldsymbol{F}\boldsymbol{n}(\boldsymbol{p}), \quad (\boldsymbol{F}, \boldsymbol{x}, \boldsymbol{p}) \in A \times S,$$

where π is a given *piecewise constant* function on S. This means that S may be written as a union of r pairwise disjoint relatively open (in $\partial\Omega$) sets S_i, $i = 1, \ldots, r$ such that $\pi(\boldsymbol{p}) = \pi_i$ on S_i where π_i are constants. We also assume that π has a $C^1(\operatorname{cl}\Omega)$ extension to $\operatorname{cl}\Omega$, denoted by the same letter. An example is an inflated balloon on whose outer and inner boundaries act gases of different, but prescribed, pressures. If condition (13.3.21) holds, then the referentially-dependent pressure loading is conservative and ψ, ϕ may be chosen to be

$$\rho_\circ \psi(\boldsymbol{F}, \boldsymbol{x}, \boldsymbol{p}) = \rho_\circ \psi_b(\boldsymbol{x}) + \pi(\boldsymbol{p}) \det \boldsymbol{F} + \tfrac{1}{3}(\operatorname{cof} \boldsymbol{F}) \nabla \pi(\boldsymbol{p}) \cdot \boldsymbol{x}, \quad \phi \equiv 0 \tag{13.3.23}$$

for every $\boldsymbol{F} \in \mathrm{Lin}^+$, $\boldsymbol{x} \in R$ and $\boldsymbol{p} \in \Omega$. This is verified by the same method as in 13.3.10.

13.3.12 Example (Currently dead surfaces force) A loading environment is said to be a ***currently dead surfaces force*** if

$$s(\boldsymbol{F}, \boldsymbol{x}, \boldsymbol{p}) = |\operatorname{cof} \boldsymbol{F}\boldsymbol{n}| \boldsymbol{t}(\boldsymbol{x}, \boldsymbol{p}) \quad \text{for every } (\boldsymbol{F}, \boldsymbol{x}, \boldsymbol{p}) \in A \times S \tag{13.3.24}$$

where $\boldsymbol{t}$ is a given vector-valued function on $R \times S$. In terms of the Cauchy stress,

$$\boldsymbol{T}(\boldsymbol{p}, t)\boldsymbol{m}(\boldsymbol{p}, t) = \boldsymbol{t}(\boldsymbol{x}, \boldsymbol{p}).$$

A currently dead surfaces force is not conservative unless $\boldsymbol{t} \equiv \boldsymbol{0}$. Indeed, one finds that the form (13.3.24) of s is incompatible with the general form Proposition 13.7.1 unless $\boldsymbol{t} \equiv \boldsymbol{0}$. Other nonconservative surface forces are described in GUO & URBANOWSKI [1963], WESOŁOWSKI [1964], and SEWELL [1967]. However, the energy stability criteria to be developed in the subsequent chapters do not apply to the nonconservative environments and they are outside the scope of the present monograph.

13.3.13 The Total Quantities For a body in a conservative loading environment, we define the ***total stress*** and the ***total Eshelby tensor*** by

$$\boldsymbol{\mathsf{S}} = S + \Sigma, \quad \boldsymbol{\mathsf{P}} = P + \Pi, \tag{13.3.25}$$

where

$$\boldsymbol{\Sigma} := \rho_\circ \partial_{\boldsymbol{F}} \psi, \quad \boldsymbol{\Pi} := \psi \mathbf{1} - \rho_\circ^{-1} \boldsymbol{F}^{\mathsf{T}} \boldsymbol{\Sigma} = \psi \mathbf{1} - \boldsymbol{F}^{\mathsf{T}} \partial_{\boldsymbol{F}} \psi. \tag{13.3.26}$$

Notice that (13.3.10) imply that

$$[\boldsymbol{\Sigma}]\boldsymbol{n} = \mathbf{0}, \quad [\boldsymbol{\Pi}]\boldsymbol{n} = \mathbf{0} \tag{13.3.27}$$

on the singular surface. If $q = q(\boldsymbol{F}, \theta, \boldsymbol{x}, \boldsymbol{p})$ is the ***total free energy***

$$q = \bar{f} + \psi, \tag{13.3.28}$$

then for equilibrium states (where there is no dynamic contribution to S), we have

$$\boldsymbol{\mathsf{S}} = \rho_\circ \partial_{\boldsymbol{F}} q, \quad \boldsymbol{\mathsf{P}} = q\mathbf{1} - \boldsymbol{F}^{\mathsf{T}} \partial_{\boldsymbol{F}} q. \tag{13.3.29}$$

Notice that $\boldsymbol{\mathsf{S}}$, $\boldsymbol{\mathsf{P}}$ differ from S, P only if the potential energy depends on $\boldsymbol{F}$. For instance, $\boldsymbol{\mathsf{S}} = S$, $\boldsymbol{\mathsf{P}} = P$ for dead loadings. Not so for the pressure environments, where the volume density of the potential energy has the special form (13.3.22) and we have

$$\boldsymbol{\Sigma} = \pi(\boldsymbol{x}) \operatorname{cof} \boldsymbol{F}, \quad \boldsymbol{\Pi} = \psi_b \mathbf{1}, \quad q = \bar{f} + \pi v + \psi_b. \tag{13.3.30}$$

For equilibrium states of fluids,

$$\boldsymbol{\mathsf{S}} = (\pi - p) \operatorname{cof} \boldsymbol{F}, \quad \boldsymbol{\mathsf{P}} = (f + pv + \psi_b)\mathbf{1}. \tag{13.3.31}$$

The equation of balance of energy takes an especially simple form when it is formulated in terms of the total stress, namely, for every process compatible with a conservative loading environment we have

$$\rho_\circ (\tfrac{1}{2} v^2 + e + \psi)^{\cdot} = \operatorname{Div}(\boldsymbol{\mathsf{S}}^{\mathsf{T}} v - \boldsymbol{q}), \quad -U\rho_\circ [\tfrac{1}{2} v^2 + e + \psi] = [\boldsymbol{\mathsf{S}}^{\mathsf{T}} v - \boldsymbol{q}] \cdot \boldsymbol{n},$$

outside of and on the singular surface, respectively. Likewise, we shall see in Sect. 14.2 that the equations of mechanical equilibrium and their consequences are most easily formulated also in terms of $\boldsymbol{\mathsf{S}}$, $\boldsymbol{\mathsf{P}}$.

13.4 The Total Canonical Free Energy

Let us define the ***total energy*** (for conservative loading environments)

$$\mathsf{U}(\sigma) = \mathsf{K}(\sigma) + \mathsf{E}(\sigma) + \mathsf{W}(\sigma),$$

and the ***total canonical free energy*** (for equilibrium heating environment of ambient temperature θ_0 and conservative loading environment with potential energy W)

$$\mathsf{P}(\sigma) = \mathsf{K}(\sigma) + \mathsf{E}(\sigma) + \mathsf{W}(\sigma) - \theta_0 \mathsf{H}(\sigma).$$

13.4.1 Proposition *If π is a process compatible with the conservative environment then*

(1) *the total canonical free energy P at the ambient temperature θ_0 is a monotonically nonincreasing function; more precisely, P is an absolutely continuous function and for a.e. time t during the process,*

$$\dot{\mathsf{P}}(t) = -\theta_0 \Gamma(t) \le 0; \tag{13.4.1}$$

(2) *in an adiabatic isolation, the total entropy H is a monotonically nondecreasing function and the total energy is constant; H is an absolutely continuous function and for a.e. time t during the process,*

$$\dot{\mathsf{H}}(t) = \Gamma(t) \ge 0. \tag{13.4.2}$$

Here $\Gamma \ge 0$ is the total production of entropy (13.2.5).

Item (2) is the famous principle of increase of entropy. Both (1) and (2) provide *natural* Liapunov functions for processes in the specific environments, i.e. functions monotonically decreasing along all processes. As will be shown in the subsequent treatment, this is a basis of the thermodynamic stability theory.

Proof To show that P is a nonincreasing function, it suffices to eliminate $\mathsf{W}(\pi)$ between (13.2.3) and (13.3.8). To obtain (13.4.1), it suffices to invoke (13.2.7). (2) is proved similarly. □

13.5 Homogeneous Null Lagrangians

The results of this section will be employed in the next section to determine the form of the potential energy of a conservative environment and in Sect. 17.5 to define the polyconvex functions.

13.5.1 Definition Let U be an open *connected* subset of Lin and $f : U \to \mathbb{R}$ a continuous function. f is said to be a ***homogeneous null Lagrangian*** if

$$f(\alpha \boldsymbol{F} + (1-\alpha)\boldsymbol{G}) = \alpha f(\boldsymbol{F}) + (1-\alpha) f(\boldsymbol{G}) \tag{13.5.1}$$

for every $\boldsymbol{F}, \boldsymbol{G} \in U$ such that $\boldsymbol{F} - \boldsymbol{G}$ is of rank at most 1 and for every α, $0 \le \alpha \le 1$ such that $\alpha \boldsymbol{F} + (1-\alpha)\boldsymbol{G} \in U$.

The reason for this terminology will become apparent in Sect. 13.6. To avoid repeated hypotheses, throughout this section, let $f : U \to \mathbb{R}$ be a continuous function defined on an open connected subset of Lin. The following two propositions will be proved simultaneously:

13.5.2 Proposition *The following conditions are equivalent:*

(1) *f is a homogeneous null Lagrangian;*

(2) *f is of class C^1 and*

$$f(\boldsymbol{F} + \boldsymbol{a} \otimes \boldsymbol{b}) = f(\boldsymbol{F}) + \partial_{\boldsymbol{F}} f(\boldsymbol{F}) \cdot (\boldsymbol{a} \otimes \boldsymbol{b}); \tag{13.5.2}$$

for every $\boldsymbol{F} \in U$, $\boldsymbol{a}, \boldsymbol{b} \in$ Vect such that $\boldsymbol{F} + \boldsymbol{a} \otimes \boldsymbol{b} \in U$;

(3) *f is of class C^2 and $\partial^2_{\boldsymbol{F}\boldsymbol{F}} f(\boldsymbol{F})[\boldsymbol{a} \otimes \boldsymbol{b}, \boldsymbol{a} \otimes \boldsymbol{b}] = 0$ for every $\boldsymbol{F} \in U$, $\boldsymbol{a}, \boldsymbol{b} \in$ Vect.*

This proposition will serve as a preparation to finding the explicit form of homogeneous null Lagrangians:

13.5.3 Proposition *Let* dim Vect $= n$. *Then* f *is a homogeneous null Lagrangian if and only if* $f(\boldsymbol{F})$ *is a linear combination, with constant coefficients, of minors of* $\boldsymbol{F}$ *in some (and hence every) fixed basis. In particular, if* dim Vect $= 3$, *then* f *is a homogeneous null Lagrangian if and only if it is of the form*

$$f(\boldsymbol{F}) = A + \boldsymbol{B} \cdot \boldsymbol{F} + \boldsymbol{C} \cdot \operatorname{cof} \boldsymbol{F} + D \det \boldsymbol{F}, \qquad \boldsymbol{F} \in U, \tag{13.5.3}$$

where A, D *are constant numbers and* $\boldsymbol{B}, \boldsymbol{C} \in$ Lin *constant tensors.*

Proof of Propositions 13.5.2 and 13.5.3 Suppose first that f is of class C^∞ and prove that (1) $\Rightarrow$ (2) $\Rightarrow$ (3) $\Rightarrow$ conclusion of Proposition 13.5.3 $\Rightarrow$ (1).

(1) $\Rightarrow$ (2): Let $\boldsymbol{F}, \boldsymbol{a}, \boldsymbol{b}$ be as in (2) and set $\boldsymbol{G} = \boldsymbol{F} + \boldsymbol{a} \otimes \boldsymbol{b}$. Since U is open, for every $\alpha \le 1$ sufficiently close to 1 (13.5.1) makes sense and differentiating it at $\alpha = 1$ we obtain (13.5.2). (2) $\Rightarrow$ (3): Let $\boldsymbol{F}, \boldsymbol{a}, \boldsymbol{b}$ be as in (3). Replacing $\boldsymbol{a}$ by $\alpha\boldsymbol{a}$, where $\alpha > 0$ is sufficiently small, in (13.5.2) and differentiating at $\alpha = 0$ we obtain the formula in (3).

(3) $\Rightarrow$ (1): Let $\boldsymbol{a}_1, \boldsymbol{u}_1 \in$ Vect and $\boldsymbol{F} \in U$. Omitting the argument $\boldsymbol{F}$, we have

$$\partial_F^2 f[\boldsymbol{a}_1 \otimes \boldsymbol{u}_1, \boldsymbol{a}_1 \otimes \boldsymbol{u}_1] = 0. \tag{13.5.4}$$

We shall repeatedly use the following consequence of the polarization identity: if $B : \text{Vect}\times\text{Vect}\to\mathbb{R}$ is a bilinear form and $B(\boldsymbol{x}, \boldsymbol{x}) = 0$ for every $\boldsymbol{x} \in$ Vect, then B is skew, i.e., $B(\boldsymbol{x}, \boldsymbol{y}) + B(\boldsymbol{y}, \boldsymbol{x}) = 0$ for every $\boldsymbol{x}, \boldsymbol{y} \in$ Vect. Fixing $\boldsymbol{a}_1$ and applying this observation to the form $(\boldsymbol{x}, \boldsymbol{y}) \mapsto \partial_F^2 f[\boldsymbol{a}_1 \otimes \boldsymbol{x}, \boldsymbol{a}_1 \otimes \boldsymbol{y}]$, one finds from (13.5.4) that

$$\partial_F^2 f[\boldsymbol{a}_1 \otimes \boldsymbol{u}_1, \boldsymbol{a}_1 \otimes \boldsymbol{u}_2] + \partial_F^2 f[\boldsymbol{a}_1 \otimes \boldsymbol{u}_2, \boldsymbol{a}_1 \otimes \boldsymbol{u}_1] = 0 \tag{13.5.5}$$

for every $\boldsymbol{u}_1, \boldsymbol{u}_2 \in$ Vect. The interchangeability of the second partial derivatives (13.5.5) gives $\partial_F^2 f[\boldsymbol{a}_1 \otimes \boldsymbol{u}_1, \boldsymbol{a}_1 \otimes \boldsymbol{u}_2] = 0$. Using the above observation again, we find that

$$\partial_F^2 f[\boldsymbol{a}_1 \otimes \boldsymbol{u}_1, \boldsymbol{a}_2 \otimes \boldsymbol{u}_2] + \partial_F^2 f[\boldsymbol{a}_2 \otimes \boldsymbol{u}_1, \boldsymbol{a}_1 \otimes \boldsymbol{u}_2] = 0. \tag{13.5.6}$$

In other words, for every fixed $\boldsymbol{u}_1, \boldsymbol{u}_2$ the expression

$$\partial_F^2 f[\boldsymbol{a}_1 \otimes \boldsymbol{u}_1, \boldsymbol{a}_2 \otimes \boldsymbol{u}_2] \tag{13.5.7}$$

is skew in $\boldsymbol{a}_1, \boldsymbol{a}_2$. Interchanging the roles of $\boldsymbol{a}_i$, and $\boldsymbol{u}_i$ in the above argument, one also finds that for every fixed $\boldsymbol{a}_1, \boldsymbol{a}_2$, the expression (13.5.7) is skew in $\boldsymbol{u}_1, \boldsymbol{u}_2$. Differentiating (13.5.6) $k - 2$ times $(2 \le k \le n + 1)$ and using the interchangeability of the partial derivatives, one deduces that for every fixed $\boldsymbol{F}$,

$$\partial_F^k f(\boldsymbol{F})[\boldsymbol{a}_1 \otimes \boldsymbol{u}_1, \ldots, \boldsymbol{a}_k \otimes \boldsymbol{u}_k]$$

is completely skew in $\boldsymbol{a}_1, \ldots, \boldsymbol{a}_k$ for every fixed $\boldsymbol{u}_1, \ldots, \boldsymbol{u}_k$ and completely skew in $\boldsymbol{u}_1, \ldots, \boldsymbol{u}_k$ for every fixed $\boldsymbol{a}_1, \ldots, \boldsymbol{a}_k$. By Proposition 1.1.6 then $\partial_F^{n+1} f(\boldsymbol{F})[\boldsymbol{A}, \ldots, \boldsymbol{A}] = 0$ for every $\boldsymbol{F} \in U, \boldsymbol{A} \in$ Lin from which $\partial_F^{n+1} f(\boldsymbol{F}) = 0$ for every $\boldsymbol{F} \in U$. Hence f is a polynomial of degree n in the components of $\boldsymbol{F}$, the n-form $\omega(\boldsymbol{A}, \ldots, \boldsymbol{A}) := \partial_F^n f(\boldsymbol{F})[\boldsymbol{A}, \ldots, \boldsymbol{A}]$ is independent of $\boldsymbol{F}$ and the integration gives

$$f(\boldsymbol{F}) = \frac{1}{n!}\omega(\boldsymbol{F}, \ldots, \boldsymbol{F}) + g(\boldsymbol{F}), \qquad \boldsymbol{F} \in U,$$

where g is a polynomial of degree $n - 1$. By Proposition 1.1.6, $\omega(\boldsymbol{F}, \ldots, \boldsymbol{F}) = c \det \boldsymbol{F}$ and hence

$$f(\boldsymbol{F}) = \frac{c}{n!} \det \boldsymbol{F} + g(\boldsymbol{F}), \qquad \boldsymbol{F} \in U.$$

Then for every $k \ge 2$,

$$\partial_F^k g(\boldsymbol{F})[\boldsymbol{a}_1 \otimes \boldsymbol{u}_1, \ldots, \boldsymbol{a}_k \otimes \boldsymbol{u}_k]$$

is completely skew in $\boldsymbol{a}_1, \ldots, \boldsymbol{a}_k$ and also in $\boldsymbol{u}_1, \ldots, \boldsymbol{u}_k$, and moreover $\partial_F^n g(\boldsymbol{F}) = 0$. Hence the $n-1$-form $\omega(\boldsymbol{A}, \ldots, \boldsymbol{A}) := \partial_F^{n-1} g(\boldsymbol{F})[\boldsymbol{A}, \ldots, \boldsymbol{A}]$ is independent of $\boldsymbol{F}$ and by Proposition 1.1.6 $\omega(\boldsymbol{A}, \ldots, \boldsymbol{A})$ is a linear combination of minors of order $n - 1$. Proceeding inductively, and using Proposition 1.1.6 in every step, it is found that f is of the asserted form.

If f is as in the conclusion of Proposition 13.5.3, then a direct verification using the form of minors (1.1.15) shows that it satisfies (13.5.1).

This proves that Propositions 13.5.2 and 13.5.3 hold under the additional assumption that f is infinitely differentiable. The proof is now completed by showing that the infinite differentiability is an automatic consequence of (13.5.1) and the assumed continuity of f. Namely we mollify f using the Friedrichs mollifier. Each mollification will satisfy (13.5.1) as a consequence of (13.5.1) for the original f. This gives us a sequence f_m, $m = 1, \dots,$ of infinitely differentiable homogeneous null Lagrangians such that $f_m \to f$ uniformly on compact sets as $m \to \infty$. By the already proved part, each f_m is a polynomial of degree n in $\boldsymbol{F}$ and hence their limit f is again a polynomial of degree n and hence infinitely differentiable. □

The following two propositions show that homogeneous null Lagrangians satisfy balance equations and the equations of equilibrium, including the jump conditions, identically for every motion. Throughout the rest of this section, let U be any open connected subset of Lin, and let us give the term motion the following meaning: A motion is any continuous and piecewise continuously differentiable function χ mapping a set $\Omega \times [t_1, t_2]$ into Eucl, where Ω is an open subset of Eucl and $[t_1, t_2] \subset \mathbb{R}$, such that outside the singular surface, f is of class C^2 and the values of $\boldsymbol{F} = \nabla\chi$, including the limiting values on the singular surface, are in U.

13.5.4 Proposition *f is a homogeneous null Lagrangian if and only if f is of class C^2 and any of the following three conditions holds (separately) for every motion χ :*

(1) $\operatorname{Div}(\partial_{\boldsymbol{F}} f) = \mathbf{0}$ *outside the singular surface;*

(2) $[\partial_{\boldsymbol{F}} f]\boldsymbol{n} = \mathbf{0}$ *on the singular surface;*

(3) $\operatorname{Div}(\partial_{\boldsymbol{F}} f)$ *is independent of* $\nabla\boldsymbol{F}$ *outside the singular surface.*

The requirement that $\operatorname{Div}(\partial_{\boldsymbol{F}} f)$ be independent of $\nabla\boldsymbol{F}$ is explained in Proposition 13.3.7(1).

Proof That a homogeneous null Lagrangian satisfies (1), (2), and (3) is a verification using the explicit form Proposition 13.5.3, and the details are suggested to the reader as an exercise. Clearly, (1) ⇒ (3) and for the proof of (3) ⇒ (1) it suffices to note that $\operatorname{Div}(\partial_{\boldsymbol{F}} f) = \partial^2_{\boldsymbol{F}} f \nabla\boldsymbol{F}$ and as the last expression is linear in $\nabla\boldsymbol{F}$, it is independent of $\nabla\boldsymbol{F}$ if and only if the divergence vanishes. Let us prove that (1) implies that f is a homogeneous null Lagrangian. Let $\boldsymbol{a}, \boldsymbol{b} =$ Vect, $\boldsymbol{F} \in U$ and let χ be a motion for which, at a particular $\boldsymbol{p}, t$, we have $\nabla\chi = \boldsymbol{F}$, $\nabla^2\chi = \boldsymbol{a} \otimes \boldsymbol{b} \otimes \boldsymbol{b}$. The requirement $\operatorname{Div}(\partial_{\boldsymbol{F}} f) = \mathbf{0}$ is easily shown to give Condition (3) of Proposition 13.5.2 and hence that f is a homogeneous null Lagrangian. Finally let us prove that (2) implies that f is a homogeneous null Lagrangian. Let $\boldsymbol{a}, \boldsymbol{b} =$ Vect, $\boldsymbol{F} \in U$ and let χ be a motion for which, at a particular $\boldsymbol{p}, t$ on the singular surface, the normal be $\boldsymbol{b}$ and $[\boldsymbol{F}] = \boldsymbol{a} \otimes \boldsymbol{b}$. Then (2) reduces to $\partial_{\boldsymbol{F}} f(\boldsymbol{F} + \boldsymbol{a} \otimes \boldsymbol{b})\boldsymbol{b} = \partial_{\boldsymbol{F}} f(\boldsymbol{F})\boldsymbol{b}$. A differentiation with respect to $\boldsymbol{a}$ and a multiplication by $\boldsymbol{a}$ gives Condition (3) of Proposition 13.5.2 again. □

13.5.5 Proposition *f is a homogeneous null Lagrangian if and only if f is of class C^2 and any of the following two conditions holds (separately) for every motion χ :*

(1) $\dot f = \operatorname{Div}((\partial_{\boldsymbol{F}} f)^{\mathsf{T}}\boldsymbol{v})$, *outside the singular surface;*

(2) $U[f] + [(\partial_{\boldsymbol{F}} f)^{\mathsf{T}}\boldsymbol{v}] \cdot \boldsymbol{n} = 0$ *on the singular surface.*

Proof f is a homogeneous null Lagrangian ⇔ (1): For any f, we have $\dot f - \operatorname{Div}((\partial_{\boldsymbol{F}} f)^{\mathsf{T}}\boldsymbol{v}) = -\boldsymbol{v} \cdot \operatorname{Div}(\partial_{\boldsymbol{F}} f)$; hence (1) holds if and only if $\operatorname{Div}(\partial_{\boldsymbol{F}} f) = \mathbf{0}$. By Proposition 13.5.4(1) this is equivalent to f being a homogeneous null Lagrangian. f is a homogeneous null Lagrangian ⇒ (2): We have, for any f,

$$U[f] + [(\partial_{\boldsymbol{F}} f)^{\mathsf{T}}\boldsymbol{v}] \cdot \boldsymbol{n} = U([f] - \langle\partial_{\boldsymbol{F}} f\rangle \cdot [\boldsymbol{F}]) + [\partial_{\boldsymbol{F}} f]\boldsymbol{n} \cdot \langle\boldsymbol{v}\rangle. \tag{13.5.8}$$

If f is a homogeneous null Lagrangian, then by Proposition 13.5.4(2) we have $[\partial_{\boldsymbol{F}} f]\boldsymbol{n} = \mathbf{0}$ and hence $\langle\partial_{\boldsymbol{F}} f\rangle\,\boldsymbol{n} = \partial_{\boldsymbol{F}} f(\boldsymbol{F}^+)\boldsymbol{n}$. Therefore, by Proposition 13.5.4(2) and Proposition 13.5.2(2) the right-hand side vanishes. (2) ⇒ f is a homogeneous null Lagrangian: If (2) holds, we can replace $\boldsymbol{v}$ by $\boldsymbol{v} + \boldsymbol{u}$ to obtain $-U[f] = [(\partial_{\boldsymbol{F}} f)^{\mathsf{T}}\boldsymbol{v}] \cdot \boldsymbol{n} + [\partial_{\boldsymbol{F}} f]\boldsymbol{n} \cdot \boldsymbol{u}$. The arbitrariness of $\boldsymbol{u}$ then implies $[\partial_{\boldsymbol{F}} f]\boldsymbol{n} = \mathbf{0}$ and hence, by Proposition 13.5.2(2), f is homogeneous null Lagrangian. □

13.5.6 Proposition *f is a homogeneous null Lagrangian if and only if for each $\boldsymbol{F} \in U$, each bounded open subset $D \subset$ Eucl, and each $\boldsymbol{u} \in C_0^\infty(D)$ such that $\boldsymbol{F} + \nabla \boldsymbol{u} \in U$, we have*

$$\int_D f(\boldsymbol{F} + \nabla \boldsymbol{u})\, dV = |D| f(\boldsymbol{F}). \tag{13.5.9}$$

Here $|D|$ is the volume of D.

Proof We shall first prove that

$$\int_D \det(\boldsymbol{F} + \nabla \boldsymbol{u})\, dV = |D| \det \boldsymbol{F}, \tag{13.5.10}$$

for any $\boldsymbol{F} \in$ Lin, $\boldsymbol{u} \in C_0^\infty(D)$, where D is as in the statement of the proposition. Clearly, if suffices to prove (13.5.10) under the additional hypothesis that D has smooth boundary; otherwise one passes from D to a bounded region with smooth boundary containing D. We use the following consequence of (1.1.5) and (1.1.14): For any $\chi \in C^\infty(\text{Vect})$ we have $\text{Div}\,((\text{cof}\,\boldsymbol{F})^{\mathsf{T}}\chi) = n \det \boldsymbol{F}$, where $\boldsymbol{F} = \nabla\chi$ and n is the dimension of Vect. This is proved by using indices and the symmetry of second partial derivatives. Setting $\chi(\boldsymbol{p}) := \boldsymbol{F}\boldsymbol{p} + \boldsymbol{u}(\boldsymbol{p})$, $\chi_0(\boldsymbol{p}) := \boldsymbol{F}\boldsymbol{p}$, $\boldsymbol{p} \in$ Vect, we therefore have

$$\begin{aligned} n \int_D \det(\boldsymbol{F} + \nabla \boldsymbol{u})\, dV &= \int_D \text{Div}\,((\text{cof}\,\boldsymbol{F})^{\mathsf{T}}\chi)\, dV = \int_{\partial D} (\text{cof}\,\boldsymbol{F})^{\mathsf{T}}\chi \cdot \boldsymbol{n}\, dA \\ &= \int_{\partial D} (\text{cof}\,\boldsymbol{F})^{\mathsf{T}}\chi_0 \cdot \boldsymbol{n}\, dA = \int_D \text{Div}\,((\text{cof}\,\boldsymbol{F})^{\mathsf{T}}\chi_0)\, dV \\ &= n \int_D \det \boldsymbol{F}\, dV = n|D| \det \boldsymbol{F}. \end{aligned}$$

This establishes (13.5.10). We now use the fact that any minor of $\boldsymbol{F}$ is a derivative of certain order of $\det \boldsymbol{F}$ with respect to (some component of) $\boldsymbol{F}$. Hence by differentiating (13.5.10) with respect to $\boldsymbol{F}$ we prove that (13.5.9) holds for any minor of $\boldsymbol{F}$. As by Proposition 13.5.3 any homogeneous null Lagrangian is a linear combination of minors, (13.5.9) holds for every homogeneous null Lagrangian.

The converse implication will be proved in Sect. 17.3. There it will be seen that (13.5.1) implies that both f and $-f$ are rank 1 convex and hence that both f and $-f$ satisfy the Legendre–Hadamard condition (see Proposition 17.3.9), which means that f satisfies condition (3) of Proposition 13.5.2 and hence it is a homogeneous null Lagrangian. □

13.6 General Null Lagrangians

13.6.1 Definition Let Ω, $R \subset$ Eucl, $U \subset$ Lin be open sets and let $f : U \times R \times \Omega \to \mathbb{R}$ be twice continuously differentiable. The function f is said to be a ***null Lagrangian*** if

$$\delta_x f := \partial_x f - \text{Div}(\partial_F f) = \mathbf{0} \text{ on } \Omega \tag{13.6.1}$$

for every twice continuously differentiable mapping $\chi : \Omega \to R$ such that $\nabla\chi(\boldsymbol{p}) \in U$ for every $\boldsymbol{p} \in \Omega$. In (13.6.1), $\partial_x f$ and $\partial_F f$ are evaluated at $(\nabla\chi(\boldsymbol{p}), \chi(\boldsymbol{p}), \boldsymbol{p})$, $\boldsymbol{p} \in \Omega$. This terminology is consistent with that of Definition 13.5.1, since, as we shall see in Proposition 13.6.2, if f is independent of $\boldsymbol{x}, \boldsymbol{p}$, i.e., if $f(\boldsymbol{F}, \boldsymbol{x}, \boldsymbol{p}) = f(\boldsymbol{F})$, then f is a null Lagrangian on Ω if and only it is a homogeneous null Lagrangian.

If we define a functional I on mappings χ from Ω to R by

$$\mathsf{I}(\chi) = \int_\Omega f(\nabla\chi, \chi, \boldsymbol{p})\, dV,$$

then the function f is a null Lagrangian if and only if $\delta\mathsf{I}(\chi, \boldsymbol{u}) = 0$ for every χ and every $\boldsymbol{u} \in C_0^\infty(\Omega)$, where $\delta\mathsf{I}$ is the first variation of I to be defined in Sect. 14.3. Thus the null Lagrangians are exactly those functions f which lead to Euler–Lagrange equations that are satisfied by *any* smooth mapping. In other words, adding a null Lagrangian to any Lagrangian has no effect on Euler–Lagrange equations. (It may have an effect on the boundary conditions, though.)

13.6.2 Proposition *If $f : U \times \mathrm{Eucl} \times \Omega \to \mathbb{R}$ is a null Lagrangian on Ω then for every $\boldsymbol{x} \in \mathrm{Eucl}$ and $\boldsymbol{p} \in \Omega$ the function $f(\cdot, \boldsymbol{x}, \boldsymbol{p})$ is a homogeneous null Lagrangian.*

Proof For general null Lagrangians the arbitrariness of $\chi(\boldsymbol{p})$, $\nabla\chi(\boldsymbol{p})$, $\nabla^2\chi(\boldsymbol{p})$ at the given point shows that (13.6.1) splits into

$$\partial_{F_{iA}}\partial_{F_{jB}} f \chi_{j,AB} = 0 \quad \text{and} \quad \partial_{F_{iA}}\partial_{x_j} f \chi_{j,A} + \partial_{F_{iA}}\partial_{p_A} f - \partial_{x_i} f = 0.$$

Fixing $\boldsymbol{p}$ and $\boldsymbol{x} = \chi$, it is observed that the first of these equations says that $f(\cdot, \boldsymbol{p}, \boldsymbol{x})$ is a null Lagrangian. Taking χ such that $\nabla^2\chi(\boldsymbol{p}) = \boldsymbol{a} \otimes \boldsymbol{b} \otimes \boldsymbol{b}$ and using Proposition 13.5.2, we see that $f(\cdot, \boldsymbol{p}, \boldsymbol{x})$ is a homogeneous null Lagrangian in the sense of Definition 13.5.1. □

We now give a characterization of the null Lagrangians as divergences of vector fields which depend on the same variables.

13.6.3 Proposition *If U, R, Ω are star-shaped, then a function $f : U \times R \times \Omega \to \mathbb{R}$ is a null Lagrangian if and only if there exists a $\boldsymbol{Q} : U \times \mathrm{Eucl} \times \Omega \to \mathrm{Vect}$ of class C^2 such that $f(\boldsymbol{F}, \boldsymbol{x}, \boldsymbol{p}) = \mathrm{Div}\, \boldsymbol{Q}(\boldsymbol{F}, \boldsymbol{x}, \boldsymbol{p})$ for every smooth χ.*

See OLVER [1986] or BALL, CURRIE & OLVER [1981] for proof.

13.6.4 The Three-Dimensional Case If $n = 3$ and U is connected, then by Propositions 13.6.2 and 13.5.3 every null Lagrangian has the form

$$f = A + \boldsymbol{B} \cdot \boldsymbol{F} + \boldsymbol{C} \cdot \mathrm{cof}\, \boldsymbol{F} + D \det \boldsymbol{F}; \tag{13.6.2}$$

the coefficients A, $\boldsymbol{B}$, $\boldsymbol{C}$, D being class C^2 functions on $R \times \Omega$. A calculation gives that

$$\delta_{\boldsymbol{x}} f = \mathrm{grad}\, A - \mathrm{Div}\, \boldsymbol{B} - (\boldsymbol{F}(\mathrm{Rot}\, \boldsymbol{C})^{\mathsf{T}} + \boldsymbol{F}(\mathrm{rot}\, \boldsymbol{B}^{\mathsf{T}}))^{\wedge} + \mathrm{cof}\, \boldsymbol{F}(\mathrm{div}\, \boldsymbol{C}^{\mathsf{T}} - \mathrm{Grad}\, D).$$

Here $(\boldsymbol{M}^{\wedge})_i = \varepsilon_{ijk} M_{jk}$ for any second order tensor $\boldsymbol{M}$; grad, div, rot are the gradient, divergence, and rotation with respect to $\boldsymbol{x}$ keeping $\boldsymbol{p}$ fixed and Grad, Div, Rot are the gradient, divergence, and rotation with respect to $\boldsymbol{p}$ keeping $\boldsymbol{x}$ fixed. The rotations are given by

$$(\mathrm{Rot}\, \boldsymbol{C})_{iA} = \varepsilon_{ABC} C_{iB,C}, \qquad (\mathrm{rot}\, \boldsymbol{D})_{Ai} = \varepsilon_{ijk} D_{Aj,k};$$

the comma followed by a subscript i denotes the derivative with respect to x_i; the comma followed by a subscript A denotes the derivative with respect to p_A. The details of a rather lengthy calculation are suggested as an exercise to the reader (use indices).

13.6.5 Proposition *Let $n = 3$ and U be connected. Then*

(1) *f is a null Lagrangian if and only if it is of the form (13.6.2) with the functions A, $\boldsymbol{B}$, $\boldsymbol{C}$, D satisfying*

$$\mathrm{grad}\, A = \mathrm{Div}\, \boldsymbol{B}, \quad (\mathrm{Rot}\, \boldsymbol{C})^{\mathsf{T}} = -\,\mathrm{rot}\, \boldsymbol{B}^{\mathsf{T}}, \quad \mathrm{Grad}\, D = \mathrm{div}\, \boldsymbol{C}^{\mathsf{T}};$$

(2) *if f is of the form (13.6.2) and*

$$A = \mathrm{Div}\, \boldsymbol{L}, \;\; \boldsymbol{B} = \mathrm{Rot}\, \boldsymbol{K} + (\mathrm{grad}\, \boldsymbol{L})^{\mathsf{T}}, \;\; \boldsymbol{C} = \mathrm{Grad}\, \boldsymbol{M} - (\mathrm{rot}\, \boldsymbol{K}^{\mathsf{T}})^{\mathsf{T}}, \;\; D = \mathrm{div}\, \boldsymbol{M},$$

where $\boldsymbol{K}$, $\boldsymbol{L}$, $\boldsymbol{M}$ are class C^2 functions on $R \times \Omega$ with values in Lin, Vect, Vect, *then f is a null Lagrangian; moreover,*

$$f = \mathrm{Div}\, \boldsymbol{Q} \quad \text{where} \quad \boldsymbol{Q} = \boldsymbol{L} + (\boldsymbol{F}^{\mathsf{T}}\boldsymbol{K})^{\wedge} + (\mathrm{cof}\, \boldsymbol{F})^{\mathsf{T}}\boldsymbol{M}; \tag{13.6.3}$$

(3) *if R, Ω are star-shaped, then f is a null Lagrangian if and only if it is of the form (13.6.2) with A, $\boldsymbol{B}$, $\boldsymbol{C}$, D given by $\boldsymbol{K}$, $\boldsymbol{L}$, $\boldsymbol{M}$ as in* (2).

Proof (1): This follows from the calculation of $\delta_{\boldsymbol{x}} f$ in 13.6.4. (2): That every f as in (2) is a null Lagrangian follows by verifying that A, $\boldsymbol{B}$, $\boldsymbol{C}$, D satisfy the relations in (1). This is straightforward using the well-known rules for the operators grad, div, rot and Grad, Div, Rot, such as rot grad = $\mathbf{0}$, div rot = $\mathbf{0}$, etc., and the commutativity of each of the grad, div, rot with each of Grad, Div, Rot (interchangeability of the second partial derivatives). The formula $f = \operatorname{Div} \boldsymbol{Q}$ with $\boldsymbol{Q}$ as in (2) is just a calculation. (3): We shall use the Poincaré lemma (e.g., ABRAHAM & MARSDEN [1978, Thm. 2.4.17(ii)] that on a star-shaped domain every closed form is exact; in particular, every rotation-free vector field is a gradient of a scalar field and every divergence-free vector field is a rotation of a vector field, and similarly for Grad, Div, Rot. Hence, if $A, \boldsymbol{B}, \boldsymbol{C}, D$ satisfy the relations in (1), we write $D = \operatorname{div} \boldsymbol{M}$ for some vector field $\boldsymbol{M} = \boldsymbol{M}(\boldsymbol{x}, \boldsymbol{p})$ on $R \times \Omega$, with $\boldsymbol{p} \in \Omega$ playing the role of a parameter. The last equation in (1) can be written as $\operatorname{div}(\operatorname{Grad} \boldsymbol{M}^{\mathsf{T}} - \boldsymbol{C}^{\mathsf{T}}) = \mathbf{0}$ and hence $\operatorname{Grad} \boldsymbol{M}^{\mathsf{T}} - \boldsymbol{C}^{\mathsf{T}} = \operatorname{rot} \boldsymbol{K}^{\mathsf{T}}$ for some tensor-field $\boldsymbol{K}$ on $R \times \Omega$, which gives $\boldsymbol{C} = \operatorname{Grad} \boldsymbol{M} - (\operatorname{rot} \boldsymbol{K}^{\mathsf{T}})^{\mathsf{T}}$. With this the middle equation in (1) gives $\operatorname{rot}(\boldsymbol{B} - \operatorname{Rot} \boldsymbol{K}) = \mathbf{0}$ from which $\boldsymbol{B} - \operatorname{Rot} \boldsymbol{K} = \operatorname{grad} \bar{\boldsymbol{L}}^{\mathsf{T}}$ for some vector-filed $\bar{\boldsymbol{L}}$ on $R \times \Omega$. This gives $\boldsymbol{B} = \operatorname{Rot} \boldsymbol{K} + \operatorname{grad} \bar{\boldsymbol{L}}^{\mathsf{T}}$ and the first equation in (1) reduces to $\operatorname{grad}(A - \operatorname{Div} \bar{\boldsymbol{L}}) = \mathbf{0}$. This may be integrated to give $A = \operatorname{Div} \bar{\boldsymbol{L}} + m$ where m is a scalar function of $\boldsymbol{p}$ only. Finally, introducing $\boldsymbol{X} = \boldsymbol{X}(\boldsymbol{p})$ such that $\operatorname{Div} \boldsymbol{X} = m$ and defining $\boldsymbol{L} = \bar{\boldsymbol{L}} + \boldsymbol{X}$ we obtain the relations asserted in (3). Moreover, a more detailed argument shows that the fields $\boldsymbol{K}, \boldsymbol{L}, \boldsymbol{M}$ may be chosen to have the asserted smoothness. □

13.6.6 Null Divergences (OLVER [1983, 1986]) A continuously differentiable function $\boldsymbol{Q} : U \times R \times \Omega \to \text{Vect}$ is said to be a ***null divergence*** if $\operatorname{Div} \boldsymbol{Q} = \mathbf{0}$ for every twice continuously differentiable mapping $\chi : \Omega \to R$, where $\boldsymbol{Q} = \boldsymbol{Q}(\nabla\chi(\boldsymbol{p}), \chi(\boldsymbol{p}), \boldsymbol{p})$, $\boldsymbol{p} \in \Omega$. This is a notion closely related to that of a null Lagrangian. A ***homogeneous null divergence*** is a null divergence that depends solely on $\boldsymbol{F} = \nabla\chi$. In analogy with (13.5.3), if $n = 3$ and U is connected, then every homogeneous null divergence $\boldsymbol{Q}$ is of the form

$$\boldsymbol{Q}(\boldsymbol{F}) = \boldsymbol{L} + (\boldsymbol{F}^{\mathsf{T}}\boldsymbol{K})^{\wedge} + (\operatorname{cof} \boldsymbol{F})^{\mathsf{T}}\boldsymbol{M}, \quad \boldsymbol{F} \in U,$$

where $\boldsymbol{L}, \boldsymbol{M} \in \text{Vect}$ and $\boldsymbol{K} \in \text{Lin}$ are constants. A general null divergence may be characterized by the condition that $\operatorname{Div} \boldsymbol{Q}$ is independent of $\nabla^2\chi$ for every twice continuously differentiable mapping χ, in a complete analogy with 13.5.4(3). Moreover, for every $\boldsymbol{x} \in R$ and $\boldsymbol{p} \in \Omega$, $\boldsymbol{Q}(\cdot, \boldsymbol{x}, \boldsymbol{p})$ is a homogeneous null divergence. We observe that if $\boldsymbol{Q}$ is as in Proposition 13.6.5(2) then $\boldsymbol{Q}(\cdot, \boldsymbol{x}, \boldsymbol{p})$ is a homogeneous null divergence, a consequence of the fact that by $f = \operatorname{Div} \boldsymbol{Q}$ necessarily $\operatorname{Div} \boldsymbol{Q}$ is independent of $\nabla^2\chi$.

To provide an additional insight into the structure of the equations in Proposition 13.6.5, let us associate with every null Lagrangian f of the form (13.6.2) a differential 3-form ω on $R \times \Omega$ by

$$\omega = A\, dp_1 \wedge dp_2 \wedge dp_3 + \tfrac{1}{2}\varepsilon_{ABC} B_{iA} dx_i \wedge dp_B \wedge dp_C + \tfrac{1}{2}\varepsilon_{ijk} C_{iA} dx_j \wedge dx_k \wedge dp_A + D dx_1 \wedge dx_2 \wedge dx_3$$

and with $\boldsymbol{Q}$ of the form $(13.6.3)_2$ a differential 2-form ζ on $R \times \Omega$ by

$$\zeta = \tfrac{1}{2}\varepsilon_{ABC} L_A dp_B \wedge dp_C + K_{iA} dx_i \wedge dp_A + \tfrac{1}{2}\varepsilon_{ijk} M_i dx_j \wedge dx_k$$

with the summation taken over repeated indices. Then the condition that f be a null Lagrangian reads $d\omega = 0$ (i.e., ω is a closed form) and the condition $f = \operatorname{Div} \boldsymbol{Q}$ reads $\omega = d\zeta$. (This follows from OLVER & SIVALOGANATHAN [1988, Lemmas 5 & 6].)

13.7 The Form of the Potential Energy

The potential energy ψ of a conservative environment satisfies Conditions (1)–(3) of Proposition 13.3.7. Let us fix $\boldsymbol{x}, \boldsymbol{p}$ and consider the function $f := \rho_0\psi(\cdot, \boldsymbol{x}, \boldsymbol{p})$. By the results of Sect. 13.5, the assertion that the divergence of $\rho_0\partial_{\boldsymbol{F}}\psi$ is independent of $\nabla\boldsymbol{F}$ and (13.3.10) are equivalent and equivalent to $\rho_0\psi$ being a homogeneous null Lagrangian. Hence Proposition 13.5.3 tells us that $\rho_0\psi$ is of the form

$$\rho_0\psi = A + \boldsymbol{B} \cdot \boldsymbol{F} + \boldsymbol{C} \cdot \operatorname{cof} \boldsymbol{F} + D \det \boldsymbol{F}, \quad \boldsymbol{F} \in \text{Lin}^+, \tag{13.7.1}$$

where $A, \boldsymbol{B}, \boldsymbol{C}, D$ are smooth functions on $R \times \Omega$. To evaluate the density of the body force and the surface tractions by (13.3.9) and (13.3.11), we have to evaluate the variational and normal derivatives of $\rho_0\psi$. The variational derivative has been calculated in Subsect. 13.6.4; here we note that

$$\rho_\circ \partial_F \psi \boldsymbol{n} = \boldsymbol{B}\boldsymbol{n} + ((\boldsymbol{C}\cdot\boldsymbol{F}^{-\mathsf{T}} + D)\mathbf{1} - \boldsymbol{F}^{-\mathsf{T}}\boldsymbol{C}^{\mathsf{T}})\operatorname{cof}\boldsymbol{F}\boldsymbol{n}.$$

To prove it, use the formula

$$\partial_F(\boldsymbol{C}\cdot\operatorname{cof}\boldsymbol{F}) = ((\boldsymbol{C}\cdot\boldsymbol{F}^{-\mathsf{T}})\mathbf{1} - \boldsymbol{F}^{-\mathsf{T}}\boldsymbol{C}^{\mathsf{T}})\operatorname{cof}\boldsymbol{F}.$$

Hence

13.7.1 Proposition *The volume density of the potential energy of a conservative loading environment is of the form* (13.7.1) *and*

$$\rho_\circ \boldsymbol{b}_0 = \operatorname{Div}\boldsymbol{B} - \operatorname{grad}A + (\boldsymbol{F}(\operatorname{rot}\boldsymbol{B}^{\mathsf{T}}) + \boldsymbol{F}(\operatorname{Rot}\boldsymbol{C})^{\mathsf{T}})^{\wedge} + \operatorname{cof}\boldsymbol{F}(\operatorname{Grad}D - \operatorname{div}\boldsymbol{C}^{\mathsf{T}}),$$

$$\boldsymbol{s} = \boldsymbol{S}\boldsymbol{n} = -\boldsymbol{B}\boldsymbol{n} - ((\boldsymbol{C}\cdot\boldsymbol{F}^{-\mathsf{T}} + D)\mathbf{1} - \boldsymbol{F}^{-\mathsf{T}}\boldsymbol{C}^{\mathsf{T}})\operatorname{cof}\boldsymbol{F}\boldsymbol{n} - \partial_{\boldsymbol{x}}\phi.$$

In terms of the Cauchy stress the last can be rewritten as

$$\boldsymbol{T}\boldsymbol{m} = -(\det\boldsymbol{F})^{-1}\boldsymbol{B}\boldsymbol{F}^{\mathsf{T}}\boldsymbol{m} - ((\boldsymbol{C}\cdot\boldsymbol{F}^{-\mathsf{T}} + D)\mathbf{1} - \boldsymbol{F}^{-\mathsf{T}}\boldsymbol{C}^{\mathsf{T}})\boldsymbol{m} - (\det\boldsymbol{F})^{-1}|\boldsymbol{F}^{\mathsf{T}}\boldsymbol{m}|\partial_{\boldsymbol{x}}\phi.$$

The interpretation of the terms in $\boldsymbol{b}_0$ involving $\boldsymbol{F}$ is unclear. The densities $\rho_\circ\psi$ derived in Examples 13.3.9–13.3.11 have the form (13.7.1) with special choices of $A, \boldsymbol{B}, \boldsymbol{C}, D$ but $\boldsymbol{b}_0$ is independent of $\boldsymbol{F}$. Also the contribution from $\boldsymbol{B}, \boldsymbol{C}$ to the expression for $\boldsymbol{s}$ do not seem to have any direct physical interpretation, in contrast to the terms $\partial_{\boldsymbol{x}}\phi$ and the term D, which arise in the dead loading and pressure environments.

13.7.2 Nonuniqueness The coefficients $A, \boldsymbol{B}, \boldsymbol{C}, D$ are not uniquely determined by the environment. Clearly, the body force $\boldsymbol{b}_0$ remains invariant under the addition of a null Lagrangian to $\rho_\circ\psi$ and any two volume densities of the potential energy giving the same body force differ by a null Lagrangian. The null Lagrangians in three dimensions have been described in Proposition 13.6.5. In particular, the body force remains invariant under the passages

$$\left.\begin{aligned} A \mapsto A' &= A + \operatorname{Div}\boldsymbol{L}, \\ \boldsymbol{B} \mapsto \boldsymbol{B}' &= \boldsymbol{B} + \operatorname{Rot}\boldsymbol{K} + (\operatorname{grad}\boldsymbol{L})^{\mathsf{T}}, \\ \boldsymbol{C} \mapsto \boldsymbol{C}' &= \boldsymbol{C} + \operatorname{Grad}\boldsymbol{M} - (\operatorname{rot}\boldsymbol{K}^{\mathsf{T}})^{\mathsf{T}}, \\ D \mapsto D' &= D + \operatorname{div}\boldsymbol{M} \end{aligned}\right\} \qquad (13.7.2)$$

where $\boldsymbol{K}, \boldsymbol{L}, \boldsymbol{M}$ are class C^2 functions on $R\times\Omega$ with values in Lin, Vect, Vect. On the other hand, an inspection of the expression for $\boldsymbol{s}$ in Proposition 13.7.1 shows that $\rho_\circ\partial_F\psi\boldsymbol{n}$ remains unchanged on S under the passages

$$A \mapsto A', \quad \boldsymbol{B} \mapsto \boldsymbol{B}' = \boldsymbol{B} + \boldsymbol{E} \text{ where } \boldsymbol{E}\boldsymbol{n} = \mathbf{0}, \quad \boldsymbol{C} \mapsto \boldsymbol{C}' = \boldsymbol{C} + \boldsymbol{G}\otimes\boldsymbol{n}, \quad D \mapsto D \qquad (13.7.3)$$

where $A', \boldsymbol{E}, \boldsymbol{G}$ are functions of $\boldsymbol{x}, \boldsymbol{p}$ on $R\times S$ with values in $\mathbb{R}$, Lin, Vect, respectively, assuming that ϕ remains unchanged for the rest of this discussion. (Indeed replacing in the expression for $\boldsymbol{s}$ in Proposition 13.7.1 the quantities $\boldsymbol{B}, \boldsymbol{C}, D$ by $\boldsymbol{B}', \boldsymbol{C}', D'$ and subtracting so obtained equation from the original equation, we are lead to

$$\boldsymbol{B}''\boldsymbol{n} + ((\boldsymbol{C}''\cdot\boldsymbol{F}^{-\mathsf{T}} + D'')\mathbf{1} - \boldsymbol{F}^{-\mathsf{T}}\boldsymbol{C}''^{\mathsf{T}})\operatorname{cof}\boldsymbol{F}\boldsymbol{n} = \mathbf{0}$$

where $\boldsymbol{B}'' = \boldsymbol{B}' - \boldsymbol{B}, \boldsymbol{C}'' = \boldsymbol{C}' - \boldsymbol{C}, D'' = D' - D$. As this must be satisfied for every $\boldsymbol{F} \in \mathrm{Lin}^+$, we replace $\boldsymbol{F}$ by $\lambda\boldsymbol{F}$, $\lambda > 0$, to see that the individual terms scale differently. As $\lambda > 0$ is arbitrary, this gives

$$\boldsymbol{B}''\boldsymbol{n} = \mathbf{0}, \qquad D\operatorname{cof}\boldsymbol{F}\boldsymbol{n} = \mathbf{0}, \qquad ((\boldsymbol{C}''\cdot\boldsymbol{F}^{-\mathsf{T}})\mathbf{1} - \boldsymbol{F}^{-\mathsf{T}}\boldsymbol{C}''^{\mathsf{T}})\operatorname{cof}\boldsymbol{F}\boldsymbol{n} = \mathbf{0}.$$

The first two equations give $\boldsymbol{E}\boldsymbol{n} = \mathbf{0}, D = 0$. The third equation and the arbitrariness of $\boldsymbol{F}$ give $\boldsymbol{C}'' = \boldsymbol{G}\otimes\boldsymbol{n}$.) The general freedom in $A, \boldsymbol{B}, \boldsymbol{C}, D$ is then obtained by (13.7.2) with $\boldsymbol{K}, \boldsymbol{L}, \boldsymbol{M}$ such that (13.7.3) hold.

As an illustration, consider the pure traction environment with $\boldsymbol{s} = \mathbf{0}, \boldsymbol{b}_0 = \mathbf{0}$. Then $\psi \equiv 0, \phi \equiv 0$ give a potential energy for which $A = D = 0, \boldsymbol{B} = \boldsymbol{C} = \mathbf{0}$. Let us show that if ψ is independent of $\boldsymbol{x}, \boldsymbol{p}$ and $\phi \equiv 0$, then ψ is a constant function. Indeed, form (13.7.3) with $A = D = 0, \boldsymbol{B} = \boldsymbol{C} = \mathbf{0}$ and from the assumed constancy of $A', \boldsymbol{B}', \boldsymbol{C}', D'$ follows that $\boldsymbol{E} = \boldsymbol{E}_0$ and $\boldsymbol{G}\otimes\boldsymbol{n} = \boldsymbol{M}_0$ are constant on $S \equiv \partial\Omega$. But then $\boldsymbol{E}_0\boldsymbol{n}(\boldsymbol{p}) = \boldsymbol{0}, \boldsymbol{p} \in \partial\Omega$ implies $\boldsymbol{E}_0 = \boldsymbol{0}$. Similarly, from $\boldsymbol{G} \otimes \boldsymbol{n} = \boldsymbol{M}_0$ follows that $\boldsymbol{M}_0^{\mathrm{T}}\boldsymbol{a} = \boldsymbol{n}(\boldsymbol{p})(\boldsymbol{G}(\boldsymbol{p})\cdot\boldsymbol{a})$ where $\boldsymbol{a} \in$ Vect is any vector. In particular, if $\boldsymbol{M}_0^{\mathrm{T}}\boldsymbol{a} \neq \boldsymbol{0}$, then $\boldsymbol{M}_0^{\mathrm{T}}\boldsymbol{a} = \boldsymbol{n}(\boldsymbol{p})(\boldsymbol{G}(\boldsymbol{p})\cdot\boldsymbol{a})$ provides a contradiction, as it equates a nonzero constant vector with a multiple of a vector $\boldsymbol{n}$, which changes its direction as $\boldsymbol{p}$ ranges on $\partial\Omega$. Hence $\boldsymbol{M}_0^{\mathrm{T}}\boldsymbol{a} = \boldsymbol{0}$ for every $\boldsymbol{a} \in$ Vect and therefore $\boldsymbol{G} \otimes \boldsymbol{n} = \boldsymbol{0}$, which by (13.7.3) means that $\boldsymbol{B}', \boldsymbol{C}', D$ vanish and A' is constant. Consequently also ψ is constant.

On the other hand, within the class of general potentials ψ, there are nonconstant potentials which give $\boldsymbol{s} \equiv \boldsymbol{0}, \boldsymbol{b}_0 \equiv \boldsymbol{0}$. Indeed, let $\pi : \mathrm{cl}\,\Omega \to \mathbb{R}$ be a continuously differentiable function such that $\pi = 0$ on $\partial\Omega$ and let $\rho_0\psi(\boldsymbol{F}, \boldsymbol{x}, \boldsymbol{p}) := \pi(\boldsymbol{p})\det\boldsymbol{F} + \frac{1}{3}(\mathrm{cof}\,\boldsymbol{F})\nabla\pi(\boldsymbol{p}) \cdot \boldsymbol{x}$. This corresponds to

$$A' = 0, \quad \boldsymbol{B}' = \boldsymbol{0}, \quad \boldsymbol{C}' = \tfrac{1}{3}\boldsymbol{x} \otimes \nabla\pi, \quad D' = \pi,$$

and to (13.7.2) with

$$\boldsymbol{L} = \boldsymbol{0}, \quad \boldsymbol{K} = \boldsymbol{0}, \quad \boldsymbol{M} = \tfrac{1}{3}\boldsymbol{x}\pi.$$

Then by the results of 13.3.11, ψ is a potential for $\boldsymbol{s} \equiv \boldsymbol{0}, \boldsymbol{b}_0 \equiv \boldsymbol{0}$.

Bibliographical Note

The use of the canonical free energy in the context of continuous bodies is due to Duhem (1911); see also Ericksen [1966], Koiter [1967], Coleman & Dill [1968, 1973] and Gurtin [1973, 1975] and the references therein. The canonical free energy is also called the availability, cf. Müller [1985] or Jou, Casas-Vázquez & Lebon [1993]. Ball & Knowles [1986] show that for materials satisfying the Clausius–Planck inequality the canonical free energy may be generalized to spatially varying temperature fields θ_0. The assertion about the maximum work 13.2.4 is classical. Most of these general assertions hold for more general heating environments which Gurtin [1973, 1975] calls passive.

The confinement condition 13.2.2, due to Ciarlet & Nečas [1987], is a nonlinear generalization of the conditions arising in Signorini's problem, see Fichera [1972], Nečas & Hlaváček [1981]. The earlier systematic studies of elastic loading environments are Guo & Urbanowski [1963], Sewell [1967] and Knops & Wilkes [1973]. However, these articles do not use the powerful method of null Lagrangians and the results are not so explicit. The use of null Lagrangians to clarify various issues about loading environments is in Edelen [1981] and Podio-Guidugli & Vergara Caffarelli [1990]. The last reference studies also surface tractions that depend on $\nabla\boldsymbol{F}$. Beatty [1987] attributes the dead loading environment to Bryan (1886–89). For the pressure environments 13.3.10 and 13.3.11, see Sewell [1967], Beatty [1970], Ball [1977] and Podio-Guidugli & Vergara Caffarelli [1990]. As has been pointed out, the format of the environments adopted above by no means exhausts all types of external conditions; for additional information, see Batra [1972], Bharatha & Levinson [1980], Spector [1980, 1982], Podio-Guidugli, Vergara Caffarelli & Virga [1987] and Ciarlet [1988, Chap. 5]. In fluid dynamics, often nonconservative body forces are considered.

The theory of null Lagrangians is due to Landers (1942), Ericksen [1962] and Edelen [1962]. See also Olver [1986] and Ball, Currie & Olver [1981], Olver & Sivaloganathan [1988] for further references and additional results. Ball, Currie & Olver [1981] provide a general theory of null Lagrangians of arbitrary order (depending on higher derivatives of the independent fields). A complete set of conditions for f to be a null Lagrangian was given by Edelen [1981]. For a closely related class of invariant functionals arising in the theory of defective crystals, see Fonseca & Parry [1992b]. See also Giaquinta, Modica & Souček [1989, 1994, 1996].

14 Equilibrium States

This chapter examines the equilibrium states of a body in a given environment. The heating conditions are trivially satisfied and the question reduces to examining the conditions of mechanical equilibrium. As a general consequence of the latter, the Eshelby conservation law for the bulk matter will be derived. Then the conditions are given a variational form: the first variations of certain functionals vanish in the equilibrium states. The origin of this fact is deep: these functionals involve the entropy; the variational principles are ultimately related to the second law of thermodynamics. That the first variations vanish is a necessary condition for an extremum, but unstable equilibrium states, admitted by the definition of an equilibrium state, do not correspond to extrema. In this chapter only the first variations will be examined. The extrema, the second variations, and their relation to the stability will be treated in the next chapter.

14.1 Equilibrium States and Dissipation of Energy

This section gives two propositions to explain why the equilibrium states have a special position among the rest states. Recall the specific and total productions of entropy γ, Γ ((9.2.9), (9.2.12) and (13.2.5)).

14.1.1 Proposition *Suppose that Ω is connected and that*

$$\gamma = 0 \quad \text{only if} \quad \boldsymbol{D} = \boldsymbol{0}, \ \boldsymbol{g} = \boldsymbol{0}. \tag{14.1.1}$$

Let $\pi = (\chi, \eta, \boldsymbol{v})$ be a process without a singular surface for which $r = 0$ and $\Gamma = 0$ identically. Then there exists an equilibrium *state $\sigma_0 = (\chi_0, \eta_0)$ and a change of frame as in* 6.1.1 *such that*

$$\chi(\boldsymbol{p}, t) = \boldsymbol{R}(t)(\chi_0(\boldsymbol{p}) - \boldsymbol{x}_0) + \boldsymbol{x}_0'(t), \qquad \eta(\boldsymbol{p}, t) = \eta_0(\boldsymbol{p}) \tag{14.1.2}$$

for every t during the process.

Proof Since $\Gamma = 0$ and $\gamma \ge 0$, it follows from (13.2.5), in which the surface integral vanishes as the singular surface is absent, that $\gamma = 0$ identically. From (14.1.1) we find

$$\boldsymbol{D} = \boldsymbol{0}, \quad \boldsymbol{g} = \boldsymbol{0} \tag{14.1.3}$$

identically. In view of $(14.1.3)_1$ and the assumption that Ω is connected we can apply Killing's theorem (see GURTIN [1981, Exercise 3, p. 76 and Theorem on p. 69]) to assert $(14.1.2)_1$. From $(14.1.3)_2$ and the connectedness of Ω we deduce that the temperature is a function of time only and from (9.2.13) that $\boldsymbol{q} = \boldsymbol{0}$. Combining $\boldsymbol{q} = \boldsymbol{0}$, $r = 0$, $\gamma = 0$ with (9.2.9), we obtain $(14.1.2)_2$. We have seen that the temperature field is spatially uniform and hence $\sigma_0 := (\chi_0, \eta_0)$ is an equilibrium state. □

14.1.2 Proposition *Suppose that Ω is connected and that* (14.1.1) *holds. Let* $\sigma = (\chi, \eta)$ *be a rest state without a singular surface and* π *a rest process at the state* σ *(that is,* $\pi = (\chi, \eta, \mathbf{0})$*). Then the following conditions are equivalent:*

(1) *σ is an equilibrium state;*

(2) *the heat flux vector vanishes over Ω;*

(3) *identically during the process π,*

$$r = 0 \text{ in } \Omega, \quad \boldsymbol{q} \cdot \boldsymbol{n} = 0 \text{ on } \partial\Omega; \tag{14.1.4}$$

(4) *the heating measure $Q(\pi, \Omega, \cdot)$ vanishes;*

(5) *the Clausius integral corresponding to the process π and the whole body Ω vanishes, i.e.,*

$$\int_{t_1}^{t_2} \Big(- \int_{\partial\Omega} \frac{\boldsymbol{q} \cdot \boldsymbol{n}}{\theta}\, dA + \int_{\Omega} \rho_\circ \frac{r}{\theta}\, dV \Big)\, dt = 0. \tag{14.1.5}$$

Items (3) and (4) say, each in its way, that the body does not exchange heat with its environment. Equations (14.1.4) are conditions of adiabatic isolation. (3) says that if an adiabatically isolated body can stay in the rest state σ for a while, then the state is necessarily an equilibrium state. Condition (4) establishes a connection between equilibrium states and the fundamental concept of heating measure.

Proof Since in π all quantities are time independent, the equation of balance of energy reduces to

$$-\operatorname{Div} \boldsymbol{q} + \rho_\circ r = 0. \tag{14.1.6}$$

(1) ⇒ (2): If σ is an equilibrium state, then $\boldsymbol{D} = \mathbf{0}$, $\boldsymbol{G} = \mathbf{0}$ and (9.2.13) gives $\boldsymbol{q} = \mathbf{0}$. (2) ⇒ (3): If $\boldsymbol{q} = \mathbf{0}$ then $(14.1.4)_1$ follows from (14.1.6), and $(14.1.4)_2$ is trivial. (3) ⇒ (4): This follows immediately from the definition of the heating measure (cf. Subsect. 7.1.5). (4) ⇒ (5): Use Lemma 7.1.9. (5) ⇒ (1): (14.1.5) gives

$$-\int_{\partial\Omega} \frac{\boldsymbol{q} \cdot \boldsymbol{n}}{\theta}\, dA + \int_{\Omega} \rho_\circ \frac{r}{\theta}\, dV = 0$$

and application of the divergence theorem and a reference to (14.1.6) justify the following computation:

$$0 = \int_{\Omega} \Big(-\operatorname{Div} \frac{\boldsymbol{q}}{\theta} + \rho_\circ \frac{r}{\theta} \Big) dV = \int_{\Omega} \Big(\frac{-\operatorname{Div} \boldsymbol{q} + \rho_\circ r}{\theta} + \frac{\boldsymbol{q} \cdot \nabla\theta}{\theta^2} \Big) dV = \int_{\Omega} \frac{\boldsymbol{q} \cdot \nabla\theta}{\theta^2}\, dV.$$

Since $\boldsymbol{q} \cdot \nabla\theta$ is nonnegative in a rest state, it follows that $\boldsymbol{q} \cdot \nabla\theta = 0$ and combining this with (14.1.1) we obtain $\nabla\theta = \mathbf{0}$. As Ω is connected, θ is uniform over Ω. □

14.2 Equilibrium States for Given Environments

In Sects. 14.2–14.4 we consider a body in a conservative environment (see 13.3.6) of ambient temperature θ_0.

14.2.1 Definition An equilibrium state $\sigma = (\chi, \eta)$ is said to be an ***equilibrium state compatible with the environment*** if its temperature equals the ambient temperature and if the rest process $\pi = (\chi, \eta, \boldsymbol{v})$ given by

$$\chi(\boldsymbol{p}, t) = \chi(\boldsymbol{p}), \quad \eta(\boldsymbol{p}, t) = \eta(\boldsymbol{p}), \quad \boldsymbol{v}(\boldsymbol{p}, t) = \mathbf{0}, \quad \boldsymbol{p} \in \Omega, \ t \in [0, \infty),$$

is a process compatible with the environment.

14.2.2 Equilibrium States and Balance of Energy To verify that the rest process π is compatible with the environment, one has to verify the balance equations with the prescribed supply terms and the boundary conditions. Since the state is static and the gradient of temperature vanishes, also the heat flux vanishes. As the external supply of heat is absent by definition, it is concluded that *the equation of balance of energy and the boundary condition for the heat flux are automatically satisfied.* One has only to check whether the temperature of the state equals the ambient temperature and whether the state is compatible with the loading environment. Using the definitions 13.3.1–13.3.3 and inserting the potentials calculated in Proposition 13.3.7, we are led to the following

14.2.3 Conditions of Equilibrium An equilibrium state $\sigma = (\chi, \eta)$ is compatible with the conservative environment if its temperature equals the ambient temperature and the following conditions (1), (2) hold:

(1) for ***all three environments:***

$$\operatorname{Div} \boldsymbol{S} + \operatorname{Div}(\rho_o \partial_{\boldsymbol{F}} \psi) - \rho_o \partial_{\boldsymbol{x}} \psi = \boldsymbol{0}, \quad [\boldsymbol{S}]\boldsymbol{n} = \boldsymbol{0}$$

outside of and on the singular surface, respectively, and

(2) for the ***displacement-traction environment:***

$$\chi = \boldsymbol{d} \text{ on } D \text{ and } \boldsymbol{Sn} = -\rho_o \partial_{\boldsymbol{F}} \psi \boldsymbol{n} - \partial_{\boldsymbol{x}} \phi \text{ on } S,$$

for the ***bilateral contact environment:***

$$\chi(\boldsymbol{p}) \in P_i \text{ and } (\boldsymbol{1} - \boldsymbol{m} \otimes \boldsymbol{m})\boldsymbol{Sn} = \boldsymbol{0} \text{ if } \boldsymbol{p} \in D_i \text{ for some } i,$$
$$\boldsymbol{Sn} = -\rho_o \partial_{\boldsymbol{F}} \psi \boldsymbol{n} - \partial_{\boldsymbol{x}} \phi \text{ on } S;$$

and for the ***unilateral contact environment:***

$$\chi(\Omega) \subset R, \quad (\boldsymbol{1} - \boldsymbol{m} \otimes \boldsymbol{m})\boldsymbol{Sn} = \boldsymbol{0} \text{ on } C$$
$$\boldsymbol{Sn} = -\rho_o \partial_{\boldsymbol{F}} \psi \boldsymbol{n} - \partial_{\boldsymbol{x}} \phi \text{ on } F.$$

Here the free part of the boundary F is defined by

$$F := \begin{cases} S & \text{for the displacement-traction and bilateral contact environments} \\ \{\boldsymbol{p} \in \partial\Omega : \chi(\boldsymbol{p}) \notin \partial R\} & \text{for the unilateral contact environment;} \end{cases}$$

and the contact part of the boundary C by

$$C = \partial\Omega \setminus F.$$

The conditions involving the force take an especially simple form when the total stress (see Subsect. 13.3.13) is used, namely, using the properties of the potentials ψ, ϕ stated in Proposition 13.3.7, one obtains the following:

(3) for ***all three environments***

$$\operatorname{Div} \mathbf{S} - \rho_o \partial_{\boldsymbol{x}} q = \boldsymbol{0}, \quad [\mathbf{S}]\boldsymbol{n} = \boldsymbol{0},$$

outside of and on the singular surface, respectively. Here $q = f + \psi$ is the total stored energy (see 13.3.13) and note that $\partial_{\boldsymbol{x}} q = \partial_{\boldsymbol{x}} \psi$;

(4) for ***all three environments***

$$\boldsymbol{S}\boldsymbol{n} = -\partial_x \phi \text{ on } F;$$

(5) for the ***bilateral contact environment***

$$(\mathbf{1} - \boldsymbol{m} \otimes \boldsymbol{m})\boldsymbol{S}\boldsymbol{n} = \mathbf{0} \text{ if } \boldsymbol{p} \in D_i \text{ for some } i,$$

for ***the unilateral contact environment***

$$(\mathbf{1} - \boldsymbol{m} \otimes \boldsymbol{m})(\boldsymbol{S}\boldsymbol{n} + \partial_x \phi) = \mathbf{0} \text{ on } C.$$

Only stable states are practically realizable. This leads to the extremum principles of Chap. 15. For concrete equilibrium states of isotropic solids and fluids the stability will be examined in Sect. 20.7 and Sect. 19.7.

14.2.4 Proposition *Let σ be an equilibrium state compatible with a conservative environment. If ρ_o is uniform over Ω, the following assertions hold:*
(1) **(*DuBois–Reymond equation*)** *Outside the singular surface,*

$$\operatorname{Div} \boldsymbol{P} - \partial_p q = \mathbf{0}. \tag{14.2.1}$$

(2) **(*Eshelby's conservation law for bulk matter*)** *If q is independent of $\boldsymbol{p}$, then outside the singular surface*

$$\operatorname{Div} \boldsymbol{P} = \mathbf{0}. \tag{14.2.2}$$

Recall that $\boldsymbol{P} = \boldsymbol{P}$ if q is independent of $\boldsymbol{F}$.

Proof If ρ_o is uniform, a calculation using (13.3.29) and (13.3.28) gives

$$\operatorname{Div} \boldsymbol{P} = \boldsymbol{F}^{\mathsf{T}}(\partial_x q - \rho_o^{-1} \operatorname{Div} \boldsymbol{S}) + \partial_p q \tag{14.2.3}$$

and a combination with the first equation in 14.2.3(3) leads to (14.2.1). □

14.2.5 *Remarks*

(1) In the calculus of variations (see, e.g., CESARI [1983, p. 31]), the one-dimensional version of (14.2.1) is called the DuBois–Reymond necessary condition; here it will be convenient to employ the terminology suggested in Proposition 14.2.4. The proof shows that (14.2.1) is actually *equivalent* to the balance of forces 14.2.3(3) outside the singular surface.

(2) Eshelby's "conservation law" for the bulk matter implies that if P is any part of Ω that does not intersect the singular surface, then under the conditions leading to (14.2.2) we have

$$\int_{\partial P} \boldsymbol{P}\boldsymbol{n}\, dA = \mathbf{0}. \tag{14.2.4}$$

The last equation for arbitrary P (possibly intersecting the singular surface) requires ***Maxwell's relation***

$$[\boldsymbol{P}]\boldsymbol{n} = \mathbf{0} \tag{14.2.5}$$

on the singular surface. Equation (14.2.5) *does not* hold for general, possibly unstable equilibrium states. In Sect. 17.3, Maxwell's relation will be shown to be a consequence of stability, i.e., of (strong) extrema of thermodynamic functionals. In Sect. 14.5, (14.2.5) will be shown to be a consequence of the *assumption* that the total stored energy be stationary on *inner variations*. Also the exchange of the roles of the jump condition for stress with that for the Eshelby tensor under the exchange of the roles of the reference and actual configurations will be explained there.

(3) (14.2.2) is the conservation law associated with the invariance of the theory under translations in the reference configurations via Noether's theorem, see 14.5.5–14.5.7. When the displacement gradient $\nabla \boldsymbol{u} = \nabla(\chi - \boldsymbol{p}) = \boldsymbol{F} - \mathbf{1}$ is used instead of $\boldsymbol{F}$, the construction leads to $\boldsymbol{P}' = \boldsymbol{P} - \rho_o^{-1}\boldsymbol{S}$. When ψ is independent of $\boldsymbol{x}$, then still $\operatorname{Div} \boldsymbol{P}' = \mathbf{0}$ by (14.2.2) and the first equation of 14.2.3(3). It is $\boldsymbol{P}'$ which is called the energy–momentum tensor in the continuum theory of dislocations and the linear elasticity, cf. ESHELBY [1975].

(4) For fluids in the reference configuration in which ρ_o is uniform, ψ is always independent of $\boldsymbol{p}$, see (13.3.22). By $(13.3.31)_2$ then (14.2.2) reduces to the well-known integral $\operatorname{grad}(f + pv + \psi_b) = \mathbf{0}$ outside the singular surface. See Sect. 19.8 for applications.

(5) Material Inhomogeneities. For materially inhomogeneous bodies (see Subsect. 9.1.4), equation (14.2.1) generalizes to

$$\mathrm{Div}(\rho_\circ \boldsymbol{P}) - \partial_p(\rho_\circ q) = \mathbf{0}.$$

Then (14.2.4) in general does not hold even in the absence of body forces. In the continuum theory of dislocations (see ESHELBY [1956, 1975] and HANYGA [1985, Chap. 4]), the integral

$$\mathsf{J} := \int_{\partial P} \boldsymbol{Pn}\, dA$$

is identified with the (pseudo-)force acting on the material inhomogeneities enclosed in P. Another case when J is nonzero arises when the fields of stress or other quantities have singularities in P so that the divergence theorem does not apply, and J is identified with the (pseudo-)force acting on the singularities. The quantity J is the famous J-integral from fracture mechanics (see Sect. 10.3 and CHEREPANOV [1967, 1968], RICE [1968], and MAUGIN [1992, Chap. 7]). These matters will not be pursued here.

14.3 Integral Functionals

This section introduces variations of total quantities and variations of states that figure in the variational forms of conditions of equilibrium to be formulated in Sect. 14.4.

14.3.1 Total Quantities For a rest state $\sigma = (\chi, \eta)$ the quantities E, H, U, F, P and W from the preceding chapter are given by

$$\left.\begin{aligned}
&\mathsf{E}(\sigma) = \int_\Omega \tilde{e}\rho_\circ\, dV, \quad \mathsf{H}(\sigma) = \int_\Omega \eta\rho_\circ\, dV, \quad \mathsf{W}(\chi) = \int_\Omega \psi\rho_\circ\, dV + \int_S \phi\, dA,\\
&\mathsf{U}(\sigma) = \mathsf{E}(\sigma) + \mathsf{W}(\chi) = \int_\Omega (\tilde{e} + \psi)\rho_\circ\, dV + \int_S \phi\, dA,\\
&\mathsf{F}(\sigma) = \mathsf{E}(\sigma) - \theta_0 \mathsf{H}(\sigma) = \int_\Omega (\tilde{e} - \theta_0\eta)\rho_\circ\, dV,\\
&\mathsf{P}(\sigma) = \mathsf{F}(\sigma) + \mathsf{W}(\chi) = \mathsf{E}(\sigma) - \theta_0\mathsf{H}(\sigma) + \mathsf{W}(\chi), \quad \text{i.e.,}\\
&\mathsf{P}(\sigma) = \int_\Omega (\tilde{e} - \theta_0\eta + \psi)\rho_\circ\, dV + \int_S \phi\, dA,
\end{aligned}\right\} \qquad (14.3.1)$$

where the specific internal energy $\tilde{e}$ is evaluated at $(\nabla\chi(\boldsymbol{p}), \eta(\boldsymbol{p}))$, the volume density of the potential energy ψ at $(\nabla\chi(\boldsymbol{p}), \chi(\boldsymbol{p}), \boldsymbol{p})$, and the surface density ϕ at $(\chi(\boldsymbol{p}), \boldsymbol{p})$. For each of the three types of the conservative environments defined in Sect. 13.3, these functionals are defined on the set of all rest states $\sigma = (\chi, \eta) \in \Sigma^\#$ such that $\chi(\Omega) \subset R$, where $R \subset \mathrm{Eucl}$ is the container for the unilateral environments and $R = \mathrm{Eucl}$ for the displacement traction and bilateral contact environment. (See Subsect. 13.1.1 for the definition of $\Sigma^\#$.) Also define for every configuration χ the total stored energy $\mathsf{M}(\chi)$ by

$$\mathsf{M}(\chi) = \int_\Omega \bar{f}(\nabla\chi(\boldsymbol{p}), \theta_0)\rho_\circ\, dV + \mathsf{W}(\chi).$$

This section adapts standard definitions of the first and second variations to E, H, U, F, P, M, and W (see, e.g., CESARI [1983, Sect. 2.3] or ALEXEEV & al. [1979, Chap. IV]). A different type of variations will be considered in Sect. 14.5.

14.3.2 Admissible States and Variations For a given conservative environment, the set of all admissible states $\Sigma^{\#\#} \subset \Sigma^\#$ is defined as follows

(1) for the ***displacement-traction environment:***

$$\Sigma^{\#\#} = \{\sigma = (\chi, \eta) \in \Sigma^{\#} : \chi = \boldsymbol{d} \text{ on } D\};$$

(2) for the ***bilateral contact environment:***

$$\Sigma^{\#\#} = \{\sigma = (\chi, \eta) \in \Sigma^{\#} : \chi(D_i) \subset P_i \text{ for each } i = 1, \ldots, s\};$$

(3) for the ***unilateral contact environment:***

$$\Sigma^{\#\#} = \{\sigma = (\chi, \eta) \in \Sigma^{\#} : \chi(\Omega) \subset R\};$$

The elements of $\Sigma^{\#\#}$ are called ***admissible states.***

Let $C^\infty := C^\infty(\text{Eucl}, \text{Vect} \times \mathbb{R})$ denote the set of all pairs $\xi = (\boldsymbol{u}, \omega)$ if infinitely differentiable functions on Eucl with $\boldsymbol{u}$ with values in Vect and ω with values in $\mathbb{R}$. For a given state $\sigma \in \Sigma^{\#\#}$ we define the space of ***admissible variations*** at σ as follows:
(4) for the ***displacement-traction environment:***

$$\text{Var}(\sigma) = \{(\boldsymbol{u}, \omega) \in C^\infty : \boldsymbol{u} = \boldsymbol{0} \text{ on } D\};$$

(5) for the ***bilateral contact environment:***

$$\text{Var}(\sigma) = \{(\boldsymbol{u}, \omega) \in C^\infty : \boldsymbol{u} \cdot \boldsymbol{m} = 0 \text{ on } D := \cup_{i=1}^{n} D_i\};$$

(6) for the ***unilateral contact environment:***

$$\text{Var}(\sigma) = \{(\boldsymbol{u}, \omega) \in C^\infty : \boldsymbol{u} \cdot \boldsymbol{m} = 0 \text{ on } C\}.$$

Here $\boldsymbol{m} = \boldsymbol{m}(\boldsymbol{p})$ is the normal to the actual boundary $\partial\chi(\Omega)$ of the body expressed as function of $\boldsymbol{p}$, and $C = \{\boldsymbol{p} \in \partial\Omega : \chi(\boldsymbol{p}) \in \partial R\}$ is the surfaces of contact. Any element $\xi = (\boldsymbol{u}, \omega)$ of $\text{Var}(\sigma)$ is called a ***admissible variation*** at σ, or briefly a ***variation.*** Note that while the states can have singular surfaces, the variations are, by definition, infinitely differentiable. To motivate (4)–(6), observe that if $\bar{\sigma}(\alpha) = (\bar{\chi}(\cdot, \alpha), \bar{\eta}(\cdot, \alpha))$, $-\varepsilon < \alpha < \varepsilon$ is any twice continuously differentiable parametric family of states such that

$$\bar{\sigma}(0) = \sigma, \qquad \frac{d}{d\alpha}\bar{\sigma}(\alpha)\bigg|_{\alpha=0} = \xi, \tag{14.3.2}$$

then the conditions on $\boldsymbol{u} = d\chi/d\alpha$ in the definitions of $\text{Var}(\sigma)$ above follow from the kinematical constrains (13.2.1) (vacuous if $R = \text{Eucl}$) and $(13.3.1)_1$, $(13.3.2)_{1,1}$ or (13.2.1) for each type of environment.

Conversely, it is expectable that for each admissible variation ξ at σ there exists a family of states as above which satisfies the kinematical constrains appropriate for each of the environments. The desired family $\bar{\sigma}(\alpha) = (\bar{\chi}(\cdot, \alpha), \bar{\eta}(\cdot, \alpha))$ is $\bar{\chi}(\boldsymbol{p}, \alpha) = \chi(\boldsymbol{p}) + \alpha\boldsymbol{u}(\boldsymbol{p})$, $\bar{\eta}(\cdot, \alpha) = \eta + \alpha\omega$ for the displacement-traction environment. For other environments, the construction of $\bar{\chi}(\cdot, \alpha)$ is not so straightforward in view of the more complicated nature of the kinematical constraints. The family $\bar{\chi}(\cdot, \alpha)$ can be sought as a solution of the problem

$$\frac{d}{d\alpha}\bar{\chi}(\boldsymbol{p}, \alpha) = \boldsymbol{v}(\bar{\chi}(\boldsymbol{p}, \alpha)), \quad -\varepsilon < \alpha < \varepsilon, \quad \bar{\chi}(\boldsymbol{p}, 0) = \chi(\boldsymbol{p}), \quad \boldsymbol{p} \in \text{cl}\,\Omega,$$

where $\boldsymbol{v} = \boldsymbol{u} \circ \chi^{-1}$. Peano's existence theorem (e.g., CODDINGTON & LEVINSON [1955]) may be invoked to guarantee the local unique solvability, with the solution defined on an interval of α that is independent of $\boldsymbol{p}$. Under mild additional assumptions, the restrictions on $\boldsymbol{u}$ in (5), (6) can be shown to guarantee that $\bar{\chi}(\cdot, \alpha)$ satisfies the kinematical constrains imposed by the environment. Such considerations are somewhat apart from the main course, and instead we directly *assume that each variation is a derivative at $\alpha = 0$ of some parametric family of states that satisfies the kinematical constrains of the environment in question.*

The definition of the set of all variations was chosen so as it be a linear space even in the presence of the unilateral constraint R. Another possibility is the nonequivalent definition in which the condition $\boldsymbol{u} \cdot \boldsymbol{m} = 0$ in (6) above is replaced by

$$\boldsymbol{u} \cdot \boldsymbol{m} \geq 0 \text{ on } C. \tag{14.3.3}$$

With this definition, the set of all variations becomes a convex cone, and the family $\bar{\sigma}(\alpha) = (\bar{\chi}(\cdot, \alpha), \bar{\eta}(\cdot, \alpha))$ as in (14.3.2) can now be defined only for $\alpha \geq 0$, $\alpha < \varepsilon$.

14.3.3 The First and Second Variations Except for W and M, each functional in 14.3.1 is of the form

$$\mathrm{Q}(\sigma) = \int_{\Omega} L(\nabla\chi, \eta, \chi, \boldsymbol{p})\, dV + \int_{S} M(\chi, \boldsymbol{p})\, dA,$$

where (a) L is a twice continuously differentiable function L defined on $\tilde{U} \times R \times \Omega$ with $\tilde{U} \subset \mathrm{InvLin} \times \mathbb{R}$ the domain of the equilibrium response functions $\tilde{e}, \tilde{\boldsymbol{S}}, \tilde{\theta}$; (b) M is a twice continuously differentiable function M defined on $R \times S$. For a given environment, we take for the domain of Q the set of all admissible states. The requirement that the functions L, M are twice continuously differentiable is to be understood that they have twice continuously differentiable extensions to a neighborhood of the closure of their domains.

The ***first variation*** $\delta\mathrm{Q}(\sigma, \cdot)$ of Q at $\sigma \in \Sigma^{**}$ is a linear functional on $\mathrm{Var}(\sigma)$ defined by

$$\delta\mathrm{Q}(\sigma, \xi) = \frac{d}{d\alpha}\mathrm{Q}(\bar{\sigma}(\alpha))\Big|_{\alpha=0}, \quad \xi \in \mathrm{Var}(\sigma),$$

where $\bar{\sigma}(\cdot)$ is any smooth parametric family of states in Σ^{**} such that (14.3.2) and the kinematical constrains of the environment hold. A differentiation under the integral sign gives

$$\delta\mathrm{Q}(\sigma, \xi) = \int_{\Omega} (\partial_F L \cdot \nabla\boldsymbol{u} + \partial_\eta L \cdot \omega + \partial_{\boldsymbol{x}} L \cdot \boldsymbol{u})\, dV + \int_{S} \partial_{\boldsymbol{x}} M \cdot \boldsymbol{u}\, dA. \tag{14.3.4}$$

The ***second variation*** $\delta^2\mathrm{Q}(\sigma, \cdot, \cdot)$ is a bilinear form on $\mathrm{Var}(\sigma)$ given by

$$\delta^2\mathrm{Q}(\sigma, \xi_1, \xi_2) = \frac{d}{d\alpha}\delta\mathrm{Q}(\bar{\sigma}_2(\alpha), \xi_1)\Big|_{\alpha=0}, \quad \xi_1, \xi_2 \in \mathrm{Var}(\sigma), \tag{14.3.5}$$

where $\bar{\sigma}_2(\cdot)$ is any smooth parametric family of states in Σ^{**} such that (14.3.2) hold with $\sigma, \bar{\sigma}, \xi$ replaced by $\sigma, \bar{\sigma}_2, \xi_2$. Again, the second variation can be computed by differentiating under the integral sign.

The domain of Q was chosen to consist of states with the degree of smoothness that allows us to state the variational form of the conditions of mechanical equilibrium in terms of classical derivatives, as opposed to the generalized derivatives of functions from Sobolev spaces. A more natural setting for the calculus of variations is that of Sobolev spaces, however. If L, M are bounded below, the domain of Q can be extended naturally to contain states with deformations from Sobolev spaces, provided one admits that Q can take infinite values. See Chap. 21 for such an approach.

The first and second variations are defined here as directional derivatives. The reader is referred to Sect. 20.3 for the discussion of the conditions under which the first and second variations are or are not the Fréchet differentials on Sobolev spaces.

14.3.4 Functionals of χ Only W and M are of the form

$$\mathrm{L}(\chi) = \int_{\Omega} L(\nabla\chi, \chi, \boldsymbol{p})\, dV + \int_{S} M(\chi, \boldsymbol{p})\, dA.$$

The notions introduced above extend without any modification to L. In particular, L is a twice continuously differentiable function defined on $U_0 \times R \times \Omega$, where $U_0 \subset \mathrm{InvLin}$ is the domain of the stored energy $\bar{f}(\cdot, \theta_0)$ at θ_0 and the domain of M is as above.

An ***admissible configuration*** χ is a configuration such that $(\chi, \eta) \in \Sigma^{**}$ for some entropy field η. This implies that $\chi(\Omega) \subset R$ (vacuous in the case of the displacement-traction and bilateral contact environments) and (a) $\chi = \boldsymbol{d}$ on D in the case of a displacement-traction environment, (b) $\chi(D_i) \subset P_i$ for every $i = 1, \ldots, s$, in the case of the bilateral contact environment. The set of all admissible configurations is denoted by Δ^{**} and we regard L as defined on Δ^{**}. If χ kinematically admissible motion (see 13.3.4), then $\chi(\cdot, t) \in \Delta^{**}$ for every t. Conversely, any appropriately regular family of admissible configurations gives rise to a kinematically admissible motion. An ***admissible variation*** (of configuration) at χ is any $\boldsymbol{u} \in C^\infty(\mathrm{cl}\,\Omega)$ such that the restrictions imposed on $\boldsymbol{u}$ in 14.3.2(4)–(6) hold. The space of all admissible variations $\boldsymbol{u}$ at $\chi \in \Delta^{**}$ is denoted by $\mathrm{Var}(\chi)$. If χ is a kinematically admissible motion of class C^∞ and $\boldsymbol{v}$ its velocity field, then $\boldsymbol{v}(\cdot, t) \in \mathrm{Var}(\chi(\cdot, t))$ for every t. The first and the second variations $\delta \mathsf{L}(\chi, \boldsymbol{u}), \delta^2 \mathsf{L}(\chi, \boldsymbol{u}_1, \boldsymbol{u}_2)$ are defined in a way analogous to that in Subsect. 14.3.3.

14.4 Variational Conditions for Thermodynamic Equilibrium

Equations 14.2.3(3)–(5) will now be shown to be equivalent to the condition that the first variations of total canonical free energy, total energy, or total entropy vanish for admissible variations subject to appropriate side conditions. The derivations in this section involve standard steps in the derivations of the *Euler–Lagrange equations* (GEL'FAND & FOMIN [1961, Chaps. I–III, VII], CESARI [1983, Chap. 2] and OLVER [1986, Chap. 4]).

14.4.1 Proposition *Consider a conservative environment.*
(1) *If $\sigma = (\chi, \eta) \in \Sigma^{**}$ is an admissible state then*

$$\delta \mathsf{H}(\sigma, \zeta) = \int_\Omega \omega \rho_\circ \, dV \tag{14.4.1}$$

and

$$\begin{aligned} \delta \mathsf{U}(\sigma, \xi) &= \int_\Omega ((\rho_\circ \partial_x q - \mathrm{Div}\, \mathbf{S}) \cdot \boldsymbol{u} + \rho_\circ \theta \cdot \omega)\, dV \\ &\quad + \int_S [\mathbf{S}]\boldsymbol{n} \cdot \boldsymbol{u}\, dA + \int_{\partial\Omega} \mathbf{S}\boldsymbol{n} \cdot \boldsymbol{u}\, dA + \int_S \partial_x \phi \cdot \boldsymbol{u}\, dA \end{aligned} \tag{14.4.2}$$

for every $\xi = (\boldsymbol{u}, \omega) \in \mathrm{Var}(\sigma)$. Here θ is the temperature field in σ.
(2) *For every $\chi \in \Delta^{**}$ and $\boldsymbol{u} \in \mathrm{Var}(\chi)$ we have*

$$\begin{aligned} \delta \mathsf{M}(\chi, \boldsymbol{u}) &= \int_\Omega (\rho_\circ \partial_x \psi - \mathrm{Div}\, \mathbf{S}) \cdot \boldsymbol{u}\, dV \\ &\quad + \int_S [\mathbf{S}]\boldsymbol{n} \cdot \boldsymbol{u}\, dA + \int_{\partial\Omega} \mathbf{S}\boldsymbol{n} \cdot \boldsymbol{u}\, dA + \int_S \partial_x \phi \cdot \boldsymbol{u}\, dA. \end{aligned} \tag{14.4.3}$$

Proof (1): Equation (14.4.1) is immediate. To prove (14.4.2), assume that $\rho_\circ = 1$. We have $\delta \mathsf{U} = \delta \mathsf{E} + \delta \mathsf{W}$. The calculation of $\delta \mathsf{W}$ is identical with the steps in the proof of Proposition 13.3.5. Further,

$$\delta \mathsf{E}(\sigma,\xi) = \int_\Omega (\partial_F \tilde{e} \cdot \nabla \boldsymbol{u} + \partial_\eta \tilde{e} \cdot \omega)\, dV = \int_\Omega (\boldsymbol{S} \cdot \nabla \boldsymbol{u} + \theta \cdot \omega)\, dV$$
$$= \int_\Omega (-\boldsymbol{u} \cdot \operatorname{Div} \boldsymbol{S} + \theta \cdot \omega)\, dV + \int_S [\boldsymbol{S}]\boldsymbol{n} \cdot \boldsymbol{u}\, dA + \int_{\partial\Omega} \boldsymbol{S}\boldsymbol{n} \cdot \boldsymbol{u}\, dA. \quad \square$$

14.4.2 Proposition *If $\sigma = (\chi, \eta)$ is an admissible state for a conservative environment, the following conditions are equivalent:*

(1) *σ is an equilibrium state compatible with the environment;*

(2) ***(principle of stationary canonical free energy)***

$$\delta \mathsf{P}(\sigma,\xi) = 0 \text{ for every } \xi \in \operatorname{Var}(\sigma);$$

(3) ***(principle of stationary energy)*** *σ has a uniform temperature equal to the ambient temperature and*

$$\delta \mathsf{U}(\sigma,\xi) = 0 \text{ for every } \xi \in \operatorname{Var}(\sigma) \text{ satisfying } \delta \mathsf{H}(\sigma,\xi) = 0;$$

(4) ***(principle of stationary entropy)*** *σ has a uniform temperature equal to the ambient temperature and*

$$\delta \mathsf{H}(\sigma,\xi) = 0 \text{ for every } \xi \in \operatorname{Var}(\sigma) \text{ satisfying } \delta \mathsf{U}(\sigma,\xi) = 0.$$

As only conditions on the first variation are present, no extrema of the functionals are implied. Without the information about the ambient temperature, Conditions (3) or (4) still imply that σ is an equilibrium state; its temperature, though, need not be the ambient temperature.

Proof Only (1) ⇒ (2) ⇒ (4) ⇒ (1) will be proved, the rest, namely (2) ⇒ (3) ⇒ (1), is analogous. (1) ⇒ (2) : Let σ be an equilibrium state and calculate

$$\delta \mathsf{P}(\sigma,\xi) = \delta \mathsf{U}(\sigma,\xi) - \theta_0 \delta \mathsf{H}(\sigma,\xi), \tag{14.4.4}$$

by (14.4.1) and (14.4.2). Using 14.2.3(3)-(5), the definitions of the admissible variations for each type of environment in 14.3.2 and the fact that the temperature of σ coincides with the ambient temperature, one finds that the coefficients in front of the variations $\boldsymbol{u}$, ω vanish and hence $\delta \mathsf{P}(\sigma,\xi) = 0$. (2) ⇒ (4) follows from (14.4.4). (4) ⇒ (1) : Assume that (4) holds. Set $\rho_0 = 1$ for convenience. By the theorem on Lagrange multipliers (see, e.g., CESARI [1983, Sect. 4.8]), there exists a $\mu \in \mathbb{R}$ such that

$$\delta \mathsf{U}(\sigma,\xi) - \mu \delta \mathsf{H}(\sigma,\xi) = 0$$

for every $\xi \in \operatorname{Var}(\sigma)$. By (14.4.2) and (14.4.1) this means that

$$\int_\Omega ((\partial_x q - \operatorname{Div} \boldsymbol{S}) \cdot \boldsymbol{u} + (\theta - \mu)\omega)\, dV + \int_S [\boldsymbol{S}]\boldsymbol{n} \cdot \boldsymbol{u}\, dA + \int_{\partial\Omega} \boldsymbol{S}\boldsymbol{n} \cdot \boldsymbol{u}\, dA + \int_S \partial_x \phi \cdot \boldsymbol{u}\, dA = 0.$$

From the arbitrariness of ω we deduce that $\theta = \mu$. This shows that the temperature is uniform over the body and hence equal to the uniform temperature which by (4) is identified with the ambient temperature. Using the freedom in $\boldsymbol{u}$ given by the definitions of the variations for each type of the environment, one obtains the conditions 14.2.3(3)–(5) of balance of forces. □

The thermal variables play a trivial role in the above variational principles: They imply only that the temperature is uniform and equal to the ambient temperature. It is therefore natural to formulate a variational principle in the mechanical variables only.

14.4.3 Proposition *Let $\chi \in \Delta^{**}$ be a configuration and $\sigma = (\chi, \eta)$ an equilibrium state of temperature θ_0 and configuration χ. Then the following conditions are equivalent:*

(1) *σ is an equilibrium state compatible with the environment;*

(2) ***(principle of stationary stored energy)***

$$\delta \mathsf{M}(\chi, \boldsymbol{u}) = 0 \text{ for every } \boldsymbol{u} \in \operatorname{Var}(\chi).$$

(Condition (2) is also called the principle of stationary potential energy.)

14.5 Spatial Description. Standard, Inner, and Outer Variations

The variations of states and deformations of the type considered in 14.3 are not always the most convenient ones. This section describes a more general class of variations within a purely mechanical context. The stationarity of the total stored energy for different subclasses of these variations leads to different, but closely related, equations which may be inequivalent if the singular surface is present.

14.5.1 Spatial Description The total quantities can be expressed in terms of the inverse deformation $\mu = \chi^{-1}$ (see Subsect. 10.3.2). Consider only the isothermal variations and the pure displacement environment defined 13.3.1 and set $\rho_o = 1$. Let χ be a deformation, denote $r \equiv \chi(\Omega)$ the actual region occupied by the body and set $\mu := \chi^{-1}$. Since, roughly, all deformations which coincide with χ on $\partial\Omega$ map Ω onto r, we consider the inverse deformations $\bar{\mu}$ defined on the domain r which coincide with μ on ∂r. If we define $\mathsf{M}^{\sharp}(\bar{\mu}) := \mathsf{M}(\bar{\mu}^{-1})$ then

$$\mathsf{M}^{\sharp}(\bar{\mu}) = \int_r (\bar{f}^{\sharp}(\bar{\mathbf{Z}}, \theta_0) + \rho\psi(\bar{\mathbf{Z}}^{-1}, x, \mu(x))\, dv$$

where $\bar{f}^{\sharp}$ is defined in Subsect. 10.3.2 and $\bar{\mathbf{Z}}$ is the inverse deformation gradient of $\bar{\mu}^{-1}$, i.e., the gradient of $\bar{\mu}$. The functional $\mathsf{M}^{\sharp}$ is of the form discussed in Sect. 14.3 with the role of Ω played by r. Let $\mathrm{Var}(\mu)$ be the class of all infinitely differentiable variations $\boldsymbol{r}$: Eucl $\to$ Vect with compact supports contained in r. The question is whether $\delta\mathsf{M}^{\sharp}(\mu, \boldsymbol{r}) = 0$ for all $\boldsymbol{r} \in \mathrm{Var}(\mu)$ is equivalent to the conditions of thermodynamic equilibrium. Calculating the variations and using the thermostatic relations in the inverse deformation gradient (10.3.1), one arrives at

$$\mathrm{Div}\, \boldsymbol{P} - \partial_p \psi = \mathbf{0}, \quad [\boldsymbol{P}]n = \mathbf{0} \tag{14.5.1}$$

outside and on the singular surface, respectively, where $\boldsymbol{P}$ is the total Eshelby tensor at χ. That is, from $\delta\mathsf{M}^{\sharp}(\mu, \boldsymbol{r}) = 0$ one does not obtain the equations of balance of forces, but the Eshelby conservation law, *including the jump condition.* From Remark 14.2.5(1) we know that (14.5.1) is equivalent to the balance of forces if the singular surface is absent, but in the presence of the singular surface, (14.5.1) may be inequivalent.

The variations of μ generally shift the position of the singular surface *in the reference configuration,* and it is exactly this kind of variations which leads to the jump condition $(14.5.1)_2$. The variational conditions of equilibrium are not invariant under the transformation $\chi \mapsto \chi^{-1}$. The standard assertions on the invariant character of the Euler–Lagrange equations known from the analytical mechanics and from the calculus of variations (e.g., GEL'FAND & FOMIN [1961, Sect. 6], CESARI [1983, Sect. 2.4D] and OLVER [1986, Sect. 4.1]) apply only to continuously differentiable solutions of the Euler–Lagrange equations.

A more symmetric and illuminative way of treating the variations of χ and μ is obtained by enlarging the class of variations.

14.5.2 General Variations Let us describe a class of variations which change the independent and dependent variables $\boldsymbol{p}$ and $\boldsymbol{x}$ simultaneously. Denote by $\mathbf{F} = \Omega \times R$ the space of pairs $(\boldsymbol{p}, \boldsymbol{x})$. Let $\mathbf{T}_\alpha$, $-\varepsilon < \alpha < \varepsilon$, be a one-parameter family of smooth transformations of $\mathbf{F}$ onto itself and write

$$\mathbf{T}_\alpha(\boldsymbol{p}, \boldsymbol{x}) = (\mathbf{p}_\alpha(\boldsymbol{p}, \boldsymbol{x}), \mathbf{x}_\alpha(\boldsymbol{p}, \boldsymbol{x})),$$

so that the point $(\boldsymbol{p}, \boldsymbol{x})$ is transformed by $\mathbf{T}_\alpha$ to the point $(\bar{\boldsymbol{p}}, \bar{\boldsymbol{x}}) \in \mathbf{F}$, where $\bar{\boldsymbol{p}} = \mathbf{p}_\alpha(\boldsymbol{p}, \boldsymbol{x})$, $\bar{\boldsymbol{x}} = \mathbf{x}_\alpha(\boldsymbol{p}, \boldsymbol{x})$. Define the ***generators of*** $\mathbf{T}_\alpha$ by

$$\dot{\mathbf{p}}(\boldsymbol{p}, \boldsymbol{x}) := \partial_\alpha \mathbf{p}_\alpha(\boldsymbol{p}, \boldsymbol{x})\big|_{\alpha=0}, \quad \dot{\mathbf{x}}(\boldsymbol{p}, \boldsymbol{x}) := \partial_\alpha \mathbf{x}_\alpha(\boldsymbol{p}, \boldsymbol{x})\big|_{\alpha=0}, \quad (\boldsymbol{p}, \boldsymbol{x}) \in \mathbf{F}.$$

Assume that

(1) $\mathbf{T}_0(\boldsymbol{p}, \boldsymbol{x}) = (\boldsymbol{p}, \boldsymbol{x})$, $(\boldsymbol{p}, \boldsymbol{x}) \in \mathbf{F}$;

(2) the function $(\boldsymbol{p}, \boldsymbol{x}, \alpha) \mapsto \mathbf{T}_\alpha(\boldsymbol{p}, \boldsymbol{x})$ is twice continuously differentiable.

The family of transformations $\mathbf{T}_\alpha$ induces a family of transformations on the class of functions $\chi : \Omega \to R$ which we now describe. Denote, for any mapping $g : M \to N$, where M, N are arbitrary sets, its graph by $\operatorname{graph} g := \{(x, g(x)) \in M \times N : x \in M\}$. The graphs of deformations χ are subsets of $\mathbf{F}$. For a fixed α, one can consider the image $\mathbf{T}_\alpha(\operatorname{graph}\chi)$ of $\operatorname{graph}\chi$ under $\mathbf{T}_\alpha$. This set need not be a graph of any mapping. However, we shall assume that χ is such that

(3) for every sufficiently small $|\alpha|$, the mapping $p \in \Omega \mapsto \bar{p} := \mathbf{p}_\alpha(p, \chi(p))$ is a diffeomorphism from Ω onto $\Omega_\alpha := \{\mathbf{p}_\alpha(p, \chi(p)) : p \in \Omega\}$.

Under this assumption, $\mathbf{T}_\alpha(\operatorname{graph}\chi)$ is a graph of some mapping which we denote by $\bar{\chi}_\alpha$, i.e., we define the transform $\bar{\chi}_\alpha : \Omega_\alpha \to R$ of a given χ under $\mathbf{T}_\alpha$ by $\operatorname{graph}\bar{\chi}_\alpha = \mathbf{T}_\alpha(\operatorname{graph}\chi)$ for a sufficiently small $|\alpha|$. In other words, we set

$$\bar{\chi}_\alpha(\bar{p}) = \mathbf{x}_\alpha(p, \chi(p)) \quad \text{where } \bar{p} = \mathbf{p}_\alpha(p, \chi(p)). \tag{14.5.2}$$

In this class, the changes of independent variables may be influenced also by the dependent variables and vice versa; notice that the domain of definition of $\bar{\chi}_\alpha$ Ω_α depends on χ.

Consider a functional M defined by

$$\mathsf{M}(\chi) := \int_\Omega f(\nabla\chi, \chi, p)\, dV \tag{14.5.3}$$

for any deformation χ (possibly with a singular surface) such that $\chi(\Omega) \subset R$. In (14.5.3), the integration is taken over the domain of definition of χ, so that

$$\mathsf{M}(\bar{\chi}_\alpha) = \int_{\Omega_\alpha} f(\bar{\nabla}\bar{\chi}_\alpha(\bar{p}), \bar{\chi}_\alpha(\bar{p}), \bar{p})\, dV(\bar{p}), \tag{14.5.4}$$

where for convenience, the gradient $\nabla\bar{\chi}_\alpha$ of $\bar{\chi}_\alpha$ is denoted by $\bar{\nabla}\bar{\chi}_\alpha$ to indicate that the differentiation is made with respect to the natural variable $\bar{p}$ of $\bar{\chi}_\alpha$. (A redundant notation.) Denote

$$\boldsymbol{S} = \partial_F f, \quad \boldsymbol{P} = f\mathbf{1} - \boldsymbol{F}^{\mathsf{T}}\boldsymbol{S},$$

which are the total stress tensor and total Eshelby tensor if f is identified with the total free energy and $\rho_0 = 1$. The differentiation of M along $\bar{\chi}_\alpha$ at $\alpha = 0$ leads to formulas that display a remarkable symmetry in $\boldsymbol{x}, \boldsymbol{p}$ and $\boldsymbol{S}$, $\boldsymbol{P}$:

14.5.3 Proposition *If the family $\mathbf{T}_\alpha$ and a deformation χ satisfy Assumptions* (1)–(3) *in* 14.5.2, *then*

$$\frac{d}{d\alpha}\mathsf{M}(\bar{\chi}_\alpha)\Big|_{\alpha=0} = \int_\Omega (\partial_x f \cdot \dot{\mathbf{x}} + \boldsymbol{S} \cdot \nabla\dot{\mathbf{x}} + \partial_p f \cdot \dot{\mathbf{p}} + \boldsymbol{P} \cdot \nabla\dot{\mathbf{p}})\, dV. \tag{14.5.5}$$

If, moreover, $\dot{\mathbf{p}}(p, \chi(p)) = \dot{\mathbf{x}}(p, \chi(p)) = \mathbf{0}$ *for every* $p \in \partial\Omega$, *then*

$$\frac{d}{d\alpha}\mathsf{M}(\bar{\chi}_\alpha)\Big|_{\alpha=0} = \int_\Omega ((\partial_x f - \operatorname{Div}\boldsymbol{S}) \cdot \dot{\mathbf{x}} + (\partial_p f - \operatorname{Div}\boldsymbol{P}) \cdot \dot{\mathbf{p}})\, dV + \int_S ([\boldsymbol{S}]\boldsymbol{n} \cdot \dot{\mathbf{x}} + [\boldsymbol{P}]\boldsymbol{n} \cdot \dot{\mathbf{p}})\, dA, \tag{14.5.6}$$

where $\partial_x f$, $\partial_p f$, $\boldsymbol{S}$, $\boldsymbol{P}$ *are evaluated at* $(\nabla\chi(p), \chi(p), p)$ *and* $\dot{\mathbf{x}}$, $\dot{\mathbf{p}}$ *at* $(p, \chi(p))$; *here* $p \in \Omega$ *is the variable of integration in the formula above.*

Proof We employ a change of variables $\bar{p} \mapsto p$ by $\bar{p} = \mathbf{p}_\alpha(p, \chi(p))$ in (14.5.4). This gives

$$\mathsf{M}(\bar{\chi}_\alpha) = \int_\Omega f(\bar{\nabla}\bar{\chi}_\alpha(q_\alpha(p)), \bar{\chi}_\alpha(q_\alpha(p)), q_\alpha(p)) \cdot \det\nabla q_\alpha(p)\, dV(p), \tag{14.5.7}$$

where $q_\alpha(p) = \mathbf{p}_\alpha(p, \chi(p))$, $p \in \Omega$. The plan is to differentiate under the integral sign, which is permissible here. The integrand is differentiated via the chain rule and our immediate goal is to evaluate the derivatives of the separate arguments of f. First,

$$\partial_\alpha q_\alpha(p)\big|_{\alpha=0} = \dot{\mathbf{p}}, \quad \partial_\alpha\bar{\chi}_\alpha(q_\alpha(p))\big|_{\alpha=0} = \dot{\mathbf{x}},$$

where $\dot{\mathbf{p}}$, $\dot{\mathbf{x}}$ are evaluated at $(p, \chi(p))$. This follows by differentiating the definition of $\bar{\chi}_\alpha$, namely

$$\bar{\chi}_\alpha(q_\alpha(p)) = \mathbf{x}_\alpha(p, \chi(p)). \tag{14.5.8}$$

Next, by differentiating (14.5.8) with respect to p and then with respect to α at $\alpha = 0$, we obtain

$$\partial_\alpha\bar{\nabla}\bar{\chi}_\alpha(q_\alpha(p))\big|_{\alpha=0} = \nabla\dot{\mathbf{x}} - \nabla\chi\nabla\dot{\mathbf{p}}.$$

Here it is understood that the differentiation with respect to α is made only after the gradient $\bar{\nabla}\bar{\chi}_\alpha$ has been calculated and the argument $q_\alpha(p)$ has been inserted, and $\nabla\dot{\mathbf{x}}$, $\nabla\dot{\mathbf{p}}$ are the gradients of the functions $p \mapsto \dot{\mathbf{x}}(p, \chi(p))$, $p \mapsto \dot{\mathbf{p}}(p, \chi(p))$. (The need of these comments arises from the well-known imperfections in the notations for derivatives and compositions of functions.) Finally,

$$\partial_\alpha \det\nabla q_\alpha(p)\big|_{\alpha=0} = \operatorname{Div}\dot{\mathbf{p}}.$$

Collecting the above formulas, one obtains (14.5.5). Performing the integration by parts on the terms containing $\nabla\dot{\mathbf{x}}$, $\nabla\dot{\mathbf{p}}$ and using that $\dot{\mathbf{x}} = \dot{\mathbf{p}} = \mathbf{0}$ on $\partial\Omega$ one obtains (14.5.6). □

14.5.4 Special Cases

(1) *Standard variations* Let $\boldsymbol{u}$ be a smooth function on Ω that vanishes near the boundary of Ω and take $\mathbf{T}_\alpha$ of the form

$$\mathbf{T}_\alpha(\boldsymbol{p}, \boldsymbol{x}) = (\boldsymbol{p}, \boldsymbol{x} + \alpha \boldsymbol{u}(\boldsymbol{p})), \quad (\boldsymbol{p}, \boldsymbol{x}) \in \mathbf{F}.$$

Hence

$$\dot{\mathbf{p}} \equiv \mathbf{0}, \quad \dot{\mathbf{x}} = \boldsymbol{u}(\boldsymbol{p}), \quad \bar{\chi}_\alpha(\boldsymbol{p}) = \chi(\boldsymbol{p}) + \alpha \boldsymbol{u}(\boldsymbol{p}), \quad (\boldsymbol{p}, \boldsymbol{x}) \in \mathbf{F}.$$

Let us call the family $\bar{\chi}_\alpha$ of the above form a ***standard variation*** of χ. Equation (14.5.6) reduces to

$$\frac{d}{d\alpha} \mathsf{M}(\bar{\chi}_\alpha)\Big|_{\alpha=0} = \int_\Omega (\partial_x f - \operatorname{Div} \mathbf{S}) \cdot \boldsymbol{u} \, dV + \int_S [\mathbf{S}]\boldsymbol{n} \cdot \boldsymbol{u} \, dA$$

established in Sect. 14.3. The requirement that $d\mathsf{M}(\bar{\chi}_\alpha)/d\alpha = 0$ at $\alpha = 0$ for every standard variation of χ leads to the referential form of the equation of balance of forces, i.e.,

$$\partial_x f - \operatorname{Div} \mathbf{S} = \mathbf{0}, \quad [\mathbf{S}]\boldsymbol{n} = \mathbf{0}, \tag{14.5.9}$$

of which the first is the ***Euler–Lagrange equation*** corresponding to the integrand (Lagrangian) f.

(2) *Outer variations* Let χ be given and let $\boldsymbol{w}$ be a smooth function on $\chi(\Omega)$ such that $\boldsymbol{w}(\chi(\boldsymbol{p})) = \mathbf{0}$ if $\boldsymbol{p} \in \partial\Omega$. Take $\mathbf{T}_\alpha$ of the form

$$\mathbf{T}_\alpha(\boldsymbol{p}, \boldsymbol{x}) = (\boldsymbol{p}, \boldsymbol{x} + \alpha \boldsymbol{w}(\boldsymbol{x})), \quad (\boldsymbol{p}, \boldsymbol{x}) \in \mathbf{F}.$$

Hence

$$\dot{\mathbf{p}} \equiv \mathbf{0}, \quad \dot{\mathbf{x}}(\boldsymbol{p}, \boldsymbol{x}) = \boldsymbol{w}(\boldsymbol{x}), \quad \bar{\chi}_\alpha(\boldsymbol{p}) = \chi(\boldsymbol{p}) + \alpha \boldsymbol{w}(\chi(\boldsymbol{p})), \quad (\boldsymbol{p}, \boldsymbol{x}) \in \mathbf{F}.$$

Let us call the family $\bar{\chi}_\alpha$ of the above form an ***outer variation*** of χ. Equation (14.5.6) reduces to

$$\frac{d}{d\alpha} \mathsf{M}(\bar{\chi}_\alpha)\Big|_{\alpha=0} = \int_\Omega (\partial_x f - \operatorname{Div} \mathbf{S}) \cdot \boldsymbol{w}(\chi(\boldsymbol{p})) \, dV + \int_S [\mathbf{S}]\boldsymbol{n} \cdot \boldsymbol{w}(\chi(\boldsymbol{p})) \, dA,$$

which is just the special case of (1) above with $\boldsymbol{u} = \boldsymbol{w} \circ \chi$. However, using the assumed invertibility of χ this can be rewritten as

$$\frac{d}{d\alpha} \mathsf{M}(\bar{\chi}_\alpha)\Big|_{\alpha=0} = \int_{\chi(\Omega)} (J^{-1}\partial_x f - \operatorname{div} \boldsymbol{T}) \cdot \boldsymbol{w} \, dv + \int_{\chi(S)} [\boldsymbol{T}]\boldsymbol{m} \cdot \boldsymbol{w} \, da$$

where $\boldsymbol{T}$ is the Cauchy stress for the state χ and $J = \det \nabla\chi$. Therefore, the requirement that $d\mathsf{M}(\bar{\chi}_\alpha)/d\alpha = 0$ at $\alpha = 0$ for every outer variation of χ leads to the spatial form of the equation of balance of forces, which is equivalent to the referential balance of forces.

(3) *Inner variations* Let χ be given and let $\boldsymbol{z}$ be a smooth function on Ω which vanishes near $\partial\Omega$. Take $\mathbf{T}_\alpha$ of the form

$$\mathbf{T}_\alpha(\boldsymbol{p}, \boldsymbol{x}) = (\boldsymbol{p} + \alpha \boldsymbol{z}(\boldsymbol{p}), \boldsymbol{x}), \quad (\boldsymbol{p}, \boldsymbol{x}) \in \mathbf{F}.$$

Hence

$$\dot{\mathbf{p}}(\boldsymbol{p}, \boldsymbol{x}) = \boldsymbol{z}(\boldsymbol{p}), \quad \dot{\mathbf{x}} \equiv \mathbf{0}, \quad \bar{\chi}_\alpha(\boldsymbol{p} + \alpha \boldsymbol{z}(\boldsymbol{p})) = \chi(\boldsymbol{p}), \quad (\boldsymbol{p}, \boldsymbol{x}) \in \mathbf{F}.$$

Let us call the family $\bar{\chi}_\alpha$ of the above form an ***inner variation*** of χ. Equation (14.5.6) reduces to

$$\frac{d}{d\alpha} \mathsf{M}(\bar{\chi}_\alpha)\Big|_{\alpha=0} = \int_\Omega (\partial_p f - \operatorname{Div} \mathbf{P}) \cdot \boldsymbol{z}(\boldsymbol{p}) \, dV + \int_S [\mathbf{S}]\boldsymbol{n} \cdot \boldsymbol{z}(\boldsymbol{p}) \, dA.$$

Therefore, the requirement that $d\mathsf{M}(\bar{\chi}_\alpha)/d\alpha = 0$ at $\alpha = 0$ for every inner variation of χ leads to the DuBois–Reymond equation (14.2.1) *including the jump condition* $[\mathbf{P}]\boldsymbol{n} = \mathbf{0}$. Inner variations are used in the classical calculus of variations in the proof of the DuBois–Reymond necessary condition; see, e.g., CESARI [1983, p. 46].

14.5.5 Variational Symmetries Let M be given by (14.5.3). A one-parameter family of smooth transformations $\mathbf{T}_\alpha$ as in 14.5.2 is said to be a ***variational symmetry*** for M if for every open subset ω of Ω such that $\operatorname{cl}\omega \subset \Omega$, every deformation χ satisfying 14.5.2(1)–(3) and every α sufficiently close to 0,

$$\int_{\tilde{\omega}_\alpha} f(\nabla\tilde{\chi}_\alpha(\bar{p}), \tilde{\chi}_\alpha(\bar{p}), \bar{p})\, dV(\bar{p}) = \int_\omega f(\nabla\chi(p), \chi(p), p)\, dV(p). \tag{14.5.10}$$

Here $\tilde{\omega}_\alpha = \{\mathbf{p}_\alpha(p, \chi(p)) : p \in \omega\}$ is the image of ω under the χ-dependent change of the independent variable p. Define a vector-valued function $\boldsymbol{Q}$ on the domain of f by

$$\boldsymbol{Q} = \boldsymbol{S}^\mathsf{T}\dot{\mathbf{x}} + \boldsymbol{P}^\mathsf{T}\dot{\mathbf{p}}$$

for each (F, x, p) in the domain of f where $\boldsymbol{Q}$, $\boldsymbol{S}$, $\boldsymbol{P}$ are evaluated at (F, x, p) and $\dot{\mathbf{x}}$, $\dot{\mathbf{p}}$ at (p, x).

14.5.6 Noether's Theorem *If* $\mathbf{T}_\alpha$ *as in* 14.5.5 *is a variational symmetry for* M *then for every deformation* χ *that is a solution of the Euler–Lagrange equations* (14.5.9) *outside the singular surface and that satisfies* 14.5.2(3) *we have*

$$\operatorname{Div} \boldsymbol{Q} = \mathbf{0}$$

outside the singular surface.

Thus the existence of a variational symmetry implies additional conservation equation satisfied by each solution of the Euler–Lagrange equations.

Proof If $\mathbf{T}_\alpha$ is a variational symmetry for M, then by (14.5.10),

$$\frac{d}{d\alpha} \int_{\tilde{\omega}_\alpha} f(\nabla\tilde{\chi}_\alpha(\bar{p}), \tilde{\chi}_\alpha(\bar{p}), \bar{p})\, dV(\bar{p}) = 0$$

at $\alpha = 0$. Calculating the derivative in the same way as in the proof of Proposition 14.5.3, it is found that if $\mathbf{T}_\alpha$ is a variational symmetry, then

$$\int_\omega (\partial_x f \cdot \dot{\mathbf{x}} + \boldsymbol{S} \cdot \nabla\dot{\mathbf{x}} + \partial_p f \cdot \dot{\mathbf{p}} + \boldsymbol{P} \cdot \nabla\dot{\mathbf{p}})\, dV = 0 \tag{14.5.11}$$

for every χ and every ω such that $\operatorname{cl}\omega \subset \Omega$. By the arbitrariness of ω the integrand must vanish. From this, using standard divergence identities,

$$0 = \partial_x f \cdot \dot{\mathbf{x}} + \boldsymbol{S} \cdot \nabla\dot{\mathbf{x}} + \partial_p f \cdot \dot{\mathbf{p}} + \boldsymbol{P} \cdot \nabla\dot{\mathbf{p}} = (\partial_x f - \operatorname{Div} \boldsymbol{S}) \cdot \dot{\mathbf{x}} + (\partial_p f - \operatorname{Div} \boldsymbol{P}) \cdot \dot{\mathbf{p}} + \operatorname{Div} \boldsymbol{Q}.$$

Therefore, if χ satisfies the Euler–Lagrange equations (14.5.9) and hence also the DuBois–Reymond equations (cf. Proposition 14.2.4), we obtain the result. □

14.5.7 *Remarks*

(1) The above is a restricted version of Noether's theorem. Possible generalizations involve

(a) a replacement of the condition that the integrals in (14.5.10) be invariant by the condition that the (standard) variation of M be invariant in an appropriate sense. This, using the idea of a null Lagrangian 13.6 means that the integrand in (14.5.11) be a null Lagrangian (rather than being identically zero) and hence a divergence of something by 13.6.3. Such transformations $\mathbf{T}_\alpha$ are called divergence symmetries,

(b) enlargement of the class of transformations $\mathbf{T}_\alpha$. One can let the new variables $\bar{p}, \bar{x}$ depend also on the gradient of χ and more generally on the higher gradients. Even more generally, one can consider the generalized symmetries of M (see OLVER [1986, Chap. 5] for a definition).

Under these generalizations, there is a one-to-one correspondence between the generalized variational symmetries and conservation laws. For a *complete* treatment of these matters, see OLVER [1986, Chaps. 4 & 5].

(2) More detailed treatments of Noether's theorem (e.g., GEL'FAND & FOMIN [1961], OVSJANNIKOV [1978, Chap. 8], and OLVER [1986, Chaps. 4 & 5]) make explicit the concrete forms of the conservation laws corresponding to concrete symmetries. We only mention that if f is independent of p, then the translations in the reference configuration

$$\mathbf{p}_\alpha(p, x) = p + \alpha r, \quad \mathbf{x}_\alpha(p, x) = x, \quad (p, x) \in \mathbf{F},$$

where $r \in \text{Vect}$ is any fixed direction, leads to $\boldsymbol{Q} = \boldsymbol{P}^\mathsf{T} r$ and hence the conservation law reads $0 = \operatorname{Div}(\boldsymbol{P}^\mathsf{T} r) = r \cdot \operatorname{Div} \boldsymbol{P}$. By the arbitrariness of r this implies the Eshelby conservation law $\operatorname{Div} \boldsymbol{P} = \mathbf{0}$.

In the Lagrangian field theory, the quantity $\boldsymbol{P}$ is known as the canonical energy–momentum tensor of the field.

Bibliographical Note

Under additional assumptions on the nature of the response of the body, such as the invertibility of the stress strain relation, linearity, etc., many further variational principles can be formulated in addition to those of Sect. 14.4. Among them the leading role is played by the principle of minimum of complementary energy, with the complementary energy expressed in terms of statically admissible fields of the referential stress tensor, see, e.g., OGDEN [1975, 1984, Subsec. 5.4.3]. The mixed principles are discussed in the just cited subsection of the book by Ogden and further in TRUESDELL & NOLL [1965, Sect. 89], WASHIZU [1968], ODEN & REDDY [1977], BERDICHEVSKY [1983], HANYGA [1985, Sect. 2.6] and SEWELL [1987, Chap. 5]. For variational principles in linear elasticity, see, GURTIN [1972, 1981] and NEČAS & HLAVÁČEK [1981].

The variations from 14.5.2 occur in, e.g., GEL'FAND & FOMIN [1961], EDELEN [1981], and in OLVER [1986, Chaps. 4 & 5] (with generalizations). The formulas for the variation of M in 14.5.3 and for $\boldsymbol{Q}$ in 14.5.5 are written in a new form; their advantage is the symmetric position of $\boldsymbol{S}$ and $\boldsymbol{P}$. Moreover, the formula for $\boldsymbol{Q}$ shows that the stress and Eshelby's tensors are the basic building blocks for all conservation laws resulting from Noether's theorem. For a description of the symmetries and conservation laws in linear elasticity, see OLVER [1984a,b, 1988].

For the states resulting from the contemporary theorems on the existence of equilibrium states in nonlinear elasticity (see Chap. 21) the standard variations cannot be applied, but the inner and outer variations can. See BALL [1983], GIAQUINTA, MODICA & SOUČEK [1989, 1994], BAUMAN, OWEN, & PHILLIPS [1991], and MÜLLER & SPECTOR [1995].)

15 Extremum Principles

Extremum principles are assertions that one or another integral functional has an extremum in an equilibrium state among competitor states subject to various constraints. The energy criterion of stability in elasticity and the principle of maximum entropy are examples. The treatment starts with a general theory of systems with Liapunov functions. This framework describes the relevance of the minima of Liapunov functions to the dynamical stability. Then the extremum principles are formulated, their simple interrelations are established, and finally it is shown that they imply the conditions of mechanical and thermal equilibrium. This chapter presents only the results for bodies of arbitrary symmetry. Specializations to fluids are found in Chap. 19 and to solids in Chaps. 20 and 21. There also the existence of solutions to the extremum problems and the *strong* extrema are discussed.

15.1 Liapunov Functions and Stability

Consider a continuous body, which up to time $t_0 < 0$ had been in an equilibrium state σ_0 compatible with the external conditions. If external conditions change between times t_0 and $t_1 = 0$ and then the original external conditions are established again and are maintained in the future, the system will leave σ_0, and one can ask whether:

(1) after a small perturbation, the state will stay near the original state in all future times, or even

(2) besides (1), the state will eventually tend to σ_0 as time goes to infinity.

If (1) holds, σ_0 is Liapunov stable; if (2) is true, σ_0 is asymptotically stable. For example, the equilibrium state of a harmonic oscillator without damping is Liapunov stable but not asymptotically stable. If the damping is present, the equilibrium state is also asymptotically stable.

15.1.1 An implementation of these ideas requires us to introduce measures of the distance between states, the class of perturbations, and realistic assumptions enabling us to prove the existence of the time evolution of the system. For continuous bodies these questions are not settled at present (the exceptions are the linearized model of Chap. 25 and some one-dimensional bodies). Since the class of admissible perturbations is infinite dimensional, some arguments from the finite dimensional case break down. We shall discuss this in connection with the strong extrema later in this section.

Let $\Pi^\natural$ be a class of processes compatible with a conservative environment 13.3.6. Let $\Sigma^\natural$ be a set of all states attainable in the processes form $\Pi^\natural$ and let σ_0 be an equilibrium state compatible with the environment. The state σ_0 is said to be ***formally***

asymptotically stable if for every $\sigma \in \Sigma^\natural$ there exists a $\pi \in \Pi^\natural$ such that $\sigma = \pi^i$ and $\pi(t)$ converges to σ_0 in the sense that

$$\mathsf{P}(\pi(t)) \to \mathsf{P}(\sigma_0), \quad \mathsf{U}(\pi(t)) \to \mathsf{U}(\sigma_0) \quad \text{as } t \to \infty. \tag{15.1.1}$$

The states from $\Sigma^\natural$ are constrained by the mechanical or thermal boundary conditions; the choice of $\Pi^\natural$ can depend on the material considered.

15.1.2 Proposition *If σ_0 is formally asymptotically stable then*
(1) *for every $\sigma \in \Sigma^\natural$,*

$$\mathsf{P}(\sigma) \geq \mathsf{P}(\sigma_0); \tag{15.1.2}$$

(2) *if additionally the heating environment is adiabatic isolation, then*

$$\mathsf{U}(\sigma) = \mathsf{U}(\sigma_0), \quad \mathsf{H}(\sigma) \leq \mathsf{H}(\sigma_0). \tag{15.1.3}$$

Assertions (1) and (2) are the principle of minimum canonical free energy and the principle of maximum entropy, respectively. They are *necessary* but not sufficient for the formal asymptotic stability.

Proof (1): Let $\pi \in \Pi^\natural$ be the process from the definition of formal asymptotic stability. By Proposition 13.4.1, $\mathsf{P}(\sigma) = \mathsf{P}(\pi^i) \geq \mathsf{P}(\pi(t))$ for every $t \geq 0$; letting $t \to \infty$ and using $(15.1.1)_1$, we obtain (15.1.2). (2): From $(15.1.1)_{1,2}$ also follows that $\mathsf{H}(\pi(t)) \to \mathsf{H}(\sigma_0)$ as $t \to \infty$; $(15.1.3)_2$ follows. Since in the processes in adiabatic isolation the total energy is conserved, $(15.1.3)_1$ follows from $(15.1.1)_2$. □

We shall now examine to what extent the strong versions of the extremum principles guarantee the stability. The ideas to be formulated below are called the second or direct or energy method in the stability theory. In the rest of this section we shall work with a pair $(\Sigma^\natural, \Pi^\natural)$ where $\Sigma^\natural$ is a set and $\Pi^\natural$ a set of functions π from $\mathbb{R}^+ = [0, \infty)$ into $\Sigma^\natural$ not necessarily identified with the objects in Subsect. 15.1.1. Two distance functions d^i, d on $\Sigma^\natural$ will be used to measure the distances of initial and the future states, respectively. By an ***equilibrium state*** σ_0 of $(\Sigma^\natural, \Pi^\natural)$ we mean a state for which a process with constant value σ_0 belongs to $\Pi^\natural$.

15.1.3 Definitions Let σ_0 be an equilibrium state of $(\Sigma^\natural, \Pi^\natural)$.
(1) σ_0 is said to be ***Liapunov stable*** if for every $\varepsilon > 0$ there exists a $\delta > 0$ such that every $\pi \in \Pi^\natural$ with $\pi^i := \pi(0)$ satisfying

$$d^i(\pi^i, \sigma_0) < \delta \tag{15.1.4}$$

also satisfies

$$d(\pi(t), \sigma_0) < \varepsilon \ \text{ for every } \ t \geq 0. \tag{15.1.5}$$

(2) $\sigma_0 \in \Sigma^\natural$ is said to be ***asymptotically stable*** if it is Liapunov stable and if there exists an $\varepsilon_0 > 0$ such that for every $\pi \in \Pi^\natural$ with $d(\pi^i, \sigma_0) \leq \varepsilon_0$ we have $d(\pi(t), \sigma_0) \to 0$ as $t \to \infty$.
(3) A function $\mathsf{V} : \Sigma^\natural \to \mathbb{R}$ is said to be a ***Liapunov function*** for $(\Sigma^\natural, \Pi^\natural)$ if for every $\pi \in \Pi^\natural$ the function $t \mapsto \mathsf{V}(\pi(t))$ is nonincreasing.
(4) If $\varepsilon' \in (0, \infty]$, then a function $\hat{\gamma} : [0, \varepsilon') \to \mathbb{R} \cup \{\infty\}$ is said to be positive-definite if $\hat{\gamma}(0) = 0$ and $\hat{\gamma}(\varepsilon) > 0$ for every $\varepsilon \in (0, \varepsilon')$.

The results from Sect. 13.4 say that the canonical free energy or minus the entropy are Liapunov functions for bodies in conservative environments. To imply Liapunov stability, the Liapunov function must have a minimum at σ_0, but the usual concept of a strict minimum does not suffice. The next two propositions introduce stronger concepts. Their relation to stability will be clarified in Propositions 15.1.8 and 15.1.9.

15.1.4 Proposition *Let $\sigma_0 \in \Sigma^\natural$ and $\mathsf{V} : \Sigma^\natural \to \mathbb{R}$. The following two conditions are equivalent:*

(1) *there exists an $\varepsilon' \in (0, +\infty]$ such that for every $\varepsilon \in (0, \varepsilon')$ there exists a $\gamma > 0$ satisfying*

$$\mathsf{V}(\sigma) - \mathsf{V}(\sigma_0) \geq \gamma \quad \text{whenever} \quad \sigma \in \Sigma^\natural \quad \text{and} \quad d(\sigma, \sigma_0) = \varepsilon; \tag{15.1.6}$$

(2) *there exists an $\varepsilon' \in (0, +\infty]$ and a positive-definite function $\hat{\gamma}$ on $[0, \varepsilon')$ such that*

$$\mathsf{V}(\sigma) - \mathsf{V}(\sigma_0) \geq \hat{\gamma}(d(\sigma, \sigma_0)) \quad \text{whenever} \quad \sigma \in \Sigma^\natural \quad \text{and} \quad d(\sigma, \sigma_0) < \varepsilon'. \tag{15.1.7}$$

This is immediate. If these equivalent conditions hold, the function V is said to have a ***strong local minimum*** at σ_0 (with respect to d). The function V is said to have a ***strong global minimum*** at σ_0 if in (1) and (2) one can set $\varepsilon' = \infty$.

15.1.5 Proposition *Let $\sigma_0 \in \Sigma^\natural$ and $\mathsf{V} : \Sigma^\natural \to \mathbb{R}$. The following two conditions are equivalent:*

(1) *there exists an $\varepsilon' \in (0, +\infty]$ such that for every $\varepsilon \in (0, \varepsilon')$ there exists a $\gamma > 0$ such that*

$$\mathsf{V}(\sigma) - \mathsf{V}(\sigma_0) \geq \gamma \quad \text{whenever} \quad \sigma \in \Sigma^\natural \quad \text{and} \quad \varepsilon \leq d(\sigma, \sigma_0) < \varepsilon'; \tag{15.1.8}$$

(2) *there exists an $\varepsilon' \in (0, +\infty]$ and a* nondecreasing *positive-definite function $\hat{\gamma}$ on $[0, \varepsilon')$ such that*

$$\mathsf{V}(\sigma) - \mathsf{V}(\sigma_0) \geq \hat{\gamma}(d(\sigma, \sigma_0)) \quad \text{whenever} \quad \sigma \in \Sigma^\natural \quad \text{and} \quad d(\sigma, \sigma_0) < \varepsilon'. \tag{15.1.9}$$

If these two equivalent conditions are satisfied, V is said to have a ***uniformly strong local minimum*** at σ_0 (with respect to d).

Proof (1) $\Rightarrow$ (2) : Let (1) hold and define, for every $\varepsilon \in [0, \varepsilon')$, the number

$$\hat{\gamma}(\varepsilon) := \inf\{\mathsf{V}(\sigma) - \mathsf{V}(\sigma_0) : \sigma \in \Sigma^\natural, \varepsilon \leq d(\sigma, \sigma_0) < \varepsilon'\}.$$

By (1), $\hat{\gamma}$ is positive-definite and nondecreasing because as ε increases, the infimum is taken over a smaller set. Clearly, (15.1.9) holds. (2) $\Rightarrow$ (1) is immediate: it suffices to set $\gamma := \hat{\gamma}(\varepsilon)$. □

For a strong minimum, the values of V on the 'sphere' $\{\sigma \in \Sigma^\natural : d(\sigma, \sigma_0) = \varepsilon\}$ are bounded away from $\mathsf{V}(\sigma_0)$ by a positive number. For a uniformly strong minimum this is true too, but in addition, if $\mathsf{V}(\sigma)$ is close to $\mathsf{V}(\sigma_0)$, then the distance between σ and σ_0 is small. The terms "potential well" and "uniform potential well" are often employed for strong and uniformly strong minima. The existence of uniformly strong minima will be examined in Sect. 19.7 for fluids and Sect. 21.5 for solids.

15.1.6 ***Remarks***

(1) For continuous functions on finite dimensional spaces the concepts of strong and uniformly strong minima coincide with the concept of strict minimum since every continuous function has a minimum on a compact set.

(2) Let $\Omega \subset$ Eucl be an open bounded set and consider the space $L^1(\Omega) \equiv L^1(\Omega, \mathbb{R}^m)$ of all m-tuples $\boldsymbol{u} = (u_1, \ldots, u_m)$ of integrable functions on Ω (see Sect. 20.2 for a definition). Let $f : \mathbb{R}^m \to \mathbb{R}^+$ be a continuous function and define $\mathsf{V} : \Sigma^\natural \to \mathbb{R}^+$ by

$$\mathsf{V}(\boldsymbol{u}) = \int_\Omega f(\boldsymbol{u}(p))\, dV, \quad \boldsymbol{u} \in \Sigma^\natural, \tag{15.1.10}$$

where $\Sigma^\natural$ consists of all $\boldsymbol{u} \in L^1(\Omega)$ for which the integral in (15.1.10) is finite, and set $d(\boldsymbol{u}, \boldsymbol{w}) = \|\boldsymbol{u} - \boldsymbol{w}\|_{L^1(\Omega)}$, where the norm $\|\cdot\|_{L^1(\Omega)}$ is defined in Subsect. 20.2.1. Suppose that $f(0) = 0$ and normalize Ω so that its volume be 1.

15.1.7 Proposition *Consider the situation above and let* $\boldsymbol{u} \in \Sigma^{\natural}$ *be defined by* $\boldsymbol{u}(\boldsymbol{p}) = 0$ *for every* $\boldsymbol{p} \in \Omega$. *Then*

(1) $\boldsymbol{u}$ *is a point of a strict minimum of* V *if and only if* $f(a) > 0$ *for every* $a \in \mathbb{R}^m$, $a \neq 0$;

(2) $\boldsymbol{u}$ *is a point of a strong minimum of* V *if and only if it is a point of uniformly strong minimum of* V; *these two equivalent conditions are satisfied if and only if there exists a positive-definite function* $C : \mathbb{R}^+ \to \mathbb{R}^+$ *such that*

$$f(a) \geq C(\varepsilon)|a| \tag{15.1.11}$$

for every $\varepsilon > 0$ *and every* $a \in \mathbb{R}^m$ *such that* $|a| \geq \varepsilon$.

Functionals like V above arise in the thermostatics of fluids; see Chap. 19. Proposition 16.2.8 gives a reasonable condition for (15.1.11) to hold.

Proof (1) is immediate. To prove (2), let us show that $\boldsymbol{u}$ is a strong minimum of V implies the existence of C as in (2), and that the existence of C implies that $\boldsymbol{u}$ is a uniformly strong minimum. Hence, let $\boldsymbol{u}$ be a point of strong minimum of V. Set $C(0) = 0$ and $C(\varepsilon) = \hat{\gamma}(\varepsilon)/\varepsilon$ for $\varepsilon > 0$, where $\hat{\gamma}$ as in Proposition 15.1.4(2). To prove (15.1.11), let ε and $a \in \mathbb{R}^m$ be as in (2) and define $\boldsymbol{w}(\boldsymbol{p}) := \chi_M(\boldsymbol{p})a$, where $M \subset \Omega$ is any measurable set with volume $\varepsilon/|a| \leq 1$, and χ_M is a characteristic function of M. Then $\mathsf{V}(\boldsymbol{w}) = \varepsilon f(a)/|a|$, $d(\boldsymbol{w}, \boldsymbol{u}) = \varepsilon$ and hence (15.1.7) reads as (15.1.11). To prove that the existence of C as in (2) implies that $\boldsymbol{u}$ is a point of uniformly strong minimum, let us first note that C may be chosen to be nondecreasing without any loss of generality. With this knowledge, the proof is completed in exactly the same way as the proof of Theorem 19.7.1, below. □

The next two propositions give two versions of Liapunov's second method.

15.1.8 Proposition *Let* σ_0 *be an equilibrium state of* $(\Sigma^{\natural}, \Pi^{\natural})$, V *a Liapunov function for* $(\Sigma^{\natural}, \Pi^{\natural})$ *and assume that*

(1) V *is continuous at* σ_0 *with respect to* d^i;

(2) V *has uniformly strong global minimum at* σ_0 *(with respect to* d*).*

Then σ_0 *is Liapunov stable.*

Proof Given $\varepsilon > 0$, the definition of uniformly strong minimum provides a $\gamma > 0$ such that (15.1.8) holds. For this γ the continuity of V with respect to d^i gives a $\delta > 0$ such that $\mathsf{V}(\sigma) - \mathsf{V}(\sigma_0) < \gamma$ whenever $d^i(\sigma, \sigma_0) < \delta$. We shall now show that with such a δ Definition 15.1.3(1) holds. Indeed, if π satisfies (15.1.4) then the above choice of δ gives $\mathsf{V}(\pi^i) - \mathsf{V}(\sigma_0) < \gamma$. Since V is nonincreasing along processes, we have $\mathsf{V}(\pi(t)) - \mathsf{V}(\sigma_0) < \gamma$ and hence the definition of the uniformly strong global minimum (see Proposition 15.1.5) shows that (15.1.5) holds. □

15.1.9 Proposition *Let* σ_0 *be an equilibrium state of* $(\Sigma^{\natural}, \Pi^{\natural})$, V *a Liapunov function for* $(\Sigma^{\natural}, \Pi^{\natural})$ *and assume that*

(1) d^i *is stronger than* d;

(2) *for every* $\pi \in \Pi^{\natural}$ *the function* $t \mapsto d(\pi(t), \sigma_0)$ *is continuous;*

(3) V *is continuous at* σ_0 *with respect to* d;

(4) V *has a strong local minimum at* σ_0 *(with respect to* d*).*

Then σ_0 *is Liapunov stable.*

Proof Let $\varepsilon > 0$ be given. The definition of strong minimum provides a $\gamma > 0$ such that (15.1.6) holds. For this γ the continuity of V with respect to d gives a $\delta_1 > 0$ such that $\mathsf{V}(\sigma) - \mathsf{V}(\sigma_0) < \gamma$ whenever $d(\sigma, \sigma_0) < \delta$. Since d^i is stronger than d, there exists a $\delta_2 > 0$ such that

$$B(\sigma_0, d^i, \delta_2) \subset B(\sigma_0, d, \varepsilon), \tag{15.1.12}$$

where $B(\sigma_0, d^i, \delta_2)$ is the d^i-ball centered at σ_0 of radius δ_2 and $B(\sigma_0, d, \varepsilon)$ the d-ball centered at σ_0 of radius ε. We now define δ by $\delta = \min\{\delta_1, \delta_2\}$ and show that with such a δ Definition 15.1.3 holds. Hence suppose that π satisfies (15.1.4). If (15.1.5) were not true, then by the continuity of the function $t \mapsto d(\pi(t), \sigma_0)$ and by the fact that $d(\pi^i, \sigma_0) < \varepsilon$ (this follows from (15.1.12)) there would exist a $t_0 > 0$ such that $d(\pi(t_0), \sigma_0) = \varepsilon$ and $d(\pi(t), \sigma_0) < \varepsilon$ for all $t \in [0, t_0)$. From $\mathsf{V}(\pi^i) \geq \mathsf{V}(\pi(t_0))$ follows

$$\mathsf{V}(\pi^i) - \mathsf{V}(\sigma_0) \geq \mathsf{V}(\pi(t_0)) - \mathsf{V}(\sigma_0) \geq \gamma. \tag{15.1.13}$$

But on the other hand from (15.1.4), from the fact that $\delta \leq \delta_1$ and δ_1 comes from the definition of continuity, we have

$$V(\pi^i) - V(\sigma_0) < \gamma. \qquad (15.1.14)$$

But (15.1.13) and (15.1.14) contradict each other, and this contradiction shows that (15.1.5) must hold. □

Let us now sketch a scheme for establishing the asymptotic stability by the energy method. The asymptotic stability needs a nontrivial dissipation. Let V be a Liapunov function for $(\Sigma^\natural, \Pi^\natural)$. A function $D : \Sigma^\natural \to [0, \infty)$ is said to be a dissipation function for V if for every $\pi \in \Pi^\natural$ the function $t \mapsto V(\pi(t))$ is absolutely continuous and

$$\dot{V}(\pi(t)) = -D(\pi(t)) \quad \text{for a.e. } t \geq 0. \qquad (15.1.15)$$

As V is nonincreasing, it follows that D is nonnegative. Proposition 13.4.1 calculated the dissipation for the Liapunov function P.

15.1.10 Proposition *Let σ_0 be an equilibrium state of $(\Sigma^\natural, \Pi^\natural)$, V a Liapunov function for $(\Sigma^\natural, \Pi^\natural)$, D a dissipation function for V and assume that σ_0 is a point of global minimum of V in the sense that $V(\sigma) \geq V(\sigma_0)$ for every $\sigma \in \Sigma^\natural$. Then*

$$\int_0^\infty D(\pi(t))\,dt \leq V(\pi^i) - V(\sigma_0) < \infty \qquad (15.1.16)$$

for every process π.

Since D depends on the state parameters of the process, the finiteness of the integral shows that $D(t)$ must be small enough on a sufficiently large set of times t, i.e., the dissipation tends to 0 in this sense. Unfortunately, (15.1.16) is insufficient to conclude that $D(\pi(t)) \to 0$ as $t \to \infty$ without extra information. Assuming that the conclusion $D(\pi(t)) \to 0$ is available, the explicit form of D allows one to infer that the process decays to some rest state. For example, in the case of $V \equiv P$ we have $D = \theta_0 \Gamma$ where Γ is the production of entropy given by (13.2.5). For materials with a linear dynamic response, Γ is a positive-definite quadratic form in $\boldsymbol{D}$ and $\boldsymbol{G}$ and one concludes that the squares of these quantities decay to 0 in the mean.

Proof We have $D(\pi(t)) \geq 0$; integrating (15.1.15) from 0 to $t > 0$ and using that $V(\pi(t)) \geq V(\sigma_0)$ we obtain

$$\int_0^t D(\pi(s))\,ds = V(\pi^i) - V(\pi(t)) \leq V(\pi^i) - V(\sigma_0).$$

As this must be true for every $t > 0$, the result follows. □

15.1.11 Proposition *Let σ_0 be an equilibrium state of $(\Sigma^\natural, \Pi^\natural)$, V a Liapunov function for $(\Sigma^\natural, \Pi^\natural)$ and assume that*

(1) *V is continuous at σ_0 with respect to d^i;*

(2) *V has a uniformly strong global minimum at σ_0;*

(3) *there exist $\varepsilon_0, \delta_0 > 0$ and a nondecreasing positive-definite function $m : [0, \delta_0) \to [0, \infty)$ such that for every $\pi \in \Pi^\natural$ satisfying $d(\pi^i, \sigma_0) \leq \varepsilon_0$, the function $t \mapsto V(\pi(t))$ is continuously differentiable, $0 \leq V(\pi(t)) - V(\sigma_0) \leq \delta_0$ and*

$$\dot{V}(\pi(t)) \leq -m(V(\pi(t)) - V(\sigma_0)) \quad \text{for every } t \geq 0. \qquad (15.1.17)$$

Then σ_0 is asymptotically stable.

(4) *If (15.1.17) holds in the strengthened form*

$$\dot{\mathsf{V}}(\pi(t)) \le -c(\mathsf{V}(\pi(t)) - \mathsf{V}(\sigma_0)) \text{ for every } t \ge 0 \tag{15.1.18}$$

where $c > 0$ is a constant independent of the process, then

$$\hat{\gamma}(d(\pi(t), \sigma_0)) \le (\mathsf{V}(\pi^i) - \mathsf{V}(\sigma_0))e^{-ct} \tag{15.1.19}$$

for every process such that $d(\pi^i, \sigma_0) \le \varepsilon_0$ and every $t \ge 0$, where $\hat{\gamma}$ is the function from Proposition 15.1.5(2).

Inequalities (15.1.18) and (15.1.19) say that the dissipation is so strong that it can be estimated by the excess Liapunov function over its value at σ_0. Proposition 15.1.11 is a standard procedure for establishing the asymptotic stability. However, we shall find no model for which (15.1.18) or (15.1.19) can be established with $\mathsf{V} \equiv \mathsf{P}$; for results of this type different Liapunov functions must be sought. For example, for the linearized equations of viscous solids with heat conduction, in Proposition 25.2.4 we shall establish (15.1.19) for a Liapunov function which is a slight, but nontrivial, modification of P. Incidentally, for these equations the asymptotic stability may be established also by the spectral methods using the theory of linear semigroups. It appears that, with a few simple exceptions like incompressible isothermal viscous fluids, all other existing proofs of the asymptotic stability use more complicated methods than merely Proposition 15.1.11. I refer to BEIRÃO DA VEIGA [1989], ANTONTSEV, KAZHIKHOV & MONAKHOV [1990], VALLI [1992], SALVI & STRAŠKRABA [1993], JIANG [1995], STRAŠKRABA [1996] for asymptotic stability results for fluids, and to RACKE & al. [1993] for one-dimensional solids.

Proof By Proposition 15.1.8, σ_0 is Liapunov stable. To complete the proof, let us first show that for every process such that $d(\pi^i, \sigma_0) \le \varepsilon_0$ we have $\mathsf{V}(\pi^i) \to \mathsf{V}(\sigma_0)$ as $t \to \infty$. Since $t \mapsto \mathsf{V}(\pi(t))$ is nondecreasing, the limit $l := \lim_{t\to\infty}(\mathsf{V}(\pi(t)) - \mathsf{V}(\sigma_0)) \ge 0$ exists. If $l > 0$, then by (15.1.17) and by the nondecreasing character of m we have $\dot{\mathsf{V}}(\pi(t)) < -m_0$ for some $m_0 > 0$ and all sufficiently large t, which contradicts the fact that $\mathsf{V}(\pi(t))$ is bounded below as a consequence of a local minimum of V at σ_0. Hence $l = 0$, which proves the claim $\mathsf{V}(\pi^i) \to \mathsf{V}(\sigma_0)$ as $t \to \infty$. By Proposition 15.1.5(2) then $d(\pi(t), \sigma_0) \to 0$ as $t \to \infty$, which shows that σ_0 is asymptotically stable. Finally, if (15.1.18) holds, then setting $f(t) = \mathsf{V}(\pi(t)) - \mathsf{V}(\sigma_0)$, $t \ge 0$, (15.1.18) can be rewritten as $(fe^{ct})^{\cdot} \le 0$ which upon integration yields $f(t) \le f(0)e^{-ct}$, i.e., $\mathsf{V}(\pi(t)) - \mathsf{V}(\sigma_0)) \le (\mathsf{V}(\pi^i) - \mathsf{V}(\sigma_0))e^{-ct}$. A combination with $(15.1.9)_1$ leads to (15.1.19). □

15.2 The Extremum Principles

This section formulates the extremum principles for bodies of unspecified material symmetry. Of them, the principle of minimum canonical free energy and the principle of maximum entropy were motivated in Subsect. 15.1.1. The following section shows that the other principles are also reasonable as they are equivalent to the principle of minimum canonical free energy or to the principle of maximum entropy.

Throughout, we consider a body Ω in a conservative environment and an equilibrium state σ_0 compatible with it. To formulate the extremum principles, one has to specify the class of competitor states on which the values of the thermodynamical functionals are compared. In view of the diversity of environments we shall deal with an abstract subset Σ_0 of the set of all rest states $\Sigma^\#$ subject to simple axioms. It is assumed that Σ_0 is contained in the domain of definition of the integral functionals in (14.3.1). The standard choice of Σ_0 is $\Sigma_0 := \Sigma^{**}$. The formalism also covers local extrema: if Σ^{**} is endowed with a distance function d, one can choose Σ_0 to consist of only those $\sigma \in \Sigma^{**}$ which satisfy $d(\sigma, \sigma_0) < \varepsilon$.

Whether or not the extremum principles formulated below are satisfied depends on the details of the response of the body and on the boundary conditions.

15.2.1 The Minimum of Canonical Free Energy *For every* $\sigma \in \Sigma_0$,

$$\mathsf{P}(\sigma) \geq \mathsf{P}(\sigma_0). \tag{15.2.1}$$

In the next section we will show that this principle is the strongest of all extremum principles formulated in this book.

Recall from Subsect. 14.3.4 the total stored energy M on $\boldsymbol{\Delta}^{**}$. Let $\boldsymbol{\Delta}_0$ be the set of all deformations $\chi \in \boldsymbol{\Delta}^{**}$ for which there exists an entropy field η such that (χ, η) is an *equilibrium state* of temperature θ_0 and $(\chi, \eta) \in \Sigma_0$.

15.2.2 Energy Criterion of Stability *For every* $\chi \in \boldsymbol{\Delta}_0$,

$$\mathsf{M}(\chi) \geq \mathsf{M}(\chi_0). \tag{15.2.2}$$

Since for states with temperature θ_0 the canonical free energy coincides with the stored energy it is clear that the principle of minimum canonical free energy implies the energy criterion of stability. The elasticity literature calls $\bar{f}$ just the "energy;" however, $\bar{f}$ is *not* the internal energy. Condition 15.2.2 is apparently weaker than 15.2.1. Interestingly, for bodies with positive specific heat at constant deformation, (15.2.2) will be shown to be equivalent to 15.2.1.

15.2.3 The Maximum of Entropy *For every* $\sigma \in \Sigma_0$ *such that* $\mathsf{U}(\sigma) = \mathsf{U}(\sigma_0)$ *one has* $\mathsf{H}(\sigma) \leq \mathsf{H}(\sigma_0)$.

15.2.4 The Minimum of Energy *For every* $\sigma \in \Sigma_0$ *such that* $\mathsf{H}(\sigma) = \mathsf{H}(\sigma_0)$ *one has* $\mathsf{U}(\sigma) \geq \mathsf{U}(\sigma_0)$.

The last two principles are dual to each other, and we shall see in the next section that under mild additional assumptions they are equivalent. Further extremum principles can be formulated for fluids, cf. Chap. 19.

The rest of this part of the book is devoted to the analysis of the extremum principles. The treatment will give a number of conditions and situations when the extremum principles *hold or fail to hold,* either directly or by implication. The range of applicability of the extremum principles is extremely wide: they govern the global stability of large engineering structures as well as purely local, small-scale phenomena like the internal or surface instabilities (such as the formation of voids, patterns, and the occurrence of surface wrinkling).

We have already encountered the minimization of energy in Sect. 2.4: the ground states satisfy the extremum principles with the choice $\Sigma_0 = \Sigma^{*}$. The energy criterion of stability for equilibrium states of a column is examined in Sect. 20.7. The stability of equilibrium states of fluids is treated in Sects. 19.3–19.5 and 19.8. The surface instabilities are examined in Sects. 20.4 and 20.6. For the extension of the extremum principles to states with microstructure, i.e., to Young measure minimizers, see Sect. 21.6.

15.3 Relationships Among the Principles

I now propose to consider the extremum principles on their own and to establish the relationships among them.

15.3.1 Proposition *Generally we have the following relations among the extremum principles:*

$$\begin{matrix}\text{Minimum of}\\ \text{Canonical Free Energy}\end{matrix} \Rightarrow \left\{\begin{matrix}\text{Energy Criterion of Stability}\\ \text{Maximum of Entropy}\\ \text{Minimum of Energy}\end{matrix}\right.$$

These are immediate implications, no reference to the details of the form of the functionals is necessary.

Proof That the principle of minimum canonical free energy implies the energy criterion of stability is immediate: the latter is the same assertion about a smaller class of competitor states. To prove that the principle of minimum canonical free energy implies the other four extremum principles, notice that the inequality $\mathsf{P}(\sigma) \geq \mathsf{P}(\sigma_0)$ can be rewritten as

$$\mathsf{U}(\sigma) - \mathsf{U}(\sigma_0) \geq \theta_0[\mathsf{H}(\sigma) - \mathsf{H}(\sigma_0)] \tag{15.3.1}$$

and from this all the implications can be read off. For instance, let us prove that the principle of minimum canonical free energy implies the extended principle of maximum entropy. Accordingly, suppose that the principle of minimum canonical free energy holds and that $\sigma \in \Sigma_0$ be a state such that $\mathsf{U}(\sigma) \leq \mathsf{U}(\sigma_0)$. Then the left-hand side of (15.3.1) is nonpositive and hence also the right-hand side is nonpositive, i.e., $\mathsf{H}(\sigma) \leq \mathsf{H}(\sigma_0)$. The other implications are proved similarly. □

Further relations hold if realistic additional assumptions are imposed.

15.3.2 Assumptions

(1) $\tilde{U} = U_1 \times (0, \infty)$ where U_1 is an open subset of InvLin.

(2) For every $\boldsymbol{F} \in U_1$ we have $\tilde{e}(\boldsymbol{F}, \eta) \to \infty$ for $\eta \to \infty$.

(3) If $(\chi, \eta) \in \Sigma_0$ and $\alpha \in (0, 1)$ then also $(\chi, \alpha\eta) \in \Sigma_0$.

(4) If $(\chi, \eta) \in \Sigma_0$ and $c \in \mathbb{R}$, $c > 0$, then also $(\chi, \eta + c) \in \Sigma_0$.

(5) Let $\sigma = (\chi, \eta) \in \Sigma_0$ be a state, $D \subset \Omega$ an open region and ω : Eucl $\to \mathbb{R}$ an infinitely differentiable function vanishing outside D such that $\int_D \omega(\boldsymbol{p})\rho_\circ dV = 0$. Then $(\chi, \eta + \omega) \in \Sigma_0$.

The hypotheses (1), (2) are compatible with the third law of thermodynamics (see Sect. 7.7).

15.3.3 Lemma *Suppose that Assumptions* 15.3.2(1)–(4) *hold. Then*

(1) *if* $\sigma = (\chi, \eta) \in \Sigma_0$ *and* H_0 *is a number satisfying*

$$0 < \mathsf{H}_0 < \mathsf{H}(\sigma), \tag{15.3.2}$$

then there exists a $\sigma' = (\chi, \eta') \in \Sigma_0$ *with the same configuration as* σ *such that*

$$\mathsf{H}(\sigma') = \mathsf{H}_0, \quad \mathsf{U}(\sigma') < \mathsf{U}(\sigma); \tag{15.3.3}$$

(2) *if* $\sigma = (\chi, \eta) \in \Sigma_0$ *and* U_0 *is a number satisfying* $\mathsf{U}_0 > \mathsf{U}(\sigma)$, *then there exists a* $\sigma' = (\chi, \eta') \in \Sigma_0$ *with the same configuration as* σ *such that* $\mathsf{U}(\sigma') = \mathsf{U}_0$, $\mathsf{H}(\sigma') > \mathsf{H}(\sigma)$.

Proof (1): By (15.3.1) the number $\alpha := \mathsf{H}(\sigma)/\mathsf{H}_0$ is in (0, 1) and we define $\sigma' = (\chi, \alpha\eta')$. By Assumptions (1), (3), $\sigma' \in \Sigma_0$ and clearly $(15.3.3)_1$ holds; moreover, as $\alpha\eta' \leq \eta$ and $\tilde{e}$ is an increasing function of η (since its derivative is the reciprocal of the positive absolute temperature), also $(15.3.3)_2$ is satisfied. The proof of (2) is similar, but Assumptions (2) and (4) of Subsect. 15.3.2 must be invoked. □

15.3.4 Proposition *Under Assumptions* 15.3.2(1)–(4) *the principle of maximum entropy is equivalent to the principle of minimum of energy.*

Proof Let us prove, e.g., that the minimum of energy implies the maximum of entropy. Accordingly, suppose the principle of minimum of energy to hold and let $\sigma \in \Sigma_0$ and $\mathsf{U}(\sigma) = \mathsf{U}(\sigma_0)$. Our goal is to prove that $\mathsf{H}(\sigma) \le \mathsf{H}(\sigma_0)$. Suppose, on the contrary, that $\mathsf{H}(\sigma) > \mathsf{H}(\sigma_0)$. Set $\mathsf{H}_0 = \mathsf{H}(\sigma_0)$ and apply Assertion (1) of Lemma 15.3.3 to obtain a state σ' for which (15.3.3) hold. In view of $(15.3.3)_1$ the state σ' satisfies the hypothesis of the principle of minimum of energy which allows us to conclude that $\mathsf{U}(\sigma') \ge \mathsf{U}(\sigma)$, a contradiction with $(15.3.3)_2$. The other implications are proved similarly. □

15.3.5 Proposition *Let the body satisfy Assumptions* 15.3.2(1)-(5) *and assume the state* σ_0 *satisfies the principle of minimum energy. Then for every* $\boldsymbol{p} \in \Omega$ *and* $\eta' > 0$ *we have*

$$\tilde{e}(\boldsymbol{F}_0(\boldsymbol{p}), \eta') \ge \tilde{e}(\boldsymbol{F}_0(\boldsymbol{p}), \eta_0(\boldsymbol{p})) + \theta_0(\eta' - \eta_0(\boldsymbol{p})). \tag{15.3.4}$$

As was already pointed out in Chap. 10, this convexity inequality is related to the positivity of the specific heat at constant deformation. Here we obtain it at every pair $(\boldsymbol{F}_0(\boldsymbol{p}), \eta_0(\boldsymbol{p}))$ that is realized in the equilibrium state considered. Proof of this inequality will be given in Chap. 17, but the condition itself is of interest already here because under it the energy criterion of stability becomes fully equivalent to the principle of minimum canonical free energy:

15.3.6 Proposition *Assume that the body satisfies Assumptions* 15.3.2(1)-(5). *Then the state satisfies the principle of minimum canonical free energy if and only if it satisfies the energy criterion and inequality* (15.3.4).

Because there appear to be no bodies violating (15.3.4), this proposition reduces the testing of whether the state satisfies the principle of minimum canonical free energy to the isothermal energy test. This is remarkable.

15.3.7 Summary *Under realistic hypotheses we have proved that the relations among the principles are as follows:*

$$\begin{matrix} \text{Minimum of} \\ \text{Canonical Free Energy} \\ \Updownarrow \\ \text{Energy Criterion of Stability} \end{matrix} \quad \Rightarrow \quad \left\{ \begin{matrix} \text{Maximum of Entropy} \\ \Updownarrow \\ \text{Minimum of Energy} \end{matrix} \right.$$

We thus see that the principles split into two pairs of mutually equivalent conditions. In Chap. 19 will be proved that for fluids all the principles are equivalent and in Chap. 17 that all the principles are equivalent for homogeneous states if the displacement is prescribed on the whole of the boundary of the body.

15.4 Extremum Principles and Variations

For the standard choice $\Sigma_0 = \Sigma^{**}$ the calculus of variations provides standard necessary conditions on the first and the second variations at the extrema which will be now described (cf., e.g., CESARI [1983, Sect. 2.3]). While the first variations are independent of the choice of independent variables, the second variations are not: With our choice of η as independent variable, H is a linear functional and hence $\delta^2\mathsf{H} = 0$ identically; if the internal energy e were to be used an independent variable then $\delta^2\mathsf{H}$ would be a nonzero functional. With the entropy as independent variable then

$$\delta^2 \mathsf{P}(\sigma_0, \xi, \xi) = \delta^2 \mathsf{U}(\sigma_0, \xi, \xi) = \delta^2 \mathsf{E}(\sigma_0, \xi, \xi) + \delta^2 \mathsf{W}(\chi_0, \boldsymbol{u}, \boldsymbol{u}), \tag{15.4.1}$$

for every $\xi = (\boldsymbol{u}, \omega) \in \mathrm{Var}(\sigma_0)$. For the conclusions about the signs of the second variations, the weak local extrema in the sense of the calculus of variations are enough.

15.4.1 Definitions

(1) If $\sigma_i = (\chi_i, \eta_i) \in \Sigma^{**}$, $i = 1, 2$ are two rest states, we define two distances between σ_1 and σ_2 by

$$\begin{aligned} \|\sigma_1 - \sigma_2\|_\infty &= \|\chi_1 - \chi_2\| + \|\nabla\chi_1 - \nabla\chi_2\| + \|\eta_1 - \eta_2\|, \\ \|\sigma_1 - \sigma_2\|_0 &= \|\chi_1 - \chi_2\| + \|\eta_1 - \eta_2\|, \end{aligned} \tag{15.4.2}$$

where, for the purpose of this definition only, $\|f\| := \sup\{|f(\boldsymbol{p})| : \boldsymbol{p} \in \Omega\}$ for any bounded function f on Ω with values in a vector space with norm $|\cdot|$. (More general distance functions will be defined in Subsect. 20.3.1.)

(2) Let $\mathsf{Q} : \Sigma^{**} \to \mathbb{R}$. A state $\sigma_0 \in \Sigma^{**}$ is said to be a point of ***weak local minimum*** (in the sense of calculus of variations) of Q if $\mathsf{Q}(\sigma_0) \le \mathsf{Q}(\sigma)$ for some $\varepsilon > 0$ and every $\sigma \in \Sigma^{**}$ satisfying $\|\sigma_0 - \sigma\|_\infty < \varepsilon$.

(3) A state $\sigma_0 \in \Sigma^{**}$ is said to be a point of ***strong local minimum*** (in the sense of calculus of variations) of Q if $\mathsf{Q}(\sigma_0) \le \mathsf{Q}(\sigma)$ for some $\varepsilon > 0$ and every $\sigma \in \Sigma^{**}$ satisfying $\|\sigma_0 - \sigma\|_0 < \varepsilon$.

(4) The weak and strong local maxima are defined analogously. If $\mathsf{L} : \boldsymbol{\Delta}^{**} \to \mathbb{R}$ is a functional of χ only, then the weak or strong local extrema are defined via the distances $\|\chi_1 - \chi_2\|_\infty$ and $\|\chi_1 - \chi_2\|_0$ defined as in (15.4.2) with the differences of entropies deleted.

15.4.2 Proposition *Let an equilibrium state $\sigma = (\chi, \eta) \in \Sigma^{**}$ compatible with a conservative environment be a point of weak local minimum of canonical free energy. Then*

$$\delta \mathsf{P}(\sigma, \xi) = 0 \text{ and } \delta^2 \mathsf{P}(\sigma, \xi, \xi) \ge 0 \text{ for every } \xi \in \mathrm{Var}(\sigma). \tag{15.4.3}$$

Moreover, in the case of the unilateral contact environment 13.3.3, *the total stress* $\boldsymbol{S}$ *in σ satisfies*

$$\boldsymbol{m} \cdot (\boldsymbol{S} N + \partial_x \phi) \le 0 \text{ on } C \tag{15.4.4}$$

where C is the contact part of $\partial\Omega$ defined in 14.2.3 *and* $\boldsymbol{m}$ *is the outer normal to $\partial\chi(\Omega)$.*

In the case of the contact environment we have $\boldsymbol{S} N + \partial_x \phi = -m \operatorname{cof} \boldsymbol{F} N$ on C where m is a scalar-valued function to be interpreted as the total pressure exerted by R on C (see 14.2.3(5)). Inequality (15.4.4) is equivalent to $m \ge 0$.

Proof Let $\xi \in \mathrm{Var}(\sigma)$ be given and let $\bar\sigma(\alpha), -\varepsilon < \alpha < \varepsilon$ be the family satisfying (14.3.2), which exists according to the assumption in 14.3.2. The function of a real variable $\alpha \mapsto \mathsf{P}(\bar\sigma(\alpha))$ then has a local minimum at $\alpha = 0$ and (15.4.3) are just the conditions that the first derivative vanish and the second derivative be nonnegative. The proof of (15.4.4) will be only sketched. Consider any smooth family $\bar\sigma(\alpha), 0 \le \alpha < \varepsilon$ of admissible states such that $\bar\sigma(0) = \sigma$. Write $\bar\sigma(\alpha) = (\bar\chi_\alpha, \bar\eta_\alpha)$, denote $\boldsymbol{u} = d\bar\chi_\alpha / d\alpha$, $\omega = d\bar\eta_\alpha / d\alpha$ at $\alpha = 0$ and define $\xi = (\boldsymbol{u}, \omega)$. Again, the function $\varphi(\alpha) := \alpha \mapsto \mathsf{P}(\bar\sigma(\alpha))$, $0 \le \alpha < \varepsilon$ has a local minimum at $\alpha = 0$ and so its first derivative at $\alpha = 0$ must be nonnegative, which gives $\delta\mathsf{P}(\sigma, \xi) \ge 0$. We have $\delta\mathsf{P} = \delta\mathsf{U} - \theta_0 \delta\mathsf{H}$ and $\delta\mathsf{U}, \delta\mathsf{H}$ are given by the formulas in Proposition 14.4.1, as is easy to check. Using these formulas, and the conditions of equilibrium 14.2.3, the inequality $\delta\mathsf{P}(\sigma, \xi) \ge 0$ reduces to

$$\int_C (\boldsymbol{S}\boldsymbol{N} + \partial_{\boldsymbol{x}}\phi) \cdot \boldsymbol{u}\, dA \geq 0. \tag{15.4.5}$$

The conditions that $\bar{\chi}_\alpha(\Omega) \subset R$ for $\alpha \geq 0$ and $\bar{\chi}_0 = \chi$ imply that

$$\boldsymbol{u} \cdot \boldsymbol{m} \leq 0 \text{ on } C. \tag{15.4.6}$$

Conversely, since essentially any $\boldsymbol{u}$ satisfying (15.4.6) can be realized on some family $\bar{\chi}_\alpha$ (see CIARLET & NEČAS [1987] for details), we see that (15.4.5) must hold for every $\boldsymbol{u}$ satisfying (15.4.6). Inequality (15.4.4) then follows by taking $\boldsymbol{u}$ such that $\boldsymbol{u} = \gamma\boldsymbol{m}$ on C with γ nonpositive functions on $\partial\Omega$. □

15.4.3 Proposition

(1) *If* $\chi \in \Delta^{**}$ *is a point of weak local minimum of* M *on* Δ^{**}, *then*

$$\delta \mathsf{M}(\chi, \boldsymbol{u}) = 0 \text{ and } \delta^2 \mathsf{M}(\chi, \boldsymbol{u}, \boldsymbol{u}) \geq 0 \text{ for every } \boldsymbol{u} \in \mathrm{Var}(\chi). \tag{15.4.7}$$

Moreover, in the case of the unilateral contact environment 13.3.3, *inequality* (15.4.4) *holds.*

(2) *If the specific heat at constant deformation is positive at every* $(\boldsymbol{F}, \eta)$ *realized in the state, then* (15.4.7) *imply* (15.4.3) *at the unique equilibrium state* σ *characterized by the temperature* θ_0 *and configuration* χ.

Proof Only (15.4.7) ⇒ (15.4.3) needs a proof. This is based on the following consequence of (10.7.3): for every $\xi = (\boldsymbol{u}, \omega) \in \mathrm{Var}(\sigma)$,

$$\delta^2 \mathsf{P}(\sigma, \xi, \xi) = \delta^2 \mathsf{M}(\chi, \boldsymbol{u}, \boldsymbol{u}) - \int_\Omega \partial_\theta^2 \bar{f} \cdot (\partial_\eta \bar{e} \cdot \omega - \partial_{\boldsymbol{F}} \partial_\eta \bar{e} \cdot \nabla \boldsymbol{u})^2 \rho_\circ\, dV. \tag{15.4.8}$$

The proof is complete if one realizes that the positivity of the specific heat implies $\partial_\theta^2 \bar{f} \leq 0$. □

Finally, the next result gives the differential consequences of the minimum of energy and the maximum entropy principles.

15.4.4 Proposition *Let* σ *be an equilibrium state compatible with the environment and let it satisfy any of the principles* 15.2.3 *or* 15.2.4 *locally in the sense that* $\Sigma_0 = \{\bar{\sigma} \in \Sigma^{**} : \|\bar{\sigma} - \sigma\|_\infty < \varepsilon\}$ *for some* $\varepsilon > 0$. *Then*

$$\delta \mathsf{U}(\sigma, \xi) = 0, \quad \delta^2 \mathsf{U}(\sigma, \xi, \xi) \geq 0$$

for every $\xi \in \mathrm{Var}(\sigma)$ *such that* $\delta \mathsf{H}(\sigma, \xi) = 0$.

15.4.5 *Remarks*

(1) The conclusions of the above propositions are the only consequences of the weak local extrema that can be drawn at this generality. In particular, they do not provide Maxwell's relation $(14.5.1)_2$ on the singular surface. This will be shown to be a consequence of the *strong* local extrema in Chap. 17.

(2) The partial converses of the above propositions are formulated in Proposition 20.3.4 under the assumption that the second variations are uniformly positive. However, this hypothesis can be satisfied only by solids; for fluids there are always nonzero variations $\xi = (\boldsymbol{u}, \omega)$ such that $\delta^2 \mathsf{P}(\sigma, \xi, \xi) = 0$ as a mere consequence of the material symmetry of the fluid.

Bibliographical Note

The material of Sect. 15.1 is standard, see, e.g., ZUBOV [1964] or HENRY [1981, Chap. 4], where also refinements are to be found. The use of Liapunov functions is also standard in the theory of ordinary differential equations. As shown in HENRY [1981], the method works almost equally efficiently for semilinear parabolic differential equations. Is is not so efficient for general evolution partial differential equations. The importance of the strong extrema in continuum thermodynamics is emphasized in COLEMAN & DILL [1968, 1973], GURTIN [1973, 1975], and BALL & MARSDEN [1984]. These papers also contain further references and some mild refinements due to the specific structure of P as a sum of a positive-definite kinetic energy plus and equilibrium canonical free energy which can have a minimum. Such refinements include, e.g., that only a rest process can emanate from an equilibrium state in which P has a local, but not necessarily strict, minimum (see GURTIN [1975, Thm. 6.3]), etc. Proposition 15.1.9 is an adaptation of a result from BALL & MARSDEN [1984]. The necessity of extremum principles for asymptotic stability, i.e., Proposition 15.1.2, is also standard. The scheme in Proposition 15.1.11 can be substantially weakened in the finite-dimensional case, but probably not too much in the infinite-dimensional case. The proof of Proposition 15.1.11 is close to HENRY [1981, Thm. 4.1.4]. The same reference also contains the definition of the asymptotic stability of orbits and processes, and LaSalle's invariance principle, a substantial generalization of the ideas presented in Proposition 15.1.11. The idea that the positivity of the specific heat enables one to reduce the verification of the minimum of canonical free energy to the verification of the minimum of the stored energy (Proposition 15.3.5) is due to ERICKSEN [1966]. The "field" formulation of the extremum principles and the equivalence of the minimum of energy with the maximum of entropy are due to COLEMAN & NOLL [1959]. DUNN & FOSDICK [1980] and MAN [1985a] pointed out that the standard formulation of the maximum entropy principle 15.2.3 does not exclude equilibrium states with negative absolute temperature (provided negative absolute temperatures are admitted by the response functions). MAN [1985a] gives a formulation of the maximum entropy principle that is free from this flaw (see also STUECKELBERG & SCHEURER [1967, 1974]). His formulation involves competitor states and equilibrium states with nonzero velocity fields. When augmented with the condition that the total linear and angular momenta of the competitor states be the same as those of the equilibrium state under consideration, this form of the maximum entropy principle also applies to states of unconfined motion such as equilbrium states of rotating stars. See also LIN [1990].

16 Convexity

Many inequalities applying to the equilibrium response functions involve either the convexity directly or concepts related to it. It appears worthwhile to collect the facts needed about convexity in a single chapter, if only so that the results can be stated in a neater notation and can be seen from a more general perspective.

The convexity enters directly into the theory of equilibrium of fluids when the internal energy is expressed as a function of the specific volume and entropy. In this case convex combinations have a direct physical significance as combinations of homogeneous phases of a heterogeneous state with the coefficients representing the relative weights of the phases. If the energy of a general body (of unspecified symmetry) is expressed as a function of the deformation gradient, then the convexity enters somewhat less directly through the notion of rank 1 convexity. The latter can be formulated as the true convexity of certain functions associated with different directions in the reference configuration (see Remark 17.3.2(3)). The basic assertions about the equilibrium of fluids or about the rank 1 convexity are just translations of some assertions about convex functions.

The main difference from the standard expositions is that here we have to deal with convex functions on nonconvex domains and with functions convex only at certain points of their domains. (On the other hand, we can afford stronger differentiability assumptions.) The reader is referred to ROCKAFELLAR [1970] or EKELAND & TÉMAM [1976] for more details on convexity and for proofs of propositions not proved below. Sections 16.1 and 16.2 form a background for all convexity properties of materials. Sections 16.3 and 16.4 are more specific as they introduce notions to be used only in the thermostatics of fluids.

16.1 Convex Sets

Throughout this chapter, let X be a finite-dimensional real vector space with inner product and D its subset.

16.1.1 Definitions If $\boldsymbol{x}, \boldsymbol{y} \in$ X, then the ***closed line segment*** $[\boldsymbol{x}, \boldsymbol{y}]$ with endpoints $\boldsymbol{x}, \boldsymbol{y}$ is the set of all $\boldsymbol{z}$ of the form $\boldsymbol{z} = \alpha\boldsymbol{x} + (1-\alpha)\boldsymbol{y}$ where $0 \le \alpha \le 1$. The ***open line segment*** $(\boldsymbol{x}, \boldsymbol{y})$ with endpoints $\boldsymbol{x}, \boldsymbol{y}$ is $[\boldsymbol{x}, \boldsymbol{y}] \setminus \{\boldsymbol{x}, \boldsymbol{y}\}$. A subset D of X is said to be ***convex*** if for every two points $\boldsymbol{x}, \boldsymbol{y}$ of D also the closed line segment with endpoints $\boldsymbol{x}, \boldsymbol{y}$ is in D. If $\boldsymbol{x}_i$, $i = 1, \dots, s$, are points of a convex set D and $\alpha_i \ge 0$ are numbers such that $\sum_{i=1}^{s} \alpha_i = 1$, then also the convex combination $\boldsymbol{x} := \sum_{i=1}^{s} \alpha_i \boldsymbol{x}_i$ is in D. For every set $K \subset$ X there exists the smallest convex set co(K) containing K; this set is called the ***convex hull*** of K; it is identical with the set of all convex combinations of points from K. A collection $\boldsymbol{x}_i$ of s points of X is said to be ***affinely independent*** if the only set of numbers $\beta_i \in \mathbb{R}$ satisfying $\sum_{i=1}^{s} \beta_i \boldsymbol{x}_i = \boldsymbol{0}$ and $\sum_{i=1}^{s} \beta_i = 0$ is $\beta_i = 0, 1 \le i \le s$. A collection of points is said to be ***affinely dependent*** if it is not affinely independent. Any collection of points containing more than $n + 1$ points is affinely dependent. A point $\boldsymbol{x}$ is said to be an ***affine combination*** of the collection $\boldsymbol{x}_i, i = 1, \ldots, s$, if $\boldsymbol{x} = \sum_{i=1}^{s} \beta_i \boldsymbol{x}_i$ for some $\beta_i \in \mathbb{R}$ such that $\sum_{i=1}^{s} \beta_i = 1$. A collection $\boldsymbol{x}_i$ is affinely dependent if and only if one of the points $\boldsymbol{x}_i$ is an affine combination of the others.

16.1.2 Carathéodory's Theorem *If $K \subset$ X then every point of* co(K) *is a convex combination of an affinely independent collection of points of K.*

A unit vector $\boldsymbol{n} \in$ X is said to be an ***exterior normal*** to D at the point $\boldsymbol{x} \in \partial D$ if $\boldsymbol{n} \cdot (z - \boldsymbol{x}) \le 0$ for every $z \in D$, i.e., if D is contained in the closed half-space $H = \{z \in \mathrm{X} : \boldsymbol{n} \cdot (z - \boldsymbol{x}) \le 0\}$. The hyperplane $\partial H = \{z \in \mathrm{X} : \boldsymbol{n} \cdot (z - \boldsymbol{x}) = 0\}$ is called a ***support hyperplane*** for D at $\boldsymbol{x}$. There can be several normals at a given $\boldsymbol{x} \in \partial D$ (e.g., if D has a corner at $\boldsymbol{x}$) and there need not exist any normal (e.g., if X \ D has a corner at $\boldsymbol{x}$).

16.1.3 Proposition *Every boundary point of a convex set D has an exterior normal. An open set D is convex if and only if every boundary point of D has an exterior normal.*

The next proposition is a version of the mean value theorem for vector-valued functions; it will be employed to prove Proposition 19.4.5 which in turn leads to the phase rule. A probability measure μ on M (see, e.g., RUDIN [1970]) is a nonnegative measure defined on a σ-algebra of subsets of M such that $\mu(M) = 1$. A μ-integrable vector-valued mapping $\boldsymbol{f} : M \to$ X is any mapping such that the scalar-valued function $\boldsymbol{f} \cdot \boldsymbol{m}$ is (absolutely) μ-integrable in the usual sense for every $\boldsymbol{m} \in$ X. We shall use the proposition in the special case when M is an open subset of Vect of Lebesgue measure 1 and μ is a restriction of the Lebesgue measure to M. The reader unfamiliar with the general notion of measure can consider this special situation.

16.1.4 Proposition *If μ is a probability measure on M and* $\boldsymbol{f}$ *a μ-integrable mapping from M to* X, *then* $\int_M \boldsymbol{f}\, d\mu \in$ co$(\boldsymbol{f}(M))$.

The point is that the integral is in co$(\boldsymbol{f}(M))$; one does not have to take the closure of co$(\boldsymbol{f}(M))$. The reader is referred to NOLL [1970, p. 3] who proved this for M locally compact, μ a Radon probability measure, and $\boldsymbol{f}$ continuous. However, inspection of his proof shows that it can be modified to provide the proof of the statement above. That X is finite-dimensional is essential.

16.2 Convex Functions

In the following definition of a convex function we admit functions f which take the values $\pm\infty$ at some points; moreover, the domain of f need not be convex.

16.2.1 Definition A function $f : D \to \bar{\mathbb{R}} := \mathbb{R} \cup \{-\infty, \infty\}$ is said to be ***convex*** at $\boldsymbol{x} \in D$ (and $\boldsymbol{x}$ is said to be a ***point of convexity*** of f) if

$$f(x) \le \sum_{i=1}^{s} \alpha_i f(x_i) \tag{16.2.1}$$

whenever $x_i \in D$, $\alpha_i \ge 0\,(i = 1, \ldots, s)$ are such that $\sum_{i=1}^{s} \alpha_i = 1$, $\sum_{i=1}^{s} \alpha_i x_i = x$ and $f(x_i) < \infty$ for every $i = 1, \ldots, s$. If D is convex, this is equivalent to saying that $f(x) \le \alpha f(y) + (1-\alpha) f(z)$ for every $y, z \in D$, $\alpha \in \mathbb{R}$ such that $0 < \alpha < 1$, $x = \alpha y + (1-\alpha) z$ and $f(y) < \infty$, $f(z) < \infty$. The function f is said to be ***strictly convex*** at x (and x is said to be a ***point of strict convexity*** of f) if the inequality (16.2.1) is strict whenever $x_i \neq x$ for all $i = 1, \ldots, s$. A function is said to be ***convex*** if it is convex at every point of its domain and ***strictly convex*** if it is strictly convex at every point of its domain. f is said to be ***concave*** at x if $-f$ is convex at x. See Fig. 16.2.1.

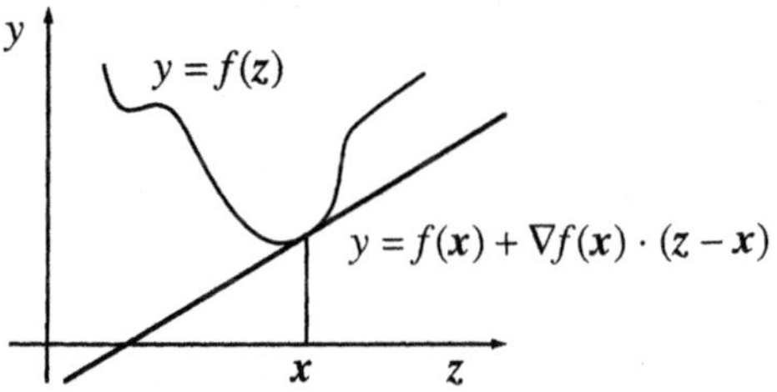

Fig. 16.2.1. A point of convexity

We shall see in Sect. 16.3 that a function f is convex in this sense if and only if it has an extension to a convex function on X, and that for functions on X the present notion of convexity coincides with that in ROCKAFELLAR [1970]. It is not required that the domain of a convex function be open. It may therefore happen that no point of the domain of f can be written as a nontrivial convex combination of other points of the domain. Consider, e.g., the unit circle $D := \{(x_1, x_2) \in \mathbb{R}^2 : x_1^2 + x_2^2 = 1\}$ in $\mathbb{R}^2$. Then *any* function f on D is convex. Nevertheless, f can still be extended to a convex function $\tilde{f}$ on $\mathbb{R}^2$. The point is that it is admitted that the extensions can take the values $\pm\infty$. One extension f is $f(x) := -\infty$ if $x_1^2 + x_2^2 < 1$, $\tilde{f}(x) := f(x)$ if $x \in D$ and $\tilde{f}(x) := \infty$ if $x_1^2 + x_2^2 > 1$. The values ∞ and $-\infty$ do not play symmetric roles in convexity: many convex functions are finite on a nonempty open subset of X and simultaneously take the value ∞ on a large set, while if a convex function has $f(x) = -\infty$ at some x, then the set of all points where it is finite has empty interior.

16.2.2 Definition If $f : D \to \bar{\mathbb{R}}$, $x \in D$, and $c \in$ X, then c is said to be a ***subgradient*** of f at x if $f(y) \ge f(x) + c \cdot (y - x)$ for every $y \in D$. The set of all subgradients of f at x is denoted by $\partial f(x)$ and called the ***subdifferential*** of f at x.

The subdifferential $\partial f(x)$ is a (possibly empty) closed convex subset of X. If $\partial f(x)$ is nonempty at some point where $f(x) > -\infty$ then $f > -\infty$ on its domain. The point of the definition of a subdifferential is that it can be nonempty even when the function is not differentiable at the given point. (Recall that in our terminology the terms "differential" and "differentiability" mean "Fréchet differential" and "Fréchet differentiability;" if f is differentiable at x then x is an interior point of the domain of f.) There is a good reason for considering nondifferentiable convex (or concave) functions. We shall see in Sect. 19.6 that the Gibbs function g of a fluid is nondifferentiable at any point (p, θ) which corresponds to the pressure and temperature of a strictly stable state of two or more coexistent phases. However, the subdifferential of $-g$ at (p, θ) is nonempty, convex, and compact.

16.2.3 Proposition *If $f : D \to \bar{\mathbb{R}}$ is finite in some neighborhood of $\boldsymbol{x} \in D$, then*
(1) *$\partial f(\boldsymbol{x})$ is convex and compact;*
(2) *f is convex at $\boldsymbol{x}$ if and only if $\partial f(\boldsymbol{x}) \neq \emptyset$;*
(3) *if f is (globally) convex on D, then f is differentiable at $\boldsymbol{x}$ if and only if $\partial f(\boldsymbol{x})$ consists of exactly one point; if this is the case, then this point is $\nabla f(\boldsymbol{x})$.*

Thus for finite-valued functions f on open sets one can take the existence of a subgradient at $\boldsymbol{x}$ as a definition of the convexity of f at $\boldsymbol{x}$. However, if D is not open or if any neighborhood of $\boldsymbol{x}$ contains points at which f is infinite then a subgradient at $\boldsymbol{x}$ need not exist even when f is convex. Consider, e.g., $f : \mathbb{R} \to \bar{\mathbb{R}}$ defined by $f(x) = \infty$ for $x < 0$, $f(0) = a$ (with $a \in \bar{\mathbb{R}}$ given) and $f(x) = 0$ for $x > 0$. Then f is convex on $\mathbb{R}$ if and only if $0 \leq a \leq \infty$; however, a subgradient at $x = 0$ exists only if $a = 0$.

The following two propositions provide convenient criteria for a differentiable function to be convex at a point or on a convex domain. They are likely to be the first tools to be tried when checking the convexity of a concrete function.

16.2.4 Proposition *Let $f : D \to \mathbb{R}$ be differentiable at $\boldsymbol{x} \in D$. Then f is convex at $\boldsymbol{x}$ if and only if*

$$f(\boldsymbol{y}) \geq f(\boldsymbol{x}) + \nabla f(\boldsymbol{x}) \cdot (\boldsymbol{y} - \boldsymbol{x}) \tag{16.2.2}$$

for every $\boldsymbol{y} \in D$. If f is convex at $\boldsymbol{x}$ and the second differential $d^2 f(\boldsymbol{x})$ exists, then it is positive semidefinite: $d^2 f(\boldsymbol{x})[\boldsymbol{u}, \boldsymbol{u}] \geq 0$ for every $\boldsymbol{u} \in X$.

16.2.5 Proposition *For a continuously differentiable function $f : D \to \mathbb{R}$ on an open convex set the following conditions are equivalent:*
(1) *f is convex;*
(2) (16.2.2) *holds for every $\boldsymbol{x}, \boldsymbol{y} \in D$;*
(3) $(\nabla f(\boldsymbol{x}) - \nabla f(\boldsymbol{y})) \cdot (\boldsymbol{x} - \boldsymbol{y}) \geq 0$ *for every $\boldsymbol{x}, \boldsymbol{y} \in D$.*
Moreover, if f is twice continuously differentiable in D, then (1), (2), *and* (3) *are also equivalent to:*
(4) *$d^2 f(\boldsymbol{x})$ is positive semidefinite at every $\boldsymbol{x} \in D$.*

A special case of (3) is that a continuously differentiable function f on an open interval in $\mathbb{R}$ is convex if and only if its first derivative is a nondecreasing function; if f is twice continuously differentiable, it is convex if and only if its second derivative is nonnegative. Note that the assumption that D be convex is essential: If f is a function defined on a union of two separated open intervals in $\mathbb{R}$, then (3), (4) above are equivalent to the assertion that the restrictions of f to the two intervals are convex, but the whole function need not be convex: One has the freedom in shifting vertically the graph of f on one of the two intervals without violating (3), (4), but destroying the convexity.

16.2.6 Proposition ("Transfer of Points of Convexity") *Let $f : D \to \mathbb{R}$ be continuously differentiable and $\boldsymbol{x}, \boldsymbol{y} \in D$. We have the following assertions:*
(1) *if $\boldsymbol{x}$ is a point of convexity of f and*

$$f(\boldsymbol{y}) = f(\boldsymbol{x}) + \nabla f(\boldsymbol{x}) \cdot (\boldsymbol{y} - \boldsymbol{x}), \tag{16.2.3}$$

then $\boldsymbol{y}$ is a point of convexity of f and

$$\nabla f(\boldsymbol{y}) = \nabla f(\boldsymbol{x}); \tag{16.2.4}$$

(2) *if x, y are points of convexity of f and* (16.2.4) *holds, then* (16.2.3) *holds; moreover, if z is any point of convexity of f on the open line segment (x, y) then $\nabla f(z) = \nabla f(x)$ and $f(z) = f(x) + \nabla f(x) \cdot (z - x)$.*

Equation (16.2.3) is an abstract form of Maxwell's relation for stable states with coexistent phases, i.e., the equality of the Gibbs functions. Equation (16.2.3) is the equality of the stresses and temperatures of the phases. The proposition will be used several times in diverse situations.

Proof (1): Combining (16.2.2) with (16.2.3) we obtain $f(w) \geq f(y) + \nabla f(x) \cdot (w - y)$ for every $w \in D$ which shows that f is convex at y. (2): Writing inequality (16.2.2) and a similar one in which the roles of x, y are exchanged, and using (16.2.4) to eliminate $\nabla f(y)$, we obtain two inequalities which differ just in the inequality signs $\geq$, $\leq$ and these two give (16.2.3). To prove the assertion about the point z, let us write $z = \alpha x + (1 - \alpha)y$, $0 < \alpha < 1$. Then $f(z) \geq f(x) + \nabla f(x) \cdot (z - x)$ and $f(z) \geq f(y) + \nabla f(x) \cdot (z - y)$. Making the convex combination of the last two inequalities with coefficients α, $1 - \alpha$, we obtain $f(z) \geq \alpha f(x) + (1 - \alpha) f(y)$ and combining this with $f(z) \leq \alpha f(x) + (1 - \alpha) f(y)$, then $f(z) = \alpha f(x) + (1 - \alpha) f(y)$. Eliminating $f(y)$ via (16.2.3) gives $f(z) = f(x) + \nabla f(x) \cdot (z - x)$. Applying (1) to the points x, z, we obtain $\nabla f(x) = \nabla f(z)$. □

16.2.7 Definition A function $f : D \to \bar{\mathbb{R}}$ is said to be ***very strictly convex*** at $x \in D$ (and x is said to be a point of a ***very strict convexity*** of f) if f has a differential at x and

$$f(y) > f(x) + \nabla f(x) \cdot (y - x)$$

for every $y \in D$ such that $y \neq x$.

The following proposition will be employed in Sect. 19.7 to establish strong minima of the canonical free energy for fluids.

16.2.8 Proposition *Let $f : D \to \mathbb{R}$ be continuously differentiable, let $w \in D$ be a point of very strict convexity of f, and let every point from some neighborhood of w be a point of convexity of f. Then there exists a nondecreasing function $C : [0, \infty) \to [0, \infty)$ with $C(r) > 0$ for $r > 0$ such that*

$$f(x) \geq f(w) + \nabla f(w) \cdot (x - w) + C(r)|w - x| \tag{16.2.5}$$

for every $r > 0$ and every $x \in D$ such that $|w - x| \geq r$.

Proof Without any loss of generality we assume that $f(w) = 0$ and $w = \mathbf{0}$. Define

$$C(r) := \inf \{(f(x) - \nabla f(\mathbf{0}) \cdot x)/|x| : x \in D, |x| \geq r\}, \quad r > 0. \tag{16.2.6}$$

The proof will be complete if we show that $C(r) > 0$ for every $r > 0$ and that C is nondecreasing. The last property follows from the fact that as r increases, the infimum in (16.2.6) is taken over a smaller set. It remains to prove that $C(r) > 0$ for every $r > 0$. By hypothesis, there exists an $R > 0$ such that every point z with $|z| \leq R$ is a point of convexity of f. From this we now derive that if $0 \leq \lambda \leq 1$ and $x \in D$ is such that $|\lambda x| \leq R$, then

$$f(\lambda x) \leq \lambda f(x). \tag{16.2.7}$$

To see this, it suffices to notice that $\lambda x = \lambda x + (1 - \lambda)\mathbf{0}$ and as λx is a point of convexity of f, we have that $f(\lambda x) \leq \lambda f(x) + (1 - \lambda) f(\mathbf{0}) = \lambda f(x)$. Let us now prove that for every $r \leq R$ we have

$$C(r) = \inf \{(f(z) - \nabla f(\mathbf{0}) \cdot z)/|z| : z \in D, |z| = r\}. \tag{16.2.8}$$

Let $x \in D$ be such that $|x| \geq r$. Define λ and z by $\lambda = r/|x|$, $z = \lambda x$. From $|x| \geq r$ we have $0 < \lambda \leq 1$ and $|z| = r \leq R$. Hence, by (16.2.7), $f(z) = f(\lambda x) \leq \lambda f(x)$. From this

$$(f(z) - \nabla f(\mathbf{0}) \cdot z)/|z| \leq (\lambda f(x) - \lambda \nabla f(\mathbf{0}) \cdot x)/(\lambda |x|) = (f(x) - \nabla f(\mathbf{0}) \cdot x)/|x|. \tag{16.2.9}$$

To summarize, for every $x \in D$ with $|x| \geq r$ there exists a $z \in D$ with $|z| = r$ such that (16.2.9) holds. This proves (16.2.8). But the infimum in (16.2.8) is strictly positive since the expression in the infimum is continuous and positive (by the hypothesis that f is very strictly convex at $\mathbf{0}$) and the infimum is taken over a compact set. □

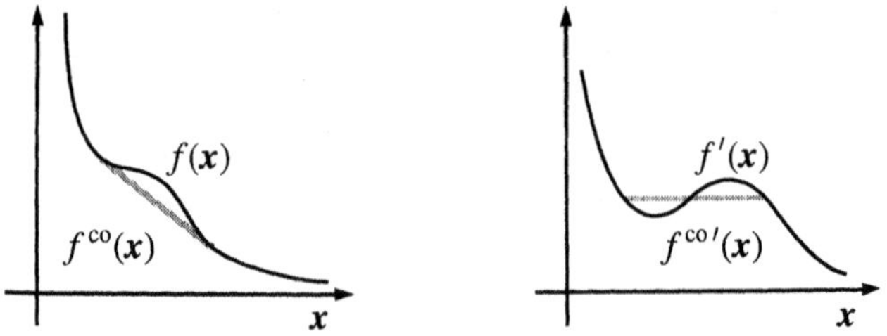

(a) (b)

Fig. 16.3.1. The functions f and f^{co} and their derivatives

16.3 The Lower Convex Hull

The results on the lower convex hull will be employed in Chap. 19 to study the energy functions of fluids admitting states of coexistent phases and in Sect. 21.6 to explain the basic properties of equilibrium of crystals.

16.3.1 Proposition *For every $f : D \to \bar{\mathbb{R}}$ there exists a unique convex function $f^{\text{co}} : \mathrm{X} \to \bar{\mathbb{R}}$ such that the following two conditions hold:*
(1) $f^{\text{co}} \le f$ *on* D;
(2) *if* $g : \mathrm{X} \to \bar{\mathbb{R}}$ *is convex and* $g \le f$ *on* D *then* $g \le f^{\text{co}}$ *on* X.
The function f^{co} has the following properties:
(3) $f^{\text{co}} = \infty$ *on* $\mathrm{X} \setminus \mathrm{co}(D)$;
(4) *if $\boldsymbol{x} \in D$ is a point of convexity of f then $f^{\text{co}}(\boldsymbol{x}) = f(\boldsymbol{x})$;*
(5) *if $f < \infty$ on D then $f^{\text{co}} < \infty$ on* $\mathrm{co}(D)$;
(6) *if f has an affine lower bound, i.e., if $f(\boldsymbol{x}) \ge \boldsymbol{c} \cdot \boldsymbol{x} + b$ for every $\boldsymbol{x} \in \mathrm{X}$ and some $b \in \mathbb{R}$, $\boldsymbol{c} \in \mathrm{X}$, then $f^{\text{co}}(\boldsymbol{x}) \ge \boldsymbol{c} \cdot \boldsymbol{x} + b$ and hence $f^{\text{co}} > -\infty$ on* X.

The function f^{co} is the largest convex function minorizing f; it is called the ***lower convex hull*** of f, see Fig. 16.3.1. It may happen that $f^{\text{co}} = -\infty$ identically, e.g., if $f(x) = -x^2$ on $\mathbb{R}$, for then $g = -\infty$ is the only convex function minorizing f. Let us briefly describe the construction of f^{co}. The convexity of f^{co} implies that (16.2.1) holds with f replaced by f^{co} for every $\boldsymbol{x} \in \mathrm{X}$ represented as a convex combination of s points $\boldsymbol{x}_i$ of X with weights α_i as described in Definition 16.2.1. If the points $\boldsymbol{x}_i$ happen to be in D then by the property that f^{co} does not exceed f we obtain $f^{\text{co}}(\boldsymbol{x}) \le \sum_{i=1}^{s} \alpha_i f(\boldsymbol{x}_i)$. The right-hand side of this inequality does not contain f^{co} and must hold for all choices of the representations $\boldsymbol{x}_i$, α_i of $\boldsymbol{x}$. This suggests that one define f^{co} as infimum of the right-hand sides of this inequality over all possible representations, with the convention that when $\boldsymbol{x} \notin \mathrm{co}(D)$, then the infimum is taken over an empty set and hence is equal to ∞. This indeed leads to the desired function. Moreover, using Carathéodory's theorem, it turns out that it suffices to consider only the representations with at most $n + 1$ points of D (here $n = \dim \mathrm{X}$), and one can assume additionally that $f(\boldsymbol{x}_i) < \infty$, i.e.,

$$f^{\text{co}}(\boldsymbol{x}) = \inf \{ \sum_{i=1}^{n+1} \beta_i f(\boldsymbol{x}_i) : \sum_{i=1}^{n+1} \beta_i \boldsymbol{x}_i = \boldsymbol{x}, \sum_{i=1}^{n+1} \beta_i = 1, \beta_i \ge 0, f(\boldsymbol{x}_i) < \infty, \boldsymbol{x}_i \in D \}. \quad (16.3.1)$$

If x is a point of convexity of f, then this equation shows that $f^{co}(x) = f(x)$. Physically, the sums entering (16.3.1) represent the total energy of unit mass of a fluid in a heterogeneous state of total volume and entropy described by the vector x. The state consists of at most $n + 1$ homogeneous phases x_i with relative weights β_i. The lower convex hull then gives the greatest lower bound of all possible total energies reachable in heterogeneous states with total volume and entropy x. We are especially interested in the situations in which the infimum in (16.3.1) is a minimum (recall minimum of energy). Let $f : D \to \mathbb{R}$ be a function and $x \in X$ a point. We say that $\{x_i, i = 1, \dots, s\} \subset D$ is a ***stable phase splitting*** of x if there exist numbers $\alpha_i > 0$ such that

$$x = \sum_{i=1}^{s} \alpha_i x_i, \quad f^{co}(x) = \sum_{i=1}^{s} \alpha_i f(x_i), \quad \sum_{i=1}^{s} \alpha_i = 1. \tag{16.3.2}$$

We say that x has a ***unique stable phase splitting*** if it has a stable phase splitting $Z = \{x_i, i = 1, \dots, s\}$ and any other stable phase splitting is a rearrangement of Z. If x has a unique stable phase splitting Z then the convex hull of Z is called the ***phase simplex*** of x. The following proposition describes the properties of the stable phase splittings. A sufficient condition for their existence will be given later.

16.3.2 Proposition *Let $f : D \to \mathbb{R}$ be a continuously differentiable function on an open convex set and $x \in D$ a point. If $\{x_i : i = 1, \dots, s\}$ is a stable phase splitting of x then f^{co} has a (Fréchet) differential $\nabla f^{co}(x)$ at x, the points x_i are points of convexity of f, and for every $i = 1, \dots, s$,*

$$\nabla f(x_i) = \nabla f^{co}(x), \quad f(x_i) = f^{co}(x) + \nabla f^{co}(x) \cdot (x_i - x). \tag{16.3.3}$$

Moreover, there exists an affinely independent set $Z \subset \{x_i : i = 1, \dots, s\}$ which is still a stable phase splitting of x.

Geometrically, relations (16.3.3) say that the affine function $\varphi(y) := f^{co}(x) + \nabla f^{co}(x) \cdot (y - x)$, $y \in X$, gives a plane that touches the graph of f at the points x_i and this graph is above that plane. Physically, when applied to the energy function of a fluid, relations (16.3.2) and (16.3.3) express the equality of pressures, temperatures, and chemical potentials of phases.

Proof Set $M := \{\xi \equiv (\gamma_1, \dots, \gamma_s, w_1, \dots, w_s) \in (0, 1)^s \times D^s\}$. If $\alpha_i > 0$ are the numbers in (16.3.2), consider $\zeta := (\alpha_1, \dots, \alpha_s, x_1, \dots, x_s) \in M$. Define functions φ, ψ, ω on M by

$$\varphi(\xi) = \sum_{i=1}^{s} \gamma_i f(w_i), \quad \psi(\xi) = \sum_{i=1}^{s} \gamma_i, \quad \omega(\xi) = \sum_{i=1}^{s} \gamma_i w_i, \quad \xi \in M.$$

It follows from the definition of a stable phase splitting that at $\zeta \in M$ φ has a minimum relative to the constraints $\psi(\xi) = 1$, $\omega(\xi) = x$. The theorem on Lagrange multipliers (see, e.g., Alexeev & al. [1979, Sect. 1.3]) asserts the existence of a $\lambda \in \mathbb{R}$ and $\mu \in X$ such that all partial derivatives of $\varphi - \lambda\psi - \mu \cdot \omega$ vanish at ζ. This gives

$$\nabla f(x_i) = \mu, \quad f(x_i) = \lambda + \mu \cdot x_i. \tag{16.3.4}$$

By $(16.3.2)_2$ then $f^{co}(x) = \lambda + \mu \cdot x = f(x_i) + \mu \cdot (x - x_i)$, $i = 1, \dots, s$. Let $c \in X$ be any subgradient of f^{co} at x. Then

$$f(y) \ge f^{co}(y) \ge f^{co}(x) + c \cdot (y - x) = f(x_i) + \mu \cdot (x - x_i) + c \cdot (y - x), \tag{16.3.5}$$

for every $y \in D$. The choice $y = x_i$ gives $(c - \mu) \cdot (x - x_i) \ge 0$, $i = 1, \dots, s$. Since

$$\sum_{i=1}^{s} \alpha_i (c - \mu) \cdot (x - x_i) = 0$$

and $\alpha_i > 0$, each of the nonnegative terms $(c - \mu) \cdot (x - x_i)$ in this sum must be 0. Then (16.3.5) can be rewritten as

$$f(y) - f(x_i) - c \cdot (y - x_i) \geq 0, \quad i = 1, \ldots, s, \quad y \in D, \tag{16.3.6}$$

and hence f is convex at x_i and $c = \nabla f(x_i)$ by Proposition 16.2.3. It follows that c as in (16.3.5) is unique and hence Proposition 16.2.3 gives that f^{co} has a differential at x and $\nabla f^{co}(x) = c = \mu$. Equations (16.3.3) then follow from $\nabla f(x_i) \equiv \mu = c$ and (16.3.4). To prove the assertion about the subset Z, note that since x is a convex combination of x_i, $i = 1, \ldots, s$, by Carathéodory's theorem x is a convex combination of an affinely independent subset Z of $\{x_i, i = 1, \ldots, s\}$. Using $(16.3.4)_2$, is is found that Z is a stable phase splitting of x. □

16.3.3 Corollary *Let $f : D \to \mathbb{R}$ be a continuously differentiable function on an open convex set. If every point of D has a stable phase splitting, then f^{co} is continuously differentiable.*

16.3.4 Proposition *Let $\dim X = n$, let $f : D \to \mathbb{R}$ be a continuously differentiable function on an open convex set and $x \in D$ a point. If x has a unique stable phase splitting $Z = \{x_i : i = 1, \ldots, s\}$, then*

(1) *Z is affinely independent and hence $s \leq n + 1$;*

(2) *if $y \in D$ is a point of convexity of f that is not in Z and $\nabla f(y) = \nabla f^{co}(x)$, then y is affinely independent of Z.*

Proof (1) If the points of Z were to form an affinely dependent set, then Proposition 16.3.2 guarantees the existence of an affinely independent stable phase splitting $Z' \subset Z$, which would then be different from Z, contradicting the uniqueness. (2) Assume that the point y in (2) is an affine combination of Z, i.e., there exist numbers β_i, $i = 1, \ldots, s$, such that

$$y = \sum_{i=1}^{s} \beta_i x_i, \quad \sum_{i=1}^{s} \beta_i = 1.$$

Set $x_{s+1} = y$, choose $\lambda > 0$ such that $\alpha_i - \lambda\beta_i > 0$, and define $\gamma_i := \alpha_i - \lambda\beta_i$, $i = 1, \ldots, s$, and $\beta_{s+1} = \lambda$. Then $\{x_i : i = 1, \ldots, s + 1\}$ is a stable phase splitting of x, with the coefficients γ_i instead of α_i in (16.3.2). This contradicts the uniqueness. □

To see that the existence of a stable phase splitting is not automatic, consider $f : \mathbb{R} \to \mathbb{R}$ given by $f(x) = x^2 e^{-x^2}$. Then the only point of convexity of f is $x = 0$ (in fact a point of very strict convexity). It is seen from the graph of f that $f^{co} \equiv 0$ and the only point having a stable phase splitting is $x = 0$. A function $f : D \to \mathbb{R}$ is said to be ***convex outside a compact set*** if D is open and convex and if there exists a compact set $K \subset D$ such that every point of $D \setminus K$ is a point of convexity of f. The proof of the following proposition will be published elsewhere.

16.3.5 Proposition *If $f : D \to \mathbb{R}$ is a continuously differentiable function that is convex outside a compact set, then every $x \in D$ has a stable phase splitting.*

16.4 The Fenchel Transformation

The Fenchel transformation will be applied to define the Gibbs function of a fluid in Sect. 19.6. We describe here the basic properties of the Fenchel transformation with special emphasis on the differentiability; also its relationship to the Legendre transformation (Sect. 10.1) is discussed in some detail.

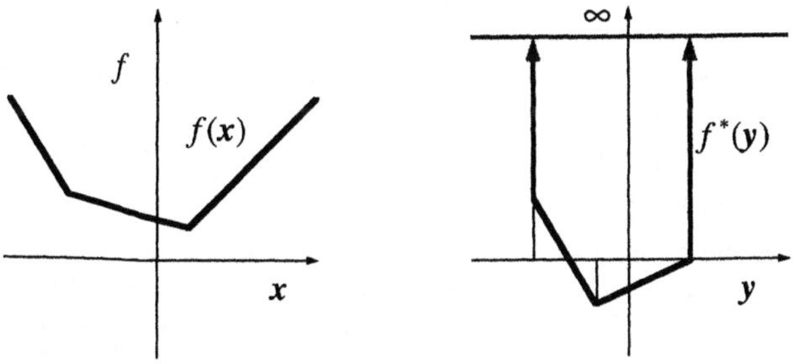

Fig. 16.4.1. The Fenchel transform of a piecewise linear function

16.4.1 Conjugate Functions For every $f : D \to \bar{\mathbb{R}}$ denote by $\tilde{f} : \mathrm{X} \to \mathbb{R}$ be the extension of f defined by $\tilde{f}(x) = f(x)$ if $x \in D$ and $\tilde{f}(x) = \infty$ if $x \notin D$. The ***conjugate function*** or the ***Fenchel transformation*** of f is the function f^* defined on X by

$$f^*(y) = \sup\{x \cdot y - \tilde{f}(x) : x \in \mathrm{X}\}, \quad y \in \mathrm{X}.$$

The function f^* can take infinite values, even when f is finite on the whole X. Since f^* is the supremum over a collection of affine functions $\{y \mapsto x \cdot y - f(x), x \in \mathrm{X}\}$, it is convex. The Fenchel transformation reverses the order: if $f \le g$ then $f^* \ge g^*$. Denote by f^{**} the conjugate of f^* :

$$f^{**}(x) = \sup\{x \cdot y - f^*(y) : y \in \mathrm{X}\}, \quad x \in \mathrm{X}.$$

Since the Fenchel transform reverses order, the double Fenchel transform preserves the order, $f \le g \Rightarrow f^{**} \le g^{**}$. The function f^{**} is convex; moreover,

$$f^{**} \le f \text{ on } D \text{ and } f^{***} = f^* \text{ on X}. \tag{16.4.1}$$

ROCKAFELLAR [1970] considers only conjugates of *convex* functions and hence his results do not apply directly to our case where f need not be convex. However, many of the results to be given below can be reduced to the situation considered in ROCKAFELLAR [1970]; in so doing, we shall use square brackets to refer to the results from ROCKAFELLAR [1970]. Thus, e.g., $(16.4.1)_2$ follows from the assertion of [Thm. 12.2] that $g^{**} = g$ whenever g is a closed convex function by setting $g = f^*$ and using that f^* is closed by [Thm. 9.4]. This theorem applies directly if f^* is finite at some $y \in \mathrm{X}$. If this is not the case, then either $f^* \equiv \infty$ identically on X or there exists an y such that $f^*(y) = -\infty$. In the former case f^* is clearly closed while in the latter case the definition of f^* implies that $f \equiv \infty$ on X and this in turn implies that $f^* \equiv -\infty$ which is again a closed function. Inequality $(16.4.1)_1$ follows from Young's inequality (16.4.2) below. Indeed, that inequality implies $\tilde{f}(x) \ge x \cdot y - f^*(y)$ for all $y \in \mathrm{X}$ and as $f^{**}(x)$ is the supremum of $x \cdot y - f^*(y)$ over all $y \in \mathrm{X}$, the result follows.

16.4.2 Example Consider a continuous, convex and piecewise linear function f of one real variable on Fig. 16.4.1. Its Fenchel transformation is again a continuous, convex and piecewise linear function f^* and the vertices of the graph of f correspond to the edges of the graph of f^* and the edges of the graph of f correspond to the vertices of f^*, in the following way. For every y which is a slope of some linear piece of f there exists a unique $c(y) \in \mathbb{R}$ such that $f(x) = xy + c(y)$ for some $x \in \mathbb{R}$. The value of f^* at such a y is precisely $-c(y)$, and so the function f^* is determined at all such y. The graph of f^* is then obtained by joining the points $(y, -c(y))$ by line segments and by setting $g = +\infty$ elsewhere. See Fig. 16.4.1. For example, if $f(x) = |x|$, then $g = f^*$ is given by $g(y) = 0$ for $y \in [-1, 1]$ and $g(y) = +\infty$ for $|y| > 1$.

16.4.3 Proposition *If $f : D \to \bar{\mathbb{R}}$, then*

(1) *for every $x, y \in \mathrm{X}$ we have the Young inequality*

$$f^*(y) \geq x \cdot y - \tilde{f}(x); \tag{16.4.2}$$

(2) *if* $x \in D$ *and* $y \in \partial f(x)$, *then* (16.4.2) *holds with the equality sign, i.e.,* $f^*(y) = y \cdot x - f(x)$; *moreover* $x \in \partial f^*(y)$;

(3) *if* (16.4.2) *holds with the equality sign for some* $x, y \in \mathrm{X}$ *and* $f(x), f^*(y)$ *are finite, then* $x \in D$ *and* $y \in \partial f(x)$.

Proof (1) follows directly from the definition. (2): If $y \in \partial f(x)$ then $y \cdot z - f(z) \leq x \cdot y - \tilde{f}(x)$ for every $z \in \mathrm{X}$ and hence, taking the supremum over al z, we obtain $f^*(y) \leq x \cdot y - \tilde{f}(x)$. A combination with (16.4.2) gives the equality. Replacing y by $z \in \mathrm{X}$ in (16.4.2) and using the equality just obtained, we get $f^*(z) \geq x \cdot z - f(x) = x \cdot (z - y) + f^*(y)$ which means that $x \in \partial f^*(y)$. The proof of (3) is similar to that of (2). □

Assertion (2) in Proposition 16.4.3 evaluates f^* at arguments that are subgradients of f at points of convexity and gives some information about the subgradients of f^*. Otherwise, formally, this is a duality similar to that for the Legendre transformation. This will be made much stricter for a special class of f below in Proposition 16.4.5. The following proposition explains the relationship of the conjugate functions to the convex hull. In particular, Assertion (2) says that f^{**} and f^{co} are almost identical under reasonable additional conditions.

16.4.4 Proposition *If* $f : D \to \bar{\mathbb{R}}$, *then*

(1) $f^* = f^{\mathrm{co}\,*}$ *and hence* $f^{**} = f^{\mathrm{co}\,**}$;

if, moreover, D is open and $-\infty < f^{\mathrm{co}} < \infty$ *on* $\mathrm{co}(D)$, *then*

(2) $f^{**} = f^{\mathrm{co}}$ *on* $\mathrm{X} \setminus \partial\,\mathrm{co}(D)$ *and* $f^{**} \leq f^{\mathrm{co}}$ *on* $\partial\,\mathrm{co}(D)$;

(3) $f^* > -\infty$ *on* X *and* $f^*(x) < \infty$ *for some* $x \in \mathrm{X}$;

(4) $\partial f^*(y) \subset \mathrm{cl}\,\mathrm{co}(D)$ *for every* $y \in \mathrm{X}$;

(5) *if* $\partial f^*(y) \neq \emptyset$ *for some* $y \in \mathrm{X}$ *then* $f^*(y) \in \mathbb{R}$.

Proof (1): From the inequality $f^{**} \leq \tilde{f}$ (see (16.4.1)) and the definition of the lower convex hull follows that $f^{**} \leq f^{\mathrm{co}} \leq \tilde{f}$. This implies $f^{***} \geq f^{\mathrm{co}\,*} \geq f^*$ and combining it with $f^{***} = f^*$ (see $(16.4.1)_2$) we obtain (1). (2): The inequality being proved in (1), we turn to the equality on $\mathrm{X} \setminus \partial\,\mathrm{co}(D)$. By [Thm. 12.2], $f^{\mathrm{co}\,**} = \mathrm{cl} f^{\mathrm{co}}$, where the closure cl is defined in [Sect. 6]. The hypothesis implies that f^{co} is a proper convex function (see [Sect. 4]) and the conclusion follows from [Thm. 7.4]. (3) says that f^* is a proper convex function. Since $f^* = f^{\mathrm{co}\,*}$ by (1), and f^{co} is proper by hypothesis, the conclusion follows from [Thm. 12.2]. (4): By the results at the beginning of [Sect. 24], range $\partial g \subset \mathrm{dom}\, g^*$ for every closed proper $g : \mathrm{X} \to \bar{\mathbb{R}}$. Applying this to $g = f^*$ we obtain $\partial f^*(y) \subset \mathrm{dom} f^{**} = \mathrm{dom} f^{\mathrm{co}\,**} \subset \mathrm{cl}\,\mathrm{co}(D)$ since $f^{\mathrm{co}\,**}$ differs from f^{co} at most on $\partial\,\mathrm{co}(D)$ and $\mathrm{dom} f^{\mathrm{co}} = \mathrm{co}(D)$. (5): This follows from the fact that if a proper function is subdifferentiable at some point, then its value is finite at that point, see [Thm. 23.4]. □

The following proposition determines the subdifferential of f^* for functions which have stable phase splittings. For any $f : D \to \bar{\mathbb{R}}$ on a convex set D, denote by G the set of all $y \in \mathrm{X}$ for which $\emptyset \neq \partial f^*(y) \subset D$ (see Proposition 16.4.4(4) for the motivation of this definition). For every $y \in \mathrm{X}$, denote further by $F(y)$ the set of all points of convexity $x \in D$ of f such that $\nabla f(x) = y$.

16.4.5 Proposition *Let* $f : D \to \mathbb{R}$ *be continuously differentiable on an open convex set and let every point of D have a stable phase splitting. Then*

(1) $\partial f^*(y) = \mathrm{co}(F(y))$ *for every* $y \in G$;

(2) $G \subset \{\nabla f^{\mathrm{co}}(x) : x \in D\}$;

(3) *if* $y \in G$, *then* f^* *is differentiable at* y *if and only if* $y = \nabla f(x)$ *for some point* $x \in D$ *of very strict convexity of* f; *if this is the case, then* $\nabla f^*(y) = x$.

Thus, by Assertion (1), to find $\partial f^*(y)$ at a point $y \in G$, one finds all points x of convexity of f which satisfy $y = \nabla f(x)$ and makes all possible convex combinations of x. This proposition will be applied to evaluate the subdifferential of $-g$, where g is the Gibbs function of a fluid at $y = (p, \theta)$. When (p, θ) are the pressure and temperature of a strictly stable heterogeneous equilibrium state, the subdifferential will be found to contain a convex combination of the specific volumes and pressures of the phases in the state, with coefficients equal to the relative amounts of the phases.

Proof (1): We have $F(y) \subset \partial f^*(y)$ by Proposition 16.4.3(2); since $\partial f^*(y)$ is convex, it follows that $\mathrm{co}(F(y)) \subset \partial f^*(y)$. Conversely, if $c \in \partial f^*(y)$, then $c \in D$ by the definition of G and $y \in \partial f^{**}(c)$ by Proposition 16.4.3(2) applied to f^*. As on D the function f^{**} coincides with f^{co} by Proposition 16.4.4(2) and the latter is continuously differentiable on D by Corollary 16.3.3, we have $y = \nabla f^{\mathrm{co}}(c)$. Since by hypothesis c has a stable phase splitting $\{x_i : i = 1, \ldots, s\}$, we have $y = \nabla f(x_i)$ by Proposition 16.3.2. But then $x_i \in F(y)$ and c is a convex combination of the points x_i, which completes the proof of (1). (2): This was proved while proving (1). (3): By the hypothesis that every point of D has a stable phase splitting we can apply Proposition 16.4.4(3) to conclude that f^* is a proper convex function and Proposition 16.4.4(5) that $\partial f^*(y) \neq \emptyset$ only if $f^*(y) \in \mathbb{R}$. Then by [Thm. 25.1] f^* is differentiable at y if and only if $\partial f^*(y)$ is a singleton. By (1), this happens if and only if $F(y)$ is a singleton, i.e., there exists exactly one point x of convexity of f such that $y = \nabla f(x)$. This is true if and only if x is a point of very strict convexity. □

The following result describes the relationship between the Fenchel and the Legendre transformations for convex functions and gives conditions for the differentiability of the Fenchel transformation. For a general mapping $m : A \to B$ and a subset C of A, denote by $m|C$ the restriction of m to C.

16.4.6 Proposition *Let $f : D \to \mathbb{R}$ be continuously differentiable, denote by S the interior of the set of all points of very strict convexity of f and set $T = \nabla f(S)$. Then*

(1) *T is open, $f^*|T$ is continuously differentiable, and every point of T is a point of very strict convexity of $f^*|T$;*

(2) *$\nabla f|S$ and $\nabla f^*|T$ are injective and inverses of one another;*

(3) *$f|S$ and $f^*|T$ are the Legendre transformations of one another.*

The set of all points of very strict convexity is not generally open: the function f described immediately before Proposition 16.3.2 has exactly one point of very strict convexity. The reader is referred to ROCKAFELLAR [1970, Sect. 26] for a different set of conditions when the Fenchel and Legendre transformations coincide globally.

Proof (1): To prove that T is open, let $y = \nabla f(x) \in T$ for some $x \in S$. Since x is a point of very strict convexity of f and every point in some neighborhood of x is a point of convexity of f, Proposition 16.2.8 is available. Let $\varepsilon > 0$ be such that the ball $B(x, \varepsilon)$ centered at x with radius ε is contained in S and set $\delta = C(\varepsilon)$, where $C(\varepsilon)$ is as in Proposition 16.2.8. We shall prove that for every $h \in B(\nabla f(x), \delta)$ there exists a $z \in B(x, \varepsilon)$ such that $h = \nabla f(z)$. Let us seek a maximum of the function $h : w \mapsto h \cdot w - f(w)$, $w \in D$, and prove that it is realized at some point $z \in B(x, \varepsilon)$. Let w be such that $|w - x| \geq \varepsilon$. Then by (16.2.5),

$$\begin{aligned} h(w) &\leq h \cdot w - f(x) - \nabla f(x) \cdot (w - x) - \delta|w - x| \\ &= h \cdot x - f(x) + (h - \nabla f(x)) \cdot (w - x) - \delta|w - x| \\ &\leq h \cdot x - f(x) + (|h - \nabla f(x)| - \delta)|w - x| < h \cdot x - f(x) = h(x). \end{aligned}$$

Hence $h(w) < h(x)$ for every $w \notin B(x, \varepsilon)$ and consequently the function h has a point z of maximum in $B(x, \varepsilon)$. The condition $\partial_w h(z) = 0$ gives $h = \nabla f(z)$ which means that $B(\nabla f(x), \delta) \subset T$. To prove that f^* is differentiable at every point of T let $y = \nabla f(x) \in T$ for some $x \in S$. By Proposition 16.4.3(2), $x \in \partial f^*(y)$, $f^*(y) = x \cdot y - f(x)$ and it suffices to prove that there is no other point in $\partial f^*(y)$. If there is another point in $\partial f^*(y)$ then, by the convexity of the subdifferential, there exists an $x' \in \partial f^*(y)$, $x' \neq x$ such that $x' \in S$. Set $z = \nabla f(x')$ so that by Proposition 16.4.3(2), $f^*(z) = x' \cdot z - f(x')$. Since x' is a subgradient of f^* at y, we have $f^*(z) \geq f^*(y) + x' \cdot (z - y)$. Inserting the above calculated values of $f^*(y), f^*(z)$, one obtains $f(x') \leq f(x) + y \cdot (x' - x)$ and combining this with $y = \nabla f(x)$ we see that x is not a point of very strict convexity of f, a contradiction. Hence f^* is differentiable at y. For a convex function, the differentiability implies the continuous differentiability on open sets. We have also proved that

$$\nabla f^*(\nabla f(\boldsymbol{x})) = \boldsymbol{x} \quad \text{for every } \boldsymbol{x} \in S. \tag{16.4.3}$$

To prove that every point of T is a point of very strict convexity of $f^*|T$, suppose, in contradiction, that $\boldsymbol{y} \in T$ is not a point of very strict convexity. Then $f^*(\boldsymbol{z}) = f^*(\boldsymbol{y}) + \nabla f^*(\boldsymbol{y}) \cdot (\boldsymbol{z} - \boldsymbol{y})$ for some $\boldsymbol{z} \in T, \boldsymbol{z} \neq \boldsymbol{y}$. Since both $\boldsymbol{y}, \boldsymbol{z}$ are points of convexity of f^*, it follows from Proposition 16.2.6 that $\nabla f^*(\boldsymbol{z}) = \nabla f^*(\boldsymbol{y})$. We have $\boldsymbol{y} = \nabla f(\boldsymbol{x}), \boldsymbol{z} = \nabla f(\boldsymbol{w})$ for some $\boldsymbol{x}, \boldsymbol{w} \in S$. Applying (16.4.3) to $\boldsymbol{x}$ and then to $\boldsymbol{w}$, we find that $\boldsymbol{x} = \nabla f^*(\nabla f(\boldsymbol{x})) = \nabla f^*(\boldsymbol{y}) = \nabla f^*(\boldsymbol{z}) = \nabla f^*(\nabla f(\boldsymbol{w})) = \boldsymbol{w}$, i.e., $\boldsymbol{x} = \boldsymbol{w}$. Hence $\boldsymbol{y} = \boldsymbol{z}$, a contradiction. (2): To prove that $\nabla f|S$, $\nabla f^*|T$ are injective, it suffices to note that $(\nabla g(\boldsymbol{x}_1) - \nabla g(\boldsymbol{x}_2)) \cdot (\boldsymbol{x}_1 - \boldsymbol{x}_2) > 0$ for every two different points of very strict convexity $\boldsymbol{x}_1, \boldsymbol{x}_2$ of a function g. The assertion about the inverses now follows from (16.4.3). (3): We have already proved that $f(\boldsymbol{x}) + f^*(\boldsymbol{y}) = \boldsymbol{x} \cdot \boldsymbol{y}$ whenever $\boldsymbol{x} \in S$, $\boldsymbol{y} \in T$ are related by $\boldsymbol{y} = \nabla f(\boldsymbol{x})$, $\boldsymbol{x} = \nabla f^*(\boldsymbol{y})$. □

16.4.7 Corollary *Let $f : D \to \mathbb{R}$ be continuously differentiable and let every point of D be a point of very strict convexity of f. Then f has the Legendre transform $g : P \to \mathbb{R}$ defined on an open set; g is continuously differentiable, every point of P is a point of very strict convexity of g and the Legendre transform of g is f.*

This is symmetric. For functions without points of convexity the Fenchel and Legendre transformations can give quite different results: Let $f : \mathbb{R}^2 \to \mathbb{R}$ be defined by $f(x) = \frac{1}{2}(x_1^2 - x_2^2)$, $x = (x_1, x_2) \in \mathbb{R}^2$. Then f has the Legendre transformation g given by $g(y) = \frac{1}{2}(y_1^2 - y_2^2)$, $y = (y_1, y_2) \in \mathbb{R}^2$ while $f^*(y) = \infty$, $y \in \mathbb{R}^2$. The Legendre transform is a direct calculation. To calculate f^*, we use that $f^* = f^{\mathrm{co}\,*}$. However, the only convex function minorizing f is equal to $-\infty$, i.e., $f^{\mathrm{co}}(x) = -\infty$, $x \in \mathbb{R}^2$. But the Fenchel transform of this f^{co} is as claimed.

Bibliographical Note

The notion of very strict convexity is due to COLEMAN [1970], who calls it superconvexity. Proposition 16.2.8 is a mild generalization of a result from the cited work. (See also COLEMAN & GREENBERG [1967], DUNN & FOSDICK [1978, 1980].) The notions of a stable phase splitting and phase simplex are due to RABIER & GRIEWANK [1992]. They also proved the existence of the stable phase splitting for lower semicontinuous functions f that satisfy $f(\boldsymbol{x})/|\boldsymbol{x}| \to \infty$ as $|\boldsymbol{x}| \to \infty$, $\boldsymbol{x} \in D$ and $f(\boldsymbol{x}) \to \infty$ as $\boldsymbol{x}$ tends to the boundary of D. Unfortunately, these assumptions cannot be satisfied by the internal energies of fluids. Indeed $\tilde{e}(v, \eta)$ is a decreasing function of v since $-\partial_v \tilde{e}(v, \eta)$ is the positive pressure. (RABIER & GRIEWANK [1992] apply their result in a different situation, where there is more hope that their assumption is satisfied.) Proposition 16.3.5, which seems to be new, gives an alternative sufficient condition, which does not contradict the known asymptotic properties of the internal energy function. The notion of a function convex outside a compact set is due to COLEMAN [1970]. The results of Sect. 16.4 on the differentiability of the conjugate function are new.

17 Constitutive Inequalities

Whether or not the extremum principles hold for a given state is a result of an interplay between the environment, the state, and the response functions. This chapter concentrates on the restrictions which the extremum principles place on the response functions. As in the preceding chapters, the discussion applies to bodies of unspecified symmetry, solids, fluids, and crystals. A condition called quasiconvexity will be isolated as a necessary condition. From the quasiconvexity will be derived a simpler condition of rank 1 convexity. The rank 1 convexity will play a key role in the proof of the Maxwell relation on the interface separating coexistent phases in Sect. 17.4. We shall also see in that section that rank 1 convexity necessarily must be violated in some states on a line segment connecting two states capable of coexistence in a stable equilibrium state with a phase transition. Also it will be shown that rank 1 convexity is necessarily violated by crystals. Rank 1 convexity implies the Legendre–Hadamard condition. At the points of the free part of the boundary the extremum principles imply a condition stronger than quasiconvexity – the quasiconvexity at the boundary, which will be also briefly discussed. The subsequent chapters will show that the validity or violation of quasiconvexity, rank 1 convexity, or the Legendre–Hadamard condition always gives important information about the material. Throughout the chapter we identify Eucl with Vect in the geometric constructions.

17.1 Quasiconvexity

For any open set $D \subset \mathrm{Eucl}$ we denote by $\mathrm{Var}(D)$ the set of all pairs $(\boldsymbol{u}, \omega)$, $\boldsymbol{u} \in C_0^\infty(D)$, $\omega \in C_0^\infty(D)$, $\boldsymbol{u}$ vector-valued, ω real-valued, with

$$\int_D \omega(\boldsymbol{p})\, dV = 0. \tag{17.1.1}$$

It is convenient to think of $(\boldsymbol{u}, \omega)$ as perturbations of states which leave the boundary of D fixed and the total entropy unchanged. The open set D, however, is in no direct way related to the reference configuration of the body Ω. This will become clear later.

17.1.1 Definitions Let $g : U \to \mathbb{R}$ be a continuous function on an open subset U of $\mathrm{InvLin} \times \mathbb{R}$ and $(\boldsymbol{F}, \eta)$ a point in U. g is said to be ***quasiconvex*** at $(\boldsymbol{F}, \eta)$ if

$$\int_D g(\boldsymbol{F} + \nabla \boldsymbol{u}, \eta + \omega)\, dV \geq |D| g(\boldsymbol{F}, \eta) \tag{17.1.2}$$

for every bounded open set $D \subset \mathrm{Eucl}$ and every $(\boldsymbol{u}, \omega) \in \mathrm{Var}(D)$ for which the argument in the integral is in U. The symbol $|D|$ stands for the volume of D. Here and at similar places below the argument $\boldsymbol{p}$ of $\nabla \boldsymbol{u}$ and ω has been omitted. The function g is said to be ***quasiconcave*** at $(\boldsymbol{F}, \eta)$ if $-g$ is quasiconvex at $(\boldsymbol{F}, \eta)$. g is said to be ***quasiconvex*** if it is quasiconvex at every point of its domain. A continuous function $h : V \to \mathbb{R}$ on an open set $V \subset \mathrm{Lin}$ is said to be ***quasiconvex*** at $\boldsymbol{F} \in V$ if

$$\int_D h(\boldsymbol{F} + \nabla \boldsymbol{u})\, dV \geq |D| h(\boldsymbol{F}) \tag{17.1.3}$$

for every bounded open set $D \subset$ Eucl and every $\boldsymbol{u} \in C_0^\infty(D)$ for which the argument in the integral is in V.

17.1.2 ***Remarks***

(1) For a fixed D, inequality (17.1.2) is the principle of minimum energy for a body with the reference region D and the internal energy g under homogeneous Dirichlet boundary conditions. Explicitly, the state of minimum energy is a homogeneous state $\sigma = (\chi, \eta)$ where $\chi(\boldsymbol{p}) = \boldsymbol{F}\boldsymbol{p}$, $\boldsymbol{p} \in D$, is a homogeneous deformation, η is the uniform field with value η, and the class of competitor states (see Sect. 15.2) is

$$\Sigma_0 = \{(\chi, \eta) + (\boldsymbol{u}, \omega) : (\boldsymbol{u}, \omega) \in \mathrm{Var}(D), (\boldsymbol{F} + \nabla \boldsymbol{u}(\boldsymbol{p}), \eta + \omega(\boldsymbol{p})) \in \tilde{U}, \boldsymbol{p} \in D\}. \tag{17.1.4}$$

The competitors $(\bar{\chi}, \bar{\eta}) = (\chi, \eta)+(\boldsymbol{u}, \omega)$ thus satisfy the Dirichlet boundary condition $\bar{\chi}(\boldsymbol{p}) = \boldsymbol{F}\boldsymbol{p} = \chi(\boldsymbol{p})$ for $\boldsymbol{p} \in \partial D$ and $\int_D \bar{\eta}\, dV = \int_D \eta\, dV$. The quasiconvexity of g at $(\boldsymbol{F}, \eta)$ means that the homogeneous state (χ, η) gives the minimum energy. A violation of this would mean that under the homogeneous Dirichlet boundary conditions the body would tend to move to an inhomogeneous state to reduce its energy. This phenomenon is sometimes called *internal buckling*, see TRUESDELL & NOLL [1965, pp. 129–130] and BIOT [1965].

(2) Let h : Lin $\to (-\infty, \infty]$ be borel measurable and bounded below, and let $1 \leq p \leq \infty$. The function h is called $W^{1,p}$***-quasiconvex*** at $\boldsymbol{F} \in$ Lin if (17.1.3) holds for every bounded open set $D \subset$ Eucl with $|\partial D| = 0$ and all $\boldsymbol{u} \in W_0^{1,p}(D)$. h is said to be $W^{1,p}$***-quasiconvex*** if it is $W^{1,p}$-quasiconvex at every $\boldsymbol{F} \in$ Lin. See BALL & MURAT [1984]. Here $W_0^{1,p}(D)$ is the Sobolev space to be defined in Sect. 20.2. $W^{1,p}$-quasiconvexity for $1 \leq p \leq \infty$ is strictly stronger than the quasiconvexity defined in Subsect. 17.1.1. Moreover, the $W^{1,p}$-quasiconvexity is stronger than the $W^{1,q}$-quasiconvexity for $1 \leq p \leq q \leq \infty$, and $W^{1,\infty}$-quasiconvexity is equivalent to the quasiconvexity defined in Subsect. 17.1.1 under simple additional conditions. In Subsect. 21.5.6 we shall give an example of a h for which the $W^{1,p}$-quasiconvexity depends dramatically on p.

(3) The definition 17.1.1 was introduced by MORREY [1952, 1966]. The term "quasiconvexity" in Definition 17.1.1 applies to the character of the dependence on $\boldsymbol{F}$; the dependence on η is convex.

(4) The definition (17.1.3) can be extended to the following situation, which is helpful for comparisons with the classical calculus of variations. Let us consider two vector spaces V and W, of possibly different dimensions n and m, respectively. Let V be an open subset of the space Lin(V, W) of all linear transformations from V to W, and let $h : V \to \mathbb{R}$ be continuous. Then h is said to be quasiconvex at $\boldsymbol{F} \in U$ if (17.1.3) holds for every bounded open subset D of V and every $\boldsymbol{u} \in C_0^\infty(D, W)$ for which the argument in the integral is in V. In this case $\nabla \boldsymbol{u}(\boldsymbol{p})$ is in Lin(V, W) and the definition makes sense. The original definition results when $V = W =$ Vect; if $V = \mathbb{R}^n$, $W = \mathbb{R}^m$, then $\nabla \boldsymbol{u}(\boldsymbol{p})$ is identified with an m by n matrix.

(5) The classical one-dimensional calculus of variations on an interval $[t_1, t_2] \subset \mathbb{R}$ deals with the integrals of the type

$$\int_{t_1}^{t_2} f(t, \chi(t), \chi'(t))\, dt$$

where $\chi : [t_1, t_2] \to \mathbb{R}^m$ and the prime denotes the differentiation with respect to $t \in [t_1, t_2]$. This corresponds to $V = \mathbb{R}$ and $W = \mathbb{R}^m$ with m arbitrary in (4) above. We shall see that in this case the quasiconvexity is equivalent to the ordinary convexity.

(6) The quasiconvexity is difficult to verify. We shall see in Sect. 17.3 that for quadratic functions, quasiconvexity is equivalent to a simpler algebraic condition, Legendre–Hadamard condition. Moreover, in Sect. 17.5 we shall give a good general sufficient condition for quasiconvexity, the polyconvexity, which is again of algebraic character. We shall also see that the gap between polyconvex and quasiconvex functions is 'relatively narrow.' See also the recapitulation in Subsect. 17.5.6 and Example 17.5.7.

We consider a body in a conservative environment with the potential energy (13.3.5). Let $\sigma = (\chi, \eta)$ be an equilibrium state compatible with the environment and let $\Sigma_0 \subset \Sigma^{**}$ be the class of competitor states as in Sect. 15.2.

17.1.3 Assumption If $D \subset \Omega$ and $(\boldsymbol{u}, \omega) \in \mathrm{Var}(D)$ are such that the values of $(\nabla\chi + \nabla\boldsymbol{u}, \eta + \omega)$ over Ω are in $\tilde{U}$, then $(\chi + \boldsymbol{u}, \eta + \omega) \in \Sigma_0$.

The following proposition is the main result of this section.

17.1.4 Proposition (Localization theorem) *Let the state* $\sigma = (\chi, \eta)$ *satisfy the principle of minimum energy with respect to the class* Σ_0, *and let Assumption* 17.1.3 *hold. Then for every* $\boldsymbol{p} \in \Omega$ *outside the singular surface,* $\tilde{e}(\cdot, \cdot)$ *is quasiconvex at* $(\boldsymbol{F}(\boldsymbol{p}), \eta(\boldsymbol{p}))$. $\tilde{e}(\cdot, \cdot)$ *is quasiconvex also at every limiting value of* $(\boldsymbol{F}, \eta)$ *on* $\partial\Omega$ *and on the singular surface.*

The proof is based on a localization procedure in which small perturbations of the state near $\boldsymbol{p}$ are magnified via a scaling argument. Scaling arguments are extremely useful in the calculus of variations and will be used frequently in the sequel. We have seen that the minimum canonical free energy implies the principle of minimum energy and the latter is equivalent to the principle of maximum entropy. Hence *every extremum principle implies quasiconvexity.*

Proof Assume $\rho_0 = 1$. Let $g(\boldsymbol{F}, \eta, \boldsymbol{x}, \boldsymbol{p}) := \tilde{e}(\boldsymbol{F}, \eta) + \psi(\boldsymbol{F}, \boldsymbol{x}, \boldsymbol{p})$, $(\boldsymbol{F}, \eta) \in U, \boldsymbol{p} \in \Omega, \boldsymbol{x} \in$ Eucl, and set $\boldsymbol{v}(\cdot) := (\nabla\chi(\cdot), \eta(\cdot))$. Let $\boldsymbol{p}$ be a fixed interior point of Ω outside the singular surface. We shall prove the quasiconvexity at $(\nabla\chi(\boldsymbol{p}), \eta(\boldsymbol{p}))$. Let $D, \boldsymbol{u}, \omega$ be as in the definition of quasiconvexity. Denote $\boldsymbol{w}(\cdot) := (\nabla\boldsymbol{u}(\cdot), \omega(\cdot))$ and assume

$$\boldsymbol{v}(\boldsymbol{p}) + \boldsymbol{w}(\boldsymbol{r}) \in \tilde{U} \quad \text{for every } \boldsymbol{r} \in \text{Eucl}. \tag{17.1.5}$$

For every $\varepsilon > 0$ define the state $\sigma_\varepsilon = (\chi_\varepsilon, \eta_\varepsilon)$ by

$$\chi_\varepsilon(\boldsymbol{q}) = \chi(\boldsymbol{q}) + \varepsilon\boldsymbol{u}(\varepsilon^{-1}(\boldsymbol{q}-\boldsymbol{p})), \quad \eta_\varepsilon(\boldsymbol{q}) = \eta(\boldsymbol{q}) + \omega(\varepsilon^{-1}(\boldsymbol{q}-\boldsymbol{p})), \qquad \boldsymbol{q} \in \Omega.$$

Since the supports of $(\boldsymbol{u}, \omega)$ are in D, the functions $(\chi_\varepsilon, \eta_\varepsilon)$ differ from (χ, η) only at points $\boldsymbol{q}$ for which $\varepsilon^{-1}(\boldsymbol{q}-\boldsymbol{p}) \in D$, i.e., at points for which

$$\boldsymbol{q} \in D_\varepsilon \equiv \varepsilon D + \boldsymbol{p} = \{\boldsymbol{q}' \in \text{Eucl} : \boldsymbol{q}' = \varepsilon\boldsymbol{r} + \boldsymbol{p} \text{ for some } \boldsymbol{r} \in D\}.$$

For every $\varepsilon > 0$ small enough, $\varepsilon D + \boldsymbol{p}$ is in Ω. We have

$$\nabla\chi_\varepsilon(\boldsymbol{q}) = \nabla\chi(\boldsymbol{q}) + \nabla\boldsymbol{u}(\varepsilon^{-1}(\boldsymbol{q}-\boldsymbol{p}))$$

and hence by (17.1.5) for all $\varepsilon > 0$ sufficiently small, $(\nabla\chi_\varepsilon(\boldsymbol{q}), \eta_\varepsilon(\boldsymbol{q}))$ is in the domain $\tilde{U}$. The values of $(\chi_\varepsilon, \eta_\varepsilon)$ on $\partial\Omega$ coincide with those of (χ, η) and by (17.1.1) the total entropy of $(\chi_\varepsilon, \eta_\varepsilon)$ coincides with that of (χ, η). We apply the principle of minimum energy to obtain

$$\int_\Omega g(\boldsymbol{v}(\boldsymbol{p}) + \boldsymbol{w}(\boldsymbol{r}), \chi(\boldsymbol{q}) + \varepsilon\boldsymbol{u}(\boldsymbol{r}), \boldsymbol{q})\, dV(\boldsymbol{q}) \ge \int_\Omega g(\boldsymbol{v}(\boldsymbol{q}), \chi(\boldsymbol{q}), \boldsymbol{q})\, dV(\boldsymbol{q}), \tag{17.1.6}$$

where here and in the next inequality the perturbations $\boldsymbol{u}$ and $\boldsymbol{w}$ are evaluated at $\boldsymbol{r} = \boldsymbol{r}(\boldsymbol{q}) = \varepsilon^{-1}(\boldsymbol{q}-\boldsymbol{p})$. We have also used the fact that the boundary contribution to the potential energy from the surface potential $\phi(\boldsymbol{F}, \boldsymbol{x}, \boldsymbol{q})$ cancels on the two sides of the inequality. Also the integrals outside $\varepsilon D + \boldsymbol{p}$ cancel in (17.1.6) and one obtains

$$\int_{D_\varepsilon} g(\boldsymbol{v}(\boldsymbol{q}) + \boldsymbol{w}(\boldsymbol{r}), \chi(\boldsymbol{q}) + \varepsilon\boldsymbol{u}(\boldsymbol{r}), \boldsymbol{q})\, dV(\boldsymbol{q}) \ge \int_{D_\varepsilon} g(\boldsymbol{v}(\boldsymbol{q}), \chi(\boldsymbol{q}), \boldsymbol{q})\, dV(\boldsymbol{q}). \tag{17.1.7}$$

The substitution $\boldsymbol{q} \mapsto \boldsymbol{r}(\boldsymbol{q}) = \varepsilon^{-1}(\boldsymbol{q}-\boldsymbol{p})$ and the division by ε^3 reduces (17.1.7) to

$$\int_D g(\boldsymbol{v}(\boldsymbol{q}) + \boldsymbol{w}(\boldsymbol{r}), \chi(\boldsymbol{q}) + \varepsilon\boldsymbol{u}(\boldsymbol{q}), \boldsymbol{q})\, dV(\boldsymbol{r}) \ge \int_D g(\boldsymbol{v}(\boldsymbol{q}), \chi(\boldsymbol{q}), \boldsymbol{q})\, dV(\boldsymbol{r})$$

where now $q = q(r) = \varepsilon r + p$. Letting $\varepsilon \to 0$, we obtain

$$\int_D g((F, \eta) + w(r), \chi(p), p)\, dV(r) \ge \int_D g(F, \eta, \chi(p), p)\, dV(r), \tag{17.1.8}$$

where $F = \nabla\chi(p)$, $\eta = \eta(p)$. By Proposition 13.7.1 the potential energy is a null Lagrangian in F and hence by Proposition 13.5.6

$$\int_D \psi(F + \nabla u(r), \chi(p), p)\, dV = |D|\psi(F, \chi(p), p). \tag{17.1.9}$$

This reduces (17.1.8) to

$$\int_D \tilde{e}(F + \nabla u(r), \eta + \omega(r))\, dV(r) \ge |D|\tilde{e}(F, \eta). \tag{17.1.10}$$

An obvious limit establishes the quasiconvexity at every limiting value of (F, η) on the singular surface and on $\partial\Omega$. □

17.1.5 *Remarks*

(1) From Proposition 17.1.4 one deduces that it is enough to check (17.1.2) in the definition of the quasiconvexity for one fixed bounded open set D_0; the validity for a general D then follows. Indeed, if (17.1.2) holds for D_0, one can apply Proposition 17.1.4 to $\Omega := D_0$ and $\chi(p) = Fp$, $\eta(p) = \eta$, $p \in D_0$.

(2) The inspection of the proof shows that the conclusion of Proposition 17.1.4 remains valid under weaker hypotheses than Assumption 17.1.3. Namely, one can replace this assumption by the following: for every $p \in \Omega$ there exist $\varepsilon > 0$, $\delta > 0$ such that if $(u, \omega) \in$ Var(Vect) satisfy $\|(u, \omega)\|_0 < \varepsilon$ and $u(q) = 0$, $\omega(q) = 0$ whenever $|q - p| > \delta$, then $(\chi + u, \eta + \omega) \in \Sigma_0$. See Definition 15.4.1(1) for $\|\cdot\|_0$. This weaker assumption corresponds to a strong *local* minimum in the sense of calculus of variations (see Definition 15.4.1(3)) with the perturbations, moreover, restricted by the additional requirement that they differ from $\sigma = (\chi, \eta)$ only on a small neighborhood of p. (The original Assumption 17.1.3 corresponds to a global minimum.) More generally, one can replace the norm $\|\cdot\|_0$ with the norm $\|\cdot\|_p$ to be defined in Subsect. 20.3.1 with p satisfying $1 \le p < \infty$. However, the weak local minimum in the sense of the calculus of variations (see Definition 17.1.3(2)) is *not enough* to obtain the conclusion of Proposition 17.1.4.

(3) By Proposition 17.1.4, the quasiconvexity of $\tilde{e}$ holds also at (F^+, η^+) and (F^-, η^-) where $(F^\pm, \eta^\pm)$ are the limiting values on the singular surface. GURTIN [1983, Assertion (ii), p. 9] gives a generalization of the quasiconvexity pertinent to the point on the singular surface, formulated in one inequality involving simultaneously both the two limits $(F^\pm, \eta^\pm)$. See also BALL & MARSDEN [1984, Remark 3, p. 259].

(4) Proposition 17.1.4 is due to MEYERS [1965] and BALL [1977]. In the more general situation when the competitors are taken from $W^{1,p}(\Omega)$, one can sometimes draw a stronger conclusion that $\tilde{e}$ be $W^{1,p}$-quasiconvex. See BALL & MURAT [1984], BALL [1987], and JAMES & SPECTOR [1992].

(5) The motivation for quasiconvexity presented above comes from the extremum principles. A different motivation for quasiconvexity (which is related) is the sequential weak lower semicontinuity (swlsc) of the integral functionals. This enables one to prove the existence of minimizers of integral functionals of the calculus of variations. See Chap. 21 and especially Subsect. 21.5.6.

The next two propositions translate quasiconvexity of $\tilde{e}$ into conditions on $\hat{\eta}$ and $\bar{f}$. Recall that throughout Part IV it is assumed that the body admits both the temperature and the entropy as independent thermal variables.

17.1.6 Proposition *Suppose that the body satisfies Assumptions* 15.3.2(1) *and* (2), *and let* $(F, \eta) \in \tilde{U}$, $(F, e) \in \hat{U}$ *be related by* $e = \tilde{e}(F, \eta)$. *Then* $\tilde{e}$ *is quasiconvex at* (F, η) *if and only if* $\hat{\eta}$ *is quasiconcave at* (F, η).

Proof This follows from Proposition 15.3.4. As has been pointed out, the quasiconvexity is the principle of minimum energy for the homogeneous state (F, η), on the set D, and for the class of competitor states (17.1.4). Similarly, the quasiconcavity of $\hat{e}$ is the principle of maximum entropy for the homogeneous state (F, e). □

A somewhat more complicated task is the translation of quasiconvexity of $\tilde{e}$ into the language of $\bar{f}$. The definition of quasiconvexity uses perturbations of the entropy field whose integral over D vanishes. We cannot expect that, say, $\bar{f}$ will be quasiconvex because the perturbations of temperature with vanishing integral make no sense as the temperature is an intensive variable.

17.1.7 Proposition *Let $(\boldsymbol{F}, \eta) \in \tilde{U}$, $(\boldsymbol{F}, \theta) \in \bar{U}$ be related by $\eta = \bar{\eta}(\boldsymbol{F}, \theta)$. Then*

(1) *if $\tilde{e}$ is quasiconvex at $(\boldsymbol{F}, \eta)$, then $\bar{f}(\cdot, \theta)$, taken as function of the deformation gradient only, is quasiconvex at $\boldsymbol{F}$ and*

$$\bar{f}(\boldsymbol{F}, \theta) \le \bar{f}(\boldsymbol{F}, \theta') + (\theta' - \theta)\bar{\eta}(\boldsymbol{F}, \theta') \tag{17.1.11}$$

for every θ' such that $(\boldsymbol{F}, \theta') \in \bar{U}$;

(2) *if*

$$\bar{f}(\boldsymbol{F}', \theta) \le \bar{f}(\boldsymbol{F}', \theta') + (\theta' - \theta)\bar{\eta}(\boldsymbol{F}', \theta') \tag{17.1.12}$$

for every $(\boldsymbol{F}', \theta') \in \bar{U}$, then $\tilde{e}$ is quasiconvex at $(\boldsymbol{F}, \eta)$, if and only if $\bar{f}(\cdot, \theta)$ is quasiconvex at $\boldsymbol{F}$;

(3) *$\tilde{e}$ is quasiconvex (globally) if and only if $\bar{f}(\cdot, \theta)$ is quasiconvex for every θ and $\bar{f}(\boldsymbol{F}, \cdot)$ concave for every $\boldsymbol{F}$.*

Assertion (3) is most suggestive: the condition of quasiconvexity of $\tilde{e}$, which is a joint restricted convexity in $\boldsymbol{F}$ and η, splits into two *separate* conditions: the concavity of $\bar{f}$ in θ and the quasiconvexity of $\bar{f}$ in $\boldsymbol{F}$. Assertions (1) and (2) are local versions of this. It is seen that (17.1.11) and (17.1.12) are related to the positivity of the specific heat at constant deformation.

Proof (1) Suppose that $\tilde{e}$ is quasiconvex at $(\boldsymbol{F}, \eta)$. To prove that $\bar{f}(\cdot, \theta)$ is quasiconvex at $\boldsymbol{F}$ let $D \subset$ Eucl be a open set and $\boldsymbol{u} \in C_0^\infty(D)$ and $\omega \in C_0^\infty(D)$ be such that

$$(\boldsymbol{F} + \nabla \boldsymbol{u}(\boldsymbol{p}), \eta + \omega(\boldsymbol{p})) \in \tilde{U}, \quad \boldsymbol{p} \in D. \tag{17.1.13}$$

(It is not assumed that the mean value of ω over D is 0.) Let $\alpha \in [0, 1)$ and define $D_\alpha^\pm$ by

$$D_\alpha^+ = \alpha^{1/3} D, \quad D_\alpha^- = (1 - \alpha)^{1/3} D + \boldsymbol{t},$$

where $\boldsymbol{t}$ is chosen so that $D_\alpha^+ \cap D_\alpha^- = \emptyset$. Define $D_\alpha := D_\alpha^+ \cup D_\alpha^-$ and $\boldsymbol{u}_\alpha, \omega_\alpha$ by

$$(\boldsymbol{u}_\alpha(\boldsymbol{p}), \omega_\alpha(\boldsymbol{p})) = \begin{cases} (\alpha^{1/3}\boldsymbol{u}(\boldsymbol{p}/\alpha^{1/3}), \omega(\boldsymbol{p}/\alpha^{1/3})) & \text{for } \boldsymbol{p} \in D_\alpha^+, \\ (\boldsymbol{0}, -\alpha(1-\alpha)^{-1}\omega((\boldsymbol{p}-\boldsymbol{t})/(1-\alpha)^{1/3})) & \text{for } \boldsymbol{p} \in D_\alpha^-, \end{cases}$$

and both these two functions are set equal to 0 outside D_α. It is not hard to see that these functions are infinitely differentiable, have supports in D_α, and that $\int_{D_\alpha} \omega_\alpha \, dV = 0$. Hence $(\boldsymbol{u}_\alpha, \omega_\alpha)$ is in $\mathrm{Var}(D_\alpha)$. Using (17.1.13), the definition of $(\boldsymbol{u}_\alpha, \omega_\alpha)$ and the fact that $\tilde{U}$ is open, one sees that for every $\alpha > 0$ sufficiently small, the values $(\boldsymbol{F} + \nabla \boldsymbol{u}_\alpha(\boldsymbol{p}), \eta + \omega_\alpha(\boldsymbol{p})), \boldsymbol{p} \in D_\alpha$, are in $\tilde{U}$. Writing the inequality from the definition of quasiconvexity of $\tilde{e}$ for $D_\alpha, \boldsymbol{u}_\alpha, \omega_\alpha$ and making the same substitutions as in the proof of Proposition 17.1.4 we obtain

$$\alpha \int_D \tilde{e}(\boldsymbol{F} + \nabla \boldsymbol{u}(\boldsymbol{r}), \eta + \omega(\boldsymbol{r})) \, dV + (1-\alpha) \int_D \tilde{e}(\boldsymbol{F}, \eta - \alpha(1-\alpha)^{-1}\omega(\boldsymbol{r})) \, dV - |D|\tilde{e}(\boldsymbol{F}, \eta) \ge 0.$$

We now consider the left-hand side of the last inequality as a function $h(\alpha)$ of $\alpha > 0$, with α sufficiently small. We have $h(0) = 0$ and it has just been proved that $h(\alpha) \ge 0$, $\alpha > 0$. It follows that the derivative $h'(0)$, which is easily seen to exist, must be nonnegative. Carrying out the differentiation and using $\partial_\eta \tilde{e}(\boldsymbol{F}, \eta) = \theta$, this gives

$$\int_D \tilde{e}(\boldsymbol{F} + \nabla \boldsymbol{u}(\boldsymbol{r}), \eta + \omega(\boldsymbol{r})) \, dV - |D|\tilde{e}(\boldsymbol{F}, \eta) - \theta \int_D \omega(\boldsymbol{r}) \, dV \ge 0. \tag{17.1.14}$$

Now let $\boldsymbol{u}, D$ be such that $(\boldsymbol{F} + \nabla \boldsymbol{u}, \theta) \in \bar{U}$. Setting $\omega(\boldsymbol{r}) = \bar{\eta}(\boldsymbol{F} + \nabla \boldsymbol{u}(\boldsymbol{r}), \theta) - \eta$ we see that (17.1.13) is satisfied. The definition of $\bar{f}$ and (17.1.14) give

$$\int_D \bar{f}(\boldsymbol{F} + \nabla \boldsymbol{u}(\boldsymbol{r}), \theta) \, dV \ge |D|\bar{f}(\boldsymbol{F}, \theta). \tag{17.1.15}$$

This is quasiconvexity of $\bar{f}(\cdot, \theta)$ at $\boldsymbol{F}$. To prove (17.1.11) it suffices to set $\boldsymbol{u} = \boldsymbol{0}$, $\omega = \bar{\eta}(\boldsymbol{F}, \theta') - \eta$ in (17.1.14) and to use the definition of $\bar{f}$ again.

(2) Let us suppose that $\bar{f}(\cdot, \theta)$ is quasiconvex at $\boldsymbol{F}$ and prove that $\tilde{e}$ is quasiconvex at $(\boldsymbol{F}, \eta)$. Let $D, \boldsymbol{u}, \omega$ be as in the definition of quasiconvexity. Set $\boldsymbol{F}' = \boldsymbol{F} + \nabla \boldsymbol{u}$, $\eta' = \eta + \omega$ and $\theta' = \tilde{\theta}(\boldsymbol{F}', \eta')$. Equation (17.1.12) gives

$$\bar{f}(\boldsymbol{F}', \theta) + \theta\eta' \le \tilde{e}(\boldsymbol{F}', \eta'). \tag{17.1.16}$$

Integrating over D, and using (17.1.1), we obtain

$$\int_D \bar{f}(\boldsymbol{F} + \nabla \boldsymbol{u}, \theta) \, dV + |D|\theta\eta \le \int_D \tilde{e}(\boldsymbol{F} + \nabla \boldsymbol{u}, \eta + \omega) \, dV.$$

Combining this with the quasiconvexity inequality (17.1.15) leads to (17.1.10).

(3) For this it suffices to realize that (17.1.12) is the inequality asserting the concavity of $\bar{f}(\boldsymbol{F}, \cdot)$ at θ. □

17.2 Quasiconvexity at the Boundary

At points $\boldsymbol{p}$ of $F \subset \partial\Omega$ of the free part of the boundary (see 14.2.3), the extremum principles lead to the quasiconvexity at the boundary, a condition stronger than quasiconvexity. Let $H \subset$ Eucl be an open half-space in Eucl with the boundary plane ∂H and outer normal $\boldsymbol{n}$, i.e.,

$$H = \{\boldsymbol{q} \in \text{Eucl} : (\boldsymbol{q} - \boldsymbol{p}) \cdot \boldsymbol{n} < 0\} \tag{17.2.1}$$

where $\boldsymbol{p} \in$ Eucl is fixed.

17.2.1 Definition Let $g : U \to \mathbb{R}$ be a continuous function on an open subset U of Lin $\times\, \mathbb{R}$ and $(\boldsymbol{F}, \eta) \in U$. g is said to be ***quasiconvex at the boundary*** with normal $\boldsymbol{n}$ at $(\boldsymbol{F}, \eta)$ if

$$\int_{D\cap H} g(\boldsymbol{F} + \nabla\boldsymbol{u}, \eta + \omega)\, dV \geq |D \cap H| g(\boldsymbol{F}, \eta) + \partial_{\boldsymbol{F}} g(\boldsymbol{F}, \eta)\boldsymbol{n} \cdot \int_{D\cap\partial H} \boldsymbol{u}\, dA \tag{17.2.2}$$

for every bounded open subset D of Eucl and every $(\boldsymbol{u}, \omega) \in \text{Var}(D)$ for which the argument in the integral is in U and $\int_{D\cap H} \omega\, dV = 0$. If f is a function of $\boldsymbol{F}$ alone, then its quasiconvexity at the boundary with normal $\boldsymbol{n}$ at $\boldsymbol{F}$ is defined by omitting η in (17.2.2) and assuming it to hold for every bounded open $D \subset$ Vect and every $\boldsymbol{u} \in C_0^\infty(D)$.

17.2.2 *Remarks*

(1) The original definition by BALL & MARSDEN [1984] is different but can be shown to be equivalent to the one above.

(2) Inequality (17.2.2) with a fixed D is a principle of minimum energy under dead loading for a homogeneous state $\sigma = (\chi, \eta)$ with deformation gradient $\boldsymbol{F}$ and homogeneous field of entropy of value η. The referential configuration of the body is $\Omega := D \cap H$, the body force is $\boldsymbol{b}_0 \equiv \boldsymbol{0}$, the subsets S, D of $\partial\Omega$ are $S = D \cap \partial H$, $D = \partial\Omega \setminus S$, and $\boldsymbol{d}(\boldsymbol{p}) = \boldsymbol{F}\boldsymbol{p}$ on D while $\boldsymbol{s}(\boldsymbol{p}) = \partial_{\boldsymbol{F}} g(\boldsymbol{F}, \eta)\boldsymbol{n}$ on S.

(3) In the situation (2), a violation of (17.2.2) would mean that the body can reduce its energy by passing from the homogeneous state σ to an inhomogeneous one, $\bar{\sigma} = (\bar{\chi}, \bar{\eta})$, which still satisfies the homogeneous displacement condition $\bar{\chi}(\boldsymbol{p}) = \boldsymbol{F}\boldsymbol{p}$ on D. If g is not quasiconvex at $(\boldsymbol{F}, \eta)$, this can be achieved by a state which satisfies the homogeneous condition $\bar{\chi}(\boldsymbol{p}) = \boldsymbol{F}\boldsymbol{p}$ on S, and the type of instability is the internal buckling discussed in Remark 17.1.2(1). However, if this is not the case, then necessarily the condition $\bar{\chi}(\boldsymbol{p}) = \boldsymbol{F}\boldsymbol{p}$ on S must be violated, i.e., the originally flat boundary of the body in the actual configuration in the state σ will become curved in $\bar{\sigma}$. Such instabilities are termed *surface instabilities,* and can result, e.g., in small-scale wrinkling of the boundary of the body. Within the linear approximation, the surface instabilities were first studied by BIOT [1965]. They will be related to the linear counterpart of the quasiconvexity at the boundary, namely the complementing and Agmon's conditions in Sects. 20.4 and 20.6.

(4) The quasiconvexity at the boundary implies the quasiconvexity. For this it suffices to choose a bounded open set $D \subset H$, to notice that (17.2.2) then reduces to (17.1.2), and use the fact that in the definition of the quasiconvexity it suffices to check (17.1.2) for just one open bounded set; see Remark 17.1.5(1). The quasiconvexity at the boundary is strictly stronger than the ordinary quasiconvexity: the function $f : \boldsymbol{F} \mapsto \det \boldsymbol{F}$ is quasiconvex (in fact even polyconvex, see Sect. 17.5) but for no $\boldsymbol{F}, \boldsymbol{n}$ is f quasiconvex at the boundary with normal $\boldsymbol{n}$ at $\boldsymbol{F}$. (See below for proof.)

17.2.3 Null Lagrangians at the Boundary A continuously differentiable function $f : U \to \mathbb{R}$ on an open subset U of Lin is said to be a ***null Lagrangian at the boundary*** with normal $\boldsymbol{n}$ if

$$\int_{D\cap H} f(\boldsymbol{F}+\nabla \boldsymbol{u})\, dV = |D\cap H| f(\boldsymbol{F}) + \partial_{\boldsymbol{F}} f(\boldsymbol{F})\boldsymbol{n}\cdot \int_{D\cap\partial H} \boldsymbol{u}\, dA \tag{17.2.3}$$

for every $\boldsymbol{F}\in U$, every bounded open subset D of Eucl and every $\boldsymbol{u}\in C_0^\infty(D)$ for which the argument in the integral is in U. Clearly, if f is a null Lagrangian at the boundary, then it is a homogeneous null Lagrangian and hence of the form (13.5.3). However, the class of null Lagrangians at the boundary is narrower.

17.2.4 Proposition *Let $f : U \to \mathbb{R}$ be a continuously differentiable function on an open connected subset U of* Lin*. Then f is a null Lagrangian at the boundary with normal $\boldsymbol{n}$ if and only if*

$$f(\boldsymbol{F}) = A + \boldsymbol{B}\cdot\boldsymbol{F} + \boldsymbol{G}\cdot \operatorname{cof}\boldsymbol{F}\boldsymbol{n}, \quad \boldsymbol{F}\in U, \tag{17.2.4}$$

where $A\in\mathbb{R}$, $\boldsymbol{B}\in$ Lin and $\boldsymbol{G}\in$ Vect are constants.

In particular, $\boldsymbol{F}\mapsto \det\boldsymbol{F}$ is not a null Lagrangian at the boundary.

Proof Let f be a null Lagrangian at the boundary with normal $\boldsymbol{n}$. Then f is a null Lagrangian and hence it is of the form (13.5.3). By Proposition 13.5.4 then $\operatorname{Div}\partial_{\boldsymbol{F}} f(\nabla\boldsymbol{\chi}) = \boldsymbol{0}$ for every sufficiently smooth $\boldsymbol{\chi}$. We now replace the $\boldsymbol{u}$ in (17.2.3) by $\boldsymbol{u}+\lambda\boldsymbol{v}$, $\boldsymbol{u},\boldsymbol{v}\in C_0^\infty(D)$, $\lambda\in\mathbb{R}$ and differentiate with respect to λ at $\lambda = 0$ to obtain

$$\int_{D\cap H} \partial_{\boldsymbol{F}} f(\boldsymbol{F}+\nabla\boldsymbol{u})\cdot\nabla\boldsymbol{v}\, dV = \partial_{\boldsymbol{F}} f(\boldsymbol{F})\boldsymbol{n}\cdot\int_{D\cap\partial H} \boldsymbol{v}\, dA. \tag{17.2.5}$$

Using that $\operatorname{Div}\partial_{\boldsymbol{F}} f(\boldsymbol{F}+\nabla\boldsymbol{u}) = \boldsymbol{0}$ and integrating by parts we obtain

$$\int_{D\cap\partial H} \partial_{\boldsymbol{F}} f(\boldsymbol{F}+\nabla\boldsymbol{u})\boldsymbol{n}\cdot\boldsymbol{v}\, dA = \partial_{\boldsymbol{F}} f(\boldsymbol{F})\boldsymbol{n}\cdot\int_{D\cap\partial H} \boldsymbol{v}\, dA$$

and as $\boldsymbol{v}$ may be chosen arbitrarily on ∂H (smooth and subject to $\boldsymbol{v} = \boldsymbol{0}$ outside D) we obtain $\partial_{\boldsymbol{F}} f(\boldsymbol{F}+\nabla\boldsymbol{u})\boldsymbol{n} = \partial_{\boldsymbol{F}} f(\boldsymbol{F})\boldsymbol{n}$ on $D\cap\partial H$. For a given point $\boldsymbol{p}\in D\cap\partial H$, the value $\nabla\boldsymbol{u}(\boldsymbol{p})$ may be chosen arbitrarily and hence

$$\partial_{\boldsymbol{F}} f(\boldsymbol{F}+\boldsymbol{H})[\boldsymbol{a}\otimes\boldsymbol{n}] = \partial_{\boldsymbol{F}} f(\boldsymbol{F})[\boldsymbol{a}\otimes\boldsymbol{n}]$$

for every $\boldsymbol{a}\in$ Vect and $\boldsymbol{H}\in$ Lin such that $\boldsymbol{F}+\boldsymbol{H}\in U$. By (13.5.3) this condition reads

$$\boldsymbol{C}\cdot\partial_{\boldsymbol{F}}\operatorname{cof}(\boldsymbol{F}+\boldsymbol{H})[\boldsymbol{a}\otimes\boldsymbol{n}] + D\operatorname{cof}(\boldsymbol{F}+\boldsymbol{H})\cdot(\boldsymbol{a}\otimes\boldsymbol{n}) = \boldsymbol{C}\cdot\partial_{\boldsymbol{F}}\operatorname{cof}\boldsymbol{F}[\boldsymbol{a}\otimes\boldsymbol{n}] + D\operatorname{cof}\boldsymbol{F}\cdot(\boldsymbol{a}\otimes\boldsymbol{n}). \tag{17.2.6}$$

Using the fact that the function $\boldsymbol{F}\mapsto\partial_{\boldsymbol{F}}\operatorname{cof}\boldsymbol{F}[\boldsymbol{a}\otimes\boldsymbol{n}]$ is linear while $\boldsymbol{F}\mapsto\operatorname{cof}\boldsymbol{F}\cdot(\boldsymbol{a}\otimes\boldsymbol{n})$ quadratic in $\boldsymbol{F}$, one sees easily that (17.2.6) splits into

$$\boldsymbol{C}\cdot\partial_{\boldsymbol{F}}\operatorname{cof}(\boldsymbol{F}+\boldsymbol{H})[\boldsymbol{a}\otimes\boldsymbol{n}] = \boldsymbol{C}\cdot\partial_{\boldsymbol{F}}\operatorname{cof}\boldsymbol{F}[\boldsymbol{a}\otimes\boldsymbol{n}], \tag{17.2.7}$$

$$D\operatorname{cof}(\boldsymbol{F}+\boldsymbol{H})\cdot(\boldsymbol{a}\otimes\boldsymbol{n}) = D\operatorname{cof}\boldsymbol{F}\cdot(\boldsymbol{a}\otimes\boldsymbol{n}). \tag{17.2.8}$$

Invoking the linearity once more, one concludes that (17.2.7) implies

$$\boldsymbol{C}\cdot\partial_{\boldsymbol{F}}\operatorname{cof}\boldsymbol{H}[\boldsymbol{a}\otimes\boldsymbol{n}] = 0 \tag{17.2.9}$$

for every $\boldsymbol{H}\in$ Lin. Setting $\boldsymbol{H} = \boldsymbol{b}\otimes\boldsymbol{c}$ and using the easily verifiable formula

$$\partial_{\boldsymbol{F}}\operatorname{cof}(\boldsymbol{b}\otimes\boldsymbol{c})[\boldsymbol{a}\otimes\boldsymbol{n}] = (\boldsymbol{b}\times\boldsymbol{a})\otimes(\boldsymbol{c}\times\boldsymbol{n}),$$

one concludes from (17.2.9) that $(\boldsymbol{b}\times\boldsymbol{a})\cdot\boldsymbol{C}(\boldsymbol{c}\times\boldsymbol{n}) = 0$ for every $\boldsymbol{a},\boldsymbol{b},\boldsymbol{c}\in$ Vect. This implies $\boldsymbol{C}(\boldsymbol{c}\times\boldsymbol{n}) = \boldsymbol{0}$ for every $\boldsymbol{c}\in$ Vect. In other words, $\boldsymbol{C}\boldsymbol{d} = \boldsymbol{0}$ for every $\boldsymbol{d}\in$ Vect such that $\boldsymbol{d}\cdot\boldsymbol{n} = 0$. This implies that $\boldsymbol{C} = \boldsymbol{G}\otimes\boldsymbol{n}$ for some $\boldsymbol{G}\in$ Vect. Similarly, we shall now show that (17.2.8) implies $D = 0$. Setting $\boldsymbol{H} = \varepsilon\boldsymbol{F}$ where $\varepsilon > 0$ is sufficiently small, (17.2.8) gives

$$(1+\varepsilon)^2 D\operatorname{cof}\boldsymbol{F}\cdot(\boldsymbol{a}\otimes\boldsymbol{n}) = D\operatorname{cof}\boldsymbol{F}\cdot(\boldsymbol{a}\otimes\boldsymbol{n})$$

from which $D\operatorname{cof}\boldsymbol{F}\cdot(\boldsymbol{a}\otimes\boldsymbol{n}) = 0$. Consequently, $D\operatorname{cof}\boldsymbol{F}\boldsymbol{n} = 0$ for every $\boldsymbol{F}\in U$, and since U is open, this can be true only if $D = 0$. This completes the proof that every null Lagrangian at the boundary is of the form (17.2.4). The converse implication is proved by checking directly that every f given by (17.2.4) satisfies (17.2.3). This is accomplished by integration by parts in a way very similar to that in the proof of Proposition 13.5.6 and the details are omitted. □

We shall now show, in a way completely parallel with Proposition 17.1.4, that the principle of minimum energy implies the quasiconvexity at the boundary at the points of the free part of the boundary F. Consider an equilibrium state $\sigma = (\chi, \eta)$ compatible with the conservative loading environment with the volume density of the potential energy ψ, see (13.3.5). Let Σ_0 be the class of competitor states as in Sect. 15.2. Let $\boldsymbol{p} \in F$ be a point such that $\partial\Omega$ is of class C^1 near $\boldsymbol{p}$ with $\boldsymbol{n}$ the outer normal at $\boldsymbol{p}$ and define $g : \tilde{U} \to \mathbb{R}$ by

$$g(\boldsymbol{F}', \eta') = \tilde{e}(\boldsymbol{F}', \eta') + \psi(\boldsymbol{F}', \chi(\boldsymbol{p}), \boldsymbol{p}), \quad (\boldsymbol{F}', \eta') \in \tilde{U}. \tag{17.2.10}$$

Recall that F is relatively open in $\partial\Omega$ and that it is permanently assumed that near every point $\boldsymbol{p}$ of $\partial\Omega$ outside the possible singular surface the deformation χ can be smoothly extended to a neighborhood of $\boldsymbol{p}$; in (17.2.10) such an extension is employed.

17.2.5 Assumption For every $\boldsymbol{p} \in F$ there exists a bounded neighborhood D of $\boldsymbol{p}$ such that
(1) $D \cap \partial\Omega \subset F$ and
(2) for every $(\boldsymbol{u}, \omega) \in \mathrm{Var}(D)$ such that the values of $(\nabla\chi + \nabla\boldsymbol{u}, \eta + \omega)$ over Ω are in $\tilde{U}$ we have $(\chi + \boldsymbol{u}, \eta + \omega) \in \Sigma_0$.

17.2.6 Proposition *Let the state $\sigma = (\chi, \eta)$ satisfy the principle of minimum energy with respect to the class Σ_0, and let Assumption* 17.2.5 *hold. Let $\boldsymbol{p} \in F$ be a point outside the singular surface of σ such that $\partial\Omega$ is of class C^1 with the normal denoted $\boldsymbol{n} = \boldsymbol{n}(\boldsymbol{p})$. Then the function g defined in* (17.2.10) *is quasiconvex at the boundary with normal $\boldsymbol{n}$ at $(\boldsymbol{F}(\boldsymbol{p}), \eta(\boldsymbol{p}))$.*

17.2.7 *Remarks*

(1) This is proved by a magnification argument very similar to the one in the proof of Proposition 17.1.4, and the reader is referred to BALL & MARSDEN [1984] for details.

(2) Proposition 17.2.6 can be used to show that it is enough the verify the inequality (17.2.2) for just one region D_0 such that $D_0 \cap \partial H \neq \emptyset$; the validity of (17.2.2) with a general D then follows. (As in Remark 17.1.5(1), apply Proposition 17.2.6 to $\Omega = D_0 \cap H$.)

(3) Proposition 17.2.6 asserts the quasiconvexity at the boundary of the sum of the internal and potential energies, in contrast to Proposition 17.1.4, where the quasiconvexity of $\tilde{e}$ was asserted. This is because the potential energy is automatically a null Lagrangian with respect to $\boldsymbol{F}$ (see (17.1.9)) but in general not quasiconvex at the boundary. However, if the potential energy is independent of $\boldsymbol{F}$, as, e.g., in the case of dead loading, the condition that $g \equiv \tilde{e} + \psi$ be quasiconvex at the boundary implies that $\tilde{e}$ is quasiconvex at the boundary.

(4) Quasiconvexity of $\tilde{e}$ at the boundary is related to the quasiconvexity at the boundary of $\bar{f}(\cdot, \theta)$ and quasiconcavity at the boundary of $\hat{\eta}$ in a way completely analogous to those described in Propositions 17.1.6 and 17.1.7.

17.3 Rank 1 Convexity and the Legendre-Hadamard Condition

The theoretically well-motivated concept of quasiconvexity has a big drawback that for concrete functions it is very difficult to verify: one has to check an integral condition. In this section we shall establish a simple consequence of quasiconvexity, rank 1 convexity. Recall from Subsect. 1.1.1 the notion of a rank 1 tensor and from Sect. 2.3 the notion of rank 1 connected tensors.

17.3.1 Definitions A subset U of InvLin × ℝ is said to be ***rank 1 convex*** if for every $(\boldsymbol{F}, \eta), (\boldsymbol{G}, \zeta) \in U$ with $\boldsymbol{G}, \boldsymbol{F}$ of rank 1 connected, the closed line segment with endpoints $(\boldsymbol{F}, \eta), (\boldsymbol{G}, \zeta)$ belongs to U. Let $g : U \to \mathbb{R}$ be a function on an open set $U \subset \text{InvLin} \times \mathbb{R}$ and $(\boldsymbol{F}, \eta) \in U$. g is said to be ***rank 1 convex*** at $(\boldsymbol{F}, \eta)$ if for each $\boldsymbol{a}, \boldsymbol{b} \in$ Vect, $\omega \in \mathbb{R}$ and $\alpha \in [0, 1]$,

$$(1-\alpha)g(\boldsymbol{F}-\alpha\boldsymbol{H}, \eta-\alpha\omega)+\alpha g(\boldsymbol{F}+(1-\alpha)\boldsymbol{H}, \eta+(1-\alpha)\omega) \geq g(\boldsymbol{F}, \eta) \qquad (17.3.1)$$

where $\boldsymbol{H} = \boldsymbol{a} \otimes \boldsymbol{b}$, provided the closed line segment with endpoints $(\boldsymbol{F} - \alpha\boldsymbol{H}, \eta - \alpha\omega)$ and $(\boldsymbol{F} + (1-\alpha)\boldsymbol{H}, \eta + (1-\alpha)\omega)$ is in U. g is said to be ***rank 1 concave*** at $(\boldsymbol{F}, \eta)$ if $-g$ is rank 1 convex at $(\boldsymbol{F}, \eta)$. g is said to be ***rank 1 convex*** if it is rank 1 convex at every point of its domain. Rank 1 convexity or concavity of a function of $\boldsymbol{F}$ alone are defined analogously.

If g is (globally) rank 1 convex, then (17.3.1) can be written in the form

$$g(\alpha(\boldsymbol{F}, \eta)+(1-\alpha)(\boldsymbol{G}, \zeta)) \leq \alpha g(\boldsymbol{F}, \eta)+(1-\alpha)g(\boldsymbol{G}, \zeta) \qquad (17.3.2)$$

for every $(\boldsymbol{F}, \eta), (\boldsymbol{G}, \zeta) \in U$ such that $\boldsymbol{F} - \boldsymbol{G}$ is of rank at most 1 and the closed line segment with endpoints $(\boldsymbol{F}, \eta), (\boldsymbol{G}, \zeta)$ in the domain of g. This should be contrasted with the convexity, which requires that U be a convex set and (17.3.2) hold for every $(\boldsymbol{F}, \eta), (\boldsymbol{G}, \zeta) \in U$. Results of Sect. 2.3, and especially Proposition 2.3.2, explain the meaning of the arguments appearing in the rank 1 convexity condition (17.3.1).

17.3.2 ***Remarks***

(1) $\text{Lin}^+ \times \mathbb{R}^{++}$ is rank 1 convex. This follows from Proposition 17.7.1. $\text{InvLin} \times \mathbb{R}^{++}$ is not rank 1 convex because $\mathbf{1}$ and $\boldsymbol{F} := \mathbf{1} - 2\boldsymbol{e} \otimes \boldsymbol{e}$, with $\boldsymbol{e}$ a unit vector, differ by a rank 1 tensor, but the line segment with endpoints $\mathbf{1}, \boldsymbol{F}$ does not lie in InvLin, since it contains $\mathbf{1} - \boldsymbol{e} \otimes \boldsymbol{e}$.

(2) In the definition of global rank 1 convexity, it is not required that the domain of g be rank 1 convex. If this requirement is incorporated, then the formulations of some of the results below simplify inasmuch as the requirements that some line segments lie in the domain can be dropped.

(3) Let g be as in 17.3.1(2) and define, for every $\boldsymbol{b} \in$ Vect, the function $g_{\boldsymbol{b}} : U_{\boldsymbol{b}} \to \mathbb{R}$ as follows. The domain $U_{\boldsymbol{b}}$ consists of all $(\boldsymbol{a}, \zeta) \in \text{Vect} \times \mathbb{R}$ such that the closed line segment with endpoints $(\boldsymbol{F}, \zeta), (\boldsymbol{F} + \boldsymbol{a} \otimes \boldsymbol{b}, \zeta)$ is in U and $g_{\boldsymbol{b}}(\boldsymbol{a}, \zeta) := g(\boldsymbol{F} + \boldsymbol{a} \otimes \boldsymbol{b}, \zeta)$ for every $(\boldsymbol{F} + \boldsymbol{a} \otimes \boldsymbol{b}, \zeta) \in U_{\boldsymbol{b}}$. Then g is rank 1 convex at $(\boldsymbol{F}, \eta)$ if and only if $g_{\boldsymbol{b}}$ is convex at $(\boldsymbol{0}, \eta)$ for every $\boldsymbol{b} \in$ Vect, $\boldsymbol{b} \neq \boldsymbol{0}$. This enables us to transfer some results on convex functions to rank 1 convex functions.

(4) The definition can clearly be generalized to the more general case where m, n are arbitrary as in Remark 17.1.2(4). In the situation pertinent to the calculus of variations on an interval (see Remark 17.1.2(5)), the rank 1 convexity reduces to the ordinary convexity.

(5) Each globally rank 1 convex (finite-valued) function is automatically continuous on its domain. See MORREY [1966; Sect. 4.4] and DACOROGNA [1985].

17.3.3 Proposition *Let $g : U \to \mathbb{R}$ be continuously differentiable and $(\boldsymbol{F}, \eta) \in U$. Then g is rank 1 convex at $(\boldsymbol{F}, \eta)$ if and only if*

$$g(\boldsymbol{F}+\boldsymbol{a}\otimes\boldsymbol{b}, \eta+\omega) \geq g(\boldsymbol{F}, \eta)+\partial_{\boldsymbol{F}} g(\boldsymbol{F}, \eta)\cdot(\boldsymbol{a}\otimes\boldsymbol{b})+\partial_\eta g(\boldsymbol{F}, \eta)\cdot\omega, \qquad (17.3.3)$$

for every $\boldsymbol{a}, \boldsymbol{b} \in$ Vect, $\omega \in \mathbb{R}$ such that the closed line segment with endpoints $(\boldsymbol{F}, \eta)$ and $(\boldsymbol{F}+\boldsymbol{a}\otimes\boldsymbol{b}, \eta+\omega)$ belongs to U. g is rank 1 convex (globally) if and only if (17.3.3) *holds at every $(\boldsymbol{F}, \eta) \in U$.*

Proof We use the form of the rank 1 convexity in Remark 17.3.2(3). The result follows from Proposition 16.2.4(2) by realizing that the derivative of $g_{\boldsymbol{b}}$ at $(\boldsymbol{0}, \eta)$ is $(\partial_{\boldsymbol{F}} g(\boldsymbol{F}, \eta)\boldsymbol{b}, \partial_\eta g(\boldsymbol{F}, \eta))$ and hence (16.2.2) takes the form (17.3.3). □

The main motivation for rank 1 convexity is Proposition 17.3.5, below. For the proof, we need a lemma. A subset P of Vect is said to be a parallelepiped if

$$P = \{p \in \text{Vect} : p = \mu_i e_i + p_0, 0 < \mu_i < 1\}, \tag{17.3.4}$$

where $\{e_i, i = 1, \ldots, n\}$ is a basis in Vect, p_0 a point, and the summation convention has been employed. A function φ on Vect is said to be P-periodic if $\varphi(p) = \varphi(p + n_i e_i)$ for every p and every n-tuple $(n_1, \ldots, n_n)$ of integers.

17.3.4 Lemma *Let $g : U \to \mathbb{R}$ be quasiconvex at $(F, \eta) \in U$.*

(1) *Let P be a parallelepiped in* Vect, *and let $(u, \omega) \in C^\infty(\text{Vect})$ be P-periodic functions such that* (a) $\int_P \omega\, dV = 0$; (b) *for every $p \in$ Vect, the closed line segment with endpoints (F, η) and $(F + \nabla u(p), \eta + \omega(p))$ is in U. Then*

$$\int_P g(F + \nabla u, \eta + \omega)\, dV \geq |P| g(F, \eta).$$

(2) *Let b be a unit vector and let $w : \mathbb{R} \to$ Vect, $\mu : \mathbb{R} \to \mathbb{R}$ be 1-periodic functions of class C^∞ such that* (a) $\int_0^1 \mu(t)\, dt = 0$; (b) *for every $t \in \mathbb{R}$, the closed line segment with endpoints (F, η) and $(F + w'(t), \eta + \mu(t))$ is in U. Then*

$$\int_0^1 g(F + w'(t) \otimes b, \eta + \mu(t))\, dt \geq g(F, \eta). \tag{17.3.5}$$

It will be clear from the proof that the lemma holds with (b) weakened to require only that the closed line segment with endpoints (F, η) and $(F + \nabla u(p), \eta + \omega(p))$ is in U for every $p \in \partial P$. On the other hand, that some condition like (b) is needed is seen as follows. Consider a function g of F only, defined on $U \subset$ Lin. If U is not connected, then by continuity, the perturbations $F + \nabla u(p)$, $u \in C_0^\infty(P)$ occurring in the definition of quasiconvexity can reach only the values of $G \in U$ from the connected component of F and hence the values of g outside this component do not influence the quasiconvexity of g at F. On the other hand, one can find simple examples of g, U such that there is a P-periodic $u \in C^\infty(\text{Vect})$ with $F + \nabla u(p)$ lying in a component of U different from that of F.

Proof (1): Let dim Vect = 3 and assume for simplicity of notation that P is the unit cube

$$P = \{p \in \text{Vect} : 0 < p \cdot e_i < 1, i = 1, 2, 3\},$$

where $\{e_i, i = 1, 2, 3\}$ an orthonormal basis (otherwise we would have to use a dual basis in some of the formulas below). Let k, m be positive integers and define the cubes

$$P_k := \{p \in \text{Vect} : 0 < p \cdot e_i < k, i = 1, 2, 3\}, \quad Q_k := \{p \in \text{Vect} : -m < p \cdot e_i < k + m, i = 1, 2, 3\},$$

$P_k \subset Q_k$. In the following proof, m will be eventually fixed while k will eventually tend to infinity. It will be convenient to think of the set $L_k = Q_k \setminus P_k$ as a *relatively* thin "boundary layer" of P_k, see Fig. 17.3.1, where $m = 1$.

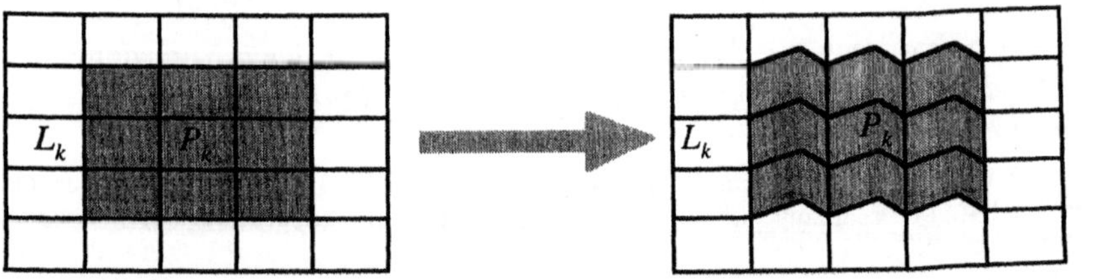

Fig. 17.3.1

Let $(\boldsymbol{u}, \omega)$ be as in the statement of the lemma. By multiplying by a suitable cutoff function, it is possible to modify $(\boldsymbol{u}, \omega)$ to obtain $(\boldsymbol{u}_k, \omega_k)$ such that $(\boldsymbol{u}_k, \omega_k) = (\boldsymbol{u}, \omega)$ on P_k, $(\boldsymbol{u}_k, \omega_k) = 0$ outside Q_k, $\int_{Q_k} \omega_k \, dV = 0$, and both $(\boldsymbol{u}_k, \omega_k)$ and their first gradients are bounded by a constant C independent of k. To avoid complications, let us sketch this construction in the case when g is a function of $\boldsymbol{F}$ only, and hence its domain is a subset of Lin. Let $\varphi : \mathbb{R} \to \mathbb{R}$ be a C^∞ function such that $\varphi(t) = 0$ for $t \le 0$, $\varphi(t) = 1$ for $t \ge m$, and $0 \le \varphi'(t) \le 2/m$ for all $t \in \mathbb{R}$. Let φ_k : Vect $\to \mathbb{R}$ be defined by

$$\varphi_k(\boldsymbol{p}) = \prod_{i=1}^{3} [\varphi(\boldsymbol{p} \cdot \boldsymbol{e}_i + m)\varphi(-\boldsymbol{p} \cdot \boldsymbol{e}_i + k + m)], \quad \boldsymbol{p} \in \text{Vect}.$$

Then $\varphi_k(\boldsymbol{p}) = 1$ on P_k, $\varphi_k(\boldsymbol{p}) = 0$ outside Q_k, and $|\nabla\varphi_k(\boldsymbol{p})| \le 12/m$ for all k and $\boldsymbol{p}$. The function $\boldsymbol{u}_k = \varphi_k \boldsymbol{u}$ satisfies the above requirements and

$$\nabla \boldsymbol{u}_k = \boldsymbol{u} \otimes \nabla\varphi_k + \varphi_k \nabla \boldsymbol{u}$$

on Vect with $\nabla \boldsymbol{u}_k = \nabla \boldsymbol{u}$ on P_k. By hypothesis, $\boldsymbol{F} + \varphi_k(\boldsymbol{p})\nabla\boldsymbol{u}(\boldsymbol{p})$ belongs to U for each $\boldsymbol{p} \in$ Vect. By increasing m, $\boldsymbol{u} \otimes \nabla\varphi_k$ may be made as small as needed, and uniformly in k and $\boldsymbol{p}$. Therefore, since U is open, we may always find an m such that $\boldsymbol{F} + \nabla\boldsymbol{u}_k(\boldsymbol{p}) \in U$ for all k and $\boldsymbol{p} \in$ Vect. Let us now return to the setting of the proposition, with the thermal variable included. We apply the quasiconvexity inequality to the domain Q_k and to the functions $(\boldsymbol{u}_k, \omega_k)$ to obtain

$$\int_{L_k} g(\boldsymbol{F} + \nabla\boldsymbol{u}_k, \eta + \omega_k)\, dV + \int_{P_k} g(\boldsymbol{F} + \nabla\boldsymbol{u}, \eta + \omega)\, dV \ge |Q_k| g(\boldsymbol{F}, \eta), \tag{17.3.6}$$

where we have also used the fact that $(\boldsymbol{u}_k, \omega_k)$ coincide with $(\boldsymbol{u}, \omega)$ on P_k. Since $(\boldsymbol{u}, \omega)$ are P-periodic, the second integral in (17.3.6) is k^3 times that over the unit cell P and taking into account that $|Q_k| = (k + 2m)^3$, (17.3.6) reduces to

$$\int_{L_k} g(\boldsymbol{F} + \nabla\boldsymbol{u}_k, \eta + \omega_k)\, dV + k^3 \int_{P} g(\boldsymbol{F} + \nabla\boldsymbol{u}, \eta + \omega)\, dV \ge (k + 2m)^3 g(\boldsymbol{F}, \eta). \tag{17.3.7}$$

Since the functions $(\boldsymbol{u}_k, \omega_k)$ and their gradients are bounded on L_k by a constant, the integrand in the first term on the right-hand side of the last inequality is also bounded on L_k by a constant independent of k. Hence the integral over L_k can be estimated by a constant times the volume of L_k, which is of order $(k + 2m)^3 - k^3 \sim 6mk^2$. We now divide (17.3.7) by k^3 and let $k \to \infty$. The contribution from the first integral divided by k^3 vanishes, and we obtain

$$\int_{P} g(\boldsymbol{F} + \nabla\boldsymbol{u}, \eta + \omega)\, dV \ge g(\boldsymbol{F}, \eta).$$

(2): Let $\{\boldsymbol{e}_i, i = 1, 2, 3\}$ be any orthonormal basis such that $\boldsymbol{e}_1 = \boldsymbol{b}$, and let P be as above. Define $\boldsymbol{u}(\boldsymbol{p}) = \boldsymbol{w}(\boldsymbol{p} \cdot \boldsymbol{b})$, $\omega(\boldsymbol{p}) = \mu(\boldsymbol{p} \cdot \boldsymbol{b})$, $\boldsymbol{p} \in$ Vect. The application of (1) completes the proof. □

17.3.5 Proposition *If $g : U \to \mathbb{R}$ is quasiconvex at $(\boldsymbol{F}, \eta)$ then it is rank 1 convex at $(\boldsymbol{F}, \eta)$.*

Combining Proposition 17.3.5 with Proposition 17.1.4 and Remark 17.1.5(2), we see that the rank 1 convexity at $(\boldsymbol{F}(\boldsymbol{p}), \eta(\boldsymbol{p}))$ is a necessary condition for a state $(\boldsymbol{\chi}, \eta)$ to be a strong local minimizer. This conclusion holds also in the more general situation of Remark 17.1.2(4). The special case of the latter in the calculus of variations on an interval (m arbitrary, $n = 1$, see Remark 17.1.2(5)) is the Weierstrass necessary condition: for every $t \in [t_1, t_2]$, $f(t, \boldsymbol{\chi}(t), \cdot)$ must be convex at $\boldsymbol{\chi}'(t)$; see CESARI [1983, pp. 31–32]. The important difference is that in this special case the convexity is equivalent to the quasiconvexity.

Proof Let α, $\boldsymbol{a}$, $\boldsymbol{b}$, ω be as in the definition of rank 1 convexity 17.3.1. We apply Lemma 17.3.4(2) to $\boldsymbol{w}(t) = \boldsymbol{a}q(t)$, $\mu(t) = \omega q'(t)$ where $q : \mathbb{R} \to \mathbb{R}$ a 1-periodic function such that

$$q'(t) = 1 - \alpha \text{ if } 0 \le t < \alpha, \quad q'(t) = -\alpha \text{ if } \alpha \le t < 1.$$

Then (17.3.5) reduces to (17.3.1). All hypotheses of Lemma 17.3.4(2) are verified except that $\boldsymbol{w}$, μ are not smooth. An obvious limiting procedure is necessary to smooth out the corners of q. The details are omitted; it is only noted that in this procedure the condition on the line segment occurring in the definition of rank 1 convexity is employed, besides it being used to satisfy the hypotheses of Lemma 17.3.4(2). □

For quadratic functions, the rank 1 convexity is equivalent to quasiconvexity, see Proposition 17.3.9 and Proposition 20.5.1. Moreover, the experience shows that the difference between global quasiconvexity and rank 1 convexity is relatively small. A longstanding problem has been the question of whether the global rank 1 convexity implies quasiconvexity in the general case, see MORREY [1966] and BALL [1977]. This has been proved to be false by ŠVERÁK [1992a] if $m \geq 3, n \geq 2$. The problem is still open if $m = n = 2$. (See also BALL [1990].)

We now apply the preceding discussion to the response functions.

17.3.6 Proposition *Suppose that the body satisfies Assumptions* 15.3.2(1) *and* (2), *and let* $(\boldsymbol{F}, \eta) \in \tilde{U}$, $(\boldsymbol{F}, e) \in \hat{U}$ *be related by* $e = \tilde{e}(\boldsymbol{F}, \eta)$. *Then* $\tilde{e}$ *is rank 1 convex at* $(\boldsymbol{F}, \eta)$ *if and only if* $\hat{\eta}$ *is rank 1 concave at* $(\boldsymbol{F}, e)$.

This proposition is proved in a manner analogous to Proposition 17.1.6. The details are omitted.

17.3.7 Proposition *Let* $(\boldsymbol{F}, \eta) \in \tilde{U}$, $(\boldsymbol{F}, \theta) \in \bar{U}$ *be related by* $\eta = \bar{\eta}(\boldsymbol{F}, \theta)$. *Then*

(1) *if* $\tilde{e}$ *is rank 1 convex at* $(\boldsymbol{F}, \eta)$, *then* $\bar{f}(\cdot, \theta)$, *taken as a function of the deformation gradient only, is rank 1 convex at* $\boldsymbol{F}$ *and* (17.1.11) *holds for every* θ' *such that* $(\boldsymbol{F}, \theta') \in \bar{U}$.

(2) *If* (17.1.12) *holds for every* $(\boldsymbol{F}', \theta') \in \bar{U}$, *then* $\tilde{e}$ *is rank 1 convex at* $(\boldsymbol{F}, \eta)$, *if and only if* $\bar{f}(\cdot, \theta)$ *is rank 1 convex at* $\boldsymbol{F}$.

(3) $\tilde{e}$ *is rank 1 convex (globally) if and only if* $\bar{f}(\cdot, \theta)$ *is rank 1 convex for every* θ *and* $\bar{f}(\boldsymbol{F}, \cdot)$ *concave for every* $\boldsymbol{F}$.

Assertion (3) again shows that the rank 1 convexity of $\tilde{e}$ splits into two *separate* conditions: concavity of $\bar{f}$ in θ for every $\boldsymbol{F}$ and the rank 1 convexity of $\bar{f}$ in $\boldsymbol{F}$ for every θ. This proposition is similar in form to Proposition 17.1.7. The proof is similar, too, and is omitted.

17.3.8 Definitions Let $g : U \to \mathbb{R}$ be a twice continuously differentiable function on an open subset U of InvLin $\times\, \mathbb{R}$ and $(\boldsymbol{F}, \eta)$ a point in U. g is said to satisfy the ***Legendre–Hadamard condition*** at $(\boldsymbol{F}, \eta)$ if for every $\boldsymbol{a}, \boldsymbol{b} \in$ Vect, $\omega \in \mathbb{R}$,

$$\partial_{\boldsymbol{F}}^2 g[\boldsymbol{a} \otimes \boldsymbol{b}, \boldsymbol{a} \otimes \boldsymbol{b}] + 2\partial_{\boldsymbol{F}}\partial_\eta g \cdot (\boldsymbol{a} \otimes \boldsymbol{b}) \cdot \omega + \partial_\eta^2 g \cdot \omega^2 \geq 0, \tag{17.3.8}$$

where the derivatives are evaluated at $(\boldsymbol{F}, \eta)$. Let h be a function on a subset V of Lin and $\boldsymbol{F}$ a point in V. h is said to satisfy the ***Legendre–Hadamard condition*** at $\boldsymbol{F}$ if for every $\boldsymbol{a}, \boldsymbol{b} \in$ Vect,

$$\partial_{\boldsymbol{F}}^2 h(\boldsymbol{F})[\boldsymbol{a} \otimes \boldsymbol{b}, \boldsymbol{a} \otimes \boldsymbol{b}] \geq 0. \tag{17.3.9}$$

h is said to satisfy the ***strong ellipticity condition*** at $\boldsymbol{F}$ if (17.3.9) is strict whenever $\boldsymbol{a} \neq 0, \boldsymbol{b} \neq 0$.

The forms of the Legendre–Hadamard and strong ellipticity conditions for objective-isotropic functions is discussed in Sect. 18.6. We shall see that fluids satisfy the Legendre–Hadamard condition in their normal regimes; in contrast to this, they can never satisfy the strong ellipticity condition.

17.3.9 Proposition *Let* $g : U \to \mathbb{R}$ *be a twice continuously differentiable function on an open subset* U *of* InvLin $\times\, \mathbb{R}$ *and* $(\boldsymbol{F}, \eta)$ *a point in* U. *Then*

(1) *if* g *is rank 1 convex at* $(\boldsymbol{F}, \eta)$, *it satisfies the Legendre–Hadamard condition at* $(\boldsymbol{F}, \eta)$;

(2) *if* U *is rank 1 convex, then* g *is globally rank 1 convex if and only if it satisfies the Legendre–Hadamard condition at every point of* U.

Proof We shall use the form of the rank 1 convexity stated in Remark 17.3.2(3). The result follows from Proposition 16.2.4 and 16.2.5 by applying them to g_b from Remark 17.3.2(3) and identifying the derivatives. □

The relationships among the Legendre–Hadamard conditions for various choices of independent thermal variables are particularly simple.

17.3.10 Proposition *Let $(F, \eta) \in \tilde{U}$, $(F, e) \in \hat{U}$, $(F, \theta) \in \bar{U}$ be related by $e = \tilde{e}(F, \eta)$ and $\eta = \bar{\eta}(F, \theta)$. Then the following conditions are equivalent:*

(1) *the function $\tilde{e}$ satisfies the Legendre–Hadamard condition at (F, η);*
(2) *the function $-\hat{\eta}$ satisfies the Legendre–Hadamard condition at (F, e);*
(3) *the function $\bar{f}(\cdot, \theta)$, taken as function of F, satisfies the Legendre–Hadamard condition at F and*

$$\partial_\theta^2 \bar{f}(F, \theta) \leq 0. \tag{17.3.10}$$

Assertion (3) shows that the single Legendre–Hadamard condition for η and e splits into two inequalities for $\bar{f}$: namely, the isothermal version of the Legendre–Hadamard condition and the positivity of the specific heat at constant deformation. In terms of the tensors of elastic constants, the Legendre–Hadamard condition for $\bar{f}$ reads $\bar{A}[a \otimes b, a \otimes b] \geq 0$, and when combined with the positivity of the specific heat at constant deformation, it reflects the full Legendre–Hadamard condition. The Legendre–Hadamard condition for $\hat{\eta}$ *implies* (with the choice $\omega = 0$) $\hat{A}[a \otimes b, a \otimes b] \geq 0$, but this condition does not imply the full Legendre–Hadamard condition, even when the specific heat is positive.

Proof It is necessary to invoke the relations (10.7.2) and (10.7.3) among the second differentials of the thermodynamic potentials. Let us prove, for instance, the equivalence (1) $\Leftrightarrow$ (3). Set $G_1 = G_2 = a \otimes b$, $v_1 = v_2 = \omega$ in (10.7.3) to obtain

$$d^2\tilde{e}[a \otimes b, \omega; a \otimes b, \omega] = \partial_F \bar{f}[a \otimes b, a \otimes b] - \partial_\theta^2 \bar{f} \cdot (d\tilde{\theta}[a \otimes b, \omega])^2 \geq 0. \tag{17.3.11}$$

Let us suppose first that $\tilde{e}$ satisfies the Legendre–Hadamard condition. Then the left-hand side is always nonnegative. Let a, b be two arbitrary vectors. Corresponding to them, there is an ω such that $d\tilde{\theta}[a \otimes b, \omega] = 0$. Then second term on the right-hand side of (17.3.11) disappears and we read off that $\bar{f}(\cdot, \theta_0)$, taken as function of F only, satisfies the Legendre–Hadamard condition. Then, setting $a = b = 0$ we find from (17.3.11) that (17.3.10) holds. Hence (3) holds. Conversely, suppose that (3) holds. Then in view of the rank 1 convexity of $\bar{f}(\cdot, \theta_0)$ and in view of (17.3.11) we conclude that the right-hand side is always nonnegative. Hence the left-hand side is also nonnegative; but this is the Legendre–Hadamard condition for $\tilde{e}$. The other equivalences are proved similarly. □

17.4 Maxwell's Relation

Throughout this section, we consider an equilibrium state $\sigma = (\chi, \eta)$, with a singular surface, of temperature θ, in a conservative environment 13.3.6. We interpret the singular surface as a (coherent) interface separating two-phases of the material. We say that σ is stable if it satisfies the principle of minimum energy. Let Assumption 17.1.3 hold. The main result of this section, Proposition 17.4.1 below, shows that the limiting values $F^\pm$ of the deformation gradient at the interface in a stable state must satisfy an *additional scalar equation* beyond the geometric compatibility and the equilibrium of forces, namely, Maxwell's relation. We shall also see that coexistent phases satisfying Maxwell's relation are necessarily accompanied by a *violation* of the rank 1 convexity at some point on the line segment connecting the states on the two sides of the singular surface.

17.4.1 Proposition (Maxwell's relation) *If σ is stable, then the limiting values $F^\pm$ of the deformation gradient at each point of the singular surface satisfy*

$$\bar{f}(F^+, \theta) = \bar{f}(F^-, \theta) + \rho_\circ^{-1}\bar{S}(F^+, \theta) \cdot (F^+ - F^-). \tag{17.4.1}$$

Further, (17.4.1) *can be given the following three equivalent forms:*

$$[f - \rho_\circ^{-1} Sn \cdot Fn] = 0, \quad [P]n = 0, \quad \text{and} \quad [\mathbf{P}]n = 0 \tag{17.4.2}$$

on the singular surface.

Proof Assume $\rho_o = 1$. By Propositions 17.1.4 and 17.1.7 the free energy function $\bar{f}(\cdot, \theta)$ is quasiconvex and hence rank 1 convex at $\boldsymbol{F}^+, \boldsymbol{F}^-$. We shall use the form of the rank 1 convexity condition in Remark 17.3.2(3). We omit the scalar arguments η, ζ occurring there, of course, and identify the function g with $\bar{f}(\cdot, \theta)$ and $\boldsymbol{F}$ with $\boldsymbol{F}^-$. Then the function $g := g_{\boldsymbol{n}}$, is convex at the points $\boldsymbol{0}$ and $\boldsymbol{a}$, where $\boldsymbol{a}$ is the amplitude: $\boldsymbol{F}^+ = \boldsymbol{F}^- + \boldsymbol{a} \otimes \boldsymbol{n}$, see (2.2.9). For every $\boldsymbol{e} \in U_{\boldsymbol{n}}$, we have $\partial_{\boldsymbol{e}} g(\boldsymbol{e}) = \bar{\boldsymbol{S}}(\boldsymbol{F}^- + \boldsymbol{e} \otimes \boldsymbol{n}, \theta)\boldsymbol{n}$ and the balance of forces on the singular surface, $[\boldsymbol{S}]\boldsymbol{n} = \boldsymbol{0}$, reads $\partial_{\boldsymbol{e}} g(\boldsymbol{0}) = \partial_{\boldsymbol{e}} g(\boldsymbol{a})$. By Proposition 16.2.6(2) then $g(\boldsymbol{a}) = g(\boldsymbol{0}) + \partial_{\boldsymbol{e}} g(\boldsymbol{0}) \cdot \boldsymbol{a}$ which gives (17.4.1). To prove $[\boldsymbol{P}]\boldsymbol{n} = \boldsymbol{0}$, note that $[\boldsymbol{S}]\boldsymbol{n} = \boldsymbol{0}$ implies $\langle \boldsymbol{S} \rangle \boldsymbol{n} = \boldsymbol{S}^+\boldsymbol{n}$. This provides $[\boldsymbol{F}^{\mathsf{T}}\boldsymbol{S}]\boldsymbol{n} = \boldsymbol{a} \cdot (\boldsymbol{S}^+\boldsymbol{n}) \cdot \boldsymbol{n}$ and hence $[\boldsymbol{P}]\boldsymbol{n} = [f]\boldsymbol{n} - \boldsymbol{a} \cdot (\boldsymbol{S}^+\boldsymbol{n})\boldsymbol{n} = \boldsymbol{0}$ by (17.4.1). From (13.3.27) we also know that the tensor $\boldsymbol{\Pi}$ does not contribute to the jump of the total Eshelby tensor; hence also $[\boldsymbol{P}]\boldsymbol{n} = \boldsymbol{0}$. □

17.4.2 ***Remarks***

(1) In the case of the calculus of variations on an interval (see Remark 17.1.2), Maxwell's relation reduces to the classical Erdman corner condition, cf. CESARI [1983, p. 33].

(2) It is worthwhile to assess the kind of extrema leading to Maxwell's relation. The above proposition refers to the global principle of minimum of energy. Combining Remark 17.1.5(2) with Proposition 17.1.4, we see that Maxwell's relation still holds under the weaker assumption that σ is a strong local minimum of energy in the sense of calculus of variations (under the constraint that the total entropy be fixed) and more generally if σ is a local extremum with respect to the metric $\|\cdot\|_p$ for $1 \le p < \infty$, as defined in Subsect. 20.3.1. However, the weak local minimum is insufficient for the validity of Maxwell's relation. It is easy to see that Maxwell's relation can be also derived from the same type of local minima of the total stored energy M. The variations of the temperature are irrelevant.

(3) Yet another derivation of Maxwell's relation was given in Subsects. 14.5.1 and 14.5.4(3), where it was shown to be a consequence of the *assumption* of the stationarity of the first variation of the total stored energy with respect to inner variations.

(4) Maxwell's relation is the generalization of the classical Maxwell's and Gibbs's equality of the chemical potentials of the stably coexisting fluid phases. Eshelby's tensor is the generalization of the scalar chemical potential of fluids to bodies of general symmetry. See ESHELBY [1970], ROBIN [1974], LARCHÉ & CAHN [1978], GRINFELD [1980], JAMES [1981], EDELEN [1981], GURTIN [1983], TRUSKINOVSKY [1983b], ŠILHAVÝ [1985c], and KAGANOVA & ROITBURD [1988]. The evolution of the idea that the phase equilibria in solids are controlled by a tensor was not straightforward, cf. GRINFELD [1991, Introduction & Sect. 2.8]. This reference also contains a treatment of incoherent phase boundaries, where Maxwell's relation is different. For an application of the Eshelby tensor in modeling phase equilibria within the theory of mixtures, see SAMOHÝL [1996a–d].

17.4.3 Thermodynamic Equilibrium at the Interface We have the following ***conditions of thermodynamic equilibrium*** at the phase interface in a stable state:

(1) ***geometrical compatibility***

$$[\boldsymbol{F}] = \boldsymbol{a} \otimes \boldsymbol{n} \quad \text{for some} \quad \boldsymbol{a} \in \text{Vect}; \tag{17.4.3}$$

(2) ***balance of forces***

$$[\boldsymbol{S}]\boldsymbol{n} = \boldsymbol{0}; \quad \text{equivalently} \quad [\boldsymbol{T}]\boldsymbol{m} = \boldsymbol{0}; \tag{17.4.4}$$

(3) ***Maxwell's relation***

$$[f - \rho_o^{-1}\boldsymbol{S}\boldsymbol{n} \cdot \boldsymbol{F}\boldsymbol{n}] = 0; \quad \text{equivalently} \quad [\boldsymbol{P}]\boldsymbol{n} = \boldsymbol{0}. \tag{17.4.5}$$

If $\boldsymbol{n} \in \text{Sph}$ and $(\boldsymbol{F}^{\pm}, \theta) \in \bar{U}$, the quadruple $\mathcal{S} = (\boldsymbol{F}^+, \boldsymbol{F}^-, \theta, \boldsymbol{n}) \equiv (\boldsymbol{F}^{\pm}, \theta, \boldsymbol{n})$ is called a ***two-phase state*** if (17.4.3)–(17.4.5) hold. An inspection that is left to the reader shows that if $(\boldsymbol{F}^{\pm}, \theta, \boldsymbol{n})$ is a two-phase state and $\boldsymbol{Q} \in \text{Orth}$, $\boldsymbol{H} \in \mathsf{G} \cap \text{Orth}^+$, then also $(\boldsymbol{Q}\boldsymbol{F}^{\pm}\boldsymbol{H}, \theta, \boldsymbol{H}^{\mathsf{T}}\boldsymbol{n})$ is a two-phase state. The following proposition describes situations when Maxwell's relation is automatically satisfied.

17.4.4 Proposition *If* $S = (F^{\pm}, \theta, n)$ *are such that* F^+, F^- *differ by a tensor of the form* $a \otimes n$ *then the following assertions hold:*

(1) *if both* $F^{\pm}$ *are the points of minimum of* $\bar{f}(\cdot, \theta)$, *then* S *is a two-phase state;*

(2) *if the balance of forces* (17.4.4) *holds and*

$$F^+ = QF^-H \text{ and either } Hn = n \text{ or } Hn = -n \tag{17.4.6}$$

for some $Q \in \text{Orth}^+$ *and* $H \in G \cap \text{Orth}^+$ *then* S *satisfies Maxwell's relation.*

(1) applies to the ground states considered in Sect. 2.3; (2) applies to twins satisfying the balance of forces; see Proposition 2.5.4(2).

Proof (1): Under the hypotheses, $\bar{f}(F^+, \theta) = \bar{f}(F^-, \theta)$, $\bar{S}(F^+, \theta) = \bar{S}(F^-, \theta) = \mathbf{0}$ and hence to the balance of forces and Maxwell's relation hold. (2): We have $\bar{S}(F^+, \theta)n \cdot F^+n = \bar{S}(QF^-H, \theta)n \cdot QF^-Hn = Q\bar{S}(F^-, \theta)Hn \cdot QF^-Hn = \bar{S}(F^-, \theta)n \cdot QF^-n$ by $(17.4.6)_2$ or $(17.4.6)_3$. Furthermore, clearly, $\bar{f}(F^+, \theta) = \bar{f}(F^-, \theta)$ and so we have arrived at $(17.4.2)_1$. □

17.4.5 Proposition *Let* $S_0 = (F_0^{\pm}, \theta_0, n_0)$ *be at two-phase state and assume that*

$$\bar{A}(F_0^+, \theta_0)[e \otimes n_0]n_0 \neq \mathbf{0} \text{ for every } e \in \text{Vect},\ e \neq \mathbf{0} \tag{17.4.7}$$

and

$$[\eta_0] + \rho_\circ^{-1}\partial_\theta S(F_0^-, \theta_0)n_0 \cdot [F_0]n_0 \neq 0, \text{ where } \eta_0^{\pm} = \bar{\eta}(F_0^{\pm}, \theta_0). \tag{17.4.8}$$

Then

(1) *there exists a neighborhood* N *of* (F_0^-, n_0) *in* $\text{Lin}^+ \times \text{Sph}$ *and two mappings* $\Phi : N \to \text{Lin}^+$, $\Theta : N \to \mathbb{R}^{++}$ *such that for every* $(F^-, n) \in N$, *the quadruple* $S = (F^+, F^-, \theta, n)$, *given by*

$$F^+ = \Phi(F^-, n), \quad \theta = \Theta(F^-, n), \tag{17.4.9}$$

is a two-phase state; each two-phase state with F^-, n *from this neighborhood may be written in this way;*

(2) *if we define* $s : N \to \text{Vect}$ *by* $s(F, n) = \bar{S}(F, \Theta(F))n$ *then for every* $(F, n) \in N$ *and every* $b \in \text{Vect}$ *we have the* ***Clapeyron equation***

$$[\eta]\, d\Theta + \rho_\circ^{-1}\, ds \cdot [F]n = 0,$$

where

$$d\Theta = \partial_F\Theta \cdot (b \otimes n), \quad ds = \partial_F s[b \otimes n],$$

and $[\eta]$, $[F]$ *denote the jumps at* $(F^{\pm}, \theta, n)$ *given by* (17.4.9);

(3) *if* $(F, n) \in N$ *and* $Q \in \text{Orth}$, $H \in G \cap \text{Orth}^+$, *then also* $(QFH, H^{\mathsf{T}}n) \in N$ *and*

$$\Phi(QFH, H^{\mathsf{T}}n) = Q\Phi(F, n)H, \quad \Theta(QFH, H^{\mathsf{T}}n) = \Theta(F, n).$$

Hence, under the nondegeneracy conditions (17.4.7), (17.4.8), each two-phase state S_0 has a neighborhood which contains an 11-dimensional manifold of coexistent states satisfying all the conditions of thermodynamic equilibrium. Note that we have $\bar{A}(F_0^{\pm}, \theta_0)[e \otimes n_0, e \otimes n_0] \geq 0$ for every $e \in \text{Vect}$ as a consequence of the Legendre–Hadamard condition at $(F^{\pm}, \theta)$ (see 17.3.9), and for solids it is realistic to assume the strict inequality for each $e \neq \mathbf{0}$, which implies (17.4.7). Nevertheless, (17.4.7) is not as general as desirable. For example, it cannot hold for fluids, merely as a consequence of their material symmetry. Elementary considerations (see Subsect. 22.1.4) show that the quantity

$$l = \theta[\eta] = [e - \rho_\circ^{-1} \boldsymbol{Sn} \cdot \boldsymbol{Fn}]$$

is the latent heat of the phase transformation, i.e., the heat that must be communicated to a unit mass to allow the transformation from $\boldsymbol{F}^+, \theta$ into $\boldsymbol{F}^-, \theta$.

Proof Let $\rho_\circ = 1$. (1): We seek $\boldsymbol{F}^+, \theta$ as functions of $\boldsymbol{F} \equiv \boldsymbol{F}^-$ and $\boldsymbol{n}$ by the implicit function theorem. To satisfy the geometrical compatibility, we set $\boldsymbol{F}^+ = \boldsymbol{F} + \boldsymbol{a} \otimes \boldsymbol{n}$ where $\boldsymbol{a} \in \text{Vect}$ is a vector to be determined from equations $(17.4.4)_1$ and $(17.4.5)_2$. These are four equations for four scalar unknowns $\boldsymbol{a}, \theta$, which we write in the form

$$\bar{\boldsymbol{S}}(\boldsymbol{F} + \boldsymbol{a} \otimes \boldsymbol{n}, \theta)\boldsymbol{n} - \bar{\boldsymbol{S}}(\boldsymbol{F}, \theta)\boldsymbol{n} = \boldsymbol{0}, \quad \bar{f}(\boldsymbol{F} + \boldsymbol{a} \otimes \boldsymbol{n}, \theta) - \bar{f}(\boldsymbol{F}, \theta) - \bar{\boldsymbol{S}}(\boldsymbol{F}, \theta)\boldsymbol{n} \cdot \boldsymbol{a} = 0. \tag{17.4.10}$$

We shall see that the Jacobi matrix J of the mapping given by the left-hand side of (17.4.10) with respect to the variables $\boldsymbol{a}, \theta$ is nonsingular at $\boldsymbol{a}_0, \theta_0$ as a consequence of the hypotheses. Indeed, using the balance of forces for $S_0 = (\boldsymbol{F}_0^\pm, \theta_0, \boldsymbol{n}_0)$, one finds that this matrix acts on $x \equiv (\boldsymbol{e}, \zeta) \in \text{Vect} \times \mathbb{R}$ by $Jx = (\boldsymbol{c}, \omega)$ where

$$\boldsymbol{c} = \bar{\boldsymbol{A}}^+[\boldsymbol{e} \otimes \boldsymbol{n}_0]\boldsymbol{n}_0 + (\partial_\theta \bar{\boldsymbol{S}}^+ \boldsymbol{n}_0 - \partial_\theta \bar{\boldsymbol{S}}^- \boldsymbol{n}_0)\zeta, \quad \omega = -([\eta] + \partial_\theta \bar{\boldsymbol{S}}^- \boldsymbol{n}_0 \cdot \boldsymbol{a}_0)\zeta,$$

where the superscripts ± indicate that the quantity is evaluated at $(\boldsymbol{F}_0^\pm, \theta_0)$. Clearly, conditions (17.4.7) and (17.4.8) imply that $Jx = 0$ can happen only if $x = 0$. (2): It suffices to insert $\boldsymbol{F}^+ = \boldsymbol{\Phi}(\boldsymbol{F}, \boldsymbol{n})$, $\boldsymbol{F}^- = \boldsymbol{F}$, $\theta = \Theta(\boldsymbol{F}, \boldsymbol{n})$ into Maxwell's relation, to differentiate with respect to $\boldsymbol{F}$ on the increment $\boldsymbol{b} \otimes \boldsymbol{n}$ and to use $\bar{\boldsymbol{S}}^+ \cdot d\boldsymbol{\Phi} = \bar{\boldsymbol{S}}^+ \boldsymbol{n} \cdot d\boldsymbol{\Phi} \boldsymbol{n}$, $\bar{\boldsymbol{S}}^- \cdot d\boldsymbol{F} = \bar{\boldsymbol{S}}^- \boldsymbol{n} \cdot d\boldsymbol{F}\boldsymbol{n}$ since on the rank 1 increments, $d\boldsymbol{\Phi} = d\boldsymbol{a} \otimes \boldsymbol{n} + d\boldsymbol{F} = (d\boldsymbol{a} + \boldsymbol{b}) \otimes \boldsymbol{n}$. (3): We use the fact that if $(\boldsymbol{F}^\pm, \theta, \boldsymbol{n})$ is a two-phase state and $\boldsymbol{Q} \in \text{Orth}$, $\boldsymbol{H} \in \mathsf{G} \cap \text{Orth}^+$, then also $(\boldsymbol{Q}\boldsymbol{F}^\pm \boldsymbol{H}, \theta, \boldsymbol{H}^\mathsf{T} \boldsymbol{n})$ is a two-phase state. This allows us to extend the functions $\boldsymbol{\Phi}$, Θ constructed above to the set $\{(\boldsymbol{QFH}, \boldsymbol{H}^\mathsf{T}) : \boldsymbol{Q} \in \text{Orth}, \boldsymbol{H} \in \mathsf{G} \cap \text{Orth}^+\}$ in such a way that the asserted relations hold. The details are left to the reader. □

17.4.6 Solid-Melt Interface Consider a solid of arbitrary symmetry in thermodynamic equilibrium with its melt. Assume that the material is globally modeled by the free energy function $\bar{f} = \bar{f}(\boldsymbol{F}, \theta)$. The occurrence of a fluid-like behavior is reflected by the existence of a region $U_f \subset \bar{U}$ such that on U_f, $\bar{f}$ has a representation

$$\bar{f}(\boldsymbol{F}, \theta) = f_f(v, \theta), \quad v = \rho_\circ^{-1} |\det \boldsymbol{F}|,$$

where f_f is a function defined on a suitable domain in $\mathbb{R}^2$. An example of a free energy function of this type is found in Subsect. 9.4.6. Let us consider a global stable state $\sigma = (\chi, \eta)$ with a connected phase interface S which divides Ω into two disjoint connected regions Ω_s, Ω_f such that $\boldsymbol{F}$ is of constant value $\boldsymbol{F}^+$ on Ω_s, while on Ω_f, the specific volume v is of constant value v^-. We have $\boldsymbol{T}^- = -p\boldsymbol{1}$ where $p = p_f(v^-, \theta) = -\partial_v f_f(v^-, \theta)$, and set $\boldsymbol{T}^+ = \bar{\boldsymbol{T}}(\boldsymbol{F}^+, \theta)$, where $\bar{\boldsymbol{T}}$ is the response function giving the equilibrium stress. The equilibrium of forces at the interface (17.4.4) reads

$$\boldsymbol{T}^+ \boldsymbol{m} = -p\boldsymbol{m}, \tag{17.4.11}$$

where $\boldsymbol{m}$ is the spatial normal to S, and Maxwell's relation takes the form

$$\bar{f}(\boldsymbol{F}^+, \theta) + pv^+ = f_f(v^-, \theta) + pv^- \tag{17.4.12}$$

where v^+ is the specific volume of the solid in the state $(\boldsymbol{F}^+, \theta)$. (The derivation of (17.4.12) is left to the reader as an exercise.) The geometric compatibility (17.4.3) plays no role in the present problem and we ignore it. By (17.4.11), at every point of S, $\boldsymbol{m}$ is an eigenvector of the constant tensor $\boldsymbol{T}^+$ with the eigenvalue $-p$. Let, for any $(\boldsymbol{F}, \theta)$, $t_1(\boldsymbol{F}, \theta) \ge t_2(\boldsymbol{F}, \theta) \ge t_3(\boldsymbol{F}, \theta)$ be the ordered triple of eigenvalues of the symmetric tensor $\bar{\boldsymbol{T}}(\boldsymbol{F}, \theta)$. There are three possibilities:

(1) at some (and hence by continuity at every) point of S, $\boldsymbol{m}$ is an eigenvector corresponding to a simple eigenvalue of $\boldsymbol{T}^+$; then $\boldsymbol{m}$ is constant and S is planar;

(2) at some (and hence by continuity at every) point of S, $\boldsymbol{m}$ is an eigenvector corresponding to a double eigenvalue of $\boldsymbol{T}^+$. Then S is a curved cylindrical surface, with axis given by the simple eigenvector of $\boldsymbol{T}^+$ and the basis an arbitrary curve in the two-dimensional eigenspace of $\boldsymbol{T}^+$;

(3) $\boldsymbol{T}^+ = -p\boldsymbol{1}$ for some $p \in \mathbb{R}$. Then S may be arbitrary.

(3) appears to be the most typical, and let us examine the locus of all triples $\boldsymbol{F}^+, v^-, \theta$ giving rise to states of the above type near a triple $\boldsymbol{F}_0^+, v_0^-, \theta_0$ that is realized in some state of the above type. We shall assume that in an appropriately chosen reference configuration, the material satisfies the following condition:

17.4.7 Assumption There exists a neighborhood N_0 of $(\boldsymbol{F}_0^+, \theta_0)$ such that for any $(\boldsymbol{F}, \theta) \in N_0$,

$$\bar{\boldsymbol{T}}(\boldsymbol{F}, \theta) = -p\mathbf{1} \quad \text{for some} \quad p \in \mathbb{R}$$

if and only if

$$\boldsymbol{F} \text{ is of the form } \boldsymbol{F} = w\boldsymbol{Q} \text{ for some } w > 0, \quad \boldsymbol{Q} \in \text{Orth}^+.$$

We assume consistently that $\boldsymbol{F}_0^+ = w_0\mathbf{1}$ and denote v_0^+ the specific volume of $\boldsymbol{F}_0^+$. Clearly, Assumption 17.4.7 is appropriate for isotropic solids, but it may be satisfied also by crystals (given that we examine a neighborhood of a state at which the stress reduces to pressure). Assumption 17.4.7 allows us to introduce functions f_s, p_s on $D_s \subset \mathbb{R}^2$ by

$$D_s = \{(v, \theta) : (w\mathbf{1}, \theta) \in N_0\}, \quad f_s(v, \theta) = \bar{f}(w\mathbf{1}, \theta), \quad \bar{\boldsymbol{T}}(w\mathbf{1}, \theta) = -p_s(v, \theta)\mathbf{1},$$

where w stands for $(\rho_\circ v)^{1/3}$ for brevity. The stress relation reads $p_s = -\partial_v f_s$, and the conditions of thermodynamic equilibrium read

$$p_s(v^+, \theta) = p_f(v^-, \theta), \quad f_s(v^+, \theta) + p_s(v^+, \theta)v^+ = f_f(v^-, \theta) + p_f(v^-, \theta)v^-. \tag{17.4.13}$$

The problem is thus formally identical with the fluid-fluid phase equilibrium. The following proposition determines that the solutions form locally a one-dimensional family.

17.4.8 Proposition *Let Assumption* 17.4.7 *hold, and let*

$$\partial_v p_s(v_0^+, \theta_0) \neq 0, \quad \partial_v p_f(v_0^-, \theta_0) \neq 0.$$

Then there exists an $\varepsilon > 0$ *and functions* $\bar{v}^+, \bar{v}^+ : (\theta_0 - \varepsilon, \theta_0 + \varepsilon) \to (0, \infty)$ *such that for every* $\theta \in (\theta_0 - \varepsilon, \theta_0 + \varepsilon)$, *the values* $v^\pm := \bar{v}^\pm(\theta)$ *satisfy* (17.4.13), *and all values sufficiently close to* v_0^+, v_0^- *may be written in this form.*

This is proved by the implicit function theorem as in the case of Proposition 17.4.5. Similar analyses apply to the cases (1), (2) above.

Let us return to the general situation.

17.4.9 Eshelby's Conservation Law *Suppose that* σ *is an equilibrium state in a conservative loading environment that satisfies the principle of minimum energy. If the volume density of the potential energy is independent of* $\boldsymbol{p}$ *and if* $\rho_\circ$ *is uniform, then*

$$\int_{\partial P} \boldsymbol{P n}\, dA = \mathbf{0}$$

for every part $P \subset \Omega$. *In particular, if the volume density potential energy vanishes, then* $\boldsymbol{\mathsf{P}}$ *can be replaced by* $\boldsymbol{P}$.

Proof The surface integrals are converted to volume integrals by Proposition 3.1.4:

$$\int_{\partial P} \boldsymbol{\mathsf{P}} \cdot \boldsymbol{n}\, dA = \int_P \operatorname{Div} \boldsymbol{\mathsf{P}}\, dV + \int_S [\boldsymbol{\mathsf{P}}] \cdot \boldsymbol{n}\, dA.$$

Outside the singular surface $\operatorname{Div} \boldsymbol{\mathsf{P}} = \mathbf{0}$ by 14.2.3(3) and on the singular surface $[\boldsymbol{\mathsf{P}}] \cdot \boldsymbol{n} = \mathbf{0}$ by Proposition 14.2.4. □

17.4.10 Proposition *Let* σ *be a stable state,* $\boldsymbol{p}$ *a point on the singular surface, let* $\boldsymbol{F}^-, \boldsymbol{F}^+$ *be the limit values of the deformation gradient at* $\boldsymbol{p}$ *and* $[\boldsymbol{F}^-, \boldsymbol{F}^+]$ *the closed line segment with endpoints* $\boldsymbol{F}^-, \boldsymbol{F}^+$. *If* $\boldsymbol{F} \in [\boldsymbol{F}^-, \boldsymbol{F}^+]$ *is a point such that*

$$\bar{\boldsymbol{S}}(\boldsymbol{F}, \theta)\boldsymbol{n} \neq \bar{\boldsymbol{S}}(\boldsymbol{F}^+, \theta)\boldsymbol{n}, \tag{17.4.14}$$

then $\bar{f}(\cdot, \theta)$ *is not rank 1 convex at* $\boldsymbol{F}$.

Proof We shall apply Proposition 16.2.6(2) to the function g defined in the proof of Proposition 17.4.1. For any $\boldsymbol{e} \in U_{\boldsymbol{n}}$, the rank 1 convexity of $\bar{f}$ at $(\boldsymbol{F}^- + \boldsymbol{e} \otimes \boldsymbol{n}, \theta)$ implies the convexity of g at $\boldsymbol{e}$. The points $\mathbf{0}$ and $\boldsymbol{a}$ are points of convexity of g and $\partial_e g(\mathbf{0}) = \partial_e g(\boldsymbol{a}) = \bar{\boldsymbol{S}}(\boldsymbol{F}^\pm, \theta)\boldsymbol{n}$. By Proposition 16.2.6(2) any point of convexity of g on the line segment $(\mathbf{0}, \boldsymbol{a})$ must satisfy $\partial_e g(\boldsymbol{e}) = \partial_e g(\mathbf{0})$ which means $\bar{\boldsymbol{S}}(\boldsymbol{F}^- + \boldsymbol{e} \otimes \boldsymbol{n}, \theta)\boldsymbol{n} = \bar{\boldsymbol{S}}(\boldsymbol{F}^+, \theta)\boldsymbol{n}$. □

17.4.11 Spinodal Region In realistic constitutive hypotheses (17.4.14) is to be expected to hold for "almost all" points $\boldsymbol{F}$ on $[\boldsymbol{F}^-, \boldsymbol{F}^+]$. (An example is the well-known "serpentine" Van der Waals isotherm, where (17.4.14) holds at all but one point on the line segment: see Fig. 19.4.1 and also the free energy function of Subsect. 9.4.6.) Hence $\bar{f}(\cdot, \theta)$ cannot be rank 1 convex and consequently quasiconvex at "almost all" points $\boldsymbol{F}$ on $[\boldsymbol{F}^-, \boldsymbol{F}^+]$. This shows first that globally rank 1 convex free energy functions cannot be used to model phase transitions. Second, the homogeneous state with the deformation gradient $\boldsymbol{F}$ satisfying (17.4.14) and temperature θ cannot satisfy any of the extremum principles formulated in Sect. 15.2 globally (absolute minimizers), for these imply quasiconvexity. For the same reason they also cannot be strong local minimizers. Consequently such homogeneous states $(\boldsymbol{F}, \theta)$ are not absolutely stable in this sense. Some of the states on the line segment can be metastable in that they are weak local minimizers. Such will be the case when they satisfy the strong ellipticity condition Definition 17.3.8. Furthermore, it can be shown that if the states $(\boldsymbol{F}^\pm, \theta)$ and the amplitude $\boldsymbol{a}$ satisfy $\partial_{\boldsymbol{F}}^2 \bar{f}(\boldsymbol{F}^\pm, \theta)[\boldsymbol{a} \otimes \boldsymbol{n}, \boldsymbol{a} \otimes \boldsymbol{n}] > 0$, which is the usual case, then there must be states $(\boldsymbol{F}, \theta)$ with $\boldsymbol{F}$ on the line segment such that $\partial_{\boldsymbol{F}}^2 \bar{f}(\boldsymbol{F}, \theta)[\boldsymbol{a} \otimes \boldsymbol{n}, \boldsymbol{a} \otimes \boldsymbol{n}] < 0$. We shall call the set of all such $\boldsymbol{F}$ the ***spinodal region*** corresponding to the two-phase state $S = (\boldsymbol{F}^\pm, \theta, \boldsymbol{n})$. In particular, if $\boldsymbol{F}^\pm$ correspond to the absolute minima of $\bar{f}(\cdot, \theta)$, e.g., if the two-phase state is a ground state in the sense of Sect. 2.3, then $\boldsymbol{S}^\pm = \boldsymbol{0}$, and the spinodal region corresponds to the potential barrier of $\bar{f}(\cdot, \theta)$ between the two minima $\boldsymbol{F}^\pm$. (The general case is reduced to this by subtracting the linear term $\boldsymbol{S}^+ \cdot (\boldsymbol{F} - \boldsymbol{F}^+)$.) The homogeneous states with the deformation gradients from the spinodal region are absolutely unstable: they violate even the weak local extrema in the extremum principles. However, it must be noted that the real existence of the spinodal states is questionable in view of their high degree of instability, and it may be the case that the response functions in the spinodal region are just conventions. Even the validity of the thermostatic relations can be questioned there (ŠILHAVÝ [1985c]). Despite this, the above shows that the phase transitions are bound to be accompanied by metastable and unstable homogeneous states. See also Sect. 22.2.

17.5 Convexity and Polyconvexity

We shall give two simple sufficient conditions for the quasiconvexity: convexity of the energy in the deformation gradient and polyconvexity. The first of them will be seen to be too restrictive as a *general* condition on the potentials, but nevertheless is often valid on large subsets of deformation gradients of bodies under *tension*. The polyconvexity will be seen to be very close to the quasiconvexity so that the latter can be replaced, at a given point, by the former without any practical difference.

17.5.1 Points of Convexity As will be seen in Proposition 17.5.3, if the assumption of convexity is combined with the principle of material frame indifference, it leads to very restrictive positivity properties of the stress. For this reason, there is a widespread belief that "convexity contradicts the frame indifference." Contrary to that, many realistic free energy functions can have a large supply of points of convexity. It is only impossible that a realistic free energy function be a globally convex function (of the deformation gradient). Relevant to this discussion is also the question of domains of the thermodynamic potentials expressed in terms of the deformation gradient: the most natural choices, Lin^+ and InvLin are not convex, see Subsect. 18.3.3. Recall the definition of a point of convexity of a function on a nonconvex set Definition 16.2.1. This definition applies without any change to the thermodynamic potentials: if $g : U \to \mathbb{R}$ is a continuous function defined on an open subset U of $\mathrm{InvLin} \times \mathbb{R}$ and $(\boldsymbol{F}, \eta) \in U$ a point of convexity, then (17.3.1) holds for every $(\boldsymbol{H}, \omega) \in \mathrm{InvLin} \times \mathbb{R}$ for which both sides of the inequality make sense, and from the general assertions about differentiable convex functions (Proposition 16.2.4), follows that an analog of Proposition 17.3.3 holds, with the only (essential) change being that (17.3.3) holds with a general $\boldsymbol{H} \in \mathrm{Lin}$ replacing $\boldsymbol{a} \otimes \boldsymbol{b}$. Also, if g is convex at $(\boldsymbol{F}, \eta)$, then the second differential $d^2 g(\boldsymbol{F}, \eta)$ is a positive semidefinite form on $\mathrm{Lin} \oplus \mathbb{R}$. A complete discussion of the points of convexity of objective-isotropic functions is given in Subsect. 18.3.

To emphasize the role of the points of convexity, consider a body Ω under the displacement-traction dead loading (Example 13.3.9), with the free energy function $\bar{f}$. Fix the temperature θ and write $\bar{f} : U \to \mathbb{R}$ for $\bar{f}(\cdot, \theta)$. The total stored energy M is given by the formula in Subsect. 14.3.1 with the potentials given in Example 13.3.9. Let $\boldsymbol{\Delta}^{**}$ be the set of admissible deformations (see 14.3.4) and $\boldsymbol{\chi}_0 \in \boldsymbol{\Delta}^{**}$ be a deformation with $\boldsymbol{F}_0 := \nabla \boldsymbol{\chi}_0$.

17.5.2 Proposition *In this situation, if $\boldsymbol{\chi}_0$ is an equilibrium deformation compatible with the environment such that for every $\boldsymbol{p} \in \Omega$ the value $\boldsymbol{F}_0(\boldsymbol{p})$ is a point of convexity of $\bar{f}$, then $\boldsymbol{\chi}_0$ is a global minimizer of M in the class of all continuous $\boldsymbol{\chi} \in \boldsymbol{\Delta}^{**}$, i.e., $\mathsf{M}(\boldsymbol{\chi}) \ge \mathsf{M}(\boldsymbol{\chi}_0)$ for every such $\boldsymbol{\chi}$.*

The global convexity of $\bar{f}$ would therefore imply a stability of every equilibrium state. This is undesirable and impossible in general, as it would exclude the barreling and buckling instabilities. See Sect. 20.7 for a discussion of these matters for a particular isotropic stored energy function. Notice also that

(1) applying this proposition to a homogeneous state of deformation gradient $\boldsymbol{F}$ and the class $\boldsymbol{\Delta}^{**}$ corresponding to the homogeneous Dirichlet boundary conditions (see Remark 17.1.5(1)), one finds that the convexity of $\bar{f}$ at $\boldsymbol{F}$ implies the quasiconvexity of $\bar{f}$ at $\boldsymbol{F}$;

(2) applying the proposition to the dead loading situation described in Remark 17.2.2(2), one finds that the convexity of $\bar{f}$ at $\boldsymbol{F}$ implies the quasiconvexity at the boundary with arbitrary normal $\boldsymbol{n}$ of $\bar{f}$ at $\boldsymbol{F}$.

Proof Denoting $\boldsymbol{F}(\boldsymbol{p}) = \nabla \boldsymbol{\chi}(\boldsymbol{p})$, the definition of a point of convexity gives $\bar{f}(\boldsymbol{F}) - \bar{f}(\boldsymbol{F}_0) \ge \boldsymbol{S}(\boldsymbol{F}_0) \cdot (\boldsymbol{F} - \boldsymbol{F}_0)$ for every $\boldsymbol{p} \in \Omega$. Integrating over Ω and using the divergence theorem and the conditions of mechanical equilibrium 14.2.3 leads to $\mathsf{M}(\boldsymbol{\chi}) - \mathsf{M}(\boldsymbol{\chi}_0) \ge 0$. □

Let us also briefly mention the nonnegativity of the second differential at a point. This requirement can be viewed as a consequence of the convexity at the point in question.

17.5.3 Proposition *Let $(\boldsymbol{F}, \eta) \in \tilde{U}$. If the second differential $d^2\tilde{e}(\boldsymbol{F}, \eta)$ is positive semidefinite, then the principal tractions at $(\boldsymbol{F}, \eta)$ satisfy*

$$t_1 + t_2 \ge 0, \quad t_2 + t_3 \ge 0, \quad t_3 + t_1 \ge 0. \tag{17.5.1}$$

In fact the above inequalities follow from the weaker requirement that $\tilde{\mathsf{B}}[\boldsymbol{W}, \boldsymbol{W}] \ge 0$ for every $\boldsymbol{W} \in$ Skew and also from the same inequality with $\tilde{\mathsf{B}}$ replaced by $\bar{\mathsf{B}}$.

This implies that either all three principal tractions are nonnegative or one of them is negative and the other two strictly positive. For isotropic states the stress reduces to a hydrostatic pressure, i.e., $\boldsymbol{T} = -p\mathbf{1}$ for some $p \in \mathbb{R}$. The positive semidefiniteness of the second differential then necessarily implies $p \le 0$, i.e., the material cannot support compressions. Hence no isotropic state of a body under tension can be a point of convexity. For fluids one can show that the convexity of $\tilde{e}(\boldsymbol{F}, \eta)$ implies that $p \equiv 0$ for all states. The result is similar to Proposition 18.3.5(2) on objective-isotropic functions, but Proposition 17.5.3 refers to a material of arbitrary symmetry. For the proof, it suffices to combine $\tilde{\boldsymbol{B}}[\boldsymbol{W}, \boldsymbol{W}] \ge 0$ with Proposition 10.5.3(4). See also TRUESDELL & NOLL [1965, p. 163].

Let us now proceed to the polyconvexity. The discussion applies both to solids and fluids, but it will turn out that the condition of polyconvexity is particularly suitable for solids, establishing the existence of the solutions to the extremum problems for solids in Chap. 21. The discussion below is based on the properties of the homogeneous null Lagrangians from Sect. 13.5.

17.5.4 Definition Let $g : U \to \mathbb{R}$ be a continuous function on an open subset U of $\mathrm{Lin} \times \mathbb{R}$ and $(\boldsymbol{F}, \eta) \in U$ a point. The function g is said to be ***polyconvex*** at $(\boldsymbol{F}, \eta)$ if there exist $\boldsymbol{A}, \boldsymbol{B} \in \mathrm{Lin}$, $C, D \in \mathbb{R}$ such that

$$\begin{aligned} g(\boldsymbol{G}, \omega) \ge\ & g(\boldsymbol{F}, \eta) + \boldsymbol{A} \cdot (\boldsymbol{G} - \boldsymbol{F}) \\ & + \boldsymbol{B} \cdot (\mathrm{cof}\, \boldsymbol{G} - \mathrm{cof}\, \boldsymbol{F}) + C(\det \boldsymbol{G} - \det \boldsymbol{F}) + D(\omega - \eta) \end{aligned} \tag{17.5.2}$$

for every $(\boldsymbol{G}, \omega) \in U$. The function g is said to be polyconvex if it is polyconvex at every point of its domain. This is equivalent to the existence of a convex function $\tilde{g} : \tilde{\tilde{U}} \to \mathbb{R}$ such that

$$g(\boldsymbol{F}, \eta) = \tilde{g}(\boldsymbol{F}, \mathrm{cof}\, \boldsymbol{F}, \det \boldsymbol{F}, \eta) \tag{17.5.3}$$

for every $(\boldsymbol{F}, \eta) \in U$. Here $\tilde{\tilde{U}}$ is the convex hull of $\{(\boldsymbol{F}, \mathrm{cof}\, \boldsymbol{F}, \det \boldsymbol{F}, \eta) : (\boldsymbol{F}, \eta) \in U\}$ in the 19-dimensional space $\mathrm{Lin} \times \mathrm{Lin} \times \mathbb{R}$. For instance, if $U = \mathrm{Lin}^+ \times (0, \infty)$, then $\tilde{\tilde{U}} = \mathrm{Lin} \times \mathrm{Lin} \times (0, \infty) \times (0, \infty)$, see BALL [1977] and BALL, CURRIE & OLVER [1981]. The polyconvexity of the function of $\boldsymbol{F}$ alone is defined analogously and also the polyconcavity.

17.5.5 Proposition *If g is polyconvex at $(\boldsymbol{F}, \eta)$, then it is quasiconvex at $(\boldsymbol{F}, \eta)$; if g is globally polyconvex then it is globally quasiconvex.*

Proof By Proposition 13.5.3, the functions $\boldsymbol{F} \mapsto \boldsymbol{F}$, $\mathrm{cof}\, \boldsymbol{F}$, $\det \boldsymbol{F}$ are homogeneous null Lagrangians and hence satisfy (13.5.9) of Proposition 13.5.6, i.e., (13.5.10) and two similar equations with $\det \boldsymbol{F}$ replaced by $\boldsymbol{F}$ and $\mathrm{cof}\, \boldsymbol{F}$. Let $(\boldsymbol{u}, \omega)$ be a perturbation in $\mathrm{Var}(D)$, where $\mathrm{Var}(D)$ is the class of perturbations occurring in the definition of quasiconvexity. It suffices to set $\boldsymbol{G} = \nabla \boldsymbol{u}$ in (17.5.2), to integrate over D, and to use the above mentioned property of the homogeneous null Lagrangians to find that (17.1.2) holds. □

17.5.6 Recapitulation Collecting the main results of the preceding exposition, we see that for finite-valued, continuous functions g,

$$\text{convexity} \Rightarrow \text{polyconvexity} \Rightarrow \text{quasiconvexity} \Rightarrow \text{rank 1 convexity}, \tag{17.5.4}$$

where these conditions may be interpreted as conditions at a given point of the domain of g, or as global conditions on the whole domain. It is easy to show that these concepts are mutually different when interpreted as conditions at the given point. In the following discussion, let us therefore speak about the global interpretation. Recall from Sect. 17.3 that for quadratic functions,

$$\text{quasiconvexity} \iff \text{rank 1 convexity.}$$

For quadratic functions in two dimensions ($n = 2$)

$$\text{polyconvexity} \iff \text{quasiconvexity} \iff \text{rank 1 convexity,} \tag{17.5.5}$$

see MORREY [1966] and HESTENES & MACSHANE [1949]. The examples due to TERPSTRA [1938] and SERRE [1983] show that for $n \geq 3$ there are quasiconvex quadratic functions that are not polyconvex. Recall from Sect. 17.3 that for $n \geq 3$, there are globally rank 1 convex functions that are not quasiconvex. Thus for $n \geq 3$, the implications in (17.5.4) are strict even in the global interpretation, and even with the choice of the domain the whole Lin. If $n = 2$, the equivalence of the global quasiconvexity and rank 1 convexity is an open problem, but the remaining implications in (17.5.4) are strict. Sects. 17.7, 17.8, and 21.6 will show that for fluids and for crystals with the symmetry group the nonorthogonal symmetry group of a Bravais lattice, the equivalences (17.5.5) hold. The following example is particularly impressive.

17.5.7 Example (ALIBERT & DACOROGNA [1992]) Let $n = 2$, $\gamma \in \mathbb{R}$ and let f_γ : Lin $\to \mathbb{R}$ be defined as

$$f_\gamma(\boldsymbol{A}) = |\boldsymbol{A}|^2(|\boldsymbol{A}|^2 - 2\gamma \det \boldsymbol{A}), \quad \boldsymbol{A} \in \text{Lin}.$$

Then

$$\begin{aligned} f_\gamma \text{ is convex} &\iff |\gamma| \leq \tfrac{2}{3}\sqrt{2}, \\ f_\gamma \text{ is polyconvex} &\iff |\gamma| \leq 1. \end{aligned}$$

There exists an $\varepsilon > 0$ such that

$$\begin{aligned} f_\gamma \text{ is quasiconvex} &\iff |\gamma| \leq 1 + \varepsilon, \\ f_\gamma \text{ is rank 1 convex} &\iff |\gamma| \leq \frac{2}{\sqrt{3}}. \end{aligned}$$

These equivalences are nontrivial. It is not known if $1 + \varepsilon = 2/\sqrt{3}$. Note that by the discussion in Subsect. 17.5.6, for $n = 2$, the examples showing the difference between polyconvexity, quasiconvexity and rank 1 convexity must be nonquadratic. Earlier examples were given by AUBERT [1987] and DACOROGNA & MARCELLINI [1987]. All these, including the one above, are in the class of Orth^+ invariant functions, i.e., functions satisfying $g(\boldsymbol{QAR}) = g(\boldsymbol{A})$ for every point $\boldsymbol{A}$ of the domain and every $\boldsymbol{Q}, \boldsymbol{R} \in \text{Orth}^+$. See also DACOROGNA & al. [1990], and ŠVERÁK [1990, 1991, 1992b].

17.5.8 Proposition *The energy function $\tilde{e}$ is polyconvex if and only if for every θ the function $\bar{f}(\cdot, \theta)$ is polyconvex and for every $\boldsymbol{F}$ the function $\bar{f}(\boldsymbol{F}, \cdot)$ is concave.*

This is in a prefect analogy with the previous results about the quasiconvexity and rank 1 convexity in terms of the temperature. The discussion of polyconvex functions continues in Sect. 18.5, where several families of isotropic polyconvex functions will be determined.

Let us give a sufficient condition for the quasiconvexity at the boundary analogous to the polyconvexity being sufficient to the quasiconvexity.

17.5.9 Proposition *Let $g : U \to \mathbb{R}$ be a continuously differentiable function on an open subset of* Lin *and* $\boldsymbol{F} \in U$, $\boldsymbol{n} \in \mathrm{Vect}$, $|\boldsymbol{n}| = 1$. *Let there exist constants* $\boldsymbol{B} \in \mathrm{Lin}$, $\boldsymbol{M} \in \mathrm{Vect}$ *such that*

$$g(\boldsymbol{G}) \ge g(\boldsymbol{F}) + \boldsymbol{B} \cdot (\boldsymbol{G} - \boldsymbol{F}) + \boldsymbol{M} \cdot (\operatorname{cof} \boldsymbol{G} - \operatorname{cof} \boldsymbol{F})\boldsymbol{n} \tag{17.5.6}$$

for every $\boldsymbol{G} \in U$. *Then* g *is quasiconvex at the boundary with normal* $\boldsymbol{n}$ *at* $\boldsymbol{F}$.

Proof Define $f(\boldsymbol{F}) = \boldsymbol{B} \cdot \boldsymbol{F} + \boldsymbol{M} \cdot \operatorname{cof} \boldsymbol{F}\boldsymbol{n}$, $\boldsymbol{F} \in \mathrm{Lin}$. Inequality (17.5.6) reads

$$g(\boldsymbol{G}) - g(\boldsymbol{F}) \ge f(\boldsymbol{G}) - f(\boldsymbol{F}) \tag{17.5.7}$$

which implies $\partial_{\boldsymbol{F}} g(\boldsymbol{F}) = \partial_{\boldsymbol{F}} f(\boldsymbol{F})$. Let $\boldsymbol{u}, D$ be as in the definition of the quasiconvexity at the boundary. Setting $\boldsymbol{G} = \boldsymbol{F} + \nabla \boldsymbol{u}$, integrating (17.5.7) over $D \cap H$ and using that f is a null Lagrangian at the boundary with normal $\boldsymbol{n}$ (see Proposition 17.2.4) we obtain (17.2.2). □

17.6 The Exchange of the Actual and Reference Configurations

We shall show that the quasiconvexity, rank 1 convexity, the Legendre–Hadamard condition and the polyconvexity are preserved under the operation of the exchange of the roles of the actual and reference configurations: Not only globally, but from point to point. Let us ignore the thermal variables, set $\rho_\circ = 1$, and consider a continuously differentiable stored energy function $f : \mathrm{Lin}^+ \to \mathbb{R}$. Then the spatial density $f^\sharp$ expressed as a function of the inverse deformation gradient $\boldsymbol{Z} = \boldsymbol{F}^{-1}$ is the function defined by

$$f^\sharp(\boldsymbol{Z}) = (\det \boldsymbol{Z}) f(\boldsymbol{Z}^{-1}), \quad \boldsymbol{Z} \in \mathrm{Lin}^+,$$

cf. Subsect. 10.3.2.

17.6.1 Lemma *Let V be a real vector space with inner product, $M \subset V \times (0, \infty)$, and $m : M \to \mathbb{R}$ a function. Define a change of variables $\phi : V \times (0, \infty) \to V \times (0, \infty)$ by $(\boldsymbol{x}, a) \mapsto \phi(\boldsymbol{x}, a) = (\boldsymbol{x}/a, 1/a)$ and a function $m^\sharp : M^\sharp \to \mathbb{R}$ by $m^\sharp = m \circ \phi^{-1}$ where $M^\sharp = \phi(M)$. Then*

(1) *ϕ maps $V \times (0, \infty)$ bijectively onto itself and $\phi^{-1} = \phi$;*

(2) *if $(\boldsymbol{p}, q) \in V \times \mathbb{R}$ is a subgradient of m at some point $(\boldsymbol{x}, a) \in M$ then $(\boldsymbol{p}^\sharp, q^\sharp)$ is a subgradient of $m^\sharp$ at $(z, b) = \phi(\boldsymbol{x}, a)$, where*

$$\boldsymbol{p}^\sharp = \boldsymbol{p}, \quad q^\sharp = m(\boldsymbol{x}, a) - \boldsymbol{p} \cdot \boldsymbol{x} - q \cdot a. \tag{17.6.1}$$

Recall that Definition 16.2.2 allows us to speak about a subgradient of a function defined on a nonconvex set with empty interior. The proof of the lemma amounts to a direct inspection, and is left to the reader.

17.6.2 Proposition *Let $f, f^\sharp$ be as above and let $\boldsymbol{F}, \boldsymbol{Z} \in \mathrm{Lin}^+$ be related by $\boldsymbol{Z} = \boldsymbol{F}^{-1}$. Then f is quasiconvex (rank 1 convex, polyconvex, satisfies the Legendre–Hadamard condition) at $\boldsymbol{F}$ if and only if $f^\sharp$ has the same property at $\boldsymbol{Z}$.*

This may be used to create, e.g., new polyconvex functions from the known ones. Note also that the convexity does not reproduce under the passage $f \mapsto f^\sharp$.

Proof First, since $(f^\sharp)^\sharp = f$, it suffices to prove only the implications in the above statement, the corresponding equivalences follow. Second, we may assume that $\boldsymbol{F} = \boldsymbol{Z} = \mathbf{1}$, otherwise we pass from f to $f'(\boldsymbol{G}) := f(\boldsymbol{G}\boldsymbol{F}^{-1})$, which simplifies the notation and allows us to use the letter $\boldsymbol{F}$ for an arbitrary element of Lin^+ and $\boldsymbol{Z}$ for $\boldsymbol{F}^{-1}$.

The proof that the quasiconvexity of f at $\mathbf{1} \Rightarrow$ quasiconvexity of $f^\sharp$ at $\mathbf{1}$: Let D be a bounded open, connected subset of Vect such that $\mathrm{int}(\mathrm{cl}(D)) = D$ and let $\boldsymbol{u} \in C_0^\infty(D)$ be such that $\mathbf{1} + \nabla \boldsymbol{u}(\boldsymbol{p}) \in \mathrm{Lin}^+$ for every $\boldsymbol{p} \in \mathrm{Eucl}$. Setting $\chi(\boldsymbol{p}) = \boldsymbol{p} + \boldsymbol{u}(\boldsymbol{p})$, $\boldsymbol{p} \in \mathrm{cl}(D)$, using $\chi(\boldsymbol{p}) = \boldsymbol{p}$ on $\partial\Omega$ and applying Ciarlet & Nečas [1987, Thm. 3] we conclude that χ is a homeomorphism from $\mathrm{cl}(D)$ onto $\mathrm{cl}(D)$, the restriction of χ to D is a class C^∞ diffeomorphism. It follows that the inverse $\mu(\boldsymbol{q}) = \chi^{-1}(\boldsymbol{p})$ is of the form $\mu(\boldsymbol{q}) = \boldsymbol{q} + \boldsymbol{v}(\boldsymbol{q})$, $\boldsymbol{q} \in D$, for some $\boldsymbol{v} \in C_0^\infty(D)$. The quasiconvexity of f at $\mathbf{1}$ provides

$$\int_D f(\nabla \mu(\boldsymbol{q}))\, dV(\boldsymbol{q}) = \int_D f(\mathbf{1} + \nabla \boldsymbol{v}(\boldsymbol{q}))\, dV(\boldsymbol{q}) \ge |D| f(\mathbf{1}). \tag{17.6.2}$$

A change of variables $q \mapsto p := \mu(q)$, $q = \chi(p)$ and the use of $\nabla\mu(\chi(p)) = (\nabla\chi(p))^{-1}$ leads to

$$\int_D f(\nabla\mu(q))\,dV(q) = \int_D f(\nabla\mu(\chi(p)))\det\nabla\chi(p)\,dV(p) = \int_D f^\sharp(\mathbf{1} + \nabla u(p))\,dV(p),$$

which reduces (17.6.2) to the quasiconvexity inequality for $f^\sharp$. The proof that the Legendre–Hadamard condition for $f \Rightarrow$ the Legendre–Hadamard condition for $f^\sharp$: Use an easily verifiable formula $\partial_F^2 f(\mathbf{1})[a \otimes b, a \otimes b] = \partial_Z^2 f(\mathbf{1})[a \otimes b, a \otimes b]$, $a, b \in \text{Vect}$, from which the conclusion follows.

For the rest of the proof, we apply Lemma 17.6.1 with the following identifications: $V = \text{Lin} \times \text{Lin}$,

$$M = \{(F, \operatorname{cof} F, \det F) \in V \times \mathbb{R} : F \in \text{Lin}^+\}, \quad x = (F, \operatorname{cof} F), \quad a = \det F,$$
$$m(F, \operatorname{cof} F, \det F) = m(x, a) := f(F), \quad F \in \text{Lin}^+,$$

so that the change of variables $\phi : (x, a) \mapsto (x/a, 1/a)$ is $(F, \operatorname{cof} F, \det F) \mapsto (\operatorname{cof} Z^{\mathrm{T}}, Z^{\mathrm{T}}, \det Z)$ where $Z = F^{-1}$ and

$$M^\sharp = \{(\operatorname{cof} Z^{\mathrm{T}}, Z^{\mathrm{T}}, \det Z) \in V \times \mathbb{R} : Z \in \text{Lin}^+\}, \quad z = (\operatorname{cof} Z^{\mathrm{T}}, Z^{\mathrm{T}}), \quad b = \det Z,$$
$$m^\sharp(\operatorname{cof} Z^{\mathrm{T}}, Z^{\mathrm{T}}, \det Z) := f^\sharp(Z), \quad Z \in \text{Lin}^+.$$

The points where the subgradients are to be examined are $x = z = (\mathbf{1}, \mathbf{1})$, $a = b = 1$.

The proof that the polyconvexity of f at $\mathbf{1} \Rightarrow$ polyconvexity of $f^\sharp$ at $\mathbf{1}$: The polyconvexity of f at $\mathbf{1}$ means that m has a subgradient at $(\mathbf{1}, \mathbf{1}, 1) \in M$, and the lemma says that then $m^\sharp$ has a subgradient at $(\mathbf{1}, \mathbf{1}, 1) \in M^\sharp$. This is equivalent to the polyconvexity of $f^\sharp$ at $\mathbf{1}$, which proves the implication. (The lemma also calculates the coefficients A, B, C, D in (17.5.2) for $f^\sharp$ in terms of those for f.) The proof that rank 1 convexity of $f \Rightarrow$ rank 1 convexity of $f^\sharp$: We apply the lemma with the change of variables and functions $m, m^\sharp$ as above, except that now M consists of triples $(F, \operatorname{cof} F, \det F)$ with F restricted to the form $F = \mathbf{1} + a \otimes b \in \text{Lin}^+$, $a, b \in \text{Vect}$. Since $Z = F^{-1} = (\mathbf{1} + a \otimes b)^{-1} = \mathbf{1} - (1 + a \cdot b)^{-1} a \otimes b$, it is seen that a tensor F differs from $\mathbf{1}$ by a rank 1 tensor if and only if Z differs from $\mathbf{1}$ by a rank 1 tensor. The rank 1 convexity of f at $\mathbf{1}$ means that m has at $(\mathbf{1}, \mathbf{1}, 1) \in M$ a subgradient (p, q) of the form $p = (S(\mathbf{1}), \mathbf{0})$, $q = 0$ where $S(\mathbf{1}) = \partial_F f(\mathbf{1})$. By the lemma then $m^\sharp$ has at $(\mathbf{1}, \mathbf{1}, 1) \in M^\sharp$ a subgradient $p^\sharp$, $q^\sharp$ where, by (17.6.1), $p^\sharp = (S(\mathbf{1}), \mathbf{0})$, $q^\sharp = f(\mathbf{1}) - S(\mathbf{1}) \cdot \mathbf{1}$. The latter means that for every $Z = \mathbf{1} + r \otimes s \in M^\sharp$,

$$f^\sharp(Z) - f^\sharp(\mathbf{1}) \geq S(\mathbf{1}) \cdot (\operatorname{cof} Z - \mathbf{1}) + (f(\mathbf{1}) - S(\mathbf{1}) \cdot \mathbf{1}) \cdot (\det Z - 1),$$

and an expansion of the right-hand side gives $f^\sharp(\mathbf{1} + r \otimes s) - f^\sharp(\mathbf{1}) \geq S(\mathbf{1})^\sharp \cdot (r \otimes s)$ where $S(\mathbf{1})^\sharp = f(\mathbf{1})\mathbf{1} - S(\mathbf{1})$ is the Eshelby tensor of the material with the stored energy function f, as was to be expected; cf. (10.3.1)). □

17.7 Constitutive Inequalities for Fluids

We here show that for fluids the quasiconvexity with respect to the deformation gradient is equivalent to rank 1 convexity and that these two concepts are equivalent to the convexity in the specific volume and entropy. In terms of the deformation gradient, the thermodynamic potentials take the form

$$\bar{f}(F, \theta) = \bar{f}(v, \theta), \quad \tilde{e}(F, \eta) = \tilde{e}(v, \eta), \tag{17.7.1}$$

where $v = \det F/\rho_\circ = 1/\rho$ is the specific volume. (The same letters are used for the functions on the right-hand sides.) The relation of the domains $\bar{U}$, $\tilde{U}$ of the functions on the left-hand sides of the above equations to the domains $\bar{V}$, $\tilde{V}$ of the functions on the right-hand sides is given by (11.4.3) and (11.4.4). We write $\tilde{U} = \tilde{U}^+ \cup \tilde{U}^-$, where $\tilde{U}^\pm = \tilde{U} \cap (\text{Lin}^\pm \times \mathbb{R})$ is the set of all pairs $(F, \eta) \in \tilde{U}$ with $\pm \det F > 0$. Similar notations apply to $\bar{U}$. An equivalent and more direct foundation of the convexity properties of the thermodynamic potentials of fluids is given in Chap. 19.

17.7.1 Proposition *$\tilde{U}^+$ is rank 1 convex if and only if $\tilde{V}$ is convex. For every $\theta > 0$, the isothermal section $\tilde{U}^+_\theta = \{F \in \mathrm{Lin}^+ : (F, \theta) \in \tilde{U}\}$ is rank 1 convex if and only if the isothermal section $\tilde{V}_\theta = \{v \in \mathbb{R}^{++} : (v, \theta) \in \tilde{V}\}$ is convex. Similar assertions apply to $\tilde{U}^-$ and $\tilde{U}^-_\theta$.*

Proof Let $\rho_o = 1$. We shall only prove that $\tilde{U}^+$ is rank 1 convex if and only if $\tilde{V}$ is convex; the proof of the rest is similar. Let $\tilde{U}^+$ be rank 1 convex, let $(v, \eta), (w, \zeta) \in \tilde{V}$ and $\alpha \in [0, 1]$. Our goal is to prove that

$$(\alpha v + (1-\alpha)w, \alpha\eta + (1-\alpha)\zeta) \in \tilde{V}. \tag{17.7.2}$$

Using (11.4.3) we see that there exist $\boldsymbol{F}, \boldsymbol{G}$ such that $(\boldsymbol{F}, \eta), (\boldsymbol{G}, \zeta) \in \tilde{U}^+$ and $v = \det \boldsymbol{F}$, $w = \det \boldsymbol{G}$. Take two vectors $\boldsymbol{a}, \boldsymbol{b}$ such that $\boldsymbol{F}^{-1}\boldsymbol{a} \cdot \boldsymbol{b} \neq 0$ and a number λ. We have $\det(\boldsymbol{F} + \lambda \boldsymbol{a} \otimes \boldsymbol{b}) = \det \boldsymbol{F} \cdot \det(\mathbf{1} + \lambda \boldsymbol{F}^{-1}\boldsymbol{a} \otimes \boldsymbol{b}) = \det \boldsymbol{F}(\mathbf{1} + \lambda \boldsymbol{F}^{-1}\boldsymbol{a} \cdot \boldsymbol{b})$ and hence, for some λ, $\det(\boldsymbol{F} + \lambda \boldsymbol{a} \otimes \boldsymbol{b}) = w$. Since $(w, \zeta) \in \tilde{V}$, using (11.4.4) we see that $(\boldsymbol{F} + \lambda \boldsymbol{a} \otimes \boldsymbol{b}, \zeta) \in \tilde{U}^+$. This means that we can choose $\boldsymbol{G}$ such that $\boldsymbol{F} - \boldsymbol{G}$ is of rank at most 1. Assuming that $\boldsymbol{G}$ is so chosen, we have

$$\det(\alpha \boldsymbol{F} + (1-\alpha)\boldsymbol{G}) = \alpha \det \boldsymbol{F} + (1-\alpha)\det \boldsymbol{G} = \alpha v + (1-\alpha)w \tag{17.7.3}$$

and since $\tilde{U}^+$ is rank 1 convex, we have $(\alpha \boldsymbol{F} + (1-\alpha)\boldsymbol{G}, \alpha\eta + (1-\alpha)\zeta) \in \tilde{U}^+$. But it then follows from (11.4.3) that (17.7.2) holds. The proof of the converse implication is similar and therefore I omit it. □

We shall assume throughout the chapter that the domains of the functions on the right of the equations (17.7.1) are convex and the fluid admits both the temperature and entropy as independent thermodynamical variables. We denote the domains of the functions on the right of the equations (17.7.1) by $\bar{U}, \tilde{U}$ whenever there is no danger of confusion.

17.7.2 Proposition *Let $(\boldsymbol{F}, \eta) \in \tilde{U}$, and $v = |\det \boldsymbol{F}|/\rho_o$. Then*

(1) *$\tilde{e}$ is quasiconvex at $(\boldsymbol{F}, \eta)$ if and only if it is rank 1 convex at $(\boldsymbol{F}, \eta)$. These two conditions are satisfied if and only if $\tilde{e}$ is convex at (v, η);*

(2) *$\tilde{e}$ satisfies the Legendre–Hadamard condition at $(\boldsymbol{F}, \eta)$ if and only if*

$$\partial_v^2 \tilde{e} \cdot \mu^2 + 2\partial_v \partial_\eta \tilde{e} \cdot \mu \cdot \omega + \partial_\eta^2 \tilde{e} \cdot \omega^2 \geq 0, \qquad \omega, \mu \in \mathbb{R} \tag{17.7.4}$$

with the derivatives evaluated at (v, η); equivalently,

$$\partial_v^2 \tilde{e} \geq 0, \quad \partial_\eta^2 \tilde{e} \geq 0, \quad \partial_v^2 \tilde{e} \cdot \partial_\eta^2 \tilde{e} - (\partial_v \partial_\eta \tilde{e})^2 \geq 0. \tag{17.7.5}$$

The terms quasiconvex and rank 1 convex apply to $\tilde{e}$ taken as function of $(\boldsymbol{F}, \eta)$ and the term convex applies to $\tilde{e}$ taken as function of (v, η). The same convention is used throughout the whole section.

Proof (1): Let $\rho_o = 1$ and assume that $\det \boldsymbol{F} > 0$; hence $v = \det \boldsymbol{F}$. The implication quasiconvexity ⇒ rank 1 convexity is a general assertion, see Chap. 17. Rank 1 convexity ⇒ convexity. Let $(u, \omega), (w, \zeta) \in \tilde{V}$ and $\alpha \in [0, 1]$ be such that $v = \alpha u + (1-\alpha)w$, $\eta = \alpha\omega + (1-\alpha)\zeta$. Our goal is to prove that

$$\tilde{e}(v, \eta) \leq \alpha \tilde{e}(u, \omega) + (1-\alpha)\tilde{e}(w, \zeta). \tag{17.7.6}$$

As in the proof of the preceding proposition there exist $\boldsymbol{F}, \boldsymbol{G}$ such that $(\boldsymbol{F}, \eta), (\boldsymbol{G}, \zeta) \in \tilde{U}$, $u = \det \boldsymbol{F}$, $w = \det \boldsymbol{G}$ and $\boldsymbol{F} - \boldsymbol{G}$ is of rank at most 1. Using (17.7.3) we have

$$\begin{aligned}
\tilde{e}(v, \eta) &= \tilde{e}\big((\alpha \det \boldsymbol{F} + (1-\alpha)\det \boldsymbol{G}), \alpha\omega + (1-\alpha)\zeta\big) \\
&= \tilde{e}\big(\det(\alpha \boldsymbol{F} + (1-\alpha)\boldsymbol{G}), \alpha\omega + (1-\alpha)\zeta\big) \\
&= \tilde{e}\big(\alpha \boldsymbol{F} + (1-\alpha)\boldsymbol{G}, \alpha\omega + (1-\alpha)\zeta\big) \leq \alpha \tilde{e}(\boldsymbol{F}, \omega) + (1-\alpha)\tilde{e}(\boldsymbol{G}, \zeta) \\
&= \alpha \tilde{e}(u, \omega) + (1-\alpha)\tilde{e}(w, \zeta).
\end{aligned}$$

Hence $\tilde{e}$ is convex. Convexity of $\tilde{e}(v, \eta)$ ⇒ quasiconvexity: For this it suffices to note that if $\tilde{e}(v, \eta)$ is convex at (v, η), then $\tilde{e}(\boldsymbol{F}, \eta)$ is polyconvex at $(\boldsymbol{F}, \eta)$ (the inequality (17.5.2) even holds with $\boldsymbol{A} = \boldsymbol{B} = \boldsymbol{0}$) and to use Proposition 17.5.5. (2): Using (1.1.10), and denoting $\mu = v(\boldsymbol{F}^{-1}\boldsymbol{a} \cdot \boldsymbol{b})$, we find

$$d^2\tilde{e}[(\boldsymbol{a} \otimes \boldsymbol{b}, \omega), (\boldsymbol{a} \otimes \boldsymbol{b}, \omega)] = \partial_v^2 \tilde{e} \cdot \mu^2 + 2\partial_v \partial_\eta \tilde{e} \cdot \mu \cdot \omega + \partial_\eta^2 \tilde{e} \cdot \omega^2 \tag{17.7.7}$$

from which the result follows. □

From (17.7.7) one can also see that a fluid can never satisfy the strong ellipticity condition: there are always vectors $\boldsymbol{a} \neq \boldsymbol{0} \neq \boldsymbol{b}$, and a number ω for which $d^2\tilde{e}[(\boldsymbol{a} \otimes \boldsymbol{b}, \omega), (\boldsymbol{a} \otimes \boldsymbol{b}, \omega)] = 0$. It suffices to take $\omega = 0$ and $\boldsymbol{a}, \boldsymbol{b}$ such that $\boldsymbol{F}^{-1}\boldsymbol{a} \cdot \boldsymbol{b} = 0$.

17.7.3 Proposition *Let* $(\boldsymbol{F}, \theta) \in \bar{U}$, *and* $v = |\det \boldsymbol{F}|/\rho_o$. *Then*

(1) $\bar{f}(\cdot, \theta)$ *is quasiconvex at* $\boldsymbol{F}$ *if and only if* $\bar{f}(\cdot, \theta)$ *is rank 1 convex at* $\boldsymbol{F}$. *These two conditions hold if and only if* $\bar{f}(\cdot, \theta)$ *is convex at* v. *When these conditions are combined with the concavity of* $\bar{f}(v, \cdot)$ *at* θ, *they are equivalent to the conditions formulated in Proposition* 17.7.2(1);

(2) $\bar{f}(\cdot, \theta)$ *satisfies the isothermal Legendre–Hadamard condition at* $\boldsymbol{F}$ *if and only if* $\bar{f}$ *satisfies* $\partial_v^2 \bar{f} \geq 0$ *at* (v, θ). *When these conditions are combined with* $\partial_\theta^2 \bar{f} \leq 0$, *at* (v, θ), *then they are equivalent to the conditions formulated in Proposition* 17.7.2(2).

The proof is analogous to the proof of the previous proposition and therefore is omitted. Let us now turn to the quasiconvexity at the boundary.

17.7.4 Proposition *Let* $h : \tilde{V} \to \mathbb{R}$ *be a continuously differentiable function on a convex subset of* $(0, \infty) \times \mathbb{R}$ *and define* $g : \tilde{U} \to \mathbb{R}$ *by* $g(\boldsymbol{F}, \eta) = h(v, \eta)$ *on* $\tilde{U}$ *given by* (11.4.4). *Let* $(\boldsymbol{F}, \eta) \in \tilde{U}$ *with* $\det \boldsymbol{F} > 0$, $\boldsymbol{n} \in \mathrm{Sph}$ *be given and set* $v := \det \boldsymbol{F}/\rho_o$. *Then*

(1) *g is quasiconvex at the boundary with normal* $\boldsymbol{n}$ *at* $(\boldsymbol{F}, \eta)$ *if and only if* (v, η) *is a point of convexity of h and* $\partial_v h(v, \eta) = 0$;

(2) *if h is independent of* η, *then g is quasiconvex at the boundary with normal* $\boldsymbol{n}$ *at* $\boldsymbol{F}$ *if and only if* v *is a point of global minimum of h; if g is quasiconvex at the boundary with normal* $\boldsymbol{n}$ *at every* $\boldsymbol{F}$ *from its domain, then g is constant.*

In particular, the function $\boldsymbol{F} \mapsto \det \boldsymbol{F}$ is not quasiconvex at the boundary; nevertheless, if is quasiconvex.

Proof (1): Let g be quasiconvex at the boundary with normal $\boldsymbol{n}$ at $(\boldsymbol{F}, \eta)$. Then g is also quasiconvex at $(\boldsymbol{F}, \eta)$ and hence (v, η) is a point of convexity of h by Proposition 17.7.2. Also the function $\boldsymbol{G} \mapsto g(\boldsymbol{G}, \eta)$ (with η fixed) is quasiconvex at the boundary with normal $\boldsymbol{n}$ at $\boldsymbol{F}$. The proof will be completed using results of Sect. 20.4 and 20.6. Namely, if $\bar{\boldsymbol{A}} := \partial_{\boldsymbol{F}}^2 g(\boldsymbol{F})$, then $\bar{\boldsymbol{A}}$ is quasiconvex at the boundary with normal $\boldsymbol{n}$ in the sense of Definition 20.4.3(6). The form of $\bar{\boldsymbol{A}}$ is that described in Subsect. 11.4.3 with $\bar{p}(\cdot) := -\partial_v h(\cdot, \eta)$. We may set $\boldsymbol{F} = \mathbf{1}$ or, equivalently, to use $\bar{\boldsymbol{B}}$ rather than $\bar{\boldsymbol{A}}$. Assuming $\boldsymbol{F} = \mathbf{1}$, we see that $\bar{\boldsymbol{A}}$ is of the form (20.6.2) with $w = 1, \lambda = -\partial_v \bar{p}(v), \mu = 0$. Then $\bar{\boldsymbol{A}}$ is quasiconvex at the boundary with normal $\boldsymbol{n}$ if and only if the inequalities (2) in Proposition 20.6.5 are satisfied, from which $p = 0$, i.e., $\partial_v h(v, \eta) = 0$. This completes the proof of the direct implication. The converse follows from $h(v', \eta') \geq h(v, \eta) + \partial_\eta h(v, \eta)(\eta' - \eta)$ for every $(v', \eta') \in \tilde{V}$ by integrating. (2): This is a direct consequence of (1). □

Note that the h in the above proposition is not to be identified with any of the thermodynamic potentials, and therefore $\partial_v h(v, \eta) = 0$ *does not* imply that the pressure vanishes. As an illustration, consider a stable equilibrium state of a fluid under (possibly) spatially-dependent pressure loading under the unilateral contact environment (see 13.3.3 and 13.3.10) and define the function $g(\boldsymbol{F}, \eta, \boldsymbol{x}, \boldsymbol{p}) = \tilde{e}(\boldsymbol{F}, \eta) + \psi(\boldsymbol{F}, \boldsymbol{x}, \boldsymbol{p})$ where ψ is given by $\psi = \rho_o^{-1} \pi \det \boldsymbol{F}$. By Proposition 17.2.6, for every $\boldsymbol{p}$ on the free part of the boundary, $g(\cdot, \boldsymbol{\chi}(\boldsymbol{p}), \boldsymbol{p})$ must be quasiconvex at the boundary with normal $\boldsymbol{n}$ at $(\nabla \boldsymbol{\chi}(\boldsymbol{p}), \eta(\boldsymbol{p}))$ (and not $\tilde{e}$). Since $g(\cdot, \boldsymbol{\chi}(\boldsymbol{p}), \boldsymbol{p})$ has a representation as in Proposition 17.7.4 with $h(v, \eta) = \bar{e}(v, \eta) + \pi v$, by that proposition necessarily $\partial_v h(v, \eta) = 0$ which means $-\tilde{p}(v, \eta) + \pi = 0$, and the latter is a consequence of the balance of forces. Moreover, (v, η) is a point of convexity of h if and only if (v, η) is a point of convexity of $\tilde{e}$.

There are additional "noncanonical" constitutive inequalities of interest. The discussion will show that the convexity of $\tilde{e}$ does not place any restriction on the signs of the specific heat at constant pressure, on the signs of the latent heats and that it also does not exclude the anomalous behavior of fluids. Recall the expressions for the specific and latent heats:

$$c_v = -\theta \partial_\theta^2 \bar{f}, \quad l_v = \theta \partial_\theta \bar{p}, \tag{17.7.8}$$

$$c_p = c_v - l_v \frac{\partial_\theta \bar{p}}{\partial_v \bar{p}}, \quad l_p = \frac{l_v}{\partial_v \bar{p}}. \tag{17.7.9}$$

17.7.5 Proposition *Let* (v, θ) *be a point in* $\bar{U}$. *For a general fluid the following three inequalities are equivalent at* (v, θ):

$$\partial_\theta \bar{p} > 0; \quad l_v > 0; \quad \partial_\theta l_v - \partial_v c_v > 0. \tag{17.7.10}$$

If the energy $\tilde{e}$ *is strictly convex at* (v, θ), *then the above equivalent inequalities are equivalent to each of the following two inequalities:*

$$l_p < 0; \quad \frac{c_p - c_v}{l_p} < 0. \tag{17.7.11}$$

The ***anomalous behavior*** occurs at (v, θ) if the group of equivalent inequalities (17.7.10) fails. This phenomenon is perfectly compatible with the convexity of the energy and the other main thermostatic inequalities. (See TRUESDELL & BHARATHA [1977] for a detailed discussion of the anomalous behavior.) Note also that the inequalities related to $(17.7.10)_{2,3}$ occur in the definition of a thermometric fluid in Sect. 7.3.

Proof The equivalence of $(17.7.10)_1$ and $(17.7.10)_2$ is seen from $(17.7.8)_2$. A computation based on (17.7.8) shows that $\partial_\theta l_v - \partial_v c_v = \partial_\theta \bar{p}$ and this shows that $(17.7.10)_3$ is equivalent to $(17.7.10)_1$ and $(17.7.10)_2$. If the energy is strictly convex, then $\partial_v \bar{p} < 0$ and hence c_p and l_p are well-defined. By $(17.7.9)_2$ the latent heats l_v and l_p are of opposite sign, which proves the equivalence of $(17.7.11)_1$ with $(17.7.10)_2$. Finally, by $(17.7.9)_1$ we have

$$\frac{c_p - c_v}{l_p} = -\frac{\partial_\theta \bar{p}}{\partial_v \bar{p}}$$

which proves the equivalence of $(17.7.11)_2$ with $(17.7.10)_1$. □

There is one additional inequality for fluids: convexity of pressure with respect to volume at fixed entropy. This is significant in the discussion of shock waves in fluids. See Sect. 24.3.

17.8 Quasiconvexity and Crystals

Let us explain that the rank 1 convexity (and hence quasiconvexity) is in conflict with the assumption that the symmetry group G of the body is as large as the nonorthogonal symmetry group $G(L)$ of some Bravais lattice (see Subsect. 9.4.5). Let $n = 3$.

17.8.1 Proposition *For every $\boldsymbol{H}$ with* $\det \boldsymbol{H} = 1$ *there exists a proper orthogonal $\boldsymbol{R}$ and vectors $\boldsymbol{b}_i, \boldsymbol{c}_i \in$ Vect with $\boldsymbol{b}_i \cdot \boldsymbol{c}_i = 0$, $i = 1, 2$, such that*

$$\boldsymbol{H} = \boldsymbol{R}(\mathbf{1} + \boldsymbol{c}_1 \otimes \boldsymbol{b}_1)(\mathbf{1} + \boldsymbol{c}_2 \otimes \boldsymbol{b}_2). \tag{17.8.1}$$

Proof Let $\boldsymbol{H} = \boldsymbol{Q}\boldsymbol{U}$ be the polar decomposition of $\boldsymbol{H}$. By the spectral decomposition in the orthonormal basis of eigenvectors of $\boldsymbol{U}$,

$$\boldsymbol{U} = v_1 \boldsymbol{e}_1 \otimes \boldsymbol{e}_1 + v_2 \boldsymbol{e}_2 \otimes \boldsymbol{e}_2 + v_3 \boldsymbol{e}_3 \otimes \boldsymbol{e}_3 = \boldsymbol{U}_1 \boldsymbol{U}_2, \tag{17.8.2}$$

where

$$\boldsymbol{U}_1 = v_1 \boldsymbol{e}_1 \otimes \boldsymbol{e}_1 + \boldsymbol{e}_2 \otimes \boldsymbol{e}_2 + v_1^{-1} \boldsymbol{e}_3 \otimes \boldsymbol{e}_3, \quad \boldsymbol{U}_2 = \boldsymbol{e}_1 \otimes \boldsymbol{e}_1 + v_2 \boldsymbol{e}_2 \otimes \boldsymbol{e}_2 + v_2^{-1} \boldsymbol{e}_3 \otimes \boldsymbol{e}_3,$$

and we have used $v_1 v_2 v_3 = 1$. Since either $v_1 \geq 1$ or $v_1^{-1} \geq 1$, we see that the decreasingly ordered triple (u_1, u_2, u_3) of eigenvalues of $\boldsymbol{U}_1$ satisfies $u_2 = 1$. Hence by Proposition 2.4.3, $\boldsymbol{U}_1$ is the right stretch tensor of some rank 1 perturbation of $\mathbf{1}$, i.e., $\mathbf{1} + \bar{\boldsymbol{c}}_1 \otimes \bar{\boldsymbol{b}}_1 = \boldsymbol{R}_1 \boldsymbol{U}_1$ for some $\bar{\boldsymbol{c}}_1, \bar{\boldsymbol{b}}_1 \in$ Vect, $\boldsymbol{R}_1 \in$ Orth$^+$; moreover, since $\det \boldsymbol{U}_1 = 1$, $\bar{\boldsymbol{c}}_1 \cdot \bar{\boldsymbol{b}}_1 = 0$. Similarly, $\mathbf{1} + \boldsymbol{c}_2 \otimes \boldsymbol{b}_2 = \boldsymbol{R}_2 \boldsymbol{U}_2$. Inserting into (17.8.2) and rearranging the rotations, it is found that (17.8.1) holds with $\boldsymbol{R} = \boldsymbol{Q}\boldsymbol{R}_1^{\mathrm{T}}\boldsymbol{R}_2^{\mathrm{T}}$, $\boldsymbol{c}_1 = \boldsymbol{R}_1 \bar{\boldsymbol{c}}_1$, $\boldsymbol{b}_1 = \boldsymbol{R}_1 \bar{\boldsymbol{b}}_1$. □

Throughout the rest of the section, let L be a Bravais lattice with with basis $\boldsymbol{e}_i$ and L^* be a Bravais lattice with with basis $\boldsymbol{e}_i^*$; moreover, let $G(L)$ be the nonorthogonal symmetry group of L.

17.8.2 ***Remark*** Let $\boldsymbol{b}, \boldsymbol{c} \in$ Vect be of the form

$$\boldsymbol{b} = \sum_{i=1}^{3} b_i \boldsymbol{e}_i, \quad \boldsymbol{c} = \sum_{i=1}^{3} c_i \boldsymbol{e}_i^*, \tag{17.8.3}$$

where b_i, c_i are rationals. If $\boldsymbol{b}_0, \boldsymbol{c}_0 \in$ Vect satisfy $\boldsymbol{b}_0 \cdot \boldsymbol{c}_0 = 0$, then there exist sequences $\boldsymbol{b}_k, \boldsymbol{c}_k \in$ Vect of vectors with rational components as in (17.8.3) such that $\boldsymbol{b}_k \to \boldsymbol{b}_0$, $\boldsymbol{c}_k \to \boldsymbol{c}_0$ as $k \to \infty$, and $\boldsymbol{b}_k \cdot \boldsymbol{c}_k = 0$ for all k.

Consider a material with the free energy $\bar{f} : \bar{U} \to \mathbb{R}$ and the symmetry group G.

17.8.3 Proposition *Let $G = G(L)$, let θ be a fixed temperature and assume that $\mathrm{Lin}^+ \times \{\theta\} \subset \bar{U}$. If $\bar{f}(\cdot, \theta)$ is rank 1 convex at a given $\boldsymbol{F} \in \mathrm{Lin}^+$, then the Cauchy stress at $\boldsymbol{F}$ is of the form $\bar{\boldsymbol{T}}(\boldsymbol{F}, \theta) = -p\mathbf{1}$ where $p \in \mathbb{R}$.*

Hence unless the crystal is a fluid, its stored energy is not globally rank 1 convex. Thus the theories of crystals with $G = G(L)$ have many internally unstable states.

Proof Recall from Sect. 1.5 that if $\boldsymbol{H} := \mathbf{1} + \boldsymbol{c} \otimes \boldsymbol{b}$ where $\boldsymbol{c} \in L, \boldsymbol{b} \in L^*, \boldsymbol{c} \cdot \boldsymbol{b} = 0$ then $\boldsymbol{H} \in G(L)$ and $\boldsymbol{H}^k = \mathbf{1} + k\boldsymbol{c} \otimes \boldsymbol{b}$, $k \in \mathbb{Z}$. Using this, it is found that the function $\psi(\lambda) = \bar{f}(\boldsymbol{F}(\mathbf{1} + \lambda\boldsymbol{c} \otimes \boldsymbol{b}), \theta)$, $\lambda \in \mathbb{R}$, is periodic of period 1. Write (17.3.3) with $\boldsymbol{a} = \lambda\boldsymbol{F}\boldsymbol{c}$:

$$\psi(\lambda) \geq \psi(0) + \lambda\bar{\boldsymbol{S}}(\boldsymbol{F}, \theta) \cdot (\boldsymbol{F}\boldsymbol{c} \otimes \boldsymbol{b}).$$

Dividing by λ, letting $\lambda \to \infty$, and using that ψ is periodic and continuous and hence bounded, one obtains $\bar{\boldsymbol{S}}(\boldsymbol{F}, \theta) \cdot (\boldsymbol{F}\boldsymbol{c} \otimes \boldsymbol{b}) \leq 0$, which implies $\bar{\boldsymbol{T}}(\boldsymbol{F}, \theta)\boldsymbol{b} \cdot \boldsymbol{c} = 0$. This implies that this equality holds for all $\boldsymbol{b}, \boldsymbol{c} \in \mathrm{Vect}$ with $\boldsymbol{b} \cdot \boldsymbol{c} = 0$ and with rational components in the expressions (17.8.3); this in turn implies that this equality holds for all $\boldsymbol{b}, \boldsymbol{c} \in \mathrm{Vect}$ with $\boldsymbol{b} \cdot \boldsymbol{c} = 0$ by the density formulated in Subsect. 17.8.2, and this finally gives the asserted form of $\bar{\boldsymbol{T}}(\boldsymbol{F}, \theta)$. □

17.8.4 Proposition *Let $\bar{f} : \mathrm{Lin}^+ \to \mathbb{R}$ be a rank 1 convex function such that*

$$\bar{f}(\boldsymbol{R}\boldsymbol{F}) = \bar{f}(\boldsymbol{F}), \quad \bar{f}(\boldsymbol{F}\boldsymbol{H}) = \bar{f}(\boldsymbol{F}) \tag{17.8.4}$$

for every $\boldsymbol{F} \in \mathrm{Lin}^+$, $\boldsymbol{R} \in \mathrm{Orth}^+$, and every $\boldsymbol{H} \in G(L)$ satisfying $\det \boldsymbol{H} = 1$. Then there exists a convex function $m : (0, \infty) \to \mathbb{R}$ such that $\bar{f}(\boldsymbol{F}) = m(\det \boldsymbol{F})$ for every $\boldsymbol{F} \in \mathrm{Lin}^+$.

Proof Let $\boldsymbol{b}, \boldsymbol{c} \in \mathrm{Vect}$ be of the form (17.8.3) where b_i, c_i are rationals satisfying $b \cdot c = 0$. The components can be written $b_i = \bar{b}_i/m$, $c_i = \bar{c}_i/m$ where $\bar{b}_i, \bar{c}_i, m$ are integers and $\bar{b} \cdot \bar{c} = 0$. Set

$$\bar{\boldsymbol{b}} = \sum_{i=1}^{3} \bar{b}_i \boldsymbol{e}_i, \quad \bar{\boldsymbol{c}} = \sum_{i=1}^{3} \bar{c}_i \boldsymbol{e}_i^*, \quad \boldsymbol{H} = \mathbf{1} + \bar{\boldsymbol{c}} \otimes \bar{\boldsymbol{b}} \in G(L).$$

Since $\bar{f}$ is rank 1 convex, the function $\phi(\lambda) := \bar{f}(\boldsymbol{F}(\mathbf{1} + \lambda\boldsymbol{c} \otimes \boldsymbol{b}))$, $\lambda \in \mathbb{R}$, is convex. For $\lambda = pm^2$, where p is an integer, we have $\phi(pm^2) = \bar{f}(\boldsymbol{F}\boldsymbol{H}^p) = \bar{f}(\boldsymbol{F}) = \phi(0)$ and so ϕ is found to be a periodic convex function. Hence it is constant. In particular,

$$\bar{f}(\boldsymbol{F}) = \bar{f}(\boldsymbol{F}(\mathbf{1} + \boldsymbol{c} \otimes \boldsymbol{b})). \tag{17.8.5}$$

Since $\bar{f}$ is rank 1 convex, it is continuous (see MORREY [1966; Sect. 4.4] and DACOROGNA [1985]). The density asserted in Subsect. 17.8.2 then extends (17.8.5) to all $\boldsymbol{c}, \boldsymbol{b} \in \mathrm{Vect}$ with $\boldsymbol{c} \cdot \boldsymbol{b} = 0$. We shall prove that if $\boldsymbol{F}_1, \boldsymbol{F}_2 \in \mathrm{Lin}^+$ are such that $\det \boldsymbol{F}_1 = \det \boldsymbol{F}_2$ then $\bar{f}(\boldsymbol{F}_1) = \bar{f}(\boldsymbol{F}_2)$. Since $\det(\boldsymbol{F}_1\boldsymbol{F}_2^{-1}) = 1$, by Proposition 17.8.1 we have $\boldsymbol{F}_1\boldsymbol{F}_2^{-1} = \boldsymbol{R}(\mathbf{1} + \bar{\boldsymbol{c}}_1 \otimes \bar{\boldsymbol{b}}_1)(\mathbf{1} + \bar{\boldsymbol{c}}_2 \otimes \bar{\boldsymbol{b}}_2)$ for some $\boldsymbol{R} \in \mathrm{Orth}^+$, and $\bar{\boldsymbol{b}}_i, \bar{\boldsymbol{c}}_i \in \mathrm{Vect}$ such that $\bar{\boldsymbol{b}}_i \cdot \bar{\boldsymbol{c}}_i = 0$, $i = 1, 2$. Hence $\boldsymbol{F}_1 = \boldsymbol{R}\boldsymbol{F}_2(\mathbf{1} + \boldsymbol{c}_1 \otimes \boldsymbol{b}_1)(\mathbf{1} + \boldsymbol{c}_2 \otimes \boldsymbol{b}_2)$, where $\boldsymbol{c}_i = \boldsymbol{F}_2^{-1}\bar{\boldsymbol{c}}_i$, $\boldsymbol{b}_i = \boldsymbol{F}_2^{\mathrm{T}}\bar{\boldsymbol{b}}_i$ and thus

$$\bar{f}(\boldsymbol{F}_1) = \bar{f}(\boldsymbol{R}\boldsymbol{F}_2(\mathbf{1} + \boldsymbol{c}_1 \otimes \boldsymbol{b}_1)(\mathbf{1} + \boldsymbol{c}_2 \otimes \boldsymbol{b}_2)).$$

Applying $(17.8.4)_1$ and (17.8.5) twice, we obtain $\bar{f}(\boldsymbol{F}_1) = \bar{f}(\boldsymbol{F}_2)$. Hence $\bar{f}(\boldsymbol{F}) = m(\det \boldsymbol{F})$ for some function m, and applying Proposition 17.7.3 we see that m must be convex. □

Bibliographical Note

The rank 1 convexity is due to Corall (1937) and Graves (1938). Lemma 17.3.4 is stated in a more general form in BALL, CURRIE & OLVER [1981]. Morrey recognized the polyconvexity as a sufficient condition for quasiconvexity. However, the realization of the importance of polyconvexity in continuum mechanics is due to BALL [1977]. He showed that the polyconvexity enables one to prove the existence of solutions to the extremum problems for solids under realistic behavior of the free energy when $\det \boldsymbol{F} \to 0$. Proposition 17.4.10 is essentially by KNOWLES & STERNBERG [1978], see also BALL [1980]. The global Eshelby conservation law is due to GURTIN [1983]. Further information about coexistent phases in solids is found in JAMES [1986a,b].

In addition to the invariance of the strong ellipticity under the exchange of the roles of the referential and actual configurations discussed in Sect. 17.6, OLVER [1993] noticed that for general variational problems the strong ellipticity is also invariant under general point transformations acting on the independent and dependent variables simultaneously, like those discussed in Sect. 14.5. See also GIAQUINTA, MODICA, & SOUČEK [1994] for the invariance of polyconvexity under the exchange of the referential and actual configurations. Proposition 17.8.3 due to ERICKSEN [1981]; see also FONSECA [1987, 1988], and CHIPOT & KINDERLEHRER [1988]; the last reference also contains Proposition 17.8.1. Additional results on the strong ellipticity and Legendre–Hadamard conditions are found in ROSAKIS [1990] and CHEN [1991], and on the quasiconvexity in PARRY [1995]. Further discussion of quasiconvexity is found in Sects. 21.3–21.6, with more references.

18 Convexity Conditions for Isotropic Functions

The constitutive inequalities of the preceding chapter gain much more concrete forms for the thermodynamic potentials of isotropic solids; the goal of this chapter is to describe them. Sects. 18.1, 18.2, and 18.3 deal with the convexity of symmetric, isotropic, and objective-isotropic functions, respectively, and each section uses the results of the preceding sections. Using the results on objective-isotropic functions, Sect. 18.5 exhibits important special classes of polyconvex functions. The final section deals with the positivity properties of the second differential, the Legendre–Hadamard condition, Baker–Ericksen inequalities, and the inequalities of Coleman & Noll and Hill.

18.1 Symmetric Convex Functions and Sets

The main result of this section says that the convexity combined with the symmetry leads to additional inequalities (18.1.4) (beyond those coming from the convexity). Throughout the chapter the dimension dim Vect $= n$ is arbitrary unless stated otherwise, the indices $i, j, k, \ldots$ range from 1 to n, and any inequality containing unspecified indices is understood to hold for all values of the indices between 1 and n.

The reader is referred to Subsect. 8.1.4 for the definitions of a symmetric set and a symmetric function. Denote by $\mathbb{D}$ the convex set

$$\mathbb{D} = \{x = (x_1, x_2 \ldots, x_n) \in \mathbb{R}^n : x_1 \geq x_2 \geq \ldots \geq x_n\}$$

and recall the notation

$$\mathbb{R}^n_{++} = \{x = (x_1, \ldots, x_n) \in \mathbb{R}^n : x_i > 0, i = 1, \ldots, n\}.$$

The basic tool for investigating symmetric convex functions is the following elementary *rearrangement inequality.*

18.1.1 Proposition *If $x, y \in \mathbb{D}$ and P is a permutation matrix then*

$$\sum_{i=1}^{n} x_i y_{n-i+1} \leq x \cdot Py \leq \sum_{i=1}^{n} x_i y_i. \tag{18.1.1}$$

See HARDY, LITTLEWOOD & PÓLYA [1952, p. 261] for an intuitive interpretation. MIRSKY [1975] gives a generalization in which P is replaced by a doubly stochastic matrix.

Proof We shall need only $(18.1.1)_2$ and only that inequality will be proved. It is enough to prove the case $x_1 > x_2 > \ldots > x_n$, the general case is obtained as a limit. Let Q be a permutation for which $x \cdot Qy$ takes the maximum value among $x \cdot Py$. It is enough to prove that $w := Qy \in \mathbb{D}$ and hence $w = y$. Suppose that $w \notin \mathbb{D}$, so that $w_i < w_j$ for a pair of indices $i, j = i + 1$. Let $z = Py$ be the permutation of

w in which the ith and jth components of w are exchanged. Then $x \cdot w - x \cdot z = (x_i - x_j) \cdot (w_i - w_j) < 0$ by $x_i - x_j > 0$ and $w_i - w_j < 0$. This means $x \cdot Qy < x \cdot Py$, which is a contradiction. □

For every $x \in \mathbb{R}^n$, denote by $G(x)$ the group of all permutation matrices P satisfying $Px = x$. Clearly $G(x)$ is nontrivial only if some of the components of x coincide.

18.1.2 Proposition *If $x \in \partial D$ is a boundary point of a symmetric set $D \subset \mathbb{R}^n$ and $n \in \mathbb{R}^n$, then the following assertions hold:*

(1) *if n is an exterior normal to D at x then*

$$(n_i - n_j) \cdot (x_i - x_j) \geq 0; \tag{18.1.2}$$

moreover, if P is a permutation matrix, then Pn is an exterior normal to D at Px;

(2) *if D is open and the exterior normal to D at x exists, then it can be chosen to satisfy $Pn = n$ for every $P \in G(x)$; if $x \in \mathbb{D}$, then the normal with this property is in $\mathbb{D}$.*

Inequality (18.1.2) is the basic constraint imposed on the normal by the symmetry of D. Analogous conditions for symmetric convex functions are (18.1.3) and (18.1.4). By (2), it is always possible to find a normal that has the same symmetry as x. Observe also that by (18.1.2), every normal at x with $x_1 > x_2 > \ldots > x_n$ is in $\mathbb{D}$.

Proof (1): Let $i \neq j$ and let $y = (y_1, \ldots, y_n)$ be the permutation of x in which the ith and jth components of x are exchanged. Then $y \in D$ and $n \cdot (y - x) \leq 0$ gives (18.1.2). The assertion about P is immediate. (2): Given an exterior normal n_0 to D at x, the normal n having the symmetry required in (2) is constructed as a mean of Pn_0 over $P \in G := G(x)$ i.e., $n := w/|w|$ where $w := \sum_{P \in G} P^{-1} n_0$ (it will be proved that $w \neq 0$). This construction automatically ensures the symmetry and that n is a normal, as is easy to check. To prove $w \neq 0$, take any interior point y of D. For any permutation matrix P also Py is an interior point of D and hence $n_0 \cdot (Py - x) < 0$. But then $w \cdot (y - x) = \sum_{P \in G} n_0 \cdot (Py - x) < 0$ and hence $w \neq 0$. If $x \in \mathbb{D}$, then the normal n automatically satisfies $(n_i - n_j) \cdot (x_i - x_j) \geq 0$ for every i, j. If i, j are such that $x_i > x_j$, then this implies $n_i \geq n_j$. In the case $x_i = x_j$ this argument does not work. However, then the matrix P which permutes i, j satisfies $Px = x$ and so $Pn = n$. But this gives $n_i = n_j$. Hence for every $i < j$ we have $n_i \geq n_j$, proving $n \in \mathbb{D}$. □

18.1.3 Proposition *An open symmetric set $D \subset \mathbb{R}^n$ is convex if and only if $D \cap \mathbb{D}$ is convex and at any $x \in \mathbb{D} \cap \partial D$ there exists an exterior normal $n \in \mathbb{D}$.*

This reduces the checking of the convexity of D to checking the convexity of the part contained in $\mathbb{D}$ (the essential part) plus checking that the essential part has normals in $\mathbb{D}$. A similar proposition for symmetric functions is below.

Proof The direct implication follows from Proposition 18.1.2(2). The converse implication: Since D is open, it suffices to prove that every boundary point has an exterior normal (Proposition 16.1.3). Let $x \in \partial D$. Then $x = Py$ where $y \in (\partial D) \cap \mathbb{D}$ and P is a suitable permutation matrix. By hypothesis there exists an exterior normal $n(y)$ to $D \cap \mathbb{D}$ at y such that $n(y) \in \mathbb{D}$. Set $n(x) = Pn(y)$. It is claimed that $n(x)$ is an exterior normal to D at x. Let $z \in D$ and let z' be its permutation in $\mathbb{D}$. Then $(z - x) \cdot n(x) = (P^{\mathrm{T}} z - y) \cdot n(y)$. By Proposition 18.1.1, $P^{\mathrm{T}} z \cdot n(y) \leq z' \cdot n(y)$ and hence $(z - x) \cdot n(x) \leq (z' - y) \cdot n(y) \leq 0$ since $z' \in D \cap \mathbb{D}$ and $n(y)$ is an exterior normal to $D \cap \mathbb{D}$ at y. □

Recall that if $f : D \to \mathbb{R}$ is a symmetric C^1 function, then $\nabla f(Px) = P\nabla f(x)$ for any $x \in D$ and any permutation matrix P. Denote $f_{,i} = \partial_{x_i} f$.

18.1.4 Proposition *If $f : D \to \mathbb{R}$ is a symmetric function on an open set, $x \in D$, and $g \in \mathbb{R}^n$, then:*

(1) *any subgradient g of f at x satisfies*

$$(g_i - g_j) \cdot (x_i - x_j) \geq 0; \tag{18.1.3}$$

moreover, if P is a permutation matrix, then Pg is a subgradient of f at Px;

(2) *if a subgradient of f at x exists then it can be chosen to satisfy $Pg = g$ for every $P \in G(x)$; if $x \in D \cap \mathbb{D}$, then the subgradient with this property is in $\mathbb{D}$.*

For f of class C^1 (18.1.3) reduces to

$$(f_{,i}(x) - f_{,j}(x)) \cdot (x_i - x_j) \geq 0. \tag{18.1.4}$$

If $x \in D \cap \mathbb{D}$ and x is a point of convexity, then $\nabla f(x) \in \mathbb{D}$.

The proof is completely analogous to that of Proposition 18.1.2. Inequalities (18.1.3) and (18.1.4) are the restrictions imposed on the subgradients by the symmetry. The same inequalities figure as a basic condition in a related theory of Schur convex functions, see MARSHALL & OLKIN [1979, Chap. 3]. The symmetric function f will eventually be the stored energy of an isotropic material expressed in principal stretches; its derivatives are the principal forces s_i (eigenvalues of the Piola–Kirchhoff stress, see Subsect. 11.1.1); (18.1.4) are called ordered-forces (O-F) inequalities in this context (see TRUESDELL & NOLL [1965, Sect. 51]). The forces are ordered in the same algebraical sense as principal stretches: if $v_1 \geq v_2 \geq \ldots \geq v_n$, then $s_1 \geq s_2 \geq \ldots \geq s_n$. This cannot hold generally, as shown by HILL [1970a,b] and OGDEN [1970a, 1972a,b]. The main case against O-F is that they exclude nearly incompressible materials. Hence also the convexity in the principal stretches cannot hold generally. On the other hand, some of the stored energies in current use, such as the Blatz–Ko special I (Subsect. 11.3.3(3)), are convex functions of the principal stretches and hence satisfy the O-F inequalities, see Example 18.1.6(4). See also the discussion at the end of Sect. 18.6.

18.1.5 Proposition *Let $f : D \to \mathbb{R}$ be a symmetric function on an open set, let ϕ be its restriction to $D \cap \mathbb{D}$ and $x \in D \cap \mathbb{D}$. Then x is a point of convexity of f if and only if x is a point of convexity of ϕ and at least one subgradient of ϕ at x is in $\mathbb{D}$. A continuously differentiable f is convex if and only if ϕ is convex and $\nabla\phi(x) \in \mathbb{D}$ for every $x \in D \cap \mathbb{D}$.*

Proof The direct implication follows from Proposition 18.1.4(2). The converse implication: Let $y \in D$ and let y' be that permutation of y which is in $D \cap \mathbb{D}$. Then $f(y) = \phi(y')$. By Proposition 18.1.1 and the hypothesis that $g \in \mathbb{D}$ then $g \cdot y \leq g \cdot y'$. Hence $f(y) - f(x) - g \cdot (y - x) \geq \phi(y') - \phi(x) - g \cdot (y' - x) \geq 0$, with the last step justified by the convexity of ϕ at x. □

18.1.6 Examples

(1) Let $m_i(x)$ be the ith largest component of $x \in \mathbb{R}^n$. That is, we define the function $m \equiv (m_1, \ldots, m_n) : \mathbb{R}^n \to \mathbb{R}^n$ by $m(x) = y$ where $y \in \mathbb{D}$ is the unique decreasing *rearrangement* (= permutation) of x (cf. MARSHALL & OLKIN [1979, p. xx]). Each m_i is symmetric and if $x \in \mathbb{D}$, then $m_i(x) = x_i$; the restrictions of m_i to $\mathbb{D}$ are linear. We shall see that m_1 is convex and m_n is concave. These assertions are consequences of a more general assertion:

(2) *If $\alpha \in \mathbb{R}^n$ then the function $f := \alpha \cdot m$ is convex on $\mathbb{R}^n$ if and only if $\alpha \in \mathbb{D}$. Also the restriction of f to $\mathbb{R}^n_{++}$ is convex if and only if $\alpha \in \mathbb{D}$. In particular, m_1 is convex and m_n concave.*

Proof of (2) We check the conditions in Proposition 18.1.5. The restriction of f to $\mathbb{D}$ is linear and hence convex. Therefore the question reduces to examining whether the restriction ϕ has a subgradient $g \in \mathbb{D}$. At the interior points x of $\mathbb{D}$, the function ϕ is differentiable and $\nabla\phi(x) = \alpha$. Hence α is a subgradient of ϕ at every $x \in \mathbb{D}$, and it is the only subgradient if x is in the interior of $\mathbb{D}$. Hence f convex implies $\alpha \in \mathbb{D}$ and conversely. The choice $\alpha = (1, 0, \ldots, 0) \in \mathbb{D}$ gives $f = m_1$ and $\alpha = (0, \ldots, 0, -1) \in \mathbb{D}$ gives $f = -m_n$ and hence $-m_n$ is convex, i.e., m_n is concave. □

(3) Other choices of interest are $\alpha = (1, \ldots, 1, 0, \ldots, 0) \in \mathbb{D}$ (with the first k entries equal to 1); this provides the convexity of the functions $m_1, m_1 + m_2, \ldots, m_1 + \cdots + m_n$.

(4) *The function $f : \mathbb{R}^n_{++} \to \mathbb{R}$ given by*

$$f(v) = (v_1^\alpha + \ldots + v_n^\alpha)/\alpha + (v_1 \cdots v_n)^{-\gamma}/\gamma,$$

where $\alpha \geq 1, \gamma > 0$ are constants, is symmetric and convex. If additionally $\alpha > 1$, then f is strictly convex. This is the function representing the generalized Blatz–Ko special I, see Subsect. 11.3.3(4), which for $\alpha = 2$ reduces to Blatz–Ko special I.
Proof of (4) Denoting $\Delta = v_1 \cdots v_n$, we find

$$f_{,ij}(v) = ((\alpha - 1)v_i^{\alpha-2} + \Delta^{-\gamma} v_i^{-2})\delta_{ij} + \gamma \Delta^{-\gamma} v_i^{-1} v_j^{-1}.$$

Hence, if $x \in \mathbb{R}^n$, then $s := \sum_{i,j=1}^n f_{,ij}(v) x_i x_j = \sum_{i=1}^n A_i x_i^2 + \gamma \Delta^{-\gamma} \xi^2$, with $A_i = (\alpha - 1) v_i^{\alpha-2} + \Delta^{-\gamma} v_i^{-2} \geq 0$ and $\xi = \sum_{i=1}^n v_i^{-1} x_i$. Hence $s \geq 0$, $x \in \mathbb{R}^n$ with the inequality strict if $\alpha > 1$ and $x \neq 0$. □

18.2 Isotropic Convex Functions and Sets

Recall from Subsect. 8.1.4 and 8.2.1 the one-to-one correspondences between the isotropic functions on Sym and symmetric functions on $\mathbb{R}^n$ and a similar correspondence for sets. The main results of this section, Proposition 18.2.4, says that this correspondence perfectly preserves the convexity properties. Let dim Vect $= n$. The following proposition plays the same role for convex functions of symmetric tensors as the rearrangement inequality Proposition 18.1.1 for convex symmetric functions.

18.2.1 Proposition *Let $\boldsymbol{A}, \boldsymbol{B}$ be in* Sym *and $a \in \mathbb{D}$, $b \in \mathbb{D}$ their ordered n-tuples of eigenvalues. Then*

$$\sum_{i=1}^n a_i b_{n-i+1} \leq \boldsymbol{A} \cdot \boldsymbol{B} \leq \sum_{i=1}^n a_i b_i. \tag{18.2.1}$$

This inequality is due to Richter (1958), see MARSHALL & OLKIN [1979, p. 248]. In the continuum mechanics literature, it appears to be first used by HILL [1968, 1970a, p. 466] to establish the result Proposition 18.2.4 on convex functions of a symmetric tensor argument. (The present author could not find an analogue of Proposition 18.2.4 in the matrix theory literature.)

Proof We shall need only $(18.2.1)_2$ and only that inequality will be proved. It is enough to consider the case $a_1 > \ldots > a_n$, $b_1 > \ldots > b_n$, the general case is obtained by a limit. Let

$$\boldsymbol{A} = \sum_{i=1}^n a_i \boldsymbol{e}_i \otimes \boldsymbol{e}_i, \quad a = (a_1, \ldots, a_n), \tag{18.2.2}$$

be the spectral decomposition of $\boldsymbol{A}$ and define $\boldsymbol{C} := \sum_{i=1}^n b_i \boldsymbol{e}_i \otimes \boldsymbol{e}_i$. We have $\boldsymbol{B} = \boldsymbol{Q}\boldsymbol{C}\boldsymbol{Q}^{\mathrm{T}}$ for some $\boldsymbol{Q} \in \mathrm{Orth}^+$ and $\boldsymbol{A} \cdot \boldsymbol{C} = a \cdot b$. It is therefore enough to prove that $\boldsymbol{A} \cdot (\boldsymbol{Q}\boldsymbol{C}\boldsymbol{Q}^{\mathrm{T}}) \leq a \cdot b$ for every $\boldsymbol{Q} \in \mathrm{Orth}^+$. The continuous function $\boldsymbol{Q} \mapsto \boldsymbol{A} \cdot (\boldsymbol{Q}\boldsymbol{C}\boldsymbol{Q}^{\mathrm{T}})$ has a maximum at some point $\boldsymbol{Q} \in \mathrm{Orth}^+$ of the compact set Orth^+. Denoting $\boldsymbol{D} = \boldsymbol{Q}\boldsymbol{C}\boldsymbol{Q}^{\mathrm{T}}$, the function $t \mapsto \boldsymbol{A} \cdot (e^{\boldsymbol{W}t} \boldsymbol{D} e^{-\boldsymbol{W}t})$ is found to have a maximum at $t = 0$ for every $\boldsymbol{W} \in$ Skew. Equating the derivative at $t = 0$ to 0 gives $\boldsymbol{A} \cdot (\boldsymbol{W}\boldsymbol{D} - \boldsymbol{D}\boldsymbol{W}) = (\boldsymbol{A}\boldsymbol{D} - \boldsymbol{D}\boldsymbol{A}) \cdot \boldsymbol{W} = 0$ for every $\boldsymbol{W} \in$ Skew. Hence $\boldsymbol{A}\boldsymbol{D} - \boldsymbol{D}\boldsymbol{A}$ is a symmetric tensor, from which one finds that $\boldsymbol{A}, \boldsymbol{D}$ commute. The spectrum of $\boldsymbol{D}$ is the same as that of $\boldsymbol{B}$ and hence $\boldsymbol{D} = \sum_{i=1}^n d_i \boldsymbol{e}_i \otimes \boldsymbol{e}_i$ where $d = (d_1, \ldots, d_n)$ is a permutation of b, $d = Pb$. Then $\boldsymbol{A} \cdot \boldsymbol{D} = a \cdot Pb \leq a \cdot b$ by Proposition 18.1.1. □

18.2.2 Proposition *Let $U \subset$ Sym, $D \subset \mathbb{R}^n$ be as in Subsect.* 8.1.4, *with U isotropic and D symmetric. Let $\boldsymbol{A} \in$ Sym be a tensor with the n-tuple of eigenvalues $a \in \mathbb{R}^n$ arranged in any order. Then $\boldsymbol{A} \in \partial U$ if and only if $a \in \partial D$; moreover any exterior normal $\boldsymbol{N} \in$ Sym to U at $\boldsymbol{A} \in \partial U$ commutes with $\boldsymbol{A}$.*

Note. Let $\boldsymbol{A}$ have the spectral decomposition (18.2.2). It does not follow that any exterior normal $\boldsymbol{N}$ at $\boldsymbol{A}$ admits a spectral decomposition

$$\boldsymbol{N} = \sum_{i=1}^n n_i \boldsymbol{e}_i \otimes \boldsymbol{e}_i; \tag{18.2.3}$$

this is true only if the eigenvalues of A are distinct; in general, one can only find a basis in which both A and N have simultaneous spectral decompositions. However, we shall prove below that if U is open, then one can always *choose* an exterior normal which admits a spectral decomposition (18.2.3).

Proof The assertion about the boundary is immediate. The commutativity: Since U is isotropic and $A \in \partial U$, we have $e^{Wt}Ae^{-Wt} \in \mathrm{cl}\, U$ for every $W \in$ Skew, $t \in \mathbb{R}$ and hence $N \cdot (e^{Wt}Ae^{-Wt} - A) \le 0$ for every t. The derivative of $t \mapsto N \cdot (e^{Wt}Ae^{-Wt} - A)$ at $t = 0$ vanishes. This gives, as in Proposition 18.2.1, that A, N commute. □

For any $A \in$ Sym denote by $G(A)$ the group of all $Q \in$ Orth such that $QAQ^{\mathsf{T}} = A$. This is the symmetry group of A. If the eigenvalues of A are distinct and $QG(A)$, then Lemma 8.1.3(2) gives that Q is symmetric and hence $Q^2 = \mathbf{1}$ by the orthogonality. Then all the eigenvectors of A are also the eigenvectors of Q and the eigenvalues of Q are ± 1. Hence in the basis of the eigenvectors of A all $Q \in G(A)$ are represented by diagonal matrices with diagonal entries ± 1. In this sense $G(A)$ is trivial if all eigenvalues of A are distinct. Only when some of the eigenvalues of A coincide does $G(A)$ contain nontrivial rotations.

18.2.3 Proposition *Let $U \subset$ Sym, $D \subset \mathbb{R}^n$ be as in Subsect.* 8.1.4, *with U isotropic and open and D symmetric and open. Let $A \in \partial U$ be a tensor with the spectral decomposition* (18.2.2). *Then the following assertions hold:*

(1) *there exists an exterior normal to U at A if and only if there exists an exterior normal to D at $a = (a_1, \ldots, a_n)$;*

(2) *the exterior normal N to U at A, if it exists, can be chosen as to satisfy $QNQ^{\mathsf{T}} = N$ for every $Q \in G(A)$; the latter condition on N is satisfied if and only if N is of the form* (18.2.3) *where $n = (n_1, \ldots, n_n)$ is an exterior normal to D at a satisfying $Pn = n$ for every $P \in G(a)$.*

Proof (1): Let N' be an exterior normal to U at A; since these tensors commute, they admit a simultaneous spectral decomposition of the form

$$A = \sum_{i=1}^{n} \bar{a}_i f_i \otimes f_i, \quad N' = \sum_{i=1}^{n} \bar{n}_i f_i \otimes f_i$$

where $\{f_i, i = 1, \ldots, n\}$ is an orthonormal basis in Vect. If $b = (b_1, \ldots, b_n) \in D$ and $B := \sum_{i=1}^n b_i f_i \otimes f_i$, then the inequality $(B - A) \cdot N' \le 0$ gives that $\bar{n} = (\bar{n}_1, \ldots, \bar{n}_n)$ is a normal to D at $\bar{a}$. Necessarily $\bar{a} := (\bar{a}_1, \ldots, \bar{a}_n)$ is some permutation of a, i.e., $a = P\bar{a}$ for some permutation matrix P. But then $n := P\bar{n}$ is an exterior normal to D at a by the symmetry of D. Conversely, let D have an exterior normal at a. Since by Proposition 18.1.2(1) D has an exterior normal at Pa for any permutation matrix P, without any loss of generality we can assume that $a \in \mathbb{D}$. Then by Proposition 18.1.2(2) there is an exterior normal n to D at a such that $n \in \mathbb{D}$ and $Pn = n$ for every $P \in G(a)$. Let us define N by (18.2.3) and show that N is an exterior normal to U at A. Let $B \in U$ and $b \in \mathbb{D}$ its ordered spectrum. Then $N \cdot B \le n \cdot b$ by Proposition 18.2.1 and hence $N \cdot (B - A) \le n \cdot (b - a) \le 0$. (2): The normal N constructed in the proof of (1) has the property stated in (2). If N is an exterior normal to U at A that satisfies the condition formulated in (2), then by this condition and the fact that A, N commute, N admits a spectral decomposition (18.2.3) and moreover $Pn = n$ for every P such that $Pa = a$ and conversely. □

18.2.4 Proposition

(1) *Let $U \subset$ Sym, $D \subset \mathbb{R}^n$ be as in Subsect.* 8.1.4, *with U isotropic and D symmetric. Then U is open and convex if and only if D is open and convex.*

(2) *Let $h : U \to \mathbb{R}$ be an isotropic function defined on an open convex subset of* Sym *and let $f : D \to \mathbb{R}$ be as in the representation theorem for isotropic functions (Proposition* 8.2.1). *Then h is convex if and only if f is convex.*

Proof (1): That the convexity of U implies the convexity of D is immediate. To prove the converse, by Proposition 16.1.3 it suffices to prove that every $\boldsymbol{A} \in \partial U$ has an exterior normal. But this follows from Proposition 18.2.3(1). The proof of (2) is similar and therefore is left to the reader. □

Item (2) of the above proposition is the main result of this section. In fact we have the following more detailed proposition.

18.2.5 Proposition *Let h, f be as in Proposition* 18.2.4(2) *and $\boldsymbol{A} \in U$ be a point with the spectral decomposition* (18.2.2) *where $a = (a_1, \ldots, a_n) \in U$. Then*

(1) *$\boldsymbol{A}$ is a point of convexity of h if and only if a is a point of convexity of f;*

(2) *any subgradient of h at $\boldsymbol{A}$ commutes with $\boldsymbol{A}$;*

(3) *if $\boldsymbol{A}$ is a point of convexity of h, then there exists a subgradient $\boldsymbol{G} \in$ Sym of h such that $\boldsymbol{Q}\boldsymbol{G}\boldsymbol{Q}^{\mathsf{T}} = \boldsymbol{G}$ for every $\boldsymbol{Q} \in \mathsf{G}(\boldsymbol{A})$; the latter condition on $\boldsymbol{G}$ holds if and only if*

$$\boldsymbol{G} = \sum_{i=1}^{n} g_i \boldsymbol{e}_i \otimes \boldsymbol{e}_i \tag{18.2.4}$$

where $g := (g_1, \ldots, g_n)$ is a subgradient of f at a satisfying $Pg = g$ for every $P \in \mathsf{G}(a)$.

Proof This is completely analogous to the proof of Proposition 18.2.2. □

18.2.6 Examples

(1) Let $\hat{a}_i(\boldsymbol{A})$ be the ith largest eigenvalue of $\boldsymbol{A} \in$ Sym, see Subsect. 1.2.1. We shall see that $\hat{a}_1$ is convex. This is a consequences of a more general assertion.

(2) *If $\alpha = (\alpha_1, \ldots, \alpha_n) \in \mathbb{R}^n$, then the function $h := \sum_{i=1}^n \alpha_i \hat{a}_i$ is convex on* Sym *if and only if $\alpha \in \mathbb{D}$. In particular, $\hat{a}_1$ is convex and $\hat{a}_n$ concave.*

Proof of (2) The function h is represented by the symmetric function f of Example 18.1.6(2). By Proposition 18.2.4(2), h is convex if and only if f is convex, and the latter is convex if and only if $\alpha \in \mathbb{D}$ by Assertion (2) in Example 18.1.6. □

Warning: We also have $h(\boldsymbol{A}) = g(\hat{a}_1(\boldsymbol{A}), \ldots, \hat{a}_n(\boldsymbol{A}))$ where $g(x) = \alpha \cdot x$, $x \in \mathbb{R}^n$. This function is linear and hence convex for every $\alpha \in \mathbb{R}^n$, proving seemingly that h is convex for every $\alpha \in \mathbb{R}^n$. This is a false proof, because Proposition 18.2.4(2) does not apply: the function g is not symmetric.

(3) Let us consider dim Vect = 2 and identify Sym with the space of all symmetric 2×2 matrices $A = [A_{ij}]$, $i, j = 1, 2$, $A_{12} = A_{21}$. Then

$$\hat{a}_1(A) = r + s,\quad \hat{a}_2(A) = r - s,\quad r = \tfrac{1}{2}(A_{11} + A_{22}),\quad s = \sqrt{\tfrac{1}{4}(A_{11} - A_{22})^2 + A_{12}^2}.$$

$a_1(A)$ is a convex function and $a_2(A)$ a concave function of $(A_{11}, A_{22}, A_{12}) \in \mathbb{R}^3$.

(4) The functions $\hat{a}_1, \hat{a}_1 + \hat{a}_2, \ldots, \hat{a}_1 + \cdots + \hat{a}_n$ are convex on Sym.

18.2.7 Maximum Tangential Stress Let us interpret $\boldsymbol{A} \in$ Sym as a stress and define, for every unit vector $\boldsymbol{n} \in$ Vect, the tangential stress to be the projection $\boldsymbol{A}\boldsymbol{n} - (\boldsymbol{A}\boldsymbol{n} \cdot \boldsymbol{n})\boldsymbol{n}$ of $\boldsymbol{A}\boldsymbol{n}$ onto the plane perpendicular to $\boldsymbol{n}$. Its magnitude is $L = |\boldsymbol{A}\boldsymbol{n} - (\boldsymbol{A}\boldsymbol{n} \cdot \boldsymbol{n})\boldsymbol{n}|$. The maximal tangential stress T_m is defined by

$$T_m(\boldsymbol{A}) = \max\{|\boldsymbol{A}\boldsymbol{n} - (\boldsymbol{A}\boldsymbol{n} \cdot \boldsymbol{n})\boldsymbol{n}| : \boldsymbol{n} \in \text{Vect}, |\boldsymbol{n}| = 1\}. \tag{18.2.5}$$

Let us show that

$$T_m(\boldsymbol{A}) = \tfrac{1}{2}(\hat{a}_1(\boldsymbol{A}) - \hat{a}_n(\boldsymbol{A})) = \tfrac{1}{2}\max\{|a_i - a_j| : 1 \le i, j \le n\}, \tag{18.2.6}$$

where $a \in \mathbb{R}^n$ are (not necessarily ordered) eigenvalues of $\boldsymbol{A}$. Equation $(18.2.6)_1$ shows that T_m : Sym$\to\mathbb{R}$ is convex (see Example 18.2.6(2)). To prove $(18.2.6)_1$, notice that this equality is invariant under the changes $\boldsymbol{A} \to \boldsymbol{A} + \alpha\boldsymbol{1}$ where $\alpha \in \mathbb{R}$. It therefore suffices to prove it for $\boldsymbol{A} \in$ Sym which satisfy additionally $\hat{a}_n(\boldsymbol{A}) = 0$. Then $a_i := \hat{a}_i(\boldsymbol{A}) \ge 0$, $i = 1, \ldots, n-1$. As $L^2 = |\boldsymbol{A}\boldsymbol{n}|^2 - |\boldsymbol{A}\boldsymbol{n} \cdot \boldsymbol{n}|^2$ and

$$|\boldsymbol{A}\boldsymbol{n}|^2 \le a_1 a_1 n_1^2 + \ldots + a_1 a_{n-1} n_{n-1}^2 = a_1(\boldsymbol{A}\boldsymbol{n} \cdot \boldsymbol{n}),$$

we have $L^2 \le f(t)$, where $f(t) := t(a_1 - t)$ and $t = (\boldsymbol{A}\boldsymbol{n} \cdot \boldsymbol{n})$. The function f has a maximum $f(a_1/2) = a_1^2/4$ and hence $L^2 \le a_1^2/4$. On the other hand, the inequality holds as the equality if $\boldsymbol{n} = (\boldsymbol{e}_1 \otimes \boldsymbol{e}_1 +$

$\boldsymbol{e}_n \otimes \boldsymbol{e}_n)/\sqrt{2}$, where $\boldsymbol{e}_i$ are the eigenvectors of $\boldsymbol{A}$. This gives $T_m(\boldsymbol{A}) = a_1/2$ for all $\boldsymbol{A}$ with $\hat{a}_n(\boldsymbol{A}) = 0$ and $(18.2.6)_1$ for a general $\boldsymbol{A}$. The equality $(18.2.6)_2$ is obvious. (Another way of seeing that T_m is convex without $(18.2.6)_1$ is to note that by (18.2.5), T_m is a maximum over a family of convex functions.)

18.2.8 Yield Criteria in Plasticity Convex sets arise naturally in the rate-independent theories of plasticity (see LUCCHESI & PODIO-GUIDUGLI [1988, 1990], LUCCHESI, OWEN & PODIO-GUIDUGLI [1992], MAUGIN [1992]). Roughly, the material behaves elastically as long as the Kirchhoff stress $\boldsymbol{A} := |\det \boldsymbol{F}|\boldsymbol{T}$ does not leave a certain set $U \subset \text{Sym}$. The plastic deformation (yield) can occur only if the Kirchhoff stress $\boldsymbol{A}$ is on the boundary of U. (See the cited works for a precise formulation). From certain stability considerations, U must be convex. Two main choices of the form of U correspond to Tresca's and von Mises's yield criteria. ***Tresca's criterion:*** $U := \{\boldsymbol{A} \in \text{Sym} : T_m(\boldsymbol{A}) < \tau\}$ where $\tau > 0$ is given. Since T_m is convex by the preceding example, U is convex. In three dimensions, U is represented in the space of principal stresses $\mathbb{R}^3$ by the hexagonal prism with axis in the direction $r = (1, 1, 1)/\sqrt{3}$ and basis a regular hexagon in the plane perpendicular to r and with sides parallel to the axes. This follows from $(18.2.6)_2$. ***Von Mises' criterion:*** $U := \{\boldsymbol{A} \in \text{Sym} : |\boldsymbol{A}^{\mathrm{D}}| < \sqrt{2}\tau\}$ where $\tau > 0$ is given, and $\boldsymbol{A}^{\mathrm{D}} = \boldsymbol{A} - (\operatorname{tr}\boldsymbol{A})\mathbf{1}/3$. The function $\boldsymbol{A} \mapsto |\boldsymbol{A}^{\mathrm{D}}|$ is convex, since it is a composition of the norm $|\cdot|$ with a linear function. We have $|\boldsymbol{A}^{\mathrm{D}}|^2 = a_1^2 + a_2^2 + a_3^2 - (a_1 + a_2 + a_3)^2/3$. The latter is the symmetric function occurring in the representation theorem. U is represented in the space of principal stresses by a cylinder $D \subset \mathbb{R}^3$ of circular basis and axis $r = (1, 1, 1)/\sqrt{3}$.

18.3 Objective-Isotropic Convex Functions

Although realistic thermodynamic potentials cannot be globally convex functions of the deformation gradient (see Sect. 17.5.1), they still can have a large supply of points of convexity. This section describes the properties of points of convexity. The results on convex functions to be derived below will be also employed to construct isotropic polyconvex functions. In contrast to the convex functions h of a symmetric-tensor argument, for functions h of a general tensor argument $\boldsymbol{F} \in \text{Lin}$, the convexity of h implies *severe* restrictions on f, see Proposition 18.3.4. This proposition also shows that it makes a difference whether one is interested in the convexity of a function defined on InvLin or in the convexity of its restriction to Lin^+. This is related to the difference between Orth and Orth^+, as Proposition 18.3.2 fully explains.

18.3.1 Lemma *Let* $\boldsymbol{S} \in \text{Sym}$, $\boldsymbol{V} \in \text{Sym}^+$ *be given and let both* $\boldsymbol{V}$ *and* $\boldsymbol{S}^2$ *have distinct eigenvalues. Suppose that the function* $(\boldsymbol{Q}', \boldsymbol{R}') \mapsto \boldsymbol{S} \cdot (\boldsymbol{Q}'\boldsymbol{V}\boldsymbol{R}'^{\mathrm{T}})$, $\boldsymbol{Q}', \boldsymbol{R}' \in \text{Orth}$ *has a local extremum at* $(\boldsymbol{Q}, \boldsymbol{R})$. *Then there exists a symmetric orthogonal tensor* $\boldsymbol{J}$ *such that* $\boldsymbol{Q} = \boldsymbol{R}\boldsymbol{J}$, $\boldsymbol{J}^2 = \mathbf{1}$, *and the tensors* $\boldsymbol{V}, \boldsymbol{J}, \bar{\boldsymbol{S}} := \boldsymbol{R}^{\mathrm{T}}\boldsymbol{S}\boldsymbol{R}$ *commute.*

Proof Let $\boldsymbol{J} = \boldsymbol{Q}^{\mathrm{T}}\boldsymbol{R}$. Writing $\boldsymbol{Q}' = \boldsymbol{Q}e^{\boldsymbol{W}t}$, $\boldsymbol{R}' = \boldsymbol{R}$, where $\boldsymbol{W} \in \text{Skew}$ and $t \in \mathbb{R}$, it is found that the hypothesis implies that the function $t \mapsto \boldsymbol{J}\bar{\boldsymbol{S}} \cdot (e^{\boldsymbol{W}t}\boldsymbol{V})$ has an extremum at $t = 0$; similarly, writing $\boldsymbol{Q}' = \boldsymbol{Q}$, $\boldsymbol{R}' = e^{\boldsymbol{W}t}\boldsymbol{R}$, it is found that $t \mapsto \boldsymbol{J}\bar{\boldsymbol{S}} \cdot (\boldsymbol{V}e^{\boldsymbol{W}t})$ has an extremum at $t = 0$. The derivatives at $t = 0$ vanish, which provides, after rearrangements, that $\boldsymbol{J}\bar{\boldsymbol{S}}\boldsymbol{V}$ and $\boldsymbol{V}\boldsymbol{J}\bar{\boldsymbol{S}}$ are symmetric:

$$\boldsymbol{J}\bar{\boldsymbol{S}}\boldsymbol{V} = \boldsymbol{V}\bar{\boldsymbol{S}}\boldsymbol{J}^{\mathrm{T}}, \quad \boldsymbol{V}\boldsymbol{J}\bar{\boldsymbol{S}} = \bar{\boldsymbol{S}}\boldsymbol{J}^{\mathrm{T}}\boldsymbol{V}. \tag{18.3.1}$$

A multiplication of $(18.3.1)_2$ by $\bar{\boldsymbol{S}}\boldsymbol{J}^{\mathrm{T}}$ from the right gives $\boldsymbol{V}\boldsymbol{J}\bar{\boldsymbol{S}}^2\boldsymbol{J}^{\mathrm{T}} = \bar{\boldsymbol{S}}\boldsymbol{J}^{\mathrm{T}}\boldsymbol{V}\bar{\boldsymbol{S}}\boldsymbol{J}^{\mathrm{T}}$ and combination with $(18.3.1)_1$ leads to $\boldsymbol{V}\boldsymbol{J}\bar{\boldsymbol{S}}^2 = \bar{\boldsymbol{S}}^2\boldsymbol{V}\boldsymbol{J}$. Since the eigenvalues of $\bar{\boldsymbol{S}}^2$ are distinct, it follows that $\boldsymbol{V}\boldsymbol{J}$ is symmetric (Lemma 8.1.3(2)), which gives $\boldsymbol{J}^{\mathrm{T}}\boldsymbol{V} = \boldsymbol{V}\boldsymbol{J}$. We have $\boldsymbol{V} = \boldsymbol{J}\boldsymbol{V}\boldsymbol{J} = (\boldsymbol{J}^2)(\boldsymbol{J}^{\mathrm{T}}\boldsymbol{V}\boldsymbol{J})$. The first bracket is an orthogonal tensor and the second a positive definite symmetric tensor; the uniqueness in the polar decomposition gives $\boldsymbol{J}^2 = \mathbf{1}$, $\boldsymbol{J}^{\mathrm{T}}\boldsymbol{V}\boldsymbol{J} = \boldsymbol{V}$. The latter can be written as $\boldsymbol{V}\boldsymbol{J} = \boldsymbol{J}\boldsymbol{V}$ and hence $\boldsymbol{J}$ is symmetric (Lemma 8.1.3(2)). Equation $(18.3.1)_2$ leads to $\boldsymbol{J}\boldsymbol{V}\bar{\boldsymbol{S}} = \bar{\boldsymbol{S}}\boldsymbol{J}\boldsymbol{V}$; writing this in matrix form in the basis where $\boldsymbol{V}, \boldsymbol{J}$ are diagonal, and taking into account that the diagonal elements of $\boldsymbol{J}$

are ±1 and the diagonal elements of $\boldsymbol{V}$ are distinct, it follows that $\bar{\boldsymbol{S}}$ is also represented by a diagonal matrix in this basis. □

The following proposition plays the same role for objective-isotropic functions as Proposition 18.1.1 played for the symmetric convex functions and Proposition 18.2.1 for the convex functions of a symmetric-tensor argument.

18.3.2 Proposition *Let* $\boldsymbol{S} \in \mathrm{Sym}$, $\boldsymbol{V} \in \mathrm{Sym}^+$ *with the ordered n-tuples of eigenvalues* $s \in \mathbb{D}$, $v \in \mathbb{D}$. *Then the following assertions hold.*

(1) *The inequality*

$$\boldsymbol{S} \cdot (\boldsymbol{Q}\boldsymbol{V}\boldsymbol{R}^{\mathsf{T}}) \le s \cdot v \tag{18.3.2}$$

holds for every $\boldsymbol{Q}, \boldsymbol{R} \in \mathrm{Orth}$ *if and only if*

$$s_i \ge 0.$$

(2) *The inequality* (18.3.2) *holds for every* $\boldsymbol{Q}, \boldsymbol{R} \in \mathrm{Orth}^+$ *if and only if*

$$s_i + s_j \ge 0, \quad i \neq j.$$

The "if" part in (1) is the trace inequality of von Neumann. The "if" part in (2) is probably new.

Proof Let us first prove the "only if" parts in the implications in (1), (2). Let $\boldsymbol{S} = \sum_{i=1}^n s_i \boldsymbol{e}_i \otimes \boldsymbol{e}_i$ be a spectral decomposition of $\boldsymbol{S}$ and set $\bar{\boldsymbol{V}} := \sum_{i=1}^n \bar{v}_i \boldsymbol{e}_i \otimes \boldsymbol{e}_i$ where $\bar{v}_1 = -v_1$ and $\bar{v}_i = v_i$ for $i > 1$. Then $\bar{\boldsymbol{V}} = \boldsymbol{Q}\boldsymbol{V}\boldsymbol{R}^{\mathsf{T}}$ for some $\boldsymbol{Q}, \boldsymbol{R} \in \mathrm{Orth}$ and $\boldsymbol{S} \cdot \bar{\boldsymbol{V}} \le s \cdot v$ reduces to $s_1 v_1 \ge 0$, which gives the "only if" part in (1). In the case of (2), define $\bar{\boldsymbol{V}}$ by the formula above with the choice $\bar{v}_1 = v_2$, $\bar{v}_2 = -v_1$ and $\bar{v}_i = v_i$ for $i > 2$. We have $\bar{\boldsymbol{V}} = \boldsymbol{Q}\boldsymbol{V}\boldsymbol{R}^{\mathsf{T}}$, where $\boldsymbol{Q}, \boldsymbol{R} \in \mathrm{Orth}^+$. (Note that this is not so for $\bar{\boldsymbol{V}}$ used in the proof of (1).) The inequality $\boldsymbol{S} \cdot \bar{\boldsymbol{V}} \le s \cdot v$ then gives $s_1 + s_2 \ge 0$, proving the "only if" part in (2).

Let us now prove the "if" parts in (1), (2). We can restrict ourselves to the case when the eigenvalues of both $\boldsymbol{S}^2$, $\boldsymbol{V}$ are distinct. Let $\boldsymbol{Q}, \boldsymbol{R} \in \mathrm{Orth}$ be the points which realize the maximum in (1) or (2). The conclusion of Lemma 18.3.1 is available; let us also use its notation. The tensors $\bar{\boldsymbol{S}}, \boldsymbol{J}, \boldsymbol{V}$ admit simultaneous spectral decompositions

$$\bar{\boldsymbol{S}} = \sum_{i=1}^n \bar{s}_i \boldsymbol{e}_i \otimes \boldsymbol{e}_i, \quad \boldsymbol{J} = \sum_{i=1}^n \sigma_i \boldsymbol{e}_i \otimes \boldsymbol{e}_i, \quad \boldsymbol{V} = \sum_{i=1}^n v_i \boldsymbol{e}_i \otimes \boldsymbol{e}_i,$$

where v_i are ordered nonincreasingly, $\sigma_i = \pm 1$ for each i and $\bar{s} = (\bar{s}_1, \ldots, \bar{s}_n)$ is a permutation of s, i.e., $\bar{s} = Ps$ for some permutation matrix. Then

$$\boldsymbol{S} \cdot (\boldsymbol{Q}\boldsymbol{V}\boldsymbol{R}^{\mathsf{T}}) = (\boldsymbol{J}\bar{\boldsymbol{S}}) \cdot \boldsymbol{V} = \sum_{i=1}^n \sigma_i \bar{s}_i v_i. \tag{18.3.3}$$

Let us now complete the proof of (1). We have $s_i \ge 0$ so that the last expression in (18.3.3) can be majorized by $\bar{s} \cdot v = Ps \cdot v \le s \cdot v$ with the last step justified by Proposition 18.1.1. This proves (1). Let us now proceed to (2). The conditions $s_i + s_j \ge 0$, $i \neq j$, imply that there can be at most one index j with s_j negative. If all s_i are nonnegative, then the proof is completed in the same way as in (1). Hence assume that exactly one s_j is negative. Since $\bar{s} = Ps$, it follows that exactly one entry of $\bar{s}$ is negative. We have $\det \boldsymbol{J} = 1$ in the case (2) and hence the number of indices i with $\sigma_i = -1$ is even. If in (18.3.3) all $\sigma_i = -1$ stand in front of $\bar{s}_i \ge 0$, then again the proof is completed as in (1). Hence assume that there is a j such that $\sigma_j = -1$, $\bar{s}_j < 0$. Since the $\sigma_i = -1$ go in pairs, there must be another index, say k, such that $\sigma_k = -1$. Then the last expression in (18.3.3) is majorized by

$$\sum \bar{s}_i v_i - \bar{s}_j v_j - \bar{s}_k v_k, \tag{18.3.4}$$

where here and in the formula below the sum is taken over all i such that $i \neq j$, $i \neq k$. Using $\bar{s}_j + \bar{s}_k \ge 0$, it is found that $-\bar{s}_j v_j - \bar{s}_k v_k \le \bar{s}_k v_j + \bar{s}_j v_k$ and hence the expression in (18.3.4) is majorized by

$$\sum \bar{s}_i v_i + \bar{s}_k v_j + \bar{s}_j v_k = Q\bar{s} \cdot v = QPs \cdot v \le s \cdot v, \tag{18.3.5}$$

where Q is the permutation exchanging the indices j, k and the last step in (18.3.5) is justified by Proposition 18.1.1. □

Let $h : U \to \mathbb{R}$ be an objective-isotropic function on an open subset U of InvLin. By the representation theorem 8.5.1 it has a representation $h(\boldsymbol{F}) = f(v)$ where v is the ordered n-tuple of the singular values of $\boldsymbol{F}$ and $f : D \to \mathbb{R}$ is a symmetric function on an open symmetric subset of $\mathbb{R}^n_{++}$. The domain U has two parts: $U^+ := U \cap \mathrm{Lin}^+$ and $U^- := U \cap \mathrm{Lin}^-$. Let us denote by h^+ the restriction of h to U^+.

18.3.3 ***Remark*** If dim Vect ≥ 2, then Lin^+ is not convex – we have $\mathrm{co}\,\mathrm{Lin}^+ = \mathrm{Lin}$. Also InvLin is not convex – it is not connected.

Proof Let first dim Vect be even. For every $\boldsymbol{F} \in \mathrm{Lin}$ we have $\det(\pm\lambda\mathbf{1} + \boldsymbol{F}) = \lambda^n + O(\lambda^{n-1})$ as $\lambda \to \infty$ and hence for sufficiently large $\lambda > 0$ we have $\pm\lambda\mathbf{1} + \boldsymbol{F} \in \mathrm{Lin}^+$. Then $\boldsymbol{F} = \frac{1}{2}((+\lambda\mathbf{1} + \boldsymbol{F}) + (-\lambda\mathbf{1} + \boldsymbol{F}))$, proving that any $\boldsymbol{F}$ is a convex combination of two elements of Lin^+. Let now dim Vect be odd, dim Vect ≥ 3. Note first that $-\mathbf{1} \in \mathrm{co}\,\mathrm{Lin}^+$. To see this, consider any orthonormal basis and define the matrices $A = \mathrm{diag}(-4, \ldots, -4, -1, 2)$, $B = \mathrm{diag}(2, \ldots, 2, -1, -4)$ so that the corresponding tensors have positive determinants. If I is the identity matrix, then $-I = \frac{1}{2}(A + B)$. For a given $\boldsymbol{F} \in \mathrm{Lin}$ we have $\det(\lambda\mathbf{1} + 2\boldsymbol{F}) > 0$ provided $\lambda > 0$ is sufficiently large. Also $-\lambda\mathbf{1} \in \mathrm{co}\,\mathrm{Lin}$ and $\boldsymbol{F} = \frac{1}{2}((\lambda\mathbf{1} + 2\boldsymbol{F}) + (-\lambda\mathbf{1}))$. □

18.3.4 Proposition *Let h, h^+, f, U, U^+, D be as explained above and assume that f is continuously differentiable. Then*

(1) *h is convex on U if and only if f is convex and $f_{,i} \geq 0$ on D for every i;*

(2) *h^+ is convex on U^+ if and only if f is convex and $f_{,i} + f_{,j} \geq 0$ on D for every $i \neq j$.*

Recall our convention that a function defined on a nonconvex set is said to be convex if it has a convex extension to the convex hull of its domain. It is unrealistic to assume that a stored energy function of an isotropic material is convex: the derivatives $f_{,i}$ are the principal forces, and realistic materials violate the inequalities in (2) in a wide range of deformation gradients corresponding to states of compression. In addition to the inequalities in (1), (2) we also have the O-F inequalities (18.1.4) as a consequence of the convexity of f. The above proposition is a consequence of the following result.

18.3.5 Proposition *Let h, h^+, f, U, U^+, D be as explained above and let $\boldsymbol{F} \in U^+$ have the n-tuple of singular values $v \in \mathbb{D}$.*

(1) *$\boldsymbol{F}$ is a point of convexity of h if and only if v is a point of convexity of f and there exists a subgradient $s = (s_1, \ldots, s_n) \in \mathbb{D}$ of f at v such that*

$$s_i \geq 0, \quad [f_{,i}(v) \geq 0]. \tag{18.3.6}$$

(2) *$\boldsymbol{F}$ is a point of convexity of h^+ if and only if v is a point of convexity of f and there exists a subgradient $s = (s_1, \ldots, s_n) \subset \mathbb{R}^n$ of f at v such that*

$$s_i + s_j \geq 0, \quad [f_{,i}(v) + f_{,j}(v) \geq 0] \quad i \neq j. \tag{18.3.7}$$

Here the square brackets give the forms of the conditions when f is continuously differentiable.

Proof (1): We have $\boldsymbol{F} = \boldsymbol{Q}\boldsymbol{V}$, the polar decomposition of $\boldsymbol{F}$ with $\boldsymbol{Q} \in \mathrm{Orth}^+$, $\boldsymbol{V} \in \mathrm{Sym}^+$, and obviously $\boldsymbol{F}$ is a point of convexity of h if and only if $\boldsymbol{V}$ is a point of convexity of h. We can therefore assume that $\boldsymbol{F} = \boldsymbol{V}$ is positive definite symmetric. Let $\boldsymbol{F}$ be a point of convexity of h and let $\boldsymbol{S}$ be a subgradient of h at $\boldsymbol{F}$. Then $h(\boldsymbol{G}) \geq h(\boldsymbol{F}) + \boldsymbol{S} \cdot (\boldsymbol{G} - \boldsymbol{F})$ for every $\boldsymbol{G} \in U$. Replacing $\boldsymbol{G}$ by $\boldsymbol{G}^{\mathrm{T}}$ and using $h(\boldsymbol{G}) = h(\boldsymbol{G}^{\mathrm{T}})$, $\boldsymbol{F}^{\mathrm{T}} = \boldsymbol{F}$ we obtain that $\boldsymbol{S}^{\mathrm{T}}$ is a subgradient of h at $\boldsymbol{F}$ from which we find that the subgradient can be chosen to be symmetric. The convexity inequality at $\boldsymbol{F} = \boldsymbol{V}$ written for $\boldsymbol{G} = \boldsymbol{Q}\boldsymbol{V}\boldsymbol{R}^{\mathrm{T}}$, $\boldsymbol{Q}, \boldsymbol{R} \in \mathrm{Orth}$, reduces to $\boldsymbol{S} \cdot (\boldsymbol{Q}\boldsymbol{V}\boldsymbol{R}^{\mathrm{T}} - \boldsymbol{V}) \leq 0$ and by Proposition 18.3.2(1) this implies that $\boldsymbol{S}$ is positive-semidefinite. As in Lemma 18.3.1 it is found that $\boldsymbol{S}$, $\boldsymbol{V}$ commute (set $\boldsymbol{R} = \boldsymbol{Q} = \mathbf{1}$ in (18.3.1)). This implies that $\boldsymbol{S}$, $\boldsymbol{V}$ admit a simultaneous spectral decompositions with the eigenvalues

$s_i, v_i, i = 1, \ldots, n$. Set $s = (s_1, \ldots, s_n)$. Let $\boldsymbol{G}$ be a tensor represented by a diagonal matrix with entries $w = (w_1, \ldots, w_n) \in D$ in the basis of eigenvectors of $\boldsymbol{V}, \boldsymbol{S}$. The convexity inequality for h with this $\boldsymbol{G}$ gives that s is a subgradient of f at v which proves the direct implication in (1). The converse implication: Let s be as in (1), and let $\bar{s}$ be the symmetrization of s given by a procedure similar to that described in Proposition 18.1.2. Then $\bar{s}_i \geq 0$ as a consequence of $s_i \geq 0$ and $P\bar{s} = \bar{s}$ for every $P \in G(v)$ and $\bar{s}$ is still a subgradient of f at v. Let $\boldsymbol{S}$ be a tensor given by a diagonal matrix with entries $\bar{s}$ in the basis of eigenvectors of $\boldsymbol{V}$. If $\boldsymbol{G} \in U$ has the n-tuple of singular values $w \in \mathbb{D}$, then by Proposition 18.3.2(1) we have $\boldsymbol{S} \cdot \boldsymbol{G} \leq s \cdot w$. Hence $h(\boldsymbol{G}) - h(\boldsymbol{F}) - \boldsymbol{S} \cdot (\boldsymbol{G} - \boldsymbol{V}) \geq h(w) - h(v) - s \cdot (w - v) \geq 0$, with the last step justified by the convexity of f at v. (2): This part of the proof is completely analogous to the preceding part; one simply uses Proposition 18.3.2(2) instead of Proposition 18.3.2(1). □

Let us now state a counterpart of Proposition 18.3.5 for functions defined on the whole of Lin.

18.3.6 Proposition *Let $f : [0, \infty)^n \to \mathbb{R}$ be a function and define $h : \mathrm{Lin} \to \mathbb{R}$ by $h(\boldsymbol{A}) = f(v(\boldsymbol{A}))$ where $v(\boldsymbol{A})$ is the n-tuple of the singular values of $\boldsymbol{A} \in \mathrm{Lin}$. Then h is convex on* Lin *if and only if f is convex and nondecreasing in each variable.*

18.3.7 Examples

(1) *If $\alpha = (\alpha_1, \ldots, \alpha_n) \in \mathbb{R}^n$, then the function $h := \sum_{i=1}^n \alpha_i \hat{v}_i$ is convex on* Lin *if and only if $\alpha \in \mathbb{D} \cap \mathbb{R}^n_{++}$. Its restriction to* Lin^+ *is convex if and only if $\alpha \in \mathbb{D}$ and $\alpha_i + \alpha_j \geq 0$ for every $i \neq j$.*

Proof We have $h(\boldsymbol{F}) = \sum_{i=1}^n \alpha_i m_i(v)$, i.e., h is represented by the restriction to $[0, \infty)^n$ of the function f of Example 18.1.6(2). By Proposition 18.3.6 we have to check when f is convex and nondecreasing in each variable. From Example 18.1.6, h is convex on $\mathbb{R}^n_{++}$ if and only if $\alpha \in \mathbb{D}$. As clearly each of the m_i is nondecreasing, f is nondecreasing if and only if $\alpha \in \mathbb{R}^n_{++}$ because the derivative of f at a point v with distinct coordinates is given by $P\alpha$ where P is a permutation matrix such that $Pv \in \mathbb{D}$. The assertion about Lin^+ is proved analogously. □

(2) The functions $\hat{v}_1, \hat{v}_1 + \hat{v}_2, \ldots, \hat{v}_1 + \cdots + \hat{v}_n$ are convex on Lin.

(3) *If $\alpha \geq 1$, then the function $h : \mathrm{Lin} \to \mathbb{R}$ defined by*

$$h(\boldsymbol{A}) = \sum_{i=1}^{n} \hat{v}_i(\boldsymbol{A})^{\alpha}, \quad \boldsymbol{A} \in \mathrm{Lin},$$

is convex.

18.4 Invertibility of the Stress Relation

We have seen that a substantial condition for the convexity of objective-isotropic scalar functions are the inequalities (18.3.6) and (18.3.7). In this section it will be shown that conditions closely related to these inequalities, namely Propositions 18.4.1(2), 18.4.2 or 18.4.3(2), arise from the global invertibility of objective-isotropic tensor-valued functions. It will be seen that the *global invertibility* of the stress-strain relation is as unrealistic as the convexity of the stored energy in $\boldsymbol{F}$. The failure of local invertibility at a given deformation gradient indicates a possibility of the bifurcation of a homogeneously deformed body under overall pure traction loading environment.

We consider an objective-isotropic tensor-valued function $\boldsymbol{G} : U \to \mathrm{Lin}$ with the representation $g : D \to \mathbb{R}^n$ as in Proposition 8.4.3. Let $\boldsymbol{G}^+ : U^+ \to \mathrm{Lin}$ be the restriction of $\boldsymbol{G}$ to $U^+ := U \cap \mathrm{Lin}^+$. Throughout this section, let for every $\boldsymbol{A} \in \mathrm{Lin}$ the symbol $|\boldsymbol{A}| \in \mathrm{Sym}$ stand for the tensor-valued absolute value of $\boldsymbol{A}$, defined by $|\boldsymbol{A}| = \sqrt{\boldsymbol{A}^{\mathrm{T}}\boldsymbol{A}}$. Similarly, if $s = (s_1, \ldots, s_n) \in \mathbb{R}^n$, we denote, in this section only, by $|s| = (|s_1|, \ldots, |s_n|)$ the n-tuple of absolute values. A mapping $\phi : M \to N$ is said to be injective if $x, y \in M$ and $\phi(x) = \phi(y)$ imply $x = y$.

18.4.1 Proposition *$\boldsymbol{G}^+$ is injective if and only if the following two conditions hold:*

(1) *if $v, w \in D$ are such that*

$$|g_i(v)| = |g_i(w)| \quad \text{for all} \quad i = 1, \ldots, n \quad \text{and} \quad \prod_{i=1}^{n} g_i(v) = \prod_{i=1}^{n} g_i(w) \tag{18.4.1}$$

then $v = w$;

(2) *for every $v \in D$ there is at most one i such that $g_i(v) = 0$.*

Thus not only the mapping $g : D \to \mathbb{R}^n$ must be injective, but essentially also the mapping $v \mapsto |g(v)|$ must be injective, which is a much stronger condition, and plainly unrealistic when $\boldsymbol{G} = \bar{\boldsymbol{S}}$ is the response function for the Piola–Kirchhoff stress of an isotropic solid. In contrast, if $\boldsymbol{G} = \bar{\boldsymbol{S}}$ corresponds to a fluid, then (1), (2) are often satisfied. Such is the case, e.g., of $\bar{\boldsymbol{S}}$ of an ideal fluid, which is injective on Lin^+.

Proof Assume that $\boldsymbol{G}^+$ is injective. Let us fix some orthonormal basis in Vect and identify tensors with matrices relative to this basis. (2): Let $v \in D$, $i \neq j$, and prove preliminarily that $g_i(v) + g_j(v) \neq 0$. Assume without any loss in generality that $i = 1, j = 2$. Let Q be a proper orthogonal matrix

$$Q = \begin{bmatrix} 0 & -1 & 0 \\ 1 & 0 & 0 \\ 0 & 0 & I \end{bmatrix} \tag{18.4.2}$$

where I is $(n-2) \times (n-2)$ identity matrix. Set $V = \operatorname{diag} v$, $W = QVQ = \operatorname{diag}(-v_2, -v_1, v_3, \dots)$ so that $\boldsymbol{G}(V) = \operatorname{diag} g(v)$, $\boldsymbol{G}(W) = Q\boldsymbol{G}(V)Q = \operatorname{diag}(-g_2(v), -g_1(v), g_3(v), \dots)$. Since clearly $V \neq W$, we must have $\boldsymbol{G}(V) \neq \boldsymbol{G}(W)$ and hence $g_1(v) \neq -g_2(v)$. In particular, for every $i \neq j$ at most one of the numbers $g_i(v)$ can be 0. (1): Let v, w be as in (1). For every i such that $g_i(v) \neq 0$ define $\sigma_i := g_i(w)/g_i(v)$ so that by $(18.4.1)_1$, $\sigma_i = \pm 1$. If $g_i(v) \neq 0$ for all $i = 1, \dots, n$, then all σ_i are uniquely defined and $\prod_{i=1}^n \sigma_i = 1$ by $(18.4.1)_2$. If the preceding is not true then $g_i(v) = 0$ for exactly one i by (2). We then set σ_i equal to $+1$ or -1 in such a way that $\prod_{i=1}^n \sigma_i = 1$. This is possible. Clearly $g_i(w) = \sigma_i g_i(v)$ for every $i = 1, \dots, n$. Set $V = \operatorname{diag} v$, $W = \operatorname{diag} w$, $Q = \operatorname{diag} \sigma$. Then $\boldsymbol{G}(QV) = Q\boldsymbol{G}(V) = \operatorname{diag} \sigma \operatorname{diag} g(v) = \operatorname{diag} g(w) = \boldsymbol{G}(W)$ and hence $QV = W$. But this is possible only if $\sigma_i = 1$ for every i and $v = w$. This completes the proof of the direct implication. Conversely, assume that (1), (2) hold and prove that $\boldsymbol{G}^+$ is injective. Hence let $\boldsymbol{F}, \bar{\boldsymbol{F}} \in U^+$ be such that

$$\boldsymbol{G}(\boldsymbol{F}) = \boldsymbol{G}(\bar{\boldsymbol{F}}) \tag{18.4.3}$$

and write $\boldsymbol{F} = \boldsymbol{R}\boldsymbol{V}$, $\bar{\boldsymbol{F}} = \bar{\boldsymbol{R}}\bar{\boldsymbol{V}}$ for the polar decompositions with $\boldsymbol{R}, \bar{\boldsymbol{R}} \in \mathrm{Orth}^+$. Let $v, \bar{v} \in D$ be the ordered principal stretches of $\boldsymbol{F}, \bar{\boldsymbol{F}}$ respectively. In the basis of eigenvectors of $\boldsymbol{V}$, the $\boldsymbol{G}(\boldsymbol{V})$ is represented by $\operatorname{diag} g(v)$ and $|\boldsymbol{G}(\boldsymbol{V})|$ by $\operatorname{diag} |g(v)|$. In the basis of eigenvectors of $\bar{\boldsymbol{V}}$, the $\boldsymbol{G}(\bar{\boldsymbol{V}})$ is represented by $\operatorname{diag} g(\bar{v})$ and $|\boldsymbol{G}(\bar{\boldsymbol{V}})|$ by $\operatorname{diag} |g(\bar{v})|$. From (18.4.3) follows that $\prod_{i=1}^n g_i(v) = \prod_{i=1}^n g_i(\bar{v})$ and $|\boldsymbol{G}(\boldsymbol{V})| = |\boldsymbol{G}(\bar{\boldsymbol{V}})|$ and from this the sets $\{|g_1(v)|, \dots, |g_n(v)|\}$ and $\{|g_1(\bar{v})|, \dots, |g_n(\bar{v})|\}$ are equal. Hence there exists a permutation matrix P such that $|g_i(v)| = |g_i(P\bar{v})|$, $i = 1, \dots, n$. By (1) then $v = P\bar{v}$ and since both v and $\bar{v}$ are ordered, it follows that $v = \bar{v}$. Furthermore, if i, j are such that $v_i \neq v_j$, then $|g_i(v)| \neq |g_j(v)|$. Indeed suppose otherwise and let w be obtained by exchanging the ith and jth entries of v. Then v and w satisfy the hypothesis of (1) and violate the conclusion of (1), a contradiction. In the bases $\{\boldsymbol{e}_i, i = 1, \dots, n\}$, $\{\bar{\boldsymbol{e}}_i, i = 1, \dots, n\}$ of eigenvectors of $\boldsymbol{V}, \bar{\boldsymbol{V}}$ the equality $|\boldsymbol{G}(\boldsymbol{V})| = |\boldsymbol{G}(\bar{\boldsymbol{V}})|$ reads

$$\sum_{i=1}^n |g_i(v)| \boldsymbol{e}_i \otimes \boldsymbol{e}_i = \sum_{i=1}^n |g_i(v)| \bar{\boldsymbol{e}}_i \otimes \bar{\boldsymbol{e}}_i$$

and the property $|g_i(v)| \neq |g_j(v)|$ if $v_i \neq v_j$ allows us to conclude that

$$\sum_{i=1}^n v_i \boldsymbol{e}_i \otimes \boldsymbol{e}_i = \sum_{i=1}^n v_i \bar{\boldsymbol{e}}_i \otimes \bar{\boldsymbol{e}}_i, \quad \text{i.e.,} \quad \boldsymbol{V} = \bar{\boldsymbol{V}}.$$

Relation (18.4.3) now reads $\boldsymbol{R}\boldsymbol{G}(\boldsymbol{V}) = \bar{\boldsymbol{R}}\boldsymbol{G}(\boldsymbol{V})$ which implies $\boldsymbol{R} g_i(v) \boldsymbol{e}_i = \bar{\boldsymbol{R}} g_i(v) \boldsymbol{e}_i$, $i = 1, \dots, n$. Since at most one of the $g_i(v)$ can be 0, it follows that $\boldsymbol{R}\boldsymbol{e}_i = \bar{\boldsymbol{R}}\boldsymbol{e}_i$ for all but possibly one i. However, since $\boldsymbol{R}, \bar{\boldsymbol{R}}$ are proper orthogonal, this allows us to conclude that $\boldsymbol{R} = \bar{\boldsymbol{R}}$. □

18.4.2 Proposition *Suppose that* $g : V \to \mathbb{R}^n$ *satisfies Conditions* (1), (2) *of Proposition* 18.4.1. *Then*

(1) $g_i(v) + g_j(v) \neq 0$ *for every* $v \in D$ *and every* $i \neq j$;

(2) $g_i(v) \neq g_j(v)$ *for every* $v \in D$ *and every* $i \neq j$ *such that* $v_i \neq v_j$.

Proof (1): Suppose that $g_i(v) = -g_j(v)$ for some $v \in D$ and some pair i, j with $i \neq j$. Since at least one of the $g_i(v), g_j(v)$ is different from 0, it follows that both of them are different from 0. Furthermore we have $v_i \neq v_j$ for otherwise we would have $g_i(v) = g_j(v) \neq -g_j(v)$ by the symmetry of g. Let w be a result of exchanging the ith and jth entries of v. Then $g_k(w) = g_k(v)$ for $k \neq i, k \neq j$ and $g_i(w) = g_j(v) = -g_i(v)$ and $g_j(w) = g_i(v) = -g_j(v)$. Hence v, w satisfy the hypothesis of Proposition 18.4.1(1) which implies $v = w$, but this is impossible because $v_i \neq v_j$. (2): Suppose that $g_i(v) = g_j(v)$

for some $v \in D$ and some pair i, j with $i \neq j$ and $v_i \neq v_j$. Let again w be a result of exchanging the ith and jth entries of v. Then $g_k(w) = g_k(v)$ for all $k = 1, \ldots, n$ and hence by Proposition 18.4.1(1), $w = v$ which implies $v_i = v_j$, which is a contradiction. □

18.4.3 Proposition *$\boldsymbol{G}$ is injective if and only if the following two conditions hold:*
(1) *if $v, w \in D$ are such that $|g_i(v)| = |g_i(w)|$, $i = 1, \ldots, n$, then $v = w$;*
(2) *$g_i(v) \neq 0$ for every $v \in D$ and every $i = 1, \ldots, n$.*

Proof Assume that $\boldsymbol{G}$ is injective. (1) is proved in a way completely similar to that in Proposition 18.4.1. Let us prove (2). Suppose that $g_i(v) = 0$ for some i and assume without loss of generality that $i = 1$. Let $V = \operatorname{diag} v$ and $Q = \operatorname{diag}(-1, 1, \ldots)$. Then $\boldsymbol{G}(QV) = Q \operatorname{diag} g(v) = \operatorname{diag} v = \boldsymbol{G}(V)$ and hence $QV = V$, a contradiction. This completes the proof of the direct implication. The proof of the converse implication is obtained by a simple modification of the corresponding portion of the proof of Proposition 18.4.1. The details are omitted. □

18.4.4 Proposition *Let D be open, g continuously differentiable and let $\boldsymbol{F} \in U$ have the principal stretches $v \in D$. Then $\partial_{\boldsymbol{F}}\boldsymbol{G}(\boldsymbol{F})$ is an invertible linear transformation from* Lin *onto* Lin *if and only if the following three conditions hold simultaneously:*
(1) *$\nabla g(v)$ is an invertible linear transformation from $\mathbb{R}^n$ to $\mathbb{R}^n$;*
(2) *$g_i(v) + g_j(v) \neq 0$ for every $i \neq j$;*
(3) *for every $i \neq j$ one has*

$$g_i(v) \neq g_j(v) \text{ if } v_i \neq v_j \text{ and } g_{i,i}(v) \neq g_{i,j}(v) \text{ if } v_i = v_j.$$

Proof This follows directly from the formula for the derivative of $\boldsymbol{G}$ in Proposition 8.4.5. □

Since the conditions under which $\boldsymbol{G} = \bar{\boldsymbol{S}}$ is injective are plainly unrealistic, we proceed to determine the set of all $\boldsymbol{F} \in \text{InvLin}$ or Lin^+ for which $\boldsymbol{G}(\boldsymbol{F})$ has a prescribed value $\boldsymbol{S}$. By the objectivity and isotropy, it is possible to restrict the attention to the case $\boldsymbol{S}$ symmetric. For a mapping $\phi : M \to N$ and $z \in N$ denote by $\phi^{-1}(z)$ the set of all $x \in M$ such that $\phi(x) = z$. For simplicity, we shall describe the set $(\boldsymbol{G}^+)^{-1}(\boldsymbol{S})$ only in the case when the eigenvalues of $\boldsymbol{S}$ are different from 0 and their absolute values are distinct. For the analysis of other cases, see OGDEN [1984, Subsect. 6.2.2].

18.4.5 Proposition *Let $\boldsymbol{S} \in \text{Sym}$ have eigenvalues s_i and consider a basis $\{\boldsymbol{e}_i, i = 1, \ldots, n\}$ of eigenvectors of $\boldsymbol{S}$. Then*
(1) *$(\boldsymbol{G}^+)^{-1}(\boldsymbol{S})$ contains all $\boldsymbol{F}$ represented by matrices $\boldsymbol{F} = \operatorname{diag}(r_1 v_1, \ldots, r_n v_n)$ where $r_i = \pm 1$,*

$$r_i g_i(v) = s_i \quad i = 1, \ldots, n, \text{ and } \prod_{i=1}^{n} r_i = 1; \tag{18.4.4}$$

(2) *if $|s_i|$, $i = 1, \ldots, n$ are distinct, then $(\boldsymbol{G}^+)^{-1}(\boldsymbol{S})$ is exactly the set of all $\boldsymbol{F} = \operatorname{diag}(r_1 v_1, \ldots, r_n v_n)$ such that* (18.4.4) *holds.*

Proof (1) is immediate. (2): Let $\boldsymbol{F}$ satisfy $\boldsymbol{G}(\boldsymbol{F}) = \boldsymbol{S}$ and let $\boldsymbol{F} = \boldsymbol{R}\boldsymbol{V}$ be the polar decomposition of $\boldsymbol{F}$ with $\boldsymbol{R} \in \text{Orth}^+$. Then $\boldsymbol{R}\boldsymbol{G}(\boldsymbol{V}) = \boldsymbol{S}$ and from this $|\boldsymbol{G}(\boldsymbol{V})| = |\boldsymbol{S}|$. Hence in the basis of eigenvectors of $\boldsymbol{S}$, the $|\boldsymbol{G}(\boldsymbol{V})|$ is represented by a diagonal matrix, namely by $\operatorname{diag}|s|$. Since $|s_i|$, $i = 1, \ldots, n$ are distinct, it follows that also the eigenvalues of the symmetric tensor $\boldsymbol{G}(\boldsymbol{V})$ are distinct and necessarily $\boldsymbol{G}(\boldsymbol{V})$ is represented by a diagonal matrix in the basis of eigenvectors of $\boldsymbol{S}$, since the eigenvalues of $|\boldsymbol{G}(\boldsymbol{V})| = |\boldsymbol{S}|$ are distinct. Since $\boldsymbol{V}$ commutes with $\boldsymbol{G}(\boldsymbol{V})$, it follows that $\boldsymbol{V}$ is represented by a diagonal matrix $\boldsymbol{V} = \operatorname{diag} v$ for some $v \in D$. But then $\boldsymbol{G}(\boldsymbol{V}) = \operatorname{diag} g(v)$. Moreover at most one eigenvalue of $\boldsymbol{G}(\boldsymbol{V})$ can be 0 as a consequence of the fact that the eigenvalues of $\boldsymbol{G}(\boldsymbol{V})$ are distinct. From $\boldsymbol{R}\boldsymbol{G}(\boldsymbol{V}) = \boldsymbol{S}$ and the fact that $\boldsymbol{S}$, $\boldsymbol{G}(\boldsymbol{V})$ are diagonal and at most one of the eigenvalues of $\boldsymbol{G}(\boldsymbol{V})$ is 0 we obtain that also $\boldsymbol{R}$ is diagonal, with entries $r_i = \pm 1$ and satisfying $(18.4.4)_2$ as a consequence of $\boldsymbol{R} \in \text{Orth}^+$. The equation $\boldsymbol{R}\boldsymbol{G}(\boldsymbol{V}) = \boldsymbol{S}$ then reads as $(18.4.4)_1$. □

18.4.6 Proposition *Let $n = 3$ and assume that g maps D bijectively onto $\mathbb{R}^3$. Let $\boldsymbol{S} \in \text{Sym}$ have the eigenvalues s_i.*
(1) *If $s_i \neq 0$, $i = 1, 2, 3$, then the set $(\boldsymbol{G}^+)^{-1}(\boldsymbol{S})$ contains at least 4 elements; namely those described in Proposition* 18.4.5(1).
(2) *If $s_i \neq 0$, $i = 1, 2, 3$, and $|s_i|$, $i = 1, 2, 3$ are distinct, then the set $(\boldsymbol{G}^+)^{-1}(\boldsymbol{S})$ contains exactly 4 elements; namely those described in Proposition* 18.4.5(2).

The conclusions of the proposition in particular apply to the stress function $\bar{\boldsymbol{S}}$ of the Blatz–Ko special I (see Subsect. 11.3.3(3)), since by the strict convexity of f and coercivity, the mapping $v \mapsto \partial_v f(v) \equiv$

$s(\nu)$ maps $(0,\infty)^3$ bijectively onto $\mathbb{R}^3$. Thus for a 'generic' $\boldsymbol{S}$ there are at least 4 distinct $\boldsymbol{F} \in \text{Lin}^+$ which lead to the stress $\boldsymbol{S}$.

18.5 Isotropic Polyconvex Functions

The results on the convex functions stated in the previous section can be used to construct a rich supply of objective-isotropic polyconvex functions. We shall be concerned only with functions on Lin^+; if such a function is a restriction of an objective-isotropic function on InvLin, then it is expressible as a symmetric function of the singular values of the deformation gradient. Recall that a function $h : \text{Lin}^+ \to \mathbb{R}$ is called polyconvex if there exists a convex function $\bar{h} : \text{Lin}\times\text{Lin}\times(0,\infty)\to\mathbb{R}$ such that $h(\boldsymbol{F}) = \bar{h}(\boldsymbol{F}, \operatorname{cof}\boldsymbol{F}, \det\boldsymbol{F})$. Without introducing a more general sufficient condition for an isotropic function to be polyconvex (see BALL [1977, Thm. 5.2]) we consider here only an important subclass. Let dim Vect = 3.

18.5.1 Proposition *Let* $m : (0,\infty)^3 \to \mathbb{R}$ *be a convex function nondecreasing in the first two variables, and* $\phi, \psi : \text{Lin} \to (0,\infty)$ *objective-isotropic convex functions. Then the function* $h : \text{Lin}^+ \to \mathbb{R}$ *given by*

$$h(\boldsymbol{F}) = m(\phi(\boldsymbol{F}), \psi(\operatorname{cof}\boldsymbol{F}), \det\boldsymbol{F}), \qquad \boldsymbol{F} \in \text{InvLin}$$

is a polyconvex function which is a restriction of an objective-isotropic function to Lin^+.

Proof The function $\bar{h} : \text{Lin}\times\text{Lin}\times(0,\infty) \to \mathbb{R}$ defined by $\bar{h}(\boldsymbol{F},\boldsymbol{H},D) = m(\phi(\boldsymbol{F}), \psi(\boldsymbol{H}), D)$ for each $(\boldsymbol{F},\boldsymbol{H},D) \subset \text{Lin}\times\text{Lin}\times(0,\infty)$, is a convex function. (This follows from the easily verifiable fact that a composition of a convex function that is nondecreasing in each variable with an n-tuple of convex functions is a convex function.) □

18.5.2 Examples

(1) Let $\alpha \ge 1, \beta \ge 1$. The functions $\phi(\boldsymbol{F}) = \operatorname{tr}\boldsymbol{V}^\alpha$, $\psi(\boldsymbol{H}) = \operatorname{tr}\boldsymbol{W}^\beta$, $\boldsymbol{F},\boldsymbol{H} \in \text{Lin}$, where $\boldsymbol{V}, \boldsymbol{W}$ are the left stretch tensors of $\boldsymbol{F}, \boldsymbol{H}$, respectively, are convex on Lin by Example 18.3.7(3). Hence, if m is as in Proposition 18.5.1, the function h given by

$$h(\boldsymbol{F}) = m(\operatorname{tr}\boldsymbol{V}^\alpha, \operatorname{tr}\operatorname{cof}\boldsymbol{V}^\beta, \det\boldsymbol{F}),$$

where $\boldsymbol{V}$ is the left stretch tensor of $\boldsymbol{F} \in \text{Lin}^+$, is polyconvex on Lin^+. In particular, $\boldsymbol{F} \mapsto \operatorname{tr}\boldsymbol{V}^\alpha$, $\boldsymbol{F} \mapsto \operatorname{tr}\operatorname{cof}\boldsymbol{V}^\beta$, $\boldsymbol{F} \mapsto \det\boldsymbol{F}^\gamma$, where either $\gamma \ge 1$ or $\gamma \le 0$, are polyconvex on Lin^+. (The function $\boldsymbol{F} \mapsto \operatorname{tr}\boldsymbol{V}^\alpha$ is even convex, of course.)

(2) The functions $\phi(\boldsymbol{F}) = \hat{\nu}_1(\boldsymbol{F})$, $\psi(\boldsymbol{H}) = \hat{\nu}_1(\boldsymbol{H})$, $\boldsymbol{F},\boldsymbol{H} \in \text{Lin}$, where $\hat{\nu}_1$ is the maximal singular value, are convex by Example 18.3.7(1) or (2) and one has $\psi(\operatorname{cof}\boldsymbol{F}) = \hat{\nu}_1(\boldsymbol{F})\hat{\nu}_2(\boldsymbol{F})$. If m is as in Proposition 18.5.1, the function h given by

$$h(\boldsymbol{F}) = m(\hat{\nu}_1(\boldsymbol{F}), \hat{\nu}_1(\boldsymbol{F})\hat{\nu}_2(\boldsymbol{F}), \det\boldsymbol{F}),$$

where $\boldsymbol{V}$ is the left stretch tensor of $\boldsymbol{F} \in \text{Lin}^+$, is polyconvex on Lin^+. In particular, $\boldsymbol{F} \mapsto \hat{\nu}_1(\boldsymbol{F})$, $\boldsymbol{F} \mapsto \hat{\nu}_1(\boldsymbol{F})\hat{\nu}_2(\boldsymbol{F})$ are polyconvex on Lin^+. (The function $\boldsymbol{F} \mapsto \hat{\nu}_1(\boldsymbol{F})$ is even convex.)

(3) *Consider the Ogden material* 11.3.1. *If* $\alpha_i \ge 1, \beta_j \ge 1$ *for each* i, j *and for each* k *and either* $\gamma_k > 0$ *or* $\gamma_k < -1$, *then the stored energy* $\bar{f}(\cdot,\theta)$ *is polyconvex on* Lin^+. *The stored energy* 11.3.3(4) *is polyconvex if* $\alpha \ge 1$ *and* $\gamma > 0$.

Proof This is just a combination of the remarks in (1), (2). □

18.6 The Second Differential of the Stored Energy

We conclude with some conditions applying to the second differential of objective-isotropic functions.

18.6.1 Proposition *Let $h : U \to \mathbb{R}$ be an objective-isotropic function of class C^2 defined on an open subset of* InvLin *with the representation $f : D \to \mathbb{R}$ as in Proposition* 8.2.1. *Let $\boldsymbol{F} \in U$ be given with the n-tuple of singular values $v \in U$. Then $\partial_F^2 h(\boldsymbol{F})[\boldsymbol{B}, \boldsymbol{B}] \geq 0$ for every $\boldsymbol{B} \in$* Lin *if and only if the following conditions hold simultaneously:*

(1) $\sum_{i,j=1}^{n} f_{,ij}(v)\lambda_i\lambda_j \geq 0$ *for every* $\lambda = (\lambda_1, \ldots, \lambda_n) \in \mathbb{R}^n$;

(2) *for every $i \neq j$,*

$$\frac{f_{,i}(v) - f_{,j}(v)}{v_i - v_j} \geq 0 \text{ if } v_i \neq v_j, \text{ and } f_{,ii}(v) - f_{,ij}(v) \geq 0 \text{ if } v_i = v_j;$$

(3) $f_{,i}(v) + f_{,j}(v) \geq 0$ *for every $i \neq j$.*

In mechanical terms, the derivatives $s_i := f_{,i}$ are the principal forces, the eigenvalues of the Piola–Kirchhoff stress $\boldsymbol{S}$. The result (3) is related to Proposition 17.5.3 which gives formally similar inequalities for the *eigenvalues of* $\boldsymbol{T}$. The latter result was derived from the positivity of the second differential for materials of arbitrary symmetry using only the principle of material frame indifference. The inequalities in (3) are stronger, as one can easily check. Items (1), (2), (3) recover inequalities (18.1.4) and (18.3.7) previously established from the convexity of h^+ at $\boldsymbol{F}$.

Proof We can assume that $\boldsymbol{F} \in \mathrm{Sym}^+ \cap U$. Setting $\boldsymbol{B} = \boldsymbol{C}$ in (8.5.5), we see that if the second differential of h at $\boldsymbol{F}$ is positive semidefinite, then (1), (2), (3) follow by taking $\boldsymbol{B}$ to be diagonal, symmetric and skew, respectively. Conversely, if (1), (2), (3) hold then the second differential is positive semidefinite since (8.5.5) with $\boldsymbol{B} = \boldsymbol{C}$ contains only nonnegative terms. □

18.6.2 Proposition *Let $h : U \to \mathbb{R}$ be an objective-isotropic function defined on an open subset of* InvLin *of class C^2 with the representation $f : D \to \mathbb{R}$ as in Proposition* 8.2.1. *Let $\boldsymbol{F} \in U$ be given with the n-tuple of singular values $v \in U$. If $\partial_F^2 h(\boldsymbol{F})[\boldsymbol{a} \otimes \boldsymbol{b}, \boldsymbol{a} \otimes \boldsymbol{b}] \geq 0$ for every $\boldsymbol{a}, \boldsymbol{b} \in$* Vect *then the following conditions hold:*

(1) $f_{,ii}(v) \geq 0$ *for every $i = 1, \ldots, n$;*

(2) *for every $i \neq j$,*

$$\frac{v_i f_{,i}(v) - v_j f_{,j}(v)}{v_i - v_j} \geq 0 \text{ if } v_i \neq v_j, \text{ and } f_{,ii}(v) - f_{,ij}(v) + f_{,i}(v)/v_i \geq 0 \text{ if } v_i = v_j,$$

$$\sqrt{f_{,ii} f_{,jj}} + f_{,ij} + \frac{f_{,i}(v) - f_{,j}(v)}{v_i - v_j} \geq 0 \text{ if } v_i \neq v_j, \quad \sqrt{f_{,ii} f_{,jj}} - f_{,ij} + \frac{f_{,i}(v) + f_{,j}(v)}{v_i + v_j} \geq 0.$$

If $n = 2$, then Conditions (1), (2) *are also sufficient.*

The inequalities in (1) are called the tension–extension inequalities. The strengthened versions of the inequalities in (2), namely

$$\frac{v_i f_{,i}(v) - v_j f_{,j}(v)}{v_i - v_j} > 0, \quad i \neq j, \quad v_i \neq v_j$$

are called the Baker–Ericksen inequalities.

Proposition 18.6.2 is due to KNOWLES & STERNBERG [1977]; see also AUBERT & TAHRAOUI [1980, 1987]. The necessary and sufficient conditions for the Legendre–Hadamard condition in 3 dimensions are more complicated: different conditions have been given in SIMPSON & SPECTOR [1983], OGDEN [1984, pp. 392–393] and ROSAKIS [1990].

Proof We can assume that $F \in \mathrm{Sym}^+ \cap U$. (1): We set $B = C = a \otimes a$ in (8.5.6). The choice $a = (0, \ldots, 1, \ldots, 0)$ with 1 in the ith position gives the desired inequality. (2): The first inequality. Take $i = 1, j = 2$. We set $B = C = a \otimes b$ in (8.5.6), where $a = (1, 0, \ldots, 0)$, $b = (0, 1, 0, \ldots, 0)$. The second inequality of (2): We set $B = C = a \otimes a$ in (8.5.5) with $a = (\sqrt[4]{f_{,22}}, \sqrt[4]{f_{,11}}, 0, \ldots, 0)$. The third inequality of (2): We set $B = C = a \otimes b$ in (8.5.5) with a as in the preceding step and with $b = (a_1, -a_2, 0, \ldots, 0)$. This completes the proof of the direct implication. For the proof of the converse implication in case dim Vect = 2, see KNOWLES & STERNBERG [1977] and AUBERT & TAHRAOUI [1980]. □

18.6.3 "A Priori" Constitutive Inequalities Although the present book pursues the point of view that the properties of the constitutive functions should be derived from the experience with the phenomena they are designed to cover, we here briefly mention the main constitutive inequalities introduced on an a priori basis. The matter was raised by TRUESDELL [1956] (see also BAKER & ERICKSEN [1954]) and has stimulated extensive research. On the whole, it led to the natural conclusion that there is no universal a priori constitutive inequality. Despite this, the research enlarged the understanding of the nonlinear response, and many methods and concepts developed are of permanent value. A detailed account of the state of the problem up to 1965 is in TRUESDELL & NOLL [1965, Sects. 51–53 and 87]. The main development since then has led to Hill's inequalities HILL [1970a,b] and to the assessment of their strength and meaning. A special case of one of the inequalities considered by Hill is the Coleman–Noll inequality.

18.6.4 Hill's Inequalities

First Family For a fixed $m \in \mathbb{R}$, the inequality

$$(\overset{\circ}{A} - mAD - mDA) \cdot D \geq 0 \tag{18.6.1}$$

is postulated to hold in every process, where A is the Kirchhoff stress (Sect. 10.4) and $\overset{\circ}{A} := \dot{G} - WA + AW$ is the Jaumann's (corotational) derivative of A along a process. (W is the spin tensor ((2.2.6).) The Coleman–Noll inequality (see TRUESDELL & NOLL [1965, eq. (52.15)]) corresponds to (18.6.1) with $m = \frac{1}{2}$; however, the original formulation (see TRUESDELL & NOLL [1965, eq. (87.7)]) does not involve time derivatives. The corotational derivative (as opposed to the material time derivative) is used in (18.6.1) to eliminate the undesirable contribution of rotations.

Second Family For a fixed $m \in \mathbb{R}$, consider the Seth–Hill strain measure $Q = \phi_m(U)$ and the corresponding conjugate stress P (Example 10.4.2(3)). It is postulated that

$$\dot{P} \cdot \dot{Q} > 0 \tag{18.6.2}$$

for every process.

In discussing the consequences, let us restrict ourselves to isotropic, isothermal hyperelastic materials. Then the stored energy $f = f(F)$ at temperature θ can be expressed as a function $\bar{f} = f(v)$ of the principal stretches. We also have the representation as in Subsect. 11.1.3 (with θ suppressed): $\bar{f}$ can be expressed as a function $\bar{f} = \bar{g}(Q)$ of Q and also as a function $\bar{f} = g(q)$ of the eigenvalues q_i of Q. The eigenvalues p_i of P can be written as functions of q, i.e., $p_i = p_i(q)$. To avoid complications with domains, it is assumed that $\bar{f}$ is defined on Lin$^+$. Then $f = f(v)$ and $g = g(q)$, $p = p(q)$ are defined on convex domains. Although we shall not need it directly, note that by using the results and notations of Subsect. 2.2.7, Sect. 10.4 and Subsect. 11.1.3, it is found that for isotropic materials,

$$\dot{P} \cdot \dot{Q} = \sum_{i=1}^{n} \dot{p}_i \dot{q}_i + \sum (p_i - p_j)(q_i - q_j)\Omega_{ij}^2 = \sum_{i=1}^{n} \dot{p}_i \dot{q}_i + \sum \frac{p_i - p_j}{q_i - q_j} \dot{Q}_{ij}^2,$$

where here and below the sums with unspecified range of indices denote the sum over all $i \neq j$, the symbol $\dot{p}_i$ is the time derivative of the eigenvalues of P, and $\dot{Q}_{ij}$ are the components of $\dot{Q}$ in the basis of eigenvectors of U.

18.6.5 Proposition *For a fixed $m \in \mathbb{R}$, the following assertions hold:*

(1) *if the material satisfies the inequality of the first family (in every process) with the given m, then it satisfies the inequality of the second family with the same m. For $m = 0, \pm 1$, the converse is also true;*

(2) *the satisfaction of* (18.6.2) *in every process is equivalent to the convexity of $\bar{g}$ in Q, which is equivalent to the convexity of g in q;*

(3) (18.6.1) *holds in every process if and only if g is convex in q and for every v and every i, j such that* $v_i \neq v_j$,

$$\frac{v_i f_{,i} - v_j f_{,j}}{v_i^2 - v_j^2} \geq m \frac{v_i f_{,i} + v_j f_{,j}}{v_i^2 + v_j^2}; \qquad (18.6.3)$$

(4) *if the equivalent conditions of* (1) *hold, then*

$$(q_i - q_j)(p_i(q) - p_j(q)) \geq 0. \qquad (18.6.4)$$

For $m = \frac{1}{2}$ (the Coleman–Noll inequality) (2) expresses the convexity of f in the principal stretches and (4) the O-F inequality. For a general m, the meaning of inequalities (18.6.3) is unclear; note, however, that for $m = 0$ (logarithmic strain), (18.6.3) are equivalent to the Baker–Ericksen inequalities. The Coleman Noll inequality was shown to lead to undesirable consequences in that it leads to lower bounds on the compressibility, discriminating nearly incompressible models. (See TRUESDELL & NOLL [1965, p. 169].) HILL [1970a] showed that his second family is inconsistent with the incompressibility constraint for any $m \neq 0$, and OGDEN [1970a] showed that Hill's second family is inconsistent with the Blatz–Ko special I materials (Subsect. 11.3.3(3)) if $m \neq 0$. If $m = 0$ then Hill's inequalities of the first and second family are equivalent and require the convexity of f in $\ln \boldsymbol{U}$ or equivalently in $(\ln v_1, \ln v_2, \ln v_3)$. This does not seem to contradict any theoretical or experimental evidence.

Proof (Sketch) (2): Inequality (18.6.2) is the assertion that the Hessian of the function $\boldsymbol{Q} \mapsto \bar{g}(\boldsymbol{Q})$ is positive semidefinite. Hence (18.6.2) holds if and only if $\bar{g}$ is convex. By Proposition 18.2.4(2) this is equivalent to the convexity of $g(q)$. (3): Suppose that (18.6.1) holds. We have $\boldsymbol{A} = \bar{\boldsymbol{A}}(\boldsymbol{V})$ where $\bar{\boldsymbol{A}} : \mathrm{Sym}^+ \to \mathrm{Sym}$ is an isotropic function with the representation (8.1.3) where $a_i(v) = v_i f_{,i}(v)$. The material derivative $\dot{\boldsymbol{A}}$ is calculated by $\dot{\boldsymbol{A}} = \partial_{\boldsymbol{V}} \bar{\boldsymbol{A}}(\boldsymbol{V})[\dot{\boldsymbol{V}}]$ and $\dot{\boldsymbol{V}}$ is related to $\boldsymbol{D}$ as in Subsect. 2.2.7. Calculating $\partial_{\boldsymbol{V}} \bar{\boldsymbol{A}}$ as shown in Proposition 8.1.9, and inserting this to (18.6.1), one is led to a quadratic inequality in $\boldsymbol{D}$. This inequality, which will not be written here, is easily shown to be equivalent to the convexity of g in q and (18.6.3). Inequality in (4) is just the Schur condition (18.1.4) coming from the convexity of g in q. □

Bibliographical Note

Proposition 18.2.4(2) is proved in DUNN & FOSDICK [1978] by a somewhat different method. The Baker–Ericksen inequalities were postulated in BAKER & ERICKSEN [1954] and then derived in HAYES [1969] from the strong ellipticity condition. A discussion of further undesirable consequences of the global convexity of the stored energy function in $\boldsymbol{F}$ is found in CIARLET [1988, Sect. 4.8] and in the references to that section. Further information on the constitutive inequalities are in PARRY [1978], DUNN [1983], OGDEN [1984], SPECTOR [1984] and BEATTY [1987, Sect. 17].

19 Thermostatics of Fluids

In terms of its applications, the thermostatics of fluids is one of the broadest areas of thermodynamics. The applications range from water and steam to the fluids of elementary particles. In the main part of this chapter we consider a fluid without the influence of body forces. The equilibrium states are then only homogeneous or heterogeneous states of several coexistent phases. The extremum principles imply strong convexity consequences on the states and on the energy and the Gibbs functions of the fluid. Sect. 19.8 extends the main results to fluids under the influence of body forces.

19.1 Preview: The Energy Function

This section describes the basic point of view and foreshadows some of the results to be proved in the subsequent treatment.

19.1.1 Global Point of View The energy function $\tilde{e} = \tilde{e}(v, \eta)$ contains all the information about the behavior of the fluid in equilibrium over the whole range of pressures and temperatures. This framework is broad enough to model stable, metastable, and unstable equilibrium states and phase transitions, i.e., heterogeneous states containing several phases. (This point of view is explained in more detail generally in Subsect. 9.4.2.) No a priori restrictions are imposed on $\tilde{e}$ beyond the thermostatic relations, the positivity of the temperature, and mild technical requirements which are far from excluding any essential feature of the fluid. The states to be considered can contain several phases from the very beginning and the treatment does not artificially separate the homogeneous states from the states of coexistent phases.

19.1.2 Convexity Properties Figure 19.1.1 shows a portion of the graph of an energy function of a hypothetical fluid. The qualitative features of the behavior of the fluid are completely reflected by the convexity/nonconvexity properties of $\tilde{e}(v, \eta)$. (The reader is referred to Sects. 16.2–16.4 for the convexity theory.) If the energy function admits stable heterogeneous states with several phases, then $\tilde{e}(v, \eta)$ cannot be globally convex but it must have many points of convexity. The convexity properties are faithfully described by the geometrical properties of the graph of $\tilde{e}(v, \eta)$, and we shall sometimes use this to make the conclusions clearer. Let us consider a stable heterogeneous state with several homogeneous phases. Analytically, the Gibbs function of the phases is the same and the points (v, η) representing the phases must be points of convexity of $\tilde{e}$. Geometrically this means that the points $(v, \eta, \tilde{e}(v, \eta))$ of the graph of $\tilde{e}$ share the same tangent plane and the rest of the graph of $\tilde{e}(v, \eta)$ is above that tangent plane. In Fig. 19.1.1, these properties are satisfied by the points A, B; other

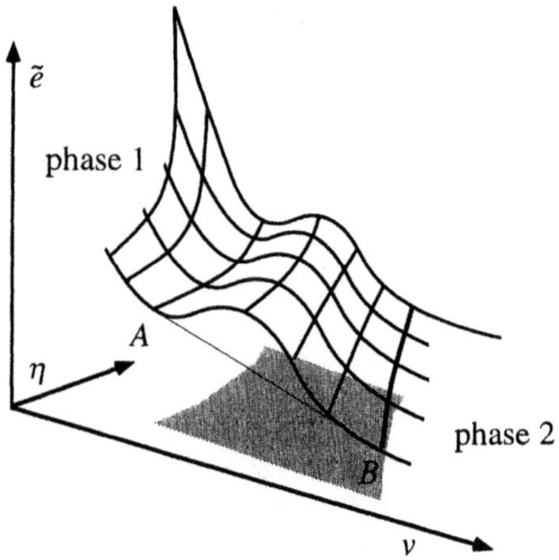

Fig. 19.1.1. An energy function, its convex hull and coexistent states

pairs of such points are easily determined by analogy. The points corresponding to different phases in a stable equilibrium must be separated each from other by regions in which $\tilde{e}$ is not convex. (The most popular example of a fluid with a nonconvex energy function is the van der Waals fluid.)

19.1.3 The Convex Hull of Energy Joining the points of the graph of $\tilde{e}$ that can coexist in a stable equilibrium state (like A, B above) by line segments, and replacing the wavy portions of $\tilde{e}$ above the line segment by the line segment itself, one obtains a convex function. This function is the greatest convex function not exceeding $\tilde{e}$, the lower convex hull $\tilde{e}^{\text{co}}$ of $\tilde{e}$. In Fig. 19.1.1, $\tilde{e}^{\text{co}}$ coincides with $\tilde{e}$ everywhere except in the shaded region of the v, η plane, above which the graph of $\tilde{e}$ is replaced by the curved surface made of the indicated segments. $\tilde{e}^{\text{co}}$ coincides with $\tilde{e}$ at points of convexity (v, η) of $\tilde{e}$. This happens at (v, η) exactly if the homogeneous state of unit mass of the fluid of total volume v and total entropy η is stable. (In the rest of this subsection we speak about the states of a unit mass of the fluid.) If the homogeneous state of total volume v and total entropy η is unstable then $\tilde{e}(v, \eta) > \tilde{e}^{\text{co}}(v, \eta)$ and the value $\tilde{e}^{\text{co}}(v, \eta)$ is the total energy of the stable *heterogeneous* state of the same total volume and total entropy. This heterogeneous state contains the two phases at the endpoints of the line segment in a definite mass fraction and they correspond to the stable phase splitting introduced in Sect. 16.3.

19.1.4 The Gibbs Function Because of the nonconvexity of $\tilde{e}(v, \eta)$, the pressure relation $p = \tilde{p}(v, \eta)$ is not globally invertible in v. Hence the Legendre transformation cannot be applied to define the Gibbs function. One has to use the Fenchel transformation instead, see Sect. 19.6. The Gibbs function defined in this way is automatically concave, and it has interesting differentiability properties. Namely, it is differentiable at (p, θ) if and only if there exists a stable homogeneous state of pressure p and temperature θ minimizing the energy. If this is not the case, then for regular energy functions (see 19.5.5 for a definition), the Gibbs function looks locally at (p, θ) either like an edge or like a vertex at the end of three joining edges, depending whether the stable heterogeneous state of pressure p and temperature θ has two or three phases.

There are two possible ways of arriving at the convexity properties of $\tilde{e}(v,\eta)$: Either by specializing the general results of Chap. 17 to fluids or by deriving them anew in the simpler context of fluids. The first possibility, foreshadowed in Sect. 17.7, shows that the general notions of quasiconvexity, rank 1 convexity, and the Legendre–Hadamard condition are comprehensive enough to explain all the properties of the thermodynamic potentials of fluids capable of phase transitions. Hence, formally, the fluids do not need a special treatment. Nevertheless, a derivation of the convexity properties starting directly from the extremum principles is much more illuminative and results in a theory of remarkable completeness and beauty. This chapter is devoted to this direct approach.

19.2 Rest States and Total Quantities

We shall use a reference configuration of the fluid Ω in which $\rho_0 \equiv 1$. Moreover, it is convenient to assume that the total mass is 1; hence the referential volume of Ω is 1. Assume that the fluid cannot leave a fixed region (canister) $R \subset$ Eucl but need not fill it completely. Accordingly, we impose the confinement condition 13.2.2 combined with the contact heating 13.2.3 of ambient temperature $\theta_0 > 0$. Assume that the volume of R is infinite (e.g., $R =$ Eucl). For the mechanical interaction we shall take the unilateral contact environment 13.3.3 with $\boldsymbol{b}_0 \equiv \mathbf{0}$ and with s as in the pressure loading 13.3.10 with a constant ambient pressure π. The ambient pressure on the free part of the boundary can be exerted by another fluid which does not mix with the examined fluid (e.g., a drop of water in the air) or a piston acting on the fluid by an invariant pressure π. For most of the treatment, we shall need just the numbers π and θ_0 from the environment.

19.2.1 Assumptions Let $\tilde{e} : \tilde{U} \to \mathbb{R}$ be the energy function of the fluid. It is assumed that

(1) $\tilde{U} = \mathbb{R}^{++} \times \mathbb{R}^{++}$, $\tilde{e}$ is continuously differentiable and bounded from below;
(2) for every $(v,\eta) \in \tilde{U}$, $\tilde{\theta}(v,\eta) := \partial_\eta \tilde{e}(v,\eta)$ is positive;
(3) for every $v > 0$, $\lim_{\eta\to\infty} \tilde{e}(v,\eta) = \infty$;
(4) if $(v(\cdot),\eta(\cdot)) : \Omega \to \tilde{U}$ are integrable functions such that $\int_\Omega \tilde{e}(v,\eta)\,dV < \infty$ and α is a positive number, then $\int_\Omega \tilde{e}(v,\eta+\alpha)\,dV < \infty$.

No other a priori assumptions are made. It follows from (2), (3) that $\tilde{e}(v,\cdot)$ is invertible; if the inverse is denoted by $\hat{\eta}(v,\cdot)$, then the domain $\hat{U}$ of $\hat{\eta}$ is of the form $\hat{U} = \{(v,e) : 0 < v < \infty, e^*(v) < e < \infty\}$, where $e^*(v) = \lim_{\eta\downarrow 0} \tilde{e}(v,\eta)$ is a finite-valued function bounded from below since $\tilde{e}$ is bounded from below. (3) is analogous to Assumption 15.3.2(2) and (4) to Assumption 15.3.2(4). The pressure function is $\tilde{p} := -\partial_v \tilde{e}$.

Further realistic properties of the response are the following: $\tilde{p} > 0$ everywhere on $\tilde{U}$; then $\tilde{e}(\cdot,\eta)$ is a decreasing function of v for each fixed $\eta > 0$. Hence $\lim_{v\to\infty} \tilde{e}(v,\eta)$ exists for each fixed η; it is also realistic to assume that this limit is independent of η and by changing the energy function by subtracting a constant, one can ensure that

$$\tilde{e}(v,\eta) \to 0 \text{ as } v \to \infty$$

for each fixed $\eta > 0$. Also

$$v\tilde{p}(v,\eta) \to 0, \quad v\partial_v \tilde{p}(v,\eta) \to 0 \text{ as } v \to \infty;$$

the last limit is a consequence of the convexity of $\tilde{p}(\cdot, \eta)$ in v at large specific volumes, which holds because then all gases are ideal gases, see Sect. 24.3. At small specific volumes,

$$\tilde{e}(v, \eta) \to \infty, \quad \tilde{p}(v, \eta)v \to \infty \text{ as } v \to 0$$

at each fixed $\eta > 0$. These properties are not used in the subsequent treatment.

19.2.2 Rest States Recall the set $\Sigma^\#$ of all rest states $\sigma = (\chi, \eta)$ introduced in Subsect. 13.1.1. For the present chapter this idea of rest states contains a superfluous information about the configuration χ and it is convenient to introduce a less detailed notion of state. Namely, we shall identify rest states with the pairs $\sigma = (v(\cdot), \eta(\cdot))$ of functions giving the specific volume and the specific entropy as functions of the reference position $\boldsymbol{p} \in \Omega$. Let us denote by $\Sigma^\#$ the set of all pairs $\sigma = (v(\cdot), \eta(\cdot))$ of measurable functions on Ω such that

(1) $(v(\boldsymbol{p}), \eta(\boldsymbol{p})) \in \tilde{U}$ for every $\boldsymbol{p} \in \Omega$;
(2) the functions $v(\cdot), \eta(\cdot)$ are integrable over Ω; and
(3) $\int_\Omega \tilde{e}(v(\boldsymbol{p}), \eta(\boldsymbol{p}))\, dV < \infty$.

Since the integrand is bounded from below, (3) is meaningful. The smoothness properties of v, η are relaxed completely. This seems to be the appropriate framework. The terms 'almost everywhere' (a.e.) and 'measurable' will always mean 'almost everywhere' and 'measurable with respect to the Lebesgue measure' (volume); the latter is denoted by V.

The state $\sigma = (v(\cdot), \eta(\cdot)) \in \Sigma^\#$ is said to be an ***equilibrium state*** if the functions $\tilde{p}(v(\cdot), \eta(\cdot))$ and $\tilde{\theta}(v(\cdot), \eta(\cdot))$ are essentially constant on Ω, i.e., if $\tilde{p}(v(\boldsymbol{p}), \eta(\boldsymbol{p})) = p_1$ and $\tilde{\theta}(v(\boldsymbol{p}), \eta(\boldsymbol{p})) = \theta_1$ for some p_1, θ_1 and a.e. $\boldsymbol{p} \in \Omega$. The numbers p_1, θ_1 are the pressure and temperature of the equilibrium state σ. An ***equilibrium state compatible with the environment*** is an equilibrium state whose pressure p_1 and temperature θ_1 coincide with the ambient pressure π and ambient temperature θ_0, respectively.

19.2.3 Heterogeneous States A state $\sigma = (v(\cdot), \eta(\cdot)) \in \Sigma^\#$ is said to be a ***heterogeneous state*** if there exists a finite number Ω_i, $i = 1, \ldots, r$, of pairwise disjoint measurable subsets of Ω, each of positive measure, such that $\bigcup_{i=1}^r \Omega_i = \Omega$, and a finite number of *distinct* points (v_i, η_i), $i = 1, \ldots, r$, of $\tilde{U}$ such that $(v(\boldsymbol{p}), \eta(\boldsymbol{p})) = (v_i, \eta_i)$ for a.e. $\boldsymbol{p} \in \Omega_i$. The number r and the set $\{(v_i, \eta_i) : i = 1, \ldots, r\}$ are uniquely determined; the former is the ***number of phases*** in σ and the latter is the ***set of phases*** of σ. The number $V(\Omega_i)$ is the amount of the phase (v_i, η_i) in the state σ. A state σ is a ***homogeneous state*** if it is a heterogeneous state with just one phase (v, η) and v, η are called the specific volume and specific entropy of the state σ.

19.2.4 Total Quantities In terms of the above notion of state, the total quantities of the fluid are expressed as follows: If $\sigma \in \Sigma^\#$, then

$$\begin{aligned} \mathsf{E}(\sigma) &= \int_\Omega \tilde{e}\rho_\circ\, dV, \quad \mathsf{H}(\sigma) = \int_\Omega \eta\rho_\circ\, dV, \quad \mathsf{V}(\sigma) = \int_\Omega v\rho_\circ\, dV, \\ \mathsf{P}(\sigma) &= \mathsf{E}(\sigma) - \theta_0\mathsf{H}(\sigma) + \pi\mathsf{V}(\sigma) = \int_\Omega (\tilde{e} - \theta_0\eta + \pi v)\rho_\circ\, dV, \end{aligned} \tag{19.2.1}$$

where $\rho_\circ$ can be omitted as $\rho_\circ = 1$. Conditions (1)–(3) imply that the integrals in the above definitions are absolutely convergent for every $\sigma \in \Sigma^\#$.

A mapping $\phi : \Omega \to \Omega$ is said to preserve the measure if it is one-to-one and onto Ω, for every subset P of Ω we have that P is measurable if and only if $\phi(P)$

is measurable, and $V(P) = V(\phi(P))$ for every measurable subset P of Ω. Two states $\sigma_1 = (v_1, \eta_1)$, $\sigma_2 = (v_2, \eta_2) \in \Sigma^\#$ are said to be ***equivalent*** if there exists a measure-preserving mapping $\phi : \Omega \to \Omega$ such that $v_1(\boldsymbol{p}) = v_2(\phi(\boldsymbol{p}))$, $\eta_1(\boldsymbol{p}) = \eta_2(\phi(\boldsymbol{p}))$ for a.e. $\boldsymbol{p} \in \Omega$. Equivalent states are identical up to a "rearrangement;" the functionals E, H, V, P have the same values on them. Two states $\sigma_1, \sigma_2 \in \Sigma^\#$ are said to be ***distinct*** if they are not equivalent. If $(V, H) \in \tilde{U}$, we denote by $\Sigma^\#(V, H)$ the set of all states $\sigma \in \Sigma^\#$ for which $\mathsf{V}(\sigma) = V$ and $\mathsf{H}(\sigma) = H$.

The following proposition deals with the lower convex hull $\tilde{e}^{\text{co}}$ of $\tilde{e}$. See Sect. 16.3 for the definition.

19.2.5 Proposition *For every* $(V, H) \in \tilde{U}$,

$$\tilde{e}^{\text{co}}(V, H) = \inf\{\mathsf{E}(\sigma) : \sigma \in \Sigma^\#(V, H)\}. \tag{19.2.2}$$

Proof We have $\tilde{e}^{\text{co}}(v(\boldsymbol{p}), \eta(\boldsymbol{p})) \le \tilde{e}(v(\boldsymbol{p}), \eta(\boldsymbol{p}))$ for every $\sigma = (v, \eta) \in \Sigma^\#(V, H)$ and every $\boldsymbol{p} \in \Omega$. Jensen's inequality (RUDIN [1970, Sect. 3.3]) for $\tilde{e}^{\text{co}}$ and the inequality $\tilde{e}^{\text{co}}(v, \eta) \le \tilde{e}(v, \eta)$ give

$$\tilde{e}^{\text{co}}(V, H) \le \int_\Omega \tilde{e}^{\text{co}}(v, \eta)\, dV \le \int_\Omega \tilde{e}(v, \eta)\, dV; \quad \text{hence } \tilde{e}^{\text{co}}(V, H) \le \inf\{\mathsf{E}(\sigma) : \sigma \in \Sigma^\#(V, H)\}.$$

To prove the opposite inequality, we employ (16.3.1), by which the value $\tilde{e}^{\text{co}}(V, H)$ is the infimum of $\sum_{i=1}^s \alpha_i \tilde{e}(v_i, \eta_i)$ over all convex combinations satisfying $V = \sum_{i=1}^s \alpha_i v_i$, $H = \sum_{i=1}^s \alpha_i \eta_i$. Each such a convex combination can be realized as a heterogeneous state $\sigma \in \Sigma^\#(V, H)$ with phases (v_i, η_i) present in the amounts α_i (recall that the total volume of Ω is 1). Hence, for every $\varepsilon > 0$ there exists a heterogeneous state σ in $\Sigma^\#(V, H)$ such that $\tilde{e}^{\text{co}}(V, H) + \varepsilon > \mathsf{E}(\sigma)$. The opposite inequality follows. □

19.3 Extremum Principles for Fluids

For the fluid in the external conditions described at the beginning of Sect. 19.2, identify the set of competitor states Σ_0 with the set of all rest states $\Sigma^\#$ from Subsect. 19.2.2, i.e., set $\Sigma_0 = \Sigma^\#$. Consider the following three extremum principles.

19.3.1 Minimum of Total Canonical Free Energy *For every* $\sigma \in \Sigma_0$, $\mathsf{P}(\sigma) \ge \mathsf{P}(\sigma_0)$.

Let us explain the relationship of the above assertion to the minimum of canonical free energy Subsect. 15.2.1 with the set of competitor states $\Sigma_1 = \Sigma^{\#\#}$, where $\Sigma^{\#\#}$ is defined in 14.3.2(3). It is claimed that the extremum principle 15.2.1 is 'essentially' equivalent to that in 19.3.1. Clearly, 19.3.1 is stronger than 15.2.1 since every state $\sigma = (\chi, \eta)$ from $\Sigma^{\#\#}$ gives rise to a state $\sigma = (v, \eta)$ from $\Sigma^\#$. That 15.2.1 is 'essentially' equivalent follows from the belief that for any sufficiently regular state $\sigma = (v, \eta) \in \Sigma^\#$ there exists a $\sigma = (\chi, \eta) \in \Sigma^{\#\#}$ leading to the specific volume v and specific entropy η. This depends on the regularity of Ω, and I refer to DACOROGNA [1981] for the existence of χ when Ω is a ball, v is continuously differentiable on $\text{cl}\,\Omega$, and χ satisfies $\chi(\boldsymbol{p}) = \boldsymbol{p}$ on $\partial\Omega$. An *a posteriori* proof of the equivalence is the following: If a state σ satisfies the principle of minimum canonical free energy formulated in Subsect. 15.2.1, then for fluids this leads to consequences on $\tilde{e}$ that enable one to prove that the state also satisfies the above stronger principle of minimum canonical free energy (and conversely).

We have $\mathsf{P}(\sigma) = \mathsf{E}(\sigma) - \theta_0 \mathsf{H}(\sigma) + \pi \mathsf{V}(\sigma)$ and therefore the validity of 19.3.1 implies the following two principles by restricting the competing states.

19.3.2 Maximum of Entropy *For every state* $\sigma \in \Sigma_0$ *satisfying* $\mathsf{E}(\sigma) = \mathsf{E}(\sigma_0)$ *and* $\mathsf{V}(\sigma) = \mathsf{V}(\sigma_0)$ *we have* $\mathsf{H}(\sigma) \le \mathsf{H}(\sigma_0)$.

This is the maximum of entropy at fixed total internal energy and fixed total volume; hence this assertion is different from the principle of maximum entropy at fixed total (i.e., internal plus potential) energy $\mathsf{U}(\sigma) = \mathsf{E}(\sigma) + \pi\mathsf{V}(\sigma)$ stated 15.2.3. In this chapter *we shall use the term 'principle of maximum entropy' for* 19.3.2. Notice that the constrains on the competing states in 19.3.2 may be interpreted as those corresponding to a rigid adiabatic container. Nevertheless, we shall see that the principle 19.3.2 is completely equivalent to 19.3.1, which we associated with a fluid whose free boundary is acted on by the pressure π.

19.3.3 Minimum of Energy *For every state* $\sigma \in \Sigma_0$ *satisfying* $\mathsf{H}(\sigma) = \mathsf{H}(\sigma_0)$ *and* $\mathsf{V}(\sigma) = \mathsf{V}(\sigma_0)$ *we have* $\mathsf{E}(\sigma) \ge \mathsf{E}(\sigma_0)$.

Again, this is the minimum of total internal energy at fixed total entropy and total volume, an assertion different from the principle of minimum (total) energy 15.2.4. Again *we shall use the term 'principle of minimum energy' exclusively for* 19.3.3 *in this chapter.*

19.3.4 Further Extremum Principles For fluids one can consider several other equivalent principles, whose obvious formulations are omitted:

(1) the minimum of total energy $\mathsf{U} = \mathsf{E} + \pi\mathsf{V}$ at fixed entropy. (This principle is related to 15.2.4 in the same way as 19.3.1 to 15.2.1.)
(2) the maximum of entropy H at fixed total energy U. (Related to 15.2.3 in the same way as 19.3.1 to 15.2.1.)
(3) the minimum of total stored energy;
(4) the minimum of volume V at fixed energy E and entropy H;
(5) the minimum of canonical free energy $\mathsf{F} = \mathsf{E} - \theta_0\mathsf{H}$ at fixed volume V.

There are also several others.

19.4 The Equivalence and Consequences of the Extremum Principles

The first result of this section shows that any solution to the extremum problems is an equilibrium state. A nice but questionable proof can be given by equating the first variations of the functionals to 0 (and on the theorem of Lagrange multipliers in the case of the constrained extremes as in 19.3.2 and 19.3.3). In the present context, where the specific volumes and entropies of the state can tend to ∞ or to 0, some care must be taken when dealing with variations and the proof below uses a different method.

19.4.1 Proposition *If the state* $\sigma \in \Sigma^{\#}$ *satisfies any of the three extremum principles* 19.3.1, 19.3.2, *and* 19.3.3, *then it is an equilibrium state. If it satisfies the principle of minimum total canonical free energy* 19.3.1, *then its pressure and temperature coincide with the ambient pressure and the ambient temperature, respectively, i.e.,* σ *is an equilibrium state compatible with the environment.*

Proof This proof is similar to, but more complicated than, the proof of Proposition 19.4.2 below and it might be a good idea to start form that proof and then to return here. (1) Let us first prove that if $\sigma = (\nu, \eta)$ satisfies the principle of minimum energy, then σ is an equilibrium state. Let us denote by $L \subset \Omega$ the set of all $\boldsymbol{p} \in \Omega$ that are simultaneously a Lebesgue point (i.e., an element of the Lebesgue set) for the functions $\nu(\cdot)$, $\eta(\cdot)$ and $\bar{e}(\nu(\cdot), \eta(\cdot))$ (see e.g., RUDIN [1970, Thm. 8.8]). Almost every $\boldsymbol{p} \in \Omega$ is such a point, i.e., $V(L) = 1$. We shall prove that if $\boldsymbol{p}_1, \boldsymbol{p}_2 \in L$, then

$$\tilde{p}(v(p_1), \eta(p_1)) = \tilde{p}(v(p_2), \eta(p_2)), \quad \tilde{\theta}(v(p_1), \eta(p_1)) = \tilde{\theta}(v(p_2), \eta(p_2)). \tag{19.4.1}$$

Let $p_1, p_2 \in L$ and let $(v_1, \eta_1), (v_2, \eta_2) \in \tilde{U}$ be such that

$$v_1 + v_2 = v(p_1) + v(p_2), \quad \eta_1 + \eta_2 = \eta(p_1) + \eta(p_2). \tag{19.4.2}$$

Let $B_i(r)$, $i = 1, 2$, be a ball centered at p_i of radius r and denote $V(r) = V(B_i(r))$, $D(r) = B_1(r) \cup B_2(r)$. Suppose that r is small enough that $D(r) \subset \Omega$. Let us define a state σ' by:

$$(v'(\cdot), \eta'(\cdot)) = \begin{cases} (v_i + \beta(r), \eta_i + \gamma(r)) & \text{on } B_i(r), \quad i = 1, 2, \\ (v, \eta) & \text{on } \Omega \setminus D(r), \end{cases}$$

where $\beta(r)$, $\gamma(r)$ are defined so as to satisfy

$$V(r)(v_1 + v_2 + 2\beta(r)) = \int_{D(r)} v \, dV, \quad V(r)(\eta_1 + \eta_2 + 2\gamma(r)) = \int_{D(r)} \eta \, dV.$$

The terms $\beta(r)$, $\gamma(r)$ ensure that σ' has the same volume and entropy as σ. (Notice that $\beta(r)$, $\gamma(r)$ vanish if σ is homogeneous.) The principle of minimum energy gives

$$\int_{D(r)} \tilde{e}(v, \eta) \, dV \le V(r)(\tilde{e}(v_1 + \beta(r), \eta_1 + \gamma(r)) + \tilde{e}(v_2 + \beta(r), \eta_2 + \gamma(r))). \tag{19.4.3}$$

Now we let r tend to 0. Standard properties of Lebesgue's points ensure that

$$\int_{B_i(r)} \tilde{e}(v, \eta) \, dV / V(r) \to \tilde{e}(v(p_i), \eta(p_i)), \quad \beta(r) \to 0, \quad \gamma(r) \to 0 \text{ as } r \to 0.$$

Inequality (19.4.3) gives

$$\tilde{e}(v_1, \eta_1) + \tilde{e}(v_2, \eta_2) \ge \tilde{e}(v(p_1), \eta(p_1)) + \tilde{e}(v(p_2), \eta(p_2)).$$

The conditions (19.4.2) imply that $v_1 = v(p_1) + \lambda$, $v_2 = v(p_2) - \lambda$, $\eta_1 = \eta(p_1) + \mu$, $\eta_2 = \eta(p_2) - \mu$ for some $\lambda, \mu \in \mathbb{R}$ and conversely, if $\lambda, \mu \in \mathbb{R}$ are small enough, then (v_1, η_1), (v_2, η_2) given by these formulas are in $\tilde{U}$ and (19.4.3) holds. But this inequality says that the left-hand side, considered as function of λ, μ has a local minimum at $(\lambda, \mu) = (0, 0)$. The condition that the first derivatives of this function with respect to (λ, μ) vanish at (0, 0) gives (19.4.1).

(2) The proof that the minimum of canonical free energy implies that the state is an equilibrium one is identical except that the constraint that the variations do not change the volume and entropy is absent. For this reason, one obtains the stronger conclusion that the pressure and temperature coincide with the ambient pressure and ambient temperature.

(3) The proof in the case of the principle of maximum entropy becomes identical with that in (1) if one uses the description of states in terms of energy rather than entropy: one identifies states with pairs $\tau = (v, e)$ of integrable functions such that $\int_\Omega \hat{\eta}(v, e) \, dV < \infty$, where $\hat{\eta}$ is the inverse of $\tilde{e}$ mentioned in Subsect. 19.2.1. The states τ are in one-to-one correspondence with the states σ. H is given by

$$\mathsf{H}(\tau) = \int_\Omega \hat{\eta}(v, e) \, dV$$

and one considers the variations of volume and energy. The details are left to the reader. □

We have seen that the minimum of total canonical free energy trivially implies the minimum internal energy and the maximum of entropy. The converse implications are established now.

19.4.2 Proposition *Let $\sigma \in \Sigma^{\#}$ be an equilibrium state compatible with the environment.*

(1) *If σ satisfies any of the three extremum principles, then it satisfies the remaining two of them; in this sense (i.e., with the assumption that the state σ is compatible with the environment) they are equivalent;*

(2) *σ satisfies the three extremum principles if and only if $(v(p), \eta(p))$ is a point of convexity of $\tilde{e}$ for a.e. $p \in \Omega$; if this is the case, then the Gibbs function $\tilde{e} - \tilde{\theta}\eta + \tilde{p}v$ is essentially constant on Ω, i.e., constant on a subset $\Omega' \subset \Omega$ of measure 1.*

We note that the differential form of the convexity of $\tilde{e}$ at $(v_0, \eta_0) \in \tilde{U}$ reads

$$\tilde{e}(v, \eta) \geq \tilde{e}(v_0, \eta_0) - \tilde{p}(v_0, \eta_0)(v - v_0) + \tilde{\theta}(v_0, \eta_0)(\eta - \eta_0) \tag{19.4.4}$$

for every $(v, \eta) \in \tilde{U}$. This in turn implies the positive-semidefiniteness of the second differential of $\tilde{e}$, see (17.7.4) and (17.7.5). The constancy of $\tilde{e} - \tilde{\theta}\eta + \tilde{p}v$ reduces to the equality of the Gibbs functions between phases of a stable heterogeneous state. Assume that σ is a heterogeneous state with r phases (v_i, η_i), $i = 1, \ldots, r$, satisfying the equivalent principles. The pressure, temperature and the Gibbs functions of the different phases must be the same, which leads to the equations

$$\tilde{p}(v_i, \eta_i) = \tilde{p}(v_j, \eta_j), \quad \tilde{\theta}(v_i, \eta_i) = \tilde{\theta}(v_j, \eta_j), \quad \tilde{g}(v_i, \eta_i) = \tilde{g}(v_j, \eta_j), \tag{19.4.5}$$

for every i, j, $1 \leq i, j \leq r$, where $\tilde{g}$ is the Gibbs function in terms of v, η : $\tilde{g}(v, \eta) = \tilde{e}(v, \eta) - \tilde{\theta}(v, \eta)\eta + \tilde{p}(v, \eta)v$. These are the ***conditions of thermodynamic equilibrium*** between phases of a heterogeneous state. The proof of the equivalence of the principles of minimum energy and maximum entropy requires the positivity of the absolute temperature.

Proof We already know that the minimum of total canonical free energy implies the minimum of energy and the maximum of entropy. The proof will be complete if we show the following: (a) the principle of minimum energy and the principle of maximum entropy are equivalent; (b) the principle of minimum energy implies that for a.e. $\boldsymbol{p} \in \Omega$ the point $(v(\boldsymbol{p}), \eta(\boldsymbol{p}))$ is a point of convexity of $\tilde{e}$; (c) if for a.e. $\boldsymbol{p} \in \Omega$ the point $(v(\boldsymbol{p}), \eta(\boldsymbol{p}))$ is a point of convexity of $\tilde{e}$ then the principle of minimum total canonical free energy holds; (d) if for a.e. $\boldsymbol{p} \in \Omega$ the point $(v(\boldsymbol{p}), \eta(\boldsymbol{p}))$ is a point of convexity of $\tilde{e}$ then $\tilde{e} - \theta\eta + \tilde{p}v$ is essentially constant on Ω.

The proof of (a) uses Assumptions 19.2.1(2),(3),(4) substantially but otherwise is sufficiently similar to the proof of Proposition 15.3.4 to be omitted.

(b): To make the proof more accessible, let us first assume that the state $\sigma = (v, \eta)$ is homogeneous (with v, η constant). Let $\boldsymbol{p}$ be a point in Ω, let $(v_1, \eta_1), (v_2, \eta_2)$ be in $\tilde{U}$, and let α be a real number such that

$$0 < \alpha < 1, \quad \alpha v_1 + (1 - \alpha)v_2 = v, \quad \alpha\eta_1 + (1 - \alpha)\eta_2 = \eta. \tag{19.4.6}$$

Let $B(r)$ be an open ball in Eucl of radius $r > 0$ centered at $\boldsymbol{p}$. Denote by $V(r)$ its volume. Let C_1 be any cone in Eucl with vertex $\boldsymbol{p}$ such that $V(C_1 \cap B(r)) = \alpha V(r)$ for any $r > 0$. Let C_2 be its complement Eucl $\setminus C_1$. Set $C_1(r) = C_1 \cap B(r)$, $C_2(r) = C_2 \cap B(r)$. Then $V(C_1(r)) = \alpha V(r)$, $V(C_2(r)) = (1 - \alpha)V(r)$, $C_1(r) \cup C_2(r) = B(r)$. Fix $r > 0$ such that $B(r) \subset \Omega$ and define a state $\sigma' = (v', \eta')$ by

$$(v'(\cdot), \eta'(\cdot)) = \begin{cases} (v_i, \eta_i) & \text{on } C_i(r), \quad i = 1, 2, \\ (v, \eta) & \text{on } \Omega \setminus B(r). \end{cases}$$

Let us check that σ satisfies the constraints $\mathsf{V}(\sigma') = \mathsf{V}(\sigma)$, $\mathsf{H}(\sigma') = \mathsf{H}(\sigma)$. It coincides with σ outside $B(r)$. Hence the contributions from $\Omega \setminus B(r)$ to the total volume and entropy in the equations $\mathsf{V}(\sigma') = \mathsf{V}(\sigma)$, $\mathsf{H}(\sigma') = \mathsf{H}(\sigma)$ cancel. In view of this it suffices to compare the contributions to these quantities from $B(r)$; they by (19.4.6). By the principle of minimum energy $\mathsf{E}(\sigma') \geq \mathsf{E}(\sigma)$. Also the contributions to the energy from the complement of $B(r)$ cancel; this gives

$$\int_{B(r)} \tilde{e}(v'(\boldsymbol{q}), \eta'(\boldsymbol{q}))\, dV(\boldsymbol{q}) \geq \int_{B(r)} \tilde{e}(v, \eta)\, dV; \tag{19.4.7}$$

the evaluation of the integral on the left-hand side gives $V(r)(\alpha\tilde{e}(v_1, \eta_1) + (1 - \alpha)\tilde{e}(v_2, \eta_2))$ and the integral on the right-hand side is $\tilde{e}(v, \eta)V(r)$. Inequality (19.4.7) is then seen to give $\alpha\tilde{e}(v_1, \eta_1) + (1 - \alpha)\tilde{e}(v_2, \eta_2) \geq \tilde{e}(v, \eta)$. That is, $\tilde{e}$ is convex at (v, η). This completes the proof of (b) when σ is homogeneous. If σ is inhomogeneous, the above argument must be refined. The convexity of $\tilde{e}$ is obtained at every $\boldsymbol{p} \in \Omega$ that is simultaneously a Lebesgue point for $v(\cdot)$, $\eta(\cdot)$ and $\tilde{e}(v(\cdot), \eta(\cdot))$. Let $(v_1, \eta_1), (v_2, \eta_2), \alpha$, be such that (19.4.6) holds with $v = v(\boldsymbol{p})$, $\eta = \eta(\boldsymbol{p})$, where $\boldsymbol{p}$ is any Lebesgue point. The definition of σ' is similar to that given above:

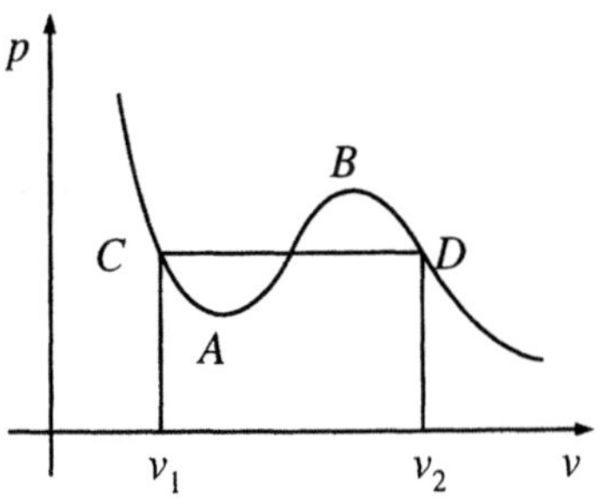

Fig. 19.4.1. A Van der Waals isotherm and Maxwell's construction

$$(v'(\cdot), \eta'(\cdot)) = \begin{cases} (v_i + \beta(r), \eta_i + \gamma(r)) & \text{on } C_i(r), \quad i = 1, 2, \\ (v, \eta) & \text{on } \Omega \setminus B(r), \end{cases}$$

where $\beta(r), \gamma(r)$ are given by

$$\beta(r) = \int_{B(r)} v\, dV/V(r) - v(\boldsymbol{p}), \quad \gamma(r) = \int_{B(r)} \eta\, dV/V(r) - \eta(\boldsymbol{p}).$$

The terms $\beta(r), \gamma(r)$ ensure that σ' has the same volume and entropy as σ. (Notice that $\beta(r), \gamma(r)$ vanish if σ is homogeneous.) The principle of minimum energy gives

$$\int_{B(r)} \tilde{e}(v, \eta)\, dV/V(r) \le \alpha \tilde{e}(v_1 + \beta(r), \eta_1 + \gamma(r)) + (1 - \alpha)\tilde{e}(v_2 + \beta(r), \eta_2 + \gamma(r)) \tag{19.4.8}$$

Now we let r tend to 0. Standard properties of Lebesgue's points ensure that

$$\int_{B(r)} \tilde{e}(v, \eta)\, dV/V(r) \to \tilde{e}(v(\boldsymbol{p}), \eta(\boldsymbol{p})), \quad \beta(r) \to 0, \quad \gamma(r) \to 0 \text{ as } r \to 0.$$

The convexity inequality then follows from (19.4.8).

(c): The state σ is an equilibrium state compatible with the environment and assume that for a.e. $\boldsymbol{p} \in \Omega$ the point $(v(\boldsymbol{p}), \eta(\boldsymbol{p}))$ is a point of convexity of $\tilde{e}$. In the state σ the pressure and temperature at a.e. point $\boldsymbol{p} \in \Omega$ are therefore equal to π and θ_0, respectively. The differential form of the convexity condition for $\tilde{e}$ at $(v(\boldsymbol{p}), \eta(\boldsymbol{p}))$ therefore reads

$$\tilde{e}(v', \eta') \ge \tilde{e}(v(\boldsymbol{p}), \eta(\boldsymbol{p})) + \theta_0(\eta' - \eta(\boldsymbol{p})) - \pi(v' - v(\boldsymbol{p})) \tag{19.4.9}$$

for a.e. $\boldsymbol{p} \in \Omega$ and every $(v', \eta') \in \tilde{U}$. If $\sigma' = (v', \eta')$ is an arbitrary state, then the integration of (19.4.9) leads to $\mathsf{P}(\sigma') \ge \mathsf{P}(\sigma)$.

(d): Let $\boldsymbol{p}, \boldsymbol{q} \in \Omega$ be points such that $(v(\boldsymbol{p}), \eta(\boldsymbol{p}))$ and $(v(\boldsymbol{q}), \eta(\boldsymbol{q}))$ are points of convexity of $\tilde{e}$. Writing (19.4.9) with $(v', \eta') = (v(\boldsymbol{q}), \eta(\boldsymbol{q}))$ we obtain

$$\tilde{e}(v(\boldsymbol{q}), \eta(\boldsymbol{q})) \ge \tilde{e}(v(\boldsymbol{p}), \eta(\boldsymbol{p})) + \theta_0(\eta(\boldsymbol{q}) - \eta(\boldsymbol{p})) - \pi(v(\boldsymbol{q}) - v(\boldsymbol{p}))$$

and exchanging the roles of $\boldsymbol{p}, \boldsymbol{q}$ we obtain an inequality with the opposite sign and hence the equality. This equality can be rewritten as the constancy of $\tilde{e} - \tilde{\theta}\eta + \tilde{p}v$. □

Let us illustrate the extremum principles and the conditions of thermodynamic equilibrium (19.4.5) on a van der Waals fluid. Let us use the temperature as independent thermal variable, and assume that for every $(\theta, v) \in \bar{U}$, $\partial_\theta^2 \bar{f} < 0$. Then by Proposition 17.7.3, a specific volume v is a point of convexity of $\bar{f}(\cdot, \theta)$ if and only if $(v, \bar{e}(v, \theta))$ is a point of convexity of $\tilde{e}$, and if this is the case, $\partial_v \bar{p}(v, \theta) \le 0$. For the van der Waals fluids at sufficiently low temperatures, the pressure has the form Fig. 19.4.1.

19.4.3 Stability Along the Isotherm and Maxwell's Construction Let us examine the convexity of $\bar{f}(\cdot, \theta)$ along the isotherm on Fig. 19.4.1. First, $\partial_v \bar{p}(v, \theta) \le 0$ is violated on the isotherm between A, B and hence the convexity of $\bar{f}(\cdot, \theta)$ fails there. However, the convexity of $\bar{f}(\cdot, \theta)$ at v requires the inequality $\bar{f}(v', \theta) \ge \bar{f}(v, \theta) - \bar{p}(v, \theta)(v' - v)$ for all v' such that $(v', \theta) \in \bar{U}$. Using this, it is not hard to see that the convexity of $\bar{f}(\cdot, \theta)$ fails everywhere on the portion C, D, where the points C, D are determined by the equality of Gibbs functions:

$$\bar{f}(v_1,\theta)+\bar{p}(v_1,\theta)v_1=\bar{f}(v_2,\theta)+\bar{p}(v_2,\theta)v_2. \quad (19.4.10)$$

For every v outside this portion, $\bar{f}(\cdot,\theta)$ is convex at v.

The homogeneous states with the values of the specific volume from the portion A, B are absolutely unstable: the principle of minimum stored energy holds locally with the incorrect (opposite) inequality sign. The states cannot be realized. The homogeneous states with the specific volume outside C, D are absolutely stable: The principle of minimum energy holds there. In the portions C, A and B, D the homogeneous states are metastable: the thermodynamic functionals have local extrema there and the states can be realized with some care.

The equations (19.4.5) can be interpreted geometrically as follows (see Fig. 19.4.1). The equality of pressures says that (v_1,p_1), (v_2,p_2) determine a horizontal line segment parallel to the v axis; there are many such segments. The equality of chemical potentials (19.4.10) says that the line segment must be such that the area of the region between the segment and the part of the isotherm below the segment equals the area of the region between the segment and the isotherm above the segment. This is the equal area rule. The line segment satisfying it is known as Maxwell's line.

The physical meaning of the lower convex hull $\tilde{e}^{\,co}$ of $\tilde{e}$ is explained in the following proposition. It will be proved simultaneously with the next proposition.

19.4.4 Proposition *If σ is a state of total volume V and entropy H that satisfies the three equivalent principles, then $\tilde{e}^{\,co}$ has a differential at (V,H) and the total energy, pressure, and temperature of σ are $\tilde{e}^{\,co}(V,H)$, $-\partial_v\tilde{e}^{\,co}(V,H)$ and $\partial_\eta\tilde{e}^{\,co}(V,H)$, respectively.*

To explain this, consider homogeneous states characterized by the variable total volume and entropy V, H. As long as (V,H) is a point of convexity of $\tilde{e}$, the state is 'stable.' When the values of (V,H) enter the region where $\tilde{e}$ is not convex, the homogeneous states with (V,H) cease to be stable. However, there is still a stable state of total volume and entropy (V,H) – a heterogeneous state with two or more phases (see Proposition 19.4.5 below). Therefore, if the fluid is in a "black box" (of volume V and entropy H) and if the matters are arranged so as to prevent the occurrence of unstable states, then the apparatus will not detect the actual energy function $\tilde{e}$ and the pressure and temperature corresponding to it, but its lower convex hull $\tilde{e}^{\,co}$ and the pressure and temperature calculated as the derivatives of $\tilde{e}^{\,co}$. Proposition 19.4.4 will be proved simultaneously with the following proposition.

19.4.5 Proposition *If σ satisfies the three equivalent principles, then there exists a heterogeneous equilibrium state $\sigma' \in \Sigma^{\#}$ with the number of phases $r \le 3$, of the same pressure and temperature as σ, such that*

$$\mathsf{V}(\sigma)=\mathsf{V}(\sigma'),\qquad \mathsf{H}(\sigma)=\mathsf{H}(\sigma'),\qquad \mathsf{E}(\sigma)=\mathsf{E}(\sigma').$$

The state σ' satisfies the three equivalent principles.

This proposition is the main step towards the phase rule. Here we encounter for the first time the number $3 = n + 1$ where $n = 2$ is the dimension of the space of variables (v, η). The proof will show that 3 enters through Carathéodory's theorem on convex combinations. The heterogeneous state σ' is a *phantom state* accompanying any state σ satisfying the extremum principles. We shall see later that if σ satisfies the principle of minimum energy strictly, then necessarily σ and σ' differ at most by a rearrangement, which gives σ' physical reality.

Proof Define a mapping $f : \Omega \to \mathbb{R}^3$ by $f(p) = (v(p), \eta(p), \tilde{e}(v(p), \eta(p)))$, $p \in \Omega$. Then $\int_\Omega f\,dV = (\mathsf{V}(\sigma), \mathsf{H}(\sigma), \mathsf{E}(\sigma))$. Since the volume of Ω is 1, Proposition 16.1.4 tells us that $\int_\Omega f\,dV \in \mathrm{co}(f(\Omega))$.

This means that there exist a positive integer s, points of $\tilde{U}$ (v_i, η_i), $i = 1, \ldots, s$ and numbers $\alpha_i > 0$, $i = 1, \ldots, s$ such that $\sum_{i=1}^{s} \alpha_i = 1$ and

$$\mathsf{E}(\sigma) = \sum_{i=1}^{s} \alpha_i \tilde{e}(v_i, \eta_i), \quad \mathsf{V}(\sigma) = \sum_{i=1}^{s} \alpha_i v_i, \quad \mathsf{H}(\sigma) = \sum_{i=1}^{s} \alpha_i \eta_i. \tag{19.4.11}$$

Let $V := \mathsf{V}(\sigma)$, $H := \mathsf{H}(\sigma)$ and consider the set $\Sigma^{\#}(V, H)$. Since σ satisfies the principle of minimum energy, the infimum in (19.2.2) is a minimum with σ the minimizer. Hence $\mathsf{E}(\sigma) = \tilde{e}^{\mathrm{co}}(V, H)$ and (19.4.11) shows that $\{(v_i, \eta_i), i = 1, \ldots, s\}$ is a stable phase splitting of the point (V, H) (relative to $\tilde{e}$, see Sect. 16.3 for the definition). By Proposition 16.3.2 the stable phase splitting can be chosen to be an affinely independent set and hence it will contain $r \le 3$ points. Assuming that, the heterogeneous state σ' with the phases $\{(\bar{v}_i, \bar{\eta}_i), i = 1, \ldots, r\}$ present in amounts α_i will satisfy all the requirements of Proposition 19.4.5. Let us now prove that the pressure and temperature of σ are given as in the assertion of Proposition 19.4.4. The existence of the differential follows from Proposition 16.3.2. Let σ'' be a combination of the states σ and σ' in which the specific volumes and entropies of the states σ and σ' are present with relative amounts $\frac{1}{2}$. Then σ'' will have the same total energy, entropy, and volume as σ and it will satisfy the principle of minimum energy. By Proposition 19.4.1 it is an equilibrium state and hence the pressure and the temperature are uniform in σ''. Since at σ' these coincide with the derivatives of $\tilde{e}^{\mathrm{co}}$ by Proposition 16.3.2, the claim is proved. □

Finally, let us mention the solvability of the principle of minimum energy.

19.4.6 Proposition *Let $(V, H) \in \tilde{U}$ be given. Then there exists a solution of the principle of minimum energy in $\Sigma^{\#}(V, H)$ (i.e., there exists a $\sigma \in \Sigma^{\#}(V, H)$ which satisfies the principle of minimum energy) if and only if (V, H) has a stable phase splitting. In particular, if $\tilde{e}$ is convex outside a compact set then the solution exists for every (V, H).*

The definition of a stable phase splitting and of a function convex outside a compact set is in Sect. 16.3.

Proof That the existence of a solution implies the existence of a stable phase splitting was proved in the course of proof of Proposition 19.4.5. Conversely, the existence of a stable phase splitting leads to a heterogeneous state σ which is a minimizer of the total energy. Finally, if $\tilde{e}$ is convex outside a compact set, then by Proposition 16.3.5 every (V, H) has a stable phase splitting. □

19.5 Strict Extremum Principles. The Phase Rule

For the phase rule and for the dynamical stability, the extremum principles with the nonstrict inequality signs as formulated in Sect. 19.3 are not strong enough. One has to pass to the strict versions. A state σ_0 is said to satisfy any of the extremum principles 19.3.1, 19.3.2 and 19.3.3 ***strictly*** if the inequality in that principle is strict for every σ distinct from σ_0.

19.5.1 Proposition *A state σ satisfies the principle of minimum energy strictly if and only if it satisfies the principle of maximum entropy strictly. If σ satisfies these two equivalent strict principles, then it is a heterogeneous state with the number of phases $r \le 3$.*

The assertion $r \le 3$ is the ***Gibbs phase rule.*** For states that only satisfy the nonstrict version of the principle of minimum energy it does not hold: one can find energy functions $\tilde{e}$ for which there can be as many coexistent phases in a state minimizing energy as one wishes.

The formal argument presented, e.g., in Fermi [1937], goes as follows. The conditions of thermodynamic equilibrium (19.4.5) between phases i, j of a heterogeneous state represent a system of

$3(r-1)$ equations for $2r$ unknowns $(v_i, \eta_i), i = 1, \ldots, r$. A calculation "the number of unknowns minus the number of equations" gives that the solutions will form a $2r - 3(r-1) = 3 - r$ parameter family; if $r > 3$, the system is overdetermined and hence has no solutions. The number $f = 3 - r$ is the ***number of degrees of freedom***. It should represent the dimension of the set of solutions in the spaces of parameters $(v_i, \eta_i), i = 1, \ldots, r$. (Let us emphasize that the absolute amounts of the phases are *not* counted as the degrees of freedom.)

Proof The equivalence of the strict principle of minimum energy with the strict principle of maximum entropy is identical with the proof of the equivalence of the corresponding nonstrict versions and is therefore omitted. To prove that the state σ is a heterogeneous state with $r \le 3$ we proceed as follows. By Proposition 19.4.5 there exists a heterogeneous state σ' with $r \le 3$ such that $\mathsf{V}(\sigma) = \mathsf{V}(\sigma')$, $\mathsf{H}(\sigma) = \mathsf{H}(\sigma')$, $\mathsf{E}(\sigma) = \mathsf{E}(\sigma')$. By the strict principle of minimum energy then necessarily σ and σ' are equivalent. But if a state σ is equivalent to a heterogeneous state σ' with $r \le 3$, then σ itself must be a heterogeneous state with $r \le 3$. □

19.5.2 Proposition *Let σ be a heterogeneous state satisfying the principle of minimum energy strictly. Let $\{(v_i, \eta_i) : i = 1, \ldots, r\}$ be its phases and p_0, θ_0, g_0 the common values of the pressure, temperature and the Gibbs function ($= e - \theta\eta + pv$) of the phases. Then $\{(v_i, \eta_i) : i = 1, \ldots, r\}$ is the unique stable phase splitting of $(\mathsf{V}(\sigma), \mathsf{H}(\sigma))$ and in particular,*

(1) *the points $\{(v_i, \eta_i) : i = 1, \ldots, r\}$ are affinely independent and each of them is a point of convexity of $\tilde{e}$;*

(2) *if $(v, \eta) \in \tilde{U}$ is any point of convexity of $\tilde{e}$ that is not in $\{(v_i, \eta_i) : i = 1, \ldots, r\}$ and its pressure, temperature, and the Gibbs function are p_0, θ_0, g_0, respectively, then (v, η) is affinely independent of $\{(v_i, \eta_i) : i = 1, \ldots, r\}$.*

The phase simplex of $(\mathsf{V}(\sigma), \mathsf{H}(\sigma))$ is $\operatorname{co}(\{(v_i, \eta_i) : i = 1, \ldots, r\}) \subset \tilde{U}$.

Proof That all the phases are points of convexity follows from Proposition 19.4.2(2). The rest follows form Proposition 16.3.4. □

19.5.3 Proposition *If the fluid admits a heterogeneous state σ with $r \ge 2$ phases that satisfies the strict principle of minimum energy, then $\tilde{e}$ is not globally convex. More precisely, if $(v_i, \eta_i), i = 1, 2$, are any two phases of σ, then every point on the open line segment with endpoints $(v_i, \eta_i), i = 1, 2$ is not a point of convexity of $\tilde{e}$.*

Proof The endpoints $(v_i, \eta_i), i = 1, 2$, are points of convexity and the pressure, temperature, and Gibbs functions at $(v_i, \eta_i), i = 1, 2$, are the same. By Proposition 16.2.6 then any point (v, η) of convexity of $\tilde{e}$ on the open line segment with endpoints $(v_i, \eta_i), i = 1, 2$, will have the same pressure, temperature, and Gibbs functions as (v_i, η_i). But such a (v, η) is affinely dependent on $(v_i, \eta_i), i = 1, 2$, and the existence of such a (v, η) contradicts the strict principle of minimum energy by Proposition 19.5.2. □

The following proposition shows that the strict versions of the three extremum principles are not equivalent: no heterogeneous state with $r \ge 2$ can satisfy the principle of minimum canonical free energy strictly.

19.5.4 Proposition *A state σ satisfies the principle of minimum total canonical free energy strictly if and only if σ is a homogeneous state of specific volume v and entropy η and (v, η) is a point of very strict convexity of $\tilde{e}$. If this is the case, the state σ satisfies the principle of minimum energy strictly.*

Proof Let us first prove that if σ satisfies the principle of minimum canonical free energy strictly, it also satisfies the principle of minimum energy strictly. Let σ' be a state distinct from σ such that $\mathsf{V}(\sigma') = \mathsf{V}(\sigma)$, $\mathsf{H}(\sigma') = \mathsf{H}(\sigma)$. Then $\mathsf{P}(\sigma') > \mathsf{P}(\sigma)$ and by the definition $\mathsf{P} = \mathsf{E} - \theta_0 \mathsf{H} + \pi \mathsf{V}$ this gives $\mathsf{E}(\sigma') > \mathsf{E}(\sigma)$, i.e., the strict principle of minimum energy. If σ satisfies the principle of minimum canonical free energy strictly, then by the just proved and by Proposition 19.5.1, $\sigma = (v, \eta)$ is a heterogeneous state with $r \le 3$. Assume that $r \ge 2$ and derive a contradiction. Take, for definiteness, $r = 2$. Then there exist disjoint measurable subsets $\Omega_i \subset \Omega, i = 1, 2$ of positive volumes $V_i > 0$, and two points $(v_i, \eta_i) \in \tilde{U}, i = 1, 2$, such that $\Omega_1 \cup \Omega_2 = \Omega$ and $(v, \eta) = (v_i, \eta_i)$ on Ω_i. Since $e - \theta\eta + pv$

is constant over Ω, i.e., $\tilde{e}(v_1, \eta_1) - \theta_0\eta_1 + \pi v_1 = \tilde{e}(v_2, \eta_2) - \theta_0\eta_2 + \pi v_2 \equiv g$ (say), we have $\mathsf{P}(\sigma) = (V_1 + V_2)g = g$. The value P is thus independent of the amounts of the two phases (recall that the total amount of mass of the fluid is fixed). Therefore, if σ' is any state which contains the two phases in different amounts, then $\mathsf{P}(\sigma') = \mathsf{P}(\sigma)$; however, σ' is not equivalent to σ, which is a contradiction with the strict principle of minimum canonical free energy. Hence necessarily σ is a homogeneous state. To prove that (v, η) is a point of very strict convexity, take any $(v', \eta') \in \tilde{U}$, $(v', \eta') \neq (v, \eta)$ and let σ' be a homogeneous state with v', η'. Then σ' is distinct from σ and hence $\mathsf{P}(\sigma') > \mathsf{P}(\sigma)$, and this inequality asserts the very strict convexity. Conversely, starting from the inequality of very strict convexity of $\tilde{e}$ at (v, η), inserting the state σ' into it and integrating over Ω, we obtain $\mathsf{P}(\sigma') > \mathsf{P}(\sigma)$. □

19.5.5 Coexistent Phases, Tangent Planes, and Phase Simplexes Propositions 19.4.1, 19.4.2 and 19.5.3 justify the description of Subsect. 19.1.2 of the graph of the energy function at phases coexistent in a stable equilibrium state. It is clear that there can be energy functions that have heterogeneous states minimizing the energy with an arbitrary number of phases. However, by Proposition 19.5.1 such heterogeneous states cannot satisfy the principle of minimum of energy *strictly*. An energy function for which for every $(V, H) \in \tilde{U}$ there exists a state of total volume and entropy (V, H) which satisfies the principle of minimum energy strictly, is called the ***regular energy function.*** One may state equivalently that every $(V, H) \in \tilde{U}$ has a unique stable phase splitting. One can then speak about phase simplexes. Every point (v, η) of convexity of $\tilde{e}$ is its own phase simplex. A point (v, η) that is not a point of convexity is a convex combination of at most three points of convexity and these are uniquely determined by (v, η). If (v, η) is a convex combination of two points of convexity, then the phase simplex of (v, η) is the closed line segment joining these two points and the phase simplexes of the endpoints are the endpoints themselves. If (v, η) is a convex combination of three points of convexity, then the phase simplex of (v, η) is the closed triangle with vertices at these points. The phase simplex of any point inside of any side of the triangle is this side, and the phase simplex of any vertex is the vertex itself. Above the triangle the graph of $\tilde{e}^{\mathrm{co}}$ has a flat portion given by the affine function uniquely determined by the values of $\tilde{e}$ at the vertices. In all cases, the vertices of the phase simplex are the stable homogeneous phases present in the unique stable heterogeneous state minimizing the total energy, and the coefficients in the affine combination are the relative amounts of the phases. Any point inside the triangle is a ***triple point*** of the fluid, these points represent all possible values of the specific volumes and entropies for all possible combinations of the three phases subject to the constraint that the total mass be 1.

19.6 The Gibbs Function

The aim of this section is to introduce the Gibbs function $\hat{g}$ when the pressure relation is not invertible and to describe the relationship between differentiability properties of $\hat{g}$ and the coexistent phases.

19.6.1 Definition Define the ***Gibbs function*** $\hat{g}$ on $\mathbb{R}^2$ by

$$\hat{g}(p, \theta) = \inf\{\tilde{e}(v, \eta) + pv - \theta\eta : (v, \eta) \in \tilde{U}\}, \quad (p, \theta) \in \mathbb{R}^2. \tag{19.6.1}$$

The function $\hat{g}$ is concave. Moreover, $\hat{g}(p, \theta) = -\tilde{e}^*(-p, \theta)$ where $\tilde{e}^* : \mathbb{R}^2 \to \bar{\mathbb{R}}$ is the conjugate of $\tilde{e}$. The second conjugate $\tilde{e}^{**}$ coincides with $\tilde{e}^{\mathrm{co}}$ everywhere on $\tilde{U}$ as a consequence of Proposition 16.4.4(2). Then by Proposition 16.4.4(1),

$$\hat{g}(p, \theta) = \inf\{\tilde{e}^{\mathrm{co}}(v, \eta) + pv - \theta\eta : (v, \eta) \in \mathbb{R}^2\}, \quad (p, \theta) \in \mathbb{R}^2.$$

The Gibbs function $\hat{g}$ is defined for all values of p, θ including the unphysical range $G := \{(p, \theta) \in \mathbb{R}^2 : \text{either } p \le 0 \text{ or } \theta \le 0\}$. One can raise the question about the meaning of $\hat{g}$ in G. Without going into details, let us state that if $p \le 0$, then $\hat{g}(p, \theta) = -\infty$. If $p > 0$ and $\theta \le 0$, then $\hat{g}(p, \theta)$ is finite, but the behavior of $\hat{g}$ on $L := \{(p, \theta) \in \mathbb{R}^2 : p > 0, \theta \le 0\}$ adds no new information: in L, $\hat{g}$ is constant on each vertical half-line $\{(p, \theta) \in \mathbb{R}^2 : \theta \le 0\}$ ($p > 0$ fixed), with value

$\hat{g}(p, \theta) = \tilde{e}(v, 0) + \tilde{p}(v, 0)v$, where $v > 0$ is determined from the condition $p = -\partial_v \tilde{e}(v, 0)$ (provided $\tilde{e}$ has a differentiable extension to the line $\eta = 0$). Also note that for any $\theta > 0$, one has

$$\hat{g}(p, \theta) \to -\infty \text{ as } p \downarrow 0.$$

19.6.2 Differentiability Any realistic smooth energy function $\tilde{e}$ leads to $\hat{g}$ which has points (p, θ) where it is not differentiable. This is not a mathematical pathology, but a reality related to phase transitions. By analogy with the case when $\hat{g}$ is obtained as a Legendre transformation of $\tilde{e}$ (under the invertibility conditions), the derivative of $\hat{g}$ with respect to p, if it exists, should be interpreted as volume; and minus the derivative of $\hat{g}$ with respect to θ should be related to entropy. Adapting the definition of the subdifferential to the present context, we denote, for a given $(p, \eta) \in \mathbb{R}^2$, by $S(p, \theta)$ the set of all pairs $(V, H) \in \mathbb{R}^2$ such that

$$\hat{g}(p', \theta') \le \hat{g}(p, \theta) + V(p' - p) - H(\theta' - \theta)$$

for every $(p', \theta') \in \mathbb{R}^2$. (The superdifferential of the concave function $\hat{g}$.) The set $S(p, \theta)$ is a (possibly empty) convex set. By Proposition 16.4.4(4), if $(V, H) \in S(p, \theta)$ then $V \ge 0$, $H \ge 0$. We shall restrict ourselves only to those pairs $(p, \theta) \in \mathbb{R}^2$ for which $S(p, \theta) \subset (0, \infty) \times (0, \infty)$. Let us denote the set of all these pairs by P. This set corresponds to the set G of Sect. 16.4. (The physical meaning of $S(p, \theta)$ is established by Proposition 19.6.5 below.) Let us further denote, for every $(p, \eta) \in \mathbb{R}^2$, by $\tilde{U}(p, \theta)$ the set of points of convexity $(v, \eta) \in \tilde{U}$ of $\tilde{e}$ such that $\tilde{p}(v, \eta) = p$, $\tilde{\theta}(v, \eta) = \theta$. Throughout the rest of this section, assume that for every $(V, H) \in \tilde{U}$ there exists a stable phase splitting.

19.6.3 Proposition *For every $(p, \theta) \in P$ we have $S(p, \theta) = \mathrm{co}(\tilde{U}(p, \theta))$.*

Proof Apply Proposition 16.4.5(2) to $f \equiv \tilde{e}$, $D \equiv (0, \infty) \times (0, \infty)$. □

This is the main result on the differentiability of $\hat{g}$. The elements (V, H) of $S(p, \theta)$ are coefficients of all planes in $\mathbb{R}^3$ which touch the graph of $\hat{g}$ at the point $(p, \theta, \hat{g}(p, \theta))$ and the rest of the graph is on or below that plane. The result shows that all coefficients of such planes coincide with the set of all convex combinations of the volume–entropy pairs of corresponding to points of convexity of $\tilde{e}$ with the pressure p and temperature θ. The function $\hat{g}$ is differentiable at (p, θ) if there is exactly one such plane. The following result gives a complete description of such points.

19.6.4 Proposition *The function $\hat{g}$ is differentiable at $(p, \theta) \in P$ if and only if $(p, \theta) = (-\partial_v \tilde{e}(v, \eta), \partial_\eta \tilde{e}(v, \eta))$ for some point (v, η) of very strict convexity of $\tilde{e}$; then $(v, \eta) := (\partial_p \hat{g}(p, \theta), -\partial_\theta \hat{g}(p, \theta))$. The mapping $(v, \eta) \mapsto (-\partial_v \tilde{e}(v, \eta), \partial_\eta \tilde{e}(v, \eta))$ establishes a one-to-one correspondence from the set of points of very strict convexity of $\tilde{e}$ onto the set of all $(p, \theta) \in P$ where $\hat{g}$ is differentiable. The inverse of this mapping is $(p, \theta) \mapsto (\partial_p \hat{g}(p, \theta), -\partial_\theta \hat{g}(p, \theta))$. On these sets, $\tilde{e}$ and $\hat{g}$ are related by the Legendre transformation.*

Proof Apply Proposition 16.4.5(3) to $f \equiv \tilde{e}$, $D \equiv (0, \infty) \times (0, \infty)$. □

Hence, if σ is a heterogeneous equilibrium state of pressure p and temperature θ satisfying the principle of minimum energy and if the number of phases of σ is $r \ge 2$, then $\hat{g}$ is not differentiable at (p, θ).

19.6.5 Proposition *Let $(p, \theta) \in P$. Then $(V, H) \in S(p, \theta)$ if and only if there exists an equilibrium state σ of pressure p and temperature θ satisfying the principle of minimum energy such that $V = \mathsf{V}(\sigma)$, $H = \mathsf{H}(\sigma)$. Each such a state then satisfies $(v(\boldsymbol{p}), \eta(\boldsymbol{p})) \in S(p, \theta)$ for a.e. $\boldsymbol{p} \in \Omega$. Among all states with the described properties is a heterogeneous equilibrium state.*

Proof If $(V, H) \in S(p, \theta)$, then by Proposition 19.6.3 there exists a positive integer s, points of convexity of $\tilde{e}$, (v_i, η_i), $i = 1, \ldots, s$, and numbers $\alpha_i > 0$, $i = 1, \ldots, s$, numbers such that

$$V = \sum_{i=1}^{s} \alpha_i v_i, \quad H = \sum_{i=1}^{s} \alpha_i \eta_i, \quad \sum_{i=1}^{s} \alpha_i = 1$$

and $\tilde{p}(v_i, \eta_i) = p$, $\tilde{\theta}(v_i, \eta_i) = \theta_0$. Then the heterogeneous state σ in which the phases (v_i, η_i) are present in amounts α_i satisfies the principle of minimum energy and is such that $V = \mathsf{V}(\sigma)$, $H = \mathsf{H}(\sigma)$. Conversely, if σ is an equilibrium state of pressure and temperature (p, θ), then by Proposition 19.4.2 for a.e. $\boldsymbol{p} \in \Omega$ the point $(v(\boldsymbol{p}), \eta(\boldsymbol{p})) \in S(p, \theta)$ is a point of convexity of $\tilde{e}$ and the pressure and temperature at $(v(\boldsymbol{p}), \eta(\boldsymbol{p}))$ are (p, θ). Then $(v(\boldsymbol{p}), \eta(\boldsymbol{p})) \in S(p, \theta)$ and as $S(p, \theta)$ is convex, $\int_\Omega (v(\boldsymbol{p}), \eta(\boldsymbol{p}))\, dV$ is also in $S(p, \theta)$ by Proposition 16.1.4. But $\int_\Omega (v(\boldsymbol{p}), \eta(\boldsymbol{p}))\, dV = (V, H)$ by hypothesis. □

19.6.6 Edges and Vertices If there exists a strictly energy-minimizing heterogeneous state of pressure and temperature (p, θ), a description of the local behavior of $\hat{g}$ near (p, θ) is available. We shall see that near (p, θ), $\hat{g}$ looks locally like a vertex at the end of three joining edges, or like an edge, or like a plane, depending on the number of phases of the strictly minimizing state. A function $f : \mathbb{R}^2 \to \mathbb{R}$ is said to be a ***concave vertex*** at (p_0, θ_0, g_0) (or a ***concave edge*** through (p_0, θ_0, g_0), or a ***plane*** through (p_0, θ_0, g_0)) if there exist $r = 3$ (or $r = 2$ or $r = 1$) affinely independent vectors $(v_i, \eta_i) \in \mathbb{R}^2$, $i = 1, \ldots, r$ such that

$$f(p, \theta) = \min\{(p - p_0)v_i - (\theta - \theta_0)\eta_i + g_0, i = 1, \ldots, r\}, \quad (p, \theta) \in \mathbb{R}^2.$$

A function $h : \mathbb{R}^2 \to \mathbb{R}$ is said to be ***locally a concave vertex (concave edge, plane)*** at (p_0, θ_0) if there exist a concave vertex (concave edge, plane) f such that

$$|f(p, \theta) - h(p, \theta)|/(|p - p_0| + |\theta - \theta_0|) \to 0 \text{ as } p \to p_0, \theta \to \theta_0.$$

19.6.7 Proposition *Let (p_0, θ_0) be given. If there exists a heterogeneous state σ of pressure and temperature (p_0, θ_0) satisfying the principle of minimum energy strictly, then $S(p, \theta)$ is exactly the phase simplex of (V, H), where V, H are the total volume and entropy of the state σ. Moreover, $\hat{g}$ is at (p_0, θ_0) locally a plane, concave edge, or a concave vertex if the number of phases of σ is $r = 1, 2, 3$, respectively.*

Of course, the coefficients (v_i, η_i) occurring in the definitions of a vertex, edge, and plane are exactly the specific volumes and entropies of the coexistent phases, i.e., the elements of the phase simplex of (V, H). The assertion about the identity of $S(p, \theta)$ with the phase simplex is a direct consequence of the definition of a phase simplex and of Proposition 19.6.3. The proof of the local behavior of $\hat{g}$ is then a consequence of the asserted form of the superdifferential $S(p, \theta)$ and of the standard differentiability properties of convex functions (cf. ROCKAFELLAR [1970, Thm. 23.4]). The details are omitted.

The results of Sect. 16.4 show that the function $\hat{g}$ depends only on the behavior of $\tilde{e}$ at its points of convexity. More precisely, if $\tilde{e}_1$, $\tilde{e}_2$ are two energy functions on $\tilde{U}$ such that (1) for every $(v, \eta) \in \tilde{U}$ the point (v, η) is a point of convexity of $\tilde{e}_1$ if and only if (v, η) is a point of convexity of $\tilde{e}_2$ and (2) $\tilde{e}_1(v, \eta) = \tilde{e}_2(v, \eta)$ for every point of convexity (v, η) for $\tilde{e}_1, \tilde{e}_2$, then the corresponding $\hat{g}_1, \hat{g}_2$ are identical. In other words, *the mechanism of Fenchel's transformation completely neglects the unstable regions of the state space.*

19.7 Strong Minima and Dynamical Stability of Equilibrium States

We have seen in Sect. 15.1 that to obtain the Liapunov stability of equilibrium states, one has to seek strong or uniformly strong extrema. We shall establish here a result on uniformly strong extrema for *homogeneous* states of fluids in the absence of external field of force.

Define a metric d on $\Sigma^{\#}$ by

$$d(\sigma_1, \sigma_2) = \int_{\Omega} |x_1(p) - x_2(p)| \, dV \tag{19.7.1}$$

for every $\sigma_1 = (v_1, \eta_1), \sigma_2 = (v_2, \eta_2) \in \Sigma^{\#}$, where $x_i(p) = (v_i(p), \eta_i(p)), i = 1, 2$, for every $p \in \Omega$ and $|\cdot|$ is the euclidean distance in $\mathbb{R}^2$. The metric d is equivalent to the metric $\tilde{d}$ given by

$$\tilde{d}(\sigma_1, \sigma_2) = \int_{\Omega} (|v_1 - v_2| + |\eta_1 - \eta_2|) \, dV.$$

The reader is referred to Sect. 16.2 for the concept of very strict convexity and to Proposition 15.1.5 for the concept of a uniformly strong local and global minima.

19.7.1 Proposition *Let $\sigma_0 = (v_0, \eta_0)$ be a homogeneous state compatible with the environment. If $\tilde{e}$ is very strictly convex at (v_0, η_0) and convex at every point sufficiently close to (v_0, η_0), then P has at σ_0 a uniformly strong global minimum on $\Sigma^{\#}$ with respect to d. That is, $\mathsf{P}(\sigma) \geq \mathsf{P}(\sigma_0)$ for every $\sigma \in \Sigma^{\#}$ and for every $\varepsilon > 0$ there exists a $\delta > 0$ such that if $\sigma \in \Sigma^{\#}$ satisfies*

$$\mathsf{P}(\sigma) - \mathsf{P}(\sigma_0) < \delta,$$

then

$$\int_{\Omega} |v - v_0| \, dV < \varepsilon, \qquad \int_{\Omega} |\eta - \eta_0| \, dV < \varepsilon.$$

That is, the knowledge that the canonical free energy of the state σ is close to that of the equilibrium state enables one to deduce that the state σ itself is also close to the equilibrium state.

Proof We shall apply the notion of a uniformly strong minimum in the form of Condition (2) of Proposition 15.1.5. Applying Proposition 16.2.8 to the function $f = \tilde{e}$ and to the point $w = (v_0, \eta_0)$ we obtain the existence of the function C occurring in that proposition. Define the function $\hat{\gamma}$: $[0, \infty) \to [0, \infty)$ by $\hat{\gamma}(n) = C(n/2)\, n/2$ for each $n \geq 0$. Then $\hat{\gamma}$ is a nondecreasing nonnegative function with $\hat{\gamma}(n) > 0$ for $n > 0$. Our aim is to prove that

$$\mathsf{P}(\sigma) - \mathsf{P}(\sigma_0) \geq \hat{\gamma}(d(\sigma, \sigma_0)) \tag{19.7.2}$$

for every state $\sigma \in \Sigma^{\#}$. From the identification $f = \tilde{e}$ and from $\nabla f(w) = (-\pi, \theta_0)$ we find that

$$\mathsf{P}(\sigma) - \mathsf{P}(\sigma_0) = \int_{\Omega} (f(x(p)) - f(w) - \nabla f(w) \cdot (x(p) - w)) \, dV = \int_{\Omega} h(x(p)) \, dV$$

where $x(p) = (v(p), \eta(p))$ and $h(x) = f(x) - f(w) - \nabla f(w) \cdot (x - w)$. Since $w = (v_0, \eta_0)$ is a point of convexity of f, the integrand h is nonnegative and hence $\mathsf{P}(\sigma) \geq \mathsf{P}(\sigma_0)$ for every $\sigma \in \Sigma^{\#}$. Define, for each $s > 0$, the regions Ω_s and Ω_s^c by

$$\Omega_s = \{p \in \Omega : |x(p) - w| \geq s\}, \quad \Omega_s^c = \{p \in \Omega : |x(p) - w| < s\};$$

hence

$$\mathsf{P}(\sigma) - \mathsf{P}(\sigma_0) = \int_{\Omega_s} h(x(p))\,dV + \int_{\Omega_s^c} h(x(p))\,dV.$$

The nonnegativeness of h, the definition of Ω_s and Proposition 16.2.8 give

$$\mathsf{P}(\sigma) - \mathsf{P}(\sigma_0) \geq \int_{\Omega_s} h(x(p))\,dV \geq C(s) \int_{\Omega_s} |x(p) - w|\,dV. \tag{19.7.3}$$

On the other hand, we have

$$d(\sigma, \sigma_0) = \int_{\Omega_s} |x(p) - w|\,dV + \int_{\Omega_s^c} |x(p) - w|\,dV \leq \int_{\Omega_s} |x(p) - w|\,dV + s,$$

where the last inequality follows from the definition of Ω_s^c and $V(\Omega) = 1$. A combination with (19.7.3) provides

$$d(\sigma, \sigma_0) \leq \frac{1}{C(s)} \int_{\Omega_s} h(x(p))\,dV + s. \tag{19.7.4}$$

Let us now set $s = d(\sigma, \sigma_0)/2$ and insert this s into (19.7.4). We obtain

$$\tfrac{1}{2} d(\sigma, \sigma_0) C(d(\sigma, \sigma_0)/2) \leq \int_{\Omega_s} h(x(p))\,dV,$$

which, using the definition of $\hat{\gamma}$, can be rewritten as (19.7.2). □

Let us now consider a fluid in an environment described at the beginning of Sect. 19.2. For dynamical states $\sigma = (\chi, \eta, v)$, the total canonical energy P is given by

$$\mathsf{P}(\sigma) = \mathsf{K}(\sigma) + \mathsf{E}(\sigma) - \theta_0 \mathsf{H}(\sigma) + \pi \mathsf{V}(\sigma),$$

where $\mathsf{K}(\sigma)$ is the kinetic energy. By Proposition 13.4.1, P is a Liapunov function for all processes compatible with the environment (of smoothness specified in Sect. 13.1). Using the general result Proposition 15.1.8 we obtain the following stability result.

19.7.2 Proposition *A state σ_0 satisfying the hypotheses of Proposition* 19.7.1 *is Liapunov stable with respect to the class of all processes in the environment in the following sense. For every $\varepsilon > 0$ there exists a $\delta > 0$ such that if $\pi = (\chi(\cdot,\cdot), \eta(\cdot,\cdot), v(\cdot,\cdot))$ is a process of the fluid which satisfies*

$$\mathsf{P}(\pi^i) - \mathsf{P}(\sigma_0) < \delta,$$

then for every $t \geq 0$ one has

$$\int_\Omega \tfrac{1}{2}\rho_\circ v(\cdot, t)^2\,dV < \varepsilon, \quad \int_\Omega |v(\cdot, t) - v_0| \rho_\circ\,dV < \varepsilon, \quad \text{and} \quad \int_\Omega |\eta(\cdot, t) - \eta_0| \rho_\circ\,dV < \varepsilon.$$

For a given ε, by choosing the initial data close enough to those of the equilibrium state one can ensure that the process does not leave the ε-neighborhood of the equilibrium state.

19.8 The Equilibrium of Fluids Under the Body Forces

Consider a fluid in the same situation as that described in the beginning of Sect. 19.2 except that now the body force b_0 is different from $\mathbf{0}$. *Assume that b_0 has a potential ψ_b (see* (13.3.21)) *and that* grad $\psi_b(x) \neq \mathbf{0}$ *for every $x \in R$.* Define, for every $l \in \mathbb{R}$, the set R_l by

$$R_l = \{x \in R : \psi_b(x) < l\},$$

i.e., the set of all points with potential below the value l.

19.8.1 Spatial Description of Rest States It is convenient to identify rest states with the triples $\sigma = (\Gamma, v, \eta)$ where $\Gamma \subset$ Eucl is the actual spatial region occupied by the fluid, and v, η are the fields of specific volume and entropy expressed as functions of the spatial position $\boldsymbol{x} \in$ Eucl. More specifically, a rest state σ is any triple $\sigma = (\Gamma, v, \eta)$ such that

(1) Γ is a connected subset of Eucl;

(2) v, η are piecewise continuously differentiable fields on Γ in the sense that Γ can be partitioned into a finite set of subregions Γ_i, $i = 1, \ldots, N$ such that each Γ_i has a Lipschitz boundary, $\Gamma_i \cap \Gamma_j = \emptyset$ for $i \neq j$,

$$\Gamma = \bigcup \{\Gamma_i, i = 1, \ldots, N\} \tag{19.8.1}$$

and v, η are continuous on each Γ_i with a smooth extension to an open set containing the closure of Γ_i;

(3) the mass of the fluid is fixed, i.e., $\int_\Gamma v^{-1}\, dv = M_0$, where M_0 is the total mass of the fluid,

(4) the values of (v, η) in Γ_i as well as their limiting values on the boundaries are in the domain $\tilde{U}$.

We denote by $\Sigma^{\#}$ the set of all triples $\sigma = (\Gamma, v, \eta)$ satisfying (1)–(4), and call its elements states. No confusion can arise from that, since it will be always clear from the context whether the referential or spatial description of rest states will be used. The regions Γ_i can represent different phases of the fluid and their boundaries the phase boundaries. Every rest state $\sigma = (\chi, \eta)$ in the sense of Subsect. 13.1.1 gives rise to a state $\sigma = (\Gamma, v', \eta')$ in the present sense, with

$$\Gamma = \chi(\Omega), \quad v'(\boldsymbol{x}) = v(\chi^{-1}(\boldsymbol{x})), \quad \eta'(\boldsymbol{x}) = \eta(\chi^{-1}(\boldsymbol{x})), \quad \boldsymbol{x} \in \Gamma. \tag{19.8.2}$$

Many pairs $\sigma = (\chi, \eta)$ lead to the same $\sigma = (\Gamma, v', \eta')$. However, not every $\sigma = (\Gamma, v', \eta')$ can be represented as in (19.8.2), because there may not exist a χ mapping Ω onto Γ in a one-to-one way with χ, χ^{-1} continuous and piecewise continuously differentiable.

A state $\sigma = (\Gamma, v, \eta)$ is said to be an ***equilibrium state*** if the field of temperature is uniform over Γ. An equilibrium state $\sigma = (\Gamma, v, \eta)$ is said to be ***compatible with the environment*** if $\Gamma \subset R$, its temperature coincides with the ambient temperature,

$$\operatorname{grad} p + \rho \operatorname{grad} \psi_b = \mathbf{0}, \quad [p] = 0 \tag{19.8.3}$$

everywhere outside and on the singular surface, respectively, and

$$\bar{p}(v, \theta_o) = \pi \ \text{ on } \ \partial\Gamma \setminus \partial R. \tag{19.8.4}$$

19.8.2 Proposition *Let $\sigma = (\Gamma, v, \eta)$ be an equilibrium state compatible with the environment. Then the Gibbs function*

$$g = e - \theta_0 \eta + pv + \psi_b = f + pv + \psi_b \tag{19.8.5}$$

is constant on every component Γ_i from 19.8.1(2).

This is the well-known integral for the equations of mechanical equilibrium for fluids. The proof amounts either to a direct verification using the thermostatic pressure relation for fluids and (19.8.3), or to the specialization of Proposition 14.2.4 (Eshelby's conservation law) to fluids. Although the direct verification is simple, it is instructive to derive the result from Proposition 14.2.4: It suffices to insert the form $(13.3.31)_2$ of $\boldsymbol{P}$ for conservative loading environments for fluids into (14.2.2) to obtain (19.8.5). We shall see that the extremum principles imply that the quantity (19.8.5) is constant over the *whole of* Γ.

19.8.3 Using the spatial description, the total thermodynamic quantities E, H, U, P and W are given by

$$\left.\begin{aligned} \mathsf{E}(\sigma) &= \int_\Gamma \tilde{e}\rho\, dv, \quad \mathsf{H}(\sigma) = \int_\Gamma \eta\rho\, dv, \quad \mathsf{W}(\sigma) = \int_\Gamma (\psi_b + \pi v)\rho\, dv, \\ \mathsf{U}(\sigma) &= \mathsf{E}(\sigma) + \mathsf{W}(\sigma) = \int_\Gamma (\tilde{e} + \psi_b + \pi v)\rho\, dv, \\ \mathsf{P}(\sigma) &= \mathsf{E}(\sigma) - \theta_0 \mathsf{H}(\sigma) + \mathsf{W}(\sigma) = \int_\Gamma (\tilde{e} - \theta_0 \eta + \psi_b + \pi v)\rho\, dV, \end{aligned}\right\} \tag{19.8.6}$$

where $\sigma = (\Gamma, v, \eta) \in \Sigma^{\#}$ and the specific internal energy $\tilde{e}$ is evaluated at $(v(\boldsymbol{x}), \eta(\boldsymbol{x}))$.

Define the set of competitor states Σ_0 by

$$\Sigma_0 = \{\sigma = (\Gamma, v, \eta) \in \Sigma^{\#} : \Gamma \subset R\}.$$

Alternatively, one can assume that the canister is completely filled by the fluid, in which case

$$\Sigma_0 = \{\sigma = (\Gamma, v, \eta) \in \Sigma^\# : \Gamma = R\}$$

The potential energy is

$$W(\sigma) = \int_\Gamma \psi_b \rho \, dv + \pi V(\sigma),$$

where $V(\sigma)$ is the total volume of the fluid (see 13.3.10). The term containing π can be omitted if the cavitation is not admitted, for in this case the value of V is constant on the set of all competitor states.

We consider the four extremum principles 15.2.1–15.2.4 with the above choice of competitor states.

19.8.4 Proposition *Let an equilibrium state σ_0 satisfy any of the four extremum principles* 15.2.1–15.2.4. *Then*

(1) *for every spatial point $(v_0, \eta_0) = (v_0(\boldsymbol{x}), \eta_0(\boldsymbol{x}))$ corresponding to $\boldsymbol{x} \in \Gamma$ one has*

$$\tilde{e}(v, \eta) \geq \tilde{e}(v_0, \eta_0) + \theta_0(\eta - \eta_0) - \tilde{p}(v_0, \eta_0)(v - v_0) \tag{19.8.7}$$

for every $(v, \eta) \in \tilde{V}$; (19.8.7) *holds also at the limiting points (v_0, η_0) on the boundary of the fluid and at the limiting values of $(v_0, \eta_0) = (v^\pm, \eta^\pm)$ on the singular surface;*

(2) *at every point of the singular surface the limiting values $(v^\pm, \eta^\pm)$ satisfy the* ***Gibbs relation***

$$\bar{f}(v^+, \theta_0) + \bar{p}(v^+, \theta_0)v^+ = \bar{f}(v^-, \theta_0) + \bar{p}(v^-, \theta_0)v^-; \tag{19.8.8}$$

(3) *the function*

$$e - \theta_0 \eta + pv + \psi_b \tag{19.8.9}$$

is uniform over the region occupied by the fluid.

We recognize (1) to assert the convexity of the internal energy with respect to the specific volume v and specific entropy η at every value (v, η) that is realized in the state. Although a direct proof of (19.8.7) is easy (cf. Sect. 19.4), the result also follows from the general theory. Our choice of the competitor states satisfies the Assumptions 15.3.2 and 17.1.3. Therefore, by the results of Chap. 17 one arrives at the quasiconvexity of the function $\tilde{e}(\boldsymbol{F}, \eta)$ and combining this with the results of Sect. 17.7 one obtains (19.8.7). The equality (2) of the chemical potentials on both sides of the phase boundary is seen to be a consequence of (19.8.7). Again, the proof is simple, but there is no need for it, because it can be obtained by specializing the Maxwell relation (see Sect. 17.4) to fluids. Concerning Assertion (3), recall that the uniformity of $e - \theta_0 \eta + pv + \psi_b$ on each region where the state parameters are continuously differentiable was established for any equilibrium state without any reference to the extremum principles. What is asserted here is the continuity of $e - \theta_0 \eta + pv + \psi_b$ across the phase boundaries. Indeed Assertion (3) follows from (2) and the constancy of $e - \theta_0 \eta + pv + \psi_b$ over every connected region where the state parameters are continuously differentiable. Recall that it is assumed that Γ is connected.

We now establish the equivalence of the extremum principles. Two results will be given.

19.8.5 Proposition *Let $\sigma_0 = (\Gamma_0, v_0, \eta_0)$ be an equilibrium state compatible with the environment. Suppose that the energy function $\tilde{e}$ is convex, that the fluid admits the pressure and temperature as independent variables in the constitutive equations and that the Gibbs function g satisfies*

$$\lim_{p \to 0} g(p, \theta_0) = -\infty. \tag{19.8.10}$$

Then the four extremum principles 15.2.1–15.2.4 *are equivalent. They hold if and only if there exists a level l_0 such that*

$$\Gamma_0 = R_{l_0}, \tag{19.8.11}$$

the quantity

$$e_0 - \theta_0 \eta_0 + p_0 v_0 + \psi_b \tag{19.8.12}$$

is constant over the fluid and the convexity inequality (19.8.7) *holds for every point of the fluid. Here e_0 and p_0 are the specific internal energy and the pressure in the state σ_0 expressed as functions of $\boldsymbol{x}$.*

The assumption that $\tilde{e}$ is convex excludes phase transitions. Below a version of the proposition will be given that is appropriate for situations when several phases are present. Equation (19.8.10) is completely realistic; let us note that the ideal gas satisfies it. Condition (19.8.11) says that the free part of the boundary of the fluid, $\partial\Gamma_0 \setminus \partial R$, coincides with some level set of the potential and that the fluid completely fills the region where the potential is smaller.

Proof If any of the extremum principles holds, then also the principle of minimum energy holds. The convexity inequality and the uniformity of the function (19.8.12) then follow from Proposition 19.8.4.

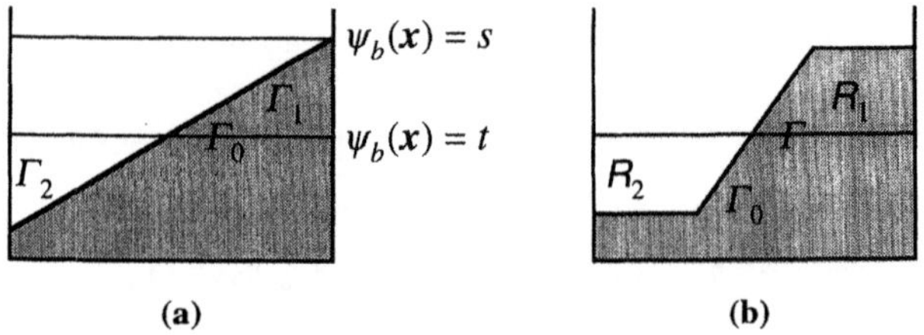

Fig. 19.8.1

Now suppose that there is no l_0 such that (19.8.11) holds. Denote by s the least value of the potential level such that $\Gamma_0 \subset R_s$. See Fig. 19.8.1(a). Then there exists a $t < s$ such that $R_t \setminus \Gamma_0$ and $\Gamma_0 \setminus R_t$ have positive volumes. In these two sets we can arbitrarily choose two regions Γ_1, Γ_2, of equal volume

$$\Gamma_1 \subset \Gamma_0 \setminus R_t, \quad \Gamma_2 \subset R_t \setminus \Gamma_0, \quad V(\Gamma_1) = V(\Gamma_2) > 0,$$

and move the fluid from Γ_1 into the set Γ_2 is such a way that the density does not change. Since the set Γ_2 is on the lower level of potential, the potential energy decreases, while the internal energy and entropy are unchanged. Formally, since the balls Γ_1 and Γ_2 have the same volumes, there exists a measure-preserving, one-to-one continuous mapping z from Γ_2 onto Γ_1. We now construct a state $\sigma = (\Gamma, \nu, \eta)$ in the following way: $\Gamma = (\Gamma_0 \setminus \Gamma_1) \cup \Gamma_2$, $\nu = \nu_0, \eta = \eta_0$ on $\Gamma_0 \setminus \Gamma_1$ and $\nu(x) = \nu_0(z(x))$, $\eta(x) = \eta_0(z(x))$ on Γ_2. Then clearly $\mathsf{H}(\sigma) = \mathsf{H}(\sigma_0)$, $\mathsf{V}(\sigma) = \mathsf{V}(\sigma_0)$ but $\mathsf{U}(\sigma) < \mathsf{U}(\sigma_0)$. This contradiction with the principle of minimum energy proves that the state σ_0 satisfies (19.8.11). This proves the necessity of the conditions enumerated in the proposition for a state to satisfy any of the extremum principles.

We shall now prove that if the state satisfies the conditions enumerated in the proposition, then it satisfies the principle of minimum canonical free energy. Let us denote by C the constant value of (19.8.12). Then

$$g(p_0, \theta_0) + \psi_b = C, \quad x \in \Gamma_0, \tag{19.8.13}$$

where the argument x has been omitted. Recall that the derivative $\hat{\nu} := \partial_p g$ is the specific volume. Hence it is positive and, consequently, g is a strictly increasing function of p. Equation (19.8.13) can be solved for p_0 to obtain the field $p_0(x)$ as a function of $\psi_b(x)$. In particular, the pressure is constant on the level sets of the potential. This serves as a motivation for the following construction of the extensions $\tilde{\nu}, \tilde{\eta}, \tilde{p}$ of the functions ν_0, η_0, p_0 outside Γ_0. Define, for each value of $\psi \in \mathbb{R}$, the number $P(\psi)$ by

$$g(P(\psi), \theta_0) + \psi = C. \tag{19.8.14}$$

In view of the assumption (19.8.10), the function $P(\cdot)$ is defined for all values of ψ. Define the extensions $\tilde{\nu}, \tilde{\eta}, \tilde{p}$ by

$$\tilde{\nu}(x) = \hat{\nu}(P(\psi_b(x)), \theta_0), \quad \tilde{\eta}(x) = -\partial_\theta g(P(\psi_b(x)), \theta_0), \quad \tilde{p}(x) = P(\psi_b(x)), \quad x \in R.$$

The functions $\tilde{\nu}, \tilde{\eta}, \tilde{p}$ coincide with ν_0, η_0, p_0 on Γ_0; moreover, we have

$$\tilde{e}(\tilde{\nu}, \tilde{\eta}) - \theta_0\tilde{\eta} + \tilde{p}\tilde{\nu} + \psi_b = C, \quad x \in R. \tag{19.8.15}$$

Now let $\sigma = (\Gamma, \nu, \eta)$ be any state of the fluid. Our goal is to prove that $\mathsf{P}(\sigma) \geq \mathsf{P}(\sigma_0)$. By the convexity of $\tilde{e}$ we have $\tilde{e}(\nu, \eta) \geq \tilde{e}(\tilde{\nu}, \tilde{\eta}) - \tilde{p}(\nu - \tilde{\nu}) + \theta_0(\eta - \tilde{\eta})$, $x \in \Gamma$, where here and in the subsequent formulas the argument x is omitted. Adding $\psi_b + \pi\nu$ to both sides and rearranging, we obtain

$$\tilde{e}(\nu, \eta) - \theta_0\eta + \psi_b + \pi\nu \geq \tilde{e}(\tilde{\nu}, \tilde{\eta}) + \tilde{p}\tilde{\nu} - \theta_0\tilde{\eta} + \psi_b - \tilde{p}\nu + \pi\nu.$$

A multiplication by ρ and the use of (19.8.15) results in

$$\rho(\tilde{e}(v,\eta) - \theta_0\eta + \psi_b) + \pi \geq C\rho + \pi - \tilde{p}.$$

We now integrate this inequality over Γ to obtain

$$\mathsf{P}(\sigma) \geq CM_0 + \int_{\Gamma} (\pi - \tilde{p}(\boldsymbol{x}))\, dv, \tag{19.8.16}$$

where M_0 is the mass of the fluid. Set $R_0 = \Gamma \cap \Gamma_0$, $R_1 = \Gamma \setminus \Gamma_0$, $R_2 = \Gamma_0 \setminus \Gamma$, so that $\Gamma_0 = R_0 \cup R_2$, $\Gamma = R_0 \cup R_1$, see Fig. 19.8.1(b). Equation (19.8.16) can be rewritten as

$$\mathsf{P}(\sigma) \geq CM_0 + \int_{R_0} (\pi - \tilde{p}(\boldsymbol{x}))\, dv + \int_{R_1} (\pi - \tilde{p}(\boldsymbol{x}))\, dv. \tag{19.8.17}$$

On the other hand, the definition of $\mathsf{P}(\sigma_0)$ gives

$$\mathsf{P}(\sigma_0) = CM_0 + \int_{R_0} (\pi - \tilde{p}(\boldsymbol{x}))\, dv + \int_{R_2} (\pi - \tilde{p}(\boldsymbol{x}))\, dv \tag{19.8.18}$$

and eliminating CM_0 between (19.8.17) and (19.8.18) we obtain

$$\mathsf{P}(\sigma) \geq \mathsf{P}(\sigma_0) + \int_{R_1} (\pi - \tilde{p}(\boldsymbol{x}))\, dv - \int_{R_2} (\pi - \tilde{p}(\boldsymbol{x}))\, dv. \tag{19.8.19}$$

As we shall see, $\tilde{p} \leq \pi$ on R_1 and $\tilde{p} \geq \pi$ on R_2. Using this, we arrive at $\mathsf{P}(\sigma) \geq \mathsf{P}(\sigma_0)$. Hence the principle of minimum canonical free energy is valid. Let us now prove that $\tilde{p} \leq \pi$ on R_1 and $\tilde{p} \geq \pi$ on R_2. We have $\tilde{p}(\boldsymbol{x}) = P(\psi_b(\boldsymbol{x}))$. By the definition of P and the increasing character of $p \mapsto g(p, \theta)$, the function P is decreasing. As a consequence of the mechanical equilibrium, we have $\tilde{p}(\boldsymbol{x}) = \pi$ on the free part of the boundary of the fluid. Hence, $\tilde{p}$ is a decreasing function of ψ_b, and we have, by (19.8.11), that $\psi(\boldsymbol{x}) \leq l_0$ on Γ_0 and $\psi(\boldsymbol{x}) \geq l_0$ on $R \setminus \Gamma_0$. This proves the claim. □

Here is the version of Proposition 19.8.5 without the assumption of convexity of $\tilde{e}$.

19.8.6 Proposition *Let $\sigma_0 = (\Gamma_0, v_0, \eta_0)$ be an equilibrium state compatible with the environment such that there exists a level l_0 such that $\Gamma_0 = R_{l_0}$, the quantity* (19.8.12) *is constant over the fluid and the convexity inequality* (19.8.7) *holds for every point of the fluid. Denote by C the constant value of* (19.8.12). *Assume that for every $z \in \mathbb{R}$ satisfying $z \leq C - l_0$ there exists at least one value $(v, \eta) \in \tilde{U}$ with the following properties:*

(1) *$\tilde{e}$ is convex at (v, η),*

(2) *$\tilde{\theta}(v, \eta) = \theta_0$,*

(3) *$\tilde{e}(v, \eta) + \tilde{p}(v, \eta)v - \theta_0\eta = z$.*

Then the state satisfies the four extremum principles 15.2.1–15.2.4.

The state σ_0 can contain several different phases.

Proof The idea is the same as that of the proof of Proposition 19.8.5, the only difference is in the construction of $\tilde{v}$, $\tilde{\eta}$. Namely we define, for every $\psi \geq l_0$, a number $z := C - l_0$. As this number satisfies $z \leq C - l_0$, the hypotheses of the proposition can be invoked to assure that there exists a $(v(\psi), \eta(\psi)) \in \tilde{U}$ such that $\tilde{e}$ is convex at $(v(\psi), \eta(\psi))$, $\tilde{\theta}(v(\psi), \eta(\psi)) = \theta_0$, and $\tilde{e}(v(\psi), \eta(\psi)) + \tilde{p}(v(\psi), \eta(\psi)) - \theta_0\eta = z$. The last equation means that $\tilde{e}(v(\psi), \eta(\psi)) + \tilde{p}(v(\psi), \eta(\psi)) - \theta_0\eta(\psi) + \psi = C$. If we define $\tilde{v}$, $\tilde{\eta}$ by $\tilde{v}(\boldsymbol{x}) = v(\psi_b(\boldsymbol{x}))$, $\tilde{\eta}(\boldsymbol{x}) = \eta(\psi_b(\boldsymbol{x}))$, the rest of the proof becomes identical with the corresponding part of the proof of Proposition 19.8.5. □

Bibliographical Note

The formulation of the principles of minimum energy and maximum entropy is due to Gibbs (1875). The notion of convexity is present implicitly in his works, dating to the same time when Weierstrass discovered the importance of convexity for the calculus of variations. The explicit use of convexity in the standard treatments of the thermodynamics of fluids is rare. They prefer to use the language of extremes of the potentials. This is based on the observation that a point $\boldsymbol{x}$ is a point of convexity of a function f if and only if $\boldsymbol{x}$ is a point of minimum of the excess function $g(\boldsymbol{y}) := f(\boldsymbol{y}) - f(\boldsymbol{x}) - \nabla f(\boldsymbol{x}) \cdot (\boldsymbol{y} - \boldsymbol{x})$. This tacit convention creates a good deal of confusion.

Sections 19.2–19.5 are influenced by COLEMAN [1970]. Proposition 19.4.5 is proved in DUNN & FOSDICK [1980] under a more special notion of state and more general constitutive hypotheses. The reader is referred to this work for formulations and proofs of the maxims of Van't Hoff and LeChatelier.

The assertion that a strict minimizer of energy must be a heterogeneous state with $r \le 3$ phases is stated without proof in COLEMAN [1970]. Within a different framework, a rigorous proof of the phase rule appears first in NOLL [1970]. See also FEINBERG [1979]. A different approach to the phase rule, based on the notion of genericity, was undertaken by MAN [1985b, 1986] and RABIER & GRIEWANK [1992]. When specialized to the present context, the results of MAN [1985b, 1986] and RABIER & GRIEWANK [1992] say that a 'typical' energy function is such that its graph does not admit a tangent plane which contains more than three points of the graph. Needless to say, the term 'typical' is rigorously defined using the notion of the residual set in an appropriate topology in the space of energy functions.

The notions of a phase splitting and a phase simplex are due to RABIER & GRIEWANK [1992]. Most of the results in Sects. 19.2–19.5 can be formulated equivalently without it, as well as without the lower convex hull. The reader can consult the papers of Coleman and Dunn & Fosdick cited above for such formulations. I have chosen to use these notions as it is my belief that they allow an efficient description the properties of the energy function. Otherwise the goal of RABIER & GRIEWANK [1992] is rather different from that of Sects. 19.2–19.5: Above, the phase simplex emerged as a consequence of the strict principle of minimum energy; in RABIER & GRIEWANK [1992] the existence of phase simplexes is studied because it is a generic property of the energy functions. Another object of RABIER & GRIEWANK [1992] is the analysis of the phase bifurcations. The paper introduces a natural notion of a phase number function, which is the minimal number of elements of the phase splitting. Phase bifurcations are defined as those points (v, η) where the phase number function is not locally constant.

The use of the convexity duality theory to define the Gibbs function and the examination of its properties in Sect. 19.6 appear to be new. The material of Sect. 19.7 and generalizations can be found in COLEMAN & GREENBERG [1967], COLEMAN [1970] and DUNN & FOSDICK [1978, 1980]. DUNN & FOSDICK [1980] and MAN [1985a] discuss the extremum principles for energy functions admitting negative temperatures. A result relevant to the lower convex hull of energy is due to DACOROGNA [1981]. In the special case of fluids it says that under mild additional conditions, for a function f of the form $f(\boldsymbol{F}) = \phi(\det \boldsymbol{F})$, $\boldsymbol{F} \in \mathrm{Lin}^+$, the function $g(\boldsymbol{F}) := f^{\mathrm{co}}(\det \boldsymbol{F})$ is the greatest quasiconvex function minorizing f, i.e., g is the lower quasiconvex hull of f.

20 A Local Approach to the Equilibrium of Solids

The local method assumes a given (possibly stressed and inhomogeneous) reference equilibrium state of a solid under given external loads, and analyzes its properties via the linearized equations. The term analysis of incremental elastic deformations is often employed for the method. The relevant questions include whether there exists an equilibrium state corresponding to loads close to those of the reference state, whether it is unique, etc. The existence of solutions to the equations of classical linear elasticity i.e., of the equations linearized about a stress-free state, is a particular case. No assumptions are made to exclude the nonuniqueness of solutions of the linearized equations. The unique solvability is related to the positive-definiteness of the second variation of the total stored energy on the variations compatible with the kinematical constraints. The bifurcation is likely to occur when the second variation is positive semidefinite on the class of all variations and equal to 0 on some nontrivial variation. This implies that the linearized equations lose the uniqueness (both zero variation and the just mentioned nontrivial variation solve the equations with the same data). In this situation, a further loading of the specimen can result in the occurrence of instabilities, such as barreling or buckling under compression, necking and formation of shear bands in tension or the formation of surface wrinkling. Under the uniform positivity of the second variation at the reference state, not only the linearized equations, but also the original nonlinear equations are solvable for data differing slightly from the data for the reference state. This is based on the implicit function theorem applied to the infinite dimensional spaces of displacements.

20.1 The Linearized Equations

We consider a body Ω in the standard heating environment (Sect. 13.2.1) of ambient temperature θ_0 and in the displacement-traction or bilateral contact loading environment 13.3.1 or 13.3.2. Let $\sigma_0 = (\chi_0, \eta_0)$ be an equilibrium state compatible with the environment. When the environment slightly changes, the linearization may be employed to seek the changing equilibrium states.

20.1.1 Linearization: The Displacement-Traction Environment Consider a family of conservative displacement-traction environments parametrized by the parameter $\tau \in (-\varepsilon, \varepsilon)$. Recall that a conservative displacement-traction environment is specified by two complementary subsets D, S of $\partial\Omega$, by the potentials ψ, ϕ and by the prescribed displacement $\boldsymbol{d}$ as in Sect. 13.3. Assume that D, S are independent of τ, but the potentials and $\boldsymbol{d}$ depend on τ, i.e., $\psi^\tau = \psi^\tau(\boldsymbol{F}, \boldsymbol{x}, \boldsymbol{p})$, $\phi^\tau = \phi^\tau(\boldsymbol{x}, \boldsymbol{p})$, $\boldsymbol{d} = \boldsymbol{d}^\tau(\boldsymbol{p})$.

Let us furthermore assume that for each τ there is an equilibrium configuration χ^τ compatible with the environment of index τ such that $\chi^0 = \chi_0$. If we set $q^\tau(\boldsymbol{F}, \boldsymbol{x}, \boldsymbol{p}) := \bar{f}(\boldsymbol{F}) + \psi^\tau(\boldsymbol{F}, \boldsymbol{x}, \boldsymbol{p})$, where $\bar{f}$ is the stored energy at the temperature θ_0, the equations of equilibrium 14.2.3(3),(4) read

$$\left.\begin{aligned} \operatorname{Div}(\rho_\circ \partial_{\boldsymbol{F}} q^\tau) - \rho_\circ \partial_{\boldsymbol{x}} q^\tau &= \boldsymbol{0} \quad \text{on } \Omega, \\ \rho_\circ \partial_{\boldsymbol{F}} q^\tau \boldsymbol{n} &= -\partial_{\boldsymbol{x}} \phi^\tau \quad \text{on } S, \\ \chi^\tau &= \boldsymbol{d}^\tau \quad \text{on } D, \end{aligned}\right\} \qquad (20.1.1)$$

where the partial derivatives of q^τ are evaluated at $(\nabla\chi^\tau(\boldsymbol{p}), \chi^\tau(\boldsymbol{p}), \boldsymbol{p})$ and the derivative of ϕ^τ at $(\chi^\tau(\boldsymbol{p}), \boldsymbol{p})$. These equations and boundary conditions may be written as

$$g(\chi^\tau, \tau) = 0 \qquad (20.1.2)$$

where for each τ, the mapping $g(\cdot, \tau)$ associates with the field χ^τ on Ω a triple of fields

$$g(\chi^\tau, \tau) = \left(\operatorname{Div}(\rho_\circ \partial_{\boldsymbol{F}} q^\tau) - \rho_\circ \partial_{\boldsymbol{x}} q^\tau,\ \chi^\tau - \boldsymbol{d}^\tau,\ \rho_\circ \partial_{\boldsymbol{F}} q^\tau \boldsymbol{n} + \partial_{\boldsymbol{x}} \phi^\tau\right)$$

on Ω, D and S, respectively. Let the superimposed dot denote the differentiation with respect to τ; furthermore, if ξ^τ is a differentiable quantity, then both $\dot{\xi}^0$ and $\partial_\tau \xi^0$ denote $\dot{\xi}^\tau$ at $\tau = 0$. The family of solutions χ^τ, $-\varepsilon < \tau < \varepsilon$, may be viewed as a curve in the space of configurations. We denote by $\boldsymbol{u} := \dot{\chi}^0$ the tangent to this curve at $\tau = 0$ and call it the ***displacement.*** The equation for $\boldsymbol{u}$ is obtained by differentiating (20.1.2) with respect to τ at $\tau = 0$, and the so obtained equations are, by definition, the ***linearized equations.*** They read,

$$\boldsymbol{L}(\chi^0) \cdot \boldsymbol{u} + \boldsymbol{M}(\chi^0) = 0,$$

where $\boldsymbol{L}(\chi^0)$, $\boldsymbol{M}(\chi^0)$ are the derivatives of g defined formally by

$$\boldsymbol{L}(\chi^0) \cdot \boldsymbol{u} := \frac{d}{d\tau} g(\chi^\tau, 0)\big|_{\tau=0}, \quad \boldsymbol{M}(\chi^0) := \frac{d}{d\tau} g(\chi^0, \tau)\big|_{\tau=0}.$$

These are inhomogeneous linear equations in $\boldsymbol{u}$. Denoting

$$q = q^0, \quad \bar{\psi} = \dot{\psi}^0, \quad \boldsymbol{d} = \boldsymbol{d}^0, \quad \bar{\phi} = \dot{\phi}^0, \qquad (20.1.3)$$

the linearizations of (20.1.1) read

$$\operatorname{Div}(\boldsymbol{S}^\flat + \rho_\circ \partial_{\boldsymbol{F}} \bar{\psi}) - \rho_\circ(\partial_{\boldsymbol{x}}^2 q\, \boldsymbol{u} + \partial_{\boldsymbol{x}} \partial_{\boldsymbol{F}} q \nabla \boldsymbol{u} + \partial_{\boldsymbol{x}} \bar{\psi}) = \boldsymbol{0} \quad \text{on } \Omega, \qquad (20.1.4)$$

$$(\boldsymbol{S}^\flat + \rho_\circ \partial_{\boldsymbol{F}} \bar{\psi})\boldsymbol{n} = -\partial_{\boldsymbol{x}}^2 \phi\, \boldsymbol{u} - \partial_{\boldsymbol{x}} \bar{\phi} \quad \text{on } S, \qquad (20.1.5)$$

$$\boldsymbol{u} = \bar{\boldsymbol{d}} \quad \text{on } D, \qquad (20.1.6)$$

where

$$\boldsymbol{S}^\flat := \rho_\circ(\partial_{\boldsymbol{F}}^2 q[\nabla \boldsymbol{u}] + \partial_{\boldsymbol{F}} \partial_{\boldsymbol{x}} q\, \boldsymbol{u}). \qquad (20.1.7)$$

The derivatives of q and $\bar{\psi}$ are evaluated at $(\nabla\chi^0(\boldsymbol{p}), \chi^0(\boldsymbol{p}), \boldsymbol{p})$ and the derivatives of ϕ and $\bar{\phi}$ at $(\chi^0(\boldsymbol{p}), \boldsymbol{p})$. (20.1.4)–(20.1.6) are inhomogeneous linear equations for $\boldsymbol{u}$ with the coefficients in front of $\boldsymbol{u}$, $\nabla \boldsymbol{u}$, $\nabla^2 \boldsymbol{u}$ depending on $\boldsymbol{p}$. The inhomogeneous terms are $\partial_{\boldsymbol{F}} \bar{\psi}$, $\partial_{\boldsymbol{x}} \bar{\psi}$, $\partial_{\boldsymbol{x}} \bar{\phi}$ and $\boldsymbol{d}$. The corresponding homogeneous equations are those obtained by setting the inhomogeneous terms equal to 0. Equations (20.1.4)–(20.1.6) simplify for specific environments:

(1) ***Dead loading*** For a family of displacement-traction environments of dead loading 13.3.9 we have $\boldsymbol{b}_0^\tau = \boldsymbol{b}_0^\tau(\boldsymbol{p})$, $\boldsymbol{s}^\tau = \boldsymbol{s}^\tau(\boldsymbol{p})$. Denoting $\bar{\boldsymbol{b}} = \dot{\boldsymbol{b}}_0^0$, $\bar{\boldsymbol{s}} = \dot{\boldsymbol{s}}^0$, one arrives at

$$\left.\begin{aligned} \operatorname{Div}(\bar{\mathsf{A}}[\nabla \boldsymbol{u}]) + \rho_{\circ}\bar{\boldsymbol{b}} &= \mathbf{0} \text{ on } \Omega, \\ \bar{\mathsf{A}}[\nabla \boldsymbol{u}]\boldsymbol{n} &= \bar{\boldsymbol{s}} \text{ on } S, \\ \boldsymbol{u} &= \bar{\boldsymbol{d}} \text{ on } D, \end{aligned}\right\} \quad (20.1.8)$$

where $\bar{\mathsf{A}} = \bar{\mathsf{A}}(\boldsymbol{p}) = \bar{\mathsf{A}}(\nabla\boldsymbol{\chi}^0(\boldsymbol{p}), \theta_0)$ is the tensor of isothermal elasticities evaluated at the state σ_0. Throughout the rest of this chapter, we shall call D and S the fixed and free parts of the boundary, respectively.

(2) *Spatially dependent pressure loading* A family of displacement-traction environments with the spatially dependent pressure loading 13.3.10 is described by a family $\psi_b^\tau = \psi_b^\tau(\boldsymbol{x})$ of the body force potentials and by the family $\pi^\tau = \pi^\tau(\boldsymbol{x})$ of ambient pressures. Denoting

$$\bar{\psi}_b = \dot{\psi}_b^0, \quad \bar{\gamma}_b = \partial_{\boldsymbol{x}}^2\psi_b^0, \quad \dot{\pi} = \dot{\pi}^0, \quad \bar{\pi} = \pi^0,$$

the linearized equations read

$$\left.\begin{aligned} \operatorname{Div}(\bar{\mathsf{A}}[\nabla \boldsymbol{u}]) - \rho_{\circ}\bar{\gamma}_b\boldsymbol{u} - \rho_{\circ}\partial_{\boldsymbol{x}}\bar{\psi}_b &= \mathbf{0} \text{ on } \Omega, \\ \bar{\mathsf{A}}[\nabla \boldsymbol{u}]\boldsymbol{n} &= \left(\bar{\pi}\boldsymbol{F}^{\mathrm{T}}\nabla\boldsymbol{u}^{\mathrm{T}} - (\dot{\pi} + \operatorname{grad}\bar{\pi}\cdot\boldsymbol{u} + \bar{\pi}(\boldsymbol{F}^{\mathrm{T}}\cdot\nabla\boldsymbol{u}))\mathbf{1}\right)\operatorname{cof}\boldsymbol{F}\boldsymbol{n} \text{ on } S, \\ \boldsymbol{u} &= \bar{\boldsymbol{d}} \text{ on } D. \end{aligned}\right\} \quad (20.1.9)$$

Here $\boldsymbol{F} \equiv \boldsymbol{F}^0$ is the deformation gradient of the deformation $\boldsymbol{\chi}^0$. To derive $(20.1.9)_2$, one writes $\bar{S}(\boldsymbol{F}^\tau, \theta_0)\boldsymbol{n} = -\pi^\tau(\boldsymbol{\chi}^\tau)\operatorname{cof}\boldsymbol{F}^\tau\boldsymbol{n}$, differentiates with respect to τ and uses (1.1.16). If $\pi^\tau(\boldsymbol{x})$ is independent of τ and $\boldsymbol{x}$, equation $(20.1.9)_2$ simplifies to

$$\bar{\mathsf{A}}[\nabla \boldsymbol{u}]\boldsymbol{n} = \bar{\pi}(\boldsymbol{F}^{\mathrm{T}}\nabla\boldsymbol{u}^{\mathrm{T}} - (\boldsymbol{F}^{\mathrm{T}}\cdot\nabla\boldsymbol{u})\mathbf{1})\operatorname{cof}\boldsymbol{F}\boldsymbol{n}.$$

20.1.2 Linearization: The Bilateral Contact Environment Consider a family of bilateral contact environments 13.3.2. It is assumed that S, D_i are independent of τ while $P_i = P_i^\tau$ may depend on τ. Conditions $(13.3.2)_{1,1}$ read $\boldsymbol{\chi}^\tau(D_i) \subset P_i^\tau$, $i = 1, \ldots, s$, and may be expressed locally as $\varphi_i^\tau(\boldsymbol{\chi}^\tau(\boldsymbol{p})) = 0$ with $\operatorname{grad}\varphi_i^\tau \neq \mathbf{0}$ on P_i^τ. The normal to P_i^τ is denoted $\boldsymbol{m}_i^\tau = \operatorname{grad}\varphi_i^\tau / |\operatorname{grad}\varphi_i^\tau|$ and we see that $\boldsymbol{m}_i^\tau = \boldsymbol{m}_i^\tau(\boldsymbol{x})$ is defined also for all points $\boldsymbol{x}$ in some neighborhood of P_i^τ. We furthermore denote

$$\boldsymbol{m}_i = \boldsymbol{m}_i^0, \quad m_i = -\partial_\tau\varphi_i^0 / |\operatorname{grad}\varphi^0|,$$

and use the notation $(20.1.3)_{1,2,4}$ and (20.1.7). The linearized equations read

$$\operatorname{Div}(\mathsf{S}^b + \rho_{\circ}\partial_{\boldsymbol{F}}\bar{\psi}) - \rho_{\circ}(\partial_{\boldsymbol{x}}^2 q\,\boldsymbol{u} + \partial_{\boldsymbol{x}}\partial_{\boldsymbol{F}}q\nabla\boldsymbol{u} + \partial_{\boldsymbol{x}}\bar{\psi}) = \mathbf{0} \text{ on } \Omega, \quad (20.1.10)$$

$$(\mathsf{S}^b + \rho_{\circ}\partial_{\boldsymbol{F}}\bar{\psi})\boldsymbol{n} = -\partial_{\boldsymbol{x}}^2\phi\,\boldsymbol{u} - \partial_{\boldsymbol{x}}\bar{\phi} \text{ on } S, \quad (20.1.11)$$

$$\boldsymbol{u}\cdot\boldsymbol{m}_i = m_i, \quad \boldsymbol{P}_i(\bar{\mathsf{A}}[\nabla\boldsymbol{u}]\boldsymbol{n} + (\operatorname{grad} m_i - \operatorname{grad}\boldsymbol{m}_i^{\mathrm{T}}\boldsymbol{u})(\boldsymbol{S}\boldsymbol{n}\cdot\boldsymbol{m}_i)) = \mathbf{0} \text{ on } D_i, \quad (20.1.12)$$

where $\boldsymbol{P}_i = \mathbf{1} - \boldsymbol{m}_i \otimes \boldsymbol{m}_i$ is the projection onto the orthogonal complement of $\boldsymbol{m}_i$, expressed as function of $\boldsymbol{p}$ via $\boldsymbol{m}_i(\boldsymbol{p}) = \boldsymbol{m}_i(\boldsymbol{\chi}^0(\boldsymbol{p}))$, S^b is given by (20.1.7), $\boldsymbol{S}$ is the stress in the state $\boldsymbol{\chi}^0$, the derivatives of q and $\bar{\psi}$ are evaluated at $(\nabla\boldsymbol{\chi}^0(\boldsymbol{p}), \boldsymbol{\chi}^0(\boldsymbol{p}), \boldsymbol{p})$, the derivatives of ϕ and $\bar{\phi}$ at $(\boldsymbol{\chi}^0(\boldsymbol{p}), \boldsymbol{p})$ and $\boldsymbol{m}_i, m_i$ together with their gradients at $\boldsymbol{\chi}^0(\boldsymbol{p})$. Equations (20.1.10)–(20.1.12) are linear in $\boldsymbol{u}$ with the coefficients in front of $\boldsymbol{u}, \nabla\boldsymbol{u}, \nabla^2\boldsymbol{u}$ depending on $\boldsymbol{p}$; the inhomogeneous terms are those containing $\bar{\psi}, \bar{\phi}, m_i$.

To derive the linearized equations, we first write the system of equations (13.3.2) of the bilateral contact environment in the form (20.1.2), where now

$$g(\boldsymbol{\chi}^\tau, \tau) = (\operatorname{Div}(\rho_{\circ}\partial_{\boldsymbol{F}}q^\tau) - \rho_{\circ}\partial_{\boldsymbol{x}}q^\tau, \rho_{\circ}\partial_{\boldsymbol{F}}q^\tau\boldsymbol{n} + \partial_{\boldsymbol{x}}\phi^\tau, \varphi_i^\tau(\boldsymbol{\chi}^\tau), \bar{\boldsymbol{P}}\bar{S}(\boldsymbol{F}^\tau)\boldsymbol{n})$$

is a quadruple of fields on Ω, S, D_i, and D_i, respectively, and

$$\bar{\boldsymbol{P}} = \mathbf{1} - \bar{\boldsymbol{m}} \otimes \bar{\boldsymbol{m}}, \quad \bar{\boldsymbol{m}}(\boldsymbol{F}) = \boldsymbol{F}^{-\mathsf{T}}\boldsymbol{n} / |\boldsymbol{F}^{-\mathsf{T}}\boldsymbol{n}|.$$

The first two members have been linearized in Subsect. 20.1.1. That the linearization of the third member results in $(20.1.12)_1$ is immediate. The explicit form of the fourth component of (20.1.2) reads $(\mathbf{1} - \bar{\boldsymbol{m}}(\boldsymbol{F}^\tau) \otimes \bar{\boldsymbol{m}}(\boldsymbol{F}^\tau))\bar{\boldsymbol{S}}(\boldsymbol{F}^\tau)\boldsymbol{n} = \mathbf{0}$ on $D := \cup_{i=1}^{s} D_i$. A differentiation with respect to τ at $\tau = 0$ gives

$$-(\partial_{\boldsymbol{F}}\bar{\boldsymbol{m}}[\nabla \boldsymbol{u}] \otimes \bar{\boldsymbol{m}}(\boldsymbol{F}) + \bar{\boldsymbol{m}}(\boldsymbol{F}) \otimes \partial_{\boldsymbol{F}}\bar{\boldsymbol{m}}[\nabla \boldsymbol{u}])\bar{\boldsymbol{S}}(\boldsymbol{F})\boldsymbol{n} + \bar{\boldsymbol{P}}\bar{\boldsymbol{A}}[\nabla \boldsymbol{u}]\boldsymbol{n} = \mathbf{0}, \tag{20.1.13}$$

where $\boldsymbol{F}$ is the deformation gradient of $\chi \equiv \chi^0$. One easily finds that

$$\partial_{\boldsymbol{F}}\bar{\boldsymbol{m}}[\boldsymbol{A}] = \bar{\boldsymbol{P}}\partial_{\boldsymbol{F}}\bar{\boldsymbol{m}}[\boldsymbol{A}] = -\bar{\boldsymbol{P}}\boldsymbol{F}^{-\mathsf{T}}\boldsymbol{A}^{\mathsf{T}}\bar{\boldsymbol{m}}, \quad \boldsymbol{A} \in \mathrm{Lin}, \tag{20.1.14}$$

with noting that $\partial_{\boldsymbol{F}}\bar{\boldsymbol{m}}[\boldsymbol{A}] \cdot \bar{\boldsymbol{m}} = 0$ as a consequence of the normalization of $\bar{\boldsymbol{m}}$. Since $\bar{\boldsymbol{S}}(\boldsymbol{F})\boldsymbol{n}$ is parallel to $\bar{\boldsymbol{m}}$ by $\bar{\boldsymbol{P}}\bar{\boldsymbol{S}}(\boldsymbol{F})\boldsymbol{n} = \mathbf{0}$, we have $\bar{\boldsymbol{m}}(\boldsymbol{F}) \otimes \partial_{\boldsymbol{F}}\bar{\boldsymbol{m}}[\nabla \boldsymbol{u}]\bar{\boldsymbol{S}}(\boldsymbol{F})\boldsymbol{n} = \mathbf{0}$ and (20.1.13) simplifies to

$$-\partial_{\boldsymbol{F}}\bar{\boldsymbol{m}}[\nabla \boldsymbol{u}]\bar{\boldsymbol{S}}(\boldsymbol{F})\boldsymbol{n} \cdot \bar{\boldsymbol{m}}(\boldsymbol{F}) + \bar{\boldsymbol{P}}\bar{\boldsymbol{A}}[\nabla \boldsymbol{u}]\boldsymbol{n} = \mathbf{0}. \tag{20.1.15}$$

A differentiation of $\boldsymbol{m}_i(\boldsymbol{x}) \cdot \boldsymbol{u}(\chi^{-1}(\boldsymbol{x})) = m_i(\boldsymbol{x})$ along P_i^τ (see $(20.1.12)_1$) gives $\operatorname{grad} m \cdot \boldsymbol{a} = (\operatorname{grad} \boldsymbol{m})\boldsymbol{a} \cdot \boldsymbol{u} + \boldsymbol{m} \cdot \nabla \boldsymbol{u}\boldsymbol{F}^{-1}\boldsymbol{a}$ for every $\boldsymbol{a} \in \mathrm{Vect}$ with $\boldsymbol{a} \cdot \boldsymbol{m} = 0$, which implies

$$\bar{\boldsymbol{P}} \operatorname{grad} m = \bar{\boldsymbol{P}} \operatorname{grad} \boldsymbol{m}^{\mathsf{T}}\boldsymbol{u} + \bar{\boldsymbol{P}}\boldsymbol{F}^{-\mathsf{T}}\nabla \boldsymbol{u}^{\mathsf{T}}\boldsymbol{m}.$$

Combining with (20.1.14), we obtain $\bar{\boldsymbol{P}}\partial_{\boldsymbol{F}}\bar{\boldsymbol{m}}[\nabla \boldsymbol{u}] = -\bar{\boldsymbol{P}}(\operatorname{grad} m - \operatorname{grad} \boldsymbol{m}^{\mathsf{T}}\boldsymbol{u})$; inserting this into (20.1.15) and using $\bar{\boldsymbol{P}} = \boldsymbol{P}_i$ on D_i, we obtain $(20.1.12)_2$.

Let us record the following special cases:

(1) ***Dead loading*** For a family of bilateral contact environments of dead loading, (20.1.10)–(20.1.12) simplify to

$$\left.\begin{aligned} &\mathrm{Div}(\bar{\boldsymbol{A}}[\nabla \boldsymbol{u}]) + \rho_\circ \bar{\boldsymbol{b}} = \mathbf{0}, \quad \text{on } \Omega, \\ \boldsymbol{u} \cdot \boldsymbol{m}_i = m_i, \quad &\boldsymbol{P}_i(\bar{\boldsymbol{A}}[\nabla \boldsymbol{u}] + (\operatorname{grad} m_i - \operatorname{grad} \boldsymbol{m}_i^{\mathsf{T}}\boldsymbol{u})(\boldsymbol{S}\boldsymbol{n} \cdot \boldsymbol{m}_i)) = \mathbf{0} \quad \text{on } D_i, \\ &\bar{\boldsymbol{A}}[\nabla \boldsymbol{u}]\boldsymbol{n} = \bar{\boldsymbol{s}}, \quad \text{on } S, \end{aligned}\right\} \tag{20.1.16}$$

where the notation of 20.1.1 has been used. If P_i^τ are planes then $\operatorname{grad} \boldsymbol{m}_i = \mathbf{0}$ and if additionally $P_i = P_i^\tau$ is independent of τ then $m_i = 0$. Equation $(20.1.16)_{2,2}$ then simplifies to

$$\boldsymbol{P}_i\bar{\boldsymbol{A}}[\nabla \boldsymbol{u}] = \mathbf{0} \quad \text{on } D_i.$$

(2) ***Spatially dependent pressure loading*** For a family of bilateral contact environments with pressure loading, the linearized equations consist of $(20.1.9)_{1,2}$ and (20.1.12).

20.1.3 The Euler–Lagrange Equations for the Second Variation The above linearization procedures have lead to inhomogeneous linear equations for $\boldsymbol{u}$. Another way to obtain linear equations for $\boldsymbol{u}$ is to take a fixed configuration χ and to consider the second variation $\delta^2\mathsf{M}(\chi, \boldsymbol{u}, \boldsymbol{u})$ as a quadratic functional in $\boldsymbol{u}$ defined on $\boldsymbol{u} \in \mathrm{Var}(\chi)$. This is an integral functional and one can consider the Euler–Lagrange equations for $\boldsymbol{u}$ derived from this Lagrangian. In other words, one demands that $\boldsymbol{u} \in \mathrm{Var}(\chi)$ is such that $\delta^2\mathsf{M}(\chi, \boldsymbol{u}, \boldsymbol{w}) = 0$ for each $\boldsymbol{w} \in \mathrm{Var}(\chi)$. Since the Lagrangian is quadratic, the resulting equations will be homogeneous linear in $\boldsymbol{u}$. Writing down the second variation and carrying out the integration by parts, one obtains

$$\mathrm{Div}\,(\rho_\circ(\partial_{\boldsymbol{F}}^2 q[\nabla \boldsymbol{u}] + \partial_{\boldsymbol{F}}\partial_{\boldsymbol{x}} q\,\boldsymbol{u})) - \rho_\circ(\partial_{\boldsymbol{x}}^2 q\,\boldsymbol{u} + \partial_{\boldsymbol{x}}\partial_{\boldsymbol{F}} q \nabla \boldsymbol{u}) = \mathbf{0} \quad \text{on } \Omega, \tag{20.1.17}$$

$$\rho_\circ(\partial_{\boldsymbol{F}}^2 q[\nabla \boldsymbol{u}] + \partial_{\boldsymbol{F}}\partial_{\boldsymbol{x}} q\,\boldsymbol{u})\boldsymbol{n} = -\partial_{\boldsymbol{x}}^2 \phi\,\boldsymbol{u} \quad \text{on } S, \tag{20.1.18}$$

$$\boldsymbol{u} = \mathbf{0} \quad \text{on } D. \tag{20.1.19}$$

These equations are identical with the homogeneous version of (20.1.4)–(20.1.6). Hence the homogeneous equations are of variational character – they are the Euler–Lagrange equations of a variational problem.

The Euler–Lagrange equations for the second variation for the bilateral contact environment are (20.1.17), (20.1.18), and

$$\boldsymbol{u}\cdot\boldsymbol{m}_i = 0, \quad \mathsf{P}_i(\partial_F^2 q[\nabla\boldsymbol{u}] + \partial_F\partial_x q\,\boldsymbol{u}) = \boldsymbol{0} \text{ on } \mathsf{D}_i.$$

These equations coincide with the homogeneous version of the linearized equations of 20.1.2 only provided $\operatorname{grad}\boldsymbol{m}_i = \boldsymbol{0}$, i.e., provided the surfaces P_i are planar. This unpleasant fact is eventually related to that the set $\operatorname{Var}(\chi)$ depends genuinely on χ for the bilateral contact environment.

20.1.4 Bifurcations Let us consider a family of displacement-traction or bilateral contact environments as in 20.1.1 and 20.1.2, with an equilibrium configuration χ^τ corresponding to the environment indexed by τ. We say that a bifurcation occurs at $\tau = \tau_0$ if there exists, for each τ in a neighborhood of τ_0, another family of equilibrium states $\bar{\chi}^\tau$ such that $\bar{\chi}^{\tau_0} = \chi^{\tau_0}$ and $\dot{\bar{\chi}}^{\tau_0} \neq \dot{\chi}^{\tau_0}$. Assuming $\tau_0 = 0$ for simplicity, and using the notation of Subsect. 20.1.1, the equations of equilibrium for the families $\chi^\tau, \bar{\chi}^\tau$ read

$$g(\chi^\tau, \tau) = 0, \quad g(\bar{\chi}^\tau, \tau) = 0;$$

differentiating these equations at 0 and subtracting the results we obtain

$$\mathsf{L}(\chi^0)\cdot\boldsymbol{u} = 0 \text{ where } \boldsymbol{u} := \dot{\bar{\chi}}^0 - \dot{\chi}^0.$$

Hence we conclude that a necessary condition for the bifurcation to occur is that the linearized equations at the point of bifurcation have a nontrivial solution $\boldsymbol{u} \neq \boldsymbol{0}$ corresponding to the homogeneous data. These equations have the form (20.1.17)–(20.1.18) for the displacement-traction environment and (20.1.17), (20.1.18) and

$$\boldsymbol{u}\cdot\boldsymbol{m}_i = 0, \quad \mathsf{P}_i(\partial_F^2 q[\nabla\boldsymbol{u}] + \partial_F\partial_x q\,\boldsymbol{u} - \operatorname{grad}\boldsymbol{m}_i^{\mathrm{T}}\boldsymbol{u}(\boldsymbol{Sn}\cdot\boldsymbol{m}_i)) = \boldsymbol{0} \text{ on } \mathsf{D}_i \qquad (20.1.20)$$

for the bilateral contact environment. Any nontrivial solution $\boldsymbol{u}$ of the homogeneous linearized equations is called an eigenmode. Unfortunately, the formal nature of the above considerations does not allow us to assert conversely that the presence of an eigenmode implies the existence of a family $\bar{\chi}^\tau$ of the solutions of the equilibrium equations.

Examples of bifurcation analyses will be given in Sects. 20.6 and 20.7. Sect. 20.6 deals with pressured isotropic states of a three-dimensional half-space. The bifurcations are surface instabilities in this case. Sect. 20.7 deals with a two-dimensional solid under a bilateral contact environment, and the bifurcations are either buckling or bulging. For the linearization about a homogeneous state, the explicit form of (20.1.17)–(20.1.18) or (20.1.17), (20.1.18) and (20.1.20) contains a limited number of scalar parameters eventually determined by the free energy. It is advantageous to consider these parameters as free, and to determine the regions of these parameters corresponding to different regimes, i.e., parabolic, hyperbolic, and elliptic, with an emphasis on the points of bifurcation. In the two-dimensional incompressible isotropic case, the classification of the regimes and solutions has been given in HILL & HUTCHINSON [1975]; in the two dimensional compressible isotropic case in OGDEN [1984, Subsect. 6.3.1]. See also OGDEN [1992b] and the literature cited in these works.

When dealing with the displacement-traction problem, we shall assume that D has a positive area. The pure traction problem, i.e., $\mathsf{D} = \emptyset$, admits a simpler and a more complete a priori analysis, for it reduces to analyzing the positive definiteness of the tensor of elasticities itself, rather than of its integral over Ω. Similarly, in this case, rather than seeking the singularities of equations (20.1.17)–(20.1.18), one seeks directly the singularities of the tensor of elasticities, i.e., those values of $\boldsymbol{F}$ for which $\bar{A}(F, \theta_0)[H] = 0$ admits solutions $H \neq 0$. I refer to OGDEN [1984, Subsect. 6.2.5; 1992b]. The global aspects of the solution χ^τ for a given loading path $b = b^\tau, d = d^\tau, s = s^\tau, \tau \in \mathbb{R}$ are studied in HEALEY & SIMPSON [1996]. (See also SPECTOR [1984].) ANTMAN [1995] gives a rigorous treatment of the bifurcations in one-dimensional models of elasticity, such as the theory of strings and rods.

20.2 Sobolev Spaces

We shall assume throughout this chapter that Ω is an open, connected, bounded subset of $\mathbb{R}^n$ = Eucl with Lipschitz boundary (see NEČAS [1967] or MORREY [1966]). Without giving a definition we note that the required smoothness of the boundary suffices for the validity of the trace theorem 20.2.4, Poincaré inequality 20.2.5 below, and other results. In our applications Ω will be the reference region of the body.

20.2.1 Definition Let p be a number satisfying $1 \le p \le \infty$. We denote by $L^p(\Omega)$ the set of all functions $f : \Omega \to \mathbb{R}$, measurable in the Lebesgue sense, such that

$$\int_\Omega |f|^p \, dV < \infty \text{ if } p < \infty \text{ and } \operatorname{sup\,ess}\{|f(\boldsymbol{p})| : \boldsymbol{p} \in \Omega\} < \infty \text{ if } p = \infty.$$

Here $\operatorname{sup\,ess}\{|f(\boldsymbol{p})| : \boldsymbol{p} \in \Omega\}$ is the infimum of all $c \in [0, \infty]$ such that $|f(\boldsymbol{p})| \le c$ for a.e. $\boldsymbol{p} \in \Omega$. The functions differing on a set of zero Lebesgue measure (zero "volume") are regarded as identical. For a $f \in L^p(\Omega)$, $\|f\|_{L^p(\Omega)}$ will denote the norm

$$\|f\|_{L^p(\Omega)} = \Big(\int_\Omega |f|^p \, dV\Big)^{1/p} \text{ if } p < \infty \text{ and } \|f\|_{L^\infty(\Omega)} = \operatorname{sup\,ess}\{|f(\boldsymbol{p})| : \boldsymbol{p} \in \Omega\}.$$

With this norm, $L^p(\Omega)$ is a Banach space, i.e., a complete linear space.

If W is a finite dimensional vector space with norm $|\cdot|$, one defines the space $L^p(\Omega, \mathrm{W})$ of functions $\boldsymbol{f}$ from $\Omega \subset$ Eucl with values in W such that $|\boldsymbol{f}(\cdot)| \in L^p(\Omega)$. For simplicity, we shall write $L^p(\Omega)$ for $L^p(\Omega, \mathrm{W})$; no confusion can arise from this.

20.2.2 Proposition (Hölder's inequality) *Let* $1 < p, q < \infty$ *satisfy*

$$\frac{1}{p} + \frac{1}{q} = 1, \quad \text{i.e.,} \quad q = p/(p-1). \tag{20.2.1}$$

If $f \in L^p(\Omega)$, $g \in L^q(\Omega)$, *then* $fg \in L^1(\Omega)$ *and*

$$\int_\Omega |fg| \, dV \le \Big(\int_\Omega |f|^p \, dV\Big)^{1/p} \Big(\int_\Omega |g|^q \, dV\Big)^{1/q}. \tag{20.2.2}$$

The index q satisfying (20.2.1) is called the index conjugate to p. When $p = 1$ or $p = \infty$, then, by definition, the conjugate index is $q = \infty$ or $q = 1$, respectively. Inequality (20.2.2) still holds if $p = 1$ provided one replaces the integral involving q by $\|g\|_{L^\infty(\Omega)}$.

20.2.3 Definition (Sobolev spaces) Let $1 \le p \le \infty$. A (scalar-valued) function $f \in L^p(\Omega)$ is said to belong to the ***Sobolev space*** $W^{1,p}(\Omega)$ if there exists a (vector-valued) function $\boldsymbol{g} \in L^p(\Omega, \mathrm{Vect})$ such that

$$\int_\Omega f \nabla\phi \, dV = -\int_\Omega \boldsymbol{g}\phi \, dV$$

for every $\phi \in C_0^\infty(\Omega)$. The function $\boldsymbol{g}$ is determined uniquely (to within a change on a set of null Lebesgue measure); it is called the ***distributional*** (or ***generalized***) ***gradient*** (or ***derivative***) of f, and denoted $\nabla f := \boldsymbol{g}$ or $Df := \boldsymbol{g}$. (Cf. Subsect. 3.8.2.) The distributional gradient coincides with the classical gradient when the latter exists and is continuous. $W^{1,p}(\Omega)$ is a Banach space with respect to the norm $\|\cdot\|_{W^{1,p}(\Omega)}$ given by

$$\|f\|_{W^{1,p}(\Omega)} = \|f\|_{L^p(\Omega)} + \|\nabla f\|_{L^p(\Omega)}.$$

If W is a finite dimensional vector space with norm $|\cdot|$, one defines the space $W^{1,p}(\Omega, \mathrm{W})$ of functions $\boldsymbol{f}$ on Ω with values in W such that "each component of $\boldsymbol{f}$" is in $W^{1,p}(\Omega)$. For simplicity, we write $W^{1,p}(\Omega)$ for $W^{1,p}(\Omega, \mathrm{W})$.

Let $C^\infty(\mathrm{cl}\,\Omega)$ be the set of all infinitely differentiable functions such that all derivatives have continuous extensions to $\mathrm{cl}\,\Omega$. The set $C^\infty(\mathrm{cl}\,\Omega)$ is dense in $W^{1,p}(\Omega)$ if $1 \le p < \infty$; accordingly, one can treat $W^{1,p}(\Omega)$ as the completion of $C^\infty(\mathrm{cl}\,\Omega)$ in the norm $\|\cdot\|_{W^{1,p}(\Omega)}$.

In contrast to functions from $L^p(\Omega)$, the functions from $W^{1,p}(\Omega)$ have well-defined boundary values. To state the result, recall that our assumption about the Lipschitz character of the boundary of Ω enables one to define the surface measure dA on $\partial\Omega$ and the corresponding surface integral. By using these concepts, one can define the space $L^p(\partial\Omega)$ in a manner completely analogous to that of $L^p(\Omega)$: $L^p(\partial\Omega)$ is the set of all A–measurable functions $f : \partial\Omega \to \mathbb{R}$ such that $\int_{\partial\Omega} |f|^p \, dA < \infty$, we endow $L^p(\partial\Omega)$ with the norm $\|\cdot\|_{L^p(\partial\Omega)}$, where

$$\|f\|_{L^p(S)} = \Big(\int_{\partial\Omega} |f|^p \, dA \Big)^{1/p}$$

for any A–measurable subset S of $\partial\Omega$.

20.2.4 Theorem (Trace theorem) *Let $1 \le p < \infty$. There exists a continuous linear operator $T : W^{1,p}(\Omega) \to L^p(\partial\Omega)$ such that*

$$(T\phi)(\boldsymbol{p}) = \phi(\boldsymbol{p}) \tag{20.2.3}$$

whenever $\phi \in C^\infty(\mathrm{cl}\,\Omega)$.

The operator T is called the ***trace operator;*** the value $T(\phi)$ is the ***trace*** of ϕ on $\partial\Omega$. When there is no danger of confusion, we simplify the notation and write ϕ instead of $T(\phi)$. The property (20.2.3) shows that this is consistent. The continuity of the trace operator means that there exists a constant c such that

$$\|f\|_{L^p(\partial\Omega)} \le c\|f\|_{W^{1,p}(\Omega)} \tag{20.2.4}$$

for every $f \in W^{1,p}(\Omega)$.

20.2.5 Theorem (Poincaré inequality) *Let $1 \le p < \infty$ and let $S \subset \partial\Omega$ be a measurable subset with $A(S) > 0$. Then there exists a constant c such that*

$$\|f\|_{L^p(\Omega)} \le c(\|\nabla f\|_{L^p(\Omega)} + \|f\|_{L^p(S)})$$

for every $f \in W^{1,p}(\Omega)$. In particular,

$$\|f\|_{L^p(\Omega)} \le c\|\nabla f\|_{L^p(\Omega)}$$

for every $f \in W^{1,p}(\Omega)$ such that $f = 0$ on S in the sense of trace.

Other topics of relevance to the treatment of the nonlinear boundary-value problems of elasticity are the Sobolev imbedding theorems. Here only a simple special case of a general result of this type will be needed. Let n be the dimension of the underlying space of which the set Ω is a subset, $\Omega \subset \mathbb{R}^n$. For bounded domains Ω the 'quality' of functions from $W^{1,p}(\Omega)$ gradually improves with increasing p; the value $p = n$ is especially remarkable because it turns out that for $p > n$, the functions f from $W^{1,p}(\Omega)$ are already continuous, while for $p \le n$ this need not be so. The phrase that a function from $W^{1,p}(\Omega)$ is continuous requires an explanation, because originally the elements f of $W^{1,p}(\Omega)$ are not exactly functions, but only classes of equivalence of functions that differ from each other only on a set of Lebesgue measure 0. The point of the imbedding theorem in question is that within the class $f \in W^{1,p}(\Omega)$ there is one function that is continuous on the closure of Ω. This continuous function in the class f, if it exists, is uniquely determined and is denoted again by f. Denote by $C^0(\mathrm{cl}\,\Omega)$ the Banach space of all continuous functions on the closure $\mathrm{cl}\,\Omega$ into $\mathbb{R}$ endowed with the norm $\|\cdot\|_\infty := \|\cdot\|_{L^\infty(\Omega)}$.

20.2.6 Theorem (Imbedding theorem) *For every $p > n$ one has $W^{1,p}(\Omega) \subset C(\mathrm{cl}\,\Omega)$ and there exists a constant $c > 0$ such that $\|f\|_\infty \le c\|f\|_{W^{1,p}(\Omega)}$ for every $f \in W^{1,p}(\Omega)$.*

The space $W^{1,p}(\Omega)$ is enough for the existence of the weak solutions of the problem of linearized elasticity. For the differentiability properties of the solutions, and for the treatment of nonlinear boundary-value problem by the implicit function theorem, it is necessary to consider Sobolev spaces of the second order. Also the following fact will be needed.

20.2.7 Proposition (Riesz's Representation Theorem) *Let H be a real Hilbert space with inner product $\langle \cdot, \cdot \rangle$. Then for every continuous linear functional F on H there exists a unique element $u \in H$ such that $F(x) = \langle x, u \rangle$ for every $x \in H$.*

20.3 The Second Variations and Extrema

It appears worthwhile to summarize some relationships between the second variation and the extrema of the nonlinear integral functionals, even though the results will not be used here. Throughout the section we consider a body Ω under displacement-traction environment of dead loadings and a heating environment of ambient temperature θ_0.

20.3.1 The Functionals Let $\tilde{e} : \tilde{U} \to \mathbb{R}$ be the internal energy as a function of the deformation gradient and entropy and let $\bar{f} : \bar{U} \to \mathbb{R}$ be the free energy as a function of the deformation gradient and temperature. Recall the set Σ^{**} of all admissible states (Subsect. 14.3.2(1)) and the set Δ^{**} of all admissible configurations (Subsect. 14.3.4). The total canonical free energy P, the total energy U, and the total entropy H are functionals on Σ^{**} and the total stored and potential energies M, W functionals on Δ^{**} defined by

$$\left.\begin{aligned} \mathsf{P}(\sigma) &= \int_\Omega (\tilde{e} - \theta_0 \eta - \boldsymbol{b}_0 \cdot \boldsymbol{\chi}) \rho_\circ \, dV - \int_S \boldsymbol{s} \cdot \boldsymbol{\chi} \, dA, \\ \mathsf{U}(\sigma) &= \int_\Omega (\tilde{e} - \boldsymbol{b}_0 \cdot \boldsymbol{\chi}) \rho_\circ \, dV - \int_S \boldsymbol{s} \cdot \boldsymbol{\chi} \, dA, \quad \mathsf{H}(\sigma) = \int_\Omega \eta \rho_\circ \, dV, \\ \mathsf{W}(\boldsymbol{\chi}) &= -\int_\Omega \boldsymbol{b}_0 \cdot \boldsymbol{\chi} \, dV - \int_S \boldsymbol{s} \cdot \boldsymbol{\chi} \, dA, \\ \mathsf{M}(\boldsymbol{\chi}) &= \int_\Omega (\bar{f}(\boldsymbol{F}, \theta_0) - \boldsymbol{b}_0 \cdot \boldsymbol{\chi}) \rho_\circ \, dV - \int_S \boldsymbol{s} \cdot \boldsymbol{\chi} \, dA. \end{aligned}\right\} \quad (20.3.1)$$

If $1 \le p \le \infty$ and $\sigma_1 = (\boldsymbol{\chi}_1, \eta_1), \sigma_2 = (\boldsymbol{\chi}_2, \eta_2) \in \Sigma^{**}$, we denote by $\|\sigma_1 - \sigma_2\|_p$ the distance between σ_1, σ_2 defined by

$$\|\sigma_1 - \sigma_2\|_p = \|\boldsymbol{\chi}_1 - \boldsymbol{\chi}_2\|_{W^{1,p}(\Omega)} + \|\eta_1 - \eta_2\|_{L^p(\Omega)}.$$

If $\boldsymbol{\chi}_1, \boldsymbol{\chi}_2 \in \Delta^{**}$, then the distance between $\boldsymbol{\chi}_1, \boldsymbol{\chi}_2$ is defined to be $\|\boldsymbol{\chi}_1 - \boldsymbol{\chi}_2\|_{W^{1,p}(\Omega)}$.

20.3.2 Variations Recall the set $\mathrm{Var}(\sigma)$ of admissible variations $\xi = (\boldsymbol{u}, \omega)$ defined in Subsect. 14.3.2(4). We introduce a wider set $\mathrm{Var}(\sigma)$ of variations with the norm $\|\cdot\|_2$ defined by

$$\mathrm{Var}(\sigma) = \{\xi = (\boldsymbol{u}, \omega) : \boldsymbol{u} \in W^{1,2}(\Omega), \omega \in L^2(\Omega), \boldsymbol{u}(\boldsymbol{p}) = \boldsymbol{0} \text{ on } D\},$$

$$\|\xi\|_2^2 = \|\boldsymbol{u}\|^2_{W^{1,2}(\Omega)} + \|\omega\|^2_{L^2(\Omega)}.$$

The space $\mathrm{Var}(\sigma)$ is a completion of $\mathrm{Var}(\sigma)$ of Subsect. 14.3.2(4) in the norm $\|\cdot\|_2$. For the variations of deformation we shall use the space

$$\mathrm{Var}(\boldsymbol{\chi}) = \{\boldsymbol{u} \in W^{1,2}(\Omega) : \boldsymbol{u}(\boldsymbol{p}) = \boldsymbol{0} \text{ on } D\} \quad (20.3.2)$$

with the norm $\|\cdot\|_{W^{1,2}(\Omega)}$. This will also be the space in which the solutions of the linearized problem will be sought in Sect. 20.8. For a given $\sigma \in \Sigma^{**}$ or $\boldsymbol{\chi} \in \Delta^{**}$ we extend the first and second variations of P, U, H, W, M to $\mathrm{Var}(\sigma)$ or $\mathrm{Var}(\boldsymbol{\chi})$ by density and continuity with respect to the norms $\|\cdot\|_2^2$, $\|\cdot\|_{W^{1,2}(\Omega)}$. This is possible: by (14.3.4) the first variation is an integral of a linear form in $\boldsymbol{u}$, $\nabla \boldsymbol{u}$, ω with the coefficients that are bounded functions of $\boldsymbol{p}$ as a consequence of the assumed smoothness of the rest states 13.1.1. Similarly the second variation is an integral of a bilinear form with bounded coefficients. With such a definition, (14.3.4) still holds with $\xi = (\boldsymbol{u}, \omega) \in \mathrm{Var}(\sigma)$.

Sect. 15.4 shows that the weak extrema imply the nonnegativity of the second variations. To establish the converse, we shall need the uniform positivity of the second variations; these cannot be satisfied, for example, by fluids. If X is a real linear space with norm $\|\cdot\|$ and Q a bilinear functional on X, we say that Q is ***uniformly positive*** on X if there exists a constant $c > 0$ such that $\mathsf{Q}(\boldsymbol{x}, \boldsymbol{x}) \geq c\|\boldsymbol{x}\|^2$ for every $\boldsymbol{x} \in \mathrm{X}$.

20.3.3 Proposition *Let $\sigma = (\chi, \eta) \in \Sigma^{**}$ be an equilibrium state compatible with the environment. Consider the following three assertions:*

(1) $\delta^2 \mathsf{P}(\sigma, \cdot, \cdot)$ *is uniformly positive on* $\mathrm{Var}(\sigma)$;

(2) $\delta^2 \mathsf{M}(\chi, \cdot, \cdot)$ *is uniformly positive on* $\mathrm{Var}(\chi)$;

(3) $\delta^2 \mathsf{U}(\sigma, \cdot, \cdot)$ *is uniformly positive on* $\{\xi \in \mathrm{Var}(\sigma) : \delta\mathsf{U}(\sigma, \xi) = 0\}$.

Then (1) $\Rightarrow$ (2), (1) $\Rightarrow$ (3). *If $\bar{c}_{\boldsymbol{F}}(\boldsymbol{F}, \theta_0) \geq c$ for every $\boldsymbol{F}$ that is realized in σ and some constant $c > 0$, then* (1) $\Leftrightarrow$ (2).

These relationships are similar to the relationships among the extremum principles discussed in Sect. 15.3. The equivalence of (1) and (2) again shows that the temperature plays a trivial role in testing the second variation of the total canonical free energy.

Proof By $(15.4.1)_1$, the implication (1) $\Rightarrow$ (3) is trivial, because Condition (3) speaks about narrower class of variations. Also the implication (1) $\Rightarrow$ (2) is trivial, since $\delta^2\mathsf{M}$ is a restriction of $\delta^2\mathsf{P}$ to variations $\xi = (\boldsymbol{u}, \omega)$ for which the variation of temperature vanishes identically. Suppose now that $\bar{c}_{\boldsymbol{F}}(\nabla\chi(\boldsymbol{p}), \theta_0) \geq c$ and prove (2) $\Rightarrow$ (1). Let (2) hold. The specific heat is $\bar{c}_{\boldsymbol{F}} = -\theta\partial_\theta^2\bar{f}$ and hence $-\partial_\theta^2\bar{f} \geq c_1/\theta_0 \equiv c_2 > 0$. Combining this with the assumption $\delta^2\mathsf{M}(\chi, \boldsymbol{u}, \boldsymbol{u}) \geq c\|\boldsymbol{u}\|^2_{W^{1,2}(\Omega)}$, we find from (15.4.8) that

$$\delta^2\mathsf{P}(\sigma, \xi, \xi) \geq c\|\boldsymbol{u}\|^2_{W^{1,2}(\Omega)} + c_2 \int_\Omega d\tilde{\theta}[\nabla\boldsymbol{u}, \omega]^2 \rho_\circ \, dV. \tag{20.3.3}$$

We have $d\tilde{\theta}[\nabla\boldsymbol{u}, \omega] = \boldsymbol{M}\cdot\nabla\boldsymbol{u} + l\omega$ where $\boldsymbol{M} = \partial_{\boldsymbol{F}}\tilde{\theta}$, $l = \partial_\eta\tilde{\theta} = \tilde{\theta}/\bar{c}_{\boldsymbol{F}}$. By $2ab \geq \alpha^{-1}a^2 + \alpha b^2$ with $\alpha = \frac{1}{2}$ we obtain

$$d\tilde{\theta}[\nabla\boldsymbol{u}, \omega]^2 = |\boldsymbol{M}\cdot\nabla\boldsymbol{u}|^2 + 2l\boldsymbol{M}\cdot\nabla\boldsymbol{u}\cdot\omega + l^2\omega^2 \geq -|\boldsymbol{M}\cdot\nabla\boldsymbol{u}|^2 + \tfrac{1}{2}l^2\omega^2.$$

Since $\nabla\chi$, η are bounded on $\mathrm{cl}\,\Omega$, the function $\boldsymbol{M}$ is bounded; by the positivity of the specific heat and the boundedness of $\nabla\chi$, η, the function $l = \tilde{\theta}/\bar{c}_{\boldsymbol{F}}$ is bounded away from 0. Therefore, there exists a $c_3 > 0$ such that $d\tilde{\theta}[\nabla\boldsymbol{u}, \omega]^2 \geq -c_3|\nabla\boldsymbol{u}|^2 + \frac{1}{2}c_3\omega^2$. Inserting this into (20.3.3), we see that

$$\delta^2\mathsf{P}(\sigma, \xi, \xi) \geq c\|\boldsymbol{u}\|^2_{W^{1,2}(\Omega)} - c_3c_2\|\nabla\boldsymbol{u}\|^2_{L^2(\Omega)} + \tfrac{1}{2}c_3c_2\|\omega\|^2_{L^2(\Omega)}.$$

Diminishing the value of c_2 to achieve

$$c\|\boldsymbol{u}\|^2_{W^{1,2}(\Omega)} - c_3c_2\|\nabla\boldsymbol{u}\|^2_{L^2(\Omega)} \geq c_4\|\boldsymbol{u}\|^2_{W^{1,2}(\Omega)},$$

with some $c_4 > 0$, we obtain Assertion (1). □

Next we are going to discuss the relationship between the positivity of the second variation and minima. Recall the definition of a weak local minimum (in the sense of calculus of variations) 15.4.1(2).

20.3.4 Proposition *Item* (1) *of Proposition* 20.3.3 *implies that* P *has a weak local minimum at* σ. *Item* (2) *implies that* M *has a weak local minimum at* χ. *Item* (3) *implies that* U *has a weak minimum at* σ *relative to* $\{\tau \in \Sigma^{**} : \mathsf{H}(\tau) = \mathsf{H}(\sigma)\}$. *Item* (3) *also implies that* H *has a weak maximum at* σ *relative to* $\{\tau \in \Sigma^{**} : \mathsf{U}(\tau) = \mathsf{U}(\sigma)\}$.

Proof These are general assertions of the calculus of variations and the proofs are omitted. □

Unfortunately, this result is only of limited interest in continuum thermodynamics, because a weak relative minimum is not sufficient for a strong minimum in the sense of definition given in 15.1.4 and 15.1.5, which, in turn, is the only version of a minimum that is relevant to the stability theory. No metric is known for which the uniform positivity of the second variation would imply a strong minimum with respect to it (in the sense of definition 15.1.4 and 15.1.5).

Let us explain this issue in a more detail. Consider the functional M only, assume $\rho_o \equiv 1$ and consider first the Dirichlet boundary conditions, i.e., $D = \partial\Omega$. Then $\mathrm{Var}(\chi) = W_0^{1,2}(\Omega)$, where $W_0^{1,2}(\Omega)$ is the space of all $\boldsymbol{u} \in W^{1,2}(\Omega)$ which satisfy $\boldsymbol{u} = \mathbf{0}$ on $\partial\Omega$ in the sense of trace. Write f for $\bar{f}(\cdot, \theta_0)$. A $\chi_0 \in \boldsymbol{\Delta}^{**}$ is said to be a local minimum in $\boldsymbol{\Delta}^{**} \cap W^{1,p}(\Omega)$ if there exists a $\delta_0 > 0$ such that $\mathsf{M}(\chi) \ge \mathsf{M}(\chi_0)$ for every $\chi \in \boldsymbol{\Delta}^{**}$ such that $\|\chi - \chi_0\|_{W^{1,p}(\Omega)} \le \delta_0$. In this situation, one would be tempted to apply the following theorem.

20.3.5 Proposition *Let* X *be a normed space with norm* $\|\cdot\|$, *let further* O *be its open subset,* $F : \mathrm{O} \to \mathbb{R}$ *a functional and* $\boldsymbol{x}_0$ *a point in* O. *Suppose that* F *has a continuous Fréchet differential* $\delta F(\boldsymbol{x}, \cdot)$ *in a neighborhood* N *of* $\boldsymbol{x}_0$ *and that the mapping* $\delta F(\cdot) : \mathrm{N} \to \mathrm{X}^*$ *has a Fréchet differential* $\delta^2 F(\boldsymbol{x}_0)$ *at* $\boldsymbol{x}_0$, *where* X^* *is the dual of* X. *If*

$$\delta F(\boldsymbol{x}_0) = \mathbf{0} \quad \text{and} \quad \delta^2 F(\boldsymbol{x}_0, \boldsymbol{h}, \boldsymbol{h}) \ge c\|\boldsymbol{h}\|^2 \tag{20.3.4}$$

for all $\boldsymbol{h} \in \mathrm{X}$ *and some* $c > 0$, *then there exists a* $\delta_0 > 0$ *such that*

$$F(\boldsymbol{x}) - F(\boldsymbol{x}_0) \ge \tfrac{1}{2}c\|\boldsymbol{x} - \boldsymbol{x}_0\|^2$$

for all $\boldsymbol{x} \in \mathrm{O}$ *such that* $\|\boldsymbol{x} - \boldsymbol{x}_0\| \le \delta_0$.

See, e.g., GEL'FAND & FOMIN [1961; Sect. 20, Thm. 2] for proof.

20.3.6 Example *Let* $1 \le p < \infty$, *let* Ω *be given and define* $\boldsymbol{d}$ *by* $\boldsymbol{d}(\boldsymbol{p}) = \boldsymbol{p}$ *for every* $\boldsymbol{p} \in \partial\Omega$. *Then there exists an infinitely differentiable function* $f : \mathrm{Lin}^+ \to \mathbb{R}$ *such that the homogeneous deformation* $\chi_0 \in \boldsymbol{\Delta}^{**}$ *with deformation gradient* $\mathbf{1}$ *satisfies*

$$\delta \mathsf{M}(\chi_0, \boldsymbol{u}) = 0, \quad \delta^2 \mathsf{M}(\chi_0, \boldsymbol{u}, \boldsymbol{u}) \ge c\|\boldsymbol{u}\|^2_{W^{1,2}(\Omega)} \tag{20.3.5}$$

for every $\boldsymbol{u} \in W_0^{1,2}(\Omega)$ *and some* $c > 0$ *and yet* χ_0 *is not a local minimum of* M *in* $\boldsymbol{\Delta}^{**} \cap W^{1,p}(\Omega)$. *Also, the function* f *can be chosen such that* χ_0 *is not a strong local minimum of* M *(see Definition* 15.4.1(3)).

To prove this assertion, note that by Remark 17.1.5(2) the local minimum of M in $\boldsymbol{\Delta}^{**} \cap W^{1,p}(\Omega)$ implies that f must be quasiconvex and hence rank 1 convex at $\mathbf{1}$, while the uniform positivity of $\delta^2 \mathsf{M}(\chi_0, \boldsymbol{u}, \boldsymbol{u})$ on $W_0^{1,2}(\Omega)$ is controlled entirely by $\partial_F^2 f(\mathbf{1})$. Since the rank 1 convexity at $\mathbf{1}$ involves the behavior of f over its entire domain, it is possible to construct an f such that $\partial_F^2 f(\mathbf{1})$ is positive definite and yet f is not rank 1 convex at $\mathbf{1}$.

The reason for this failure is that f is not quasiconvex. For the classical calculus of variations on an interval (see 17.1.2(5)) the quasiconvexity is equivalent to the ordinary convexity, and there it is known that the positivity of the second variation and a strengthened form of the convexity along the extremal imply the strong local minimum (see CESARI [1983, Sect. 2.11] and BALL & MARSDEN [1984]). In more spatial dimensions, a partial analog of the just mentioned fact is a result of SIVALOGANATHAN [1989]. It applies to the Dirichlet boundary conditions and says that if the stored energy function is polyconvex in a strengthened sense, then the integral functional has a strong local minimum in the sense of the calculus of variations on perturbations with sufficiently small support. For further results in this direction, see ZHANG [1991].

For the non-Dirichlet boundary conditions, i.e., for $S \neq \emptyset$, in addition to the ordinary quasi-convexity, also the quasiconvexity at the boundary (see Sect. 17.1) figures as a necessary condition for the minimum in $W^{1,p}(\Omega)$, $1 \leq p < \infty$. An example by BALL & MARSDEN [1984] shows that in $n > 1$ space dimensions the uniform positivity of the second variation and the quasiconvexity (in fact, even polyconvexity) of f do not imply a local minimum in the Sobolev space $W^{1,p}(\Omega)$. In that example, the integrand f is not quasiconvex at the boundary.

It is instructive to see why Proposition 20.3.5 cannot be applied to M and the space X = $W_0^{1,p}(\Omega)$, $1 \leq p < \infty$. The reason is different when $1 \leq p \leq 2$ and when $p > 2$. If $1 \leq p \leq 2$, then except for essentially quadratic energy functions the second variation δ^2M is not a Fréchet differential with respect to the norm $\|\cdot\|_{W^{1,p}(\Omega)}$, see below. On the other hand, if $p > 2$, then the condition $(20.3.4)_2$ cannot be satisfied. Let us first give a theorem on failure of second Fréchet differentiability in a version that does not require any growth conditions on the integrand.

20.3.7 Proposition *Let f : $\mathrm{Lin}^+ \to \mathbb{R}$ be a function of class C^3, let Ω be an open subset of* Eucl *and let p satisfy $1 \leq p \leq 2$. Suppose that there exists a $\chi_0 \in \Delta^{\#\#}$ such that* M *is twice Fréchet differentiable at χ_0 in the sense that*

$$\mathsf{M}(\chi_0 + \boldsymbol{u}) = \mathsf{M}(\chi_0) + \delta\mathsf{M}(\chi_0, \boldsymbol{u}) + \tfrac{1}{2}\delta^2\mathsf{M}(\chi_0, \boldsymbol{u}, \boldsymbol{u}) + \mathsf{R}(\boldsymbol{u}),$$

for every $\boldsymbol{u} \in W_0^{1,2}(\Omega)$, where

$$\mathsf{R}(\boldsymbol{u})/\|\boldsymbol{u}\|^2_{W^{1,p}(\Omega)} \to 0 \ \text{ as } \ \|\boldsymbol{u}\|_{W^{1,p}(\Omega)} \to 0.$$

Then f is of the form

$$f(\boldsymbol{F}) = c_0 + \boldsymbol{A}_0 \cdot \boldsymbol{F} + \tfrac{1}{2}\mathsf{A}_0[\boldsymbol{F}, \boldsymbol{F}] + p_0 \det \boldsymbol{F}, \quad \boldsymbol{H} \in \mathrm{Lin}^+,$$

where $c_0, p_0 \in \mathbb{R}$, $\boldsymbol{A}_0 \in$ Lin, and A_0 is a constant fourth order tensor.

The above energy function is essentially quadratic–the terms linear in $\boldsymbol{F}$ and $\det \boldsymbol{F}$ do not contribute to the values of M when the boundary of the body is held fixed. For spaces $W^{1,p}$ with $p > 2$, there are conditions which ensure that the second variation of an integral functional is the second Fréchet differential, see KRASNOSELSKII & al. [1976]. However, there the uniform positivity of the second variation cannot be satisfied:

20.3.8 Proposition *Let $\chi_0 \in \Delta^{\#\#}$, $p > 2$, and $c \geq 0$ be such that*

$$\delta^2\mathsf{M}(\chi_0, \boldsymbol{u}, \boldsymbol{u}) \geq c\|\boldsymbol{u}\|^2_{W^{1,p}(\Omega)},$$

for every $\boldsymbol{u} \in W_0^{1,p}(\Omega)$. Then $c = 0$.

The proofs of the last two propositions are omitted.

20.4 Positivity of the Second Variation (Necessary Conditions)

The second variation of the stored energy M (see $(20.3.1)_3$) at a given configuration χ_0 is a quadratic functional $\mathsf{J}(\cdot) = \delta\mathsf{M}(\chi_0, \cdot)$ given by

$$\mathsf{J}(\boldsymbol{u}) = \int_\Omega \bar{\boldsymbol{A}}[\nabla \boldsymbol{u}, \nabla \boldsymbol{u}]\, dV, \tag{20.4.1}$$

defined on the space of admissible variations $\mathrm{Var}(\chi)$. In the rest of this chapter, we define the spaces of admissible variations to be the closures in $W^{1,2}(\Omega)$ of the spaces of variations defined in 14.3.2; thus for the displacement-traction loading environment, $\mathrm{Var}(\chi)$ is given by (20.3.2); for the bilateral contact environment,

$$\mathrm{Var}(\chi) = \{\boldsymbol{u} \in W^{1,2}(\Omega) : \boldsymbol{u}(\boldsymbol{p}) \cdot \boldsymbol{m}(\boldsymbol{p}) = \boldsymbol{0} \ \text{ on } \ D := \cup_{i=1}^{s} D_i\}$$

where $\boldsymbol{m}(\boldsymbol{p})$ is the normal to $\partial\chi(\Omega)$ expressed as function of $\boldsymbol{p} \in D$. In (20.4.1), $\bar{\mathbf{A}}$ is a function defined on Ω with values in the fourth-order tensors, the tensor of isothermal elasticities at $\sigma_0 = (\chi_0, \theta_0)$. For the results of this section this interpretation of $\bar{\mathbf{A}}$ is immaterial. They apply to any function $\bar{\mathbf{A}}$ defined on Ω which possesses the major symmetry. In Chap. 25 the concepts of the present section will be applied to $\bar{\mathbf{A}}$ identified with the elasticities evaluated at static, but not necessarily an equilibrium state. *Throughout Sects.* 20.4–20.6 *it is assumed that* $\bar{\mathbf{A}}$ *is continuous on* cl Ω *and that it has the major symmetry; moreover, it is assumed that* S *is relatively open in* $\partial\Omega$.

20.4.1 Definitions The second variation J is said to be

(1) ***positive semidefinite*** (on $\operatorname{Var}(\chi)$) if for every $\boldsymbol{u} \in \operatorname{Var}(\chi)$,

$$\mathsf{J}(\boldsymbol{u}) \geq 0; \tag{20.4.2}$$

(2) ***strictly positive*** (on $\operatorname{Var}(\chi)$) if (20.4.2) holds for every $\boldsymbol{u} \in \operatorname{Var}(\chi)$ and the equality holds only if $\boldsymbol{u} = \mathbf{0}$.

(3) ***uniformly positive*** (on $\operatorname{Var}(\chi)$) if for every $\boldsymbol{u} \in \operatorname{Var}(\chi)$ and some $c > 0$,

$$\mathsf{J}(\boldsymbol{u}) \geq c\|\boldsymbol{u}\|^2_{W^{1,2}(\Omega)}; \tag{20.4.3}$$

(4) ***coercive*** (on $\operatorname{Var}(\chi)$) if for every $\boldsymbol{u} \in \operatorname{Var}(\chi)$ and some $c > 0$, $d > 0$,

$$\mathsf{J}(\boldsymbol{u}) \geq c\|\boldsymbol{u}\|^2_{W^{1,2}(\Omega)} - d\|\boldsymbol{u}\|^2_{L^2(\Omega)};$$

(5) ***weakly bounded below*** (on $\operatorname{Var}(\chi)$) if for every $\boldsymbol{u} \in \operatorname{Var}(\chi)$ and some $d > 0$,

$$\mathsf{J}(\boldsymbol{u}) \geq -d\|\boldsymbol{u}\|^2_{L^2(\Omega)}.$$

Uniform positivity will be seen in Sect. 20.8 to ensure the existence of solutions of the linearized static problem, and in Sect. 25.5 the asymptotic stability of the linear time evolution. For the existence of the linear time evolution, mere coercivity suffices (Sect. 25.4). The uniqueness of the initial value problem in the linearized dynamics holds under weak boundedness below (Sect. 25.3). Here we only note that the strict positivity of J implies that the only class C^2 solution of the linearized equations (20.1.8) with $\bar{\boldsymbol{b}} \equiv \mathbf{0}, \bar{\boldsymbol{s}} \equiv \mathbf{0}, \bar{\boldsymbol{d}} \equiv \mathbf{0}$ is $\boldsymbol{u} = \mathbf{0}$ identically. Indeed, it suffices to multiply these equations by $\boldsymbol{u}$, to integrate over Ω and to use the divergence theorem and the boundary conditions to find that $\mathsf{J}(\boldsymbol{u}) = 0$. The strict positivity of J then implies $\boldsymbol{u} = \mathbf{0}$.

Throughout Sect. 20.4–20.6 the concepts of Definition 20.4.1(1)–(5) are applied with the choice $\operatorname{Var}(\chi)$ for the displacement-traction environment as specified in (20.3.2) without further explicit mention.

Clearly, the second variation is positive semidefinite if the tensor $\bar{\mathbf{A}}$ is positive semidefinite at every $\boldsymbol{p} \in \Omega$. This trivial condition is far from necessary, and, moreover, cannot be satisfied generally. We shall see in Sect. 20.6 that for isotropic states under compression, $\bar{\mathbf{A}}$ is *not* positive semidefinite, and yet under concrete circumstances J is uniformly positive. Of course the case when $\bar{\mathbf{A}}$ is positive semidefinite will figure as a particular case in the necessary and sufficient conditions. The techniques we shall use are the Fourier transformation, for which we need the complex formalism of Subsect. 1.3.1, and scaling arguments which enable us to infer much information about the problems 20.4.1 on a general domain Ω from the much easily tractable problems on the whole space or a half-space.

20.4.2 Problems in the Half-Space Let $\boldsymbol{p} \in S$, denote $\bar{\mathbf{A}} = \bar{\mathbf{A}}(\boldsymbol{p})$, $\boldsymbol{n} = \boldsymbol{n}(\boldsymbol{p})$ and define the open half-space H with boundary plane ∂H of normal $\boldsymbol{n}$ by

$$H = \{\boldsymbol{p} \in \mathrm{Vect} : \boldsymbol{p} \cdot \boldsymbol{n} < 0\}, \quad \partial H = \{\boldsymbol{p} \in \mathrm{Vect} : \boldsymbol{p} \cdot \boldsymbol{n} = 0\}. \tag{20.4.4}$$

We shall be interested in infinitely differentiable real vector-valued functions $\boldsymbol{u} : H \to$ Vect of the form of ***exponentials,*** i.e., in functions of the form

$$\boldsymbol{u}(\boldsymbol{p}) = z(-\boldsymbol{p} \cdot \boldsymbol{n}) \exp(\mathrm{i}\boldsymbol{k} \cdot \boldsymbol{p}) + z^*(-\boldsymbol{p} \cdot \boldsymbol{n}) \exp(-\mathrm{i}\boldsymbol{k} \cdot \boldsymbol{p}), \tag{20.4.5}$$

where z is an infinitely differentiable function mapping $[0, \infty)$ into $\mathrm{Vect}^{\mathbb{C}}$ and $\boldsymbol{k}$ is a constant vector such that $\boldsymbol{k} \cdot \boldsymbol{n} = 0$. The exponential $\boldsymbol{u}$ is said to be ***bounded*** if there exists a constant C such that $|\boldsymbol{u}(\boldsymbol{p})| < C$ for every $\boldsymbol{p} \in H$. We shall consider the following two problems.

(1) Find all solutions $\boldsymbol{u} \in C^\infty(\mathrm{cl}\, H)$ of the form of bounded exponentials satisfying

$$\mathrm{Div}\, \bar{\mathbf{A}}[\nabla \boldsymbol{u}] = \mathbf{0} \ \text{ in } \ H, \quad \bar{\mathbf{A}}[\nabla \boldsymbol{u}]\boldsymbol{n} = \mathbf{0} \ \text{ on } \ \partial H. \tag{20.4.6}$$

This problem is obtained by blowing up the situation at the boundary point $\boldsymbol{p} \in S$ of the original problem (20.1.8), and it is recalled that the main goal in examining the problem (20.1.8) is to determine whether there exists a nontrivial solution to the trivial data (i.e., nonuniqueness). The existence of a decaying (as $\boldsymbol{p} \cdot \boldsymbol{n} \to -\infty$) solution $\boldsymbol{u} \neq \mathbf{0}$ satisfying (20.4.6) means that the magnified problem does not possess the uniqueness, which also sheds light on the nonuniqueness of the original problem. The nontrivial decaying solutions to (20.4.6) are called ***surface instabilities.***

(2) Let $\alpha \neq 0$, $\alpha \in \mathbb{R}$ be a constant. Find all solutions $\boldsymbol{u} \in C^\infty(\mathrm{cl}\, H)$ of the form of bounded exponentials satisfying

$$\mathrm{Div}\, \bar{\mathbf{A}}[\nabla \boldsymbol{u}] = \alpha^2 \boldsymbol{u} \ \text{ in } \ H, \quad \bar{\mathbf{A}}[\nabla \boldsymbol{u}]\boldsymbol{n} = \mathbf{0} \ \text{ on } \ \partial H. \tag{20.4.7}$$

If (20.4.7) has a bounded exponential solution $\boldsymbol{u}$ for some $\alpha \neq 0$ then it has a bounded exponential solution for every $\alpha \neq 0$. The required solution for $\beta \neq 0$ is $\bar{\boldsymbol{w}}(\boldsymbol{p}) = \boldsymbol{u}(\beta \boldsymbol{p}/\alpha)$. We shall see that when (20.4.7) possesses a decaying solution $\boldsymbol{u} \neq 0$, then the total stored energy is negative at this $\boldsymbol{u}$. When properly scaled and slightly modified away from the boundary, this solution eventually leads to a variation $\boldsymbol{u} \in \mathrm{Var}(\chi)$ of the original problem on which $\mathrm{J}(\boldsymbol{u}) < 0$. Another interpretation of a decaying solution to (20.4.7) is in terms of the operator of linear elasticity $L : D(L) \subset L^2(H) \to L^2(H)$ on a half-space, where

$$L\boldsymbol{u} = -\mathrm{Div}\, \bar{\mathbf{A}}[\nabla \boldsymbol{u}], \quad \boldsymbol{u} \in D(L) := \{\boldsymbol{u} \in W^{2,2}(H) : \bar{\mathbf{A}}[\nabla \boldsymbol{u}] \cdot \boldsymbol{n} = \mathbf{0} \ \text{ on } \ \partial H\}.$$

Any decaying nontrivial solution $\boldsymbol{u}$ to (20.4.7) is a generalized eigenvector of L with *negative* eigenvalue $-\alpha^2$. Here the modifier "generalized" means that that $\boldsymbol{u}$ is not in $L^2(H)$, as required from ordinary eigenvectors. Nevertheless, the existence of a generalized eigenvector with the eigenvalue $-\alpha^2$ still implies that the spectrum of L contains $-\alpha^2$. Notice that the above mentioned scaling also shows that if the spectrum of L contains some negative number, then it contains the whole negative semi-axis $(-\infty, 0)$. Yet another interpretation is that if $\boldsymbol{u}$ satisfies (20.4.7), then the time-dependent displacement $\boldsymbol{w}(\boldsymbol{p}, t) := \boldsymbol{u}(\boldsymbol{p})e^{ct}, \boldsymbol{p} \in H, t \in \mathbb{R}$, where $c = |\alpha|/\sqrt{\rho_\circ}$, satisfies the linearized dynamical equations

$$\rho_\circ \ddot{\boldsymbol{w}} = \mathrm{Div}\, \bar{\mathbf{A}}[\nabla \boldsymbol{w}] \ \text{ in } \ H, \quad \bar{\mathbf{A}}[\nabla \boldsymbol{w}]\boldsymbol{n} = \mathbf{0} \ \text{ on } \ \partial H$$

and $\boldsymbol{w}$ grows exponentially in time, and decays as $\boldsymbol{p} \cdot \boldsymbol{n} \to -\infty$. Agmon's condition (below) is equivalent to saying that such displacements do not exist. This is related to *surface waves* (see Subsect. 23.3.7 for the definition). However, this interpretation should not be taken literally, because the propagation of waves involves adiabatic elasticities, while here, dealing with the static problems, we have the isothermal elasticities.

We shall now introduce pointwise properties of the tensor of elasticities which will figure as necessary, sufficient, and exceptionally as necessary and sufficient conditions for the positivity properties of the second variation 20.4.1. Some of them have already been encountered in the previous chapters. Define the ***isothermal acoustic tensor*** $\bar{\boldsymbol{Q}}(\boldsymbol{K})$ in the 'direction' $\boldsymbol{K} \in \mathrm{Vect}^{\mathbb{C}}$ by $\bar{\boldsymbol{Q}}(\boldsymbol{K})\boldsymbol{a} = \bar{\boldsymbol{A}}[\boldsymbol{a} \otimes \boldsymbol{K}]\boldsymbol{K}$ for every $\boldsymbol{a} \in \mathrm{Vect}^{\mathbb{C}}$.

20.4.3 Definitions The elasticity tensor $\bar{\boldsymbol{A}}$ is said to satisfy

(1) the ***Legendre–Hadamard condition*** if

$$\bar{\boldsymbol{A}}[\boldsymbol{a} \otimes \boldsymbol{b}, \boldsymbol{a} \otimes \boldsymbol{b}] \geq 0 \tag{20.4.8}$$

for every $\boldsymbol{a}, \boldsymbol{b} \in \mathrm{Vect}$;

(2) the ***strong ellipticity condition*** if it satisfies (20.4.8) and the inequality is strict whenever $\boldsymbol{a} \neq \boldsymbol{0}, \boldsymbol{b} \neq \boldsymbol{0}$; the ***ellipticity condition*** if $\bar{\boldsymbol{Q}}(\boldsymbol{n})$ is invertible for every $\boldsymbol{n} \in \mathrm{Sph}$;

(3) the ***complementing condition*** with normal $\boldsymbol{n} \in \mathrm{Sph}$ if the only bounded exponential solution of (20.4.6) is $\boldsymbol{u} \equiv \boldsymbol{0}$;

(4) ***Agmon's condition*** with normal $\boldsymbol{n} \in \mathrm{Sph}$ if the only bounded exponential solution of (20.4.7) with $\alpha \neq 0$ is $\boldsymbol{u} \equiv \boldsymbol{0}$;

(5) the ***strong complementing condition*** with normal $\boldsymbol{n} \in \mathrm{Sph}$ if it satisfies both the complementing and Agmon's conditions.

(6) $\bar{\boldsymbol{A}}$ is said to be ***quasiconvex at the boundary*** with normal $\boldsymbol{n} \in \mathrm{Sph}$ if for every $\boldsymbol{u} \in C_0^\infty(\mathrm{Vect})$,

$$\int_H \bar{\boldsymbol{A}}[\nabla \boldsymbol{u}, \nabla \boldsymbol{u}]\, dV \geq 0. \tag{20.4.9}$$

(7) $\bar{\boldsymbol{A}}$ is said to be ***uniformly quasiconvex at the boundary*** with normal $\boldsymbol{n} \in \mathrm{Sph}$ if the inequality (20.4.9) holds for every $\boldsymbol{u} \in C_0^\infty(\mathrm{Vect})$ with the right-hand side replaced by $c\|\boldsymbol{u}\|^2_{W^{1,2}(\Omega)}$ where $c > 0$ is a constant independent of $\boldsymbol{u}$

20.4.4 ***Remarks***

(1) The idea behind the choice of the test functions $\boldsymbol{u} \in C_0^\infty(\mathrm{Vect})$ in Definition 20.4.3(6) is that the restriction of $\boldsymbol{u}$ to H takes arbitrary values on ∂H, but zero values outside some compact region in H. A more illuminative equivalent form is to demand that (20.4.9) hold for every $\boldsymbol{u} \in C_0^\infty(B)$, where B is a ball centered at $\boldsymbol{0} \in \mathrm{Vect}$ of radius 1. Clearly, the latter requirement is weaker or equivalent to Definition 20.4.3(6). Conversely, if (20.4.9) holds for every $\boldsymbol{u} \in C_0^\infty(B)$, then the validity of (20.4.9) for every $\boldsymbol{u}$ as in Definition 20.4.3(6) is proved by an obvious scaling of $\boldsymbol{u}$ to reach that the support of the scaled $\boldsymbol{u}$ is in B. Also, B can be equivalently replaced by a cube with one face being the free part of the boundary and the remaining five faces fixed, etc.

(2) We shall see that the nonnegativity of the second variation implies Agmon's condition at every $\boldsymbol{p} \in S$. Note that it is clear that the complementing condition is invariant under the change $\bar{\boldsymbol{A}} \to -\bar{\boldsymbol{A}}$, while Agmon's condition and the strong complementing conditions are not. Hence the complementing condition typically implies that some combinations of the coefficients of $\bar{\boldsymbol{A}}$ are different from 0, while Agmon's condition and the strong complementing condition typically result in inequalities on the coefficients.

(3) If $\bar{\boldsymbol{A}}$ satisfies the Legendre–Hadamard condition then $\bar{\boldsymbol{A}}[\boldsymbol{a} \otimes \boldsymbol{b}, \boldsymbol{a}^* \otimes \boldsymbol{b}] \geq 0$ for every $\boldsymbol{a} \in \mathrm{Vect}^{\mathbb{C}}$, $\boldsymbol{b} \in \mathrm{Vect}$, but in general it is not true that $\bar{\boldsymbol{A}}[\boldsymbol{a} \otimes \boldsymbol{b}, \boldsymbol{a}^* \otimes \boldsymbol{b}^*] \geq 0$ for every $\boldsymbol{a}, \boldsymbol{b} \in \mathrm{Vect}^{\mathbb{C}}$.

20.4.5 Algebraic Form of Problems 20.4.2 Equations (20.4.6) and (20.4.7) are *ordinary* differential equations. Calculating the derivatives of $\boldsymbol{u}$ in (20.4.5) one finds that (20.4.7) read

$$\begin{gathered}\bar{\mathsf{A}}[\ddot{z}\otimes\boldsymbol{n}]\boldsymbol{n}-\mathrm{i}\bar{\mathsf{A}}[\dot{z}\otimes\boldsymbol{n}]\boldsymbol{k}-\mathrm{i}\bar{\mathsf{A}}[\dot{z}\otimes\boldsymbol{k}]\boldsymbol{n}-\bar{\mathsf{A}}[z\otimes\boldsymbol{k}]\boldsymbol{k}=\alpha^2 z \ \text{ for } \ s\in[0,\infty),\\ \bar{\mathsf{A}}[\dot{z}\otimes\boldsymbol{n}-\mathrm{i}z\otimes\boldsymbol{k}]\boldsymbol{n}=\boldsymbol{0} \ \text{ for } \ s=0,\end{gathered} \tag{20.4.10}$$

and (20.4.6) give (20.4.10) with $\boldsymbol{0}$ on the right-hand side of $(20.4.10)_1$. Equivalently,

$$\begin{gathered}\boldsymbol{A}\ddot{z}-\mathrm{i}(\boldsymbol{B}+\boldsymbol{B}^{\mathrm{T}})\dot{z}-\boldsymbol{C}z=\boldsymbol{0} \ \text{ for } \ s\in[0,\infty),\\ \boldsymbol{A}\dot{z}-\mathrm{i}\boldsymbol{B}^{\mathrm{T}}z=\boldsymbol{0} \ \text{ for } \ s=0,\end{gathered} \tag{20.4.11}$$

where $\boldsymbol{A}, \boldsymbol{B}, \boldsymbol{C}$ are the second order tensors defined by

$$\boldsymbol{Aa}=\bar{\mathsf{A}}[\boldsymbol{a}\otimes\boldsymbol{n}]\boldsymbol{n},\quad \boldsymbol{Ba}=\bar{\mathsf{A}}[\boldsymbol{a}\otimes\boldsymbol{n}]\boldsymbol{k},\quad \boldsymbol{Ca}=\bar{\mathsf{A}}[\boldsymbol{a}\otimes\boldsymbol{k}]\boldsymbol{k}-\alpha^2\boldsymbol{a},$$

for every $\boldsymbol{a}\in$ Vect. The problems 20.4.2 have been addressed for a general $\bar{\mathsf{A}}$; in particular, it may happen that $\boldsymbol{A}$ is not invertible in which case $(20.4.11)_1$ may be of the second order in some components of z and of the first or zeroth order in others. If $\boldsymbol{A}$ is invertible, then $(20.4.11)_1$ is of the second order in all components of z and may be put in the form

$$\ddot{z}+\mathrm{i}\bar{\boldsymbol{D}}\dot{z}-\bar{\boldsymbol{C}}z=\boldsymbol{0} \tag{20.4.12}$$

where $\bar{\boldsymbol{D}}, \bar{\boldsymbol{C}}$ are some second-order tensors. By introducing $\boldsymbol{w} = \dot{z}$ one can reduce (20.4.12) to the first order system

$$\dot{\boldsymbol{x}}=\boldsymbol{L}\boldsymbol{x} \ \text{ where } \ \boldsymbol{x}=\begin{bmatrix}\boldsymbol{w}\\ z\end{bmatrix},\quad \boldsymbol{L}=\begin{bmatrix}-\mathrm{i}\bar{\boldsymbol{D}} & \bar{\boldsymbol{C}}\\ \boldsymbol{1} & \boldsymbol{0}\end{bmatrix}. \tag{20.4.13}$$

Standard methods of ordinary linear differential equations with constant coefficients may be used to determine a general solution. The matrix $\boldsymbol{L}$ need not be diagonalizable, i.e., $\boldsymbol{L}$ need not have $2n$ linearly independent eigenvectors. This may happen even in isotropic states of a hyperelastic body (see Proof of Proposition 20.6.2). Thus generally one has to determine the Jordan decomposition of $\boldsymbol{L}$ to find the general solution of (20.4.13). The eigenvectors of $\mathrm{i}\boldsymbol{L}$ lead to particular solutions of (20.4.11) in the form

$$z(s)=\boldsymbol{a}\exp(-\mathrm{i}\omega s),\qquad s\ge 0.$$

Then z will solve $(20.4.10)_1$ if and only if

$$(\bar{\boldsymbol{Q}}(\boldsymbol{K})+\alpha^2\boldsymbol{1})\boldsymbol{a}=\boldsymbol{0} \ \text{ where } \ \boldsymbol{K}=\omega\boldsymbol{n}+\boldsymbol{k}. \tag{20.4.14}$$

Given $\boldsymbol{k}\in$ Vect, $\boldsymbol{n}\in$ Sph, $\alpha\in\mathbb{R}$, a pair $(\omega,\boldsymbol{a})\in\mathbb{C}\times\mathrm{Vect}^{\mathbb{C}}$ with $\boldsymbol{a}\neq\boldsymbol{0}$ is called a ***characteristic pair*** if it satisfies (20.4.14). The frequency ω satisfies the characteristic equation

$$\det(\bar{\boldsymbol{Q}}(\boldsymbol{K})+\alpha^2\boldsymbol{1})=0 \ \text{ where } \ \boldsymbol{K}=\omega\boldsymbol{n}+\boldsymbol{k}. \tag{20.4.15}$$

For each fixed $\boldsymbol{k}\in$ Vect, $\boldsymbol{n}\in$ Sph, $\alpha\in\mathbb{R}$, the characteristic polynomial

$$p(\omega):=\det(\bar{\boldsymbol{Q}}(\boldsymbol{K})+\alpha^2\boldsymbol{1})$$

is of degree at most $2n$ in ω. It is of degree exactly $2n$ when $\boldsymbol{A}$ is invertible and in particular when $\bar{\mathsf{A}}$ is strongly elliptic. Taking the complex conjugate of (20.4.14) and using that $\bar{\boldsymbol{A}}$ has real elements, one sees that the characteristic pairs go in pairs: if $(\omega, \boldsymbol{a})$ is a characteristic pair, then $(\omega^*, \boldsymbol{a}^*)$ is again a characteristic pair. Moreover, if $\bar{\boldsymbol{A}}$ is strongly elliptic, then $\operatorname{Im} \omega \neq 0$ for every characteristic pair $(\omega, \boldsymbol{a})$ and hence $(\omega^*, \boldsymbol{a}^*)$ is different from $(\omega, \boldsymbol{a})$. To prove the last assertions, multiply (20.4.14) by $\boldsymbol{a}^*$ to obtain

$$\bar{\boldsymbol{A}}[\boldsymbol{a} \otimes \boldsymbol{K}, \boldsymbol{a}^* \otimes \boldsymbol{K}] + \alpha^2 |\boldsymbol{a}|^2 = 0. \qquad (20.4.16)$$

The reality of ω would mean that $\boldsymbol{K} \in \mathrm{Vect}$ and as $\boldsymbol{K} \neq \boldsymbol{0}$ by $\boldsymbol{k} \neq \boldsymbol{0}, \boldsymbol{k} \cdot \boldsymbol{n} = 0$, and $\boldsymbol{a} \neq \boldsymbol{0}$, the strong ellipticity says that $\bar{\boldsymbol{A}}[\boldsymbol{a} \otimes \boldsymbol{K}, \boldsymbol{a}^* \otimes \boldsymbol{K}] > 0$ which contradicts (20.4.16). If $\bar{\boldsymbol{A}}$ satisfies the strong ellipticity condition and $\boldsymbol{L}$ is diagonalizable, there will be n characteristic pairs $(\omega_j, \boldsymbol{a}_j), j = 1, \ldots, n$, with $\operatorname{Im} \omega_j < 0$ and $\boldsymbol{a}_j$ linearly independent. In this case every bounded solution of (20.4.10) is of the form

$$z(s) = \sum_{j=1}^{n} \xi_j \boldsymbol{a}_j \exp(-\mathrm{i}\omega_j s), \quad s \geq 0, \qquad (20.4.17)$$

where $(\boldsymbol{a}_j, \omega_j)$ are fixed characteristic pairs with $\operatorname{Im} \omega_j < 0$ and $\xi_j \in \mathbb{C}$. Equation $(20.4.10)_2$ gives

$$\sum_{j=1}^{n} \xi_j \bar{\boldsymbol{A}}[\boldsymbol{a}_j \otimes \boldsymbol{K}_j]\boldsymbol{n} = \boldsymbol{0}, \qquad (20.4.18)$$

where $\boldsymbol{K}_j := (\omega_j \boldsymbol{n} + \boldsymbol{k}), j = 1, \ldots, n$. Equations (20.4.18) form a system of n scalar equations for the n complex numbers ξ_j, and the bounded nontrivial solutions of the problems 20.4.2 will not exist if and only if the determinant of this system is nonzero. This is the final algebraic form of the complementing and Agmon's conditions when $\boldsymbol{L}$ is diagonalizable. When $\boldsymbol{L}$ has a nontrivial Jordan structure, the consideration are similar but more complicated, see the proof of Proposition 20.6.2.

We shall now show that one can extend (20.4.9) to functions periodic in the directions perpendicular to $\boldsymbol{n}$. Let $\boldsymbol{n}$ be given and $\boldsymbol{e}_1, \cdots, \boldsymbol{e}_{n-1} \in \mathrm{Vect}$ vectors such that $\{\boldsymbol{n}, \boldsymbol{e}_1, \cdots, \boldsymbol{e}_{n-1}\}$ is an orthogonal, but not necessarily orthonormal, basis. A semiprism of axis $\boldsymbol{n}$ and rectangular base $\{\boldsymbol{e}_1, \cdots, \boldsymbol{e}_{n-1}\}$ is the subset P of Vect given by

$$P = \{\boldsymbol{p} \in \mathrm{Vect} : 0 < \boldsymbol{p} \cdot \boldsymbol{e}_j < 1, j = 1, \ldots, n-1, \ \boldsymbol{p} \cdot \boldsymbol{n} < 0\} \subset H.$$

Denote by $W^{1,2}_{\mathbb{P}}(P)$ the set of all $\boldsymbol{u} \in W^{1,2}(P)$ which have an extension $\tilde{\boldsymbol{u}}$ onto H which is periodic in the sense that $\tilde{\boldsymbol{u}}(\boldsymbol{p} + p_1 \boldsymbol{e}_1 + \cdots + p_{n-1}\boldsymbol{e}_{n-1}) = \tilde{\boldsymbol{u}}(\boldsymbol{p})$ for every $\boldsymbol{p} \in H$ and every set of integers $p_1, \cdots, p_{n-1}$ and which is differentiable in the sense that $\tilde{\boldsymbol{u}}$ has generalized derivatives that are square integrable on every compact subset of H. Note that the periodicity condition applies only to the direction perpendicular to $\boldsymbol{n}$. A vector $\boldsymbol{k}$ is said to be a period of $\{\boldsymbol{e}_1, \cdots, \boldsymbol{e}_{n-1}\}$ if it is of the form $\boldsymbol{k} = 2\pi(p_1 \boldsymbol{e}_1 + \cdots + p_{n-1}\boldsymbol{e}_{n-1})$, where $p_1, \cdots, p_{n-1}$ is any set of integers. If $\boldsymbol{k}$ is a period of $\{\boldsymbol{e}_1, \cdots, \boldsymbol{e}_{n-1}\}$, then any exponential $\boldsymbol{u}$ of the form (20.4.5) with $z \in W^{1,2}(0, \infty)$ is in $W^{1,2}_{\mathbb{P}}(P)$. Item (1) of the following proposition shows that the Dirichlet boundary conditions on the side faces of the prism can be replaced with periodic boundary conditions.

20.4.6 Proposition *Let P be as above. A tensor $\bar{\boldsymbol{A}}$ is quasiconvex at the boundary with normal $\boldsymbol{n}$ if and only if for every $\boldsymbol{u} \in W^{1,2}_{\mathbb{P}}(P)$,*

$$\int_P \bar{\boldsymbol{A}}[\nabla \boldsymbol{u}, \nabla \boldsymbol{u}]\, dV \geq 0. \qquad (20.4.19)$$

This is proved in a way similar to that in the proof of Lemma 17.3.4. The details are omitted. The following result establishes universal conditions which the nonnegativity of J places on $\bar{\boldsymbol{A}}$ at the interior points of Ω and at the points of the free part of the boundary.

20.4.7 Proposition *Let* J *be weakly bounded below (in particular, let* J *be nonnegative). Then*

(1) *for every* $\boldsymbol{p} \in \Omega$, $\bar{\boldsymbol{A}}(\boldsymbol{p})$ *satisfies the Legendre–Hadamard condition;*

(2) *for every* $\boldsymbol{p} \in S$ *at which* $\partial\Omega$ *is of class* C^1, $\bar{\boldsymbol{A}}(\boldsymbol{p})$ *is quasiconvex at the boundary with normal* $\boldsymbol{n}(\boldsymbol{p})$.

Combining this with the fact that the nonnegativity of J is a consequence of the weak local minimum of M, we see that (1), (2) are necessary conditions for the weak local minimum.

Proof (1): This is Proposition 17.3.9 applied to the quadratic function $g(\boldsymbol{F}) = \bar{\boldsymbol{A}}[\boldsymbol{F}, \boldsymbol{F}]$. (2): Assume that $\boldsymbol{p} = \boldsymbol{0}$ and that S is locally flat at $\boldsymbol{p}$, i.e., that there exists a neighborhood N of $\boldsymbol{p}$ such that $\Omega \cap N = H \cap N$. Let $\boldsymbol{u} \in C_0^\infty(\text{Vect})$ and define $\boldsymbol{u}_\alpha(\boldsymbol{p}) = \boldsymbol{u}(\alpha\boldsymbol{p}), \boldsymbol{p} \in \text{Vect}, \alpha > 0$. For α sufficiently large, $\boldsymbol{u}_\alpha$ will be different from $\boldsymbol{0}$ only on a small ball around $\boldsymbol{0}$, and hence will satisfy $\boldsymbol{u}_\alpha = \boldsymbol{0}$ on D. Writing the condition of Definition 20.4.1(5) for $\boldsymbol{u}_\alpha$, we obtain,

$$\int_\Omega (\alpha^2 \bar{\boldsymbol{A}}(\boldsymbol{p})[\nabla \boldsymbol{u}(\alpha\boldsymbol{p}), \nabla \boldsymbol{u}(\alpha\boldsymbol{p})] + d|\boldsymbol{u}(\alpha\boldsymbol{p})|^2)\, dV \geq 0.$$

Making the substitution $\boldsymbol{q} = \alpha\boldsymbol{p}$, dividing by α^2, and letting $\alpha \to \infty$ we obtain (20.4.9). If the boundary in not locally flat, then one can locally flatten it. See SIMPSON & SPECTOR [1987]. □

The next goal is to examine the quasiconvexity at the boundary. We shall first reduce the nonnegativity of the quadratic form in (20.4.9) to the nonnegativity of a quadratic form on a half-line. Let us fix $\boldsymbol{k} \in \text{Vect}$ such that $\boldsymbol{k} \cdot \boldsymbol{n} = 0$ and define $Q : W^{1,2}((0, \infty), \text{Vect}^{\mathbb{C}}) \to \mathbb{R}$ by

$$Q(z) = \int_0^\infty \bar{\boldsymbol{A}}[\dot{z} \otimes \boldsymbol{n} - \mathrm{i}z \otimes \boldsymbol{k}, \dot{z}^* \otimes \boldsymbol{n} + \mathrm{i}z^* \otimes \boldsymbol{k}]\, ds, \qquad z \in W^{1,2}((0, \infty), \text{Vect}^{\mathbb{C}}).$$

The Euler–Lagrange equations corresponding to Q are exactly (20.4.10).

20.4.8 Proposition *Let* $\boldsymbol{n}$ *be given. Then*

(1) $\bar{\boldsymbol{A}}$ *is quasiconvex at the boundary with normal* $\boldsymbol{n}$ *if and only if for every* $\boldsymbol{k} \in \text{Vect}$ *with* $\boldsymbol{k} \cdot \boldsymbol{n} = 0$ *the form* Q *is nonnegative on* $W^{1,2}((0, \infty), \text{Vect}^{\mathbb{C}})$;

(2) *if* $\bar{\boldsymbol{A}}$ *is quasiconvex at the boundary with normal* $\boldsymbol{n}$, *then*

$$\bar{\boldsymbol{A}}[\boldsymbol{a} \otimes \boldsymbol{K}, \boldsymbol{a}^* \otimes \boldsymbol{K}^*] \geq 0 \tag{20.4.20}$$

for every $\boldsymbol{a} \in \text{Vect}^{\mathbb{C}}$ *and every* $\boldsymbol{K} \in \text{Vect}^{\mathbb{C}}$ *of the form* $\boldsymbol{K} = \omega\boldsymbol{n} + \boldsymbol{k}$ *where* $\omega \in \mathbb{C}$ *and* $\boldsymbol{k} \in \text{Vect}$ *such that* $\boldsymbol{k} \cdot \boldsymbol{n} = 0$ *(cf. Remark* 20.4.4(3)).

Proof (1): Let $\bar{\boldsymbol{A}}$ be quasiconvex at the boundary with normal $\boldsymbol{n}$, let $\boldsymbol{k}$ be as above and let furthermore $z \in W^{1,2}((0, \infty), \text{Vect}^{\mathbb{C}})$. Let $\{\boldsymbol{e}_1, \cdots, \boldsymbol{e}_{n-1}\}$ be any basis such that $\boldsymbol{k}$ is a period with respect to it, and define $\boldsymbol{u}$ by (20.4.5). Then $\boldsymbol{u} \in W_{\mathbb{P}}^{1,2}(P)$ and (20.4.9) applied to this $\boldsymbol{u}$ gives $Q(z) \geq 0$. Conversely, let Q be nonnegative for every $\boldsymbol{k}$ as above and $\boldsymbol{u} \in C_0^\infty(\text{Vect})$. We have

$$\boldsymbol{u}(\boldsymbol{p}) = \int_{\partial H} \phi(-\boldsymbol{p} \cdot \boldsymbol{n}, \boldsymbol{k}) \exp(\mathrm{i}\boldsymbol{k} \cdot \boldsymbol{p})\, dA(\boldsymbol{k}),$$

where, for each s, $\phi(s, \cdot)$ is the Fourier transform of $\boldsymbol{p}' \mapsto \boldsymbol{u}(s\boldsymbol{n} + \boldsymbol{p}'), \boldsymbol{p}' \in \partial H$. Then

$$\nabla \boldsymbol{u}(\boldsymbol{p}) = \int_H \{-\dot{\phi}(-\boldsymbol{p} \cdot \boldsymbol{n}, \boldsymbol{k}) \otimes \boldsymbol{n} + \mathrm{i}\phi(-\boldsymbol{p} \cdot \boldsymbol{n}, \boldsymbol{k}) \otimes \boldsymbol{k}\} \exp(\mathrm{i}\boldsymbol{k} \cdot \boldsymbol{p})\, dA(\boldsymbol{k}),$$

where $\dot{\phi}$ is the derivative of $\phi(-\boldsymbol{p}\cdot\boldsymbol{n},\boldsymbol{k})$ with respect to its first variable. By Parseval's equality,

$$\int_H \bar{\mathsf{A}}[\nabla \boldsymbol{u}, \nabla \boldsymbol{u}^*]\, dV = \frac{1}{(2\pi)^{n-1}} \int Q(\phi(\cdot, \boldsymbol{k}))\, dA(\boldsymbol{k}).$$

(2): Assume, without loss of generality, that $\operatorname{Im}\omega < 0$ and $\boldsymbol{k} \neq \boldsymbol{0}$ and define z by

$$z = \boldsymbol{a}\exp(-\mathrm{i}\omega s), \quad s \geq 0. \tag{20.4.21}$$

Then $z \in W^{1,2}((0,\infty), \mathrm{Vect}^{\mathbb{C}})$ so that, by (1), $Q(z) \geq 0$. Evaluating Q gives the result. □

The following proposition is the main result of this section.

20.4.9 Proposition (SIMPSON & SPECTOR [1989]) *Let* $\boldsymbol{n}$ *be given. Then* $\bar{\mathsf{A}}$ *is quasiconvex at the boundary with normal* $\boldsymbol{n}$ *if and only if the following three conditions hold:*

(1) $\bar{\mathsf{A}}$ *satisfies the Legendre–Hadamard condition;*
(2) $\bar{\mathsf{A}}$ *satisfies Agmon's condition with normal* $\boldsymbol{n}$;
(3) *if* $\bar{\mathsf{A}}[\boldsymbol{a}\otimes\boldsymbol{n}, \boldsymbol{a}\otimes\boldsymbol{n}] = 0$ *for some* $\boldsymbol{a} \in \mathrm{Vect}$, *then* $\bar{\mathsf{A}}[\boldsymbol{a}\otimes\boldsymbol{n}] = \boldsymbol{0}$.

This also gives precise necessary and sufficient conditions for the second variation to be positive semidefinite on any finite prism whose five faces are held fixed; the sixth face, with normal $\boldsymbol{n}$, is free and the tensor $\bar{\mathsf{A}}$ is constant. The same holds also for a half-ball whose curved surface is fixed and whose flat surface, with normal $\boldsymbol{n}$, is free, and generally for any Ω such that $\mathcal{S}$ is contained in a single plane. When, say, two opposite faces of a prism are held fixed and the rest has prescribed surface traction, then (1), (2), (3) are necessary, but not sufficient conditions for the positive semidefiniteness of the second variation.

Note that if $\bar{\mathsf{A}}$ satisfies the Legendre–Hadamard condition and $\bar{\mathsf{A}}[\boldsymbol{a}\otimes\boldsymbol{n}, \boldsymbol{a}\otimes\boldsymbol{n}] = 0$ for some $\boldsymbol{a} \in \mathrm{Vect}$, $\boldsymbol{n} \in \mathrm{Sph}$, then $\bar{\mathsf{A}}[\boldsymbol{a}\otimes\boldsymbol{n}]\boldsymbol{n} = \boldsymbol{0}$. Indeed, if $\lambda \in \mathbb{R}$, $\boldsymbol{b} \in \mathrm{Vect}$, then $\bar{\mathsf{A}}[(\boldsymbol{a}+\lambda\boldsymbol{b})\otimes\boldsymbol{n}, (\boldsymbol{a}+\lambda\boldsymbol{b})\otimes\boldsymbol{n}]$ is nonnegative for all λ and equal to 0 for $\lambda = 0$. Hence the first derivative vanishes which implies $\bar{\mathsf{A}}[\boldsymbol{a}\otimes\boldsymbol{n}]\boldsymbol{n} = \boldsymbol{0}$. However, Condition (3) above requires more, namely, $\bar{\mathsf{A}}[\boldsymbol{a}\otimes\boldsymbol{n}] = \boldsymbol{0}$.

20.5 Positivity of the Second Variation (Sufficient Conditions)

Let us now review some known sufficient conditions for various positivity properties of the second variation. Let us first consider two special cases.

20.5.1 Proposition (Van Hove) *Let* $\bar{\mathsf{A}}$ *be constant and strongly elliptic and let* $\partial\Omega$ *be Lipschitz. Then* J *is uniformly positive on* $\mathrm{Var}(\chi) \equiv W_0^{1,2}(\Omega)$.

Here $W_0^{1,2}(\Omega)$ is the set of all functions $\boldsymbol{u}$ from $W^{1,2}(\Omega)$ with $\boldsymbol{u} = 0$ on $\partial\Omega$ in the sense of trace.

Proof By density it is enough to prove (20.4.3) only for every $\boldsymbol{u} \in C_0^\infty(\Omega)$. Let $\tilde{\boldsymbol{u}} = \tilde{\boldsymbol{u}}(\boldsymbol{k})$ be the Fourier transform of $\boldsymbol{u} \in C_0^\infty(\Omega)$ and let $\boldsymbol{v}$ be the Fourier transform of $\nabla\boldsymbol{u}$. Then, since $\boldsymbol{u}$ vanishes near the boundary of Ω, we have $\boldsymbol{v}(\boldsymbol{k}) = \mathrm{i}\tilde{\boldsymbol{u}}(\boldsymbol{k})\otimes\boldsymbol{k}$, $\boldsymbol{k} \in \mathrm{Vect}$ and Parseval's equality gives

$$\int_\Omega \bar{\mathsf{A}}[\nabla\boldsymbol{u}, \nabla\boldsymbol{u}]\, dV = \int \bar{\mathsf{A}}[\tilde{\boldsymbol{u}}\otimes\boldsymbol{k}, \tilde{\boldsymbol{u}}^*\otimes\boldsymbol{k}]\, dV(\boldsymbol{k}) \geq c\int |\boldsymbol{k}|^2|\tilde{\boldsymbol{u}}|^2\, dV(\boldsymbol{k}) = c\int_\Omega |\nabla\boldsymbol{u}|^2\, dV,$$

since $\bar{\mathsf{A}}[\boldsymbol{a}\otimes\boldsymbol{k}, \boldsymbol{a}^*\otimes\boldsymbol{k}] \geq c|\boldsymbol{a}|^2|\boldsymbol{k}|^2$ for every $\boldsymbol{a} \in \mathrm{Vect}^{\mathbb{C}}$, $\boldsymbol{k} \in \mathrm{Vect}$ and some $c > 0$ as a consequence of the strong ellipticity. (The integration range of the middle two integrals is Vect.) □

20.5.2 Proposition (Gårding's Inequality) *Suppose that* $\bar{\boldsymbol{A}}$ *is strongly elliptic at every point of* $\operatorname{cl}\Omega$ *and let* $\partial\Omega$ *be Lipschitz. Then* J *is coercive on* $\operatorname{Var}(\boldsymbol{\chi}) \equiv W_0^{1,2}(\Omega)$.

This is proved by applying the partition of unity to divide Ω into parts on which $\bar{\boldsymbol{A}}$ can be assumed to be constant with a prescribed degree of precision. Then Proposition 20.5.1 is invoked to complete the proof. Alternatively, when $\partial\Omega$ and *D*, *S* satisfy the hypotheses of the following proposition, then Proposition 20.5.2 is its consequence.

20.5.3 Proposition *Let* $\partial\Omega$ *be of class* C^∞, *let* $\bar{\boldsymbol{A}}$ *be of class* C^∞ *on* $\operatorname{cl}\Omega$, *and let D, S be relatively open in* $\partial\Omega$. *A necessary and sufficient condition for the second variation* J *to be coercive is that*

(1) *for every* $\boldsymbol{p} \in \Omega$ *the tensor* $\bar{\boldsymbol{A}}$ *is strongly elliptic and*

(2) *for every* $\boldsymbol{p} \in S$ *the tensor* $\bar{\boldsymbol{A}}$ *satisfies the strong complementing condition.*

This is a complete answer, valid, however, only for the case when $\partial\Omega$ is of class C^∞ and *D*, *S* are relatively open in $\partial\Omega$. The latter condition is very restrictive: *D* and *S* cannot touch, i.e., their distance must be positive. An example of such a *D* and *S* are the inner and outer boundaries of a spherical shell, respectively. I refer to SIMPSON & SPECTOR [1987] for proof of Proposition 20.5.3.

Let us finally consider the uniform positivity of the second variation.

20.5.4 Proposition *Let* $\partial\Omega$ *be of class* C^∞, *let* $\bar{\boldsymbol{A}}$ *be of class* C^∞ *on* $\operatorname{cl}\Omega$, *and let D, S be relatively open in* $\partial\Omega$ *and* J *be strictly positive. A necessary and sufficient condition for the second variation* J *to be uniformly positive is that*

(1) *for every* $\boldsymbol{p} \in \Omega$ *the tensor* $\bar{\boldsymbol{A}}$ *is strongly elliptic and*

(2) *for every* $\boldsymbol{p} \in S$ *the tensor* $\bar{\boldsymbol{A}}$ *satisfies the complementing condition.*

I refer to SIMPSON & SPECTOR [1987] for proof. Unfortunately, there does not seem to be any tool to verify that J is strictly positive. Note that in the calculus of variations on an interval under Dirichlet boundary conditions, such a condition is available: the strong form of the Legendre condition along an extremal and the Jacobi condition (nonexistence of a conjugate point) imply that the second variation is strictly positive (cf. CESARI [1983, Subsect. 2.12.ii]). The existing generalizations, due to Carathéodory (1929) and Weyl (1935), in the multidimensional case, are not so natural (See MORREY [1966, Sect. 1.6].)

20.6 The Second Variation for Stressed Isotropic States

In this section we consider a linearization about an isotropic state $\boldsymbol{\chi}$ of an isotropic body Ω, where $\boldsymbol{\chi}(\boldsymbol{p}) = w\boldsymbol{p}$, $\boldsymbol{p} \in \Omega$ with $w > 0$ a constant. Setting $w = 1$ without any loss of generality we have the following form of the tensor of isothermal elasticities (see Subsect. 11.2.1):

$$\begin{aligned} \bar{\boldsymbol{A}}[\boldsymbol{H}] &= (\lambda - p)(\operatorname{tr}\boldsymbol{E})\mathbf{1} + (2\mu + p)\boldsymbol{E} - p\boldsymbol{W} \\ &= (\lambda - p)(\operatorname{tr}\boldsymbol{H})\mathbf{1} + \mu\boldsymbol{H} + (\mu + p)\boldsymbol{H}^{\mathrm{T}}, \end{aligned} \tag{20.6.1}$$

where λ, μ are the isothermal Lamé moduli, p is the pressure, $\boldsymbol{T} = -p\mathbf{1}$, and $\boldsymbol{E}$, $\boldsymbol{W}$ are the symmetric and the skew parts of $\boldsymbol{H} \in \operatorname{Lin}$. Assuming that λ, μ, p are uniform over Ω, the equation of equilibrium of forces takes the form

$$(\lambda + \mu)\nabla \operatorname{Div}\boldsymbol{u} + \mu\Delta\boldsymbol{u} = \mathbf{0}$$

where we have assumed that the increment of the body force vanishes; notice that the pressure does not intervene. The boundary condition $\bar{\boldsymbol{A}}[\nabla \boldsymbol{u}]\boldsymbol{n} = \boldsymbol{0}$ reads

$$(\lambda - p)(\operatorname{Div} \boldsymbol{u})\boldsymbol{n} + \mu(\nabla \boldsymbol{u})\boldsymbol{n} + (\mu + p)(\nabla \boldsymbol{u}^{\mathrm{T}})\boldsymbol{n} = \boldsymbol{0}.$$

The second variation of the stored energy is an integral of the quadratic form

$$\bar{\boldsymbol{A}}[\boldsymbol{H}, \boldsymbol{H}] = (\lambda - p)(\operatorname{tr} \boldsymbol{E})^2 + (2\mu + p)|\boldsymbol{E}|^2 - p|\boldsymbol{W}|^2. \qquad (20.6.2)$$

The goal of the section is to examine the properties introduced in Sect. 20.4 for the above $\bar{\boldsymbol{A}}$; specifically, the positivity properties of the second variation at χ and the uniqueness of the corresponding equations on a half-space.

Irrespective of the signs of λ, μ, the term $-p|\boldsymbol{W}|^2$ contributes a negative amount to $\bar{\boldsymbol{A}}[\boldsymbol{H}, \boldsymbol{H}]$ when $p > 0$, i.e., when the body is in the state of compression at the given point. Hence, in this case, $\bar{\boldsymbol{A}}$ is not pointwise positive definite, but despite this, the second variation can still be positive definite, and in fact we shall determine the range of p when this occurs (see Proposition 20.6.5 below.)

There are three groups of conditions on $\bar{\boldsymbol{A}}$ of increasing strength: **(1)**: ellipticity, strong ellipticity, and the Legendre–Hadamard condition; **(2)**: complementing condition, Agmon's condition and quasiconvexity at the boundary, **(3)**: invertibility and the pointwise positive semidefiniteness of $\bar{\boldsymbol{A}}$. Roughly, these conditions govern the behavior of the body under **(1)** the Dirichlet boundary conditions (a cube with fixed boundary), **(2)** mixed boundary conditions (a half-space whose boundary plane is free of tractions, but with deformations decaying at infinity), and **(3)** the Neumann boundary conditions (a cube with all faces free of tractions), respectively.

Proceeding to the discussion of the group **(1)**, note that the isothermal acoustic tensor of an isotropic state is

$$\bar{\boldsymbol{Q}}(\boldsymbol{K}) = \sigma \boldsymbol{K} \otimes \boldsymbol{K} + \mu K^2 \mathbf{1}, \quad \boldsymbol{K} \in \operatorname{Vect}^{\mathbb{C}},$$

where $\sigma = \lambda + \mu$. Moreover, for every $\alpha \in \mathbb{R}$,

$$\det(\bar{\boldsymbol{Q}}(\boldsymbol{K}) + \alpha^2 \mathbf{1}) = (\mu \boldsymbol{K}^2 + \alpha^2)^{n-1}((\lambda + 2\mu)\boldsymbol{K}^2 + \alpha^2) \qquad (20.6.3)$$

and, in particular, if $\boldsymbol{n} \in \operatorname{Sph}$, then

$$\bar{\boldsymbol{Q}}(\boldsymbol{n}) = \sigma \boldsymbol{n} \otimes \boldsymbol{n} + \mu \mathbf{1}, \quad \det(\sigma \boldsymbol{n} \otimes \boldsymbol{n} + \mu \mathbf{1}) = \mu^{n-1}(\lambda + 2\mu). \qquad (20.6.4)$$

20.6.1 Proposition *The elasticity tensor of an isotropic state $\bar{\boldsymbol{A}}$ satisfies*

(1) *the Legendre-Hadamard condition if and only if*

$$\lambda + 2\mu \geq 0, \quad \mu \geq 0; \qquad (20.6.5)$$

(2) *the strong ellipticity condition if and only if the two inequalities in* (20.6.5) *are strict;*

(3) *the ellipticity condition if and only if*

$$\lambda + 2\mu \neq 0, \quad \mu \neq 0.$$

Notice that these conditions do not involve p.

Proof (1),(2): Inserting $\boldsymbol{H} = \boldsymbol{a} \otimes \boldsymbol{b}$, $\boldsymbol{a}, \boldsymbol{b} \in \operatorname{Vect}$, into $\bar{\boldsymbol{A}}$, one obtains

$$\bar{\boldsymbol{A}}[\boldsymbol{a} \otimes \boldsymbol{b}, \boldsymbol{a} \otimes \boldsymbol{b}] = (\lambda + \mu)(\boldsymbol{a} \cdot \boldsymbol{b})^2 + \mu a^2 b^2.$$

The choice of first $a = b$ and then $a \perp b$ shows that the material satisfies (20.6.5). The converse implication is established with the help of Schwarz's inequality. The strict versions are treated similarly. (3): See (20.6.4). □

A somewhat more complicated is the examination of the properties in the group (2). The exponential solutions of the problems of Subsect. 20.4.2 are governed by

$$(\sigma \boldsymbol{n} \otimes \boldsymbol{n} + \mu \mathbf{1})\ddot{z} - \mathrm{i}\sigma(\boldsymbol{n} \otimes \boldsymbol{k} + \boldsymbol{k} \otimes \boldsymbol{n})\dot{z} - (\sigma \boldsymbol{k} \otimes \boldsymbol{k} + (\mu k^2 + \alpha^2)\mathbf{1})z = \mathbf{0} \qquad (20.6.6)$$

for $s \in [0, \infty)$, and

$$\sigma(\dot{z} \cdot \boldsymbol{n})\boldsymbol{n} + \mu\dot{z} - \mathrm{i}((\lambda - p)(z \cdot \boldsymbol{k})\boldsymbol{n} + (\mu + p)(z \cdot \boldsymbol{n})\boldsymbol{k}) = \mathbf{0} \text{ for } s = 0. \qquad (20.6.7)$$

The system (20.6.6) is exactly of order 2 in every component of z if and only if $\sigma\boldsymbol{n}\otimes\boldsymbol{n}+\mu\mathbf{1}$ is invertible, which by Proposition 20.6.1(3) holds if and only if $\mu \neq 0$, $\lambda+2\mu \neq 0$. If one or both of these inequalities fails, then the system is of mixed order, and namely, if $\mu = 0$, $\lambda \equiv \lambda+2\mu \neq 0$, then (20.6.6) is of order 2 in the component $z \cdot \boldsymbol{n}$ and of order ≤ 1 in $(\mathbf{1} - \boldsymbol{n} \otimes \boldsymbol{n})z$; if $\mu \neq 0$, $\lambda + 2\mu = 0$, then (20.6.6) is of order ≤ 1 in $z \cdot \boldsymbol{n}$ and of order 2 in $(\mathbf{1} - \boldsymbol{n} \otimes \boldsymbol{n})z$; if both $\mu = 0$, $\lambda \equiv \lambda + 2\mu = 0$, then the system is of order ≤ 1 in all components of z. If $\mu \neq 0$, $\lambda + 2\mu \neq 0$, then the system may be put in the form with an explicitly calculated second derivative $\ddot{z}$ by using

$$(\sigma \boldsymbol{n} \otimes \boldsymbol{n} + \mu \mathbf{1})^{-1} = \mu^{-1}\mathbf{1} - \frac{\sigma}{\mu(\lambda + 2\mu)}\boldsymbol{n} \otimes \boldsymbol{n},$$

with the result

$$\mu\ddot{z} - \mathrm{i}(\sigma \boldsymbol{k} \otimes \boldsymbol{n} + \frac{\sigma\mu}{\lambda + 2\mu}\boldsymbol{n} \otimes \boldsymbol{k})\dot{z} - (\sigma \boldsymbol{k} \otimes \boldsymbol{k} + (\mu k^2 + \alpha^2)(\mathbf{1} - \frac{\sigma}{\lambda + 2\mu}\boldsymbol{n} \otimes \boldsymbol{n}))z = \mathbf{0}.$$

The following proposition may be interpreted as the specification of the values of λ, μ, p at which the surface instabilities occur.

20.6.2 Proposition *Let $n \geq 2$ and $\boldsymbol{n} \in$ Sph. The elasticity tensor of an isotropic state $\bar{\boldsymbol{A}}$ satisfies the complementing condition with normal $\boldsymbol{n}$ if and only if*

$$\lambda + 2\mu \neq 0, \quad \mu \neq 0, \quad 2\mu + p \neq 0, \quad \text{and} \quad p(\lambda + 3\mu) \neq 2\mu(\lambda + \mu). \qquad (20.6.8)$$

In particular, if $p = 0$, then (20.6.8) *reduce to*

$$\lambda + 2\mu \neq 0, \quad \mu \neq 0, \quad \text{and} \quad \lambda + \mu \neq 0.$$

Proof Throughout the proof, let $z_n = z \cdot \boldsymbol{n}$, $z_k = z \cdot \boldsymbol{k}$ and let $z^{\perp}$ denote the projection of z onto the orthogonal complement of $\{\boldsymbol{n}, \boldsymbol{k}\}$. Suppose that $\bar{\boldsymbol{A}}$ satisfies the complementing condition. The necessity of $(20.6.8)_2$: Suppose for contradiction that $\mu = 0$. Then (20.6.6), (20.6.7) with $\alpha = 0$ read

$$\begin{aligned} &\sigma(\ddot{z} \cdot \boldsymbol{n})\boldsymbol{n} - \mathrm{i}\sigma(\boldsymbol{n}(\dot{z} \cdot \boldsymbol{k}) + \boldsymbol{k}(\dot{z} \cdot \boldsymbol{n})) - \sigma\boldsymbol{k}(z \cdot \boldsymbol{k}) = \mathbf{0} \text{ for } s \geq 0, \\ &(\lambda(\dot{z} \cdot \boldsymbol{n}) - \mathrm{i}(\lambda - p)(z \cdot \boldsymbol{k}))\boldsymbol{n} - \mathrm{i}p(z \cdot \boldsymbol{n})\boldsymbol{k} = \mathbf{0} \text{ for } s = 0. \end{aligned} \qquad (20.6.9)$$

These contain only z_n, z_k; $z^{\perp}$ is absent and can therefore be chosen arbitrarily. Equations (20.6.9) can be rewritten as

$$\sigma\ddot{z}_n - \mathrm{i}\sigma\dot{z}_k = 0, \quad \sigma\dot{z}_n - \mathrm{i}\sigma z_k = 0 \text{ for } s \geq 0, \quad \lambda\dot{z}_n - \mathrm{i}(\lambda - p)z_k = 0, \quad pz_n = 0 \text{ for } s = 0,$$

and any function of the form $z = w\boldsymbol{n} - \mathrm{i}\dot{w}\boldsymbol{k}/k^2$, where w is any smooth function on $[0, \infty)$ with values in $\mathbb{C}$, with $w = \dot{w} = 0$ at $s = 0$, satisfies them. This proves that the complementing condition implies $\mu \neq 0$. The necessity of $(20.6.8)_1$: Suppose for contradiction that $\lambda + 2\mu = 0$. Splitting (20.6.6) into the components in the direction $\boldsymbol{n}, \boldsymbol{k}$, and $\{\boldsymbol{n}, \boldsymbol{k}\}^{\perp}$, we obtain

$$\mu\dot{z}_k + \mathrm{i}\mu k^2 z_n = 0, \quad \mu\ddot{z}_k + \mathrm{i}\mu k^2 \dot{z}_n = 0, \quad \mu\ddot{z}^{\perp} - \mu k^2 z^{\perp} = \mathbf{0} \text{ for } s \ge 0, \tag{20.6.10}$$

$$(\lambda - p)z_k = 0, \quad \mu\dot{z}_k - \mathrm{i}(\mu + p)k^2 z_n = 0, \quad \mu\dot{z}^{\perp} = \mathbf{0} \text{ for } s = 0 \tag{20.6.11}$$

Moreover, $(20.6.10)_1$ implies $(20.6.10)_2$. This implies that every function z of the form $z = w\boldsymbol{k} + \mathrm{i}\dot{w}\boldsymbol{n}$, where w is any smooth function on $[0, \infty)$ with values in $\mathbb{C}$, with $w = \dot{w} = 0$ at $s = 0$, satisfies (20.6.6). This function may be chosen bounded; if it vanishes in some neighborhood of 0; z satisfies the boundary condition (20.6.7). The complementing condition thus fails and $(20.6.8)_1$ is necessary.

To complete the proof, we shall now show that granted $(20.6.8)_{1,2}$, $\bar{\boldsymbol{A}}$ satisfies the complementing condition if and only if $(20.6.8)_{3,4}$ hold. We shall first determine the characteristic pairs. By $(20.6.8)_{1,2}$ and (20.6.3), $\det\bar{\mathbf{Q}}(\boldsymbol{K}) = 0$ if and only if $\boldsymbol{K}^2 = 0$ where $\boldsymbol{K} = \omega\boldsymbol{n} + \boldsymbol{k}$. From this, $\omega = \pm\mathrm{i}|\boldsymbol{k}|$ and the only ω with negative imaginary part is $\omega = -\mathrm{i}|\boldsymbol{k}|$. Equation (20.4.14) reduces to $\sigma(\boldsymbol{a} \cdot \boldsymbol{K}) = 0$. If $\sigma = 0$, then $\boldsymbol{a}$ is arbitrary; if $\sigma \neq 0$, then there are only $n - 1$ characteristic pairs with $\omega = -\mathrm{i}|\boldsymbol{k}|$ and with linearly independent amplitudes. That is, the matrix $\boldsymbol{L}$ from (20.4.13) is not diagonalizable and $\boldsymbol{L}$ has nontrivial Jordan blocks. Moreover, as $\boldsymbol{K}^2 = 0$, we have $\bar{\mathbf{Q}}(\boldsymbol{K})\boldsymbol{K} = 0$, i.e., $(\omega, \boldsymbol{K})$ is a characteristic pair. We join the cases $\sigma = 0$, $\sigma \neq 0$ in a single treatment and seek, using standard methods, the solution of (20.6.6) in the form

$$z = (\boldsymbol{c} + \xi\boldsymbol{K}s)e^{-\mathrm{i}\omega s} \tag{20.6.12}$$

where $\omega = -\mathrm{i}|\boldsymbol{k}|$, $\boldsymbol{K} = \omega\boldsymbol{n} + \boldsymbol{k}$, $\xi \in \mathbb{C}$, $\boldsymbol{c} \in \mathrm{Vect}^{\mathbb{C}}$. Inserting this z into $(20.4.10)_1$ and invoking (20.6.1), we find, after a lengthy but otherwise straightforward calculation, that this equation is equivalent to

$$\sigma(\boldsymbol{c} \cdot \boldsymbol{K}) + \xi|\boldsymbol{k}|(\lambda + 3\mu) = 0. \tag{20.6.13}$$

Using $\sigma = \lambda + \mu$ one finds that at least one of the numbers σ, $\lambda + 3\mu$ is nonzero as a consequence of $(20.6.8)_{1,2}$. The functions z as in (20.6.12) are parametrized by $\boldsymbol{c} \in \mathrm{Vect}^{\mathbb{C}}$, $\xi \in \mathbb{C}$ with the nontrivial scalar constraint (20.6.13) and so they form an n-dimensional family; for dimensional reasons, z is the general bounded solution of $(20.4.10)_1$. Inserting z into the boundary condition $(20.4.10)_2$, we obtain

$$-\mathrm{i}((\lambda - p)(\boldsymbol{c} \cdot \boldsymbol{K})\boldsymbol{n} + \mu\omega\boldsymbol{c} + (\mu + p)(\boldsymbol{c} \cdot \boldsymbol{n})\boldsymbol{K}) + \xi((\lambda + \mu)\omega\boldsymbol{n} + \mu\boldsymbol{K}) = \mathbf{0}. \tag{20.6.14}$$

Hence any bounded solution of (20.6.6), (20.6.7) is of the form (20.6.12) where $\boldsymbol{c}$, ξ satisfy the system (20.6.13) and (20.6.14). Since $\mu \neq 0$, we deduce from (20.6.14) that $\boldsymbol{c}$ is a linear combination of $\boldsymbol{n}$, $\boldsymbol{K}$. Writing $\boldsymbol{c} = \alpha\boldsymbol{n} + \beta\boldsymbol{K}$, $\alpha, \beta \in \mathbb{C}$, and inserting this form of $\boldsymbol{c}$ into (20.6.13) and (20.6.14), we arrive at

$$\begin{bmatrix} -\mathrm{i}(\lambda - p + \mu) & 0 & \sigma \\ -\mathrm{i}(\mu + p) & -\mathrm{i}(2\mu + p)\omega & \mu \\ -\mathrm{i}\sigma & 0 & \lambda + 3\mu \end{bmatrix} \begin{bmatrix} \alpha \\ \beta \\ \xi \end{bmatrix} = 0 \tag{20.6.15}$$

The determinant of this system is

$$D = \omega(2\mu + p)(p(\lambda + 3\mu) - 2\mu(\lambda + \mu)).$$

Hence the system (20.6.6), (20.6.7) has a nontrivial bounded solution if and only if $D = 0$ or, put otherwise, $\bar{\boldsymbol{A}}$ satisfies the complementing condition if and only if $D \neq 0$. But $D \neq 0$ is equivalent to $(20.6.8)_{3,4}$. □

20.6.3 Instabilities The violation of any of the inequalities (20.6.8) leads to a nontrivial bounded exponential solution of the problem 20.4.2(1). Let us view the changes of λ, μ, p as a result of the change of the basic homogeneous state χ of an isotropic material filling the half-space H. At the values of λ, μ, p which violate any of the inequalities in (20.6.8), the linearized equations lose the uniqueness. A further change of external forces will result in an inhomogeneous deformation; the solution of the linearized equations gives the direction of the change of the deformation. The nontrivial bounded exponential solutions of the problem (20.4.2)(1) are called surface instabilities. The proof of Proposition 20.6.2 essentially determines them; nevertheless they withstand a brief recapitulation. Each surface instability is of the form

$$\boldsymbol{u}(\boldsymbol{p}) = z(-\boldsymbol{p} \cdot \boldsymbol{n})\exp(\mathrm{i}\boldsymbol{k} \cdot \boldsymbol{p}) + z^*(-\boldsymbol{p} \cdot \boldsymbol{n})\exp(-\mathrm{i}\boldsymbol{k} \cdot \boldsymbol{p}), \quad \boldsymbol{n} \cdot \boldsymbol{p} \le 0$$

where z is as follows:

(1) if $\lambda + 2\mu = 0$ and $\mu \neq 0$, then

$$z(s) = w\boldsymbol{k} + \mathrm{i}\dot{w}\boldsymbol{n},$$

where $w : [0, \infty) \to \mathbb{C}$ is any smooth bounded function with bounded derivative satisfying $w = 0,\ \dot{w} = 0$ at $s = 0$;

(2) if $\mu = 0$, then

$$z(s) = w\boldsymbol{n} - \mathrm{i}\dot{w}\boldsymbol{k}/\boldsymbol{k}^2 + \text{ any function with values in } \{\boldsymbol{n}, \boldsymbol{k}\}^{\perp},$$

where $w : [0, \infty) \to \mathbb{C}$ is any smooth bounded function with bounded derivative satisfying $w = 0,\ \dot{w} = 0$ at $s = 0$;

(3) if $p = -2\mu$, then

$$z(s) = (-\mathrm{i}|\boldsymbol{k}|\boldsymbol{n} + \boldsymbol{k})e^{-|\boldsymbol{k}|s};$$

(4) if $p = 2\mu(\lambda + \mu)/(\lambda + 3\mu)$, then

$$z(s) = \left[\mu(\mathrm{i}|\boldsymbol{k}|\boldsymbol{n} + \boldsymbol{k}) + p(\mathrm{i}|\boldsymbol{k}|\boldsymbol{n} - \boldsymbol{k})|\boldsymbol{k}|s\right] e^{-|\boldsymbol{k}|s};$$

in all the above formulas, $\boldsymbol{k}$ is orthogonal to $\boldsymbol{n}$ but otherwise arbitrary.

The instabilities in (1), (2) are associated with the failure of the ellipticity condition; only (3), (4) are the surface instabilities in the proper sense. The forms of z in (3), (4) were found by solving (20.6.15) for the given values of λ, μ, p. The following proposition is a preparation for the discussion of the quasiconvexity at the boundary of $\bar{\boldsymbol{A}}$; it also enables one to evaluate the Korn constant for a half-space (see Proposition 20.6.7(2)).

20.6.4 Proposition *Let $\gamma, \delta \in \mathbb{R}$ and define the tensor $\bar{\boldsymbol{B}}$ by*

$$\bar{\boldsymbol{B}}[\boldsymbol{H}] = (\delta - \gamma)(\operatorname{tr}\boldsymbol{H})\boldsymbol{1} + \boldsymbol{H} + (\gamma - 1)\boldsymbol{H}^{\mathrm{T}}, \quad \boldsymbol{H} \in \mathrm{Lin}. \tag{20.6.16}$$

Then

(1) *$\bar{\boldsymbol{B}}$ satisfies the Legendre–Hadamard condition if and only if*

$$\delta \geq 0;$$

(2) *if $\delta > 0$ and $\boldsymbol{n} \in$ Sph, then $\bar{\boldsymbol{B}}$ satisfies Agmon's condition with normal $\boldsymbol{n}$ if and only if*

$$0 \leq \gamma \leq 4\delta/(\delta + 1); \tag{20.6.17}$$

(3) *if $\boldsymbol{n} \in$ Sph, then $\bar{\boldsymbol{B}}$ is quasiconvex at the boundary with normal $\boldsymbol{n}$ if and only if*

$$\delta \geq 0, \quad 0 \leq \gamma \leq 4\delta/(\delta + 1). \tag{20.6.18}$$

Proof (1): This is completely analogous to the proof of Proposition 20.6.1(1). (2): Let $\delta > 0$, and prove that $\bar{\boldsymbol{B}}$ satisfies Agmon's condition if and only if (20.6.17) hold. Let us consider separately the case $\delta \neq 1$ and $\delta = 1$. ***Case I,*** $\delta \neq 1$. Let $\alpha \in \mathbb{R}$, $\alpha \neq 0$, $\boldsymbol{k} \in$ Vect, $\boldsymbol{k} \cdot \boldsymbol{n} = 0$, and let $\boldsymbol{u}$ be a solution of (20.4.7) (with $\bar{\boldsymbol{A}}$ replaced by $\bar{\boldsymbol{B}}$) of the form (20.4.5). We proceed as in the proof of Proposition 20.6.2: We first determine the characteristic frequencies from (20.4.15). By (20.6.3) this equation takes the form

$$\det((\boldsymbol{K}^2 + \alpha^2)\boldsymbol{1} + (\delta - 1)\boldsymbol{K} \otimes \boldsymbol{K}) = (\boldsymbol{K}^2 + \alpha^2)^{n-1}(\delta\boldsymbol{K}^2 + \alpha^2) = 0,$$

and hence either $(\boldsymbol{K}^2 + \alpha^2) = 0$ or $\delta\boldsymbol{K}^2 + \alpha^2 = 0$ from which the roots ω are $\pm\mathrm{i}\sqrt{\boldsymbol{k}^2 + \alpha^2}$ and $\pm\mathrm{i}\sqrt{\boldsymbol{k}^2 + \alpha^2/\delta}$. We consider only the roots with the negative imaginary parts

$$\omega_1 = -\mathrm{i}\sqrt{\boldsymbol{k}^2 + \alpha^2} \quad\text{and}\quad \omega_2 = -\mathrm{i}\sqrt{\boldsymbol{k}^2 + \alpha^2/\delta}.$$

The characteristic equation (20.4.14) for the present $\bar{\boldsymbol{B}}$ reads

$$(\delta-1)\boldsymbol{K}(\boldsymbol{a}\cdot\boldsymbol{K})+(\boldsymbol{K}^2+\alpha^2)\boldsymbol{a}=\boldsymbol{0},\quad \boldsymbol{K}:=\omega\boldsymbol{n}+\boldsymbol{k},\quad \omega=\omega_1 \text{ or } \omega_2 \tag{20.6.19}$$

and setting $\boldsymbol{K}_i=\omega_i\boldsymbol{n}+\boldsymbol{k}$, $i=1,2$, the set of all characteristic pairs is $(\omega_1,\boldsymbol{a}_1)$, $(\omega_2,\boldsymbol{a}_2)$, where

$$\boldsymbol{a}_1\cdot\boldsymbol{K}_1=0,\quad \boldsymbol{a}_1\neq\boldsymbol{0},\quad\text{and}\quad \boldsymbol{a}_2=\xi_2\boldsymbol{K}_2\quad\text{for some }\ \xi_2\in\mathbb{C},\quad \xi_2\neq 0. \tag{20.6.20}$$

If $\boldsymbol{b}_j$, $j=1,\ldots,n-1$ are linearly independent vectors such that $\boldsymbol{b}_j\cdot\boldsymbol{K}_1=0$, then one easily checks that the system $(\omega_1,\boldsymbol{b}_j)$, $j=1,\ldots,n-1$, $(\omega_2,\boldsymbol{K}_2)$ forms n characteristic pairs with linearly independent characteristic vectors $\boldsymbol{b}_j$, $j=1,\ldots,n-1$, $\boldsymbol{K}_2$. For dimensional reasons, a general bounded solution of (20.6.6), (20.6.7) is of the form (20.4.17) with some $\omega_j\in\mathbb{C}$, $\boldsymbol{a}_j\in\mathrm{Vect}^{\mathbb{C}}$. From this, a general bounded solution of (20.6.6), (20.6.7) is of the form

$$z(s)=\boldsymbol{a}_1\exp(-\mathrm{i}\omega_1 s)+\boldsymbol{a}_2\exp(-\mathrm{i}\omega_2 s),\quad s\geq 0,$$

where

$$\boldsymbol{a}_1\cdot\boldsymbol{K}_1=0,\quad \boldsymbol{a}_2=\xi_2\boldsymbol{K}_2,\quad \boldsymbol{K}_1=\omega_1\boldsymbol{n}+\boldsymbol{k},\quad \boldsymbol{K}_2=\omega_2\boldsymbol{n}+\boldsymbol{k}, \tag{20.6.21}$$

with some $\xi_2\in\mathbb{C}$. We now insert this form of z into the boundary condition $(20.4.10)_2$ to obtain

$$\omega_1\boldsymbol{a}_1+(\gamma-1)(\boldsymbol{a}_1\cdot\boldsymbol{n})\boldsymbol{K}_1+(\delta-\gamma)(\boldsymbol{a}_2\cdot\boldsymbol{K}_2)\boldsymbol{n}+\omega_2\boldsymbol{a}_2+(\gamma-1)(\boldsymbol{a}_2\cdot\boldsymbol{n})\boldsymbol{K}_2=\boldsymbol{0} \tag{20.6.22}$$

where we have used $(20.6.21)_{1,3,4}$. Since $\omega_1\neq 0$ and $\boldsymbol{K}_1,\boldsymbol{K}_2,\boldsymbol{a}_2$ are linear combinations of $\boldsymbol{n}$ and $\boldsymbol{k}$, we see that also $\boldsymbol{a}_1$ is a linear combination of $\boldsymbol{n},\boldsymbol{k}$. From $\boldsymbol{a}_1\cdot\boldsymbol{K}_1=0$ then $\boldsymbol{a}_1=\xi_1(\boldsymbol{k}^2\boldsymbol{n}-\omega_1\boldsymbol{k})$ for some $\xi_1\in\mathbb{C}$. Inserting these expressions into (20.6.22) one obtains

$$\begin{bmatrix}\omega_1\boldsymbol{k}^2\gamma & -(\alpha^2+\gamma\boldsymbol{k}^2)\\ \alpha^2+\gamma\boldsymbol{k}^2 & \omega_2\gamma\end{bmatrix}\begin{bmatrix}\xi_1\\ \xi_2\end{bmatrix}=0.$$

Hence the existence of a nontrivial bounded exponential solution to (20.4.7) is equivalent to the nullity of the determinant of the above system. Put otherwise, Agmon's condition holds if and only if the determinant of the system is nonzero for every $\alpha\neq 0$. If $\gamma=0$, then the determinant is $\neq 0$, there is no bounded exponential solution, and (20.6.17) holds. Assume that $\gamma\neq 0$. Then the determinant is

$$D=|\boldsymbol{k}|^4\gamma^2(1+x)^{1/2}(1+x/\delta)^{1/2}(f(x)^{1/2}-1)$$

where

$$f(x):=\frac{\delta(\gamma+x)^4}{\gamma^4(1+x)(\delta+x)},\qquad x:=\alpha^2/\boldsymbol{k}^2.$$

Agmon's condition is equivalent to the existence of an $x>0$ such that $f(x)=1$. Clearly, $f(0)=1$ and $f(x)\to\infty$ as $x\to\infty$. Hence the absence of a positive x with $f(x)=1$ is equivalent to $f(x)>1$ for all $x>0$. ***Proof that Agmon's condition implies*** (20.6.17). If $\gamma<0$, then $f(-\gamma)=0$ and hence $f(x)=1$ for some positive x, which is a violation of Agmon's condition; consequently, Agmon's condition implies $(20.6.17)_1$. Furthermore, as $f(x)>1$ for $x>0$, necessarily $f'(0)\geq 0$. A calculation gives $f'(x)=g(x)h(x)$ where g is a function such that $g(x)>0$ for all $x\geq 0$ and $h(x)=4(1+x)(\delta+x)-(\gamma+x)(\delta+1+2x)$. The condition $h(0)\geq 0$ is equivalent to $(20.6.17)_2$. ***Proof that*** (20.6.17) ***implies Agmon's condition.*** The preceding part of the proof shows that if (20.6.17) hold, then $f'(0)\geq 0$, $h(0)\geq 0$. Further, it is easily shown that (20.6.17) imply that $h'(x)>0$ for every $x>0$; hence $h(x)>0$ for every $x>0$. From this $f'(x)>0$ for every $x>0$ and using $f(0)=1$ we conclude that $f(x)>1$ for all $x>0$ which implies $D>0$ for all $x>0$. The nontrivial decaying solutions are excluded and Agmon's condition follows. This completes the proof in Case I. ***Case II,*** $\delta=1$. Equation (20.6.19) simplifies to

$$(\boldsymbol{K}^2+\alpha^2)\boldsymbol{a}=\boldsymbol{0},\quad \boldsymbol{K}:=\omega\boldsymbol{n}+\boldsymbol{k},$$

and from $\boldsymbol{a}\neq\boldsymbol{0}$ necessarily $\omega=-i\sqrt{\boldsymbol{k}^2+\alpha^2}$. Hence $z=\boldsymbol{a}\exp(-\mathrm{i}\omega s)$ with $\boldsymbol{a}\in\mathrm{Vect}^{\mathbb{C}}$ arbitrary. The boundary condition reads

$$(1-\gamma)(\boldsymbol{a}\cdot\boldsymbol{K})\boldsymbol{n}+\omega\boldsymbol{a}+(\gamma-1)(\boldsymbol{a}\cdot\boldsymbol{n})\boldsymbol{K}=\boldsymbol{0}.$$

From this we deduce firstly that $\boldsymbol{a}$ is a linear combination of $\boldsymbol{n},\boldsymbol{k}$, namely, $\boldsymbol{a}=\xi_1\boldsymbol{n}+\xi_2\boldsymbol{k}$, and secondly, that ξ_1,ξ_2 satisfy

$$\begin{bmatrix}\omega & (1-\gamma)\boldsymbol{k}^2\\ \gamma-1 & \omega\end{bmatrix}\begin{bmatrix}\xi_1\\ \xi_2\end{bmatrix}=0.$$

The determinant is $D = ((\gamma - 1)^2 - 1)k^2 - \alpha^2$. The condition that $D \neq 0$ for all $\alpha \neq 0$ is equivalent to $(\gamma - 1)^2 \le 1$. This in turn is equivalent to $0 \le \gamma \le 2$, which is identical with (20.6.17) with $\delta = 1$. This completes the proof of (2).

(3): By Proposition 20.4.9, the quasiconvexity at the boundary is equivalent to Conditions (1)–(3) of that proposition. Assume that $\bar{\boldsymbol{B}}$ is quasiconvex at the boundary. Then by Proposition 20.4.9(1), the Legendre–Hadamard condition holds, which by (1) of the present proposition is equivalent to $(20.6.18)_1$. Let us now prove $(20.6.18)_{2,3}$. By Proposition 20.4.9(2), Agmon's condition holds. If $\delta > 0$, then by Item (2) of the present proposition, $(20.6.18)_{2,3}$ hold. If $\delta = 0$, we derive $(20.6.18)_{2,3}$ from Proposition 20.4.9(3): Setting $\boldsymbol{a} = \boldsymbol{n}$ and using $\delta = 0$ we find $\bar{\boldsymbol{B}}[\boldsymbol{a} \otimes \boldsymbol{n}, \boldsymbol{a} \otimes \boldsymbol{n}] = 0$ and so by Proposition 20.4.9(3), $\bar{\boldsymbol{B}}[\boldsymbol{a} \otimes \boldsymbol{n}] = \boldsymbol{0}$, which means $\gamma(\boldsymbol{n} \otimes \boldsymbol{n} - \boldsymbol{1}) = \boldsymbol{0}$ from which $\gamma = 0$ and $(20.6.18)_{2,3}$ hold again. Conversely, suppose that (20.6.18) hold and prove the quasiconvexity at the boundary. If $\delta = 0$ then $\gamma = 0$ by $(20.6.18)_{2,3}$ and $\bar{\boldsymbol{B}}[\boldsymbol{H}] = \boldsymbol{H} - \boldsymbol{H}^{\mathrm{T}} = 2\boldsymbol{W}$, where $\boldsymbol{W}$ is the skew part of $\boldsymbol{H}$. Then $\bar{\boldsymbol{B}}[\boldsymbol{H}, \boldsymbol{H}] = 2|\boldsymbol{W}|^2 \ge 0$. Consequently $\bar{\boldsymbol{B}}$ is positive semidefinite and hence also quasiconvex at the boundary. If $\delta > 0$, we verify Conditions (1)–(3) of Proposition 20.4.9. By $(20.6.18)_1$ and Item (1) of the present proposition Condition (1) of Proposition 20.4.9 holds and by Item(2) of the present proposition Condition (2) of Proposition 20.4.9 also holds. Finally, to verify Proposition 20.4.9(3), let $\boldsymbol{a}$ satisfy

$$\bar{\boldsymbol{B}}[\boldsymbol{a} \otimes \boldsymbol{n}, \boldsymbol{a} \otimes \boldsymbol{n}] = 0 \tag{20.6.23}$$

This implies $0 = (\delta - 1)|\boldsymbol{a} \cdot \boldsymbol{n}|^2 + |\boldsymbol{a}|^2 \ge \delta |\boldsymbol{a} \cdot \boldsymbol{n}|^2 \ge 0$ by Schwartz's inequality. From this $|\boldsymbol{a} \cdot \boldsymbol{n}| = 0$ which in combination with the original equation gives $|\boldsymbol{a}| = 0$ and hence

$$\bar{\boldsymbol{B}}[\boldsymbol{a} \otimes \boldsymbol{n}] = \boldsymbol{0}. \quad \square$$

20.6.5 Proposition *If* $\boldsymbol{n} \in \mathrm{Sph}$, *then the tensor of elasticities of an isotropic state* $\bar{\boldsymbol{A}}$ *is quasiconvex at the boundary with normal* $\boldsymbol{n}$ *if and only if one of the following two possibilities occurs:*

(1) $$\mu > 0, \quad \lambda + 2\mu \ge 0, \quad -2\mu \le p \le 2\mu(\lambda + \mu)/(\lambda + 3\mu);$$

(2) $$\mu = 0, \quad \lambda \ge 0, \quad p = 0.$$

In particular, the last two inequalities in (1) determine an interval of pressures for which the second variation is nonnegative for problems on H. Notice that the inequalities in (1) are in one-to-one correspondence with those in Proposition 20.6.2. This shows that when the positivity of the second variation on H is lost at some point of the given loading path then the complementing condition fails as well and the surface instabilities occur at this point. By adding a positive quantity 2μ to the numerator of the last term in (1) we obtain $p < 2\mu$, hence the last two inequalities in (1) imply

$$-2\mu \le p \le 2\mu.$$

The strict version of these are well-known necessary conditions for the absence of surface instabilities in the incompressible case (see, e.g., OGDEN [1992b]). Here in the compressible case the upper bound on p is stricter; it approaches the upper bound of the incompressible case when $\lambda \to \infty$.

Proof If $\bar{\boldsymbol{A}}$ is quasiconvex at the boundary, then it is quasiconvex and from the Legendre-Hadamard condition follows $\mu \ge 0$. Assume first that $\mu > 0$. It is found that then $\bar{\boldsymbol{A}} = \mu \bar{\boldsymbol{B}}$ where $\bar{\boldsymbol{B}}$ is given by (20.6.16) with $\gamma = 2 + p/\mu$, $\delta = 2 + \lambda/\mu$. With this identification, inequalities (20.6.18) become equivalent to the inequalities of Assertion (1) of the present proposition. If $\mu = 0$, then an analysis analogous to, but simpler than, that in the proof of Proposition 20.6.4 gives the quasiconvexity at the boundary is equivalent to the inequalities in Assertion (2). □

Let us proceed to the pointwise positive definiteness properties of (20.6.2).

20.6.6 Proposition *Let $n = 3$ and let $\bar{\boldsymbol{A}}$ be the elasticity tensor of an isotropic state. Then*

(1) *$\bar{\boldsymbol{A}}$ is positive semidefinite on* Lin *if and only if*

$$3\lambda + 2\mu - 2p \geq 0, \quad 2\mu + p \geq 0, \quad p \leq 0; \tag{20.6.24}$$

$\bar{\boldsymbol{A}}$ is positive definite on Lin *if and only if the three inequalities in* (20.6.24) *are strict; there exists a constant $c > 0$ such that $\bar{\boldsymbol{A}}[\boldsymbol{H}, \boldsymbol{H}] \geq c|\boldsymbol{E}|^2$ for every $\boldsymbol{H} \in$* Lin *(here $\boldsymbol{E}$ is the symmetric part of $\boldsymbol{H}$) if and only if the first two inequalities in* (20.6.24) *are strict.*

(2) *If $p = 0$, then $\bar{\boldsymbol{A}}[\boldsymbol{E}, \boldsymbol{E}] > 0$ for every $\boldsymbol{E} \in$* Sym, *$\boldsymbol{E} \neq \boldsymbol{0}$ if and only if*

$$3\lambda + 2\mu > 0, \quad \mu > 0. \tag{20.6.25}$$

Proof The proofs of these assertions are based on the decomposition of a general second-order tensor $\boldsymbol{H} \in \mathrm{Lin}$ of the form $\boldsymbol{H} = \frac{1}{3}(\mathrm{tr}\,\boldsymbol{E})\boldsymbol{1} + \boldsymbol{E}^{\mathrm{D}} + \boldsymbol{W}$ where $\boldsymbol{E}$ is the symmetric part of $\boldsymbol{H}$, $\boldsymbol{W}$ the skew part, and $\boldsymbol{E}^{\mathrm{D}}$ the deviatoric part of $\boldsymbol{E}$. One finds that

$$\bar{\boldsymbol{A}}[\boldsymbol{H}, \boldsymbol{H}] = (\lambda + \tfrac{2}{3}\mu - \tfrac{2}{3}p)(\mathrm{tr}\,\boldsymbol{E})^2 + (2\mu + p)|\boldsymbol{E}^{\mathrm{D}}|^2 - p|\boldsymbol{W}|^2, \quad \boldsymbol{H} \in \mathrm{Lin}. \tag{20.6.26}$$

Now $\mathrm{tr}\,\boldsymbol{E}$, $\boldsymbol{E}^{\mathrm{D}}$ and $\boldsymbol{W}$ are independent and from this the positivity properties (1), (2) can be easily deduced. The details are omitted. □

Remark. Another possibility of proving the above proposition is to find the eigenvalues of the linear transformation $\boldsymbol{H} \mapsto \bar{\boldsymbol{A}}[\boldsymbol{H}]$ on Lin. These are: (1) $\lambda + \frac{2}{3}\mu - \frac{2}{3}p$ with the eigenvector $\boldsymbol{1} \in \mathrm{Lin}$; (2) $2\mu + p$ with the eigenspace the space of all symmetric traceless tensors; (3) $-p$ with the eigenspace Skew.

Let us specialize the results to a natural state:

Table 20.6.1. Properties of the isotropic elasticity tensor ($p = 0$)

positive definiteness on Sym ⇓	⇔	$3\lambda + 2\mu > 0$,	$\mu > 0$
uniform quasiconvexity at the boundary ⇓	⇔	$\lambda + \mu > 0$,	$\mu > 0$
strong ellipticity	⇔	$\lambda + 2\mu > 0$,	$\mu > 0$

As mentioned above, if $p = 0$, $\bar{\boldsymbol{A}}$ can never be strictly positive definite on Lin, however, by combining the assumption that $\bar{\boldsymbol{A}}$ is strictly positive definite on Sym (which is a consistent assumption) with Korn's inequality, one can prove that the second variation is uniformly positive. More precisely, let Ω, D be as in Proposition 20.6.7(3) (below), suppose that $\bar{\boldsymbol{A}} = \bar{\boldsymbol{A}}(\boldsymbol{p})$ depends continuously on $\boldsymbol{p}$ over $\mathrm{cl}\,\Omega$ and that it is of the form (20.6.1) with $p \equiv 0$ and with λ, μ satisfying (20.6.25) at every $\boldsymbol{p} \in \mathrm{cl}\,\Omega$. Then J is uniformly positive on the space of all $\boldsymbol{u} \in W^{1,2}(\Omega)$ satisfying $\boldsymbol{u} = \boldsymbol{0}$ on D in the sense of trace. Indeed we have $\bar{\boldsymbol{A}}[\boldsymbol{H}] = \bar{\boldsymbol{A}}[\boldsymbol{E}]$ and by Item (2) of Proposition 20.6.6 there exists a $k > 0$ such that

$$\bar{\boldsymbol{A}}[\boldsymbol{H}, \boldsymbol{H}] \geq k|\boldsymbol{E}|$$

for every $\boldsymbol{p} \in \mathrm{cl}\,\Omega$ and every $\boldsymbol{H} \in \mathrm{Lin}$. Setting $\boldsymbol{H} = \nabla\boldsymbol{u}$ and combining with (20.6.29) we obtain the assertion.

20.6.7 Proposition (Korn's Inequalities)

(1) *For every $\boldsymbol{u} \in W^{1,2}(\mathrm{Vect})$,*

$$\int |\boldsymbol{E}|^2 \, dV \geq \tfrac{1}{2} \int |\nabla\boldsymbol{u}|^2 \, dV, \tag{20.6.27}$$

where the integration takes place over Vect *and where here and in* (2), (3), *below,* $\boldsymbol{E} = \frac{1}{2}(\nabla \boldsymbol{u} + \nabla \boldsymbol{u}^{\mathrm{T}})$.

(2) *if H is a half-space* (*see* (20.4.4)), *then for every* $\boldsymbol{u} \in W^{1,2}(H)$,

$$\int_H |\boldsymbol{E}|^2 \, dV \geq \tfrac{1}{4} \int_H |\nabla \boldsymbol{u}|^2 \, dV; \tag{20.6.28}$$

(3) *if* $\Omega \subset$ Vect *is a bounded region with Lipschitz boundary and* $D \subset \partial\Omega$ *a set of positive area, then there exists a* $c > 0$ *such that*

$$\int_\Omega |\boldsymbol{E}|^2 \, dV \geq c \int_\Omega |\nabla \boldsymbol{u}|^2 \, dV \tag{20.6.29}$$

for every $\boldsymbol{u} \in W^{1,2}(\Omega)$ *satisfying* $\boldsymbol{u} = \boldsymbol{0}$ *on D in the sense of trace.*

The value $\frac{1}{2}$ in (20.6.27) is optimal, i.e., it cannot be replaced by any larger number without violating (20.6.27) for some $\boldsymbol{u} \in W^{1,2}(\text{Vect})$; also the value $\frac{1}{4}$ in (20.6.28) is optimal. One also deduces from (1) that if $D = \partial\Omega$ in (3), then (20.6.29) holds with $c = \frac{1}{2}$. From (2) one deduces that if $\partial\Omega \setminus D$ contains at least one point $\boldsymbol{p}$ such that $\partial\Omega$ is of class C^1 at $\boldsymbol{p}$, then necessarily $c \leq \frac{1}{4}$; if there exists a plane P in Vect such that $\emptyset \neq \partial\Omega \setminus D \subset P$, then (20.6.29) holds with $c = \frac{1}{4}$ and this value is optimal, see the proof below. Inequalities (20.6.27)–(20.6.29) can be stated equivalently as

$$\int |\boldsymbol{E}|^2 \, dV \geq c' \int |\boldsymbol{W}|^2 \, dV, \tag{20.6.30}$$

where $c' = c/(1-c)$, $\boldsymbol{W} = \frac{1}{2}(\nabla \boldsymbol{u} - \nabla \boldsymbol{u}^{\mathrm{T}})$. In particular, $c' = 1$ and $c' = \frac{1}{3}$ in the cases of (20.6.27) and (20.6.28), respectively.

Proof (1): It suffices to verify (20.6.27) only when $\boldsymbol{u}$: Vect $\to$ Vect is infinitely differentiable with compact support, which is contained in some cube C. Inequality (20.6.27) with $\frac{1}{2}$ replaced by a $c > 0$ means that

$$\int_C \bar{\boldsymbol{A}}[\nabla \boldsymbol{u}, \nabla \boldsymbol{u}] \, dV \geq 0 \tag{20.6.31}$$

for every $\boldsymbol{u} \in W_0^{1,2}(C)$, where the tensor $\bar{\boldsymbol{A}}$ is given by

$$\bar{\boldsymbol{A}}[\boldsymbol{H}] = (1 - 2c)\boldsymbol{H} + \boldsymbol{H}^{\mathrm{T}}, \quad \boldsymbol{H} \in \text{Lin}. \tag{20.6.32}$$

This means that $\bar{\boldsymbol{A}}$ is quasiconvex, and this is equivalent to the Legendre–Hadamard condition by Proposition 20.5.1. The latter is equivalent to $c \leq \frac{1}{2}$, as is easily verified. This proof also shows that the value $c = \frac{1}{2}$ is optimal. (2): Again, (20.6.28) is equivalent to (20.6.31) with $\bar{\boldsymbol{A}}$ from (20.6.32) with $c = \frac{1}{4}$ and with C replaced by H. But this is the quasiconvexity at the boundary with normal $\boldsymbol{n}$. Writing $\bar{\boldsymbol{A}} = (1 - 2c)\bar{\boldsymbol{B}}$, one finds that $\bar{\boldsymbol{B}}$ is of the form (20.6.16) with $\delta = \gamma = 2(1 - c)/(1 - 2c)$. As $1 - 2c > 0$, we see that also $\bar{\boldsymbol{B}}$ must be quasiconvex at the boundary, and hence (20.6.18) must hold. But these inequalities are found to be equivalent to $c \leq \frac{1}{4}$. (3): See, e.g., NEČAS & HLAVÁČEK [1981, Chap. 6]. □

Propositions 20.6.5 and 20.6.1 can be now combined with Propositions 20.4.9, 20.5.1, 20.5.2, 20.5.3, and 20.5.4. One obtains various necessary, sufficient, and sometimes necessary and sufficient conditions for the second variation of the stored energy in an isotropic state to be uniformly positive or coercive. The straightforward detailed formulation of the corresponding results is omitted. Propositions 20.6.5 and 20.6.1 also give a complete list of conditions that are independent of the shape of the body. A further examination of, for instance, the positive semidefiniteness of J requires detailed information about Ω, S, D. A convenient hypothetical strategy is to check the inequalities in Propositions 20.6.5 and 20.6.1. If any of them is violated, then certainly J is not positive semidefinite. If all of them are satisfied, then J still need not be positive semidefinite, and it will be positive semidefinite if and only if the spectrum of the associated eigenvalue problem is contained in the nonnegative half-axis. The spectrum of the associated eigenvalue problem is necessarily discrete by the standard compact imbedding results if Ω is bounded. Note that if λ, μ, p are uniform over Ω and the free part of the boundary is entirely contained in some fixed plane, then the inequalities of Proposition 20.6.5, or their appropriate strict versions, give necessary and sufficient conditions for the positive semidefiniteness or the uniform positivity of J, respectively.

20.7 Stability and Bifurcation for a Column

We here outline an explicitly solvable, two-dimensional bifurcation model of a body in a bilateral contact environment 13.3.2.

20.7.1 Formulation of the Problem Let $n = 2$ and identify Vect with $\mathbb{R}^2$ with the standard basis $\boldsymbol{e}_i$, $i = 1, 2$. Consider a rectangular elastic solid $\Omega = (-R, R) \times (0, H)$ of width $2R$ and height H. Let $D_1 := (-R, R) \times \{0\}$, $D_2 := (-R, R) \times \{H\}$ be its lower and upper ends, respectively, and D their union. Denote by $S := \partial\Omega \setminus D$ the union of the sides. Suppose that the reference configuration Ω is stress-free, $\rho_\circ = 1$, that the processes are isothermal, and that the body is composed of a Blatz–Ko special I with the stored energy $\bar{f}$ in 11.3.3(3). Alternatively, this is a Hadamard material 11.3.3(2) with $h(\Delta) = \Delta^{-\gamma}/\gamma$. For this material,

$$\bar{f}(\boldsymbol{F}) = \tfrac{1}{2}|\boldsymbol{F}|^2 + (\det \boldsymbol{F})^{-\gamma}/\gamma = f(v_1, v_2), \quad \bar{\boldsymbol{S}}(\boldsymbol{F}) = \boldsymbol{F} - (\det \boldsymbol{F})^{-\gamma}\boldsymbol{F}^{-\mathrm{T}},$$
$$\bar{\mathsf{A}}[\boldsymbol{H}] = \boldsymbol{H} + (\det \boldsymbol{F})^{-\gamma}(\gamma \operatorname{tr}(\boldsymbol{F}^{-1}\boldsymbol{H})\boldsymbol{F}^{-\mathrm{T}} + \boldsymbol{F}^{-\mathrm{T}}\boldsymbol{H}^{\mathrm{T}}\boldsymbol{F}^{-\mathrm{T}}), \quad \boldsymbol{H} \in \mathrm{Lin}.$$

Let $A > 0$ be given and let P_i, $i = 1, 2$, be two parallel lines in $\mathbb{R}^2$ at separation AH. Identify P_1 with the x_1 axis and P_2 with the line $x_2 = AH$. We are looking for a deformation χ such that

$$\operatorname{Div} \boldsymbol{S} = \boldsymbol{0} \text{ on } \Omega, \tag{20.7.1}$$

$$\chi(\boldsymbol{p}) \in P_i \text{ and } (\boldsymbol{1} - \boldsymbol{e}_2 \otimes \boldsymbol{e}_2)\boldsymbol{S}(\boldsymbol{p})\boldsymbol{e}_2 = \boldsymbol{0} \text{ if } \boldsymbol{p} \in D_i, \quad i = 1, 2, \tag{20.7.2}$$

$$\boldsymbol{S}(\boldsymbol{p})\boldsymbol{e}_1 = \boldsymbol{0} \text{ if } \boldsymbol{p} \in S. \tag{20.7.3}$$

Since $\boldsymbol{b}_0 \equiv \boldsymbol{0}$, $\boldsymbol{s} \equiv \boldsymbol{0}$, the environment is conservative with the potential energy $\mathsf{W} \equiv 0$. For a fixed $A > 0$, assume that the solution of (20.7.1)–(20.7.3) is of the form $\chi = \boldsymbol{F}\boldsymbol{p}$, $\boldsymbol{p} \in \Omega$ where $\boldsymbol{F} = \operatorname{diag}(v_1, v_2) = \text{const}$. Then

$$\boldsymbol{S} = \operatorname{diag}(s_1, s_2) \text{ where } s_i = v_i - (v_1 v_2)^{-\gamma}/v_i, \quad i = 1, 2. \tag{20.7.4}$$

Equation (20.7.1) and and $(20.7.2)_2$ are trivially satisfied, $(20.7.2)_1$ requires that $v_2 = A$ and (20.7.3) that $s_1 = 0$, which gives $v_1 = v_2^{-\delta} = A^{-\delta}$ where $\delta = \gamma/(2+\gamma)$. Hence for every $A > 0$ the deformation χ defined by

$$\begin{gathered} \chi(\boldsymbol{p}) = \boldsymbol{F}\boldsymbol{p}, \quad \boldsymbol{p} \in \Omega, \text{ where} \\ \boldsymbol{F} = \operatorname{diag}(v_1, v_2), \; v_1 = A^{-\delta}, \; v_2 = A, \; \delta = \gamma/(2+\gamma) \end{gathered} \tag{20.7.5}$$

satisfies (20.7.1)–(20.7.3). As A varies, we have a family of problems with a family of trivial solutions χ and our goal is to examine the stability of and bifurcations at the state χ.

For a given $A > 0$, the sets Δ^{**} and $\mathrm{Var}(\chi)$ of admissible deformations and admissible variations, respectively, are defined by

$$\Delta^{**} = \{\bar{\chi} \in W^{1,1}(\Omega) : \det \nabla\bar{\chi} > 0 \text{ a.e. on } \Omega \text{ and } \bar{\chi}(D_i) \subset P_i, i = 1, 2\},$$
$$\mathrm{Var}(\chi) = \{u \in W^{1,2}(\Omega) : u \cdot e_2 = 0 \text{ on } D\}.$$

The total stored energy is

$$\mathsf{M}(\bar{\chi}) = \int_\Omega \bar{f}(\nabla\bar{\chi})\, dV,$$

and $\delta\mathsf{M}(\chi, \cdot), \delta^2\mathsf{M}(\chi, \cdot, \cdot)$ are linear and bilinear functionals on $\mathrm{Var}(\chi)$ given by the formulas identical with those for the smooth variations.

For $A \geq 1$ no bifurcation takes place and the trivial solution is stable:

20.7.2 Proposition *Let $A \geq 1$ and let χ be the trivial homogeneous solution* (20.7.5). *Then*

(1) $s_2 \geq 0$ *and F is a point of convexity of $\bar{f}$;*

(2) χ *is a global minimizer of* M *on* Δ^{**}, *i.e.,*

$$\mathsf{M}(\bar{\chi}) \geq \mathsf{M}(\chi) \text{ for every } \bar{\chi} \in \Delta^{**}; \tag{20.7.6}$$

moreover, the equality holds in (20.7.6) *only if $\bar{\chi} = \chi + me_1$ for some $m \in \mathbb{R}$.*

Proof (1): From (20.7.4) and $s_1 = 0$ we obtain $v_1^2 = (v_1v_2)^{-\gamma}$ and hence $s_2 = (v_2^2 - v_1^2)/v_2 \geq 0$ as $v_2 \geq v_1$ by (20.7.5). By Example 18.1.6(4), f is convex on $(0, \infty)^2$ and as $s_1 \geq 0, s_2 \geq 0$ also $\bar{f}$ is convex at F by Proposition 18.3.5(1). (2): Let $\bar{\chi} \in \Delta^{**}$. The convexity of $\bar{f}$ at F gives $\bar{f}(\nabla\bar{\chi}) \geq \bar{f}(F) + \bar{S}(F) \cdot (\nabla\bar{\chi} - F)$ for every $p \in \Omega$. Integrating this inequality over Ω we obtain

$$\mathsf{M}(\bar{\chi}) \geq \mathsf{M}(\chi) + \int_\Omega \bar{S}(F) \cdot (\nabla\bar{\chi} - \nabla\chi)\, dV = \mathsf{M}(\chi) + \int_{\partial\Omega} \bar{S}(F)n \cdot (\bar{\chi} - \chi)\, dA.$$

By the boundary conditions (20.7.2), the surface integral vanishes, which proves (20.7.6). Since F can be shown to be a point of very strict convexity of $\bar{f}$, one can prove the uniqueness assertion as well. □

The situation is different when $A < 1$. We shall use the linearized equations to determine the values of A at which the bifurcation occurs.

20.7.3 Linearization For the present problem, the linearized equations (20.1.16) read

$$\mathrm{Div}(\bar{\mathbf{A}}[\nabla u]) = \mathbf{0} \text{ on } \Omega,$$
$$u \cdot e_2 = 0 \text{ and } (\mathbf{1} - e_2 \otimes e_2)\bar{\mathbf{A}}[\nabla u]e_2 = \mathbf{0} \text{ on } D,$$
$$\mathbf{A}[\nabla u]n = \mathbf{0} \text{ on } S,$$

where $\bar{\mathbf{A}}$ is the referential tensor of elasticities at $F \in \mathrm{Lin}^+$. Inserting the value of $\bar{\mathbf{A}}$ calculated above and using the identities

$$\mathrm{Div}\left(\mathrm{tr}(F^{-1}\nabla u)F^{-\mathrm{T}}\right) = F^{-\mathrm{T}}\nabla\,\mathrm{Div}(F^{-1}u),$$
$$\mathrm{Div}(F^{-\mathrm{T}}\nabla u^{\mathrm{T}}F^{-\mathrm{T}}) = F^{-\mathrm{T}}\nabla\,\mathrm{Div}(F^{-1}u),$$

the linearized problem may be written as

$$\Delta u + (\det F)^{-\gamma}(\gamma + 1)F^{-\mathrm{T}}\nabla\,\mathrm{Div}(F^{-1}u) = \mathbf{0} \text{ on } \Omega,$$
$$u_2 = 0, \quad v_1v_2u_{1,2} + (v_1v_2)^{-\gamma}u_{2,1} = 0 \text{ on } D,$$
$$v_1u_{,1} + (v_1v_2)^{-\gamma}(\gamma\,\mathrm{Div}(F^{-1}u)e_1 + F^{-\mathrm{T}}\nabla u_1) = \mathbf{0} \text{ on } S,$$

where $\boldsymbol{u} = (u_1, u_2)$, $\boldsymbol{u}_{,i} = \partial \boldsymbol{u}/\partial p_i$ and $\Delta \boldsymbol{u} = \operatorname{Div} \nabla \boldsymbol{u}$ is the Laplacian. Introducing $\boldsymbol{w} = \boldsymbol{F}^{-1}\boldsymbol{u}$, the linearized equations are finally written as

$$\boldsymbol{F}^2 \Delta \boldsymbol{w} + (v_1 v_2)^{-\gamma}(\gamma + 1)\nabla \operatorname{Div} \boldsymbol{w} = \boldsymbol{0} \text{ on } \Omega, \tag{20.7.7}$$

$$w_2 = 0, \quad v_1^2 w_{1,2} + (v_1 v_2)^{-\gamma} w_{2,1} = 0 \text{ on } D, \tag{20.7.8}$$

$$\boldsymbol{F}^2 \boldsymbol{w}_{,1} + (v_1 v_2)^{-\gamma}(\gamma(\operatorname{Div} \boldsymbol{w})\boldsymbol{e}_1 + \nabla w_1) = \boldsymbol{0} \text{ on } S. \tag{20.7.9}$$

20.7.4 Bifurcations We now follow BURGESS & LEVINSON [1972] in a special case to determine the critical values $A < 1$ at which the bifurcations may occur, i.e., at which (20.7.7)–(20.7.9) has a nontrivial solution. It can be shown that the solution can be sought in the separated form

$$w_1 = f_1(\omega_n p_1) \cos \omega_n p_2, \quad w_2 = f_2(\omega_n p_1) \sin \omega_n p_2,$$

where $\omega_n = n\pi/H$, $n = 1, \ldots$ Inserting this into (20.7.7), one finds a system of two second-order equations with constant coefficients

$$\begin{aligned} v_1^2(f_1'' - f_1) + (v_1 v_2)^{-\gamma}(\gamma + 1)(f_1'' + f_2') &= 0, \\ v_2^2(f_2'' - f_2) - (v_1 v_2)^{-\gamma}(\gamma + 1)(f_1' + f_2) &= 0. \end{aligned} \tag{20.7.10}$$

The boundary conditions (20.7.8) are automatically satisfied and (20.7.9) read as

$$\begin{aligned} (v_1^2 + (v_1 v_2)^{-\gamma}(\gamma + 1))f_1' + \gamma(v_1 v_2)^{-\gamma} f_2 &= 0, \\ v_2^2 f_2' - (v_1 v_2)^{-\gamma} f_1 &= 0. \end{aligned} \tag{20.7.11}$$

The characteristic frequencies of (20.7.10) are $\omega = \pm 1, \pm K$ where

$$K = [(1 + v_1^2 B)/(1 + v_2^2 B)]^{\frac{1}{2}}, \quad B = (\gamma + 1)(v_1 v_2)^{-\gamma - 2}. \tag{20.7.12}$$

The general solution of (20.7.10) is

$$f(t) = c_1^+(1, -1)e^t + c_1^-(1, 1)e^{-t} + c_2^+(Kv_2^2/v_1^2, -1)e^{Kt} + c_2^-(Kv_2^2/v_1^2, 1)e^{-Kt};$$

in components,

$$f_1 = c_1^+ e^t + c_1^- e^{-t} + v_2^2 K/v_1^2 (c_2^+ e^{Kt} + c_2^- e^{-Kt}),$$

$$f_2 = (c_1^- e^{-t} - c_1^+ e^t) + c_2^- e^{-Kt} - c_2^+ e^{Kt},$$

where $c_1^\pm$, $c_2^\pm$ are constants. This solution f is now inserted into the boundary conditions (20.7.9). This gives, after rearrangements,

$$\begin{aligned} \alpha \operatorname{ch} t(c_1^+ - c_1^-) + \beta \operatorname{ch}(Kt)(c_2^+ - c_2^-) &= 0, \\ \beta \operatorname{sh} t(c_1^+ - c_1^-) + \delta \operatorname{sh}(Kt)(c_2^+ - c_2^-) &= 0, \\ \alpha \operatorname{sh} t(c_1^+ + c_1^-) + \beta \operatorname{sh}(Kt)(c_2^+ + c_2^-) &= 0, \\ \beta \operatorname{ch} t(c_1^+ + c_1^-) + \delta \operatorname{ch}(Kt)(c_2^+ + c_2^-) &= 0, \end{aligned}$$

where $t = \omega_n R$ and $\alpha = 2v_1^2$, $\beta = v_1^2 + v_2^2$, $\delta = 2v_2^2 K$. This system has a nontrivial solution if and only if its determinant vanishes, which happens if and only if either

$$(v_1^2 + v_2^2)^2 = 4K(v_1 v_2)^2 \tanh\left(\frac{n\pi R}{H}\right) \coth\left(\frac{Kn\pi R}{H}\right) \tag{20.7.13}$$

or

$$(v_1^2 + v_2^2)^2 = 4K(v_1 v_2)^2 \coth\left(\frac{n\pi R}{H}\right) \tanh\left(\frac{Kn\pi R}{H}\right) \qquad (20.7.14)$$

where K is as above and $v_2 = A$, $v_1 = A^{-\delta}$. The corresponding solutions can be written explicitly, and it turns out that (20.7.14) corresponds to a $\boldsymbol{u}$ that is symmetric about the x_2 axis (bulging) while (20.7.13) corresponds to an asymmetric $\boldsymbol{u}$ (buckling).

Thus, for a given A, the linearized equations have a nonzero solution if and only if equation (20.7.13) or (20.7.14) holds with some $n = 1, \ldots$ This gives a sequence of values $A_n < 1$ at which a bifurcation occurs. One is especially interested in the largest value of $A < 1$ for which one of the two equations holds. This gives the first critical A as A diminishes form 1 as a result of the compression. It can be proved that the largest $A < 1$ corresponding to buckling tends to 1 as $H \to \infty$ at a fixed R, while the largest $A < 1$ corresponding to bulging is smaller than the A for buckling at large H. This means that for sufficiently slender blocks, the buckling occurs first and almost immediately. The reader is referred to BURGESS & LEVINSON [1972] for the numerically determined graphs of the critical A as a function of H/R.

20.7.5 The Second Variation There is a Korn inequality for variations $\boldsymbol{u} \in \mathrm{Var}(\chi)$, of the previous form (20.6.29). Combining this with the fact that the Lamé moduli at $\boldsymbol{F} = \mathbf{1}$ satisfy (20.6.25), one concludes that $\delta^2 \mathsf{M}(\chi, \cdot, \cdot)$ is uniformly positive in the stress-free state χ corresponding to $A = 1$. This can be used to prove that $\delta^2 \mathsf{M}(\chi, \cdot, \cdot)$ is uniformly positive also at the states χ with $A > 1$. Since, as A changes, the coefficients of the quadratic form $\delta^2 \mathsf{M}(\chi, \cdot, \cdot)$ change continuously, $\delta^2 \mathsf{M}(\chi, \cdot)$ is uniformly positive also for all $A < 1$ sufficiently close to 1. Hence there is always an interval $(1 - \varepsilon, \infty)$, $\varepsilon > 0$, of A where the trivial solution is *infinitesimally stable* and consequently is a weak local minimizer of M. It is conceivable that $\delta^2 \mathsf{M}(\chi, \cdot)$ is uniformly positive until the first bifurcation occurs, where $\delta^2 \mathsf{M}(\chi, \cdot)$ definitely ceases to be uniformly positive, as $\delta^2 \mathsf{M}(\chi, \boldsymbol{u}_0, \boldsymbol{u}_0) = 0$ for the $\boldsymbol{u}_0 \neq \mathbf{0}$ coming from the solution above.

Notice that the Blatz–Ko special I stored energy is polyconvex and hence quasiconvex. We thus see a bifurcation and a loss of stability in the elliptic region of the material. This kind of instability is much finer than the violation of the Legendre–Hadamard condition in the spinodal region associated with the phase transition. The above bifurcation analysis is formal in the sense that it is not obvious that at every $A < 1$ at which the system (20.7.7)–(20.7.9) has a nontrivial solution, there is a family of nontrivial solutions of the original nonlinear problem bifurcating from this point.

20.8 Existence in Linearized Elasticity

In this section we shall establish the existence of the solution of the linearized equations for the displacement-traction environment of dead loadings (20.1.8) with $\bar{\boldsymbol{d}} = \mathbf{0}$. Any pair $(\bar{\boldsymbol{b}}, \bar{\boldsymbol{s}})$ such that $\bar{\boldsymbol{b}} \in L^2(\Omega)$ and $\bar{\boldsymbol{s}} \in L^2(S)$ is called the ***data.*** The displacement $\boldsymbol{u} \in \mathrm{Var}(\chi)$ (see (20.3.2)) is said to be a ***weak solution*** corresponding to the data $\bar{\boldsymbol{b}}, \bar{\boldsymbol{s}}$ if

$$\int_\Omega (-\bar{\boldsymbol{A}}[\nabla \boldsymbol{u}, \nabla \boldsymbol{v}] + \rho_0 \bar{\boldsymbol{b}} \cdot \boldsymbol{v})\, dV + \int_S \bar{\boldsymbol{s}} \cdot \boldsymbol{v}\, dA = 0 \quad \text{for every } \boldsymbol{v} \in \mathrm{Var}(\chi). \qquad (20.8.1)$$

The values of $\boldsymbol{v}$ in the surface integral are understood in the sense of trace (Theorem 20.2.4). This can be equivalently restated as

$$\boldsymbol{u} \in \mathrm{Var}(\chi), \quad \delta\mathsf{G}(\boldsymbol{u}, \boldsymbol{v}) = 0 \text{ for every } \boldsymbol{v} \in \mathrm{Var}(\chi),$$

where

$$\mathsf{G}(\boldsymbol{u}) = \int_{\Omega} (\tfrac{1}{2}\bar{\mathbf{A}}[\nabla\boldsymbol{u}, \nabla\boldsymbol{u}] - \rho_o\bar{\boldsymbol{b}} \cdot \boldsymbol{u})\, dV - \int_{S} \bar{\boldsymbol{s}} \cdot \boldsymbol{u}\, dA \tag{20.8.2}$$

on $\mathrm{Var}(\chi)$. The quadratic part of G is a half of the second variation J at χ of the stored energy $\mathsf{M}_0(\chi, \cdot, \cdot)$ (without the potential energy term),

$$\mathsf{M}_0(\chi) = \int_{\Omega} \bar{f}(\nabla\chi, \theta_0)\rho_o\, dV.$$

It is assumed that χ gives rise to a tensor of elasticities $\bar{\mathbf{A}}$ that is continuous on $\mathrm{cl}\,\Omega$; moreover, assume that ρ_o is bounded on Ω and D has a positive area.

20.8.1 Proposition *If* J *is uniformly positive then for every data* $\bar{\boldsymbol{b}}, \bar{\boldsymbol{s}}$ *there exists a unique weak solution* $\boldsymbol{u} \in \mathrm{Var}(\chi)$ *corresponding to* $\bar{\boldsymbol{b}}, \bar{\boldsymbol{s}}$. *The solution minimizes the functional* G *over* $\mathrm{Var}(\chi)$ *and this solution is the only minimizer of* G *over* $\mathrm{Var}(\chi)$. *Moreover there exists a constant* $c > 0$ *such that*

$$\|\boldsymbol{u}\|_{W^{1,2}(\Omega)} \le c(\|\bar{\boldsymbol{b}}\|_{L^2(\Omega)} + \|\bar{\boldsymbol{s}}\|_{L^2(S)}). \tag{20.8.3}$$

Proof The hypotheses imply that the form $\langle\cdot,\cdot\rangle$ on $\mathrm{Var}(\chi)$, defined by

$$\langle \boldsymbol{u}, \boldsymbol{v}\rangle = \int_{\Omega} \bar{\mathbf{A}}[\nabla\boldsymbol{u}, \nabla\boldsymbol{v}]\, dV, \quad \boldsymbol{u}, \boldsymbol{v} \in \mathrm{Var}(\chi),$$

is a scalar product and

$$c_1\|\boldsymbol{u}\|^2_{W^{1,2}(\Omega)} \le \langle \boldsymbol{u}, \boldsymbol{u}\rangle \le c_2\|\boldsymbol{u}\|^2_{W^{1,2}(\Omega)} \tag{20.8.4}$$

for all $\boldsymbol{u} \in \mathrm{Var}(\chi)$ and some positive constants c_1, c_2. Hence the norm derived form $\langle\cdot,\cdot\rangle$ is equivalent to the Sobolev norm $\|\cdot\|_{W^{1,2}(\Omega)}$. The functional F defined on $\mathrm{Var}(\chi)$ by

$$F(\boldsymbol{v}) = \int_{\Omega} \rho_o\bar{\boldsymbol{b}} \cdot \boldsymbol{v}\, dV + \int_{S} \bar{\boldsymbol{s}} \cdot \boldsymbol{v}\, dA$$

is continuous on $\mathrm{Var}(\chi)$ endowed with the scalar product $\langle\cdot,\cdot\rangle$. This follows from the continuity of F with respect to $\|\cdot\|_{W^{1,2}(\Omega)}$ and from the equivalence of the latter to the norm derived from $\langle\cdot,\cdot\rangle$. Hence by the Riesz representation theorem 20.2.7 there exists a unique element $\boldsymbol{u} \in \mathrm{Var}(\chi)$ such that

$$\langle \boldsymbol{u}, \boldsymbol{v}\rangle = F(\boldsymbol{v}) \tag{20.8.5}$$

for every $\boldsymbol{v} \in \mathrm{Var}(\chi)$. In view of the definition of $\langle\cdot,\cdot\rangle$, equation (20.8.5) is exactly the assertion that $\boldsymbol{u}$ is a weak solution corresponding to $\bar{\boldsymbol{b}}, \bar{\boldsymbol{s}}$. To prove the uniqueness, it suffices to prove that if $\bar{\boldsymbol{b}} = \mathbf{0}, \bar{\boldsymbol{s}} = \mathbf{0}$, then also the weak solution $\boldsymbol{u}$ is $\mathbf{0}$. If $\boldsymbol{u}$ is a weak solution corresponding to $\bar{\boldsymbol{b}} = \mathbf{0}, \bar{\boldsymbol{s}} = \mathbf{0}$, then $\int_{\Omega} \bar{\mathbf{A}}[\nabla\boldsymbol{u}, \nabla\boldsymbol{v}]\, dV = 0$ for every $\boldsymbol{v} \in \mathrm{Var}(\chi)$. Hence $\int_{\Omega} \bar{\mathbf{A}}[\nabla\boldsymbol{u}, \nabla\boldsymbol{u}]\, dV = 0$ and consequently $\boldsymbol{u} = \mathbf{0}$ by the uniform positivity of J. To prove that the solution is a minimum of G, let $\boldsymbol{v} \in \mathrm{Var}(\chi)$. Using (20.8.5) one finds

$$\mathsf{G}(\boldsymbol{u} + \boldsymbol{v}) = \tfrac{1}{2}\langle \boldsymbol{u}, \boldsymbol{u}\rangle + \langle \boldsymbol{u}, \boldsymbol{v}\rangle + \tfrac{1}{2}\langle \boldsymbol{v}, \boldsymbol{v}\rangle - F(\boldsymbol{u}) - F(\boldsymbol{v}) = \mathsf{G}(\boldsymbol{u}) + \tfrac{1}{2}\langle \boldsymbol{v}, \boldsymbol{v}\rangle$$

and as $\langle \boldsymbol{v}, \boldsymbol{v}\rangle \ge 0$, we see that $\boldsymbol{u}$ is a minimum. Conversely, if $\boldsymbol{u}$ is a minimum of G, then for any $\lambda \in \mathbb{R}$ and $\boldsymbol{v} \in \mathrm{Var}(\chi)$, $\mathsf{G}(\boldsymbol{u} + \lambda\boldsymbol{v}) - \mathsf{G}(\boldsymbol{u}) \ge 0$. By a calculation similar to the one above, this gives $\lambda(\langle \boldsymbol{u}, \boldsymbol{v}\rangle - F(\boldsymbol{v})) + \lambda^2\langle \boldsymbol{v}, \boldsymbol{v}\rangle \ge 0$ for any λ. Taking $\lambda > 0$, dividing by λ and letting $\lambda \to 0$ then gives $\langle \boldsymbol{u}, \boldsymbol{v}\rangle - F(\boldsymbol{v}) = 0$ and this is the definition of the weak solution. Let us finally prove (20.8.3). Let $\boldsymbol{u}$ be a weak solution corresponding to $\bar{\boldsymbol{b}}, \bar{\boldsymbol{s}}$. The boundedness of ρ_o by, say, c_3 and Hölder's inequality applied to the volume and surface integrals in the definition of F lead to

$$|F(\boldsymbol{u})| \le c_3\|\bar{\boldsymbol{b}}\|_{L^2(\Omega)}\|\boldsymbol{u}\|_{L^2(\Omega)} + \|\bar{\boldsymbol{s}}\|_{L^2(S)}\|\boldsymbol{u}\|_{L^2(S)}$$

for every $\boldsymbol{u} \in \mathrm{Var}(\chi)$. Combining this with (20.2.4) of the trace theorem and with $\|\boldsymbol{u}\|_{L^2(\Omega)} \le \|\boldsymbol{u}\|_{W^{1,2}(\Omega)}$ one finds that there exist constants c_4, c_5 such that

$$|F(u)| \le c_4 \|\bar{\boldsymbol{b}}\|_{L^2(\Omega)} \|\boldsymbol{u}\|_{W^{1,2}(\Omega)} + c_5 \|\bar{\boldsymbol{s}}\|_{L^2(S)} \|\boldsymbol{u}\|_{W^{1,2}(\Omega)}.$$

Setting $\boldsymbol{v} = \boldsymbol{u}$ in (20.8.5), using $(20.8.4)_1$ and the inequality above, one finds

$$c_1 \|\boldsymbol{u}\|^2_{W^{1,2}(\Omega)} \le c_6 (\|\bar{\boldsymbol{b}}\|_{L^2(\Omega)} + \|\bar{\boldsymbol{s}}\|_{L^2(S)}) \|\boldsymbol{u}\|_{W^{1,2}(\Omega)}.$$

Hence (20.8.3) holds with $c = c_6/c_1$. □

The following theorem presents a regularity result for the pure displacement problem.

20.8.2 Theorem *Suppose that* $D = \partial\Omega$, *that* $\partial\Omega$ *is of class* C^∞ *and* $\bar{\boldsymbol{A}}$ *is of class* C^∞ *on* $\mathrm{cl}\,\Omega$. *Let* $p \ge 2$, $\bar{\boldsymbol{b}} \in L^p(\Omega)$, *and let* J *be uniformly positive. The unique weak solution* $\boldsymbol{u}$ *of the linearized problem corresponding to the data* $\bar{\boldsymbol{b}}$ *satisfies* $\boldsymbol{u} \in W^{2,p}(\Omega)$ *and one has*

$$\mathrm{Div}\, \bar{\boldsymbol{A}}[\nabla \boldsymbol{u}] + \rho_0 \bar{\boldsymbol{b}} = \boldsymbol{0} \qquad (20.8.6)$$

for almost every point $\boldsymbol{p} \in \Omega$.

The solution possesses more differentiability than a priori assumed in the definition of the weak solution. For functions $\boldsymbol{u}$ from $W^{2,p}(\Omega)$ the divergence in (20.8.6) is an integrable function. The proof of (20.8.2) is technical; it is omitted. I refer to NEČAS [1967] for the treatment of the regularity of solutions of elliptic systems; to CIARLET [1988, Thm. 6.3-6] for a sketch of the proof and to VALENT [1988] and HEALEY & SIMPSON [1996] for related results. The result 20.8.2 is needed in Sect. 20.9 in dealing with the solutions to the nonlinear equations by the implicit function theorem.

20.9 Existence Via the Implicit Function Theorem

In this section it will be shown how the solvability of the linear problem can be used to obtain the information about the solvability of the nonlinear boundary-value problem of elasticity. Only the pure displacement problem will be considered. (For the pure traction problem, see CHILLINGWORTH, MARSDEN & WAN [1982, 1983] and VALENT [1988]; for a mixed problem, however, with *S*, *D* that do not touch, see HEALEY & SIMPSON [1996].) The idea is to solve first the linearized equations in $W^{1,2}(\Omega)$. Next it is shown that the solutions are regular (see Proposition 20.8.2): also the second derivatives are integrable. Finally, the implicit function theorem applied to the infinite dimensional spaces of displacements is invoked to show that for loads sufficiently close to the loads of the known reference state, the solution exists, is unique, and depends continuously on these loads.

20.9.1 Local Inversion Theorem *Let* X, Y *be Banach spaces,* O *an open subset of* X, $f : \mathrm{O} \to \mathrm{Y}$ *a continuously Fréchet differentiable mapping and* $\boldsymbol{x}_0 \in \mathrm{O}$. *If the derivative* $\partial_x f(\boldsymbol{x}_0) : \mathrm{X} \to \mathrm{Y}$ *has a bounded inverse* $(\partial_x f(\boldsymbol{x}_0))^{-1} : \mathrm{Y} \to \mathrm{X}$, *then there exists a neighborhood* M *of* $\boldsymbol{x}_0$ *in* X *and a neighborhood* N *of* $y_0 : f(\boldsymbol{x}_0)$ *in* Y *such that for every* $\boldsymbol{y} \in \mathrm{N}$ *there exists exactly one solution* $\boldsymbol{x} \in \mathrm{M}$ *of the equation* $f(\boldsymbol{x}) = \boldsymbol{y}$. *The restriction of the function* f *on* X *is thus an invertible mapping and the inverse is continuously differentiable.*

See DIEUDONNÉ [1960, Sect. X.2] for proof.

Let $p > 3$ and denote by O^p the set of all mappings $\chi : \Omega \to \mathrm{Eucl}$ such that $\chi \in W^{2,p}(\Omega)$ and $\nabla\chi(\boldsymbol{p}) \in \mathrm{Lin}^+$ for every $\boldsymbol{p} \in \Omega$. If $\chi \in W^{2,p}(\Omega)$ with $p > 3$, then $\nabla\chi \in W^{1,p}(\Omega)$, and the imbedding Theorem 20.2.6 implies that $\nabla\chi$ is a continuous function on the closure $\mathrm{cl}\,\Omega$ of the reference domain.

20.9.2 Proposition *Let $\partial\Omega$ be of class C^∞. For every $p > 3$ the set O^p is an open subset of $W^{2,p}(\Omega)$.*

Proof We have to prove that for every $\chi_0 \in \mathrm{O}^p$ there exists a $\delta > 0$ such that if χ is in $W^{2,p}(\Omega)$ and $\|\chi - \chi_0\|_{W^{2,p}(\Omega)} < \delta$, then the deformation gradient $\nabla\chi$ has a positive determinant everywhere. The set Lin^+ is an open subset of Lin : if $\boldsymbol{F}_0$ has a positive determinant, then also all $\boldsymbol{F}$ close to $\boldsymbol{F}_0$ have positive determinant, because the determinant is a positive function. The function $\nabla\chi_0$ is continuous on $\mathrm{cl}\,\Omega$ a explained above, and therefore its range $\nabla\chi_0(\mathrm{cl}\,\Omega)$ is a compact set, which, by the definition of O^p is contained in Lin^+. Since Lin^+ is open, there exists an $\varepsilon > 0$ such that the ε-neighborhood N of $\nabla\chi_0(\mathrm{cl}\,\Omega)$ is still in Lin^+. Let c be the constant from the imbedding Theorem 20.2.6 for $W^{1,p}(\Omega)$. It is now claimed that with δ defined by $\delta = \varepsilon/c$ the desired assertion stated at the beginning of the proof holds. Indeed, let $\|\chi - \chi_0\|_{W^{2,p}(\Omega)} < \delta = \varepsilon/c$. Then also $\|\nabla\chi - \nabla\chi_0\|_{W^{1,p}(\Omega)} < \delta = \varepsilon/c$ and by the inequality from the imbedding theorem consequently $\|\chi - \chi_0\|_\infty \le c \cdot \|\nabla\chi - \nabla\chi_0\|_{W^{1,p}(\Omega)} \le \varepsilon$. But as the ε-neighborhood of the range of $\nabla\chi_0$ is still in Lin^+, the assertion follows. □

20.9.3 Theorem *Let $\partial\Omega$ be of class C^∞, $p > 3$ and assume that $\bar{S} : \mathrm{Lin}^+ \to \mathrm{Lin}$ is of class C^2. Then for every $\chi \in \mathrm{O}^p$ the function $\bar{S} \circ \nabla\chi$ is in $W^{1,p}(\Omega)$, and its generalized divergence $\mathrm{Div}(\bar{S} \circ \nabla\chi)$ is in $L^p(\Omega)$. The nonlinear mapping $\mathrm{A} : \mathrm{O}^p \to L^p(\Omega)$ defined by $\mathrm{A}(\chi) = \mathrm{Div}(\bar{S} \circ \nabla\chi)$ is Fréchet continuously differentiable and its differential $\mathrm{A}'(\chi, \cdot) : W^{2,p}(\Omega) \to L^2(\Omega)$ is given by $\mathrm{A}'(\chi, \boldsymbol{u}) = \mathrm{Div}(\bar{\boldsymbol{A}}(\nabla\chi)[\nabla\boldsymbol{u}])$ for every $\boldsymbol{u} \in W^{2,p}(\Omega)$.*

This is a technical result, and the reader is referred to Valent [1988] and Ciarlet [1988] for proof. Here we restrict ourselves to the explanation of some points.

(1) To prove that the composite mapping $\bar{S} \circ \nabla\chi$ is in $W^{1,p}(\Omega)$ one has to check first that $\bar{S}\circ\nabla\chi$ in $L^p(\Omega)$, i.e., that its p's power is integrable. This can be seen from the already mentioned fact that by the imbedding theorem $\nabla\chi$ is bounded and continuous on $\mathrm{cl}\,\Omega$. Therefore also the compound mapping is bounded and continuous. Next it is necessary to verify that the generalized derivative of $\bar{S} \circ \nabla\chi$ belongs to $L^p(\Omega)$. If χ is twice continuously differentiable on Ω, but not necessarily with bounded derivatives, then the generalized derivatives necessarily coincide with the classical derivatives calculated by the chain rule, hence in this case

$$\nabla(\bar{S} \circ \nabla\chi) = (\partial_F \bar{S} \circ \nabla\chi) \cdot \nabla^2\chi. \tag{20.9.1}$$

The position of the indices in this tensorial expression is determined by the chain rule, but it is unimportant at this moment. It is the structure of the expression for the derivative that needs to be examined. We have to check that the above function is in $L^p(\Omega)$. We again invoke the imbedding theorem to see that the factor $\partial_F \bar{S} \circ \nabla\chi$ is bounded and continuous, because $\nabla\chi$ is so. The second factor $\nabla^2\chi$ is in $L^p(\Omega)$ by the definition of the space $W^{1,p}(\Omega)$. Hence the left-hand side of (20.9.1), being a product of a bounded continuous function with a function from $L^p(\Omega)$ is again in $L^p(\Omega)$. This partially explains the usefulness of the assumption $p > 3$.

(2) Next one has to verify that the expression on the right-hand side of the equation giving the claimed value of the derivative belongs to L^p. This is done in much the same way as in (1), and again the condition $p > 3$ is used in an essential way.

(3) Finally it is necessary to verify that the expression claimed as the derivative of the nonlinear mapping really approximates the mapping itself in the way required in the definition of the derivative. This can be done by using the fact that the derivative is clearly a Gateaux derivative, is linear, and is continuous as an operator-valued function.

20.9.4 Theorem *Let $\partial\Omega$ be of class C^∞, $p > 3$ and assume that $\bar{S} : \mathrm{Lin}^+ \to \mathrm{Lin}$ is of class C^2. Let $\chi_0 \in \mathrm{O}^p$ be an equilibrium state corresponding to the body force b_0. If the second variation of the stored energy J is uniformly positive on $W_0^{1,2}(\Omega)$, then there exist constants $\delta_1 > 0, \delta_2 > 0$ such that for every $b \in L^p(\Omega)$ satisfying $\|b - b_0\|_{L^p(\Omega)} \leq \delta_1$ there exists a unique $\chi \in \mathrm{O}^p$ such that*

$$\mathrm{Div}\, \bar{S}(\nabla\chi) + \rho_0 b = 0 \quad \text{and} \quad \chi(p) = \chi_0(p) \quad \text{on } \partial\Omega.$$

The unique solution depends continuously on b in the sense that two body forces b_1, b_2 with $\|b_1 - b_2\|_{L^p(\Omega)}$ small lead to solutions χ_1, χ_2 with $\|\chi_1 - \chi_2\|_{W^{2,p}(\Omega)}$ small.

Bibliographical Note

Biot (in 1939; see his works cited in BIOT [1965]) was the first to see the equations of the linearized elasticity free from the traditional crude ellipticity assumptions (such as those listed in (20.6.25)), and analyzed instabilities by seeking nontrivial solutions to the linearized equations with trivial boundary data. His approach was subsequently extended by Hill and others. The systematic development of linearized elasticity is due to Green, Rivlin & Shiled (1952); the earliest book treatments are GREEN & ZERNA [1954, Chaps. IV & V], BIOT [1965], and TRUESDELL & NOLL [1965, Sects. 68–70].

An exposition of linear elasticity theory at an unstressed state is found in GURTIN [1972]. For a treatment with an emphasis on the existence of solutions, see NEČAS & HLAVÁČEK [1981]; for the existence and regularity of solutions, see FICHERA [1972].

For a relationship between the positivity of the second variation and the local uniqueness and continuous dependence, see GURTIN & SPECTOR [1979], SPECTOR [1980, 1982] and the references therein. The complementing condition is also called the Lopatinsky–Shapiro or the covering condition (see LOPATINSKY [1953], SHAPIRO [1951], AGMON, DOUGLIS & NIRENBERG [1964], and HÖRMANDER [1985, Chap. 20]). In these works, it figures as a necessary condition for the solutions of the elliptic systems to have integral representations in terms of boundary values (such as in terms of s in elasticity) via kernels. Such representations allow for detailed estimates of the solutions. In the elasticity literature, the complementing condition is first used in THOMPSON [1969], who also gave necessary and sufficient conditions for its validity for states of uniaxial extension. A further examination of the complementing condition is in SIMPSON & SPECTOR [1985, 1987, 1989]. Sects. 20.4 and 20.5 follow in part their work.

SIMPSON & SPECTOR [1987, Proposition 4.1 and Sect. 9] summarize the properties of the operator $L : u \to -\mathrm{Div}\,\bar{A}[\nabla u]$ viewed as an unbounded self-adjoint operator in L^2. The existing regularity theory requires that the boundary conditions be smooth functions of position, which excludes abrupt changes from Dirichlet's to Neumann's conditions. The results therefore apply only to the case when Ω is smooth and S, D do not touch, as in Proposition 20.5.3. For certain aspects of the linear elasticity on nonsmooth domains, see GRISVARD [1989; 1992, Chap. 4].

The optimal value of Korn's constant (20.6.28) for the half-space and the upper bound for c in the general case appear to be new. The standard proofs of Korn's inequality in the general case (such as the one in NEČAS & HLAVÁČEK [1981]) use the theorem on negative norms to derive first the coercivity of strains. Next a contradiction argument is invoked to pass from the coercivity of strains to Korn's inequality; this argument does not provide any information about the value of Korn's constant. In NITSCHE [1981] the extension operators are used to reduce the case of a bounded domain to the case of the whole space. This gives some control over the value of Korn's constant. Korn's inequality holds also when the requirement that a part of the boundary is held fixed is replaced by the requirement that the mean value of the infinitesimal rotation tensor W vanishes. In this case the estimates of the value of Korn's constant are available. This type of Korn's inequality is useful when dealing with the pure traction problem and so is not needed in the present monograph. The special case of Proposition 20.6.5 with $p = 0$ has been established by SIMPSON & SPECTOR [1987] (see also GRINFELD [1991, Subsect. 5.4.3]). In contrast to the well-known conditions on the first and third lines of Table 20.6.6, the inequality $\lambda + \mu > 0$ seems to be relatively unknown. The surface and shear instabilities in the case of two-dimensional bodies admit a much fuller treatment, and I refer to NOWINSKI [1969], REDDY [1982, 1983] and OGDEN [1984, 1992b] for the treatment and further references. For the bifurcation analysis of compressible balls under different boundary conditions, including the Dirichlet ones (internal instabilities), see WESOŁOWSKI [1964].

There is a large literature on the column problem with different materials, in both the compressible and incompressible cases. I refer to WESOŁOWSKI [1964], BURGESS & LEVINSON [1972], SIMPSON & SPECTOR [1984a,b], and HILL & HUTCHINSON [1975] for further references. The theory of elastic stability (see, e.g., THOMPSON & HUNT [1973, 1984]) has developed perturbation methods to determine also the post-buckling behavior. Unfortunately, the analytical difficulties can be overcome only in discrete models (governed by a finite number of scalar parameters) or in models that are essentially one- or two-dimensional, such as plates and rods, or in the three-dimensional models with the perturbations restricted so as to effectively lower the dimension (e.g., by considering radial perturbations). THOMPSON & HUNT [1973] contains also hints on the correlation of the theory with experiment. For a comprehensive account of a rigorous theory of bifurcations within lower-dimensional nonlinear theories of elasticity, see ANTMAN [1995].

The existence theorems in nonlinear elasticity based on the implicit function theorem were originated by Stoppelli (1954–1957). The reader is referred to VALENT [1988] and CIARLET [1988] for more detailed expositions and further references. That the positivity of the second variation does not have any immediate implications for the stability of the state was first realized by KOITER [1976]. A result on the failure of differentiability related to Proposition 20.3.7 is given in VALENT [1988]. His result says, in the present terminology, that if the stress field is a differentiable function from $W_0^{1,p}$ to L^q, then the stress response is linear. This result does not catch the term with determinant, as he assumes a differentiability of stress rather than of the divergence of stress, which is what is done in the variational calculus setting in Sect. 20.3. See also MARTINI [1979].

21 Direct Methods in Equilibrium Theory

This chapter treats the equilibrium states by the direct methods of the calculus of variations. Given a class of admissible rest states Σ_0 satisfying various constraints (such as the boundary conditions), one seeks a state σ_0 for which the total canonical free energy P takes the least possible value on Σ_0. This state need not exist, depending on the free energy $\bar{f}$ and on the class Σ_0. However, if the total energy is bounded from below on Σ_0, then $\mathsf{P}_0 := \inf\{\mathsf{P}(\sigma) : \sigma \in \Sigma_0\}$ is finite and one can always find a sequence $\sigma_k \in \Sigma_0$ such that $\mathsf{P}(\sigma_k) \to \mathsf{P}_0$ as $k \to \infty$, the minimizing sequence. Under the condition of coercivity, one can find a subsequence, still denoted by σ_k, which converges weakly to some state σ_0 and we assume that $\sigma_0 \in \Sigma_0$. (To guarantee this inclusion, one has to admit states with a lower degree of smoothness than, say, continuous differentiability, see Sect. 21.2.) Since σ_k converges weakly to σ_0 and $\mathsf{P}(\sigma_0)$ approaches P_0, a natural question is whether $\mathsf{P}_0 = \mathsf{P}(\sigma_0)$. This will be the case if P is sequentially weakly lower semicontinuous (swlsc), i.e., if for every sequence σ_k in Σ_0 converging weakly to some $\sigma \in \Sigma_0$, we have

$$\liminf_{k\to\infty} \mathsf{P}(\sigma_k) \geq \mathsf{P}(\sigma).$$

Then the state σ_0 will be a global minimizer of P on Σ_0, i.e., it will satisfy the principle of minimum canonical free energy 15.2.1. The swlsc is essentially equivalent to the quasiconvexity and/or polyconvexity of $\bar{f}$ ("elliptic free energies", see Subsect. 21.5.6). The experience indicates that the global polyconvexity of $\bar{f}$ is an appropriate assumption for rubber-like materials. On the other hand, Proposition 17.4.10 says that $\bar{f}$ cannot be globally quasiconvex if the body admits a stable equilibrium state with two or more coexistent phases, and hence P cannot be swlsc. For instance, the wells structure of $\bar{f}$ at the temperature of the martensitic transformation is incompatible with the quasiconvexity (see Subsects. 2.3.4 and 2.3.5). Also, $\bar{f}$ cannot be quasiconvex if the symmetry group of the material is as large as the nonorthogonal symmetry group of a Bravais lattice, see Sect. 17.8. In these cases, it may happen that $\mathsf{P}(\sigma_0) > \mathsf{P}_0$ and one cannot expect the existence of a global minimizer of P on Σ_0. One is thus led to examining minimizing sequences σ_k rather than minimizing states.

It often happens that the states σ_k exhibit a microstructure which consists in intensive small-scale spatial oscillations of the deformation gradient. For example, in striving to achieve the minimum energy, a crystal may exhibit a fine mixture of twinned phases, and to reach the lower bound P_0, the mixture must get 'infinitely fine,' cf. Subsect. 2.3.8. These fine oscillations are not reflected in the limiting macroscopic state σ_0, which describes only the mean deformation. In particular, since the microstructure is not taken into account in evaluating $\mathsf{P}(\sigma_0)$, we have

$\mathsf{P}(\sigma_0) > \mathsf{P}_0\,(\approx \mathsf{P}(\sigma_k))$. A convenient device to describe the asymptotic distribution of oscillations is the Young measure associated with the sequence σ_k, 'the Young measure limit of σ_k' as $k \to \infty$. Young measures represent an enlargement of the state space by 'states with microstructure,' and in this sense one speaks about Young measure minimizers of energy.

In the absence of a unified theory, the chapter treats the elliptic and nonelliptic cases separately. The first three sections introduce tools common to both. The theory of equilibrium of elliptic (rubber-like) materials is then presented in Sects. 21.4 and 21.5. The nonelliptic case is not so complete at present. We shall restrict ourselves to the calculation of the minimum energy of crystals and to the basic properties of the Young measure minimizers in Sect. 21.6. Recall that Sect. 2.3 treated another aspect of the minimization problem for nonelliptic energy functions, the structure of ground states in the presence of several wells of $\bar{f}$.

21.1 Weak Convergence and Young Measures

This section describes the main ingredients of the direct methods of the calculus of variations, the weak convergence, the extraction of a weakly convergent subsequence from a bounded sequence, the Young measure, and the lower semicontinuity of integral functionals. Throughout the section we consider the space $L^p(\Omega) \equiv L^p(\Omega, \mathrm{W})$ of functions f from an open set $\Omega \subset$ Eucl with values in W, where W is a finite-dimensional vector space with norm $|\cdot|$, either Vect or Lin. Similar notation is used for Sobolev spaces.

21.1.1 Definitions Let $1 \le p \le \infty$ and $f, f_k \in L^p(\Omega)$, $(k = 1, 2, \ldots)$.

(1) f_k is said to ***converge strongly*** to f in $L^p(\Omega)$ if $\|f_k - f\|_{L^p(\Omega)} \to 0$ as $k \to \infty$; this fact is denoted

$$f_k \to f \text{ in } L^p(\Omega) \text{ as } k \to \infty.$$

(2) If $1 \le p < \infty$, f_k is said to ***converge weakly*** to f in $L^p(\Omega)$ if $\int_\Omega f_k \phi \, dV \to \int_\Omega f \phi \, dV$ as $k \to \infty$ for every $\phi \in L^{p'}(\Omega)$ (here $p' = p/(p-1)$ is the conjugate exponent); this fact is denoted

$$f_k \rightharpoonup f \text{ in } L^p(\Omega) \text{ as } k \to \infty.$$

(3) If $p = \infty$, f_k is said to ***converge weak**** to f in $L^\infty(\Omega)$ if $\int_\Omega f_k \phi \, dV \to \int_\Omega f \phi \, dV$ as $k \to \infty$ for every $\phi \in L^1(\Omega)$; this fact is denoted

$$f_k \overset{*}{\rightharpoonup} f \text{ in } L^\infty(\Omega) \text{ as } k \to \infty.$$

The strong and weak convergences in $W^{1,p}$ are defined by requiring that both f_k and ∇f_k converge strongly or weakly in $L^p(\Omega)$, respectively. All the properties of the strong or weak convergence in $L^p(\Omega)$ have their counterparts in the properties of the strong or weak convergence in $W^{1,p}(\Omega)$.

21.1.2 Examples (Weak Convergence and Periodic Structures)

(1) Let $P \subset$ Vect be a parallelepiped (see Sect. 17.3), let $1 \le p \le \infty$, and let f be a P-periodic function such that $f \in L^p(D)$ for every bounded open subset of Vect. Define $f_k(\boldsymbol{p}) = f(k\boldsymbol{p})$ for $\boldsymbol{p} \in$ Vect and $k = 1, 2 \ldots$ If Ω is any open subset of Vect, then the restrictions of f_k, f to Ω satisfy

$$f_k \rightharpoonup \bar{f} \text{ in } L^p(\Omega) \;\; (\overset{*}{\rightharpoonup} \text{ if } p = \infty) \quad \text{as } k \to \infty,$$

where $\bar{f}$ is a constant function of value

$$\bar{f} = |P|^{-1} \int_P f(p)\, dV(p),$$

the mean value of f over the period P. The reader is referred to BALL & MURAT [1984, Appendix] for proof. Here we only note that in the special case when P is a unit cube and the P-periodic function f is a trigonometric polynomial, i.e., a linear combination of the exponentials $\exp(2\pi i \boldsymbol{k} \cdot \boldsymbol{p})$ where $\boldsymbol{k} \in$ Vect has integral components relative to a basis of vectors along the sides of P, then the assertion follows from the Riemann–Lebesgue localization lemma.

(2) Consider a periodic pairwise homogeneous deformation (see Sect. 2.3). Let α satisfy $0 < \alpha < 1$ and let $f : \mathbb{R} \to \mathbb{R}$ be a 1-periodic continuous and piecewise continuously differentiable function defined by (2.3.9) (with the period set equal to 1). Let $\boldsymbol{F} \in \mathrm{Lin}^+$, $\boldsymbol{a}, \boldsymbol{n} \in$ Vect, with $\boldsymbol{n}$ a unit vector. Define mappings $\bar{\chi}, \chi_k, \chi$ from Vect into itself by

$$\bar{\chi}(\boldsymbol{p}) = \boldsymbol{Fp} + \boldsymbol{a} f(\boldsymbol{p} \cdot \boldsymbol{n}), \quad \chi_k(\boldsymbol{p}) = k^{-1}\bar{\chi}(k\boldsymbol{p}) = \boldsymbol{Fp} + \boldsymbol{a}k^{-1} f(k\boldsymbol{p} \cdot \boldsymbol{n}), \quad \chi(\boldsymbol{p}) = \boldsymbol{Fp},$$

for every $\boldsymbol{p} \in$ Vect, $k = 1, \dots$ If Ω is any open subset of Vect and $1 \le p \le \infty$, then the restrictions of χ_k, χ to Ω satisfy

$$\chi_k \rightharpoonup \chi \ \text{ in } W^{1,p}(\Omega) \ (\rightharpoonup^* \text{ if } p = \infty) \quad \text{as } k \to \infty.$$

For k large, χ_k represents a finely layered deformation consisting of periodically repeating pairs of plates orthogonal to $\boldsymbol{n}$, of width α/k and $(1-\alpha)/k$. On each of these plates, the deformation is homogeneous with deformation gradients $\boldsymbol{F} + (1-\alpha)\boldsymbol{a} \otimes \boldsymbol{n}$ and $\boldsymbol{F} - \alpha \boldsymbol{a} \otimes \boldsymbol{n}$, respectively. The weak limit is a (macroscopically) homogeneous deformation with deformation gradient $\boldsymbol{F}$, which can be viewed as the mean deformation gradient over one period of the structure. More generally, we have

(3) Let $P \subset$ Vect be a parallelepiped, let $1 \le p \le \infty$, let $\bar{\chi}$: Vect $\to$ Vect be such that $\bar{\chi} \in W^{1,p}(D)$ for every bounded open subset D of Vect and assume that $\nabla\bar{\chi}$ is P-periodic. Define

$$\chi_k(\boldsymbol{p}) = k^{-1}\bar{\chi}(k\boldsymbol{p}), \quad \chi(\boldsymbol{p}) = \boldsymbol{Fp} \ \text{ for } \boldsymbol{p} \in \text{Vect}, \ k = 1, 2, \dots, \ \text{ where } \boldsymbol{F} = |P|^{-1} \int_P \nabla\bar{\chi}(\boldsymbol{p})\, dV.$$

If Ω is a bounded open subset of Vect then

$$\chi_k \rightharpoonup \chi \ \text{ in } W^{1,p}(\Omega) \ (\rightharpoonup^* \text{ if } p = \infty) \quad \text{as } k \to \infty.$$

For k large, χ_k is a fine periodic structure with the deformation gradient $\boldsymbol{F}_k(\boldsymbol{p}) = \bar{\boldsymbol{F}}(k\boldsymbol{p})$ where $\bar{\boldsymbol{F}} = \nabla\bar{\chi}$. The deformation gradient $\boldsymbol{F}_k$ is thus periodic of period P/k, and it repeats k^n times the scaled "motif" $\bar{\boldsymbol{F}}$. Macroscopically, χ_k converges to a homogeneous deformation with the deformation gradient equal to the mean deformation over one period.

(4) In the situation of Subsect. 2.3.8, the approximate ground state χ_ε describing the austenite/martensite interface satisfies

$$\chi_\varepsilon \rightharpoonup \bar{\chi}_0 \ \text{ in } W^{1,p}(\Omega), \ \text{ where } \bar{\chi}_0(\boldsymbol{p}) = \begin{cases} \boldsymbol{Gp} & \text{if } \boldsymbol{p} \subset \Omega_0, \\ \bar{\boldsymbol{F}}\boldsymbol{p} & \text{if } \boldsymbol{p} \in \Omega \setminus \Omega_0, \end{cases} \quad 1 \le p \le \infty;$$

here $\bar{\boldsymbol{F}}$ is the mean deformation gradient of the periodic deformation χ_0. Thus, macroscopically, the state χ_ε looks like a pairwise homogeneous state $\bar{\chi}_0$, and the austenite/martensite interface becomes sharp. However, the total stored energy of the body in χ_ε does not converge to that in the state $\bar{\chi}_0$.

The weak convergence is not preserved by nonlinear mappings. That is, if $f_k \rightharpoonup f$ in $L^p(\Omega, \mathrm{W})$ as $k \to \infty$ and if $G \in C(\mathrm{W})$ is a continuous function on W, it does not follow that the composition, $G \circ f_k$, converges weakly to $G \circ f$ as $k \to \infty$. Here $G \circ f(\boldsymbol{p}) = G(f(\boldsymbol{p}))$ for every $\boldsymbol{p} \in \Omega$ and $C(\mathrm{W})$ is the space of all continuous functions on W. The reason of this failure is that the values of f_k at a given point $\boldsymbol{p} \in \Omega$ may oscillate rapidly as $k \to \infty$, and while these oscillations may cancel in f_k to yield the weak limit f, this need not be so for, say, f_k^2. A convenient device to take into account the oscillations is the Young measure to be now introduced. A probability measure ν on W is a nonnegative Borel measure on W such that $\nu(\mathrm{W}) = 1$. Recall from Sect. 7.1 the notion of the support of a measure and assume that ν is a probability measure with compact support. If G is a continuous function on W, one defines the mean value $\bar{G}$ of G with respect to ν by

$$\bar{G} = \int_{W} G(A)\, d\nu(A),$$

where we have denoted the integration variable by $A \in W$. The following proposition associates with every weakly convergent sequence f_k and every $p \in \Omega$ a probability measure ν_p on W which describes the asymptotic distribution of the values of $f_k(p)$ at the point p as $k \to \infty$.

21.1.3 Proposition *Let $f_k, f \in L^\infty(\Omega, W)$, $k = 1 \ldots$, be such that $f_k \rightharpoonup^* f$ in $L^\infty(\Omega)$ as $k \to \infty$. Then there exists a subsequence of f_k, denoted again by f_k, and a family of probability measures $\{\nu_p : p \in \Omega\}$ on W, with compact support contained in some fixed compact set $K \subset W$ that is independent of p, such that for every $G \in C(W)$,*

$$G \circ f_k \rightharpoonup^* \bar{G} \text{ in } L^\infty(\Omega) \text{ as } k \to \infty,$$

where

$$\bar{G}(p) = \int_{W} G(A)\, d\nu_p(A), \quad p \in \Omega. \tag{21.1.1}$$

The reader is referred to BALL [1989] for proof, references, and refinements. It is a part of the assertion that $\bar{G}$ is a bounded measurable function for every $G \in C(W)$. The family $\{\nu_p : p \in \Omega\}$ is called the ***Young measure*** associated with the sequence f_k; alternatively, $\{\nu_p : p \in \Omega\}$ may be called the ***Young measure limit*** of f_k. In particular the weak limit f can be reconstructed from the Young measure by

$$f(p) = \int_{W} A\, d\nu_p(A), \quad p \in \Omega,$$

which is a special case of (21.1.1) with $G(A) = A$ identically. Note that if $f_k(p) \in K \subset W$ for all k and a.e. $p \in \Omega$, where K is a fixed compact subset, then $\operatorname{supp} \nu_p \subset K$ for a.e. $p \in \Omega$; however, it does not follow that the weak limit satisfies $f(p) \in K$ for a.e. $p \in \Omega$. We shall use the Young measures associated with deformation gradients $F_k = \nabla \chi_k$ of a weakly convergent sequence of deformations χ_k; in this case W = Lin, for every $p \in \Omega$, ν_p is a probability measure on Lin, and the integration variable in (21.1.1) is the second-order tensor $A \in$ Lin. We call $\{\nu_p : p \in \Omega\}$ the ***Young measure limit of deformation gradients*** of χ_k. The gradient F of the weak limit χ of χ_k can be reconstructed from

$$\nabla \chi(p) = F(p) = \int_{\text{Lin}} A\, d\nu_p(A), \quad p \in \Omega, \tag{21.1.2}$$

by integration. The Young measure $\{\nu_p : p \in \Omega\}$ is said to be ***homogeneous*** if ν_p is independent of $p \in \Omega$.

21.1.4 Examples

(1) Let $f_k, f, \bar{f}$ be W-valued functions as in Example 21.1.2(1) with $p = \infty$, and let Ω be a bounded open subset of Vect. By the cited example, $f_k \rightharpoonup^* \bar{f}$ in $L^\infty(\Omega)$. Let us show that the Young measure limit of f_k is the homogeneous measure $\nu_p = \bar{\nu}$,

$$\bar{\nu}(M) = |P|^{-1} V(\{p \in P : f(p) \in M\}), \tag{21.1.3}$$

for every Borel subset M of W, where V is the Lebesgue measure in Vect. The measure ν_p is thus the distribution measure of the function f on P. The asserted form of ν_p can be deduced form the assertion in Example 21.1.2(1): We have to prove that for every $G \in C(W)$ the sequence $G \circ f_k$ converges weak* to $\bar{G}$ where

$$\bar{G}(\boldsymbol{p}) = \int_{\mathrm{W}} G(\boldsymbol{F})\, d\nu_{\boldsymbol{p}}(\boldsymbol{F}) = \int_{\mathrm{W}} G(\boldsymbol{F})\, d\bar{\nu}(\boldsymbol{F}) = |P|^{-1} \int_{P} G(f(\boldsymbol{p}))\, dV(\boldsymbol{p})$$

(the last equality is proved in the same way as in the proof of Lemma 7.1.9). But applying Example 21.1.2(1) to the sequence of P-periodic functions $F_k = G \circ f_k$, $F = G \circ f$ we see that $F_k \rightharpoonup^* \bar{F}$, where

$$\bar{F} = |P|^{-1} \int_{P} G(f(\boldsymbol{p}))\, dV(\boldsymbol{p})$$

and the assertion follows.

(2) For χ_k a sequence of scaled periodic two-phase deformations as in 21.1.2(2), one easily finds, applying item (1), that the Young measure limit of deformation gradients of χ_k is the homogeneous measure

$$\nu_{\boldsymbol{p}} = (1-\alpha)\delta_{\boldsymbol{F}-\alpha\boldsymbol{a}\otimes\boldsymbol{n}} + \alpha\delta_{\boldsymbol{F}+(1-\alpha)\boldsymbol{a}\otimes\boldsymbol{n}} \quad \text{for every } \boldsymbol{p} \in \Omega,$$

a combination of Dirac's measures on Lin at $\boldsymbol{F} - \alpha\boldsymbol{a}\otimes\boldsymbol{n}$ and $\boldsymbol{F} + (1-\alpha)\boldsymbol{a}\otimes\boldsymbol{n}$. More generally,

(3) If χ_k is the sequence of scaled P-periodic deformations with the "motif" $\bar{\chi}$ as in Example 21.1.2(3), then the Young measure limit of deformation gradients of χ_k is the homogeneous measure $\nu_{\boldsymbol{p}} = \bar{\nu}$, where $\bar{\nu}$ is given by (21.1.3) with f identified with $\bar{\boldsymbol{F}}$, the deformation gradient of $\bar{\chi}$. Thus the "motif" $\bar{\boldsymbol{F}}$ is encoded in the distribution measure $\bar{\nu}$.

(4) In the situation of Subsect. 2.3.8, the Young measure limit of deformation gradients of the sequence of the approximate ground states χ_ε is

$$\nu_p = \begin{cases} \delta_G & \text{if } \boldsymbol{p} \in \Omega_0, \\ (1-\alpha)\delta_{\boldsymbol{F}^+} + \alpha\delta_{\boldsymbol{F}^-} & \text{if } \boldsymbol{p} \in \Omega \setminus \Omega_0. \end{cases}$$

The Young measure limit is inhomogeneous in this case.

(5) Let $f_k, f \in L^\infty(\Omega)$ be a sequence such that $f_k(\boldsymbol{p}) \to f(\boldsymbol{p})$ for a.e. $\boldsymbol{p} \in \Omega$ as $k \to \infty$. This implies that $f_k \rightharpoonup^* f$ in $L^\infty(\Omega)$, and the Young measure associated with this sequence is

$$\nu_{\boldsymbol{p}} = \delta_{f(\boldsymbol{p})},$$

the Dirac measure in W concentrated at the value $f(\boldsymbol{p})$. Thus in the case of strong convergence, the phenomenon of distribution of values of f_k is absent.

The rest of this section is devoted to proving a simple lower semicontinuity theorem that will be useful when treating the polyconvex free energy functions of rubber-like materials.

21.1.5 Proposition *Let $f, f_k \in L^p(\Omega)$, $(k = 1, 2, \ldots)$ and $f_k \to f$ in $L^p(\Omega)$. Then there exists a subsequence f_{k_m} such that $f_{k_m}(\boldsymbol{p}) \to f(\boldsymbol{p})$ for almost every $\boldsymbol{p} \in \Omega$.*

21.1.6 Theorem (Fatou's lemma) *Let $\phi_k \ge 0$ be a sequence of measurable functions on Ω. Then*

$$\int_{\Omega} \liminf_{k\to\infty} \phi_k\, dV \le \liminf_{k\to\infty} \int_{\Omega} \phi_k\, dV. \tag{21.1.4}$$

The importance of the weak convergence is due to the following property.

21.1.7 Theorem *Let $f_k \in L^p(\Omega)$, $p > 1$. Suppose that there exists a $C < \infty$ such that*

$$\|f_k\|_{L^p(\Omega)} \le C, \quad k = 1, 2, \ldots$$

Then there exists a subsequence f_{k_m} of f_k, $k_1 < k_2 < \ldots < k_m < \ldots$ and $f \in L^p(\Omega)$ such that

$$f_{k_m} \rightharpoonup f \ \text{ in } L^p(\Omega) \ \text{ as } m \to \infty.$$

Every sequence bounded in $L^p(\Omega)$ contains a subsequence which converges weakly to some element of $L^p(\Omega)$. The following result deals with the relationship of the weak convergence to the strong convergence.

21.1.8 Theorem (Mazur's lemma) *Let $f, f_k \in L^p(\Omega)$, $(k = 1, 2, \ldots)$. If $f_k \rightharpoonup f$ in $L^p(\Omega)$, then there exists a sequence of convex combinations a_k of f_k,*

$$a_k = \sum_{m\geq k}^{K_k} \lambda_{k,m} f_k, \quad K_k > 0, \quad \sum_{m\geq k}^{K_k} \lambda_{k,m} = 1, \quad \lambda_{k,m} \geq 0, \tag{21.1.5}$$

such that $a_k \to f$ in $L^p(\Omega)$.

By passing to an appropriate sequence of convex combinations, one can pass from weak convergence to strong convergence.

21.1.9 Proposition *Let $g : D \to \mathbb{R}$ be a continuous convex function on a convex domain $D \subset W$, bounded below, $p > 1$ and $f, f_k \in L^p(\Omega)$, $k = 1, 2, \ldots$ a sequence of functions such that $f(\boldsymbol{p}), f_k(\boldsymbol{p}) \in D$ for all k and a.e. $\boldsymbol{p} \in \Omega$. If $f_k \rightharpoonup f$ in $L^p(\Omega)$, then*

$$\int_\Omega g(f(\boldsymbol{p}))\, dV \leq \liminf_{k\to\infty} \int_\Omega g(f_k(\boldsymbol{p}))\, dV. \tag{21.1.6}$$

This is lower semicontinuity of the integrals of convex integrands. The above is the simplest version of the results of this kind. For more general versions, the reader is referred to treatises on the calculus of variations. The result also holds when f is a multicomponent function with the components in the spaces $L^p(\Omega)$ with different p for different components. We shall employ the result in this generalized version.

Proof Suppose that (21.1.6) does not hold and denote the *lim inf* on the right-hand side of (21.1.6) by l so that

$$l < \int_\Omega g(f(\boldsymbol{p}))\, dV. \tag{21.1.7}$$

By the definition of *lim inf* there exists a subsequence of f_k, again denoted by f_k, such that $\lim_{k\to\infty} \int_\Omega g(f_k(\boldsymbol{p}))\, dV = l$. Let $\varepsilon > 0$ be arbitrary. Then for all sufficiently large k

$$\int_\Omega g(f_k(\boldsymbol{p}))\, dV < l + \varepsilon. \tag{21.1.8}$$

Since f_k converges weakly to f, Mazur's lemma 21.1.8 tells us that there exists a sequence of convex combinations a_k as in $(21.1.5)_1$ which converges strongly to f_k. In view of Proposition 21.1.5 then a subsequence of a_k, again denoted a_k, converges to f a.e. on Ω. Then also $g(f_k(\cdot))$ converges to $g(f_k(\cdot))$ a.e. on Ω. Fatou's lemma 21.1.6 tells us that

$$\int_\Omega g(f(\boldsymbol{p}))\, dV \leq \liminf_{k\to\infty} \int_\Omega g(a_k(\boldsymbol{p}))\, dV. \tag{21.1.9}$$

Since g is convex, we have, in view of $(21.1.5)_1$,

$$g(a_k(\boldsymbol{p})) \leq \sum_{m\geq k}^{K_k} \lambda_{k,m} g(f_k(\boldsymbol{p}))$$

and hence in view of (21.1.8)

$$\int_\Omega g(a_k(\boldsymbol{p}))\, dV \leq \sum_{m\geq k}^{K_k} \lambda_{k,m} \int_\Omega g(f_k(\boldsymbol{p}))\, dV \leq l + \varepsilon \tag{21.1.10}$$

where we have also used $(21.1.5)_3$. From (21.1.10), $\liminf_{k\to\infty} \int_\Omega g(a_k(\boldsymbol{p}))\, dV \leq l+\varepsilon$. As $\varepsilon > 0$ is arbitrary, we have $\liminf_{k\to\infty} \int_\Omega g(a_k(\boldsymbol{p}))\, dV \leq l$ which in conjunction with (21.1.9) gives $\int_\Omega g(f(\boldsymbol{p}))\, dV \leq l$ and this inequality contradicts (21.1.7). □

21.2 Deformations from Sobolev Spaces

The existence theorems to be described in Sects. 21.3–21.5 are based on the classes of deformations

$$\mathcal{A}_{pq}(\Omega) = \{\chi \in W^{1,p}(\Omega) : \operatorname{cof} \nabla\chi \in L^q(\Omega)\},$$
$$\mathcal{A}_{pq}^+(\Omega) = \{\chi \in \mathcal{A}_{pq}(\Omega) : \det \nabla\chi > 0 \text{ a.e. in } \Omega\}$$

where, as an integral part of the definition, p, q satisfy

$$p \geq n-1 \ (= 2 \text{ if } n = 3), \quad \text{and} \quad q \geq p/(p-1), \tag{21.2.1}$$

as motivated by Proposition 21.3.1. Observe that by Hölder's inequality, the condition $p \geq n-1$ ensures that all minors of a $\chi \in \mathcal{A}_{pq}$ of order $\leq n-1$ are integrable on Ω; $q \geq p/(p-1)$ then ensures that the unique minor of order n, $\det \nabla\chi$, is also integrable on Ω. If $p \geq n$, then $\mathcal{A}_{pq}(\Omega) = W^{1,p}(\Omega)$ for each q such that $p/(p-1) \leq q \leq p/(n-1)$.

The purpose of this section is to describe informally the properties of deformations from $W^{1,p}(\Omega)$ and from the above classes. Recall that the elements of $W^{1,p}(\Omega)$ (and hence also of $\mathcal{A}_{pq}(\Omega)$ and $\mathcal{A}_{pq}^+(\Omega)$) are only equivalence classes of mappings differing on sets of null Lebesgue measure, and one can destroy, e.g., injectivity by changing χ on a set of measure 0. Moreover, these spaces may contain nonsmooth, even discontinuous elements; we shall see that these are not just mathematical artifacts. A mapping $\chi \in W^{1,p}(\Omega)$, $1 \leq p \leq \infty$, is said to be ***orientation-preserving*** if $\det \nabla\chi(\boldsymbol{p}) > 0$ for a.e. $\boldsymbol{p} \in \Omega$. Let us first consider a special class of deformations.

21.2.1 Radial Deformations Let $\Omega = B(\mathbf{0}, R)$ be a ball of radius $R > 0$ centered at $\mathbf{0}$. A function $\chi : B(\mathbf{0}, R) \to \text{Vect}$ is said to be a ***radial deformation*** if

$$\chi(\boldsymbol{p}) = \rho(r)\boldsymbol{p}/r, \quad r := |\boldsymbol{p}|, \quad \boldsymbol{p} \in B(\mathbf{0}, R), \quad \boldsymbol{p} \neq \mathbf{0},$$

where $\rho : (0, R) \to \mathbb{R}$ is a function. Formally, assuming that ρ is differentiable and denoting $\boldsymbol{F} := \nabla\chi$, one finds that

$$\boldsymbol{F} = \frac{\rho}{r}\mathbf{1} + \Big(\dot\rho - \frac{\rho}{r}\Big)\frac{\boldsymbol{p}}{r} \otimes \frac{\boldsymbol{p}}{r}, \quad \operatorname{cof} \boldsymbol{F} = \frac{\rho\dot\rho}{r}\mathbf{1} - \Big(\frac{\rho\dot\rho}{r} - \frac{\rho^2}{r^2}\Big)\frac{\boldsymbol{p}}{r} \otimes \frac{\boldsymbol{p}}{r}, \quad \det \boldsymbol{F} = \rho^2\dot\rho/r^2$$

for every $\boldsymbol{p} \in B(\mathbf{0}, R)$. Proceeding rigorously, it can be shown that a radial deformation χ is in $W^{1,p}(B(\mathbf{0}, R))$ $(1 \leq p < \infty)$ if and only if ρ is absolutely continuous on every closed subinterval of $(0, R)$ and

$$\int_0^R (|\dot\rho(r)|^p + |\rho(r)/r|^p) r^2 \, dr < \infty. \tag{21.2.2}$$

(See Ball [1977, 1982 p. 566].) In this case the distributional gradient $\boldsymbol{F} := \nabla\chi$ is given by the formula above for a.e. $\boldsymbol{p} \in B(\mathbf{0}, R)$; if $1 \leq q < \infty$, then $\operatorname{cof} \boldsymbol{F} \in L^q(B(\mathbf{0}, R))$ if and only if

$$\int_0^R (|\dot\rho(r)|^{2q} + |\rho(r)/r|^{2q}) r^2 \, dr < \infty. \tag{21.2.3}$$

Let us consider the orientation-preserving radial deformations with $\rho(r) \geq 0$ for every $r \in (0, R)$. The condition $\det \boldsymbol{F} > 0$ a.e. on $B(\mathbf{0}, R)$ is equivalent to $\dot\rho(r) > 0$ for a.e. $r \in (0, R)$. The function χ is injective on $B(\mathbf{0}, R) \setminus \{\mathbf{0}\}$ if and only if ρ is strictly monotone in the sense that $\rho(r) < \rho(s)$ for every $r, s \in (0, R)$ such that $r < s$. Under the monotonicity assumption, ρ can be extended to a continuous function on $[0, R]$, and we denote its values at $0, R$ by $\rho(0) \geq 0$, $\rho(R) > 0$, respectively. If $\rho(0) > 0$, then χ is a one-to-one mapping from $B(\mathbf{0}, R)\backslash\{\mathbf{0}\}$ onto $\Omega' = \{\boldsymbol{x} \in \text{Vect} : \rho(0) < |\boldsymbol{x}| < \rho(R)\}$, while if $\rho(0) = 0$, then χ is a one-to-one mapping from $B(\mathbf{0}, R)\backslash\{\mathbf{0}\}$ onto $B(\mathbf{0}, \rho(R))\backslash\{\mathbf{0}\}$, and this mapping can be extended to a continuous one-to-one mapping from $B(\mathbf{0}, R)$ onto $B(\mathbf{0}, \rho(R))$ by setting $\chi(\mathbf{0}) = \mathbf{0}$. In the case $\rho(0) > 0$, the mapping χ is discontinuous at $\boldsymbol{p} = \mathbf{0}$ as 'it expands the point $\boldsymbol{p} = \mathbf{0}$ to a sphere centered at $\mathbf{0}$ of radius $\rho(0)$.' That is, χ creates a hole of radius $\rho(0)$ in the deformed configuration. Let us refer to the radial deformations with ρ nonnegative, monotone and with $\rho(0) > 0$ to as the cavitating radial deformations. Let us now show that there are cavitating radial deformations $\chi \in W^{1,p}(B(\mathbf{0}, R))$ if and only if $1 \leq p < 3$, and that there are no cavitating radial deformations in $A^+_{pq}(B(\mathbf{0}, R))$. To see that for $p \geq 3$ there are no cavitating radial deformations in $W^{1,p}(B(\mathbf{0}, R))$, it suffices to note that (21.2.2) implies that for $p \geq 3$ necessarily $\rho(0) = 0$, for otherwise the integral would diverge. On the other hand, if $1 \leq p < 3$, then $\rho = (Dr^3 + E)^{1/3}$ where $D \geq 0, E > 0$, leads to a cavitating radial deformation. To see that there are no cavitating radial deformations in the classes $A^+_{pq}(B(\mathbf{0}, R))$, it suffices to note that if $\rho(0) > 0$, then $p < 3$ from the above and similarly $q < 3/2$ from (21.2.3). Combining with $q \geq p/(p-1)$ one obtains a contradiction. Let us now return to $W^{1,p}(\Omega)$, $A_{pq}(\Omega)$ and $A^+_{pq}(\Omega)$ with a general Ω.

21.2.2 Pointwise Values The imbedding theorem 20.2.6 assures that for $p > 3$, every $\chi \in W^{1,p}(\Omega)$ has a representative that is continuous on $\operatorname{cl}\Omega$. For $p \leq 3$ there are χ which cannot be represented by a continuous function. We checked that on radial deformations $\chi \in W^{1,p}(B(\mathbf{0}, R))$: for $p \geq 3$ they can be extended continuously by $\chi(\mathbf{0}) = \mathbf{0}$, while the cavitating radial deformations for $1 \leq p < 3$ cannot be changed to continuous ones by modifying χ on a set of null Lebesgue measure. Thus, if $p > 3$, there is no doubt about the pointwise values of the elements of $W^{1,p}(\Omega)$ and hence also of $A_{pq}(\Omega)$ and $A^+_{pq}(\Omega)$. For $p \leq 3$, the situation is more complicated. While the results on fine properties of weakly differentiable functions provide a number of items to the effect that the elements if $W^{1,p}(\Omega)$ have representatives which are continuous and/or differentiable outside small sets, we here restrict ourselves to orientation-preserving mappings $\chi \in W^{1,p}(\Omega)$. First, if $p = 3$ (generally $p = n$) and $\chi \in W^{1,n}(\Omega)$ is orientation-preserving, then a result of VODOPYANOV & GOLDSTEIN [1976] says that χ has a continuous representative; hence also in this case there is no doubt about the pointwise values of such mappings. For $A^+_{pq}(\Omega)$, $2 < p < 3$, ŠVERÁK [1988] showed, using the topological degree and a number of results on fine properties of weakly differentiable functions, that each element of $\chi \in A^+_{pq}(\Omega)$ has a representative $\tilde{\chi}$ that is continuous outside a set $S \subset \Omega$ of Hausdorff dimension $3 - p$. Moreover, he showed that it is possible to define in a natural way a set function $\boldsymbol{p} \mapsto F(\boldsymbol{p}, \chi)$ such that $F(\boldsymbol{p}, \chi) = \{\tilde{\chi}(\boldsymbol{p})\}$ for each $\boldsymbol{p} \in \Omega \setminus S$ and, for each $\boldsymbol{p} \in S$, $F(\boldsymbol{p}, \chi)$ is a compact connected set describing the singularity of the deformation at $\boldsymbol{p}$. This set function is then a substitute of the value $\chi(\boldsymbol{p})$. (See also the cited work of Vodopyanov & Goldstein, TANG [1988], GIAQUINTA, MODICA & SOUČEK [1989, 1992, 1994, 1996], MÜLLER, SPECTOR, & TANG [1994], MÜLLER, TANG, & YAN [1994], and MÜLLER & SPECTOR [1995]).

21.2.3 Injectivity and the Range of Mappings Another natural requirement is that χ be one-to-one. Since the elements $\chi \in W^{1,p}(\Omega)$ are only equivalence classes, one has to deal with appropriate generalizations of injectivity, or with a suitably chosen representative in the equivalence class. Moreover, one cannot be too cavalier in imposing the invertibility assumptions freely, for only the properties stable under the weak convergence of minimizing sequences are useful for the existence theory. One generalization of injectivity is injectivity a.e. An equivalence class χ is said to be ***injective a.e.*** if for each representative of χ there exists a subset E of Ω of Lebesgue measure 0 such the representative is injective on $\Omega \setminus E$. CIARLET & NEČAS [1987] showed that within the class of orientation-preserving mappings $\chi \in W^{1,p}(\Omega)$ with $p > 3$, $(p > n)$ the injectivity a.e. is equivalent to the requirement that the continuous representative satisfies

$$\int_\Omega \det \nabla\chi \, dV \le |\chi(\Omega)|. \tag{21.2.4}$$

Moreover, they showed that (21.2.4) is stable under the weak limit of minimizing sequences in $W^{1,p}(\Omega)$. That (21.2.4) is equivalent to the injectivity a.e. follows from the *area formula*: if χ is the continuous representative of a class from $W^{1,p}(\Omega)$, $p > n$, and S a Lebesgue measurable subset of Ω, then $\chi(S)$ is Lebesgue measurable and

$$\int_S |\det \nabla\chi| \, dV = \int_{\mathrm{Vect}} N(\boldsymbol{x}, \chi, S) \, dv(\boldsymbol{x}) \tag{21.2.5}$$

where, for each $\boldsymbol{x} \in$ Vect, $N(\boldsymbol{x}, \chi, S)$ is the number (possibly 0 or ∞) of all $\boldsymbol{p} \in S$ such that $\chi(\boldsymbol{p}) = \boldsymbol{x}$. It is a part of the assertion that $\boldsymbol{x} \mapsto N(\boldsymbol{x}, \chi, S)$ is a measurable function on Vect. (For the proof of (21.2.5), see MARCUS & MIZEL [1973].) Hence, if χ satisfies (21.2.4) and $\det \nabla\chi > 0$, then

$$|\chi(\Omega)| \le \int_{\mathrm{Vect}} N(\boldsymbol{x}, \chi, \Omega) \, dv(\boldsymbol{x}) = \int_\Omega |\det \nabla\chi| \, dV \le |\chi(\Omega)|$$

from which $N(\boldsymbol{x}, \chi, \Omega) = 1$ for a.e. $\boldsymbol{x} \in \chi(\Omega)$. Consequently, if E is defined by $E := \{\boldsymbol{p} \in \Omega : N(\chi(\boldsymbol{p}), \chi, \Omega) > 1\}$, then by (21.2.5) E has null Lebesgue measure and χ is injective on $\Omega \setminus E$. The converse implication is proved similarly. The condition (21.2.4) was extended to $A^+_{pq}(\Omega)$ with $p > 2$ by TANG [1988], using the definition of the range of χ of ŠVERÁK [1988], and shown to be stable under weak convergence.

Formula (21.2.5) also shows that the continuous representative of a $\chi \in W^{1,p}(\Omega)$, $p > n$, satisfies Lusin's (N) condition: a mapping (not an equivalence class) χ from Ω to Vect is said to satisfy Lusin's (N) condition if for any $A \subset \Omega$ with $|A| = 0$ also $|\chi(A)| = 0$. There are examples of mappings in $\chi \in W^{1,p}(\Omega, \mathrm{Vect})$, $p \le n$, violating Condition (N). Hence additional hypotheses must be imposed on χ if $p \le n$. For instance, if $\chi \in W^{1,n}(\Omega, \mathrm{Vect})$ is orientation-preserving, then the continuous representative of χ, guaranteed by the result of VODOPYANOV & GOLDSTEIN [1976] mentioned above, satisfies (N) (see MARTIO & ZIEMER [1992]). Generally, the area formula can hold only if χ satisfies Condition (N). There is a number of results about the area formula at various levels of generality, starting from diffeomorphisms, when $N(\boldsymbol{x}, \chi, \Omega) = 1$ for every $\boldsymbol{x} \in \chi(\Omega)$, with the intermediate cases of Lipschitz continuous mappings, mappings from $W^{1,p}(\Omega)$, $p > n$, up to the case when χ is only approximately differentiable a.e. In this case, one has to redefine the range of χ and

$N(x, \chi, S)$ to render (21.2.5) valid. Namely, it turns out that if $\chi : \Omega \to \mathrm{Vect}$ is approximately differentiable at a.e. point $p \in \Omega$ and if Ω_0 is the set of all Lebesgue points $p \in \Omega$ where the approximate differential $\mathrm{ap}\nabla\chi(p)$ exists, then $\Omega \setminus \Omega_0$ is a set of null Lebesgue measure, and

$$\int_S |\det \mathrm{ap}\nabla\chi|\, dV = \int_{\mathrm{Vect}} \tilde{N}(x, \chi, S)\, dv(x)$$

for every measurable subset S of Ω, where $\tilde{N}(x, \chi, S)$ is the number of all $p \in \Omega_0$ such that $\chi^*(p) = x$, and χ^* is the precise representative of χ, defined by

$$\chi^*(p) = \lim_{r \to 0} \frac{1}{\omega_n r^n} \int_{B(p,r)} \chi\, dV,$$

at those points p of Ω at which the limit exists, cf. Subsect. 3.8.2. (See GIAQUINTA, MODICA & SOUČEK [1992] for the proof of this version of the area formula). The set $\{x \in \mathrm{Vect} : \tilde{N}(x, \chi, S) \geq 1\}$ may be called the essential image of S under χ; it depends only on the equivalence class, and it is Lebesgue measurable. (In contrast to this, easy examples, such as mappings with values 0, 1 and different from 0 only on some null set, show that the set-theoretic image of a measurable set S under a given representative of χ need not be measurable; also, the set-theoretic images of a given measurable set S under two representatives of a given class may differ by a set of positive Lebesgue measure.) In particular, $\{x \in \mathrm{Vect} : \tilde{N}(x, \chi, \Omega) \geq 1\} = \chi^*(\Omega_0)$ is the essential range of χ, i.e., within the context of deformations, the current spatial configuration of the body.

The area formula and Condition (N) are constant sources of inspiration. In addition to the references given above, I refer to RADÓ & REICHELDERFER [1955], FEDERER [1969], BOJARSKI & IWANIEC [1983], EVANS & GARIEPY [1992], MALÝ [1994b], GIAQUINTA, MODICA & SOUČEK [1994, 1996], MALÝ & MARTIO [1995], and MÜLLER & SPECTOR [1995].

In the special case of the Dirichlet problem, BALL [1981] established a result on invertibility of a different type. Namely, if $p > 3$, $q > 3$ and $\varepsilon(D) = D^{-s}$ for $s > 2q/(q-3)$ in Assumption 21.4.1 below, then the displacement problem ($S = \emptyset$) with d a restriction of a homeomorphism on $\mathrm{cl}\,\Omega$, has a solution χ_0 that is a diffeomorphism of Ω onto $\chi(\Omega)$. (See also ŠVERÁK [1988].)

21.2.4 Cavitation The cavitation is the spontaneous formation of holes in certain elastomers under tensile loading conditions. It was first observed by GENT & LINDLEY [1958]; they also gave a first theoretical background for the phenomenon. The possibility of modeling the cavitation within the contemporary nonlinear elasticity theory was realized by BALL [1982], who also provided a mathematical theory of radially symmetric cavities. The analysis was subsequently extended in a number of papers; the reader is referred to JAMES & SPECTOR [1991, 1992] and MÜLLER & SPECTOR [1995] for references. As mentioned above, the classes $A^+_{pq}(\Omega)$ do not allow for cavitation, and within the Sobolev spaces setting, only the spaces $W^{1,p}(\Omega)$ with $p < n$ allow for discontinuous deformations. MÜLLER & SPECTOR [1995] provide an existence theory for materials that allow for cavitation within the classes

$$C^+_p := \{\chi \in W^{1,p}(\Omega) : \det \nabla\chi > 0 \text{ a.e. on } \Omega \text{ and } \chi \text{ satisfies (INV)}\}$$

where $2 < p < 3$ if $n = 3$, and (INV) will be described below. The authors point out that the injectivity a.e. is not efficient enough in the spaces $W^{1,p}$ with $p < 3$; in particular, it is not stable under weak convergence of minimizing sequences (see also MÜLLER, SPECTOR & TANG [1994]). Following VODOPYANOV & GOLDSTEIN [1976], the authors define a stronger condition of invertibility, (INV), based on the topological degree, which is stable under the weak convergence. Roughly speaking, this condition requires that the cavities in one part of the body are not filled by material from other parts of the body. It turns out that such mappings have a number of additional regularity properties, and in particular, each map from this class has a representative which is continuous outside a set of Hausdorff dimension $3-p$, and this representative satisfies Condition (N). The existence theory in the classes C_p^+ needs a term in the expression for the total energy that is proportional to the perimeter of the essential range of χ; this term penalizes the unlimited creation of holes.

21.2.5 Cartesian Currents and Weak Diffeomorphisms A different approach to the geometry and kinematics of deformations for nonlinear elasticity was developed by GIAQUINTA, MODICA & SOUČEK [1989, 1994, 1996]. The approach is based on the the notion of integer rectifiable current (see FEDERER [1969]). Using this notion, they interpret the deformation as, roughly, the graph of the mapping χ in Vect×Vect. More precisely, they associate with each graph of a smooth mapping a current and show that the deformation gradient, its cofactor, and determinant are naturally associated with the tangent space to the graph of χ. The closure of the set of all graphs of smooth diffeomorphisms under certain weak convergences of currents is the set of all weak diffeomorphisms. The set of all weak diffeomorphisms has good closure properties under weak convergence as a consequence of the Federer–Fleming closure theorem for integer rectifiable currents. Thus the set of all weak diffeomorphisms is well suited for the direct methods of the calculus of variations. Moreover, the notion of boundary of a current is used to distinguish the cases when the deformation exhibits fracture (in a generalized sense; such as the formation of cavities), and the approach also permits a detailed discussion of the notion of weak invertibility and of other desirable properties of deformations.

21.3 Weak Convergence of Determinant and Cofactor

This section describes the behavior of $\operatorname{cof} \boldsymbol{F}$ and $\det \boldsymbol{F}$ under the weak convergence in $W^{1,p}(\Omega)$. The results will be employed in the proofs of the existence of the solutions to the extremum problems.

21.3.1 Proposition *Let $n = 3$ and let p, q satisfy* (21.2.1). *If $\chi \in A_{pq}(\Omega)$ then* $\det \nabla\chi \in L^1(\Omega)$ *and*

$$\int_\Omega (\operatorname{cof} \nabla\chi)_{iA}\varphi\, dV = -\tfrac{1}{2}\varepsilon_{ijk}\varepsilon_{ABC}\int_\Omega \chi_k\chi_{j,B}\varphi_{,C}\, dV,$$

$$\int_\Omega (\det \nabla\chi)\varphi\, dV = -\tfrac{1}{3}\int_\Omega (\operatorname{cof} \nabla\chi)_{iA}\chi_i\varphi_{,A}\, dV$$

for every $\varphi \in C_0^\infty(\Omega)$. If $p \ge 3$ then these identities hold for every $\chi \in W^{1,p}(\Omega)$.

Recall that $A_{pq}(\Omega) = W^{1,p}(\Omega)$ if $p \ge n$ and $p/(p-1) \le q \le p/(n-1)$. A remarkable feature of these formulas is that the integrals on their right-hand sides contain one less derivative of χ than their left-hand sides. If χ is twice continuously differentiable on the closure of Ω, the assertion of the proposition is obtained from (1.1.19) and (1.1.22) by integration by parts. Then the density of $C^\infty(\mathrm{cl}\,\Omega)$ in $W^{1,p}(\Omega)$ is employed to extend the validity of the first formula to $W^{1,p}(\Omega)$. The limiting procedure involves a product of two functions from $L^p(\Omega)$ and for this the assumption $p \geq 2$ is needed. A similar procedure gives the second formula if $p \geq 3$ without the hypothesis that $\mathrm{cof}\,\nabla\chi \in L^q(\Omega)$. For the more general case with this hypothesis, see BALL [1977, pp. 369–370] or CIARLET [1988, pp. 366–368]. The above result enables one to use Rellich's lemma to prove the following.

21.3.2 Theorem *Let $n = 3$ and let p, q, r satisfy* (21.2.1) *and $r \geq 1$. Let $\chi_k, \chi \in A_{pq}$ satisfy*

$$\chi_k \rightharpoonup \chi \text{ in } W^{1,p}(\Omega), \quad \mathrm{cof}\,\nabla\chi_k \rightharpoonup \boldsymbol{A} \text{ in } L^q(\Omega), \quad \det\nabla\chi_k \rightharpoonup D \text{ in } L^r(\Omega)$$

for some functions $\boldsymbol{A} \in L^q(\Omega)$, $D \in L^r$. Then $\boldsymbol{A} = \mathrm{cof}\,\nabla\chi$ and $D = \det\nabla\chi$.

This fact is crucial: cof and det of the deformation gradient have sequentially weakly closed graphs. It can be also proved that the only mappings with this property are affine combinations of $\boldsymbol{F}$, $\mathrm{cof}\,\boldsymbol{F}$ and $\det\boldsymbol{F}$. The mappings $\boldsymbol{F}$, $\mathrm{cof}\,\boldsymbol{F}$ and $\det\boldsymbol{F}$ are also weakly continuous, which leads to Item (2) of the following proposition.

21.3.3 Proposition *Let $n = 3$, $\chi_k \rightharpoonup^* \chi$ in $W^{1,\infty}(\Omega)$ as $k \to \infty$, let $\{\nu_p : \boldsymbol{p} \in \Omega\}$ be the Young measure limit of deformation gradients of χ_k and set $\boldsymbol{F} = \nabla\chi$. Then*
(1) *if $f : \mathrm{Lin} \to \mathbb{R}$ is a quasiconvex function, then*

$$f(\boldsymbol{F}(\boldsymbol{p})) \leq \int_{\mathrm{Lin}} f(\boldsymbol{A})\, d\nu_p(\boldsymbol{A}) \quad \text{for a.e. } \boldsymbol{p} \in \Omega;$$

(2) **(*minors relations*)** *in particular,*

$$\boldsymbol{F}(\boldsymbol{p}) = \int_{\mathrm{Lin}} \boldsymbol{A}\, d\nu_p(\boldsymbol{A}), \quad \mathrm{cof}\,\boldsymbol{F}(\boldsymbol{p}) = \int_{\mathrm{Lin}} \mathrm{cof}\,\boldsymbol{A}\, d\nu_p(\boldsymbol{A}), \quad \det\boldsymbol{F}(\boldsymbol{p}) = \int_{\mathrm{Lin}} \det\boldsymbol{A}\, d\nu_p(\boldsymbol{A}),$$

for a.e. $\boldsymbol{p} \in \Omega$.

(1) follows from the crucial fact, to be explained in more detail in Subsect. 21.5.6, that integrals of quasiconvex functions are sequentially weakly lower semicontinuous with respect to weak* convergence in $W^{1,\infty}(\Omega)$. For the proof of (1), see, e.g., DIPERNA [1985] or TARTAR [1983]. These results are also related to a deep problem of describing the set of all Young measure limits of deformation gradients: Evaluating $\boldsymbol{F}$ via (21.1.2) to eliminate χ, we see that the above two items place obvious constraints on $\{\nu_p : \boldsymbol{p} \in \Omega\}$. KINDERLEHRER & PEDREGAL [1991] show that every parametric measure satisfying (1) is a Young measure limit of gradients of some sequence $\chi_k \in W^{1,\infty}(\Omega)$. For restrictions on the Young measure limits of gradients of more algebraic type, see BHATTACHARYA & al. [1994].

21.3.4 Weak Continuity and Integrability of Jacobians Let n be arbitrary and consider the mapping $\chi \mapsto J(\chi) := \det\nabla\chi$ from $W^{1,p}(\Omega)$ to $L^q(\Omega)$ for different values of p, q.

(1) Let $p > n$. Hölder's inequality implies that $J(\chi) \in L^{p/n}(\Omega)$ whenever $\chi \in W^{1,p}(\Omega)$ and using Proposition 21.3.1 one can prove that if $\chi_k \rightharpoonup \chi$ in $W^{1,p}(\Omega)$, then $J(\chi_k) \rightharpoonup J(\chi)$ in $L^{p/n}(\Omega)$.

(2) Let $p = n$. Hölder's inequality implies that $J(\chi) \in L^1(\Omega)$ whenever $\chi \in W^{1,n}(\Omega)$. Using Proposition 21.3.1 one can prove that if $\chi_k \rightharpoonup \chi$ in $W^{1,n}(\Omega)$, then $J(\chi_k) \rightharpoonup {}^*J(\chi)$ in the sense measures, i.e.,

$$\int_\Omega \phi J(\chi_k)\, dV \to \int_\Omega \phi J(\chi)\, dV \quad \text{as } k \to \infty$$

for every continuous function ϕ on Vect with compact support in Ω. This, as well as (1), is due to RESHETNYAK [1968]. BALL & MURAT [1984] show that it is not true that $J(\chi_k) \rightharpoonup J(\chi)$ in $L^1(\Omega)$.

(3) For mappings $\chi \in W^{1,n}(\Omega)$ with $J(\chi) \ge 0$, the Jacobian has a surprising higher integrability: MÜLLER [1989] showed that then $J(\chi)\ln(2 + J(\chi)) \in L^1(K)$ for every compact subset K of Ω. Also if $\chi_k \rightharpoonup \chi$ in $W^{1,n}(\Omega)$, $J(\chi_k) \ge 0$, then $J(\chi_k) \rightharpoonup {}^*J(\chi)$ in $L^1(K)$ for every compact $K \subset \Omega$.

(4) Let $p < n$. If $\chi \in W^{1,p}(\Omega)$ then generally $J(\chi)$ need not be an integrable function. However, for $p \ge n^2/(n+1)$ one can define a weak Jacobian as a distribution by $\operatorname{Det} \nabla\chi = \operatorname{Div}((\operatorname{cof} \nabla\chi)^{\mathsf T}\chi)/n$ (in the distributional sense) for every $\chi \in W^{1,p}(\Omega)$, see BALL [1977]. If $\chi \in W^{1,n}(\Omega)$ then $\operatorname{Det}\nabla\chi = \det\nabla\chi$. See BALL & MURAT [1984, Sect. 7] for the continuity properties of $\operatorname{Det}\nabla\chi$.

For refinements and further references on the above material, see MÜLLER [1990], IWANIEC & SBORDONE [1992], IWANIEC & LUTOBORSKI [1993], and GRECO & IWANIEC [1994].

21.4 States of Rubber-Like Bodies

We consider a rubber-like body in the conservative environment of ambient temperature θ_0. Suppose, for simplicity, the displacement-traction loading environment 13.3.1 with dead loadings 13.3.9 specified by the displacement $\boldsymbol d$ on D, surface traction $\boldsymbol s$ on S and a body force $\boldsymbol b_0$. Suppose that Ω has Lipschitz boundary, that D, S are area-measurable and that the area of D is positive. Let $\rho_\circ = 1$. Throughout the rest of the chapter, let $\boldsymbol F^\# := (\boldsymbol F, \operatorname{cof}\boldsymbol F, \det \boldsymbol F)$ for every $\boldsymbol F \in \mathrm{Lin}$. The entropy is used as independent variable and the following general assumptions are made about the energy function $\tilde e : \tilde U \to \mathbb{R}$.

21.4.1 Assumptions

(1) $\tilde U = \mathrm{Lin}^+ \times \mathbb{R}^{++}$.

(2) $\tilde e$ is polyconvex in the sense that there exists a convex function $\tilde e^0 : \mathrm{Lin} \times \mathrm{Lin} \times \mathbb{R}^{++} \times \mathbb{R}^+$ such that $\tilde e(\boldsymbol F, \eta) = \tilde e^0(\boldsymbol F^\#, \eta)$ for every $(\boldsymbol F, \eta) \in \tilde U$. Moreover, $\tilde e^0$ is continuous on its domain and twice continuously differentiable in its interior, the inequality $\partial_\eta \tilde e^0 > 0$ holds on $\mathrm{Lin} \times \mathrm{Lin} \times \mathbb{R}^{++} \times \mathbb{R}^{++}$ and we denote by $\tilde\theta^0$ the derivative $\partial_\eta \tilde e^0$.

(3) For every $(\boldsymbol F, \boldsymbol A, D) \subset \mathrm{Lin} \times \mathrm{Lin} \times \mathbb{R}^{++}$, the range of $\tilde\theta^o(\boldsymbol F, \boldsymbol A, D, \cdot) : \mathbb{R}^{++} \to \mathbb{R}^{++}$ is exactly $\mathbb{R}^{++}$; moreover $\tilde\theta^o(\boldsymbol F, \boldsymbol A, D, \cdot)$ is a strictly increasing function.

(4) There exist constants p, q, r, s, with $p \ge 2$, $q \ge p/(p-1)$, $r > 1$, $s > 1$, $c_1, c_2 > 0$, and an increasing function $\varepsilon : \mathbb{R}^{++} \to \mathbb{R}^{++}$ such that $\varepsilon(D) \to \infty$ for $D \to 0$ and

$$\tilde e^0(\boldsymbol F, \boldsymbol A, D, \eta) \ge c_1 + c_2 b(\boldsymbol F, \boldsymbol A, D, \eta)$$

for every $(\boldsymbol F, \boldsymbol A, D, \eta) \in \mathrm{Lin} \times \mathrm{Lin} \times \mathbb{R}^{++} \times \mathbb{R}^{++}$, where

$$b(\boldsymbol F, \boldsymbol A, D, \eta) = |\boldsymbol F|^p + |\boldsymbol A|^q + D^r + \eta^s + \varepsilon(D). \tag{21.4.1}$$

(5) $\boldsymbol b_0$ and $\boldsymbol s$ are bounded on their domains.

It is Assumption 21.4.1(4) which excludes fluids since for them $\tilde e^0$ cannot be bounded below by the powers of magnitudes of $\boldsymbol F$ and $\operatorname{cof}\boldsymbol F$; all others can be satisfied. We denote by $\mathrm Z := L^p(\Omega)\times L^q(\Omega)\times L^r(\Omega)\times L^s(\Omega)$ the Banach space of quadruples $(\boldsymbol F, \boldsymbol A, D, \eta)$ of functions with norm

$$\|(\boldsymbol F, \boldsymbol A, D, \eta)\|_{\mathrm Z} := \|\boldsymbol F\|_{L^p(\Omega)} + \|\boldsymbol A\|_{L^q(\Omega)} + \|D\|_{L^r(\Omega)} + \|\eta\|_{L^s(\Omega)}.$$

21.4.2 The State Space Let p, q, r, s be as in 21.4.1(4). We denote by $\Sigma^{\#\#}$ the set of all fields $\sigma = (\chi, \eta)$ on Ω with values in Eucl $\times$ $\mathbb{R}$ such that

(1) $(\nabla\chi^{\#}, \eta) \in Z$;

(2) $\chi = d$ on D in the sense of trace;

(3) $\det \nabla\chi > 0$, $\eta \geq 0$ a.e. on Ω and

(4)

$$\int_\Omega \tilde{e}(\nabla\chi, \eta)\, dV < \infty. \tag{21.4.2}$$

It follows from Assumption 21.4.1(4) that the integrand in (21.4.2) is bounded from below and hence it is absolutely integrable, and the integral is either finite or ∞. The set $\Sigma^{\#\#}$ contains the set $\Sigma^{\#\#}$ from 14.3.2(1) as a *proper* subset. We shall consider the extremum principles with the competitors from $\Sigma^{\#\#}$. Notice that among the states in $\Sigma^{\#\#}$ we have also included those for which $\eta = 0$ on a set of nonzero volume. Such states correspond to the zero absolute temperature. Let us also denote by $\Delta^{\#\#}$ the set of all χ such that there exists an η such that $(\chi, \eta) \in \Sigma^{\#\#}$.

The functionals U, P, H, M, W(σ) = W(χ) given by (20.3.1) originally on $\Sigma^{\#\#}$ in the sense of Sect. 20.3 are now extended to $\Sigma^{\#\#}$ or $\Delta^{\#\#}$ in the present sense by the same formulas (20.3.1). All of them take finite values on $\Sigma^{\#\#}$. To see that, note first that H, W are restrictions, to $\Sigma^{\#\#}$, of continuous linear functionals on $W^{1,p}(\Omega) \times L^s(\Omega)$, given by the same formulas. The continuity follows by a simple application of Hölder's inequality in the case of H and of Assumption 21.4.1(5) and the trace theorem in the case of W. The finiteness of U follows from the finiteness of W and 21.4.2(4). The finiteness of P follows from P = U$-\theta_0$H. The finiteness of M follows from the fact that M(χ) = P(χ, η) for an appropriate η such that $(\chi, \eta) \in \Sigma^{\#\#}$. Now some estimates of U, P, W on $\Sigma^{\#\#}$ will be presented.

21.4.3 Proposition

(1) *For every $\mu > 0$ there exists a $c \in \mathbb{R}$ such that*

$$\mathsf{W}(\chi) \geq c - \mu \int_\Omega |F|^p\, dV \tag{21.4.3}$$

for every $\sigma = (\chi, \eta) \in \Sigma^{\#\#}$ (with $F = \nabla\chi$).

(2) *There exist constants c_0, c_1, with $c_1 > 0$ such that*

$$\mathsf{U}(\sigma) \geq c_0 + c_1 \int_\Omega b(F^{\#}, \eta)\, dV, \qquad \mathsf{P}(\sigma) \geq c_0 + c_1 \int_\Omega b(F^{\#}, \eta)\, dV$$

for every $\sigma = (\chi, \eta) \in \Sigma^{\#\#}$ (b is given by (21.4.1)).

Proof (1): Since the area of D is positive, by Poincaré's inequality, the norm $\chi \mapsto \|\chi\|_{W^{1,p}(\Omega)}$ is equivalent to the norm $\chi \mapsto \|F\|_{L^p(\Omega)}$. Since W is a continuous linear functional on $W^{1,p}(\Omega)$ we have W$(\chi) \geq -c_1 \|F\|_{L^p(\Omega)}$ for all $\chi \in W^{1,p}(\Omega)$ and some $c_1 > 0$. For every $\mu > 0$ there exists a $c_2 \in \mathbb{R}$ such that $|t| \leq \mu|t|^p + c_2$ for every $t \in \mathbb{R}$. Hence W$(\chi) \geq -c_1\mu\|F\|^p_{L^p(\Omega)} - c_1 c_2$, which is (21.4.3). (2): By 21.4.1(4) and (1),

$$\mathsf{U}(\sigma) = \mathsf{W}(\chi) + \int_\Omega e\, dV \geq -\mu \int_\Omega |F|^p\, dV + c + c_1 + c_2 \int_\Omega b(F^{\#}, \eta)\, dV$$

for every $\mu > 0$, $\sigma = (\chi, \eta) \in \Sigma^{\#\#}$. It suffices to choose $\mu > 0$ small enough to satisfy $c_2 - \mu > 0$. The first estimate of the proposition then follows. To obtain the second, let us prove that

$$\tilde{e}^0(F^{\#}, \eta) - \theta_0 \eta \geq c_1 + c_2 b(F^{\#}, \eta) \tag{21.4.4}$$

for some c_1, c_2 with $c_2 > 0$ and every $(\boldsymbol{F}^{\#}, \eta) \in \text{Lin} \times \text{Lin} \times \mathbb{R}^{++} \times \mathbb{R}^{++}$. To see this, note that by Assumption 21.4.1(4)

$$\tilde{e}^0(\boldsymbol{F}^{\#}, \eta) - \theta_0\eta \geq c_0 + c_1 b(\boldsymbol{F}^{\#}, \eta) - \theta_0\eta \geq c_0 + \tfrac{1}{2}c_1\eta^s - \theta_0\eta + \tfrac{1}{2}c_1 b(\boldsymbol{F}^{\#}, \eta).$$

We have $\frac{1}{2}c_1\eta^s - \theta_0\eta \geq c_2$ for some $c_2 \in \mathbb{R}$ and all $\eta \geq 0$. Inequality (21.4.4) then follows. The rest of the proof is identical with the proof of the first estimate. □

21.4.4 Lemma *Let $\sigma' = (\chi, \eta') \in \Sigma^{**}$, $\theta > 0$, and define $\eta : \Omega \to \mathbb{R}^{++}$ by*

$$\eta(\boldsymbol{p}) = \bar{\eta}(\nabla\chi, \theta), \quad \boldsymbol{p} \in \Omega, \tag{21.4.5}$$

where $\bar{\eta}$ gives the entropy as a function of the deformation gradient and temperature. Then $\sigma = (\chi, \eta) \in \Sigma^{**}$.

Proof Denote $\boldsymbol{F} = \nabla\chi$. It is necessary to verify Conditions (1) and (4) of 21.4.2. The convexity of $\tilde{e}^0$ with respect to η and (21.4.4) yield

$$\tilde{e}^0(\boldsymbol{F}^{\#}, \eta') - \theta\eta' \geq \tilde{e}^0(\boldsymbol{F}^{\#}, \eta) - \theta\eta \geq c_1 + c_2 b(\boldsymbol{F}^{\#}, \eta). \tag{21.4.6}$$

As part of the definition of Σ^{**} we have

$$\int_\Omega \eta'^s\, dV < \infty, \qquad \int_\Omega |\tilde{e}^0(\boldsymbol{F}^{\#}, \eta')|\, dV < \infty. \tag{21.4.7}$$

From $(21.4.7)_1$ and Hölder's inequality we have

$$\int_\Omega \eta'\, dV < \infty. \tag{21.4.8}$$

Hence the integral of the left-hand-side of (21.4.6) is finite and the inequality shows that the integral of $b(\boldsymbol{F}^{\#}, \eta)$ is finite which implies that Condition (1) of 21.4.2 holds. Writing the convexity inequality of $\tilde{e}^0$ in the form

$$\tilde{e}^0(\boldsymbol{F}^{\#}, \eta') - \theta(\eta' - \eta) \geq \tilde{e}^0(\boldsymbol{F}^{\#}, \eta), \tag{21.4.9}$$

we see from $(21.4.7)_2$ and (21.4.8) that the left-hand side of (21.4.9) has finite integral and hence also that $\tilde{e}^0(\boldsymbol{F}^{\#}, \eta)$ is integrable. □

21.4.5 Lemma *Let $\sigma = (\chi, \eta) \in \Sigma^{**}$. Then the functions*

$$\theta \mapsto \int_\Omega \bar{\eta}(\nabla\chi, \theta)\, dV, \quad \theta \mapsto \int_\Omega \bar{e}(\nabla\chi, \theta)\, dV, \quad \theta > 0,$$

are increasing and continuous. The range of the first function is $\mathbb{R}^{++}$ and the second satisfies

$$\lim_{\theta\to\infty} \int_\Omega \bar{e}(\nabla\chi, \theta)\, dV = \infty. \tag{21.4.10}$$

Proof This follows from the monotone convergence theorem: Notice first that Assumption 21.4.1(3) implies that the function $\theta \mapsto \bar{\eta}(\boldsymbol{F}, \theta)$ is an increasing function, since it is the inverse of the function $\eta \mapsto \bar{\eta}^o(\boldsymbol{F}, \text{cof}\,\boldsymbol{F}, \det \boldsymbol{F}, \eta)$, and the latter is strictly increasing. The continuity of $\theta \mapsto \int_\Omega \bar{\eta}(\nabla\chi, \theta)\, dV$: By Lemma 21.4.4 $\bar{\eta}(\nabla\chi, \theta)$ is absolutely integrable, ≥ 0, and as $\theta \mapsto \bar{\eta}(\nabla\chi, \theta)$ is an increasing function, the monotone convergence theorem yields the continuity. Further, since the range of $\bar{\theta}_0(\boldsymbol{F}, \boldsymbol{A}, D, \cdot)$ is $\mathbb{R}^{++}$ for every $(\boldsymbol{F}, \boldsymbol{A}, D)$ (cf. Assumption 21.4.1(3)), we see from the monotonicity of $\bar{\eta}(\boldsymbol{F}, \cdot)$ that

$$\lim_{\theta\to 0+} \bar{\eta}(\boldsymbol{F}, \theta) = 0, \quad \lim_{\theta\to\infty} \bar{\eta}(\boldsymbol{F}, \theta) = \infty.$$

The monotone convergence theorem then yields

$$\lim_{\theta\to 0+} \int_\Omega \bar{\eta}(\nabla\chi, \theta)\, dV = 0, \qquad \lim_{\theta\to\infty} \int_\Omega \bar{\eta}(\nabla\chi, \theta)\, dV = \infty$$

and the continuity of this function implies the assertion about the range. The proofs of the assertions about $\theta \mapsto \int_\Omega \bar{e}(\nabla\chi, \theta)\, dV$ are similar. □

21.4.6 Proposition *Let* $\sigma' = (\chi, \eta') \in \Sigma^{**}$ *be a state. Then*

(1) *there exists an equilibrium state* $\sigma = (\chi, \eta) \in \Sigma^{**}$ *with the same deformation as* σ' *and of temperature equal to the ambient temperature* θ_0*; this state satisfies*

$$\mathsf{P}(\sigma) \le \mathsf{P}(\sigma'). \tag{21.4.11}$$

(2) *If* $\mathsf{U}_0 \in \mathbb{R}$ *is such that* $\mathsf{U}_0 \ge \mathsf{U}(\sigma')$ *then there exists an equilibrium state* $\sigma = (\chi, \eta) \in \Sigma^{**}$ *with the same deformation as* σ' *and such that* $\mathsf{U}(\sigma) = \mathsf{U}_0$*; this state satisfies*

$$\mathsf{H}(\sigma) \ge \mathsf{H}(\sigma'). \tag{21.4.12}$$

(3) *If* $\mathsf{H}_0 \in \mathbb{R}$ *is such that* $0 < \mathsf{H}_0 \le \mathsf{H}(\sigma')$ *then there exists an equilibrium state* $\sigma = (\chi, \eta) \in \Sigma^{**}$ *with the same deformation as* σ' *and such that* $\mathsf{H}(\sigma) = \mathsf{H}_0$*; this state satisfies*

$$\mathsf{U}(\sigma) \le \mathsf{U}(\sigma').$$

That is, according to (1), by passing from the state σ' to the equilibrium state σ with the same deformation and with the temperature equal to the ambient temperature, the canonical free energy is generally diminished. Assertions (2), (3) have similar interpretations.

Proof (1) It suffices to define the state σ by $\sigma = (\chi, \eta)$ with η given by (21.4.5) (with $\theta = \theta_0$). The integration of (21.4.9) gives (21.4.11). (2) Consider a family $\{\sigma_\theta, \theta > 0\}$ of equilibrium states given by $\sigma_\theta = (\chi, \eta_\theta)$ with $\eta_\theta(\boldsymbol{p}) = \bar{\eta}(\nabla\chi, \theta)$. By Lemma 21.4.4 then $\sigma_\theta \in \Sigma^{**}$ and by Lemma 21.4.5 the functions

$$\theta \mapsto \int_\Omega \bar{\eta}(\nabla\chi, \theta)\, dV, \quad \theta > 0, \qquad \theta \mapsto \int_\Omega \bar{e}(\nabla\chi, \theta)\, dV, \quad \theta > 0,$$

are increasing continuous functions. In view of (21.4.10) and the hypothesis of the assertion then $\mathsf{U}(\sigma_{\theta'}) = \mathsf{U}(\sigma') \le \mathsf{U}_0$ for some $\theta' > 0$. Since the energy is an increasing function of the temperature and $\mathsf{U}_0 \ge \mathsf{U}(\sigma')$, there exists a $\theta \ge \theta'$ such that $\mathsf{U}(\sigma_\theta) = \mathsf{U}_0$. Integrating the convexity inequality (21.4.9) and using the constraint $\mathsf{U}(\sigma_{\theta'}) = \mathsf{U}(\sigma')$ we obtain $\mathsf{H}(\sigma_{\theta'}) \ge \mathsf{H}(\sigma')$ and since the entropy is an increasing function of the temperature, we obtain from $\theta \ge \theta'$ inequality (21.4.12). (3) is proved similarly. □

21.5 Existence of Solutions to Extremum Problems for Rubber-Like Bodies

The following theorem asserts the existence of a solution, in Σ^{**}, to the problem of minimum canonical free energy.

21.5.1 Theorem *Let the body satisfy Assumptions* 21.4.1 *and let* $\Sigma^{**} \neq \emptyset$*. Then*

(1) *there exists a* $\sigma_0 = (\chi_0, \eta_0) \in \Sigma^{**}$ *such that* $\mathsf{P}(\sigma) \ge \mathsf{P}(\sigma_0)$ *for every* $\sigma \in \Sigma^{**}$*; the temperature field of* σ_0 *satisfies* $\theta(\boldsymbol{p}) = \theta_0$ *for a.e.* $\boldsymbol{p} \in \Omega$*;*

(2) *the deformation* χ_0 *satisfies* $\mathsf{M}(\chi) \ge \mathsf{M}(\chi_0)$ *for every* $\chi \in \Delta^{**}$*.*

(3) *Conversely, if* $\chi_0 \in \Delta^{**}$ *is such that* $\mathsf{M}(\chi) \ge \mathsf{M}(\chi_0)$ *for every* $\chi \in \Delta^{**}$*, and* $\eta_0 := \bar{\eta}(\nabla\chi_0, \theta_0)$*, then the state* $\sigma_0 := (\chi_0, \eta_0)$ *satisfies* $\mathsf{P}(\sigma) \ge \mathsf{P}(\sigma_0)$ *for every* $\sigma \in \Sigma^{**}$*.*

This theorem asserts the existence, but does not assert the uniqueness, as the uniqueness is not to be expected on physical grounds. In addition to the buckling discussed in Sect. 20.7, the reader is referred to TRUESDELL & NOLL [1965], OGDEN [1984],

BEATTY [1987], CIARLET [1988] and ANTMAN [1995] for numerous examples, mathematical or intuitive, where the uniqueness fails.

Proof (1): By $\Sigma^{\#\#} \neq \emptyset$ we have $\mathsf{P}_0 := \inf\{\mathsf{P}(\sigma) : \sigma \in \Sigma^{\#\#}\} < \infty$ and by 21.4.3(2), $\mathsf{P}_0 > -\infty$. Hence, by the definition of *infimum* there exists a sequence $\sigma_k = (\chi_k, \eta_k) \in \Sigma^{\#\#}$ such that

$$\mathsf{P}(\sigma_1) \geq \mathsf{P}(\sigma_2) \geq \cdots \geq \mathsf{P}(\sigma_k) \geq \cdots \tag{21.5.1}$$

and $\lim_{k\to\infty} \mathsf{P}(\sigma_k) = \mathsf{P}_0$. (A minimizing sequence.) By (21.5.1) there exists a C such that $\mathsf{P}(\sigma_k) \leq C$, $k = 1, 2, \ldots$ It then follows from the coercivity (see Proposition 21.4.3(2)) that the sequence $(\boldsymbol{F}_k^{\#}, \eta_k)$ is bounded in Z. By Poincaré's inequality $(\chi_k, \boldsymbol{F}_k^{\#}, \eta_k)$ is bounded in $L^p(\Omega)\times$Z. By Theorem 21.1.7 there exists a subsequence of σ_k, again denoted σ_k, such that

$$(\chi_k, \boldsymbol{F}_k^{\#}, \eta_k) \rightharpoonup (\chi, \boldsymbol{F}, \boldsymbol{A}, D, \eta') \text{ in } L^p(\Omega)\times \mathrm{Z}, \tag{21.5.2}$$

where $\chi \in L^p(\Omega)$ and $(\boldsymbol{F}, \boldsymbol{A}, D, \eta') \in$ Z are some elements. From the definitions of the weak convergence and of the generalized derivatives it follows that $\boldsymbol{F} = \nabla\chi$. By Proposition 21.3.2, $\boldsymbol{A} = \operatorname{cof}\nabla\chi$, $D = \det\nabla\chi$ and hence $(\boldsymbol{F}, \boldsymbol{A}, D) = \boldsymbol{F}^{\#}$. Relations (21.5.2) then imply

$$(\boldsymbol{F}_k^{\#}, \eta_k) \rightharpoonup (\boldsymbol{F}^{\#}, \eta') \text{ in Z}. \tag{21.5.3}$$

We assert that $\sigma' := (\chi, \eta')$ is in $\Sigma^{\#\#}$. We have to verify Conditions (1)–(4) of 21.4.2. Condition (1) follows directly from the construction. Condition (2): By the trace theorem, $\chi_k \rightharpoonup \chi$ in $L^{p'}(D)$ and as $\chi_k = \boldsymbol{d}$ on D, it follows that $\chi = \boldsymbol{d}$ on D. Condition (3): Since $\det\nabla\chi_k > 0$, $\eta_k \geq 0$ a.e. in Ω, we have

$$\det\nabla\chi \geq 0, \quad \eta' \geq 0 \text{ a.e. in } \Omega. \tag{21.5.4}$$

It follows from (21.5.1) and estimate 21.4.3(2) that $\int_\Omega \varepsilon(\det\nabla\chi_k)\,dV < C < \infty$ and hence Fatou's lemma gives

$$\int_\Omega \tilde\varepsilon(\det\nabla\chi)\,dV \leq C, \tag{21.5.5}$$

where $\tilde\varepsilon(D) = \varepsilon(D)$ for $D > 0$ and $\tilde\varepsilon(0) = \infty$. Inequality (21.5.5) implies

$$\det\nabla\chi > 0 \text{ a.e. on } \Omega \tag{21.5.6}$$

for otherwise the integral in (21.5.5) would be equal to ∞. Condition (4): The finiteness of the integral (21.4.2) follows from the fact that the sequence $\mathsf{P}(\sigma_k)$ is bounded and $\mathsf{U} \geq \mathsf{P}$. Relations (21.5.3) and Proposition 21.1.9 yield

$$\int_\Omega \tilde e^0(\boldsymbol{F}^{\#}, \eta')\,dV \leq \liminf_{k\to\infty} \int_\Omega \tilde e^0(\boldsymbol{F}_k^{\#}, \eta_k)\,dV. \tag{21.5.7}$$

Since H, W are continuous linear functionals,

$$\mathsf{H}(\sigma_k) \to \mathsf{H}(\sigma'), \quad \mathsf{W}(\chi_k) \to \mathsf{W}(\chi). \tag{21.5.8}$$

Relations (21.5.7) and (21.5.8) yield $\mathsf{P}(\sigma') \leq \liminf_{k\to\infty} \mathsf{P}(\sigma_k)$ and since the sequence on the right-hand side converges to P_0, one has $\mathsf{P}(\sigma') \leq \mathsf{P}_0$. Define now η_0 by $\eta_0(\boldsymbol{p}) = \bar\eta(\nabla\chi, \theta_0)$, $\boldsymbol{p} \in \Omega$, and set $\sigma_0 = (\chi, \eta_0)$. This is the equilibrium state about which Proposition 21.4.6(1) speaks; one has $\sigma_0 \in \tilde\Omega$, $\eta_0 > 0$ a.e. and $\mathsf{P}(\sigma') \geq \mathsf{P}(\sigma_0)$. Since P_0 is already the infimum, we have $\mathsf{P}(\sigma_0) = \mathsf{P}_0$. We therefore have a minimizer which is an equilibrium state. The same convexity argument implies that *every* minimizer is an equilibrium state.

(2): Let $\chi \in \boldsymbol{\Delta}^{\#\#}$ and define η by $\eta(\boldsymbol{p}) = \bar\eta(\boldsymbol{F}(\boldsymbol{p}), \theta_0)$, for every $\boldsymbol{p} \in \Omega$, where $\boldsymbol{F} = \nabla\chi$ and $\bar\eta$ is the function giving the entropy as a function of the deformation gradient and temperature. We have $\chi, \eta' \in \Sigma^{\#\#}$ for some η' and hence Lemma 21.4.4 tells us that the pair $\sigma := (\chi, \eta)$ is in $\Sigma^{\#\#}$. For the state σ we have $\mathsf{P}(\sigma) = \mathsf{M}(\chi)$ and, as σ_0 is a minimum of P on $\Sigma^{\#\#}$, we have $\mathsf{M}(\chi) = \mathsf{P}(\sigma) \geq \mathsf{P}(\sigma_0) = \mathsf{M}(\sigma_0)$. The proof of (2) is complete.

(3): We have

$$\tilde e(\boldsymbol{F}, \eta) - \theta_0\eta = \bar f(\boldsymbol{F}, \theta_0) + \tilde e(\boldsymbol{F}, \eta) - \tilde e(\boldsymbol{F}, \bar\eta(\boldsymbol{F}, \theta_0)) - \theta_0(\eta - \bar\eta(\boldsymbol{F}, \theta_0)).$$

By the convexity of $\tilde e$ with respect to η the term $\tilde e(\boldsymbol{F}, \eta) - \tilde e(\boldsymbol{F}, \bar\eta(\boldsymbol{F}, \theta_0)) - \theta_0(\eta - \bar\eta(\boldsymbol{F}, \theta_0))$ is nonnegative. Hence $\tilde e(\boldsymbol{F}, \eta) - \theta_0\eta \geq \bar f(\boldsymbol{F}, \theta_0)$. It follows that for every $\sigma = (\chi, \eta) \in \Sigma^{\#\#}$ we have $\mathsf{P}(\sigma) \geq \mathsf{M}(\chi)$. Hence, if χ_0 is a minimum of M on $\boldsymbol{\Delta}^{\#\#}$, we have $\mathsf{P}(\sigma) \geq \mathsf{M}(\chi) \geq \mathsf{M}(\chi_0) = \mathsf{P}(\sigma_0)$ for every $\sigma \in \Sigma^{\#\#}$. □

The next result gives a solution to the problem of maximum entropy.

21.5.2 Theorem *Let the body satisfy Assumptions* 21.4.1 *and let* U_0 *be a given number. If* $\{\sigma \in \Sigma^{\#\#} : \mathsf{U}(\sigma) = \mathsf{U}_0\} \neq \emptyset$, *then*

(1) *there exists an equilibrium state* $\sigma_0 = (\chi_0, \eta_0) \in \Sigma^{\#\#}$ *such that* $\mathsf{U}(\sigma_0) = \mathsf{U}_0$ *and* $\mathsf{H}(\sigma_0) \geq \mathsf{H}(\sigma)$ *for every* $\sigma \in \Sigma^{\#\#}$ *with* $\mathsf{U}(\sigma) = \mathsf{U}_0$;

(2) σ_0 *also satisfies* $\mathsf{U}(\sigma) \geq \mathsf{U}(\sigma_0)$ *for every* $\sigma \in \Sigma^{\#\#}$ *such that* $\mathsf{H}(\sigma) = \mathsf{H}(\sigma_0)$.

Proof (1): We have for every σ from the set $A = \{\sigma \in \Sigma^{\#\#} : \mathsf{U}(\sigma) = \mathsf{U}_0\}$ the estimate 21.4.3(2) for U and from this estimate and the fact that U is equal to U_0 on the set $\{\sigma \in \Sigma^{\#\#} : \mathsf{U}(\sigma) = \mathsf{U}_0\}$ we conclude that there exists a C such that $\int_\Omega \eta^s \, dV < C$ for every $\sigma = (\chi, \eta) \in \Sigma^{\#\#}$ such that $\mathsf{U}(\sigma) = \mathsf{U}_0$. (Here s is the exponent occurring in 21.4.1(4).) Hölder's inequality implies that $\int_\Omega \eta \, dV < C'$ for some C' and every $\sigma \in \Sigma^{\#\#}$ such that $\mathsf{U}(\sigma) = \mathsf{U}_0$. Hence

$$0 < \sup\{\mathsf{H}(\sigma) : \sigma \in \Sigma^{\#\#}, \mathsf{U}(\sigma) = \mathsf{U}_0\} < \infty.$$

Denote this supremum by S_0. Let $\sigma_k \in A$ be the maximizing sequence for H,

$$\mathsf{U}(\sigma_k) = \mathsf{U}_0, \quad \lim_{k\to\infty} \mathsf{H}(\sigma_k) = \mathsf{S}_0. \tag{21.5.9}$$

Denote $\boldsymbol{F}_k = \nabla\chi$. From $(21.5.9)_1$ and the coercivity of U (see Proposition 21.4.3(2)) we find that $(\chi_k, \boldsymbol{F}_k^{\#}, \eta_k)$ is a bounded sequence in $L^p(\Omega) \times \mathrm{Z}$. As in the proof of Theorem 21.5.1, by passing to a subsequence, we have (21.5.2) and hence (21.5.7). Moreover, $\det \nabla\chi_0 > 0$ a.e. on Ω. We define $\sigma_0' = (\chi_0, \eta_0')$. A reference to Proposition 21.1.9 gives, as in the proof of Theorem 21.5.1,

$$\mathsf{U}(\sigma_0') \leq \liminf_{k\to\infty} \mathsf{U}(\sigma_k) = \mathsf{U}_0, \quad \mathsf{H}(\sigma_0') = \lim_{k\to\infty} \mathsf{H}(\sigma_k) = \mathsf{S}_0. \tag{21.5.10}$$

From Proposition 21.4.6(2) we learn that there exists an equilibrium state $\sigma_0 = (\chi_0, \eta_0) \in \Sigma^{\#\#}$ such that

$$\mathsf{U}(\sigma_0) = \mathsf{U}_0, \quad \mathsf{H}(\sigma_0) \geq \mathsf{H}(\sigma_0'). \tag{21.5.11}$$

In view of (21.5.10), $(21.5.11)_2$, and the fact that S_0 is already the supremum, then necessarily

$$\mathsf{H}(\sigma_0) = \mathsf{S}_0. \tag{21.5.12}$$

But $(21.5.11)_1$ and (21.5.12) show that the equilibrium state is a solution to the problem of maximum entropy. (2): The proof is the same as that of Proposition 15.3.4. □

The next result gives a solution to the problem of minimum energy.

21.5.3 Theorem *Let the body satisfy Assumptions* 21.4.1 *and let* $\mathsf{H}_0 > 0$ *be a given number. If* $\{\sigma \in \Sigma^{\#\#} : \mathsf{H}(\sigma) = \mathsf{H}_0\} \neq \emptyset$, *then*

(1) *there exists an equilibrium state* $\sigma_0 = (\chi_0, \eta_0) \in \Sigma^{\#\#}$ *such that* $\mathsf{H}(\sigma_0) = \mathsf{H}_0$ *and* $\mathsf{U}(\sigma_0) \leq \mathsf{U}(\sigma)$ *for every* $\sigma \in \Sigma^{\#\#}$ *with* $\mathsf{H}(\sigma) = \mathsf{H}_0$;

(2) σ_0 *also satisfies* $\mathsf{H}(\sigma_0) \geq \mathsf{H}(\sigma)$ *for every* $\sigma \in \Sigma^{\#\#}$ *with* $\mathsf{U}(\sigma) = \mathsf{U}(\sigma_0)$.

The proof is similar to that of the preceding theorem. It is omitted. The following theorem shows that if the canonical free energy has a strict minimum in $\Sigma^{\#\#}$, then it has a strong minimum with respect to a metric d defined on $\Sigma^{\#\#}$ by

$$d(\sigma_1, \sigma_2) = \|(\boldsymbol{F}_1^{\#}, \eta_1) - (\boldsymbol{F}_2^{\#}, \eta_2)\|_{\mathrm{Z}},$$

$\sigma_1 = (\chi_1, \eta_1), \sigma_2 = (\chi_2, \eta_2) \in \Sigma^{\#\#}$, where $\boldsymbol{F}_1 = \nabla\chi_1, \boldsymbol{F}_2 = \nabla\chi_2$ and $\|\cdot\|_{\mathrm{Z}}$ is defined in 21.4.1.

We shall also consider $\boldsymbol{\Delta}^{\#\#}$ endowed with the metric d_0 given by

$$d_0(\chi_1, \chi_2) = \|\boldsymbol{F}_1 - \boldsymbol{F}_2\|_{L^p(\Omega)} + \|\operatorname{cof} \boldsymbol{F}_1 - \operatorname{cof} \boldsymbol{F}_2\|_{L^q(\Omega)} + \|\det \boldsymbol{F}_1 - \det \boldsymbol{F}_2\|_{L^r(\Omega)}.$$

21.5.4 Theorem *Let the body satisfy Assumptions* 21.4.1 *with* 21.4.1(2) *strengthened to assure that* $\tilde{e}^0$ *is strictly convex. Let* $\sigma_0 = (\chi_0, \eta_0) \in \Sigma^{**}$ *be an equilibrium state compatible with the environment such that the stored energy* M *has a strict local minimum at* χ_0 *with respect to* d_0 *on* Δ^{**} *in the sense that* $\mathsf{M}(\chi) > \mathsf{M}(\chi_0)$ *for every* $\chi \in \Delta^{**}$ *such that* $\chi \neq \chi_0$ *and* $d_0(\chi, \chi_0) \le \varepsilon_0'$ *for some* $\varepsilon_0' > 0$. *Then* P *has a uniformly strong local minimum (see Proposition* 15.1.5) *at* σ_0 *with respect to d.*

Theorem 21.5.4 is an adaptation, to nonisothermal situations, of the result by BALL & MARSDEN [1984], and the reader is referred there for proof. If the set of processes compatible with the environment satisfies hypotheses of Proposition 15.1.9, then Theorem 21.5.4 provides the Liapunov stability of the state σ_0. However, the present knowledge of the evolution equations of viscous materials with heat conduction in more than one space dimension does not allow one to verify the hypotheses of Proposition 15.1.9.

21.5.5 ***Remarks***

(1) No attempt has been made to make the hypotheses minimal. Note, e.g., that for Theorem 21.5.1 it would suffice to assume that the stored energy $\bar{f}(\cdot, \theta)$ is polyconvex and that the growth conditions involve only this function. However, Theorems 21.5.2 and 21.5.3 need the assumptions as formulated above.

(2) Finer questions concern the choice of the spaces of deformations discussed in Sect. 21.2. In addition to the remarks there, it is necessary to mention the coercivity conditions (see 21.4.1(4)). They control the quality of minimizers. We have seen in Subsect. 21.2.1 that the above choice, i.e., 21.4.1(4), does not allow for cavitation. Still within the choices of spaces without cavitation, the assumptions 21.4.1(4) can be substantially refined. The reader is referred to BALL & MURAT [1984] for optimal coercivity conditions, as well as to the relevant papers cited in Sect. 21.2. The theories that allow for cavitation must have $p < 3$ and moreover the requirement that $\operatorname{cof} \nabla\chi \in L^q(\Omega)$ must be relaxed, as in BALL [1981], JAMES & SPECTOR [1991, 1992] and MÜLLER & SPECTOR [1995], and the term with $\boldsymbol{A} = \operatorname{cof} \boldsymbol{F}$ in (21.4.1) must be deleted. However, MÜLLER & SPECTOR [1995] penalize the minimizers by including a surface energy term that is proportional to the area of $\partial\chi(\Omega)$ (which is not the image of $\partial\Omega$ if the cavitation is present).

(3) Another point is the form of the equations of equilibrium. Of course, the classical forms of these, discussed in Sect. 14.2, cannot be expected, and one has to pass to a weak form (the equations in the sense of distributions). But even there one has to be cautious, because, e.g., in the presence of cavitation or other types of defects, it is not to be expected that the referential and spatial forms are equivalent. Reasonable assumptions are known which guarantee the spatial weak form of the equations of equilibrium of forces and the referential weak form of the Eshelby conservation law. These are derived by using the outer and inner variations, respectively; see Subsect. 14.5.2 and the references given there.

21.5.6 Sequential Weak Lower Semicontinuity The above existence theorems, as well as those in the references cited so far, do not directly use the sequential weak lower semicontinuity (swlsc) of the minimized functionals mentioned in the introduction to this chapter. The swlsc is present only indirectly through the swlsc of the integral of the extension $\tilde{e}^0(\boldsymbol{F}, \boldsymbol{A}, D, \eta)$ of the energy $\tilde{e}(\boldsymbol{F}, \eta)$. However, the question of the swlsc of the integrals of quasiconvex functions on the Sobolev spaces $W^{1,p}(\Omega)$ is of independent interest. Consider, for simplicity, functionals of the form

$$I(\chi) = \int_\Omega f(\nabla\chi)\, dV, \quad \chi \in W^{1,p}(\Omega), \tag{21.5.13}$$

where Ω is a bounded open set, $f : \mathrm{Lin} \to \mathbb{R} \cup \{\infty\}$ a continuous function, bounded below, and $1 \le p \le \infty$. The functional I is said to be swlsc on $W^{1,p}(\Omega)$ if

$$\liminf_{k\to\infty} I(\chi_k) \ge I(\chi) \tag{21.5.14}$$

for every sequence $\chi_k \in W^{1,p}(\Omega)$ and every $\chi \in W^{1,p}(\Omega)$ such that $\chi_k \rightharpoonup \chi$ in $W^{1,p}(\Omega)$ ($\rightharpoonup^*$ if $p = \infty$). The basic fact is that the quasiconvexity of f in the form Definition 17.1.1 is a necessary, and under technical hypotheses also sufficient, condition for the swlsc of I. In the case of the calculus of variations on an interval (see Remark 17.1.2(4), (5)), quasiconvexity is equivalent to convexity, and the equivalence of the convexity to the swlsc in this special case is a basic discovery of Tonelli (1922). For the general case, when m, n in Remark 17.1.2(4) are arbitrary, the necessity of the quasiconvexity for swlsc is due to Morrey (1952), see MORREY [1966], where the notion of quasiconvexity is that defined in Definition 17.1.1 (see also MEYERS [1965], BALL, CURRIE & OLVER [1981]). More precisely, BALL & MURAT [1984] proved that the swlsc of I on $W^{1,p}(\Omega)$ implies the stronger condition of $W^{1,p}$-quasiconvexity of f defined in Remark 17.1.4(2). These necessity results of Morrey and Ball & Murat are actually simple and involve the testing of (21.5.14) on the sequence χ_k of scaled periodic patterns from Example 21.1.2(3). Morrey also proved that the quasiconvexity is sufficient for the swlsc on $W^{1,\infty}(\Omega)$. If $1 \le p < \infty$, the sufficency of the quasiconvexity for the swlsc is harder, and requires additional conditions on f. In the early result of Morrey (1952), these conditions were fairly restrictive. A specialization of the main result of ACERBI & FUSCO [1984] to the present case of f independent of $\boldsymbol{p}$ and $\boldsymbol{x}$ says that if f satisfies the growth condition

$$0 \le f(\boldsymbol{F}) \le a + c|\boldsymbol{F}|^p \tag{21.5.15}$$

for some $a, c \in \mathbb{R}$ and all $\boldsymbol{F}$, then I is swlsc on $W^{1,p}(\Omega)$ if and only if f is quasiconvex (in the sense of Definition 17.1.1). Unfortunately, (21.5.15) rules out the energy functions of nonlinear elasticity as it is inconsistent with $f(\boldsymbol{F}) \to \infty$ for $\det \boldsymbol{F} \to 0$; for this reason it was necessary to pass from the quasiconvexity to the stronger, but still realistic, assumption of polyconvexity in the treatment above. There is a large literature on the subject of swlsc, and the state up to 1989 is described in the monograph DACOROGNA [1989]. More recent relevant results and literature may be found in MALÝ [1993, 1994], FUSCO & HUTCHINSON [1995], and FONSECA & MALÝ [1996].

It is worth mentioning that the $W^{1,p}$-quasiconvexity depends dramatically on p. Consider the integral

$$\int_\Omega g(\det \nabla\chi)\, dV, \quad \chi \in W^{1,p}(\Omega),$$

where Ω is a bounded open set and $g : \mathbb{R} \to \mathbb{R}$ a continuous function, bounded below. The results of Sect. 17.7 imply that the function $h(\boldsymbol{F}) := g(\det \boldsymbol{F})$ is quasiconvex if and only if g is convex. However, by a result of BALL & MURAT [1984, Thm. 4.1], for $p \ge n$, h is $W^{1,p}$-quasiconvex if and only if g is convex while for $1 \le p < n$, h is $W^{1,p}$-quasiconvex if and only if g is constant. In particular, $\boldsymbol{F} \mapsto |\det \boldsymbol{F}|$ is $W^{1,p}$-quasiconvex only for $p \ge n$ and

$$\int_\Omega |\det \nabla\chi|\, dV,$$

is swlsc on $W^{1,p}(\Omega)$ if $p \ge n$ but not swlsc if $p < n$. While the polyconvexity unconditionally implies quasiconvexity (see Sect. 17.5), the same is true for $W^{1,p}$-quasiconvexity only if $p \ge n$ (see ACERBI, BUTTAZZO & FUSCO [1983]).

21.6 Minimum Energy in Crystals and Young Measure Minimizers

The existence of solutions to the extremum principles in the preceding sections was established because the stored energy was assumed to be coercive and polyconvex. We have already pointed out that the coercivity assumption 21.4.1(4) trivially cannot be satisfied by fluids. However, for fluids the existence of equilibrium states is trivial for other reasons. Another case when both the coercivity and the polyconvexity fail are crystals with the symmetry group the nonorthogonal symmetry group of some Bravais lattice L. In this section we shall briefly review some aspects of the theory of equilibrium of such materials; some of the considerations apply to more general materials.

Consider a bounded body Ω with Lipschitz boundary at some fixed temperature θ. Denote by $\bar{f} : \mathrm{Lin}^+ \to \mathbb{R}$ the stored energy function at θ and by G the symmetry group of the body. We shall use some or all of the following assumptions.

21.6.1 Assumptions

(1) $\bar{f}(\boldsymbol{F}) \to \infty$ as $\det \boldsymbol{F} \to 0$;

(2) $\bar{f}(\boldsymbol{F}) \geq 0$ for each $\boldsymbol{F} \in \mathrm{Lin}^+$ and $\bar{f}(\mathbf{1}) = 0$;

(3) $\mathsf{G} = \mathsf{G}(\mathsf{L})$ for some Bravais lattice $\mathsf{L} \subset \mathrm{Vect}$ (see Sect. 1.5).

Recall that (3) implies that $\bar{f}$ cannot be coercive: we have $\bar{f}(\boldsymbol{H}) = \bar{f}(\mathbf{1})$ for every $\boldsymbol{H} \in \mathsf{G}(\mathsf{L})$ and $\mathsf{G}(\mathsf{L})$ is unbounded. Also recall that by Proposition 17.8.3, under (3), the stored energy cannot be rank 1 convex unless it depends only on $\det \boldsymbol{F}$; the crystals are in this sense internally unstable. FONSECA [1987] has shown that under the pure traction dead-loading conditions, the global infimum of the total stored energy satisfying (1)–(3) is $-\infty$ unless the body forces and surface tractions vanish identically. We shall therefore consider the Dirichlet boundary conditions combined with zero body forces. Specifically, let $\boldsymbol{d} : \Omega \to \mathrm{Vect}$ be a Lipschitz continuous function and define $\boldsymbol{\Delta}^{**} = \boldsymbol{\Delta}^{**}(\boldsymbol{d})$ by

$$\boldsymbol{\Delta}^{**}(\boldsymbol{d}) = \{\boldsymbol{\chi} \in W^{1,\infty}(\Omega) : \det \nabla\boldsymbol{\chi} > 0 \text{ a.e. and } \boldsymbol{\chi} = \boldsymbol{d} \text{ on } \partial\Omega\}.$$

The total stored energy is

$$\mathsf{M}(\boldsymbol{\chi}) \equiv \mathsf{M}(\boldsymbol{\chi}, \Omega) = \int_\Omega \bar{f}(\nabla\boldsymbol{\chi})\, dV, \quad \boldsymbol{\chi} \in \boldsymbol{\Delta}^{**}(\boldsymbol{d}).$$

Our first concern will be to calculate the minimum total stored energy, i.e.,

$$E_\Omega(\boldsymbol{d}) := \inf\{\mathsf{M}(\boldsymbol{\chi}, \Omega) : \boldsymbol{\chi} \in \boldsymbol{\Delta}^{**}(\boldsymbol{d})\}. \tag{21.6.1}$$

We start with the homogeneous boundary conditions, when $\boldsymbol{d}(\boldsymbol{p}) = \boldsymbol{F}\boldsymbol{p}$ for some fixed $\boldsymbol{F} \in \mathrm{Lin}^+$, and write $E_\Omega(\boldsymbol{F}) \equiv E_\Omega(\boldsymbol{d})$, $\boldsymbol{\Delta}_\Omega^{**}(\boldsymbol{F}) \equiv \boldsymbol{\Delta}^{**}(\boldsymbol{d})$. Clearly, $E_\Omega(\boldsymbol{F}) \leq |\Omega| \bar{f}(\boldsymbol{F})$.

Define the ***subenergy*** $h : (0, \infty) \to \mathbb{R}$ of the body by

$$h(v) = \inf\{\bar{f}(\boldsymbol{F}) : \boldsymbol{F} \in \mathrm{Lin}^+, \det \boldsymbol{F} = v\}, \quad v \in (0, \infty).$$

Let furthermore $h^{\mathrm{co}} : (0, \infty) \to \mathbb{R}$ be the lower convex hull of h, see Sect. 16.3. Note that under Assumption 21.6.1(2), $0 \leq h^{\mathrm{co}}(v) \leq h(v) < \infty$ for every $v > 0$ and hence h^{co} is continuous. Proposition 21.6.3, below, provides an explicit calculation of the minimum energy under the homogeneous boundary conditions. The following lemma describes preliminary steps needed to prove the result.

21.6.2 Lemma *Let Assumptions* 21.6.1(1)–(3) *hold.*

(1) *There exists a function* $\tilde{f} : \mathrm{Lin}^+ \to \mathbb{R}$ *such that*

$$E_\Omega(\boldsymbol{F}) = |\Omega| \tilde{f}(\boldsymbol{F})$$

for every $\boldsymbol{F} \in \mathrm{Lin}^+$ *and every* $\Omega \subset \mathrm{Vect}$ *such that* $\partial\Omega$ *has a three-dimensional measure* 0.

(2) $\tilde{f}$ *is rank 1 convex and* $\tilde{f}(\boldsymbol{QFH}) = \tilde{f}(\boldsymbol{F})$ *for every* $\boldsymbol{F} \in \mathrm{Lin}^+$, $\boldsymbol{Q} \in \mathrm{Orth}^+$, *and* $\boldsymbol{H} \in \mathsf{G}(\mathsf{L})$ *with* $\det \boldsymbol{H} = 1$.

By (1), the minimum energy per unit mass of a crystal subject to homogeneous boundary conditions is independent of its shape.

Proof (1): Let $\{\boldsymbol{e}_i, i = 1, 2, 3\}$ be an orthonormal basis in Vect, let

$$C = \{\boldsymbol{p} \in \mathrm{Vect} : 0 \leq \boldsymbol{p} \cdot \boldsymbol{e}_i \leq 1, i = 1, 2, 3\}$$

and define $\tilde{f}$ by $\tilde{f}(\boldsymbol{F}) = E_C(\boldsymbol{F})$. Let $\Omega \subset \mathrm{Vect}$ be bounded and open with $|\partial\Omega| = 0$, and let $\boldsymbol{u} \in W_0^{1,\infty}(\Omega)$ be such that $\det(\boldsymbol{F} + \nabla\boldsymbol{u}(\boldsymbol{p})) > 0$ for a.e. $\boldsymbol{p} \in \Omega$. By Vitali's covering theorem there exists a finite or countable disjoint sequence $\boldsymbol{p}_i + \varepsilon_i \operatorname{cl}\Omega$ of subsets of C, where $\boldsymbol{p}_i \in \mathrm{Vect}$ and $0 < \varepsilon_i < 1$, such that

$$|C \setminus \bigcup_i^\infty (p_i + \varepsilon_i \operatorname{cl} \Omega)| = 0.$$

Since $|\partial\Omega| = 0$, we have $|\Omega| \sum_i \varepsilon_i^3 = 1 = |C|$. Define

$$v(p) = \varepsilon_i u((p - p_i)/\varepsilon_i) \text{ if } p \in p_i + \varepsilon_i \operatorname{cl} \Omega \text{ and } v(p) = 0 \text{ otherwise.}$$

Clearly, $v \in W_0^{1,\infty}(C)$ and $\det(F + \nabla v(p)) > 0$ for a.e. $p \in C$. Hence

$$E_C(F) \le \int_C \bar{f}(F + \nabla v)\, dV = \sum_i \int_{p_i + \varepsilon_i \Omega} \bar{f}(F + \nabla u((p - p_i)/\varepsilon_i))\, dV(p)$$

$$= \sum_i \varepsilon_i^3 \int_\Omega \bar{f}(F + \nabla u(p))\, dV(p) = \frac{1}{|\Omega|} \int_\Omega \bar{f}(F + \nabla u(p))\, dV(p).$$

Taking the infimum over all u, we obtain $\bar{f}(F) \equiv E_C(F) \le E_\Omega(F)/|\Omega|$ and, in the same way, $E_\Omega(F)/|\Omega| \le \bar{f}(F) \equiv E_C(F)$. (2): Let $F \in \mathrm{Lin}^+$, and let $\alpha, 0 < \alpha < 1, a, b \in$ Vect be as in the definition of rank 1 convexity 17.3.1, i.e., let the closed line segment with endpoints $F - \alpha a \otimes b$ and $F + (1 - \alpha) a \otimes b$ be in Lin^+. Here it suffices to assume that $F - \alpha a \otimes b$, $F + (1 - \alpha) a \otimes b \in \mathrm{Lin}^+$; that the line segment with these endpoints is in Lin^+ follows from the fact that Lin^+ is rank 1 convex. Assume without any loss of generality that $|b| = 1$. Let $\{e_i, i = 1, 2, 3\}$ be an orthonormal basis in Vect such that $e_1 = b$. Let

$$P = \{p \in \text{Vect} : 0 < p \cdot e_i < 1 : i = 1, 2, 3\},$$
$$P_\alpha = \{p \in P : 0 < p \cdot e_1 < \alpha\}, \quad P_{1-\alpha} = \{p \in P : \alpha < p \cdot e_1 < 1\}.$$

Let $v \in W_0^{1,\infty}(P_\alpha)$, $w \in W_0^{1,\infty}(P_{1-\alpha})$ be such that $\det(F + (1 - \alpha) a \otimes b + \nabla v(p)) > 0$ for a.e. $p \in P_\alpha$, and $\det(F - \alpha a \otimes b + \nabla w(p)) > 0$ for a.e. $p \in P_{1-\alpha}$. Set $v = 0$ on $P_{1-\alpha}$ and $w = 0$ on P_α, and let us extend v, w to P-periodic functions on Vect. Let m, k be positive integers and define P_k, Q_k, L_k as in the proof of Lemma 17.3.4. Let furthermore φ_k be the cutoff function defined in that proof, and $g : \mathbb{R} \to \mathbb{R}$ a 1-periodic function such that

$$g'(s) = 1 - \alpha \text{ if } 0 \le s < \alpha, \quad g'(s) = -\alpha \text{ if } \alpha \le s < 1.$$

Define mappings u, z_k by

$$u(p) = a g(p \cdot b), \quad z_k(p) = \varphi_k(p) u(p) + \chi_k(p)(v(p) + w(p)), \quad p \in \text{Vect},$$

where χ_k is the characteristic function of P_k. Then for every k,

$$z_k \in W_0^{1,\infty}(Q_k), \quad z_k = u + v + w \text{ on } P_k,$$

$$\nabla z_k(p) = (1 - \alpha) a \otimes b + \nabla v(p) \text{ on } P_\alpha, \text{ and } \nabla z_k(p) = -\alpha a \otimes b + \nabla w(p) \text{ on } P_{1-\alpha}.$$

Moreover, if m is sufficiently large, then $\det(F + \nabla z_k(p)) > 0$ for all k and a.e. $p \in Q_k$ (see Proof of Lemma 17.3.4). If $\chi(p) := Fp + z_k(p)$, $p \in Q_k$, then $\chi \in\in \Delta_{Q_k}^{\#\#}(F)$. The application of Assertion (1) to $\Omega = Q_k$, the definition of $E_{Q_k}(F)$, and the periodicity of z_k on P_k therefore give

$$(k + 2m)^3 \bar{f}(F) \le \int_{Q_k} \bar{f}(F + \nabla z_k)\, dV = \int_{L_k} \bar{f}(F + \nabla z_k)\, dV + k^3 \int_P \bar{f}(F + \nabla(u + v + w))\, dV,$$

where $(k + 2m)^3$ is the volume of Q_k. Dividing by k^3, letting $k \to \infty$, and using the fact that $\int_{Q_k} f(F + \nabla z_k)\, dV / k^3 \to 0$ (see Proof of Lemma 17.3.4), we obtain

$$\bar{f}(F) \le \int_P \bar{f}(F + \nabla(u + v + w))\, dV = \int_{P_\alpha} \bar{f}(F + (1 - \alpha) a \otimes b + \nabla v(p))\, dV + \int_{P_{1-\alpha}} \bar{f}(F - \alpha a \otimes b + \nabla w(p))\, dV.$$

Taking the infimum over all v, w as above, using (1), and noticing that $|P_\alpha| = \alpha$, $|P_{1-\alpha}| = 1 - \alpha$, we obtain

$$\bar{f}(F) \le \alpha \bar{f}(F + (1 - \alpha) a \otimes b) + (1 - \alpha) \bar{f}(F - \alpha a \otimes b).$$

Finally, the objectivity and the symmetry of $\bar{f}$ under $G(L)$ are consequence of the same properties of $\bar{f}$, and the easy proof is left to the reader. □

21.6.3 Proposition *If Assumptions* 21.6.1(1)–(3) *hold then for every* $\boldsymbol{F} \in \mathrm{Lin}^+$,

$$E_\Omega(\boldsymbol{F}) = |\Omega| h^{\mathrm{co}}(\det \boldsymbol{F}).$$

Here $|\Omega|$ *is the volume of* Ω.

Proof Since $\bar{f}$ is rank 1 convex, Proposition 17.8.4 tells us that $\bar{f}$ is of the form $\bar{f}(\boldsymbol{F}) = m(\det \boldsymbol{F})$ $\boldsymbol{F} \in \mathrm{Lin}^+$, where $m : (0, \infty) \to \mathbb{R}$ is a convex function. To prove that $m = h^{\mathrm{co}}$ notice that we have $\bar{f}(\boldsymbol{F}) = m(\det \boldsymbol{F}) \le \bar{f}(\boldsymbol{F})$ for every $\boldsymbol{F} \in \mathrm{Lin}^+$ and the definition of the subenergy h gives $m(\det \boldsymbol{F}) \le h(\det \boldsymbol{F})$. To complete the proof, it suffices to prove that if g is a convex function such that $g \le h$, then $g \le m$. Any such a function satisfies $g(\det \boldsymbol{F}) \le h(\det \boldsymbol{F}) \le \bar{f}(\boldsymbol{F})$. The function $\boldsymbol{F} \mapsto g(\det \boldsymbol{F})$ is polyconvex and hence

$$|P| g(\det \boldsymbol{F}) \le \int_P g(\det \nabla \chi)\, dV \le \int_P \bar{f}(\det \nabla \chi)\, dV$$

for every $\chi = \boldsymbol{F}\boldsymbol{p} + \boldsymbol{u}$ with $\boldsymbol{u} \in W_0^{1,\infty}(P)$ and any parallelepiped P. Taking the infimum over χ and using Lemma 21.6.2 we obtain $g(\det \boldsymbol{F}) \le \bar{f}(\boldsymbol{F}) = m(\det \boldsymbol{F})$. □

21.6.4 Proposition *If Assumptions* 21.6.1(1)–(3) *hold then*

(1) *the function* $\bar{f} : \mathrm{Lin}^+ \to \mathbb{R}$ *given by*

$$\bar{f}(\boldsymbol{F}) = h^{\mathrm{co}}(\det \boldsymbol{F}), \quad \boldsymbol{F} \in \mathrm{Lin}^+,$$

is the greatest quasiconvex function on Lin^+ *such that* $\bar{f} \le \bar{f}$;

(2) *the set of all* $\boldsymbol{F} \in \mathrm{Lin}^+$ *where* $\bar{f}$ *is quasiconvex coincides with the set*

$$Q = \{\boldsymbol{F} \in \mathrm{Lin}^+ : \bar{f}(\boldsymbol{F}) = h^{\mathrm{co}}(\det \boldsymbol{F})\};$$

(3) *if* $\boldsymbol{F} \in Q$ *and* h^{co} *is differentiable at* $v = \det \boldsymbol{F}$, *then the Piola–Kirchhoff stress* $\bar{\boldsymbol{S}} = \partial_{\boldsymbol{F}} \bar{f}$ *and the Eshelby tensor* $\bar{\boldsymbol{P}} = \bar{f}\mathbf{1} - \boldsymbol{F}^{\mathrm{T}}\bar{\boldsymbol{S}}$ *are given by*

$$\bar{\boldsymbol{S}}(\boldsymbol{F}) = -p \operatorname{cof} \boldsymbol{F}, \quad \bar{\boldsymbol{P}}(\boldsymbol{F}) = g(v)\mathbf{1},$$

where

$$p = -\partial_v h^{\mathrm{co}}(v) \ \text{ and } \ g(v) := h^{\mathrm{co}}(v) + pv.$$

Proof We only sketch the main points of the proof; the details may be filled by inspecting the proofs of Lemma 21.6.2 and Proposition 21.6.3. (1): Recall that h^{co} is finite and nonnegative on $(0, \infty)$ and hence continuous; consequently also $\bar{f}$ is continuous on Lin^+. That $\bar{f}$ is quasiconvex follows from the result of Subsect. 17.7 saying that any convex function of $\det \boldsymbol{F}$ is a quasiconvex function of $\boldsymbol{F}$. That $\bar{f}$ is the largest quasiconvex function $\le \bar{f}$ follows from the definition of $E_\Omega(\boldsymbol{F})$ and item (1) of Lemma 21.6.2. (2): if $\bar{f}$ is quasiconvex at $\boldsymbol{F}$, then the definition of $E_\Omega(\boldsymbol{F})$ and item (1) of Lemma 21.6.2 imply that $\bar{f}(\boldsymbol{F}) = \bar{f}(\boldsymbol{F})$ and hence $\boldsymbol{F} \in Q$. Conversely, if $\boldsymbol{F} \in Q$, then the quasiconvexity of $\bar{f}$, the equality $\bar{f}(\boldsymbol{F}) = \bar{f}(\boldsymbol{F})$ and the inequality $\bar{f}(\boldsymbol{A}) \ge \bar{f}(\boldsymbol{A})$ for every $\boldsymbol{A} \in \mathrm{Lin}^+$ imply that $\bar{f}$ is quasiconvex at $\boldsymbol{F}$. (3): Immediate. □

21.6.5 The Lower Quasiconvex Hull Generally, given a function f on Lin, the definition of quasiconvexity implies that if $m : \mathrm{Lin} \to \mathbb{R}$ is a quasiconvex function such that $m \le f$ on Lin, then necessarily $m \le Qf$, where Qf is given by

$$Qf(\boldsymbol{F}) = \inf \{ |P|^{-1} \int_P f(\boldsymbol{F} + \nabla \boldsymbol{u})\, dV : \boldsymbol{u} \in C_0^\infty(P)\}, \quad \boldsymbol{F} \in \mathrm{Lin}, \tag{21.6.2}$$

where P is a parallelepiped in Vect. (Alternatively, one may replace the infimum over $\boldsymbol{u} \in C_0^\infty(P)$ by the infimum over all P-periodic $\boldsymbol{u} \in C^\infty(\mathrm{Vect})$, cf. Lemma 17.3.4.) This suggests that the greatest quasiconvex function $\le f$, i.e., the lower quasiconvex hull of f, should be given by (21.6.2). This is indeed the case when f is continuous, finite-valued, and defined on Lin. (See Dacorogna [1982, 1985], Acerbi & Fusco [1984].) Proposition 21.6.4 shows that (21.6.2) is also valid for the lower quasiconvex hull of $\bar{f}$ in the special case of crystals. (The condition that f be defined and continuous on Lin is not needed because of other features of $\bar{f}$.)

The formula (21.6.2) is meaningful also when χ is a function of n scalar variables with values in an m-dimensional space, as described in Remark 17.1.2(4). This includes as special cases the scalar-valued functions χ of n variables and vector-valued functions χ of a scalar variable. In both these special cases, under fairly general conditions, the lower quasiconvex hull of f reduces to the lower convex hull of f as defined in Chap. 16. (See EKELAND & TEMAM [1975].) Another case when the calculation of the lower quasiconvex hull reduces to the lower convex hull is when $f(\boldsymbol{F}) = g(\det \boldsymbol{F})$. Then, under mild conditions, the lower quasiconvex hull $\tilde{f}$ of f is $\tilde{f}(\boldsymbol{F}) = g^{\mathrm{co}}(\det \boldsymbol{F})$, where g^{co} is the lower convex hull of g. See DACOROGNA [1981]; the proof can be also completed by using that $f(\boldsymbol{F}) = g(\det \boldsymbol{F})$ is quasiconvex in $\boldsymbol{F}$ if and only if g is convex (see Sect. 17.7) and modifying the proofs above in this section. This is the reason of the importance of the lower convex hull of energy of a fluid (expressed as a function of v, η) explained in Chap. 19.

Similarly, one can consider a lower rank 1 convex hull Rf and the lower polyconvex hull Pf of f; in addition, we have defined the lower convex hull f^{co} of f. Since convexity $\Rightarrow$ polyconvexity $\Rightarrow$ quasiconvexity $\Rightarrow$ rank 1 convexity, we have $f^{\mathrm{co}} \le Pf \le Qf \le Rf$. An explicit calculation of Qf is possible only exceptionally (see KOHN & STRANG [1983, 1986]), which leads to seeking optimal lower bounds for Qf; the reader is referred to FIROOZYE [1991] and the references therein. Similar methods are used to obtain optimal lower bounds for the effective conductivity and elasticity tensors of composites, see MURAT & TARTAR [1985], LURIE & CHERKAEV [1984, 1986], and FRANCFORT & MURAT [1986]. The theory of homogenization deals with the overall response of composite materials, like layered or fibered materials, matrix-inclusion composites, etc. I refer to CHRISTENSEN [1979], KOHN & STRANG [1986], MILTON [1990], and DAL MASO [1993].

21.6.6 Relaxation Related to the topics of quasiconvexity and swlsc is the relaxation. Namely, if I as in (21.5.13) is not swlsc on $W^{1,p}(\Omega)$, it is natural to consider the *relaxed functional*, i.e., the greatest swlsc functional which does not exceed I. In particular, one may define a functional I^* by

$$I^*(\chi) = \inf \{ \liminf_{k\to\infty} I(\chi_k) \}$$

where the infimum is taken over all sequences $\chi_k \in W^{1,p}(\Omega)$ such that $\chi_k \rightharpoonup \chi$ in $W^{1,p}(\Omega)$ ($\rightharpoonup^*$ if $p = \infty$), where, in general, f is not quasiconvex. If f satisfies the growth condition (21.5.15), then by DACOROGNA [1981, 1982] and ACERBI & FUSCO [1984],

$$I^*(\chi) = \int_\Omega Qf(\nabla\chi)\, dV, \quad \chi \in W^{1,p}(\Omega), \tag{21.6.3}$$

where Qf is the lower quasiconvex hull of f. The relaxation in the spaces $W^{1,p}(\Omega)$ where p is below the growth exponent r of f is more complicated, and new phenomena arise even when dealing with the relaxation of integrals of polyconvex functions. The reader is referred to the literature cited at the end of this paragraph. However, it may be useful to give an example illustrating the idea behind these works. Namely, following MARCELLINI [1986], one may believe that the formula (21.5.13) gives the correct value if I only provided χ is smooth, or restricted by other requirements depending on the nature of the concrete problem, while for general functions $\chi \in W^{1,p}(\Omega)$ it is given by (21.6.3) with the sequences χ_k consisting of smooth functions, possibly restricted by other requirements. An impressive example by MARCELLINI [1986] is the evaluation of

$$I(\chi) := \int_\Omega |\det \nabla\chi|\, dV$$

where Ω is the unit ball $B(\mathbf{0}, 1) \subset$ Vect and $\chi(\boldsymbol{p}) = \boldsymbol{p}/|\boldsymbol{p}|$ for $\boldsymbol{p} \in B(\mathbf{0}, 1), \boldsymbol{p} \neq \mathbf{0}$. Using the criteria from Subsect. 21.2.1, it is found that $\chi \in W^{1,p}(\Omega)$ for every p such that $1 \le p < 3$ if $n = 3$ and $1 \le p < n$ for a general n. Moreover, the distributional gradient of χ is given by the formula in Subsect. 21.2.1 with $\rho \equiv 1$, from which $\det \nabla\chi = 0$ for a.e. $\boldsymbol{p} \in B(\mathbf{0}, 1)$. Hence, $I(\chi) = 0$ by (21.5.13). However, if I^* is defined by (21.6.3) where the infimum is taken over all sequences $\chi_k \in C^1(\Omega, \text{Vect})$ such that $\chi_k \rightharpoonup \chi$ in $W^{1,p}(\Omega)$, then $I^*(\chi) = \omega_n$, where ω_n is the Lebesgue measure of the unit ball in the n-dimensional Euclidean space. The inequality $I^*(\chi) > I(\chi)$ is explained by the obvious singularity of χ at $\boldsymbol{p} = \mathbf{0}$, which is not reflected in the gradient of χ. In addition to the literature cited above, the reader is referred to DACOROGNA [1989], BUTTAZZO [1989], DAL MASO [1993], FONSECA & MÜLLER [1993], MALÝ [1993], AMAR & DE CICCO [1994], FONSECA & MALÝ [1996] and the references therein, and to GIAQUINTA, MODICA & SOUČEK [1989, 1994, 1996] for the relaxation within the framework of elasticity based on the theory of currents.

Proposition 21.6.3 evaluates $E_\Omega(\boldsymbol{F})$ for the homogeneous boundary condition. For a general $\boldsymbol{d}$, CHIPOT & KINDERLEHRER [1988, Thm. 4.1] show that under Assumptions 21.6.1(1)–(3) and certain technical conditions,

$$E_\Omega(\boldsymbol{d}) = \inf\{\int_\Omega h^{\mathrm{co}}(\det\nabla\boldsymbol{\chi})\,dV : \boldsymbol{\chi}\in\boldsymbol{\Delta}^{**}(\boldsymbol{d})\}.$$

In general, there is not a state $\boldsymbol{\chi}$ on which the minimum over $\boldsymbol{\Delta}^{**}(\boldsymbol{d})$ is attained. One has, however, minimizing sequences, i.e., sequences $\boldsymbol{\chi}_k\in\boldsymbol{\Delta}^{**}(\boldsymbol{d})$ such that

$$\mathsf{M}(\boldsymbol{\chi}_k,\Omega)\to E_\Omega(\boldsymbol{d}) \;\text{ as }\; k\to\infty \tag{21.6.4}$$

as a consequence of the definition of $E_\Omega(\boldsymbol{d})$. In the case of rubber-like materials, the coercivity of $\bar f$ enabled us to conclude that the minimizing sequence was bounded in the appropriate norms, and consequently, we were able to extract a weakly convergent subsequence. Here the coercivity is not satisfied and thus the boundedness of the minimizing sequences must be assumed, and we shall employ a strong form of this assumption. Namely, we shall assume the existence of the minimizing sequence which converges weak* to some $\boldsymbol{\chi}\in\boldsymbol{\Delta}^{**}(\boldsymbol{d})$ in $W^{1,\infty}(\Omega)$. Since the sequence is weakly convergent, we may associate the Young measure limit of deformation gradients with it (see Proposition 21.1.3). More precisely, we lay down the following definition, which applies to a general material.

21.6.7 Definition A ***Young measure minimizer*** $\{\nu_p : \boldsymbol{p}\in\Omega\}$ for $\boldsymbol{\Delta}^{**}(\boldsymbol{d})$ is a Young measure limit of gradients of a sequence $\boldsymbol{\chi}_k\in\boldsymbol{\Delta}^{**}(\boldsymbol{d})$ such that $\boldsymbol{\chi}_k \rightharpoonup^* \boldsymbol{\chi}$ in $W^{1,\infty}(\Omega)$ for some $\boldsymbol{\chi}\in\boldsymbol{\Delta}^{**}(\boldsymbol{d})$ and $\mathsf{M}(\boldsymbol{\chi}_k,\Omega)\to E_\Omega(\boldsymbol{d})$ as $k\to\infty$. A Young measure minimizer $\{\nu_p : \boldsymbol{p}\in\Omega\}$ for $\boldsymbol{\Delta}^{**}(\boldsymbol{d})$ is said to be ***strongly supported*** in Lin^+ if the sequence $\boldsymbol{\chi}_k$ above may be chosen such that $\nabla\boldsymbol{\chi}_k(\boldsymbol{p})\in K$ for all k and a.e. $\boldsymbol{p}\in\Omega$, where K is a fixed compact subset of Lin^+. It then follows that $\operatorname{supp}\nu_p\subset K$ for a.e. point $\boldsymbol{p}\in\Omega$ (with respect to the Lebesgue measure).

If $\{\nu_p : \boldsymbol{p}\in\Omega\}$ is a Young measure minimizer for $\boldsymbol{\Delta}^{**}(\boldsymbol{d})$ and G a continuous function on Lin, one may define the mean value field of G in $\{\nu_p : \boldsymbol{p}\in\Omega\}$ by

$$\bar G(\boldsymbol{p}) = \int_{\mathrm{Lin}} G(\boldsymbol{A})\,d\nu_p(\boldsymbol{A}), \quad \boldsymbol{p}\in\Omega; \tag{21.6.5}$$

moreover, $\bar G$ is the weak limit of $G\circ\nabla\boldsymbol{\chi}_k$ as a direct consequence of the definition of the Young measure. The definition (21.6.5) may be also applied to any continuous function G on Lin^+ which can be extended to a continuous function $\tilde G : \mathrm{Lin}\to\mathbb{R}\cup\{\infty\}$ that is bounded from below. This in particular includes $\bar f$. If G is defined and continuous on Lin^+, then the integral in (21.6.5) may not exist. However, if $\{\nu_p : \boldsymbol{p}\in\Omega\}$ is strongly supported in Lin^+, then $\bar G$ in (21.6.5) is well defined and gives a bounded measurable function of $\boldsymbol{p}$. For instance, the Piola–Kirchhoff stress $\bar{\boldsymbol{S}} = \partial_{\boldsymbol{F}}\bar f$ and the Eshelby tensor $\bar{\boldsymbol{P}} = \bar f\boldsymbol{1} - \boldsymbol{F}^{\mathrm{T}}\bar{\boldsymbol{S}}$ are continuous on Lin^+, but they cannot be extended to a continuous function on Lin, for the stress may become infinite for $\det\boldsymbol{F}\to 0$. Thus, for a Young measure minimizer that is strongly supported in Lin^+, the mean Piola-Kirchhoff stress, the mean Eshelby tensor, and the mean density of the stored energy are

$$\bar{\bar{\boldsymbol{S}}}(\boldsymbol{p}) = \int_{\mathrm{Lin}} \bar{\boldsymbol{S}}(\boldsymbol{A})\,d\nu_p(\boldsymbol{A}), \quad \bar{\bar{\boldsymbol{P}}}(\boldsymbol{p}) = \int_{\mathrm{Lin}} \bar{\boldsymbol{P}}(\boldsymbol{A})\,d\nu_p(\boldsymbol{A}), \quad \bar{\bar f}(\boldsymbol{p}) = \int_{\mathrm{Lin}} \bar f(\boldsymbol{A})\,d\nu_p(\boldsymbol{A}), \quad \boldsymbol{p}\in\Omega.$$

21.6.8 Change of Reference Configuration Consider a change of the reference configuration of the body Ω of the form $\Omega \mapsto \Omega' = z(\Omega)$, $\boldsymbol{p} \mapsto \boldsymbol{p}' = z(\boldsymbol{p})$ as in Subsect. 9.4, where $z : \Omega \to \Omega'$ is a mapping which can be extended to an orientation-preserving diffeomorphism from an open set containing $\mathrm{cl}\,\Omega$ to an open set containing $\mathrm{cl}\,\Omega'$. Set $\boldsymbol{H}(\boldsymbol{p}) = \nabla z(\boldsymbol{p})$. Let $\chi_k \rightharpoonup^* \chi$ in $W^{1,\infty}(\Omega)$ and let $\{\nu_{\boldsymbol{p}}, \boldsymbol{p} \in \Omega\}$ be the Young measure limit of deformation gradients of χ_k. In the new reference configuration, the deformations χ_k, χ are described by mappings $\chi_k' = \chi_k \circ z^{-1}$, $\chi' = \chi \circ z^{-1}$ where $\circ$ denotes the composition and z^{-1} is the inverse function of z. It can be proved that $\chi_k' \rightharpoonup^* \chi'$ in $W^{1,\infty}(\Omega')$ and the Young measure limit $\{\nu'_{\boldsymbol{p}'}, \boldsymbol{p}' \in \Omega'\}$ of deformation gradients of χ_k' is

$$\nu'_{\boldsymbol{p}'}(M') = \nu_{\boldsymbol{p}}(M), \quad \text{where } \boldsymbol{p} = z^{-1}(\boldsymbol{p}') \text{ and } M = \{\boldsymbol{F}'\boldsymbol{H}(\boldsymbol{p}) : \boldsymbol{F}' \in M'\}$$

for a.e. $\boldsymbol{p}' \in \Omega'$, and every Borel set $M' \subset \mathrm{Lin}$. Put otherwise, if $G \in C(\mathrm{W})$, then

$$\int_{\mathrm{Lin}} G(\boldsymbol{A}')\, d\nu'_{\boldsymbol{p}'}(\boldsymbol{A}') = \int_{\mathrm{Lin}} G(\boldsymbol{A}\boldsymbol{H}^{-1}(\boldsymbol{p}))\, d\nu_{\boldsymbol{p}}(\boldsymbol{A})$$

for a.e. $\boldsymbol{p}' \in \Omega'$, where $\boldsymbol{p} = z^{-1}(\boldsymbol{p}')$. This follows from the rule for the change of the deformation gradient under the change of the reference configuration $\boldsymbol{F}_k' := \nabla\chi_k' = (\boldsymbol{F}_k \circ z^{-1})\boldsymbol{H}^{-1}$. Consequently,

$$\int_{\Omega'} \int_{\mathrm{Lin}} G(\boldsymbol{A}')\, d\nu'_{\boldsymbol{p}'}(\boldsymbol{A}')\, dV(\boldsymbol{p}') = \int_{\Omega} \int_{\mathrm{Lin}} G(\boldsymbol{A}\boldsymbol{H}^{-1}(\boldsymbol{p}))\, \det \boldsymbol{H}(\boldsymbol{p})\, d\nu_{\boldsymbol{p}}(\boldsymbol{A})\, dV(\boldsymbol{p}).$$

The following proposition gives two versions of the principle of minimum total stored energy for 'generalized states' described by Young measures. Moreover, it is also shown that the mean stress and mean Eshelby tensor satisfy the balance equation and the Eshelby conservation law in the sense of distributions.

21.6.9 Proposition *Let $\{\nu_{\boldsymbol{p}} : \boldsymbol{p} \in \Omega\}$ be a Young measure minimizer for $\Delta^{**}(\boldsymbol{d})$ which is strongly supported in* Lin^+. *Then the following assertions hold:*

(1) *Let z be an orientation-preserving diffeomorphism from an open set containing $\mathrm{cl}\,\Omega$ to an open set containing $\mathrm{cl}\,\Omega$ such that $z(\Omega) = \Omega$ and $z(\boldsymbol{p}) = \boldsymbol{p}$ for every $\boldsymbol{p} \in \partial\Omega$. Then*

$$\int_{\Omega} \int_{\mathrm{Lin}} \bar f(\boldsymbol{A}\boldsymbol{H}^{-1}(\boldsymbol{p}))\, \det \boldsymbol{H}(\boldsymbol{p})\, d\nu_{\boldsymbol{p}}(\boldsymbol{A})\, dV(\boldsymbol{p}) \ge \int_{\Omega} \int_{\mathrm{Lin}} \bar f(\boldsymbol{A})\, d\nu_{\boldsymbol{p}}(\boldsymbol{A})\, dV(\boldsymbol{p})$$

where $\boldsymbol{H} = \nabla z$.

(2) *There exists an $\varepsilon > 0$ such that for every $\boldsymbol{u} \in W_0^{1,\infty}(\Omega, \mathrm{Vect})$ with $\|\boldsymbol{u}\|_{W^{1,\infty}(\Omega)} < \varepsilon$,*

$$\int_{\Omega} \int_{\mathrm{Lin}} \bar f(\boldsymbol{A} + \nabla\boldsymbol{u}(\boldsymbol{p}))\, d\nu_{\boldsymbol{p}}(\boldsymbol{A})\, dV(\boldsymbol{p}) \ge \int_{\Omega} \int_{\mathrm{Lin}} \bar f(\boldsymbol{A})\, d\nu_{\boldsymbol{p}}(\boldsymbol{A})\, dV(\boldsymbol{p}).$$

(3) *The mean Piola–Kirchhoff stress and the mean Eshelby tensor satisfy*

$$\mathrm{Div}\,\bar{\bar{\boldsymbol{S}}} = \boldsymbol{0}, \quad \mathrm{Div}\,\bar{\bar{\boldsymbol{P}}} = \boldsymbol{0} \text{ in } \Omega$$

in the sense of distributions, i.e.,

$$\int_{\Omega} \bar{\bar{\boldsymbol{S}}} \cdot \nabla\boldsymbol{u}\, dV = 0, \quad \int_{\Omega} \bar{\bar{\boldsymbol{P}}} \cdot \nabla\boldsymbol{u}\, dV = 0 \text{ for every } \boldsymbol{u} \in C_0^\infty(\Omega). \tag{21.6.6}$$

Proof (1): Let χ_k be the minimizing sequence occurring in the definition of the Y. M. minimizer, and let χ_k' be the sequence constructed in Subsect. 21.6.8. By the hypotheses of (1), $\Omega' = \Omega$ and

$\chi_k' \in \Delta^{\#\#}(d)$. Since the values of $\nabla\chi_k$ are contained in some fixed compact subset K of Lin^+, the values of $\nabla\chi_k'$ are contained in some fixed compact subset K' of Lin^+. Consequently,

$$\lim_{k\to\infty} \mathsf{M}(\chi_k', \Omega) = \int_\Omega \int_{\mathrm{Lin}} \bar f(A')\, d\nu_{p'}'\, dV(p') = \int_\Omega \int_{\mathrm{Lin}} G(AH^{-1}(p)) \det H(p)\, d\nu_p(A)\, dV(p),$$

where $\nu_{p'}'$ is as in Subsect. 21.6.8. On the other hand, since $\chi_k' \in \Delta^{\#\#}(d)$, necessarily $\mathsf{M}(\chi_k', \Omega) \ge E_\Omega(d)$ for each k and the assertion follows. (2): This is proved similarly, one only considers the sequence $\chi_k' = \chi_k + u$. The hypotheses imply that $\chi_k' \in \Delta^{\#\#}(d)$ provided $\varepsilon > 0$ is chosen such that the set $K' := \{A \in \mathrm{Lin} : \mathrm{dist}(A, K) < \varepsilon\} \subset \mathrm{Lin}^+$, and such an ε exists since K is a compact subset of Lin^+. (3): Let $u \in C_0^\infty(\Omega)$. To obtain $(21.6.6)_1$, we apply (2) with u replaced by λu where $\lambda \in \mathbb{R}$ is sufficiently close to 0 to obtain

$$\int_\Omega \int_{\mathrm{Lin}} \bar f(A + \lambda \nabla u(p)))\, d\nu_p(A)\, dV(p) \ge \int_\Omega \int_{\mathrm{Lin}} \bar f(A)\, d\nu_p(A)\, dV(p).$$

Thus the function of λ given by the left-hand side of the last equation has a local minimum at $\lambda = 0$. Its derivative exists as a consequence of the hypothesis that $\{\nu_p, p \in \Omega\}$ is strongly supported in Lin^+ by the standard theorem on the derivative of the integral with respect to a parameter. Moreover, by this theorem, the derivative is given by formally differentiating under the integral sign. Performing the differentiation and using the definition of $\bar S$, one obtains $(21.6.6)_1$. Equation $(21.6.6)_2$ is proved similarly. We apply (1) with the choice $z^{-1} := p + \lambda u(p)$ where $\lambda \in \mathbb{R}$ is sufficiently close to 0 to obtain

$$\int_\Omega \int_{\mathrm{Lin}} \bar f(A(1 + \lambda \nabla u(p))) \det(1 + \lambda \nabla u(p))^{-1}\, d\nu_p(A)\, dV(p) \ge \int_\Omega \int_{\mathrm{Lin}} \bar f(A)\, d\nu_p(A)\, dV(p)$$

and differentiate as above. □

For the materials of the crystal symmetry, the following specific results hold.

21.6.10 Proposition *Let Assumptions* 21.6.1(1)–(3) *hold, and let* $\{\nu_p, p \in \Omega\}$ *be a Young measure minimizer for* $\Delta^{\#\#}(d)$ *which is strongly supported in* Lin^+. *Then*

(1) $\bar f$ *is* $d\nu_p$ *integrable and*

$$\int_{\mathrm{Lin}} \bar f(A)\, d\nu_p(A) = h^{\mathrm{co}}(\det F(p)) \quad \text{for a.e. } p \in \Omega,$$

where $F := \nabla\chi$;

(2) *for a.e. point* $p \in \Omega$ *(with respect to the Lebesgue measure)*

$$\operatorname{supp} \nu_p \subset Q,$$

where Q *is the set introduced in Proposition* 21.6.4;

(3) *if and* h^{co} *is continuously differentiable, then*

$$p := -\partial_v h^{\mathrm{co}}(\det F(p)) \quad \text{and} \quad g := h^{\mathrm{co}}(\det F(p)) + p \det F(p)$$

are constant on Ω *and*

$$\bar{\bar S} = -p \operatorname{cof} F(p), \quad \bar{\bar P} = g\mathbf{1}$$

for a.e. $p \in \Omega$.

I refer to CHIPOT & KINDERLEHRER [1988] for proof. The last two propositions say that under mild additional conditions, the Young measure minimizers satisfy appropriate generalizations of most of the main properties of the classical minimizers. Thus, e.g., using the interpretation of Q from Proposition 21.6.4, we see that Item (2) of the above proposition may be viewed as a generalization of Proposition 17.1.4 that the stored energy function must be quasiconvex at all values of deformation gradient realized in a classical minimizer. The reader is referred to CHIPOT & KINDERLEHRER

[1988], JAMES & KINDERLEHRER [1989], and KINDERLEHRER & PEDREGAL [1991] for a number of further results of this nature.

Another major case of minimization of a nonelliptic stored energy function is the two-well problem. The objective stored energy function is assumed to have two compatible or incompatible wells (see Sect. 2.3 for the terminology), and one seeks the Young measure minimizers giving the absolute minimum of the total stored energy. We saw in Sect. 2.3 that the wells structure of this type arise at the temperature of the martensitic transformation in solids. The reader is referred to BALL & JAMES [1987, 1992] for details of the physical motivation and for the mathematical treatment; for further results, see BHATTACHARYA [1992, 1993], BHATTACHARYA & al. [1994], RUDDOCK [1994], and the references therein.

Bibliographical Note

The existence theory for polyconvex stored energy functions is due to BALL [1977]. The reader is referred to BALL, CURRIE & OLVER [1981] for further developments. ZHANG [1991] discusses the relationships between the energy minimizers and the solutions obtained by the implicit function theorem as outlined in Sect. 20.9. LIN [1990] studies the existence of solutions of the entropy maximum principle for steady states of bodies with nonzero velocity fields.

The existence theory for problems with unilateral constraints, and the relevant literature, may be found in CIARLET [1988]. Additional remarks and the existence theory for lower-dimensional models of nonlinear elasticity may be found in ANTMAN [1995]. The treatment above in this section is based on CHIPOT & KINDERLEHRER [1988] and FONSECA [1988]. For a theory of equilibrium and for the relaxation in defective crystals, see the bibliographical note to Chap. 9.

Part V

Dynamics

22 Dynamical Thermoelastic and Adiabatic Theories

This part deals with aspects of behavior of thermoelastic bodies with heat conduction and viscosity in dynamical situations. The governing equations are the balance equations of Sects. 3.4–3.7 combined with the constitutive equations of Sect. 9.1; moreover, the constitutive equations must be compatible with the Clausius–Duhem inequality for smooth processes, see Sect. 9.2. In their full generality, the constitutive equations allow for a dissipation by the heat conduction and viscosity, but they also include as special cases the thermoelastic materials 9.1.3(1), where the heat conduction is present and the viscosity is absent, and the idealized dissipationless ('adiabatic') materials 9.1.3(2), where both the heat conduction and viscosity are absent. The behavior and interpretation of solutions strongly depend on whether the equations describe a material in which the viscosity and heat conduction are really present or whether they describe one of the two subcases mentioned above. Moreover, there are also situations when one can pass to isothermal or isentropic dynamic theories, thus omitting formally the thermal phenomena altogether. The latter two theories are often referred to by the common name 'elastodynamics'. Their basic equations are formally identical, but for a given material, the isothermal elastodynamics uses the stress expressed as a function of the deformation gradient at a given fixed temperature, whereas the isentropic dynamics uses the stress expressed as a function of the deformation gradient at a fixed entropy.

The present chapter gives a review of the basic equations of the thermoelastic, isothermal, adiabatic, and isentropic theories, with the emphasis on moving singular surfaces. Unlike the equilibrium case, where the singular surface can represent only a phase boundary, in the dynamical situation it may represent either an evolving phase boundary (melting, crystallization) or a shock wave. The constitutive differences between the two classes of singular surfaces will be described, and then the propagating phase boundaries will be treated in some detail. The propagating phase boundaries can be consistently described only if the model is capable of supporting the change of phase by bringing the necessary balance of latent heat to the phase boundary. Within the above simplified theories, this condition is met only by the thermoelasticity and the isothermal elastodynamics. The wave phenomena (including the shock waves) are consistently modeled by the thermoelasticity and adiabatic or isentropic dynamic theories, but they cannot be adequately modeled by the isothermal elastodynamics (e.g., it gives wrong values of the speed of sound). Note that although the present chapter uses the referential description, the results apply both to solids and fluids.

Chaps. 23 and 24 treat in more detail the wave motions within the adiabatic dynamic theories, and Chap. 25 treats the bodies in which both the heat conduction and viscosity are really present.

22.1 Equations of Dynamic Thermoelasticity

22.1.1 The Constitutive and Balance Equations For a thermoelastic material 9.1.3(1) the constitutive equations reduce to

$$f = \bar{f}(\boldsymbol{F}, \theta), \quad \boldsymbol{S} = \bar{\boldsymbol{S}}(\boldsymbol{F}, \theta), \quad \eta = \bar{\eta}(\boldsymbol{F}, \theta), \quad \boldsymbol{q} = \bar{\boldsymbol{q}}(\boldsymbol{F}, \theta, \boldsymbol{G}), \tag{22.1.1}$$

with

$$\bar{\boldsymbol{S}} = \rho_\circ \partial_{\boldsymbol{F}} \bar{f}, \quad \bar{\eta} = -\partial_\theta \bar{f}, \quad \bar{\boldsymbol{q}}(\boldsymbol{F}, \theta, \boldsymbol{G}) \cdot \boldsymbol{G} \le 0 \tag{22.1.2}$$

identically for $(\boldsymbol{F}, \theta) \in \bar{U} \subset \mathrm{InvLin} \times \mathbb{R}$, $\boldsymbol{G} \in \mathrm{Vect}$. In the absence of the body supply of heat r, the balance equations take the form

$$\rho_\circ \ddot{\boldsymbol{\chi}} = \operatorname{Div} \boldsymbol{S} + \rho_\circ \boldsymbol{b}, \quad \rho_\circ (e + \tfrac{1}{2} v^2)^{\cdot} = \operatorname{Div}(\boldsymbol{S}^{\mathrm{T}} \boldsymbol{v} - \boldsymbol{q}) + \rho_\circ \boldsymbol{b} \cdot \boldsymbol{v} \tag{22.1.3}$$

outside the singular surface, and

$$-U\rho_\circ [\boldsymbol{v}] = [\boldsymbol{S}]\boldsymbol{n}, \quad -U\rho_\circ [e + \tfrac{1}{2} v^2] = [\boldsymbol{S}^{\mathrm{T}} \boldsymbol{v} - \boldsymbol{q}] \cdot \boldsymbol{n} \tag{22.1.4}$$

on the singular surface. The reduced equations of balance of energy (3.6.5) read

$$\rho_\circ \dot{e} = \boldsymbol{S} \cdot \dot{\boldsymbol{F}} - \operatorname{Div} \boldsymbol{q}, \quad U\rho_\circ [e] = \langle \boldsymbol{S} \rangle \cdot [\boldsymbol{F}] + [\boldsymbol{q}] \cdot \boldsymbol{n} \tag{22.1.5}$$

outside and on the singular surface, respectively. As a consequence of the Hadamard lemma we have

$$[\boldsymbol{F}] = \boldsymbol{a} \otimes \boldsymbol{n}, \quad [\boldsymbol{v}] = -U\boldsymbol{a}, \quad \rho_\circ U^2 [\boldsymbol{F}]\boldsymbol{n} = [\boldsymbol{S}]\boldsymbol{n}, \tag{22.1.6}$$

where $\boldsymbol{a}$ is the amplitude of the jump. The Clausius–Duhem inequality on the singular surface reads

$$-U\rho_\circ [\eta] \ge -\left[\frac{\boldsymbol{q}}{\theta}\right] \cdot \boldsymbol{n}, \tag{22.1.7}$$

with the reduced forms (see Sect. 3.7)

$$-U\rho_\circ \langle \theta \rangle [\eta] \ge -U(\rho_\circ [e] - \langle \boldsymbol{S} \rangle \cdot [\boldsymbol{F}]) + [\theta] \langle \boldsymbol{q}/\theta \rangle \cdot \boldsymbol{n}, \tag{22.1.8}$$

or equivalently

$$U(\rho_\circ [f] - \langle \boldsymbol{S} \rangle \cdot [\boldsymbol{F}] + \rho_\circ \langle \eta \rangle [\theta]) - [\theta] \langle \boldsymbol{q}/\theta \rangle \cdot \boldsymbol{n} \ge 0. \tag{22.1.9}$$

For a prescribed body force $\boldsymbol{b}$ and specified boundary conditions the equations (22.1.1)–(22.1.6) form a closed evolutionary system, with the same number of unknown functions as the number of evolution equations. The initial value problem is to find a solution satisfying the equations, the appropriate boundary conditions, and the prescribed initial data, namely, $\boldsymbol{\chi}(\cdot, 0) = \boldsymbol{\chi}^i(\cdot)$, $\dot{\boldsymbol{\chi}}(\cdot, 0) = \boldsymbol{v}^i(\cdot)$, $\theta(\cdot, 0) = \theta^i(\cdot)$, where $\boldsymbol{\chi}^i, \boldsymbol{v}^i, \theta^i$ are given functions of the referential position, subject to appropriate restrictions. The available proofs of existence are either for the one-dimensional case or only short-time existence theorems. Of particular physical importance is the Riemann problem, wherein the body is assumed to fill the whole space, $\Omega = \mathrm{Eucl}$, no boundary conditions are prescribed, and the initial data are such that the velocity, deformation gradient and temperature (or entropy) are assumed to be constant on two complementary half-spaces.

The reader is referred to Bibliographical Note to Chap. 25 for relevant literature on the existence theory for equations of thermoelasticity.

22.1.2 The Entropy Inequality and Admissibility Outside the singular surface, the Clausius–Duhem inequality is a consequence of (22.1.2), but on the singular surface, inequality (22.1.7) is an independent requirement. Only the processes which satisfy (22.1.7) are considered as physically reasonable: the entropy inequality is an admissibility criterion for jumps. In certain *special cases,* this entropy admissibility criterion enables one to prove the uniqueness for the initial-value problem, while without (22.1.7) the uniqueness fails. However, matters are not so quite simple: (22.1.7) is not strong enough to secure the uniqueness in general, and stronger admissibility criteria must used to guarantee the physical soundness and the mathematical uniqueness of the solutions. (For shocks, some criteria will be reviewed in Sects. 23.8 and 24.4.) When the system allows phase boundaries, the nonuniqueness is so large that the system of equations admits extra kinetic equations for the dynamics of phase boundaries. This may be viewed as an extremely strong admissibility criterion. See Sect. 22.2.

As mentioned in the introduction to this chapter, within the framework of the above constitutive and balance equations, the singular surfaces can describe both the propagating phase boundaries and shock waves. However, the distinction between the two classes is not just a matter of interpretation. Let us now describe the properties of propagating phase boundaries and the associated consequences on the response functions; the framework for the shock waves will be described in Sect. 22.3.

22.1.3 Coexistent Phases and Moving Phase Boundaries We shall take the point of view that a jump describes a moving phase boundary provided the values of the state parameters $\boldsymbol{F}^{\pm}, \theta^{\pm}$ at the singular surface are close to some static values $\boldsymbol{F}_0^{\pm}, \theta$ realized in some stable equilibrium state with two coexistent phases. As a consequence to be proved below, the phase boundaries move subsonically with respect to the states on both sides of the jump; more precisely a moving phase boundary can be connected to a stable equilibrium state of two phases by a continuous family of phase boundaries of speeds monotonically tending to zero, while for shocks this is impossible: the shock waves move with the speeds of order of the speed of sound on both sides of the jump. Secondly, the existence of a stable equilibrium state with the two coexistent phases implies that the free energy has to violate the Legendre–Hadamard condition in the spinodal region, see Sect. 17.4. In the one-dimensional case this means that the stress–strain relation cannot be monotonic. This implies that the *evolution* equations change their type from hyperbolic or parabolic to *elliptic* in the spinodal region. The systems are thus formally of mixed hyperbolic–elliptic type. Thirdly, the dynamic changes of phase are often accompanied by the latent heat, i.e., the heat that must be communicated to the substance to convert it from one phase to another. A consistent theory therefore requires a nontrivial heat conduction to bring the heat to the interface.

Suppose the interface separating the stable equilibrium phases has a normal $\boldsymbol{n}$ and that $\boldsymbol{F}_0^+\boldsymbol{n} \neq \boldsymbol{F}_0^-\boldsymbol{n}$. Maxwell's relation and the balance of forces at the interface (see Sect. 17.4) read

$$\rho_0 \bar{f}(\boldsymbol{F}_0^+, \theta) = \rho_0 \bar{f}(\boldsymbol{F}_0^-, \theta) + \boldsymbol{S}_0^+ \cdot (\boldsymbol{F}_0^+ - \boldsymbol{F}_0^-), \quad \boldsymbol{S}_0^+\boldsymbol{n} = \boldsymbol{S}_0^-\boldsymbol{n}, \tag{22.1.10}$$

where $\boldsymbol{S}_0^{\pm} = \bar{\boldsymbol{S}}(\boldsymbol{F}_0^{\pm}, \theta)$. To demonstrate the subsonic character of propagating phase boundaries, let us show that their speed is arbitrarily small provided the values $\boldsymbol{F}^{\pm}, \theta^{\pm}$ are close to the equilibrium values $\boldsymbol{F}_0^{\pm}, \theta_0^{\pm}$. Hence, suppose that we have a family of processes, indexed by $k = 1, \ldots,$ such that each contains a discontinuity with $\boldsymbol{F}^{\pm} = \boldsymbol{F}_k^{\pm}, \theta^{\pm} = \theta_k^{\pm}, U = U_k \neq 0$, and moreover $\boldsymbol{F}_k^{\pm} \to \boldsymbol{F}_0^{\pm}, \theta_k^{\pm} \to \theta$ as $k \to \infty$. Writing $(22.1.6)_3$ for the jump k, letting $k \to \infty$ and using $[\boldsymbol{F}_k]\boldsymbol{n} \to [\boldsymbol{F}_0]\boldsymbol{n} \neq \boldsymbol{0}$, $[\boldsymbol{S}_k]\boldsymbol{n} \to [\boldsymbol{S}_0]\boldsymbol{n} = \boldsymbol{0}$ we find that $U_k \to 0$. There is no analog of this for shocks in the hyperbolic region of the equations. Indeed in Proposition 23.6.2 we shall see that for a family of shocks of amplitudes tending to $\boldsymbol{0}$ the speed will tend to one of the speeds of sound at the given point (sonic character of shocks).

22.1.4 Latent Heat For a general jump as in Subsect. 22.1.1, the quantity $[\boldsymbol{q}] \cdot \boldsymbol{n}$ is the heat that must be communicated to unit area of the singular surface per unit time to support the change of the state across the jump, and $\rho_\circ U$ is the mass that passes through the unit referential area of the singular surface per unit time; see Sect. 3.4. Hence the quantity

$$l = \frac{[\boldsymbol{q}] \cdot \boldsymbol{n}}{\rho_\circ U} = [e] + \rho_\circ^{-1} \langle \boldsymbol{S} \rangle \cdot [\boldsymbol{F}]$$

is the heat that must be communicated to the singular surface to support the process in which unit mass passes through it and changes its state from + to −. For a sequence of jumps as above the corresponding $l = l_k$ then satisfy $l_k \to l_0$, where

$$l_0 = e_0^+ - e_0^- - \rho_\circ^{-1} \boldsymbol{S}_0^+ \boldsymbol{n} \cdot (\boldsymbol{F}_0^+ - \boldsymbol{F}_0^-)\boldsymbol{n} = \theta(\eta_0^+ - \eta_0^-), \qquad (22.1.11)$$

with the last equation obtained with the use of (22.1.10). The quantity l_0 is the (quasi-static) ***latent heat of phase transformation***, the amount of heat that must be delivered to the interface to change unit mass of phase + into the phase −. Each member of the family describes a propagating phase boundary, in which the phase labeled + is changed to a phase labeled − if $U_k > 0$. There are phase transitions in which $l_0 \neq 0$, but there are also ones for which $l_0 = 0$. The latter happens when the phases $(\boldsymbol{F}^+, \theta)$ and $(\boldsymbol{F}^-, \theta)$ are related by a symmetry transformation of the material, like different variants in crystals (cf. Subsect. 2.3.5). We see that if $l_0 \neq 0$, then necessarily the material must be capable of conducting heat to support dynamic changes of phase with $(\boldsymbol{F}^{\pm}, \theta^{\pm})$ close to $(\boldsymbol{F}_0^{\pm}, \theta)$.

22.2 Extra Conditions for Evolving Phase Boundaries

22.2.1 A Kinetic Equation for a Phase Boundary Experience with the initial-value problem for materials that exhibit states with coexistent phases shows a massive nonuniqueness in the initial-value problem, even when the entropy inequality is in force as an admissibility criterion, see JAMES [1980], SHEARER [1986] and ABEYARATNE & KNOWLES [1992]. It appears that in the general, nonisothermal case, one can impose two additional kinetic equations on a phase boundary. Observe first that the system of equations of Sect. 22.1, including the Clausius–Duhem inequality, admits as a solution any process in which (1) (χ, θ) are independent of time, (2) (χ, θ) is pairwise homogeneous in the sense that the pair $(\boldsymbol{F}(\boldsymbol{p}), \theta(\boldsymbol{p}))$ takes only two values $(\boldsymbol{F}^{\pm}, \theta^{\pm})$, and (3) the equation of balance of forces $(22.1.4)_1$ holds at the interface. If $\theta^+ \neq \theta^-$, then this state is not even an equilibrium state; if $\theta^+ = \theta^-$ but the Maxwell relation $[f] - \rho_\circ^{-1} \langle \boldsymbol{S} \rangle \cdot [\boldsymbol{F}] = 0$ is violated, then the state is unstable. Thus the mere requirement that the only time-independent processes admitted by the model are stable the equilibrium states calls for additional restrictions on the class of admissible processes; it will be seen that the imposition of

$$[f]-\rho_\circ^{-1}\langle \boldsymbol{S}\rangle\cdot[\boldsymbol{F}]=0,\quad [\theta]=0 \tag{22.2.1}$$

everywhere on the phase boundary is a consistent choice; one which leads to a dissipationless phase boundary. More generally, let us rewrite the entropy inequality (22.1.9) in the form $\boldsymbol{J}\cdot\boldsymbol{X}\geq 0$, where the thermodynamic force $\boldsymbol{X}\in\mathbb{R}^2$ and the thermodynamic flux $\boldsymbol{J}\in\mathbb{R}^2$ are defined by

$$\boldsymbol{X}=([f]-\rho_\circ^{-1}\langle \boldsymbol{S}\rangle\cdot[\boldsymbol{F}]+\langle\eta\rangle[\theta],[\theta]),\qquad \boldsymbol{J}=(U,-\rho_\circ^{-1}\langle \boldsymbol{q}/\theta\rangle\cdot\boldsymbol{n});$$

the quantity $[f]-\rho_\circ^{-1}\langle \boldsymbol{S}\rangle\cdot[\boldsymbol{F}]$ is called the driving traction of the phase transition. Note that in a stable static state with two coexistent phases $\boldsymbol{X}=\boldsymbol{0}$ as a consequence of the Maxwell relation. An appeal to the general prescription for a formulation of the kinetic equations (see Sect. 12.2) leads to a pair of independent constitutive equations

$$\boldsymbol{J}=\boldsymbol{L}\boldsymbol{X}\ \text{ where }\ \boldsymbol{L}=\begin{bmatrix}\alpha & \beta\\ \beta & \delta\end{bmatrix}$$

is a given positive-definite symmetric matrix of kinetic coefficients. In components,

$$\begin{aligned}U&=\alpha([f]-\rho_\circ^{-1}\langle \boldsymbol{S}\rangle\cdot[\boldsymbol{F}]+\langle\eta\rangle[\theta])+\beta[\theta],\\ -\rho_\circ^{-1}\langle \boldsymbol{q}/\theta\rangle\cdot\boldsymbol{n}&=\beta([f]-\rho_\circ^{-1}\langle \boldsymbol{S}\rangle\cdot[\boldsymbol{F}]+\langle\eta\rangle[\theta])+\delta[\theta].\end{aligned} \tag{22.2.2}$$

If the equilibrium conditions prevail on the singular surface, i.e., if $U=0$, $\boldsymbol{q}^{\pm}=\boldsymbol{0}$ then the kinetic equation implies $\boldsymbol{X}=\boldsymbol{0}$, i.e.,

$$[f]-\rho_\circ^{-1}\langle \boldsymbol{S}\rangle\cdot[\boldsymbol{F}]=0,\quad [\theta]=0.$$

There are some special or limiting cases considered previously in the literature. Thus, in a completely isothermal theory, when θ is a specified spacetime constant (see Subsect. 22.2.4 below), the kinetic relation reduces to that of ABEYARATNE & KNOWLES [1987–1994], TRUSKINOVSKY [1987] and GURTIN & STRUTHERS [1990], namely

$$U=\alpha([f]-\rho_\circ^{-1}\langle \boldsymbol{S}\rangle\cdot[\boldsymbol{F}]),\quad \alpha>0,\ \text{ or }\ [f]-\rho_\circ^{-1}\langle \boldsymbol{S}\rangle\cdot[\boldsymbol{F}]=\alpha^{-1}U, \tag{22.2.3}$$

Also a nonlinear analog of $(22.2.3)_2$, namely

$$[f]-\rho_\circ^{-1}\langle \boldsymbol{S}\rangle\cdot[\boldsymbol{F}]=\phi(U),\quad \phi(0)=0,\quad \phi(U)\geq 0,$$

with ϕ a prescribed scalar-valued function, is considered by ABEYARATNE & KNOWLES [1987–1994]. These authors also show that with the linear kinetic equation, the Riemann problem is solvable in some special cases, which indicates the consistency of the kinetic equation with the rest of the constitutive and balance structure for isothermal thermoelasticity. In the nonisothermal case, they impose the following relations at the interface:

$$[f]-\rho_\circ^{-1}\langle \boldsymbol{S}\rangle\cdot[\boldsymbol{F}]=\alpha^{-1}U,\quad [\theta]=0. \tag{22.2.4}$$

However, the mathematical experience with the extra conditions at the interface is only limited, with one significant exception.

22.2.2 The Stefan Problem The above setting of propagating phase boundaries within the complete thermomechanical context may be considered as a broad generalization of the Stefan problem (see, e.g., RUBINSTEIN [1971], MEIRMANOV [1992]). This is a problem of determining an unknown phase boundary between a solid and its melt in which the mechanical part of the phenomenon is completely neglected. In the case of the isotropic solid phase, the only material parameters are the heat conductivity coefficients $\kappa_s, \kappa_f > 0$ of the solid and fluid phases, their specific heats $c_s, c_f > 0$, the melting temperature $\theta_m > 0$, and the latent heat l of the phase transformation at θ_m. Assume that $\kappa_s, \kappa_f, c_s, c_f, \theta_m, l$ are absolute constants; the generalization to the case when $\kappa_s, \kappa_f, c_s, c_f$ are functions of θ is straightforward; set $\rho_\circ = 1$. The only unknown is the temperature $\theta = \theta(\boldsymbol{p}, t)$ which is assumed continuous and piecewise continuously differentiable, with the singular surface S_t, which divides Ω into two time-dependent regions $\Omega_t^\pm$, not necessarily connected, such that $\Omega_t^\pm = \{\boldsymbol{p} \in \Omega : \pm(\theta(\boldsymbol{p}, t) - \theta_m) < 0\}$. These represent the solid and fluid phases, respectively. The only relevant equation is the balance of energy. In the absence of the external sources of heat, it reads

$$c\dot{\theta} = -\operatorname{Div} \boldsymbol{q}, \quad lU = [\boldsymbol{q}] \cdot \boldsymbol{n}$$

outside and on the singular surface, respectively, where $U, \boldsymbol{n}$ are the speed of and the normal to S_t and

$$\boldsymbol{q} = -\kappa_s \nabla\theta \ \text{ in } \ \Omega_t^+, \quad \boldsymbol{q} = -\kappa_l \nabla\theta \ \text{ in } \ \Omega_t^-.$$

Appropriate boundary conditions must be prescribed on $\partial\Omega$. Thus, as part of the formulation, we have the ***interfacial Stefan condition*** $[\theta] = 0$ on S_t; the conditions (22.2.1) may be considered as a generalization of this to the full thermomechanical setting and the kinetic equations as further extensions. The large body of experimental and mathematical experience with the Stefan problem has shown that the interfacial Stefan condition is an appropriate assumption; in particular, it leads to physically natural existence and uniqueness results.

22.2.3 The Structure of the Interface: Gradient Theory The constitutive equations (22.2.2) or (22.2.3) may be viewed as admissibility criteria for propagating phase boundaries. A different approach to the admissibility of phase boundaries is to view the interface as an idealization of a region of finite width in which the deformation gradient, temperature, and the related quantities change rapidly but smoothly. (The problem is similar to the problem of the structure of shocks to be considered in Sect. 23.8 and 24.5.) The detailed structure of the interface is governed by separate constitutive equations and in the obvious idealization the interface is modeled as an essentially one-dimensional process in the range of the spatial variables from $-\infty$ to ∞. The values of $\boldsymbol{F}^\pm, \theta^\pm$, etc. in a phase boundary are identified consistently with the limiting values of the respective quantities at $\pm\infty$ in the smooth model. To admit a smooth equilibrium solution with nonconstant deformation gradient (static interface), the stress must depend on higher gradients of deformation, i.e., the interface cannot be modeled by the thermoelastic material with heat conduction and viscosity. Such broader models go back to van der Waals (1893) and Korteweg (1901), with important contributions by Cahn and Hilliard (1958). In the simplest case of fluids, the free energy is given by

$$\bar{f}(\rho, \operatorname{grad}\rho, \theta) = \bar{f}_0(\rho, \theta) + \varepsilon(\rho, \theta)|\operatorname{grad}\rho|^2,$$

where ρ is the density, and $\varepsilon > 0$ a constant. Another possibility is to use the Ginzbug–Landau equations, see TRUSKINOVSKY [1991, Sect. 4]. More importantly, these models require also a significant modification of the structure of the *balance equations*, otherwise they are incompatible with the Clausius–Duhem inequality. I refer to DUNN & SERRIN [1985] and the references therein for such modifications. Another way to reconcile the higher-order models with the usual structure of the balance equations and the Clausius–Duhem inequality has been suggested in ŠILHAVÝ [1984,1985c].

Apart from this interpretational problem, the main conclusion is that these models impose a relation among the parameters of the sharp phase boundaries similar to the kinetic relation considered above. (See ABEYARATNE & KNOWLES [1991b].)

There is a large literature on the gradient theory of phase transformations: CAHN & HILLIARD [1958, 1971], CAHN [1961, 1962], TRUSKINOVSKY [1982, 1983a, 1985, 1987, 1991], SLEMROD [1983, 1984, 1989], HAGAN & SLEMROD [1983], SHEARER [1982, 1988], HAGAN & SERRIN [1986], MODICA [1987], STERNBERG [1988], FONSECA [1989a,b], FONSECA & TARTAR [1989], KOHN & STERNBERG [1989], BRONSARD & REITICH [1993] and the references therein.

In addition to the shock waves or propagating phase boundaries, there are also rapidly moving, nearly adiabatic shocks that involve change in state, such as liquefaction shocks, see THOMPSON & al. [1986], DETTLEFF & al. [1979], and THOMPSON & MEIER [1989].

Further conditions for the interface, such as the Gibbs–Thompson condition, and the related motion by curvature, are discussed by GURTIN [1995] (see also TRUSKINOVSKY [1991]).

22.2.4 Isothermal Approximation (Isothermal Elastodynamics) If the processes are slow, one can pass to the *isothermal dynamic theory,* with no thermal variables. The constitutive equations reduce to

$$f = \bar{f}(\boldsymbol{F}), \quad \boldsymbol{S} = \bar{\boldsymbol{S}}(\boldsymbol{F}), \quad \text{with } \bar{\boldsymbol{S}} = \rho_\circ \partial_{\boldsymbol{F}} \bar{f} \tag{22.2.5}$$

on $\bar{U} \subset \mathrm{InvLin}$. Here $\bar{f}(\boldsymbol{F}) = \bar{f}(\boldsymbol{F}, \theta)$ is the stored energy of the material at the given prescribed temperature θ. The equation of balance of linear momentum reads

$$\rho_\circ \ddot{\boldsymbol{\chi}} = \mathrm{Div}\, \boldsymbol{S} + \rho_\circ \boldsymbol{b}, \quad -U\rho_\circ [\boldsymbol{v}] = [\boldsymbol{S}]\boldsymbol{n}, \tag{22.2.6}$$

outside and on the singular surface. The equation of balance of energy is irrelevant; it only serves to determine the amount of heat r that must be communicated to the body to keep the process isothermal. A combination of the Clausius–Duhem inequality with the equation of balance of energy and the assumption that the process is isothermal leads to the following form of the "entropy inequality"

$$\rho_\circ (f + \tfrac{1}{2}\boldsymbol{v}^2)^{\boldsymbol{\cdot}} = \mathrm{Div}(\boldsymbol{S}^{\mathrm{T}}\boldsymbol{v}) + \rho_\circ \boldsymbol{b} \cdot \boldsymbol{v} \tag{22.2.7}$$

outside the singular surface and

$$-U\rho_\circ [f + \tfrac{1}{2}\boldsymbol{v}^2] \le [\boldsymbol{S}^{\mathrm{T}}\boldsymbol{v}] \cdot \boldsymbol{n} \tag{22.2.8}$$

on the singular surface, with the reduced forms

$$\rho_\circ \dot{f} = \boldsymbol{S} \cdot \dot{\boldsymbol{F}}, \quad U\rho_\circ [f] \le \tfrac{1}{2}(\boldsymbol{S}^+ + \boldsymbol{S}^-)\boldsymbol{n} \cdot [\boldsymbol{F}]\boldsymbol{n}. \tag{22.2.9}$$

Equation (22.2.7) is a consequence of $(22.2.6)_1$ outside the singular surface. Loosely speaking, the isothermal theory gives a good approximation to the full thermoelastic theory of Sect. 22.1 only for slow processes. For fast processes, such as the wave motions, it leads to wrong predictions; in particular, it gives the speeds of sound based on the isothermal tensor of elasticities, whereas the correct values require the adiabatic tensor of elasticities.

22.3 Adiabatic and Isentropic Dynamics; Shock Waves

The experience has shown that the shock waves can be modeled adequately in a theory which neglects both the viscosity and heat conduction. I call this theory the "adiabatic dynamics." The explicit mechanisms of dissipation are neglected, but this does not mean the absence of dissipation altogether – the energy is dissipated at the singular surfaces, as we shall see. An appropriate assumption is that the *internal* energy function be elliptic, i.e., that is satisfy the Legendre–Hadamard or even the strong ellipticity conditions so that the phase boundaries are excluded from the outset. The evolution equations are then hyperbolic. The entropy of a material point is conserved until the shock wave passes through it, whereby the entropy increases. To simplify the treatment, it is often assumed that the entropy is conserved altogether, even at shocks, it being tacitly assumed that heat is taken away (by an unspecified mechanism) to keep the shock isentropic. This results in a decrease of the total (internal plus kinetic) energy. I call this theory an "isentropic theory" and its equations are formally entirely identical with those of the isothermal dynamic of Subsect. 22.2.4, with, however, different response functions.

22.3.1 The Constitutive and Balance Equations For an adiabatic material 9.1.3(2) the constitutive equations reduce to

$$e = \tilde{e}(\boldsymbol{F}, \eta), \quad \boldsymbol{S} = \tilde{\boldsymbol{S}}(\boldsymbol{F}, \eta), \quad \theta = \tilde{\theta}(\boldsymbol{F}, \eta), \quad \boldsymbol{q} = \boldsymbol{0},$$

with $\tilde{\boldsymbol{S}}, \tilde{\theta}$ given by $\tilde{\boldsymbol{S}} = \rho_\circ \partial_{\boldsymbol{F}} \tilde{e}, \tilde{\theta} = \rho_\circ \partial_\eta \tilde{e}$ on $\tilde{U} \subset \text{InvLin} \times \mathbb{R}$. All the constitutive information is contained in $\tilde{e}$. The Gibbs equation

$$\rho_\circ \dot{e} = \boldsymbol{S} \cdot \dot{\boldsymbol{F}} + \rho_\circ \theta \dot{\eta} \tag{22.3.1}$$

holds outside the singular surface. The production of entropy, computed in Proposition 9.2.3(2), vanishes identically outside the singular surface. In the absence of the body supply of heat r, the balance equations take the form

$$\rho_\circ \ddot{\boldsymbol{\chi}} = \text{Div}\, \boldsymbol{S} + \rho_\circ \boldsymbol{b}, \quad \rho_\circ (e + \tfrac{1}{2} v^2)^{\cdot} = \text{Div}(\boldsymbol{S}^{\mathrm{T}} \boldsymbol{v}) + \rho_\circ \boldsymbol{b} \cdot \boldsymbol{v} \tag{22.3.2}$$

outside the singular surface and

$$-U\rho_\circ [\boldsymbol{v}] = [\boldsymbol{S}]\boldsymbol{n}, \quad -U\rho_\circ [e + \tfrac{1}{2} v^2] = [\boldsymbol{S}^{\mathrm{T}} \boldsymbol{v}] \cdot \boldsymbol{n} \tag{22.3.3}$$

on the singular surface. The reduced equations of balance of energy (3.6.5) read

$$\rho_\circ \dot{e} = \boldsymbol{S} \cdot \dot{\boldsymbol{F}}, \quad U\rho_\circ [e] = \langle \boldsymbol{S} \rangle \boldsymbol{n} \cdot [\boldsymbol{F}]\boldsymbol{n} \tag{22.3.4}$$

outside and on the singular surface, respectively. Outside the singular surface, the explicit forms of $(22.3.2)_1$ and $(22.3.4)_1$ are

$$\rho_\circ \ddot{\boldsymbol{\chi}} = \tilde{\boldsymbol{A}}[\nabla^2 \boldsymbol{\chi}] + \tilde{\boldsymbol{M}} \nabla \eta + \rho_\circ \boldsymbol{b}, \quad \dot{\eta} = 0, \tag{22.3.5}$$

where $\tilde{\boldsymbol{A}}$ is the tensor of adiabatic elasticities and $\tilde{\boldsymbol{M}}$ the stress entropy tensor (see Sect. 10.7).

22.3.2 Conservation and Nonconservation of Entropy Combining $(22.3.4)_1$ with (22.3.1) we obtain

$$\dot{\eta} = 0 \tag{22.3.6}$$

outside the singular surface. The evolution equations (22.3.2) thus have an additional conservation law for smooth processes. The evolution equations admitting an independent additional conservation law are exceptional and the additional conservation law greatly simplifies matters. On the singular surface the Clausius–Duhem inequality with $\boldsymbol{q} = \boldsymbol{0}$, $r = 0$ is

$$-U\rho_o[\eta] \geq 0. \tag{22.3.7}$$

Again, the jump condition (22.3.7) is *not* a consequence of the jump conditions (22.3.3). Our convention is that the velocity of the shock is always nonnegative; if $U > 0$, then (22.3.7) gives $\eta^+ \leq \eta^-$; that is, the entropy at a given material point can only increase or remain constant when the singular surface passes through the point. This is an additional condition which must be postulated independently of the evolution equations, and it serves as the simplest, and most important, admissibility criterion (see Sect. 23.8).

22.3.3 The Initial-Value Problem The initial-value problem is to find a solution satisfying the equations, the boundary conditions, and the prescribed initial data, namely,

$$\boldsymbol{\chi}(\cdot, 0) = \boldsymbol{\chi}^i(\cdot), \quad \dot{\boldsymbol{\chi}}(\cdot, 0) = \boldsymbol{v}^i(\cdot), \quad \eta(\cdot, 0) = \eta^i(\cdot)$$

where $\boldsymbol{\chi}^i, \boldsymbol{v}^i, \eta^i$ are given. Of importance is that when the material is genuinely nonlinear (see Sect. 23.1), then for a large class of smooth initial data $\boldsymbol{\chi}^i, \boldsymbol{v}^i, \eta^i$ the process will remain smooth only up to some time. Starting from that time it will have a singular surface on which the velocity and the deformation gradient will have jump discontinuities. (See. e.g., LAX [1964], MACCAMY & MIZEL [1967], DAFERMOS & HSIAO [1986] and HRUSA & MESSAOUDI [1990].) Thus the singular surfaces are typical, not exceptional, and the existence of solutions can be guaranteed only in the class of processes with singular surfaces. On the other hand, in the class of piecewise smooth processes there are initial data for which there exist several processes starting from them – uniqueness is lost. (We shall see in Subsect. 23.9.3 that in the case of a genuinely nonlinear, strictly hyperbolic material there is a large class of almost equilibrium discontinuous initial data from which two different solutions will emanate; see also PERICAK-SPECTOR & SPECTOR [1988] for an example of an entirely different type.) Of particular interest is the variant of the initial-value problem called the Riemann problem; I refer to Sect. 23.9 for a detailed formulation for the hyperbolic systems. The Riemann problem is often, although not always, explicitly solvable by tailoring the solutions from pieces of constant solutions and simple waves (see Sect. 23.5) separated by shock waves. The nonuniqueness is then seen explicitly. In Sect. 23.9 a solution of the Riemann problem will be given for nearly equilibrium initial data in the strictly hyperbolic genuinely nonlinear case. The method of proof will clearly show that the nonuniqueness arises without the admissibility criteria.

22.3.4 Isentropic Approximation (Elastodymanics) There are situations where one can proceed with the simplification of the equations of adiabatic dynamics further and neglect even the jumps of entropy, i.e., one can assume that $\eta(\boldsymbol{p}, t) = \eta(\boldsymbol{p})$. The only relevant variable in the constitutive equations is $\boldsymbol{F}$,

$$\tilde{e} = \tilde{e}(\boldsymbol{F}, \boldsymbol{p}) = \tilde{e}(\boldsymbol{F}, \eta(\boldsymbol{p})), \quad \boldsymbol{S} = \tilde{\boldsymbol{S}}(\boldsymbol{F}, \boldsymbol{p}) = \tilde{\boldsymbol{S}}(\boldsymbol{F}, \eta(\boldsymbol{p})) \text{ with } \tilde{\boldsymbol{S}} = \rho_o \partial_{\boldsymbol{F}} \tilde{e}.$$

A further simplification comes when $\eta(\boldsymbol{p}) = \eta$ is uniform over Ω, in which case

$$\tilde{e} = \tilde{e}(\boldsymbol{F}) \quad \boldsymbol{S} = \tilde{\boldsymbol{S}}(\boldsymbol{F}) \text{ with } \tilde{\boldsymbol{S}} = \rho_o \partial_{\boldsymbol{F}} \tilde{e}. \tag{22.3.8}$$

The only relevant evolution equation is the equation of balance of linear momentum

$$\rho_o \ddot{\boldsymbol{\chi}} = \operatorname{Div} \boldsymbol{S} + \rho_o \boldsymbol{b}, \quad -U\rho_o[\boldsymbol{v}] = [\boldsymbol{S}] \cdot \boldsymbol{n}. \tag{22.3.9}$$

As a consequence of (22.3.9) one obtains that outside the singular surface the equation of balance of energy holds

$$\rho_{\circ}(e + \tfrac{1}{2}v^2)^{\cdot} = \mathrm{Div}(S^{\mathsf{T}}v) + \rho_{\circ}b \cdot v. \tag{22.3.10}$$

The imposition of the conservation of energy at the singular surface,

$$-U\rho_{\circ}[e + \tfrac{1}{2}v^2] = [S^{\mathsf{T}}v]n,$$

would lead to an overdetermined system of jump equations at the singular surface. Rather, it is conceivable that the total energy decreases in shocks,

$$-U\rho_{\circ}[e + \tfrac{1}{2}v^2] \le [S^{\mathsf{T}}v]n, \tag{22.3.11}$$

while at regular points one has the equation of balance of energy (22.3.10). Inequality (22.3.11) can be motivated as follows. During a shock the energy is dissipated and heat is produced which increases the entropy: $\eta^- > \eta^+$. However, if the jumps in entropy are neglected, in order to satisfy the constraint $\eta^- = \eta^+$, the heat produced in the shock must be taken away from the body: the body must be cooled, and this decreases the total energy of the body. In the isentropic approximation, the inequality (22.3.11) plays the role of the admissibility criterion, in complete analogy with the entropy condition in the case of the adiabatic evolution. Note that the equations are formally identical with those of the isothermal dynamics, provided one identifies the stored energy (i.e., the isothermal free energy) with the internal energy at a fixed entropy. However, for a given material the stored energy and the internal energies at a fixed entropy are distinct, so the equations are distinct, too. The difference is in the applicability theories: The isothermal theory is good for slow processes with moving phase boundaries; it is incapable of describing the shocks adequately. The isentropic dynamics is good for fast processes and shocks. A deeper motivation for the isentropic approximation and (22.3.11) follows from the results of Sects. 23.6 and 23.8, where it will be shown that the jump in entropy is of the third order or higher of the jump in the deformation gradient (when the jumps are weak).

The dynamical counterpart of the Eshelby conservation law is

$$\rho_{\circ}(F^{\mathsf{T}}v)^{\cdot} + \mathrm{Div}\left[\rho_{\circ}(e - \tfrac{1}{2}v^2)\mathbf{1} - F^{\mathsf{T}}S\right] = \mathbf{0}$$

provided $\tilde{e}$ does not depend explicitly on p and $b = \mathbf{0}$ identically or equivalently

$$(F^{\mathsf{T}}v)^{\cdot} + \mathrm{Div}(\tilde{P} - \tfrac{1}{2}v^2\mathbf{1}) = \mathbf{0} \quad \text{where} \quad \tilde{P} = e - \rho_{\circ}^{-1}F^{\mathsf{T}}S.$$

Moreover, under the ellipticity condition and simple additional hypotheses, the equation of balance of energy, the equations of balance of linear and angular momenta, and the above Eshelby conservation law are the only linearly independent conservation laws (to within null divergences). The reader is referred to ROZHDESTVENSKII [1957], SUHUBI [1989], and the references therein.

The exposition of shock waves is much simpler in the isentropic approximation; however, the main line of exposition in the following two chapters deals with the general system of equations in which the entropy can change in jumps. The isentropic case is described in detail in HANYGA [1985, Chap. 3].

22.4 Equations in the Form of a First-Order System

22.4.1 First Order Systems of Conservation Laws The treatment of the shock waves is most systematic if the basic equations can be put in the form of a first-order system of conservation laws,

$$\dot{\boldsymbol{P}} + \mathrm{DIV}\,\boldsymbol{G} = 0, \tag{22.4.1}$$

or in indices, $\dot{P}_\alpha + (G_{\alpha,A})_{,A} = 0$ for an m–tuple of unknown functions $w = (w_1, \ldots, w_m) = w(z, t)$, with $z = (z_1, \ldots, z_n)$ standing either for $\boldsymbol{p}$ or $\boldsymbol{x}$, and with $\boldsymbol{P} = (P_1, \ldots, P_m) = \boldsymbol{P}(w)$, $\boldsymbol{G} = G_{\alpha,A} = \boldsymbol{G}(w)$ specified functions. DIV denotes the divergence with respect to z. One has to admit discontinuous solutions with moving singular surfaces $S_t \subset \Omega$. The equations must be given weak or integral forms as discussed in Sect. 3.3. The kinematics of the singular surface is identical with that described in Sect. 2.1; in particular, the singular surface has a well-defined normal $\boldsymbol{n}$ and speed of propagation U. The jump conditions read

$$U[\boldsymbol{P}] = [\boldsymbol{G}]\boldsymbol{n}. \tag{22.4.2}$$

In this section we examine the possibility of writing the equations of adiabatic or isentropic dynamics in the reference configuration in such a form. The resulting system is not equivalent to the original second-order system if the number of spatial dimensions is $n > 1$; this system is mentioned because in the case $n = 1$ it is equivalent and widely used.

22.4.2 First-Order Systems in Adiabatic Dynamics The equations from Subsect. 22.3 imply the equations

$$\rho_\circ \dot{\boldsymbol{v}} = \mathrm{Div}\,\boldsymbol{S}, \quad \dot{\boldsymbol{F}} = \mathrm{Div}(\boldsymbol{v} \otimes \boldsymbol{1}), \quad \rho_\circ (e + \tfrac{1}{2}\boldsymbol{v}^2)^{\cdot} = \mathrm{Div}(\boldsymbol{S}^{\mathrm{T}}\boldsymbol{v}), \tag{22.4.3}$$

where $\boldsymbol{v} \otimes \boldsymbol{1}$ is a tensor of order 3 with components $(\boldsymbol{v} \otimes \boldsymbol{1})_{iAB} = v_i \delta_{AB}$ and the jump conditions

$$-U\rho_\circ[\boldsymbol{v}] = [\boldsymbol{S}]\boldsymbol{n}, \quad -U[\boldsymbol{F}] = [\boldsymbol{v}] \otimes \boldsymbol{n}, \quad -U\rho_\circ[e + \tfrac{1}{2}\boldsymbol{v}^2] = [\boldsymbol{S}^{\mathrm{T}}\boldsymbol{v}]\boldsymbol{n}. \tag{22.4.4}$$

They are of the form (22.4.1) and (22.4.2) with $m = n^2 + n + 1$, where n is the number of spatial dimensions. As in the case of the original system, the system (22.4.3) has an independent additional conservation law (22.3.6), (22.3.7).

In indices, $(22.4.3)_2$ reads $\dot{F}_{iA} = v_{i,A}$. This is the interchangeability of the material time derivative with the referential gradient; $(22.4.4)_2$ is a parallel kinematical compatibility condition that was shown in Sect. 2.2 to be a consequence of the Hadamard lemma. It is important to realize that the above equations (22.4.3) and (22.4.4) are consequences of the equations of adiabatic dynamics, but they are not equivalent to them unless the number of spatial dimensions is 1. In more than one spatial dimension we have the following phenomenon. From the interchangeability of the second partial derivatives of χ follows that if $\boldsymbol{F}$ is a deformation gradient of some motion, then

$$F_{iA,B} = F_{iB,A}. \tag{22.4.5}$$

But from (22.4.3) the motion χ has disappeared and as a consequence of $(22.4.3)_2$ we obtain only that $\dot{F}_{iA,B} = \dot{F}_{iB,A}$. Hence if the initial value of $\boldsymbol{F}$ satisfies (22.4.5), then $\boldsymbol{F}$ will satisfy it forever. If this additional restriction is neglected when discussing the characteristic equation of the system, one arrives at solutions of (22.4.3) which do not satisfy (22.4.5). These give unphysical characteristic speeds with unphysical characteristic vectors. A parallel remark concerns the jump conditions: On the singular surface we have from the Hadamard lemma for χ (see (2.2.9)) that $[F_{iA}]n_B = [F_{iB}]n_A$, i.e., that $[F_{iA}]n_B$ is symmetric in the indices A, B. But $(22.4.4)_2$ implies only that $U[F_{iA}]n_B = U[F_{iB}]n_A$; hence, if the singular surface is immobile, this places no restriction at all.

22.4.3 The One-Dimensional Case In the one-dimensional case the system (22.4.3) reads:

$$\rho_\circ w_t = S_p, \quad F_t = w_p, \quad \rho_\circ (e + \tfrac{1}{2}w^2)_t = (Sw)_p, \tag{22.4.6}$$

where the subscript t denotes the differentiation with respect to time, the subscript p denotes the differentiation with respect to the referential position $p \in \mathbb{R}$, and w is the velocity. The jump conditions are

$$-U\rho_\circ[w] = [S], \quad -U[F] = [w], \quad -U\rho_\circ[e + \tfrac{1}{2}w^2] = [Sw]. \tag{22.4.7}$$

As above, for smooth processes $(22.4.6)_2$ can be replaced by $\dot{\eta}_t = 0$, which is convenient for the computation of the characteristic speeds. The Piola–Kirchhoff stress S coincides with the Cauchy stress T, and the deformation gradient is a number different from 0. The constitutive equations read

$S = \tilde{S}(F, \eta)$, $e = \tilde{e}(F, \eta)$. In the one-dimensional case the first-order system is completely equivalent to the original system with the balance of linear momentum in the form of the second-order equation.

In the isentropic case one deletes the equation of balance of energy from the list of the evolution equations, the entropy in the constitutive equations is replaced by $\boldsymbol{p}$, and the jump condition for the entropy on the singular surface is replaced by the decrease of energy (22.3.11) on the singular surface. The system admits the additional conservation law – the equation of balance of energy (smooth processes only).

Bibliographical Note

For a consistent theory of sharp evolving phase boundaries with interfacial structure, such as surface tension, see GURTIN [1993a,b; 1994, 1995] and CERMELLI & GURTIN [1994]. The approach in these papers also includes incoherent phase boundaries, whose kinematics essentially allows that the deformation function χ be discontinuous, but such that no cavitation occurs, i.e., the phases can slip at the interface. (See also PFENNING & WILLIAMS [1993], TRUSKINOVSKY [1983a,b] and GRINFELD [1991].) A further examination of the kinetic equation for phase boundary is found in LE FLOCH [1993] and TRUSKINOVSKY [1994]. The approach proposed in ABEYARATNE & KNOWLES [1987–1994] also contains an initiation criterion for the appearance of a new phase boundary that is outside the scope of this volume. Within the framework of first-order systems of conservation laws, DAFERMOS [1986] calls restrictions of the type (22.4.5) involutions and develops a theory of them.

23 Waves in the Referential Description

This chapter discusses elastic waves: Sect. 23.3 the plane, surface and acceleration waves, Sect. 23.5 the centered waves and Sects. 23.6–23.8 the shock waves. The exposition starts with the characteristic equation, which is important for all kinds of waves. Attention is concentrated especially on shock waves; here thermodynamics provides the entropy admissibility criterion. Some other proposed admissibility criteria are briefly reviewed and their relationship to the entropy criterion is established under additional assumptions. The Riemann problem is formulated and solved for small data in the strictly hyperbolic, genuinely nonlinear case. The nonuniqueness of solutions is explained.

23.1 The Characteristic Equation

Throughout the chapter, we consider an adiabatic material governed by the constitutive and balance equations 22.3.1. It will be convenient to use single letters $\boldsymbol{f}, \boldsymbol{g}$ for the pairs $(\boldsymbol{F}, \eta)$, $(\boldsymbol{G}, \zeta)$ from the domain $\tilde{U}$. Recall that Sph is the unit sphere in Vect and set $D := \tilde{U} \times \mathrm{Sph}$. Let n be the dimension of Vect as always.

23.1.1 The Acoustic Tensor Let $\boldsymbol{f} \in \tilde{U}$ and let $\boldsymbol{n}$ be a vector of arbitrary length. The (referential) ***acoustic tensor*** at $\boldsymbol{f}$ in the direction $\boldsymbol{n}$ is a second-order tensor $\boldsymbol{Q}(\boldsymbol{n}) \equiv \boldsymbol{Q}(\boldsymbol{f}, \boldsymbol{n})$ defined by

$$\boldsymbol{Q}(\boldsymbol{n})\boldsymbol{e} = \rho_{\circ}^{-1}\tilde{\boldsymbol{A}}[\boldsymbol{e} \otimes \boldsymbol{n}]\boldsymbol{n}, \quad \boldsymbol{e} \in \mathrm{Vect},$$

where $\tilde{\boldsymbol{A}}$ is the referential tensor of adiabatic elasticities at $\boldsymbol{f}$. The acoustic tensor is symmetric; its components are

$$\boldsymbol{Q}(\boldsymbol{n})_{ij} = \rho_{\circ}^{-1}\tilde{\boldsymbol{A}}_{iAjB} n_A n_B.$$

Define also $\boldsymbol{D} := \rho_{\circ}^{-1}\partial_\eta \tilde{\boldsymbol{S}}$, which is a multiple of the stress-entropy tensor $\tilde{\boldsymbol{M}}$ (see Sect. 10.7 for the definitions of $\tilde{\boldsymbol{A}}$, $\tilde{\boldsymbol{B}}$ and $\tilde{\boldsymbol{M}}$).

The ***spatial acoustic tensor*** $\boldsymbol{q}(\boldsymbol{m}) = \boldsymbol{q}(\boldsymbol{f}, \boldsymbol{m})$ at $\boldsymbol{f} \in \tilde{U}$ in the direction $\boldsymbol{m} \in \mathrm{Vect}$ is defined by

$$\boldsymbol{q}(\boldsymbol{m})\boldsymbol{e} = \rho^{-1}\tilde{\boldsymbol{B}}[\boldsymbol{e} \otimes \boldsymbol{m}]\boldsymbol{m}, \quad \boldsymbol{e} \in \mathrm{Vect},$$

where $\tilde{\boldsymbol{B}}$ is the spatial tensor of adiabatic elasticities at $\boldsymbol{f}$, and $\rho = \rho_{\circ}/|\det \boldsymbol{F}|$ is the actual density. Using the analog of Proposition 10.5.3(1) for the adiabatic elasticities, one finds that

$$\boldsymbol{q}(\boldsymbol{f}, \boldsymbol{m}) = \boldsymbol{Q}(\boldsymbol{f}, \boldsymbol{F}^{\mathsf{T}}\boldsymbol{m}) \tag{23.1.1}$$

for every $\boldsymbol{m} \in \mathrm{Vect}$. Note also that in a state with deformation gradient $\boldsymbol{F}$ the referential and spatial *unit* normals $\boldsymbol{n}, \boldsymbol{m}$ to a material surface are related by $\boldsymbol{m} = \boldsymbol{F}^{-\mathsf{T}}\boldsymbol{n}/|\boldsymbol{F}^{-\mathsf{T}}\boldsymbol{n}|$; then

$$\boldsymbol{q}(\boldsymbol{m}) = |\boldsymbol{F}^{-\mathsf{T}}\boldsymbol{n}|^{-2}\boldsymbol{Q}(\boldsymbol{n}).$$

23.1.2 The Characteristic Equation The subsequent sections will show that the characteristic equation (23.1.2), below, or its simplified version (23.1.3), arise in connection with the propagation of plane, acceleration, centered, and shock waves. With all these waves one can associate the direction $\boldsymbol{n} \in \mathrm{Sph}$ and the speed C of propagation, and the amplitudes of disturbances $\boldsymbol{e} \in \mathrm{Vect}$, $b \in \mathbb{R}$ of deformation and entropy. The equations of balance of momenta and energy say that the wave can exist (propagate) only if the parameters $\boldsymbol{n}, C, \boldsymbol{e}, b$ satisfy the characteristic equation.

Following the definition of the characteristic equation for a general system (e.g., OLVER [1986, Sect. 2.6]), we take a 'direction' $(\boldsymbol{n}, -C) \in \mathrm{Vect} \times \mathbb{R}$ in the space of independent variables of the system, a vector $\xi = (\boldsymbol{e}, b) \in \mathrm{Vect} \times \mathbb{R}$ in the space of unknowns, and make the replacements

$$\dot{\boldsymbol{\chi}} \mapsto -C\boldsymbol{e}, \ \ddot{\boldsymbol{\chi}} \mapsto C^2\boldsymbol{e}, \ \nabla\boldsymbol{\chi} \mapsto \boldsymbol{e} \otimes \boldsymbol{n}, \ \nabla^2\boldsymbol{\chi} \mapsto \boldsymbol{e} \otimes \boldsymbol{n} \otimes \boldsymbol{n},$$

$$\dot{\eta} \mapsto -Cb, \ \nabla\eta \mapsto b\boldsymbol{n}$$

in the explicit form of the equations (22.3.5) with $\boldsymbol{b} = \boldsymbol{0}$. This results in the ***characteristic equation***

$$C^2\boldsymbol{e} = \mathbf{Q}(\boldsymbol{n})\boldsymbol{e} + \mathbf{D}\boldsymbol{n}b, \quad Cb = 0. \tag{23.1.2}$$

In indices, the first of (23.1.2) reads $C^2 a_i = \tilde{\mathbf{A}}_{iAjB} n_A n_B a_j + \mathbf{D}_{iA} n_A b$. Given $\boldsymbol{f} \in \tilde{U}$ and $\boldsymbol{n} \in \mathrm{Sph}$, (23.1.2) is a system of equations for unknowns $\boldsymbol{e}$, b, and C. For a fixed $\boldsymbol{n}$, the characteristic equation is an *eigenvalue problem* with C (or C^2) playing the role of eigenvalue and $\xi = (\boldsymbol{e}, b)$ that of eigenvector.

The relevance of the characteristic equation to the plane and acceleration waves is established in Sect. 23.3 for materials of arbitrary symmetry. The implications of the material symmetry on the form of the characteristic equation are briefly described in Sect. 23.4. The remaining sections of this chapter apply and extend the material associated with the characteristic equation to centered and shock waves.

Equation $(23.1.2)_2$ and the occurrence of b are due to the equation of balance of energy and the occurrence of entropy in the constitutive equations. From Sect. 22.3 we know that the entropy is conserved during the regular parts of the process and hence in this sense both the equation of balance of energy and the entropy are 'redundant.' In the same sense $(23.1.2)_2$ and b are also 'redundant.' In fact the latter two disappear when dealing with the isentropic approximation of Subsect. 22.3.4 and with the propagation of plane waves, wherein the characteristic equation reduces to

$$C^2\boldsymbol{e} = \mathbf{Q}(\boldsymbol{n})\boldsymbol{e}. \tag{23.1.3}$$

23.1.3 Definitions A vector $\xi = (\boldsymbol{e}, b) \neq 0$ is said to be a ***characteristic vector*** at $(\boldsymbol{f}, \boldsymbol{n}) \in D$ corresponding to a characteristic speed $C \geq 0$ if (23.1.2) holds. Then C and ξ are referred to as the ***characteristic speed*** and the ***characteristic vector*** at $(\boldsymbol{f}, \boldsymbol{n})$, respectively. If C is a characteristic speed at $(\boldsymbol{f}, \boldsymbol{n})$ then the set of all $\xi = (\boldsymbol{e}, b)$ satisfying (23.1.2) is called the ***characteristic space*** at $(\boldsymbol{f}, \boldsymbol{n})$ and its dimension the ***multiplicity*** of C. There are two kinds of the characteristic vectors:

23.1.4 *Remark* *A* $\xi = (\boldsymbol{e}, b) \in \mathrm{Vect} \times \mathbb{R}$, $\xi \neq 0$, *is a characteristic vector at* $(\boldsymbol{f}, \boldsymbol{n}) \in D$ *corresponding to the characteristic speed* $C \geq 0$ *if and only if one of the following two assertions holds:*

(1) $\xi = (\boldsymbol{e}, 0)$ *and* $C^2\boldsymbol{e} = \mathbf{Q}(\boldsymbol{n})\boldsymbol{e}$, *or*

(2) $\xi = (\boldsymbol{e}, b)$ *with* $b \neq 0$, $\mathbf{Q}(\boldsymbol{n})\boldsymbol{e} + \mathbf{D}\boldsymbol{n}b = 0$ *and* $C = 0$.

This is immediate. The characteristic vectors $\xi = (\boldsymbol{e}, b)$ with $b \neq 0$ play a somewhat singular role; only the characteristic vectors $\xi = (\boldsymbol{e}, 0)$ corresponding to positive speeds are related to true waves, and we shall be concerned mainly with them.

23.1.5 Definitions A pair $(C, \boldsymbol{e}) \in \mathbb{R} \times \text{Vect}$ is said to be a ***characteristic pair*** at $(f, \boldsymbol{n})$ if $\boldsymbol{e} \neq \boldsymbol{0}$, $C \geq 0$, and $C^2\boldsymbol{e} = \boldsymbol{Q}(\boldsymbol{n})\boldsymbol{e}$. Then $\boldsymbol{e}$ and C are referred to as the ***characteristic amplitude*** and the ***the characteristic speed*** at $(f, \boldsymbol{n})$, respectively. (By Remark 23.1.4 the term characteristic speed introduced in connection with the characteristic amplitudes is consistent with the term characteristic speed introduced in connection with the characteristic vectors.) A characteristic pair $(C, \boldsymbol{e})$ at $(f, \boldsymbol{n})$ is said to be ***simple*** if C^2 is a simple eigenvalue of $\boldsymbol{Q}(\boldsymbol{n})$, i.e., if the dimension of the space of all solutions $\boldsymbol{e}$ of $C^2\boldsymbol{e} = \boldsymbol{Q}(\boldsymbol{n})\boldsymbol{e}$ is 1. The material is said to be ***hyperbolic*** at $(f, \boldsymbol{n}) \in D$ if there are $n + 1$ linearly independent characteristic vectors at $(f, \boldsymbol{n})$. The material is said to be ***strictly hyperbolic*** at $(f, \boldsymbol{n})$ if there are $n + 1$ mutually different characteristic speeds at $(\boldsymbol{F}, \boldsymbol{n})$.

The hyperbolicity and strict hyperbolicity are too restrictive as permanent assumptions since, as we shall see, the hyperbolicity implies the adiabatic Legendre–Hadamard condition, and we know that this condition excludes phase transitions. Hence for realistic materials there are regions in $\tilde{U}$ where the hyperbolicity fails. We shall also see that the strict hyperbolicity fails in isotropic states. For instance, in three spatial dimensions, every isotropic state of a solid has a characteristic speed of multiplicity ≥ 2; every fluid has only one nonzero characteristic speed and $C = 0$ is a characteristic speed of multiplicity ≥ 3.

23.1.6 Proposition

(1) *The material is hyperbolic at* $(f, \boldsymbol{n}) \in D$ *if and only if* $\boldsymbol{D}\boldsymbol{n}$ *is in the range of* $\boldsymbol{Q}(\boldsymbol{n})$ *and the adiabatic Legendre–Hadamard condition* $\tilde{\boldsymbol{A}}(f)[\boldsymbol{e} \otimes \boldsymbol{n}, \boldsymbol{e} \otimes \boldsymbol{n}] \geq 0$ *holds for every* $\boldsymbol{e} \in \text{Vect}$;

(2) *if the material is hyperbolic at* $(f, \boldsymbol{n})$, *then* 0 *is a characteristic speed at* $(f, \boldsymbol{n})$; *corresponding to this characteristic speed there is a characteristic vector of the form* $\xi = (\boldsymbol{e}, 1) \in \text{Vect} \times \mathbb{R}$.

Proof Assume that the material is hyperbolic at $(f, \boldsymbol{n})$. Denote the $n+1$ linearly independent characteristic vectors by $\xi_i = (\boldsymbol{e}_i, b_i)$ and the corresponding characteristic speeds by C_i, $i = 0, \dots, n$. Since $\dim(\text{Vect} \times \mathbb{R}) = n + 1$, there must be at least one ξ_i with $b_i \neq 0$; let the corresponding index be $i = 0$. Then $C_0 = 0$ by $(23.1.2)_2$, proving (2). Set $b_0 = 1$. The characteristic equation for ξ_0 reads $\boldsymbol{Q}(\boldsymbol{n})\boldsymbol{e}_0 + \boldsymbol{D}\boldsymbol{n} = \boldsymbol{0}$, proving the assertion about the range in (1). To prove the Legendre–Hadamard condition, we have to show that $\boldsymbol{Q}(\boldsymbol{n})$ has n linearly independent eigenvectors with nonnegative eigenvalues. Let p be the number of C_i equal to 0 and assume that the corresponding indices are $0, \dots, p-1$. The characteristic equations for the ξ_i, $i = 0, \dots, p-1$ read $\boldsymbol{Q}(\boldsymbol{n})\boldsymbol{e}_i + b_i\boldsymbol{D}\boldsymbol{n} = \boldsymbol{0}$ and as $b_0 = 1$, by passing from ξ_i to $\xi_i - b_i\xi_0$, $i = 1, \dots, p-1$, if necessary, we can assume that $b_1 = \cdots = b_{p-1} = 0$. But then the characteristic equations read $\boldsymbol{Q}(\boldsymbol{n})\boldsymbol{e}_i = \boldsymbol{0}$, so that $\boldsymbol{e}_1, \dots, \boldsymbol{e}_{p-1}$ are $p - 1$ linearly independent eigenvectors corresponding to eigenvalue 0. Since for $i = p, \dots, n$ we have $C_i \neq 0$, by Remark 23.1.4(1) the characteristic vectors corresponding to C_i, $i = p, \dots, n$ are of the form $\xi_i = (\boldsymbol{e}_i, 0)$ and $\boldsymbol{e}_i$ are the eigenvectors of $\boldsymbol{Q}(\boldsymbol{n})$ with the eigenvalues C_i^2. Hence the linearly independent eigenvectors a_i, $i = p, \dots, n$ have positive eigenvalues. To summarize, $\boldsymbol{Q}(\boldsymbol{n})$ has $(p-1)+(n-p+1) = n$ linearly independent eigenvectors with nonnegative eigenvalues, which proves the direct implication in (1). To prove the converse implication, observe that the Legendre–Hadamard condition gives that $\boldsymbol{Q}(\boldsymbol{n})$ has n linearly independent eigenvectors $\boldsymbol{e}_i$, $i = 1, \dots, n$ with nonnegative eigenvalues λ_i. Then $\xi_i = (\boldsymbol{e}_i, 0)$ are n linearly independent characteristic vectors with the characteristic speeds $C_i = \sqrt{\lambda_i}$. To find the $n+1$st characteristic vector, one uses the assertion about the range to find a vector $\boldsymbol{e} \in \text{Vect}$ such that $\boldsymbol{Q}(\boldsymbol{n})\boldsymbol{e} = -\boldsymbol{D}\boldsymbol{n}$. Then $\xi_0 = (\boldsymbol{e}, 1)$ is a characteristic vector corresponding to $C = 0$. □

23.1.7 Proposition

(1) *The material is strictly hyperbolic at* $(f, \boldsymbol{n})$ *if and only if* $\boldsymbol{Q}(\boldsymbol{n})$ *has n distinct positive eigenvalues; the strict hyperbolicity implies the adiabatic strong ellipticity condition* $\tilde{\boldsymbol{A}}(f)[\boldsymbol{e} \otimes \boldsymbol{n}, \boldsymbol{e} \otimes \boldsymbol{n}] > 0$ *for every* $\boldsymbol{e} \in \text{Vect}$, $\boldsymbol{e} \neq \boldsymbol{0}$;

(2) *if the material is strictly hyperbolic at* $(\boldsymbol{f}, \boldsymbol{n})$ *then it is strictly hyperbolic at every* $(\boldsymbol{g}, \boldsymbol{m})$ *sufficiently close to* $(\boldsymbol{f}, \boldsymbol{n})$.

Proof The considerations in the proof of Proposition 23.1.6 show that if the material is strictly hyperbolic at $(\boldsymbol{f}, \boldsymbol{n})$, then $\boldsymbol{Q}(\boldsymbol{n})$ has n distinct positive eigenvalues, which proves the direct implication in (1). To prove the converse implication, it suffices to note that if $\boldsymbol{Q}(\boldsymbol{n})$ has n distinct positive eigenvalues, then it is an invertible tensor; hence $\boldsymbol{Dn}$ is in its range. Proposition 23.1.6(1) gives that the material is hyperbolic at $(\boldsymbol{f}, \boldsymbol{n})$; and as the characteristic speeds are distinct the material is strictly hyperbolic at $(\boldsymbol{f}, \boldsymbol{n})$. To prove (2), it suffices to apply (1) and to note that if $\boldsymbol{Q}(\boldsymbol{f}, \boldsymbol{n})$ has n distinct positive eigenvectors, then the same is true for $\boldsymbol{Q}(\boldsymbol{g}, \boldsymbol{m})$ with $(\boldsymbol{g}, \boldsymbol{m})$ from some neighborhood of $(\boldsymbol{f}, \boldsymbol{n})$; see Proposition 1.2.2. □

23.2 Characteristic Fields. Genuine Nonlinearity

23.2.1 Definition A ***characteristic field*** in the direction $\boldsymbol{n} \in \mathrm{Sph}$ is any pair $(\tilde{C}(\cdot, \boldsymbol{n}), \tilde{\boldsymbol{e}}(\cdot, \boldsymbol{n}))$ of continuously differentiable functions on an open subset N of $\tilde{U}$ such that $(\tilde{C}(\boldsymbol{f}, \boldsymbol{n}), \tilde{\boldsymbol{e}}(\boldsymbol{f}, \boldsymbol{n}))$ is a characteristic pair at $(\boldsymbol{f}, \boldsymbol{n})$ for every $\boldsymbol{f} \in N$. The characteristic field is said to be ***simple*** if $(\tilde{C}(\boldsymbol{f}, \boldsymbol{n}), \tilde{\boldsymbol{e}}(\boldsymbol{f}, \boldsymbol{n}))$ is a simple characteristic pair at $(\boldsymbol{f}, \boldsymbol{n})$ for every $\boldsymbol{f} \in N$.

For the rest of the chapter assume that $\tilde{e}$ is of class C^5. (This will be shown in Proposition 23.7.6 to lead to the Hugoniot curves of class C^3, and the latter smoothness is needed for the equivalence of the admissibility criteria; for various parts, a lower degree of smoothness of $\tilde{e}$ suffices.)

23.2.2 Proposition *Any simple characteristic pair* $(C, \boldsymbol{e})$ *at* $(\boldsymbol{f}, \boldsymbol{n}) \in D$ *with* $C > 0$ *can be extended to a simple characteristic field* $(\tilde{C}(\cdot, \boldsymbol{n}), \tilde{\boldsymbol{e}}(\cdot, \boldsymbol{n}))$ *on a neighborhood* N *of* $\boldsymbol{f}$ *(i.e.,* $\tilde{C}(\boldsymbol{f}, \boldsymbol{n}) = C, \tilde{\boldsymbol{e}}(\boldsymbol{f}, \boldsymbol{n}) = \boldsymbol{e}$*). Moreover,* N *can be chosen so that for some* $\varepsilon > 0$, $\tilde{C}(\boldsymbol{g}, \boldsymbol{n})$ *is the only characteristic speed at* $(\boldsymbol{g}, \boldsymbol{n})$ *in the interval* $(C - \varepsilon, C + \varepsilon)$ *for any* $\boldsymbol{g} \in N$.

The characteristic field $(\tilde{C}, \tilde{\boldsymbol{e}})$ is the ***extension of the characteristic pair*** $(C, \boldsymbol{e})$ to a neighborhood of $\boldsymbol{f}$. For N small, the speed of the extension is uniquely determined and the amplitude $\tilde{\boldsymbol{e}}$ is determined uniquely to within a nonzero scalar multiple depending on $\boldsymbol{f}$.

Proof We apply Proposition 1.2.2 to $\boldsymbol{M}(\boldsymbol{x}) = \boldsymbol{Q}(\boldsymbol{f}, \boldsymbol{n})$ with the identification $\boldsymbol{x} = \boldsymbol{f}, \boldsymbol{y} = \boldsymbol{g}$. Since $(C, \boldsymbol{e})$ is a simple characteristic pair at $(\boldsymbol{f}, \boldsymbol{n})$ and $\boldsymbol{Q}(\boldsymbol{f}, \boldsymbol{n})$ is symmetric, it follows that C^2 is a simple root of the characteristic polynomial of $\boldsymbol{Q}(\boldsymbol{f}, \boldsymbol{n})$. □

23.2.3 Definition A characteristic pair $(C, \boldsymbol{e})$ at $(\boldsymbol{f}, \boldsymbol{n}) \in D$ is said to be ***genuinely nonlinear*** (or ***linearly degenerate***) if $\partial_{\boldsymbol{F}}^3 \tilde{e}[\boldsymbol{e} \otimes \boldsymbol{n}, \boldsymbol{e} \otimes \boldsymbol{n}, \boldsymbol{e} \otimes \boldsymbol{n}] \neq 0$ (or $= 0$). A characteristic field $(\tilde{C}(\cdot, \boldsymbol{n}), \tilde{\boldsymbol{e}}(\cdot, \boldsymbol{n}))$ on N is said to be ***genuinely nonlinear*** (or ***linearly degenerate***) if $(\tilde{C}(\boldsymbol{f}, \boldsymbol{n}), \tilde{\boldsymbol{e}}(\boldsymbol{f}, \boldsymbol{n}))$ is a genuinely nonlinear (or linearly degenerate) characteristic pair for every $\boldsymbol{f} \in N$.

With this definition we encounter a restriction on the *third* derivative of a thermodynamic potential. A linear material has a quadratic $\tilde{e}$ and hence $\partial_{\boldsymbol{F}}^3 \tilde{e}$ vanishes. On the other hand, if $\partial_{\boldsymbol{F}}^3 \tilde{e}[\boldsymbol{e} \otimes \boldsymbol{b}, \boldsymbol{e} \otimes \boldsymbol{b}, \boldsymbol{e} \otimes \boldsymbol{b}] = 0$ for every $\boldsymbol{e}, \boldsymbol{b} \in \mathrm{Vect}$ on a connected region $(\boldsymbol{F}, \eta) \in V \subset \tilde{U}$, then a simple modification of the proof of Proposition 13.5.3 shows that for every η, the function $\tilde{e}(\cdot, \eta)$ must be a null Lagrangian with respect to $\boldsymbol{F}$ plus a function quadratic in $\boldsymbol{F}$, i.e.,

$$\tilde{e}(\boldsymbol{F}, \eta) = A(\eta) + \boldsymbol{B}(\eta) \cdot \boldsymbol{F} + \boldsymbol{C}(\eta) \cdot \operatorname{cof} \boldsymbol{F} + D(\eta) \det \boldsymbol{F} + \mathbf{A}(\eta)[\boldsymbol{F}, \boldsymbol{F}]$$

for every $(\boldsymbol{F}, \eta) \in V$, where $A, \boldsymbol{B}, \boldsymbol{C}, D, \mathbf{A}$ are functions of η, A, D scalar-valued, $\boldsymbol{B}, \boldsymbol{C}$ with values in Lin and $\mathbf{A}$ with values in the space of all fourth-order tensors. The genuine nonlinearity, however, requires that $\partial_F^3\tilde{e}$ be nonzero on specific arguments. For fluids the genuine nonlinearity results in a simple condition on the pressure function, see Proposition 23.4.4 below. The reason for the specific form of the definition is explained in Proposition 23.7.4 which shows that the genuine nonlinearity implies that locally the entropy of a shock is a strictly increasing function of the speed. This is important for establishing the equivalence of the admissibility criteria for small shocks in Proposition 23.8.11. The following proposition is a characterization of the genuine nonlinearity in terms of the characteristic speeds.

23.2.4 Proposition *Let $(\tilde{C}(\cdot, \boldsymbol{n}), \tilde{e}(\cdot, \boldsymbol{n}))$ be a characteristic field on N. Then for every $f \in \tilde{U}$,*

$$e^2\partial_F\tilde{C}^2(f, \boldsymbol{n}) \cdot (\boldsymbol{e} \otimes \boldsymbol{n}) = \partial_F^3\tilde{e}(f)[\boldsymbol{e} \otimes \boldsymbol{n}, \boldsymbol{e} \otimes \boldsymbol{n}, \boldsymbol{e} \otimes \boldsymbol{n}]. \tag{23.2.1}$$

Hence, the pair $(\tilde{C}(f, \boldsymbol{n}), \tilde{e}(f, \boldsymbol{n}))$ is genuinely nonlinear at f if and only if

$$\partial_F\tilde{C}^2(f, \boldsymbol{n}) \cdot (\boldsymbol{e} \otimes \boldsymbol{n}) \neq 0. \tag{23.2.2}$$

If $(\tilde{C}(\cdot, \boldsymbol{n}), \tilde{e}(\cdot, \boldsymbol{n}))$ is a genuinely nonlinear characteristic field with $\tilde{C}(f, \boldsymbol{n}) > 0$ everywhere on its domain, then one can choose a normalization of $\tilde{e}$ such that

$$\partial_F\tilde{C}(f, \boldsymbol{n}) \cdot (\boldsymbol{e} \otimes \boldsymbol{n}) = 1 \tag{23.2.3}$$

for every f from the domain. (A characteristic field satisfying (23.2.3) is called a normalized genuinely nonlinear characteristic field.)

Proof Suppose, for this proof only, that $\boldsymbol{e}$ is normalized to 1 and that $\rho_o = 1$. Then $C^2 = \boldsymbol{Q}(\boldsymbol{n})\boldsymbol{e} \cdot \boldsymbol{e}$. A differentiation with respect to $\boldsymbol{F}$ in the direction $\boldsymbol{L} = \boldsymbol{e} \otimes \boldsymbol{n}$ and the definition of the acoustic tensor give

$$\partial_F C^2 \cdot (\boldsymbol{e} \otimes \boldsymbol{n}) = \partial_F^3\tilde{e}[\boldsymbol{e} \otimes \boldsymbol{n}, \boldsymbol{e} \otimes \boldsymbol{n}, \boldsymbol{e} \otimes \boldsymbol{n}] + 2\boldsymbol{Q}(\boldsymbol{n})\boldsymbol{e} \cdot \partial_F\boldsymbol{e}[\boldsymbol{e} \otimes \boldsymbol{n}].$$

The last term vanishes since with the help of $\boldsymbol{Q}(\boldsymbol{n})\boldsymbol{e} = C^2\boldsymbol{e}$ it can be rewritten as $2C^2\boldsymbol{e} \cdot \partial_F\boldsymbol{e}[\boldsymbol{e} \otimes \boldsymbol{n}]$ and $\boldsymbol{e} \cdot \partial_F\boldsymbol{e}[\boldsymbol{e} \otimes \boldsymbol{n}] = 0$ by the normalization condition $\boldsymbol{e} \cdot \boldsymbol{e} = 1$. □

23.3 Plane, Surface, and Acceleration Waves

Elastic plane waves will be discussed here in the linear approximation and the acceleration waves in the context of exact equations. Suppose that the body Ω extends over the whole space Eucl which we identify with Vect, that $n = 3$ and that external forces and supplies of heat are absent. Consider a homogeneous reference state characterized by the deformation gradient $\boldsymbol{F}_0$ and the density of entropy η_0; it is not assumed that the stress vanishes in this state. Write $\boldsymbol{u}$ for the displacement relative to the reference state and $\bar{\eta}$ for the difference of the actual entropy from the reference entropy:

$$\boldsymbol{\chi}(\boldsymbol{p}, t) = \boldsymbol{F}_0\boldsymbol{p} + \boldsymbol{u}(\boldsymbol{p}, t), \quad \eta(\boldsymbol{p}, t) = \eta_0 + \bar{\eta}(\boldsymbol{p}, t).$$

The linearized equations about the reference state read

$$\rho_o\ddot{\boldsymbol{u}} = \mathrm{Div}(\tilde{\boldsymbol{A}}[\nabla\boldsymbol{u}] + \tilde{\boldsymbol{M}}\nabla\bar{\eta}), \quad \rho_o\dot{\bar{\eta}} = 0 \tag{23.3.1}$$

where the adiabatic elasticities $\tilde{\boldsymbol{A}}$ and the stress entropy tensor $\boldsymbol{D}$ are evaluated at $(\boldsymbol{F}_0, \eta_0)$.

23.3.1 Plane Waves The wave solutions of the linearized equations will be assumed to be of the form

$$\boldsymbol{u}(\boldsymbol{p}, t) = \boldsymbol{e}\cos P, \quad \bar{\eta}(\boldsymbol{p}, t) = b\sin P, \tag{23.3.2}$$

where P is the phase, given by the following three equivalent expressions,

$$P = 2\pi\Lambda^{-1}(\boldsymbol{n}\cdot\boldsymbol{p} - Ct) = \boldsymbol{K}\cdot\boldsymbol{p} - \omega t = \omega(\boldsymbol{S}\cdot\boldsymbol{p} - t)$$

where all symbols except for $\boldsymbol{p}$, t are constants and moreover $C > 0$. The following terminology and notation are employed for the quantities associated with a plane wave:

$\boldsymbol{e}, b$ amplitudes,
C referential phase speed of propagation,
Λ referential wavelength,
$\boldsymbol{n}$ referential direction of propagation,
$\boldsymbol{K} = 2\pi\boldsymbol{n}/\Lambda$ referential wavevector,
$\omega = 2\pi C/\Lambda$ cyclic (angular) frequency,
$\boldsymbol{C} = C\boldsymbol{n}$ referential phase velocity,
$\boldsymbol{S} = \boldsymbol{K}/\omega = \boldsymbol{C}/|\boldsymbol{C}|^2$ referential slowness,
$\nu = \omega/2\pi$ linear frequency,
$T = 2\pi/\omega$ period.

In the *reference* configuration given by the constant deformation gradient $\boldsymbol{F}_0$, the material point $\boldsymbol{p}$ has the position $\boldsymbol{x} = \boldsymbol{F}_0$. The phase P can be written as

$$P = 2\pi\lambda^{-1}(\boldsymbol{m}\cdot\boldsymbol{x} - ct) = \boldsymbol{k}\cdot\boldsymbol{x} - \omega t = \omega(\boldsymbol{s}\cdot\boldsymbol{x} - t)$$

where the *spatial* quantities associated with a plane wave are referred to as follows

$c = C/|\boldsymbol{F}_0^{-\mathrm{T}}\boldsymbol{n}|$ phase speed of propagation,
$\lambda = \Lambda/|\boldsymbol{F}_0^{-\mathrm{T}}\boldsymbol{n}|$ wavelength,
$\boldsymbol{m} = \boldsymbol{F}_0^{-\mathrm{T}}\boldsymbol{n}/|\boldsymbol{F}_0^{-\mathrm{T}}\boldsymbol{n}|$ direction of propagation,
$\boldsymbol{k} = 2\pi\boldsymbol{m}/\lambda$ wavevector,
$\boldsymbol{c} = c\boldsymbol{m}$ phase velocity,
$\boldsymbol{s} = \boldsymbol{k}/\omega = \boldsymbol{c}/|\boldsymbol{c}|^2$ slowness;

hence $\omega = 2\pi c/\lambda$ and $|\boldsymbol{F}_0^{-\mathrm{T}}\boldsymbol{n}|\cdot|\boldsymbol{F}_0^{\mathrm{T}}\boldsymbol{m}| = 1$.

Inserting (23.3.2) into (23.3.1), one finds that the $\boldsymbol{e}$ and C satisfy the characteristic equation (23.1.2); moreover, as it is assumed that $C > 0$, necessarily $b = 0$ and the amplitude $\boldsymbol{e}$ must satisfy the characteristic equation. Hence, a plane wave satisfies (23.3.1) if and only if $b = 0$ and $(C, \boldsymbol{e})$ is a characteristic pair at $(\boldsymbol{F}_0, \eta_0, \boldsymbol{n})$ with $C > 0$. The following are equivalent forms of the characteristic equation:

$$C^2\boldsymbol{e} = \boldsymbol{Q}(\boldsymbol{n})\boldsymbol{e}, \quad \omega^2\boldsymbol{e} = \boldsymbol{Q}(\boldsymbol{K})\boldsymbol{e}, \quad \boldsymbol{e} = \boldsymbol{Q}(\boldsymbol{S})\boldsymbol{e},$$
$$c^2\boldsymbol{e} = \boldsymbol{q}(\boldsymbol{m})\boldsymbol{e}, \quad \omega^2\boldsymbol{e} = \boldsymbol{q}(\boldsymbol{k})\boldsymbol{e}, \quad \boldsymbol{e} = \boldsymbol{q}(\boldsymbol{s})\boldsymbol{e}.$$

The ***characteristic polynomial*** at $(\boldsymbol{F}_0, \eta_0)$ is the bicubic polynomial p in the variables $\boldsymbol{K}$, C given by

$$p(\boldsymbol{K}, C) = \det(C^2\boldsymbol{1} - \boldsymbol{Q}(\boldsymbol{K})).$$

23.3.2 Slowness and Wave Surfaces; Group Velocity Corresponding to a given slowness $\boldsymbol{S} \in \mathrm{Vect}$ there exists a nonzero $\boldsymbol{e}$ solving the characteristic equation $\boldsymbol{e} = \boldsymbol{Q}(\boldsymbol{S})\boldsymbol{e}$ if and only if $\det(\boldsymbol{1} - \boldsymbol{Q}(\boldsymbol{S})) = 0$. The set

$$S := \{\mathbf{S} \in \text{Vect} : \det(\mathbf{1} - \mathbf{Q}(\mathbf{S})) = 0\}$$

is called the ***(referential) slowness surface*** at $(\boldsymbol{F}_0, \eta_0)$. The slowness surface is a sextic surface and as $\mathbf{Q}$ depends on $\mathbf{S}$ quadratically, S is a union of three sheets which can generally intersect. It is also clear that S is centrally symmetric with respect to the origin in Vect, i.e., if $\mathbf{S} \in S$ then also $-\mathbf{S} \in S$; moreover $\mathbf{0} \notin S$. Without any assumption on $\tilde{\mathbf{A}}$, S may be empty. Proposition 23.3.3, below, ensures a local surface-like form of S near the slownesses corresponding to *simple* characteristic amplitudes $\boldsymbol{e}$.

The phase velocity $\boldsymbol{C}$ describes the direction and speed of propagation of the displacement $\boldsymbol{u}$; it will be seen in Subsect. 23.3.6 that the energy propagates generally in the direction of the group velocity. Let $(C, \boldsymbol{e})$ be a characteristic pair at $(\boldsymbol{F}_0, \eta_0, \boldsymbol{n})$ with $C > 0$. The ***referential group velocity*** $\mathbf{V}$ of $(C, \boldsymbol{e}, \boldsymbol{n})$ is

$$\mathbf{V} := \tilde{\mathbf{A}}[\boldsymbol{e} \otimes \boldsymbol{n}]^{\mathsf{T}}\boldsymbol{e}/(\rho_{\circ} C e^2).$$

From $C^2\boldsymbol{e} = \mathbf{Q}(\boldsymbol{n})\boldsymbol{e}$ we obtain

$$\mathbf{V} \cdot \mathbf{C} = C^2, \quad \mathbf{V} \cdot \mathbf{S} = 1, \quad |\mathbf{V}| \geq C. \tag{23.3.3}$$

The ***wave (front) surface*** V is defined by

$$V = \{\mathbf{V} \in \text{Vect} : \mathbf{V} \text{ is the group velocity of some characteristic pair } (C, \boldsymbol{e})\}.$$

The slowness and wave surfaces are the basic characteristics of the material from the acoustic point of view. I refer to PAYTON [1983, Chaps. 1 & 2], CHADWICK [1989, Part I] for detailed descriptions and classifications of the slowness surfaces of transversely isotropic media. See also RACKE [1992, Chap. 11] for cubic media, and for the role of the slowness surface for the decay rates of solutions.

Recall that a subset K of a vector space X is called a cone if for every $\boldsymbol{x} \in K$ and every $\lambda > 0$ also $\lambda\boldsymbol{x} \in K$. A conical neighborhood of a point $\boldsymbol{x}$ is a neighborhood of $\boldsymbol{x}$ that is also a cone. If X, Y are vector spaces and $s \in \mathbb{R}$, a function $f : K \to$ Y is said to be positively homogeneous of degree s if $K \subset$ X is a cone and $f(\lambda\boldsymbol{x}) = \lambda^s f(\boldsymbol{x})$ for every $\boldsymbol{x} \in K$ and every $\lambda > 0$.

23.3.3 Proposition *If $(C, \boldsymbol{e})$ is a simple characteristic pair at $(\boldsymbol{F}_0, \eta_0, \boldsymbol{n}) \in D$ with $C > 0$ and with the wavevector, frequency and slowness $\mathbf{K}$, ω, $\mathbf{S}$, respectively, then there exists a conical neighborhood M of $\mathbf{K}$ in* Vect *and class C^∞ functions $\tilde{\omega} : M \to (0, \infty)$, $\tilde{\boldsymbol{e}} : M \to$* Vect, *positively homogeneous of degree* 1, 0, *respectively, such that*

(1) *$S \cap M$ is the set of all solutions $\mathbf{S}'$ of $\tilde{\omega}(\mathbf{S}') = 1$;*

(2) *$\mathbf{V} = \partial_{\mathbf{S}}\tilde{\omega}(\mathbf{S})$ and hence $\mathbf{V}$ is normal to S at $\mathbf{S}$;*

(3) *$\tilde{\omega}(\boldsymbol{K}) = \omega$, $\tilde{\boldsymbol{e}}(\boldsymbol{K}) = \boldsymbol{e}$ and*

$$\tilde{\omega}(\boldsymbol{L})^2\tilde{\boldsymbol{e}}(\boldsymbol{L}) = \mathbf{Q}(\boldsymbol{L})\tilde{\boldsymbol{e}}(\boldsymbol{L}) \quad \text{for every } \boldsymbol{L} \in M;$$

(4) $\partial_{\mathbf{S}} \det(\mathbf{1} - \mathbf{Q}(\mathbf{S})) = -2\mathbf{V} \operatorname{tr}[\operatorname{cof}(\mathbf{1} - \mathbf{Q}(\mathbf{S}))] \neq \mathbf{0}$.

Hence, locally near $\mathbf{S}$, S is a class C^∞ surface by (1) and (4). Since $\partial_{\boldsymbol{K}}\tilde{\omega}$ is positively homogeneous of degree 0, (2) also implies $\mathbf{V} = \partial_{\boldsymbol{K}}\tilde{\omega}(\boldsymbol{K})$. A standard argument (e.g., DIEULESAINT & ROYER [1980, Subsect. 1.1.3]) shows that $\partial_{\boldsymbol{K}}\tilde{\omega}(\boldsymbol{K})$ is the velocity of propagation of signals in the form of spatially concentrated wave groups.

Proof (3): The existence of $\tilde{\omega}$, $\tilde{\boldsymbol{e}}$ satisfying (3) follows from Proposition 1.2.2 in much the same way as in the proof of Proposition 23.2.2, and the details are omitted. It is only noted that the conical form of M and the homogeneity of $\tilde{\omega}$ follow from the form of the characteristic equation. (2): Assume, as we may, that $\tilde{e}$ is normalized, i.e., $|\tilde{e}(L)| = 1$, $L \in M$ and set $\rho_\circ = 1$. A differentiation of $\tilde{\omega}(L)^2\tilde{e}(L) = Q(L)\tilde{e}(L)$ in the direction $H \in$ Vect provides

$$\begin{aligned}\omega\partial_K\omega \cdot H &= \tilde{A}[\tilde{e}\otimes H, \tilde{e}\otimes K] + \tilde{A}[\partial_K\tilde{e}[H]\otimes K, \tilde{e}\otimes K] \\ &= \tilde{A}[\tilde{e}\otimes K]\cdot(\tilde{e}\otimes H) + Q(K)\tilde{e}\cdot\partial_K\tilde{e}[H] \\ &= \tilde{A}[\tilde{e}\otimes K]^{\mathsf{T}}\tilde{e}\cdot H + \omega^2\tilde{e}\cdot\partial_K\tilde{e}[H] = \tilde{A}[\tilde{e}\otimes K]^{\mathsf{T}}\tilde{e}\cdot H = \omega V\cdot H\end{aligned}$$

since $\tilde{e}\cdot\partial_K\tilde{e}[H] = 0$ by normalization. This implies $V = \partial_K\tilde{\omega}(K)$ and hence also (2). (1): This follows immediately from the definition of the slowness $S' = L/\tilde{\omega}(L)$ and the homogeneity of $\tilde{\omega}$. (4): By (1) and (2), $\partial_S \det(\mathbf{1} - Q(S)) = \alpha V$ for some $\alpha \in \mathbb{R}$. Hence, differentiating the determinant in the direction $M \in$ Vect, we obtain

$$\alpha V\cdot M = -\operatorname{cof}(\mathbf{1} - Q(S))\cdot\partial_S Q(S)[M].$$

Setting $M = S$ and using $V\cdot S = 1$, $\partial_S Q(S)[S] = 2Q(S)$ and $\operatorname{cof}(\mathbf{1} - Q(S))\cdot(\mathbf{1} - Q(S)) = 3\det(\mathbf{1} - Q(S)) = 0$, we obtain

$$\alpha = -2\operatorname{cof}(\mathbf{1} - Q(S))\cdot Q(S) = -2\operatorname{tr}\operatorname{cof}(\mathbf{1} - Q(S)).$$

This proves the equation in (4). Using the fact that (C, e) is a simple characteristic pair, we find that $\operatorname{tr}\operatorname{cof}(\mathbf{1}-Q(S)) \neq 0$ in the same way as in the proof of Proposition 1.2.2, and hence $\partial_S \det(\mathbf{1}-Q(S)) \neq 0$. □

For a smooth surface S of a vector space X with inner product, the polar reciprocal surface is defined as the set T of all $y \in$ X such that y is normal to S at some point $x \in S$ and $x\cdot y = 1$. The passage from S to its polar reciprocal is an analog of the Legendre transformation for surfaces. From the fact that the group velocities are normals to the slowness surface and satisfy $(23.3.3)_2$ we see that *the wave surface is the polar reciprocal of the slowness surface,* at least when all characteristic pairs are simple. Assertions (1), (2) of the following proposition are therefore duals of Assertions (1), (2) of Proposition 23.3.3.

23.3.4 Proposition *Let V be a group velocity of a simple characteristic pair (C, e) of slowness S, and let $\tilde{\omega} : M \to \mathbb{R}$ be as in Proposition* 23.3.3. *Then there exists a conical neighborhood N of V and a C^∞ function $\tilde{\pi} : N \to \mathbb{R}$, positively homogeneous of degree* 1, *such that*

(1) *$V \cap N$ is the set of all solutions V' of $\tilde{\pi}(V') = 1$;*

(2) *$S = \partial_V\tilde{\pi}(V)$ and hence S is normal to V at V;*

(3) *if $S' \in S \cap M$ and $V' \in V \cap N$, then*

$$S' = \partial_V\tilde{\pi}(V') \text{ if and only if } V' = \partial_S\tilde{\omega}(S').$$

Proof (1): There exists a neighborhood N of V such that $V \cap N$ is a class C^∞ surface. This follows from the fact that $\tilde{\omega}$ maps $S \cap M$ into V and that its rank at S is $2 = 3 - 1$. Let us prove that S is normal to this surface. Let $S(t) : (-\varepsilon, \varepsilon) \to S$ be any smooth curve such that $S(0) = S$. Then $\tilde{\omega}(S(t)) = 1$ and $V(t) := \partial_K\tilde{\omega}(S(t))$ is a smooth curve in V, with $V(0) = V$. If n is a normal to V at V, then $n\cdot\dot{V}(0) = 0$. On the other hand by $(23.3.3)_2$, the equation $\tilde{\omega}(S(t)) = 1$ can be written as $\tilde{\omega}(S(t)) = V(t)S(t)$. Differentiating this at $t = 0$, we obtain $\partial_K\tilde{\omega}(S)\cdot\dot{S}(0) = \dot{V}(0)\cdot S + V\cdot\dot{S}(0)$, which implies $\dot{V}(0)\cdot S = 0$. As $\dot{V}(0)$ may be chosen arbitrarily, we have $M\cdot n = 0$ for every $M \in$ Vect such that $M\cdot S = 0$. This implies $n = \alpha S$ for some $\alpha \in \mathbb{R}$. From $V\cdot S = 1$ follows that if N is chosen small enough, then for every $V' \in V \cap N$ the ray $\{\lambda V' : \lambda > 0\}$ intersects $V \cap N$ just at V'. We change the notation and denote by N the union of all the mentioned rays, and define $\tilde{\pi}(\lambda V') = \lambda$ for every $\lambda > 0$ and $V' \in V \cap N$. This gives a well defined function on N which clearly satisfies (1). (2): From the fact that $\tilde{\pi}$ is constant on $V \cap N$ and from the normality of S to $V \cap N$ at V we have $\partial_V\tilde{\pi}(V) = \alpha S$ for some $\alpha \in \mathbb{R}$. On the other hand, from the homogeneity we obtain $\partial_V\tilde{\pi}(V)\cdot V = \tilde{\pi}(V) = 1$ which means that $\alpha S\cdot V = 1$. By $(23.3.3)_2$ then $\alpha = 1$. This completes the proof of (2). Assertion (3) is just a simple consequence whose proof is left to the reader. □

23.3.5 Proposition *Let* $\boldsymbol{V}$ *be a group velocity of a simple characteristic pair* $(C, \boldsymbol{e})$ *of slowness* $\mathbf{S}$, *and let* $\tilde{\pi} : M \to \mathbb{R}$ *be as in Proposition 23.3.4. For every* $t > 0$ *define a subset* S_t *of* Vect *by* $S_t = t \cdot (V \cap N) = \{\boldsymbol{p} \in \text{Vect} : \boldsymbol{p}/t \in V \cap N\}$. *Then* S_t *is a moving smooth surface and*

(1) *for every* $t > 0$, S_t *is the set of all solutions* $\boldsymbol{V}'$ *of* $\tilde{\pi}(\boldsymbol{V}') = t$;

(2) *the normal* $\boldsymbol{n}$ *and speed* U *of* S_t *at any* $(\boldsymbol{p}, t)$ *coincide with the normal and speed of the characteristic pair of which* $\boldsymbol{p}/t \in V$ *is the group velocity;*

(3) S_t *is a characteristic surface, i.e., its normal and speed satisfy the characteristic equation* $p(\boldsymbol{n}, U) = 0$ *at any point* $(\boldsymbol{p}, t)$ *on* S_t.

Generally, the disturbances propagate along the characteristic surfaces (see Subsect. 23.3.8). In the present case, the surface S_1 (and hence also V) may be identified with the wave front at time $t = 1$ of the wave caused by a point disturbance at the origin $\boldsymbol{0} \in \text{Vect}$ at time $t = 0$.

Proof (1) is an immediate consequence of Proposition 23.3.4(1). (2): Let us apply the definitions in Subsect. 2.1.2 to the present situation. The surface S_t can be expressed as the set of all solutions $(\boldsymbol{p}, t)$ of $\phi(\boldsymbol{p}, t) = 0$ where ϕ is of the form $\phi(\boldsymbol{p}, t) = \tilde{\pi}(\boldsymbol{p}) - t$. If $\bar{\boldsymbol{n}}$, C, $\mathbf{S}'$ are the normal, velocity and slowness of the characteristic pair of which $\boldsymbol{p}/t$ is the group velocity, then $\nabla\phi = \mathbf{S}$, $|\nabla\phi| = |\mathbf{S}| = 1/C$, by Proposition 23.3.4 and hence $\boldsymbol{n} = \bar{\boldsymbol{n}}$, $U = C$ by Subsect. 2.1.2. (3) is a consequence of (2). □

23.3.6 The Dynamical Quantities Consider a plane wave of wavevector $\boldsymbol{K}$, cyclic frequency ω and amplitude $\boldsymbol{e}$. We have

$$\boldsymbol{v} = \omega\boldsymbol{e}\sin P, \quad \boldsymbol{F} = \boldsymbol{F}_0 - \boldsymbol{e}\otimes\boldsymbol{K}\sin P, \quad \nabla\boldsymbol{u} = -\boldsymbol{e}\otimes\boldsymbol{K}\sin P, \quad \omega\nabla\boldsymbol{u} = -\boldsymbol{v}\otimes\boldsymbol{K}.$$

Inserting these expressions into the approximate constitutive equations, we obtain

$$\rho_\circ\tilde{e} = \rho_\circ e_0 + \boldsymbol{S}_0\cdot\nabla\boldsymbol{u} + \tfrac{1}{2}\rho_\circ\tilde{\boldsymbol{A}}[\nabla\boldsymbol{u}, \nabla\boldsymbol{u}] + o(|\nabla\boldsymbol{u}|^2),$$
$$\tilde{\boldsymbol{S}} = \boldsymbol{S}_0 + \tilde{\boldsymbol{A}}[\nabla\boldsymbol{u}] + o(|\nabla\boldsymbol{u}|);$$

hence it is meaningful to compute the energy to within the quadratic terms and the momentum only to within linear terms. We identify the reference configuration with the configuration about which the linearization is made, set $e_0 = 0$, and write ρ for $\rho_\circ$. The internal and kinetic energies per unit volume are

$$\rho e = -\boldsymbol{S}_0\boldsymbol{K}\cdot\boldsymbol{e}\sin P + \tfrac{1}{2}\rho\omega^2\sin^2 P = \boldsymbol{S}_0\cdot\nabla\boldsymbol{u} + \tfrac{1}{2}\rho v^2, \quad \tfrac{1}{2}\rho\omega^2\sin^2 P = \tfrac{1}{2}\rho v^2,$$

respectively, and the total energy ρu per unit volume is

$$\rho u = \boldsymbol{S}_0\cdot\nabla\boldsymbol{u} + \rho v^2.$$

The flux of energy $\boldsymbol{w}$ is defined as $\boldsymbol{w} = -\boldsymbol{S}^{\mathsf{T}}\boldsymbol{v}$; it is given by

$$\boldsymbol{w} = -\boldsymbol{S}^{\mathsf{T}}\boldsymbol{v} = -(\boldsymbol{S}_0 + \tilde{\boldsymbol{A}}[\nabla\boldsymbol{u}])^{\mathsf{T}}\boldsymbol{e}\omega\sin P = -\boldsymbol{S}_0^{\mathsf{T}}\boldsymbol{v} + \omega\tilde{\boldsymbol{A}}[\boldsymbol{e}\otimes\boldsymbol{K}]^{\mathsf{T}}\boldsymbol{e}\sin^2 P$$
$$= -\boldsymbol{S}_0^{\mathsf{T}}\boldsymbol{v} + \rho\boldsymbol{V}\omega^2\boldsymbol{e}^2\sin^2 P = -\boldsymbol{S}_0^{\mathsf{T}}\boldsymbol{v} + \rho v^2\boldsymbol{V}.$$

The momentum per unit volume and its flux $\boldsymbol{z}$ are

$$\rho\boldsymbol{v} = \rho\boldsymbol{e}\omega\sin P, \quad \boldsymbol{z} = -\boldsymbol{S} = -\boldsymbol{S}_0 + \tilde{\boldsymbol{A}}[\boldsymbol{e}\otimes\boldsymbol{K}]\sin P.$$

We have the relations

$$\boldsymbol{w}\cdot\boldsymbol{K} = \omega\rho u, \quad \boldsymbol{z}\boldsymbol{K} = -\boldsymbol{S}_0\boldsymbol{K} + \omega\rho\boldsymbol{v}.$$

The mean time values of the total energy and of the flux of energy are

$$\overline{\rho u} = \tfrac{1}{2}\rho\omega^2 e^2 = \overline{\rho v^2}, \quad \overline{w} = \tfrac{1}{2}\rho\omega^2 e^2 \boldsymbol{V} = \overline{\rho u}\cdot \boldsymbol{V}.$$

Hence the energy propagates in the direction of the referential group velocity, as claimed in Subsect. 23.3.2. The mean values of momentum and of the flux of momentum are identically zero.

23.3.7 Surface Waves A different characteristic equation arises for surface waves. Let $\boldsymbol{n} \in$ Vect be a unit vector and let H be a half-space

$$H = \{\boldsymbol{p} \in \text{Vect} : \boldsymbol{p}\cdot\boldsymbol{n} < 0\}, \quad \partial H = \{\boldsymbol{p} \in \text{Vect} : \boldsymbol{p}\cdot\boldsymbol{n} = 0\}.$$

Consider the displacement field $\boldsymbol{u} : H \times \mathbb{R} \to \text{Vect}$ of the form

$$\boldsymbol{u}(\boldsymbol{p}, t) = \boldsymbol{z}(-\boldsymbol{p}\cdot\boldsymbol{n})\exp(\mathrm{i}(\boldsymbol{k}\cdot\boldsymbol{p} - \omega t)) + \boldsymbol{z}^*(-\boldsymbol{p}\cdot\boldsymbol{n})\exp(-\mathrm{i}(\boldsymbol{k}\cdot\boldsymbol{p} - \omega t))$$

with given $\omega \in \mathbb{R}$, $\boldsymbol{k} \in$ Vect such that $\boldsymbol{k}\cdot\boldsymbol{n} = 0$ and with $\boldsymbol{z} : [0, \infty) \to$ Vect such that $\boldsymbol{z}(s) \to \boldsymbol{0}$ as $s \to \infty$. The displacement field of this form is said to be a surface wave if $\boldsymbol{u}$ solves the linearized equations (23.3.1) with $\bar{\eta} \equiv \boldsymbol{0}$ on H and satisfies the boundary condition

$$\tilde{\mathsf{A}}[\nabla \boldsymbol{u}]\boldsymbol{n} = \boldsymbol{0} \text{ on } \partial H.$$

This leads to

$$\tilde{\mathsf{A}}[\ddot{\boldsymbol{z}} \otimes \boldsymbol{n}]\boldsymbol{n} - \mathrm{i}\tilde{\mathsf{A}}[\dot{\boldsymbol{z}} \otimes \boldsymbol{n}]\boldsymbol{k} - \mathrm{i}\tilde{\mathsf{A}}[\dot{\boldsymbol{z}} \otimes \boldsymbol{k}]\boldsymbol{n} - \tilde{\mathsf{A}}[\boldsymbol{z} \otimes \boldsymbol{k}]\boldsymbol{k} = -\omega^2\rho_\circ \boldsymbol{z} \text{ for } s \in [0, \infty),$$
$$\tilde{\mathsf{A}}[\dot{\boldsymbol{z}} \otimes \boldsymbol{n} - \mathrm{i}\boldsymbol{z} \otimes \boldsymbol{k}]\boldsymbol{n} = \boldsymbol{0} \text{ for } s = 0.$$

In the absence of degeneracies, a general solution $\boldsymbol{z}$ of these equations is a linear combination of exponentials,

$$\boldsymbol{z}(s) = \sum_j \gamma_j \boldsymbol{e}_j \exp(-\mathrm{i}p_j s), \quad s \ge 0, \tag{23.3.4}$$

where $\boldsymbol{e}_j \in \text{Vect}^{\mathbb{C}}$, $p_j \in \mathbb{C}$, $\gamma_j \in \mathbb{C}$, $|\boldsymbol{e}_j| = 1$, and j ranges over $1 \le j \le 6$. Equations (20.4.10) give

$$\tilde{\mathsf{A}}[\boldsymbol{e}_j \otimes \boldsymbol{K}_j]\boldsymbol{K}_j = \rho_\circ\omega^2 \boldsymbol{e}_j, \quad \sum_j \tilde{\mathsf{A}}[\gamma_j \boldsymbol{e}_j \otimes \boldsymbol{K}_j]\boldsymbol{n} = \boldsymbol{0}, \tag{23.3.5}$$

where $\boldsymbol{K}_j := (p_j\boldsymbol{n} + \boldsymbol{k})$. The eigenvalue problem $(23.3.5)_1$ determines the possible values of p_j and the corresponding $\boldsymbol{e}_j$. Next one considers only those p_j in (23.3.4) for which $\operatorname{Im} p_j < 0$ as to satisfy the decay condition on $\boldsymbol{z}$, and $(23.3.5)_2$ is then used to determine the relations among $\boldsymbol{e}_j$. The form of the problem is identical to that considered in connection with Agmon's condition in Sect. 20.4. The reader is referred to CHADWICK & SMITH [1977], DIEULESAINT & ROYER [1980] and CHADWICK [1989, 1993] for more details on the surface waves and for further references.

23.3.8 Acceleration Waves The solution is said to have a ***weak singularity*** along a moving singular surface S_t if $\boldsymbol{\chi}$, $\boldsymbol{v}$, $\boldsymbol{F}$, η are continuous and the first spacetime gradients of $\boldsymbol{v}$, $\boldsymbol{F}$, η have jump discontinuities across S_t. These spacetime gradients are $\dot{\boldsymbol{v}}$, $\nabla\boldsymbol{v}$, $\dot{\boldsymbol{F}}$, $\nabla\boldsymbol{F}$, $\dot{\eta}$, $\nabla\eta$, or in other words, $\ddot{\boldsymbol{\chi}}$, $\nabla\dot{\boldsymbol{\chi}}$, $\nabla\dot{\boldsymbol{\chi}}$, $\nabla^2\boldsymbol{\chi}$, $\dot{\eta}$, $\nabla\eta$. Because the waves of this type have a discontinuity in the acceleration, they are called the ***acceleration waves***. A double application of the Hadamard lemma 2.1.6 gives the existence of $\boldsymbol{e}$, $\boldsymbol{b}$ defined on the singular surface, such that

$$[\dot{v}] = eU^2, \quad [\nabla v] = [\dot{F}] = -Ue \otimes n, \quad [\nabla F] = e \otimes n \otimes n,$$
$$[\dot{\eta}] = -Ub, \quad [\nabla \eta] = bn.$$

Writing down the equation $\dot{\eta} = 0$ and the equation of balance of linear momentum with explicitly computed dependencies on ∇F and $\nabla\eta$, for both sides of the singular surface and subtracting the results, one finds that the amplitudes e, b and the speed U satisfy the characteristic equation (23.1.2) with C replaced by U. Remark 23.1.4 applies, hence either $U = 0$ or, if it is not the case, the amplitude e is one of the characteristic vectors of the acoustic tensor and the speed of propagation U is equal to the speed of sound corresponding to the amplitude. See BOWEN & WANG [1970, 1971], CHADWICK & OGDEN [1971a,b] and TRUESDELL & NOLL [1965, Sects. 71–78] for more details.

23.4 The Characteristic Equation and Material Symmetry

For a discussion of the consequences of the material symmetry on the characteristic equation, the spatial form $c^2 e = \mathsf{q}(m)e$ is more convenient than the referential form, which is explained by the following observation.

23.4.1 Proposition *For every $f \in \tilde{U}$ and $m \in$ Vect the spatial acoustic tensor $\mathsf{q}(m)$ at f satisfies*

$$\mathsf{q}(Qm) = Q\mathsf{q}(m)Q^{\mathrm{T}} \tag{23.4.1}$$

for every $Q \in G_F$, where $G_F = FGF^{-1} \cap \mathrm{Orth}$ is the local symmetry group at f, cf. Definition 9.4.4.

This follows immediately from Proposition 10.5.3(5). The transformation law of the referential acoustic tensor is more complicated than (23.4.1).

For the rest of this section, let $n = 3$. We define the ***symmetry group $G(\mathsf{q})$ of the acoustic tensor*** at f as the set of all $Q \in$ Orth such that (23.4.1) holds for every $m \in$ Vect. Since $\mathsf{q}(m) = \mathsf{q}(-m)$ we have that if $Q \in G(\mathsf{q})$ then also $-Q \in G(\mathsf{q})$. Note that the last property need not hold for G_F. Proposition 23.4.1 says that $G_F \subset G(\mathsf{q})$. Thus, e.g., by Proposition 11.1.4, for an isotropic material $G(\mathsf{q}) =$ Orth if f is an isotropic state with all three principal stretches of the same value, $G(\mathsf{q})$ contains the transversely isotropic group if two principal stretches coincide and the third is different and $G(\mathsf{q})$ contains the orthotropic group if the three principal stretches are distinct. Similarly, for a crystal material with $G = G(L)$ (see Subsect. 9.4.5), $G(\mathsf{q})$ contains the point group of the deformed lattice FL (see 1.5) etc.

An $m \in$ Sph is called an ***axis of symmetry*** of the acoustic tensor at f if there exists a proper orthogonal tensor $R \in G(\mathsf{q})$ such that $Rm = m$ and $R \neq 1$. Then necessarily R is a rotation about m through some angle ω, $0 < \omega < 2\pi$, i.e., R is given by the right-hand side of (1.3.6). If additionally R is a rotation of order k about m, (see 1.3.5) then k is called the ***order of the axis of symmetry m***. A plane $P \subset$ Vect through $\mathbf{0}$ with normal m is called a ***plane of symmetry*** of the acoustic tensor at f if the reflection $Q = 1 - 2m \otimes m$ on P is in $G(\mathsf{q})$. Below we abbreviate and use the terms 'axis (or plane) of symmetry' to mean 'axis (or plane) of symmetry of the acoustic tensor at f.'

Items (2)–(4) of the following proposition outline the relationships between the elements of symmetry and the characteristic amplitudes.

23.4.2 Proposition *Let $f \in \tilde{U}$ and let q be the acoustic tensor at f. Then*

(1) *there exists an $m \in$ Sph such that m is an eigenvector of $\mathsf{q}(m)$;*

(2) *if $m \in$ Sph is an axis of symmetry then m is an eigenvector of $\mathsf{q}(m)$ (and hence the two other eigenvectors may be chosen in the plane orthogonal to m);*

(3) *if $m \in$ Sph is an axis of symmetry that is not of order 2 then $\mathsf{q}(m)$ has at most two distinct eigenvalues c_1, c_2 and the corresponding eigenspaces are the line spanned by m and the plane perpendicular to m; consequently*

$$\mathsf{q}(m) = c_1 m \otimes m + c_2(1 - m \otimes m);$$

(4) *if P is a plane of symmetry of normal $\boldsymbol{r} \in$ Sph and $\boldsymbol{m}$ is in P, then $\boldsymbol{r}$ is an eigenvector of $\boldsymbol{q}(\boldsymbol{m})$ and the two other eigenvectors lie in P.*

Proof (1): The quartic form $K(\boldsymbol{m}) := \boldsymbol{q}(\boldsymbol{m})\boldsymbol{m} \cdot \boldsymbol{m}$, $\boldsymbol{m} \in$ Vect has an extremum on Sph at some $\boldsymbol{m} \in$ Sph, i.e., an extremum relative to the constraint $\boldsymbol{m}^2 = 1$. By the Lagrange multipliers, there is a $\lambda \in \mathbb{R}$ such that $\partial_{\boldsymbol{m}} K(\boldsymbol{m}) = 4\lambda \boldsymbol{m}$. Since $\partial_{\boldsymbol{m}} K(\boldsymbol{m}) = 4\boldsymbol{q}(\boldsymbol{m})\boldsymbol{m}$, we have $\boldsymbol{q}(\boldsymbol{m})\boldsymbol{m} = \lambda \boldsymbol{m}$. (2): Let $\boldsymbol{R}$ be the rotation associated with $\boldsymbol{m}$. Then from $\boldsymbol{R}\boldsymbol{m} = \boldsymbol{m}$ and (23.4.1) follows that $\boldsymbol{R}\boldsymbol{q}(\boldsymbol{m})\boldsymbol{m} = \boldsymbol{q}(\boldsymbol{m})\boldsymbol{m}$. As the only vectors invariant under $\boldsymbol{R}$ are multiples of $\boldsymbol{m}$, it follows that $\boldsymbol{q}(\boldsymbol{m})\boldsymbol{m} = \alpha \boldsymbol{m}$. (3) By (2), $\boldsymbol{m}$ is an eigenvector of $\boldsymbol{q}$ and hence there exists an eigenvector $\boldsymbol{e}$ in P. From $\boldsymbol{q}(\boldsymbol{m})\boldsymbol{e} = \lambda \boldsymbol{e}$ and (23.4.1) follows that $\boldsymbol{q}(\boldsymbol{m})\boldsymbol{R}\boldsymbol{e} = \lambda \boldsymbol{R}\boldsymbol{e}$, i.e., $\boldsymbol{b} := \boldsymbol{R}\boldsymbol{e} \neq \boldsymbol{e}$ is another eigenvector in P with the same eigenvalue. As any other vector in P is a linear combination of $\boldsymbol{e}, \boldsymbol{b}$, the result follows. (4): If P is a plane of symmetry with normal $\boldsymbol{r}$ and $\boldsymbol{R} = \boldsymbol{1} - 2\boldsymbol{r} \otimes \boldsymbol{r}$ the associated reflection, then $\boldsymbol{R}\boldsymbol{m} = \boldsymbol{m}$ and hence (23.4.1) gives $\boldsymbol{R}\boldsymbol{q}(\boldsymbol{m}) = \boldsymbol{q}(\boldsymbol{m})\boldsymbol{R}$. If now $\boldsymbol{b}$ is any vector in P, the last equation implies $\boldsymbol{R}\boldsymbol{q}(\boldsymbol{m})\boldsymbol{b} = \boldsymbol{q}(\boldsymbol{m})\boldsymbol{b}$ since $\boldsymbol{R}\boldsymbol{b} = \boldsymbol{b}$. As the only vectors invariant under $\boldsymbol{R}$ are the elements of P, it follows that $\boldsymbol{q}(\boldsymbol{m})\boldsymbol{b} \in P$, i.e., P is invariant under $\boldsymbol{q}(\boldsymbol{m})$. Hence invariant is also the orthogonal complement of P, i.e., the line spanned by $\boldsymbol{m}$. □

For further details on the relationship between the wave propagation and the elements of symmetry, for other general features, and for the forms of the acoustic tensor for crystal classes at stress-free states, see FEDOROV [1968].

23.4.3 The Characteristic Equation for Fluids Using the results of Subsect. 11.4.3, one finds that for fluids,

$$\boldsymbol{Q}(\boldsymbol{n}) = -v^2 \partial_v \tilde{p}\, \boldsymbol{F}^{-\mathsf{T}} \boldsymbol{n} \otimes \boldsymbol{F}^{-\mathsf{T}} \boldsymbol{n}, \quad \boldsymbol{D} = -v \partial_\eta \tilde{p}\, \boldsymbol{F}^{-\mathsf{T}}, \quad \boldsymbol{q}(\boldsymbol{m}) = -v^2 \partial_v \tilde{p}\, \boldsymbol{m} \otimes \boldsymbol{m}.$$

The characteristic equation reads

$$C^2 \boldsymbol{e} = -v^2 \partial_v \tilde{p}\, (\boldsymbol{F}^{-\mathsf{T}} \boldsymbol{n} \otimes \boldsymbol{F}^{-\mathsf{T}} \boldsymbol{n})\boldsymbol{e} - v \partial_\eta \tilde{p}\, \boldsymbol{F}^{-\mathsf{T}} \boldsymbol{n} b, \quad C b = 0.$$

We use Proposition 23.1.6(1) to clarify the conditions under which the fluid is hyperbolic. We have

$$\boldsymbol{Q}(\boldsymbol{n})\boldsymbol{e} \cdot \boldsymbol{e} = -v^2 \partial_v \tilde{p}\, (\boldsymbol{F}^{-\mathsf{T}} \boldsymbol{n} \cdot \boldsymbol{e})^2$$

and hence the adiabatic Legendre–Hadamard condition is satisfied if and only if $\partial_v \tilde{p} \le 0$. The range of $\boldsymbol{Q}(\boldsymbol{n})$ is $\{\boldsymbol{0}\}$ if $\partial_v \tilde{p} = 0$ and the line spanned by $\boldsymbol{m} := \boldsymbol{F}^{-\mathsf{T}} \boldsymbol{n} / |\boldsymbol{F}^{-\mathsf{T}} \boldsymbol{n}|$ if $\partial_v \tilde{p} \neq 0$. Hence $\boldsymbol{D}\boldsymbol{n}$ is in the range of $\boldsymbol{Q}(\boldsymbol{n})$ if and only if $\partial_v \tilde{p} \neq 0$ whenever $\partial_\eta \tilde{p} \neq 0$. Assume now that $\partial_v \tilde{p} < 0$ and determine the characteristic vectors and speeds. Any vector $\xi = (\boldsymbol{e}, b)$ such that

$$v \partial_v \tilde{p}\, |\boldsymbol{F}^{-\mathsf{T}} \boldsymbol{n}|\, (\boldsymbol{m} \cdot \boldsymbol{e}) + \partial_\eta \tilde{p}\, b = 0$$

is a characteristic vector corresponding to the characteristic speed $C = 0$. There are three linearly independent vectors satisfying the last equation: two vectors $\xi = (\boldsymbol{e}, 0)$ with $\boldsymbol{e} \perp \boldsymbol{m}$ and a vector $\xi = (\boldsymbol{m}, b)$ where b satisfies

$$v \partial_v \tilde{p}\, |\boldsymbol{F}^{-\mathsf{T}} \boldsymbol{n}| + \partial_\eta \tilde{p}\, b = 0.$$

The only characteristic field with positive characteristic speed is

$$C = C(f, \boldsymbol{n}) = \sqrt{-v^2 \partial_v \tilde{p}}\, |\boldsymbol{F}^{-\mathsf{T}} \boldsymbol{n}|, \quad \boldsymbol{e} = \boldsymbol{e}(f, \boldsymbol{n}) = \boldsymbol{F}^{-\mathsf{T}} \boldsymbol{n} \tag{23.4.2}$$

and one has

$$\partial_{\boldsymbol{F}} C^2 = (2v^2 \partial_v \tilde{p}\, (\boldsymbol{F}^{-\mathsf{T}} \boldsymbol{n} \otimes \boldsymbol{F}^{-\mathsf{T}} \boldsymbol{n}) - |\boldsymbol{F}^{-\mathsf{T}} \boldsymbol{n}|^2 v \partial_v (v^2 \partial_v \tilde{p}) \boldsymbol{1}) \boldsymbol{F}^{-\mathsf{T}};$$

consequently,

$$\partial_{\boldsymbol{F}} C^2 \cdot (\boldsymbol{e} \otimes \boldsymbol{n}) = -v^3 \partial_v^2 \tilde{p}\, |\boldsymbol{F}^{-\mathsf{T}} \boldsymbol{n}|^4.$$

Hence this characteristic field is genuinely nonlinear if and only if $\partial_v^2 \tilde{p} \neq 0$. This means that the point v is not an inflection point of the function $\tilde{p}(\cdot, \eta)$. The speed C is the referential speed of the propagation of the wave, the spatial speed (of sound) is $c = C / |\boldsymbol{F}^{\mathsf{T}} \boldsymbol{n}| = \sqrt{-v^2 \partial_v \tilde{p}} = \sqrt{\partial_\rho \bar{p}}$. The considerations above prove the following summary information.

23.4.4 Proposition *Consider a fluid and let* $f = (F, \eta)$ *be in* $\tilde{U}$, $n \in$ Sph *with the corresponding specific volume* v. *Then the following assertions hold (with the derivatives of* $\tilde{p}$ *evaluated at* (v, η)*)*:

(1) *the fluid is hyperbolic at* (f, n) *if and only if* $\partial_v \tilde{p} \le 0$ *and the inequality is strict if* $\partial_\eta \tilde{p} \ne 0$;

(2) *if* $\partial_v \tilde{p} < 0$, *then* (23.4.2) *gives a simple characteristic pair;*

(3) *the characteristic pair from* (23.4.2) *is genuinely nonlinear if and only if* $\partial_v \tilde{p} < 0$ *and* $\partial_v^2 \tilde{p} \ne 0$;

(4) *if* $n > 1$, *the fluid is never strictly hyperbolic; if* $\partial_v \tilde{p} < 0$, *then the multiplicity of the speed* $C = 0$ *is* n.

23.4.5 The Acoustic Tensor of Isotropic Solids Let the response of the isotropic solid be given by the internal energy expressed as a function of the principal stretches v_i and entropy: $e = \tilde{e}(v_1, \ldots, v_n, \eta)$. It is enough to calculate the acoustic tensor for states $f = (F, \eta) \in \tilde{U}$ with $F \in \mathrm{Sym}^+$. Using Proposition 8.5.4 to calculate $\tilde{A}$, one finds that the components of $Q(n)$ in the basis of eigenvectors of $F \in \mathrm{Sym}^+$ are

$$Q(n)_{ii} = \tilde{e}_{,ii} n_i^2 + \sum_{j \ne i}^{n} K_{ij} n_j^2, \qquad Q(n)_{ij} = (\tilde{e}_{,ij} + M_{ij}) n_i n_j, \qquad i \ne j,$$

where n_i are the components of n in the basis of eigenvectors of F and

$$K_{ij} = \frac{v_i \tilde{e}_{,i} - v_j \tilde{e}_{,j}}{v_i^2 - v_j^2}, \qquad M_{ij} = \frac{v_j \tilde{e}_{,i} - v_i \tilde{e}_{,j}}{v_i^2 - v_j^2} \quad \text{if } v_i \ne v_j,$$

$$K_{ij} = \tfrac{1}{2}(\tilde{e}_{,ii} - \tilde{e}_{,ij} + \tilde{e}_{i}/v_i), \qquad M_{ij} = \tfrac{1}{2}(\tilde{e}_{,ii} - \tilde{e}_{,ij} - \tilde{e}_{i}/v_i), \quad \text{if } v_i = v_j.$$

The spatial acoustic tensor is

$$q(m)_{ii} = \tilde{e}_{,ii} v_i^2 m_i^2 + \sum_{j \ne i}^{n} K_{ij} v_j^2 m_j^2, \qquad q(m)_{ij} = (\tilde{e}_{,ij} + M_{ij}) v_i v_j m_i m_j, \qquad i \ne j.$$

For a general material, the eigenvectors of the acoustic tensor are in no direct way related to the principal directions of strain at F and to n or m. Only special cases are solvable:

(1) ***Principal waves*** A principal wave is a wave whose direction of propagation m is parallel to one of the eigenvectors e_i of the left stretch tensor V. Without any loss of generality, one can assume that $m = e_1$. Then $q(m)$ is represented by a diagonal matrix $q = \mathrm{diag}(c_1, c_2, c_3)$ with entries

$$c_1 = \tilde{e}_{,11} v_1^2, \qquad c_2 = K_{21} v_1^2, \qquad c_3 = K_{31} v_1^2,$$

and e_1, e_2, e_3 are characteristic amplitudes with the speeds of propagation c_1, c_2, c_3, respectively. These are Ericksen's formulas for the speeds of propagation of principal waves. See BOWEN & WANG [1970, 1971], CHADWICK & OGDEN [1971a,b] and TRUESDELL & NOLL [1965, Sect. 74] for more details about the principal waves.

(2) ***Transversely isotropic states*** If the state f has two identical principal stretches and the third different from them, then the symmetry group of the acoustic tensor contains the transversely isotropic group described 11.1.4(2). The form of the transversely isotropic acoustic tensor at the stress-free state and the corresponding explicit solution of the characteristic equation are described in a number of works, e.g., SYNGE [1956a,b], FEDOROV [1968, Chap. 6], DIEULESAINT & ROYER [1980, pp. 193–197] and CHADWICK [1989, Part I]. The form of the acoustic tensor at a stressed transversely isotopic state can be derived from these results by applying Proposition 10.5.4. Namely, one finds that

$$q(m) = q_{\circ}(m) + (Tm \cdot m)\mathbf{1}$$

where $q_{\circ}(m)$ is the acoustic tensor based on the reduced tensor $\tilde{B}_{\circ}$ of adiabatic elasticities. As $\tilde{B}_{\circ}$ has all the symmetries of the classical elasticity tensor at the stress-free state, $q_{\circ}(m)$ is of the form discussed in the cited works, and the term $(Tm \cdot m)\mathbf{1}$ represents a correction.

(3) ***Isotropic states*** The longitudinal and two transversal waves (see, e.g., GURTIN [1981, Sect. 35]). The strict hyperbolicity fails: the speeds of transversal waves are degenerate.

(4) ***Hadamard–Green materials*** By definition, the Hadamard–Green material is an isotropic material in which for every $(F, \eta) \in \tilde{U} \cap (\mathrm{Sym}^+ \times \mathbb{R})$ and every $m \in$ Sph, the vector m is an eigenvector of $q(m)$. It turns out (see, e.g., OGDEN [1970b]) that a necessary and sufficient condition for that is that its internal energy function be of the form

$$\tilde{e}(F, \eta) = a \,\mathrm{tr}\, B + b \,\mathrm{tr}\,\mathrm{cof}\, B + h(\det F, \eta), \qquad (F, \eta) \in \tilde{U}, \quad \text{where } B = FF^{\mathrm{T}},$$

where $a \equiv a(\eta)$, $b \equiv b(\eta)$ are arbitrary functions of entropy and $h \equiv h(D, \eta)$ is a function of $D = \det \boldsymbol{F}$ and η. This is proved using the expressions for $\boldsymbol{q}$ above.

23.5 Centered Waves

Another type of process related closely to the characteristic equation are centered waves. Let $\Omega = \text{Vect}$.

23.5.1 Centered Processes Let a, b be such that $-\infty \le a < b \le \infty$ and consider a spacetime wedge $W = \{(\boldsymbol{p}, t) : a < \boldsymbol{p} \cdot \boldsymbol{n}/t < b, 0 < t\}$. A process $(\boldsymbol{\chi}, \eta)$ defined on W is said to be a ***centered process*** if it is of the form

$$\boldsymbol{\chi}(\boldsymbol{p}, t) = (\boldsymbol{p} \cdot \boldsymbol{n})\boldsymbol{\psi}(\boldsymbol{p} \cdot \boldsymbol{n}/t) + \boldsymbol{F}_0\boldsymbol{p}, \quad \eta(\boldsymbol{p}, t) = \zeta(\boldsymbol{p} \cdot \boldsymbol{n}/t), \tag{23.5.1}$$

for some unit vector $\boldsymbol{n}$, some $\boldsymbol{F}_0 \in \text{InvLin}$, where $\boldsymbol{\psi}$ is a continuous and piecewise C^2 function defined on $\mathbb{R}$ and ζ is a piecewise continuously differentiable function defined on $\mathbb{R}$. It follows that $\boldsymbol{\chi}$ is continuous and piecewise C^2 and η piecewise continuously differentiable on $\text{Vect} \times \mathbb{R}^{++}$. For a processes of this form,

$$\boldsymbol{F}(\boldsymbol{p}, t) = \boldsymbol{\phi}(z) \otimes \boldsymbol{n} + \boldsymbol{F}_0, \quad \boldsymbol{v}(\boldsymbol{p}, t) = -z^2\boldsymbol{\psi}'(z), \quad \dot{\eta}(\boldsymbol{p}, t) = -z\zeta'(z)/t,$$

where $z = \boldsymbol{p} \cdot \boldsymbol{n}/t$, $\boldsymbol{\phi}(z) = \boldsymbol{\psi}(z) + z\boldsymbol{\psi}'(z)$ and the prime denotes differentiation with respect to z. It is assumed that $\boldsymbol{f}(z) := (\boldsymbol{\phi}(z) \otimes \boldsymbol{n} + \boldsymbol{F}_0, \zeta(z))$ is in $\tilde{U}$ for every $z \in (a, b)$. The deformation gradient, velocity, and entropy are thus constant along spacetime surfaces given by $\boldsymbol{p} \cdot \boldsymbol{n} - zt = 0$ with z fixed. The solution thus makes a spacetime "fan" with the center at $\boldsymbol{p} \cdot \boldsymbol{n} = 0, t = 0$. Moreover, if $(\boldsymbol{\phi}, \zeta)$ are discontinuous at z, then the process $(\boldsymbol{\chi}, \eta)$ has a singular surface S_t given by $\boldsymbol{p} \cdot \boldsymbol{n} - zt = 0$. If $z \neq 0$ then this singular surface moves with the speed $U = |z|$ in the referential direction $\operatorname{sgn} z\, \boldsymbol{n}$. If $z = 0$ then the singular surface is immobile. It can be shown that the velocity, deformation gradient, and entropy expressed as functions of spatial positions $\boldsymbol{x}$ and time t have the forms

$$\boldsymbol{v}(\boldsymbol{x}, t) = \boldsymbol{w}_0(\boldsymbol{x} \cdot \boldsymbol{m}/t), \quad \boldsymbol{F}(\boldsymbol{x}, t) = \boldsymbol{\varphi}_0(\boldsymbol{x} \cdot \boldsymbol{m}/t), \quad \eta(\boldsymbol{x}, t) = \eta_0(\boldsymbol{x} \cdot \boldsymbol{m}/t),$$

where $\boldsymbol{m} = (\operatorname{cof} \boldsymbol{F}_0\boldsymbol{n})/|\operatorname{cof} \boldsymbol{F}_0\boldsymbol{n}|$ and the functions $\boldsymbol{w}_0, \boldsymbol{\varphi}_0, \eta_0$ are expressible through $\boldsymbol{\psi}, \zeta$.

23.5.2 The Balance Equations for a Centered Process The process $(\boldsymbol{\chi}, \eta)$ satisfies the balance equations and the Clausius–Duhem inequality with $\boldsymbol{b} \equiv \boldsymbol{0}$, $r \equiv 0$ if and only if the following assertions hold:

$$z^2\boldsymbol{\phi}' = \boldsymbol{Q}(\boldsymbol{f}(z), \boldsymbol{n})\boldsymbol{\phi}', \quad \zeta' = 0 \tag{23.5.2}$$

at every point $z \in \mathbb{R}$ where $(\boldsymbol{\phi}, \zeta)$ are differentiable,

$$-z[\boldsymbol{v}] = [\boldsymbol{S}]\boldsymbol{n}, \quad \rho_{\circ}[e] = \langle \boldsymbol{S} \rangle \boldsymbol{n} \cdot [\boldsymbol{F}]\boldsymbol{n}, \quad -\operatorname{sgn} z\, [\eta] \ge 0 \tag{23.5.3}$$

at every point $z \neq 0$ of discontinuity of $(\boldsymbol{\phi}, \zeta)$ and

$$[\boldsymbol{S}]\boldsymbol{n} = \boldsymbol{0} \tag{23.5.4}$$

at $z = 0$ if $(\boldsymbol{\phi}, \zeta)$ is discontinuous at $z = 0$. We use the obvious notation concerning the jumps of quantities given by the constitutive equations in terms of $\boldsymbol{\phi}, \zeta$. Equations

(23.5.3) and (23.5.4) will be discussed in Sects. 23.6–23.8; here we shall concentrate on (23.5.2).

A centered process (χ, η) on W satisfying the balance equations is said to be a ***centered wave*** if χ is twice continuously differentiable, η is continuously differentiable and $\boldsymbol{\phi}'(z) \neq 0$ for every $z \in (a, b)$. In the case of a fluid the term rarefaction wave is employed.

23.5.3 Proposition *Let (χ, η) be a centered wave on W. Then*

(1) *$(|z|, \boldsymbol{\phi}'(z))$ is a characteristic pair at $(\boldsymbol{f}(z), \boldsymbol{n})$ for every $z \in (a, b)$; if $z \neq 0$, then it is genuinely nonlinear;*

(2) *if either $a > 0$ or $b < 0$ and if for every $z \in (a, b)$ the characteristic pair $(|z|, \boldsymbol{\phi}'(z))$ is simple, then there exists a normalized (see (23.2.3)) genuinely nonlinear characteristic field $(\tilde{C}(\cdot, \boldsymbol{n}), \tilde{\boldsymbol{e}}(\cdot, \boldsymbol{n}))$ on an open set N containing $\{\boldsymbol{f}(z) : a \le z \le b\}$ such that*

$$\boldsymbol{\phi}'(z) = \pm\tilde{\boldsymbol{e}}(\boldsymbol{f}(z), \boldsymbol{n}), \quad z = \pm\tilde{C}(\boldsymbol{f}(z), \boldsymbol{n}) \tag{23.5.5}$$

for every $z \in (a, b)$ with the + sign holding if $a > 0$ and − if $b < 0$.

The necessity of the genuine nonlinearity shows that a centered wave is a purely nonlinear phenomenon; nontrivial centered waves do not exist, say, for the linear wave equation.

Proof (1): That $(|z|, \boldsymbol{\phi}'(z))$ is a characteristic pair at $(\boldsymbol{f}(z), \boldsymbol{n})$ follows from $(23.5.2)_1$. Let $z \in (a, b)$ be fixed, $z \neq 0$. Differentiating $(23.5.2)_1$ with respect to z at this value of z and multiplying the resulting equation scalarly by $\boldsymbol{\phi}'(z)$, we obtain, as in the proof of Proposition 23.2.4, $2z|\boldsymbol{\phi}'|^2 = \partial^3_{\boldsymbol{F}}\tilde{e}(\boldsymbol{f})[\boldsymbol{\phi}'\otimes\boldsymbol{n}, \boldsymbol{\phi}'\otimes\boldsymbol{n}, \boldsymbol{\phi}'\otimes\boldsymbol{n}]$ where $\boldsymbol{f} = \boldsymbol{f}(z)$ and $\boldsymbol{\phi}' \equiv \boldsymbol{\phi}'(z)$. Since $2z|\boldsymbol{\phi}'|^2 \neq 0$, we see that $(|z|, \boldsymbol{\phi}'(z))$ is genuinely nonlinear. (2): Using the hypotheses of (2) and Proposition 23.2.2, for every $z \in (a, b)$ there exists an extension of the characteristic pair $(z, \boldsymbol{\phi}'(z))$ to a neighborhood of $\boldsymbol{f}(z)$. Since this is true for every $z \in (a, b)$, one can find a single characteristic field $(\tilde{C}(\cdot, \boldsymbol{n}), \tilde{\boldsymbol{e}}(\cdot, \boldsymbol{n}))$ on $N \supset \{\boldsymbol{f}(z) : a \le z \le b\}$ such that $\boldsymbol{\phi}'(z) = \alpha(z)\tilde{\boldsymbol{e}}(\boldsymbol{f}(z), \boldsymbol{n})$ and $z = \pm\tilde{C}(\boldsymbol{f}(z), \boldsymbol{n})$ for every $z \in (a, b)$ and some $\alpha(z) \neq 0$. Under the hypotheses of (2) Item (1) implies that $(\tilde{C}(\boldsymbol{f}(z), \boldsymbol{n}), \tilde{\boldsymbol{e}}(\boldsymbol{f}(z), \boldsymbol{n}))$ is a genuinely nonlinear characteristic pair for every $z \in (a, b)$. It follows from continuity that $(\tilde{C}(\boldsymbol{g}, \boldsymbol{n}), \tilde{\boldsymbol{e}}(\boldsymbol{g}, \boldsymbol{n}))$ is genuinely nonlinear also for every $\boldsymbol{g}$ from a neighborhood of $\{\boldsymbol{f}(z) : a \le z \le b\}$. Then the restriction of $(\tilde{C}(\cdot, \boldsymbol{n}), \tilde{\boldsymbol{e}}(\cdot, \boldsymbol{n}))$ to this neighborhood is a genuinely nonlinear characteristic pair. We can normalize it (see (23.2.3)). Differentiation of $z = \pm\tilde{C}(\boldsymbol{f}(z), \boldsymbol{n})$ gives $1 = \pm\partial_{\boldsymbol{F}}\tilde{C}(\boldsymbol{f}, \boldsymbol{n}) \cdot (\boldsymbol{\phi}' \otimes \boldsymbol{n}) = \pm\alpha\partial_{\boldsymbol{F}}\tilde{C}(\boldsymbol{f}, \boldsymbol{n}) \cdot (\tilde{\boldsymbol{e}}(\boldsymbol{f}, \boldsymbol{n}) \otimes \boldsymbol{n}) = \pm\alpha$ which implies that $\alpha = \pm 1$. This proves (23.5.5). □

The following proposition says that the equations (23.5.5) have solutions. In view of the future applications of this proposition, the parameter z is shifted to the origin of $\mathbb{R}$.

23.5.4 Proposition *Let $(\tilde{C}(\cdot, \boldsymbol{n}), \tilde{\boldsymbol{e}}(\cdot, \boldsymbol{n}))$ be a normalized genuinely nonlinear characteristic field on $N \subset \tilde{U}$ and $\boldsymbol{f} = (\boldsymbol{F}, \eta) \in N$. Then there exists a function $\bar{\boldsymbol{g}} : (-\varepsilon, \varepsilon) \to N$, $\varepsilon > 0$, of class C^3, of the form $\bar{\boldsymbol{g}}(\cdot) = (\boldsymbol{\phi}(\cdot) \otimes \boldsymbol{n} + \boldsymbol{F}, \eta)$, such that $\bar{\boldsymbol{g}}(0) = \boldsymbol{f}$ and*

$$\boldsymbol{\phi}'(\tau) = \tilde{\boldsymbol{e}}(\bar{\boldsymbol{g}}(\tau), \eta, \boldsymbol{n}), \quad \tau + \tilde{C}(\boldsymbol{f}, \boldsymbol{n}) = \tilde{C}(\bar{\boldsymbol{g}}(\tau), \boldsymbol{n}), \tag{23.5.6}$$

for every $\tau \in (-\varepsilon, \varepsilon)$.

Proof The existence of the solution of $(23.5.6)_1$ with the initial condition follows from the existence theorem for ordinary differential equations. This solution is of class C^3 because from the assumption that $\tilde{e}$ is of class C^5 follows that $\tilde{\boldsymbol{e}}$ is of class C^3. To prove $(23.5.6)_2$ one calculates the derivative of $\tilde{C}(\bar{\boldsymbol{g}}(\tau), \boldsymbol{n})$ by the chain rule and uses the normalization (23.2.3) to find that $d\tilde{C}(\bar{\boldsymbol{g}}(\tau), \boldsymbol{n})/d\tau = 1$. Then $\tilde{C}(\bar{\boldsymbol{g}}(\tau), \boldsymbol{n}) = c + \tau$ and the initial condition that $\tilde{C}(\bar{\boldsymbol{g}}(0), \boldsymbol{n}) = \tilde{C}(\boldsymbol{f}, \boldsymbol{n})$ gives the result. □

23.6 Discontinuities

Discontinuities are regions in which the velocity, deformation gradient, and entropy have jump discontinuities. They are governed by the equations

$$-U\rho_\circ[\boldsymbol{v}] = [\boldsymbol{S}]\boldsymbol{n}, \quad -U[\boldsymbol{F}] = [\boldsymbol{v}] \otimes \boldsymbol{n}, \quad -U\rho_\circ[e + \tfrac{1}{2}\boldsymbol{v}^2] = [\boldsymbol{S}^{\mathrm{T}}\boldsymbol{v}]\boldsymbol{n},$$
$$[\boldsymbol{F}] = \boldsymbol{a} \otimes \boldsymbol{n}, \quad [\boldsymbol{v}] = -U\boldsymbol{a}, \tag{23.6.1}$$

with the consequences

$$\rho_\circ U^2[\boldsymbol{F}]\boldsymbol{n} = [\boldsymbol{S}]\boldsymbol{n}, \quad \rho_\circ U[e] = U\,\langle \boldsymbol{S}\rangle\,\boldsymbol{n} \cdot [\boldsymbol{F}]\boldsymbol{n}. \tag{23.6.2}$$

The jump condition for entropy has not been included; it will be considered separately in Sect. 23.8. The quantities in the jump equations are related by the constitutive equations; at a given spacetime point, they are completely determined by U, $\boldsymbol{v}^\pm$ and $\boldsymbol{f}^\pm = (\boldsymbol{F}^\pm, \eta^\pm)$. We will call any collection $\mathcal{S} = (U, \boldsymbol{v}^\pm, \boldsymbol{f}^\pm, \boldsymbol{n})$ satisfying (23.6.1) a ***discontinuity*** or a ***jump.*** We now fix $\boldsymbol{f} = (\boldsymbol{F}, \eta)$, $\boldsymbol{n}$ and determine the set of all deformation gradient–entropy pairs $\boldsymbol{g} = (\boldsymbol{G}, \zeta)$ which can be connected with $\boldsymbol{f}$ by a discontinuity.

23.6.1 Hugoniot's Set The ***Hugoniot set*** $H(\boldsymbol{f}, \boldsymbol{n})$ at $(\boldsymbol{f}, \boldsymbol{n}) \in D$ is the set of all $\boldsymbol{g} = (\boldsymbol{G}, \zeta) \in \tilde{U}$ of the form $\boldsymbol{G} = \boldsymbol{F} + \boldsymbol{a} \otimes \boldsymbol{n}$, ($\boldsymbol{a} \in$ Vect) for which there exists a $U \in \mathbb{R}$ such that

$$\rho_\circ U^2(\boldsymbol{G} - \boldsymbol{F})\boldsymbol{n} = (\tilde{\boldsymbol{S}}(\boldsymbol{g}) - \tilde{\boldsymbol{S}}(\boldsymbol{f}))\,\boldsymbol{n}, \tag{23.6.3}$$

$$\rho_\circ U\,(\tilde{e}(\boldsymbol{g}) - \tilde{e}(\boldsymbol{f})) = \tfrac{1}{2}U\,(\tilde{\boldsymbol{S}}(\boldsymbol{g}) + \tilde{\boldsymbol{S}}(\boldsymbol{f}))\,\boldsymbol{n} \cdot (\boldsymbol{G} - \boldsymbol{F})\boldsymbol{n}. \tag{23.6.4}$$

These are ***Hugoniot's equations.*** Equivalently, (23.6.4) reads

$$\rho_\circ^2 U^2\,(\tilde{e}(\boldsymbol{g}) - \tilde{e}(\boldsymbol{f})) = \tfrac{1}{2}\,(|\,\tilde{\boldsymbol{S}}(\boldsymbol{g})\boldsymbol{n}\,|^2 - |\,\tilde{\boldsymbol{S}}(\boldsymbol{f})\boldsymbol{n}\,|^2). \tag{23.6.5}$$

The definition of Hugoniot's set using the deformation gradients and entropies has been adopted. By $\boldsymbol{G} = \boldsymbol{F} + \boldsymbol{a} \otimes \boldsymbol{n}$ with $\boldsymbol{F}, \boldsymbol{n}$ fixed, the Hugoniot set is a priori confined to a four-dimensional subspace of Lin $\times\ \mathbb{R}$. The point $\boldsymbol{f}$ is in $H(\boldsymbol{f}, \boldsymbol{n})$. The velocities play a trivial role in the system of jump equations; hence we shall often identify discontinuities with quadruples $\mathcal{S} = (U, \boldsymbol{f}^\pm, \boldsymbol{n})$ such that $\boldsymbol{f}^+ \in H(\boldsymbol{f}^-, \boldsymbol{n})$.

For every $\boldsymbol{g} \in H(\boldsymbol{f}, \boldsymbol{n})$, $\boldsymbol{g} \neq \boldsymbol{f}$, the speed U in (23.6.3), (23.6.4) can be chosen to be nonnegative; with this choice, it is unique and we denote it by $U = \hat{U}_{\boldsymbol{f},\boldsymbol{n}}(\boldsymbol{g}) \geq 0$. To see this uniqueness, note that if $\boldsymbol{G} \neq \boldsymbol{F}$, then $(\boldsymbol{G} - \boldsymbol{F})\boldsymbol{n} \neq \boldsymbol{0}$ and U can be calculated from (23.6.3). If $\boldsymbol{g} \neq \boldsymbol{f}$ is such that $\boldsymbol{G} = \boldsymbol{F}$, then $\zeta \neq \eta$ and since the energy is an increasing function of entropy, we have $\tilde{e}(\boldsymbol{g}) \neq \tilde{e}(\boldsymbol{f})$. Using again $\boldsymbol{G} = \boldsymbol{F}$, the right-hand side of (23.6.4) vanishes and it follows that $U = 0$. *For* $\boldsymbol{g} = \boldsymbol{f}$ *the speed* U *is undefined.*

The solution of the Hugoniot equations is a bifurcation problem: for every $U \in \mathbb{R}$ we have the trivial solution $\boldsymbol{g} = \boldsymbol{f}$; at certain values of U, a nontrivial set of solutions is bifurcating. As we shall see, this bifurcation occurs exactly at the values of U equal to some characteristic speed at $(\boldsymbol{f}, \boldsymbol{n})$ and the tangent to the set of the nontrivial solutions is one of the characteristic vectors. The first observations about the relationship of the Hugoniot set to the characteristic equation are made in Propositions 23.6.2 and 23.6.3. The first of them shows that for all $\boldsymbol{g} \in H(\boldsymbol{f}, \boldsymbol{n})$ sufficiently close to $\boldsymbol{f}$, $\boldsymbol{g} \neq \boldsymbol{f}$, the value $\hat{U}_{\boldsymbol{f},\boldsymbol{n}}(\boldsymbol{g})$ is close to one of the characteristic speeds at $(\boldsymbol{f}, \boldsymbol{n})$.

23.6.2 Proposition *Let $\sigma(\boldsymbol{f}, \boldsymbol{n}) := \{C_1, \ldots, C_p\}$ be the set of all characteristic speeds at $(\boldsymbol{f}, \boldsymbol{n}) \in D$. Then for every $\varepsilon > 0$ there exists a neighborhood N of $\boldsymbol{f}$ such that for every $\boldsymbol{g} \in H(\boldsymbol{f}, \boldsymbol{n}) \cap N$ the distance of $\hat{U}_{f,n}(\boldsymbol{g})$ from the set $\sigma(\boldsymbol{f}, \boldsymbol{n})$ does not exceed ε; i.e., there exists a k, $1 \le k \le p$, such that $|\hat{U}_{f,n}(\boldsymbol{g}) - C_k| \le \varepsilon$.*

Proof Set $\rho_o = 1$. Let $\boldsymbol{f} = (\boldsymbol{F}, \eta)$, let $X = \text{Vect} \times \mathbb{R}$, let O be an open subset of X of all $\boldsymbol{z} = (\boldsymbol{A}, \zeta)$ such that $(\boldsymbol{F} + \boldsymbol{A} \otimes \boldsymbol{n}, \zeta) \in \tilde{U}$, let ϕ, F be functions from O to X defined by

$$\phi(\boldsymbol{z}) = (\boldsymbol{A}, \tilde{e}(\boldsymbol{g})), \quad F(\boldsymbol{z}) = (\tilde{\boldsymbol{S}}(\boldsymbol{g})\boldsymbol{n}, \tfrac{1}{2} |\tilde{\boldsymbol{S}}(\boldsymbol{g})\boldsymbol{n}|^2), \quad \boldsymbol{z} = (\boldsymbol{A}, \zeta) \in O, \tag{23.6.6}$$

where $\boldsymbol{g} = (\boldsymbol{F} + \boldsymbol{A} \otimes \boldsymbol{n}, \zeta)$. Let $\boldsymbol{x} = (\boldsymbol{0}, \eta)$. The system (23.6.3), (23.6.5) takes the form

$$w(\phi(\boldsymbol{z}) - \phi(\boldsymbol{x})) = F(\boldsymbol{z}) - F(\boldsymbol{x}) \tag{23.6.7}$$

where $w = U^2$. We have

$$\nabla\phi(\boldsymbol{x}) = \begin{bmatrix} \boldsymbol{1} & 0 \\ \boldsymbol{s} & \tilde{\theta} \end{bmatrix}, \quad \nabla F(\boldsymbol{x}) = \begin{bmatrix} \boldsymbol{Q} & \boldsymbol{d} \\ \boldsymbol{Qs} & \boldsymbol{s} \cdot \boldsymbol{d} \end{bmatrix}, \tag{23.6.8}$$

where $\boldsymbol{s} = \tilde{\boldsymbol{S}}(\boldsymbol{f})\boldsymbol{n}$, $\boldsymbol{d} = \boldsymbol{Dn}$, $\tilde{\theta} = \tilde{\theta}(\boldsymbol{f})$ and $\boldsymbol{Q} = \boldsymbol{Q}(\boldsymbol{f}, \boldsymbol{n})$. Notice that $\nabla\phi(\boldsymbol{x})$ is an invertible linear transformation and hence ϕ is a diffeomorphism in some neighborhood of $\boldsymbol{x}$. Moreover, it is easily checked that $\xi = (\boldsymbol{e}, b)$ is a characteristic vector corresponding to the characteristic speed C at $(\boldsymbol{f}, \boldsymbol{n})$ if and only if $C^2 \nabla\phi(\boldsymbol{x})\xi = \nabla F(\boldsymbol{x})\xi$. Let us first observe that $\hat{U}_{f,n}(\cdot)$ is bounded in some neighborhood of $\boldsymbol{f}$ on $H(\boldsymbol{f}, \boldsymbol{n})$. This amounts to checking that the w occurring in (23.6.7) is bounded when $\boldsymbol{z}$ is taken from some neighborhood of $\boldsymbol{x}$. We have $w(\phi(\boldsymbol{z}) - \phi(\boldsymbol{x})) = \nabla F(\boldsymbol{x})(\boldsymbol{z} - \boldsymbol{x}) + o(|\boldsymbol{z} - \boldsymbol{x}|)$. Multiplying scalarly by $\phi(\boldsymbol{z}) - \phi(\boldsymbol{x})$, dividing by $|\phi(\boldsymbol{z}) - \phi(\boldsymbol{x})|^2$, taking the absolute value of the resulting equation and making straightforward estimates, we obtain

$$|w| \le |\nabla F(\boldsymbol{x})| \cdot |\boldsymbol{z} - \boldsymbol{x}| / |\phi(\boldsymbol{z}) - \phi(\boldsymbol{x})| + o(|\boldsymbol{z} - \boldsymbol{x}|) / |\phi(\boldsymbol{z}) - \phi(\boldsymbol{x})|.$$

The proof of the boundedness of w is then completed by noting that $|\boldsymbol{z} - \boldsymbol{x}| / |\phi(\boldsymbol{z}) - \phi(\boldsymbol{x})|$ is bounded in a neighborhood of $\boldsymbol{x}$ as a consequence of the fact that ϕ is a diffeomorphism: there exist constants $c > 0$, $d > 0$ such that $c < |\phi(\boldsymbol{z}) - \phi(\boldsymbol{x})| / |\boldsymbol{z} - \boldsymbol{x}| < d$ in some neighborhood of $\boldsymbol{x}$. Suppose that there exists a sequence $\boldsymbol{z}_n$, $n = 1, \ldots,$ such that $\boldsymbol{z}_n \to \boldsymbol{x}$ and yet the corresponding w_n as in (23.6.7) is such that $|w_n - C_k^2| \ge \varepsilon > 0$ for all $n = 1, \ldots,$ all $k = 1, \ldots, p$, and some $\varepsilon > 0$. Since w_n is bounded and $(\boldsymbol{z}_n - \boldsymbol{x}) / |\boldsymbol{z}_n - \boldsymbol{x}|$ has norm 1, for some subsequence $w_n \to w_0 \ge 0$ and $(\boldsymbol{z}_n - \boldsymbol{x}) / |\boldsymbol{z}_n - \boldsymbol{x}| \to \xi \neq 0$. A division of (23.6.7) (written for $\boldsymbol{z}_n$, w_n) by $|\boldsymbol{z}_n - \boldsymbol{x}|$ and a limit give $w_0 \nabla\phi(\boldsymbol{x})\xi = \nabla F(\boldsymbol{x})\xi$ so that w_0 is a square of a characteristic speed at $(\boldsymbol{f}, \boldsymbol{n})$, which means that $w_0 = C_k^2$ for some k, $1 \le k \le p$, and $|w_n - C_k^2| \ge \varepsilon > 0$ is violated for this particular k. This contradiction completes the proof. (Alternatively, one can use the implicit function theorem to show that for values of w close to some w_0 different from the characteristic speeds, only the trivial solutions $\boldsymbol{z} = \boldsymbol{x}$ of (23.6.7) exist in a neighborhood of $\boldsymbol{x}$.) □

23.6.3 Proposition *Let $(\boldsymbol{f}, \boldsymbol{n}) \in D$ and let $\boldsymbol{g}(\cdot) = (\boldsymbol{G}(\cdot), \zeta(\cdot))$ be a continuously differentiable function on $(-\varepsilon, \varepsilon)$, $\varepsilon > 0$, with values in $H(\boldsymbol{f}, \boldsymbol{n})$, such that $\boldsymbol{g}(0) = \boldsymbol{f}$, $\boldsymbol{g}(\tau) \neq \boldsymbol{f}$ for every $\tau \neq 0$ and $\dot{\boldsymbol{g}}(0) \neq \boldsymbol{0}$. Then $u_0 := \lim_{\tau \to 0} \hat{U}_{f,n}(\boldsymbol{g}(\tau))$ exists, $\dot{\boldsymbol{g}}(0)$ is of the form $\dot{\boldsymbol{g}}(0) = (\boldsymbol{e} \otimes \boldsymbol{n}, b)$ and $\xi := (\boldsymbol{e}, b)$ is a characteristic vector corresponding to the characteristic speed u_0 at $(\boldsymbol{f}, \boldsymbol{n})$.*

Proof We have $\boldsymbol{g}(\tau) = (\boldsymbol{A}(\tau) \otimes \boldsymbol{n}, \zeta(\tau))$ and setting $\boldsymbol{y}(\tau) = (\boldsymbol{A}(\tau), \zeta(\tau))$, we see that for every τ, (23.6.7) holds with $\boldsymbol{z} = \boldsymbol{y}(\tau)$, $w = \hat{U}_{f,n}(\boldsymbol{g}(\tau))^2$. The function $\hat{U}_{f,n}(\boldsymbol{g}(\cdot))$ is bounded by the preceding proposition and so if τ_n is a sequence such that $\tau_n \to 0$, then for some subsequence, again denoted by τ_n, $\hat{U}_{f,n}(\boldsymbol{g}(\tau_n)) \to C \ge 0$. Writing down (23.6.7) for $\boldsymbol{z}_n$, and using that $w_n \to C^2$, dividing by $\tau_n \neq 0$ and using the differentiability of $\boldsymbol{y}$, ϕ, F, it is found that $C^2 \nabla\phi(\boldsymbol{x})\dot{\boldsymbol{y}}(0) = \nabla F(\boldsymbol{x})\dot{\boldsymbol{y}}(0)$. As $\dot{\boldsymbol{y}}(0) \neq 0$, we see that $\dot{\boldsymbol{y}}(0)$ is a characteristic vector corresponding to the characteristic speed C. Since $\dot{\boldsymbol{y}}(0)$ is the same for each subsequence, so also is C (being the eigenvalue for $\dot{\boldsymbol{y}}(0)$). Hence the limit $\lim_{n\to\infty} \hat{U}_{f,n}(\boldsymbol{g}(\tau_n))$ is the same for every subsequence, which implies that the limit $u_0 := \lim_{\tau\to 0} \hat{U}_{f,n}(\boldsymbol{g}(\tau))$ exists, is equal to C and the conclusion is proved. □

We have seen that for discontinuities with $\boldsymbol{g}$ close to $\boldsymbol{f}$, the speed is close to one of the characteristic speeds at $\boldsymbol{f}$. When the speed of the discontinuity is exactly the characteristic speed at $\boldsymbol{f}$, one speaks about a contact discontinuity. Various definitions of contact discontinuities exist in the literature; the one below does not mention the characteristic speeds but the proposition that follows establishes this relation.

23.6.4 Definition A discontinuity $S = (U, \boldsymbol{f}^{\pm}, \boldsymbol{n})$ is said to be a ***contact discontinuity*** if there exists a continuously differentiable function $\boldsymbol{g} : [\tau^+, \tau^-] \to H(\boldsymbol{f}^-, \boldsymbol{n})$ such that $\boldsymbol{f}^{\pm} = \boldsymbol{g}(\tau^{\pm})$, $\boldsymbol{g}(\tau) \neq \boldsymbol{f}^-$ for every $\tau \neq \tau^-$, $\dot{\boldsymbol{g}}(\tau) \neq 0$ for all τ, and $\hat{U}_{\boldsymbol{f}^-, \boldsymbol{n}}(\boldsymbol{g}(\tau)) = U$ for all $\tau \in [\tau^+, \tau^-)$.

The special role of the entropy in the system of equations of adiabatic dynamics causes these equations to always have contact discontinuities with $U = 0$, see Proposition 23.6.7. On the other hand, the occurrence of the contact discontinuities with $U > 0$ is rather exceptional: they exist only if the material has a linearly degenerate characteristic field, see Proposition 23.7.8.

23.6.5 Proposition *Let $S = (U, \boldsymbol{f}^{\pm}, \boldsymbol{n})$ be a contact discontinuity and let $\boldsymbol{g}$ be the function occurring in Definition 23.6.4. Its derivative is of the form $\dot{\boldsymbol{g}}(\tau) = (\boldsymbol{e}(\tau) \otimes \boldsymbol{n}, b(\tau))$ and $(\boldsymbol{e}(\tau), b(\tau))$ is a characteristic vector at $(\boldsymbol{g}(\tau), \boldsymbol{n})$ with the characteristic speed U for all τ.*

Proof It suffices to differentiate (23.6.7) along $z = \boldsymbol{y}(\tau)$, where $\boldsymbol{y}$ is defined in the proof of Proposition 23.6.3, and to use that $w = U^2$ is constant. □

23.6.6 Shocks and Rest Discontinuities We divide the Hugoniot set into two parts

$$H(\boldsymbol{f}, \boldsymbol{n}) = H_s(\boldsymbol{f}, \boldsymbol{n}) \cup H_r(\boldsymbol{f}, \boldsymbol{n})$$

where $H_s(\boldsymbol{f}, \boldsymbol{n})$ consists of the point $\boldsymbol{f}$ and of all $\boldsymbol{g} \in H(\boldsymbol{f}, \boldsymbol{n})$ for which $\hat{U}_{\boldsymbol{f}, \boldsymbol{n}}(\boldsymbol{g}) \neq 0$; the discontinuities with $U \neq 0, \boldsymbol{f}^+ \neq \boldsymbol{f}^-$ are called ***shocks.*** The set $H_r(\boldsymbol{f}, \boldsymbol{n})$ consists of the point $\boldsymbol{f}$ and of all $\boldsymbol{g} \in H(\boldsymbol{f}, \boldsymbol{n})$ for which $\hat{U}_{\boldsymbol{f}, \boldsymbol{n}}(\boldsymbol{g}) = 0$; the discontinuities with $U = 0, \boldsymbol{f}^+ \neq \boldsymbol{f}^-$ are called ***rest discontinuities.*** Apart from the point $\boldsymbol{f}$, which is common for both $H_s(\boldsymbol{f}, \boldsymbol{n})$ and $H_r(\boldsymbol{f}, \boldsymbol{n})$, these sets are disjoint. The set $H_s(\boldsymbol{f}, \boldsymbol{n})$ is called the shock set. For $\boldsymbol{g}$ in $H_s(\boldsymbol{f}, \boldsymbol{n})$ the nonzero speed U can be canceled in (23.6.4) with the result

$$\rho_\circ(\tilde{e}(\boldsymbol{g}) - \tilde{e}(\boldsymbol{f})) = \tfrac{1}{2}(\tilde{\boldsymbol{S}}(\boldsymbol{g}) + \tilde{\boldsymbol{S}}(\boldsymbol{f}))\boldsymbol{n} \cdot (\boldsymbol{G} - \boldsymbol{F})\boldsymbol{n}. \tag{23.6.9}$$

On $H_r(\boldsymbol{f}, \boldsymbol{n})$, (23.6.4) is trivially satisfied while (23.6.3) reduces to

$$\tilde{\boldsymbol{S}}(\boldsymbol{g})\boldsymbol{n} = \tilde{\boldsymbol{S}}(\boldsymbol{f})\boldsymbol{n}. \tag{23.6.10}$$

We thus see that the rest discontinuities are subject only to the equation of equilibrium of forces (and of course to the geometrical constraint that $\boldsymbol{G}$ and $\boldsymbol{F}$ differ by a rank 1 tensor).

23.6.7 Proposition *Let $\boldsymbol{f} = (\boldsymbol{F}, \eta) \in \tilde{\mathcal{U}}$ and assume that $C = 0$ is a characteristic speed of multiplicity 1 with the characteristic vector of the form $(\boldsymbol{e}, 1)$. Then there exists a continuously differentiable function $\boldsymbol{g}(\cdot)$ defined on $(-\varepsilon, \varepsilon)$, $\varepsilon > 0$, with values in $H_r(\boldsymbol{f}, \boldsymbol{n})$ such that $\boldsymbol{g}(0) = \boldsymbol{f}$, $\dot{\boldsymbol{g}}(0) = (\boldsymbol{e} \otimes \boldsymbol{n}, 1)$. The intersection of $H_r(\boldsymbol{f}, \boldsymbol{n})$ with some neighborhood of $\boldsymbol{f}$ is precisely the range of $\boldsymbol{g}$. For every τ, the discontinuity $S = (0, \boldsymbol{g}(\tau), \boldsymbol{f}, \boldsymbol{n})$ is a contact discontinuity.*

That is, $H_r(f, n)$ is locally a smooth curve consisting entirely of contact discontinuities. The hypothesis of the proposition is satisfied, e.g., if the material is strictly hyperbolic at (f, n).

Proof The elements $g = (A \otimes n + F, \zeta)$ of the set $H_r(f, n)$ satisfy $\tilde{S}(A \otimes n + F, \zeta)n - \tilde{S}(f)n = 0$. We shall apply the implicit function theorem to the function $\Phi : (a, \zeta) \mapsto \tilde{S}(A \otimes n + F, \zeta)n - \tilde{S}(f)n$ at the point $(a, \zeta) = (0, \eta)$. We have $\partial_a \Phi(A, \eta) = Q(f, n)$ and under the hypothesis of the proposition, $Q(f, n)$ is an invertible transformation (for otherwise the occurrence of eigenvectors of $Q(f, n)$ with eigenvalue 0 would increase the multiplicity of the characteristic speed 0). Then, by the implicit function theorem, locally the solutions of $\Phi(A, \zeta) = 0$ are expressible as a function of ζ, and this leads directly to the existence of $g(\cdot)$. Since the speed along $g(\tau)$ is 0 and hence constant, it follows that each point on $g(\cdot)$ is a contact discontinuity. □

23.7 The Shock Set

We shall derive a local form of the set $H_s(f, n)$ for the case in which all positive characteristic speeds at (f, n) are distinct. Then locally $H_s(f, n)$ consists of m curves that intersect at f. In view of this, we are now going to examine the properties of curves in $H_s(f, n)$ assuming that they are given. Their existence, however, is proved only later, in Proposition 23.7.6.

23.7.1 Curves in the Shock Set In the subsequent three propositions we assume that $(f, n) \in D$ is a given point and we consider a function $g(\cdot) = (G(\cdot), \zeta(\cdot))$ of class C^3 defined on $(-\varepsilon, \varepsilon)$, $\varepsilon > 0$, with values in $H_s(f, n)$, such that

$$g(0) = f \quad \text{and} \quad g(\tau) \neq f \quad \text{for} \quad \tau \neq 0.$$

The speed $u(\tau) := \hat{U}_{f,n}(g(\tau))$ is defined for all $\tau \neq 0$; this formula cannot be used for $\tau = 0$ since $\hat{U}_{f,n}(f)$ is undefined. However, by Proposition 23.6.3 the value $u(0)$ can be defined as $u(0) := \lim_{\tau \to 0} u(\tau)$ and the resulting function is continuous. We assume that $u : (-\varepsilon, \varepsilon) \to \mathbb{R}^{++}$ is of class C^3. As Proposition 23.7.6 shows, this is a natural assumption. We also write $G(\tau) = F + a(\tau) \otimes n$, suppose that $\dot{a}(0) \neq 0$ and denote $e := \dot{a}(0)$, $u_0 = u(0)$.

23.7.2 Proposition *Let $g(\cdot)$ be as in 23.7.1. Then (u_0, e) is a characteristic pair at (f, n). The derivatives of $g(\cdot)$ at $\tau = 0$ satisfy:*

$$\dot{\zeta} = 0, \quad \ddot{\zeta} = 0, \quad \dddot{\zeta} = 2u_0 \dot{u} e^2 \tag{23.7.1}$$

and

$$4u_0 \dot{u} e^2 = \partial_F^3 \tilde{e}[e \otimes n, e \otimes n, e \otimes n]. \tag{23.7.2}$$

Equations (23.7.1) show that the jump in entropy is of the third or higher order of the magnitude in the jump in the deformation gradient.

Proof Assume $\rho_0 = 1$. Introduce the notations $s(\tau) = \tilde{S}(g(\tau))n$, $e(\tau) = \tilde{e}(g(\tau))$, $s_0 = \tilde{S}(f)n$, $e_0 = \tilde{e}(f)$. Relations (23.6.3), (23.6.9) read

$$u^2 a = s - s_0, \quad e - e_0 = \tfrac{1}{2}(s + s_0) \cdot a. \tag{23.7.3}$$

Differentiating $(23.7.3)_2$ with respect to τ and using the Gibbs equation in the form $\dot{e} = s \cdot \dot{a} + \theta \dot{\zeta}$ we obtain $s \cdot \dot{a} + \theta \dot{\zeta} = \frac{1}{2}\dot{s} \cdot a + \frac{1}{2}s \cdot \dot{a} + \frac{1}{2}s_0 \cdot \dot{a}$, and upon rearrangement,

$$\theta \dot{\zeta} = \tfrac{1}{2}\dot{s} \cdot a - \tfrac{1}{2}s \cdot \dot{a} + \tfrac{1}{2}s_0 \cdot \dot{a}. \tag{23.7.4}$$

Setting $\tau = 0$ and using $a(0) = 0$, $s(0) = s_0$ leads to $(23.7.1)_1$. Differentiation of (23.7.4) gives

$$\theta\dot{\zeta} + \theta\ddot{\zeta} = \tfrac{1}{2}\ddot{s}\cdot a - \tfrac{1}{2}s\cdot \ddot{a} + \tfrac{1}{2}s_0\cdot \ddot{a}. \tag{23.7.5}$$

Setting $\tau = 0$ and using $\dot{\zeta}(0) = 0$, $a(0) = \mathbf{0}$, $s(0) = s_0$ provides $(23.7.1)_2$. Differentiation of (23.7.5) at $\tau = 0$ yields

$$\ddot{\zeta}(0) = \tfrac{1}{2}\ddot{s}\cdot \dot{a} - \tfrac{1}{2}\dot{s}\cdot \ddot{a}. \tag{23.7.6}$$

On the other hand, differentiation of $(23.7.3)_1$ gives

$$2u\dot{u}a + u^2\dot{a} = \dot{s}; \tag{23.7.7}$$

differentiating once more and setting $\tau = 0$ we obtain

$$4u_0\dot{u}\dot{a} + u_0^2\ddot{a} = \ddot{s}. \tag{23.7.8}$$

Inserting this value of $\ddot{s}$ into (23.7.6),

$$\dddot{\zeta}(0) = 2u_0\dot{u}\dot{a}\cdot\dot{a} + \tfrac{1}{2}u_0^2\ddot{a}\cdot\dot{a} - \tfrac{1}{2}\dot{s}\cdot\ddot{a}. \tag{23.7.9}$$

At $\tau = 0$, (23.7.7) gives

$$u_0^2\dot{a} = \dot{s}; \tag{23.7.10}$$

using this, the right-hand side of (23.7.9) reduces to $2u_0\dot{u}|\dot{a}|^2$ and $(23.7.1)_3$ follows. By the definition of s and the chain rule we have, at $\tau = 0$, that $\dot{s} = Q(n)\dot{a} + D\dot{\zeta} = Q(n)\dot{a}$ (since $\dot{\zeta} = 0$). A combination with (23.7.10) gives $u_0^2\dot{a} = Q(n)\dot{a}$; as $e \equiv \dot{a} \neq \mathbf{0}$ (recall that $\dot{G} \neq \mathbf{0}$) this shows that e is a characteristic amplitude and u_0 a characteristic speed. A scalar multiplication of (23.7.8) by e gives

$$4u_0\dot{u}e^2 + u_0^2\ddot{a}\cdot e = \ddot{s}\cdot e. \tag{23.7.11}$$

The evaluation of $\ddot{s}$ at $\tau = 0$ using the chain rule gives, with the help of $(23.7.1)_2$,

$$\ddot{s}\cdot e = Q(n)\ddot{a}\cdot e + \partial_F^3\bar{e}[e\otimes n, e\otimes n, e\otimes n]$$

and so (23.7.11) reduces to

$$4u_0\dot{u}e^2 + u_0^2\ddot{a}\cdot e = Q(n)\ddot{a}\cdot e + \partial_F^3\bar{e}[e\otimes n, e\otimes n, e\otimes n]. \tag{23.7.12}$$

By the symmetry of the acoustic tensor we have $Q(n)\ddot{a}\cdot e = Q(n)e\cdot\ddot{a} = u_0^2 e\cdot\ddot{a}$ and (23.7.12) leads to (23.7.2). □

23.7.3 Proposition *Let $g(\cdot)$ be as in 23.7.1, assume that the pair (u_0, e) is a simple characteristic pair at (f, n) and let $(\tilde{C}(\cdot, n), \tilde{e}(\cdot, n))$ be an extension of (u_0, e) to a neighborhood of f (see Proposition 23.2.2). The derivatives $\dot{u}$, $\dot{\tilde{C}}$ of the functions $u(\cdot)$ and $\tau \mapsto \tilde{C}(g(\tau), n)$ at $\tau = 0$ satisfy*

$$\dot{u} = \dot{\tilde{C}}/2. \tag{23.7.13}$$

Hence $\tilde{C}(g(\tau), n)$ changes faster than $u(\tau)$.

Proof By Proposition 23.7.2, e is a characteristic amplitude corresponding to u_0; moreover $u_0 > 0$ by the hypothesis. Let us consider the derivative $\dot{S}$ of the function $\tau \mapsto S(\tau) := \tilde{C}(g(\tau), n)^2$ at $\tau = 0$. We have $\dot{S} = \partial_F\tilde{C}^2\cdot\dot{G} + \partial_\eta\tilde{C}^2\cdot\dot{\zeta} = \partial_F\tilde{C}^2\cdot\dot{G}$ since $\dot{\zeta} = 0$ by $(23.7.1)_1$. Now $\dot{G} = e\otimes n$ and (23.2.1) gives $\dot{S} = \partial_F^3\bar{e}[e\otimes n, e\otimes n, e\otimes n]/e^2$. Replacing $\dot{S}$ in the last equation by $2\tilde{u}_0\cdot\dot{\tilde{C}}$ (which is equal to $\dot{S}$) and comparing the resulting equation with (23.7.2), we obtain (23.7.13). □

23.7.4 Proposition *Let $g(\cdot)$ be as in 23.7.1; assume that (u_0, e) is a genuinely non-linear characteristic pair at (f, n). Then $\dot{u} \neq 0, \dddot{\zeta} \neq 0$ at $\tau = 0$ and $\dot{u}, \dddot{\zeta}$ are of the same sign. By passing from $\tau \mapsto g(\tau)$ to $\tau \mapsto g(-\tau)$ if necessary, one can achieve that u and ζ are strictly increasing functions of τ near $\tau = 0$.*

This shows that if the material is genuinely nonlinear, then the parametrization with $\dot{G}(0) \equiv e\otimes n \neq \mathbf{0}$ is such that $\dot{u} \neq 0$ and hence also $\dot{\tilde{C}}_i \neq 0$. The jump in entropy is of *exactly* the third order in the magnitude of the jump in the deformation gradient. This is of basic importance for establishing the equivalence of the admissibility criteria for small jumps.

Proof A combination of equations (23.7.1) and (23.7.2) with the definition of the genuine nonlinearity shows that $\dot{u} \neq 0, \dddot{\zeta}(0) \neq 0$ and that these two derivatives are of the same sign. □

The following proposition is a preparation for the proof of the existence of the curve $\boldsymbol{g}(\cdot)$ in $H_s(\boldsymbol{f}, \boldsymbol{n})$.

23.7.5 Proposition *Let* X *be a finite-dimensional real vector space with inner product, let* O *be an open subset of* X *and let* F, ϕ *be functions of class* C^r, $r \geq 2$, *from* O *to* X, *with* ϕ *a diffeomorphism. Let* $\boldsymbol{x} \in$ O *and let* $\boldsymbol{u} \in$ X *be an eigenvector of* $\nabla F(\boldsymbol{x})(\nabla\phi(\boldsymbol{x}))^{-1}$ *with a simple eigenvalue* $\lambda \in \mathbb{R}$. *Then there exist functions* $\boldsymbol{y}, u$ *on* $(-\varepsilon, \varepsilon)$, $\varepsilon > 0$, *with values in* O *and* $\mathbb{R}$, *respectively, of class* C^{r-1}, *such that*

(1) $\boldsymbol{y}(0) = \boldsymbol{x}$, $u(0) = \lambda$, $\dot{\boldsymbol{y}}(0) = \boldsymbol{u} \neq \boldsymbol{0}$ *and* $\boldsymbol{y}(\tau) \neq \boldsymbol{x}$ *if* $\tau \neq 0$;

(2) *for every* $\tau \in (-\varepsilon, \varepsilon)$,

$$u(\tau)(\phi(\boldsymbol{y}(\tau)) - \phi(\boldsymbol{x})) = F(\boldsymbol{y}(\tau)) - F(\boldsymbol{x});$$

(3) *there exists a* $\delta > 0$ *and a neighborhood* O′ *of* $\boldsymbol{x}$ *such that the only solution* (z, w) *of*

$$w(\phi(z) - \phi(\boldsymbol{x})) = F(z) - F(\boldsymbol{x}) \tag{23.7.14}$$

in $\mathrm{O}' \times (\lambda - \delta, \lambda + \delta)$ *is* $z = \boldsymbol{y}(\tau)$, $w = u(\tau)$ *for some* $\tau \in (-\varepsilon, \varepsilon)$.

Proof The change of variables $\boldsymbol{x} \mapsto \boldsymbol{x}' = \phi(\boldsymbol{x})$ reduces the general situation of the proposition to the one in which $\phi(z) = z$ identically, and hence it suffices to prove the assertion only in this special case. Hence assume that $\phi(z) = z$ identically. Then $\boldsymbol{u}$ is a simple eigenvector of $\nabla F(\boldsymbol{x})$ and λ the eigenvalue. Let $\mathrm{O}' \subset \mathrm{O}$ be any ball centered at $\boldsymbol{x}$ and define the linear transformation $\boldsymbol{M}(y)$ from X to X by

$$\boldsymbol{M}(y) = \int_0^1 \nabla F(\boldsymbol{x} + \sigma(y - \boldsymbol{x}))\, d\sigma, \quad y \in \mathrm{O}'.$$

Then $\boldsymbol{M}$ is of class C^{r-1} in O′ and $\boldsymbol{M}(\boldsymbol{x}) = \nabla F(\boldsymbol{x})$. Notice that $\boldsymbol{M}(z)(z - \boldsymbol{x}) = F(z) - F(\boldsymbol{x})$ for every $z \in \mathrm{O}'$. Hence a pair $(z, w) \in \mathrm{O}' \times \mathbb{R}$ satisfies (23.7.14) if and only if $z - \boldsymbol{x}$ is an eigenvector of $\boldsymbol{M}(z)$ corresponding to the eigenvalue w. Since $\boldsymbol{u}$ is an eigenvector of $\boldsymbol{M}(\boldsymbol{x})$ corresponding to the simple eigenvalue λ, Proposition 1.2.2 says that there exist functions $\tilde{\boldsymbol{u}}, \tilde{\lambda}$ defined in a neighborhood of $\boldsymbol{x}$, of class C^{r-1}, such that $\boldsymbol{M}(\tilde{\boldsymbol{u}}(y)) = \tilde{\lambda}(y)\tilde{\boldsymbol{u}}(y)$; moreover, there exists a $\delta > 0$ such that $\tilde{\lambda}(z)$ is the only eigenvalue of $\boldsymbol{M}(z)$ in $(\lambda - \delta, \lambda + \delta)$. Let us denote the neighborhood for which this holds again by O′. Let n be the dimension of X and let $\boldsymbol{u}_2, \ldots, \boldsymbol{u}_n$ be fixed vectors in X such that for every $\boldsymbol{y}$ from a neighborhood of $\boldsymbol{x}$ (again denoted by O′), the set $\tilde{\boldsymbol{u}}(y), \boldsymbol{u}_2, \ldots, \boldsymbol{u}_n$ is a basis in X. Write, for uniformity of notation, $\tilde{\boldsymbol{u}}_1(y) := \tilde{\boldsymbol{u}}(y)$ and $\tilde{\boldsymbol{u}}_i(y) := \boldsymbol{u}_i$, $i = 2, \ldots, n$. For every $z \in \mathrm{O}'$ there exist numbers $\alpha_i(z)$ such that

$$z - \boldsymbol{x} = \sum_{i=1}^{n} \alpha_i(z)\tilde{\boldsymbol{u}}_i(z).$$

Clearly $\alpha_i(\boldsymbol{x}) = 0$, $i = 1, \ldots, n$. The values $\alpha_i(z)$ are determined by Cramer's rule in terms of $z, \tilde{\boldsymbol{u}}_i(z)$ and, as these functions are all of class C^{r-1}, it follows that also the functions α_i are of class C^{r-1} on O′. Define the mapping $\boldsymbol{\alpha} : \mathrm{O}' \to \mathbb{R}^n$ by $\boldsymbol{\alpha}(z) = (\alpha_1(z), \ldots, \alpha_n(z))$. It is readily found that the derivative (Jacobian matrix) of $\boldsymbol{\alpha}$ at $\boldsymbol{x}$ maps X onto $\mathbb{R}^n$. By the inverse function theorem there exists a neighborhood of $\boldsymbol{x}$, again denoted by O′, such that $\boldsymbol{\alpha}$ is injective on O′ and its inverse $\boldsymbol{\beta}$ is defined on a neighborhood of $(0, \ldots, 0) \in \mathbb{R}^n$ which can be chosen to be of the form $(-\varepsilon, \varepsilon) \times \ldots \times (-\varepsilon, \varepsilon)$ with $\varepsilon > 0$. Moreover $\boldsymbol{\beta}$ is of class C^{r-1}. Hence

$$\boldsymbol{\beta}(\tau_1, \ldots, \tau_n) - \boldsymbol{x} = \sum_{i=1}^{n} \tau_i \tilde{\boldsymbol{u}}_i(\boldsymbol{\beta}(\tau_1, \ldots, \tau_n))$$

for every $(\tau_1, \ldots, \tau_n) \in (-\varepsilon, \varepsilon)\times\ldots\times(-\varepsilon, \varepsilon)$. Let us now define $\boldsymbol{y} : (-\varepsilon, \varepsilon) \to \mathrm{O}'$ by $\boldsymbol{y}(\tau) = \boldsymbol{\beta}(\tau, 0, \ldots, 0)$. Then

$$y(\tau) - x = \tau \tilde{u}_1(y(\tau)) = \tau \tilde{u}(y(\tau)), \quad \tau \in (-\varepsilon, \varepsilon) \tag{23.7.15}$$

and combining it with $M(z)\tilde{u}(z) = \tilde{\lambda}(z)\tilde{u}(z)$, we obtain $M(y(\tau))(y(\tau) - x) = \tilde{\lambda}(y(\tau))(y(\tau) - x)$. By the remark following the introduction of M this implies Assertion (2) with $u(\tau) = \tilde{\lambda}(y(\tau))$. We have $y(0) = x$ by (23.7.15). A differentiation of (23.7.15) at $\tau = 0$ gives that $\dot{y}(0) = u(y(0)) = u \neq 0$. Hence y is nonconstant near 0 and diminishing the value of ε if necessary, we can achieve that $y(\tau) \neq x$ if $\tau \neq 0$. This proves (1). To prove (3), let (z, w) be a pair satisfying (23.7.14) with $z \in O'$, $w \in (\lambda-\delta, \lambda+\delta)$. Then $z - x$ is an eigenvector of $M(z)$ corresponding to the eigenvalue w and as $\tilde{\lambda}(z)$ is the only eigenvalue of $M(z)$ in $(\lambda - \delta, \lambda + \delta)$ and $\tilde{u}(z)$ is the only eigenvector corresponding to this eigenvalue, we have $w = \tilde{\lambda}(z)$, and $z - x = \gamma\tilde{u}(z)$ for some $\gamma \in \mathbb{R}$. But it follows that $\alpha_1(z) = \gamma$, $\alpha_i(z) = 0$, $i = 2, \dots, n$. Then $z = y(\gamma)$. □

23.7.6 Proposition *Let (C, e) be a simple characteristic pair at $(f, n) \in D$ with $C > 0$. Then there exists a function $g(\cdot) : (-\varepsilon, \varepsilon) \to H_s(f, n)$, $\varepsilon > 0$, of class C^3 such that $g(0) = f$ and $\dot{g}(0) = (e \otimes n, 0)$; moreover, the function u defined by $u(\tau) := \hat{U}_{f,n}(g(\tau))$ for $\tau \neq 0$ and by $u(0) = C$ is of class C^3. If ε is chosen small enough, then there exists a $\delta > 0$ and a neighborhood N of f such that every $h \in H_s(f, n) \cap N$ with $|\hat{U}_{f,n}(h) - C| < \delta$ is of the form $h = g(\tau)$ for some τ.*

Proof Set $\rho_o = 1$. We shall use the Hugoniot equations in the form (23.6.7) with the identifications of ϕ, F as in the proof of Proposition 23.6.2. We are going to apply Proposition 23.7.5. Then $\nabla\phi(x)$, $\nabla F(x)$ are given by (23.6.8), from which

$$\nabla F(x)(\nabla\phi(x))^{-1} = \begin{bmatrix} Q - \tilde{\theta}^{-1} d \otimes s, & \tilde{\theta}^{-1} d \\ Qs - \tilde{\theta}^{-1}(s \cdot d)s, & \tilde{\theta}^{-1} s \cdot d \end{bmatrix}.$$

Let us check that C^2 is a simple root of the characteristic polynomial of $M := \nabla F(x)(\nabla\phi(x))^{-1}$. Notice the following facts. If $e \in$ Vect is an eigenvector of Q with eigenvalue $\mu \in \mathbb{R}$, then $\xi = (e, s \cdot e)$ is an eigenvector of M with eigenvalue μ. If $\xi = (e, b)$ is an eigenvector of M with eigenvalue $\mu \neq 0$, then it is of the form $\xi = (e, s \cdot e)$ with e an eigenvector of Q with eigenvalue μ. Since (C, e) is a simple characteristic pair, it follows from these facts that C^2 is an eigenvector of M with C^2 the eigenvalue and as $C > 0$, there is no other linearly independent eigenvector of M with the eigenvalue C^2. Hence the only way the value C^2 can fail to be a simple root of the characteristic polynomial is if the Jordan block of M corresponding to the eigenvalue C^2 is two-dimensional, i.e., if there exists a vector $\omega \in \text{Vect} \times \mathbb{R}$ such that $(e, e \cdot s) = (M - C^2)\omega$. Now Q is symmetric and hence it has n linearly independent eigenvectors e and hence M has n linearly independent eigenvectors $(e, e \cdot s)$. Their span coincides with the set of all $\xi \in \text{Vect} \times \mathbb{R}$ satisfying $\xi \cdot \beta = 0$ where $\beta = (s, -1)$. The vector ω must be linearly independent of the eigenvectors of M and hence $\omega \cdot \beta \neq 0$. However, multiplying $(e, e \cdot s) = (M - C^2)\omega$ scalarly by β and using that $M^{\mathrm{T}}\beta = 0$, we obtain $0 = (e, e \cdot s) \cdot \beta = (M - C^2)\omega \cdot \beta = \omega \cdot (M^{\mathrm{T}} - C^2)\beta = -C^2\omega \cdot \beta$, i.e., $C^2\omega \cdot \beta = 0$, which is a contradiction because $C^2 > 0$ and $\omega \cdot \beta \neq 0$. We have thus verified the main hypothesis of Proposition 23.7.5 that λ is a simple root of the characteristic polynomial of M. Further, recall our permanent smoothness assumption that $\tilde{e}$ is of class C^5. It then follows that F, ϕ are of class C^4 and hence the index r occurring in Proposition 23.7.5 is $r = 4$. All the hypotheses of Proposition 23.7.5 have been checked and its conclusion is just the assertion of the present proposition. □

We can now describe the local form of the Hugoniot set under the strict hyperbolicity assumption.

23.7.7 Proposition *If the material is strictly hyperbolic at $(f, n) \in D$, then there exists a neighborhood N of f such that $H(f, n) \cap N$ consists of exactly $n + 1$ smooth curves H_i, $i = 0, 1, \dots, n$ which pairwise intersect only at f. For $i = 0$ the curve lies in $H_r(f, n) \cap N$ and its tangent at f is $(e_0 \otimes n, 1)$ where $(e_0, 1)$ is a characteristic vector corresponding to the characteristic speed 0; the curve H_0 can be parametrized by a function g_0 of the type described in Proposition 23.6.7. For each $i = 1, \dots, n$, the the* *tangent of H_i at f is $(e_i \otimes n, 0)$ where e_i is the characteristic amplitude corresponding to $C_i > 0$; the curve H_i can be parametrized by a function g_i of the type described in Proposition 23.7.6 with the associated function u_i.*

Proof Let C_i, $i = 0, 1, \ldots, n$ be the characteristic speeds, ordered increasingly; by Proposition 23.1.6(2) necessarily $C_0 = 0$. The corresponding characteristic vectors are of the form $\xi_0 = (e_0, 1)$, $\xi_i = (e_i, 0)$, $i = 1, \ldots, n$. By Proposition 23.6.2 for every $\delta > 0$ there exists a neighborhood N of f such that $\{\hat{U}_{f,n}(g) : g \in N\} \subset \bigcup_{i=1}^{n} (C_i - \delta, C_i + \delta) \cup [0, \delta)$. Let us prove that N can be chosen in such a way that $\{\hat{U}_{f,n}(g) : g \in N\} \subset \bigcup_{i=1}^{n} (C_i - \delta, C_i + \delta) \cup \{0\}$. Suppose that this is not true. Then there exists a sequence $g_k = (a_k \otimes n + F, b_k)$, $k = 1, \ldots,$ such that $\hat{U}_{f,n}(g_k) > 0$ for every k and $\hat{U}_{f,n}(g_k) \to 0$. There exists a subsequence, denoted again g_k, such that $(a_k, b_k)/|(a_k, b_k)| \to (b, c) \neq 0$. Since $\hat{U}_{f,n}(g_k) > 0$ we have $g_k \in H_s(f, n)$ and so (23.6.3) and (23.6.9) apply to $g = g_k$. However, dividing these equations by $|(a_k, b_k)|$ and taking the limit, we find that

$$Q(f, n)b + Dnc = 0, \quad \tilde{S}(f)n \cdot b + \tilde{\theta}(f)c = \tilde{S}(f)n \cdot b,$$

from which it follows that $c = 0$ and that b is a characteristic amplitude corresponding to characteristic speed 0; in other words, $(b, 0)$ is a characteristic vector corresponding to the characteristic speed 0. Since $(e_0, 1)$ is another linearly independent characteristic vector corresponding to the characteristic speed 0, we have a contradiction with the strict hyperbolicity. This proves that $\{\hat{U}_{f,n}(g) : g \in N\} \subset \bigcup_{i=1}^{n} (C_i - \delta, C_i + \delta) \cup \{0\}$ if N is chosen small enough. Further, if $\delta > 0$ is chosen small enough, then the intervals $(C_i - \delta, C_i + \delta)$, $i = 1, \ldots, n$ do not overlap. Diminishing $\delta > 0$ further if necessary, then by Proposition 23.7.6 for every $i = 1, \ldots, n$, the set $\{g \in H_s(f, n) : |\hat{U}_{f,n}(g) - C_i| < \delta\}$ can be parametrized by the function $g(\cdot)$ described there. Similarly, by Proposition 23.6.7, the set $\{g \in H_s(f, n) : \hat{U}_{f,n}(g) = 0\}$ can be parametrized by the function $g(\cdot)$ described there. □

The above proposition describes only the local structure of $H_s(f, n)$. The curves H_i are (parts of) the ***Hugoniot curves*** centered at f corresponding to the direction n. Under additional assumptions not to be formulated here, they can be continued outside H. There are, however, model systems of conservation laws for which the curves H_i, H_j, $i \neq j$, join somewhere away from f so that topologically $H_s(f, n)$ contains a closed curve which intersects itself at the point f, looking like the symbol ∞. The reader is referred to HANYGA [1985] for a detailed analysis of the global structure of the Hugoniot set of strictly hyperbolic systems of conservation laws and of the *isentropic* elasticity. Let us also briefly mention the case in which there is a degeneracy of some characteristic speeds. If C is a characteristic speed of multiplicity $p > 1$, it can, but need not, happen that $H_s(f, n)$ contains a surface of dimension p through the point f. For instance, if f is an isotropic state then $H_s(f, n)$ contains an $(n-1)$-dimensional surface; this reflects the multiplicity of the transverse waves. It will be seen in Chap. 24 that for fluids satisfying the assumptions stated there, there is a single global Hugoniot curve through a given f which does not intersect itself and which does not return back to f. Moreover, $H_r(f, n)$ is a three-dimensional surface in $\mathrm{Lin} \times \mathbb{R}$ in this case.

Let us conclude this section with a proposition about contact discontinuities of the type of shocks. The simple proof is omitted.

23.7.8 Proposition

(1) *Let $(U, \bar{g}, f, n)$ be a contact discontinuity with $\bar{g} \in H_s(f, n)$. Let g be the function occurring in Definition 23.6.4 and write $\dot{g}(\tau) = (e(\tau) \otimes n, b(\tau))$ for its derivative. Then $b \equiv 0$ identically and $(U, e(\tau))$ is a linearly degenerate characteristic pair at $g(\tau)$ for every $\tau \in [\tau^+, \tau^-]$.*

(2) *If $(\tilde{C}(\cdot, n), \tilde{e}(\cdot, n))$ is a linearly degenerate characteristic field, then there exists a function $g : (-\varepsilon, \varepsilon) \to H_s(f, n)$ such that $g(0) = f$, $\dot{g}(0) \neq 0$ and $(\tilde{C}(f, n), g(\tau), f, n)$*

is a contact discontinuity for every $\tau \in (-\varepsilon, \varepsilon)$. If the derivative of g is written $\dot{g}(\tau) = (e(\tau) \otimes n, b(\tau))$, then $b \equiv 0$ and $e(\tau) = \tilde{e}(g(\tau), n)$.

Hence, if every characteristic pair of the material is genuinely nonlinear, then the material does not admit contact discontinuities with positive speed.

23.8 The Shock Admissibility Criteria

The purpose of the admissibility criteria is to select physically admissible discontinuities from the inadmissible ones. There is a large literature on the subject, and different criteria are proposed to cover different situations. The criteria considered below coincide for small shocks in strictly hyperbolic and genuinely nonlinear materials. They do not coincide when the equations are not strictly hyperbolic or when they even change the type from hyperbolic to parabolic, and some of the admissibility criteria are designed exactly for these situations. The reader is referred to the selected literature at the end of this chapter for these cases. It is felt that the admissibility criteria should be helpful to prove the uniqueness or facilitate the proofs of the existence of the solutions, and such is indeed the case in special circumstances.

Recall that a discontinuity is a shock if $U > 0$ and $f^+ \neq f^-$. Here and in what follows, $S = (U, f^\pm, n)$ and $f^\pm = (F^\pm, \eta^\pm)$. Recall that the state labeled with the + sign is in front of the shock and the state labeled with the – sign behind of the shock. The admissibility criteria to be formulated below deal only with shocks; they impose no restrictions on the rest discontinuities.

23.8.1 The Entropy Criterion The shock S is admissible according to the entropy criterion if $\eta^+ \leq \eta^-$. A process (χ, η) is said to satisfy the entropy criterion if every shock on its singular surface satisfies the entropy criterion.

Under some assumptions about the material, DAFERMOS [1979a,b] showed that if there exists a smooth process satisfying particular initial conditions, then there is no other process (even discontinuous) satisfying the same initial conditions and the entropy criterion. For the one-dimensional isentropic dynamics, DIPERNA [1979] proved the uniqueness in the class of piecewise Lipschitz solutions satisfying the entropy criterion. (He also cites earlier results.)

The role of the entropy criterion has long been recognized in fluid dynamics, as documented in COURANT & FRIEDRICHS [1948], ROZHDESTVENSKII & JANENKO [1978], and LANDAU & LIFSCHITZ [1978]. The interpretation behind the increase of entropy across shocks is that even a small amount of viscosity and heat conduction produces strong dissipation in the region of the shock, converting a substantial amount of mechanical energy into heat. This provides a motivation for the next two criteria. It will be seen that the magnitudes of the coefficients of viscosity and heat conduction influence the *width of the region of shock, but not the amount of dissipation in the shock,* which remains the same even when the viscosity and heat conduction tend to zero.

23.8.2 A Family of Dissipative Materials To formulate the next two criteria, let us associate with the adiabatic material described by response functions 22.3.1 a family of viscous materials with heat conduction. We first express the response of the original adiabatic material in terms of temperature:

$$e = \bar{e}(F, \theta), \quad \eta = \bar{\eta}(F, \theta), \quad S = \bar{S}(F, \theta), \quad \bar{f}(F, \theta) = \bar{e}(F, \theta) - \theta\bar{\eta}(F, \theta).$$

Now let

$$\bar{f}^* = \bar{f}^*(F, \theta), \; \bar{\eta}^* = \bar{\eta}^*(F, \theta), \; \bar{S}^* = \bar{S}^*(F, \theta, G, \dot{F}), \; \bar{q}^* = \bar{q}^*(F, \theta, G, \dot{F})$$

be any thermodynamically compatible material whose equilibrium response coincides with that of $\bar{e}, \bar{S}$, i.e., $\bar{f}^*(F, \theta) = \bar{f}(F, \theta)$, $\bar{S}^*(F, \theta, \mathbf{0}, \mathbf{0}) = \bar{S}(F, \theta)$. Assume that its production of entropy (9.2.12) satisfies $\gamma^* \geq c(|D|^2 + |G|^2)$ in every process, where $c > 0$ is a constant independent of $\dot{F}, G$. (A strong dissipator.) Finally consider a family of materials with the response functions

$$\bar{f}^\varepsilon(F, \theta) = \bar{f}^*(F, \theta),\ \ \bar{\eta}^\varepsilon(F, \theta) = \bar{\eta}^*(F, \theta),$$
$$\bar{S}^\varepsilon(F, \theta, G, \dot{F}) = \bar{S}^*(F, \theta, \varepsilon G, \varepsilon \dot{F}),\ \ \bar{q}^\varepsilon(F, \theta, G, \dot{F}) = \bar{q}^*(F, \theta, \varepsilon G, \varepsilon \dot{F}),$$

where $\varepsilon > 0$ is a parameter. Each member of the family is thermodynamically compatible and its equilibrium response coincides with that of $\bar{e}, \bar{S}$. As $\varepsilon \to 0$, the dissipation becomes weaker and weaker. Physically there are no bodies without dissipation; mathematically the viscous perturbation changes the character of the evolution equations from hyperbolic to parabolic, for which there is a larger spectrum of methods, including the proofs of the existence of solutions in various special cases. Let us denote by Ω_ε the body Ω with the response functions $\bar{f}^\varepsilon$, etc. In this situation we say that the adiabatic material Ω is imbedded in the family Ω_ε of viscous materials with heat conduction.

23.8.3 The Viscosity Criterion A process (χ, η) of Ω is said to satisfy the viscosity criterion if there exists a family of smooth processes $(\chi^\varepsilon, \eta^\varepsilon)$ of Ω_ε such that $\eta^\varepsilon \to \eta$, $q^\varepsilon/\theta^\varepsilon \to \mathbf{0}$ on $\Omega \times (t_1, t_2)$ in the sense of distributions, i.e.,

$$\int_{t_1}^{t_2}\int_\Omega \rho_0 \eta^\varepsilon \varphi \, dV\, dt \to \int_{t_1}^{t_2}\int_\Omega \rho_0 \eta \varphi \, dV\, dt, \qquad \int_{t_1}^{t_2}\int_\Omega q^\varepsilon \varphi/\theta^\varepsilon \, dV\, dt \to \mathbf{0}$$

as $\varepsilon \to 0$ for every $\varphi \in C_0^\infty(\Omega \times (t_1, t_2))$.

Whether or not a given process satisfies the viscosity criterion depends, at least in principle, on the family Ω_ε. Ideally, the information that $\eta^\varepsilon \to \eta$, $q^\varepsilon/\theta^\varepsilon \to \mathbf{0}$ should come from the balance laws for the material Ω_ε; such is indeed the case in special circumstances. Note that in the limit $\varepsilon \to 0$, the dissipation generally does not tend to zero, rather, it tends to a measure concentrated at shocks.

23.8.4 Proposition *If a process satisfies the viscosity criterion, then it satisfies the entropy criterion.*

Proof Let $(\chi^\varepsilon, \eta^\varepsilon)$ be the family of processes of Ω_ε that converges to (χ, η). The Clausius–Duhem inequality for $(\chi^\varepsilon, \eta^\varepsilon)$ reads

$$\rho_0 \dot{\eta}^\varepsilon \geq -\mathrm{Div}(q^\varepsilon/\theta^\varepsilon) \tag{23.8.1}$$

on $\Omega \times (t_1, t_2)$. Let φ be a nonnegative function in $C_0^\infty(\Omega \times (t_1, t_2))$. Multiply (23.8.1) by φ and integrate over $\Omega \times (t_1, t_2)$. The use of the integration by parts and of the divergence theorem (see Sect. 3.1–3.3) gives the *weak form* of the Clausius–Duhem inequality:

$$-\int_{t_1}^{t_2}\int_\Omega \rho_0 \eta^\varepsilon \dot{\varphi} \, dV\, dt \geq \int_{t_1}^{t_2}\int_\Omega q^\varepsilon \cdot \nabla\varphi/\theta^\varepsilon \, dV\, dt.$$

Letting $\varepsilon \to 0$ and using the assumed convergence, one obtains $-\int_{t_1}^{t_2}\int_\Omega \rho_0 \eta \dot{\varphi} \, dV\, dt \geq 0$. The integral can be transformed back to the form

$$\int_{t_1}^{t_2}\int_\Omega \rho_0 \dot{\eta} \varphi \, dV\, dt - \int_{t_1}^{t_2}\int_{S_t} \rho_0 U[\eta] \varphi \, dA\, dt \geq 0.$$

As this must be satisfied by every smooth nonnegative function, one obtains $\dot{\eta} \geq 0$ outside the singular surface and $-\rho_0 U[\eta] \geq 0$ on the singular surface. □

23.8.5 ***Remark*** Let $\boldsymbol{n} \in$ Vect be a unit vector and $U \in \mathbb{R}$, $U \neq 0$. Assume that $\boldsymbol{\chi} : \text{Vect} \times \mathbb{R} \to \text{Vect}$ is a smooth motion with $\det \boldsymbol{F} > 0$ and with $\boldsymbol{F}, \boldsymbol{v}$ of the form

$$\boldsymbol{F}(\boldsymbol{p}, t) = \boldsymbol{\phi}(s), \quad \boldsymbol{v}(\boldsymbol{p}, t) = \boldsymbol{\psi}(s),$$

where $s = \boldsymbol{p} \cdot \boldsymbol{n} - Ut$ and $\boldsymbol{\phi}, \boldsymbol{\psi}$ are continuously differentiable functions on $\mathbb{R}$. Then there exists an $\boldsymbol{F}_0 \in \text{Lin}^+$, $\boldsymbol{V} \in$ Vect, and a twice continuously differentiable function $\boldsymbol{u} : \mathbb{R} \to \text{Vect}$ such that

$$\boldsymbol{\chi}(\boldsymbol{p}, t) = \boldsymbol{F}_0\boldsymbol{p} + \boldsymbol{u}(s) + \boldsymbol{V}t \text{ for every } \boldsymbol{p} \in \text{Vect},\ t \in \mathbb{R}. \tag{23.8.2}$$

Proof Let $\boldsymbol{h} : \mathbb{R} \to \text{Vect}$ be any primitive function to $\boldsymbol{\psi}$, i.e., $\boldsymbol{h}' = \boldsymbol{\psi}$ on $\mathbb{R}$. Then

$$(\boldsymbol{\chi} + U^{-1}\boldsymbol{h}(\boldsymbol{p} \cdot \boldsymbol{n} - Ut))^{\cdot} = \boldsymbol{0} \text{ and hence } \boldsymbol{\chi}(\boldsymbol{p}, t) + U^{-1}\boldsymbol{h}(\boldsymbol{p} \cdot \boldsymbol{n} - Ut) = \boldsymbol{g}(\boldsymbol{p})$$

for some $\boldsymbol{g} : \text{Vect} \to \text{Vect}$ and all $\boldsymbol{p} \in$ Vect, $t \in \mathbb{R}$. A differentiation with respect to $\boldsymbol{p}$ gives

$$\boldsymbol{\phi}(\boldsymbol{p} \cdot \boldsymbol{n} - Ut) + U^{-1}\boldsymbol{\psi}(\boldsymbol{p} \cdot \boldsymbol{n} - Ut) \otimes \boldsymbol{n} = \nabla \boldsymbol{g}(\boldsymbol{p}) \tag{23.8.3}$$

from which $\nabla \boldsymbol{g}(\boldsymbol{p}) := \boldsymbol{F}_1$ is found to be constant and $\boldsymbol{g}(\boldsymbol{p}) = \boldsymbol{F}_1\boldsymbol{p} + \boldsymbol{C}$, with $\boldsymbol{C} \in$ Vect. Recalling that all values of $\boldsymbol{\phi}$ are in Lin^+, from (23.8.3) we see that $\boldsymbol{F}_1$ differs at most by a rank 1 tensor from some tensor with positive determinant. Hence, writing $\boldsymbol{F}_1 = \boldsymbol{F}_0 - U^{-1}\boldsymbol{V} \otimes \boldsymbol{n}$ with $\boldsymbol{F}_0 \in \text{Lin}^+$, $\boldsymbol{V} \in$ Vect, we see that (23.8.2) holds with $\boldsymbol{u}(s) = U^{-1}(-\boldsymbol{h}(s) + \boldsymbol{V}s) + \boldsymbol{C}$. □

23.8.6 The Viscosity Profile Consider the family Ω_ε described in 23.8.2 and assume that the referential region Ω is the whole space, which we identify with Vect. We shall consider processes of the traveling wave form

$$\boldsymbol{\chi}(\boldsymbol{p}, t) = \boldsymbol{F}_0\boldsymbol{p} + \varepsilon\boldsymbol{u}(s), \quad \theta(\boldsymbol{p}, t) = \vartheta(s), \quad \boldsymbol{p} \in \text{Vect}, \quad t \in \mathbb{R},$$

where $s = (\boldsymbol{p} \cdot \boldsymbol{n} - Ut)/\varepsilon$, the functions $\boldsymbol{u}, \vartheta$ are to be determined, and $\boldsymbol{F}_0, \boldsymbol{n}, U$ are constants, with $|\boldsymbol{n}| = 1$. Then

$$\boldsymbol{F} = \boldsymbol{F}_0 + \boldsymbol{w} \otimes \boldsymbol{n}, \quad \dot{\boldsymbol{F}} = -\varepsilon^{-1}U\boldsymbol{w}' \otimes \boldsymbol{n}, \quad \boldsymbol{v} = -U\boldsymbol{w}, \quad \dot{\boldsymbol{v}} = \varepsilon^{-1}U^2\boldsymbol{w}',$$
$$\dot{\theta} = -\varepsilon^{-1}U\vartheta', \quad \boldsymbol{G} = \varepsilon^{-1}\vartheta'\boldsymbol{n},$$

where $\boldsymbol{w} = \boldsymbol{u}'$ and the prime denotes the derivative with respect to s. We are looking for processes for which

$$\boldsymbol{F} \to \boldsymbol{F}^+, \ \boldsymbol{v} \to \boldsymbol{v}^+, \ \eta \to \eta^+, \ \theta \to \theta^+, \ \dot{\boldsymbol{F}} \to \boldsymbol{0}, \ \boldsymbol{G} \to \boldsymbol{0} \text{ as } s \to \pm\infty, \tag{23.8.4}$$

where $(U, \boldsymbol{v}^\pm, \boldsymbol{f}^\pm, \boldsymbol{n})$ is a jump. The tensor $\boldsymbol{F}_0$ must then differ by a tensor of rank one from $\boldsymbol{F}^\pm$, and we write $\boldsymbol{F}^\pm = \boldsymbol{F}_0 + \boldsymbol{A}^\pm \otimes \boldsymbol{n}$. The requirements (23.8.4) reduce to

$$\boldsymbol{w} \to \boldsymbol{A}^\pm, \ \boldsymbol{w}' \to \boldsymbol{0}, \ \vartheta \to \theta^\pm, \ \vartheta' \to 0 \text{ as } s \to \pm\infty. \tag{23.8.5}$$

If one inserts these values into the equations of balance of linear momentum and energy for Ω_ε, it turns out that the scale factors ε cancel and one ends up with equations that are independent of ε. Hence, the intensity of viscosity and heat conduction, measured by ε, does not affect the shape of the viscosity profile. It affects only its width. The balance equations and the Clausius–Duhem inequality read:

$$\begin{aligned} \rho_\circ U^2\boldsymbol{w}' &= \bar{\boldsymbol{S}}^*\boldsymbol{n}', \\ U\rho_\circ(\tfrac{1}{2}U^2\boldsymbol{w}^2 + \bar{e}^*)' &= (U\bar{\boldsymbol{S}}^*\boldsymbol{n} \cdot \boldsymbol{w} + \bar{\boldsymbol{q}}^* \cdot \boldsymbol{n})', \\ -U\rho_\circ\bar{\eta}^{*\prime} &\le -(\bar{\boldsymbol{q}}^* \cdot \boldsymbol{n}/\vartheta)', \end{aligned} \tag{23.8.6}$$

with $\bar{e}^*$, $\bar{\eta}^*$ evaluated at $(\boldsymbol{F}_0 + \boldsymbol{w} \otimes \boldsymbol{n}, \vartheta)$ and $\bar{\boldsymbol{S}}^*, \bar{\boldsymbol{q}}^*$ at $(\boldsymbol{F}_0 + \boldsymbol{w} \otimes \boldsymbol{n}, \vartheta, \vartheta'\boldsymbol{n}, -U\boldsymbol{w}' \otimes \boldsymbol{n})$. One can integrate from $-\infty$ to s to obtain

$$\begin{aligned} \rho_\circ U^2(\boldsymbol{w} - \boldsymbol{w}^-) &= \bar{\boldsymbol{S}}^*\boldsymbol{n} - \boldsymbol{S}^-\boldsymbol{n}, \\ U\rho_\circ(\bar{e}^* - e^-) &= U\boldsymbol{S}^-\boldsymbol{n} \cdot (\boldsymbol{w} - \boldsymbol{w}^-) + \tfrac{1}{2}\rho_\circ U^3(\boldsymbol{w} - \boldsymbol{w}^-)^2 + \boldsymbol{q}^* \cdot \boldsymbol{n}. \end{aligned} \tag{23.8.7}$$

Alternatively, one can write the last equation in the form

$$\begin{aligned} U\rho_\circ(\bar{e}^* - e^-) &= \tfrac{1}{2}U(\bar{\boldsymbol{S}}^*\boldsymbol{n} + \boldsymbol{S}^-\boldsymbol{n}) \cdot (\boldsymbol{w} - \boldsymbol{w}^-) + \boldsymbol{q}^* \cdot \boldsymbol{n} \quad \text{or} \\ U^2\rho_\circ^2(\bar{e}^* - e^-) &= \tfrac{1}{2}(|\bar{\boldsymbol{S}}^*\boldsymbol{n}|^2 - |\boldsymbol{S}^-\boldsymbol{n}|^2) + \rho_\circ U\boldsymbol{q}^* \cdot \boldsymbol{n}. \end{aligned}$$

(23.8.7) is a system of ordinary differential equations of the first order for $\boldsymbol{w}, \vartheta$ with $\boldsymbol{w}', \vartheta'$ contained in the arguments of $\bar{\boldsymbol{S}}^*, \bar{\boldsymbol{q}}^*$. A process χ, θ of the traveling wave form satisfying these equations and (23.8.4) will be referred to as the ***viscosity profile*** or ***shock layer*** connecting the states of the shock $(U, \boldsymbol{v}^\pm, \boldsymbol{f}^\pm, \boldsymbol{n})$. One can eliminate $\boldsymbol{q}^*$ between the second and third of (23.8.6) to show that the quantity

$$\Phi = -\rho_\circ(\eta - \eta^-) + (\rho_\circ(e - e^-) - \boldsymbol{S}^-(\boldsymbol{w} - \boldsymbol{w}^-) - \tfrac{1}{2}U^2(\boldsymbol{w} - \boldsymbol{w}^-)^2)/\vartheta,$$

decreases along solutions, $\Phi' \le 0$; it is the Liapunov function for the system. In particular,

$$\Phi(\boldsymbol{w}^-, \theta^-) \ge \Phi(\boldsymbol{w}, \vartheta). \tag{23.8.8}$$

One can let s tend to ∞ in the last inequality and in (23.8.7) to obtain Hugoniot's equations (23.6.3), (23.6.4) and the entropy condition $\eta^+ \le \eta^-$.

23.8.7 The Viscosity Profile Criterion Let the adiabatic material Ω be imbedded in the family Ω_ε. A shock S is said to satisfy the viscosity profile criterion if its states can be connected through the viscosity profile. A process (χ, η) is said to satisfy the viscosity profile criterion if every shock on its singular surface satisfies the viscosity profile criterion.

Equation (23.8.8) implies the following.

23.8.8 Proposition *If a shock satisfies the viscosity profile criterion, then it satisfies also the entropy criterion.*

The converse can be established for various one-dimensional models, for which the reader is referred to the literature at the end of this chapter.

23.8.9 Lax's Criterion The shock S is said to be admissible according to Lax's criterion if there exists a characteristic field $(\tilde{C}(\cdot, \boldsymbol{n}), \tilde{\boldsymbol{e}}(\cdot, \boldsymbol{n}))$ in the direction $\boldsymbol{n}$ such that

$$\tilde{C}(\boldsymbol{f}^+, \boldsymbol{n}) \le U \le \tilde{C}(\boldsymbol{f}^-, \boldsymbol{n}). \tag{23.8.9}$$

A process (χ, η) is said to satisfy Lax's criterion if every shock on its singular surface satisfies Lax's criterion.

The inequalities (23.7.5) say that the shock is subsonic relative to the speed of sound behind the shock and supersonic relative to the speed in front of the shock. Riemann (1860) noticed that compressible shocks in gases have this property. SMOLLER [1983, Sect. 15D] gives a motivation by an elementary calculation of the number of conditions on the singular surface to make the solution formally determined. The inequalities (23.8.9) apply to general hyperbolic systems with possible degeneracies (as in isotropic states), but it may be the case that its spirit is preserved only when the degeneracies do not occur and, moreover, (23.8.9) are strict. In any case, we shall establish the relation of Lax's criterion to the entropy criterion only in the absence of degeneracies. It also appears that the apparatus developed in SMOLLER [1983, Chaps. 22–24] can be applied to show that at least for imbeddings Ω_ε satisfying additional assumptions, if a small shock in a strictly hyperbolic material satisfies Lax's criterion, then it satisfies the viscosity profile criterion. The reader is in particular referred to Section 24.A of the cited work for a discussion of the relationship of Lax's inequalities to the dimensions of the stable and unstable manifolds at the equilibrium points $\boldsymbol{f}^+, \boldsymbol{f}^-$ of the system of shock layer equations from Subsect. 23.8.6. We shall give such a proof only for fluids (in Sect. 24.5). See also MALEK-MADANI [1981] for a proof in a special case within the isentropic approach.

23.8.10 Liu's Criterion The shock S is said to be admissible according to Liu's criterion if there exists a continuously differentiable function $\boldsymbol{g} : [\tau^+, \tau^-] \to H(\boldsymbol{f}^-, \boldsymbol{n})$ such that $\boldsymbol{g}(\tau^\pm) = \boldsymbol{f}^\pm$ and $U \le \hat{U}_{\boldsymbol{f}^-, \boldsymbol{n}}(\boldsymbol{g}(\tau))$ for every $\tau \in (\tau^+, \tau^-)$. A process (χ, η) is said to satisfy Liu's criterion if every shock on its singular surface satisfies Liu's criterion.

This formulation does not presuppose any a priori structure of the Hugoniot set. However, if the Hugoniot set centered at $\boldsymbol{f}^-$ consists of disjoint open curves, then necessarily the function $\boldsymbol{g}$ occurring in the above formulation will coincide with a part of one of the Hugoniot curves in the Hugoniot set (see Proof of Proposition 23.8.11 below.) Liu's criterion was designed as a strengthening of Lax's criterion in the case of hyperbolic but not necessarily genuinely nonlinear systems. It may be also viewed as a generalization of Oleinik's E-condition from the case of a single conservation law to systems. In the case of *isentropic fluid dynamics* the Liu criterion unconditionally implies Lax's criterion. Liu showed that his criterion implies the existence of a purely viscous shock (without heat conduction) in the one-dimensional case. PEGO [1986] gave an example of a one-dimensional fluid with a nonconvex pressure function $\tilde{p}(\cdot, \eta)$ which admits a shock satisfying Liu's criterion for which the viscous profile with heat conduction does not exist.

The following two propositions establish the equivalence of the entropy, Lax's, and Liu's criteria in the genuinely nonlinear case. The first of them deals with a single characteristic pair, while the second, which reads more nicely, is formulated under the assumption of strict hyperbolicity. The difference between them is that the former applies, e.g., to fluids, while the latter does not if $n > 1$.

23.8.11 Proposition *Let $\boldsymbol{f} \in \tilde{U}, \boldsymbol{n} \in \mathrm{Sph}$ and let $(C, \boldsymbol{e})$ be a genuinely nonlinear simple characteristic pair at $(\boldsymbol{f}, \boldsymbol{n})$. Then there exists a $\delta > 0$ and a neighborhood N of $\boldsymbol{f}$ such that for every shock $(U, \boldsymbol{f}^\pm, \boldsymbol{n})$ with $\boldsymbol{f}^- = \boldsymbol{f}$, $|U - C| < \delta$ and $\boldsymbol{f}^+ \in N$ the entropy criterion, Lax's criterion, and Liu's criterion are equivalent.*

In Chap. 24 it will be shown that for fluids whose pressure is a convex function of the specific volume at fixed entropy, the three criteria are equivalent for shocks of arbitrary magnitude.

Proof Since $(C, \boldsymbol{e})$ is simple, Proposition 23.2.2 applies and says that $(C, \boldsymbol{e})$ has an extension $(\tilde{C}(\cdot, \boldsymbol{n}), \tilde{\boldsymbol{e}}(\cdot, \boldsymbol{n}))$ to some neighborhood of $\boldsymbol{f}$. Again since $(C, \boldsymbol{e})$ is simple, Proposition 23.7.6 applies to the pair $(C, \boldsymbol{e})$, which gives us a neighborhood N, which can be assumed to be contained in the N from the preceding step, the values δ, ε and the functions $\boldsymbol{g}$, u with the properties stated there. Since

$(C, \boldsymbol{e})$ is a genuinely nonlinear characteristic pair, Proposition 23.7.4 tells us that we can change the parametrization to achieve that $\dot{u} > 0$ and $\ddot{\zeta} > 0$ at $\tau = 0$ and that ζ, u are increasing functions of τ near 0. By Proposition 23.7.3 we have that $\dot{u} = \dot{\tilde{C}}/2 < \dot{\tilde{C}}$ at $\tau = 0$ and hence by continuity $\dot{u} < \dot{\tilde{C}}$ for all τ sufficiently close to 0. Diminishing the value of ε and the size of N if necessary, we can achieve that $\dot{u} < \dot{\tilde{C}}$ on $(-\varepsilon, \varepsilon)$, the functions u, ζ are increasing on the whole interval $(-\varepsilon, \varepsilon)$ and the conclusion of Proposition 23.7.6 still holds. This is the final choice of N and of δ. We shall now prove that with this choice the conclusion of the present proposition holds. Let $(U, \boldsymbol{f}^{\pm}, \boldsymbol{n})$ be a shock with $\boldsymbol{f}^{-} = \boldsymbol{f}$, $|U - C| < \delta$ and $\boldsymbol{f}^{+} \in N$. By Proposition 23.7.6 then $\boldsymbol{f}^{+} = \boldsymbol{g}(\tau^{+})$ for some τ^{+} in $(-\varepsilon, \varepsilon)$ with $\tau^{+} \neq 0$. Set $\tau^{-} = 0$.

Let us now prove that if S satisfies the entropy criterion, the it satisfies Liu's and Lax's criteria. Hence assume that $\eta^{+} \le \eta^{-}$. Since ζ is an increasing function and $\eta^{\pm} = \zeta(\tau^{\pm})$, it follows that $\tau^{+} < \tau^{-} = 0$. Since u is an increasing function, it then follows that $U = u(\tau^{+}) \le u(\tau)$ for every τ, $\tau^{+} \le \tau \le \tau^{-}$. This proves that the function $\boldsymbol{g}$, restricted to $[\tau^{+}, \tau^{-}]$ has the properties of the $\boldsymbol{g}$ occurring in Liu's criterion. Hence the entropy criterion implies Liu's criterion. Integrating $\dot{u} < \dot{\tilde{C}}$ from τ^{+} to τ^{-} and using $\tilde{C}(\tau^{-}) = u(\tau^{-})$ we obtain $\tilde{C}(\tau^{+}) < u(\tau^{+})$, which means that $\tilde{C}(\boldsymbol{f}^{+}, \boldsymbol{n}) < U$. Since u is an increasing function, it follows that $U = u(\tau^{+}) < u(\tau^{-}) = \tilde{C}(\boldsymbol{f}^{-}, \boldsymbol{n})$.

Conversely, if Liu's criterion holds then necessarily the $\boldsymbol{g}$ from that criterion is the $\boldsymbol{g}$ from above, restricted to $[\tau^{+}, \tau^{-}]$ (up to a reparametrization). This is true since there are no other points $\boldsymbol{h}$ in $H_s(\boldsymbol{f}, \boldsymbol{n})$ with $\boldsymbol{h} \in N$ and $\hat{U}_{\boldsymbol{f},\boldsymbol{n}}(\boldsymbol{h})$ in $(C - \delta, C + \delta)$ but those in the range of $\boldsymbol{g}$. With this identification, Liu's criterion says that $U = u(\tau^{+}) \le u(\tau)$ for all τ on the segment with endpoints $\tau^{\pm}$ and in particular, $u(\tau^{+}) \le u(\tau^{-})$. Since u is increasing, it follows that $\tau^{+} \le \tau^{-}$ and since ζ is increasing, $\zeta(\tau^{+}) \le \zeta(\tau^{-})$, i.e., the entropy criterion holds. If Lax's criterion holds, then the inequality $(23.8.9)_2$ alone and the increasing character of u suffice to establish the validity of the entropy criterion. □

23.8.12 Proposition *Suppose that the material is strictly hyperbolic at $(\boldsymbol{f}, \boldsymbol{n})$ and that all the n characteristic pairs at $(\boldsymbol{f}, \boldsymbol{n})$ are genuinely nonlinear. Then there exists a neighborhood N of $\boldsymbol{f}$ such that for every shock $(U, \boldsymbol{f}^{\pm}, \boldsymbol{n})$ with $\boldsymbol{f}^{-} = \boldsymbol{f}, \boldsymbol{f}^{+} \in N$ the entropy criterion, Lax's criterion and Liu's criterion are equivalent.*

This is just a combination of Proposition 23.8.11 with Proposition 23.7.7.

For a process (χ, η), a time t during the process, and a part P of the body Ω, the production of entropy $\Gamma(\chi, \eta, P, t)$ at time t in P during (χ, η) is defined by

$$\Gamma(\chi, \eta, P, t) = \lim_{\varepsilon \downarrow 0} \frac{1}{\varepsilon}\Big(\int_P \eta(\boldsymbol{p}, t + \varepsilon)\rho_\circ \, dV - \int_P \eta(\boldsymbol{p}, t)\rho_\circ \, dV\Big).$$

The limit exists and is given by (see (13.2.5))

$$\Gamma(\chi, \eta, P, t) = -\int_{S_t \cap P} U\rho_\circ[\eta] \, dA.$$

23.8.13 The Entropy Rate Criterion A process (χ, η) defined on a time interval $[t_1, t_2]$ is said to satisfy the entropy rate criterion if for every $t \in (t_1, t_2)$, and for every part P the value $\Gamma(\chi, \eta, P, t)$ takes the maximum value among all processes $(\bar{\chi}, \bar{\eta})$ defined on $[t_1, t + \delta)$ which coincide with (χ, η) on the interval $[t_1, t]$ and satisfy the balance equations and the constitutive equations.

The entropy rate criterion uses only intrinsic concepts associated with the adiabatic material and does not refer to any artificial device like the family Ω_ε. It also does not refer to shocks, and so it may be formulated for processes of more complicated structure, such as those resulting from a hypothetical existence theorem. In the one-dimensional isothermal case DAFERMOS [1974] has shown that the viscosity profile criterion, the entropy rate criterion, and Liu's criterion are equivalent. The reader is referred to HATTORI [1986a,b], HSIAO [1980], DAFERMOS [1989], and PENCE [1992] for more recent developments.

23.9 The Riemann Problem

Suppose that Ω = Vect. Let (χ, η) be a centered process as in Subsect. 23.5.1 defined on the half-spacetime W = Vect $\times \mathbb{R}^{++}$, i.e., $a = -\infty, b = \infty$ in the notation of Subsect. 23.5.1. If we assume that there exist numbers $U_L < U_R$ such that on $(-\infty, U_L]$ and on $[U_R, \infty)$ the functions ϕ, ζ are constant, then the limits of $f(p, t) := (F(p, t), v(p, t)), \eta(p, t)$ as $t \downarrow 0$ exist for every p such that $p \cdot n \neq 0$. We denote them consistently by $f_0(p), v_0(p)$, respectively.

23.9.1 Data Let $(f_L, v_L), (f_R, v_R) \in \tilde{U} \times$ Vect be given and let n be a unit vector such that F_L, F_R differ by a tensor of the form $a \otimes n$. The Riemann problem is to find a centered process (χ, η) such that

$$(f_0(p), v_0(p)) = \begin{cases} (f_L, v_L) & \text{for } p \cdot n < 0, \\ (f_R, v_R) & \text{for } p \cdot n > 0. \end{cases}$$

A physical situation corresponding to the special choice $v_L = v_R = \mathbf{0}$ is that at time $t = 0$ the left half of the body is in a homogeneous state characterized by (F_L, η_L) and the right half in the state (F_R, η_R). This is generally a state in which the balance of forces is violated at the interface $p \cdot n = 0$ and some artificial device must be introduced to keep the body in this state. In the case of a fluid this can be a membrane separating the two regions. When the device is removed at time $t = 0$, the body can no longer remain in the original state and the state of the body will evolve. This process will solve the Riemann problem.

23.9.2 Proposition *Let the material be strictly hyperbolic at (f_R, n) and let every characteristic pair at (f_R, n) be genuinely nonlinear. Then there exists a neighborhood M of (f_R, v_R) such that for every $(f_L, v_L) \in M$ with $F_R - F_L$ a rank one tensor there exists a solution of the Riemann problem with data $(f_L, v_L), (f_R, v_R)$. This solution is a centered process satisfying the entropy, Lax, and Liu criteria and consists of at most $2n + 2$ constant states separated by shocks, centered waves and at most one contact discontinuity. There is precisely one solution of this kind in a neighborhood of (f_R, v_R).*

Proof Only a sketch of the proof will be given; the details can be completed using the results of the preceding sections. By Proposition 23.1.6 one of the characteristic speeds is 0 and let $C_i, i = 1, \ldots, n$ be the positive characteristic speeds ordered increasingly. The characteristic vector corresponding to $C_0 := 0$ can be chosen of the form $\xi_0 = (e_0, 1)$ with $e_0 \in$ Vect. The characteristic vectors corresponding to $C_i, i = 1, \ldots, n$, are of the form $\xi_i = (e_i, 0)$ with e_i the characteristic amplitude. The characteristic pairs $(C_i, e_i), i = 1, \ldots, n$ are all simple and genuinely nonlinear by the hypothesis. By Proposition 23.2.2 they can be extended to genuinely nonlinear, normalized simple characteristic fields $(\tilde{C}_i(\cdot, n), \tilde{e}_i(\cdot, n))$ defined on a common neighborhood of $f_R := (F_R, \eta_R)$. Under these conditions, we apply Propositions 23.7.6 and 23.5.4 to $(\tilde{C}(\cdot, n), \tilde{e}(\cdot, n)) \equiv (\tilde{C}_i(\cdot, n), \tilde{e}_i(\cdot, n))$ and the point $f \equiv f_R$ for every $i = 1, \ldots, n$. Proposition 23.7.6 gives functions g, u, defined in a neighborhood of 0 in $\mathbb{R}$, which we denote by g_i, u_i, with the property that $g_i(\tau)$ solves the Hugoniot equations with the speed $U = u_i(\tau)$. Moreover, if we write $g_i(\tau) = (a_i(\tau) \otimes n + F_R, \zeta_i(\tau))$, then $a_i(0) = \mathbf{0}, \dot{a}_i(0) = e_i$. Proposition 23.5.4 gives a function $\bar{g}$ defined in a neighborhood of 0 in $\mathbb{R}$, which we denote by $\bar{g}_i$, with the property that if we write $\bar{g}_i(\tau) = (\phi_i(\tau) \otimes n + F_R, \zeta_i(\tau))$, then (23.5.6) holds. This gives a centered wave with a velocity $v_i(\tau)$ which satisfies $v_i'(\tau) = -z\phi_i'(\tau)$, and we choose the integration constant as to give $v_i(0) = \mathbf{0}$. Let us now define functions H_i, H_{-i} by

$$\boldsymbol{H}_{\pm i}(\tau) = \begin{cases} (\boldsymbol{a}_i(\tau), \zeta_i(\tau), \mp u_i(\tau)\boldsymbol{a}_i(\tau) + \boldsymbol{v}_R) & \text{if } \tau \le 0, \\ (\boldsymbol{\phi}_i(\tau), \zeta_i(\tau), \mp \boldsymbol{v}_i(\tau) + \boldsymbol{v}_R) & \text{if } \tau \ge 0 \end{cases} \tag{23.9.1}$$

and write $\boldsymbol{T}_{\pm i}(\boldsymbol{0}, \eta_R, \boldsymbol{v}_R, \tau)$ for $\boldsymbol{H}_{\pm i}(\pm\tau)$. Furthermore, under the hypothesis of the proposition the set of all rest discontinuities $H_r(\boldsymbol{f}_R, \boldsymbol{n})$ is locally a curve which can be parametrized by a function $\boldsymbol{g}_0$: $(-\varepsilon, \varepsilon) \to H_r(\boldsymbol{f}_R, \boldsymbol{n})$, $\boldsymbol{g}_0 = (\boldsymbol{a}_0 \otimes \boldsymbol{n} + \boldsymbol{F}_R, \zeta_0)$ such that $\dot{\boldsymbol{a}}(0) = \boldsymbol{e}_0$, $\dot{\zeta}(0) = 1$ and $\boldsymbol{g}_0(0) = \boldsymbol{f}_R$. (See Proposition 23.6.7.) Write $\boldsymbol{T}_0(\boldsymbol{0}, \eta_R, \boldsymbol{v}_R, \tau)$ for $(\boldsymbol{a}_0(\tau), \zeta_0(\tau), \boldsymbol{v}_R)$. The same construction of $\boldsymbol{H}_{\pm i}$, $\boldsymbol{g}_0$ can be given for every $\boldsymbol{f} = (\boldsymbol{a} \otimes \boldsymbol{n} + \boldsymbol{F}_R, \eta)$ from some neighborhood of $\boldsymbol{f}_R$, and every $\boldsymbol{v}$ from some neighborhood of $\boldsymbol{v}_R$, since the hypothesis of the proposition implies that the material is strictly hyperbolic at every $(\boldsymbol{f}, \boldsymbol{n})$ with $\boldsymbol{f}$ from some neighborhood of $\boldsymbol{f}_R$ and all characteristic pairs are genuinely nonlinear. (See Proposition 23.1.6. Write $\boldsymbol{T}_{\pm i}(\boldsymbol{a}, \zeta, \boldsymbol{v}, \tau)$, $\boldsymbol{T}_0(\boldsymbol{a}, \zeta, \boldsymbol{v}, \tau)$ for the values of $\boldsymbol{H}_{\pm i}(\tau)$, $\boldsymbol{g}_0(\tau)$ constructed on the basis of $(\boldsymbol{f}, \boldsymbol{v}) = (\boldsymbol{a} \otimes \boldsymbol{n} + \boldsymbol{F}_R, \zeta, \boldsymbol{v})$. We have obtained in this way a collection of mappings $\boldsymbol{T}_i(\cdot, \tau)$, $i = 0, \pm 1, \dots, \pm n$, defined, for each sufficiently small τ, in a neighborhood M of $(\boldsymbol{0}, \eta_R, \boldsymbol{v}_R) \in$ Vect $\times$ $\mathbb{R}$ $\times$ Vect with values in Vect $\times$ $\mathbb{R}$ $\times$ Vect. The latter set is a $2n + 1$-dimensional space. The mappings satisfy $\boldsymbol{T}_i(\boldsymbol{0}, \eta_R, \boldsymbol{v}_R, 0) = (\boldsymbol{0}, \eta_R, \boldsymbol{v}_R)$, $i = 0, \pm 1, \dots, \pm n$. Let us now define $\boldsymbol{T}(\tau, \cdot) = \boldsymbol{T}_{-n}(\tau_{-n}) \circ \dots \circ \boldsymbol{T}_{-1}(\tau_{-1}) \circ \boldsymbol{T}_0(\tau_0) \circ \boldsymbol{T}_1(\tau_1) \circ \dots \circ \boldsymbol{T}_n(\tau_n)$, $\tau = (\tau_{-n}, \dots, \tau_n) \in (-\varepsilon, \varepsilon)^{(2n+1)}$ and write $\boldsymbol{U}(\tau)$ for the value of $\boldsymbol{T}(\tau)$ at $(\boldsymbol{0}, \eta_R, \boldsymbol{v}_R)$. The mapping $\boldsymbol{U}$ maps $(-\varepsilon, \varepsilon)^{(2n+1)}$ to Vect$\times\mathbb{R}\times$Vect, satisfies $\boldsymbol{U}(0, \dots, 0) = (\boldsymbol{0}, \eta_R, \boldsymbol{v}_R)$, is continuously differentiable, and its linear part $\boldsymbol{L}(\cdot)$ at $(0, \dots, 0)$ is a linear transformation from $\mathbb{R}^{2n+1}$ to Vect $\times$ $\mathbb{R}$ $\times$ Vect, which maps a $\tau = (\tau_{-n}, \dots, \tau_0, \dots, \tau_n) \in \mathbb{R}^{2n+1}$ to an element $\boldsymbol{L}(\tau)$ of the form

$$\boldsymbol{L}(\tau) = \sum_{i=1}^{n} \tau_i(\boldsymbol{e}_i, 0, -C_i\boldsymbol{e}_i) + \tau_0(\boldsymbol{e}_0, 1, \boldsymbol{0}) - \sum_{i=1}^{n} \tau_{-i}(\boldsymbol{e}_i, 0, C_i\boldsymbol{e}_i).$$

It is clear that $\boldsymbol{L}$ maps $\mathbb{R}^{2n+1}$ onto Vect $\times$ $\mathbb{R}$ $\times$ Vect, which is a $2n + 1$-dimensional space. By the inverse mapping theorem then for all $(\boldsymbol{f}_L, \boldsymbol{v}_L)$ in Vect $\times$ $\mathbb{R}$ $\times$ Vect, sufficiently close to $(\boldsymbol{f}_R, \boldsymbol{v}_R)$ there exists a $\tau \in (-\varepsilon, \varepsilon)^{(2n+1)}$ such that $U(\tau) = (\boldsymbol{f}_L, \boldsymbol{v}_L)$. Having this τ, one can construct the corresponding centered process. The details are omitted. It is, however, noted that the number of shocks in the process is the sum of the number of indices $i > 0$ with $\tau_i < 0$ with the number of indices $i < 0$ with $\tau_i > 0$. The contact discontinuity is present if and only if $\tau_0 \neq 0$; if this is the case, then the solution will contain a rest discontinuity.

□

23.9.3 ***Remark on Nonuniqueness*** *Under the conditions of Proposition* 23.9.2 *every neighborhood of* $(\boldsymbol{f}_R, \boldsymbol{v}_R)$ *contains a* $(\boldsymbol{f}_L, \boldsymbol{v}_L)$ *for which there exist two different processes satisfying the initial condition* 23.9.1. *One of them is the process asserted by Proposition* 23.9.2 *and the other consists entirely of constant states separated by shocks and possibly one contact discontinuity.*

Both the two processes satisfy the balance equations (including the jump conditions); the second process need not satisfy the entropy criterion.

Proof By the proof of the preceding theorem, for every neighborhood of $(\boldsymbol{F}_R, \eta_R, \boldsymbol{v}_R)$ there exist an $(\boldsymbol{F}_L, \eta_L, \boldsymbol{v}_L)$ such that the process asserted by Proposition 23.9.2 contains at least one centered wave: It is enough to choose any $\tau \in (-\varepsilon, \varepsilon)^{(2n+1)}$ where at least one of the τ_i with $i > 0$ is positive and to set $(\boldsymbol{f}_L, \boldsymbol{v}_L) = \boldsymbol{U}(\tau)$. Now repeat the construction of $\boldsymbol{U}$ in the proof of Proposition 23.9.2 with (23.9.1) replaced by

$$\boldsymbol{H}_{\pm i}(\tau) = (\boldsymbol{a}_i(\tau), \zeta_i(\tau), \mp u_i(\tau)\boldsymbol{a}_i(\tau) + \boldsymbol{v}_R), \quad \tau \in (-\varepsilon, \varepsilon).$$

The mapping $\bar{\boldsymbol{U}}$ obtained in this way will have the same properties as $\boldsymbol{U}$, in particular, the inverse function theorem can be applied to it, and so for some $\bar{\tau}$, we have $\bar{\boldsymbol{U}}(\bar{\tau}) = (\boldsymbol{f}_L, \boldsymbol{v}_L)$, which will give rise to a process satisfying the Riemann data and consisting entirely constant states separated by shocks possibly a contact discontinuity. The process asserted by Proposition 23.9.2 contains a centered wave, while this one does not, so these processes are different. □

Bibliographical Note

The calculation of the acoustic tensor of an isotropic solid in Sect. 23.4 is based on the papers BOWEN & WANG [1970, 1971] and CHADWICK & OGDEN [1971a,b].

A primitive example of an admissibility criterion is the statement that only compression shocks are admissible. According to DAFERMOS [1989] the requirement that the entropy be increased by a shock goes back to Burton (1893), Weber (1901), Rayleigh (1908). Further elaborations are due to Jouguet (1901, 1904) and Zemplen (1905). The formulation of the entropy criterion within the framework of general systems of quasilinear equations is due to LAX [1971].

Proposition 23.7.2 is an adaptation, to the mixed second-order-first-order system of adiabatic dynamics, of the classical argument by LAX [1957]. This paper also contains the formulation of the Lax admissibility criterion, within the context of the hyperbolic system of conservation laws (with the mathematical entropy equal to minus the physical entropy, and with the assumption of convexity of it). That paper also shows that for genuinely nonlinear strictly hyperbolic systems, the entropy criterion is equivalent to the Lax criterion for small shocks. The criterion of Subsect. 23.8.10 is first formulated in T.-P. LIU [1975]. The criterion of Subsect. 23.8.13 is due to DAFERMOS [1973], who has analyzed it from the point of view of the uniqueness of the solutions of the equations with shocks. For other admissibility criteria, see, e.g., SLEMROD [1983], HAGAN & SERRIN [1986]. The proof of Proposition 23.9.2 is an adaptation of the proof in SMOLLER [1983, Chap. 17, C]. The function $\boldsymbol{M}$ introduced in the proof of Proposition 23.7.5 is from SMOLLER [1983, Chap. 17, B], who attributes it to CONLON [1980]. The proof is, however, different and the result stronger inasmuch as it does not rely on the hyperbolicity. Alternatively, Proposition 23.9.2 can be proved using the bifurcation theorem from simple eigenvalues, see CRANDALL & RABINOWITZ [1971, Thm. 1.7]. For an application of the bifurcation theory to a nonstrictly hyperbolic model system, see SHEARER & al. [1987]. Additional information about the entropy criteria in the isentropic approximation for solids in three dimensions can be found in MALEK-MADANI [1981].

A mathematical theory for global existence and uniqueness for the equations of three dimensional adiabatic or isentropic dynamics is missing. These equations fall within the framework of quasilinear systems of equations. I refer to ROZHDESTVENSKII & JANENKO [1978] and to HANYGA [1985, Chap. 3] for more information about the global existence/nonexistence and local existence results. Further information about nonlinear wave propagation may be found in WHITHAM [1974] and about the balance equations in the presence of singularities in KOSIŃSKI [1986].

24 Adiabatic Fluid Dynamics

For fluids many assertions of Chap. 23 gain much more definite forms and it is the purpose of this chapter to derive them. The spatial description is a natural setting. The equations are of first order in all choices of variables, in contrast to when the motion is used as the unknown. Under the classical hypotheses about the fluid, the equivalence of the entropy criterion to the Lax–Riemann inequalities and Liu's criterion is established and the existence of the shock layer connecting the states of the jump is proved.

24.1 The Equations of Fluid Dynamics

24.1.1 Balance Equations The flow an adiabatic fluid in n dimensions is described by the triple $(\rho, \boldsymbol{v}, \eta)$ of functions of $\boldsymbol{x} \in r_t$, $t \in [t_1, t_2]$, where $r_t \subset$ Eucl is the actual region occupied by the fluid at time t. The balance equations read

$$\left.\begin{aligned} \frac{\partial \rho}{\partial t} + \operatorname{div}(\rho \boldsymbol{v}) &= 0, \\ \frac{\partial \rho \boldsymbol{v}}{\partial t} + \operatorname{div}(\rho \boldsymbol{v} \otimes \boldsymbol{v} + p\mathbf{1}) &= \mathbf{0}, \\ \frac{\partial \rho(e + \frac{1}{2}\boldsymbol{v}^2)}{\partial t} + \operatorname{div}(\rho(e + \tfrac{1}{2}\boldsymbol{v}^2)\boldsymbol{v} + p\boldsymbol{v}) &= 0, \end{aligned}\right\} \qquad (24.1.1)$$

outside the singular surface and

$$\left.\begin{aligned} -u[\rho] + [\rho \boldsymbol{v}] \cdot \boldsymbol{m} &= 0, \\ -u[\rho \boldsymbol{v}] + [\rho \boldsymbol{v} \otimes \boldsymbol{v} + p\mathbf{1}]\boldsymbol{m} &= \mathbf{0}, \\ -u[\rho(e + \tfrac{1}{2}\boldsymbol{v}^2)] + [\rho(e + \tfrac{1}{2}\boldsymbol{v}^2)\boldsymbol{v} + p\boldsymbol{v}] \cdot \boldsymbol{m} &= 0, \end{aligned}\right\} \qquad (24.1.2)$$

on the singular surface s_t moving with the spatial speed u in the direction $\boldsymbol{m}$. The constitutive information is fully provided by $\tilde{e} : \tilde{U} \to \mathbb{R}$ with $\tilde{U} \subset (0, \infty) \times \mathbb{R}$, and by the constitutive equations

$$e = \tilde{e}(v, \eta), \quad p = \tilde{p}(v, \eta) = -\partial_v \tilde{e}(v, \eta),$$

where $v = 1/\rho > 0$ is the specific volume and η the entropy. The Clausius–Duhem inequality reduces to

$$\frac{\partial \rho\eta}{\partial t} + \operatorname{div}(\rho\eta\boldsymbol{v}) = 0, \quad -u[\rho\eta] + [\rho\eta\boldsymbol{v}] \cdot \boldsymbol{m} \ge 0 \qquad (24.1.3)$$

outside and on the singular surface, respectively. Equation $(24.1.3)_1$ expresses the constancy of the specific entropy along trajectories. Outside the singular surface the equation of balance of energy $(24.1.1)_3$ can be replaced by $(24.1.3)_1$. Note that the expression $\rho(e+\frac{1}{2}\boldsymbol{v}^2)\boldsymbol{v}+p\boldsymbol{v}$ in $(24.1.1)_3$ is the transport of the total specific enthalpy $h = e+\frac{1}{2}\boldsymbol{v}^2+pv$. (24.1.1) is a system of $n+2$ conservation laws for the unknowns $\rho, \boldsymbol{v}, \eta$. It is of the form (22.4.1) with $m = n+2$, $w = (\rho, \boldsymbol{v}, \eta)$, $\boldsymbol{z} = \boldsymbol{x}$.

24.1.2 The Characteristic Equation Since for the considerations about the shock waves the specific volume is more convenient than the density we regard (24.1.1) as a system for the unknowns $(v, \boldsymbol{v}, \eta)$. The characteristic equation is obtained in a way similar to that in 23.1.2: In the explicit form of the system (24.1.1) with $\rho = v^{-1}$ we make the replacements

$$\partial v/\partial t \to -r\alpha, \quad \partial \boldsymbol{v}/\partial t \to -r\boldsymbol{b}, \quad \partial \eta/\partial t \to -r\gamma,$$
$$\operatorname{grad} v \to \alpha\boldsymbol{m}, \quad \operatorname{grad} \boldsymbol{v} \to \boldsymbol{b}\otimes\boldsymbol{m}, \quad \operatorname{grad}\eta \to \gamma\boldsymbol{m},$$

where $\xi = (\alpha, \boldsymbol{b}, \gamma) \in \mathbb{R}\times\mathrm{Vect}\times\mathbb{R}$ is the ***characteristic vector***, $r \in \mathbb{R}$ the ***characteristic speed*** and $\boldsymbol{m} \in \mathrm{Vect}$, $|\boldsymbol{m}| = 1$, the spatial direction. (Some work is saved if (24.1.3) is used instead of $(24.1.1)_3$.) For a given state $(v, \boldsymbol{v}, \eta)$ the ***characteristic equation*** reads

$$\left.\begin{aligned}\rho(-r+\boldsymbol{v}\cdot\boldsymbol{m})\alpha - \boldsymbol{m}\cdot\boldsymbol{b} &= 0,\\ \rho(-r+\boldsymbol{v}\cdot\boldsymbol{m})\boldsymbol{b} + \partial_v\tilde{p}\cdot\alpha\boldsymbol{m} + \partial_\eta\tilde{p}\cdot\gamma\boldsymbol{m} &= \boldsymbol{0},\\ (-r+\boldsymbol{v}\cdot\boldsymbol{m})\gamma &= 0.\end{aligned}\right\} \tag{24.1.4}$$

Equating the determinant of (24.1.4) to 0 shows that there are at most three distinct characteristic speeds r_0, r_+, r_-. They are real if and only if $\partial_v\tilde{p} \le 0$ and if this is the case then these characteristic speeds and the corresponding characteristic vectors are

$$r_0 = \boldsymbol{v}\cdot\boldsymbol{m} \text{ and } \xi_b = (0, \boldsymbol{b}, 0) \text{ or } \xi_0 = (\partial_\eta\tilde{p}, \boldsymbol{0}, -\partial_v\tilde{p}), \tag{24.1.5}$$
$$r_\pm = \boldsymbol{v}\cdot\boldsymbol{m} \pm c \text{ and } \xi_\pm = (v, \mp c\boldsymbol{m}, 0), \tag{24.1.6}$$

where $c = \sqrt{-v^2\partial_v\tilde{p}}$ is the speed of sound and $\boldsymbol{b}$ any nonzero vector orthogonal to $\boldsymbol{m}$. In three dimensions, there are two linearly independent vectors $\boldsymbol{b}$, in two dimensions – one, and in one dimension none.

If $\xi = (\alpha, \boldsymbol{b}, \gamma)$ is a characteristic vector at $(v, \boldsymbol{v}, \eta)$ with the characteristic speed $r \in \mathbb{R}$ and $\boldsymbol{m}$ the direction, then

$$(\bar{v}, \bar{\boldsymbol{v}}, \bar{\eta})(\boldsymbol{x}, t) := (v, \boldsymbol{v}, \eta) + (\alpha, \boldsymbol{b}, \gamma)\sin(2\pi l^{-1}(\boldsymbol{x}\cdot\boldsymbol{m} - rt)), \quad \boldsymbol{x}\in\mathrm{Vect},\ t\in\mathbb{R},$$

is a solution of the linearizations of equations (24.1.1) about $(v, \boldsymbol{v}, \eta)$. Here l is an arbitrary wavelength, and r is the spatial speed of propagation of the plane wave. Clearly for $r_\pm$, $\xi_\pm$ these are really longitudinal waves propagating with the speed of sound in the direction $\pm\boldsymbol{m}$ relative to the basic motion of the fluid. The speed r_0 gives a solution in a relative rest with respect to the basic motion of the fluid. The amplitudes ξ_b correspond to transversal disturbances of the velocity field; such disturbances are not accompanied by changes of density and entropy. The amplitude $(24.1.6)_3$ corresponds to transversal disturbances of the density and entropy field without disturbances of the velocity; they are subject to the condition that the pressure be unchanged.

If for a given $(\rho, \boldsymbol{v}, \eta, \boldsymbol{m})$ all the characteristic speeds are real, the system is said to be hyperbolic at the given point; if the characteristic speeds are distinct, the system is said to be strictly hyperbolic at the given point.

24.1.3 Proposition *For a given* $\omega = (\rho, \boldsymbol{v}, \eta, \boldsymbol{m})$ *the following assertions hold:*
(1) *the system* (24.1.1) *is hyperbolic at* ω *if and only if* $\partial_v\tilde{p} \le 0$ *at* (v, η);

(2) *if $\partial_v \tilde{p} < 0$, there are exactly three distinct characteristic speeds r_0, r_+, r_- as in* (24.1.5) *and* (24.1.6), *with the multiplicities n,* 1, 1, *respectively; for $n > 1$ the system is never strictly hyperbolic. If $\partial_v \tilde{p} < 0$ then for $n = 1$ the system is strictly hyperbolic;*

(3) *the characteristic speeds $r_\pm$ are genuinely nonlinear, i.e., $\xi_\pm \cdot \partial_w r_\pm \neq 0$ where $w = (v, \boldsymbol{v}, \eta)$ if and only if $\partial_v^2 \tilde{p} \neq 0$.*

The conditions $\partial_v \tilde{p} < 0, \partial_v^2 \tilde{p} \neq 0$ were already encountered in Chap. 23 when describing the fluid referentially. Of the assertions above, only that about the genuine nonlinearity needs a proof. One finds that

$$\partial_w r_\pm \cdot \xi_\pm = \mp v^3 \partial_v^2 \tilde{p}/(2c). \quad \square$$

24.2 Shock Waves in Fluids

In the spatial description, the shock waves are the jump discontinuities in the spatial density, velocity, and entropy.

24.2.1 The Rankine–Hugoniot Conditions The reduced forms (3.5.5), (3.6.6), and (3.7.3) of the jump conditions for fluids are

$$j[\boldsymbol{v}] - [p]\boldsymbol{m} = \boldsymbol{0}, \quad j[e] - \langle p \rangle [\boldsymbol{v}] \cdot \boldsymbol{m} = 0, \quad -j[\eta] \geq 0, \tag{24.2.1}$$

where j is the flux of mass through the singular surface, see Proposition 3.4.2. In this section it is admitted that j can be of arbitrary sign. Recall the following consequences of $(3.4.3)_2$ and (2.2.13):

$$u = \boldsymbol{v}^\pm \cdot \boldsymbol{m} + j v^\pm, \quad j[v] + [\boldsymbol{v}] \cdot \boldsymbol{m} = 0 \tag{24.2.2}$$

where v is the specific volume. Eliminating $[\boldsymbol{v}]$ from (24.2.1) by $(24.2.2)_2$, one obtains

$$j^2[v] + [p] = 0, \quad j([e] + \langle p \rangle [v]) = 0, \quad -j[\eta] \geq 0. \tag{24.2.3}$$

These equations will be referred to as the Rankine–Hugoniot conditions. If $j \neq 0$, $(24.2.3)_2$ leads to Hugoniot's relation

$$[e] + \langle p \rangle [v] = 0. \tag{24.2.4}$$

Let $\tilde{Z}$ be a subset of the $n + 2$-dimensional space $\mathbb{R} \times \text{Vect} \times \mathbb{R}$ consisting of triples $\boldsymbol{p} = (v, \boldsymbol{v}, \eta)$ with $(v, \eta) \in \tilde{U}$. If $\boldsymbol{p}^\pm = (v^\pm, \boldsymbol{v}^\pm, \eta^\pm) \in \tilde{Z}$, a collection $(\boldsymbol{p}^+, \boldsymbol{p}^-, \boldsymbol{m})$ is called a ***jump*** or ***discontinuity*** if it satisfies (24.1.2) with some $u \in \mathbb{R}$. If $\boldsymbol{p}^+ \neq \boldsymbol{p}^-$, the speed u is uniquely determined, and so also is the flux of mass j. A jump is called a $\pm$ ***shock*** if $\pm j > 0$. The + shocks move in the positive direction $\boldsymbol{m}$ with a positive speed relative to the state in front of and behind the shock and the state + is in front of the singular surface. A jump is called a ***contact discontinuity*** if $j = 0$. The contact discontinuities are immobile relative to the states on both sides of the singular surface, and the jump in velocity is tangential to the singular surface. Equation (24.2.3) shows that the balance of energy and the entropy jump conditions are trivially satisfied by contact discontinuities, and the only restriction is the balance of forces $[p] = 0$.

24.2.2 Hugoniot's Adiabat Let $\boldsymbol{p} = (v, \boldsymbol{v}, \eta) = (v^+, \boldsymbol{v}^+, \eta^+) \in \tilde{Z}$ and $\boldsymbol{m} \in$ Sph. The ***Hugoniot's set*** $H(\boldsymbol{p}, \boldsymbol{m})$ at $(\boldsymbol{p}, \boldsymbol{m})$ is the set of all $\boldsymbol{q} = (v^-, \boldsymbol{v}^-, \eta^-) \in \tilde{Z}$ for which there exists a $u \in \mathbb{R}$ such that (24.1.2) hold. For $\boldsymbol{q} \neq \boldsymbol{p}$ the value u is uniquely determined and it is denoted by $\hat{u}_{p,m}(\boldsymbol{q})$. We divide Hugoniot's set into three parts, $H(\boldsymbol{p}, \boldsymbol{m}) = H^\circ(\boldsymbol{p}, \boldsymbol{m}) \cup H^+(\boldsymbol{p}, \boldsymbol{m}) \cup H^-(\boldsymbol{p}, \boldsymbol{m})$, where $H^\circ(\boldsymbol{p}, \boldsymbol{m})$ consists of $\boldsymbol{p}$ and of all $\boldsymbol{q}$ for which $(\boldsymbol{q}, \boldsymbol{p}, \boldsymbol{m})$ is a contact discontinuity and $H^\pm(\boldsymbol{p}, \boldsymbol{m})$ consists of $\boldsymbol{p}$ and of all $\boldsymbol{q}$ for which $(\boldsymbol{q}, \boldsymbol{p}, \boldsymbol{m})$ is a $\pm$ shock. By (24.2.4) the elements $\boldsymbol{q} \in H^\pm(\boldsymbol{p}, \boldsymbol{m})$ must satisfy Hugoniot's equation

$$\tilde{e}(w, \zeta) - \tilde{e}(v, \eta) + \tfrac{1}{2}(\tilde{p}(w, \zeta) + \tilde{p}(v, \eta))(w - v) = 0. \tag{24.2.5}$$

Hugoniot's adiabat centered at (v, η) is the set $A(v, \eta)$ of all pairs $(w, \zeta) \in \tilde{U}$ satisfying (24.2.5). Let us observe that if $(w, \zeta) \in A(v, \eta)$ and $(w, \zeta) \neq (v, \eta)$, then $w \neq v$. Suppose for contradiction that $w = v$. Then (24.2.5) reduces to $\tilde{e}(v, \zeta) = \tilde{e}(v, \eta)$ and since $\tilde{e}$ is strictly increasing in η, this implies $\zeta = \eta$, i.e., $(w, \zeta) = (v, \eta)$. This contradiction proves the claim. Moreover, if $\boldsymbol{q} \in H^\pm(\boldsymbol{p}, \boldsymbol{m})$ and $\boldsymbol{q} \neq \boldsymbol{p}$ then

$$(\tilde{p}(v, \eta) - \tilde{p}(w, \zeta))/(w - v) > 0. \tag{24.2.6}$$

This follows from $w \neq v$ and $(24.2.3)_1$. Conversely, if $(w, \zeta) \in A(v, \eta)$, $(w, \zeta) \neq (v, \eta)$ and (24.2.6) holds, then the values (w, ζ) can be connected to (v, η) by a jump, as we shall prove later. (The following section will give assumptions which guarantee that (24.2.6) is an automatic consequence of (24.2.5).) Let $J(w, \zeta, v, \eta)$ be defined for every $(w, \zeta) \in A(v, \eta)$ by

$$J(w, \zeta, v, \eta) = \begin{cases} \sqrt{(\tilde{p}(v, \eta) - \tilde{p}(w, \zeta))/(w - v)} & \text{if } (w, \zeta) \neq (v, \eta), \\ \sqrt{-\partial_v \tilde{p}(v, \eta)} & \text{if } (w, \zeta) = (v, \eta), \end{cases} \tag{24.2.7}$$

where it is understood that in the case $(w, \zeta) \neq (v, \eta)$ the quantity $J(w, \zeta, v, \eta)$ is defined only provided (24.2.6) holds. Under additional assumptions it will be shown (see Proposition 24.3.7) that the function $J(w, \zeta, v, \eta)$ is continuous along Hugoniot's adiabat.

24.2.3 Proposition *Let* $\boldsymbol{p} = (v, \boldsymbol{v}, \eta) \in \tilde{Z}$ *and* $\boldsymbol{m}$ *be given. Then*

$$H^\circ(\boldsymbol{p}, \boldsymbol{m}) = \{(w, \boldsymbol{w}, \zeta) : (\boldsymbol{w} - \boldsymbol{v}) \cdot \boldsymbol{m} = 0, \tilde{p}(w, \zeta) = \tilde{p}(v, \eta)\}, \tag{24.2.8}$$

$$H^\pm(\boldsymbol{p}, \boldsymbol{m}) = \{(w, \boldsymbol{w}, \zeta) : (w, \zeta) \in A(v, \eta), \boldsymbol{w} = \boldsymbol{v} \pm \boldsymbol{m} J(w, \zeta, v, \eta)(v - w)\} \tag{24.2.9}$$

(in the last formula only those triples $(w, \boldsymbol{w}, \zeta)$ *are admitted for which* $J(w, \zeta, v, \eta)$ *is defined) and*

$$\hat{u}_{p,m}(\boldsymbol{q}) = \boldsymbol{v} \cdot \boldsymbol{m}, \quad \hat{u}_{p,m}(\boldsymbol{q}) = \boldsymbol{v} \cdot \boldsymbol{m} \pm v J(w, \zeta, v, \eta)$$

on $H^\circ(\boldsymbol{p}, \boldsymbol{m})$, $H^\pm(\boldsymbol{p}, \boldsymbol{m})$, *respectively.*

Thus the set $H^\circ(\boldsymbol{p}, \boldsymbol{m})$ is essentially an n-dimensional manifold in the $n + 2$-dimensional space $\mathbb{R} \times \text{Vect} \times \mathbb{R}$. The sets $H^\pm(\boldsymbol{p}, \boldsymbol{m})$ are curves in $\mathbb{R} \times \text{Vect} \times \mathbb{R}$; the form of these curves is completely determined by the form of Hugoniot's adiabat $A(v, \eta)$. The tangents to Hugoniot's curves $H^\pm(\boldsymbol{p}, \boldsymbol{m})$ at $(v, \boldsymbol{v}, \eta)$ are the characteristic vectors $\boldsymbol{\xi}_\pm(\boldsymbol{p}, \boldsymbol{m})$ defined in (24.1.6).

Proof Let $(\boldsymbol{q}, \boldsymbol{p}, \boldsymbol{m})$ be a contact discontinuity. Then $j = 0$ by the definition and $(24.2.1)_1$ implies $\tilde{p}(w, \zeta) = \tilde{p}(v, \eta)$ while $(24.2.2)_1$ with $j = 0$ implies $(\boldsymbol{w} - \boldsymbol{v}) \cdot \boldsymbol{m} = 0$. This shows that the inclusion $\subset$ holds in (24.2.9). To prove the opposite inclusion, we use $(\boldsymbol{w} - \boldsymbol{v}) \cdot \boldsymbol{m} = 0$ to define $u := \boldsymbol{w} \cdot \boldsymbol{m} = \boldsymbol{v} \cdot \boldsymbol{m}$ and check that with this choice (24.1.2) hold. Indeed, we have $j = 0$ from $(24.2.2)_1$; moreover, $(24.2.3)_{1,2}$ hold, with $(24.2.3)_1$ a consequence of $\tilde{p}(w, \zeta) = \tilde{p}(v, \eta)$. With the help of (3.4.5) then (24.1.2) are established. This proves the formula for H°. Let $(\boldsymbol{q}, \boldsymbol{p}, \boldsymbol{m})$ be a + shock. We have proved above that $(w, \zeta) \in A(v, \eta)$ and (24.2.6) holds. Then $J(w, \zeta, v, \eta)$ is defined and coincides with j by $(24.2.3)_1$ and the assumption that $j > 0$. From $(24.2.1)_1$ and $(24.2.2)_2$ we obtain $\boldsymbol{w} = \boldsymbol{v} \pm \boldsymbol{m}J(w, \zeta, v, \eta)(v - w)$. This proves that the inclusion $\subset$ holds in (24.2.8). To prove the converse inclusion, we identify $j \equiv J(w, \zeta, v, \eta)$ and find from $\boldsymbol{w} = \boldsymbol{v} \pm \boldsymbol{m}j(v - w)$ that $\boldsymbol{w} \cdot \boldsymbol{m} + jw = \boldsymbol{v} \cdot \boldsymbol{m} + jv$ which enables us define u to be the common value of the expressions on the two sides of the last equality. With this identification, it is found that (24.1.2) hold in the following way. We have $(24.2.3)_1$ from the identification $j = J(w, \zeta, v, \eta)$ and $(24.2.3)_2$ from the definition of Hugoniot's adiabat. Using these equations and (3.4.5), we are led to (24.1.2) This proves the part of the proposition dealing with the equalities between the sets. The expressions for the speeds are now evident in view of what has been said above. □

Since the passage from $\boldsymbol{m}$ to $-\boldsymbol{m}$ changes a + shock into a – shock and conversely, *from now on we restrict ourselves to + shocks, which we shall call just shocks.*

24.3 Hugoniot's Adiabat

Under the assumptions to be formulated below, Propositions 24.3.5–24.3.7 assert that Hugoniot's adiabat can be parametrized by the entropy and determine the properties of the parametrization.

24.3.1 Assumptions

(1) $\tilde{U} = (0, \infty) \times (a, \infty)$, where a is either finite or $-\infty$, and $\tilde{e}$ is of class C^3 on $\tilde{U}$;
(2) throughout $\tilde{U}$,

$$\tilde{p} > 0, \quad \partial_v \tilde{p} < 0, \quad \partial_v^2 \tilde{p} > 0, \quad \partial_\eta \tilde{p} > 0; \tag{24.3.1}$$

(3) for every $\eta \in (a, \infty)$,

$$\lim_{v \to \infty} \tilde{p}(v, \eta) = 0, \quad \lim_{v \to 0} \tilde{p}(v, \eta) = \infty.$$

By the pressure relation, $(24.3.1)_2$ expresses the very strict convexity of $\tilde{e}$ with respect to v. We have encountered the convexity of $\tilde{e}$ in v, η as a consequence of the extremum principles. Here the convexity of $\tilde{e}$ in η is not needed, the fluid can have negative specific heat at constant volume in this section. Equation $(24.3.1)_3$ expresses the very strict convexity of the pressure $\tilde{p}$ with respect to the specific volume. See Fig. 24.3.1(a). A related inequality $\partial_v^2 \tilde{p} \neq 0$ is equivalent to genuine nonlinearity (see Proposition 23.4.4 and 24.1.3). Under this condition the equivalence of the shock admissibility criteria was established for shocks of small magnitude. Under the stronger Assumptions (1)–(3), the admissibility criteria are equivalent for shocks of arbitrary magnitude. Inequality $(24.3.1)_4$ is the increase of the pressure with respect to the increase of entropy (or equivalently with respect to the increase of temperature).

24.3.2 Pressure as Independent Variable At several places, it will be convenient to use the pressure and entropy as independent variables. By $(24.3.1)_2$ and 24.3.1(3), $\tilde{p}$ is invertible in v; we denote by $\bar{v} : \bar{U} \to (0, \infty)$ its inverse and by $\bar{h} : \bar{U} \to \mathbb{R}$ the enthalpy $h = e + pv$ expressed as function of (p, η), see Sect. 11.4. Assumptions 24.3.1(1)–(3) are equivalent to

(1′) $\bar{U} = (0, \infty) \times (a, \infty)$ where a is the same number as in 24.3.1(1);
(2′) throughout $\bar{U}$,

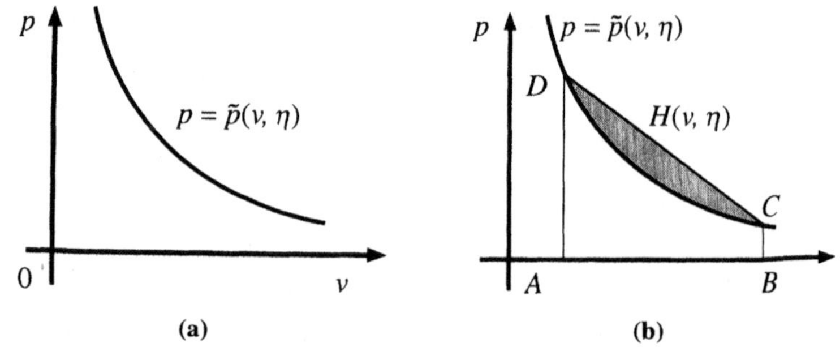

Fig. 24.3.1

$$\bar{v} > 0, \quad \partial_p \bar{v} < 0, \quad \partial_p^2 \bar{v} > 0, \quad \partial_\eta \bar{v} > 0; \tag{24.3.2}$$

(3′) for every $\eta \in (a, \infty)$,

$$\lim_{p \to \infty} \bar{v}(p, \eta) = 0, \quad \lim_{p \to 0} \bar{v}(p, \eta) = \infty.$$

This can be proved by simple manipulations with the derivatives, but a preferable proof amounts to visualizing the curves $p = \tilde{p}(v, \eta)$ in the p–v plane. Note that Assertions **(1′)**–**(3′)** are obtained by interchanging the positions of v and p.

Throughout the rest of this section it is assumed that 24.3.1(1)–(3) or equivalently 24.3.2(1′)–(3′) hold.

24.3.3 Hugoniot's Equations Suppose that $(v_0, \eta_0) \in (0, \infty) \times (a, \infty)$ is given and denote $e_0 = \tilde{e}(v_0, \eta_0)$, $p_0 = \tilde{p}(v_0, \eta_0)$, $h_0 = \bar{h}(p_0, \eta_0)$. The equation of Hugoniot's adiabat reads

$$H(v, \eta) := \tilde{e}(v, \eta) - e_0 + \tfrac{1}{2}(\tilde{p}(v, \eta) + p_0)(v - v_0) = 0. \tag{24.3.3}$$

This is completely equivalent to

$$K(p, \eta) := \bar{h}(p, \eta) - h_0 + \tfrac{1}{2}(\bar{v}(p, \eta) + v_0)(p_0 - p) = H(\bar{v}(p, \eta), \eta) = 0. \tag{24.3.4}$$

It will turn out that for $\eta < \eta_0$ it is convenient to solve (24.3.3) while for $\eta > \eta_0$ one chooses (24.3.4). We have the following formulas for the derivatives:

$$\partial_v H = \tfrac{1}{2}(p_0 - \tilde{p} + \partial_v \tilde{p} \cdot (v - v_0)), \quad \partial_\eta H = \tilde{\theta} + \tfrac{1}{2}\partial_\eta \tilde{p} \cdot (v - v_0) \tag{24.3.5}$$

$$\partial_v^2 H = \tfrac{1}{2}(v - v_0) \cdot \partial_v^2 \tilde{p}, \tag{24.3.6}$$

$$\partial_p K = \tfrac{1}{2}(\bar{v} - v_0 + \partial_p \bar{v} \cdot (p_0 - p)), \quad \partial_\eta K = \bar{\theta} + \tfrac{1}{2}\partial_\eta \bar{v} \cdot (p_0 - p), \tag{24.3.7}$$

$$\partial_p^2 K = \tfrac{1}{2}(p_0 - p) \cdot \partial_p^2 \bar{v}. \tag{24.3.8}$$

If $v > v_0$, then $H(v, \eta_0)$ is the area between the graph of $\tilde{p}(\cdot, \eta_0)$ and its chord through the points (v_0, p_0), $(v, \tilde{p}(v, \eta_0))$, see Fig. 24.3.1(b). Indeed, using the pressure relation, one finds that

$$H(v, \eta_0) = \int_{v_0}^{v} \Big(\frac{\tilde{p}(v, \eta_0) - \tilde{p}(v_0, \eta_0)}{v - v_0}(v' - v_0) + \tilde{p}(v_0, \eta_0)\Big)\, dv' - \int_{v_0}^{v} \tilde{p}(v', \eta_0)\, dv',$$

and the first integral is the area of the quadrangle $ABCD$, while the second integral is the area below the graph. It is intuitively clear that for $v \to \infty$ this area tends to infinity.

24.3.4 Lemma

(1) *For every* $\eta \in (a, \infty)$,

$$\lim_{v\to\infty} \partial_v H(v, \eta) = \tfrac{1}{2}p_0 \quad \text{and} \quad \lim_{p\to\infty} \partial_p K(p, \eta) = -\tfrac{1}{2}v_0; \tag{24.3.9}$$

(2) *if* $\eta < \eta_0$, *then* $H(\cdot, \eta)$ *is strictly increasing on* $(0, \infty)$ *and*

$$H(v_0, \eta) < 0, \qquad \lim_{v\to\infty} H(v, \eta) = \infty; \tag{24.3.10}$$

(3) *if* $\eta > \eta_0$, *then*

$$H(v_0, \eta) > 0, \qquad \lim_{v\to 0} H(v, \eta) = -\infty. \tag{24.3.11}$$

Note that there is not a complete symmetry between Assertions (2) and (3).

Proof (1): Note that $(24.3.9)_1$ is a consequence of $(24.3.5)_1$ and the following assertions:

$$\lim_{v\to\infty} \partial_v \tilde{p}(v, \eta) = 0, \qquad \lim_{v\to\infty} v\partial_v \tilde{p}(v, \eta) = 0. \tag{24.3.12}$$

Equation $(24.3.12)_1$ is an obvious consequence of $(24.3.12)_2$. To prove the latter, observe that since $\tilde{p}$ is convex in v, we have $\tilde{p}(\tfrac{1}{2}v, \eta) - \tilde{p}(v, \eta) \ge \partial_v \tilde{p}(v, \eta)(\tfrac{1}{2}v - v) = -\tfrac{1}{2}v\partial_v\tilde{p}(v, \eta)$. Since $-\tfrac{1}{2}v\partial_v\tilde{p}(v, \eta) \ge 0$ by $(24.3.1)_2$ and $\tilde{p}(\tfrac{1}{2}v, \eta) - \tilde{p}(v, \eta) \to 0$ as $v \to \infty$ by Assumption 24.3.1(3), $(24.3.12)_2$ follows. Equation $(24.3.9)_2$ is proved analogously. (2): Let $\eta < \eta_0$. Since $\tilde{p}(\cdot, \eta)$ is convex, we have $\tilde{p}(v_0, \eta) \ge \tilde{p}(v, \eta) + \partial_v\tilde{p}(v, \eta)(v_0 - v)$ and combining this with $(24.3.5)_1$ we obtain $\partial_v H(v, \eta) \ge \tfrac{1}{2}(\tilde{p}_0 - \tilde{p}(v_0, \eta))$. As by $\partial_\eta \tilde{p} > 0$ the function $\tilde{p}(v_0, \cdot)$ is increasing, it follows that

$$\partial_v H(v, \eta) > 0. \tag{24.3.13}$$

Hence $H(\cdot, \eta)$ is strictly increasing. To prove $(24.3.10)_1$, it suffices to note that $H(v_0, \eta) = \tilde{e}(v_0, \eta) - e_0$ and that $\tilde{e}(v_0, \eta)$ is an increasing function of η. To prove $(24.3.10)_2$, notice that by (24.3.6) and $(24.3.1)_3$ the second derivative of $H(\cdot, \eta)$ is positive on (v_0, ∞) and hence $H(\cdot, \eta)$ is convex on (v_0, ∞). Therefore, if $v > v_1 > v_0$, then

$$H(v, \eta) \ge H(v_1, \eta) + \partial_v H(v_1, \eta)(v - v_1). \tag{24.3.14}$$

By $(24.3.9)_1$, $\partial_v H(v_1, \eta) > \tfrac{1}{4}p_0 > 0$ if v_1 is sufficiently large, and (24.3.14) then gives $H(v, \eta) \ge H(v_1, \eta) + \tfrac{1}{4}p_0(v - v_1)$. Fixing v_1 and letting $v \to \infty$ we obtain the result from $p_0 > 0$. (3): Let $\eta > \eta_0$. Equation $(24.3.11)_1$ follows again from the increasing character of $\tilde{e}(v_0, \cdot)$. Equation $(24.3.11)_2$ is equivalent to

$$\lim_{p\to\infty} K(p, \eta) = -\infty, \tag{24.3.15}$$

since we have (24.3.4) and at constant η, $v \to 0$ if and only if $p \to \infty$. By (24.3.8) and $(24.3.2)_3$ the second derivative of $K(\cdot, \eta)$ is negative and hence $K(\cdot, \eta)$ is concave on (p_0, ∞). Therefore, if $p > p_1 > p_0$, then $K(p, \eta) \le K(p_1, \eta) + \partial_p K(p_1, \eta)(p - p_1)$. If p_1 is large enough, then $\partial_p K(p_1, \eta) < -\tfrac{1}{4}v_0 < 0$ by $(24.3.9)_2$ and we obtain the inequality $K(p, \eta) \le K(p_1, \eta) - \tfrac{1}{4}v_0(p - p_1)$. Letting $p \to \infty$ gives $(24.3.11)_2$. □

24.3.5 Proposition *For every* $\eta \in (a, \infty)$ *there exists exactly one* $v = \hat{v}(\eta) > 0$ *such that* $(\hat{v}(\eta), \eta) \in A(v_0, \eta_0)$. *Moreover,*

$$\hat{v}(\eta) > v_0 \ \text{ if } \ \eta < \eta_0 \ \text{ and } \ \hat{v}(\eta) < v_0 \ \text{ if } \ \eta > \eta_0. \tag{24.3.16}$$

Proof Since for every η the function $H(\cdot, \eta)$ is continuous, the existence of at least one v satisfying $H(v, \eta) = 0$ follows from Lemma 24.3.4(2),(3). To prove the uniqueness of v and (24.3.16), we distinguish the cases $\eta < \eta_0$ and $\eta > \eta_0$.

Case 1: $\eta < \eta_0$. In this case $H(\cdot, \eta)$ is strictly increasing by Lemma 24.3.4(2); this immediately implies the uniqueness of the solution of $H(v, \eta) = 0$. Also, again since $H(\cdot, \eta)$ is strictly increasing and $H(v_0, \eta) < 0$ by Lemma 24.3.4(2), we have $v > v_0$, i.e., the first assertion of (24.3.16). This completes the proof of uniqueness and of (24.3.16) in this case.

Case 2: $\eta > \eta_0$. Let (v, η) solve $H(v, \eta) = 0$, i.e., $(v, \eta) \in A(v_0, \eta_0)$. Then $(v_0, \eta_0) \in A(v, \eta)$ and applying the already proved *Case 1* to the adiabat $A(v, \eta)$, we obtain $v < v_0$. Hence, in the present case, any solution of $H(v, \eta) = 0$ is in the interval $(0, v_0)$. Let us now prove the uniqueness of this solution. In this case the function $H(\cdot, \eta)$ is not strictly increasing, and the idea is to use the concavity of $H(\cdot, \eta)$. Hence, let us first check that $H(\cdot, \eta)$ is strictly concave on $(0, v_0)$. The second derivative $\partial_v^2 H$ is given by (24.3.6) and it is negative for $v < v_0$ by $(24.3.1)_3$. Suppose now that there are two roots v_1, v_2, $0 < v_1 < v_2 < v_0$, of $H(v, \eta) = 0$,

$$H(v_1, \eta) = 0, \quad H(v_2, \eta) = 0. \tag{24.3.17}$$

From the concavity of $H(\cdot, \eta)$ it then follows that $H(v, \eta) \geq 0$ for each $v \in (v_1, v_2)$. From this and $(24.3.17)_2$ then

$$\partial_v H(v_2, \eta) \leq 0. \tag{24.3.18}$$

Now $\partial_v^2 H < 0$ on $(0, v_0)$ and hence the first derivative $\partial_v H$ is a decreasing function on $(0, v_0)$ which in conjunction with (24.3.18) implies that $\partial_v H \leq 0$ on (v_2, v_0). This in turn implies that $H(\cdot, \eta)$ is decreasing on (v_2, v_0) and combining this with $(24.3.17)_2$ we obtain $H(v_0, \eta) \leq 0$, in contradiction with $(24.3.11)_1$. This contradiction proves the uniqueness. This completes the proof of the uniqueness and of (24.3.16) in this case. Note also for future use that the strict concavity of $H(\cdot, \eta)$ and Lemma 24.3.4(3) give that in *Case 2*,

$$\partial_v H(\hat{v}(\eta), \eta) > 0. \quad \square \tag{24.3.19}$$

24.3.6 Proposition *$\hat{v}$ is continuous on $(0, \infty)$ and continuously differentiable on (a, ∞) except at η_0; its derivative $\hat{v}'$ satisfies*

$$(\tilde{p}(v_0, \eta_0) - \tilde{p} - (v_0 - \hat{v})\partial_v \tilde{p}) \cdot \hat{v}' + 2\tilde{\theta} + \partial_\eta \tilde{p} \cdot (\hat{v} - v_0) = 0 \tag{24.3.20}$$

and

$$\lim_{\eta \to \eta_0} \frac{d\hat{v}(\eta)}{d\eta} = -\infty. \tag{24.3.21}$$

Moreover, $\hat{v}$ is strictly decreasing on $(\eta_0 - \varepsilon, \infty)$ and the function $\eta \mapsto \hat{p}(\eta) := \tilde{p}(\hat{v}(\eta), \eta)$ is strictly increasing on $(\eta_0 - \varepsilon, \infty)$ for some $\varepsilon > 0$.

Proof Granted the differentiability of $\hat{v}$, equation (24.3.20) is obtained by differentiating the equation $H(\hat{v}(\eta), \eta) = 0$. Now the expression in front of $\hat{v}'$ in that formula is exactly $\partial_v H(\hat{v}(\eta), \eta)$. That derivative is positive for every $\eta \neq \eta_0$: In the case $\eta < \eta_0$, it suffices to consult (24.3.13). In the case $\eta > \eta_0$, see (24.3.19). By the positivity of that coefficient, the implicit function theorem implies the differentiability of $\hat{v}(\eta)$ at $\eta \neq \eta_0$ and justifies the formal calculation leading to (24.3.20). The decreasing character of $\hat{v}$ on $(\eta_0 - \varepsilon, \infty)$ follows from (24.3.20) and from the fact that the coefficient in front of $\hat{v}'$ in that equation is positive everywhere along the adiabat except at η_0. This was proved in the proof of Lemma 24.3.5. On the other hand, the absolute term $2\tilde{\theta} + \partial_\eta \tilde{p}(\hat{v} - v_0)$ in that equation is positive for v sufficiently close to v_0 in view of the positivity of the absolute temperature and for all $v > v_0$ in view of $(24.3.1)_3$. The negativity of $\hat{v}'$ then follows. A limit gives (24.3.21). The increasing character of $\hat{p}$ follows by applying the chain rule to $\hat{p} = \tilde{p}(\hat{v}(\eta), \eta)$ and invoking the known signs of the resulting terms. $\square$

Figure 24.3.2 depicts a graph of a Hugoniot adiabat consistent with the above proposition.

24.3.7 Proposition *If $(v, \eta) \in A(v_0, \eta_0)$, $v \neq v_0$, then $(p_0 - \tilde{p}(v, \eta))/(v - v_0) > 0$. Hence (24.2.7) gives a well-defined positive-valued function J; the function $\eta \mapsto J(\hat{v}(\eta), \eta, v_0, \eta_0)$ is increasing and continuous on (a, ∞) and continuously differentiable everywhere on (a, ∞) except at η_0.*

Proof If $(v, \eta) \in A(v_0, \eta_0)$ and $v < v_0$, then by (24.3.16) $\eta > \eta_0$ and hence $\tilde{p}(v, \eta) > \tilde{p}(v, \eta_0) > \tilde{p}(v_0, \eta_0)$ since $\tilde{p}$ is increasing in η and decreasing in v. It follows that the fraction is positive. If $(v, \eta) \in A(v_0, \eta_0)$ and $v > v_0$, then by (24.3.16) $\eta < \eta_0$ and hence $\tilde{p}(v, \eta) < \tilde{p}(v, \eta_0) < \tilde{p}(v_0, \eta_0)$; the positivity of the fraction follows again. The continuity of J is clear everywhere except at $(v, \eta) = (v_0, \eta_0)$. The continuity at (v_0, η_0) is proved as follows. From (24.3.21) and the mean value theorem follows that

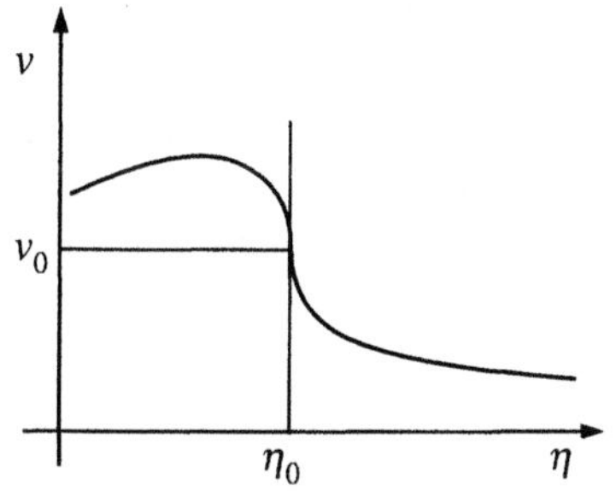

Fig. 24.3.2

$$\lim_{\eta\to\eta_0} \frac{\hat{v}(\eta) - v_0}{\eta - \eta_0} = -\infty \text{ and hence } \lim_{\eta\to\eta_0} \frac{\eta - \eta_0}{\hat{v}(\eta) - v_0} = 0. \tag{24.3.22}$$

Applying the Taylor expansion with remainder to $\tilde{p}$ at (v_0, η_0) we find that

$$\tilde{p}(v, \eta) - p_0 = \partial_v \tilde{p}(v_0, \eta_0)(v - v_0) + \partial_\eta \tilde{p}(v_0, \eta_0) + o(v - v_0, \eta - \eta_0) \tag{24.3.23}$$

where $o(v - v_0, \eta - \eta_0)/(|v - v_0| + |\eta - \eta_0|) \to 0$ as $(v, \eta) \to (v_0, \eta_0)$. Setting $v = \hat{v}(\eta)$ in (24.3.23), dividing by $\hat{v}(\eta) - v_0$, letting $\eta \to \eta_0$ and using $(24.3.22)_2$ we obtain

$$\lim_{\eta\to\eta_0} \frac{\tilde{p}(\hat{v}(\eta), \eta) - p_0}{\hat{v}(\eta) - v_0} = \partial_v \tilde{p}(v_0, \eta_0)$$

from which the continuity follows immediately. The differentiability and the increasing character of the function $\eta \mapsto J(\hat{v}(\eta), \eta, v_0, \eta_0)$ is equivalent to proving the same properties for its square. The differentiability follows from the differentiability of $\hat{v}$. Further, omitting the obvious arguments of the functions, we have

$$(\hat{v} - v_0)J^2 = \tilde{p}(v_0, \eta_0) - \tilde{p}, \quad (\hat{v} - v_0)\frac{dJ^2}{d\eta} + J^2\hat{v}' + \partial_v \tilde{p} \cdot \hat{v}' + \partial_\eta \tilde{p} = 0. \tag{24.3.24}$$

Multiplying $(24.3.24)_2$ by $\hat{v} - v_0$ and eliminating J^2 by $(24.3.24)_1$ we obtain

$$(\hat{v} - v_0)^2 \frac{dJ^2}{d\eta} + (\tilde{p}(v_0, \eta_0) - \tilde{p} + (\hat{v} - v_0)\partial_v \tilde{p})\hat{v}' + (\hat{v} - v_0)\partial_\eta \tilde{p} = 0$$

and a combination with (24.3.20) gives $(\hat{v} - v_0)^2 dJ^2/d\eta - 2\tilde{\theta} - 0$. Hence the derivative is positive. □

24.3.8 *Remark* By Proposition 24.3.6 one can also express the entropy as function of the specific volume near the point (v_0, η_0), i.e., $\eta = \hat{\eta}(v)$, where $\hat{\eta}$ is the inverse of $\hat{v}$ near the point η_0 defined also by the requirement that

$$H(v, \hat{\eta}(v)) = 0 \tag{24.3.25}$$

for every $v \in (v_0 - \delta, v_0 + \delta)$, where $\delta > 0$. If $\tilde{e}$ is of class C^5, then $\hat{\eta}$ is of class C^3. A differentiation of (24.3.25) gives

$$\partial_v H + \partial_\eta H \cdot \hat{\eta}' = 0, \quad \partial_v^2 H + 2\partial_v \partial_\eta H \cdot \hat{\eta}' + \partial_\eta^2 H \cdot (\hat{\eta}')^2 + \partial_\eta H \cdot \hat{\eta}'' = 0,$$

where the prime denotes differentiation with respect to v. At $v = v_0$ these equations give $\hat{\eta}'(v_0) = 0$, $\hat{\eta}''(v_0) = 0$, and one more differentiation with respect to v at $v = v_0$ gives $\partial_v^3 H + \partial_\eta H \cdot \hat{\eta}''' = 0$. One finds that $\partial_v H = \frac{1}{2}(p_0 - \tilde{p} + \partial_v \tilde{p}(v - v_0))$, $\partial_v^2 H = \frac{1}{2}\partial_v^2 \tilde{p}(v - v_0)$ and at (v_0, η_0), $\partial_v^3 H = \frac{1}{2}\partial_v^2 \tilde{p}$, $\partial_\eta H = \tilde{\theta}$. Using this, we can summarize that

$$\hat{\eta}'(v_0) = 0, \quad \hat{\eta}''(v_0) = 0, \quad \hat{\eta}'''(v_0) = -\tfrac{1}{2}\partial_v^2 \tilde{p}(v_0, \eta_0)/\tilde{\theta}(v_0, \eta_0) < 0.$$

From this one concludes that *the jump in entropy is of the third order of magnitude of the jump in specific volume, in agreement with the general result Proposition* 23.7.2.

24.4 The Equivalence of the Admissibility Criteria

In this section we restate the Lax's and Liu's admissibility criteria of Sect. 23.8 in the variables appropriate for fluids and establish their equivalence. The existence of the shock layer will be considered separately in the following section. Denote, for every $\boldsymbol{p} = (v, \boldsymbol{v}, \eta)$, $\boldsymbol{m} \in \mathrm{Sph}$, by $r(\boldsymbol{p}, \boldsymbol{m})$ the spatial speed of sound given by

$$r(\boldsymbol{p}, \boldsymbol{m}) = \boldsymbol{v} \cdot \boldsymbol{m} + c(v, \eta), \tag{24.4.1}$$

where $c(v, \eta) = \sqrt{-v^2 \partial_v \tilde{p}(v, \eta)}$ is the speed of sound relative to the moving material points of the fluid, cf. (24.1.6). Recall our convention to call shocks the discontinuities with $j > 0$. Let $\boldsymbol{p}^{\pm} = (v^{\pm}, \boldsymbol{v}^{\pm}, \eta^{\pm})$ and let $(\boldsymbol{p}^+, \boldsymbol{p}^-, \boldsymbol{m})$ be a shock of speed u. Consider the Hugoniot curve $H^+(\boldsymbol{p}^-, \boldsymbol{m})$ at $(\boldsymbol{p}^-, \boldsymbol{m})$, see Proposition 24.2.3. Then for any $\boldsymbol{q} = (w, \boldsymbol{w}, \zeta) \in H^+(\boldsymbol{p}^-, \boldsymbol{m})$ the speed of the shock $(\boldsymbol{q}, \boldsymbol{p}^-, \boldsymbol{m})$ is

$$\hat{u}_{\boldsymbol{p}^-, \boldsymbol{m}}(\boldsymbol{q}) = \boldsymbol{v}^- \cdot \boldsymbol{m} + v^- J(w, \zeta, v^-, \eta^-) \tag{24.4.2}$$

and in particular the speed of the shock $(\boldsymbol{p}^+, \boldsymbol{p}^-, \boldsymbol{m})$ is

$$u = \hat{u}_{\boldsymbol{p}^-, \boldsymbol{m}}(\boldsymbol{p}^+) = \boldsymbol{v}^- \cdot \boldsymbol{m} + v^- J(v^+, \eta^+, v^-, \eta^-), \tag{24.4.3}$$

cf. $(24.2.2)_1$.

The following result gives three conditions equivalent to the entropy criterion for shocks of arbitrary magnitude.

24.4.1 Proposition *If Assumptions* 24.3.1 *hold and* $(\boldsymbol{p}^+, \boldsymbol{p}^-, \boldsymbol{m})$ *is a shock of speed* u, *then the following conditions are equivalent:*
(1) $\eta^- > \eta^+$;
(2) $v^- < v^+$;
(3) $r(\boldsymbol{p}^+, \boldsymbol{m}) < u$;
(4) $u < r(\boldsymbol{p}^-, \boldsymbol{m})$.

In particular, under the Assumptions 24.3.1 only the compression shocks are admissible. Another equivalent condition is that the speed of the shock be supersonic relative to the state in front of the shock and subsonic relative to the state behind the shock, $r(\boldsymbol{p}^+, \boldsymbol{m}) < u < r(\boldsymbol{p}^-, \boldsymbol{m})$, the Lax–Riemann shock inequalities, cf. Sect. 23.8.

Proof (1) $\Leftrightarrow$ (2) : By Proposition 24.3.5 $\eta^- > \eta^+$ is equivalent to $v^+ > v^-$. (1) $\Leftrightarrow$ (3) : By Proposition 24.3.7 the flux $J(v, \eta, v^+, \eta^+)$ is an increasing function of η on $A(v^+, \eta^+)$. Hence $\eta^- > \eta^+$ is equivalent to $J(v^-, \eta^-, v^+, \eta^+) > J(v^+, \eta^+, v^+, \eta^+) = \sqrt{-\partial_v \tilde{p}(v^+, \eta^+)}$ (see (24.2.7)) which is equivalent to $v^+ J(v^-, \eta^-, v^+, \eta^+) > c(v^+, \eta^+)$. By (24.4.1) and (24.4.3) this is equivalent to the inequality in (3). The proof of (1) $\Leftrightarrow$ (4) is similar. □

24.4.2 Liu's Criterion The formulation of Liu's criterion depends on the fact that the Hugoniot set is an open curve, so that the concept of "between" points on the Hugoniot set is meaningful. The shock $(\boldsymbol{p}^+, \boldsymbol{p}^-, \boldsymbol{m})$ is said to satisfy Liu's criterion if its speed u satisfies $u \le \hat{u}_{\boldsymbol{p}^-, \boldsymbol{m}}(\boldsymbol{q})$ for every point $\boldsymbol{q}$ on $H^+(\boldsymbol{p}^-, \boldsymbol{m})$ between $\boldsymbol{p}^-$ and $\boldsymbol{p}^+$.

24.4.3 Proposition *If Assumptions* 24.3.1 *hold, then the entropy and Liu's criteria are equivalent.*

Proof Let the shock satisfy Liu's criterion. Let the point $\boldsymbol{q}$ occurring in the formulation of the criterion converge to the point $\boldsymbol{p}^-$ along the Hugoniot set $H^+(\boldsymbol{p}^-, \boldsymbol{m})$. Then $\hat{u}_{\boldsymbol{p}^-, \boldsymbol{m}}(\boldsymbol{q})$ converges to $\boldsymbol{v}^- \cdot \boldsymbol{m} + c(v^-, \eta^-) = r(\boldsymbol{p}^-, \boldsymbol{m})$. The criterion gives $u \le r(\boldsymbol{p}^-, \boldsymbol{m})$, i.e., that Condition (4) of

Proposition 24.4.1 holds. Hence also the entropy criterion holds. Conversely, suppose that the entropy criterion holds. Then $\eta^+ < \eta^-$. Let $q = (w, \mathbf{w}, \zeta)$ be a point on $H^+(p^-, m)$ between p^- and p^+. Then $(v, \eta) \in A(v^-, \eta^-)$, and as the points of the Hugoniot adiabat are in one-to-one correspondence with the entropies (Proposition 24.3.5), one has $\eta \in [\eta^+, \eta^-]$ because the point q is between p^+ and p^-. Moreover, as $J(w, \zeta, v^-, \eta^-)$ is an increasing function of entropy along the Hugoniot adiabat (Proposition 24.3.7), we have $J(v^+, \eta^+, v^-, \eta^-) \le J(w, \zeta, v^-, \eta^-)$, which implies $\mathbf{v}^- \cdot \mathbf{m} + v^- J(v^+, \eta^+, v^-, \eta^-) \le \mathbf{v}^- \cdot \mathbf{m} + v^- J(w, \zeta, v^-, \eta^-)$. By (24.4.2) this can be rewritten as $u = \hat{u}_{p^-,m}(p^+) \le \hat{u}_{p^-,m}(q)$. This is the inequality of Liu's criterion. □

Liu's criterion is stronger than the entropy and the Lax criteria if the fluid violates Assumption 24.3.1, i.e., when essentially the genuine nonlinearity fails. In this situation, the entropy and the Lax criteria do not to secure the uniqueness of the Riemann problem, while Liu's criterion does, at least under additional assumptions. See T.-P. Liu [1975, 1981]. For basic information about the systems of conservation laws that violate the genuine nonlinearity, see Wendroff [1972], Rozhdestvenskii & Janenko [1978], and the references therein.

24.5 Shock Layers in Fluids

In this section we shall complete the discussion of the admissibility criteria by showing that under natural additional conditions, every shock satisfying the entropy criterion can be visualized as a limit of a structured shock layer in a viscous and heat conducting fluid when the viscosity and conductivity of heat tend to zero.

24.5.1 Shock Layers in a Viscous Fluid with Heat Conduction Assume that the adiabatic fluid admits the temperature as independent variable and let $f = \bar{f}(v, \theta)$ be its free energy defined on $\bar{U}$, the image of $\tilde{U} = (0, +\infty) \times (a, \infty)$ under the passage from η to θ. We shall now consider a fluid with viscosity and heat conduction given by the constitutive equations

$$\mathbf{T} = -\bar{p}(v, \theta)\mathbf{1} + \mathbf{T}_d, \quad \mathbf{T}_d = \lambda(v, \theta)(\operatorname{tr}\mathbf{D})\mathbf{1} + 2\mu(v, \theta)\mathbf{D},$$
$$\mathbf{h} = -\kappa(v, \theta)\operatorname{grad}\theta,$$

and the rest of the response coinciding with that of the original adiabatic fluid. Here λ, μ are the coefficients of viscosity and κ the coefficient of heat conduction. The second law requires that $3\lambda + 2\mu \ge 0$, $\mu \ge 0$, $\kappa \ge 0$ and hence $\lambda + 2\mu \ge 0$. Here we assume

$$\bar{\mu} := \lambda + 2\mu > 0, \quad \kappa > 0, \tag{24.5.1}$$

and that $\bar{\mu}$, κ are continuously differentiable functions of (v, θ) on $\bar{U}$.

We seek solutions of the form

$$\phi = \phi(\mathbf{x} \cdot \mathbf{m} - ut) \tag{24.5.2}$$

$\mathbf{x} \in \text{Vect}$, $t \in \mathbb{R}$, where $\mathbf{m}$ is a constant unit vector, u a constant number and ϕ stands for ρ, $\mathbf{v}$, p, θ, $\mathbf{T}_d$, e, $\mathbf{h}$. The balance equations and the Clausius–Duhem inequality read

$$\begin{aligned}
&(-u\rho + \rho\mathbf{v} \cdot \mathbf{m})' = 0,\\
&(-u\rho\mathbf{v} + \rho\mathbf{v}(\mathbf{v} \cdot \mathbf{m}) + p\mathbf{m} - \mathbf{T}_d\mathbf{m})' = \mathbf{0},\\
&(-u\rho(e + \tfrac{1}{2}\mathbf{v}^2) + \rho(e + \tfrac{1}{2}\mathbf{v}^2)\mathbf{v} \cdot \mathbf{m} + p\mathbf{v} \cdot \mathbf{m} - (\mathbf{T}_d\mathbf{m}) \cdot \mathbf{v} + \mathbf{h} \cdot \mathbf{m})' = 0,\\
&(-u\rho\eta + \rho\eta\mathbf{v} \cdot \mathbf{m} + \mathbf{h} \cdot \mathbf{m}/\theta)' \ge 0,
\end{aligned} \tag{24.5.3}$$

where the prime denotes differentiation with respect to $s = \boldsymbol{x} \cdot \boldsymbol{m} - ut$. We have $\boldsymbol{D} = \frac{1}{2}(\boldsymbol{v}' \otimes \boldsymbol{m} + \boldsymbol{m} \otimes \boldsymbol{v}')$, $\operatorname{grad}\theta = \theta'\boldsymbol{m}$ and the constitutive equations give

$$\boldsymbol{T}_d\boldsymbol{m} = \bar{\mu}(\boldsymbol{v}' \cdot \boldsymbol{m})\boldsymbol{m}, \quad \boldsymbol{h} \cdot \boldsymbol{m} = -\kappa\theta'.$$

It is convenient to use the specific volume $v = 1/\rho$ rather than the density. Let $\boldsymbol{p}^\pm = (v^\pm, \boldsymbol{v}^\pm, \eta^\pm)$ be given and consider a shock $(\boldsymbol{p}^+, \boldsymbol{p}^-, \boldsymbol{m})$ with speed u; denote $\theta^\pm = \tilde{\theta}(v^\pm, \eta^\pm)$. A ***shock layer*** connecting the states of the shock $(\boldsymbol{p}^+, \boldsymbol{p}^-, \boldsymbol{m})$ is any solution $(\boldsymbol{v}(\cdot), v(\cdot), \theta(\cdot))$ of the form (24.5.2) such that

$$(\boldsymbol{v}(s), v(s), \theta(s)) \to (\boldsymbol{v}^\pm, v^\pm, \theta^\pm), \quad (\boldsymbol{v}'(s), v'(s), \theta'(s)) \to (\boldsymbol{0}, 0, 0) \text{ as } s \to \pm\infty.$$

24.5.2 Assumptions

(1) $\tilde{c}_v > 0$ on $\tilde{U}$;

(2) $\bar{\mu}$, κ are continuously differentiable and (24.5.1) hold on $\tilde{U}$;

(3) the set

$$U^\sharp = \{(v, \tilde{p}(v, \eta)) : (v, \eta) \in \tilde{U}\} \tag{24.5.4}$$

is convex.

24.5.3 Proposition *If Assumptions* 24.3.1 *and* 24.5.2 *hold, then a shock* $(\boldsymbol{p}^+, \boldsymbol{p}^-, \boldsymbol{m})$ *satisfies the entropy criterion if and only if there exists a shock layer connecting the states of the jump. Moreover, the temperature is decreasing and the specific volume is increasing along the shock layer in the direction of increasing* s.

Proof Suppose that there is a shock layer connecting the two states. The integration of the balance equations and of the Clausius–Duhem inequality gives the jump conditions for the shock and the condition that

$$-u\rho^-\eta^- + \rho^-\eta^-\boldsymbol{v}^- \cdot \boldsymbol{m} \le -u\rho^+\eta^+ + \rho^+\eta^+\boldsymbol{v}^+ \cdot \boldsymbol{m}.$$

The latter says that the shock satisfies the entropy criterion. □

A number of preliminary results are needed to prove the converse implication in the proposition. Throughout the rest of the section, suppose that Assumptions 24.3.1 and 24.5.2 hold and consider a shock $(v^+, \boldsymbol{v}^+, \eta^+; v^-, \boldsymbol{v}^-, \eta^-, \boldsymbol{m})$ satisfying the entropy criterion: $\eta^- > \eta^+$. Our goal is to find the shock layer connecting the states of the shock. We first use the Galilean invariance to simplify the equations.

24.5.4 Shock Layer Equations By (24.2.4), the tangential component of velocity is continuous across the singular surface, i.e., $\boldsymbol{v}^+ - (\boldsymbol{v}^+ \cdot \boldsymbol{m})\boldsymbol{m} = \boldsymbol{v}^- - (\boldsymbol{v}^- \cdot \boldsymbol{m})\boldsymbol{m}$. Denoting this common value by $\boldsymbol{c}$, we define $\boldsymbol{w} = \boldsymbol{v}^+ - \boldsymbol{c} - u\boldsymbol{m}$ and arrive at the equations

$$\begin{aligned} (\rho\boldsymbol{w} \cdot \boldsymbol{m})' &= 0, \\ (\rho\boldsymbol{w}(\boldsymbol{w} \cdot \boldsymbol{m}) + p\boldsymbol{m} - \boldsymbol{T}_d\boldsymbol{m})' &= \boldsymbol{0}, \\ (\rho(e + \tfrac{1}{2}w^2)\boldsymbol{w} \cdot \boldsymbol{m} + p\boldsymbol{w} \cdot \boldsymbol{m} - \boldsymbol{T}_d\boldsymbol{m} \cdot \boldsymbol{w} + \boldsymbol{h} \cdot \boldsymbol{m})' &= 0, \\ (\rho\eta\boldsymbol{w} \cdot \boldsymbol{m} + \boldsymbol{h} \cdot \boldsymbol{m}/\theta)' &\ge 0, \end{aligned} \tag{24.5.5}$$

with the constitutive equations

$$\boldsymbol{T}_d\boldsymbol{m} = \bar{\mu}(\boldsymbol{w}' \cdot \boldsymbol{m})\boldsymbol{m}, \quad \boldsymbol{h} \cdot \boldsymbol{m} = -\kappa\theta'. \tag{24.5.6}$$

Furthermore, the limits $\boldsymbol{w}^\pm$ of $\boldsymbol{w}$ at $\pm\infty$ are easily found to satisfy $\boldsymbol{w}^\pm = -jv^\pm\boldsymbol{m}$, where j is the flux of mass through the singular surface, see (24.2.2). Integrating $(24.5.5)_2$ from $-\infty$ to s and using that $\boldsymbol{w}^-$, $\boldsymbol{T}_d\boldsymbol{m}$ are parallel with $\boldsymbol{m}$, (see $(24.5.6)_1$) we conclude that $\boldsymbol{w}$ is parallel with $\boldsymbol{m}$ for all s, i.e., $\boldsymbol{w} = w\boldsymbol{m}$ where w is a scalar-valued function of s. In terms of w, equations (24.5.5) read

$$\begin{aligned}(\rho w)' &= 0,\\ (\rho w^2 + p - \bar{\mu} w')' &= 0,\\ \big(\rho(e + \tfrac{1}{2}w^2)w + pw - \bar{\mu} w w' - \kappa\theta'\big)' &= 0,\end{aligned} \tag{24.5.7}$$

where the constitutive equations have also been used. From the first equation follows that $m := -\rho w$ is constant and we also have $w = -mv$, where v is the specific volume. At $\pm\infty$, $w^{\pm} = -mv^{\pm}$ and a comparison with $w^{\pm} = -jv^{\pm}\boldsymbol{m}$ gives $m = j$. Hence

$$m = \sqrt{(p^- - p^+)/(v^- - v^+)}.$$

We assume throughout that m is so chosen. An elimination of w in (24.5.7) via $w = -mv$ and an integration of the resulting equations from $-\infty$ to s gives

$$\begin{aligned} m^2 v + p + \bar{\mu} v' &= m^2 v^- + p^-,\\ m(e + \tfrac{1}{2}m^2 v^2) + mpv + \bar{\mu} v' v m^2 + \kappa\theta' &= m(e^- + \tfrac{1}{2}m^2 v^{-2}) + mp^- v^-.\end{aligned}$$

Finally, one can eliminate the term $\bar{\mu} v'$ in the second equation by using the first equation. A rearrangement gives the ***shock layer equations***

$$v' = \bar{\mu}(v,\theta)^{-1} M(v,\theta), \quad \theta' = \kappa(v,\theta)^{-1} L(v,\theta), \tag{24.5.8}$$

where

$$\begin{aligned} M(v,\theta) &= m^{-1}(m^2(v^{\pm} - v) + p^{\pm} - \bar{p}(v,\theta)),\\ L(v,\theta) &= m(e^{\pm} - \bar{e}(v,\theta) + \tfrac{1}{2}m^2(v^{\pm} - v)^2 + p^{\pm}(v^{\pm} - v)).\end{aligned} \tag{24.5.9}$$

Here, the equations with the + superscripts give the same L, M as those with the − superscripts, since $e^+ - e^- + \frac{1}{2}(p^+ + p^-)(v^+ - v^-) = 0$ and $m^2(v^+ - v^-) + p^+ - p^- = 0$. Also note that

$$L(v,\theta) = m(e^{\pm} - \bar{e}(v,\theta) + \tfrac{1}{2}(\bar{p}(v,\theta) + p^{\pm})(v^{\pm} - v) - \tfrac{1}{2}mM(v,\theta)(v - v^{\pm})). \tag{24.5.10}$$

The system (24.5.8) can be written as $(v', \theta') = (X, Y)$, where $(X, Y) = (\bar{\mu}^{-1}M, \kappa^{-1}L)$ is viewed as a vector field on $\bar{U}$. We are looking for the orbit (v, θ), defined on $\mathbb{R}$, such that $(v(s), \theta(s)) \to (v^{\pm}, \theta^{\pm})$ as $s \to \pm\infty$. The idea of the proof is as follows. We shall see that the points $(v^{\pm}, \theta^{\pm})$ are the only critical points of (X, Y), i.e., the only points where (X, Y) vanishes. The solution (v, θ) is an orbit connecting these equilibrium points. The linearization of (X, Y) at $(v^{\pm}, \theta^{\pm})$ will show that, thanks to the entropy criterion, the point (v^+, θ^+) is a stable point of (X, Y) while (v^-, θ^-) is a saddle point. The former means that all solutions starting in some neighborhood of (v^+, θ^+) are defined for all positive times and approach (v^+, θ^+) as the time $s \to \infty$; the latter means that near (v^-, θ^-) there are stable and unstable one-dimensional manifolds consisting of points which are attracted or repelled. Any solution starting at the unstable manifold can be defined for all negative times s, and as $s \to -\infty$, this solution tends to (v^-, θ^-). This unstable manifold is divided into two connected arcs by the point (v^-, θ^-). We shall find a positively invariant region $R \subset \bar{U}$ such that R contains exactly one of the two arcs of the unstable manifold at (v^-, θ^-) and such that (v^+, θ^+) is a boundary point of R. Then any solution which starts at any point of this arc is defined for all s negative (by the properties of the unstable manifold) and approaches (v^-, θ^-) as $s \to -\infty$. By the properties of the positively invariant region, this solution can be defined for all positive s and must approach (v^+, θ^+) as $s \to \infty$ since (v^+, θ^+) is the only stable point. We now turn to details.

24.5.5 Proposition

(1) *We have* $L(v^\pm, \theta^\pm) = M(v^\pm, \theta^\pm) = 0$, *and* $(v^\pm, \theta^\pm)$ *are the only points where* L, M *vanish simultaneously.*

(2) *On* $\bar{U}$,

$$\partial_v M = -m^{-1}(m^2 + \partial_v \bar{p}), \quad \partial_\theta M = -m^{-1}\partial_\theta \bar{p}, \tag{24.5.11}$$

$$\partial_v L = -m^2 M - m\theta\partial_\theta \bar{p}, \quad \partial_\theta L = -m\partial_\theta \bar{e} = -mc_v. \tag{24.5.12}$$

(3) L, M *are decreasing functions of* θ *for every constant* v.

Proof (1): The equations follow from (24.5.9). Let us now prove that the points $(v^\pm, \theta^\pm)$ are the only points where L, M vanish simultaneously. Let $L(v, \theta) = M(v, \theta) = 0$ and denote $\eta = \bar{\eta}(v, \theta)$. The identity (24.5.10) with the minus sign gives

$$e^- - \bar{e}(v, \theta) + \tfrac{1}{2}(\bar{p}(v, \theta) + p^-)(v^- - v) = 0, \tag{24.5.13}$$

which shows that (v, η) is on the Hugoniot adiabat A^- centered at (v^-, η^-). If $v = v^-$, (24.5.13) gives $\bar{e}(v^-, \theta^-) = \bar{e}(v^-, \theta)$ and since $\bar{e}(v, \cdot)$ is an increasing function, this implies $\theta = \theta^-$, i.e., $(v, \theta) = (v^-, \theta^-)$. If $v \neq v^-$, then by $(24.5.9)_1$ with the $-$ sign the equation $M(v, \theta) = 0$ gives $m = J(v, \eta, v^-, \eta^-)$. But the function $J(v, \eta, v^-, \eta^-)$ is monotonically increasing along A^- and the equation $M(v^+, \theta^+) = 0$ shows that the value m of $J(v, \eta, v^-, \eta^-)$ is already taken by (v^+, η^+). From this $(v, \eta) = (v^+, \eta^+)$. (2): Let us prove $(24.5.12)_1$, the rest is immediate. We have $\partial_v L = m(-\partial_v \bar{e} - p^- - m^2(v^- - v))$. Now we use the identity

$$\partial_v \bar{e} = \partial_v(\bar{e} - \theta\bar{\eta}) + \theta\partial_v \bar{\eta} = \partial_v \bar{f} - \theta\partial_v\partial_\theta \bar{f} = -\bar{p} + \theta\partial_\theta \bar{p}.$$

This allows us to rearrange

$$\partial_v L = m(\bar{p} - \theta\partial_\theta \bar{p} - p^- - m^2(v^- - v)) = m(-mM(v, \theta) - \theta\partial_\theta \bar{p}).$$

(3): This follows from $(24.5.11)_2$, $(24.5.12)_2$ and the fact that $c_v > 0$ and

$$\partial_\theta \bar{p} = \partial_\eta \bar{p} \cdot \partial_\theta \bar{\eta} > 0 \tag{24.5.14}$$

since $\partial_\eta \bar{p} > 0$ and $\partial_\theta \bar{\eta} = c_v/\theta > 0$. □

Let A^+ and A^- be the Hugoniot adiabats centered at (v^+, η^+) and (v^-, η^-), respectively. By Proposition 24.3.5 each of the $A^\pm$ can be parametrized by the entropy, i.e., $v = \hat{v}^\pm(\eta)$ on $A^\pm$, and moreover $v^\pm = \hat{v}^\mp(\eta^\pm)$. Since $\eta^- > \eta^+$, Proposition 24.3.5 tells us that $v^- < v^+$. By Proposition 24.3.7, the function $J(v, \eta, v^\pm, \eta^\pm)$ of (24.2.7) monotonically increases along $A^\pm$ as η increases. This fact will be used several times in the sequel. Let $B^\pm = \{(\hat{v}^\pm(\eta), \eta) : \eta \in (\eta^+, \eta^-)\}$ be the portions of $A^\pm$ between η^- and η^+ and let $C^\pm$ be the images of $B^\pm$ in the (v, θ)-space,

$$C^\pm = \{(\hat{v}^\pm(\eta), \tilde{\theta}(\hat{v}^\pm(\eta), \eta)) : \eta \in (\eta^+, \eta^-)\}.$$

24.5.6 Proposition

(1) *If* $(v, \eta) \in B^+$, *then*

$$-\partial_v \bar{p}(v^+, \eta^+) < J^2(v, \eta, v^+, \eta^+) < m^2; \tag{24.5.15}$$

if $(v, \theta) \in C^+$ *and* $v < v^+$, *then* $M(v, \theta) > 0$, $L(v, \theta) > 0$.

(2) *If* $(v, \eta) \in B^-$, *then*

$$m^2 < J^2(v, \eta, v^-, \eta^-) < -\partial_v \bar{p}(v^-, \eta^-); \tag{24.5.16}$$

if $(v, \theta) \in C^-$ *and* $v > v^-$, *then* $M(v, \theta) > 0$, $L(v, \theta) < 0$.

Consequently,

$$-\partial_v \bar{p}(v^+, \eta^+) < m^2 < -\partial_v \bar{p}(v^-, \eta^-). \tag{24.5.17}$$

Proof Inequalities (24.5.15) and (24.5.16) follow form the monotone character of the dependence of the functions $J^2(v, \eta, v^\pm, \eta^\pm)$ along $B^\pm$ and from the fact that at the endpoints of $B^\pm$ the values of

$J^2(v, \eta, v^{\pm}, \eta^{\pm})$ are exactly the bounds occurring in (24.5.15) and (24.5.16). Inequality (24.5.17) is a consequence. The inequality $M(v, \theta) > 0$ in (1) is just a restatement of $(24.5.15)_2$ in terms of (v, θ). By the identity (24.5.10) with the plus sign we have $L(v, \theta) = -\frac{1}{2}mM(v, \theta)(v - v^+)$ on C^+ and the result $L(v, \theta) > 0$ follows from $M(v, \theta) > 0$ and $v < v^+$. Inequality $M(v, \theta) > 0$ in (2) is a restatement of $(24.5.16)_1$. By the identity (24.5.10) with the minus sign we have $L(v, \theta) = -\frac{1}{2}mM(v, \theta)(v - v^-)$ on C^- and the result $L(v, \theta) < 0$ follows from $M(v, \theta) > 0$ and $v > v^-$. □

24.5.7 Proposition

(1) *For every* $v \in [v^-, v^+]$ *there exists exactly one* $m(v)$ *and exactly one* $l(v)$ *such that*

$$M(v, m(v)) = 0, \quad L(v, l(v)) = 0;$$

(2) *everywhere on* $R := \{(v, \theta) \in \bar{U} : v \in (v^-, v^+), l(v) < \theta < m(v)\}$,

$$L < 0, \quad M > 0 \tag{24.5.18}$$

and also $M(v, l(v)) > 0$, $L(v, m(v)) < 0$ *for every* $v \in (v^-, v^+)$;

(3) *for every* $v \in (v^-, v^+)$,

$$l(v) < m(v), \quad M(v, l(v)) > 0, \quad L(v, m(v)) < 0; \tag{24.5.19}$$

(4) *the functions* l *and* m *are continuously differentiable on* $[v^-, v^+]$, *and for every* $v \in [v^-, v^+]$,

$$\dot{l}(v) < 0, \tag{24.5.20}$$

where a superimposed dot denotes the derivative with respect to v.

Proof (1): Let us first prove the existence and uniqueness of $m(v)$. The points $(v^{\pm}, p^{\pm})$ belong to the set $U^{\sharp}$ defined in (24.5.4), and by Assumption 24.5.2 this set is convex. Hence the whole line segment with endpoints $(v^{\pm}, p^{\pm})$ is in $U^{\sharp}$. Returning to the variables v, θ, this means that for every $v \in [v^-, v^+]$ there exist at least one θ such that $M(v, \theta) = 0$. This proves the existence. The uniqueness follows from the fact that M is a strictly increasing function of θ for each fixed v. Let us now turn to the proof of the existence of $l(v)$. In view of Proposition 24.5.6 for every $v \in [v^-, v^+]$ there exist θ_1, θ_2 such that $L(v, \theta_1) < 0 < L(v, \theta_2)$, namely the temperatures corresponding to the adiabats $C^{\pm}$. The existence of $l(v)$ such that $L(v, l(v)) = 0$ then follows. The uniqueness follows from the fact that L is a strictly decreasing function of θ. (2): Inequality $(24.5.18)_1$ follows from the fact that the region R is situated above the curve consisting of those (v, θ) for which $L(v, \theta) = 0$ and from the fact that L is strictly decreasing in θ. Inequality $(24.5.18)_2$ follows from the fact that the region R is situated below the curve consisting of those (v, θ) for which $M(v, \theta) = 0$ and from the fact that M is strictly decreasing in θ. (3): By Proposition 24.5.6(2) we have: if $(v, \theta) \in C^-$, then $M(v, \theta) > 0$ and $L(v, \theta) < 0$. Since both L, M are strictly decreasing in θ, it follows that $m(v) > \theta > l(v)$ where θ is such that $(v, \theta) \in C^-$. This gives $(24.5.19)_1$. For the same reason, $(24.5.19)_{2,3}$ hold. (4): The differentiability is a simple application of the implicit function theorem; the details are omitted. We furthermore have

$$\dot{l}(v^-) = -\partial_v L / \partial_\theta L = -(mM(v, \theta) + \theta \partial_\theta \bar{p}) / c_v.$$

By $(24.5.19)_2$ and (24.5.14) we have $M(v, \theta) > 0$; also $\theta \partial_\theta \bar{p} > 0$ and $c_v > 0$. Inequality (24.5.20) follows. □

By (2) of the preceding proposition, on the region R the vector field (X, Y) points to the right ($M > 0$) and downwards ($L < 0$). We shall now show that no solution of the shock layer equations starting in R can ever leave it.

24.5.8 Proposition *The region* R *is positively invariant with respect to the solutions of the shock layer equations, i.e., any solution* $(v(\cdot), \theta(\cdot))$ *on* $[0, \varepsilon)$ *with initial value* $(v(0), \theta(0))$ *in* R *can be extended to a solution on* $[0, +\infty)$ *and the solution does not leave* R.

Proof The boundary of R is the union $\partial R = \{(v^+, \theta^+)\} \cup \{(v^-, \theta^-)\} \cup L \cup M$, of the critical points $(v^\pm, \theta^\pm)$ with the following two arcs:

$$M = \{(v, m(v)) : v \in (v^-, v^+)\}, \quad L = \{(v, l(v)) : v \in (v^-, v^+)\}.$$

From the definitions of l, m it follows that the vector field (X, Y) is vertical on M, namely $(X, Y) = (0, \kappa^{-1}L)$ with $L < 0$. Hence, on M the vector field points into the region R and so no solution can leave R across M. Similarly, on L the vector field is horizontal, namely $(X, Y) = (\bar{\mu}^{-1}M, 0)$ with $M > 0$. Since the function l is decreasing (see Proposition 24.5.7(4)), it follows that on L, too, the vector field points into the region. Since the solution cannot leave R through the critical points, it follows that any local solution starting in R will remain in R for all s for which the solution is defined. Since the only way how a local solution to an ordinary differential equation cannot be prolonged to all positive s is that the solution leaves the domain of the equation, we see from the preceding that every solution starting at a point of R can be defined for all $s > 0$. □

We now want to examine the character of the equilibrium points $(v^\pm, \theta^\pm)$. With this in mind, we calculate the matrix of derivatives of (X, Y) at these points,

$$D^\pm = \begin{bmatrix} \partial_v X & \partial_\theta X \\ \partial_v Y & \partial_\theta Y \end{bmatrix},$$

where the derivatives are calculated at $(v^\pm, \theta^\pm)$. By the expressions in Proposition 24.5.5 it is found that

$$D^\pm = \begin{bmatrix} -\bar{\mu}^{-1}m^{-1}(m^2 + \partial_v \bar{p}) & -\bar{\mu}^{-1}m^{-1}\partial_\theta \bar{p} \\ -\kappa^{-1}m\theta\partial_\theta \bar{p} & -\kappa^{-1}mc_v \end{bmatrix},$$

If $P = \operatorname{diag}\left(1, m^{-1}\sqrt{\bar{\mu}^{-1}\theta^{-1}\kappa}\right)$, then

$$PD^\pm P^{-1} = \begin{bmatrix} -\bar{\mu}^{-1}m^{-1}(m^2 + \partial_v \bar{p}) & -\sqrt{\bar{\mu}^{-1}\kappa^{-1}\theta}\,\partial_\theta \bar{p} \\ -\sqrt{\bar{\mu}^{-1}\kappa^{-1}\theta}\,\partial_\theta \bar{p} & -\kappa^{-1}mc_v \end{bmatrix},$$

is a symmetric matrix and hence the eigenvalues of $D^\pm$ are real. Further, a calculation using (11.4.5) gives

$$\det D^\pm = c_v(m^2 + \partial_v \bar{p})/\bar{\mu}\kappa, \quad \operatorname{tr} D^\pm = -\bar{\mu}^{-1}m^{-1}(m^2 + \partial_v \bar{p}) - \kappa^{-1}mc_v. \qquad (24.5.21)$$

24.5.9 Proposition

(1) *At* (v^-, θ^-) *we have* $\det D^- < 0$; *hence one eigenvalue of* D^- *is positive and the other negative. The point* (v^-, θ^-) *is a saddle point of the shock layer equations.*

(2) *At* (v^+, θ^+) *we have* $\det D^+ > 0$ *and* $\operatorname{tr} D^+ < 0$; *hence both the eigenvalues of* D^+ *are negative and* (v^+, θ^+) *is a stable node of the shock layer equations.*

Proof In view of $(24.5.21)_1$ the assertions $\pm\det D^\pm > 0$ are equivalent to the inequalities (24.5.17). Let us now prove the assertion about the trace in (2). We have $m^2 + \partial_v \bar{p} > 0$, which in view of (11.4.5) can be rewritten as $m^2 + \partial_v \bar{p} - \theta(\partial_\theta \bar{p})^2/c_v > 0$. This implies $m^2 + \partial_v \bar{p} > 0$ and the assertion follows from $(24.5.21)_2$ and the inequalities $m > 0$, $\kappa > 0$, $\bar{\mu} > 0$, $c_v > 0$. □

Let us consider the unstable manifold at (v^-, θ^-) corresponding to the unique positive eigenvalue $\omega > 0$ of D^-. The slope of the unstable manifold is given by the ratio of the components of the eigenvector of D^- corresponding to ω. We have

$$D^- - \omega I = \begin{bmatrix} -\bar{\mu}^{-1}m^{-1}(m^2 + \partial_v \bar{p}) - \omega & -\bar{\mu}^{-1}m^{-1}\partial_\theta \bar{p} \\ -\kappa^{-1}m\theta\partial_\theta \bar{p} & -\kappa^{-1}mc_v - \omega \end{bmatrix},$$

and so the eigenvector is $(\bar{\mu}^{-1}m^{-1}\partial_\theta\bar{p}, -\bar{\mu}^{-1}m^{-1}(m^2+\partial_v\bar{p})-\omega)$ which is a multiple of the other form of the eigenvector $(\kappa^{-1}mc_v+\omega, -\kappa^{-1}m\theta\partial_\theta\bar{p})$. The slope s is therefore

$$s = -\frac{\bar{\mu}^{-1}m^{-1}(m^2+\partial_v\bar{p})+\omega}{\bar{\mu}^{-1}m^{-1}\partial_\theta\bar{p}} = -\frac{\kappa^{-1}m\theta\partial_\theta\bar{p}}{\kappa^{-1}mc_v+\omega}. \tag{24.5.22}$$

24.5.10 Proposition *We have*

$$\dot{l}(v^-) < s < \dot{m}(v^-). \tag{24.5.23}$$

Proof Differentiation of the relation $L(v, l(v)) = 0$ and the use of (24.5.12) with $M = 0$ gives

$$\dot{l}(v^-) = -\partial_v L/\partial_\theta L = -\theta\partial_\theta\bar{p}/c_v.$$

A combination with $(24.5.22)_2$ and some calculation shows that $(24.5.23)_1$ holds since $\omega > 0$. Differentiation of the relation $M(v, m(v)) = 0$ and the use of (24.5.11) gives

$$\dot{m}(v^-) = -\partial_v M/\partial_\theta M = -m^{-1}(m^2+\partial_v\bar{p})/(m^{-1}\partial_\theta\bar{p}) = -(m^2+\partial_v\bar{p})/\partial_\theta\bar{p}.$$

A combination with $(24.5.22)_1$ and a calculation shows that $(24.5.23)_2$ holds since $\omega > 0$. □

Completion of the Proof of Proposition 24.5.3 Inequality (24.5.23) shows that the unstable manifold U through the point (v^-, θ^-) contains a nonempty arc U such that $U \subset R$. Let (v_0, θ_0) be any point of U. From the properties of the unstable manifold, there exists a solution to the system of shock layer equations $(v(\cdot), \theta(\cdot))$ defined on $(-\infty, 0]$ such that $(v(s), \theta(s)) \to (v^-, \theta^-)$ as $s \to -\infty$ and $(v(0), \theta(0)) = (v_0, \theta_0)$. The point (v_0, θ_0) is in R and R is positively invariant. Hence the solution $(v(\cdot), \theta(\cdot))$ can be continued to all positive values s. We have thus the solution to the shock layer equations defined for all $s \in \mathbb{R}$. Since $M > 0$, $L < 0$ on R, we see that the solution satisfies $v' > 0, \theta' < 0$ and hence the limits $(v(s), \theta(s))$ as $s \to +\infty$ exist, and must lie in the bounded set cl R. At the limiting point the vector field must vanish and as $(v^\pm, \theta^\pm)$ are the only points where the vector field vanishes, we see that the limit must be one of these two points. However, the possibility that the solution converges to (v^-, θ^-) is excluded since $v(s)$ is increasing. This proves that $(v(s), \theta(s)) \to (v^+, \theta^+)$ as $s \to +\infty$. Hence the constructed solution is a shock layer. □

Bibliographical Note

I deviate slightly from the common terminology, where Hugoniot's adiabat means the set of all solutions of Hugoniot's equation in the (p, v)-plane. Results of the type of Proposition 24.4.1 and its consequences are known as Bethe-Weyl's theorems, with the important contribution from Zemplén. The Assumptions 24.3.1, under which the theorem is proved here, are standard except for the assumed form of the domain $\tilde{U}$ in (1) and the first limit in (3). These more restrictive assumptions allow us to obtain the global parametrization of the Hugoniot adiabat by entropy. Inequalities $(24.3.1)_{2,3,4}$ are used in COURANT & FRIEDRICHS [1948] (with an obvious misprint in the inequality sign in their (65.02), as a comparison with their (2.06) shows), ROZHDESTVENSKII & JANENKO [1978], and SMOLLER [1983], to mention prominent examples. The original paper WEYL [1949], completed in 1944, contains an equivalent form stated by using the pressure and specific volume as independent variables. Weyl also mentions that his analysis would not be affected if $(24.3.1)_4$ is replaced by $\partial_\eta\tilde{p} < 0$. (LANDAU & LIFSCHITZ [1978] do not even mention any condition on $\partial_\eta\tilde{p}$.) The existence of the shock layers is proved by essentially the same method as in the classical paper GILBARG [1951]. Gilbarg also proves the uniqueness of the shock layer, and the limiting behavior as the viscosity or the heat conduction tend separately to zero. The stability character of the critical points $(v^\pm, \theta^\pm)$ in Proposition 24.5.9 was first realized by WEYL [1949]. In ideal gases, the shock layers were considered by Becker (1921). For the shock profiles and viscosity-type perturbations in the more general class of systems of conservation laws, see MAJDA & PEGO [1985].

25 Dissipation of Energy in Solids

This chapter examines the properties of a linearized system of equations of a viscous solid with heat conduction. The linearization is first carried out about an inhomogeneous equilibrium state. The linearized model is shown to have a well-defined time evolution under natural hypotheses, with the solution depending continuously on the initial data. The Liapunov functions are determined, and the solutions are shown to decay to the basic equilibrium state if the second variation of the stored energy is uniformly positive.

Linearization about a nonequilibrium static state of nonuniform temperature is also considered. The equations are more complicated. In general, the Liapunov functions for the linearization about an equilibrium state are no longer Liapunov functions in this case. Only for a linearization about a slightly nonequilibrium state can some of the results of the preceding sections be extended; in particular, the Liapunov functions are exhibited, and the asymptotic stability is established.

25.1 Review of Basic Equations

25.1.1 The Full System of Equations Consider a viscous body with heat conduction occupying the reference region Ω. In the reference description the process is determined by the evolution of the deformation and temperature (χ, θ). These are the unknown functions. The complete equations governing the evolution are

$$\rho_0 \ddot{\chi} = \operatorname{Div} \boldsymbol{S} + \rho_0 \boldsymbol{b}, \quad \rho_0 (\tfrac{1}{2} v^2 + e)^{\dot{}} = \operatorname{Div}(\boldsymbol{S}^{\mathrm{T}} \boldsymbol{v}) + \rho_0 \boldsymbol{b} \cdot \boldsymbol{v} + \rho_0 r \tag{25.1.1}$$

and

$$\left.\begin{aligned} e = \bar{e}(\boldsymbol{F}, \theta), \quad \eta = \bar{\eta}(\boldsymbol{F}, \theta), \quad f = \bar{f}(\boldsymbol{F}, \theta), \\ \boldsymbol{S} = \boldsymbol{S}_e + \boldsymbol{S}_d, \quad \boldsymbol{S}_e = \bar{\boldsymbol{S}}_e(\boldsymbol{F}, \theta), \quad \boldsymbol{S}_d = \bar{\boldsymbol{S}}_d(\boldsymbol{F}, \theta, \dot{\boldsymbol{F}}, \boldsymbol{G}), \\ \boldsymbol{q} = \bar{\boldsymbol{q}}(\boldsymbol{F}, \theta, \dot{\boldsymbol{F}}, \boldsymbol{G}), \end{aligned}\right\} \tag{25.1.2}$$

where $\boldsymbol{F}, \boldsymbol{v}, \boldsymbol{G}$ are the deformation gradient, velocity, and referential gradient of temperature. For convenience the nonequilibrium quantities $\boldsymbol{S}_d, \boldsymbol{q}$ are expressed as functions of $\dot{\boldsymbol{F}}$ rather than of $\boldsymbol{D}$. It is assumed throughout that the response functions are frame-indifferent and satisfy the Clausius–Duhem inequality.

25.1.2 Alternative Forms of Balance of Energy One can replace $(25.1.1)_2$ with one of the following: the reduced equation of balance of energy

$$\rho_0 \dot{e} = \boldsymbol{S} \cdot \dot{\boldsymbol{F}} - \operatorname{Div} \boldsymbol{q} + \rho_0 r,$$

the Clausius–Duhem inequality with the explicit production of entropy (combine (9.2.9) with $(9.2.12)_1$)

$$\rho_o \dot{\eta} = -\mathrm{Div}(\boldsymbol{q}/\theta) + \boldsymbol{S}_d \dot{\boldsymbol{F}}/\theta - \boldsymbol{q} \cdot \boldsymbol{G}/\theta^2 + \rho_o r/\theta,$$

or with

$$\rho_o \theta \dot{\eta} = -\mathrm{Div}\, \boldsymbol{q} + \boldsymbol{S}_d \cdot \dot{\boldsymbol{F}} + \rho_o r.$$

Computing the time-derivative of the entropy,

$$\rho_o \bar{c}_F \dot{\theta} = -\mathrm{Div}\, \boldsymbol{q} + \boldsymbol{S}_d \cdot \dot{\boldsymbol{F}} + \theta \bar{\boldsymbol{M}} \cdot \dot{\boldsymbol{F}} + \rho_o r,$$

where $\bar{c}_F$ is the specific heat at constant deformation and $\bar{\boldsymbol{M}}$ is the referential stress-temperature tensor, $\bar{\boldsymbol{M}} = \partial_\theta \bar{\boldsymbol{S}}_e(\boldsymbol{F}, \theta)$.

25.1.3 The Initial-Boundary-Value Problem The time evolution (χ, θ) is expected to be completely determined when the external body force and the external supply of heat are known functions of position and time, $\boldsymbol{b} = \boldsymbol{b}(\boldsymbol{p}, t)$, $r = r(\boldsymbol{p}, t)$, the boundary conditions are prescribed, and the initial conditions are given. The boundary conditions are postulated in the form

$$\chi = \boldsymbol{d} \text{ on } D, \quad \boldsymbol{Sn} = \boldsymbol{s} \text{ on } S, \quad \theta = \theta_0 \text{ on } T, \quad \boldsymbol{q} \cdot \boldsymbol{n} = q \text{ on } Q, \tag{25.1.3}$$

where D, S and T, Q are pairs of complementary subsets of $\partial\Omega$ and $\boldsymbol{d}, \boldsymbol{s}, \theta_0, q$ are prescribed functions of $\boldsymbol{p}$, t. The initial conditions read

$$\chi(\boldsymbol{p}, 0) = \chi^i(\boldsymbol{p}), \quad \boldsymbol{v}(\boldsymbol{p}, 0) = \boldsymbol{v}^i(\boldsymbol{p}), \quad \theta(\boldsymbol{p}, 0) = \theta^i(\boldsymbol{p}), \quad \boldsymbol{p} \in \Omega, \tag{25.1.4}$$

where $\chi^i, \boldsymbol{v}^i, \theta^i$ are given. Equations (25.1.1), (25.1.2), (25.1.3) and (25.1.4) form a complete system of equations. At the present time there is no mathematical theory for this system with general nonlinear constitutive equations. Partial results are known only in one spatial dimension. (A brief review of the literature is at the end of this chapter.) Here we restrict ourselves to the properties of the linearized equations in an arbitrary number of dimensions, for which a complete theory can be constructed using standard means.

25.1.4 Linearization Assume that the functions $\boldsymbol{b}, \boldsymbol{d}, \boldsymbol{s}, \theta_0$ are independent of time and the same for all processes, that $q = 0$, $r = 0$ identically and that θ_0 is uniform and time independent on T. This means that the body is in a standard heating environment and under dead loadings (see Subsects. 13.2.1 and 13.3.9). Suppose further that there is an equilibrium reference state $\sigma_0 = (\chi_0, \tilde{\theta}_0)$ compatible with these conditions. We shall linearize (25.1.1), (25.1.2), (25.1.3) and (25.1.4) about σ_0. (The linearization about nonequilibrium rest states is described in Sect. 25.6.) For σ_0, the equation of energy is automatically satisfied with $r = 0$ (see Sects. 14.1 and 14.2), the equality $\tilde{\theta}_0 = \theta_0$ is part of the definition, and the equation of mechanical equilibrium reads $\mathrm{Div}\, \boldsymbol{S}_0 + \rho_o \boldsymbol{b} = \boldsymbol{0}$, with $\boldsymbol{S}_0$ the stress in the state σ_0 given by the exact constitutive equations (25.1.2). Let us write θ_0 for $\tilde{\theta}_0$, denote by $\boldsymbol{u}(\boldsymbol{p}, t) := \chi(\boldsymbol{p}, t) - \chi_0(\boldsymbol{p})$ the displacement from the reference deformation, and by $\vartheta(\boldsymbol{p}, t) := \theta(\boldsymbol{p}, t) - \theta_0$ the difference of the actual temperature from the reference temperature. In the rest of this chapter assume that χ_0 has a continuously differentiable extension to an open set containing the closure of Ω, and that this extension is one-to-one, with the inverse continuously differentiable. The linearization is done in a way similar to that in Sect. 20.1. The resulting system is

$$\left.\begin{aligned}
&\rho_\circ \ddot{\boldsymbol u} = \operatorname{Div}(\bar{\boldsymbol A}[\nabla \boldsymbol u] + \boldsymbol V[\nabla \dot{\boldsymbol u}] + \bar{\boldsymbol M}\vartheta + \boldsymbol D[\nabla\vartheta]),\\
&\rho_\circ \bar c_F \dot\vartheta = \operatorname{Div}(\boldsymbol K[\nabla\vartheta] + \boldsymbol E[\nabla\dot{\boldsymbol u}]) + \theta_0 \bar{\boldsymbol M}\cdot\nabla\dot{\boldsymbol u},\\
&\boldsymbol u = \boldsymbol 0 \text{ on } D, \quad \vartheta = 0 \text{ on } T,\\
&(\bar{\boldsymbol A}[\nabla \boldsymbol u] + \boldsymbol V[\nabla \dot{\boldsymbol u}] + \bar{\boldsymbol M}\vartheta + \boldsymbol D[\nabla\vartheta])\boldsymbol n = \boldsymbol 0 \text{ on } S,\\
&(\boldsymbol K[\nabla\vartheta] + \boldsymbol E[\nabla\dot{\boldsymbol u}])\cdot \boldsymbol n = 0 \text{ on } Q.
\end{aligned}\right\} \quad (25.1.5)$$

Here $\bar c_F$, $\bar{\boldsymbol A}$, $\bar{\boldsymbol M}$ are the specific heat at constant deformation, the referential tensor of isothermal elasticities and the stress–temperature coefficient evaluated at $(\boldsymbol F_0(\boldsymbol p), \theta_0)$, $\boldsymbol F_0 = \nabla\chi_0$; generally $\bar c_F$, $\bar{\boldsymbol A}$, $\bar{\boldsymbol M}$ are functions of $\boldsymbol p$. Further, $\boldsymbol V, \boldsymbol D, \boldsymbol E, \boldsymbol K$ are referential kinetic coefficients defined by

$$\boldsymbol V = \partial_{\dot F}\bar{\boldsymbol S}_d, \quad \boldsymbol D = \partial_G \bar{\boldsymbol S}_d, \quad \boldsymbol E = -\partial_{\dot F}\bar{\boldsymbol q}, \quad \boldsymbol K = -\partial_G \bar{\boldsymbol q},$$

with the derivatives evaluated at $(\boldsymbol F_0(\boldsymbol p), \theta_0, \boldsymbol 0, \boldsymbol 0)$; hence also the kinetic coefficients may be functions of $\boldsymbol p$. The referential kinetic coefficients are perfect analogs of the spatial kinetic coefficients of Chap. 12 and have analogous properties.

25.1.5 Conservation Laws Equations (25.1.5) can be given the form:

$$\begin{aligned}
&\rho_\circ \ddot{\boldsymbol u} = \operatorname{Div}\boldsymbol S^b, \quad \rho_\circ\theta_0\dot\eta^b = -\operatorname{Div}\boldsymbol q^b\\
&\boldsymbol u = \boldsymbol 0 \text{ on } D, \ \vartheta = 0 \text{ on } T, \ \boldsymbol S^b\boldsymbol n = \boldsymbol 0 \text{ on } S, \ \boldsymbol q^b\cdot\boldsymbol n = 0 \text{ on } Q,
\end{aligned} \quad (25.1.6)$$

where

$$\rho_\circ\eta^b = \rho_\circ \bar c_F\vartheta/\theta_0 - \bar{\boldsymbol M}\cdot\nabla\boldsymbol u,$$

$$\boldsymbol S^b = \bar{\boldsymbol A}[\nabla\boldsymbol u] + \bar{\boldsymbol M}\vartheta + \boldsymbol V[\nabla\dot{\boldsymbol u}] + \boldsymbol D[\boldsymbol G], \quad \boldsymbol q^b = -\boldsymbol K[\nabla\vartheta] - \boldsymbol E[\nabla\dot{\boldsymbol u}]$$

are the linearized entropy, stress, and heat flux, without the absolute terms in η^b, $\boldsymbol S^b$, cf. Sect. 10.9. In calculating the Liapunov functions, we shall also make extensive use of the density of the canonical free energy f_0,

$$\rho_\circ f_0^b = \tfrac12\bar{\boldsymbol A}[\nabla\boldsymbol u, \nabla\boldsymbol u] + \tfrac12\rho_\circ\bar c_F\vartheta^2/\theta_0,$$

see the discussion in Subsect. 10.9.2.

25.1.6 The Dissipation The kinetic coefficients $\boldsymbol V, \boldsymbol D, \boldsymbol E, \boldsymbol K$ satisfy the residual dissipation inequality (cf. Chap. 10)

$$\boldsymbol S_d^b\cdot\nabla\dot{\boldsymbol u} - \boldsymbol q^b\cdot\nabla\vartheta/\theta_0 \ge 0. \quad (25.1.7)$$

If we denote by β the left-hand side of (25.1.7) then

$$\beta(\boldsymbol H, \boldsymbol G) = \boldsymbol V[\boldsymbol H, \boldsymbol H] + \boldsymbol D[\boldsymbol H]\cdot\boldsymbol G + \boldsymbol E[\boldsymbol G]\cdot\boldsymbol H/\theta_0 + \boldsymbol K\boldsymbol G\cdot\boldsymbol G/\theta_0. \quad (25.1.8)$$

The explicit dependence of β on $\boldsymbol p$ through the dependence of the kinetic coefficients on $\boldsymbol p$ is suppressed in this notation. Let us denote,

$$B := \int_\Omega \beta(\boldsymbol H, \boldsymbol G)\, dV. \quad (25.1.9)$$

The material is said to be ***uniformly dissipative*** if there exists a constant $c > 0$ such that for every $\boldsymbol p \in \Omega$, every $\boldsymbol G \in$ Vect and every $\boldsymbol H \in$ Lin

$$\beta(\boldsymbol H, \boldsymbol G) \ge c(|\boldsymbol D|^2 + |\boldsymbol G|^2) \text{ where } \boldsymbol D = \tfrac12(\boldsymbol M + \boldsymbol M^{\mathrm T}), \ \boldsymbol M := \boldsymbol H\boldsymbol F_0(\boldsymbol p)^{-1} \quad (25.1.10)$$

(and $\boldsymbol F_0(\boldsymbol p) = \nabla\chi_0(\boldsymbol p)$). Note that the assumption $\beta(\boldsymbol H, \boldsymbol G) \ge c(|\boldsymbol H|^2 + |\boldsymbol G|^2)$ for all $\boldsymbol H \in$ Lin would be inconsistent with the frame indifference because $\beta(\boldsymbol H, \boldsymbol 0) = 0$ whenever $\boldsymbol H\boldsymbol F_0(\boldsymbol p)^{-1}$ is skew.

Table 25.1.1. Properties of the linearized dynamics

property of solutions	implied by the property of coefficients	
	stored energy	dissipation
uniqueness	$\delta^2 \mathrm{M} \geq -d\|\boldsymbol{u}\|^2_{L^2}$	$\beta \geq 0$
existence, continuous dependence	$\delta^2 \mathrm{M} \geq -d\|\boldsymbol{u}\|^2_{L^2} + c_1\|\boldsymbol{u}\|^2_{W^{1,2}}$	$\beta \geq c(\|\boldsymbol{D}\|^2 + \|\boldsymbol{G}\|^2)$
asymptotic stability	$\delta^2 \mathrm{M} \geq c_1\|\boldsymbol{u}\|^2_{W^{1,2}}$	$\beta \geq c(\|\boldsymbol{D}\|^2 + \|\boldsymbol{G}\|^2)$

25.1.7 Lemma *If the material is uniformly dissipative and D has positive area, then there exists a constant $m > 0$ such that*

$$\int_\Omega \beta(\boldsymbol{H}, \boldsymbol{G})\, dV \geq m \int_\Omega (|\nabla \dot{\boldsymbol{u}}|^2 + |\boldsymbol{G}|^2)\, dV \tag{25.1.11}$$

for every process $\boldsymbol{u}, \vartheta$ such that $\boldsymbol{u} = \boldsymbol{0}$ on D.

Proof The idea is to pass to the reference configuration Ω_0 given by the deformation χ_0. Let $\boldsymbol{u} : \Omega \to$ Vect be a sufficiently smooth displacement field, $\boldsymbol{H} = \nabla \dot{\boldsymbol{u}}$ and $\boldsymbol{M}, \boldsymbol{D}$ be as in (25.1.10). Let $\bar{\boldsymbol{u}} : \Omega_0 \to$ Vect be given by $\bar{\boldsymbol{u}} \circ \chi_0 = \boldsymbol{u}$. Then $\boldsymbol{M} = \nabla \dot{\bar{\boldsymbol{u}}}$ and $\boldsymbol{D}$ is a symmetric part of $\boldsymbol{M}$. Hence, by (25.1.10),

$$\int_\Omega \beta(\boldsymbol{H}, \boldsymbol{G})\, dV \geq \int_\Omega c(|\boldsymbol{D}|^2 + |\boldsymbol{G}|^2)\, dV = c\Big(\int_{\Omega_0} |\boldsymbol{D}|^2 J_0^{-1}\, dV_0 + \int_\Omega |\boldsymbol{G}|^2\, dV\Big)$$

where $J_0 = \det \boldsymbol{F}_0$ and dV_0 is the element of volume in Ω_0. By using that χ_0 and its inverse are continuously differentiable up to the boundary, we have that $J_0, J_0^{-1}, |\boldsymbol{F}_0|, |\boldsymbol{F}_0^{-1}|$ are bounded above and below by positive constants independent of position, so that

$$\int_{\Omega_0} |\boldsymbol{D}|^2 J_0^{-1}\, dV_0 \geq c_1 \int_{\Omega_0} |\boldsymbol{D}|^2\, dV_0 \geq c_2 \int_{\Omega_0} |\nabla \dot{\bar{\boldsymbol{u}}}|^2\, dV_0 \geq c_3 \int_\Omega |\nabla \dot{\boldsymbol{u}}|^2\, dV_0,$$

where the intermediate step uses Korn's inequality Proposition 20.6.7(3). A combination of the last two inequalities gives (25.1.11). □

To avoid repeated hypotheses, for the rest of the chapter we lay down the following assumptions.

25.1.8 Assumptions
(1) Ω is a nonempty, bounded open subset of Eucl with Lipschitz boundary;
(2) *D*, *S* and *T*, *Q* are pairs of complementary subsets of $\partial\Omega$, *D*, *T* have positive areas and *S*, *Q* are relatively open in $\partial\Omega$;
(3) the coefficients ρ_o, $\bar{c}_F$, $\bar{\mathbf{A}}$, $\bar{\mathbf{M}}$, $\mathbf{V}$, $\mathbf{D}$, $\mathbf{E}$, $\mathbf{K}$ have continuous extensions to the closure of Ω;
(4) the functions ρ_o, $\bar{c}_F$ are bounded above and below by positive constants.

Also it is assumed that the material is thermodynamically compatible, i.e., the kinetic coefficients satisfy the dissipation inequality (25.1.7). For the existence of the time evolution and its stability, however, a stronger version of dissipativity will be needed.

25.1.9 Program and Review The subsequent sections discuss the uniqueness, existence, continuous dependence and asymptotic stability of the linearized equations. Each of these properties requires different hypotheses, but all of them are stated in terms of the second variation of the stored energy and in terms of the dissipation.

The dissipation has been considered above. Let M be the total stored energy of the original nonlinear model, including the potential energy of the dead loadings by $\boldsymbol{b}_0$, s as above, see Subsect. 14.3.1. The second variation of the total stored energy at χ_0 is

$$\mathsf{J}(\boldsymbol{u}) \equiv \delta^2 \mathsf{M}(\chi_0, \boldsymbol{u}, \boldsymbol{u}) = \int_\Omega \bar{\boldsymbol{A}}[\nabla \boldsymbol{u}, \nabla \boldsymbol{u}]\, dV$$

for any $\boldsymbol{u} \in W^{1,2}(\Omega)$ satisfying $\boldsymbol{u} = \boldsymbol{0}$ on D. Various properties of J to be used below have been discussed in Sect. 20.4. Table 25.1.8 summarizes the properties of the linearized dynamics; the constants c_1, c occurring there are positive and the constant d arbitrary.

The existence of solutions is established by the techniques of the linear semigroup theory. The uniqueness and asymptotic stability can be established by elementary means. For a reader who does not want to enter the existence, the next two sections are devoted to calculating the Liapunov functions which lead to the uniqueness and asymptotic stability directly.

25.2 Liapunov Functions

The Liapunov functions derived here apply to the linearization about an equilibrium state, but we shall see in Sect. 25.6 that the same Liapunov functions also work for linearization about slightly nonequilibrium states to be defined there. This is interesting, in the author's opinion, because some of the texts on nonequilibrium thermodynamics claim that when passing to the nonequilibrium states, the canonical free energy ceases to be a Liapunov function, and that the production of entropy is the Liapunov function instead.

25.2.1 The Energy Identities Consider the linearized equations about an equilibrium state. We have the following identity

$$\rho_o(\tfrac{1}{2}\dot{\boldsymbol{u}}^2 + f_0^b)^{\cdot} = \operatorname{Div}(\boldsymbol{S}^{b\mathsf{T}}\dot{\boldsymbol{u}} - \boldsymbol{q}^b\vartheta/\theta_0) - \beta, \tag{25.2.1}$$

and from the residual dissipation inequality (25.1.7)

$$\rho_o(\tfrac{1}{2}\dot{\boldsymbol{u}}^2 + f_0^b)^{\cdot} \le \operatorname{Div}(\boldsymbol{S}^{b\mathsf{T}}\dot{\boldsymbol{u}} - \boldsymbol{q}^b\vartheta/\theta_0), \tag{25.2.2}$$

where $\beta = \beta(\nabla\dot{\boldsymbol{u}}, \nabla\vartheta)$ is defined in Subsect. 25.1.6. Note that the equation of balance of energy $(25.1.6)_2$ is linear, while the identities (25.2.1) and (25.2.2) are quadratic. The identities (25.2.1) and (25.2.2) will be seen to play a major role in establishing the existence, uniqueness, and stability of the linear time evolution. Note also for future reference the following useful identity

$$(\rho_o\boldsymbol{u}\cdot\boldsymbol{v})^{\cdot} = \operatorname{Div}(\boldsymbol{S}^{b\mathsf{T}}\boldsymbol{u}) - \bar{\boldsymbol{A}}[\nabla\boldsymbol{u}, \nabla\boldsymbol{u}] + \rho_o\boldsymbol{v}^2 - \bar{\boldsymbol{M}}\cdot\nabla\boldsymbol{u}\cdot\vartheta - \boldsymbol{S}_d^b\cdot\nabla\boldsymbol{u}, \tag{25.2.3}$$

where $\boldsymbol{S}_d^b = \boldsymbol{V}[\nabla\boldsymbol{v}] + \boldsymbol{D}[\boldsymbol{G}]$. The next step is to integrate (25.2.1), (25.2.2) over Ω, to use (25.2.3) and the boundary conditions to obtain the following observation.

25.2.2 Proposition *For every $\alpha \in \mathbb{R}$, every process $(\boldsymbol{u}, \boldsymbol{v}, \vartheta)$ satisfying the linearized equations and the boundary conditions we have*

$$\begin{aligned}\frac{d}{dt}\int_\Omega (\tfrac{1}{2}\dot{\boldsymbol{u}}^2 + f_0^b + \alpha\boldsymbol{u}\cdot\boldsymbol{v})\rho_o\, dV &= -\alpha\int_\Omega (\boldsymbol{V}[\nabla\boldsymbol{v}]\cdot\nabla\boldsymbol{u} + \boldsymbol{D}[\nabla\vartheta]\cdot\nabla\boldsymbol{u})\, dV \\ &\quad - \alpha\int_\Omega (\bar{\boldsymbol{A}}[\nabla\boldsymbol{u}, \nabla\boldsymbol{u}] - \rho_o\boldsymbol{v}^2 + \bar{\boldsymbol{M}}\cdot\nabla\boldsymbol{u}\cdot\vartheta)\, dV - \int_\Omega \beta\, dV;\end{aligned} \tag{25.2.4}$$

for $\alpha = 0$ this reduces to

$$\frac{d}{dt}\int_{\Omega}(\tfrac{1}{2}\dot{\boldsymbol{u}}^2 + f_0^{\flat})\rho_{\circ}\, dV = -\int_{\Omega}\beta\, dV. \tag{25.2.5}$$

This is just a computation and is left to the reader. Let us now use these identities to determine the Liapunov functions. The right-hand side of (25.2.5) is nonpositive and hence the total canonical free energy (i.e., the integral on the left-hand side of (25.2.5)) is a Liapunov function:

$$\frac{d}{dt}\int_{\Omega}(\tfrac{1}{2}\dot{\boldsymbol{u}}^2 + f_0^{\flat})\rho_{\circ}\, dV \le 0. \tag{25.2.6}$$

This is not surprising. We shall see in Sect. 25.5 that this Liapunov function is not suitable for establishing the exponential decay of the solutions to the reference state. For this, the integral on the left-hand side of (25.2.4) with $\alpha > 0$, α small, will turn out to be appropriate. The following two propositions establish the properties of the Liapunov functions of this type. To state the first, and also for other purposes, let us introduce the norm $\|\cdot\|_H$ on the triples $x = (\boldsymbol{u}, \boldsymbol{v}, \vartheta)$ (the corresponding Hilbert space will be introduced in Sect. 25.4) by

$$\|x\|_H^2 = \|\boldsymbol{u}\|_{W^{1,2}(\Omega)}^2 + \|\boldsymbol{v}\|_{L^2(\Omega)}^2 + \|\vartheta\|_{L^2(\Omega)}^2. \tag{25.2.7}$$

25.2.3 Proposition *If* $\mathsf{J} = \delta^2\mathsf{M}(\chi_0, \cdot)$ *is uniformly positive (see Definition* 20.4.1(3)) *in the reference state, then for every sufficiently small* $\alpha > 0$ *there exist constants* $c_0 = c_0(\alpha) > 0$, $c_1 = c_1(\alpha) > 0$ *such that*

$$c_0\|x\|_H^2 \le \int_{\Omega}(\tfrac{1}{2}\dot{\boldsymbol{u}}^2 + f_0^{\flat} + \alpha\boldsymbol{u}\cdot\boldsymbol{v})\rho_{\circ}\, dV \le c_1\|x\|_H^2 \tag{25.2.8}$$

for every process.

This says, among other things, that the indicated function is nonnegative and bounded below by the indicated norms.

Proof Let $\alpha > 0$. Denoting the integral in (25.2.8) by V and using $\alpha\boldsymbol{u}\cdot\boldsymbol{v} \ge -\frac{1}{2}\alpha(\boldsymbol{u}^2 + \boldsymbol{v}^2)$, we find

$$V \ge \tfrac{1}{2}\int_{\Omega}(\bar{\boldsymbol{A}}[\nabla\boldsymbol{u}, \nabla\boldsymbol{u}] - \alpha\rho_{\circ}\boldsymbol{u}^2 + \rho_{\circ}(1-\alpha)\boldsymbol{v}^2 + \rho_{\circ}\bar{c}_F\vartheta^2/\theta_0)\, dV.$$

By the uniform positivity of J and the Poincaré inequality 20.2.5, if α is sufficiently small, we have

$$\int_{\Omega}(\bar{\boldsymbol{A}}[\nabla\boldsymbol{u}, \nabla\boldsymbol{u}] - \alpha\rho_{\circ}\boldsymbol{u}^2)\, dV \ge k_1\|\boldsymbol{u}\|_{W^{1,2}(\Omega)}^2$$

for some $k_1 > 0$ and all $\boldsymbol{u}$ with $\boldsymbol{u} = \boldsymbol{0}$ on D. Further,

$$\int_{\Omega}(\rho_{\circ}(1-\alpha)\boldsymbol{v}^2 + \rho_{\circ}\bar{c}_F\vartheta^2/\theta_0)\, dV \ge k_2(\|\boldsymbol{v}\|_{L^2(\Omega)}^2 + \|\vartheta\|_{L^2(\Omega)}^2)$$

for some k_2 if α is small. This proves $(25.2.8)_1$. To establish $(25.2.8)_2$, one simply bounds the coefficients in the bilinear expressions occurring in the integrand by their maximum values, and uses the inequality $ab \le \frac{1}{2}(a^2 + b^2)$. The details are left to the reader. □

25.2.4 Proposition *If* J *is uniformly positive (see Definition* 20.4.1(3)) *and the material is uniformly dissipative in the reference state, then for every sufficiently small* $\alpha > 0$ *there exists a constant* $c = c(\alpha) > 0$ *such that*

$$\frac{d}{dt}\int_{\Omega}(\tfrac{1}{2}\dot{\boldsymbol{u}}^2 + f_0^{\flat} + \alpha\boldsymbol{u}\cdot\boldsymbol{v})\rho_{\circ}\, dV \le -c\int_{\Omega}(\tfrac{1}{2}\dot{\boldsymbol{u}}^2 + f_0^{\flat} + \alpha\boldsymbol{u}\cdot\boldsymbol{v})\rho_{\circ}\, dV \tag{25.2.9}$$

for every process.

This Liapunov function is not only decreasing, but its rate of decrease is bounded by the negative of the function itself.

Proof Let V be the integral on the left-hand side of (25.2.9). From the right-hand side of (25.2.4) we see that it is necessary to bound the mixed terms like $\boldsymbol{V}[\nabla \boldsymbol{v}, \nabla \boldsymbol{u}]$. Each of them is first bounded by the maximum K of the norms of $\boldsymbol{V}, \boldsymbol{D}, \bar{\boldsymbol{M}}$ over Ω times the product of the norms of the appropriate pairs of the expressions $\nabla \boldsymbol{v}, \nabla \boldsymbol{u}, \nabla\vartheta$, and ϑ. For instance, the term $K|\nabla \boldsymbol{v}| \cdot |\nabla \boldsymbol{u}|$ comes from an estimation of $\boldsymbol{V}[\nabla \boldsymbol{v}, \nabla \boldsymbol{u}]$. Then the inequality $ab \le \frac{1}{2}\mu a^2 + \frac{1}{2}b^2/\mu$ is applied to each of these three products to obtain,

$$-(\boldsymbol{V}[\nabla \boldsymbol{v}, \nabla \boldsymbol{u}] + \boldsymbol{D}[\nabla\vartheta, \nabla \boldsymbol{u}] + \vartheta \bar{\boldsymbol{M}} \cdot \nabla \boldsymbol{u}) \le \tfrac{1}{2}K\mu(|\nabla \boldsymbol{v}|^2 + |\vartheta|^2 + |\nabla\vartheta|^2) + \tfrac{3}{2}K\mu^{-1}|\nabla \boldsymbol{u}|^2,$$

where μ is any positive number to be chosen later. Equation (25.2.4) leads to

$$\begin{aligned}\dot{V} \le -B + \alpha \int_\Omega (\tfrac{1}{2}K\mu(|\nabla \boldsymbol{v}|^2 + |\vartheta|^2 + |\nabla\vartheta|^2) + \rho_\circ \boldsymbol{v}^2)\, dV \\ - \alpha \int_\Omega (\bar{\boldsymbol{A}}[\nabla \boldsymbol{u}, \nabla \boldsymbol{u}] - \tfrac{3}{2}K\mu^{-1}|\nabla \boldsymbol{u}|^2)\, dV,\end{aligned} \tag{25.2.10}$$

where B is defined in (25.1.9). Let us now use the freedom in $\mu > 0$ and α to estimate the three integrals on the right-hand side of (25.2.10). By the uniform positivity of J it is possible first to choose μ large enough so that

$$\int_\Omega (\bar{\boldsymbol{A}}[\nabla \boldsymbol{u}, \nabla \boldsymbol{u}] - \tfrac{3}{2}K\mu^{-1}|\nabla \boldsymbol{u}|^2)\, dV \ge c_1 \|\boldsymbol{u}\|^2_{W^{1,2}(\Omega)}$$

for some positive constant c_1 and all $\boldsymbol{u}$. This reduces (25.2.10) to

$$\dot{V} \le -B + \alpha \int_\Omega (\tfrac{1}{2}K\mu(|\nabla \boldsymbol{v}|^2 + |\vartheta|^2 + |\nabla\vartheta|^2) + \rho_\circ \boldsymbol{v}^2)\, dV - \alpha c_1 \|\boldsymbol{u}\|^2_{W^{1,2}(\Omega)}.$$

Next, with this choice of μ it is possible to take $\alpha > 0$ small enough to ensure that

$$-B + \alpha \int_\Omega (\tfrac{1}{2}K\mu(|\nabla \boldsymbol{v}|^2 + |\vartheta|^2 + |\nabla\vartheta|^2) + \rho_\circ \boldsymbol{v}^2)\, dV \le -c_2(\|\boldsymbol{v}\|^2_{W^{1,2}(\Omega)} + \|\vartheta\|^2_{W^{1,2}(\Omega)})$$

for some positive constant c_2 and all processes. This is possible in view of Lemma 25.1.7 and Poincaré's inequality. From this we obtain

$$\dot{V} \le -c_2\|\boldsymbol{v}\|^2_{W^{1,2}(\Omega)} - c_2\|\vartheta\|^2_{W^{1,2}(\Omega)} - \alpha c_1 \|\boldsymbol{u}\|^2_{W^{1,2}(\Omega)}.$$

Using

$$\|\boldsymbol{v}\|^2_{W^{1,2}(\Omega)} \ge \|\boldsymbol{v}\|^2_{L^2(\Omega)}, \quad \|\vartheta\|^2_{W^{1,2}(\Omega)} \ge \|\vartheta\|^2_{L^2(\Omega)},$$

we can summarize that the expression on the left-hand side of (25.2.9) is bounded by $-c_3\|x\|^2_H$. Combining, this with Proposition 25.2.3, we obtain the result. □

25.3 Uniqueness

As a simple application of the identity (25.2.2) let us prove the uniqueness for the linearization about an equilibrium state.

25.3.1 Proposition *Let* J *be weakly bounded below (see Definition* 20.4.1(5)). *If* $\boldsymbol{u}, \vartheta$ *is a classical solution to the linearized equations such that* $\boldsymbol{u}(\cdot, 0) = \boldsymbol{0}, \dot{\boldsymbol{u}}(\cdot, 0) = \boldsymbol{0}, \vartheta(\cdot, 0) = 0$ *then* $\boldsymbol{u} = \boldsymbol{0}, \vartheta = 0$ *identically.*

Proof Let $T > 0$ be a fixed time, and $t \le T$. An integration of (25.2.5) over the interval $[0, t]$ gives

$$\int_\Omega (\tfrac{1}{2}\boldsymbol{v}^2 + f_0^\flat(\cdot, t))\rho_\circ\, dV \le \int_\Omega f_0^\flat(\cdot, 0)\rho_\circ\, dV = 0.$$

A reference to the definition of $f_0^\flat$, the uniform positiveness of $\bar{c}_F, \rho_\circ$, and the weak boundedness below of the second variation of the stored energy give

$$0 \geq \int_{\Omega} (\tfrac{1}{2}v^2 + f_0^{\flat}(\cdot, t))\rho_o \, dV \geq c \int_{\Omega} (|\dot{u}|^2 + |\vartheta|^2) \, dV - d \int_{\Omega} |u|^2 \, dV$$

for every process and fixed positive constants c, d. Denoting $f(t) := \int_{\Omega} (|\dot{u}|^2 + |\vartheta|^2) \, dV$, we have

$$f(t) \leq c_1 \int_{\Omega} |u|^2 \, dV \tag{25.3.1}$$

with $c_1 = d/c > 0$. As $u(\cdot, 0) = \mathbf{0}$, Hölder's inequality implies

$$|u(\cdot, t)|^2 \leq \left(\int_0^t |\dot{u}| \, d\tau\right)^2 \leq t \int_0^t |\dot{u}|^2 \, d\tau \leq T \int_0^t |\dot{u}|^2 \, d\tau$$

Integration over Ω gives $\int_{\Omega} |u(\cdot, t)|^2 \, dV \leq T \int_0^t \int_{\Omega} |\dot{u}|^2 \, dV \, d\tau \leq T \int_0^t f(\tau) \, d\tau$ since from the definition of f we have $f(t) \geq \int_{\Omega} |\dot{u}|^2 \, dV$. A combination with (25.3.1) provides

$$f(t) \leq c_2 \int_0^t f(\tau) \, d\tau$$

for every process and a fixed positive constant c_2. Denoting by g the indefinite integral of f, this can be rewritten as (observe that $g \geq 0$) the differential inequality $\dot{g}(t) \leq cg(t)$ or $(ge^{-ct})^{\cdot} \leq 0$. Upon integration with the initial condition $g(0) = 0$ this gives $g(t)e^{-ct} \leq 0$ and hence $g = 0$ identically. Since f is nonnegative and g its indefinite integral, it follows that $f = 0$ identically, from which $\dot{u} = \mathbf{0}$, $\vartheta = 0$ and consequently also $u = \mathbf{0}$. □

25.4 The Existence of the Linear Time Evolution

The object describing the evolution of linear systems of equations is a continuous semigroup.

25.4.1 Definition A ***strongly continuous semigroup*** on a Hilbert space H is a family $\{U(t), t \geq 0\}$ of bounded linear operators from H into H such that the following conditions hold:

(1) semigroup property:

$$U(s+t) = U(s)U(t), \qquad s, t \geq 0;$$

(2) $U(0)$ is the identity operator I on H;

(3) strong continuity: for every $x \in H$, $\lim_{t \downarrow 0} U(t)x = x$, i.e.,

$$\lim_{t \downarrow 0} \|U(t)x - x\| = 0.$$

The ***infinitesimal generator*** A of the semigroup $\{U(t), t \geq 0\}$ is the (generally unbounded) linear operator given by

$$Ax = \lim_{t \downarrow 0} \frac{U(t)x - x}{t} \tag{25.4.1}$$

defined on the set $D(A)$ of all $x \in H$ for which the limit (25.4.1) exists. It is a densely defined, closed linear operator, and if two strongly continuous semigroups have the same infinitesimal generator, then they are the same. It is possible to prove that if $x_0 \in D(A)$ and $x(t) = U(t)x_0$, $t \geq 0$, then the function $x(\cdot)$ is continuously differentiable (as a mapping from $\mathbb{R}^+$ into H) and

$$\frac{d}{dt}x(t) = Ax(t), \qquad t \geq 0, \qquad x(0) = x_0. \tag{25.4.2}$$

Hence the semigroup $\{U(t), t \geq 0\}$ can be viewed as giving solutions to (25.4.2), and in analogy with the case when $H = \mathbb{R}$ one can write $x(t) = e^{At}x_0$ and $U(t) = e^{At}$.

The following theorem gives a sufficient condition for an operator to generate a strongly continuous semigroup. Such operators determine a well-defined time evolution.

25.4.2 Theorem *Let A be a linear operator defined on a domain $D(A)$ in a Hilbert space H with inner product $\langle \cdot, \cdot \rangle$. Suppose that*
(1) $\langle Ax, x\rangle \leq \beta \langle x, x\rangle$ *for all $x \in D(A)$ and some $\beta \in \mathbb{R}$;*
(2) *there exists a $\lambda_0 > \beta$ such that $(\lambda_0 I - A)D(A) = H$.*
Then $D(A)$ is dense in H and A is a generator of a strongly continuous semigroup $\{U(t), t \geq 0\}$ on H that satisfies

$$\|U(t)x\| \leq e^{\beta t}\|x\| \tag{25.4.3}$$

for every $x \in H$ and $t \geq 0$.

Inequality (25.4.3) is an abstract version of the continuous dependence of the solution on the initial data.

Proof This is a consequence of the Lummer–Phillips theorem, as stated in PAZY [1983, Thm. 4.3(a), p. 14]. To show this, assume that A satisfies (1), (2). Let us denote by A_1 a linear operator in H defined on $D(A_1) = D(A)$ by $A_1 = A - \beta I$. Let us show that A_1 satisfies the conditions of PAZY [1983, Thm. 4.3(a), p. 14]. We have to prove that (i) $D(A_1) = D(A)$ is dense in H, (ii) $\langle A_1 x, x\rangle \leq 0$ for every $x \in D(A_1) = D(A)$, and (iii) there exists a $\lambda_1 > 0$ such that $(\lambda_1 I - A_1)D(A_1) = H$. Setting $\lambda_1 = \lambda_0 - \beta$, one sees immediately that (ii), (iii) are just restatements of (1), (2). Let us now prove (i), i.e., that $D(A_1) = D(A)$ is dense in H. As a consequence of (1) we have

$$\langle(\lambda_1 I - A_1)x, x\rangle \geq \lambda_1 \|x\|^2$$

for every $x \in D(A)$. Assume that $D(A)$ is not dense in H. Then there exists a $y_0 \in H$, $y_0 \neq 0$ such that y_0 is orthogonal to $D(A)$, i.e., $\langle y_0, x\rangle = 0$ for every $x \in D(A)$. By (iii) there exists an $x_0 \in D(A)$ such that $y_0 = (\lambda_1 I - A_1)x_0$. But then

$$0 = \langle y_0, x_0\rangle = \langle(\lambda_1 I - A_1)x_0, x_0\rangle \geq \lambda_1 \|x_0\|^2$$

and hence $x_0 = 0$. This gives $y_0 = 0$, in contradiction with the hypothesis. Hence A_1 satisfies the hypotheses of PAZY [1983, Thm. 4.3(a), p. 14]. By that theorem, A_1 is a generator of a strongly continuous semigroup $\{U_1(t), t \geq 0\}$ that satisfies $\|U_1(t)x\| \leq \|x\|$ for every $x \in H$. But then the semigroup $\{U(t), t \geq 0\}$ defined by $U(t) = e^{\beta t}U_1(t)$ is easily shown to have the generator $A = A_1 + \beta I$, and moreover it satisfies (25.4.3) as a consequence of $\|U_1(t)x\| \leq \|x\|$. □

Throughout this section, assume that J is coercive in the reference state (see Definition 20.4.1(4)) and that the material is uniformly dissipative there. It is convenient to seek the triples $(\boldsymbol{u}, \boldsymbol{v}, \vartheta)$ satisfying the system

$$\dot{\boldsymbol{u}} = \boldsymbol{v}, \quad \rho_\circ \dot{\boldsymbol{v}} = \operatorname{Div}(\bar{\mathbf{A}}[\nabla \boldsymbol{u}] + \mathbf{V}[\nabla \boldsymbol{v}] + \bar{\mathbf{M}}\vartheta + \mathbf{D}[\nabla \vartheta]),$$
$$\rho_\circ \bar{c}_F \dot{\vartheta} = \operatorname{Div}(\mathbf{K}[\nabla \vartheta] + \mathbf{E}[\nabla \boldsymbol{v}]) + \theta_0 \bar{\mathbf{M}} \cdot \nabla \boldsymbol{v},$$

which is of the first order in time. The equations can be more compactly written as $\dot{x} = Ax$ where $x = (\boldsymbol{u}, \boldsymbol{v}, \vartheta)$ and A is a linear operator defined formally by $A(\boldsymbol{u}, \boldsymbol{v}, \vartheta) = (\boldsymbol{v}, \boldsymbol{f}, g)$ where

$$\begin{aligned} \boldsymbol{f} &= \rho_\circ^{-1} \operatorname{Div}(\bar{\mathbf{A}}[\nabla \boldsymbol{u}] + \mathbf{V}[\nabla \boldsymbol{v}] + \bar{\mathbf{M}}\vartheta + \mathbf{D}[\nabla \vartheta]), \\ g &= \rho_\circ^{-1}\bar{c}_F^{-1}(\operatorname{Div}(\mathbf{K}[\nabla \vartheta] + \mathbf{E}[\nabla \boldsymbol{v}]) + \theta_0 \bar{\mathbf{M}} \cdot \nabla \boldsymbol{v}). \end{aligned} \tag{25.4.4}$$

Our immediate goal is to give a mathematical formulation to these evolution laws. With a view to applying Theorem 25.4.2, this first requires a definition of a Hilbert space of elements x.

25.4.3 Hilbert Spaces Let H be the set of all $x = (\boldsymbol{u}, \boldsymbol{v}, \vartheta) \in W^{1,2}(\Omega)\times L^2(\Omega)\times L^2(\Omega)$ such that $\boldsymbol{u} = \boldsymbol{0}$ on D in the sense of trace. Define a norm $\|\cdot\|$ on H by

$$\|x\|^2 = \tfrac{1}{2}\int_\Omega (\bar{\boldsymbol{A}}[\nabla\boldsymbol{u}, \nabla\boldsymbol{u}] + \rho_o\boldsymbol{v}^2 + \rho_o\bar{c}_F\vartheta^2/\theta_0 + 2\,d|\boldsymbol{u}|^2)\,dV \tag{25.4.5}$$

derived from the inner product

$$\langle x, x'\rangle := \tfrac{1}{2}\int_\Omega (\bar{\boldsymbol{A}}[\nabla\boldsymbol{u}, \nabla\boldsymbol{u}'] + \rho_o\boldsymbol{v}\cdot\boldsymbol{v}' + \rho_o\bar{c}_F\vartheta\vartheta'/\theta_0 + 2\,d\boldsymbol{u}\cdot\boldsymbol{u}')\,dV,$$

for every $x = (\boldsymbol{u}, \boldsymbol{v}, \vartheta)$, $x' = (\boldsymbol{u}', \boldsymbol{v}', \vartheta') \in H$. Here d is the constant in the definition of coercivity 20.4.1(4). By the coercivity and the uniform positivity of ρ_o, $\bar{c}_F$ this is really an inner product. Lemma 25.4.4 (below) shows that H is complete under this inner product; hence it is really a Hilbert space. If $x \in H$, the value $\|x\|^2$ is the sum of the canonical free energy, the kinetic energy, and the term with d in the right-hand side of (25.4.5).

Let us also denote by H_1 the space of all $x = (\boldsymbol{u}, \boldsymbol{v}, \vartheta) \in W^{1,2}(\Omega)\times W^{1,2}(\Omega)\times W^{1,2}(\Omega)$ which satisfy $\boldsymbol{u} = \boldsymbol{0}$, $\boldsymbol{v} = \boldsymbol{0}$ on D and $\vartheta = 0$ on T in the sense of trace. This is a Hilbert space with respect to the inner product on the underlying Sobolev spaces. Its natural norm is denoted by $\|\cdot\|_{H_1}$.

25.4.4 Lemma *The norm $\|\cdot\|$ on H is equivalent to the norm $\|\cdot\|_H$, given by (25.2.7), i.e., there exist constants $k_1 > 0$, $k_2 > 0$ such that $k_1\|x\| \le \|x\|_H \le k_2\|x\|$ for every $x = (\boldsymbol{u}, \boldsymbol{v}, \vartheta) \in H$.*

Proof Inequality $k_1\|x\| \le \|x\|_H$ follows from the boundedness of the coefficients in the definition of $\|\cdot\|$; recall that the coefficients are assumed to be continuous on $\bar{\Omega}$. Inequality $\|x\|_H \le k_2\|x\|$ follows from the uniform positiveness of ρ_o, $\bar{c}_F$, and the coercivity of J. □

25.4.5 Definition Define a linear operator $A : D(A) \to H$. The domain $D(A)$ consists of all $x = (\boldsymbol{u}, \boldsymbol{v}, \vartheta) \in H_1$ for which there exist $\boldsymbol{f}$, $g \in L^2(\Omega)$ such that

$$\int_\Omega (\bar{\boldsymbol{A}}[\nabla\boldsymbol{u}] + \boldsymbol{V}[\nabla\boldsymbol{v}] + \boldsymbol{D}[\nabla\vartheta] + \boldsymbol{M}\vartheta)\cdot\nabla\boldsymbol{w}\,dV = -\int_\Omega \rho_o\boldsymbol{f}\cdot\boldsymbol{w},\,dV \tag{25.4.6}$$

$$\int_\Omega ((-\boldsymbol{K}[\nabla\vartheta - \boldsymbol{E}[\nabla\boldsymbol{v}])\nabla\varphi + \theta_o\bar{\boldsymbol{M}}\cdot\nabla\boldsymbol{v}\cdot\varphi)\,dV = \int_\Omega \rho_o\bar{c}_F g\varphi\,dV \tag{25.4.7}$$

for every $\boldsymbol{w}$, $\varphi \in W^{1,2}(\Omega)$ satisfying

$$\boldsymbol{w} = \boldsymbol{0} \text{ on } D, \quad \varphi = 0 \text{ on } T. \tag{25.4.8}$$

The set of all $\boldsymbol{w}$, φ from the indicated Sobolev spaces satisfying (25.4.8) is dense in $L^2(\Omega)$. Hence $\boldsymbol{f}$, g are uniquely determined and we set $A(\boldsymbol{u}, \boldsymbol{v}, \vartheta) = (\boldsymbol{v}, \boldsymbol{f}, g)$. This is indeed an element of H: We have $x = (\boldsymbol{u}, \boldsymbol{v}, \vartheta) \in H_1$ as part of the definition of the domain, hence $\boldsymbol{v} \in W^{1,2}(\Omega)$ and $\boldsymbol{v}$ satisfies the displacement boundary condition, so the conditions imposed on the first component of an element belonging to H are satisfied. The second and the third components of $A(\boldsymbol{u}, \boldsymbol{v}, \vartheta)$ are $\boldsymbol{f}$, g and the only condition on them is that they belong to $L^2(\Omega)$, and this is again satisfied.

We shall prove that A is a generator of a semigroup, an implementation of the time evolution of the linearized dynamics, by verifying (1), (2) of Theorem 25.4.2. A series of preliminary results is needed. Let us first show that on "nice" functions, A coincides with the A defined formally by (25.4.4).

25.4.6 Proposition *Let $x = (\boldsymbol{u}, \boldsymbol{v}, \vartheta) \in C^\infty(\mathrm{cl}\,\Omega)$. Then $x \in D(A)$ if and only if*

$$\boldsymbol{u} = \boldsymbol{v} = \boldsymbol{0} \text{ on } D, \quad \vartheta = 0 \text{ on } T,$$
$$(\bar{\boldsymbol{A}}[\nabla \boldsymbol{u}] + \boldsymbol{V}[\nabla \boldsymbol{v}] + \bar{\boldsymbol{M}}\vartheta + \boldsymbol{D}[\nabla\vartheta])\boldsymbol{n} = \boldsymbol{0} \text{ on } S,$$
$$(\boldsymbol{K}[\nabla\vartheta] + \boldsymbol{E}[\nabla \boldsymbol{v}]) \cdot \boldsymbol{n} = \boldsymbol{0} \text{ on } Q.$$

If this is the case, then $Ax = (\boldsymbol{v}, \boldsymbol{f}, g)$ where $\boldsymbol{f}, g$ are given by (25.4.4).

Proof Suppose that $x = (\boldsymbol{u}, \boldsymbol{v}, \vartheta)$ is infinitely differentiable on $\mathrm{cl}\,\Omega$ and in the domain of A. By the definition of $D(A)$ the boundary conditions $\boldsymbol{u} = \boldsymbol{0}, \boldsymbol{v} = \boldsymbol{0}$ on D and $\vartheta = 0$ on T hold. Further, there exist $\boldsymbol{f}, g$ as in the definition of A. Apply Green's theorem to (25.4.7) to obtain

$$-\int_\Omega \rho_0 \boldsymbol{f} \cdot \boldsymbol{w}\, dV = \int_{\partial\Omega} (\bar{\boldsymbol{A}}[\nabla \boldsymbol{u}] + \boldsymbol{V}[\nabla \boldsymbol{v}] + \boldsymbol{D}[\nabla\vartheta] + \bar{\boldsymbol{M}}\vartheta)\boldsymbol{n} \cdot \boldsymbol{w}\, dV$$
$$- \int_\Omega \mathrm{Div}\,(\bar{\boldsymbol{A}}[\nabla \boldsymbol{u}] + \boldsymbol{V}[\nabla \boldsymbol{v}] + \boldsymbol{D}[\nabla\vartheta] + \bar{\boldsymbol{M}}\vartheta) \cdot \boldsymbol{w}\, dV$$

and an analogous relation from (25.4.7). These identities must hold for every $\boldsymbol{w}, \varphi$ satisfying (25.4.8). If one first chooses these functions to vanish on the whole boundary, the surface integrals vanish and the equalities then say that $\boldsymbol{f}, g$ equal the expressions in (25.4.4). With this knowledge, the volume integrals cancel and the surface integrals must be equal to 0 for every $\boldsymbol{w}, \varphi$ satisfying (25.4.8). The integration is effective only over S and Q. Thanks to the assumptions that S, Q are relatively open in $\partial\Omega$, one can choose the functions $\boldsymbol{w}, \varphi$ to satisfy (25.4.8) on D and T and to have essentially arbitrary prescribed values on S and Q (subject only, say, to continuity). It then follows that the stress and heat-flux part of the boundary conditions must hold. The converse implication is obtained by reversing the proof above. □

25.4.7 Bilinear Forms Let us introduce bilinear forms S, D, C on H_1 in the following way. If $x = (\boldsymbol{u}, \boldsymbol{v}, \vartheta), y = (\boldsymbol{u}', \boldsymbol{v}', \vartheta') \in H_1$, we set

$$S(x, y) = \tfrac{1}{2} \int_\Omega (\bar{\boldsymbol{A}}[\nabla \boldsymbol{v}, \nabla \boldsymbol{u}'] - \bar{\boldsymbol{A}}[\nabla \boldsymbol{u}, \nabla \boldsymbol{v}'] - \vartheta\bar{\boldsymbol{M}} \cdot \nabla \boldsymbol{v}' + \vartheta'\bar{\boldsymbol{M}} \cdot \nabla \boldsymbol{v})\, dV,$$
$$D(x, y) = \tfrac{1}{2} \int_\Omega (\boldsymbol{V}[\nabla \boldsymbol{v}, \nabla \boldsymbol{v}'] + \boldsymbol{D}[\nabla\vartheta, \nabla \boldsymbol{v}'] + \boldsymbol{E}[\nabla \boldsymbol{v}, \nabla\vartheta']/\theta_0 + \boldsymbol{K}[\nabla\vartheta, \nabla\vartheta']/\theta_0)\, dV.$$

and

$$C(x, y) = S(x, y) - D(x, y) + d \int_\Omega \boldsymbol{v} \cdot \boldsymbol{u}'\, dV.$$

We have $S(x, y) = -S(y, x)$, for every $x, y \in H_1$ and

$$\langle Ax, y\rangle = C(x, y), \quad x \in D(A),\ y \in H_1. \tag{25.4.9}$$

The last formula is a direct computation which is left to the reader. If $x = y$, (25.4.9) reduces to

$$\langle Ax, x\rangle = -D(x, x) + d \int_\Omega \boldsymbol{v} \cdot \boldsymbol{u}\, dV \tag{25.4.10}$$

for every $x \in D(A)$. This formula has the following meaning. We have that $\|x\|^2 = \langle x, x\rangle$ is essentially the energy in the state x. During the process, therefore, the time derivative of energy is $(\|x\|^2)^\cdot = \langle x, x, \rangle^\cdot = 2\langle \dot{x}, x\rangle = 2\langle Ax, x\rangle$. Hence $\langle Ax, x\rangle$ is one half of the time derivative of energy and (25.4.10) says that this time derivative is expressible in terms of the dissipation $D(x, x)$. Equation (25.4.10) reduces to (25.2.5) if $d = 0$.

We can now verify Condition (1) of Theorem 25.4.2. By $D(x, x) \ge 0$ and (25.4.10) it suffices to verify that $\int_\Omega \boldsymbol{v} \cdot \boldsymbol{u}\, dV \le c_1 \|x\|^2$ for some $c_1 > 0$. We have

$$\int_\Omega \boldsymbol{v}\cdot\boldsymbol{u}\,dV \le \tfrac{1}{2}\int_\Omega(\boldsymbol{u}^2+\boldsymbol{v}^2)\,dV \le c_2\int_\Omega|\nabla\boldsymbol{u}|^2\,dV+\tfrac{1}{2}\int_\Omega \boldsymbol{v}^2\,dV \le c_3\|x\|^2 \qquad (25.4.11)$$

by the Poincaré inequality and Lemma 25.4.4. Condition (1) then holds with $\beta = c_3$. □

25.4.8 Proposition *Let $x \in H_1$. Then $x \in D(A)$ if and only if there exists a $z \in H$ such that*

$$C(x, y) = \langle z, y\rangle \quad \text{for every } y \in H_1. \qquad (25.4.12)$$

If this is the case, then $z = Ax$.

Proof If $x \in D(A)$, it suffices to set $z = Ax$ and to observe that (25.4.12) reduces to (25.4.9). Assume now that there exists a $z \in H$ such that (25.4.12) holds. Write $x = (\boldsymbol{u}, \boldsymbol{v}, \vartheta)$, $y = (\boldsymbol{u}', \boldsymbol{v}', \vartheta')$, $z = (\boldsymbol{u}'', \boldsymbol{v}'', \vartheta'')$. Next write down (25.4.12) explicitly using the definition of C. Since $y = (\boldsymbol{u}', \boldsymbol{v}', \vartheta')$ is arbitrary, one can split the so-obtained equation into three equations containing $\boldsymbol{u}', \boldsymbol{v}', \vartheta'$, respectively. These read:

$$\left.\begin{aligned}
-\int_\Omega \bar{\boldsymbol{A}}[\nabla\boldsymbol{u},\nabla\boldsymbol{v}'] + \boldsymbol{V}[\nabla\boldsymbol{v},\nabla\boldsymbol{v}'] + \boldsymbol{D}[\nabla\vartheta]\cdot\nabla\boldsymbol{v}' + \vartheta\bar{\boldsymbol{M}}\cdot\nabla\boldsymbol{v}'\,dV &= \int_\Omega \rho_0\boldsymbol{v}''\cdot\boldsymbol{v}'\,dV,\\
\int_\Omega \bar{\boldsymbol{A}}[\nabla\boldsymbol{v},\nabla\boldsymbol{u}'] + 2d\boldsymbol{v}\cdot\boldsymbol{u}'\,dV &= \int_\Omega \bar{\boldsymbol{A}}[\nabla\boldsymbol{u}'',\nabla\boldsymbol{u}'] + 2d\boldsymbol{u}''\cdot\boldsymbol{u}'\,dV,\\
\int_\Omega \vartheta'\bar{\boldsymbol{M}}\cdot\nabla\boldsymbol{v} - \boldsymbol{E}[\nabla\boldsymbol{v}]\cdot\nabla\vartheta'/\theta_0 - \boldsymbol{K}\nabla\vartheta\cdot\nabla\vartheta'/\theta_0\,dV &= \int_\Omega \rho_0\bar{c}_F\vartheta''\vartheta'/\theta_0\,dV.
\end{aligned}\right\} \qquad (25.4.13)$$

A comparison of $(25.4.13)_{1,3}$ with the definition of A shows that $x \in D(A)$ and $Ax = (\boldsymbol{v}, \boldsymbol{v}'', \vartheta'')$. That $\boldsymbol{v} = \boldsymbol{u}''$ follows from $(25.4.13)_2$ by rearranging it as

$$\int_\Omega \bar{\boldsymbol{A}}[\nabla\boldsymbol{v}-\nabla\boldsymbol{u}'',\nabla\boldsymbol{u}'] + 2d(\boldsymbol{v}-\boldsymbol{u}'')\cdot\boldsymbol{u}'\,dV = 0,$$

then choosing $\boldsymbol{u}' = \boldsymbol{v} - \boldsymbol{u}''$ and invoking the coercivity of J. □

25.4.9 Lemma *Let $\lambda \ge 0$ and define a bilinear form C_λ on H_1 by $C_\lambda(x, y) = \lambda\langle x, y\rangle - C(x, y)$, $x, y \in H_1$. This is a bounded bilinear form on H_1; i.e., there exists a $c_1 = c_1(\lambda)$ such that*

$$|C_\lambda(x, y)| \le c_1\|x\|_{H_1}\|y\|_{H_1}, \quad x, y \in H_1. \qquad (25.4.14)$$

If λ is sufficiently large, then there exists a $c_2 = c_2(\lambda) > 0$ such that

$$C_\lambda(x, x) \ge c_2\|x\|^2_{H_1} \qquad (25.4.15)$$

for every $x \in H_1$.

Proof The boundedness is clear from Subsect. 25.4.7 and from the boundedness of the coefficients occurring in the expressions for S, D. Inequality (25.4.15) is proved as follows. We have

$$C_\lambda(x, x) = \lambda\|x\|^2 - C(x, x) = \lambda\|x\|^2 + D(x, x) - d\int_\Omega \boldsymbol{v}\cdot\boldsymbol{u}\,dV.$$

Next,

$$-d\int_\Omega \boldsymbol{v}\cdot\boldsymbol{u}\,dV \ge -\tfrac{1}{2}d\int_\Omega(\boldsymbol{u}^2+\boldsymbol{v}^2)\,dV \ge -c_3\int_\Omega|\nabla\boldsymbol{u}|^2\,dV - \tfrac{1}{2}d\int_\Omega \boldsymbol{v}^2\,dV \ge -c_4\|x\|^2 \qquad (25.4.16)$$

by the Poincaré inequality and Lemma 25.4.4. Further observe that

$$D(x, x) = \tfrac{1}{2}\int_\Omega \beta(\nabla\boldsymbol{u}, \nabla\vartheta)\,dV$$

where β is given by (25.1.8). By the uniform dissipativity and Lemma 25.1.7 then

$$D(x, x) \ge c_5\int_\Omega(|\nabla\boldsymbol{v}|^2 + |\nabla\vartheta|^2)\,dV \ge c_6\|\boldsymbol{v}\|^2_{W^{1,2}(\Omega)} + c_6\|\vartheta\|^2_{W^{1,2}(\Omega)}.$$

Hence

$$C_\lambda(x, x) \ge \lambda\|x\|^2 - c_4\|x\|^2 + c_6\|\boldsymbol{v}\|^2_{W^{1,2}(\Omega)} + c_6\|\vartheta\|^2_{W^{1,2}(\Omega)}.$$

It now suffices to take $\lambda > c_4$ to obtain (25.4.15). □

25.4.10 Proposition (The Lax–Milgram lemma) *Let $B : H_1 \times H_1 \to \mathbb{R}$ be a bilinear form on a Hilbert space H_1 such that there exist constants $c_1, c_2 > 0$ with $|B(u, v)| \le c_1\|u\| \cdot \|v\|$ for every $u, v \in H_1$ and $B(u, u) \ge c_2\|u\|^2$ for every $u \in H_1$. Then for every continuous linear functional F on H_1 there exists exactly one $u \in H_1$ such that $B(u, v) = F(v)$ for every $v \in H_1$.*

25.4.11 Lemma *For every λ sufficiently large, $(\lambda I - A)D(A) = H$.*

Proof Let λ be large enough so that (25.4.15) holds. Let $\bar{x} = (\bar{\boldsymbol{u}}, \bar{\boldsymbol{v}}, \bar{\vartheta})$ be in H. The linear form $x' \mapsto \langle \bar{x}, x' \rangle$ is a bounded linear form on H_1 and hence by the Lax–Milgram lemma applied to the bilinear form $B \equiv C_\lambda$ there exists a unique element $x \in H_1$ such that $C_\lambda(x, x') = \langle \bar{x}, x' \rangle$ for every $x' \in H_1$. This can be rewritten as $C(x, x') = \langle z, x' \rangle$ where $z = \lambda x - \bar{x}$. Proposition 25.4.8 yields $x \in D(A)$ and $z = Ax$. By $z = \lambda x - \bar{x}$ the last equation reads $\lambda x - Ax = \bar{x}$. □

The preceding lemma shows that Condition (2) of Theorem 25.4.2 is also satisfied and hence A is a generator of a semigroup. We have thus proved

25.4.12 Theorem *Suppose that* J *is coercive in the reference state (see Definition* 20.4.1(4)) *and the material uniformly dissipative there. The operator A defined in Subsect.* 25.4.5 *generates a strongly continuous semigroup $\{U(t), t \ge 0\}$ in H that satisfies* (25.4.3).

This semigroup gives an abstract solution to the equations of the linearized dynamics. Equation (25.4.3) proves the continuous dependence of solutions on initial data. If J is negative on some variations, it can happen that the solution blows up exponentially, i.e., the evolution need not be stable.

25.5 Asymptotic Stability

The purpose of this section is to show that if the second variation of the stored energy is uniformly positive in the reference state, then the solutions tend exponentially to the equilibrium solution. Hence, in the linearized case the energy criterion of stability really works.

25.5.1 Proposition *Let* J *be uniformly positive in the reference state (see Definition* 20.4.1(3)) *and let the material be uniformly dissipative there. Then there exist constants $K > 0$, $C > 0$ such that for any solution $x(t) = (\boldsymbol{u}(\cdot, t), \boldsymbol{v}(\cdot, t), \vartheta(\cdot, t))$ with the initial data $x_0 = (\boldsymbol{u}_0, \boldsymbol{v}_0, \vartheta_0)$ and any $t \ge 0$ one has*

$$\|x(t)\|_H \le K\|x_0\|_H e^{-Ct}. \tag{25.5.1}$$

Proof An appeal to Proposition 25.2.4 shows that there exists an $\alpha > 0$ such that if one denotes

$$V(t) = \int_\Omega (\tfrac{1}{2}\dot{\boldsymbol{u}}^2 + f_0^\flat + \alpha \boldsymbol{u} \cdot \boldsymbol{v})\rho_\circ \, dV, \tag{25.5.2}$$

then $\dot{V}(t) \le -cV(t)$, for some $c > 0$. By Proposition 25.2.3, $V \ge 0$. The integration leads to $V(t) \le V(0)e^{-ct}$. By Proposition 25.2.3, $c_0\|x(t)\|_H^2 \le V(t) \le c_1\|x(t)\|_H^2$, which gives (25.5.1) with $K = \sqrt{c_1/c_2}$ and $C = c/2$. □

The existence of the Liapunov function V as in (25.5.2) can be given a functional analytic form. Namely, it is possible to introduce a new scalar product on H which will decay exponentially along any solution. In addition to the assumptions of the preceding section, let J be uniformly positive. We set $d = 0$ in the definition of $\|\cdot\|$ (see Subsect. 25.4.3) and all of the assertions of the preceding section continue to hold. For a fixed $\alpha > 0$ we introduce the bilinear form $\langle\!\langle \cdot, \cdot \rangle\!\rangle$ on H by

$$\langle x, x' \rangle = \frac{1}{2} \int_{\Omega} (\bar{\boldsymbol{A}}[\nabla \boldsymbol{u}, \nabla \boldsymbol{u}'] + \rho_0 \boldsymbol{v} \cdot \boldsymbol{v}' + \rho_0 \bar{c}_{\boldsymbol{F}} \vartheta \vartheta' / \theta_0 + \rho_0 \alpha (\boldsymbol{u} \cdot \boldsymbol{v}' + \boldsymbol{u}' \cdot \boldsymbol{v})) \, dV,$$

for every $x = (\boldsymbol{u}, \boldsymbol{v}, \vartheta)$, $x' = (\boldsymbol{u}', \boldsymbol{v}', \vartheta') \in H$. The value of α will be determined later. If $x = (\boldsymbol{u}, \boldsymbol{v}, \vartheta)$, then $|x|^2 = \langle x, x \rangle$ is exactly V from (25.5.2). Proposition 25.2.3 tells us that for every sufficiently small $\alpha > 0$ the form $\langle \cdot, \cdot \rangle$ is a scalar product on H and the norm $|\cdot|$ induced by it is equivalent to $\| \cdot \|$ and to $\| \cdot \|_H$. Furthermore, by (25.2.4), for every solution $x(t) = e^{At} x_0$, $t \geq 0$, $x_0 \in D(A)$, we have

$$\frac{d}{dt} |x|^2 = 2 \langle Ax, x \rangle, \quad t \geq 0.$$

Combining this with Proposition 25.2.4, it is found that for every $\alpha > 0$ sufficiently small there exists a constant $c > 0$ such that $\langle Ax, x \rangle \leq -c|x|^2$ for every $x \in D(A)$. Let α be so chosen and set $C = c/2$. An integration of

$$\frac{d}{dt} |x(t)|^2 = 2 \langle A e^{At} x_0, e^{At} x_0 \rangle \leq -2C |x(t)|^2$$

gives that the conclusion of Proposition 25.5.1 holds with (25.5.1) replaced by $|x(t)| \leq e^{-Ct} |x_0|$.

25.6 The Linearization About Nonequilibrium States

The purpose of this section is to show how the theory is modified when the linearization is carried out about a nonequilibrium rest state (χ_0, θ_0), where $\theta_0(\cdot)$ is a field of temperature.

25.6.1 The Linearization About a General State The linearization of the equations of the thermomechanics about the general rest state (χ_0, θ_0) is done in a way similar to that of Subsect. 25.1.4. The resulting system is

$$\rho_0 \ddot{\boldsymbol{u}} = \operatorname{Div} (\bar{\boldsymbol{A}}[\nabla \boldsymbol{u}] + \boldsymbol{V}[\nabla \dot{\boldsymbol{u}}] + \bar{\boldsymbol{M}} \vartheta + \boldsymbol{G}[\nabla \boldsymbol{u}] + \boldsymbol{F} \vartheta + \boldsymbol{D}[\nabla \vartheta]), \tag{25.6.1}$$

$$\rho_0 \bar{c}_{\boldsymbol{F}} \dot{\vartheta} = -\operatorname{Div} (\boldsymbol{P} \vartheta + \boldsymbol{Q}[\nabla \boldsymbol{u}] - \boldsymbol{K}[\nabla \vartheta] - \boldsymbol{E}[\nabla \dot{\boldsymbol{u}}]) + \theta_0 \bar{\boldsymbol{M}} \cdot \nabla \dot{\boldsymbol{u}}, \tag{25.6.2}$$

where in addition to the coefficients of Subsect. 25.1.4 there are also the following coefficients:

$$\boldsymbol{F} = \partial_\theta \bar{\boldsymbol{S}}_d, \quad \boldsymbol{G} = \partial_{\boldsymbol{F}} \bar{\boldsymbol{S}}_d, \quad \boldsymbol{P} = \partial_\theta \bar{\boldsymbol{q}}, \quad \boldsymbol{Q} = \partial_{\boldsymbol{F}} \bar{\boldsymbol{q}}. \tag{25.6.3}$$

The quantities $\bar{c}_{\boldsymbol{F}}$, $\bar{\boldsymbol{A}}$, $\bar{\boldsymbol{M}}$ are evaluated at $(\boldsymbol{F}_0(\boldsymbol{p}), \theta_0(\boldsymbol{p}))$, the quantities $\boldsymbol{V}$, $\boldsymbol{D}$, $\boldsymbol{E}$, $\boldsymbol{K}$ and the partial derivatives of the response functions are evaluated at $(\boldsymbol{F}_0(\boldsymbol{p}), \theta_0(\boldsymbol{p}), \boldsymbol{0}, \boldsymbol{G}_0)$, where $\boldsymbol{F}_0 = \nabla \chi_0$, $\boldsymbol{G}_0 = \nabla \theta_0$. All these functions figure as time independent coefficients which depend at most on $\boldsymbol{p}$ through their arguments $\boldsymbol{F}_0(\boldsymbol{p})$, $\theta_0(\boldsymbol{p})$, $\boldsymbol{0}$, $\boldsymbol{G}_0$.

We retain the definitions of $f^\flat$, $e^\flat$, $f_0^\flat$, $\eta^\flat$, $\boldsymbol{S}_e^\flat$, $\boldsymbol{S}_d^\flat$, but it must be observed that θ_0 is a field, and in the previous treatment the canonical free energy was introduced only in the case where θ_0 was uniform. The response of the stress is written

$$\boldsymbol{S}^\flat = \boldsymbol{S}_e^\flat + \boldsymbol{S}_d^\flat + \boldsymbol{S}_f^\flat, \quad \boldsymbol{S}_e^\flat = \bar{\boldsymbol{A}}[\nabla \boldsymbol{u}] + \bar{\boldsymbol{M}} \vartheta, \quad \boldsymbol{S}_d^\flat = \boldsymbol{V}[\nabla \dot{\boldsymbol{u}}] + \boldsymbol{D}[\boldsymbol{G}], \quad \boldsymbol{S}_f^\flat = \boldsymbol{G}[\nabla \boldsymbol{u}] + \boldsymbol{F} \vartheta,$$

where $\boldsymbol{S}_e^\flat$ is the linearized expression for the equilibrium stress, $\boldsymbol{S}_d^\flat$ for the part of the dynamic stress that depends on the gradients of velocity and temperature and $\boldsymbol{S}_f^\flat$ the part of the dynamic stress depending on the gradient of displacement and on the excess temperature. It will be seen below that each of these three parts behaves in a different way. The heat flux is written as

$$\boldsymbol{q}^\flat = \boldsymbol{q}_d^\flat + \boldsymbol{q}_f^\flat, \quad \boldsymbol{q}_d^\flat = -\boldsymbol{K}[\nabla \vartheta] - \boldsymbol{E}[\nabla \dot{\boldsymbol{u}}], \quad \boldsymbol{q}_f^\flat = \boldsymbol{P} \vartheta + \boldsymbol{Q}[\nabla \boldsymbol{u}],$$

where again $\boldsymbol{q}^\flat$ is the part of the linearized heat flux that depends on the gradient of temperature and the gradient of velocity, while $\boldsymbol{q}_f^\flat$, is the part of the heat flux that depends on the excess temperature and displacement gradient.

Equations (25.6.1) and (25.6.2) reduce to those of Sect. 25.1 when the linearization is carried out about an equilibrium state. Indeed, then $\boldsymbol{S}_f^\flat = \boldsymbol{0}$, $\boldsymbol{q}_f^\flat = \boldsymbol{0}$ identically, since if $\nabla \theta_0 = \boldsymbol{0}$ then the partial derivatives occurring in the definitions (25.6.3) are evaluated at the point $(\boldsymbol{F}_0(\boldsymbol{p}), \theta_0, \boldsymbol{0}, \boldsymbol{0})$, and it suffices to recall that $\bar{\boldsymbol{S}}_d(\boldsymbol{F}, \theta, \boldsymbol{0}, \boldsymbol{0}) = \boldsymbol{0}$, $\bar{\boldsymbol{q}}(\boldsymbol{F}, \theta, \boldsymbol{0}, \boldsymbol{0}) = \boldsymbol{0}$ to realize that all the quantities in (25.6.3) vanish.

25.6.2 The Energy Identities For the linearization about a nonequilibrium state, the energy identity reads

$$\rho_o(\tfrac{1}{2}\dot{u}^2 + f_0^b)^{\cdot} = \mathrm{Div}(S^{bT}\dot{u} - q^b\vartheta/\theta_0) - \beta - \zeta \tag{25.6.4}$$

where $\beta = \beta(\nabla\dot{u}, \nabla\vartheta)$ is given by (25.1.8) and $\zeta = \zeta(\nabla u, \nabla\dot{u}, \vartheta, \nabla\vartheta)$ by

$$\zeta = q_d^b\vartheta \cdot \nabla\theta_0/\theta_0^2 - q_f^b \cdot \nabla\vartheta/\theta_0 + q_f^b\vartheta \cdot \nabla\theta_0/\theta_0^2 + S_f^b \cdot \nabla u. \tag{25.6.5}$$

Unfortunately the right-hand side of (25.2.1) can have arbitrary sign. This is so because the thermodynamics imposes no restriction on the signs of the coefficients ***F***, ***G***, ***P***, ***Q*** in the expressions for S_f^b, q_f^b. Even when the reference state is only slightly nonequilibrium in the sense specified below, the expression on the right-hand side can be an arbitrarily large positive quantity. Nevertheless, this identity will be shown below to lead to Liapunov functions with suitable properties, but finer methods than the mere positive definiteness of the production of entropy must be employed. Identity (25.2.3) generalizes to

$$(\rho_o u \cdot v)^{\cdot} = \mathrm{Div}(S^{bT}u) - \bar{A}[\nabla u, \nabla u] + \rho_o v^2 - \vartheta\bar{M} \cdot \nabla u - S_d^b \cdot \nabla u - S_f^b \cdot \nabla u. \tag{25.6.6}$$

25.6.3 Proposition *For every $\alpha \in \mathbb{R}$ and every process (u, v, ϑ) satisfying the equations of linearized dynamics and the boundary conditions we have*

$$\begin{aligned}\frac{d}{dt}\int_\Omega(\tfrac{1}{2}\dot{u}^2 + f_0^b + \alpha u \cdot v)\rho_o\,dV &= -\alpha\int_\Omega(S_d^b \cdot \nabla u + S_f^b \cdot \nabla u)\,dV \\ &\quad - \alpha\int_\Omega(\bar{A}[\nabla u, \nabla u] - \rho_o v^2 + \bar{M} \cdot \nabla u\vartheta)\,dV - \int_\Omega(\beta + \zeta)\,dV.\end{aligned} \tag{25.6.7}$$

Most of the results of the previous sections remain valid if the reference state differs only slightly from the equilibrium state. Consider an equilibrium state of the body with exact constitutive equations with the deformation $\chi^\#$ and temperature $\theta^\#$. A state (χ_0, θ_0) is said to be a ***slightly nonequilibrium state of tolerance*** $\delta > 0$ if

$$|\nabla\chi_0(p) - \nabla\chi^\#(p)| \le \delta, \quad |\theta_0(p) - \theta^\#| \le \delta, \quad |\nabla\theta_0(p)| \le \delta \text{ for every } p \in \Omega.$$

25.6.4 Proposition *Let $(\chi^\#, \theta^\#)$ be an equilibrium state in which the material is uniformly dissipative. Then for every $\varepsilon > 0$ there exists a $c > 0$ and $\delta > 0$ such that the kinetic coefficients at any slightly nonequilibrium state of tolerance δ satisfy*

$$\beta(\nabla u, \nabla\vartheta) \ge c(|\nabla u|^2 + |\nabla\vartheta|^2), \tag{25.6.8}$$

$$\zeta(\nabla u, \nabla\dot{u}, \vartheta, \nabla\vartheta) \le \varepsilon(|\nabla u|^2 + |\nabla v|^2 + |\vartheta|^2 + |\nabla\vartheta|^2) \tag{25.6.9}$$

for every process (u, v, ϑ).

Proof Inequality (25.6.8): For a linearization about any state the coefficients ***V***, ***D***, ***E***, ***K*** occurring in the expressions for the quantities S_d^b, q_d^b are continuous functions of the deformation gradient $F(p)$, temperature $\theta(p)$, and gradient of temperature $\nabla\theta(p)$. For the equilibrium state $(\chi^\#, \theta^\#)$ they satisfy the strict dissipativity with a constant $c' > 0$. It then follows from a uniform continuity argument that, for deformation gradients, temperatures, and gradients of temperature close to those of the equilibrium state, the inequality of strict dissipativity holds with possibly smaller constant, say $c'/2$. But this means that (25.6.8) holds with $c = c'/2$. Inequality (25.6.9) follows by a similar continuity argument. The quantity ζ vanishes identically for all ∇u, $\nabla\dot{u}$, ϑ, $\nabla\vartheta$ when it is evaluated in the equilibrium state $(\chi^\#, \theta^\#)$, as a consequence of the fact that all the coefficients of the quadratic form ζ vanish. These coefficients are continuous functions of the deformation gradient, temperature, and the gradient of temperature, which leads to the assertion. □

25.6.5 Proposition *Let $\sigma^\# = (\chi^\#, \theta^\#)$ be an equilibrium state in which* J *is uniformly positive (see Definition 20.4.1(3)) and in which the material is uniformly dissipative. Then there exist $c_1 > 0$, $c_2 > 0$, $\delta > 0$ and $\alpha > 0$ such that for the linearization about a slightly nonequilibrium reference state of tolerance δ from the state $\sigma^\#$ one has*

$$c_1\|x\|_H^2 \le \int_\Omega(\tfrac{1}{2}\dot{u}^2 + f_0^b + \alpha u \cdot v)\rho_o\,dV \le c_2\|x\|_H^2, \quad (x = (u, v, \vartheta)) \tag{25.6.10}$$

and

$$\frac{d}{dt}\int_\Omega(\tfrac{1}{2}\dot{u}^2 + f_0^b + \alpha u \cdot v)\rho_o\,dV \le -c(\alpha)\int_\Omega(\tfrac{1}{2}\dot{u}^2 + f_0^b + \alpha u \cdot v)\rho_o\,dV \tag{25.6.11}$$

for every process.

Proof The proof is based on the identity (25.6.7). Since at $\sigma^{\#}$ the second variation of energy is uniformly positive, one can invoke Propositions 25.2.3 and 25.2.4 to find $\alpha > 0$, α small, such that the conclusions of these propositions hold for the linearization about the state $\sigma^{\#}$. Next it is shown that (25.6.10) holds for every nonequilibrium state of sufficiently small tolerance. This is possible in view of the following facts. (1) the state $\sigma^{\#}$ satisfies such an estimate; (2) the coefficients occurring in the expression for the integral (25.6.10) are continuous functions of the deformation gradient and temperature. Hence if they are uniformly close to those in the equilibrium state $\sigma^{\#}$, the inequality still holds, possibly with slightly different constants. Next Proposition 25.6.4 is invoked to find that if the tolerance is diminished (if necessary), then the material coefficients $\boldsymbol{V}, \boldsymbol{D}, \boldsymbol{E}, \boldsymbol{K}$ corresponding to a linearization about any state σ of small tolerance from $\sigma^{\#}$ will satisfy the conclusions (25.6.8) with some prescribed constant $c > 0$. Knowing that, it is possible to reduce α (if necessary) so that the integral $\int_\Omega \beta(\nabla \boldsymbol{u}, \nabla\vartheta)\, dV$ on the right-hand side of the identity (25.6.7) can be estimated to provide

$$\frac{d}{dt}\int_\Omega (\tfrac{1}{2}\dot{\boldsymbol{u}}^2 + f_0^b + \alpha \boldsymbol{u}\cdot\boldsymbol{v})\rho_\circ\, dV \le -c\int_\Omega (\tfrac{1}{2}\dot{\boldsymbol{u}}^2 + f_0^b + \alpha\boldsymbol{u}\cdot\boldsymbol{v})\rho_\circ\, dV - \int_\Omega \zeta\, dV \tag{25.6.12}$$

This is done by steps completely analogous to those in the proof of Proposition 25.2.4. Finally Proposition 25.6.4 is invoked to find a final tolerance for the slightly nonequilibrium state in such a way that the second integral in (25.6.12) bounded by

$$\varepsilon\int_\Omega (|\nabla\boldsymbol{u}|^2 + |\nabla\boldsymbol{v}|^2 + |\vartheta|^2 + |\nabla\vartheta|^2)\, dV$$

to obtain

$$\begin{aligned}\frac{d}{dt}\int_\Omega (\tfrac{1}{2}\dot{\boldsymbol{u}}^2 + f_0^b + \alpha\boldsymbol{u}\cdot\boldsymbol{v})\rho_\circ\, dV \le &- c\int_\Omega (\tfrac{1}{2}\dot{\boldsymbol{u}}^2 + f_0^b + \alpha\boldsymbol{u}\cdot\boldsymbol{v})\rho_\circ\, dV \\ &+ \varepsilon\int_\Omega (|\nabla\boldsymbol{u}|^2 + |\nabla\boldsymbol{v}|^2 + |\vartheta|^2 + |\nabla\vartheta|^2)\, dV.\end{aligned} \tag{25.6.13}$$

Hence, for sufficiently small ε, which can be chosen independently of the reference state (and of course of the process), the right-hand side is bounded by

$$-\frac{c}{2}\int_\Omega (\tfrac{1}{2}\dot{\boldsymbol{u}}^2 + f_0^b + \alpha\boldsymbol{u}\cdot\boldsymbol{v})\rho_\circ\, dV,$$

which is the result. □

25.6.6 Proposition *Let $\sigma^{\#} = (\chi^{\#}, \theta^{\#})$ be an equilibrium state in which* J *is uniformly positive (see Definition* 20.4.1(3)) *and in which the material is uniformly dissipative. Then there exists a $\delta > 0$ such that the linearization about a slightly nonequilibrium state of tolerance δ has the property of uniqueness and the solutions decay exponentially to the reference state.*

That is, if the equilibrium state $\sigma^{\#}$ is exponentially asymptotically stable, then all nonequilibrium states sufficiently close to the equilibrium state are also exponentially asymptotically stable.

Proof By continuity, the energy is coercive for any state sufficiently close to $\sigma^{\#}$. Also the inequality of strict dissipativity holds for all reference states sufficiently close to $\sigma^{\#}$. An appeal to formula (25.6.7), Proposition 25.6.4, and the proof of Proposition 25.6.5 then shows that for a linearization about states sufficiently close to $\sigma^{\#}$ one still has

$$\frac{d}{dt}\int_\Omega (\tfrac{1}{2}\boldsymbol{v}^2 + f_0^b(\cdot, t))\rho_\circ\, dV \le 0.$$

With this knowledge, it is now possible to complete the proof in the same way as in Proposition 25.3.1. To prove the exponential decay, use Proposition 25.6.5 and repeat the proof above. □

Bibliographical Note

Section 25.2 is partly based on unpublished work of C. E. Beevers & Šilhavý (1980); see also Beevers & Šilhavý [1988]. There is a large literature on the existence theory for purely thermoelastic bodies or thermoelastic bodies with viscosity in one spatial dimension. For isothermal elastic one-dimensional bodies with viscosity, see Greenberg, MacCamy & Mizel [1968], Dafermos [1969], Greenberg [1969], Greenberg & MacCamy [1970], Andrews [1980], Pego [1987]; for one-dimensional thermoelasticity with heat conduction, see Slemrod [1981], Hrusa & Tarabek

[1989], RACKE & al. [1993], and for one-dimensional thermoviscoelasticity, see DAFERMOS [1982], DAFERMOS & HSIAO [1982], JIANG [1993, 1995]. In the three dimensional case, there is also an existence and asymptotic stability result by POTIER-FERRY [1982], which deals with an isothermal, nonlinearly elastic body with viscosity, but only with prescribed displacement on the whole boundary. The reader is also referred to the review by RACKE [1992] for further references. The dynamical system generated by the equations of three-dimensional linear thermoelasticity can have a nontrivial attractor consisting in isothermal vibrations of the body. The reader is referred to DAFERMOS [1968] for details. Classical treatises, such as BOLEY & WEINER [1960], CARSLAW & JAEGER [1959], and CARLSON [1972] contain the applications of the theory.

References

ABEYARATNE, R., & KNOWLES, J. K.:

[1987] *Non-elliptic elastic materials and the modeling of elastic-plastic behavior for finite deformation.* J. Mech. Phys. Solids **35,** 343–365

[1988] *On the dissipative response due to discontinuous strains in bars of unstable elastic material.* Int. J. Solids Structures **24,** 1021–1044

[1990] *On the driving traction acting on a surface of strain discontinuity in a continuum.* J. Mech. Phys. Solids **38,** 345–360

[1991a] *Kinetic relations and the propagation of phase boundaries in solids.* Arch. Rational Mech. Anal. **114,** 119–154

[1991b] *Implication of viscosity and strain-gradient effects for the kinetics of propagating phase boundaries in solids.* Siam J. Appl. Math. **38,** 345–360

[1992] *On the propagation of maximally dissipative phase boundaries in solids.* Quart. Appl. Math. **50,** 149–172

[1994] *Dynamics of propagating phase boundaries: Thermoelastic solids with heat conduction.* Arch. Rational Mech. Anal. **126,** 203–230

ABRAHAM, R., & MARSDEN, J. E.:

[1978] *Foundations of mechanics.* The Benjamin/Cummings Publishing Co., London

ACERBI, E., BUTTAZZO, G., & FUSCO, N.:

[1983] *Semicontinuity and relaxation for integrals depending on vector-valued functions.* J. Math. Pures Appl. **62,** 371–378

ACERBI, E., & FUSCO, N.:

[1984] *Semicontinuity problems in the calculus of variations.* Arch. Rational Mech. Anal. **83,** 125–145

AGMON, S., DOUGLIS, A., & NIRENBERG, L.:

[1964] *Estimates near the boundary for solutions of elliptic partial differential equations satisfying general boundary conditions II.* Comm. Pure Appl. Math. **17,** 35–92

ALEXEEV, V. M., TICHOMIROV, V. M., & FOMIN, C. B.:

[1979] *Optimal Control.* Nauka, Moscow

ALIBERT, J.-J., & DACOROGNA, B.:

[1992] *An example of a quasiconvex function that is not polyconvex in two dimensions.* Arch. Rational Mech. Anal. **117,** 155–166

AMAR, M., & DE CICCO, V.:

[1994] *Relaxation of quasi-convex integrals of arbitrary order.* Proc. Royal Soc. Edinburgh **124A,** 927–946

ANDREWS, G.:

[1980] *On the existence of solutions to the equation $u_{tt} = u_{xxt} + \sigma(u_x)_x$.* J. Diff. Equations **35,** 200–231

ANGENENT, S. B., & GURTIN, M. E.:

[1989] *Multiphase thermomechanics with interfacial structure, 2. Evolution of an isothermal interface.* Arch. Rational Mech. Anal. **108,** 323–391

ANTMAN, S. S.:

[1970] *Existence of solutions of the equilibrium equations for nonlinearly elastic rings and arches.* Indiana Univ. Math. J. **20,** 281–302

[1971] *Existence and non-uniqueness of axisymmetric equilibrium states of nonlinearly elastic shells.* Arch. Rational Mech. Anal. **40,** 329–372

[1972] *The theory of rods.* In *Handbuch der Physik, VIa/2,* Springer, Berlin

[1995] *Nonlinear problems of elasticity.* Springer, New York

ANTMAN, S. S., & OSBORN, J. E.:
[1979] *The principle of virtual work and integral laws of motion.* Arch. Rational Mech. Anal. **69,** 231–262

ANTONTSEV, S. N., KAZHIKHOV, A. V., & MONAKHOV, V. N.:
[1990] *Boundary value problems in mechanics of inhomogeneous fluids.* North-Holland, Amsterdam

ARENS, R.:
[1963] *An axiomatic basis for classical thermodynamics.* J. Math. Anal. Appl. **6,** 207

AUBERT, G.:
[1987] *On a counterexample of a rank 1 convex function which is not polyconvex in the case $N = 2$.* Proc. Royal Soc. Edinburgh **106A,** 237–240

AUBERT, G., & TAHRAOUI, R.:
[1980] *Sur la faible fermeture de certains ensembles de contrainte en élasticité non-linéaire plane.* C. R. Acad. Sci. Paris **290,** 537–540
[1987] *Sur la faible fermeture de certains ensembles de contrainte en elasticite non-lineaire plane.* Arch. Rational Mech. Anal. **97,** 33–58

BAKER, M., & ERICKSEN, J. E.:
[1954] *Inequalities restricting the form of the stress-deformation relations for isotropic elastic solids and Reiner-Rivlin fluids.* J. Wash. Acad. Sci. **44,** 33–35

BALL, J. M.:
[1977] *Convexity conditions and existence theorems in nonlinear elasticity.* Arch. Rational Mech. Anal. **63,** 337–403
[1980] *Strict convexity, strong ellipticity, and regularity in the calculus of variations.* Math. Proc. Camb. Phil. Soc. **87,** 501–513
[1981] *Global invertibility of Sobolev functions and the interpenetration of matter.* Proc. Royal Soc. Edinburgh **88A,** 315–328
[1982] *Discontinuous equilibrium solutions and cavitation of non-linear elasticity.* Phil. Trans. Royal Soc. London **A306,** 557–611
[1983] *Minimizers and the Euler-Lagrange equations.* In *Proc. of ISIMM Conference, Paris,* Springer, Berlin
[1984] *Differentiability properties of symmetric and isotropic functions.* Duke Math. J. **51,** 699–728
[1987] *Does rank-one convexity imply quasiconvexity?.* In *Metastability and incompletely posed problems,* S. S. Antman, J. L. Ericksen, D. Kinderlehrer, & I. Müller (ed.), pp. 17–32. Springer, Berlin
[1989] *A version of the fundamental theorem for Young measures.* In *Partial differential equations and continuum models of phase transitions,* M. Rascle, D. Serre, & M. Slemrod (ed.), pp. 207–215. Springer, Berlin
[1990] *Sets of gradients with no rank-one connections.* J. Math. pures appl. **69,** 241–259

BALL, J. M., CURRIE, J. C., & OLVER, P. J.:
[1981] *Null lagrangians, weak continuity and variational problems of any order.* J. Funct. Anal. **41,** 135–174

BALL, J. M., & JAMES, R. D.:
[1987] *Fine phase mixtures as minimizers of energy.* Arch. Rational Mech. Anal. **100,** 13–52
[1992] *Proposed experimental tests of a theory of fine microstructure and the two-well problem.* Phil. Trans. Royal Soc. Lond. **338,** 389–450

BALL, J. M., & KNOWLES, G.:
[1986] *Liapunov functions for thermomechanics with spatially varying boundary temperature.* Arch. Rational Mech. Anal. **92,** 193–204

BALL, J. M., & MARSDEN, J. E.:
[1984] *Quasiconvexity at the boundary, positivity of the second variation and elastic stability.* Arch. Rational Mech. Anal. **86,** 251–277

BALL, J. M., & MURAT, F.:
[1984] *$W^{1,p}$-quasiconvexity and variational problems for multiple integrals.* J. Funct. Anal. **58,** 225–253

BATRA, R. C.:
[1972] *On non-classical boundary conditions.* Arch. Rational Mech. Anal. **48,** 163–191

BAUMAN, P., OWEN, N. C., & PHILLIPS, D.:
[1991] *Maximum principles and a priori estimates for a class of problems in nonlinear elasticity.* Ann. Inst. H. Poincaré, Analyse non linéaire **8,** 119–157

BEATTY, M. F.:
[1970] *Stability of hyperelastic bodies subject to hydrostatic loading.* Nonlinear. Mech. **5,** 367–383
[1987] *Topics in finite elasticity: hyperelasticity of rubber, elastomers, and biological tissues–with examples.* Appl. Mech. Rev. **40,** 1699–1734
BEATTY, M. F., & STALNAKER, D. O.:
[1986] *The Poisson function of finite elasticity.* J. Appl. Mech. **53,** 807–813
BEEVERS, C. E., & ŠILHAVÝ, M.:
[1988] *The asymptotic behavior of classical solutions to the mixed initial-boundary value problem in finite thermo-viscoelasticity.* Quart. Appl. Math. **46,** 319–329
BEIRÃO DA VEIGA, H.:
[1989] *Long time behavior for one-dimensional motion of a general barotropic viscous fluid.* Arch. Rational Mech. Anal. **108,** 141–160
BERDICHEVSKY, V. L.:
[1983] *Variational principles of continuum mechanics.* Nauka, Moscow
BESSELING, J. F., & VAN DER GIESSEN, E.:
[1993] *Mathematical modelling of inelastic deformation.* Chapman & Hall, London
BHARATHA, S., & LEVINSON, M.:
[1980] *A theory of elastic foundations.* Arch. Rational Mech. Anal. **74,** 249
BHATTACHARYA, K.:
[1991] *Wedge-like microstructure in martensites.* Acta Metall. **39,** 2431–2444
[1992] *Self-accommodation in martensite.* Arch. Rational Mech. Anal. **120,** 201–244
[1993] *Comparison of the geometrically nonlinear and linear theories of martensitic transformation.* Continuum Mech. Thermodyn. **5,** 205–242
BHATTACHARYA, K., FIROOZYE, N. B., JAMES, R. D., & KOHN, R. V.:
[1994] *Restrictions on microstructure.* Proc. Royal Soc. Edinburgh **124A,** 843–878
BIOT, M. A.:
[1965] *Mechanics of incremental deformations.* Wiley, New York
BISHOP, R. L., & GOLDBERG, S. I.:
[1968] *Tensor analysis on manifolds.* Macmillan, New York
BLATZ, P. J., & KO, W. L.:
[1962] *Application of finite elasticity to the deformation of the rubbery materials.* Trans. Soc. Rheology **6,** 223–251
BOJARSKI, B., IWANIEC, T.:
[1983] *Analytical foundations of the theory of quasiconformal mappings in* $\mathbb{R}^n$. Ann. Acad. Sci. Fen. Ser. A. I. **8,** 157–324
BOLEY, B. A., & WEINER, J. H.:
[1960] *Theory of thermal stresses.* Wiley, New York
BOURBAKI, N.:
[1966] *Elements of Mathematics. General Topology.* Addison Wesley, New York
BOWEN, R. M.:
[1967] *Toward a thermodynamics and mechanics of mixtures.* Arch. Rational Mech. Anal. **24,** 370–403
[1976] *Theory of mixtures.* In *Continuum Physics, vol. III,* A. C. Eringen (ed.), Academic Press, New York
BOWEN, R. M., & WANG, C.-C.:
[1970] *Acceleration waves in inhomogeneous isotropic elastic bodies.* Arch. Rational Mech. Anal. **38,** 13–45
[1971] *Acceleration waves in inhomogeneous isotropic elastic bodies.* Arch. Rational Mech. Anal. **40,** 403
BOWLES, J. S., & MACKENZIE, J. K.:
[1954] *The crystallography of martensitic transformations I & II.* Acta Met. **2,** 129–127 & 138–147
BRADLEY, C. J., & CRACKNELL, A. P.:
[1972] *The mathematical theory of symmetry in solids.* Clarendon Press, Oxford
BRANDON, D., & HRUSA, W. J.:
[1988] *Construction of a class of integral models for heat flow in materials with memory.* J. Integral Equations and Applications **1,** 175–201
BRONSARD, L., & REITICH, F.:
[1993] *On the phase boundary motion and singular limit of a vector-valued Ginzburgh–Landau equation.* Arch. Rational Mech. Anal. **124,** 355

BURGESS, I. W., & LEVINSON, M.:
[1972] *The instability of slightly compressible rectangular rubberlike solids under biaxial loadings.* Int. J. Solids Structures **8,** 133–148
BUTTAZZO, G.:
[1989] *Semicontinuity, relaxation and integral representation in the calculus of variations. Pitman Res. Notes Mathematics,* vol. **207**, Longman, Harlow
CAHN, J. W.:
[1961] *On spinodal decomposition.* Act. Metall. **9,** 795–801
[1962] *On spinodal decomposition in cubic crystals.* Act. Metall. **10,** 179–183
CAHN, J. W., & HILLIARD, J. E.:
[1958] *Free energy of a nonuniform system. I. Interfacial free energy.* J. Chem. Phys. **28,** 258–267
[1971] *Spinodal decomposition: a reprise.* Act. Metall. **19,** 151–161
CARATHÉODORY, C.:
[1909] *Untersuchungen über die Grundlagen der Thermodynamik.* Math. Ann. **67,** 355–386
CARLSON, D. E.:
[1972] *Linear thermoelasticity.* In *Handbuch der Physik, VIa/2,* Springer, Berlin
CARLSON, D. E., & HOGER, A.:
[1986] *The derivative of a tensor-valued function of a tensor.* Quart. Appl. Math. **44,** 409–423
CARRASSI, M., & MORRO, A.:
[1972] *A modified Navier-Stokes equation and its consequences on sound dispersion.* Nuovo Cimento **9B,** 321
CARSLAW, H. S., & JAEGER, J. C.:
[1959] *Conduction of heat in solids.* Oxford University Press, Oxford
CERMELLI, P., & GURTIN, M. E.:
[1994] *The dynamics of solid-solid phase transitions. 2. Incoherent interfaces.* Arch. Rational Mech. Anal. **127,** 41–99
CESARI, L.:
[1983] *Optimization–theory and applications.* Springer, New York
CHADWICK, P.:
[1975] *Applications of an energy-momentum tensor in non-linear elastostatics.* J. Elasticity **5,** 249–258
[1989] *Wave propagation in transversely isotropic media, I, II, III.* Proc. Royal Soc. Lond. **A422,** 23–121
[1993] *Wave propagation in incompressible transversely isotropic media, I.* Proc. Royal Ir. Acad. **93A,** 231–253
CHADWICK, P., & OGDEN, R. W.:
[1971a] *On the definition of elastic moduli.* Arch. Rational Mech. Anal. **44,** 41–53
[1971b] *A theorem of tensor calculus and its application to isotropic elasticity.* Arch. Rational Mech. Anal. **44,** 54–68
CHADWICK, P., & SMITH, G. D.:
[1977] *Foundations of the theory of surface waves in anisotropic materials.* In *Advances in applied mechanics, vol. 17,* C-S. Yih (ed.), pp. 303–376. Academic Press, New York
CHEN, Y.-C.:
[1991] *On strong ellipticity and the Legendre-Hadamard condition.* Arch. Rational Mech. Anal. **113,** 165–175
CHEREPANOV, G. P.:
[1967] *Crack propagation in continuous media.* Prikl. Math. Mekh. **31,** 476–488
[1968] *Cracks in solids.* Int. J. Solids Structures **4,** 811–831
CHILLINGWORTH, D. R. J., MARSDEN, J. E., & WAN, Y. H.:
[1982] *Symmetry and bifurcation in three-dimensional elasticity, Part I.* Arch. Rational Mech. Anal. **80,** 295–331
[1983] *Symmetry and bifurcation in three-dimensional elasticity, Part II.* Arch. Rational Mech. Anal. **83,** 363–395
CHIPOT, M., & KINDERLEHRER, D.:
[1988] *Equilibrium configurations of crystals.* Arch. Rational Mech. Anal. **103,** 237–277
CHRISTENSEN, R. M.:
[1979] *Mechanics of composite materials.* Wiley, New York
CIARLET, P. G.:
[1988] *Mathematical elasticity.* North-Holland, Amsterdam

CIARLET, P. G., & NEČAS, J.:
[1987] *Injectivity and self-contact in nonlinear elasticity.* Arch. Rational Mech. Anal. **97,** 171–188

CODDINGTON, E. A., & LEVINSON, N.:
[1955] *Theory of ordinary differential equations.* McGraw-Hill, New York

COLEMAN, B. D.:
[1964a] *Thermodynamics of materials with memory.* Arch. Rational Mech. Anal. **17,** 1–46
[1964b] *On thermodynamics, strain impulses, and viscoelasticity.* Arch. Rational Mech. Anal. **17,** 230–254
[1970] *On the stability of equilibrium states of general fluids.* Arch. Rational Mech. Anal. **36,** 1–32

COLEMAN, B. D., & DILL, E. H.:
[1968] *On the stability of certain motions of incompressible materials with memory.* Arch. Rational Mech. Anal. **30,** 197–224
[1971] *On the thermodynamics of electromagnetic fields in materials with memory.* Arch. Rational Mech. Anal. **41,** 132–162
[1973] *On thermodynamics and the stability of motions of materials with memory.* Arch. Rational Mech. Anal. **51,** 1–53

COLEMAN, B. D., FABRIZIO, M., & OWEN, D.:
[1982] *On the thermodynamics of second sound in dielectric crystals.* Arch. Rational Mech. Anal. **80,** 135–158
[1986] *Thermodynamics and the constitutive relations for second sound in crystals.* In *New Perspectives in Thermodynamics,* J. Serrin (ed.), pp. 171–184. Springer, Berlin

COLEMAN, B. D., & GREENBERG, J. M.:
[1967] *Thermodynamics and the stability of fluid motion.* Arch. Rational Mech. Anal. **25,** 321–341

COLEMAN, B. D., HRUSA, W. J., & OWEN, D. R.:
[1986] *Stability of equilibrium for a nonlinear hyperbolic system describing heat propagation by second sound in solids.* Arch. Rational Mech. Anal. **94,** 267–289

COLEMAN, B. D., & LAI, P.-H.:
[1994] *Waves of discontinuity and sinusoidal waves in the theory of second sound is solids.* Arch. Rational Mech. Anal. **126,** 1–20

COLEMAN, B. D., & NOLL, W.:
[1959] *On the thermostatics of continuous media.* Arch. Rational Mech. Anal. **4,** 97–128
[1960] *An approximation theorem for functionals with applications in continuum mechanics.* Arch. Rational Mech. Anal. **6,** 355–370
[1961] *Foundations of linear viscoelasticity.* Rev. Mod. Phys. **33,** 239–249
[1963] *The thermodynamics of elastic materials with heat conduction and viscosity.* Arch. Rational Mech. Anal. **13,** 167–178

COLEMAN, B. D., & OWEN, D.:
[1974] *A mathematical foundation for thermodynamics.* Arch. Rational Mech. Anal. **54,** 1–104
[1975] *On thermodynamics and elastic-plastic materials.* Arch. Rational Mech. Anal. **59,** 25–51
[1977] *On the thermodynamics of semi-systems with restrictions on the accessibility of states.* Arch. Rational Mech. Anal. **66,** 173–181

COLEMAN, B. D., OWEN, D., & SERRIN, J.:
[1981] *The second law of thermodynamics for systems with approximate cycles.* Arch. Rational Mech. Anal. **77,** 103–142

CONLON, J.:
[1980] *A theorem in ordinary differential equations with application to hyperbolic conservation laws.* Advances in Mathematics **35,** 1–18

COOPER, J. L. B.:
[1967] *The foundations of thermodynamics.* J. Math. Anal. Appl. **17,** 172–193

COURANT, R., & FRIEDRICHS, K. O.:
[1948] *Supersonic flow and shock waves.* Interscience Publishers, Inc., New York, London

CRANDALL, M. G., & RABINOWITZ, P. H.:
[1971] *Bifurcation from simple eigenvalues.* J. Funct. Anal. **8,** 321–340

DACOROGNA, B.:
[1981] *A relaxation theorem and its application to the equilibrium of gases.* Arch. Rational Mech. Anal. **77,** 359–386
[1982] *Quasiconvexity and relaxation of nonconvex problems in the calculus of variations.* J. Functional. Anal. **46,** 102–118

[1985] *Remarques sur les notions des polyconvexité, quasi-convexité et convexité de rang 1.* J. Math. Purres et Appl. **64,** 403–438
[1989] *Direct methods in the calculus of variations.* Springer, Berlin

DACOROGNA, B., DOUCHET, J., GANGBO, W., & RAPPAZ, J.:
[1990] *Some examples of rank one convex functions in dimension two.* Proc. Royal Soc. Edinburgh **114A,** 135–150

DACOROGNA, B., & FONSECA, I.:
[1992] *A minimization problem involving variation of the domain.* Comm. Pure Appl. Math. **45,** 871–897

DACOROGNA, B., & MARCELLINI, P.:
[1987] *A counterexample in the vectorial calculus of variations.* In *Material Instabilities in Continuum Mechanics,* J. M. Ball (ed.), pp. 77–83. Calrendon Press, Oxford

DAFERMOS, C.:
[1968] *On the existence and asymptotic stability of solutions to the equations of linear thermoelasticity.* Arch. Rational Mech. Anal. **29,** 241–271
[1969] *The mixed initial boundary value problem for the equations of nonlinear one dimensional viscoelasticity.* J. Diff. Equations **6,** 71–86
[1973] *The entropy rate admissibility criterion for solutions of hyperbolic systems of conservation laws.* J. Diff. Equations **14,** 202–212
[1974] *The entropy rate admissibility criterion in thermoelasticity.* Atti Accad. Naz. Lincei, Rend. Cl. Sci. Fis. Mat. Natur. (8) **57,** 113–119
[1979a] *The second law of thermodynamics and stability.* Arch. Rational Mech. Anal. **70,** 167–179
[1979b] *Stability of motions of thermoelastic fluids.* J. Thermal Stresses **2,** 127–134
[1982] *Global smooth solutions to the initial-boundary value problem for the equations of one-dimensional nonlinear thermoviscoelasticity.* SIAM J. Math. Anal. **13,** 397–408
[1986] *Quasilinear hyperbolic systems with involutions.* Arch. Rational Mech. Anal. **94,** 373–389
[1989] *Admissible wave fans in non-linear hyperbolic systems.* Arch. Rational Mech. Anal. **106,** 243–260

DAFERMOS, C., & HSIAO, L.:
[1982] *Global smooth thermomechanical processes in one dimensional nonlinear thermoviscoelasticity.* J. Nonlinear Analysis **6,** 435–454
[1986] *Development of singularities in solutions of the equations of nonlinear thermoelasticity.* Quart. Appl. Math. **44,** 463–474

DAL MASO, G.:
[1993] *An introduction to Γ-convergence.* Birkhäuser, Boston

DAVINI, C.:
[1986] *A proposal for a continuum theory of defective crystals.* Arch. Rational Mech. Anal. **96,** 295–317

DAVINI, C., & PARRY, G.:
[1989] *On defect-preserving deformations in crystals.* Int. J. Plasticity **5,** 337–369
[1991] *A complete list of invariants for defective crystals.* Proc. Royal Soc. Lond. **A432,** 341–365
[1992] *Uniform rearrangement in defective crystals.* Proc. Royal Soc. Lond. **A439,** 247–256

DAY, W. A.:
[1969] *A theory of thermodynamics for materials with memory.* Arch. Rational Mech. Anal. **34,** 85–96
[1972a] *The thermodynamics of simple materials with fading memory.* Springer, Berlin
[1972b] *A condition equivalent to the existence of entropy in classical thermodynamics.* Arch. Rational Mech. Anal. **49,** 151–171

DAY, W. A., & ŠILHAVÝ, M.:
[1977] *Efficiency and the existence of entropy in classical thermodynamics.* Arch. Rational Mech. Anal. **66,** 73–81

DE FIGUEIREDO, D. G.:
[1963] *The coerciveness problem for forms over vector valued functions.* Comm. Pure Appl. Math. **16,** 63–94

DE GROOT, S., & MAZUR, P.:
[1962] *Non-equilibrium thermodynamics.* North-Holland, Amsterdam

DEL PIERO, G., & OWEN, D. R.:
[1993] *Structured deformations of continua.* Arch. Rational Mech. Anal. **124,** 99–155

[1995] *Integral-gradient formulae for structured deformations.* Arch. Rational Mech. Anal. **131,** 121–138

DETTLEFF, G., THOMPSON, P. A., MEIER, G. E. A., & SPECKMANN, H. D.:

[1979] *An experimental study of liquefaction shock waves.* J. Fluid Mech. **95,** 279–304

DIEUDONNÉ, J.:

[1960] *Foundations of modern analysis.* Academic Press, New York

DIEULESAINT, E., & ROYER, D.:

[1980] *Elastic waves in solids. Applications to signal processing.* Wiley, Chichester

DIPERNA, R. J.:

[1979] *Uniqueness of solutions to hyperbolic conservation laws.* Indiana Univ. Math. J. **28,** 137–188

[1985] *Compensated compactness and general systems of conservation laws.* Trans. A. M. S. **292,** 383–420

DONOGHUE, JR, W. F.:

[1974] *Monotone matrix functions and analytic continuation.* Springer, Berlin

DUNN, J. E.:

[1983] *Certain a priori inequalities and a peculiar elastic material.* Quarterly J. Mech. Appl. Math. **36,** 351–363

DUNN, J. E., & FOSDICK, R. L.:

[1974] *Thermodynamics, stability, and boundedness of fluids of complexity 2 and fluids of second grade.* Arch. Rational Mech. Anal. **56,** 191–252

[1978] *Lemmas in thermodynamic stability: a priori constitutive inequalities and induced global bounds on thermodynamic states.* Arch. Rational Mech. Anal. **67,** 183–224

[1980] *The morphology and stability of material phases.* Arch. Rational Mech. Anal. **74,** 1–99

DUNN, J. E., & SERRIN, J.:

[1985] *On the thermodynamics of interstitial working.* Arch. Rational Mech. Anal. **88,** 95–133

ECKART, C.:

[1940] *Thermodynamics of irreversible processes, I, II, & III.* Phys. Rev. **58,** 267, 269, 919

[1948] *Thermodynamics of irreversible processes, IV.* Phys. Rev. **73,** 373–382

EDELEN, D. G. B.:

[1962] *The null set of the Euler–Lagrange operator.* Arch. Rational Mech. Anal. **11,** 117–121

[1973] *On the existence of the symmetry relations and dissipation potentials.* Arch. Rational Mech. Anal. **51,** 218–227

[1981] *Aspects of variational arguments in the theory of elasticity: Facts and folklore.* Int. J. Solids Structures **17,** 729–740

EKELAND, I., & TEMAM, R.:

[1976] *Convex analysis and variational problems.* North-Holland, Amsterdam

ERICKSEN, J. L.:

[1962] *Nilpotent energies in liquid crystal theory.* Arch. Rational Mech. Anal. **10,** 189–196

[1966] *A thermo-kinetic view in elastic stability theory.* Int. J. Solids Structures **2,** 573–580

[1970] *Nonlinear elasticity of diatomic crystals.* Int. J. Solids Structures **6,** 951

[1973] *Loading devices and stability of equilibrium.* Nonlinear Elasticity, Academic Press, New York 161–173

[1979] *On the symmetry of deformable crystals.* Arch. Rational Mech. Anal. **72,** 1–13

[1980] *Some phase transitions in crystals.* Arch. Rational Mech. Anal. **73,** 99–124

[1981] *Some simpler cases of the Gibbs phenomenon for thermoelastic solids.* J. Thermal Stresses **4,** 13–30

[1985] *Some surface defects in unstressed thermoelastic solids.* Arch. Rational Mech. Anal. **88,** 337–345

[1989] *Weak martensitic transformations in Bravais lattices.* Arch. Rational Mech. Anal. **107,** 23–36

[1992] *Bifurcation and martensitic transformations in Bravais lattices.* J. Elasticity **28,** 55–78

ESHELBY, J. D.:

[1951] *The force on an elastic singularity.* Phil. Trans. Royal Soc. London **A244,** 87–112

[1956] *Continuum theory of defects.* In *Solid State Physics, vol. 3,* D. Turnbull (ed.), pp. 79–144. Academic Press, New York

[1970] *Energy relations and the energy–momentum tensor in continuum mechanics.* In *Inelastic behavior of solids,* M. Kanninen, W. Adler, A. Rosenfield, & R. Jaffee (ed.), pp. 77–115. Academic Press, New York

[1975] *The elastic energy-momentum tensor.* J. Elasticity **5,** 321–335

EVANS, L. C., & GARIEPY, R. F.:

[1992] *Measure theory and fine properties of functions.* CRC Press, Inc., Boca Raton

FEDERER, H.:
[1969] *Geometric measure theory.* Springer, New York

FEDOROV, F. I.:
[1968] *The theory of elastic waves in crystals.* Plenum Press, New York

FEINBERG, M.:
[1979] *On Gibbs's phase rule.* Arch. Rational Mech. Anal. **70,** 219–234

FEINBERG, M., & LAVINE, R.:
[1983] *Thermodynamics based on the Hahn–Banach theorem: the Clausius inequality.* Arch. Rational Mech. Anal. **82,** 203–293

FERMI, E.:
[1937] *Thermodynamics.* Prentice-Hall, New York

FICHERA, G.:
[1972] *Existence theorems in elasticity.* In *Handbuch der Physik, VIa/2,* Springer, Berlin

FIROOZYE, N. B.:
[1991] *Optimal use of the translation method and relaxations of variational problems.* Comm. Pure Appl. Math. **44,** 643–678

FONSECA, I.:
[1987] *Variational methods for elastic crystals.* Arch. Rational Mech. Anal. **97,** 189–220
[1988] *The lower quasiconvex envelope of the stored energy functions for an elastic crystal.* J. Math. pures et appl. **67,** 175–195
[1989a] *Interfacial energy and the Maxwell rule.* Arch. Rational Mech. Anal. **106,** 63–95
[1989b] *Phase transitions in elastic solid materials.* Arch. Rational Mech. Anal. **107,** 195–223

FONSECA, I., & MALÝ, J.:
[1996] *Relaxation of multiple integrals in Sobolev spaces below the growth exponent for the energy density.* (To appear)

FONSECA, I., & MÜLLER, S.:
[1993] *Relaxation of quasiconvex functionals in $BV(\Omega, \mathbb{R}^p)$ for integrands $f(x, u, \nabla u)$.* Arch. Rational Mech. Anal. **123,** 1–49

FONSECA, I., & PARRY, G.:
[1992a] *Equilibrium configurations of defective crystals.* Arch. Rational Mech. Anal. **120,** 245–283
[1992b] *On a class of invariant functionals.* Proc. Royal Soc. Lond. **A436,** 317–329

FONSECA, I., & TARTAR, L.:
[1989] *The gradient theory of phase transitions for systems with two potential wells.* Proc. Royal Soc. Edinburgh **A111,** 89–102

FOSDICK, R. L., & HERTOG, B.:
[1990] *Material symmetry and crystals.* Arch. Rational Mech. Anal. **110,** 43–79

FOSDICK, R. L., & SERRIN, J.:
[1975] *Global properties of continuum thermodynamic processes.* Arch. Rational Mech. Anal. **59,**

FRANCFORT, G. A. & MURAT, F.:
[1986] *Homogenization and optimal bounds in linear elasticity.* Arch. Rational Mech. Anal. **94,** 307–334

FUSCO, N., & HUTCHINSON, J. E.:
[1995] *A direct proof for lower semicontinuity of polyconvex functionals.* Manuscripta Math. **87,** 35–50

GEL'FAND, I. M., & FOMIN, S. V.:
[1963] *Calculus of variations.* Prentice-Hall, Englewood Cliffs, NJ

GENT, A. N., & LINDLEY, P. B.:
[1958] *Internal rupture of bonded rubber cylinders in tension.* Proc. Royal Soc. London **A249,** 195–205

GIAQUINTA, M., MODICA, G., & SOUČEK, J.:
[1989] *Cartesian currents, weak diffeomorphisms and existence theorems in non-linear elasticity.* Arch. Rational Mech. Anal. **106,** 97–160
[1992] *Area and the Area Formula.* Rend. Sem. Mat. Fis. Milano **62,** 53–87
[1994] *A weak approach to finite elasticity.* Calc. Var. **2,** 65–100
[1996] *Calculus of variations and cartesian currents.* (To appear)

GILBARG, D.:
[1951] *The existence and limit behavior of the one-dimensional shock layer.* Am. J. Math. **73,** 256–274

GIUSTI, E.:
[1984] *Minimal surfaces and functions of bounded variation.* Birkhäuser, Boston

GRECO, L., & IWANIEC, T.:
[1994] *New inequalities for the Jacobian.* Ann. Inst. H. Poincaré, Analyse non linéaire **11,** 17–35
GREEN, A. E., & NAGHDI, P. M.:
[1978] *The second law of thermodynamics and cyclic processes.* J. Appl. Mech. **45,** 487–492
GREEN, A. E., & RIVLIN, R. S.:
[1964a] *On Cauchy's equations of motion.* ZAMP **15,** 290–292
[1964b] *Theories of elasticity with stress multipoles.* Arch. Rational Mech. Anal. **17,** 85–112
GREEN, A. E., & ZERNA, W.:
[1954] *Theoretical elasticity.* Clarendon Press, Oxford
GREENBERG, J. M.:
[1969] *On the existence, uniqueness and stability of equation* $\rho_0\chi_{tt} = E(\chi_x)\chi_{xx} + \lambda\chi_{xxt}$. J. Math. Anal. Appl. **25,** 575–591
GREENBERG, J. M., & MACCAMY, R. C.:
[1970] *On the exponential stability of solutions of* $E(u_x)u_{xx} + \lambda u_{xtx} = \rho u_{tt}$. J. Math. Anal. Appl. **31,** 406–417
GREENBERG, J. M., MACCAMY, R. C., & MIZEL, V. J.:
[1968] *On the existence, uniqueness and stability of solutions of the equation* $\sigma'(u_x)u_{xx} + \lambda u_{xxt} = \rho_0 u_{tt}$. J. Math. Mech. **17,** 707–728
GRINFELD, M.:
[1980] *Conditions for thermodynamic phase equilibrium in a nonlinear elastic material.* Dokl. AN SSSR **251,** 824–827
[1991] *Thermodynamic methods in the theory of heterogeneous systems.* Longman, Harlow
GRISVARD, P.:
[1989] *Singularités en élasticité.* Arch. Rational Mech. Anal. **107,** 157–180
[1992] *Singularities in boundary value problems.* Masson, Paris
GUO, Z.-H.:
[1984] *Rates of stretch tensors.* J. Elasticity **14,** 263–267
GUO, Z.-H., LEHMANN, TH., LIANG, H., & & MAN, C.-S.:
[1992] *Twirl tensors and the tensor equation* $\mathbf{AX} - \mathbf{XA} = \mathbf{C}$. J. Elasticity **27,** 227–245
GUO, Z.-H., & MAN, C.-S.:
[1992] *Conjugate stress and tensor equation* $\sum_{r=1}^{m} U^{m-r}XU^{r-1} = C$. Int. J. Solids Structures **29,** 2063–2076
GUO, Z.-H., & URBANOWSKI, W.:
[1963] *Stability of non-conservative systems in the theory of elasticity of finite deformations.* Archives of Mechanics **15,** 309–321
GURTIN, M. E.:
[1972] *The linear theory of elasticity.* In *Handbuch der Physik, VIa/2,* Springer, Berlin
[1973] *Thermodynamics and the energy criterion for stability.* Arch. Rational Mech. Anal. **52,** 93–103
[1975] *Thermodynamics and stability.* Arch. Rational Mech. Anal. **59,** 63–96
[1981] *An introduction to continuum mechanics.* Academic Press, Boston
[1983] *Two-phase deformations of elastic solids.* Arch. Rational Mech. Anal. **84,** 1–29
[1988] *Multiphase thermomechanics with interfacial structure, 1. Heat conduction and the capillary balance law.* Arch. Rational Mech. Anal. **103,** 195–221
[1993a] *Thermomechanics of evolving phase boundaries in the plane.* Clarendon Press, Oxford
[1993b] *The dynamics of solid-solid phase transitions. 1. Coherent interfaces.* Arch. Rational Mech. Anal. **123,** 305–335
[1994] *The characterization of configurational forces.* Arch. Rational Mech. Anal. **126,** 387–394
[1995] *The nature of configurational forces.* Arch. Rational Mech. Anal. **131,** 67–100
GURTIN, M. E., & HRUSA, W. J.:
[1991] *On the thermodynamics of viscoelastic materials of single-integral type.* Quart. Appl. Math. **49,** 67–85
GURTIN, M. E., & MARTINS, L. C.:
[1976] *Cauchy's theorem in classical physics.* Arch. Rational Mech. Anal. **60,** 305–324
GURTIN, M. E., & PIPKIN, A. C.:
[1968] *A general theory of heat conduction with finite wave speeds.* Arch. Rational Mech. Anal. **31,** 113–126
GURTIN, M. E., & SPECTOR, S. J.:
[1979] *On stability and uniqueness in finite elasticity.* Arch. Rational Mech. Anal. **70,** 153–165

GURTIN, M. E., & STRUTHERS, A.:
[1990] *Multiphase thermomechanics with interfacial structure, 3. Evolving phase boundaries in the presence of bulk deformation.* Arch. Rational Mech. Anal. **112,** 97–160
HAGAN, R., & SERRIN, J.:
[1986] *Dynamic changes of phase in a van der Waals fluid.* In *New Perspectives in Thermodynamics,* J. Serrin (ed.), pp. 241–260. Springer, Berlin
HAGAN, R., & SLEMROD, M.:
[1983] *The viscosity-capillarity criterion for shocks and phase transitions.* Arch. Rational Mech. Anal. **83,** 333–361
HALMOS, P. R.:
[1958] *Finite-dimensional vector spaces, 2nd ed.* Van Nostrand, Princeton
[1967] *A Hilbert space problem book.* Van Nostrand, Princeton
HANYGA, A.:
[1985] *Mathematical theory of non-linear elasticity.* Polish Scientific Publishers–Ellis Horwood Ltd., Warsaw–Chicester
HARDY, G. H., LITTLEWOOD, J. E., & POLYA, G.:
[1952] *Inequalities.* Cambridge University Press, Cambridge
HATTORI, H.:
[1986a] *The Riemann problem for a van der Waals fluid with the entropy rate admissibility criterion–isothermal case.* Arch. Rational Mech. Anal. **92,** 247–263
[1986b] *The Riemann problem for a van der Waals fluid with the entropy rate admissibility criterion–nonisothermal case.* J. Diff. Equations **65,** 158–174
HAYES, M.:
[1969] *Static implications of the strong-ellipticity condition.* Arch. Rational Mech. Anal. **33,** 181–191
HEAD, A. K., HOWISON, S. D., OCKENDON, J. R., & TIGHE, S. P.:
[1993] *An equilibrium theory of dislocation continua.* SIAM Review **35,** 580–609
HEALEY, T. J., & SIMPSON, H. C.:
[1996] *Global continuation in nonlinear elasticity.* Arch. Rational Mech. Anal. (To appear)
HENRY, D.:
[1981] *Geometric theory of semilinear parabolic equations.* Springer, Berlin
HESTENES, M. R., & MACSHANE, E. J.:
[1949] *A theorem on quadratic forms and its applications in the calculus of variations.* Trans. A. M. S. **47,** 501–512
HILL, R.:
[1957] *On uniqueness and stability in the theory of finite elastic strain.* J. Mech. Phys. Solids **5,** 229–241
[1958] *A general theory of uniqueness and stability in elastic-plastic solids.* J. Mech. Phys. Solids **6,** 236–249
[1968] *Constitutive inequalities for simple materials.* J. Mech. Phys. Solids **16,** 229–242
[1970a] *Constitutive inequalities for isotropic elastic solids under finite strain.* Proc. Royal Soc. Lond. **A 314,** 457–472
[1970b] *Some questions of the behaviour of isotropic solids under small deformation superimposed on finite deformations.* In *Problems in mechanics,* L. I. Sedov (ed.), pp. 459–466. Leningrad
[1981] *Invariance relations in thermoelasticity with generalized variables.* Math. Proc. Camb. Phil. Soc. **90,** 373–384
HILL, R., & HUTCHINSON, J. W.:
[1975] *Bifurcation phenomena in the plane tension test.* J. Mech. Phys. Solids **23,** 239–264
HOGER, A.:
[1987] *The stress conjugate to the logarithmic strain.* Int. J. Solids Structures **23,** 1645–1656
HOGER, A., & CARLSON, D. E.:
[1984] *Determination of the stretch and rotation in the polar decomposition of the deformation gradient.* Quart. Appl. Math. **42,** 113–117
HÖRMANDER, L.:
[1985] *The analysis of partial differential operators III. Pseudo-differential operators.* Springer, Berlin
HORN, R. A., & JOHNSON, CH. R.:
[1986] *Matrix analysis.* Cambridge University Press, Cambridge
HRUSA, W. J., & MESSAOUDI, S. A.:
[1990] *On formation of singularities in one-dimensional nonlinear thermoelasticity.* Arch. Rational Mech. Anal. **111,** 135–151

HRUSA, W., & TARABEK, M. A.:
[1989] *On smooth solutions of the Cauchy problem in one dimensional nonlinear thermoelasticity.* Quart. Appl. Math. **47,** 631–644
HSIAO, L.:
[1980] *The entropy rate admissibility criterion in gas dynamics.* J. Diff. Equations **38,** 226–238
IWANIEC, T., & LUTOBORSKI, C.:
[1993] *Integral estimates for null-Lagrangians.* Arch. Rational Mech. Anal. **125,** 25–79
IWANIEC, T., & SBORDONE, C.:
[1992] *On the integrability of the Jacobian under minimal hypotheses.* Arch. Rational Mech. Anal. **119,** 129–143
JAMES, R. D.:
[1980] *The propagation of phase boundaries in elastic bars.* Arch. Rational Mech. Anal. **73,** 125–158
[1981] *Finite deformation by mechanical twinning.* Arch. Rational Mech. Anal. **77,** 143–176
[1986a] *Phase transformations and non-elliptic free energy functions.* In *New Perspectives in Thermodynamics,* J. Serrin (ed.), pp. 223–239. Springer, Berlin
[1986b] *Displacive phase transformations in solids.* J. Mech. Phys. Solids **34,** 359–384
JAMES, R. D., & KINDERLEHRER, D.:
[1989] *Theory of diffusionless phase transitions.* In *Partial differential equations and continuum models of phase transitions,* M. Rascle, D. Serre, & M. Slemrod (ed.), pp. 51–84. Springer, Berlin
JAMES, R. D., & SPECTOR, S.:
[1991] *The formation of filamentary voids in solids.* J. Mech. Phys. Solids **39,** 783–813
[1992] *Remarks on $W^{1,p}$ quasiconvexity, interpenetration of matter, and function spaces in elasticity.* Ann. Inst. H. Poincaré, Analyse non linéaire **9,** 263–280
JIANG, S.:
[1993] *Global large solutions to initial boundary value problems in one-dimensional nonlinear thermoviscoelasticity.* Quart. Appl. Math. **51,** 731–744
[1995] *Asymptotic stability of shearing flows of incompressible thermoviscous fluids.* Nonlinear Anal. **24,** 1207–1220
JOU, D., & CASAS-VÁZQUEZ, J.:
[1992] *Possible experiment to check the reality of a nonequilibrium temperature.* Phys. Rev. A **45,** 8371–8373
JOU, D., CASAS-VÁZQUEZ, J., & LEBON, G.:
[1988] *Extended irreversible thermodynamics.* Rep. Prog. Phys. **51,** 1105–1179
[1993] *Extended irreversible thermodynamics.* Springer, Berlin
KAGANOVA, I. M., & ROITBURD, A. L.:
[1988] *Equilibrium between elastically-interacting phases.* Zh. Eks. Teor. Fiz. **94,** 156–173
KALMAN, R. E., FALB, P. L., & ARBIB, M. A.:
[1969] *Topics in mathematical systems theory.* McGraw-Hill, New York
KHACHATURYAN, A. G.:
[1967] *Some questions concerning the theory of phase transformations in solids.* Sov. Phys. Solid State **8,** 2163–2168
[1983] *Theory of structural transformations in solids.* J. Wiley, New York
KHACHATURYAN, A. G., & SHATALOV, G. A.:
[1969] *Theory of macroscopic periodicity for a phase transition in the solid state.* Sov. JETP **29,** 557–561
KINDERLEHRER, D.:
[1988] *Remarks about equilibrium configurations of crystals.* In *Material instabilities in continuum mechanics,* J. M. Ball (ed.), pp. 217–241. Clarendon Press, Oxford
KINDERLEHRER, D., & PEDREGAL, P.:
[1991] *Characterizations of Young measures generated by gradients.* Arch. Rational Mech. Anal. **115,** 329–366
KNOPS, R. J., & WILKES, E. W.:
[1973] *Theory of elastic stability.* In *Handbuch der Physik, VIa/3,* Springer, Berlin
KNOWLES, J. K.:
[1979] *On a dissipation associated with equilibrium shocks in finite elasticity.* J. Elasticity **9,** 131–158
KNOWLES, J. K., & STERNBERG, E.:
[1975] *On the failure of ellipticity of the equations of nonlinear elastostatics for a special material.* J. Elasticity **5,** 341–361

[1977] *On the failure of ellipticity of the equations for finite elastostatic plane strain.* Arch. Rational Mech. Anal. **63,** 321–326

[1978] *On the failure of ellipticity and emergence of discontinuous gradients in plane finite elastostatics.* J. Elasticity **8,** 329–379

KOHN, R. V., & STERNBERG, P.:

[1989] *Local minimizers and singular perturbations.* Proc. Royal Soc. Edinburgh **A111,** 69–84

KOHN, R. V., & STRANG, G.:

[1983] *Explicit relaxation of a variational problem in optimal design.* Bull. AMS **9,** 211–214

[1986] *Optimal design and relaxation of variational problems, I, II, III.* Comm. Pure Appl. Math. **39,** 113–137, 139–182, 353–377

KOITER, W.T.:

[1967] *On the thermodynamic background of elastic stability theory.* Report, Technological University, Delft

[1976] *A basic open problem in the theory of elastic stability.* Springer Lecture Notes in Mathematics **503,** 366–373

KOLASIŃSKA, G.:

[1986] *Correlation and prediction of VLE and LLE by empirical EOS.* Fluid Phase Equilibria **27,** 289–308

KOSIŃSKI, W.:

[1986] *Field singularities and wave analysis in continuum mechanics.* PWN, Wasraw

KRASNOSELSKII, M. A., ZABREIKO, P. P., PUSTYLNIK, E. I., & SOBOLEVSKII, P. E.:

[1976] *Integral operators in spaces of summable functions.* Noordhoff International Publishing, Leyden

KRATOCHVÍL, J., & ŠILHAVÝ, M.:

[1982] *On thermodynamics of non-equilibrium processes.* J. Non-Equilib. Thermodyn. **7,** 339–354

KRAWIETZ, A.:

[1986] *Materialtheorie, mathematische Beschreibung des phanomenologischen termo-dynamischen Verhaltens.* Springer, Berlin

LANDAU, L. D., & LIFSCHITZ, E. M.:

[1978] *Fluid mechanics.* Pergamon Press, Oxford

LARCHÉ, F. C., & CAHN, J. W.:

[1978] *Thermochemical equilibrium of multiphase solids under stress.* Act. Metall. **26,** 1579–1589

LAX, P. D.:

[1957] *Hyperbolic systems of conservation laws.* Comm. Pure Appl. Math. **10,** 537

[1964] *Development of singularities of solutions of nonlinear hyperbolic partial differential equations.* J. Math. Phys. **5,** 611–613

[1971] *Shock waves and entropy.* In *Contributions to nonlinear functional analysis,* E. H. Zarantonello (ed.), pp. 603–634. Academic Press, New York

LEBON, G.:

[1978] *Derivation of generalized Fourier and Navier-Stokes equations based on thermodynamics of irreversible processes.* Bull. Acad. Royal Sci. Belgique **64,** 45

[1987] *An extended thermodynamic description of linear viscoelasticity.* In *Kinetic theory and extended thermodynamics,* I. Müller & T. Ruggeri (ed.), pp. 197–208. Pitagora Editrice, Bologna

LE FLOCH, P.:

[1993] *Propagating phase boundaries.* Arch. Rational Mech. Anal. **123,** 153–197

LEHMAN, T., & GUO, Z.-H.:

[1991] *The conjugacy between Cauchy stress and logarithm of the left stretch tensor.* European J. Mech A. Solids **10,** 395–404

LIN, P.:

[1990] *Maximization of entropy for an elastic body free of surface traction.* Arch. Rational Mech. Anal. **112,** 161–191

LIU, I-SHIH:

[1972] *Method of Lagrange multipliers for exploitation of the entropy principle.* Arch. Rational Mech. Anal. **46,** 131–148

LIU, I-SHIH, & MÜLLER, I.:

[1983] *Extended thermodynamics of classical and degenerate gases.* Arch. Rational Mech. Anal. **83,** 285

LIU, T.-P.:

[1975] *The Riemann problem for general conservation laws.* J. Diff. Equations **18,** 218–234

[1981] *Admissible solutions of hyperbolic systems of conservation laws.* Mem. American Math. Soc., **240**
[1985] *Nonlinear stability of shock waves for viscous conservation laws.* Mem. American Math. Soc., **328**

LIU, Z., & ZHENG, S.:
[1993] *Exponential stability of the semigroup associated with a thermoelastic system.* Quart. Appl. Math. **51,** 535–545

LOPATINSKY, YA. B.:
[1953] *On a method of reducing boundary problems for a system of partial differential equations of elliptic type to regular integral equations.* Ukrain. Math. Z. **5,** 123–151

LUCCHESI, M.:
[1993] *Free-energy functions for elastic-plastic material elements.* Quart. Appl. Math. **51,** 299–318

LUCCHESI, M., OWEN, D. R., & PODIO-GUIDUGLI, P.:
[1992] *Materials with elastic range: A theory with a view towards applications. Part III.* Arch. Rational Mech. Anal. **117,** 53–96

LUCCHESI, M., & PODIO-GUIDUGLI, P.:
[1988] *Materials with elastic range: A theory with a view towards applications. Part I.* Arch. Rational Mech. Anal. **102,** 23–43
[1990] *Materials with elastic range: A theory with a view towards applications. Part II.* Arch. Rational Mech. Anal. **110,** 9–42

LUCCHESI, M., & ŠILHAVÝ, M.:
[1993] *Thermoplastic materials with combined hardening.* Int. J. Plasticity **9,** 291–315

LURIE, K. A., & CHERKAEV, A. V.:
[1984] *Exact estimates of conductivity of composites formed by two isotropically conducting media taken in prescribed proportion.* Proc. Royal Soc. Edinburgh **A99,** 71
[1986] *Exact estimates of the conductivity of a binary mixture of isotropic components.* Proc. Royal Soc. Edinburgh **A04,** 21

MACCAMY, R. C., & MIZEL, V. J.:
[1967] *Existence and non-existence in the large of solutions of quasilinear wave equations.* Arch. Rational Mech. Anal. **25,** 299–320

MAJDA, A., & PEGO, R.:
[1985] *Stable viscosity matrices for systems of conservation laws.* J. Diff. Equations **56,** 229–262

MALEK-MADANI, R.:
[1981] *Energy Criteria for Finite Hyperelasticity.* Arch. Rational Mech. Anal. **77,** 177–188

MALÝ, J.:
[1993] *Weak lower semicontinuity of polyconvex integrals.* Proc. Royal Soc. Edinburgh **123A,** 681–691
[1994a] *Lower semicontinuity of quasiconvex integrals.* Manuscripta Math. **85,** 419–428
[1994b] *The area formula for $W^{1,n}$-mappings.* Comment. Math. Univ. Carolinae **35,** 291–298

MALÝ, J., & MARTIO, O.:
[1995] *Lusin's condition* (N) *and mappings of the class $W^{1,n}$.* J. reine angew. Math. **458,** 19–36

MAN, C.-S.:
[1980] *Proof that the Clausius-Serrin inequality is equivalent to a strengthened Kelvin's law.* Atti Accad. Naz. Lincei, Rend. Cl. Sci. Fis. Mat. Natur. (8) **69,** 399–499
[1985a] *Dynamic admissible states, negative absolute temperature, and the entropy maximum principle.* Arch. Rational Mech. Anal. **90,** 263–289
[1985b] *Material stability, the Gibbs conjecture and the first phase rule for substances.* Arch. Rational Mech. Anal. **91,** 1–53
[1986] *Genericity and the Gibbs conjecture on the maximum number of coexistent phases.* In *New Perspectives in Thermodynamics,* J. Serrin (ed.), pp. 57–168. Springer, Berlin
[1989] *Thermodynamics based on a work inequality.* Arch. Rational Mech. Anal. **106,** 1–62
[1994] *Remarks on the continuity of the scalar coefficients in the representation* $\mathbf{H}(\mathbf{A}) = \alpha\mathbf{I} + \beta\mathbf{A} + \gamma\mathbf{A}^2$ *for isotropic tensor functions.* J. Elasticity **34,** 229–238
[1995] *Remarks on global and local versions of the second law of thermomechanics.* In *Thermoelastic problems and the thermodynamics of continua,* L. M. Brock (ed.), ASME, New York

MAN, C.-S., & CARLSON, D. E.:
[1994] *On the traction problem of dead loading in linear elasticity with initial stress.* Arch. Rational Mech. Anal. **128,** 223–247

MAN, C.-S., & GUO ZHONG-HENG:
[1993] *A basis-free formula for time rate of Hill's strain tensors.* Int. J. Solids Structures **30,** 2819–2842
MARCELLINI, P.:
[1986] *On the definition and the lower semicontinuity of certain quasiconvex integrals.* Ann. Inst. H. Poincaré, Analyse non linéaire **3,** 391–409
MARCUS, M., & MIZEL, V.:
[1973] *Transformations by functions in Sobolev spaces and lower semicontinuity for parametric variational problems.* Bull. AMS **79,** 790–795
MARSDEN, J. E., & HUGHES, T. J. R.:
[1983] *Mathematical foundations of elasticity.* Prentice-Hall, Englewood Cliffs
MARSHALL, A. W., & OLKIN, I.:
[1979] *Inequalities: theory of majorization and its applications.* Academic Press, New York
MARTINI, R.:
[1979] *On the Fréchet differentiability of certain energy functionals.* Proc. Kon. Ned. Akad. Wet. **B82,** 42–45
MARTIO, O., & ZIEMER, W.:
[1992] *Lusin's condition* (N) *and mappings with non-negative Jacobians.* Mich. Math. J. **39,** 495–508
MAUGIN, G. A.:
[1992] *The thermomechanics of plasticity and fracture.* Cambridge University Press, Cambridge
MCLELLAN, A. G.:
[1980] *The classical thermodynamics of deformable materials.* Cambridge University Press, Cambridge
MEIRMANOV, A. M.:
[1992] *The Stefan problem.* Walter de Gruyter, Berlin
MEIXNER, J.:
[1943] *Zur Thermodynamik der irreversiblen Prozessen in Gasen mit chemisch reagierenden, dissozierenden und anregbaren Komponenten.* Ann. Phys. **43,** 244–270
MENDOZA, E.:
[1961] *Sketch for a history of early thermodynamics.* Physics Today **14,** 32–42
MEYERS, N.:
[1965] *Quasi-convexity and lower semicontinuity of multiple variational integrals of any order.* Trans. American Math. Soc. **119,** 125–149
MILTON, G. W.:
[1990] *On characterizing the set of possible effective tensors of composites: variational method and the translation method.* Comm. Pure Appl. Math. **43,** 63–125
MIRSKY, L.:
[1975] *A trace inequality of John von Neumann.* Monatshefte für Mathematik **79,** 303–306
MODICA, L.:
[1987] *Gradient theory of phase transitions and minimal interface criterion.* Arch. Rational Mech. Anal. **98,** 123–142
MONLEÓN PRADAS, M.:
[1994] *On baricentres of heating measures and thermodynamical inequalities.* Arch. Rational Mech. Anal. **125,** 311–340
MORREY, JR, C. B.:
[1952] *Quasi-convexity and the lower semicontinuity of multiple integrals.* Pacific J. Math. **2,** 25–53
[1966] *Multiple integrals in the calculus of variations.* Springer, New York
MORRO, A.:
[1987] *Hyperbolic models of heat conduction.* In *Kinetic theory and extended thermodynamics,* I. Müller & T. Ruggeri (ed.), pp. 233–243. Pitagora Editrice, Bologna
MÜLLER, I.:
[1967] *Zum Paradox der Wärmeleitungstheorie.* Z. Phys. **198,** 329
[1985] *Thermodynamics.* Pitman, Boston
MÜLLER, I., & RUGGERI, T.:
[1993] *Extended Thermodynamics.* Springer, Berlin
MÜLLER, S.:
[1989] *A surprising higher integrability property of mappings with positive determinant.* Bull. AMS **21,** 245–248
[1990] *Higher integrability of determinant and weak convergence in L^1.* J. reine angew. math. **412,** 20–34

MÜLLER, S., & SPECTOR, S.:
[1995] *An existence theory for nonlinear elasticity that allows for cavitation.* Arch. Rational Mech. Anal. **131,** 1–66
MÜLLER, S., & SPECTOR, S., & TANG, Q.:
[1993] *Invertibility and a topological property of Sobolev maps (Preprint)*
MÜLLER, S., TANG, Q., & YAN, B. S.:
[1994] *On a new class of elastic deformations not allowing for cavitation.* Ann. Inst. H. Poincaré, Analyse non linéaire **11,** 217–243
MURAT, F., & TARTAR, L.:
[1985] *Calcul des variations et homogeneisation.* In *Les Méthodes d'Homogeneisation: Théorie et Applications en Physique,* pp. 319. Paris
MURDOCH, I.:
[1978] *Radial vibrations of a gravitating sphere with a material boundary.* Q. J. Mech. Appl. Math. **31,** 531–542
MUSGRAVE, M. J. P.:
[1970] *Crystal acoustics.* Holden-Day, San Francisco
NEČAS, J.:
[1967] *Les méthodes directes en théorie des equations elliptiques.* Masson, Paris
NEČAS, J., & HLAVÁČEK, I.:
[1981] *Mathematical theory of elastic and elastic-plastic bodies: An introduction.* Elsevier Scientific Publishing, Amsterdam
NITSCHE, J. A.:
[1981] *On Korn's second inequality.* RAIRO Numerical Analysis **15,** 237–248
NOLL, W.:
[1958] *A mathematical theory of the mechanical behaviour of continuous media.* Arch. Rational Mech. Anal. **2,** 197–226
[1959] *The foundations of classical mechanics in the light of recent advances in continuum mechanics.* In *The Axiomatic Method, with Special Reference to Geometry and Physics,* P. Suppes (ed.), pp. 266–281. North-Holland, Amsterdam
[1963] *La mécanique classique, basée sur un axiome d'objectivité.* In *La Methode Axiomatique dans les Mecaniques Classique et Nouvelles. Colloque international, Paris 1959,* pp. 47–56. Gauthier–Villars, Paris
[1970] *On certain convex sets of measures and on phases of reacting mixtures.* Arch. Rational Mech. Anal. **38,** 1–12
[1972] *A new mathematical theory of simple materials.* Arch. Rational Mech. Anal. **48,** 1–50
[1973] *Lectures on the foundations of continuum mechanics and thermodynamics.* Arch. Rational Mech. Anal. **52,** 62–92
NOWINSKI, J. L.:
[1969] *Surface instability of a half-space under high two-dimensional compression.* J. Franklin Institute **288,** 357–376
ODEN, J. T., & REDDY, J. N.:
[1976] *Variational methods in theoretical mechanics.* Springer, Berlin
OGDEN, R. W.:
[1970a] *Compressible isotropic elastic solids under finite strain–constitutive inequalities.* Quart J. Mech. Appl. Math. **23,** 457–468
[1970b] *Waves in isotropic materials of Hadamard, Green or harmonic type.* J. Mech. Phys. Solids **18,** 149–164
[1972a] *Large deformation isotropic elasticity–on the correlation of theory and experiment for incompressible rubberlike solids.* Proc. Royal Soc. London **A 326,** 565–584
[1972b] *Large deformation isotropic elasticity–on the correlation of theory and experiment for compressible rubberlike solids.* Proc. Royal Soc. London **A 328,** 567–583
[1975] *A note on variational theorems in non-linear elastostatics.* Math. Proc. Camb. Phil. Soc. **77,** 609–615
[1979] *On the anisotropy of compressibility of rubber-like solids.* J. Phys. D: Appl. Phys. **12,** 465–479
[1984] *Non-linear elastic deformations.* Ellis Horwood Ltd., Chichester
[1986] *Recent advances in the phenomenological theory of rubber elasticity.* Rubber Chemistry and Technology **59,** 361–383
[1987] *Aspects of the phenomenological theory of rubber thermoelasticity.* Polymer **28,** 379–385
[1992a] *On the thermoelastic modeling of rubberlike solids.* J. Thermal Stresses **15,** 533–557

[1992b] *Nonlinear elasticity: Incremental equations and bifurcation phenomena.* Nonlinear equations in the applied sciences **2,** 437–468

OLVER, P. J.:

[1983] *Conservation laws and null divergences.* Math. Proc. Camb. Phil. Soc. **94,** 529–540

[1984a] *Conservation laws in elasticity I. General results.* Arch. Rational Mech. Anal. **85,** 119–129

[1984b] *Conservation laws in elasticity II. Linear homogeneous isotropic elastostatics.* Arch. Rational Mech. Anal. **85,** 131–160

[1985] *Conservation laws and null divergences, II. Non-negative divergences.* Math. Proc. Camb. Phil. Soc. **97,** 511–514

[1986] *Applications of Lie groups to differential equations.* Springer, New York

[1988a] *Canonical elastic moduli.* J. Elasticity **19,** 189–212

[1988b] *Conservation laws in elasticity III. Planar linear anisotropic elastostatics.* Arch. Rational Mech. Anal. **102,** 167–181

[1988c] *The equivalence problem and canonical forms for quadratic Lagrangians.* Adv. Appl. Math. **9,** 226–257

[1993] *Equivalence and the Cartan form.* Acta Appl. Math. **31,** 99–136

OLVER, P. J., & SIVALOGANATHAN, J.:

[1988] *The structure of null Lagrangians.* Nonlinearity **1,** 389–398

ÖNCÜ, T. S. , & MOODIE, T. B.:

[1992] *On the constitutive relations for second sound in elastic solids.* Arch. Rational Mech. Anal. **121,** 87–99

OVSJANNIKOV, L. V.:

[1978] *Group analysis of differential equations.* Nauka, Moscow

OWEN, D.:

[1982] *Global and local versions of the second law of thermodynamics.* In *Categories in Continuum Physics,* F. W. Lawvere & S. H. Schanuel (ed.), pp. 100–114. Springer Lecture Notes in Mathematics **1174**, Berlin

[1984] *A first course in the mathematical foundations of thermodynamics.* Springer, New York

PARRY, G. P.:

[1976] *On the elasticity of monatomic crystals.* Math. Proc. Camb. Phil. Soc. **80,** 189–211

[1978] *On the relative strengths of intrinsic stability criteria.* Quart. J. Mechanics and Appl. Math. **31,** 1–7

[1995] *On the planar rank-one convexity condition.* Proc. Royal Soc. Edinburgh **125A,** 247–264

PARTINGTON, J. R.:

[1949] *An advanced treatise on physical chemistry, Vol. 1.* Longmans, Green & Co., London

PAYTON, R. G.:

[1983] *Elastic wave propagation in transversely isotropic media.* Nijhoff, The Hague

PAZY, A.:

[1983] *Semigroups of linear operators and applications to partial differential equations.* Springer, New York

PEGO, R. L.:

[1986] *Nonexistence of a shock layer in gas dynamics with a nonconvex equation of state.* Arch. Rational Mech. Anal. **94,** 165–178

[1987] *Phase transitions in one-dimensional nonlinear viscoelasticity: Admissibility and stability.* Arch. Rational Mech. Anal. **97,** 353–394

PENCE, T. J.:

[1992] *On the mechanical dissipation of solutions to the Riemann problem for impact involving a two-phase elastic material.* Arch. Rational Mech. Anal. **117,** 1–52

PERICAK-SPECTOR, K. A., & SPECTOR, S. J.:

[1988] *Non-uniqueness for a hyperbolic system: Cavitation in nonlinear elastodynamics.* Arch. Rational Mech. Anal. **101,** 293–317

PERZYNA, P., & KOSIŃSKI, W.:

[1973] *A mathematical theory of materials.* Bull. Acad. Polon. Sci., Serie Sci. Techn. **21,** 647–654

PFENNING, N. M., & WILLIAMS, W. O.:

[1993] *On nonmaterial surfaces with structure.* Quart. Appl. Math. **51,** 559–576

PITTERI, M.:

[1982] *Classical thermodynamics of homogeneous systems based upon Carnot's general axiom.* Arch. Rational Mech. Anal. **80,** 333–385

[1984] *Reconciliation of local and global symmetries in crystals.* J. Elasticity **14,** 175–190

[1985a] *On* $\nu + 1$ *lattices.* J. Elasticity **15,** 3–25

[1985b] *On the kinematics of mechanical twinning in crystals.* Arch. Rational Mech. Anal. **88,** 25–57

PODIO-GUIDUGLI, P., & VERGARA CAFFARELLI, G.:

[1990] *Surface interaction potentials in elasticity.* Arch. Rational Mech. Anal. **109,** 343–383

[1991] *Extreme elastic deformations.* Arch. Rational Mech. Anal. **115,** 311–328

PODIO-GUIDUGLI, P., VERGARA CAFFARELLI, G., & VIRGA, E.:

[1987] *The role of ellipticity and normality assumptions in formulating live-boundary conditions in elasticity.* Quart. Appl. Math. **44,** 659–664

PODIO-GUIDUGLI, P., & VIRGA, E. G.:

[1987] *Transversely isotropic elasticity tensors.* Proc. Royal Soc. Lond. **A411,** 85–93

POTIER-FERRY, M.:

[1982] *On the mathematical foundations of elastic stability theory I.* Arch. Rational Mech. Anal. **78,** 55–72

PUCCI, P., & SERRIN, J.:

[1994] *On the derivation of Hamilton's equations.* Arch. Rational Mech. Anal. **125,** 297–310

RABIER, P. J., & GRIEWANK, A.:

[1992] *Generic aspects of convexification with applications to thermodynamic equilibrium.* Arch. Rational Mech. Anal. **118,** 349–397

RACKE, R.:

[1992] *Mathematical aspects in nonlinear thermoelasticity.* Preprint, Bonn

[1992] *Lectures on nonlinear evolution equations.* Vieweg & Sohn, Braunschweig

RACKE, R., SHIBATA, Y., & ZHENG, S.:

[1993] *Global solvability and exponential stability in one-dimensional nonlinear thermoelasticity.* Quart. Appl. Math. **51,** 751–763

RADÓ, T. & REICHELDERFER, P.:

[1955] *Continuous transformations in analysis.* Springer, Berlin

REDDY, B. D.:

[1982] *Surface instabilities on an equibiaxially stretched half-space.* Math. Proc. Camb. Phil. Soc. **91,** 491–501

[1983] *The occurrence of surface instabilities and shear bands in plane-strain deformation of an elastic half-space.* Quart. J. Mech. Appl. Math. **36,** 337–350

RENARDY, M., HRUSA W. J., & NOHEL, J. A.:

[1987] *Mathematical problems in viscoelasticity.* Longman, New York

RESHETNYAK, YU. G.:

[1967] *On the stability of conformal mappings in multidimensional spaces.* Siberian. Math. J. **8,** 91–114

[1968] *Stability theorems for mappings with bounded excursion.* Siberian Math. J. **9,** 499–512

RICE, J. M.:

[1968] *A path-independent integral and the approximate analysis of strain concentration by notches and cracks.* ASME Trans. J. Appl. Mech. **37,** 728–737

RICOU, M.:

[1986] *The laws of thermodynamics for non-cyclic processes.* In *New Perspectives in Thermodynamics,* J. Serrin (ed.), pp. 85–97. Springer, Berlin

[1988] *A general first law for thermodynamics.* Arch. Rational Mech. Anal. **101,** 365–385

ROBIN, P.-Y. F.:

[1974] *Thermodynamic equilibrium across a coherent interface in a stressed crystal.* Am. Min. **59,** 1286–1298

ROCKAFELLAR, R. T.:

[1970] *Convex analysis.* Princeton University Press, Princeton

ROITBURD, A. L.:

[1978] *Martensitic transformation as a typical phase transformation in solids.* Solid state Physics **33,** 317–390

ROSAKIS, P.:

[1990] *Ellipticity and deformations with discontinuous gradients in finite elastostatics.* Arch. Rational Mech. Anal. **109,** 1–37

ROZHDESTVENSKII, B. L.:

[1957] *On systems of quasilinear equations.* Doklady AN SSSR **115,** 454–457

ROZHDESTVENSKII, B. L., & JANENKO, N. N.:

[1978] *Systems of quasilinear equations and their application in gas dynamics. Second edition.* Nauka, Moscow

RUBINSTEIN, L. I.:

[1971] *The Stefan problem.* AMS, Providence

RUDDOCK, G.:
[1994] *A microstructure of martensite which is not a minimiser of energy: the X-interface.* Arch. Rational Mech. Anal. **127,** 1–39

RUDIN, W.:
[1970] *Real and complex analysis.* McGraw-Hill, New York
[1973] *Functional analysis.* McGraw-Hill, New York

RUMER, J. B., & RYVKIN, M. S.:
[1977] *Thermodynamics, statistical physics, and kinetics.* Nauka, Moscow

RYCHLEWSKI, J.:
[1984] *Elastic energy decompositions and limit criteria.* Adv. in Mech. **7,** 51–80
[1995] *Unconventional approach to linear elasticity.* Archives of Mechanics **47,** 149–171

RYCHLEWSKI, J., & ZHANG, J. M.:
[1989] *Anisotropy degree of elastic materials.* Archives of Mechanics **41,** 697–715
[1991] *On representation of tensor functions: a review.* Adv. in Mech. **14,** 75–94

SALVI, R., & STRAŠKRABA, I.:
[1993] *Global existence for viscous compressible fluids and their behavior as $t \to \infty$.* J. Fac. Sci. Univ. Tokyo **40,** 17–51

SAMOHÝL, I.:
[1982] *Racionální termodynamika chemicky reagujících směsí.* Academia, Praha
[1996a] *Thermodynamics of the mixture of reacting and non-reacting fluids with heat conduction, diffusion and viscosity.* (To appear)
[1996b] *Eshelby law in mixtures of non-fluids.* (To appear)
[1996c] *The thermodynamics of non-reacting mixtures of any symmetry with heat conduction, diffusion and viscosity.* (To appear)
[1996d] *The thermodynamics of reacting mixtures of any symmetry with heat conduction, diffusion and viscosity.* (To appear)

SAWYERS, K.:
[1986] *Comments on the paper* Determination of the stretch and rotation in the polar decomposition of the deformation gradient *by A. Hoger and D. E. Carlson.* Quart. Appl. Math. **46,** 309–311

SCHEIDLER, M.:
[1991] *Time rates of generalized strain tensors. Part I: Component formulas.* Mech. Mater. **11,** 199–210

SCHWARZENBERGER, R. L. E.:
[1972] *Classification of crystal lattices.* Proc. Camb. Phil. Soc. **72,** 325–349

SERRE, D.:
[1983] *Formes quadratiques et calcul des variations.* J. Math. purres appl. **62,** 177–196

SERRIN, J.:
[1959] *The derivation of stress deformation relations for a Stokesian fluid.* J. Math. Mech. **8,** 459–468
[1972] *On the axioms of classical mechanics (an unpublished paper)*
[1978] *The concepts of thermodynamics.* In *Contemporary developments in continuum mechanics and partial differential equations,* G. M. de la Penha, & L. A. Medeiros (ed.), pp. 411–451. North-Holland, Amsterdam
[1979] *Conceptual analysis of the classical second laws of thermodynamics.* Arch. Rational Mech. Anal. **70,** 355–371
[1986] *An outline of thermodynamical structure.* In *New Perspectives in Thermodynamics,* J. Serrin (ed.), pp. 3–32. Springer, Berlin
[1990] *The equations of continuum mechanics as a consequence of invariance and thermodynamical principles. Lecture, Conference on Nolinear analysis and partial differential equations, Rutgers University, 1990*
[1995] *On the elementary thermodynamics of quasi-static systems and other remarks.* In *Thermoelastic problems and the thermodynamics of continua,* L. M. Brock (ed.), ASME, New York

SETH, B. R.:
[1964] *Generalized strain measure with applications to physical problems.* In *Second-Order Effects in Elasticity, Plasticity, and Fluid Dynamics,* M. Reiner & D. Abir (ed.), pp. 162–172. Pergamon Press, Oxford

SEWELL, M. J.:
[1967] *On configuration-dependent loading.* Arch. Rational Mech. Anal. **23,** 327–351
[1987] *Maximum and minimum principles.* Cambridge University Press, Cambridge

SHAPIRO, Z. YA.:
[1951] *The first boundary problem for elliptic systems of differential equations.* Math. Sb. **28,** 55–78

SHEARER, M.:
[1982] *The Riemann problem for a class of conservation laws of mixed type.* J. Diff. Equations **46,** 426–443
[1986] *Non-uniqueness of admissible solutions of Riemann initial-value problems for a system of conservation laws of mixed type.* Arch. Rational Mech. Anal. **93,** 45–59
[1988] *Dynamic phase transitions in a van der Waals gas.* Quart. Appl. Math. **46,** 631–636

SHEARER, M., SCHAEFFER, D. G., MARCHESIN, D., & PAES-LEME, P. L.:
[1987] *Solution of the Riemann problem for a prototype* 2×2 *system of non-strictly hyperbolic conservation laws.* Arch. Rational Mech. Anal. **97,** 299–320

ŠILHAVÝ, M.:
[1977] *A note on the existence of entropy in classical thermodynamics.* Archives of Mechanics **29,** 289–298
[1980] *On measures, convex cones, and foundations of thermodynamics. Parts I & II.* Czech. J. Phys. **B30,** 841–861 & 961–991
[1982] *On the second law of thermodynamics. Parts I & II.* Czech. J. Phys. **B32,** 987–1010 & 1073–1099
[1983] *On the Clausius inequality.* Arch. Rational Mech. Anal. **81,** 221–243. A preliminary announcement appeared in Abstracts of EUROMECH 111, Marienbad, September 26–28, 1978, p. 68
[1984] *An admissibility criterion for shocks and propagating phase boundaries via thermodynamics of non-simple materials, I & II.* J. Non- Equilib. Thermodyn **9,** 177–186 & 187–200
[1985a] *Actions with the conservation property.* Aplikace matematiky **30,** 140–153
[1985b] *The existence of the flux vector and the divergence theorem for general Cauchy fluxes.* Arch. Rational Mech. Anal. **90,** 195–212
[1985c] *Phase transitions in non-simple bodies.* Arch. Rational Mech. Anal. **88,** 135–161
[1987] *On the concepts of mass and linear momentum in Galilean thermodynamics.* Czech. J. Phys. **B37,** 133–157
[1989] *Mass, internal energy, and Cauchy's equations of motion in frame-indifferent thermodynamics.* Arch. Rational Mech. Anal. **107,** 1–22
[1991] *Cauchy's stress theorem and tensor fields with divergences in* L^p. Arch. Rational Mech. Anal. **116,** 223–255
[1992] *Energy principles and the equations of motion in Galilean thermomechanics.* Czech. J. Phys. **42,** 363–374

SIMPSON, H. C., & SPECTOR, S.:
[1983] *On copositive matrices and strong ellipticity for isotropic elastic materials.* Arch. Rational Mech. Anal. **84,** 55–68
[1984a] *On barelling for a special material in finite elasticity.* Quart. Appl. Math. **42,** 99–110
[1984b] *On barelling instabilities in finite elasticity.* J. Elasticity **14,** 103–125
[1985] *On failure of the complementing condition and nonuniqueness in linear elastostatics.* J. Elasticity **15,** 229–231
[1987] *On the positivity of the second variation in finite elasticity.* Arch. Rational Mech. Anal. **98,** 1–30
[1989] *Necessary conditions at the boundary for minimizers in finite elasticity.* Arch. Rational Mech. Anal. **107,** 106–125

SIVALOGANATHAN, J.:
[1989] *The generalised Hamilton-Jacobi inequality and the stability of equilibria in nonlinear elasticity.* Arch. Rational Mech. Anal. **107,** 347–369

SLEMROD, M.:
[1981] *Global existence, uniqueness and asymptotic stability of classical smooth solutions in one-dimensional non-linear thermoelasticity.* Arch. Rational Mech. Anal. **76,** 97–133
[1983] *Admissibility criteria for propagating phase boundaries in a van der Waals fluid.* Arch. Rational Mech. Anal. **81,** 301–315
[1984] *Dynamic phase transitions in a van der Waals fluid.* J. Diff. Equations **54,** 1–23
[1989] *A limiting "viscosity" approach to the Riemann problem for materials exhibiting change of phase.* Arch. Rational Mech. Anal. **105,** 327–365

SMOLLER, J.:
[1983] *Shock waves and reaction-diffusion equations.* Springer, New York

SPECTOR, S. J.:

[1980] *On uniqueness in finite elasticity with general loading.* J. Elasticity **10,** 145–161

[1982] *On uniqueness for the traction problem in finite elasticity.* J. Elasticity **12,** 367–383

[1984] *On the absence of bifurcation for elastic bars in uniaxial tension.* Arch. Rational Mech. Anal. **85,** 171–199

SPENCER, A. J. M.:

[1971] *Theory of invariants.* In *Continuum Physics, vol. I,* A. C. Eringen (ed.), Academic Press, New York

STERNBERG, P.:

[1988] *The effect of a singular perturbation on nonconvex variational problems.* Arch. Rational Mech. Anal. **101,** 209–260

STRAŠKRABA, I.:

[1996] *Asymptotic development of vacuums for 1-D Navier-Stokes equations.* Nonlinear World (To appear)

STUECKELBERG DE BREIDENBACH, E. C. G., & SCHEURER, P. B.:

[1967] *Phenomenological thermodynamics V: the 2nd law applied to extensive functionals with the use of Lagrange multipliers.* Helv. Phys. Acta **40,** 887–906

[1974] *Thermocinétique Phenoménologique Galiléene.* Birkhäuser, Basel

SUHUBI, E. S.:

[1989] *Conservation laws in nonlinear elastodynamics.* Int. J. Engng Sci. **27,** 441–453

ŠVERÁK, V.:

[1988] *Regularity properties of deformations with finite energy.* Arch. Rational Mech. Anal. **98,** 105–127

[1990] *Examples of rank-one convex functions.* Proc. Royal Soc. Edinburgh **114A,** 237–242

[1991] *Quasiconvex functions with subquadratic growth.* Proc. Royal Soc. London **A433,** 723–725

[1992a] *Rank-One convexity does not imply quasiconvexity.* Proc. Royal Soc. Edinburgh **120,** 185–189

[1992b] *New examples of quasiconvex functions.* Arch. Rational Mech. Anal. **119,** 293–300

SYLVESTER, J.:

[1985] *On the differentiability of* $\mathbf{O}(n)$ *invariant functions of symmetric matrices.* Duke Math. J. **52,** 475–483

SYNGE, J. L.:

[1956a] *Elastic waves in anisotropic media.* J. Math. Phys. **35,** 323–334

[1956b] *Flux of energy for elastic waves in anisotropic media.* Proc. Royal Irish Academy **58,** 13–21

TANG, Q.:

[1988] *Almost-everywhere injectivity in nonlinear elasticity.* Proc. Royal Soc. Edinburgh **109A,** 79–95

TARTAR, L.:

[1983] *The compensated compactness method applied to systems of conservation laws.* In *Systems of nonlinear partial differential equations,* J. M. Ball (ed.), pp. 263–285. Reidel, Dordrecht

TERPSTRA, F. J.:

[1938] *Die Darstellung biquadratischer Formen als Summen von Quadraten mit Anwendung auf die Variationsrechnung.* Math. Ann. **116,** 166–180

THOMPSON, J. L.:

[1969] *Some existence problems for the traction boundary value problem of linearized elastostatics.* Arch. Rational Mech. Anal. **32,** 369–399

THOMPSON, J. M. T., & HUNT, G. W.:

[1973] *A general theory of elastic stability.* Willey, London

[1984] *Elastic instability phenomena.* Willey, London

THOMPSON, P. A., CAROFANO, G. C. & YOON, G. K.:

[1986] *Shock waves and phase changes in a large-heat-capacity fluid emerging from a tube.* J. Fluid Mech. **166,** 57–92

THOMPSON, P. A., MEIER, G. E. A.:

[1989] *Adiabatic phase changes, fast and slow.* In *Partial differential equations and continuum models of phase transitions,* M. Rascle, D. Serre, & M. Slemrod (ed.), pp. 146–174. Springer, Berlin

TING, T. C. T.:

[1985] *Determination of* $C^{1/2}$, $C^{-1/2}$ *and more general isotropic tensor functions.* J. Elasticity **15,** 319–323

TRELOAR, L. R. G.:
[1975] *The physics of rubber elasticity.* Oxford University Press, Oxford
TRUESDELL, C.:
[1956] *Das ungelöste Hauptproblem der endlichen Elasticitätstheorie.* Z. angew. Math. Mech. **36,** 97–103
[1973] *The efficiency of a homogeneous heat engine.* Journal of Mathematical and Physical Sciences (Madras) **7,** 349–371
[1979] *Absolute temperatures as a consequence of Carnot's general axiom.* Archive for History of Exact Sciences **20,** 357–380
[1980] *The tragicomical history of thermodynamics, 1822–1854.* Springer, New York
[1984] *Rational thermodynamics, 2nd edition.* Springer, New York
[1986] *What did Gibbs and Carathéodory leave us about thermodynamics.* In *New Perspectives in Thermodynamics,* J. Serrin (ed.), pp. 101–124. Springer, Berlin
TRUESDELL, C., & BHARATHA, S.:
[1977] *The concepts and logics of classical thermodynamics as a theory of heat engines.* Springer, New York
TRUESDELL, C., & NOLL, W.:
[1965] *The non-linear field theories of mechanics.* In *Handbuch der Physik III/3,* Springer, Berlin
TRUESDELL, C., & TOUPIN, R. A.:
[1960] *Classical field theories of mechanics.* In *Handbuch der Physik III/1,* Springer, Berlin
TRUSKINOVSKY, L.:
[1982] *Equilibrium interface boundaries.* Dokl. Akad. Nauk SSSR **275,** 306–310
[1983a] *Critical nuclei in the van der Waals model.* Dokl. Akad. Nauk SSSR **269,** 587–592
[1983b] *On the tensor of chemical potential.* Geochimija **12,** 1730–1744
[1985] *Structure of an isothermal phase jump.* Dokl. Akad. Nauk SSSR **285,** 309–315
[1987] *Dynamics of non-equilibrium phase boundaries in a heat conducting nonlinear elastic medium.* J. Appl. Math. Mech **51,** 777–784
[1991] *Kinks versus shocks.* In *Shock Induced Transitions and Phase Structures in General Media,* R. Fosdick, E. Dunn, & M. Slemrod (ed.), Springer, Berlin
[1994] *Transition to detonation in dynamic phase changes.* Arch. Rational Mech. Anal. **125,** 375–397
VALENT, T.:
[1988] *Boundary value problems in finite elasticity.* Springer, Berlin
VALLI, A.:
[1992] *Mathematical results for compressible flow.* Dipartimento di Matematica, Universita degli studi di Trento, 365
VODOPYANOV, S. K., & GOLDSHTEIN, V. M.:
[1977] *Quasiconformal mappings and spaces of functions with generalized derivatives.* Siberian Math. J. **17,** 515–531
VOL'PERT, A. I.:
[1967] *The space BV and quasilinear equations.* Math. Sb. **73(115),** 255–302
VOL'PERT, A. I., & HUDJAEV, S. I.:
[1985] *Analysis in classes of discontinuous functions and equations of mathematical physics.* Martinus Nijhoff Publishers, The Hague
VUKALOVICH, M. P., & NOVIKOV, I. I.:
[1948] *Equations of state of real gases.* Gosudarstvennoe Energeticheskoe Izdatelstvo, Moscow
WALPOLE, L. J.:
[1984] *Fourth-rank tensors of the thirty-two crystal classes: multiplication tables.* Proc. Royal Soc. Lond. **A391,** 149–179
WALTER, J.:
[1978] *On the definition of the absolute temperature — a reconciliation of the classical method with that of Carathéodory.* Proc. Royal Soc. Edinburgh **82A,** 87–94
WANG, C.-C.:
[1984] *Appendix 7A in* C. Truesdell's Rational Thermodynamics, 2nd ed. Springer, Berlin
WANG, W. B., & DUAN, Z. P.:
[1991] *On the invariant representation of spin tensors with applications.* Int. J. Solids Structures **27,** 329–341
WANG, C.-C., & TRUESDELL, C.:
[1973] *Introduction to rational elasticity.* Noordhoff, Leyden
WASHIZU, K.:
[1968] *Variational principles in elasticity and plasticity.* Pergamon Press, Oxford

WECHSLER, M. S., LIEBERMAN, D. S., & READ, T. A.:
[1953] *On the theory of the formation of martensite.* Trans. AIME J. Metals **197,** 1503–1515
WENDROFF, B.:
[1972] *The Riemann problem for materials with nonconvex equations of state. I: Isentropic flow.* J. Math. Anal. Appl. **38,** 454–466
WESOŁOWSKI, Z.:
[1964] *Stability of a full elastic sphere uniformly loaded on the surface.* Archives of Mechanics **16,** 1131–1151
WEYL, H.:
[1949] *Shock waves in arbitrary fluids.* Comm. Pure and Applied Math. **2,** 103–122
WHAPLES, G.:
[1952] *Carathéodory's temperature equations.* J. Rational Mech. Anal. **1,** 302–307
WHITHAM, G. B.:
[1974] *Linear and nonlinear waves.* Wiley, New York
WILLEMS, J. C.:
[1972] *Dissipative dynamical systems, I, II.* Arch. Rational Mech. Anal. **45,** 321–351
ZADEH, L. A., & DESOER, CH. A.:
[1969] *Linear systems theory. The state space approach.* McGraw-Hill, New York
ZANZOTTO, G.:
[1988] *Twinning in mineral and metals: remarks on the comparison of a thermoelastic theory with some available experimental results: Note I & II.* Atti Accad. Naz. Lincei, Rend. Cl. Sci. Fis. Mat. Natur. (8) **82,** 725–742 & 743–756
[1992] *On the material symmetry group of elastic crystals and the Born rule.* Arch. Rational Mech. Anal. **121,** 1–36
ZHANG, K.:
[1991] *Energy minimizers in nonlinear elastostatics and the implicit function theorem.* Arch. Rational Mech. Anal. **114,** 95–117
ZHONG, H. Y., & DEL PIERO, G.:
[1987] *On the completeness of the crystallographic symmetries in the description of the symmetries of the elastic tensor.* Preprint
ZIEGLER, H.:
[1968] *A possible generalization of Onsager's theory.* In *Proc. IUTAM Symposium on Irreversible Aspects of Continuum Mechanics,* H. Parkus & L. I. Sedov (ed.), pp. 411–424. Springer, Berlin
ZIEMER, W.:
[1983] *Cauchy flux and sets of finite perimeter.* Arch. Rational Mech. Anal. **84,** 189–201
ZUBOV, V. I.:
[1964] *Methods of A. M. Lyapunov and their application.* Noordhoff, Groningen

Subject Index

Springer-Verlag and the Environment

We at Springer-Verlag firmly believe that an international science publisher has a special obligation to the environment, and our corporate policies consistently reflect this conviction.

We also expect our business partners – paper mills, printers, packaging manufacturers, etc. – to commit themselves to using environmentally friendly materials and production processes.

The paper in this book is made from low- or no-chlorine pulp and is acid free, in conformance with international standards for paper permanency.

9 783540 583783

GPSR Compliance

The European Union's (EU) General Product Safety Regulation (GPSR) is a set of rules that requires consumer products to be safe and our obligations to ensure this.

If you have any concerns about our products, you can contact us on ProductSafety@springernature.com

In case Publisher is established outside the EU, the EU authorized representative is:

Springer Nature Customer Service Center GmbH
Europaplatz 3
69115 Heidelberg, Germany

Batch number: 10682187

Printed by Printforce, the Netherlands